BIOLOGY

Concepts and Investigations
BIOLOGY

FOURTH EDITION

MARIËLLE HOEFNAGELS
THE UNIVERSITY OF OKLAHOMA

MEDIA CONTRIBUTIONS BY
MATTHEW S. TAYLOR

Mc
Graw
Hill
Education

BIOLOGY: CONCEPTS AND INVESTIGATIONS, FOURTH EDITION

Published by McGraw-Hill Education, 2 Penn Plaza, New York, NY 10121. Copyright © 2018 by McGraw-Hill Education. All rights reserved. Printed in the United States of America. Previous editions © 2015, 2012, and 2009. No part of this publication may be reproduced or distributed in any form or by any means, or stored in a database or retrieval system, without the prior written consent of McGraw-Hill Education, including, but not limited to, in any network or other electronic storage or transmission, or broadcast for distance learning.

Some ancillaries, including electronic and print components, may not be available to customers outside the United States.

This book is printed on acid-free paper.

1 2 3 4 5 6 7 8 9 LWI 21 20 19 18 17

ISBN 978-0-07-802420-7
MHID 0-07-802420-X

Chief Product Officer, SVP Products & Markets: *G. Scott Virkler*
Vice President, General Manager, Products & Markets: *Marty Lange*
Vice President, Content Design & Delivery: *Betsy Whalen*
Director, Biology: *Lynn Breithaupt*
Brand Manager: *Michelle Vogler*
Director of Product Development: *Rose Koos*
Product Developer: *Anne Winch*
Marketing Manager: *Britney Ross*
Digital Product Analyst: *Christine Carlson*
Director, Content Design & Delivery: *Linda Avenarius*
Program Manager: *Angela R. FitzPatrick*
Content Project Manager (print): *Peggy Selle*
Content Project Manager (assessment): *Emily Windelborn*
Buyer: *Laura Fuller*
Designer: *Tara McDermott*
Content Licensing Specialists: *Lori Hancock/Deanna Dausener*
Cover Image: *(front) ©mustamin/Getty Images RF*
Compositor: MPS Limited
Typeface: *10/12 STIX MathJax Main*
Printer: *LSC Communications*

All credits appearing on page or at the end of the book are considered to be an extension of the copyright page.
Leaf: *©image broker/Alamy RF*
Flames: *©Don Farrall/Getty Images RF*

Library of Congress Cataloging-in-Publication Data

Names: Hoefnagels, Mariëlle.
Title: Biology : concepts and investigations.
Description: Fourth edition / Mariëlle Hoefnagels, The University of Oklahoma;
   media contributions, Matthew S. Taylor. | New York, NY : McGraw-Hill Education, [2018] | Includes index.
Identifiers: LCCN 2016047958 | ISBN 9780078024207 (alk. paper)
Subjects: LCSH: Biology–Textbooks.
Classification: LCC QH307.2 .H64 2018 | DDC 570–dc23 LC record available at https://lccn.loc.gov/2016047958

The Internet addresses listed in the text were accurate at the time of publication. The inclusion of a website does not indicate an endorsement by the authors or McGraw-Hill Education, and McGraw-Hill Education does not guarantee the accuracy of the information presented at these sites.

www.mhhe.com

Brief Contents

About the Author

©Davenport Photos

Mariëlle Hoefnagels is an associate professor in the Department of Biology and the Department of Microbiology and Plant Biology at the University of Oklahoma, where she teaches courses in introductory biology, mycology, and science writing. She has received the University of Oklahoma General Education Teaching Award and the Longmire Prize (the Teaching Scholars Award from the College of Arts and Sciences). She has also been awarded honorary memberships in several student honor societies.

Dr. Hoefnagels received her B.S. in environmental science from the University of California at Riverside, her M.S. in soil science from North Carolina State University, and her Ph.D. in plant pathology from Oregon State University. Her dissertation work focused on the use of bacterial biological control agents to reduce the spread of fungal pathogens on seeds. In addition to authoring *Biology: Concepts and Investigations* and *Biology: The Essentials*, her recent publications have focused on creating investigative teaching laboratories and methods for teaching experimental design in beginning and advanced biology classes. She frequently gives presentations on study skills and related topics to student groups.

Preface

Vision and Change in Undergraduate Biology Education: A Call to Action encourages instructors to improve student engagement and learning in introductory biology courses. The central idea of the original *Vision and Change* report—and of the conferences and reports that followed—is that we need to turn away from teaching methods that reward students who memorize and regurgitate superficial knowledge. Instead, we need to emphasize deeper learning that requires students to understand and apply course content. This idea is precisely what I have tried to achieve since I started teaching at the University of Oklahoma in 1997, and it has been a guiding principle in the creation of my books and digital material as well.

This edition retains what users have always loved about this book: the art program, readable narrative, handy study tips, Investigating Life essays, and tutorial animations. We also supply a variety of supplements that make teaching easier, including eye-catching PowerPoint® lectures with integrated clicker questions that assess conceptual understanding. As you examine this new edition, however, I hope you will see an even stronger emphasis on connections and the "big picture." Our most prominent new feature, Survey the Landscape, shows how each chapter's content fits into the unit's overall emphasis. Students often struggle to connect new topics to what they have learned previously; Survey the Landscape is designed to help them keep an eye on the big picture. These new figures, which appear in each chapter opener, can be integrated with the Pull It Together figure in every chapter's summary to help students see the "forest" *and* the "trees."

Many other changes to this book reflect the growing numbers of instructors and students who are embracing digital textbooks. Much of our work for the fourth edition was behind-the-scenes adjustments that make the narrative and art more digital-friendly. Moreover, SmartBook® user data from thousands of students using the third edition helped us to identify passages that needed clarification. The user data also guided us as we created a carefully selected array of digital Learning Resources to accompany many probes in SmartBook. In addition, many chapters have bonus features for ebook users, including new digital-only miniglossaries and figures.

I agree with the *Vision and Change* report's call for instructors to embrace active learning techniques, but I also believe that one set of tools and techniques does not work in every classroom. For that reason, my team and I are proud to create a package that gives you the flexibility to teach introductory biology in a way that works best for you. The following sections illustrate the features and resources for this edition that can help you meet your teaching goals.

I hope that you and your students enjoy this text and that it helps cultivate an understanding of, and deep appreciation for, biology.

Mariëlle Hoefnagels
The University of Oklahoma

Author's Guide *To Using This Textbook*

This guide lists the main features of each chapter and describes some of the ways that I use them in my own classes.

The Learning Outline introduces the chapter's main headings and helps students keep the big picture in mind.

Each heading is a complete sentence that summarizes the most important idea of the section.

The gradual change in leaf colors as a chapter unfolds indicates where the student is in the chapter's big picture.

Students can also flip to the end of the chapter before starting to read; the chapter summary and Pull It Together concept map can serve as a review or provide a preview of what's to come.

Concept maps help students see the big picture.

New Survey the Landscape concept maps at the start of each chapter illustrate how the pieces of the entire unit fit together. These new figures integrate with the existing Pull It Together concept maps in the chapter summary.

After spending class time discussing the key points in constructing concept maps, I have my students draw concept maps of their own.

Is It Easier Being Green?

Food is expensive. It would be much cheaper and easier if we could feed ourselves using photosynthesis. Imagine the benefit of being photosynthetic: You could make your own food, free of charge, simply by sitting outside in the sun.

Of course, your body would have to have some new adaptations for photosynthesis to work. Your skin would have to be green, for starters. You might even have skin flaps that capture extra sunlight. You wouldn't eat, so you would need another way to acquire essential minerals; perhaps your feet would grow rootlike extensions that would absorb water and nutrients from soil.

Maybe photosynthetic cows, pigs, and chickens—or pets such as dogs and cats—would be a better idea. Feed-free animals would be a commercial and environmental triumph, costing less to own and generating less waste than the animals we raise now.

Fortunately or unfortunately, scientists will probably never be able to create photosynthetic people, chickens, or pooches. Mammals and birds move, breathe, pump blood, and maintain high body temperatures. All of this activity would likely require energy beyond what photosynthesis alone could supply.

Some animals, however, have adopted the "green" lifestyle by harboring live-in photosynthetic partners (see section 5.7). The closest to a true plant–animal hybrid is probably the sea slug *Elysia chlorotica*, a solar-powered mollusk with chloroplasts (photosynthetic organelles) in the cells lining its digestive tract. The chloroplasts come from algae in the slug's diet. As the animal grazes, it punctures the algal cells and discards everything but the chloroplasts, which migrate into the animal's cells. Light passes through the slug's skin and strikes the food-producing chloroplasts. Once its "solar panels" are in place, the animal may not eat again for months!

Perhaps the most famous animals to "farm" photosynthetic partners are corals. Inside the coral are single-celled protists called dinoflagellates, which use the sun's energy to feed the coral. In exchange, the animals provide a home for the protists. Sometimes, however, the partners break up. Corals under stress sometimes expel their dinoflagellates, or the protists may leave on their own. The reef then turns white. The coral animals eventually die, endangering the entire reef ecosystem. Pollution, disease, shading, excessively warm water, and ultraviolet radiation all trigger coral bleaching. Biologists predict that global climate change will only make this problem worse.

Corals and sea slugs are not the only animals whose lives depend on photosynthesis. Yours does, too, as you will learn in the next two chapters.

LEARNING OUTLINE

5.1 Life Depends on Photosynthesis

5.2 Sunlight Is the Energy Source for Photosynthesis

5.3 Photosynthesis Occurs in Two Stages

5.4 The Light Reactions Begin Photosynthesis

5.5 The Carbon Reactions Produce Carbohydrates

5.6 C₃, C₄, and CAM Plants Use Different Carbon Fixation Pathways

5.7 Investigating Life: Solar-Powered Salamanders

SURVEY THE LANDSCAPE
Science, Chemistry, and Cells

Photosynthetic cells use carbon dioxide, water, and sunlight to build carbohydrates. These molecules, which make up much of a plant's body, can also be used to store energy and generate ATP.

For more details, study the Pull It Together feature in the chapter summary.

PULL IT TOGETHER

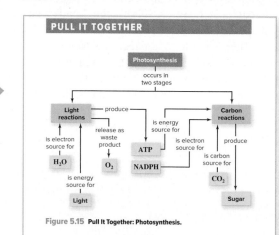

Figure 5.15 Pull It Together: Photosynthesis.

Refer to figure 5.15 and the chapter content to answer the following questions.

1. Review the Survey the Landscape figure in the chapter introduction, and then add *enzymes, cells, molecules,* and *respiration* to the Pull It Together concept map.

2. How would you incorporate the Calvin cycle, rubisco, C₃ plants, C₄ plants, and CAM plants into this concept map?

3. One possible connecting phrase in the concept map is "*Chlorophyll* reflects _____ wavelengths of *light*." Fill in the blank and explain your answer.

4. Build another small concept map showing the relationships among the terms *chloroplast, stroma, grana, thylakoid, photosystem,* and *chlorophyll.*

5. Besides respiration, what happens to the sugar produced in photosynthesis?

Learn How to Learn study tips help students develop their study skills.

Each chapter has one Learn How to Learn study tip, and a complete list is in Appendix F.

I present a *Study Minute* in class each week, with examples of how to use these study tips.

Investigating Life describes a real experiment focusing on an evolutionary topic related to each chapter's content.

Each case concludes with critical thinking questions that can be used as an in-class group activity. The studies touch on concepts found in other units; you can encourage students to draw a concept map illustrating the relationships between ideas. You might also use the case as a basis for discussion of the nature of science.

Connect interactive and test bank questions focus on the Investigating Life cases. Questions assess students' understanding of the science behind the Investigating Life case and their ability to integrate those concepts with information from other units.

LEARN HOW TO LEARN
See What's Coming

Start by reviewing the Survey the Landscape figure at the start of each chapter to see how the material fits with the rest of the unit. Then check out the Learning Outline. Each heading is a complete sentence that summarizes the most important idea of the section. Read through these statements before you start each chapter. You can also flip to the end of the chapter before you start to read; the chapter summary and Pull It Together concept map can provide a preview of what's to come.

INVESTIGATING LIFE

5.7 Solar-Powered Salamanders

This chapter's opening essay described two examples of solar-powered animals that live in the ocean: sea slugs and corals. But marine invertebrates are not the only animals with this unusual lifestyle. The eggs of spotted salamanders have live-in algae of their own.

Spotted salamanders are amphibians that live throughout the forests of North America. On rainy evenings in the spring, these animals mate in temporary ponds, where the females lay masses of fertilized eggs (figure 5.12). Each egg contains a tiny embryo and is surrounded by a thick jelly layer. Then, something unusual happens: Microscopic green algae somehow find each egg and enter the jelly layer. The algae reproduce and carry out photosynthesis in the protective confines of their new homes.

Biologists have known since 1986 that the green algae boost the O_2 concentration inside the salamander eggs, a real benefit to embryos that cannot yet breathe on their own. But do algae also feed the embryos a steady diet of sugars? Erin Graham, Robert Sanders, and two other researchers at Temple University wanted to learn more.

The team gathered algae-infected eggs from the wild and incubated them for nearly 2 hours in a solution containing CO_2 that was "tagged" with a radioactive isotope of carbon. They knew that only algae—not salamander eggs or embryos—would be able to use CO_2 directly in photosynthesis. After the incubation period, they rinsed off the excess solution and measured the amount of radioactive carbon in each egg and embryo. ① *radioactive isotopes, section 2.1C*

The researchers reasoned that any radioactive carbon in a salamander embryo could come from one of two sources. The carbon might simply diffuse in from the solution, without any help from the algae. Alternatively, the algae might use the tagged carbon to produce sugars in photosynthesis, then

Figure 5.12 Spawning. A spotted salamander lays eggs in a pool of water.
©George Grall/National Geographic Creative

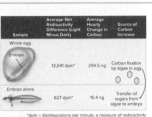

Sample	Average Net Radioactivity Difference (Light Minus Dark)	Average Hourly Change in Carbon	Source of Carbon Increase
Whole egg (Embryo)	12,041 dpm*	294.5 ng	Carbon fixation by algae in egg
Embryo alone	627 dpm*	15.4 ng	Transfer of sugars from algae to embryo

*dpm = disintegrations per minute, a measure of radioactivity

Figure 5.13 Thanks for the Snack. Using a radioactive isotope of carbon, researchers measured the amount of carbon transferred from egg-dwelling green algae to salamander embryos.

transfer some of the radioactive sugar to the embryos. ① *diffusion, section 4.5A*

To differentiate between these two possibilities, the team incubated some eggs in the light, allowing both diffusion and photosynthesis to occur. A second set of eggs was incubated in total darkness. Photosynthesis is not possible in the dark, but diffusion continues. Subtracting the amount of radioactive carbon in dark-treated embryos from the amount in light-treated embryos should therefore reveal the effect of photosynthesis.

After the experiment was complete, radioactivity measurements revealed that eggs and embryos incubated in the light incorporated more radioactive carbon than did their dark-treated counterparts, a sure sign that the algae were sharing their carbon with their tiny hosts (figure 5.13). This sugar supplement can help a developing embryo survive.

It is not surprising that animals would take on photosynthetic partners, but what's in it for the algae? They probably benefit from the partnership as well. A developing embryo releases CO_2 in respiration (see chapter 6). Perhaps this extra shot of CO_2 makes photosynthesis more efficient for the algae, completing the exchange of materials between two allies from different kingdoms of life.

Source: Graham, Erin R., Scott A. Fay, Adam Davey, and Robert W. Sanders. 2013. Intracapsular algae provide fixed carbon to developing embryos of the salamander *Ambystoma maculatum. Journal of Experimental Biology*, vol. 216, pages 452–459.

5.7 MASTERING CONCEPTS

1. On average, what percentage of the 294.5 ng of carbon that the algae fix each hour is transferred to the embryo? Refer to figure 5.13.

2. Identify a standardized variable, an independent variable, and a dependent variable in the experiment.

CHAPTER SUMMARY

5.1 Life Depends on Photosynthesis

- **Autotrophs** produce their own organic compounds from inorganic starting materials such as CO_2 and water. **Heterotrophs** rely on organic molecules produced by other organisms.

A. Photosynthesis Builds Carbohydrates Out of Carbon Dioxide and Water

- **Photosynthesis** converts kinetic energy in light to potential energy in the covalent bonds of carbohydrates, according to the following chemical equation:

$$6CO_2 + 6H_2O \xrightarrow{\text{light energy}} C_6H_{12}O_6 + 6O_2$$

B. Plants Use Carbohydrates in Many Ways

- Plants use glucose and other sugars to grow, generate **ATP,** nourish nonphotosynthetic plant parts, and produce cellulose and many other biochemicals. Most store excess carbohydrates as starch or sucrose.

C. The Evolution of Photosynthesis Changed Planet Earth

- Before photosynthesis evolved, organisms were heterotrophs. The first autotrophs made new food sources available.
- Over billions of years, oxygen produced in photosynthesis changed Earth's climate and the history of life.

5.2 Sunlight Is the Energy Source for Photosynthesis

A. What Is Light?

- Visible light is a small part of the **electromagnetic spectrum.**
- **Photons** move in waves. The shorter the **wavelength,** the more kinetic energy per photon. Visible light occurs in a spectrum of colors representing different wavelengths.

B. Photosynthetic Pigments Capture Light Energy

- **Chlorophyll *a*** is the primary photosynthetic pigment in plants. **Accessory pigments** absorb wavelengths of light that chlorophyll *a* cannot absorb.

C. Chloroplasts Are the Sites of Photosynthesis

- Plants exchange gases with the environment through pores called **stomata.**
- Leaf **mesophyll** cells contain abundant **chloroplasts.**
- A chloroplast contains a gelatinous fluid called the **stroma.** The fluid surrounds the **grana,** which are stacks of pancake-shaped **thylakoid** membranes. Photosynthetic pigments are embedded in the thylakoid membranes, which enclose the **thylakoid space.**
- A **photosystem** consists of proteins, antenna pigments, and a **reaction center.**

5.3 Photosynthesis Occurs in Two Stages

- The **light reactions** of photosynthesis produce ATP and **NADPH;** these molecules provide energy and electrons for the sugar-producing **carbon reactions** (figure 5.14).

5.4 The Light Reactions Begin Photosynthesis

A. Light Striking Photosystem II Provides the Energy to Produce ATP

- Photosystem II captures light energy and sends electrons from reactive chlorophyll *a* along the **electron transport chain.**
- Electrons from chlorophyll are replaced with electrons from water. O_2 is the waste product.
- The energy released in the electron transport chain drives the active transport of protons (H^+) into the thylakoid space. The protons diffuse out through channels in **ATP synthase.** This movement powers the phosphorylation of ADP to ATP.
- The coupling of the proton gradient and ATP formation is called **chemiosmotic phosphorylation.**

B. Electrons from Photosystem I Reduce NADP⁺ to NADPH

- Light striking photosystem I re-energizes the electrons, which pass to an enzyme that uses them to reduce NADP⁺. The product of this reaction is NADPH.

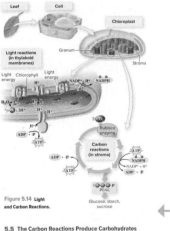

Figure 5.14 Light and Carbon Reactions.

Labels: Leaf, Cell, Chloroplast, Granum, Stroma, Light reactions (in thylakoid membranes), Light energy, Chlorophyll, Light energy, H_2O, O_2, $2H^+$, NADP⁺, H^+, NADPH, H^+, Rubisco enzyme, ADP + P, ATP, Carbon reactions (in stroma), ADP + P, ATP, ATP, NADP⁺ + H^+, ADP + P, PGAL, Glucose, starch, sucrose

5.5 The Carbon Reactions Produce Carbohydrates

- The carbon reactions use energy from ATP and electrons from NADPH in **carbon fixation** reactions that add CO_2 to organic compounds.
- In the **Calvin cycle, rubisco** catalyzes the reaction of CO_2 with **ribulose bisphosphate (RuBP)** to yield two molecules of PGA. These are converted to PGAL, the immediate product of photosynthesis. PGAL later becomes glucose and other carbohydrates.

5.6 C₃, C₄, and CAM Plants Use Different Carbon Fixation Pathways

- The Calvin cycle is also called the C_3 **pathway.** Most plant species are C_3 plants, which use only this pathway to fix carbon.
- **Photorespiration** wastes carbon and energy when rubisco reacts with O_2 instead of CO_2.
- The C_4 **pathway** reduces photorespiration by separating two carbon fixation reactions into different cells. In mesophyll cells, CO_2 is fixed as a four-carbon molecule, which moves to a **bundle-sheath cell** and liberates CO_2 to be fixed again in the Calvin cycle.
- In the **CAM pathway,** desert plants such as cacti open their stomata and take in CO_2 at night, storing the fixed carbon in vacuoles. During the day, they split off CO_2 and fix it in chloroplasts in the same cells.

5.7 Investigating Life: Solar-Powered Salamanders

- The eggs of spotted salamanders contain cells of green algae, which provide O_2 and carbon to the animal's embryo.

The Chapter Summary highlights key points and terminology from the chapter.

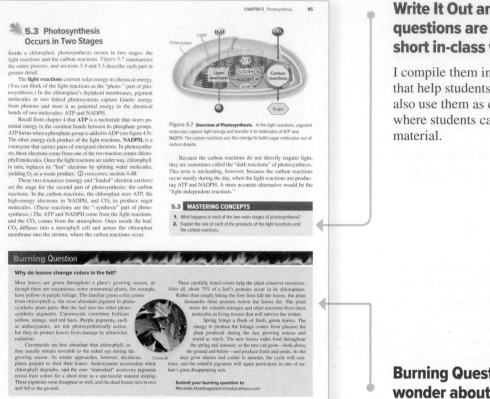

Write It Out and Mastering Concepts questions are useful for student review or as short in-class writing assignments.

I compile them into a list of *Guided Reading Questions* that help students focus on material I cover in class. I also use them as discussion questions in Action Centers, where students can come for additional help with course material.

Burning Questions cover topics that students wonder about.

I ask my students to write down a Burning Question on the first day of class. I answer all of them during the semester, whenever a relevant topic comes up in class.

Figure It Out questions reinforce chapter concepts and typically have numeric answers (supporting student math skills).

Students can work on these in small groups, in class, or in Action Center. Most could easily be used as clicker questions as well.

Figure It Out

If you could expose plants to just one wavelength of light at a time, would a wavelength of 300 nm, 450 nm, or 600 nm produce the highest photosynthetic rate?

Answer: 450 nm

Animated Tutorials *Explain Complicated Topics*

Animated tutorials guide students through complicated topics, using illustrations and examples from the book.

We created these tutorials to walk students through the most difficult material, step by step. Each tutorial places the topic in context, explains one or more concepts and related figures taken directly from the Hoefnagels text, and returns to the big picture at the end. You can assign the tutorials with accompanying critical thinking questions from the interactive question banks, or you can use the tutorials embedded in PowerPoint® slides in your presentations.

Your students can review the tutorials through SmartBook. Topics are listed below.

Organization of Life
Scientific Method and Interpreting a Graph
Chemical Bonding
Dehydration Synthesis and Hydrolysis
Protein Structure
Anatomy of a Cell Membrane
ATP
Enzymes
Reaction Energetics
Osmosis
Cell Structure
Overview of Photosynthesis
Light Reactions
The Calvin Cycle
Overview of Respiration
Mitochondrial Electron Transport Chain
Fermentation
Protein Synthesis
Overview of DNA Replication
Stages of Mitosis
Stages of Meiosis
Comparison of Mitosis and Meiosis
Crossing Over
Nondisjunction
Homologous Chromosomes
Constructing and Interpreting a Punnett Square
DNA Profiling
Mechanisms of Evolution

Genetic Variation: The Basis of Natural Selection
Understanding the Hardy–Weinberg Equation
Evidence for Evolution
Evidence for Human Evolution
Radiometric Dating
Reading an Evolutionary Tree
Origin of Life
Endosymbiont Theory
Viral Replication
Lytic and Lysogenic Cycles
Replication of HIV
Prokaryote Diversity
Protist Diversity
Plant Diversity
Moss Reproductive Cycle
Fern Reproductive Cycle
Conifer Reproductive Cycle
Sexual Reproduction in Angiosperms
Basidiomycete Reproductive Cycle
Diversity of Fungi
Animal Diversity
Overview of Plant Tissues
Phloem Sap Transport
Water Movement Through the Xylem
Alternation of Generations
Fruit Development
Auxin and Phototropism
Overview of Animal Tissues

Organ System Interactions
Example of Negative Feedback
Action Potential
The Synapse
Overview of the Senses
Sense of Vision
Sense of Hearing
Cell Responses to Hormones
Role of ATP in Muscle Contraction
The Heartbeat
Respiratory Surfaces
Digestion and Food Molecules
Nephron Function
Adaptive Immunity
Allergies
Oogenesis
Human Male Reproductive System
Human Female Reproductive System
Ovarian and Menstrual Cycles
Proximate and Ultimate Behaviors
Population Growth Models
Biomagnification
Water Cycle
Nitrogen Cycle
Phosphorus Cycle
Carbon Cycle
Earth's Climate and Biomes
CO_2 and Earth's Average Temperature
Threats to Biodiversity

Mc Graw Hill Education **connect®**
Required=Results

©Getty Images/iStockphoto

McGraw-Hill Connect®
Learn Without Limits

Connect is a teaching and learning platform that is proven to deliver better results for students and instructors.

Connect empowers students by continually adapting to deliver precisely what they need, when they need it, and how they need it, so your class time is more engaging and effective.

> 73% of instructors who use **Connect** require it; instructor satisfaction **increases** by 28% when **Connect** is required.

Analytics ———

Connect Insight®

Connect Insight is Connect's new one-of-a-kind visual analytics dashboard—now available for both instructors and students—that provides at-a-glance information regarding student performance, which is immediately actionable. By presenting assignment, assessment, and topical performance results together with a time metric that is easily visible for aggregate or individual results, Connect Insight gives the user the ability to take a just-in-time approach to teaching and learning, which was never before available. Connect Insight presents data that empowers students and helps instructors improve class performance in a way that is efficient and effective.

Mobile ———

Connect's new, intuitive mobile interface gives students and instructors flexible and convenient, anytime–anywhere access to all components of the Connect platform.

Connect's Impact on Retention Rates, Pass Rates, and Average Exam Scores

Graph showing values:
Retention Rates: 70.1% (without Connect), 89.9% (with Connect)
Course Pass Rates: 72.5% (without Connect), 85.2% (with Connect)
Average Exam Scores: 71.0% (without Connect), 80.1% (with Connect)

without Connect | with Connect

> Using **Connect** improves retention rates by **19.8%**, passing rates by **12.7%**, and exam scores by **9.1%**.

Impact on Final Course Grade Distribution

	without Connect	with Connect
A	22.9%	31.0%
B	27.4%	34.3%
C	22.9%	18.7%
D	11.5%	6.1%
F	15.4%	9.9%

> Students can view their results for any **Connect** course.

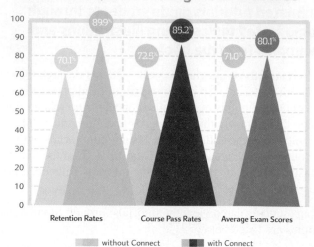

Adaptive

THE **ADAPTIVE** **READING EXPERIENCE** DESIGNED TO TRANSFORM THE WAY STUDENTS READ

More students earn **A's** and **B's** when they use McGraw-Hill Education **Adaptive** products.

SmartBook®

Proven to help students improve grades and study more efficiently, SmartBook contains the same content within the print book, but actively tailors that content to the needs of the individual. SmartBook's adaptive technology provides precise, personalized instruction on what the student should do next, guiding the student to master and remember key concepts, targeting gaps in knowledge and offering customized feedback, and driving the student toward comprehension and retention of the subject matter. Available on tablets, SmartBook puts learning at the student's fingertips—anywhere, anytime.

Over **8 billion questions** have been answered, making McGraw-Hill Education products more intelligent, reliable, and precise.

www.mheducation.com

STUDENTS WANT **SMARTBOOK®**

95% of students reported **SmartBook** to be a more effective way of reading material.

100% of students want to use the Practice Quiz feature available within **SmartBook** to help them study.

100% of students reported having reliable access to off-campus wifi.

90% of students say they would purchase **SmartBook** over print alone.

95% of students reported that **SmartBook** would impact their study skills in a positive way.

Acknowledgments

It takes an army of people to make a textbook, and while I don't work with everyone directly, I greatly appreciate the contributions of each person who makes it possible.

Matt Taylor has worked alongside me at every stage, from first draft to finished product; in addition, he has seamlessly integrated the book's approach into our digital assets. His careful work and insights are invaluable.

I appreciate the help of my colleagues at The University of Oklahoma, including Doug Gaffin, Ben Holt, Michael Markham, and J. P. Masly. Helpful colleagues from other institutions include Marjorie Weber.

I am grateful to former OU student Emily North, who spent many hours scrutinizing art and page layouts. In addition, conversations with students in my classes spark many good ideas.

My team at McGraw-Hill is wonderful. Thank you to Managing Director Lynn Breithaupt and Executive Brand Manager Michelle Vogler, who help us create the best book and digital resources possible. Product Developer Anne Winch retains her amazing ability to juggle an ever-increasing slate of tasks, all while remaining both responsive and funny. Chris Loewenberg, Britney Ross, and Jenna Paleski contribute energy and great ideas to the marketing side. Emily Tietz continues to provide outstanding service in photo selections. I also appreciate Angie Fitzpatrick, April Southwood, and Peggy Selle for their impressive skills at the interface between us and the production team. Also among the talented folks at McGraw-Hill are Emily Windelborn, Tara McDermott, Lori Hancock, Lorraine Buczek, and Jane Peden, who has made life easier in countless ways. Thanks to all of you for all you do.

My family and friends continue to encourage me. Thank you to my parents and sister for their pride and support. I also thank my friends Kelly Damphousse, Ben and Angie Holt, Michael Markham and Kristi Isacksen, Karen and Bruce Renfroe, Ingo and Andrea Schlupp, Clarke and Robin Stroud, Matt Taylor and Elise Knowlton, and Mark Walvoord. Cats Sidecar, Smudge, and Snorkels were worthy companions in my office as well. Finally, my husband, Doug Gaffin, is always there for me, helping in countless large and small ways. I could not do this work without him.

Changes by Chapter

UNIT 1 Science, Chemistry, and Cells

- **Chapter 1 (The Scientific Study of Life):** Updated figure 1.3 to use memories as examples of emergent properties; improved figure 1.12 to more clearly define independent and dependent variables; within table 1.2, clarified definitions of *independent variable, dependent variable,* and *standardized variable.* Added several learning tools to SmartBook®: a figure depicting a simple evolutionary tree in section 1.2; a figure showing the taxonomic hierarchy in section 1.2; a miniglossary of scientific knowledge in section 1.3; a figure describing how to interpret graphs in the chapter summary.
- **Chapter 2 (The Chemistry of Life):** In table 2.1, clarified that the mass of an electron is not exactly zero; added figure 2.3, which compares a neutral hydrogen atom with H^+; improved description of electron orbitals and energy shells; improved explanation of polar covalent bonds in water molecules in narrative and in figure 2.10; clarified definition of *hydrophobic*; updated illustration of cellulose in figure 2.18 and illustration

of triglycerides in figure 2.25; added figure 2.26, which shows the difference between solid and liquid lipids; wrote new Investigating Life about chemical defenses in tawny crazy ants.
- **Chapter 3 (Cells):** Based on SmartBook user data, clarified that high surface area to volume improves cell transport efficiency; improved Apply It Now box to teach students that most of their cells are not their own; replaced Burning Question with a new box on artificial cells; clarified distinction between *cytoplasm* and *cytosol* in narrative and several figures; in figure 3.10, improved accuracy of double bond in unsaturated fatty acid; adjusted figures 3.15 and 3.18 for accuracy; clarified the function of plasmodesmata. Added a new learning tool to SmartBook: a table comparing mitochondria and chloroplasts in section 3.4.
- **Chapter 4 (The Energy of Life):** In chapter opening essay, added how human evolutionary history affects our food cravings; improved illustration of potential and kinetic energy in figure 4.1; based on SmartBook user data, clarified definition of heat; updated figure 4.10 to clearly show uses of ATP; improved explanations of

negative and positive feedback; in figure 4.18, corrected position of cell membranes in plant cells in hypertonic surroundings; wrote new Apply It Now box on boosting metabolism; added aquaporins to Burning Question on headaches, to improve the connection to the chapter; wrote new Investigating Life on membrane proteins in electric fish; improved clarity of summary figure 4.24. Added a new learning tool to SmartBook: in the chapter summary, a miniglossary defining metabolism terms.

- **Chapter 5 (Photosynthesis):** Based on SmartBook user data, changed "oxidized" to "consumed" in section 5.1C to emphasize that the first organisms obtained C from their surroundings; in narrative and figure 5.8, clarified that the photosynthetic electron transport chain includes the entire pathway from photosystem II to the formation of NADPH; improved figure 5.9 to show the fates of the carbohydrates produced in the carbon reactions; clarified that rubisco is not involved in the C_4 pathway; based on SmartBook user data, explained why C_4 plants have greater water use efficiency than C_3 plants; clarified the distinction between C_4 and CAM plants; wrote new Investigating Life on carbon translocation from algae to salamander embryos. Added new learning tools to SmartBook: a miniglossary of leaf anatomy in section 5.2; a miniglossary of light reactions in section 5.4.

- **Chapter 6 (Respiration and Fermentation):** Changed chapter title to "Respiration and Fermentation" to better match the title of chapter 5; improved consistency of use of the terms *hydrogen ion gradient, H+ gradient,* and *proton gradient*; updated Burning Question on diet pills. Made several changes based on SmartBook user data: clarified the paragraph debunking the myth that lactic acid causes muscle soreness after intense exercise; modified figure 6.9 to show how glycerol and some amino acids enter metabolism and to show that nitrogen is stripped from amino acids and eliminated as waste; improved explanation of why aerobic respiration must have evolved after O_2-generating photosynthesis. Added several learning tools to SmartBook: a table in section 6.3 showing where the reactions of respiration occur; a miniglossary of mitochondrion anatomy in section 6.3; a miniglossary of aerobic respiration in section 6.5.

UNIT 2   DNA, Inheritance, and Biotechnology

- **Chapter 7 (DNA Structure and Gene Function):** Clarified in section 7.3 that DNA must be "unpacked" for the cell to use its genetic information; based on SmartBook user data, explained that terminator sequence is part of DNA, not RNA; improved definition of promoter in narrative and position of promoter in figure 7.10; extended "cookbook" analogy to the participants in translation; improved description of the *lac* operon; added information about epigenetics; made an explicit connection between transcription factors and signal transduction; clarified bold-faced terms related to mutations; reworked figure 7.24 (Investigating Life) to show an evolutionary tree of FOXP2 protein changes; added summary figure 7.26 to show three types of RNA. Added a new learning tool to SmartBook: a miniglossary of protein synthesis in the chapter summary.

- **Chapter 8 (DNA Replication, Binary Fission, and Mitosis):** Made many small clarifications to the narrative describing chromosome structure and the events of cell division; improved figure 8.8 to include a yarn analogy of DNA compaction; based on SmartBook user data, clarified that a compacted chromosome is unavailable for transcription and improved the definition of semiconservative replication; added information about epigenetics in relation to cancer; improved the list of ways that cancer cells differ from normal cells; updated figure showing ways to reduce cancer risk. Added new learning tools to SmartBook: miniglossary of cell division in section 8.1; table comparing binary fission and mitotic cell division in chapter summary.

- **Chapter 9 (Sexual Reproduction and Meiosis):** In chapter opening essay, added possible implications of fetal screening on human evolution; increased relevance of box on mules by explaining why mules are desirable; clarified the use of the word *align* in talking about the events of meiosis and the origin of genetic variation; revised box on multiple births to focus on their rising incidence; improved connection between problems during meiosis and abnormalities in chromosome number and structure; added *aneuploid cell* as a contrast to *polyploid cell.* Added several learning tools to SmartBook: a miniglossary of variability in meiosis in section 9.5; a table comparing asexual and sexual reproduction in section 9.6; a miniglossary of chromosome abnormalities in section 9.7.

- **Chapter 10 (Patterns of Inheritance):** Reworked allele designations for all figures and narrative relevant to yellow and green peas; improved connection between proteins and traits; replaced figure 10.14 to show how genotypic ratios differ in crosses between linked and unlinked genes; based on SmartBook user data, clarified explanation of product rule; related epigenetics to environmental effects on gene expression; based on SmartBook user data, explained why males cannot be symptomless carriers of X-linked traits; revised Investigating Life to incorporate information about next-generation Bt cotton; improved lightbulb analogy of dominance relationships in figure 10.34. Added several new learning tools to SmartBook: a miniglossary of tracking inheritance in section 10.3; a miniglossary of gene linkage in section 10.5; a miniglossary of dominance relationships in section 10.6; a miniglossary of modes of inheritance in section 10.8.

- **Chapter 11 (DNA Technology):** Added CRISPR as an example of a new DNA technology to chapter opening essay; based on SmartBook user data, clarified that scientists use DNA to reveal species relationships; updated figure 11.5 to show modern DNA sequencing; added figure 11.6, which shows similarity between a gene of humans and a homologous gene in other species; improved application of DNA profiling in figure 11.9; revised description of somatic cell nuclear transfer; based on SmartBook user data, clarified that pseudogenes are noncoding DNA and that gene therapy provides supplemental DNA (not replacement DNA); wrote new Investigative Life on gene transfer between GMOs and their wild relatives. Added a new learning tool to SmartBook: a table listing some additional uses of DNA analysis in section 11.2.

UNIT 3 The Evolution of Life

- **Chapter 12 (The Forces of Evolutionary Change):** Updated chapter opening essay to mention CDIFF and CRE, two of the three most dangerous antibiotic-resistant bacteria; improved figure 12.8 to connect natural selection with mutations in DNA; reworked Burning Question to explain why there is no such thing as a "pinnacle of evolution"; improved narrative and figure explaining Hardy–Weinberg equilibrium; clarified descriptions of genetic drift and nonrandom mating; based on SmartBook user data, clarified the effect of gene flow on genetic diversity; reworked Pull It Together (figure 12.24) to explain how each mechanism of evolution affects allele frequencies. Added new learning tools to SmartBook: a miniglossary of populations and evolution in section 12.1; a miniglossary of evolutionary mechanisms in section 12.7.
- **Chapter 13 (Evidence of Evolution):** In figure 13.2, replaced *Tertiary Period* with *Paleogene* and *Neogene*; added mole eyes as an example of a vestigial structure in figure 13.11; expanded the list of vestigial structures in the narrative; based on Smart-Book user data, improved figure 13.17 to better show how mutations in enhancers affect gene expression; in Investigating Life, clarified evidence that *Najash* was terrestrial. Added several learning tools to SmartBook: a miniglossary of fossil aging terms in section 13.2; a miniglossary of comparative anatomy terms in section 13.4; a figure showing all five lines of evidence for evolution in the chapter summary.
- **Chapter 14 (Speciation and Extinction):** Reorganized section on gradualism and punctuated equilibrium for clarity; added figure 14.17, which distinguishes ancestral and derived features; improved figure 14.18 to more clearly explain the anatomy of a phylogenetic tree; revised Burning Question to explain how each condition boosts the evolution rate; wrote new Investigating Life on ecological interactions that boost speciation rates. Added several learning tools to SmartBook: a miniglossary of macroevolution in section 14.1; a miniglossary of reproductive barriers in section 14.2; a miniglossary of speciation patterns in section 14.3.
- **Chapter 15 (The Origin and History of Life):** Improved illustrations of primary and secondary endosymbiosis in figure 15.9; rearranged section on human evolution for clarity; based on SmartBook user data, clarified distinction between "early" and "recent" *Homo;* replaced Investigating Life with an essay about the human and chimpanzee genome sequencing projects. Added a new learning tool to SmartBook: a table summarizing biodiversity changes over time in section 15.3.

UNIT 4 The Diversity of Life

- **Chapter 16 (Viruses):** Revised chapter opening essay to include the most recent Ebola outbreak; reworked several headings to improve clarity; reorganized paragraphs on viral envelope for clarity; rewrote the passage on latent animal viruses; based on SmartBook user data, explained which cells are infected by herpes simplex virus type I (cold sores); improved figure 16.5, which shows HIV replication; updated Investigating Life to include newer data. Added a new learning tool to SmartBook: a miniglossary of viral infections in section 16.3.
- **Chapter 17 (Bacteria and Archaea):** Based on SmartBook user data, improved figure 17.5 and clarified description of Gram-positive and Gram-negative cells; differentiated between exotoxins and endotoxins; wrote new Investigating Life on antibiotic-resistant bacteria in pig farms. Added several learning tools to section 17.2 of SmartBook: a miniglossary of prokaryote anatomy; a miniglossary of prokaryote classification; a miniglossary of gene transfer.
- **Chapter 18 (Protists):** Based on SmartBook user data, clarified differences between feeding and reproductive stages for plasmodial slime molds; corrected figure showing overall tree of life. Added new learning tools to SmartBook: a miniglossary of types of algae in section 18.2; a table summarizing the life cycles of plasmodial and cellular slime molds in section 18.3.
- **Chapter 19 (Plants):** Improved organization of introduction to section 19.1; clarified description of alternation of generations; reworked description of double fertilization; improved description of ovules; added paragraph about gluten sensitivity. Added a new learning tool to SmartBook: a miniglossary of plant reproduction in section 19.1.
- **Chapter 20 (Fungi):** Based on SmartBook user data, clarified description of figure 20.2; revised Apply It Now: Fungi and Human Health to focus on infection prevention; improved distinction between endophytes and mycorrhizal fungi. Added new learning tools to SmartBook: a miniglossary of fungal anatomy in section 20.1; a miniglossary of fungal interactions in section 20.7.
- **Chapter 21 (Animals):** Wrote new chapter opening essay to emphasize the uses of animal products in everyday objects; reorganized section 21.1 for clarity; improved definitions of *ectoderm* and *endoderm*; based on SmartBook user data, clarified that both indirect and direct development may occur in mollusks; also based on SmartBook user data, rearranged section on echinoderm defenses to put related content together; clarified description of amniotic eggs and the term *amniote*; reworked and simplified the Investigating Life section. Added new learning tools to SmartBook: a miniglossary of animal clades in section 21.1; a miniglossary of arthropod diversity in section 21.8.

UNIT 5 Plant Life

- **Chapter 22 (Plant Form and Function):** Added art of ground tissue cell types in figure 22.4; clarified the description of cells in phloem tissue; defined primary and secondary growth earlier in the chapter; improved illustrations of stem and root cross sections in figures 22.9 and 22.13; based on SmartBook user data, clarified distinction between monocots and eudicots in leaf cross sections; wrote new Apply It Now box on the topic of fire-resistant trees and shrubs; based on SmartBook user data, reworked figure 22.17 to better illustrate the definition of *bark*. Added a new learning tool to SmartBook: a miniglossary of plant anatomy in section 22.3.

- **Chapter 23 (Plant Nutrition and Transport):** Revised Apply It Now box on fertilizers; reworked some titles so the words *xylem* and *phloem* are more prominent in the chapter's main headers; clarified explanation of *sink* in description of pressure flow theory. Added a new learning tool to SmartBook: a miniglossary of plant transport in the chapter summary.

- **Chapter 24 (Reproduction and Development of Flowering Plants):** Clarified relationship between *carpel* and *ovary*; improved illustration of mature monocot and eudicot seeds in figure 24.9; based on SmartBook user data, adjusted labeling on figure depicting corn and bean seed germination; wrote new Apply It Now box on plants that attack caterpillars; improved figures 24.22 and 24.23, which illustrate photoperiod's role in flowering. Added new learning tools to SmartBook: a miniglossary of angiosperm reproduction in section 24.2; a miniglossary of plant tropisms in the chapter summary.

UNIT 6 Animal Life

- **Chapter 25 (Animal Tissues and Organ Systems):** Wrote new chapter opener on physiological changes that happen as a person runs a marathon; revised box on organ donation to focus on artificial organs; revised many glossary terms for consistency; developed new figure 25.14, which summarizes organ system interactions. Added a new learning tool to SmartBook: a miniglossary of animal anatomy and physiology in section 25.1.

- **Chapter 26 (The Nervous System):** Added narrative and glossary definitions for *membrane potential*; clarified the significance of the "all-or-none" nature of an action potential; reworked the explanation of graded potentials and action potentials; distinguished between *action potential* and *neural impulse*; added figure 26.7, which illustrates how a neural impulse is similar to a line of firecrackers exploding; improved explanation of how the sympathetic nervous system can have both instantaneous effects and prolonged effects (via adrenal hormones); wrote new Burning Question on whether neurons communicate at the speed of light; replaced Investigating Life with an essay on a grasshopper mouse's reaction to a scorpion sting. Based on SmartBook user data, added new learning tools to SmartBook: a table of action potential events in section 26.3; a miniglossary of nervous system communication in section 26.3.

- **Chapter 27 (The Senses):** Wrote new Investigating Life on taste detection in whales. Based on SmartBook user data, added new learning tools to SmartBook: a miniglossary of the visual information pathway in section 27.4; a miniglossary of the auditory information pathway in section 27.5.

- **Chapter 28 (The Endocrine System):** Based on SmartBook user data, clarified in the introduction to section 28.2 that hormone receptors may be on the cell surface, in the cytoplasm, or in the nucleus; improved the explanation of the overall role of the hypothalamus and pituitary; reworked section on adrenal hormones and their regulation. Added new learning tools to SmartBook: a miniglossary of hormones and responses in section 28.2; a chapter summary table comparing the origins and functions of many hormones.

- **Chapter 29 (The Skeletal and Muscular Systems):** Added micrograph to illustration of skeletal muscle organization; improved figure 29.23, which summarizes the relationship between muscles and bones. Added several new learning tools to SmartBook: a miniglossary of the skeletal system in section 29.3; a table summarizing the steps of muscle contraction in section 29.4; a miniglossary of the muscular system in the chapter summary.

- **Chapter 30 (The Circulatory System):** Reworked introduction to section 30.1 for clarity; improved passage describing red blood cells; clarified description of blood clotting; added information on the possible effects of overexercising. Added new learning tools to SmartBook: a miniglossary of the heartbeat in section 30.4; a miniglossary of blood vessels in section 30.5.

- **Chapter 31 (The Respiratory System):** Wrote new chapter opener on competitive breath-holding; based on SmartBook user data, clarified the features that all respiratory surfaces have in common; revised the description of red blood cells' role in carrying O_2 and CO_2; improved section on the functions of CO_2 and blood pH in regulating the breathing rate; updated figure 31.17 to compare and contrast external and internal respiration. Added a new learning tool to SmartBook: a miniglossary of breathing in the chapter summary.

- **Chapter 32 (Digestion and Nutrition):** Reworked figure 32.9 to emphasize the types of teeth; added information on essential amino acids and essential fatty acids; wrote new Burning Question on maximizing the nutrient content of food; added new table on the calorie content of various beverages; replaced Investigating Life essay with one about the evolutionary cost of a sweet tooth; reworked Pull It Together (figure 32.22) to better cover the chapter's content. Added a new learning tool to SmartBook: a miniglossary of digestive fluids in section 32.3.

- **Chapter 33 (Regulation of Temperature and Body Fluids):** Clarified the process of urination in the human urinary system. Added a new learning tool to SmartBook: a miniglossary of temperature homeostasis in section 33.1.

- **Chapter 34 (The Immune System):** Reworked figure 34.4, which provides an overview of innate defenses; clarified the roles of white blood cells in innate defenses; improved the passage about cytotoxic T cells; made many small improvements to the narrative about adaptive immunity; based on SmartBook user data, clarified the role of MHC proteins; updated information about SCID. Added a new learning tool to SmartBook: a miniglossary of adaptive immunity in section 34.3.

- **Chapter 35 (Animal Reproduction and Development):** Wrote new chapter opener about intersex conditions; combined contraception and sexually transmitted diseases in a new section on reproductive health; clarified the events of fertilization and prenatal development; improved explanation of the placenta's structure and function; wrote a new box on male pregnancy; made a new summary figure showing the stages of human development. Added a new learning tool to SmartBook: a miniglossary on embryonic support structures in section 35.5.

UNIT 7 The Ecology of Life

- **Chapter 36 (Animal Behavior):** Based on SmartBook user data, improved figure 36.1 to differentiate between proximate and ultimate causes of behavior; clarified the definition of *search image*; combined multiple figures to make figure 36.11, which shows many types of defenses against predation. Added a new learning tool to SmartBook: a miniglossary of innate and learned behaviors in section 36.2.
- **Chapter 37 (Populations):** Revised population numbers and graphs in figures 37.6 and 37.8 for accuracy and clarity; updated information about China's former one-child policy. Added a new learning tool to SmartBook: a miniglossary of population growth in section 37.3.
- **Chapter 38 (Communities and Ecosystems):** Wrote new chapter opening essay about sustainable meat production and home gardening; improved explanation of coevolution in crossbills; updated information about mercury in tuna in figure 38.17. Added new learning tool to SmartBook: a miniglossary of diversity and succession in section 38.2.

- **Chapter 39 (Biomes):** Based on SmartBook user data, revised figure 39.4 to show a convection cell from the perspective of Earth's surface; improved explanation and illustration of El Niño in the Apply It Now box; updated Investigating Life text and figure 39.26 for clarity. Added new learning tools to SmartBook: a miniglossary of lake zones in section 39.4; a miniglossary of ocean zones in section 39.5.
- **Chapter 40 (Preserving Biodiversity):** Updated data on endangered species; added inset map to figure 40.8 showing the location of the Great Pacific Garbage Patch; revised section on global climate change to include information from the recent Paris conference; mentioned the sources of methane and N_2O in the atmosphere; corrected the amount of CO_2 released by human activities; updated photo and information about the ozone hole in figure 40.12; improved figure 40.15 to show how the Arctic sea ice minimum has changed over time; updated Pull It Together (figure 40.27) to better illustrate connections between the topics. Added a new learning tool to SmartBook: a miniglossary of pollution in section 40.3.

Contents

UNIT 2 DNA, Inheritance, and Biotechnology

7 | DNA Structure and Gene Function 120

©Dr. Gopal Murti/Science Source

8 | DNA Replication, Binary Fission, and Mitosis 146

9 | Sexual Reproduction and Meiosis 166

10 | Patterns of Inheritance 186

11 | DNA Technology 216

UNIT 3   The Evolution of Life

12 | The Forces of Evolutionary Change 236

©Chris Ryan/Getty Images RF

13 | Evidence of Evolution 260

UNIT 4   The Diversity of Life

16 | Viruses 330

©Yuriy Dyachyshyn/AFP/Getty Images

35 | Animal Reproduction and Development 700

UNIT 7 The Ecology of Life

36 | Animal Behavior 728

©Gerald Hinde/Gallo Images/ Getty Images RF

37 | Populations 748

BIOLOGY

The Scientific Study of Life

Undersea World. A coral reef in the Red Sea is home to countless marine species. The prickly animal in the center is a "crown of thorns" sea star.

©Franco Banfi/WaterF/age fotostock

LEARN HOW TO LEARN
Real Learning Takes Time

You got good at basketball, running, dancing, art, music, or video games by putting in lots of practice. Likewise, you will need to commit time to your biology course if you hope to do well. To get started, look for the "Learn How to Learn" tip in each chapter of this textbook. Each hint is designed to help you use your study time productively.

With practice, you'll discover that all concepts in biology are connected. The *Survey the Landscape* figure in every chapter highlights each chapter's place in the "landscape" of the entire unit. Use it, along with the more detailed *Pull It Together* concept map in the chapter summary, to see how each chapter's content fits into the unit's big picture.

Life Is Everywhere

Welcome to biology, the scientific study of life. Living organisms surround us. You are alive, and so are your friends, your pets, and the plants in your home and yard. Bacteria thrive on and in your body. Any food you ate today was (until recently, anyway) alive. And the news is full of biology-related discoveries about fossils, new cancer treatments, genetics, global climate change, and the environment.

Stories such as these enjoy frequent media coverage because this is an exciting time to study biology. Not only is the field changing rapidly, but its new discoveries and applications might change your life. DNA technology has brought us genetically engineered bacteria that can manufacture life-saving drugs—and genetically engineered plants that produce their own pesticides. This same technology may one day enable physicians to routinely cure hemophilia, cystic fibrosis, and other genetic diseases by replacing faulty DNA with a functional "patch."

Biology also includes the study of nonhuman life. We exist only because of our interactions with other species, which provide food, oxygen, clean water, clothing, shelter, and other necessities. Even species that do not directly "serve" us are essential to the ecosystems that sustain all life. Human activities, however, are pushing many ecosystems dangerously out of balance.

Consider the "crown of thorns" sea star shown here. These animals are notorious for their arsenal of sharp, venomous spines, which may cause painful wounds. At low population densities, their coral-eating habits help maintain reef biodiversity. Sometimes, however, huge numbers of sea stars destroy entire patches of coral. What causes these infestations? Many researchers point to nutrient-polluted runoff from nearby farms and cities. The nutrient influx triggers a population explosion of algae, which sea star larvae eat as they develop into adults. Removing the adults from an infested reef is dangerous and labor-intensive, but help is coming from an unusual source: Underwater robots have been programmed to seek out the crown of thorns and deliver a lethal injection.

The list of biology-related topics goes on and on: global climate change, stem cell therapies, new cancer treatments, infectious disease, improved crop plants, synthetic life, infertility treatment, endangered species, DNA fingerprinting, biofuels, pollution, the history of life, and more. This book will bring you a taste of what we know about life and help you make sense of the science-related news you see every day. Chapter 1 begins your journey by introducing the scope of biology and explaining how science teaches us what we know about life.

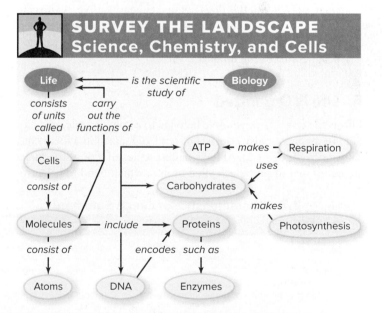

SURVEY THE LANDSCAPE
Science, Chemistry, and Cells

Organisms from all three branches of life share a unique combination of characteristics. Biologists are scientists who use evidence to test hypotheses about life.

For more details, study the **Pull It Together** feature in the chapter summary.

1.1 What Is Life?

Biology is the scientific study of life. The second half of this chapter explores the meaning of the term *scientific,* but first we will consider the question "What is life?" We all have an intuitive sense of what life is. If we see a rabbit on a rock, we know that the rabbit is alive and the rock is not. But it is difficult to state just what makes the rabbit alive. Likewise, in the instant after an individual dies, we may wonder what invisible essence has transformed the living into the dead.

One way to define life is to list its basic components. The **cell** is the basic unit of life; every **organism,** or living individual, consists of one or more cells. Every cell has an outer membrane that separates it from its surroundings. This membrane encloses the water and other chemicals that carry out the cell's functions. One of those biochemicals, deoxyribonucleic acid (DNA), is the informational molecule of life (**figure 1.1**). Cells use genetic instructions—as encoded in DNA—to produce proteins, which enable cells to carry out their functions in tissues, organs, and organ systems.

A list of life's biochemicals, however, provides an unsatisfying definition of life. After all, placing DNA, water, proteins, and a membrane in a test tube does not create life. And a crushed insect still contains all of the biochemicals that it had immediately before it died.

In the absence of a concise definition, scientists have settled on five qualities that, in combination, constitute life. **Table 1.1** summarizes them, and the rest of section 1.1 describes each one in more detail. An organism is a collection of structures that function together and exhibit all of these qualities. Note, however, that each trait in table 1.1 may also occur in nonliving objects. A rock crystal is highly organized, but it is not alive. A fork placed in a pot of boiling water absorbs heat energy and passes it to the hand that grabs it, but this does not make the fork alive. A fire can "reproduce" and grow very rapidly, but it lacks most of the other characteristics of life. It is the *combination* of these five characteristics that makes life unique.

A. Life Is Organized

Just as the city where you live belongs to a county, state, and nation, living matter also consists of parts organized in a hierarchical pattern (**figure 1.2**). At the smallest scale, all living structures are composed of particles called **atoms,** which bond together to

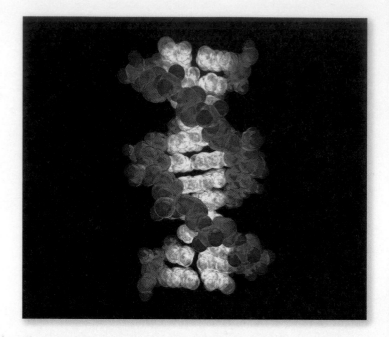

Figure 1.1 Informational Molecule of Life. All cells contain DNA, a series of "recipes" for proteins that each cell can make.
©Scott Camazine/123RF

form **molecules.** These molecules can form **organelles,** which are compartments that carry out specialized functions in cells (note that not all cells contain organelles). Many organisms consist of single cells. In multicellular organisms such as the tree illustrated in figure 1.2, however, the cells are organized into specialized **tissues** that make up **organs.** Multiple organs are linked into an individual's **organ systems.**

We have now reached the level of the organism, which may consist of just one cell or of many cells organized into tissues, organs, and organ systems. Organization in the living world extends beyond the level of the individual organism as well. A **population** includes members of the same species occupying the same place at the same time. A **community** includes the populations of different species in a region, and an **ecosystem** includes both the living and nonliving components of an area. Finally, the **biosphere** consists of all parts of the planet that can support life.

Biological organization is apparent in all life. Humans, eels, and evergreens, although outwardly very different, are all organized into specialized cells, tissues, organs, and organ systems.

TABLE **1.1**   Characteristics of Life: A Summary

Characteristic	Example
Organization	Atoms make up molecules, which make up cells, which make up tissues, and so on.
Energy use	A kitten uses the energy from its mother's milk to fuel its own growth.
Maintenance of internal constancy (homeostasis)	Your kidneys regulate your body's water balance by adjusting the concentration of your urine.
Reproduction, growth, and development	An acorn germinates, develops into an oak seedling, and, at maturity, reproduces sexually to produce its own acorns.
Evolution	Increasing numbers of bacteria survive treatment with antibiotic drugs.

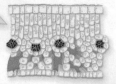

ORGANELLE
A membrane-bounded structure that has a specific function within a cell.
Example: Chloroplast

CELL
The fundamental unit of life. Multicellular organisms consist of many cells; unicellular organisms consist of one cell.
Example: Leaf cell

TISSUE
A collection of specialized cells that function in a coordinated fashion. (Multicellular life only.)
Example: Epidermis of leaf

MOLECULE
A group of joined atoms.
Example: DNA

ORGAN
A structure consisting of tissues organized to interact and carry out specific functions. (Multicellular life only.)
Example: Leaf

ATOM
The smallest chemical unit of a type of pure substance (element).
Example: Carbon atom

ORGANISM
A single living individual.
Example: One acacia tree

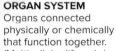

ORGAN SYSTEM
Organs connected physically or chemically that function together. (Multicellular life only.)
Example: Aboveground part of a plant

POPULATION
A group of the same species of organism living in the same place and time.
Example: Multiple acacia trees

COMMUNITY
All populations that occupy the same region.
Example: All populations in a savanna

ECOSYSTEM
The living and nonliving components of an area.
Example: The savanna

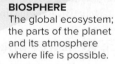

BIOSPHERE
The global ecosystem; the parts of the planet and its atmosphere where life is possible.

Figure 1.2 Life's Organizational Hierarchy. This diagram applies life's organizational hierarchy to a multicellular organism (an acacia tree). Green arrows represent the organizational hierarchy up to the level of the organism; blue arrows represent levels that include multiple organisms.

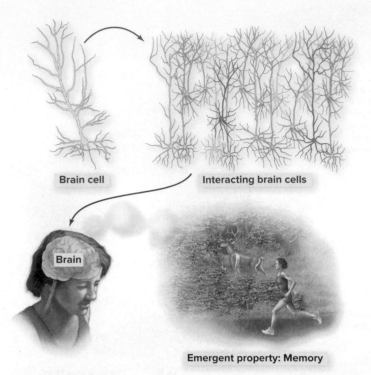

Brain cell | Interacting brain cells

Brain

Emergent property: Memory

Figure 1.3 **An Emergent Property—from Cells to Memories.** Highly branched cells interact to form a complex network in the brain. Memories, consciousness, and other qualities of the mind emerge only when these cells interact in a certain way.

Single-celled bacteria, although less complex than animals or plants, still contain DNA, proteins, and other molecules that interact in highly organized ways.

An organism, however, is more than a collection of successively smaller parts. **Emergent properties** are new functions that arise from physical and chemical interactions among a system's components, much as flour, sugar, butter, and chocolate can become brownies—something not evident from the parts themselves. **Figure 1.3** shows another example of emergent properties: the thoughts and memories produced by interactions among the neurons in a person's brain. For an emergent property, the whole is greater than the sum of the parts.

Emergent properties explain why structural organization is closely tied to function. Disrupt a structure, and its function ceases. Brain damage, for instance, disturbs the interactions between brain cells and can interfere with memory, coordination, and other brain functions. Likewise, if a function is interrupted, the corresponding structure eventually breaks down, much as unused muscles begin to waste away. Biological function and form are interdependent.

B. Life Requires Energy

Inside each cell, countless chemical reactions sustain life. These reactions, collectively called metabolism, allow organisms to acquire and use energy and nutrients to build new structures, repair old ones, and reproduce.

Biologists divide organisms into broad categories, based on their source of energy and raw materials (**figure 1.4**). **Producers,** also called autotrophs, make their own food by extracting energy

and nutrients from nonliving sources. The most familiar producers are the plants and microbes that capture light energy from the sun, but some bacteria can derive chemical energy from rocks. **Consumers,** in contrast, obtain energy and nutrients by eating other organisms, living or dead; consumers are also called heterotrophs. You are a consumer, relying on energy and atoms from food to stay alive. **Decomposers** are heterotrophs that absorb energy and nutrients from wastes or dead organisms. These organisms, which include fungi and some bacteria, recycle nutrients to the nonliving environment.

Within an ecosystem, organisms are linked into elaborate food webs, beginning with producers and continuing through several levels of consumers (including decomposers). But energy transfers are never 100% efficient; some energy is always lost to the surroundings in the form of heat (see figure 1.4). Because no organism can use it as an energy source, heat represents a permanent loss from the cycle of life. All ecosystems therefore depend on a continuous stream of energy from an outside source, usually the sun. ⓘ *food webs,* section 38.3A

C. Life Maintains Internal Constancy

The conditions inside cells must remain within a constant range, even if the surrounding environment changes. For example, a cell must maintain a certain temperature; it will die if it becomes too

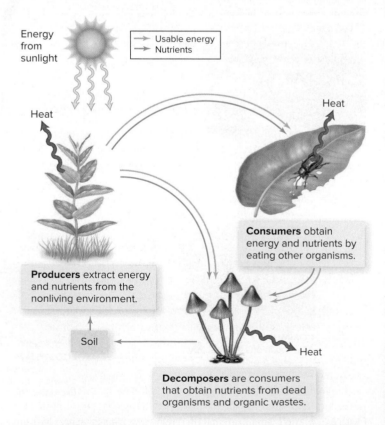

Energy from sunlight

→ Usable energy
→ Nutrients

Heat

Heat

Producers extract energy and nutrients from the nonliving environment.

Consumers obtain energy and nutrients by eating other organisms.

Soil

Heat

Decomposers are consumers that obtain nutrients from dead organisms and organic wastes.

Figure 1.4 **Life Is Connected.** All organisms extract energy and nutrients from the nonliving environment or from other organisms. Decomposers recycle nutrients back to the nonliving environment. At every stage along the way, heat is lost to the surroundings.

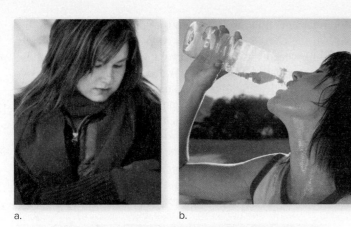

a.   b.

Figure 1.5 Temperature Homeostasis. (a) Shivering and (b) sweating are responses that maintain body temperature within an optimal range.
(a): ©Design Pics/Kristy-Anne Glubish RF; (b): ©John Rowley/Getty Images RF

a.   b.

Figure 1.6 Asexual and Sexual Reproduction. (a) Identical plantlets develop along the runners of a wild strawberry plant. (b) Two swans protect their offspring, the products of sexual reproduction.
(a): ©Dorling Kindersley/Getty Images; (b): ©Jadranko Markoc/flickr/Getty Images RF

hot or too cold. The cell must also take in nutrients, excrete wastes, and regulate its many chemical reactions to prevent a shortage or surplus of essential substances. **Homeostasis** is this state of internal constancy.

Because cells maintain homeostasis by counteracting changes as they occur, organisms must be able to sense and react to stimuli. To illustrate this idea, consider the mechanisms that help maintain your internal temperature at about 37°C (**figure 1.5**). When you go outside on a cold day, you may begin to shiver; heat from these muscle movements warms the body. In severe cold, your lips and fingertips may turn blue as your circulatory system sends blood away from your body's surface. Conversely, on a hot day, sweat evaporating from your skin helps cool your body.

D. Life Reproduces, Grows, and Develops

Organisms reproduce, making other individuals that are similar to themselves (**figure 1.6**). Reproduction transmits DNA from generation to generation; this genetic information defines the inherited characteristics of the offspring.

Reproduction occurs in two basic ways: asexually and sexually. In **asexual reproduction,** genetic information comes from only one parent, and all offspring are virtually identical (figure 1.6a). One-celled organisms such as bacteria reproduce asexually by doubling and then dividing the contents of the cell. Many multicellular organisms also reproduce asexually. A strawberry plant, for instance, produces "runners" that sprout leaves and roots, forming new plants that are identical to the parent. Fungi produce countless asexual spores, visible as the green, white, or black powder on moldy bread or cheese. Some animals, including sponges, reproduce asexually when a fragment of the parent animal detaches and develops into a new individual.

In **sexual reproduction,** genetic material from two parents unites to form an offspring, which has a new combination of inherited traits (figure 1.6b). By mixing genes at each generation, sexual reproduction results in tremendous diversity in a population. Genetic diversity, in turn, enhances the chance that some individuals will survive even if conditions change. Sexual reproduction is therefore a very successful strategy, especially in an

environment where conditions change frequently; it is extremely common among plants, animals, and fungi.

If each offspring is to reproduce, it must grow and develop to adulthood. Each young swan in figure 1.6b, for example, started as a single fertilized egg cell. That cell divided over and over, developing into an embryo. Continued cell division and specialization yielded the newly hatched swans, which will eventually mature into adults that can also reproduce—just like their parents.

E. Life Evolves

One of the most intriguing questions in biology is how organisms become so well suited to their environments. A beaver's enormous front teeth, which never stop growing, are ideal for gnawing wood. Tubular flowers have exactly the right shapes for the beaks of their hummingbird pollinators. Some organisms have color patterns that enable them to fade into the background (**figure 1.7**).

These examples, and countless others, illustrate adaptations. An **adaptation** is an inherited characteristic or behavior that enables an organism to survive and reproduce successfully in its environment.

Where do these adaptive traits come from? The answer lies in natural selection. The simplest way to think of natural selection is

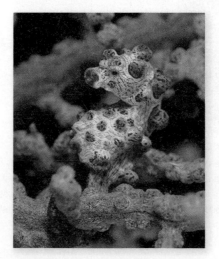

Figure 1.7 Hiding in Plain Sight. This pygmy seahorse is barely visible in its coral habitat, thanks to its unique body shape, skin color, and texture.
©Mark Webster/Getty Images

to consider two facts. First, populations produce many more off-spring than will survive to reproduce; these organisms must compete for limited resources such as food and habitat. A single mature oak tree may produce thousands of acorns in one season, but only a few are likely to germinate, develop, and reproduce. The rest die. Second, no organism is exactly the same as any other. Genetic mutations—changes in an organism's DNA sequence—generate variability in all organisms, even those that reproduce asexually.

Of all the offspring in a population, which will outcompete the others and live long enough to reproduce? The answer is those with the best adaptations to the current environment; conversely, the poorest competitors are most likely to die before reproducing. A good definition of **natural selection,** then, is the enhanced reproductive success of certain individuals from a population based on inherited characteristics (**figure 1.8**). The same principle applies to all populations. In general, individuals with the best combinations of genes survive and reproduce, while those with less suitable characteristics fail to do so. Over many generations, individuals with adaptive traits make up most or all of the population.

But the environment is constantly changing. Continents shift, sea levels rise and fall, climates warm and cool. What happens to a population when the selective forces that drive natural selection change? Only some organisms survive: those with the "best" traits in the *new* environment. Features that may once have been rare become more common as the reproductive success of individuals with those traits improves. Notice, however, that this outcome depends on variability within the population. If no individual can reproduce in the new environment, the species may go extinct.

Natural selection is one mechanism of **evolution,** which is a change in the genetic makeup of a population over multiple generations. Although evolution can also occur in other ways, natural selection is the mechanism that selects for adaptations. Charles

Darwin became famous in the 1860s after the publication of his book *On the Origin of Species by Means of Natural Selection,* which introduced the theory of evolution by natural selection; another naturalist, Alfred Russel Wallace, independently developed the same idea at around the same time.

Evolution is the single most powerful idea in biology. As unit 3 describes in detail, evolution has been operating since life began, and it explains the current diversity of life. In fact, the similarities among existing organisms strongly suggest that all species descend from a common ancestor. Evolution has molded the life that has populated the planet since the first cells formed almost 4 billion years ago, and it continues to act today.

1.1 | MASTERING CONCEPTS

1. Does any nonliving object possess all of the characteristics of life? Explain your answer.
2. List the levels of life's organizational hierarchy from smallest to largest, starting with atoms and ending with the biosphere.
3. The bacteria in figure 1.8 reproduce asexually, yet they are evolving. What is their source of genetic variation?

1.2 The Tree of Life Includes Three Main Branches

Biologists have been studying life for centuries, documenting the existence of everything from bacteria to blue whales. An enduring problem has been how to organize the ever-growing list of known organisms into meaningful categories. **Taxonomy** is the science of naming and classifying organisms.

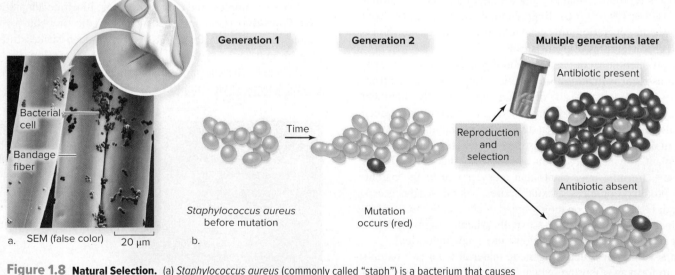

Generation 1   **Generation 2**   **Multiple generations later**

Time

Reproduction and selection

Antibiotic present

Antibiotic absent

Staphylococcus aureus before mutation

Mutation occurs (red)

Bacterial cell

Bandage fiber

a.   SEM (false color)   20 µm   b.

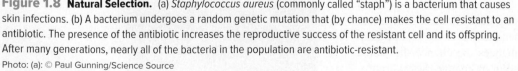

Figure 1.8 Natural Selection. (a) *Staphylococcus aureus* (commonly called "staph") is a bacterium that causes skin infections. (b) A bacterium undergoes a random genetic mutation that (by chance) makes the cell resistant to an antibiotic. The presence of the antibiotic increases the reproductive success of the resistant cell and its offspring. After many generations, nearly all of the bacteria in the population are antibiotic-resistant.

Photo: (a): © Paul Gunning/Science Source

The basic unit of classification is the **species,** which designates a distinctive "type" of organism. Closely related species are grouped into the same **genus.** Together, the genus and a specific descriptor denote the unique, two-word scientific name of each species. A human, for example, is *Homo sapiens.* (Note that scientific names are always italicized and that the genus—but not the specific descriptor—is capitalized.) Scientific names help taxonomists and other biologists communicate with one another.

But taxonomy involves more than simply naming species. Taxonomists also strive to classify organisms according to what we know about evolutionary relationships; that is, how recently one type of organism shared an ancestor with another type. The more recently they diverged from a shared ancestor, the more closely related we presume the two types of organisms to be. Researchers infer these relationships by comparing anatomical, behavioral, cellular, genetic, and biochemical characteristics.

Section 14.6 describes the taxonomic hierarchy in more detail. For now, it is enough to know that genetic evidence suggests that all species fall into one of three **domains,** the broadest (most inclusive) taxonomic category. **Figure 1.9** depicts the three domains: Bacteria, Archaea, and Eukarya. The species in domains Bacteria and Archaea are superficially similar to one another; all are prokaryotes,

meaning that their DNA is free in the cell and not confined to an organelle called a nucleus. Major differences in DNA sequences separate these two domains from each other. The third domain, Eukarya, contains all species of eukaryotes, which are unicellular or multicellular organisms whose cells contain a nucleus.

The species in each domain are further subdivided into **kingdoms;** figure 1.9 shows the kingdoms within domain Eukarya. Three of these kingdoms—Animalia, Fungi, and Plantae—are familiar to most people. Within each one, organisms share the same general strategy for acquiring energy. The plant kingdom contains autotrophs, whereas fungi and animals are consumers that differ in the details of how they obtain food. But the fourth group of eukaryotes, the Protista, contains a huge collection of unrelated species. Protista is a convenient but artificial "none of the above" category for the many species of eukaryotes that are not plants, fungi, or animals.

1.2 MASTERING CONCEPTS

1. What are the goals of taxonomy?
2. How are domains related to kingdoms?
3. List and describe the four main groups of eukaryotes.

Figure 1.9 Life's Diversity. The three domains of life arose from a hypothetical common ancestor, shown at the base of the evolutionary tree.

Photos: (Bacteria): ©Heather Davies/Getty Images RF; (Archaea): ©Eye of Science/Science Source; (Protista): ©Melba Photo Agency/PunchStock RF; (Animalia): USDA/ARS/Scott Bauer; (Fungi): ©Corbis RF; (Plantae) USDA/Keith Weller

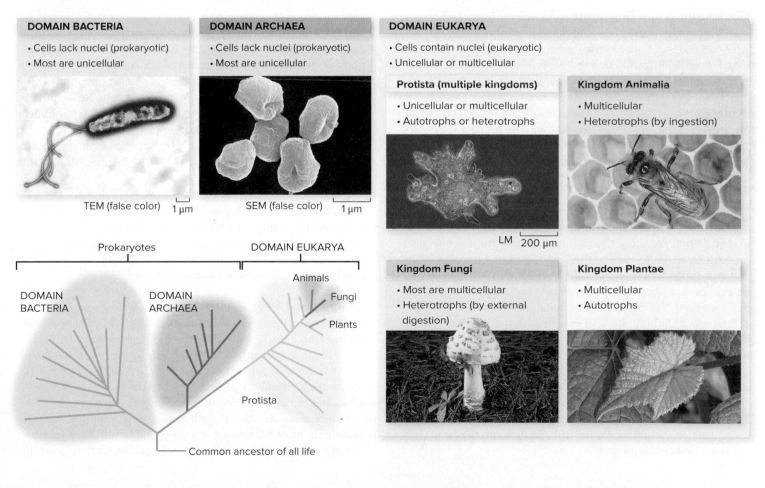

DOMAIN BACTERIA
- Cells lack nuclei (prokaryotic)
- Most are unicellular

TEM (false color) 1 μm

DOMAIN ARCHAEA
- Cells lack nuclei (prokaryotic)
- Most are unicellular

SEM (false color) 1 μm

DOMAIN EUKARYA
- Cells contain nuclei (eukaryotic)
- Unicellular or multicellular

Protista (multiple kingdoms)
- Unicellular or multicellular
- Autotrophs or heterotrophs

LM 200 μm

Kingdom Animalia
- Multicellular
- Heterotrophs (by ingestion)

Kingdom Fungi
- Most are multicellular
- Heterotrophs (by external digestion)

Kingdom Plantae
- Multicellular
- Autotrophs

Prokaryotes

DOMAIN EUKARYA

DOMAIN BACTERIA

DOMAIN ARCHAEA

Animals

Fungi

Plants

Protista

Common ancestor of all life

1.3 Scientists Study the Natural World

The idea of biology as a "rapidly changing field" may seem strange if you think of science as a collection of facts. After all, the parts of a frog are the same now as they were 50 or 100 years ago. But memorizing frog anatomy is not the same as thinking scientifically. Scientists use evidence to answer questions about the natural world. If you compare a frog to a snake, for instance, can you determine how the frog can live in water and on land, whereas the snake survives in the desert? Understanding anatomy simply gives you the vocabulary you need to ask these and other interesting questions about life.

A. The Scientific Method Has Multiple Interrelated Parts

Scientific knowledge arises from application of the **scientific method,** which is a general way of using evidence to answer questions and test ideas (**figure 1.10**). Although this diagram may give the impression that science is a tedious, step-by-step process, that is not at all true. Instead, science combines thinking, detective work, communicating with other scientists, learning from mistakes, and noticing connections between seemingly unrelated events. The resulting insights have taught us everything we know about the natural world.

Observations and Questions   The scientific method begins with observations and questions. The observations may rely on what we can see, hear, touch, taste, or smell, or they may be based on existing knowledge and experimental results. Often, a great leap forward happens when one person makes connections between previously unrelated observations. Charles Darwin, for example, developed the idea of natural selection by combining his understanding of Earth's long history with his detailed observations of organisms. Another great advance occurred decades later, when biologists realized that mutations in DNA generate the variation that Darwin saw but could not explain.

Hypothesis and Prediction   A **hypothesis** is a tentative explanation for one or more observations. The hypothesis is the essential "unit" of scientific inquiry. To be useful, the hypothesis must be testable—that is, there must be a way to collect data that can support or reject it. Interestingly, a hypothesis cannot be *proven* true, because future discoveries may contradict today's results. Nevertheless, a hypothesis becomes widely accepted when multiple lines of evidence support it, no credible data refute it, and plausible alternative hypotheses have been rejected.

A hypothesis is a general statement that should lead to specific **predictions.** Often, the prediction is written as an if–then statement. As a simple example, suppose you hypothesize that your lawn mower stopped working because it ran out of gas. A reasonable prediction would be "If I put fuel into the tank, then my lawn mower should start."

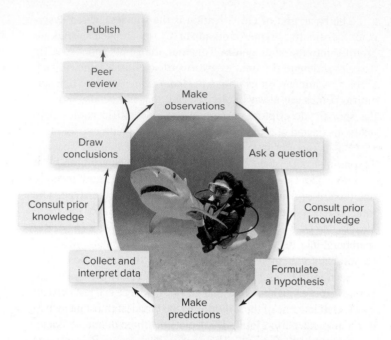

Figure 1.10 **Scientific Inquiry.** This researcher studies tiger sharks; her observations could lead to questions and testable hypotheses. Additional data, combined with prior findings, can help support or reject each hypothesis. Peer review determines w1hether the results are publishable.
Photo: © Jeff Rotman/Getty Images RF

Data Collection   Investigators draw conclusions based on data (**figure 1.11**). The data may come from careful observations of the natural world, an approach called discovery science. The National Audubon Society's annual Christmas Bird Count is a case in point: For more than a century, thousands of "citizen scientists" have documented the ups and downs of hundreds of bird species nationwide. Another way to gather data is to carry out an experiment to test a hypothesis under controlled conditions (section 1.3B explores experimental design in more detail).

Discovery and experimentation work hand in hand. As just one example, consider the well-known connection between cigarettes and lung cancer. In the late 1940s, scientists showed that smokers are far more likely than nonsmokers to develop cancer. Since that time, countless laboratory experiments have revealed how the chemicals in tobacco damage living cells.

Analysis and Peer Review   After collecting and interpreting data, investigators decide whether the evidence supports or falsifies the hypothesis. Often, the most interesting results are those that are unexpected, because they provide new observations that force scientists to rethink their hypotheses; figure 1.10 shows this feedback loop. Science advances as new information arises and explanations continue to improve.

Once a scientist has enough evidence to support or reject a hypothesis, he or she may write a paper and submit it for publication in a scientific journal. The journal's editors send the paper to

a.

b.

Figure 1.11 Different Types of Science. (a) Scientists track the number of migratory birds that visit a wildlife refuge each year—an example of discovery science. (b) Controlled experiments can help food scientists objectively compare different techniques for roasting or brewing coffee.
(a): U.S. Fish & Wildlife Service/J&K Hollingsworth; (b): ©Corbis RF

anonymous reviewers who are knowledgeable about the research topic. In a process called **peer review,** these scientists independently evaluate the validity of the methods, data, and conclusions. Overall, peer review ensures that journal articles—the tangible products of the global scientific conversation—are of high quality.

B. An Experimental Design Is a Careful Plan

Scientists test many hypotheses with the help of experiments. An **experiment** is an investigation carried out in controlled conditions. This section considers a real study that tested the hypothesis that a new vaccine protects against a deadly virus. The virus, called rotavirus, causes severe diarrhea and takes the lives of hundreds of thousands of young children each year. An effective, inexpensive vaccine would prevent many childhood deaths.

Sample Size   One of the most important decisions that an investigator makes in designing an experiment is the **sample size,** which is the number of individuals assigned to each treatment. The sample size in the rotavirus study, for example, was approximately 100 infants per treatment. In general, the larger the sample size, the more credible the results of a study.

Variables   A systematic consideration of variables is also important in experimental design (table 1.2). A **variable** is a changeable element of an experiment, and there are several types. The **independent variable** is the factor that an investigator directly manipulates to determine whether it causes another variable to change. In the rotavirus study, the independent variable was the dose of the vaccine. The **dependent variable** is any response that might *depend on* the value of the independent variable, such as the number of children with rotavirus-related illness during the study period.

A **standardized variable** is anything that the investigator holds constant for all subjects in the experiment, ensuring the best chance of detecting the effect of the independent variable. Because rotavirus infection is most common among very young children, the test of the new vaccine included only infants younger than 12 weeks. Furthermore, vaccines work best in people with healthy immune systems, so the study excluded infants who were ill or had weak immunity. Age and health were therefore among the study's standardized variables.

TABLE 1.2   Types of Variables in an Experiment: A Summary

Type of Variable	Definition	Example
Independent variable	A variable that an investigator manipulates to determine whether it influences the dependent variable	Dose of vaccine
Dependent variable	A variable that an investigator measures to determine whether it is affected by the independent variable	Number of children with illness caused by rotavirus
Standardized variable	Any variable that an investigator intentionally holds constant for all subjects in an experiment, including the control group	Age and health of children in study

Controls   Well-designed experiments compare one or more groups undergoing treatment to a group of "normal" individuals. The experimental **control** is the untreated group, and it is important because it provides a basis for comparison in measuring the effect of the independent variable. Ideally, the only difference between the control and any other experimental group is the one factor being tested.

Experimental controls may take several forms. Sometimes, the control group simply receives a "zero" value for the independent variable. If a gardener wants to test a new fertilizer in her garden, she may give some plants a lot of fertilizer, others only a little, and still others—the control plants—none. In medical research, a control group might receive a **placebo,** an inert substance that resembles the treatment given to the experimental group. The control infants in the rotavirus study received a placebo that contained all components of the vaccine except the active ingredient.

The investigators in the rotavirus study used a double-blind design, in which neither the researchers nor the participants knew who received the vaccine and who received the placebo. Double-blind studies help avoid bias in medical research. The investigators break the "code" of who received which treatment only after the experiment is complete and the data are tabulated.

Statistical Analysis   Once an experiment is complete, the investigator compiles the data and decides whether the results support the hypothesis (**figure 1.12**). The researchers in the rotavirus study concluded that the vaccine was effective, but only after applying a statistical test. All statistical tests consider both variation and sample size to estimate the probability that the results arose purely by chance. If this probability is low, then the results are considered **statistically significant.** Appendix B shows how scientists use error bars and other notation to illustrate statistical significance in graphs.

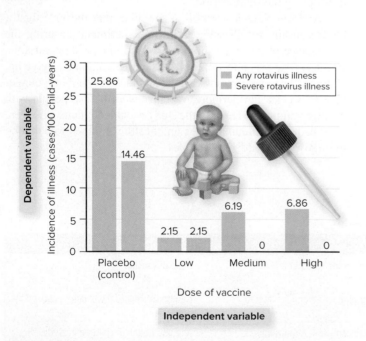

Figure 1.12 **Vaccine Test.** This graph shows that the rotavirus vaccine is more effective than a placebo. (The statistical analysis is not shown.)

C. Theories Are Comprehensive Explanations

Outside of science, the word *theory* is often used to describe an opinion or a hunch. For instance, immediately after a plane crash, experts offer their "theories" about the cause of the disaster. These tentative explanations are really untested hypotheses.

In science, the word *theory* has a distinct meaning. Like a hypothesis, a **theory** is an explanation for a natural phenomenon, but a theory is typically broader in scope than a hypothesis. For example, the germ theory—the idea that some microorganisms cause human disease—is the foundation for medical microbiology. Individual hypotheses relating to the germ theory are much narrower, such as the suggestion that rotavirus causes illness. Not all theories are as "large" as the germ theory, but they generally encompass multiple hypotheses. Note also that the germ theory does not imply that *all* microbes make us sick or that all illnesses have microbial causes. But it does explain many types of human disease.

A second difference between a hypothesis and a theory is acceptance and evidence. A hypothesis is tentative, whereas theories reflect broader agreement. This is not to imply that theories are not testable; in fact, the opposite is true. Every scientific theory is *potentially* falsifiable, meaning that a particular set of observations could prove the theory wrong. The germ theory remains widely accepted because many observations support it and no reliable tests have disproved it. The same is true for the theory of evolution and many other scientific theories.

Another quality of a scientific theory is its predictive power. A good theory not only ties together many existing observations but also suggests predictions about phenomena that have yet to be observed. Both Charles Darwin and naturalist Alfred Russel Wallace used the theory of evolution by natural selection to predict the existence of a moth that could pollinate orchid flowers with unusually long nectar tubes (**figure 1.13**). Decades later, scientists discovered the long-tongued insect (see section 1.4). This finding was consistent with evolutionary theory, but the theory would have been weakened if subsequent research had not supported its predictions.

What is the relationship between facts and a theory? One definition of the word *fact* is "a repeatable observation that everyone can agree on." It is a fact that a dropped pencil falls toward the ground; no reasonable person disagrees with that statement. Gravity is a *fact*; gravitational *theory* explains the forces that cause pencils and other objects to fall.

Biologists also consider biological evolution to be a fact. Yet the phrase "theory of evolution" persists because evolution is both a fact *and* a theory. Like gravity, evolution is a *fact*. No one can dispute that antibiotics drive evolutionary change in bacteria (see figure 1.8). On a broader scale, the combined evidence for genetic change over time is so persuasive and comes from so many different fields of study that to deny the existence of evolution is unrealistic. Evolutionary *theory* explains how life has diversified since its origin. Note that biologists do not understand everything about how evolution works. Many questions about life's history remain, but the debates swirl around *how,* not *whether,* evolution occurs.

Figure 1.13 Prediction Confirmed. When Charles Darwin saw this Madagascar orchid, he predicted that its pollinator would have long, thin mouthparts that could reach the bottom of the elongated nectar tube. He was right; the unknown pollinator turned out to be a moth with an extraordinarily long tongue.

©Kjell Sandved/Alamy

D. Scientific Inquiry Has Limitations

Scientific inquiry is neither foolproof nor always easy to implement. One problem is that experimental evidence may lead to multiple interpretations, and even the most carefully designed experiment can fail to provide a definitive answer (see the Apply It Now box in this section). Consider the observation that animals fed large doses of vitamin E live longer than similar animals that do not ingest the vitamin. So, does vitamin E slow aging?

Possibly, but excess vitamin E also causes weight loss, and other research has connected weight loss with longevity. Does vitamin E extend life, or does weight loss? The experiment alone does not distinguish between these possibilities.

Another limitation is that researchers may misinterpret observations or experimental results. For example, centuries ago, scientists sterilized a bottle of broth, corked the bottle shut, and observed bacteria in the broth a few days later. They concluded that life arose directly from the broth. The correct explanation, however, was that the cork did not keep airborne bacteria out. Although scientists may make mistakes in the short term, science is self-correcting in the long run because it remains open to new data and new interpretations.

A related problem is that the scientific community may be slow to accept new evidence that suggests unexpected conclusions. Every investigator should try to keep an open mind about observations, not allowing biases or expectations to cloud interpretation of the results. But it is human nature to be cautious in accepting an observation that does not fit what we think we know. The careful demonstration that life does not arise from broth surprised many people who believed that mice sprang from moldy grain and that flies came from rotted beef. More recently, it took many years to set aside the common belief that stress caused ulcers. Today, we know that a bacterium causes most ulcers.

Although science is a powerful tool for answering questions about the natural world, it cannot answer questions of beauty, morality, ethics, or the meaning of life (see this section's Burning Question). Nor can we directly study some phenomena that occurred long ago and left little physical evidence. Consider the many experiments that have attempted to recreate the chemical reactions that might have produced life on early Earth. Although the experiments produce interesting results and reveal ways that these early events may have occurred, we cannot know if they accurately reflect conditions at the dawn of life.

Burning Question

Why am I here?

The Burning Questions featured in each chapter of this book came from students. On the first day of class, I always ask students to turn in a "burning question"—anything they have always wondered about biology. I answer most of the questions as the relevant topics come up during the semester.

Why not answer all of the questions? It is because at least one student often asks something like "Why am I here?" or "What is the meaning of life?" Such puzzles have fascinated humans throughout the ages, but they are among the many questions that we cannot approach scientifically. Biology can explain how you developed after a sperm from your father fertilized an egg cell from your mother. But no one can develop a testable hypothesis about life's meaning or the purpose of human existence. Science must remain silent on such questions.

Instead, other ways of knowing must satisfy our curiosity about "why." Philosophers, for example, can help us see how others have

considered these questions. Religion may also provide the meaning that many people seek. Part of the value of higher education is to help you acquire the tools you need to find your own life's purpose.

In the meantime, this book's Burning Questions can help you discover the answers to many questions—asked by students just like you—about human health, environmental quality, life's diversity, and the rest of the biological world.

©Getty Images/Photodisc RF

Submit your burning question to
Marielle.Hoefnagels@mheducation.com

Apply It **Now**

It's Hard to Know What's Bad for You

You have probably heard reports that a food previously considered healthy is actually bad for you, or vice versa. These conflicting reports may tempt you to mistakenly conclude that scientific studies are no better than guesswork.

Instead, the problem lies in the fact that some questions are extremely hard to answer. Take, for example, the controversy surrounding the artificial sweetener called saccharin. In 1977, the U.S. Food and Drug Administration (FDA) proposed a ban on saccharin, based on a handful of studies suggesting that the sweetener caused bladder cancer in rats. Congress opted to require warning labels on products containing saccharin. In 1991, the FDA withdrew its proposed ban, and in 1998, saccharin was rated as "not classifiable as to its carcinogenicity to humans." Two years later, legislation removed the warning label requirement.

This tangled history raises an important issue. Why can't science reply "yes" or "no" to the seemingly simple question of whether saccharin is bad for you? To understand the answer, consider one of the studies that prompted the FDA to propose the ban on saccharin in the first place. Researchers divided 200 rats into two groups. The control animals ate standard rodent chow, whereas the experimental group got the same food supplemented with saccharin. At reproductive maturity the animals were bred, and the researchers fed the offspring the same dose of saccharin throughout their lives as well. To measure the incidence of cancer, they counted the tumors in both generations of rats for 24 months or until the rats died, whichever came first. **Figure 1.A** summarizes the results.

At first glance, the conclusion seems inescapable: Saccharin causes cancer in male lab rats. But closer study reveals several hidden complexities that make the data hard to interpret. First, the dose of saccharin was huge: 5% of the rats' diets, for life. The equivalent dose in humans would require drinking hundreds of cans of saccharin-sweetened soda every day. In addition, the experimental rats weighed much less than the

control rats by the end of the study, suggesting that high doses of the sweetener are toxic. Rather than causing cancer directly, the saccharin may have simply made the animals more susceptible to disease. Moreover, follow-up studies using other animals were inconclusive.

Perhaps the scientists should have studied the saccharin–cancer connection in humans instead. Unfortunately, however, documenting a link between any food and cancer in people is extremely difficult. One strategy might be to measure the incidence of cancer in saccharin users versus nonusers. But with so many other possible causes of cancer—smoking, poor diet, exposure to job-related chemicals, genetic predisposition—it is difficult to separate out just the effects of saccharin.

So what are we to make of the mixed news reports? It is hard to know, but one thing is certain: No matter what the headlines say, one study, especially a small one, cannot reveal the whole story.

©Photodisc/Getty Images RF

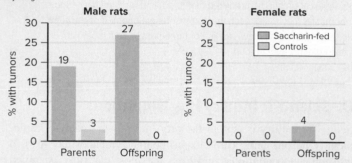

Figure 1.A **The Saccharin Scare.** These graphs summarize the results of one study examining the link between saccharin and bladder cancer in rats. Sample sizes ranged from 36 to 49 rats per treatment.

Source: Data adapted from Office of Technology Assessment Report, October 1977, *Cancer Testing Technology and Saccharin,* page 52.

E. Biology Continues to Advance

Science is just one of many ways to investigate the world, but its strength is its openness to new information. Theories change to accommodate new knowledge. The history of science is full of long-established ideas that changed as we learned more about nature, often thanks to new technology. People thought that Earth was flat and at the center of the universe before inventions and data analysis revealed otherwise. Similarly, biologists thought all organisms were plants or animals until microscopes unveiled a world of life invisible to our eyes.

Technology is the practical application of scientific knowledge. Science and technology are therefore intimately related. For example, thanks to centuries of scientific inquiry, we understand many of the differences between humans and bacteria. We can exploit these differences to invent new antibiotic drugs that kill germs without harming our own bodies. These antibiotics, in turn, can be useful tools that help biologists learn even more

about bacterial cells. The new scientific discoveries spawn new technologies, and so on.

Biology is changing rapidly because technology has expanded our ability to spy on living cells, compare DNA sequences, track wildlife, and make many other types of observations. Scientists can now answer questions about the natural world that previous generations could never have imagined.

1.3 MASTERING CONCEPTS

1. Identify the elements of the experiment summarized in the Apply It Now box.
2. What is a statistically significant result?
3. What is the difference between a hypothesis and a theory, and why are some theories regarded as facts?
4. What are some limitations of scientific inquiry?
5. Compare and contrast *science* and *technology.*

1.4 The Orchid and the Moth

Each chapter of this book ends with a section that examines how biologists use systematic, scientific observations to solve a different evolutionary puzzle from life's long history. This first installment of Investigating Life revisits the story of the orchid plant pictured in figure 1.13.

In a book on orchids published in 1862, Charles Darwin speculated about which type of insect might pollinate the unusual flowers of the *Angraecum sesquipedale* orchid, a species that lives on Madagascar (an island off the coast of Africa). As described in section 1.3C, the flowers have unusually long nectar tubes (also called nectaries). Darwin observed nectaries "eleven and a half inches long, with only the lower inch and a half filled with very sweet nectar." Darwin found it "surprising that any insect should be able to reach the nectar; our English sphinxes [moths] have probosces as long as their bodies; but in Madagascar there must be moths with probosces capable of extension to a length of between ten and eleven inches!"

Charles Darwin
©Richard Milner Archives

Alfred Russel Wallace picked up the story in a book published in 1895. According to Wallace, "There is a Madagascar orchid—the *Angraecum sesquipedale*—with an immensely long and deep nectary. How did such an extraordinary organ come to be developed?" He went on to summarize how natural selection could explain this unusual flower. He wrote: "The pollen of this flower can only be removed by the base of the proboscis of some very large moths, when trying to get at the nectar at the bottom of the vessel. The moths with the longest probosces would do this most effectually; they would be rewarded for their long tongues by getting the most nectar; whilst on the other hand, the flowers with the deepest nectaries would be best fertilized by the largest moths preferring them. Consequently, the deepest nectaried orchids and the longest tongued moths would each confer on the other an advantage in the battle of life."

Alfred Russel Wallace
©Hulton Archive/Getty Images

At that time, the pollinator had not yet been discovered. However, as Wallace wrote, moths with very long tongues were known to exist: "I have carefully measured the proboscis of a specimen . . . from South America . . . and find it to be nine inches and a quarter long! One from tropical Africa . . . is seven inches and a half. A species having a proboscis two or three inches longer could reach the nectar in the largest flowers of *Angraecum sesquipedale*. . . . That such a moth exists in Madagascar may be safely predicted; and naturalists who visit that island should search for it with as much confidence as astronomers searched for the planet Neptune—and I venture to predict they will be equally successful!"

Figure 1.14 Found at Last. More than 40 years after Darwin predicted its existence, scientists finally discovered the sphinx moth that pollinates the Madagascar orchid.
©The Natural History Museum/Alamy

A taxonomic publication from 1903 finally validated Darwin's and Wallace's predictions. The authors described a moth species, *Xanthopan morgani*, with a 225-millimeter (8-inch) tongue (figure 1.14). Given the correspondence between lengths of the orchid's nectary and the moth's tongue, the authors concluded that "*Xanthopan morgani* can do for *Angraecum* what is necessary [for pollination]; we do not believe that there exists in Madagascar a moth with a longer tongue. . . ."

This story not only illustrates how theories lead to testable predictions but also reflects the collaborative nature of science. Darwin and Wallace asked a simple question: Why are these nectar tubes so long? Other biologists cataloging the world's insect species finally solved the puzzle, decades after Darwin first raised the question of the mysterious Madagascar orchid.

Sources: Darwin, C. R. 1862. *On the Various Contrivances by Which British and Foreign Orchids Are Fertilised by Insects, and on the Good Effects of Intercrossing.* London: John Murray, pages 197–198.

Rothschild, W., and K. Jordan. 1903. A revision of the lepidopterous family Sphingidae. *Novitates Zoologicae*, vol. 9, supplement part 1, page 32.

Wallace, Alfred Russel. 1895. *Natural Selection and Tropical Nature: Essays on Descriptive and Theoretical Biology.* London: MacMillan and Co., pages 146–148.

1.4 MASTERING CONCEPTS

1. What observations led Darwin and Wallace to predict the existence of a long-tongued moth in Madagascar?
2. How does this story illustrate discovery science?

CHAPTER SUMMARY

1.1 What Is Life?

- A combination of characteristics distinguishes life: organization, energy use, internal constancy, reproduction and development, and evolution.

A. Life Is Organized (figure 1.15)

- **Atoms** form **molecules.** These molecules form the **organelles** inside many **cells.** An **organism** consists of one or more cells. In most multicellular organisms, cells form **tissues** and then **organs** and **organ systems.**

- Whether unicellular or multicellular, multiple individuals of the same species make up **populations;** multiple populations form **communities. Ecosystems** include living communities plus their nonliving environment. The **biosphere** is composed of all of the world's ecosystems.

- **Emergent properties** arise from interactions among the parts of an organism.

B. Life Requires Energy

- Life requires energy to maintain its organization and functions. **Producers** make their own food, using energy and nutrients extracted from the nonliving environment. **Consumers** eat other organisms, living or dead. **Decomposers** recycle nutrients to the nonliving environment.

- Because of heat losses, all ecosystems require constant energy input from an outside source, usually the sun.

C. Life Maintains Internal Constancy

- Organisms must maintain **homeostasis,** an internal state of constancy in changing environmental conditions.

D. Life Reproduces, Grows, and Develops

- Organisms reproduce asexually, sexually, or both. **Asexual reproduction** yields virtually identical copies of one parent, whereas **sexual reproduction** generates tremendous genetic diversity by combining and scrambling DNA from two parents.

E. Life Evolves

- In **natural selection,** environmental conditions select for organisms with inherited traits (**adaptations**) that increase the chance of survival and reproduction.

- **Evolution** through natural selection explains how common ancestry unites all species, producing diverse organisms with many similarities.

1.2 The Tree of Life Includes Three Main Branches

- **Taxonomy** is the science of classification. Biologists classify types of organisms, or **species,** according to probable evolutionary relationships. A **genus,** for example, consists of closely related species.
- The two broadest taxonomic levels are **domain** and **kingdom.**
- The three domains of life are Archaea, Bacteria, and Eukarya. Within each domain, mode of nutrition and other features distinguish the kingdoms.

1.3 Scientists Study the Natural World

A. The Scientific Method Has Multiple Interrelated Parts

- Scientific inquiry, which uses the **scientific method,** is a way of using evidence to evaluate ideas about the natural world.

- A scientist makes observations, raises questions, and uses reason to construct a testable explanation, or **hypothesis.** Specific **predictions** follow from a scientific hypothesis.

- After collecting data and making conclusions based on the evidence, the investigator may seek to publish scientific results. **Peer review** ensures that published studies meet high standards for quality.

B. An Experimental Design Is a Careful Plan

- An **experiment** is a test of a hypothesis carried out in controlled conditions.
- The larger the **sample size,** the more credible the results of an experiment.
- **Variables** are changeable elements in an experiment. The **independent variable** is the factor that the investigator manipulates. The **dependent variable** is what the investigator measures to determine the outcome of the experiment. **Standardized variables** are held constant for all subjects, including the **control** group (often a set of subjects receiving either no treatment or a **placebo**).
- Double-blind experiments minimize bias.
- **Statistically significant** results are unlikely to be due to chance.

C. Theories Are Comprehensive Explanations

- A **theory** is more widely accepted and broader in scope than a hypothesis.
- The acceptance of scientific ideas may change as new evidence accumulates.

D. Scientific Inquiry Has Limitations

- The scientific method does not always yield a complete explanation, or it may produce ambiguous results. Science cannot answer all questions—only those for which it is possible to develop testable hypotheses.

E. Biology Continues to Advance

- **Technology** is the practical application of scientific knowledge. Advances in science lead to new technologies, and vice versa.

1.4 Investigating Life: The Orchid and the Moth

- Charles Darwin and Alfred Russel Wallace knew of an orchid in Madagascar with an extremely long nectar tube. They predicted that the orchid's pollinator would be a moth with an equally long tongue.
- Years later, other scientists discovered the moth, illustrating the predictive power of evolutionary theory.

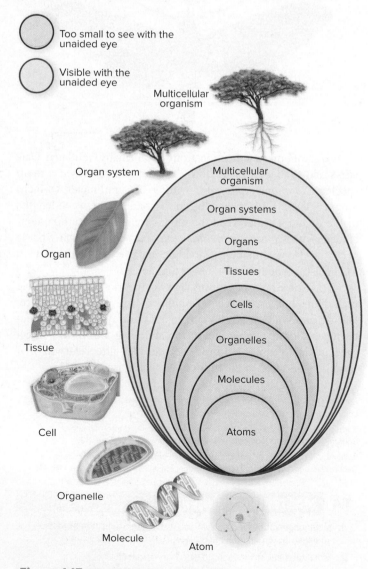

Too small to see with the unaided eye

Visible with the unaided eye

Multicellular organism

Organ system

Organ

Tissue

Cell

Organelle

Molecule

Atom

Multicellular organism

Organ systems

Organs

Tissues

Cells

Organelles

Molecules

Atoms

Figure 1.15 Life Is Organized: A Summary.

MULTIPLE CHOICE QUESTIONS

1. Which of the following is smaller than an organelle?
   a. An organ
   b. A molecule
   c. A cell
   d. A tissue

2. All of the following are characteristics of life EXCEPT
   a. evolution.
   b. reproduction.
   c. homeostasis.
   d. multicellularity.

3. The concentration of salts in blood remains relatively steady, regardless of a person's diet. This situation best illustrates
   a. homeostasis.
   b. life's organizational hierarchy.
   c. autotrophy.
   d. evolution.

4. Because plants extract nutrients from soil and use sunlight as an energy source, they are considered to be
   a. autotrophs.
   b. consumers.
   c. heterotrophs.
   d. decomposers.

5. Evolution through natural selection will occur most rapidly for populations of plants that
   a. are already well adapted to the environment.
   b. live in an unchanging environment.
   c. are in the same genus.
   d. reproduce sexually and live in an unstable environment.

6. *Homo sapiens* is the scientific name for humans. It consists of the
   a. domain name followed by the genus name.
   b. kingdom name followed by the family name.
   c. order name followed by the class name.
   d. genus name followed by a specific descriptor.

7. In an experiment to test the effect of temperature on the rate of bacterial reproduction, temperature would be the
   a. standardized variable.
   b. independent variable.
   c. dependent variable.
   d. control variable.

8. A scientist has just observed a new phenomenon and wonders how it happens. What is the next step in his or her discovery of the answer?
   a. Observe
   b. Hypothesize
   c. Experiment
   d. Peer review

9. Can a theory be proven wrong?
   a. No, theories are exactly the same as facts.
   b. No, because there is no good way to test a theory.
   c. Yes, a new observation or interpretation of data could disprove a theory.
   d. Yes, theories are exactly the same as hypotheses.

10. Which of the following statements is false?
    a. Emergent properties are functions that arise from the interactions between an organism's parts.
    b. Two of the three domains contain prokaryotic organisms.
    c. In a double-blind experiment, neither the researcher nor the subjects know which subject is assigned to which treatment.
    d. For a scientific study to be considered valid, the researchers must conduct experiments.

Answers to these questions are in appendix A.

WRITE IT OUT

1. Describe each of the five characteristics of life, and list several nonliving things that possess at least two of these characteristics.

2. Imagine two related species of single-celled protists living together in a pond. Write the organizational hierarchy of this ecosystem, starting with "atom" and ending with "ecosystem." Give an example of a structure at each level.

3. Why is a cell, and not an atom or a molecule, considered the basic unit of life?

4. Think of an analogy that will help you remember the differences between populations, communities, and ecosystems.

5. Other than the examples given in the text, name an example of emergent properties from everyday life.

6. Draw and explain the relationship between producers and consumers (including decomposers).

7. How does a home's air conditioning system illustrate homeostasis?

8. Explain why populations of organisms are typically well adapted to their environment.

9. How are the members of the three domains similar? How are they different?

10. List each step of the scientific method and explain why it is important.

11. Give two examples of questions that you cannot answer using the scientific method. Explain your reason for choosing each example.

12. Design an experiment to test the following hypothesis: "Eating chocolate causes zits." Include sample size, independent variable, dependent variable, the most important variables to standardize, and an experimental control.

PULL IT TOGETHER

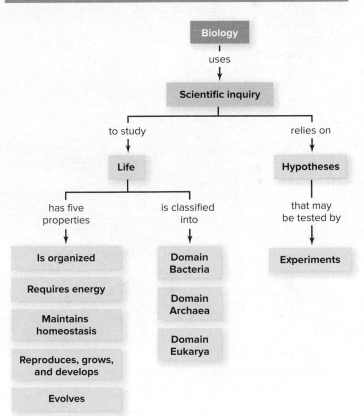

Figure 1.16 Pull It Together: The Scientific Study of Life.

Refer to **figure 1.16** and the chapter content to answer the following questions.

1. Review the Survey the Landscape figure in the chapter introduction, paying special attention to the units of life and their components. Connect these structures to the Pull It Together concept map.

2. What are the elements of a controlled experiment?

3. Give an example of each of the five properties of life.

The Chemistry of Life

Eat Me and You're Dead. The skin of the Sierra newt contains a potent toxin that protects the animal from predators.

©Brand X Pictures/PunchStock RF

LEARN HOW TO LEARN
Organize Your Time, and Don't Try to Cram

Get a calendar and study the syllabus for every class you are taking. Write each due date in your calendar. Include homework assignments, quizzes, and exams, and add new dates as you learn them. Then, block out time well before each due date to work on each task. Success comes much more easily if you take a steady pace instead of waiting until the last minute.

Chemical Warfare

The thought of chemistry often brings to mind laboratory benches, white lab coats, and flasks filled with bubbling solutions. It may be hard for you to connect that image with living organisms, yet chemistry and biology are intimately connected.

What does the complexity of life have to do with chemistry? A living body is a mixture of thousands of different types of interacting chemicals. Some of these chemicals are familiar to anyone who has read a nutrition label: sugars, proteins, and fats. Others, like DNA and water, are frequently in the news. But the vast majority are substances you may never have thought of as chemicals, like the hair on your head, the hormones in your bloodstream, or the red pigment in your blood cells.

Chemistry is not just limited to animals; all organisms, plants included, are made of chemicals. Some of these substances are the same as those in your own body, but others are unique to the plant kingdom. Some of the most interesting chemicals are those that plants deploy in their never-ending fight against herbivores and competitors. Whereas humans and other animals defend themselves by using their wits and ability to escape, many plants use chemical warfare.

Black walnut trees, for example, release a chemical that inhibits the growth of many other plants. Landscapers therefore often find it hard to establish new plants around black walnut trees. How does it benefit the trees to be such bad neighbors? By putting up a chemical "Keep Out" sign, they are reducing competition for vital resources like light, water, and soil nutrients.

Plants also produce chemicals that limit damage from animals that might otherwise dine on their foliage, flowers, fruits, or seeds. Consider the hot, spicy taste of a chili pepper; most mammals can't tolerate it, so they leave the peppers alone (see section 24.8).

Of course, plants don't have an exclusive hold on the chemical warfare market. Chemists who study nature's chemical weapons have spent decades identifying antibiotics produced by bacteria and fungi. Manufactured as drugs, these antibiotics have saved millions of lives. Other toxins occur in organisms ranging from mushrooms to sponges to snails to frogs and salamanders. Some of these chemicals protect their owners against competitors, predators, and disease. The race is on to determine whether they may also find new uses in medicine or other applications.

The majority of a cell consists of water and four classes of organic compounds, and defensive chemicals make up a relatively small proportion of any organism's body. This chapter describes these basic constituents of life. But we return to the topic of chemical warfare—this time among ants—in this chapter's Investigating Life essay (section 2.6).

SURVEY THE LANDSCAPE
Science, Chemistry, and Cells

All life is composed of chemical substances, including not only water but also DNA, proteins, and other organic molecules. Each of these substances is composed of atoms.

For more details, study the Pull It Together feature in the chapter summary.

🍁 2.1 Atoms Make Up All Matter

If you have ever touched a plant in a restaurant to see if it's fake, you know that we all have an intuitive sense of what life is made of. Most living leaves feel moist and pliable; a fake one is dry and stiff. But what does chemistry tell us about the composition of life?

Your desk, your book, your body, your sandwich, a plastic plant—indeed, all objects in the universe, including life on Earth—are composed of matter and energy. **Matter** is any material that takes up space, such as organisms, rocks, the oceans, and gases in the atmosphere. This chapter and the next concentrate on the building blocks that make up living matter. Physicists define **energy,** on the other hand, as the ability to do work. In this context, *work* means moving matter. Heat, light, and chemical bonds are all forms of energy; chapters 4, 5, and 6 discuss the energy of life in detail.

A. Elements Are Fundamental Types of Matter

The matter that makes up every object in the universe consists of one or more elements. A chemical **element** is a pure substance that cannot be broken down by chemical means into other substances. Examples of elements include oxygen (O), carbon (C), nitrogen (N), sodium (Na), and hydrogen (H).

Scientists had already noticed patterns in the chemical behavior of the elements by the mid-1800s, and several had proposed schemes for organizing the elements into categories. Nineteenth-century Russian chemist Dmitry Mendeleyev invented the chart we still use today. This chart, called the **periodic table,** arranges the elements in such a way that their chemical properties repeat in each vertical column. **Figure 2.1** illustrates an abbreviated periodic table, emphasizing the elements that make up organisms.

About 25 elements are essential to life. Of these, the **bulk elements** are required in the largest amounts because they make up the vast majority of every living cell. The four most abundant bulk elements in life are carbon, hydrogen, oxygen, and nitrogen. Other

bulk elements include phosphorus (P), sulfur (S), sodium (Na), magnesium (Mg), potassium (K), and calcium (Ca). **Trace elements,** such as iron (Fe) and zinc (Zn), are required in small amounts.

A person whose diet is deficient in any essential element can become ill or die. The thyroid gland, for example, requires the trace element iodine (I). If the diet does not supply enough iodine, the thyroid may become enlarged, forming a growth called a *goiter* in the neck. Similarly, red blood cells require iron (Fe) to carry oxygen to the body's tissues. An iron-poor diet can cause anemia, which is a decline in the number of red blood cells.

B. Atoms Are Particles of Elements

An **atom** is the smallest possible "piece" of an element that retains the characteristics of the element. An atom is composed of three types of subatomic particles (**figure 2.2** and **table 2.1**). **Protons,** which carry a positive charge, and **neutrons,** which are uncharged, together form a central **nucleus.** Negatively charged **electrons** surround the nucleus. Compared with a proton or neutron, an electron is essentially weightless.

For simplicity, most illustrations of atoms show the electrons closely hugging the nucleus. In reality, however, if the nucleus of a hydrogen atom were the size of a small marble, the entire atom would have a diameter slightly longer than a football field! Thus, most of an atom's mass is concentrated in the nucleus, while the electron cloud occupies virtually all of its volume.

How can this electron cloud, which is mostly empty space, account for the solid "feel" of the objects in our world? The fact that the electrons are in constant motion helps explain this paradox. A good analogy is a ceiling fan. When the fan is not spinning, it is easy to move your hand between two blades. But when the fan is on, the rotating blades essentially form a solid disk.

Each element has a unique **atomic number,** the number of protons in the nucleus. Hydrogen, the simplest type of atom, has an atomic number of 1. In contrast, an atom of uranium has

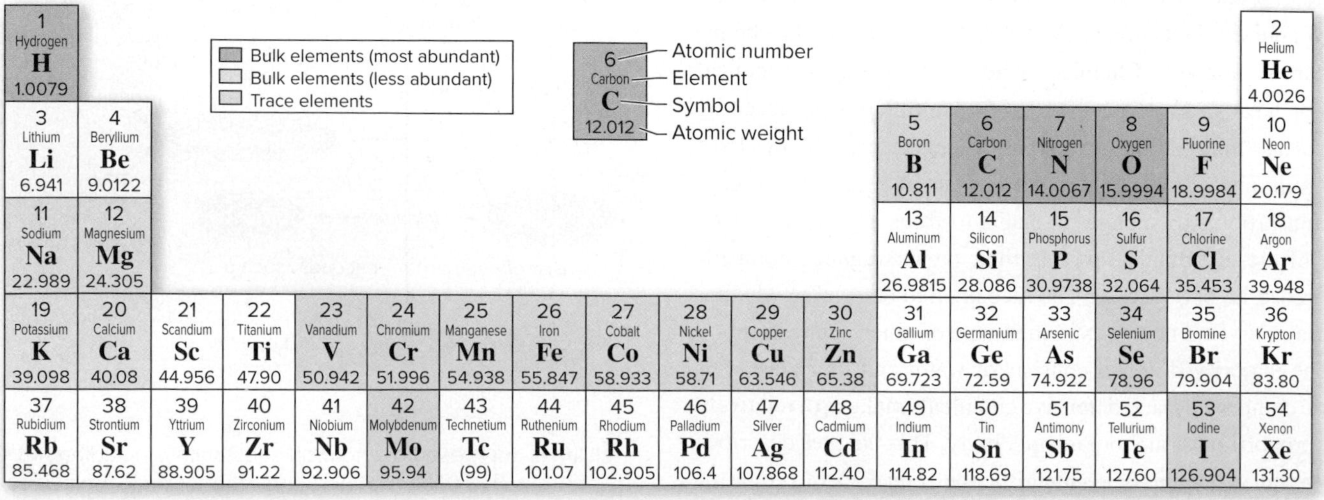

Figure 2.1 The Periodic Table of Elements. This abbreviated periodic table shows the first 54 elements, each with a unique atomic number and a symbol. A complete periodic table appears in appendix D.

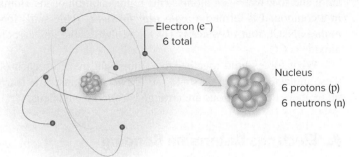

Figure 2.2 Atom Anatomy. The nucleus at the center of an atom is made of protons and neutrons. A cloud of electrons surrounds the nucleus. This example has six protons, so it is a carbon atom.

TABLE **2.1** Types of Subatomic Particles

Particle	Charge	Mass	Location
Electron	Negative (−)	~0	Surrounding nucleus
Neutron	None	1	Nucleus
Proton	Positive (+)	1	Nucleus

92 protons. Elements are arranged sequentially in the periodic table by atomic number, which appears above each element's symbol (see figure 2.1).

When the number of protons equals the number of electrons, the atom is electrically neutral; that is, it has no net charge. An **ion** is an atom (or group of atoms) that has gained or lost electrons and therefore has a net negative or positive charge. One common positively charged ion (cation) is hydrogen, H^+ (**figure 2.3**); others include sodium (Na^+) and potassium (K^+). Negatively charged ions (anions) include hydroxide (OH^-) and chloride (Cl^-). Ions participate in many biological processes, including the transmission of messages in the nervous system. They also form ionic bonds, discussed in section 2.2. ⓘ *action potential, section 26.3*

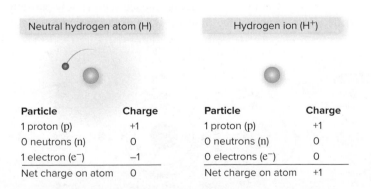

Figure 2.3 Two Forms of Hydrogen. A neutral (uncharged) atom of hydrogen consists of one proton and one electron, whose charges balance one another. A hydrogen ion (H^+) has a net positive charge because it has lost its electron.

C. Isotopes Have Different Numbers of Neutrons

An atom's **mass number** is the total number of protons and neutrons in its nucleus. Because neutrons and protons have the same mass (see table 2.1), subtracting the atomic number from the mass number yields the number of neutrons in an atom.

All atoms of an element have the same number of protons but not necessarily the same number of neutrons. An **isotope** is any of these different forms of a single element (**figure 2.4**). For example, carbon has three isotopes, designated ^{12}C (six neutrons), ^{13}C (seven neutrons), and ^{14}C (eight neutrons). The superscript denotes the mass number of each isotope.

Figure It Out

The most abundant isotope of iron (Fe) has a mass number of 56. Using the information in figure 2.1, how many neutrons are in each atom of ^{56}Fe?

Answer: 30

Often one isotope of an element is very abundant, and others are rare. For example, about 99% of carbon isotopes are ^{12}C, and only 1% are ^{13}C or ^{14}C. An element's **atomic weight,** which is the average mass of all atoms of an element, is typically close to the mass number of the most abundant isotope.

Many of the known isotopes are unstable and **radioactive,** which means they emit energy as rays or particles when they break down into more stable forms. Every radioactive isotope has a characteristic half-life, which is the time it takes for half of the atoms in a sample to emit radiation, or "decay" to a different, more stable form. Scientists have determined the half-life of each radioactive isotope experimentally. Depending on the isotope, the half-life might range from a fraction of a second to millions or even billions of years. Physicists use large samples of isotopes and precise measurements to calculate the longest half-lives.

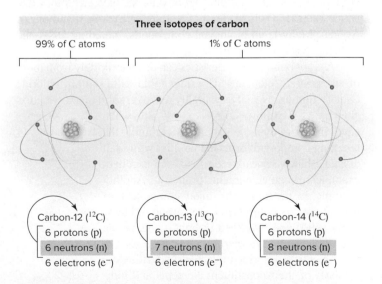

Figure 2.4 Three Isotopes of Carbon. Carbon's atomic number is 6, so its nucleus always contains six protons. These three carbon isotopes, however, have different numbers of neutrons.

Radioactive isotopes have many uses in medicine and science, ranging from detecting broken bones to determining the ages of fossils. But the same properties that make radioactive isotopes useful can also make them dangerous. Exposure to excessive radiation can lead to radiation sickness, and radiation-induced mutations of a cell's DNA can cause cancer (see chapter 8). The lead-containing "bib" that a dentist places on your chest during mouth X-rays protects you from radiation. ⓘ *radiometric dating,* section 13.2C

2.1 | MASTERING CONCEPTS

1. Which four elements do organisms require in the largest amounts?
2. Where in an atom are protons, neutrons, and electrons located?
3. What does an element's atomic number indicate?
4. What is the relationship between the mass of ^{12}C and the atomic weight of carbon (12.012)?
5. How are the isotopes of an element different from one another?

2.2 Chemical Bonds Link Atoms

Like all organisms, you are composed mostly of carbon, hydrogen, oxygen, and nitrogen atoms. But the arrangement of these atoms is not random. Instead, your atoms are organized into molecules (see figure 1.2). A **molecule** is two or more chemically joined atoms.

Some molecules, such as the gases hydrogen (H_2), oxygen (O_2), and nitrogen (N_2), consist of two atoms of the same element. More often, however, the elements in a molecule are different. A **compound** is a molecule composed of two or more different elements. Carbon monoxide (CO), for example, is a compound consisting of one carbon and one oxygen atom. Likewise, water (H_2O) is made of two atoms of hydrogen and one of oxygen. Many large biological compounds, including DNA and proteins, consist of tens of thousands of atoms.

A compound's characteristics can differ strikingly from those of its separate elements. Consider table salt, sodium chloride (NaCl). Sodium is a silvery, highly reactive, solid metal, whereas chlorine is a yellow, corrosive gas. But when equal numbers of these two atoms combine, the resulting compound forms the familiar white salt crystals that we sprinkle on food—an excellent example of an emergent property. Another example is methane, the main component of natural gas. Its components are carbon (a black, sooty solid) and hydrogen (a light, combustible gas). ⓘ *emergent properties,* section 1.1A

Scientists describe molecules by writing the symbols of their constituent elements and indicating the number of atoms of each element as subscripts. For example, methane is written CH_4, which denotes a molecule with one carbon

atom and four hydrogen atoms. This representation of the atoms in a compound is termed a *molecular formula.* Table salt's formula is NaCl, that of water is H_2O, and that of the gas carbon dioxide is CO_2.

What forces hold together the atoms that make up each of these molecules? To understand the answer, we must first learn more about how electrons are arranged around the nucleus.

A. Electrons Determine Bonding

Electrons occupy distinct energetic regions around the nucleus. They are constantly in motion, so it is impossible to determine the exact location of any electron at any given moment. Instead, chemists use the term **orbitals** to describe the most likely location for an electron relative to its nucleus. Each orbital can hold up to two electrons. Consequently, the more electrons in an atom, the more orbitals they occupy.

We can envision an atom's orbitals as being grouped into a series of concentric **energy shells,** each having a higher energy level than the one inside it (**figure 2.5**). The number of orbitals in each shell determines the number of electrons the shell can hold. The lowest energy shell, for example, contains just one orbital and thus holds up to two electrons. The next two shells each contain four orbitals and therefore hold as many as eight electrons each.

Electrons occupy the lowest energy level available to them, starting with the innermost one. As each energy shell fills, any additional electrons must reside in higher energy shells. For example, hydrogen has only one electron in the lowest energy orbital, and helium has two. Carbon has six electrons; two occupy the lowest energy orbital, and four are in the next energy shell. Oxygen, with eight electrons total, has two electrons in the lowest energy orbital and six at the next higher energy level.

An atom's **valence shell** is its outermost occupied energy shell. Atoms are most stable when their valence shells are full. The gases helium (He) and neon (Ne), for example, are inert—that is, they are chemically unreactive. Because their outermost shells are full, they exist in nature without combining with other atoms.

For most atoms, however, the valence shell is only partially filled. Look again at the first few rows of the periodic table in

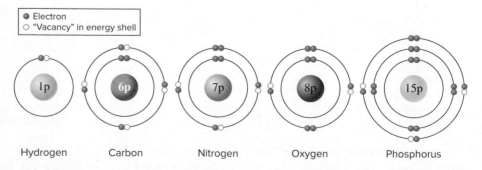

● Electron
○ "Vacancy" in energy shell

Hydrogen　　Carbon　　Nitrogen　　Oxygen　　Phosphorus

Figure 2.5 **Energy Shells.** Shown here are models of the most common atoms in organisms. Each pair of electrons represents one orbital.

figure 2.1. Lithium (Li), which occupies the first column, has one valence electron, as do all other elements directly below it. Beryllium (Be) has two, as does magnesium (Mg). Working across the table, note that boron (B) has three valence electrons, carbon has four, nitrogen has five, oxygen has six, and fluorine (F) has seven.

An atom of each of these elements will become most stable if its valence-shell "vacancies" fill. As you will soon see, atoms may donate, steal, or share electrons to arrive at exactly the right number. The exact "strategy" that an atom uses depends in part on its **electronegativity,** which measures the atom's ability to attract electrons on a scale of 0 to 4 (figure 2.6). Oxygen, for example, has high electronegativity compared to sodium. Elements with high electronegativity tend to strip electrons away from those with lower values. Elements with moderate electronegativity often share electrons.

Whether electrons are stolen or shared, the transfer of electrons from one atom to another creates a **chemical bond,** an attractive force that holds atoms together. The remainder of this section describes three types of chemical bonds that are important in biology.

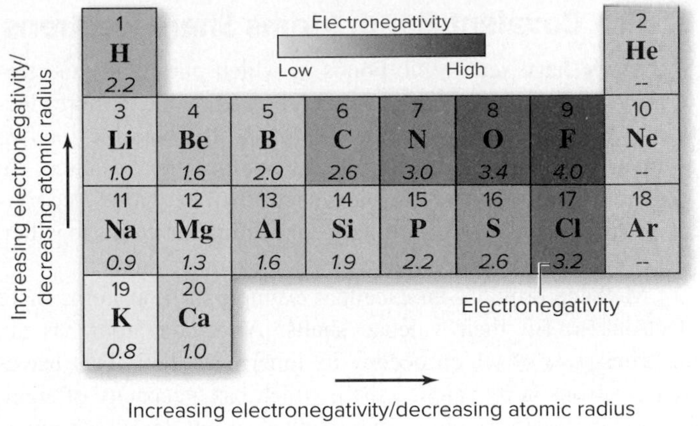

B. In an Ionic Bond, One Atom Transfers Electrons to Another Atom

Sometimes, two atoms have such different electronegativities that one actually takes one or more of its partner's electrons. Recall that an atom is most stable if its valence shell is full. The most electronegative atoms, such as chlorine (Cl), are usually those whose valence shells have only one "vacancy." Likewise, sodium (Na) and other weakly electronegative atoms have only one electron in the outermost shell. Neither chlorine nor sodium would benefit from sharing. Instead, sodium is most stable if it simply releases its extra electron to chlorine, which needs this "scrap" electron to complete its own valence shell (figure 2.7).

Figure 2.6 Unequal Attraction. Atoms vary widely in their electronegativity, which is the ability to attract electrons. Note that the most electronegative atoms pull electrons close to the nucleus, decreasing the overall atomic radius.

An ion is an atom that has lost or gained electrons. The atom that has lost electrons—such as the sodium atom in figure 2.7—is an ion carrying a positive charge. Conversely, the one that has gained electrons—chlorine in this case—acquires a negative charge. An **ionic bond** results from the electrical attraction between two ions with opposite charges. In general, such bonds form between an atom whose outermost shell is almost empty and one whose valence shell is nearly full.

The ions in figure 2.7 have bonded ionically to form NaCl. In NaCl, the most stable configuration of Na^+ and Cl^- is a three-dimensional crystal. Ionic bonds in crystals are strong, as demonstrated by the stability of the salt in your shaker. Those same crystals, however, dissolve when you stir them into water. As described in section 2.3, water molecules pull ionic bonds apart.

Figure 2.7 Table Salt, an Ionically Bonded Molecule. (a) A sodium atom (Na) can donate the single electron in its valence shell to a chlorine atom (Cl), which has seven electrons in its outermost shell. After the transfer, the valence shells of both atoms are full. The resulting ions (Na^+ and Cl^-) form the compound sodium chloride (NaCl). (b) In a salt crystal, Na^+ and Cl^- ions occur in a repeating pattern.

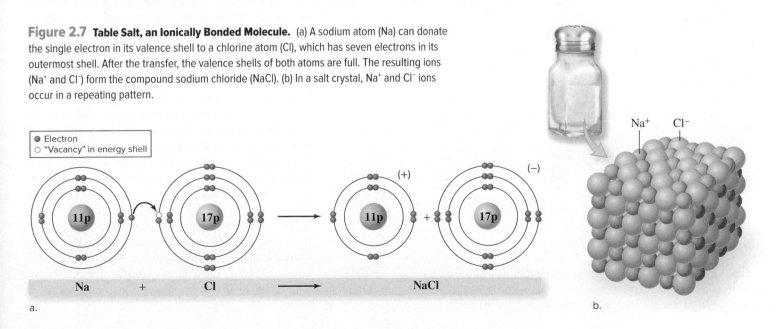

C. In a Covalent Bond, Atoms Share Electrons

So far, we have seen ionic bonds in which one highly electronegative atom fills its outermost shell by taking one or more electrons from another atom. However, it is also possible for two atoms to fill their outermost shells by pooling their resources. In a **covalent bond,** two atoms share electrons. The shared electrons travel around both nuclei, strongly connecting the atoms together. Most of the bonds in biological molecules are covalent.

Methane provides an excellent example of how atoms share electrons to fill their valence shells. A carbon atom has six electrons, two of which occupy its innermost shell. That leaves four electrons in its valence shell, which has a capacity of eight. Carbon therefore requires four more electrons to fill its outermost shell. A carbon atom can attain the stable eight-electron configuration by sharing electrons with four hydrogen atoms, each of which has one electron in its only shell. The resulting molecule is methane, CH_4 (**figure 2.8a**). Figure 2.8b shows how oxygen and hydrogen form covalent bonds as they combine to produce water.

Figure It Out

Use the information in figure 2.1 to predict the number of covalent bonds that nitrogen (N) forms.

Answer: 3

Covalent bonds are usually depicted as lines between the interacting atoms, with each line representing one bond. Each single bond contains two electrons, one from each atom. Atoms can also share two pairs of electrons, forming a double covalent bond.

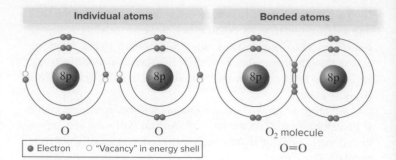

● Electron   ○ "Vacancy" in energy shell

Figure 2.9 Double Bond. A pair of atoms can share two pairs of electrons, forming a double covalent bond. This example shows the double bond joining the two oxygen atoms in O_2.

The O_2 molecule, for example, has one double bond (**figure 2.9**). The greater the number of shared electrons, the stronger the bond. A triple covalent bond (three shared pairs of electrons) is therefore extremely strong. The atoms in nitrogen gas, N_2, are joined by a triple bond.

Covalent bonding means "sharing," but the partnership is not necessarily equal. A **polar covalent bond** is a lopsided union in which one nucleus exerts a much stronger pull on the shared electrons than does the other nucleus. Polar bonds form whenever a highly electronegative atom such as oxygen shares electrons—unequally—with a less electronegative partner such as carbon or hydrogen. Like a battery, a polar covalent bond has a positive end and a negative end.

Polar covalent bonds are critical to biology. As described in section 2.2D, they are responsible for hydrogen bonds, which in turn help define not only the unique properties of water (see section 2.3) but also the shapes of DNA and proteins (section 2.5).

In contrast, a **nonpolar covalent bond** is a "bipartisan" union in which both atoms exert approximately equal pull on their shared electrons. A bond between two atoms of the same element, such as a carbon–carbon bond, is nonpolar; after all, a bond between two identical atoms must be electrically balanced. H_2, N_2, and O_2 are all nonpolar molecules. Carbon and hydrogen atoms have similar electronegativity. A carbon–hydrogen bond is therefore also nonpolar.

Ionic bonds, polar covalent bonds, and nonpolar covalent bonds represent points along a continuum. If one atom is so electronegative that it rips electrons from another atom's valence shell, an ionic bond forms. If one atom tugs at shared electrons much more than the other, a covalent bond is polar. And two atoms of similar electronegativity share electrons equally in nonpolar covalent bonds. Notice that the bond type depends on the *difference* in electronegativity, so the same element can participate in different types of bonds. Oxygen, for example, forms nonpolar bonds with itself (as in O_2) and polar bonds with hydrogen (as in H_2O).

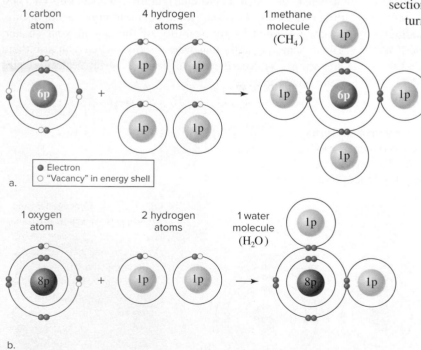

● Electron   ○ "Vacancy" in energy shell

Figure 2.8 Covalent Bonds. (a) In methane (CH_4), one carbon atom and four hydrogen atoms complete their outermost shells by sharing electrons. (b) A water molecule (H_2O) has covalent bonds between one oxygen atom and two hydrogen atoms.

TABLE **2.2** **Chemical Bonds: A Summary**

Type	Chemical Basis	Strength	Example
Ionic bond	One atom donates one or more electrons to another atom; electronegativity difference between atoms is very large (> 1.7). The resulting oppositely charged ions attract each other.	Strong but breaks easily in water	Sodium chloride (NaCl)
Covalent bond	Two atoms share pairs of electrons.	Strong	
Nonpolar	Electronegativity difference between atoms is small (< 0.4).		H–H bond in H_2 molecule
Polar	Electronegativity difference between atoms is moderate or large (0.4 to 1.7).		O–H bond within water molecule
Hydrogen bond	An atom with a partial negative charge attracts a hydrogen atom with a partial positive charge. Hydrogen bonds form between adjacent molecules or between different parts of a large molecule.	Weak	Attraction between adjacent water molecules

D. Partial Charges on Polar Molecules Create Hydrogen Bonds

When a covalent bond is polar, the negatively charged electrons spend more time around the nucleus of the more electronegative atom than around its partner. The "electron-hogging" atom therefore has a partial negative charge (written as "δ^-"), and the less electronegative partner has an electron "deficit" and a partial positive charge (δ^+).

In a **hydrogen bond,** opposite partial charges on *adjacent molecules*—or within a single large molecule—attract each other. The name comes from the fact that the atom with the partial positive charge is always hydrogen. The atom with the partial negative charge, on the other hand, is a highly electronegative atom such as oxygen or nitrogen.

Water provides the simplest illustration of hydrogen bonds (figure 2.10). Each water molecule has a "boomerang" shape, owing to the oxygen atom's two pairs of unshared valence electrons (see figure 2.8). Moreover, the O atom is joined to the two H atoms by polar covalent bonds, with the oxygen atom's nucleus attracting the shared electrons more strongly than do the hydrogen nuclei.

Each hydrogen atom in a water molecule therefore has a partial positive charge, which attracts the partial negative charge of the oxygen atom on an adjacent molecule. This attraction is the hydrogen bond. The partial charges on O and H, plus the bent shape, cause water molecules to stick to one another and to some other substances. (This slight stickiness is another example of an emergent property, because it arises from interactions between O and H.)

Hydrogen bonds are relatively weak compared with ionic and covalent bonds. In 1 second, the hydrogen bonds between one water molecule and its nearest neighbors break and re-form some 500 billion times. Even though hydrogen bonds are weak, they account for many of water's unusual characteristics—the subject of section 2.3. In addition, the collective strength of multiple hydrogen bonds helps stabilize some large molecules, including proteins and DNA (see section 2.5).

Table 2.2 summarizes the properties of ionic, covalent, and hydrogen bonds.

2.2 MASTERING CONCEPTS

1. How are atoms, molecules, and compounds related?
2. How does the number of valence electrons determine an atom's tendency to form bonds?
3. Explain how electronegativity differences between atoms result in each type of chemical bond.

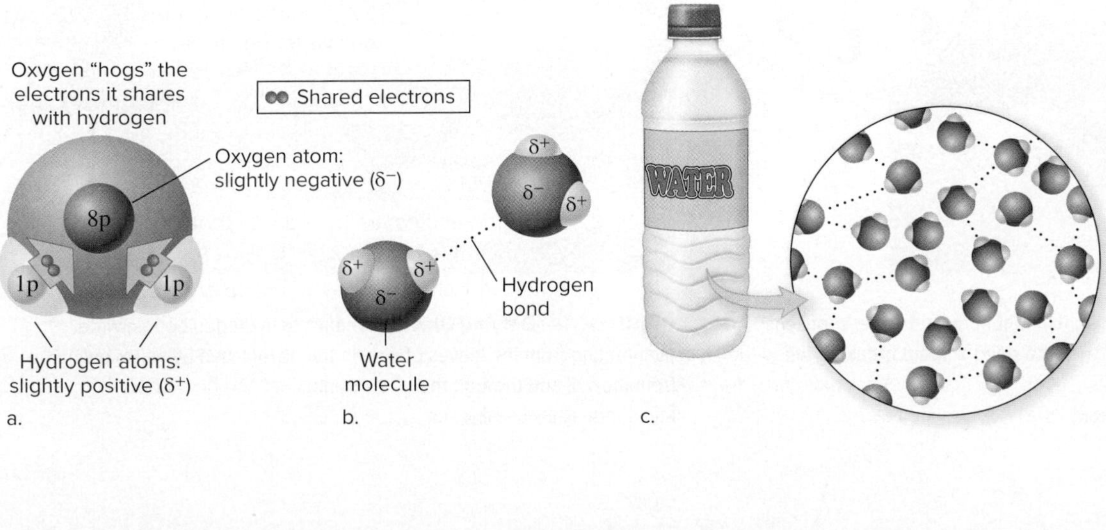

Oxygen "hogs" the electrons it shares with hydrogen

Oxygen atom: slightly negative (δ^-)

Hydrogen atoms: slightly positive (δ^+)

a.

Shared electrons

Water molecule

Hydrogen bond

b.

c.

Figure 2.10 Hydrogen Bonds in Water. (a) Water's two covalent bonds are polar; the oxygen atom attracts electrons more strongly than do the hydrogen atoms. The O atom therefore bears a partial negative charge (δ^-), and the H atoms carry partial positive charges (δ^+). (b) The hydrogen bond is the attraction between partial charges on adjacent molecules. (c) In liquid water, many molecules stick to one another with hydrogen bonds.

2.3 Water Is Essential to Life

Although water may seem to be a rather ordinary fluid, it is anything but. The tiny, three-atom water molecule has extraordinary properties that make it essential to all organisms, which explains why the search for life on other planets begins with the search for water. Indeed, life on Earth began in water, and for at least the first 3 billion years of life's history on Earth, all life was aquatic (see chapter 15). It was not until some 475 million years ago, when plants and fungi colonized land, that life could survive without being surrounded by water. Even now, terrestrial organisms cannot live without it. This section explains some of the properties that make water central to biology.

A. Water Is Cohesive and Adhesive

Hydrogen bonds contribute to a property of water called **cohesion**—the tendency of water molecules to stick together. Without cohesion, water would evaporate instantly from most locations on Earth's surface. Cohesion also contributes to the observation that you can sometimes fill a glass so full that water is above the rim, yet it doesn't flow over the side unless disturbed.

This tendency of a liquid to hold together at its surface is called surface tension, and not all liquids exhibit it. Water has high surface tension because it is cohesive. At the boundary between water and air, the water molecules form hydrogen bonds with neighbors to their sides and below them in the liquid. These bonds tend to hold the surface molecules together, creating a thin "skin" that is strong enough to support small animals without breaking through (**figure 2.11**).

A related property of water is **adhesion,** the tendency to form hydrogen bonds with substances other than water. For example, when water soaks into a paper towel, it is adhering to the molecules that make up the paper.

Both adhesion and cohesion are at work when water seemingly defies gravity as it moves from a plant's roots to its highest leaves (**figure 2.12**). This movement depends upon cohesion of water within the plant's conducting tubes. Water entering roots is drawn up through these tubes as water molecules evaporate from leaf cells. Adhesion to the walls of the conducting tubes also helps lift water to the topmost leaves of trees. ⓘ *water uptake,* section 23.2A

B. Many Substances Dissolve in Water

Another reason that water is vital to life is that it can dissolve a wide variety of chemicals. To illustrate this process, picture the slow disappearance of table salt as it dissolves in water. Although the salt crystals seem to vanish, the sodium and chloride ions remain. Water molecules surround each ion individually, separating them from one another (**figure 2.13**).

In this example, water is a **solvent:** a chemical in which other substances, called **solutes,** dissolve. A **solution** consists of one or more solutes dissolved in a liquid solvent. In a so-called aqueous solution, water is the solvent. But not all solutions are aqueous. According to the rule "Like dissolves like," polar solvents such as water dissolve polar molecules; similarly, nonpolar solvents dissolve nonpolar substances.

Scientists divide chemicals into two categories, based on their affinity for water. **Hydrophilic** substances are either polar or charged, so they readily dissolve in water (the term literally means "water-loving"). Examples include sugar, salt, and ions.

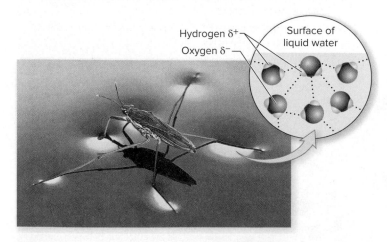

Hydrogen δ⁺
Oxygen δ⁻
Surface of liquid water

Figure 2.11 Running on Water. A lightweight body and water-repellent legs allow this water strider to "skate" across a pond without breaking the water's surface tension.

Photo: ©Herman Eisenbeiss/Science Source

1 Water evaporates through pores in leaves.

2 Evaporating molecules pull water up stem.

3 Water molecules are pulled into roots.

Figure 2.12 Defying Gravity. Thanks to hydrogen bonds, water evaporating from the leaves of a palm tree is replaced by water pulled up from the soil and through the tree's trunk.

Photo: ©Getty Images/flickr RF

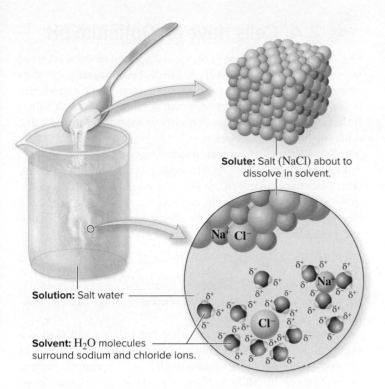

Figure 2.13 **Dissolving Salt.** As salt crystals dissolve, polar water molecules surround each sodium and chloride ion.

Electrolytes are ions in the body's fluids, and the salty taste of sweat illustrates water's ability to dissolve them. Sports drinks replace not only water but also sodium, potassium, magnesium, and calcium ions that are lost in perspiration during vigorous exercise. Electrolytes are essential to many processes, including heart and nerve function.

Not every substance, however, is water-soluble. Nonpolar molecules are called **hydrophobic** ("water-fearing") because they do not dissolve in, or form hydrogen bonds with, water. Butter and oil are hydrophobic because they are made mostly of carbon and hydrogen, which form nonpolar bonds with each other. This is why water alone will not remove grease from hands, dishes, or clothes. Detergents contain molecules that attract both water and fats, so they can dislodge greasy substances and carry the mess down the drain with the wastewater.

C. Water Regulates Temperature

Another unusual property of water is its ability to resist temperature changes. When molecules absorb energy, they move faster. Water's hydrogen bonds tend to counteract this molecular movement; as a result, more heat is needed to raise water's temperature than is required for most other liquids, including alcohols. Because an organism's fluids are aqueous solutions, the same effect holds: An organism may encounter considerable heat before its body temperature becomes dangerously high. Likewise, the body cools slowly in cold temperatures.

At a global scale, water's resistance to temperature change explains why coastal climates tend to be mild. People living along the California coast have good weather year-round because the Pacific Ocean's steady temperature helps keep winters warm and summers cool. Far away from the ocean, in the central United States, winters are much colder and summers are much hotter. These differences in local climate contribute to the unique ecosystems that occur in each region. (Chapter 39 describes climate in more detail.)

Hydrogen bonds also mean that a lot of heat is required to evaporate water. **Evaporation** is the conversion of a liquid into a vapor. When sweat evaporates from skin, individual water molecules break away from the liquid droplet and float into the atmosphere. Surface molecules must absorb energy to escape, and when they do, heat energy is removed from those that remain, drawing heat out of the body—an important part of the mechanism that regulates body temperature.

D. Water Expands As It Freezes

Water's unusual tendency to expand upon freezing also affects life. In liquid water, hydrogen bonds are constantly forming and breaking, and the water molecules are relatively close together. But in an ice crystal, the hydrogen bonds are stable, and the molecules are "locked" into roughly hexagonal shapes. Therefore, the less-dense ice floats on the surface of the denser liquid water below (**figure 2.14**). This characteristic benefits aquatic organisms.

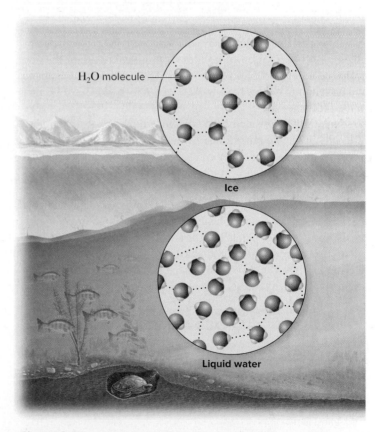

Figure 2.14 **Ice Floats.** The water molecules in ice form hexagons stabilized by hydrogen bonds. Ice is therefore less dense than—and floats on top of—the liquid water in this lake.

When the air temperature drops, a small amount of water freezes at the pond's surface. This solid cap of ice retains heat in the water below. If ice were to become denser upon freezing, it would sink to the bottom. The lake would then gradually turn to ice from the bottom up, entrapping the organisms that live there.

The formation of ice crystals inside cells, however, can be deadly. The expansion of ice inside a frozen cell can rupture the delicate outer membrane, killing the cell. How, then, do organisms survive in extremely cold weather? Mammals have thick layers of insulating fur and fat that help their bodies stay warm. Icefishes that live in the cold waters surrounding Antarctica have a different adaptation: They produce antifreeze chemicals that prevent their cells from freezing solid.

E. Water Participates in Life's Chemical Reactions

Life exists because of thousands of simultaneous chemical reactions. In a **chemical reaction,** two or more molecules "swap" their atoms to yield different molecules; that is, some chemical bonds break and new ones form. Chemists depict reactions as equations with the **reactants,** or starting materials, to the left of an arrow; the **products,** or results of the reaction, are listed to the right.

Consider what happens when the methane in natural gas burns inside a heater, gas oven, or stove:

$$CH_4 + 2O_2 \longrightarrow CO_2 + 2H_2O$$
$$\text{methane} + \text{oxygen} \longrightarrow \text{carbon dioxide} + \text{water}$$

In words, this equation says that one methane molecule combines with two oxygen molecules to produce a carbon dioxide molecule and two molecules of water. The bonds of the methane and oxygen molecules have broken, and new bonds have formed in the products.

Note that each side of the equation shows the same number of atoms of each element: one carbon, four hydrogens, and four oxygens. Atoms are neither created nor destroyed in a chemical reaction; rather, they are simply rearranged.

Nearly all of life's chemical reactions occur in the watery solution that fills and bathes cells. Moreover, water is either a reactant in or a product of many of these reactions. In photosynthesis, for example, plants use the sun's energy to assemble food out of just two reactants: carbon dioxide and water (see chapter 5). Section 2.5 describes two other water-related reactions, hydrolysis and dehydration synthesis, that are vital to life.

2.3   MASTERING CONCEPTS

1. How are cohesion and adhesion important to life?
2. Distinguish between a solute and a solvent and between a hydrophilic and a hydrophobic molecule.
3. How does water help an organism regulate its body temperature?
4. How does the density difference between ice and liquid water affect life?
5. What happens in a chemical reaction?
6. How does water participate in the chemistry of life?

2.4 Cells Have an Optimum pH

One of the most important substances dissolved in water is one of the simplest: H^+ ions. Each H^+ is a hydrogen atom stripped of its electron; in other words, it is simply a proton. But its simplicity belies its enormous effects on living systems. Too much or too little H^+ can ruin the shapes of critical molecules inside cells, rendering them nonfunctional.

One source of H^+ is pure water. At any time, about one in a million water molecules spontaneously breaks into two pieces, with one of the hydrogen atoms separating from the rest of the molecule. The highly electronegative oxygen atom keeps the electron from the breakaway hydrogen atom. The result is one hydrogen ion (H^+) and one hydroxide ion (OH^-):

$$H_2O \longrightarrow H^+ + OH^-$$

In pure water, the number of hydrogen ions exactly equals the number of hydroxide ions. A **neutral** solution likewise contains as much H^+ as it does OH^-.

Some substances, however, alter this balance. An **acid** is a chemical that adds H^+ to a solution, making the concentration of H^+ ions exceed the concentration of OH^- ions. Examples include hydrochloric acid (HCl), sulfuric acid (H_2SO_4), and sour foods such as vinegar and lemon juice. Adding hydrochloric acid to pure water releases H^+ ions into the solution:

$$HCl \longrightarrow H^+ + Cl^-$$

Because no OH^- ions were added at the same time, the balance of H^+ to OH^- skews toward extra H^+.

A **base** is the opposite of an acid: It makes the concentration of OH^- ions exceed the concentration of H^+ ions. Bases work in one of two ways. They come apart to directly add OH^- ions to the solution, or they absorb H^+ ions. Either way, the result is the same: The balance between H^+ and OH^- shifts toward OH^-. Two common household bases are baking soda and sodium hydroxide (NaOH), an ingredient in oven and drain cleaners. When NaOH dissolves in water, it releases OH^- into solution:

$$NaOH \longrightarrow Na^+ + OH^-$$

What happens if a person mixes an acid with a base? The acid releases protons, while the base either absorbs the H^+ or releases OH^-. Acids and bases therefore neutralize each other.

Both acids and bases are important in everyday life. In the environment, some air pollutants return to Earth as acid precipitation. The acidic rainfall kills plants and aquatic life, and it damages buildings and outdoor sculptures. On a more positive note, the tart flavors of yogurt and sour cream come from acid-producing bacteria, and your stomach produces strong acids that kill microbes and activate enzymes that begin the digestion of food. At the opposite end of the spectrum, drugs called antacids contain bases that neutralize excess acid, relieving an upset stomach. In addition, many household cleaners are strong bases, which break down grease without corroding metal. ⓘ *acid deposition,* section 40.3B; *stomach acid,* section 32.3B

A. The pH Scale Expresses Acidity or Alkalinity

Scientists use the **pH scale** to measure how acidic or basic a solution is. The pH scale ranges from 0 to 14, with 7 representing a neutral solution such as pure water (figure 2.15). An acidic solution has a pH lower than 7, whereas an **alkaline,** or basic, solution has a pH greater than 7. Note that the higher the H^+ concentration of a solution, the lower its pH. Thus, 0 represents a strongly acidic solution and 14 represents an extremely basic one (low H^+ concentration).

Each unit on the pH scale represents a 10-fold change in H^+ concentration. A solution with a pH of 4 is therefore 10 times more acidic than one with a pH of 5, and it is 100 times more acidic than one with a pH of 6.

All species have characteristic pH requirements. Some organisms, such as the bacteria that cause ulcers in human stomachs, are adapted to low-pH environments. In contrast, the normal pH of human blood is 7.35 to 7.45. Extremely shallow breathing or kidney failure can cause the blood's pH to drop below 7. Vomiting, hyperventilating, or taking some types of alkaloid drugs, on the other hand, can raise the blood's pH above 7.8. Straying too far from the normal pH can cause death by destroying the shapes of critical proteins (see section 2.5B).

B. Buffers Regulate pH

Maintaining the correct pH of body fluids is critical, yet organisms frequently encounter conditions that could alter their internal pH. They can maintain homeostasis because cells contain **buffers,** pairs of weak acids and bases that resist pH changes.

Hydrochloric acid is a strong acid because it releases all of its H^+ when dissolved in water. As you can see in figure 2.15, the pH of pure HCl is 0. A weak acid, in contrast, does not release all of its H^+ into solution. An example is carbonic acid, H_2CO_3, which forms one part of the human body's pH buffer:

$$H_2CO_3 \rightleftharpoons H^+ + HCO_3^-$$
$$\text{carbonic acid} \qquad \text{bicarbonate}$$

The dual arrow indicates that the reaction can proceed in either direction, depending on the pH of the fluid. If a base removes H^+ from the solution, the reaction moves to the right to produce more H^+, restoring acidity. Alternatively, if an acid contributes H^+ to the solution, the reaction proceeds to the left and consumes the excess H^+. This action keeps the pH of the solution relatively constant. Carbonic acid is just one of several buffers that maintain the pH of blood at about 7.4.

2.4 MASTERING CONCEPTS

1. How do acids and bases affect a solution's H^+ concentration?
2. How do the values of 0, 7, and 14 relate to the pH scale?
3. How do buffers regulate the pH of a fluid?

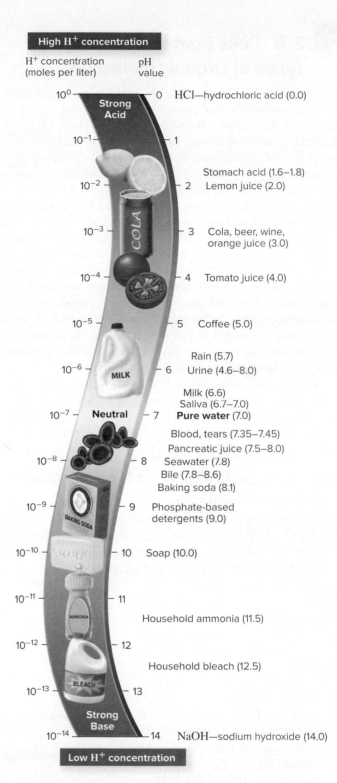

High H^+ concentration

H^+ concentration (moles per liter) — pH value

H^+ conc.	pH	
10^0	0	Strong Acid — HCl—hydrochloric acid (0.0)
10^{-1}	1	
10^{-2}	2	Stomach acid (1.6–1.8) / Lemon juice (2.0)
10^{-3}	3	Cola, beer, wine, orange juice (3.0)
10^{-4}	4	Tomato juice (4.0)
10^{-5}	5	Coffee (5.0)
10^{-6}	6	Rain (5.7) / Urine (4.6–8.0)
10^{-7}	Neutral 7	Milk (6.6) / Saliva (6.7–7.0) / **Pure water (7.0)**
10^{-8}	8	Blood, tears (7.35–7.45) / Pancreatic juice (7.5–8.0) / Seawater (7.8) / Bile (7.8–8.6) / Baking soda (8.1)
10^{-9}	9	Phosphate-based detergents (9.0)
10^{-10}	10	Soap (10.0)
10^{-11}	11	Household ammonia (11.5)
10^{-12}	12	Household bleach (12.5)
10^{-13}	13	
10^{-14}	14	Strong Base — NaOH—sodium hydroxide (14.0)

Low H^+ concentration

Figure 2.15 **The pH Scale.** A neutral solution has a pH of 7. The higher the H^+ concentration, the more acidic the solution (pH < 7). The lower the H^+ concentration, the more basic (alkaline) the solution (pH > 7).

Figure It Out

Flask A contains 100 mL of a solution with pH 5. After you add 100 mL of solution from Flask B, the pH rises to 7. What was the pH in Flask B?

2.5 Cells Contain Four Major Types of Organic Molecules

Organisms are composed mostly of water and **organic molecules,** chemical compounds that contain both carbon and hydrogen. (The first of this section's Burning Questions explains the use of the term *organic* in describing food.) As you will see later in this unit, plants and other autotrophs can produce all the organic molecules they require, whereas heterotrophs—including humans—must obtain their organic building blocks from food.

Life uses a tremendous variety of organic compounds. Organic molecules consisting almost entirely of carbon and hydrogen are called *hydrocarbons;* methane (CH_4) is the simplest example. Because a carbon atom forms four covalent bonds, however, this element can assemble into much more complex molecules, including long chains, intricate branches, and rings. Many organic compounds also include other essential elements, such as oxygen, nitrogen, phosphorus, or sulfur. A peek ahead at the molecules illustrated in this section reveals the diversity of shapes and sizes of organic molecules. Without carbon's versatility, organic chemistry—and life—would be impossible.

All organisms, from bacteria to plants to people, consist largely of the same four types of organic molecules: carbohydrates, proteins, nucleic acids, and lipids. This unity in life's chemistry is powerful evidence that all species inherited the same basic chemical structures and processes from a common ancestor.

A. Large Organic Molecules Are Composed of Smaller Subunits

Proteins, nucleic acids, and some carbohydrates all share a property in common with one another: They are **polymers,** which are chains of small molecular subunits called **monomers.** A polymer is made of monomers that are linked together, just as a train is made of individual railcars.

Railcars include two connectors, which enable one to hook to another in a long train. Similarly, organic molecules have small groups of atoms, called functional groups, that serve the same coupling function. **Figure 2.16** shows four of the most common examples: hydroxyl, carboxyl, amino, and phosphate groups. As you study this section, you will see that these functional groups participate in the reactions that create life's large organic molecules.

Cells use a chemical reaction called **dehydration synthesis** (also called a condensation reaction) to link monomers into polymers (**figure 2.17a**). In this reaction, a protein called an enzyme removes an –OH (hydroxyl group) from one molecule and a hydrogen atom

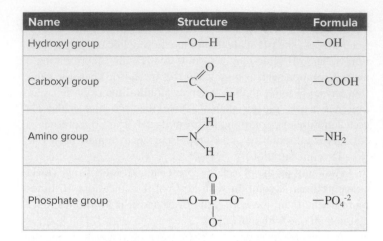

Name	Structure	Formula
Hydroxyl group	—O—H	—OH
Carboxyl group	(structure)	—COOH
Amino group	(structure)	—NH_2
Phosphate group	(structure)	—PO_4^{-2}

Figure 2.16 Functional Groups. Each of these groups of atoms occurs in one or more types of organic molecules. Look through the illustrations in the rest of section 2.5 to find examples of each type.

from another, forming H_2O and a new covalent bond between the two smaller components. (The term *dehydration* means that water is lost.) By repeating this reaction many times, cells can build extremely large polymers consisting of thousands of monomers.

The reverse reaction, called **hydrolysis,** breaks the covalent bonds that link monomers (figure 2.17b). In hydrolysis, enzymes use atoms from water to add a hydroxyl group to one molecule and a hydrogen atom to another (*hydrolysis* means "breaking with water"). Hydrolysis happens in your body when digestive enzymes in your stomach and intestines break down the proteins and other polymers in food.

The rest of this section takes a closer look at carbohydrates, proteins, nucleic acids, and lipids.

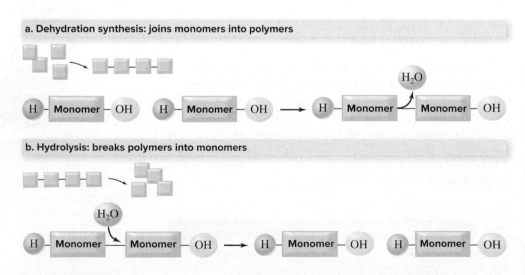

a. Dehydration synthesis: joins monomers into polymers

b. Hydrolysis: breaks polymers into monomers

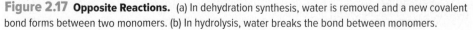

Figure 2.17 Opposite Reactions. (a) In dehydration synthesis, water is removed and a new covalent bond forms between two monomers. (b) In hydrolysis, water breaks the bond between monomers.

B. Carbohydrates Include Simple Sugars and Polysaccharides

Although they range from sweet to starchy, foods such as candy, sugary fruits, cereal, potatoes, pasta, and bread all share a common characteristic. They are rich in **carbohydrates,** organic molecules that consist of carbon, hydrogen, and oxygen, often in the proportion 1:2:1.

Carbohydrates are the simplest of the four main types of organic compounds, mostly because just a few monomers account for the most common types in cells. The two main groups of carbohydrates are simple sugars and complex carbohydrates.

Sugars (Simple Carbohydrates)
The smallest carbohydrates, the **monosaccharides,** usually contain five or six carbon atoms (**figure 2.18a**). Monosaccharides with the same number of carbon atoms can differ from one another by how their atoms are bonded. For example, glucose (blood sugar) and fructose (fruit sugar) are both six-carbon monosaccharides with the molecular formula $C_6H_{12}O_6$, but their chemical structures differ.

A **disaccharide** ("two sugars") is two monosaccharides joined by dehydration synthesis. Figure 2.18b shows how sucrose (table sugar) forms when a molecule of glucose bonds to a molecule of fructose. Lactose, or milk sugar, is also a disaccharide.

Together, the sweet-tasting monosaccharides and disaccharides are called sugars, or simple carbohydrates. Their function in cells is to provide a ready source of energy, which is released when their bonds are broken (see chapter 6). Sugarcane sap and sugar beet roots contain abundant sucrose, which the plants use to fuel growth. The disaccharide maltose provides energy in sprouting seeds; beer brewers also use it to promote fermentation.

Complex Carbohydrates
Chains of monosaccharides are collectively called complex carbohydrates. **Oligosaccharides** consist of 3 to 100 monomers. Such a carbohydrate chain sometimes attaches to a protein, forming a glycoprotein ("sugar protein"). Among other functions, glycoproteins on cell surfaces are important in immunity. For example, a person's blood type—A, B, AB, or O—refers to the combination of glycoproteins attached to the surface of his or her red blood cells. A transfusion of the "wrong" blood type can trigger a harmful immune reaction. ⓘ *blood type,* section 30.2B

Polysaccharides ("many sugars") are huge molecules consisting of hundreds or thousands of monosaccharide monomers (figure 2.18c). The most common polysaccharides are cellulose, chitin, starch, and glycogen. All are long chains of glucose, but they differ from one another by the orientation of the bonds that link the monomers.

Cellulose forms part of plant cell walls. Multiple cellulose molecules, held together along their length by hydrogen bonds, align side by side to form strong fibrils. Although it is the most common organic compound in nature, humans cannot digest it. Yet cellulose is an important component of the human diet, making up much of what nutrition labels refer to as "fiber." Cotton fibers, wood, and paper consist largely of cellulose. ⓘ *plant cell wall,* section 3.6B

Chitin is the second most common polysaccharide in nature. The cell walls of fungi contain chitin, as do the flexible

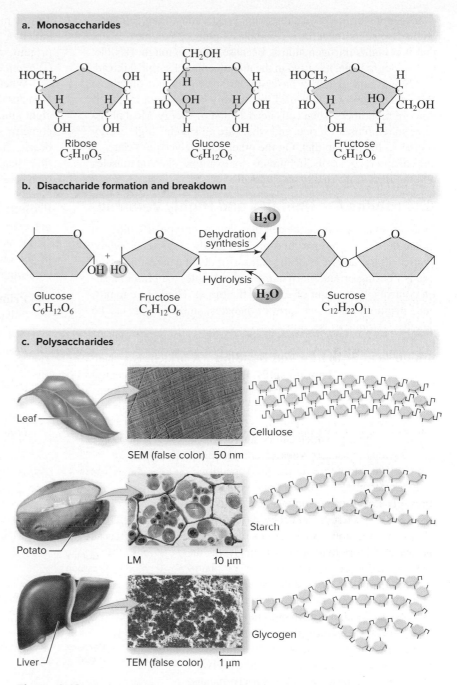

Figure 2.18 Carbohydrates. (a) Monosaccharides such as ribose, glucose, and fructose each consist of a single ring. (b) Disaccharides form by dehydration synthesis. In this example, glucose and fructose bond to form sucrose. (c) Polysaccharides such as cellulose, starch, and glycogen are long chains of monosaccharides. All of the examples shown here are composed of glucose monomers.

Photos: (c, cellulose): ©BioPhoto Associates/Science Source; (c, starch): ©Dr. Keith Wheeler/Science Source; (c, glycogen): ©Marshall Sklar/Science Source

exoskeletons of insects, spiders, and crustaceans. Like cellulose, chitin also supports cells. It resembles a glucose polymer, except that it contains nitrogen atoms. Because chitin is tough, flexible, and biodegradable, it is used in the manufacture of surgical thread.

Starch and glycogen have similar structures and functions. Both act as storage molecules that readily break down into their glucose monomers when cells need a burst of energy. Most plants store starch. Potatoes, rice, and wheat are all starchy, high-energy staples in the human diet. On the other hand, glycogen occurs in animal and fungal cells. In humans, for example, skeletal muscles and the liver store energy as glycogen.

C. Proteins Are Complex and Highly Versatile

Proteins do more jobs in the cell than any other type of biological molecule. Cells produce thousands of kinds of proteins, which control all the activities of life; illness or death can result if even one is missing or faulty. To name one example, the protein insulin controls the amount of sugar in the blood. The failure to produce insulin leads to one form of diabetes, an illness that can be deadly. ⓘ *diabetes*, section 28.4D

Amino Acid Structure and Bonding   A **protein** is a chain of monomers called **amino acids.** Each amino acid has a central carbon atom bonded to four other atoms or groups of atoms (**figure 2.19a**). One is a hydrogen atom; another is a carboxyl group; a third is an amino group ($-NH_2$); and the fourth is a side chain, or **R group,** which can be any of 20 chemical groups.

Organisms use 20 types of amino acids; figure 2.19a shows three of them. (Appendix E includes a complete set of amino acid structures.) The R groups distinguish the amino acids from one another, and they have diverse chemical structures. An R group may be as simple as the lone hydrogen atom in glycine or as complex as the two rings of tryptophan. Some R groups are acidic or basic; some are strongly hydrophilic or hydrophobic.

Just as the 26 letters in our alphabet combine to form a nearly infinite number of words in many languages, mixing and matching the 20 amino acids gives rise to an endless diversity of unique proteins. This variety means that proteins have a seemingly limitless array of structures and functions.

The dehydration synthesis reaction connects amino acids to each other; a **peptide bond** is the resulting covalent bond that links each amino acid to its neighbor (figure 2.19b). Two linked amino acids form a dipeptide; three form a tripeptide. Chains with fewer than 100 amino acids are peptides; examples include the nine-amino-acid-long hormone oxytocin, which plays a role in pair bonding and is sometimes called the "love hormone." Finally, chains with

100 or more amino acids are **polypeptides.** A polypeptide is called a *protein* once it folds into its functional shape; a protein may consist of one or more polypeptide chains. ⓘ *peptide hormones*, section 28.2A

Where do the amino acids in your own proteins come from? Humans can synthesize most of them from scratch. However, eight amino acids are considered "essential" because they must come from protein-rich foods such as meat, fish, dairy products, beans, and tofu. Digestive enzymes catalyze the hydrolysis reactions that release amino acids from proteins in food. The body then uses these monomers to build its own polypeptides.

Protein Folding   Unlike polysaccharides, most proteins do not exist as long chains inside cells. Instead, the polypeptide chain folds into a unique three-dimensional structure determined by the order and kinds of amino acids. Biologists describe the conformation of a protein at four levels (**figure 2.20**):

- **Primary (1°) structure:** The amino acid sequence of a polypeptide chain. This sequence determines all subsequent structural levels.

- **Secondary (2°) structure:** A "substructure" with a defined shape, resulting from hydrogen bonds between parts of the polypeptide. These interactions fold the chain of amino acids into coils, sheets, and loops. Figure 2.20 shows two common shapes: an alpha helix and a beta sheet. Each protein can have multiple areas of secondary structure.

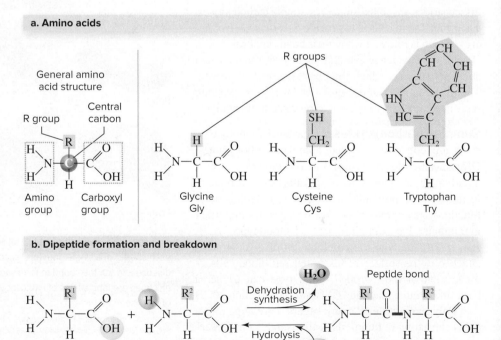

a. Amino acids

General amino acid structure

R group | Central carbon

Amino group | Carboxyl group

R groups

Glycine Gly | Cysteine Cys | Tryptophan Try

b. Dipeptide formation and breakdown

Amino acid | Amino acid

Dehydration synthesis → | Hydrolysis ← | H_2O | Peptide bond

Dipeptide

Figure 2.19 **Amino Acids.** (a) An amino acid is composed of an amino group, a carboxyl group, and one of 20 R groups attached to a central carbon atom. Three examples appear here. (b) A peptide bond forms by dehydration synthesis, joining two amino acids together. Hydrolysis breaks the peptide bond.

- **Tertiary (3°) structure:** The overall shape of a polypeptide, arising primarily through interactions between R groups and water. Inside a cell, water molecules surround each polypeptide. The hydrophobic R groups move away from water toward the protein's interior. In addition, hydrogen bonds and ionic bonds form between the peptide backbone and some R groups. Covalent bonds between sulfur atoms in some R groups further stabilize the structure. These disulfide bridges are abundant in structural proteins such as keratin, which forms hair, scales, beaks, feathers, wool, and hooves.

- **Quaternary (4°) structure:** The shape arising from interactions between multiple polypeptide subunits of the same protein. The protein in figure 2.20 consists of two polypeptides; similarly, the oxygen-toting blood protein hemoglobin is composed of four polypeptide chains. Only proteins consisting of multiple polypeptides have quaternary structure.

As detailed in chapter 7, an organism's genetic code specifies the amino acid sequence of each protein. A genetic mutation may therefore change a protein's primary structure. The protein's

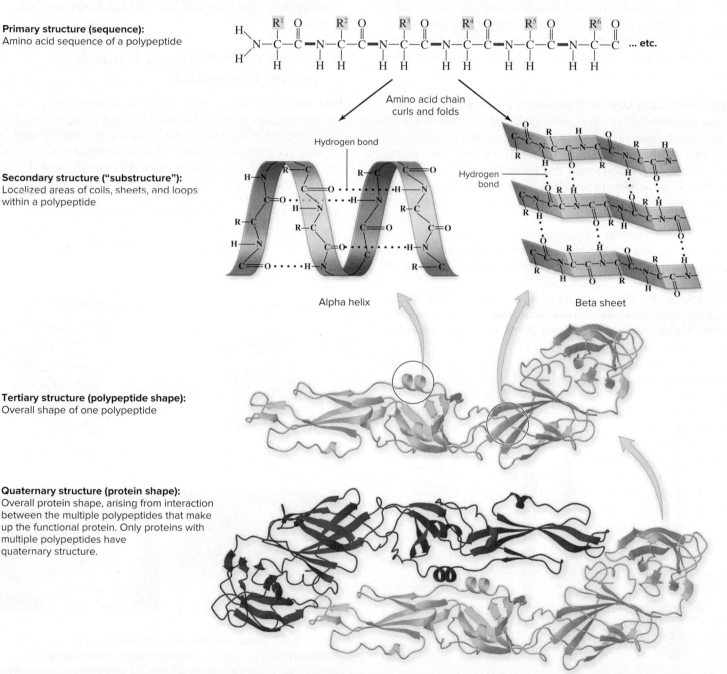

Primary structure (sequence):
Amino acid sequence of a polypeptide

Amino acid chain curls and folds

Hydrogen bond

Secondary structure ("substructure"):
Localized areas of coils, sheets, and loops within a polypeptide

Hydrogen bond

Alpha helix

Beta sheet

Tertiary structure (polypeptide shape):
Overall shape of one polypeptide

Quaternary structure (protein shape):
Overall protein shape, arising from interaction between the multiple polypeptides that make up the functional protein. Only proteins with multiple polypeptides have quaternary structure.

Figure 2.20 **Four Levels of Protein Structure.** The amino acid sequence of a polypeptide forms the primary structure, while hydrogen bonds create secondary structures such as a helix or sheet. The tertiary structure is the overall three-dimensional shape of a protein. The interaction of multiple polypeptides forms the protein's quaternary structure.

secondary, tertiary, and quaternary structures all depend upon the primary structure. Genetic mutations are often harmful because they result in misfolded, nonfunctional proteins.

Many biologists devote their careers to deducing protein structures, in part because the research has so many practical applications. Misfolded infectious proteins called prions, for instance, cause mad cow disease. Knowledge of protein structure can also aid in the treatment of infectious disease. If scientists can determine the shape of a protein unique to the organism that causes malaria, for example, they may be able to use that information to create effective new drugs with few side effects. Some consumer products also exploit protein shape. "Permanent wave" solutions and hair straighteners break disulfide bridges in keratin. The bonds return once the hair is in the desired conformation. (i) *prions,* section 16.6B

Denaturation: Loss of Function It is impossible to overstate the importance of a protein's shape in determining its function. Examine **figure 2.21**, which illustrates the major categories of protein function: structural support, contraction, transport, storage, enzymes, and antibodies. Notice the great variety of protein shapes, reflecting their different jobs in the cell. A digestive enzyme, for example, has a groove that holds a food molecule in just the right way to break the nutrient apart. Muscle proteins form long, aligned fibers that slide past one another, shortening their length to contract the muscle. Membrane channels have pores that admit some molecules but not others into a cell.

Proteins are therefore vulnerable to conditions that alter their shapes. Heat, excessive salt, or the wrong pH can **denature** a protein, changing its shape so that it can no longer carry out its function. As an example, consider what happens to an egg as it cooks (**figure 2.22**). Heat disrupts the hydrogen bonds that maintain protein shape. The proteins unfold, then clump and refold randomly as the once-clear egg protein turns solid white. Typically, a denatured protein will not "renature." That is, there is no way to uncook an egg.

Humans prevent microbes from spoiling food by denaturing proteins. These proteins are not necessarily those in the food itself. Rather, when we heat foods or preserve them in salt or vinegar, we are denaturing *microbial* proteins. Without functional proteins, the microbes die, and the food's shelf life is extended.

D. Nucleic Acids Store and Transmit Genetic Information

How does a cell "know" which amino acids to string together to form a particular protein? The answer is that each protein's primary structure is encoded in the sequence of a **nucleic acid,** a polymer consisting of monomers called nucleotides. Cells contain two types of nucleic acids, **deoxyribonucleic acid (DNA)** and **ribonucleic acid (RNA).**

Each **nucleotide** monomer consists of three components (**figure 2.23a**). At the center is a five-carbon sugar—ribose in RNA and deoxyribose in DNA. Attached to one of the sugar's carbon atoms is at least one phosphate group. Attached to the

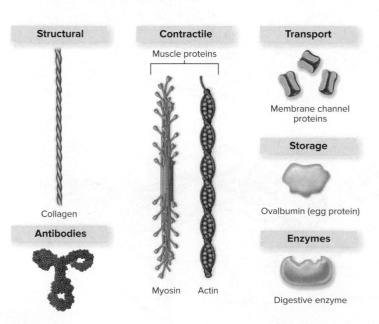

Structural
Collagen
Antibodies

Contractile
Muscle proteins
Myosin   Actin

Transport
Membrane channel proteins
Storage
Ovalbumin (egg protein)
Enzymes
Digestive enzyme

Figure 2.21 **Protein Diversity.** The function of a protein is a direct consequence of its shape. Shown here are a few of the thousands of types of known proteins.

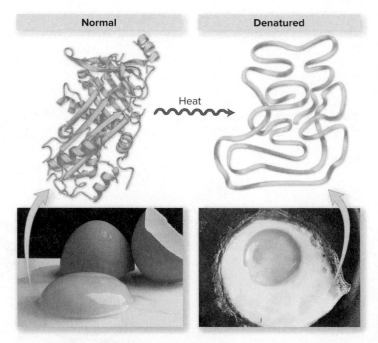

Normal   Denatured
Heat

Figure 2.22 **Denatured Proteins.** Heat from a frying pan causes the proteins in an egg to denature. The protein chains unravel, ruining their overall shape. The new arrangement causes the proteins to clump and refold at random. The clear, fluid, raw egg white becomes a rubbery solid as the proteins denature.

Photos: (both): ©Ingram Publishing RF

Burning Question

What does it mean when food is "organic" or "natural"?

The word *organic* has multiple meanings. To a chemist, an organic compound contains carbon and hydrogen. Chemically, all food is therefore organic. To a farmer or consumer, however, organic foods are produced according to a defined set of standards.

The U.S. Department of Agriculture (USDA) certifies crops as organically grown if the farmer did not apply pesticides (with few exceptions), petroleum-based fertilizers, or sewage sludge. Organically raised cows, pigs, and chickens cannot receive growth hormones or antibiotics, and they must have access to the outdoors and eat organic food. In addition, food labeled "organic" cannot be genetically engineered or treated with ionizing radiation.

A natural food may or may not be organic. The term *natural* refers to the way in which foods are processed, not how they are grown. Standards for what constitutes a natural food are fuzzy. The USDA specifies that meat and poultry labeled as natural cannot contain artificial ingredients or added color, but no such standards exist for other foods.

©Masterfile RF

Submit your burning question to
Marielle.Hoefnagels@mheducation.com

a. Nucleotides and nitrogenous bases

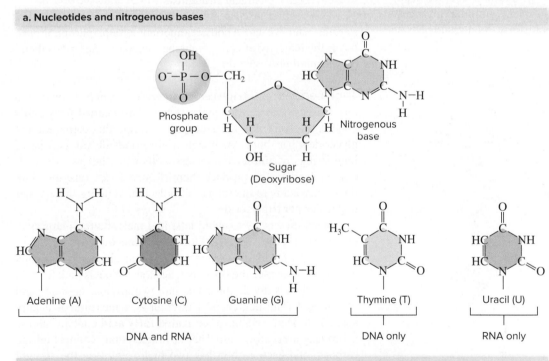

Adenine (A) Cytosine (C) Guanine (G)   Thymine (T)   Uracil (U)

DNA and RNA   DNA only   RNA only

Figure 2.23 Nucleotides. (a) A nucleotide consists of a sugar, one or more phosphate groups, and one of several nitrogenous bases. In DNA, the sugar is deoxyribose, whereas RNA nucleotides contain ribose. In addition, the base thymine appears only in DNA; uracil is only in RNA. (b) Dehydration synthesis joins two nucleotides, and hydrolysis breaks them apart.

b. Nucleic acid formation and breakdown

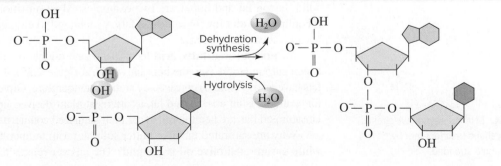

opposite side of the sugar is a nitrogenous base: adenine (A), guanine (G), thymine (T), cytosine (C), or uracil (U). DNA contains A, C, G, and T, whereas RNA contains A, C, G, and U.

Dehydration synthesis links nucleotides together (figure 2.23b). In this reaction, a covalent bond forms between the sugar of one nucleotide and the phosphate group of its neighbor.

A DNA polymer is a double helix that resembles a spiral staircase. Alternating sugars and phosphates form the rails of the staircase, and nitrogenous bases form the rungs (**figure 2.24**). Hydrogen bonds between the bases hold the two strands of nucleotides together: A with T, C with G. The two strands are therefore complementary to, or "opposites" of, each other. Because of complementary base pairing, one strand of DNA contains the information for the other, providing a mechanism for the molecule to replicate. ⓘ *DNA replication,* section 8.2

DNA's main function is to store genetic information; its sequence of nucleotides "tells" a cell which amino acids to string together to form each protein. (This process is described in detail in chapter 7.) Every organism inherits DNA from its parents (or parent, in the case of asexual reproduction). Slight changes in DNA from generation to generation, coupled with natural selection, account for many of the evolutionary changes that have occurred throughout life's history. As a result, DNA

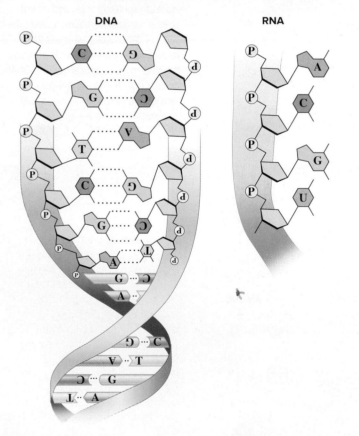

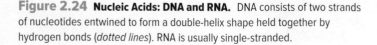

Figure 2.24 **Nucleic Acids: DNA and RNA.** DNA consists of two strands of nucleotides entwined to form a double-helix shape held together by hydrogen bonds (*dotted lines*). RNA is usually single-stranded.

and protein sequences reveal important information about how species are related to one another. ⓘ *molecular evidence for evolution,* section 13.6

Unlike DNA, RNA is typically single-stranded (see figure 2.24). One function of RNA is to enable cells to use the protein-encoding information in DNA (see chapter 7). In addition, a modified RNA nucleotide, adenosine triphosphate (ATP), carries the energy that cells use in many biological functions. ⓘ *ATP,* section 4.3

If DNA encodes only protein, where do the rest of the molecules in cells come from? The answer relates to the diverse functions of proteins: Some of them synthesize the carbohydrates, nucleic acids, and lipids that are essential to a cell's function.

E. Lipids Are Hydrophobic and Energy-Rich

Lipids are organic compounds with one property in common: They do not dissolve in water. They are hydrophobic because they contain large areas dominated by nonpolar carbon–carbon and carbon–hydrogen bonds. Moreover, lipids are not polymers consisting of long chains of monomers. Instead, they have extremely diverse chemical structures.

This section discusses three groups of lipids: triglycerides, steroids, and waxes. Another important group, phospholipids, forms the majority of cell membranes; chapter 3 describes them. ⓘ *phospholipids,* section 3.3

Triglycerides   A **triglyceride** (more commonly known as a fat) consists of three long hydrocarbon chains called **fatty acids** bonded to **glycerol,** a three-carbon molecule that forms the triglyceride's backbone. Although triglycerides do not consist of long strings of similar monomers, cells nevertheless use dehydration synthesis to produce them (**figure 2.25**). Enzymes link three fatty acids to one glycerol molecule, yielding three water molecules per triglyceride.

Many dieters try to avoid fats. Red meat, butter, margarine, oil, cream, cheese, lard, fried foods, and chocolate are all examples of high-fat foods. (A high fat content characterizes many unhealthy foods; see the Burning Question, "What is junk food?") Nutrition labels divide these fats into two groups: saturated and unsaturated. The degree of saturation is a measure of a fatty acid's hydrogen content. A **saturated fatty acid** contains all the hydrogens it possibly can. That is, single bonds connect all the carbons, and each carbon has two hydrogens (see the straight chains in figure 2.25). Animal fats are saturated and tend to be solid; bacon fat and butter are two examples. Most nutritionists recommend a diet low in saturated fats, which tend to clog arteries and cause heart disease.

An **unsaturated fatty acid** has at least one double bond between carbon atoms (see the bent fatty acid in figure 2.25). These fats have an oily (liquid) consistency at room temperature. Olive oil, for example, is an unsaturated fat, as are most plant-derived lipids. Unsaturated fats are healthier than are their saturated counterparts.

Why are saturated fats like butter solid at room temperature, while unsaturated olive oil is a liquid? The answer relates to the

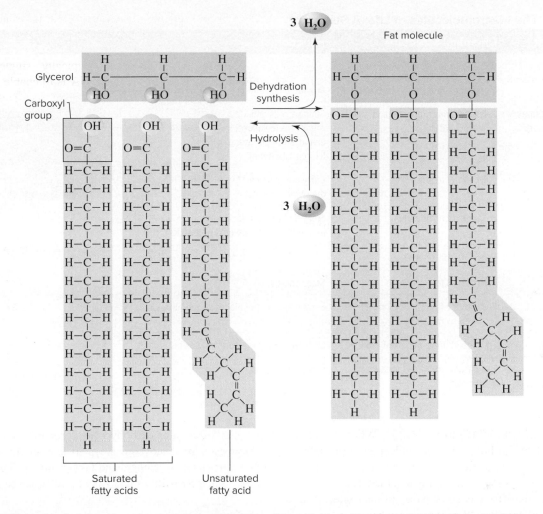

Figure 2.25 Fat Molecule. A triglyceride, or fat molecule, consists of three fatty acids bonded to glycerol. Saturated fatty acid chains contain only single carbon–carbon bonds. In unsaturated fatty acids, one or more double bonds bend the chains.

Burning Question

What is junk food?

Your favorite potato chips contain carbo-hydrates, protein, and fat—three of the four main types of organic molecules in cells. If you must obtain these mole-cules from your diet, why are chips con-sidered a junk food?

In general, junk foods like chips and candy are high in fat or sugar (or both) but low in protein and complex carbohydrates. They also typically have few vitamins and minerals. Junk foods there-fore are high in calories but deliver little nutritional value.

Many junk foods also contain chemical additives. One common ingredient in packaged cookies, pies, and other baked goods is par-tially hydrogenated vegetable oil, a type of chemically processed fat. Partial hydrogenation causes fats to remain solid at room temperature; it also produces trans fats, which have been linked to several diseases (see section 2.5E).

Some junk foods also contain artificial colors, flavor enhancers, artificial flavors, and preservatives that make food look or taste more appealing without adding nutritional value. One example is monoso-dium glutamate (MSG). This chemical consists of an amino acid and a sodium atom connected by an ionic bond. It enhances the flavor of many packaged snacks and fast foods, imparting a savory taste. More-over, preservatives such as BHA and BHT increase the shelf life of many junk foods. These chemicals prevent oxygen from interacting with fat, so it takes longer for the food to become stale.

Potato chips, pizza, fries, candy bars, snack cakes, and other junk foods are hard to resist because they tap into our desire to eat sweet, salty, and fatty foods. These snacks are tasty, appealing, easily available, and often cheap. But for a more nutritious diet, reach for whole grains, fresh fruits, and vegetables instead. ⓘ *healthful diet,* section 32.4

Submit your burning question to
Marielle.Hoefnagels@mheducation.com

TABLE **2.3**   **The Macromolecules of Life: A Summary**

Type of Molecule	Chemical Structure	Function(s)
Carbohydrates		
Simple sugars	Monosaccharides and disaccharides	Provide quick energy
Complex carbohydrates (cellulose, chitin, starch, glycogen)	Polysaccharides (polymers of monosaccharides)	Support cells and organisms (cellulose, chitin); store energy (starch, glycogen)
Proteins	Polymers of amino acids	Carry out nearly all the work of the cell
Nucleic acids (DNA, RNA)	Polymers of nucleotides	Store and use genetic information and transmit it to the next generation
Lipids	Diverse; hydrophobic	
Triglycerides (fats)	Glycerol + 3 fatty acids	Store energy
Phospholipids	Glycerol + 2 fatty acids + phosphate group (see chapter 3)	Form major part of biological membranes
Steroids	Four fused rings, mostly of C and H	Stabilize animal membranes; sex hormones
Waxes	Fatty acids + other hydrocarbons or alcohols	Provide waterproofing

Carbohydrates (starch); lipids

Proteins; lipids

Carbohydrates (cellulose)

©Ingram Publishing/Alamy RF

shapes of the fatty acid "tails" (figure 2.26a, b). When the tails are straight, as they are in butter, the fat molecules form tight, dense stacks. When double bonds cause kinks in the tails, as in olive oil, the molecules cannot pack tightly together. The fat molecules in butter are like flat pieces of paper in neat piles. The molecules in olive oil are more like crumpled paper.

Food chemists have discovered how to turn vegetable oils into solid fats such as margarine, shortening, and peanut butter. A technique called partial hydrogenation adds hydrogen to the oil to solidify it—in essence, partially saturating a formerly unsaturated fat. One byproduct of this process is **trans fats,** which are unsaturated fats whose fatty acid tails are straight, not kinked (figure 2.26c). Trans fats are common in fast foods, fried foods, and many snack products, and they raise the risk of heart disease even more than saturated fats. A healthy diet should be as low as possible in trans fats.

Despite their unhealthful reputation, fats and oils are vital to life. Fat is an excellent energy source, providing more than twice as much energy as equal weights of carbohydrate or protein. Animals must have dietary fat for growth; this requirement explains why human milk is rich in lipids, which fuel the brain's rapid growth during the first 2 years of life. Fats also slow digestion, and they are required for the use of some vitamins and minerals. Nutrition experts therefore recommend that people eat nuts and salmon, which contain polyunsaturated omega-3 fatty acids and other "good fats."

Fat-storing cells aggregate as adipose tissue in animals. The adipose tissue that forms most of the fat in human adults is important because it cushions organs and helps to retain body heat. Excess body fat, however, is associated with diabetes, heart disease, and an elevated risk of cancer.

As different as carbohydrates, proteins, and lipids are, food chemists have discovered ways to use all three substances to make artificial sweeteners and fat substitutes. The Apply It Now box, "Sugar Substitutes and Fake Fats," describes how they do it.

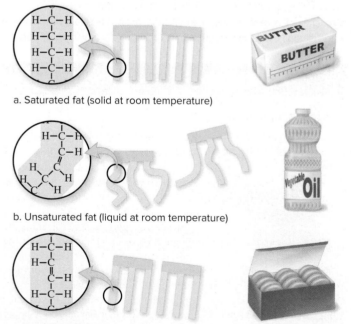

a. Saturated fat (solid at room temperature)

b. Unsaturated fat (liquid at room temperature)

c. Trans fat (solid at room temperature)

Figure 2.26 Solid or Liquid? (a) The straight fatty acid chains of saturated fats stack neatly, easily forming solids. (b) Double bonds in liquid unsaturated fats create kinks that prevent the fat molecules from packing tightly. (c) The fatty acids in a trans fat contain double bonds yet remain straight, so the fat remains a solid.

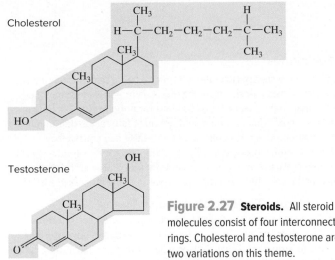

Cholesterol

Testosterone

Figure 2.27 Steroids. All steroid molecules consist of four interconnected rings. Cholesterol and testosterone are two variations on this theme.

Figure 2.28 A Waxy Nest. Beeswax makes up this honeycomb.
©Peter Arnold/Digital Vision/Getty Images RF

Steroids   Steroids are lipids that have four interconnected carbon rings. Vitamin D and cortisone are examples of steroids, as is cholesterol (figure 2.27). Cholesterol is a key part of animal cell membranes. In addition, animal cells use cholesterol as a starting material to make other lipids, including the sex hormones testosterone and estrogen. ⓘ *steroid hormones,* section 28.2B

Although cholesterol is essential, an unhealthy diet can easily contribute to cholesterol levels that are too high, increasing the risk of cardiovascular disease (see the Apply It Now box, "Bad and Good Cholesterol"). Because saturated fats stimulate the liver to produce more cholesterol, it is important to limit dietary intake of both saturated fats and cholesterol. ⓘ *cardiovascular disease,* section 30.5B

Waxes   Waxes are fatty acids combined with alcohols or other hydrocarbons, usually forming a stiff, water-repellent material. The waxy compartments of a honeycomb may hold pollen, honey, or larval bees (figure 2.28). In other species, waxes keep fur, feathers, leaves, fruits, and stems waterproof. Jojoba oil, used in cosmetics and shampoos, is unusual in that it is a liquid wax.

Table 2.3 reviews the characteristics of the four major types of organic molecules in life.

2.5 MASTERING CONCEPTS

1. Distinguish between hydrolysis and dehydration synthesis.
2. Compare and contrast the structures of polysaccharides, proteins, and nucleic acids.
3. List examples of carbohydrates, proteins, nucleic acids, and lipids, and name the function of each.
4. What is the significance of a protein's shape, and how can that shape be destroyed?
5. What are some differences between RNA and DNA?
6. What are the components of a triglyceride?

Apply It **Now**

Bad and Good Cholesterol

You may know someone who is concerned because a blood test has revealed higher-than-normal cholesterol levels. To understand the problem, you need to know a bit more about this molecule.

Cholesterol is not water-soluble, so it travels in the bloodstream encased in proteins. The resulting packets of cholesterol and protein are called lipoproteins, and they occur in low- and high-density varieties.

Low-density lipoprotein (LDL) particles carry cholesterol to the arteries. Excess LDL cholesterol that does not enter cells may accumulate inside blood vessels, blocking blood flow to the heart. Less blood flow means low oxygen delivery, which can damage the heart's muscle. Too much LDL cholesterol (commonly called "bad cholesterol") therefore increases the risk of heart attack.

In contrast, high-density lipoproteins (HDL) carry cholesterol away from the heart and to the liver, which removes it from the bloodstream. High levels of HDL cholesterol ("good cholesterol") promote heart health. Fortunately, it is possible to raise your HDL level. Exercising, losing weight, avoiding trans fats, and replacing meat and dairy with unsaturated dietary fats are among the most effective strategies.

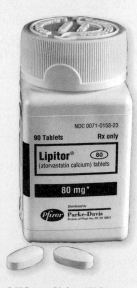

Cholesterol-lowering statin drugs like Lipitor can reduce the risk of heart attack by lowering LDL levels, although they typically have little effect on HDL. These drugs block a liver enzyme that normally helps produce cholesterol; they also boost the liver's ability to get rid of LDL.

©Jill Braaten/McGraw-Hill Education

Apply It **Now**

Sugar Substitutes and Fake Fats

Many weight-conscious people turn to artificial sweeteners and fat substitutes to cut calories while still enjoying their favorite foods. Chemically, how do these sugar and fat replacements compare with the real thing?

Artificial Sweeteners

Table sugar delivers about 4 Calories per gram. (As described in chapter 4, a nutritional Calorie—with a capital *C*—is a measure of energy that represents 1000 calories.) Consuming artificial sweeteners reduces calorie intake, but not always because the additives are truly calorie-free. Instead, most are hundreds of times sweeter-tasting than sugar, so a tiny amount of artificial sweetener achieves the same effect as a teaspoon of sugar. **Figure 2.A** shows a few popular artificial sweeteners. They include

- **Sucralose** (sold as Splenda): This sweetener is a close relative of sucrose, except that three chlorine (Cl) atoms replace three of sucrose's hydroxyl groups. Sucralose is about 600 times sweeter than sugar, and the body digests little if any of it, so it is virtually calorie-free.

- **Aspartame** (sold as NutraSweet and Equal): Surprisingly, aspartame's chemical structure does not resemble sugar. Instead, it consists of two amino acids, phenylalanine and aspartic acid. Aspartame delivers about 4 Calories per gram, but it is about 200 times sweeter than sugar.

- **Saccharin** (sold as Sweet'n Low and Sugar Twin): This sweetener, which has only 1/32 of a Calorie per gram, consists of a double-ring structure that includes nitrogen and sulfur. (Saccharin's eventful history as a food additive is the topic of the Apply It Now box in chapter 1.)

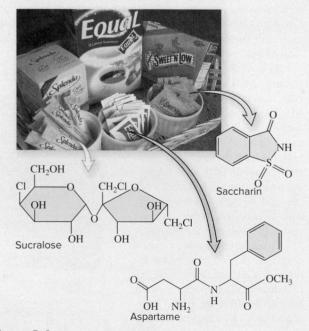

Figure 2.A **Three Artificial Sweeteners.**

Photo: ©Jill Braaten/McGraw-Hill Education

Fat Substitutes

Because fat is so energy-rich (about 9 Calories per gram), cutting fat is a quick way to trim calories from the diet. It is important to remember, however, that some dietary fat is essential for good health. Fat aids in the absorption of some vitamins and provides fatty acids that human bodies cannot produce. Fats also lend foods taste and consistency.

Fat substitutes are chemically diverse. The most common ones are based on carbohydrates, proteins, or even fats, and a careful reading of nutrition labels will reveal their presence in many processed foods.

- **Carbohydrate-based fat substitutes:** Modified food starches, dextrins, guar gum, pectin, and cellulose gels are all derived from polysaccharides, and they all mimic fat's "mouth feel" by absorbing water to form a gel. Depending on whether they are indigestible (cellulose) or digestible (starches), these fat substitutes deliver 0 to 4 Calories per gram. They cannot be used to fry foods.

- **Protein-based fat substitutes:** These food additives are derived from egg whites or whey (the watery part of milk). When ground into "microparticles," these proteins mimic fat's texture as they slide by each other in the mouth. Protein-based fat substitutes deliver about 4 Calories per gram, and they cannot be used in frying.

- **Fat-based fat substitute:** Olestra (marketed as Olean) is a hybrid molecule that combines a central sucrose molecule with six to eight fatty acids (**figure 2.B**). Its chief advantage is that it tastes and behaves like fat—even for frying. Olestra is currently approved only for savory snacks such as chips. It is indigestible and calorie-free, but some people have expressed concern that olestra removes fat-soluble vitamins as it passes through the digestive tract. Others have publicized its reputed laxative properties. Most people, however, do not experience problems after eating small quantities of olestra.

Sugar and fat substitutes can be useful for people who cannot—or do not wish to—eat much of the real thing. But nutritionists warn that these food additives should not take the place of a healthy diet and moderate eating habits.

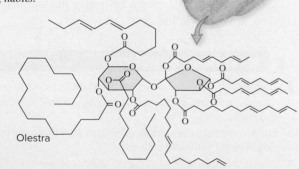

Figure 2.B **Olestra, a Fat Substitute.**

Photo: ©Jill Braaten/McGraw-Hill Education

INVESTIGATING LIFE

2.6 Chemical Warfare on a Tiny Battlefield

Chemical warfare may be illegal among humans, but in other species it is extremely common. Fungi and plants produce toxic chemicals that ward off hungry animals. Soil bacteria secrete chemicals that prevent other microbes from colonizing nearby. And many of us have firsthand experience with animal chemical defenses in the form of a sting from a jellyfish, bee, or fire ant.

Fire ants are native to South America but are famous for invading the southeastern United States, a process that started in the 1930s. So far the conquest has been easy, thanks to a potent venom that kills most other insects. But now, another invasive South American ant species is giving the fire ants a run for their money. Since the 2000s, tawny crazy ants have been outcompeting fire ants in large sections of the southern United States. How do these newcomers survive a fire ant's killer sting?

Tawny crazy ants have venom of their own, but their success against fire ants does not stem from a superior poison. Rather, crazy ants use their venom in a unique defensive behavior. Immediately after contact with fire ant venom, a crazy ant bends its abdomen underneath its body and squirts its own venom on its mouthparts. It then runs its front legs through its mouthparts and grooms itself, smearing its venom over its entire body. Somehow this unusual behavior protects the crazy ant from the fire ant's attack.

Researchers Edward LeBrun, Nathan Jones, and Lawrence Gilbert from the University of Texas at Austin hypothesized that crazy ant venom makes fire ant venom less poisonous. To test their hypothesis, they set up an experiment in the lab. They separated crazy ants into two groups. In one group, the researchers used nail polish to plug the venom gland at the rear of each ant's abdomen. The other crazy ants received nail polish on a different part of their abdomen; these ants formed a control group.

LeBrun and his colleagues then put two fire ants and one crazy ant together in a vial to encourage aggression between the two species. Once fire ant venom touched the crazy ant, the researchers removed the fire ants and monitored the crazy ant for 8 hours. Fewer than half of the crazy ants (48%) with plugged venom glands survived the attack, whereas almost all (98%) of the control ants survived (**figure 2.29a**). The team considered this strong evidence that crazy ant venom detoxifies fire ant venom.

Crazy ant venom is a cocktail of chemicals, and the biologists wondered which one counteracts the fire ant's attack. They knew that the crazy ant venom smelled strongly of a small molecule called formic acid, so they tested whether formic acid alone can do the job. The researchers applied fire ant venom to crazy ants, followed immediately by formic acid or by a control solution of mostly water. Most ants (95%) treated with formic acid survived, whereas less than 20% of those treated with control solution survived (figure 2.29b). The scientists therefore concluded that formic acid is most likely the "secret weapon" that deactivates fire ant venom.

These experiments did not answer *how* the crazy ant's formic acid disables the fire ant's attack, but the researchers speculated about a possible mechanism. Fire ant venom contains enzymes that may help the toxic compounds enter and destroy the victim's cells. All enzymes function best at a specific pH (see section 4.4). Formic acid may decrease the pH so much that the fire ant's enzymes cannot function, indirectly protecting crazy ants from the poison.

What is clear is that defensive chemicals enable tawny crazy ants to invade their rivals' nests and steal their food with little danger. This advantage has allowed crazy ants to displace fire ants in parts of their newly adopted country. It's chemical warfare, and crazy ants have the antidote to the fire ants' best weapons.

Source: LeBrun, Edward G., Nathan T. Jones, and Lawrence E. Gilbert. 2014. Chemical warfare among invaders: A detoxification interaction facilitates an ant invasion. *Science*, vol. 343, pages 1014–1017.

2.6 MASTERING CONCEPTS

1. How did the researchers determine that tawny crazy ants use formic acid in their own venom to detoxify fire ant venom?

2. Propose an experiment to test the hypothesis that formic acid's low pH denatures the enzymes in fire ant venom.

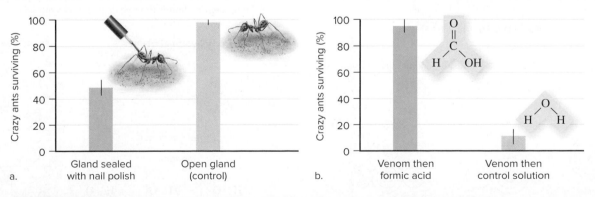

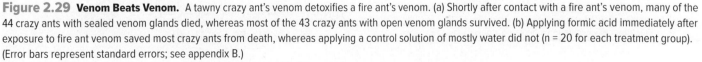

Figure 2.29 Venom Beats Venom. A tawny crazy ant's venom detoxifies a fire ant's venom. (a) Shortly after contact with a fire ant's venom, many of the 44 crazy ants with sealed venom glands died, whereas most of the 43 crazy ants with open venom glands survived. (b) Applying formic acid immediately after exposure to fire ant venom saved most crazy ants from death, whereas applying a control solution of mostly water did not (n = 20 for each treatment group). (Error bars represent standard errors; see appendix B.)

CHAPTER SUMMARY

2.1 Atoms Make Up All Matter

- All substances contain **matter** and **energy,** the ability to do work.
- All matter can be broken down into pure substances called **elements.**

A. Elements Are Fundamental Types of Matter
- **Bulk elements** are essential to life in large quantities. The most abundant are C, H, O, and N. **Trace elements** are required in smaller amounts.

B. Atoms Are Particles of Elements
- An **atom** is the smallest unit of an element. Positively charged **protons** and neutral **neutrons** form the **nucleus.** The negatively charged, much smaller **electrons** surround the nucleus.
- Elements are organized in the **periodic table** according to **atomic number** (the number of protons).
- An **ion** is an atom that gains or loses electrons.

C. Isotopes Have Different Numbers of Neutrons
- **Isotopes** of an element differ by the number of neutrons. A **radioactive isotope** is unstable.
- An element's **atomic weight** reflects the average **mass number** of all isotopes, weighted by the proportions in which they naturally occur.

2.2 Chemical Bonds Link Atoms

- A **molecule** is two or more atoms joined together; if they are of different elements, the molecule is called a **compound.**

A. Electrons Determine Bonding
- Electrons move constantly; they are most likely to occur in volumes of space called **orbitals.** Orbitals are grouped into **energy shells.**
- An atom's tendency to fill its **valence shell** with electrons drives it to form **chemical bonds** with other atoms.
- The more **electronegative** an atom, the more strongly it attracts electrons.

B. In an Ionic Bond, One Atom Transfers Electrons to Another Atom
- An **ionic bond** is an attraction between two oppositely charged ions, which form when one highly electronegative atom strips one or more electrons from another atom (**figure 2.30**).

C. In a Covalent Bond, Atoms Share Electrons
- **Covalent bonds** form between atoms that can fill their valence shells by sharing one or more pairs of electrons (see figure 2.30).
- Atoms in a **nonpolar covalent bond** share electrons equally. If one atom is more electronegative than the other, a **polar covalent bond** forms.

D. Partial Charges on Polar Molecules Create Hydrogen Bonds
- **Hydrogen bonds** result from the attraction between opposite partial charges on adjacent molecules or between oppositely charged parts of a large molecule.

2.3 Water Is Essential to Life

A. Water Is Cohesive and Adhesive
- Water molecules stick to each other (**cohesion**) and to other substances (**adhesion**).

B. Many Substances Dissolve in Water
- A **solution** consists of a **solute** dissolved in a **solvent.**
- Water dissolves **hydrophilic** (polar and charged) substances but not **hydrophobic** (nonpolar) substances.

C. Water Regulates Temperature
- Water helps regulate temperature in organisms because it resists both temperature change and **evaporation.**
- Large bodies of water help keep coastal climates mild.

D. Water Expands As It Freezes
- Ice floats because it is less dense than liquid water.

E. Water Participates in Life's Chemical Reactions
- In a **chemical reaction,** the **products** are different from the **reactants.**
- Most biochemical reactions occur in a watery solution.

2.4 Cells Have an Optimum pH

- In pure water, the concentrations of H^+ and OH^- are equal, so the solution is **neutral.** An **acid** adds H^+ to a solution, and a **base** adds OH^- or removes H^+.

A. The pH Scale Expresses Acidity or Alkalinity
- The **pH scale** measures H^+ concentration. Pure water has a pH of 7, acidic solutions have a pH below 7, and an **alkaline** solution has a pH between 7 and 14.

B. Buffers Regulate pH
- **Buffers** consist of weak acid–base pairs that maintain the optimal pH ranges of body fluids by releasing or consuming H^+.

2.5 Cells Contain Four Major Types of Organic Molecules

A. Large Organic Molecules Are Composed of Smaller Subunits
- Many **organic molecules** consist of small subunits called **monomers,** which link together to form **polymers** (**figure 2.31**). **Dehydration synthesis** is the chemical reaction that joins monomers together, releasing a water molecule.
- The **hydrolysis** reaction uses water to break polymers into monomers.

B. Carbohydrates Include Simple Sugars and Polysaccharides
- **Carbohydrates** consist of carbon, hydrogen, and oxygen in the proportions 1:2:1.
- **Monosaccharides** are single-ring sugars such as glucose. Two bonded monosaccharides form a **disaccharide.** These simple sugars provide quick energy.
- Complex carbohydrates include **oligosaccharides** (chains of 3 to 100 monosaccharides) and **polysaccharides** (chains of hundreds of monosaccharides). Polysaccharides provide support and store energy.

C. Proteins Are Complex and Highly Versatile
- **Proteins** consist of **amino acids,** which join into **polypeptides** by forming **peptide bonds** through dehydration synthesis.

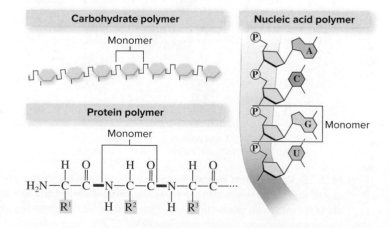

Figure 2.31 Monomers and Polymers.

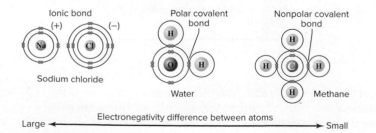

Figure 2.30 Electronegativity and Bonds.

- A protein's three-dimensional shape is vital to its function. A **denatured** protein has a ruined shape.
- Proteins have a great variety of functions, participating in all the work of a cell.

D. Nucleic Acids Store and Transmit Genetic Information

- **Nucleic acids,** including **DNA** and **RNA,** are polymers consisting of **nucleotides.** DNA carries genetic information and transmits it from generation to generation. RNA helps the cell use DNA's information to make proteins.

E. Lipids Are Hydrophobic and Energy-Rich

- **Lipids** are diverse hydrophobic compounds consisting mainly of carbon and hydrogen.
- **Triglycerides** (fats) consist of **glycerol** and three **fatty acids,** which may be **saturated** (straight chains with no double bonds) or **unsaturated** (typically kinked chains with at least one double bond). Many processed foods contain unhealthy **trans fats,** unsaturated fats with straight chains.
- Fats store energy, slow digestion, cushion organs, and preserve body heat.
- **Steroids,** including cholesterol and sex hormones, are lipids consisting of four fused rings.
- **Waxes** are hard, waterproof coverings made of fatty acids combined with other molecules.

2.6 Investigating Life: Chemical Warfare on a Tiny Battlefield

- The venom of a tawny crazy ant contains formic acid, a chemical that helps deactivate the poisons in the venom of its competitor, the fire ant.

MULTIPLE CHOICE QUESTIONS

1. A hydrogen ion (H^+) has _____ neutron(s), _____ proton(s), and _____ electron(s).
   a. 1; 0; 2
   b. 0; 1; 0
   c. 0; 0; 1
   d. 1; 1; 0

2. How many valence electrons does a neutral atom of magnesium have?
   a. 0
   b. 6
   c. 2
   d. 8

3. A *covalent bond* forms when
   a. electrons are present in a valence shell.
   b. a valence electron is removed from one atom and added to another.
   c. a pair of valence electrons is shared between two atoms.
   d. the electronegativity of one atom is much greater than that of another atom.

4. Water dissolves salts because it
   a. is hydrophobic, and salts are also hydrophobic.
   b. forms covalent bonds with the atoms of the salt crystal.
   c. has partial positive and negative charges.
   d. evaporates quickly at room temperature.

5. A hydrogen bond is distinct from ionic and covalent bonds in that it
   a. is a weak attraction between two molecules.
   b. forms only between two hydrogen atoms.
   c. is considerably stronger than the other two types of bonds.
   d. occurs more commonly in lipids than in other types of molecules.

6. _____ are monomers that form polymers called _____.
   a. Nucleotides; nucleic acids
   b. Amino acids; nucleic acids
   c. Monoglycerides; triglycerides
   d. Carbohydrates; monosaccharides

Answers to these questions are in appendix A.

WRITE IT OUT

1. The vitamin biotin contains 10 atoms of carbon, 16 of hydrogen, 3 of oxygen, 2 of nitrogen, and 1 of sulfur. What is its molecular formula?

2. Distinguish between nonpolar covalent bonds, polar covalent bonds, and ionic bonds.

3. How does electronegativity explain whether a covalent bond is polar or nonpolar?

4. Can nonpolar molecules such as CH_4 participate in hydrogen bonds? Why or why not?

5. Define *solute, solvent,* and *solution.*

6. Give an example from everyday life of each of the following properties of water: cohesion, adhesion, ability to dissolve solutes, resistance to temperature change.

7. Draw from memory a diagram showing the interactions among a few water molecules.

8. How do hydrogen ions relate to the pH scale?

9. Sketch a monosaccharide, an amino acid, a nucleotide, a glycerol molecule, and a fatty acid. Then show how those smaller molecules form carbohydrates, proteins, nucleic acids, or fats.

10. Describe what occurs when a chemical reaction removes a water molecule from two adjacent monomers.

11. You eat a sandwich made of starchy bread, ham, and cheese. What types of chemicals are in it?

PULL IT TOGETHER

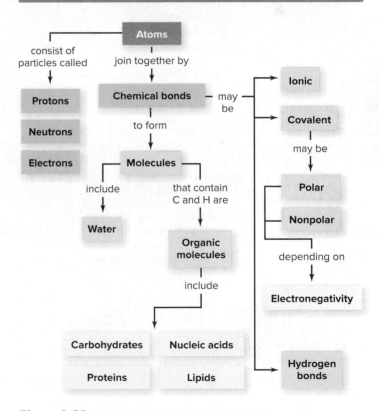

Figure 2.32 Pull It Together: The Chemistry of Life.

Refer to figure 2.32 and the chapter content to answer the following questions.

1. Compare the Survey the Landscape figure in the chapter introduction with the Pull It Together concept map. Are molecules alive? Can life exist without molecules? Why?

2. How do ions and isotopes fit into this concept map?

3. Besides water, what other molecules are essential to life?

4. Add *monomers, polymers, dehydration synthesis,* and *hydrolysis* to this concept map.

3 Cells

Think Pink. For decades, the Komen Race for the Cure has raised money for the fight against breast cancer. The color pink symbolizes support for women with this disease.

©The Columbian, Janet L. Mathews/AP Images

LEARN HOW TO LEARN
Interpreting Images from Microscopes

Any photo taken through a microscope should include information that can help you interpret what you see. First, read the caption and labels so that you know what you are looking at—usually an organ, a tissue, or an individual cell. Then study the scale bar and estimate the size of the image. (For a review of metric units, consult appendix C.) Finally, check whether a light microscope (LM), scanning electron microscope (SEM), or transmission electron microscope (TEM) was used to create the image. Note that stains and false colors are often added to emphasize the most important features.

Cancer Cells: A Tale of Two Drugs

"You have cancer." Physicians deliver this frightening diagnosis to millions of patients each year. The term *cancer* refers to a family of diseases with one feature in common: A person's own cells multiply out of control. The abnormal cells form tumors that may invade nearby tissues and spread to other parts of the body. The illness can turn deadly if the cancerous cells destroy normal tissues or interfere with vital body functions. ⓘ *cancer,* section 8.6

Not long ago, cancer patients had few options. Surgery became a common treatment in the 1800s, after anesthesia was invented. In the early 1900s, physicians discovered that radiation can kill cancer cells. The first drugs designed to treat cancer were developed in the mid-1900s. While these strategies saved countless lives, the side effects were often devastating, and medical professionals have continued to search for better tools.

Cancer research and cell biology are intimately related, since the goal of the treatment is to kill cancer cells while leaving normal cells alone. Biologists must therefore scrutinize the structural and chemical differences between cancer cells and normal cells. Their findings have yielded spectacular new treatments that exploit some of these differences. Research into the chemical signals that control cell division has been especially fruitful, producing new drugs that inhibit only abnormal cells.

One example is trastuzumab (Herceptin). This drug, which was created to treat some forms of breast cancer, got its name from its target: HER2, a receptor protein on the surface of breast cells. The HER2 receptor binds to a molecule that stimulates the cell to divide. A normal breast cell has thousands of HER2 receptors, but cells of one form of breast cancer have many millions. Cells with too many HER2 receptors divide and spread rapidly. Herceptin prevents this by binding to HER2 receptors.

Another successful drug is imatinib (Gleevec), which treats some forms of leukemia and gastrointestinal cancers. Leukemia is a disease in which the body produces many abnormal white blood cells. In a type of leukemia called chronic myeloid leukemia, a genetic mutation causes cells to produce an abnormal protein. This protein prompts the body to produce cancerous cells. Gleevec blocks the protein, selectively slowing division of the abnormal cells without harming normal cells.

Both drugs took decades to develop. Each interferes with a feature that is unique to the cancer cells, producing fewer side effects than older treatments that destroy healthy cells along with their cancerous targets. These drugs owe their success to cell biologists who painstakingly documented the structures in and on cells. These cellular parts are the subject of this chapter.

SURVEY THE LANDSCAPE
Science, Chemistry, and Cells

All life is organized into fundamental units called cells. The details of cell structure vary from organism to organism, but all cells share features that enable them to carry out life's functions.

For more details, study the Pull It Together feature in the chapter summary.

3.1 Cells Are the Units of Life

A human, a plant, a mushroom, and a bacterium appear to have little in common other than being alive. However, on a microscopic level, these organisms share many similarities. For example, all organisms consist of one or more microscopic structures called **cells,** the smallest units of life that can function independently. Within cells, highly coordinated biochemical activities carry out the basic functions of life. This chapter introduces the cell, and the chapters that follow delve into the cellular events that make life possible.

A. Simple Lenses Revealed the First Glimpses of Cells

The study of cells began in 1660, when English physicist Robert Hooke melted strands of spun glass to create lenses. He focused on bee stingers, fish scales, fly legs, feathers, and any type of insect he could hold still. When he used a lens to look at cork, which is bark from a type of oak tree, it appeared to be divided into little boxes, left by cells that were once alive. Hooke called these units "cells" because they looked like the cubicles (Latin, *cellae*) where monks studied and prayed. Although Hooke did not realize the significance of his observation, he was the first person to see the outlines of cells. His discovery initiated a new field of science, now called cell biology.

In 1673, Antony van Leeuwenhoek of Holland improved lenses further (**figure 3.1**). One of his first objects of study was tartar scraped from his own teeth, and his words best describe what he saw there:

> *To my great surprise, I found that it contained many very small animalcules, the motions of which were very pleasing to behold. The motion of these little creatures, one among another, may be likened to that of a great number of gnats or flies disporting in the air.*

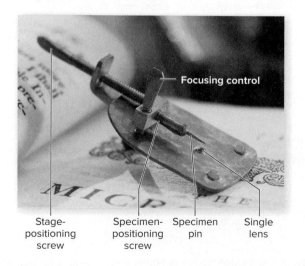

Figure 3.1 Early Microscope. Antony van Leeuwenhoek made many simple microscopes like this one. The object he was studying would have been at the tip of the specimen pin.

Focusing control

Stage-positioning screw | Specimen-positioning screw | Specimen pin | Single lens

©Tetra Images/Alamy RF

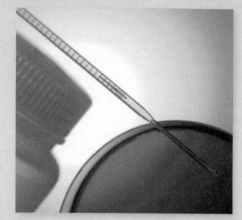
Leeuwenhoek opened a vast new world to the human eye and mind. He viewed bacteria and protists that people hadn't known existed. He also described microscopic parts of larger organisms, including human red blood cells and sperm. However, he failed to see the single-celled "animalcules" reproduce. He therefore perpetuated the idea of spontaneous generation, which suggested that life arises from nonliving matter or from nothing.

B. The Cell Theory Emerges

In the nineteenth century, more powerful microscopes with improved magnification and illumination revealed details of structures inside cells. In the early 1830s, Scottish surgeon Robert Brown noted a roughly circular object in cells from orchid plants. He saw the structure in every cell, then identified it in cells of a variety of organisms. He named it the "nucleus," a term that stuck. Soon microscopists distinguished the translucent, moving material that made up the rest of the cell, calling it the cytoplasm.

In 1839, German biologists Mathias J. Schleiden and Theodor Schwann proposed a new theory, based on many observations made with microscopes. Schleiden first noted that cells are the basic units of plants, and then Schwann compared animal cells to plant cells. After observing similarities in many different plant and animal cells, Schleiden and Schwann formulated the **cell theory,** which originally had two main components: All organisms are made of one or more cells, and the cell is the fundamental unit of all life.

German physiologist Rudolf Virchow added a third component to the cell theory in 1855, when he proposed that all cells come from preexisting cells (see this chapter's Burning Question). This idea contradicted spontaneous generation. When the French chemist and microbiologist Louis Pasteur finally disproved spontaneous generation in 1859, he provided additional evidence in support of the cell theory.

The existence of cells is an undisputed fact, yet the cell theory is still evolving (table 3.1). For 150 years after its formulation, biologists focused on documenting the parts of a cell and the process of cell division. Since the discovery of DNA's structure and function in the 1950s, however, the cell theory has focused on the role of genetic information in dictating what happens inside cells. Modern cell theory therefore adds the ideas that all cells have the same basic chemical composition (chapter 2), use energy (chapters 4, 5, and 6), and contain DNA that is duplicated and passed on as each cell divides (chapters 7, 8, and 9).

Like any scientific theory, the cell theory is *potentially* falsifiable—yet many lines of evidence support each of its components, making it one of the most powerful ideas in biology.

TABLE 3.1 The Cell Theory: A Summary

Components of early cell theory
All organisms are made of one or more cells.
The cell is the fundamental unit of life.
All cells come from preexisting cells.

Additional ideas in modern cell theory
All cells have the same basic chemical composition.
All cells use energy.
All cells contain DNA that is duplicated and passed on as each cell divides.

50 μm
LM

©Don Rubbelke/
McGraw-Hill Education

C. Microscopes Magnify Cell Structures

The unaided eye can see objects that are larger than about 0.2 mm (figure 3.2). Cells are typically smaller than this lower limit of human vision, so studying life at the cellular and molecular levels requires magnification. Cell biologists use a variety of microscopes to produce different types of images. As you will see, some microscopes show full-color structures and processes inside living cells. Others can greatly magnify cell structures, but with two significant drawbacks: They require that cells be killed, and they produce only black-and-white images. This section describes several types of microscopes; figure 3.2 provides a sense of the size of objects that each can reveal.

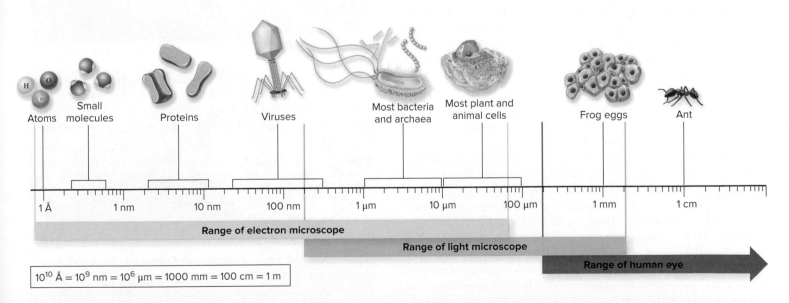

$$10^{10}\ \text{Å} = 10^9\ \text{nm} = 10^6\ \mu\text{m} = 1000\ \text{mm} = 100\ \text{cm} = 1\ \text{m}$$

Figure 3.2 Ranges of Light and Electron Microscopes. Biologists use light microscopes and electron microscopes to view a world too small to see with the unaided eye. This illustration uses the metric system to measure size (see appendix C). Thanks to the overlapping capabilities of the different microscopes, we can visualize objects ranging in size from large molecules to entire cells.

Light Microscopes   Light microscopes are ideal for generating true-color views of living or preserved cells. Because light must pass through an object to reveal its internal features, however, the specimens must be transparent or thinly sliced to generate a good image.

Two types of light microscopes are the compound microscope and the confocal microscope (**figure 3.3a, b**). A compound scope, the type you are likely to use in a biology lab course, uses two or more lenses to focus visible light through a specimen. The most powerful ones can magnify up to 1600 times and resolve objects that are 200 nanometers apart. A confocal microscope enhances resolution by focusing white or laser light through a lens to the object. The image then passes through a pinhole. The result is a scan of highly focused light on one tiny part of the specimen at a time. Computers can integrate multiple confocal images of specimens exposed to fluorescent dyes to produce spectacular three-dimensional peeks at living structures.

Transmission and Scanning Electron Microscopes

Light microscopes are useful, but their main disadvantage is that many cell structures are too small to see using light. Electron microscopes provide much greater magnification and resolution. Instead of using light, these microscopes use electrons.

A transmission electron microscope (TEM) sends a beam of electrons through a very thin slice of a specimen, using a magnetic field to focus the beam. The microscope translates the contrasts in electron transmission into a high-resolution, two-dimensional image that shows the internal features of the object (figure 3.3c). TEMs can magnify up to 50 million times and resolve objects less than 1 angstrom (10^{-10} meters) apart.

A scanning electron microscope (SEM) scans a beam of electrons over the surface of a metal-coated, three-dimensional specimen. Its images have lower resolution than those of the TEM; in SEM, the maximum magnification is about 250,000 times, and the resolution limit is 1 to 5 nanometers. SEM's chief advantage is its ability to reveal textures on a specimen's external surface (figure 3.3d).

Both TEM and SEM provide much greater magnification and resolution than light microscopes. Nevertheless, they do have limitations. First, they are extremely expensive to build, operate, and maintain. Second, electron microscopy normally requires that a specimen be killed, chemically fixed, and placed in a vacuum. These treatments can distort natural structures. Light microscopy, in contrast, allows an investigator to view living organisms. Third, unlike light microscopes, all images from electron microscopes are black-and-white, although artists often add false color to highlight specific objects in electron micrographs. (In this book, each photo taken through a microscope is tagged with the magnification and the type of microscope; the addition of false color is also noted where applicable.)

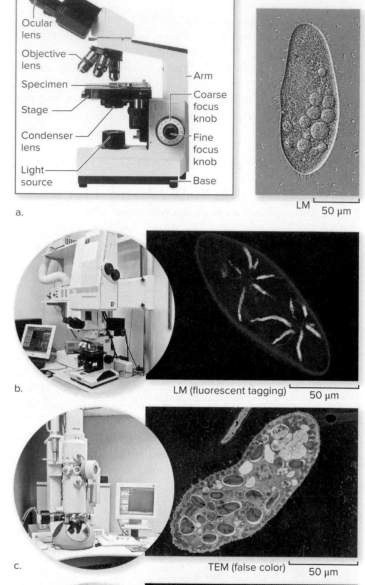

Figure 3.3 Four Types of Microscopes Compared. These photographs show four types of microscopes, along with sample images of a protist called *Paramecium*. (a) Compound light microscope. (b) Confocal microscope. (c) Transmission electron microscope. (d) Scanning electron microscope.

D. All Cells Have Features in Common

Microscopes and other tools clearly reveal that although cells can appear very different, they all share some features that reflect their shared evolutionary history (see figure 1.9). For example, all cells contain DNA, the cell's genetic information. They also contain RNA, which participates in the production of proteins (see chapter 7). These proteins, in turn, are essential to life because they carry out all of the cell's work, from orchestrating reproduction to processing energy to regulating what enters and leaves the cell.

Since all cells require proteins, they also contain **ribosomes,** which are the structures that manufacture proteins. Each cell is also surrounded by a lipid-rich **cell membrane** (also called the plasma membrane) that forms a boundary between the cell and its environment (see section 3.3). The membrane encloses the **cytoplasm,** which includes all cell contents (except the nucleus, in cells that have one). The **cytosol** is the fluid portion of the cytoplasm.

One other feature common to nearly all cells is small size, typically less than 0.1 millimeter in diameter (see figure 3.2). Why so tiny? The answer is that nutrients, water, oxygen, carbon dioxide, and waste products enter or leave a cell through its surface. Each cell must have abundant surface area to accommodate these exchanges efficiently. As an object grows, however, its volume increases much more quickly than its surface area. **Figure 3.4a** illustrates this principle for a series of cubes, but the same applies to cells: Small size maximizes the ratio of surface area to volume.

Figure It Out

For a cube 5 centimeters on each side, calculate the ratio of surface area to volume.

Answer: 1.2

Cells avoid surface area limitations in several ways. Nerve cells may be long (up to a meter or so), but they are also extremely slender, so the ratio of surface area to volume remains high. The flattened shape of a red blood cell maximizes its ability to carry oxygen, and the many microscopic extensions of an amoeba's membrane provide a large surface area for absorbing oxygen and capturing food (figure 3.4b). A transportation system that quickly circulates materials throughout the cell also helps.

The concept of surface area is everywhere in biology; many structures illustrate the principle that a large surface area maximizes contact with the environment. For example, a pine tree's pollen grains have extensions that enable them to float on air currents; root hairs have tremendous surface area for absorbing water; the broad, flat leaves of plants maximize exposure to light; a fish's feathery gills absorb oxygen from water; a jackrabbit's enormous ears help the animal lose excess body heat in the desert air—the list goes on and on. Conversely, low surface areas minimize the exchange of materials or heat with the environment. A hibernating animal, for example, conserves warmth by tucking its limbs close to its body; a cactus plant produces few if any leaves, reducing water loss in its dry habitat.

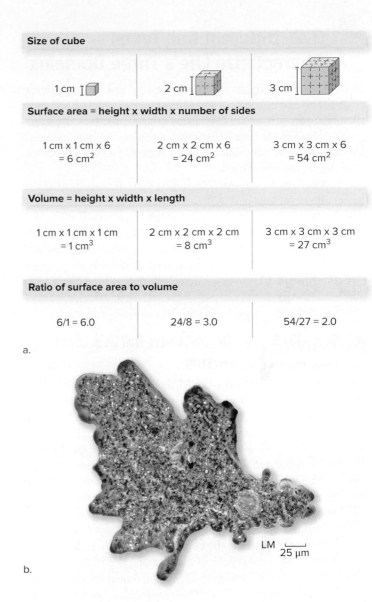

Size of cube

1 cm   2 cm   3 cm

Surface area = height x width x number of sides

| 1 cm x 1 cm x 6 = 6 cm² | 2 cm x 2 cm x 6 = 24 cm² | 3 cm x 3 cm x 6 = 54 cm² |

Volume = height x width x length

| 1 cm x 1 cm x 1 cm = 1 cm³ | 2 cm x 2 cm x 2 cm = 8 cm³ | 3 cm x 3 cm x 3 cm = 27 cm³ |

Ratio of surface area to volume

| 6/1 = 6.0 | 24/8 = 3.0 | 54/27 = 2.0 |

a.

b.

LM 25 μm

Figure 3.4 Surface Area and Volume. (a) This simple example shows that smaller objects have more surface area *relative to their volume* than do larger objects with the same overall shape. (b) The membrane of this amoeba is highly folded, producing a large surface area relative to the cell's volume.

(b): ©Roland Birke/Getty Images

3.1 MASTERING CONCEPTS

1. Why are cells, not atoms, the basic units of life?
2. How have microscopes advanced the study of cells?
3. What are the original components of the cell theory, and what parts of the theory came later?
4. Rank the three main types of microscopes from lowest to highest potential magnification.
5. Which molecules and structures occur in all cells?
6. Describe adaptations that increase the ratio of surface area to volume in cells.

3.2 Different Cell Types Characterize Life's Three Domains

Until recently, biologists organized life into just two categories: prokaryotic and eukaryotic. **Prokaryotes,** the simplest and most ancient forms of life, are organisms whose cells lack a nucleus (*pro* = before; *karyon* = kernel, referring to the nucleus). **Eukaryotes** have cells that contain a nucleus and other membranous organelles (*eu* = true).

In 1977, however, microbiologist Carl Woese studied key molecules in many cell types. He detected differences suggesting that prokaryotes actually include two forms of life that are distantly related to each other. Biologists subsequently divided life into three domains: Bacteria, Archaea, and Eukarya (**figure 3.5**). This section describes them briefly; chapters 17 through 21 cover prokaryotic and eukaryotic life in much more detail.

A. Domain Bacteria Contains Earth's Most Abundant Organisms

Bacteria are the most abundant and diverse organisms on Earth, perhaps because they have existed longer than any other group. Some species, such as *Streptococcus* and *Escherichia coli*, can cause illnesses, but most are not harmful. In fact, the bacteria living on your skin and inside your intestinal tract are essential for good health. (These microbes are surprisingly numerous; see this chapter's Apply It Now box.) Bacteria are also very valuable in research, food and beverage processing, and pharmaceutical production. In ecosystems, bacteria play critical roles as decomposers and producers.

Bacterial cells are structurally simple (**figure 3.6a**). The **nucleoid** is the area where the cell's circular DNA molecule congregates. Unlike a eukaryotic cell's nucleus, the bacterial nucleoid is not bounded by a membrane. Located near the DNA in the cytoplasm are the enzymes, RNA molecules, and ribosomes needed to produce the cell's proteins.

Apply It **Now**

Most of Your Cells Are Not Your Own

Many people are surprised to learn that nonhuman cells vastly outnumber the body's own cells. Microbiologists estimate that the number of bacteria living in and on a typical human is *10 times* the number of human cells! Although some of these bacteria can cause disease, most exist harmlessly on the skin and in the mouth and intestines. These inconspicuous guests also can help extract nutrients from food and prevent disease.

So how many *human* cells make up a person's body? For adults, estimates range from about 10 trillion to 100 trillion. No one knows for sure, because counting living cells is very difficult. After all, the number of cells changes throughout life. A child's growth comes from cell division that adds new cells, not from the expansion of existing ones. Moreover, new cells arise as old cells die, so a "true" count is a moving target. Also, no one has found a good way to count them all. Cells come in so many different shapes that it is hard to extrapolate from a small sample to the whole body.

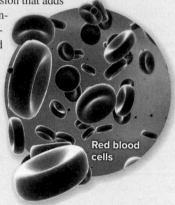

Red blood cells

©MedicalRF.com/Getty Images RF

A rigid **cell wall** surrounds the cell membrane of most bacteria, protecting the cell and preventing it from bursting if it absorbs too much water. This wall also gives the cell its shape: usually rod-shaped, round, or spiral (figure 3.6b, c, d). Many antibiotic drugs, including penicillin, halt bacterial infection by

	Cell Type	Nucleus	Membrane-Bounded Organelles	Membrane Chemistry	Typical Cell Size
Domain Bacteria	Prokaryotic	Absent	Absent	Fatty acids	1-10 μm
Domain Archaea	Prokaryotic	Absent	Absent	Nonfatty acid lipids	1-10 μm
Domain Eukarya	Eukaryotic	Present	Present	Fatty acids	10-100 μm

Common ancestor

Figure 3.5 The Three Domains of Life. Biologists distinguish domains Bacteria, Archaea, and Eukarya based on unique features of cell structure and biochemistry. The small evolutionary tree shows that archaea are the closest relatives of the eukaryotes.

Figure 3.6 Anatomy of a Bacterium. (a) Bacterial cells lack internal compartments. (b) Rod-shaped cells of *E. coli* inhabit human intestines. (c) Spherical *Staphylococcus aureus* cells cause "staph" infections that range from mild to deadly. (d) Corkscrew-shaped *Campylobacter* cells often cause diarrhea.
(b): ©Steve Gschmeissner/Getty Images RF; (c): ©David McCarthy/Science Source; (d): CDC/Melissa Brower

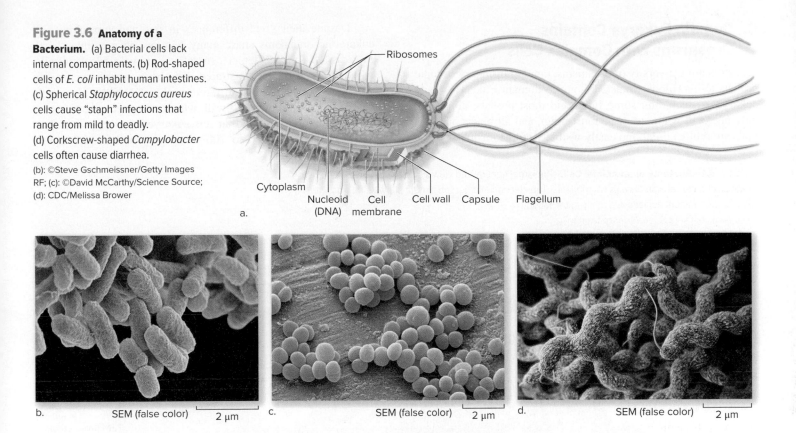

a.

b. SEM (false color) 2 μm   c. SEM (false color) 2 μm   d. SEM (false color) 2 μm

interfering with the microorganism's ability to construct its protective cell wall. In some bacteria, polysaccharides on the cell wall form a capsule that adds protection or enables the cell to attach to surfaces.

Many bacteria can swim in fluids. **Flagella** (singular: flagellum) are tail-like appendages that enable these cells to move. One or more flagella are anchored in the cell wall and underlying cell membrane. Bacterial flagella rotate like a propeller, moving the cell forward or backward.

B. Domain Archaea Includes Prokaryotes with Unique Biochemistry

Archaean cells resemble bacterial cells in some ways. Like bacteria, they are smaller than most eukaryotic cells, and they lack a nucleus and other organelles. Most have cell walls; flagella are also common. And like bacteria, most archaea are one-celled organisms. Because of these similarities, Woese first named his newly recognized group Archaebacteria.

The name later changed to Archaea after it became obvious that the resemblance to bacteria was only superficial. Archaea have their own domain because they build their cells out of biochemicals that are different from those in either bacteria or eukaryotes. Their cell membranes, cell walls, and flagella are all chemically unique. Their ribosomes, however, share similarities with those of both bacteria and eukaryotes, and key DNA sequences suggest that archaea are actually the closest relatives of eukaryotes.

The first members of Archaea to be described were methanogens, microbes that use carbon dioxide and hydrogen from the environment to produce methane (figure 3.7). Archaea subsequently became famous as "extremophiles" because scientists discovered many of them in habitats that are extremely hot, acidic, or salty. This characterization is somewhat misleading, however, because bacteria also occupy the same environments. Moreover, researchers have now discovered archaea in a variety of moderate habitats, including soil, swamps, rice paddies, oceans, and even the human mouth.

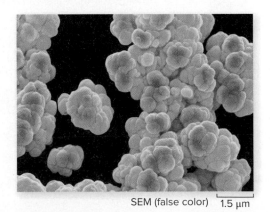

SEM (false color) 1.5 μm

Figure 3.7 An Archaeon. *Methanosarcina* is an archaeon that generates methane in habitats lacking oxygen.
©Power and Syred/Science Source

C. Domain Eukarya Contains Organisms with Complex Cells

An astonishing diversity of organisms, ranging from microscopic protists to enormous whales, belong to domain Eukarya. Many eukaryotes, including some fungi and most protists, consist of only one cell. A few protists, most fungi, and all plants and animals are multicellular and easily visible with the unaided eye.

Despite their great differences in size and appearance, all eukaryotic organisms share many features on a cellular level. **Figures 3.8** and **3.9** depict generalized animal and plant cells. Although both of the illustrated cells have many structures in common, there are some differences. Most notably, plant cells have chloroplasts and a cell wall, which animal cells lack.

One obvious feature that sets eukaryotic cells apart is their large size, typically 10 to 100 times greater than prokaryotic

Figure 3.8 **Anatomy of an Animal Cell.** The large, generalized view shows the relative sizes and locations of a typical animal cell's components. The electron micrograph at right shows a rat's pancreas cell with a prominent nucleus and many mitochondria.
Photo: ©Biophoto Associates/Science Source

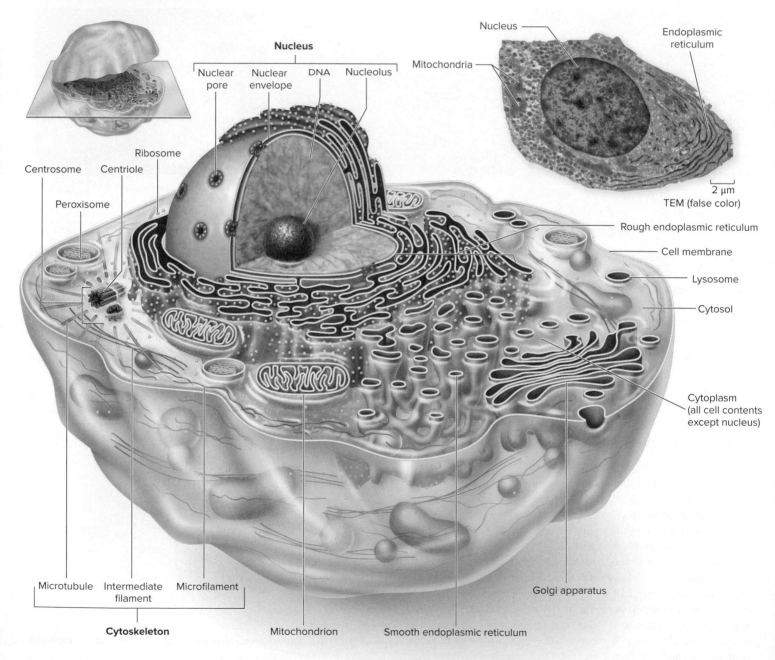

TEM (false color)

2 μm

cells. The other main difference is that the cytoplasm of a eukaryotic cell is divided into **organelles** ("little organs"), compartments that carry out specialized functions. An elaborate system of internal membranes creates these compartments.

Sections 3.4, 3.5, and 3.6 describe the structure of the eukaryotic cell in greater detail, and the illustrated table at the end of the chapter summarizes the functions of the eukaryotic organelles (see table 3.3).

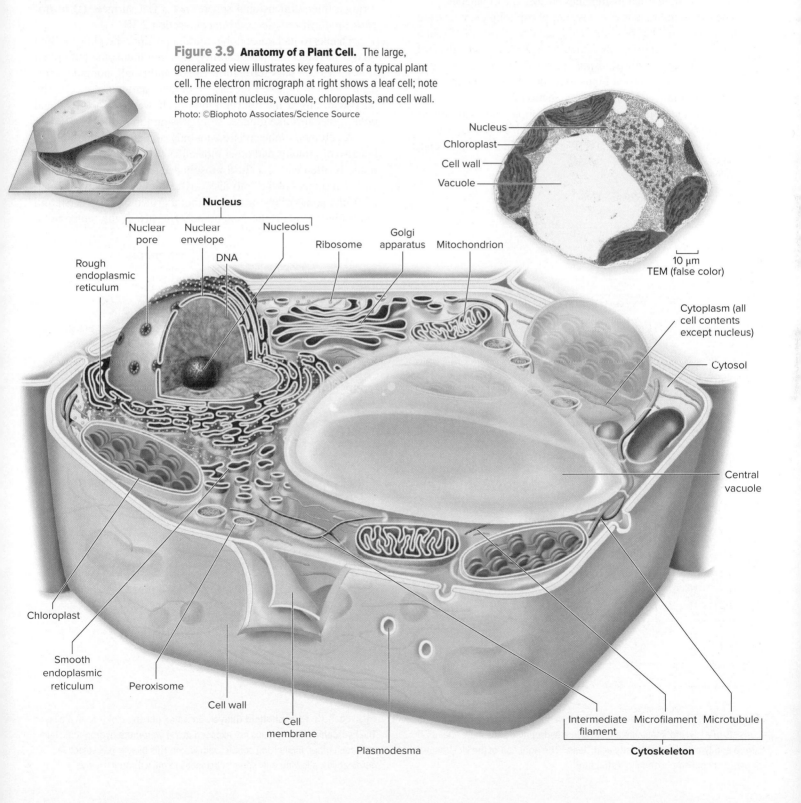

Figure 3.9 Anatomy of a Plant Cell. The large, generalized view illustrates key features of a typical plant cell. The electron micrograph at right shows a leaf cell; note the prominent nucleus, vacuole, chloroplasts, and cell wall.
Photo: ©Biophoto Associates/Science Source

3.3 A Membrane Separates Each Cell from Its Surroundings

A cell membrane is one feature common to all cells. The membrane separates the cytoplasm from the cell's surroundings. The cell's surface also transports substances into and out of the cell (see chapter 4), and it receives and responds to external stimuli. Inside a eukaryotic cell, internal membranes enclose the organelles.

The cell membrane is composed of phospholipids, which are organic molecules that resemble triglycerides (**figure 3.10**). In a triglyceride, three fatty acids attach to a three-carbon glycerol molecule. But in a **phospholipid,** glycerol bonds to only two fatty acids; the third carbon binds to a phosphate group attached to additional atoms. ⓘ *triglycerides,* section 2.5E

This chemical structure gives phospholipids unusual properties in water. The phosphate "head" end, with its polar covalent bonds, is attracted to water; that is, it is hydrophilic. The other end,

consisting of two fatty acid "tails," is hydrophobic. In water, these molecules spontaneously arrange themselves into a **phospholipid bilayer:** a double layer of phospholipids (**figure 3.11**). In some ways, this bilayer resembles a cheese sandwich. The hydrophilic head groups (the "bread" of the sandwich) are exposed to the watery medium outside and inside the cell, whereas the hydrophobic tails face each other on the inside of the sandwich, like cheese between the bread slices. Unlike a sandwich, however, the bilayer forms a three-dimensional sphere, not a flat surface. ⓘ *hydrophilic and hydrophobic substances,* section 2.3B

Thanks to its hydrophobic middle portion, the phospholipid bilayer has selective permeability, meaning that some but not all substances can pass through it. Lipids and small, nonpolar molecules such as O_2 and CO_2 pass freely into and out of a cell. The fatty acid tails at the bilayer's interior, however, block ions and polar molecules like glucose from passing through.

A cell membrane consists not only of a phospholipid bilayer but also of proteins and other molecules (**figure 3.12**). The membrane is often called a **fluid mosaic** because many of the molecules (the pieces of the "mosaic") drift laterally within the bilayer, a bit like pickpockets moving within a crowd of people. Steroid molecules maintain the membrane's fluidity as the temperature

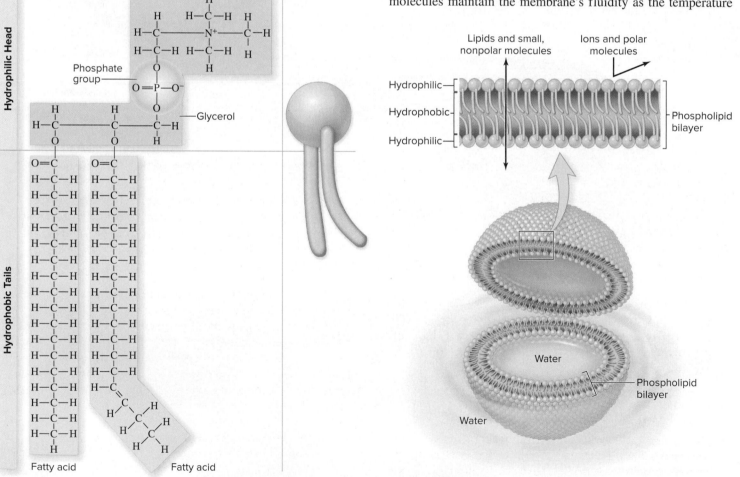

Figure 3.10 Membrane Phospholipid. A phospholipid molecule consists of a glycerol molecule attached to a hydrophilic phosphate "head" group and two hydrophobic fatty acid "tails." The right half of the illustration shows a simplified phospholipid structure.

Figure 3.11 Phospholipid Bilayer. In water, phospholipids form a bilayer. The hydrophilic head groups are exposed to the water; the hydrophobic tails face each other, minimizing contact with water. This bilayer has selective permeability, allowing only some substances to pass through freely.

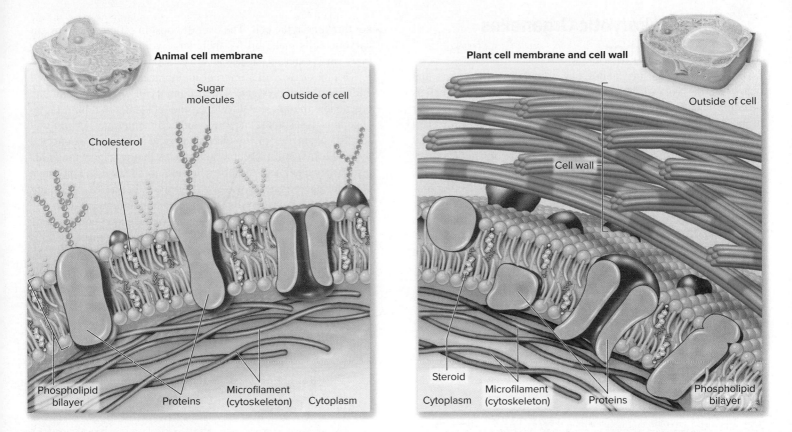

Figure 3.12 Anatomy of a Cell Membrane. The cell membrane is a "fluid mosaic" of proteins embedded in a phospholipid bilayer. Steroid molecules, such as the cholesterol in animal membranes, add fluidity. The outer face of the animal cell membrane also features carbohydrate (sugar) molecules linked to proteins. A rigid wall of cellulose fibers surrounds each plant cell.

fluctuates. Both animal and plant cell membranes contain steroids; the cholesterol in animal membranes is the most familiar example. ⓘ *steroids,* section 2.5E

Whereas phospholipids and steroids provide the membrane's structure, proteins are especially important to its function. As you can see in figure 3.12, some of the proteins extend through the phospholipid bilayer, whereas others face only the inside or outside of the cell. Cells have multiple types of membrane proteins:

- **Transport proteins:** Transport proteins embedded in the phospholipid bilayer create passageways through which ions, glucose, and other polar substances pass into or out of the cell. Section 4.5 describes membrane transport in more detail.

- **Enzymes:** These proteins facilitate chemical reactions that otherwise would proceed too slowly to sustain life. (Not all enzymes, however, are associated with membranes.) ⓘ *enzymes,* section 4.4

- **Recognition proteins:** Carbohydrates attached to cell surface proteins serve as "name tags" that help the body's immune system recognize its own cells. The immune system attacks cells with unfamiliar surface molecules, which is why transplant recipients often reject donated organs. Surface proteins also distinctively mark specialized cells within an individual, so a bone cell's surface is different from that of a nerve cell or a muscle cell.

- **Adhesion proteins:** These membrane proteins enable cells to stick to one another.

- **Receptor proteins:** Receptor proteins bind to molecules outside the cell and trigger an internal response, a process called signal transduction. For example, when a hormone binds to a receptor, the resulting chain reaction produces the hormone's effects on the cell (see section 28.2).

Understanding membrane proteins is a vital part of medicine, in part because at least half of all drugs bind to them. One example is omeprazole (Prilosec). This drug relieves heartburn by blocking some of the transport proteins that pump hydrogen ions into the stomach. Another is the antidepressant drug fluoxetine (Prozac), which prevents receptors on brain cell surfaces from absorbing a mood-altering biochemical called serotonin.

3.3 MASTERING CONCEPTS

1. Chemically, how is a phospholipid different from a triglyceride?
2. How does the chemical structure of phospholipids enable them to form a bilayer in water?
3. Where in the cell do phospholipid bilayers occur?
4. What are some functions of membrane proteins?

3.4 Eukaryotic Organelles Divide Labor

In eukaryotic cells, organelles have specialized functions that carry out the work of the cell. If you think of a eukaryotic cell as a home, each organelle is analogous to a room. For example, your kitchen, bathroom, and bedroom each hold unique items that suit the uses of those rooms. Likewise, each organelle has distinct sets of proteins and other molecules that fit the organelle's function. The "walls" of these cellular compartments are membranes, often intricately folded and studded with enzymes and other proteins. These folds provide tremendous surface area where many of the cell's chemical reactions occur.

In general, organelles keep related biochemicals and structures close together. This arrangement saves energy because high concentrations of each chemical occur only in certain organelles,

not throughout the cell. The overall boost in efficiency helps eukaryotic cells make up for their large volume and correspondingly small external surface area (see section 3.1D).

Many of the cell's internal membranes form a coordinated **endomembrane system,** which consists of several interacting organelles: the nuclear envelope, endoplasmic reticulum, Golgi apparatus, lysosomes, vacuoles, and cell membrane. As you will see, the organelles of the endomembrane system are connected by **vesicles:** small, membranous spheres that transport materials inside the cell. These "bubbles" of membrane, which can pinch off from one organelle, travel within the cell, and fuse with another, are also considered part of the endomembrane system.

Interactions between the organelles of the endomembrane system enable cells to produce, package, and release complex mixtures of biochemicals. This section focuses on each step involved in the production and secretion of one such mixture: milk (**figure 3.13**).

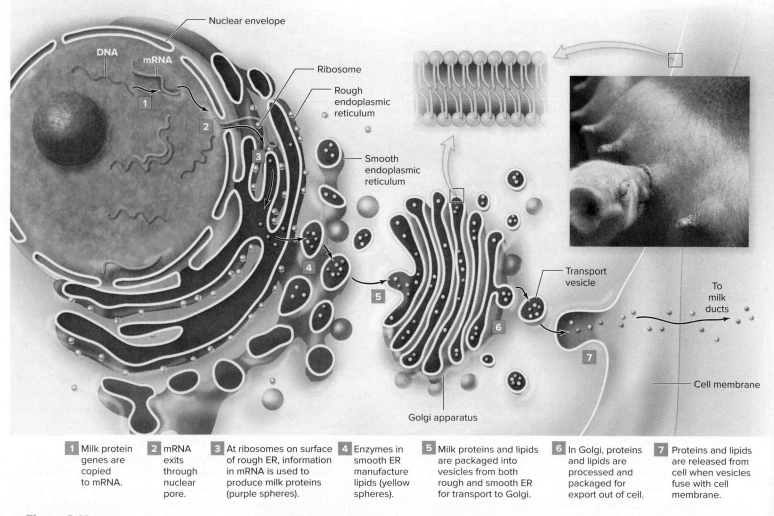

| **1** Milk protein genes are copied to mRNA. | **2** mRNA exits through nuclear pore. | **3** At ribosomes on surface of rough ER, information in mRNA is used to produce milk proteins (purple spheres). | **4** Enzymes in smooth ER manufacture lipids (yellow spheres). | **5** Milk proteins and lipids are packaged into vesicles from both rough and smooth ER for transport to Golgi. | **6** In Golgi, proteins and lipids are processed and packaged for export out of cell. | **7** Proteins and lipids are released from cell when vesicles fuse with cell membrane. |

Figure 3.13   Making Milk. Several types of organelles work together to produce and secrete milk from a cell in a mammary gland; the numbers (*1*) through (*7*) indicate each organelle's role. Note that membranes enclose each organelle and the entire cell. (The inset shows a piglet suckling from a sow.)

Photo: ©Tim Flach/The Image Bank/Getty Images

A. The Nucleus, Endoplasmic Reticulum, and Golgi Interact to Secrete Substances

Special cells in the mammary glands of female mammals produce milk, which contains proteins, fats, carbohydrates, and water in a proportion ideal for the development of a newborn. Human milk is rich in lipids, which the rapidly growing baby's nervous system requires. (Cow's milk contains a higher proportion of protein, better suited to a calf's rapid muscle growth.) Milk also contains calcium, potassium, and antibodies that help jump-start the infant's immunity to disease.

The milk-producing cells of the mammary glands are dormant most of the time, but they undergo a burst of productivity shortly after the female gives birth. How do each cell's organelles work together to manufacture milk?

The Nucleus
The process of milk production and secretion begins in the **nucleus** (see figure 3.13, step 1), the organelle that contains most of a eukaryotic cell's DNA. DNA is an informational molecule that specifies the "recipe" for every protein a cell can make (such as milk protein and enzymes required to synthesize carbohydrates and lipids). The cell copies the genes encoding these proteins into another nucleic acid, messenger RNA (mRNA).

The mRNA molecules exit the nucleus through **nuclear pores,** which are holes in the double-membrane **nuclear envelope** that separates the nucleus from the cytoplasm (figure 3.13, step 2, and **figure 3.14**). Nuclear pores are highly specialized channels composed of dozens of types of proteins. Traffic through the nuclear pores is busy, with millions of regulatory proteins entering and mRNA molecules leaving each minute.

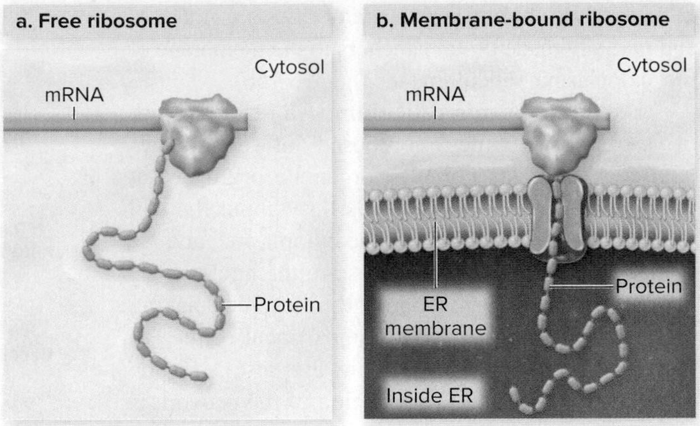

Figure 3.15 Ribosomes. (a) Free ribosomes produce proteins used in the cell's cytosol. (b) Proteins produced by ribosomes attached to the rough ER's membrane are typically used in specialized organelles or in the cell membrane; they may also be secreted outside the cell.

Also inside the nucleus is the **nucleolus,** a dense spot that assembles the components of ribosomes. These ribosomal subunits leave the nucleus through the nuclear pores, and they come together in the cytoplasm to form complete ribosomes.

The Endoplasmic Reticulum and Golgi Apparatus
The remainder of the cell, between the nucleus and the cell membrane, is the cytoplasm. The cytoplasm includes the cytosol, a watery mixture of ions, enzymes, RNA, and other dissolved substances. Organelles are also part of the cytoplasm, as are arrays of protein rods and tubules called the cytoskeleton (see section 3.5).

Once in the cytoplasm, mRNA coming from the nucleus binds to a ribosome, which manufactures proteins (see figure 3.13, step 3). Free-floating ribosomes produce proteins that remain in the cell's cytosol (**figure 3.15a**). But many proteins are destined for organelles, for the cell membrane, or for secretion (in milk, for example).

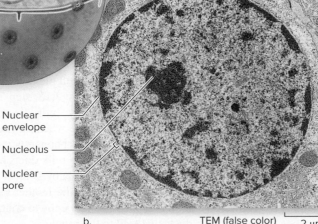

Nuclear envelope

DNA

Nuclear pore

Nucleolus

a.

Nuclear envelope

Nucleolus

Nuclear pore

b.   TEM (false color)   2 μm

Figure 3.14 The Nucleus. (a) The nucleus contains DNA and is surrounded by two membrane layers, which make up the nuclear envelope. Large pores in the nuclear envelope allow proteins to enter and mRNA molecules to leave the nucleus. (b) This transmission electron micrograph shows the nuclear envelope and nucleolus.

(b): ©Dr. David M. Phillips/Science Source

a. Free ribosome   Cytosol   mRNA   Protein

b. Membrane-bound ribosome   Cytosol   mRNA   ER membrane   Inside ER   Protein

In those cases, the entire complex of ribosome, mRNA, and partially made protein anchors to the membrane of the endoplasmic reticulum (see figure 3.15b).

The **endoplasmic reticulum (ER)** is a network of sacs and tubules composed of membranes (figure 3.16). This complex organelle originates at the nuclear envelope and winds throughout the cell (*endoplasmic* means "within the cytoplasm," and *reticulum* means "network"). Close to the nucleus, the membrane surface is studded with ribosomes making proteins that enter the inner compartment of the ER (see figure 3.15b); these proteins are destined to be secreted from the cell. This section of the network is called the **rough ER** because the ribosomes give these membranes a roughened appearance.

Adjacent to the rough ER, a section of the network called the **smooth ER** synthesizes lipids—such as those that will end up in the milk—and other membrane components (see figure 3.13, step 4, and figure 3.16). The smooth ER also houses enzymes that detoxify drugs and poisons. In muscle cells, a specialized type of smooth ER stores and delivers the calcium ions required for muscle contraction. ⓘ *muscle function*, section 29.4

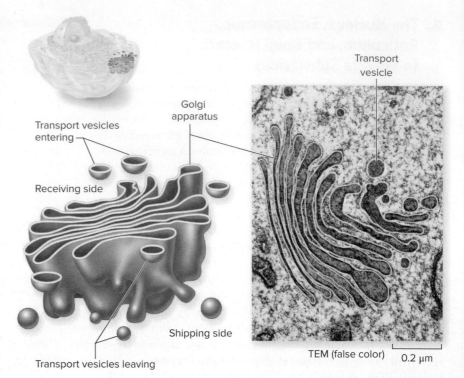

Figure 3.17 **The Golgi Apparatus.** The Golgi apparatus consists of a series of flattened sacs, plus transport vesicles that deliver and remove materials. Proteins are sorted and processed as they move through the Golgi apparatus on their way to the cell surface or to a lysosome.

Photo: ©Biophoto Associates/Science Source

The lipids and proteins made by the ER exit the organelle in vesicles. A loaded transport vesicle pinches off from the tubular endings of the ER membrane (see figure 3.13, step 5) and takes its contents to the Golgi apparatus, the next stop in the production line. The **Golgi apparatus** is a stack of flat, membrane-enclosed sacs that functions as a processing center (figure 3.17). Proteins from the ER pass through the series of Golgi sacs, where they complete their intricate folding and become functional (see figure 3.13, step 6). Enzymes in the Golgi apparatus also manufacture and attach carbohydrates to proteins or lipids, forming the "name tags" recognized by the immune system.

The Golgi apparatus sorts and packages materials into vesicles, which move toward the cell membrane. Some of the proteins received from the ER will become membrane surface proteins; other substances (such as milk protein and fat) are packaged for secretion from the cell. In the production of milk, these vesicles fuse with the cell membrane and release the proteins outside the cell (see figure 3.13, step 7). The fat droplets stay suspended in the watery milk because they retain a layer of surrounding membrane when they leave the cell.

This entire process happens simultaneously in countless specialized cells lining the milk ducts of the mother's breast, beginning shortly after a baby's birth. When the infant suckles, hormones released in the mother's body stimulate muscles surrounding balls of these cells to contract, squeezing milk into the ducts that lead to the nipple.

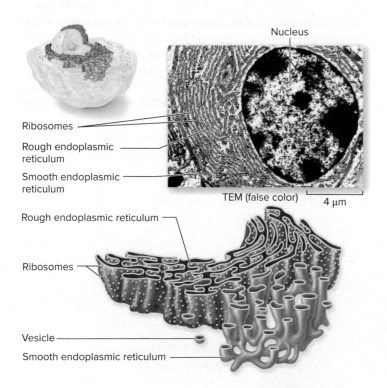

Figure 3.16 **Rough and Smooth Endoplasmic Reticulum.** The endoplasmic reticulum is a network of membranes extending from the nuclear envelope. Ribosomes dot the surface of the rough ER, giving it a "rough" appearance. The smooth ER is a series of interconnecting tubules and is the site for lipid production and other metabolic processes.

Photo: ©Prof. J. L. Kemeny/ISM/Phototake

B. Lysosomes, Vacuoles, and Peroxisomes Are Cellular Digestion Centers

Besides producing molecules for export, eukaryotic cells also break down molecules in specialized compartments. All of these "digestion center" organelles are sacs surrounded by a single membrane.

Lysosomes **Lysosomes** are organelles containing enzymes that dismantle and recycle food particles, captured bacteria, worn-out organelles, and debris (**figure 3.18**). They are so named because their enzymes lyse, or cut apart, their substrates.

The enzymes inside lysosomes originate in the rough ER. The Golgi apparatus detects these enzymes by recognizing a sugar attached to them, then packages them into vesicles that eventually become lysosomes. The lysosomes, in turn, fuse with transport vesicles carrying debris from outside or from within the cell. The lysosome's enzymes break down the large organic molecules into smaller subunits by hydrolysis, releasing them into the cytosol for the cell to use.

What keeps a lysosome from digesting the entire cell? The lysosome's membrane maintains the pH of the organelle's interior at about 4.8, much more acidic than the neutral pH of the rest of the cytoplasm. If one lysosome were to burst, the liberated enzymes would no longer be at their optimum pH, so they could not digest the rest of the cell. Nevertheless, a cell injured by extreme cold, heat, or another physical stress may initiate its own death by bursting all of its lysosomes at once. ⓘ *pH*, section 2.4; *cell death*, section 8.7

Some cells have more lysosomes than others. White blood cells, for example, have many lysosomes because these cells engulf and dispose of debris and bacteria. Liver cells require many lysosomes to process cholesterol.

Malfunctioning lysosomes can cause illness. In Tay-Sachs disease, for example, a defective lysosomal enzyme allows a lipid to accumulate to toxic levels in nerve cells of the brain. The nervous system deteriorates, and an affected person eventually becomes unable to see, hear, or move. In the most severe forms of the illness, death usually occurs by age 5.

Vacuoles Most plant cells lack lysosomes, but they do have an organelle that serves a similar function. In mature plant cells, the large central **vacuole** contains a watery solution of enzymes that degrade and recycle molecules and organelles (see figure 3.8).

The vacuole also has other roles. Most of the growth of a plant cell comes from an increase in the volume of its vacuole. In some plant cells, the vacuole occupies up to 90% of the cell's volume (**figure 3.19**). As the vacuole acquires water, it exerts pressure (called turgor pressure) against the cell membrane. Turgor pressure helps plants stay rigid and upright.

Besides water and enzymes, the vacuole also contains a variety of salts, sugars, and weak acids. Therefore, the pH of the vacuole's solution is usually somewhat acidic. In citrus fruits, the solution is very acidic, producing the tart taste of lemons and oranges. Water-soluble pigments also reside in the vacuole, producing blue, purple, and magenta colors in some leaves, flowers, and fruits.

Some protists have vacuoles, although their function is different from that in plants. The contractile vacuole in *Paramecium*, for example, pumps excess water out of the cell. In *Amoeba*, food vacuoles digest nutrients that the cell has engulfed.

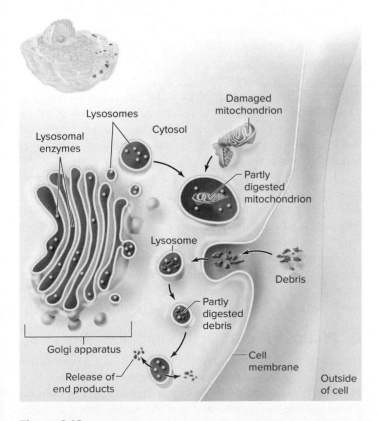

Figure 3.18 Lysosomes. Lysosomes contain enzymes that dismantle damaged organelles and other debris, then release the nutrients for the cell to use.

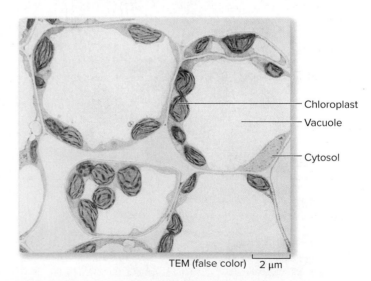

TEM (false color)   2 μm

Figure 3.19 Vacuole. Much of the volume of a spinach leaf cell is occupied by the large central vacuole. The rest of each cell's contents (including numerous chloroplasts) is pushed to the edges of the cell.

©Biophoto Associates/Science Source

Peroxisomes All eukaryotic cells contain **peroxisomes,** organelles that contain several types of enzymes that dispose of toxic substances. Although they resemble lysosomes in size and function, peroxisomes originate at the ER (not the Golgi) and contain different enzymes. In some peroxisomes, the concentration of enzymes reaches such high levels that the proteins condense into easily recognized crystals (**figure 3.20**).

The peroxisome's name comes from one of its roles in protecting the cell. Some chemical reactions in the peroxisome produce hydrogen peroxide, H_2O_2. This highly reactive compound can produce oxygen free radicals that can damage the cell. To counteract the free-radical buildup, peroxisomes contain an enzyme that detoxifies H_2O_2 and produces harmless water molecules in its place.

Peroxisomes in liver and kidney cells help dismantle toxins from the blood. Peroxisomes also break down fatty acids and produce cholesterol and some other lipids. In a disease called adrenoleukodystrophy (ALD), a faulty peroxisomal enzyme causes fatty acids to accumulate to toxic levels in the brain, causing severe brain damage and eventually death.

C. Mitochondria Extract Energy from Nutrients

Growth, cell division, protein production, secretion, and many chemical reactions in the cytoplasm all require a steady supply of energy. **Mitochondria** (singular: mitochondrion) are organelles that use a process called cellular respiration to extract this needed energy from food (see chapter 6). With the exception of a few types of protists, all eukaryotic cells have mitochondria.

A mitochondrion has two membrane layers: an outer membrane and an intricately folded inner membrane that encloses the mitochondrial matrix (**figure 3.21**). Within the matrix is DNA

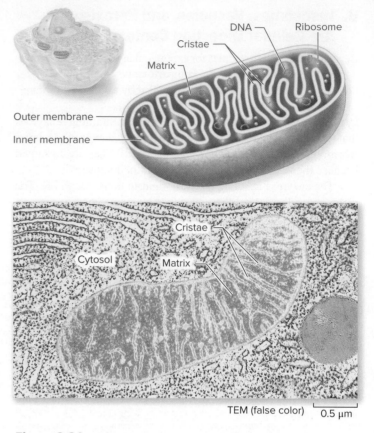

Figure 3.21 **Mitochondria.** Cellular respiration occurs inside mitochondria. Each mitochondrion contains a highly folded inner membrane, where many of the reactions of cellular respiration occur.

Photo: ©Bill Longcore/Science Source

that encodes proteins essential for mitochondrial function; ribosomes occupy the matrix as well. **Cristae** are the folds of the inner membrane. The cristae add tremendous surface area to the inner membrane, which houses the enzymes that catalyze the reactions of cellular respiration.

In most mammals, mitochondria are inherited from the female parent only. (This is because the mitochondria in a sperm cell degenerate after fertilization.) Mitochondrial DNA is therefore useful for tracking inheritance through female lines in a family. For the same reason, genetic mutations that cause defective mitochondria also pass only from mother to offspring. Mitochondrial illnesses are most serious when they affect the muscles or brain, because these energy-hungry organs depend on the functioning of many thousands of mitochondria in every cell.

D. Photosynthesis Occurs in Chloroplasts

Plants and many protists carry out photosynthesis, a process that uses energy from sunlight to produce glucose and other food molecules (see chapter 5). These nutrients sustain not only the photosynthetic organisms but also the consumers (including humans) that eat them.

The **chloroplast** (**figure 3.22**) is the site of photosynthesis in eukaryotes. Each chloroplast contains multiple membrane layers. Two outer membrane layers enclose an enzyme-rich fluid called

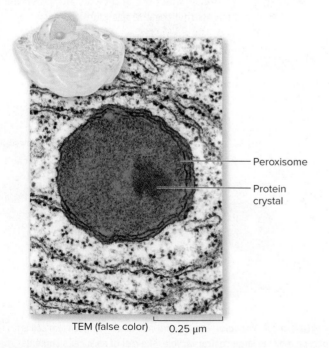
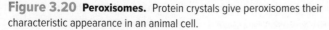

Figure 3.20 **Peroxisomes.** Protein crystals give peroxisomes their characteristic appearance in an animal cell.

Photo: ©Don W. Fawcett/Science Source

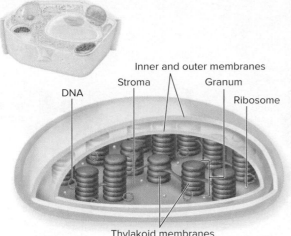

DNA
Stroma
Inner and outer membranes
Granum
Ribosome
Thylakoid membranes

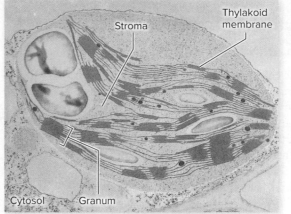

Figure 3.22
Chloroplasts.
Photosynthesis occurs inside chloroplasts. Each chloroplast contains stacks of thylakoids that form the grana within the inner compartment, the stroma. Enzymes and light-harvesting pigments embedded in the membranes of the thylakoids convert the energy in sunlight to chemical energy.

Photo: ©Biophoto Associates/Science Source

Stroma
Thylakoid membrane
Cytosol
Granum
TEM (false color) 1 µm

the stroma. Within the stroma is a third membrane system folded into flattened sacs called thylakoids, which are stacked like pancakes to form structures called grana. Photosynthetic pigments such as chlorophyll are embedded in the thylakoid membranes.

A chloroplast is one representative of a larger category of plant organelles called plastids. Some plastids synthesize lipid-soluble red, orange, and yellow carotenoid pigments, such as those found in carrots and ripe tomatoes. Plastids that assemble starch molecules are important in cells specialized for food storage, such as those in potatoes and corn kernels. Interestingly, any plastid can convert into any other type. As a tomato ripens, for example, its green chloroplasts change into plastids that store red carotenoid pigments.

Like mitochondria, all plastids (including chloroplasts) contain DNA and ribosomes. The genetic material encodes proteins unique to plastid structure and function, including some of the enzymes required for photosynthesis.

The similarities between chloroplasts and mitochondria—both have their own DNA and ribosomes, and both are surrounded by double membranes—provide clues to the origin of eukaryotic cells, an event that occurred at least 1.5 billion years ago. According to the endosymbiosis theory, some ancient organism engulfed bacterial cells. Rather than digesting them as food, the host cells kept them on as partners: mitochondria and chloroplasts. The structures and genetic sequences of today's bacteria, mitochondria, and chloroplasts supply powerful evidence for this theory. ⓘ *endosymbiosis,* section 15.2A

Organelles divide a cell's work, just as the rooms in a house contain related items: Pots and dishes are in the kitchen, whereas blankets and pillows are in the bedroom. But highly specialized buildings also exist. A restaurant, for example, has an enormous kitchen and no bedrooms at all. Likewise, cells can also have specialized functions (figure 3.23). For example, a heart muscle cell is roughly cylindrical when compared with a neuron, which produces extensions that touch adjacent nerve cells. A leaf cell is packed with chloroplasts. The protective epidermis of an onion, on the other hand, is dry and tough; because it forms underground, it lacks chloroplasts. In each case, the mix of organelles inside each cell determines its functions. Keep these specialized structures and functions in mind as you study cell processes throughout this book.

3.4 MASTERING CONCEPTS

1. How do organelles contribute to efficiency in eukaryotic cells?
2. Which parts of a cell interact to produce and secrete a complex substance such as milk?
3. What is the function of the nucleus and its contents?
4. Which organelles are the cell's "recycling centers"?
5. Which organelle houses the reactions that extract chemical energy from nutrient molecules?
6. How are the functions of plastids essential to the life of a plant cell?
7. Describe how form fits function for three organelles.

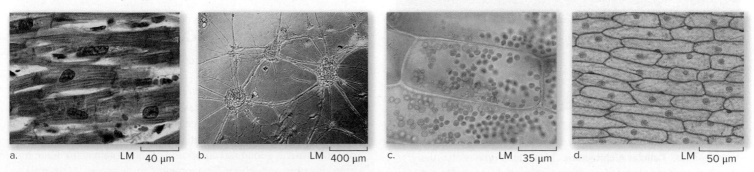

a. LM 40 µm b. LM 400 µm c. LM 35 µm d. LM 50 µm

Figure 3.23 Specialized Cells. (a) Muscle cells in the heart look and behave differently than (b) the highly branched neurons that form the nervous system. (c) A plant's leaf cells contain chloroplasts, whereas (d) the cells making up the outer skin of an onion do not.

(a): ©Corbis RF; (b): ©Francois Paquet-Durand/Science Source; (c): ©Ed Reschke/Photolibrary/Getty Images; (d): ©Ted Kinsman/Science Source

3.5 The Cytoskeleton Supports Eukaryotic Cells

The cytosol of a eukaryotic cell contains a **cytoskeleton**, an intricate network of protein "tracks" and tubules. The cytoskeleton is a structural framework with many functions. It is a transportation system, and it provides the physical support necessary to maintain the cell's characteristic three-dimensional shape (figure 3.24). It aids in cell division and helps connect cells to one another. The cytoskeleton also enables cells—or parts of a cell—to move.

Given the cytoskeleton's many functions, it is not surprising that defects can cause disease. For example, people with Duchenne muscular dystrophy lack a protein called dystrophin, part of the cytoskeleton in muscle cells. Without dystrophin, muscles—including those in the heart—degenerate. A faulty version of another cytoskeleton protein, ankyrin, causes a genetic disease of blood. In healthy red blood cells, the cytoskeleton maintains a concave disk shape. When ankyrin is missing, red blood cells are small, fragile, and misshapen, greatly reducing their ability to carry oxygen.

Injuries can also cause serious damage to the cytoskeleton. When a person suffers a strong blow to the head, such as in a fall or an auto accident, the cells that make up the brain can become stretched and distorted. The resulting damage to the cytoskeleton can trigger a chain reaction that ends with the death of the affected brain cells. People with such injuries may recover, but many die, lapse into a coma, or suffer from permanent disabilities (see the opening essay for chapter 26).

SEM (false color) 10 μm

Figure 3.24 Cellular Architecture. Thanks to its cytoskeleton, this white blood cell can produce long, thin extensions that reach out to and engulf "foreign" substances, including bacteria.
©Dennis Kunkel Microscopy, Inc./Phototake

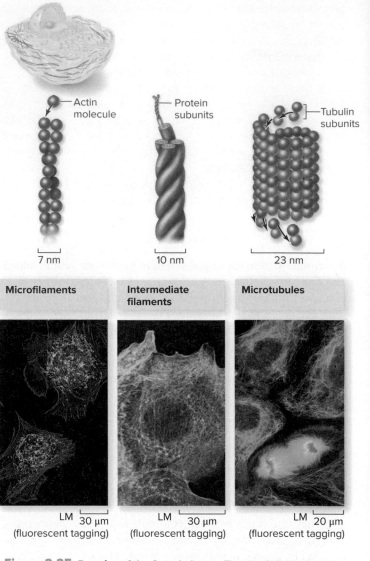

Figure 3.25 Proteins of the Cytoskeleton. The cytoskeleton consists of three sizes of protein filaments, arranged in this figure from smallest to largest diameter. Arrows indicate protein subunits being added to or removed from microfilaments and microtubules. The photos show actin microfilaments (*left, colored red*), intermediate filaments (*center, colored green*), and microtubules (*right, colored green*).
Photos: (microfilaments): ©Science Source; (intermediate): ©Phanie/Alamy; (microtubules): ©Jennifer Waters/Science Source

A. Proteins Form the Cytoskeleton

The cytoskeleton includes three major components: microfilaments, intermediate filaments, and microtubules (figure 3.25). They are distinguished by protein type, diameter, and how they aggregate into larger structures. Other proteins connect these components to one another, creating an intricate meshwork.

The thinnest component of the cytoskeleton is the **microfilament**, a long rod composed of the protein actin. Each microfilament is only about 7 nanometers in diameter. Actin microfilament networks are part of nearly all eukaryotic cells. Muscle contraction, for example, relies on actin filaments and another protein, myosin. Microfilaments also provide strength for cells to survive stretching and compression, and they help to anchor one cell to another (see section 3.6). ⓘ *sliding filaments,* section 29.4B

Intermediate filaments are so named because their 10-nanometer diameters are intermediate between those of microfilaments and microtubules. Unlike the other components of the cytoskeleton, which consist of a single protein type, intermediate filaments are made of a variety of proteins. They maintain a cell's shape by forming an internal scaffold in the cytosol and resisting mechanical stress. Intermediate filaments also help bind some cells together (see section 3.6).

A **microtubule** is composed of a protein called tubulin, assembled into a hollow tube that is 23 nanometers in diameter. The cell can change the length of a microtubule rapidly by adding or removing tubulin molecules. Microtubules have many functions in eukaryotic cells. For example, they form a type of "trackway" along which substances move within a cell. Specialized motor proteins "walk" along the tracks, toting an organelle, a vesicle, or other cargo. In addition, chapter 8 describes how microtubules split a cell's duplicated chromosomes apart during cell division.

B. Cilia and Flagella Help Cells Move

In animal cells, structures called **centrosomes** organize the microtubules. (Plants typically lack centrosomes and assemble microtubules at sites scattered throughout the cell.) The centrosome contains two centrioles, which are visible in figure 3.8. The centrioles form the basis of structures called basal bodies, which in turn give rise to the extensions that enable some cells to move: cilia and flagella (**figure 3.26**).

Cilia are short, numerous extensions resembling a fringe. Some protists, such as the *Paramecium* in figure 3.3, have thousands of cilia that enable the cells to "swim" in water. In the human respiratory tract, coordinated movement of cilia sets up a wave that propels particles up and out; other cilia can move an egg cell through the female reproductive tract. ⓘ *ciliates,* section 18.4C

Unlike cilia, flagella occur singly or in pairs, and a flagellum is much longer than a cilium. Flagella are more like tails, and their whiplike movement propels cells. Sperm cells in many species (including humans) have prominent flagella. A man whose sperm cells have defective flagella is infertile because the sperm are unable to swim to the egg cell.

Figure 3.26 shows the internal microtubules underlying the functions of both cilia and flagella. Inside each appendage's shaft, a protein called dynein links a ring of outer microtubule pairs to a central pair, a little like a wheel. Dynein molecules shift in a way that slides adjacent microtubules against each other. This movement bends the appendage from side to side. (The bacterial flagellum has a different structure.)

3.5 | MASTERING CONCEPTS

1. What are some functions of the cytoskeleton?
2. What are the main components of the cytoskeleton?
3. Why are cilia and flagella important?

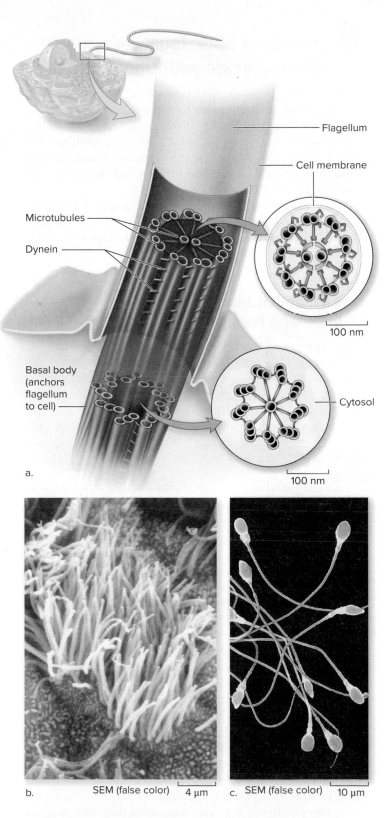

a.

b.   SEM (false color) 4 μm   c. SEM (false color) 10 μm

Figure 3.26 Microtubules Move Cells. (a) Cilia and eukaryotic flagella contain microtubules, as does the anchoring basal body. (b) These cilia help eliminate dust and other foreign particles from the human respiratory tract. (c) The flagella on human sperm cells enable them to swim.

(b): ©D.W. Fawcett/Science Source; (c): ©Dr. Tony Brain/Science Source

3.6 Cells Stick Together and Communicate with One Another

So far, this chapter has described individual cells. But multicellular organisms, including plants and animals, are made of many cells that work together. How do these cells adhere to one another so that your body—or that of a plant—doesn't disintegrate in a heavy rain? Also, how do cells in direct contact with one another communicate to coordinate development and respond to the environment?

This section describes how the cells of animal and plant tissues stick together and how neighboring cells share signals. Table 3.2 lists the types of cell–cell connections for animals and plants.

A. Animal Cell Junctions Occur in Several Forms

Unlike plants and fungi, animal cells lack cell walls. Instead, many animal cells secrete a complex extracellular matrix that holds them together and coordinates many aspects of cellular life. In these tissues, cells are not in direct contact with one another. ⓘ *extracellular matrix,* section 25.2

In other tissues, however, the membranes of adjacent cells directly connect to one another via several types of junctions (**figure 3.27**): tight junctions, anchoring junctions, and gap junctions.

A **tight junction** fuses animal cells together, forming an impermeable barrier between them. Proteins anchored in membranes connect to actin in the cytoskeleton and join cells into sheets, such as those lining the inside of the digestive tract and the tubules of the kidneys. These connections allow the body to control where biochemicals move, since fluids cannot leak between the joined cells. For example, tight junctions prevent stomach acid from seeping into the tissues surrounding the stomach. Likewise, tight junctions create the "blood–brain barrier," densely packed cells that prevent many harmful substances from entering the brain. However, this barrier readily admits lipid-soluble drugs such as heroin and cocaine across its cell membranes, accounting for the rapid action of these drugs. ⓘ *blood–brain barrier,* section 26.6D

A second type of cell junction, an **anchoring (or adhering) junction,** connects an animal cell to its neighbors or to the extracellular matrix, somewhat like a rivet. Proteins at each anchoring junction span the cell membrane and link to each cell's cytoskeleton. These junctions hold skin cells in place by anchoring them to one another and to the extracellular matrix.

A **gap junction** is a protein channel that links the cytoplasm of adjacent animal cells, allowing exchange of ions and small molecules. Gap junctions link heart muscle cells to one another, allowing groups of cells to contract together. Similarly, the muscle cells that line the digestive tract coordinate their contractions to propel food along its journey, courtesy of countless gap junctions.

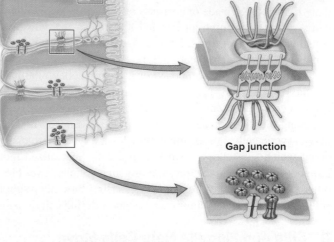

Figure 3.27 Animal Cell Connections. Tight junctions fuse neighboring cell membranes, anchoring junctions form "spot welds," and gap junctions allow small molecules to move from cell to cell.

B. Cell Walls Are Strong, Flexible, and Porous

Cell walls surround the cell membranes of nearly all bacteria, archaea, fungi, algae, and plants. But *cell wall* is a misleading term: It is not just a barrier that outlines the cell. Cell walls impart shape, regulate cell volume, prevent bursting when a cell takes in too much water, and interact with other molecules to help determine how a cell in a complex organism specializes. In plants, for example, a given cell may become a root, shoot, or leaf, depending on which cell walls it touches.

Many materials may make up a cell wall. Bacterial cell walls, for example, are composed of peptidoglycan, whereas those of fungi contain chitin. Much of the plant cell wall consists of cellulose molecules aligned into microfibrils, which in turn aggregate and twist to form larger fibrils (**figure 3.28**). This fibrous organization imparts great strength. Other molecules, including the polysaccharides hemicellulose and pectin, glue adjacent cells together and add strength and flexibility. Plant cell walls also contain glycoproteins, enzymes, and many other proteins. ⓘ *cellulose,* section 2.5B

A plant cell secretes many of the components of its wall from the inside. The oldest layer of a cell wall is therefore on the exterior of the cell, and the newer layers hug the cell membrane. The region where adjacent cell walls meet is called the middle lamella.

Some cells have rigid secondary cell walls beneath the initial, more flexible primary one (see figure 3.28b). The secondary cell

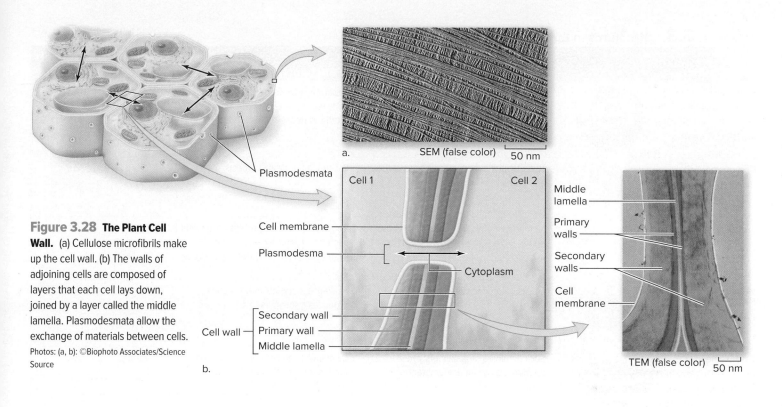

Figure 3.28 The Plant Cell Wall. (a) Cellulose microfibrils make up the cell wall. (b) The walls of adjoining cells are composed of layers that each cell lays down, joined by a layer called the middle lamella. Plasmodesmata allow the exchange of materials between cells.

Photos: (a, b): ©Biophoto Associates/Science Source

wall typically supports plant cells that are no longer growing. For example, the tough texture of walnut shells, wood, apple cores, and other durable plant parts comes from their thick secondary cell walls. In addition to cellulose and other polysaccharides, the secondary cell wall includes lignin. This strong, rigid polymer is so complex that only a few types of organisms can break it down.

How do plant cells communicate with their neighbors through the wall? **Plasmodesmata** (singular: plasmodesma) are channels that connect adjacent cells. They are essentially "tunnels" in the cell wall, and materials can move from one cell to another via a thin strand of cytoplasm that passes through each channel (see figure 3.28). Plasmodesmata are especially plentiful in parts of plants that conduct water or nutrients and in cells that secrete oils and nectars.

Like the gap junctions of animal cells, plasmodesmata enable cell-to-cell communication and coordination of function within a plant. Plasmodesmata, however, may also play a role in the spread of disease within a plant; viruses use them as conduits to pass from cell to cell.

We have now completed our tour of the eukaryotic cell. Table 3.3 summarizes the structures and functions of the main organelles and other structures in the cells of both plants and animals.

3.6 MASTERING CONCEPTS

1. What are the three types of junctions that link cells in animals?
2. Describe the functions and chemical composition of a plant cell wall.
3. What are plasmodesmata? To which type of animal cell junction are plasmodesmata most similar?

TABLE **3.2** Intercellular Junctions: A Summary

Type	Function(s)	Example of Location
Animal cells		
Tight junctions	Close the spaces between animal cells by fusing cell membranes	Cells in inner lining of small intestine
Anchoring (adhering) junctions	Connect adjacent animal cell membranes in one spot; connect cells to the extracellular matrix	Cells in outer skin layer
Gap junctions	Form channels between animal cells, allowing exchange of substances	Muscle cells in heart and digestive tract
Plant cells		
Plasmodesmata	Allow substances to move between plant cells	Plant cell walls

TABLE **3.3** **Structures in Eukaryotic Cells: A Summary**

Structure		Description	Function(s)	Plant Cells?	Animal Cells?
Nucleus		Perforated sac containing DNA, proteins, and RNA; surrounded by double membrane	Separates DNA from rest of cell; site of first step in protein synthesis; nucleolus produces ribosomal subunits	Yes	Yes
Ribosome		Two globular subunits composed of RNA and protein	Location of protein synthesis	Yes	Yes
Endoplasmic reticulum (ER)		Membrane network studded with ribosomes (rough ER) or lacking ribosomes (smooth ER)	Rough ER produces proteins destined for secretion from the cell; smooth ER synthesizes lipids and detoxifies drugs and poisons	Yes	Yes
Golgi apparatus		Stacks of flat, membranous sacs	Packages materials to be secreted; produces lysosomes	Yes	Yes
Lysosome		Sac containing digestive enzymes; surrounded by single membrane	Dismantles and recycles components of food, debris, captured bacteria, and worn-out organelles	Rarely	Yes
Central vacuole		Sac containing enzymes, acids, water-soluble pigments, and other solutes; surrounded by single membrane	Produces turgor pressure; recycles cell contents; contains pigments	Yes	No
Peroxisome		Sac containing enzymes, often forming visible protein crystals; surrounded by single membrane	Disposes of toxins; breaks down fatty acids; eliminates hydrogen peroxide	Yes	Yes
Chloroplast		Two membranes enclosing stacks of membrane sacs, which contain photosynthetic pigments and enzymes; contains DNA and ribosomes	Produces food (sugars) by photosynthesis	Yes	No
Mitochondrion		Two membranes; inner membrane is folded into enzyme-studded cristae; contains DNA and ribosomes	Releases energy from food by cellular respiration	Yes	Yes
Cytoskeleton		Network of protein filaments and tubules	Transports organelles within cell; maintains cell shape; structural basis for flagella/cilia; connects adjacent cells	Yes	Yes
Cell wall		Porous barrier of cellulose and other substances (in plants)	Protects cell; provides shape; connects adjacent cells	Yes	No

INVESTIGATING LIFE

🍁 3.7 The Tiniest Compass

Submarine captains steer their vessels through the dark ocean. They could wander in all directions in search of their destination, but that approach wastes time and fuel. Instead, they rely on navigation systems to find their way. Surprisingly, massive ships have something in common with marine bacteria: Compasses guide some of these tiny vessels as well.

Until recently, biologists thought that prokaryotic cells lacked any internal membranes. But microscopes revealed that some bacteria have small lipid bilayer spheres in their cytoplasm. Scientists found high concentrations of magnetic iron crystals within these membrane bubbles and aptly named them "magnetosomes" (figure 3.29). What is the function of these structures?

When scientists found magnetosomes, they already knew that Earth's magnetic field leaves the planet from the southern hemisphere, circles far into space, and returns to Earth in the northern hemisphere. In most parts of the ocean, the magnetic field lines are roughly vertical. Experiments on magnetic bacteria collected from oceans revealed that the magnetosomes align with magnetic field lines and that the bacteria swim either against or with the field.

These studies showed how bacteria respond to magnetism; they did not explain why orienting to magnetic fields is adaptive. A team of four researchers led by Richard Frankel at California Polytechnic State University aimed to answer this question.

The observation that the bacteria do not always swim in the same direction along the magnetic field lines led them to hypothesize that another factor must influence bacterium movement. One clue was that these bacteria cannot survive if oxygen levels are too high or too low. So Frankel and his colleagues devised an experiment to test whether magnetism and oxygen concentration jointly guided bacterial movement.

The scientists put the bacteria in a solution. They then drew the mixture into narrow glass tubes and sealed one end. Within

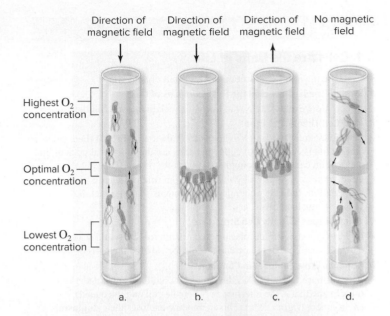

Figure 3.30 **Magnetic Orientation.** (a) Bacteria with magnetosomes turn toward magnetic fields and (b) move in straight lines toward their optimal oxygen concentration. (c) Switching the direction of the magnetic field rotates the bacteria. (d) Without a magnetic field, bacteria move toward an optimal O_2 concentration but do not take a direct path. Small arrows in (a) and (d) indicate the direction of bacterial movement within each tube.

each tube, the dissolved oxygen concentration was lowest at the sealed end and increased toward the open end. When the team produced a magnetic field across one of the tubes, all of the bacteria turned toward the field. Some then swam forward, while others moved backward. They aggregated in a distinct band in the center of the tube, at their optimal oxygen concentration (figure 3.30). The scientists then switched the direction of the magnetic field. All of the bacteria turned 180°, but none migrated out of the band in the center of the tube.

These results indicate that magnetic fields influence the direction that magnetosome-containing bacteria face, helping the cells follow a straight line through the water. Since the dissolved oxygen concentration decreases with depth, and since Earth's magnetic field runs almost vertically through the water column, bacterial cells use magnetism to find the shortest path toward or away from oxygen. Decreasing the swimming distance saves energy for other cellular tasks, such as reproduction.

Scientists used powerful microscopes and clever experiments to reveal how some bacteria avoid getting lost at sea. Lipid-enclosed magnetosomes guide them like compasses through the deep unknown.

Source: Frankel, Richard B., Dennis A. Bazylinski, Mark S. Johnson, and Barry L. Taylor. 1997. Magneto-aerotaxis in marine coccoid bacteria. *Biophysical Journal,* vol. 73, pages 994–1000.

3.7 MASTERING CONCEPTS

1. How did the researchers determine that both magnetism and oxygen guided bacteria movements?

2. How do magnetosomes help bacteria save energy?

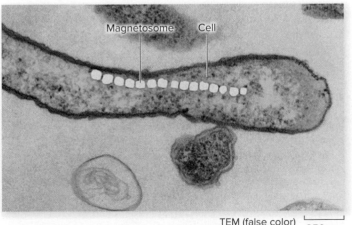

Magnetosome   Cell

TEM (false color) └── 250 nm

Figure 3.29 **Magnetosomes.** This bacterial cell contains a row of magnetosomes, which are lipid spheres containing iron crystals that align with Earth's magnetic field.

©Dennis Kunkel Microscopy, Inc./Phototake

CHAPTER SUMMARY

3.1 Cells Are the Units of Life

- **Cells** are the microscopic components of all organisms.

A. Simple Lenses Revealed the First Glimpses of Cells

- Robert Hooke and Antony van Leeuwenhoek pioneered cell biology.

B. The Cell Theory Emerges

- Schleiden, Schwann, and Virchow's formulation of the **cell theory** states that all life is composed of cells, that cells are the functional units of life, and that all cells come from preexisting cells.
- Contemporary cell biology focuses on the role of genetic information, the cell's chemical components, and the metabolic processes inside cells.

C. Microscopes Magnify Cell Structures

- Light microscopes, transmission electron microscopes, and scanning electron microscopes are essential tools for viewing the parts of a cell.

D. All Cells Have Features in Common

- All cells have DNA, RNA, **ribosomes** that build proteins, and a **cell membrane** that is the interface between the cell and the outside environment (**figure 3.31**). This membrane encloses the **cytoplasm,** which includes a fluid portion called the **cytosol.**
- Complex cells also have specialized compartments called **organelles.**
- The surface area of a cell must be large relative to its volume.

3.2 Different Cell Types Characterize Life's Three Domains

- **Eukaryotic** cells have a nucleus and other organelles; **prokaryotic** cells lack these structures. Prokaryotic cells include bacteria and archaea.

A. Domain Bacteria Contains Earth's Most Abundant Organisms

- Bacterial cells are structurally simple, but they are abundant and diverse. Most have a **cell wall** and one or more **flagella.** DNA occurs in an area called the **nucleoid.**

B. Domain Archaea Includes Prokaryotes with Unique Biochemistry

- Archaea share some characteristics with bacteria and eukaryotes but also have unique structures and chemistry.

C. Domain Eukarya Contains Organisms with Complex Cells

- Eukaryotic cells include those of protists, plants, fungi, and animals. Most eukaryotic cells are larger than prokaryotic cells.

	Prokaryote	Eukaryote
Nucleus	No	Yes
Membrane-bounded organelles	No	Yes

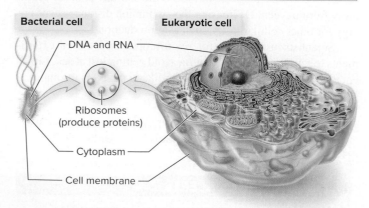

Figure 3.31 Cell Features: A Summary.

3.3 A Membrane Separates Each Cell from Its Surroundings

- A biological membrane consists of a **phospholipid bilayer** embedded with movable proteins and steroid molecules, forming a **fluid mosaic.**
- Membrane proteins carry out a variety of functions.

3.4 Eukaryotic Organelles Divide Labor

- The **endomembrane system** includes the nuclear envelope, endoplasmic reticulum, Golgi apparatus, lysosomes, vacuoles, cell membrane, and **vesicles** that transport materials within cells.

A. The Nucleus, Endoplasmic Reticulum, and Golgi Interact to Secrete Substances

- A eukaryotic cell houses DNA in a **nucleus. Nuclear pores** allow the exchange of materials through the two-layered **nuclear envelope;** assembly of the ribosome's subunits occurs in the **nucleolus.**
- The **smooth endoplasmic reticulum, rough endoplasmic reticulum,** and **Golgi apparatus** work together to synthesize, store, transport, and release molecules.

B. Lysosomes, Vacuoles, and Peroxisomes Are Cellular Digestion Centers

- A eukaryotic cell degrades wastes and digests nutrients in **lysosomes.**
- In plants, a watery **vacuole** degrades wastes, exerts turgor pressure, and stores acids and pigments.
- **Peroxisomes** help digest fatty acids and detoxify many substances.

C. Mitochondria Extract Energy from Nutrients

- **Mitochondria** house the reactions of cellular respiration. The **cristae** (folds) of the inner mitochondrial membrane add surface area.

D. Photosynthesis Occurs in Chloroplasts

- In the cells of plants and algae, **chloroplasts** use light energy to make food.

3.5 The Cytoskeleton Supports Eukaryotic Cells

- The **cytoskeleton** is a network of protein rods and tubules that provides cells with form, support, and the ability to move.

A. Proteins Form the Cytoskeleton

- **Microfilaments,** the thinnest components of the cytoskeleton, are composed of the protein actin. **Intermediate filaments** consist of various proteins that strengthen the cytoskeleton. **Microtubules** are made of tubulin subunits. They form an internal trackway and move chromosomes.
- **Centrosomes** organize microtubules in animal cells.

B. Cilia and Flagella Help Cells Move

- **Cilia** are short, numerous extensions; flagella are less numerous but much longer. Both cilia and flagella aid in the movement of cells or materials.

3.6 Cells Stick Together and Communicate with One Another

A. Animal Cell Junctions Occur in Several Forms

- **Tight junctions** create a seal between adjacent cells. **Anchoring junctions** secure cells in place. **Gap junctions** allow adjacent cells to exchange materials.

B. Cell Walls Are Strong, Flexible, and Porous

- Cell walls provide protection and shape. Plant cell walls consist of cellulose fibrils connected by hemicellulose, pectin, and various proteins.
- **Plasmodesmata** are channels extending through the walls of adjacent plant cells.

3.7 Investigating Life: The Tiniest Compass

- Some bacteria contain magnetosomes, specialized structures that help the cells navigate vertically in the water column. This adaptation boosts their efficiency in finding the optimal concentration of oxygen.

MULTIPLE CHOICE QUESTIONS

1. Which of the following is NOT a feature found in all cells?
   a. Proteins
   b. Ribosomes
   c. Cell wall
   d. Cell membrane

2. One property that distinguishes cells in domain Eukarya from those in domain Bacteria is the presence of
   a. a cell wall.
   b. DNA.
   c. flagella.
   d. membranous organelles.

3. What chemical property of phospholipids is key to the formation of the cell membrane?
   a. The positively charged nitrogen atom
   b. The covalent bond between the phosphate and the glycerol
   c. The kink in the fatty acid tail
   d. The hydrophilic head and hydrophobic tails

4. Which of the following organelles are associated with the job of cellular digestion?
   a. Lysosomes and peroxisomes
   b. Golgi apparatus and vesicles
   c. Nucleus and nucleolus
   d. Smooth and rough endoplasmic reticulum

5. Within a single cell, which of the following is physically the smallest?
   a. Nuclear envelope
   b. Phospholipid molecule
   c. Cell membrane
   d. Mitochondrion

6. A human nerve cell that has an abnormal shape most likely has a defective
   a. cell wall.
   b. cytoskeleton.
   c. nucleus.
   d. ribosome.

7. What type of cellular junction prevents stomach acid from leaking into the abdomen and digesting internal organs?
   a. Plasmodesmata
   b. Anchoring junctions
   c. Tight junctions
   d. Gap junctions

Answers to these questions are in appendix A.

WRITE IT OUT

1. How does the formation of the cell theory illustrate the process of science?

2. List the features that all cells share; then name three structures found in eukaryotic cells but not in bacteria or archaea.

3. Chapter 1 explains emergent properties and describes the characteristics of life. Use this information to explain why life is an emergent property that appears at the level of the cell.

4. If a eukaryotic cell is like a house, how is a prokaryotic cell like an efficiency (one-room) apartment?

5. List three structural differences between plant and animal cells. Explain how each structural difference reflects a functional difference between plants and animals.

6. Suppose you find a sample of cells at a crime scene. What criteria might you use to determine if the cells are from prokaryotes, plants, or animals?

7. Rank the following in order from smallest to largest: ant, prokaryotic cell, actin molecule, microtubule, nitrogen atom. What type of microscope (if any) would you need if you wanted to see each?

8. Which cell in figure 3.31 has the highest ratio of surface area to volume? Explain your answer.

9. What advantages does compartmentalization confer on a large cell?

10. List the chemicals that make up cell membranes.

11. Imagine that you could engineer a cell that exchanges gases efficiently with the environment and quickly metabolizes sugars. Describe your cell's size and shape. What organelles would be abundant?

12. One way to understand cell function is to compare the parts of a cell to the parts of a factory. For example, the Golgi apparatus would be analogous to the factory's shipping department. How would the other cell parts fit into this analogy?

13. Imagine that you found a cell that releases many proteins into the bloodstream. What organelles might be especially active in this cell? What would each of these organelles be doing?

14. How does the cytoskeleton interact with other structures in eukaryotic cells?

15. How do plant cells form cell walls?

16. Describe how animal cells use junctions in different ways.

PULL IT TOGETHER

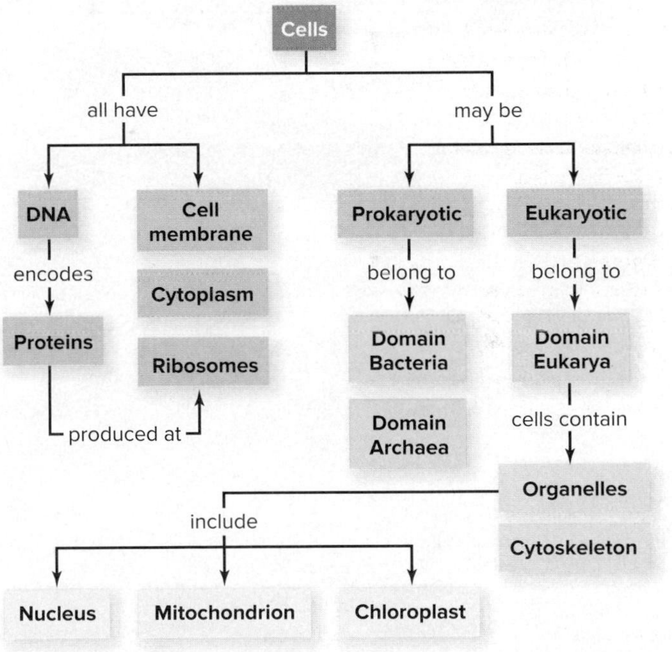

Figure 3.32 Pull It Together: Cells.

Refer to figure 3.32 and the chapter content to answer the following questions.

1. Review the Survey the Landscape figure in the chapter introduction, and then add *molecules, atoms, carbohydrates,* and *enzymes* to the Pull It Together concept map.

2. How might you connect the terms *proteins* and *cytoskeleton*?

3. Add the three main components of the cytoskeleton to this map.

4. In what ways are domains Bacteria and Archaea different?

5. Add *chloroplast, lysosome,* and *vacuole* to this concept map.

6. Which cell types have a cell wall?

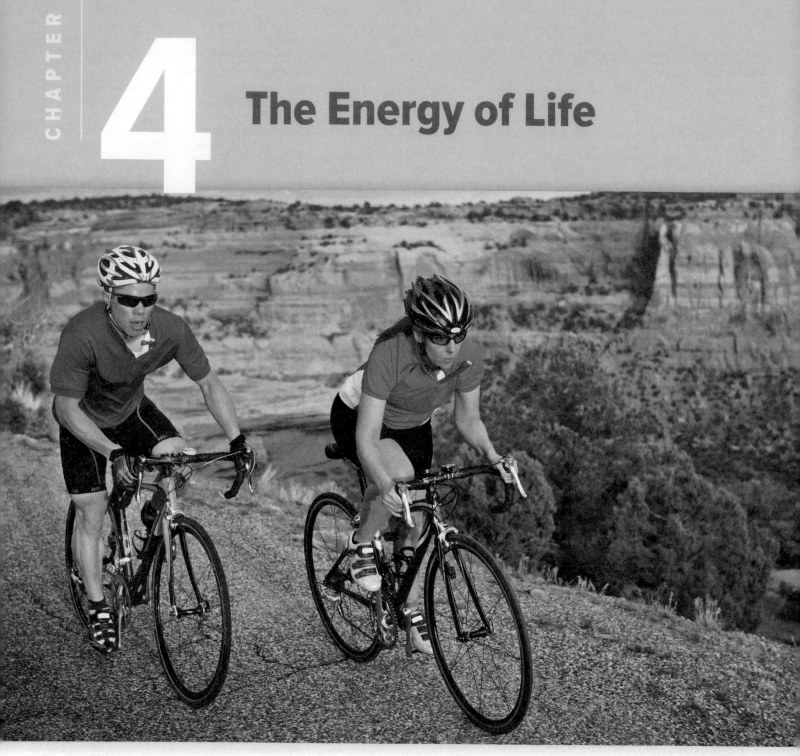

CHAPTER

4 The Energy of Life

Exercise on Wheels. Riding a bicycle up a hill takes energy, which comes from metabolic reactions inside cells.
©Tyler Stableford/The Image Bank/Getty Images

LEARN HOW TO LEARN
Focus on Understanding, Not Memorizing

When you are learning the language of biology, be sure to concentrate on how each new term fits with the others. Are you studying multiple components of a complex system? Different steps in a process? The levels of a hierarchy? As you study, always make sure you understand how each part relates to the whole.

Whole-Body Metabolism: Energy on an Organismal Level

"I wish I had your metabolism!" Perhaps you have overheard a calorie-counting friend make a similar comment to someone who stays slim on a diet of fattening foods.

In that context, the word *metabolism* means how fast a person burns food. But biochemists define metabolism as all of the chemical reactions that build and break down molecules within any cell. How are these two meanings related?

Interlocking networks of metabolic reactions supply the energy that every cell needs to stay alive. In humans, teams of metabolizing cells perform specialized functions such as digestion, muscle movement, hormone production, and countless other activities. It all takes a reliable energy supply—food, which we "burn" at an ever-changing rate.

Minimally, a body needs energy to maintain heartbeat, temperature, breathing, brain activity, and other basic life requirements. For an adult human male, the average energy use is 1750 Calories in 24 hours; for a female, 1450 Calories. These numbers do not include the energy required for physical activity or digestion, so the number of Calories needed to get through a day generally exceeds the minimum requirement.

Of course, these averages mask the fact that people have different metabolic rates. Age, sex, weight, and activity level all influence metabolic rates, as does body fat composition. All other things being equal, a person with the most lean tissue (muscle, nerve, liver, and kidney) will have the highest metabolic rate, because lean tissue consumes more energy than relatively inactive fat tissue. A thyroid hormone, thyroxine, also influences energy expenditure.

So why *do* some people gain weight? The simple answer is that if you eat more Calories than you spend, you gain weight. A typical food plan for an adult includes 2000 Calories per day. By comparison, a single fast-food meal of a burger, large fries, and a chocolate shake may contain nearly 2500 Calories. Evolution has primed us to crave sugary, fatty foods, which were hard to come by early in human history but are ubiquitous now. It is little wonder that Americans are facing an obesity epidemic.

On the other hand, if you eat fewer Calories than you spend, you slim down. This is why reducing caloric intake and exercising are two basic weight-loss recommendations. Unfortunately, metabolic differences make it difficult to translate this simple energy balance into a one-size-fits-all weight-loss plan.

This chapter describes the fundamentals of metabolism, including how cells organize, regulate, and fuel the chemical reactions that sustain life.

SURVEY THE LANDSCAPE
Science, Chemistry, and Cells

All cells require energy in the form of ATP to carry out their chemical reactions, acquire resources, and power their other activities. Enzymes are proteins that speed these reactions.

For more details, study the Pull It Together feature in the chapter summary.

4.1 All Cells Capture and Use Energy

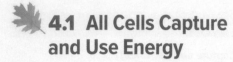

You're running late. You overslept, you have no time for breakfast, and you have a full morning of classes. You rummage through your cupboard and find something called an "energy bar"—just what you need to get through the morning. But what is energy?

A. Energy Allows Cells to Do Life's Work

Physicists define **energy** as the ability to do work—that is, to move matter. This idea, as abstract as it sounds, is fundamental to biology. Life depends on rearranging atoms and trafficking substances across membranes in precise ways. These intricate movements represent work, and they require energy.

Although it may seem strange to think of a "working" cell, all organisms do tremendous amounts of work on a microscopic scale. For example, a plant cell assembles glucose molecules into long cellulose fibers, moves ions across its membranes, and performs thousands of other tasks simultaneously. A gazelle grazes on a plant's tissues to acquire energy that will enable it to do its own cellular work. A crocodile eats that gazelle for the same reason.

The total amount of energy in any object is the sum of energy's two forms: potential and kinetic (**figure 4.1** and **table 4.1**). **Potential energy** is stored energy available to do work. A bicyclist at the top of a hill illustrates potential energy, as does a compressed spring. The covalent bonds of molecules, such as the ingredients in your energy bar, contain a form of potential energy called chemical energy. A chemical gradient is another form of potential energy (see section 4.5).

Kinetic energy is the energy of motion; any moving object possesses this form of energy. The bicyclist coasting down the hill in figure 4.1 demonstrates kinetic energy, as do moving pistons, a rolling bus, and contracting muscles. Light and sound are other types of kinetic energy. Inside a cell, each molecule also has kinetic energy. In fact, all of the chemical reactions that sustain life rely on collisions between moving molecules. The colder an object feels, the slower the movement of its atoms and molecules; this is why many cells die if conditions are too chilly.

Calories are units used to measure energy. One **calorie** (cal) is the amount of energy required to raise the temperature of 1 gram of water from 14.5°C to 15.5°C. The energy content of food, however, is usually measured in **kilocalories** (kcal), each of which equals 1000 calories. (In nutrition, 1 food Calorie—with a capital *C*—is actually a kilocalorie.) A typical energy bar, for example, contains 240 kcal of potential energy stored in the chemical bonds of its carbohydrates, proteins, and fats.

TABLE **4.1**   Examples of Energy in Biology

Type of Energy	Examples
Potential energy	Chemical energy (stored in bonds)
	Concentration gradient across a membrane
Kinetic energy	Light
	Sound
	Moving objects
	Thermal energy (heat)

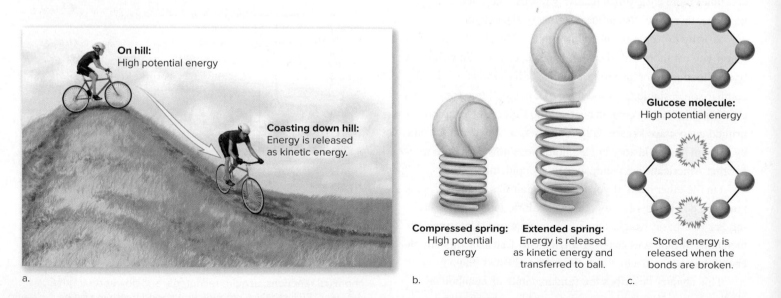

a.

b.

c.

Figure 4.1 Potential and Kinetic Energy. (a) A bicyclist at the top of a hill has potential energy, which is converted to kinetic energy as he coasts down. (b) The potential energy in a compressed spring is released as kinetic energy when the spring is released. (c) Chemical energy is a form of potential energy, which is released when a molecule's bonds break.

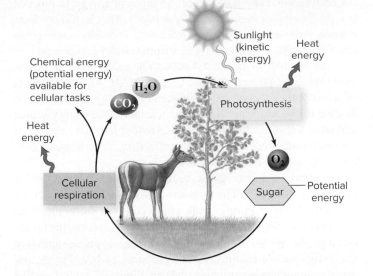

Figure 4.2 **Energy Conversions.** In photosynthesis, plants transform the kinetic energy in sunlight into potential energy stored in the chemical bonds of sugars and other organic molecules. Respiration, in turn, releases this potential energy. Heat energy is lost to the environment at every step along the way.

B. The Laws of Thermodynamics Describe Energy Transfer

Thermodynamics is the study of energy transformations. The first and second laws of thermodynamics describe the energy conversions vital for life, as well as those that occur in the nonliving world. They apply to all energy transformations—gasoline combustion in a car's engine, a burning chunk of wood, or a cell breaking down glucose.

The **first law of thermodynamics** is the law of energy conservation. It states that energy cannot be created or destroyed, although energy can be converted to other forms. This means that the total amount of energy in the universe does not change.

Living cells constantly convert energy from one form to another (**figure 4.2**). The most important energy transformations are photosynthesis and cellular respiration. In photosynthesis, plants and some microbes use carbon dioxide, water, and the kinetic energy in sunlight to produce sugars that are assembled into glucose and other carbohydrates. These molecules contain potential energy in their chemical bonds. During cellular respiration, the energy-rich glucose molecules change back to carbon dioxide and water, liberating the energy necessary to power life. Cells translate some of the potential energy in glucose into the kinetic energy of molecular motion and use that kinetic energy to do work.

Most organisms obtain energy from the sun, either directly through photosynthesis or indirectly by consuming other organisms. Even the potential energy in fossil fuels originated as solar energy. However, a few types of microbes can extract potential energy from the chemical bonds of inorganic chemicals; they use the energy to produce organic compounds, which are nutrients for their cells (and for the organisms that consume them).

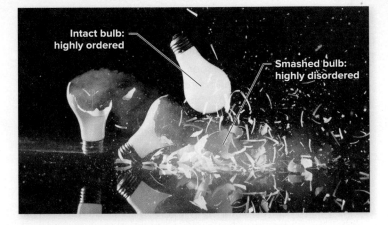

Figure 4.3 **Entropy.** In an instant, a highly organized lightbulb is transformed into broken glass and metal fragments. Entropy has irreversibly increased; no matter how many times you drop the glass and metal, the pieces will not arrange themselves back into a lightbulb.
©Ryan McVay/Getty Images

The **second law of thermodynamics** states that all energy transformations are inefficient because every reaction loses some energy to the surroundings as heat (see figure 4.2). If you eat your energy bar on the way to your first class, your cells can use the potential energy in its chemical bonds to make proteins, divide, or do other forms of work. According to the second law of thermodynamics, however, you will lose some energy as heat with every chemical reaction. This process is irreversible; cells cannot use energy that has been converted to heat.

Heat energy is disordered because it results from random molecular movements. Because heat is disordered, and all energy eventually becomes heat, it follows that all energy transformations must head toward increasing disorder. **Entropy** is a measure of this randomness. In general, the more disordered a system is, the higher its entropy (**figure 4.3**).

Because organisms are highly organized, they may seem to defy the second law of thermodynamics. But organisms are not isolated from their surroundings. Instead, a constant stream of incoming energy and matter allows organisms to maintain their organization and stay alive, using the information in DNA. In other words, organisms can increase in complexity *as long as something else decreases in complexity by a greater amount*. Ultimately, life remains ordered and complex because the sun is constantly supplying energy to Earth. But the entropy of the universe as a whole, including the sun, is increasing.

The ideas in this chapter and the two that follow describe how organisms acquire and use the energy they need to sustain life.

4.1 | MASTERING CONCEPTS

1. What are some examples of the "work" of a cell?
2. Describe how your body has both potential and kinetic energy.
3. What are some energy conversions that occur in cells?
4. Why does the amount of entropy in the universe always increase?

4.2 Networks of Chemical Reactions Sustain Life

The number of chemical reactions occurring in even the simplest cell is staggering. Thousands of reactants and products form interlocking pathways that resemble complicated road maps.

The word **metabolism** encompasses all of these chemical reactions in cells, including those that build new molecules and those that break down existing ones. Each reaction rearranges atoms into new compounds, and each reaction either absorbs or releases energy. Digesting your morning energy bar and using its carbohydrates to fuel muscle movement are part of your metabolism. Photosynthesis and respiration are part of the metabolism of the grass under your feet as you hurry to class.

A. Chemical Reactions Absorb or Release Energy

Biologists group metabolic reactions into two categories based on energy requirements: endergonic and exergonic (**figure 4.4**).

An **endergonic reaction** requires an input of energy to proceed (the prefix *end-* or *endo-* means "put into"). That is, the products contain more energy than the reactants. Typically, endergonic reactions build complex molecules from simpler components.

The top half of figure 4.4 shows one example of an endergonic reaction: photosynthesis. Glucose ($C_6H_{12}O_6$), the product of photosynthesis, contains more potential energy than do carbon dioxide (CO_2) and water (H_2O), the reactants. The energy source that powers this reaction is sunlight. Another familiar example of an endergonic process is muscle contraction, since muscles contract only with an input of energy.

In contrast, an **exergonic reaction** releases energy (*ex-* or *exo-* means "out of"). The products contain less energy than the reactants. Such reactions break large, complex molecules into their smaller, simpler components. Cellular respiration, the breakdown of glucose to carbon dioxide and water, is an example (see the bottom half of figure 4.4). The products, carbon dioxide and water, contain less energy than glucose. Similarly, your digestive system breaks down an energy bar's protein into individual amino acids. This process also releases energy.

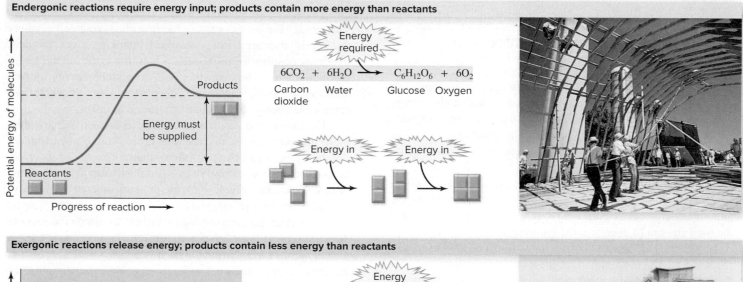

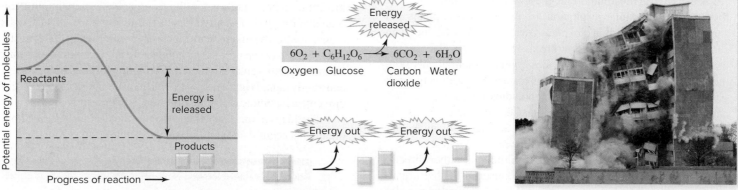

Figure 4.4 Endergonic and Exergonic Reactions. Endergonic reactions require an input of energy to build complex molecules from small components, like building a barn out of wooden boards. Exergonic reactions release energy by dismantling complex molecules. Likewise, as an old building collapses into dust and chunks of concrete and steel, it releases energy in the form of sound and heat.

Photos: (raising): ©Blair Seitz/Science Source; (demolition): ©Image Source/Corbis RF

Both graphs in figure 4.4 feature a "hump," which represents the amount of potential energy needed to start the reaction in the first place. Note that even exergonic reactions require this initial energy input. Overall, however, exergonic reactions release more energy than the amount they need to get started.

What happens to the energy released in an exergonic reaction? According to the second law of thermodynamics, some is lost to the environment as heat; entropy always increases. But some of the energy can be used to do work. For example, the cell may use the energy to form bonds or to power endergonic reactions. As we shall see, life's biochemistry is full of endergonic reactions that proceed only at the expense of exergonic ones.

Figure It Out

Complete these sentences by selecting the correct term from each pair of words in parentheses:

In an endergonic reaction, products contain (*less/more*) energy than reactants. Entropy in the universe (*decreases/increases*) through this reaction.

Answers: more; increases

B. Linked Oxidation and Reduction Reactions Form Electron Transport Chains

Electrons can carry energy. Most energy transformations in organisms occur in **oxidation–reduction ("redox") reactions,** which transfer energized electrons from one molecule to another. ⓘ *electrons,* section 2.1B

A redox reaction is similar to a person presenting a gift to a friend (**figure 4.5**). **Oxidation** means the loss of electrons—and a corresponding loss of energy—from a molecule, an atom, or an ion. In figure 4.5, the electron donor molecule being oxidized is analogous to the gift-giver. Conversely, **reduction** means a gain of electrons (and their energy); the electron acceptor being reduced is analogous to the woman receiving the package.

Each redox reaction links an exergonic process with an endergonic one. The "oxidation" half is exergonic, since energized electrons are removed from the electron donor. That is, the electron donor has more potential energy before it is oxidized than it does after the reaction is complete. On the other hand, the

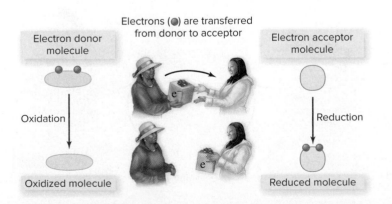

Figure 4.5 Redox Reaction. An electron donor molecule loses electrons and is therefore being oxidized. The molecule that accepts the electrons is being reduced.

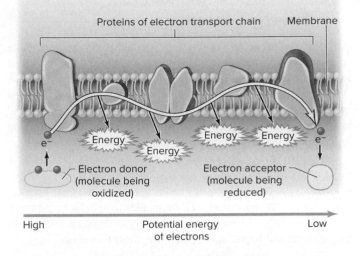

Figure 4.6 Electron Transport Chain. An electron donor molecule transfers an electron to the first protein in an electron transport chain. When the recipient passes the electron to its neighbor, energy is released. The electron continues along the chain, releasing energy at each step, until it reaches a final electron acceptor.

"reduction" half is endergonic. The acceptor molecule has gained the energy-rich electrons, so it ends up with more potential energy than it had before the reaction started.

Oxidations and reductions occur simultaneously because electrons removed from one molecule during oxidation must join another molecule and reduce it. That is, if one molecule is reduced (gains electrons), then another must be oxidized (lose electrons). Continuing the analogy from figure 4.5, a person can only give a gift if a recipient is willing to accept it. Of course, the woman receiving the gift is free to either keep the package or pass it on to someone else.

Likewise, some molecules act as electron carriers, hanging on to the electrons they receive for a short time before delivering them to another molecule. Groups of proteins that are electron-shuttling "specialists" often align in membranes. In an **electron transport chain,** each protein accepts an electron from the molecule before it and passes it to the next, like a basketball team passing a ball from one player to another (**figure 4.6**). As a result, each protein in the chain is first reduced and then oxidized. Small amounts of energy are released at each step, and the cell uses this energy in other reactions. As you will see in chapters 5 and 6, both photosynthesis and respiration rely on electron transport chains to harvest energy.

4.2 MASTERING CONCEPTS

1. What is metabolism on a cellular level?
2. Distinguish between endergonic and exergonic reactions.
3. Review figure 4.6. As electrons pass from the first to the second protein in the electron transport chain, which protein is oxidized and which is reduced? Explain your answer.

4.3 ATP Is Cellular Energy Currency

All cells contain a maze of interlocking chemical reactions, some releasing energy and others absorbing it. The covalent bonds of **adenosine triphosphate,** a molecule more commonly known as **ATP,** temporarily store much of the released energy. ATP holds energy released in exergonic reactions—such as the digestion of an energy bar—just long enough to power muscle contraction and all other endergonic reactions.

In eukaryotic cells, organelles called mitochondria produce most of a cell's ATP. As you will see in chapter 6, a mitochondrion uses the potential energy in the bonds of one glucose molecule to generate dozens of ATP molecules in cellular respiration. Not surprisingly, the most energy-hungry cells, such as those in the muscles and brain, also contain the most mitochondria.

A. Coupled Reactions Release and Store Energy in ATP

ATP is a type of nucleotide (**figure 4.7**). Its components are the nitrogen-containing base adenine, the five-carbon sugar ribose, and three phosphate groups (PO_4). Each phosphate group has a negatively charged oxygen atom. The negative charges on neighboring phosphate groups repel one another, making the molecule unstable. It therefore releases energy when the covalent bonds between the phosphates break. ⓘ *nucleotides,* section 2.5D

All cells depend on the potential energy in ATP to power their activities. When a cell requires energy for an endergonic reaction, it "spends" ATP by removing the endmost phosphate group (**figure 4.8**). The products of this exergonic hydrolysis reaction are adenosine *di*phosphate (ADP, in which only two phosphate groups remain attached to ribose), the liberated phosphate group, and a burst of energy:

$$ATP + H_2O \longrightarrow ADP + Ⓟ + energy$$

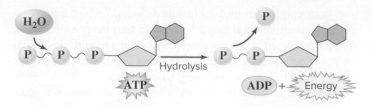

Figure 4.8 ATP Hydrolysis. Removing the endmost phosphate group of ATP yields ADP and a free phosphate group. The cell uses the released energy to do work.

In the reverse situation, energy can be temporarily stored by adding a phosphate to ADP, forming ATP and water:

$$ADP + Ⓟ + energy \longrightarrow ATP + H_2O$$

The energy for this endergonic reaction comes from molecules that are broken down in other reactions, such as those in cellular respiration.

These reactions are fundamental to biology because ATP is the "go-between" that links endergonic to exergonic reactions. **Coupled reactions,** as their name implies, are simultaneous reactions in which one provides the energy that drives the other (**figure 4.9**). Cells couple the hydrolysis of ATP to endergonic reactions. The ATP hydrolysis reaction drives the endergonic one, which does work or synthesizes new molecules.

How does this coupling work? A cell uses ATP as an energy source by **phosphorylating** (transferring its phosphate group to) another molecule. This transfer may have either of two effects (**figure 4.10**). In one scenario, the presence of the phosphate may energize the target molecule, making it more likely to bond with other molecules. ATP fuels endergonic reactions in this way. The other possible consequence of phosphorylation is a change in the shape of the target molecule. For example, adding phosphate can force a protein to take a different shape; removing phosphate returns the protein to its original form. Muscle contraction is the large-scale effect of millions of small molecules changing shape, and then changing back again, in a coordinated way. ATP hydrolysis provides the energy.

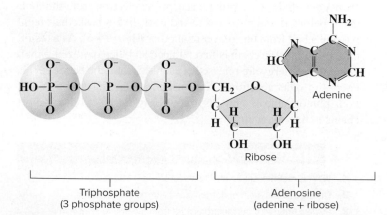

Triphosphate
(3 phosphate groups)

Adenosine
(adenine + ribose)

Figure 4.7 ATP's Chemical Structure. ATP is a nucleotide consisting of adenine, ribose, and three phosphate groups.

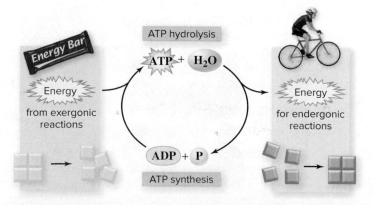

Figure 4.9 Coupled Reactions. Cells use ATP hydrolysis, an exergonic reaction, to fuel endergonic reactions. The cell regenerates ATP in other exergonic reactions, such as cellular respiration.

a. ATP energizes target molecule, making it more likely to bond with other molecules.

E.g., ATP provides the energy to build large molecules out of small subunits

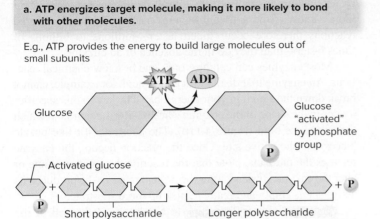

Glucose

ATP → ADP

Glucose "activated" by phosphate group

P

Activated glucose

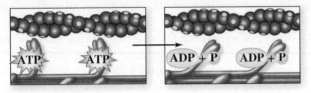

P +

→ + P

Short polysaccharide   Longer polysaccharide

b. ATP donates a phosphate group that changes the shape of the target molecule.

E.g., ATP binding changes shape of proteins involved in muscle contraction

ATP   ATP   →   ADP + P   ADP + P

Figure 4.10 ATP Use. When ATP donates a phosphate group to a molecule, the recipient may (a) be more likely to bond or (b) change its shape in a useful way.

ATP is sometimes described as energy "currency." Just as you can use money to purchase a variety of products, all cells use ATP in many chemical reactions to do different kinds of work. Besides muscle contraction, other examples of jobs that require ATP include transporting substances across cell membranes, moving chromosomes during cell division, and synthesizing the large molecules that make up cells.

ATP is also analogous to a fully charged rechargeable battery. A full battery represents a versatile source of potential energy that can provide power to many types of electronic devices. Although a dead battery is no longer useful as an energy source, you can recharge a spent battery to restore its utility. Likewise, the cell can use respiration to rebuild its pool of ATP.

B. ATP Represents Short-Term Energy Storage

Organisms require huge amounts of ATP. A typical human cell uses the equivalent of 2 billion ATP molecules a minute just to stay alive. Organisms recycle ATP at a furious pace, adding phosphate groups to ADP to reconstitute ATP, using the ATP to drive reactions, and turning over the entire supply every minute or so. If you ran out of ATP, you would die instantly.

Apply It **Now**

Boosting Your Metabolism

Metabolism describes all the chemical reactions in a cell. Because our cells always lose energy as heat, they require constant energy input to continue fueling their reactions. So the familiar definition of metabolism—how fast a person burns calories in food—relates to the rate at which cellular reactions are occurring. What can you do to make your cells use the energy in food more quickly?

Exercise speeds up the body's energy metabolism in several ways. Immediately after exercise, cells work to rebuild ATP and other energy reserves, so caloric demands are high. Also, body temperature remains elevated for hours after exercise, speeding chemical reactions and contributing to increased metabolism. Regular exercise also increases the size of muscle cells, which require more energy than fat cells even when at rest.

Caffeine may also accelerate metabolism. Although caffeine contains zero calories, many people can attest to the "energy boost" that it provides. Caffeine increases the release of fatty acids into the blood and raises the heart rate, giving cells quick access to energy reserves. However, studies have shown that getting too little sleep (a side effect of excess caffeine) disturbs normal metabolism.

Finally, cellular metabolism slows when you go for a few hours without food. One way to keep your metabolism high is therefore to maintain your blood sugar level by eating multiple small, healthy meals throughout the day.

©Corbis RF

Even though ATP is essential to life, cells do not stockpile it in large quantities. ATP's high-energy phosphate bonds make the molecule too unstable for long-term storage. Instead, cells store energy-rich molecules such as fats, starch, and glycogen. When ATP supplies run low, cells divert some of their lipid and carbohydrate reserves to the metabolic pathways of cellular respiration. This process soon produces additional ATP. (Boosting your body's overall metabolic rate speeds these reactions; see this chapter's Apply It Now box.)

4.3 MASTERING CONCEPTS

1. What are the main parts of an ATP molecule?
2. How does ATP hydrolysis supply energy for cellular functions?
3. Describe the relationships among endergonic reactions, ATP hydrolysis, and cellular respiration.

4.4 Enzymes Speed Biochemical Reactions

Enzymes are among the most important of all biological molecules. An **enzyme** is an organic molecule that catalyzes (speeds up) a chemical reaction without being consumed. Most enzymes are proteins, although some are made of RNA.

Many of the cell's organelles, including mitochondria, chloroplasts, lysosomes, and peroxisomes, are specialized sacs of enzymes. Enzymes copy DNA, build proteins, digest food, recycle a cell's worn-out parts, and catalyze oxidation–reduction reactions, just to name a few of their jobs. Without enzymes, all of these reactions would proceed far too slowly to support life. ⓘ *organelles,* section 3.4

A. Enzymes Bring Reactants Together

Enzymes speed reactions by lowering the **activation energy,** the amount of energy required to start a reaction (**figure 4.11a**). Even exergonic reactions, which ultimately release energy, require an initial "kick" to get started (see figure 4.4). The enzyme brings reactants (also called substrates) into contact with one another, so that less energy is required for the reaction to proceed. Just as it is

easier to climb a small hill than a tall mountain, reactions occur more rapidly if the activation energy is low. Enzyme-catalyzed reactions therefore occur much faster—millions to billions of times faster—than they do in the absence of an enzyme.

Most enzymes can catalyze only one or a few chemical reactions. An enzyme that dismantles a fatty acid, for example, cannot break down the starch in your energy bar. The key to this specificity lies in the shape of the enzyme's **active site,** the region to which the substrates bind (figure 4.11b). The substrates fit like puzzle pieces into the active site. Once the reaction occurs, the enzyme releases the products. Note that the reaction does not consume or alter the enzyme. Instead, after the protein releases the products, its active site is empty and ready to pick up more substrate.

Enzymes are very sensitive to conditions in the cell. If the pH or the salt concentration is too high or too low, an enzyme can become denatured and stop working. Temperature is also important (**figure 4.12**). Enzyme action generally speeds up as the temperature climbs because reactants have more kinetic energy at higher temperatures. If it gets too hot, however, the enzyme rapidly denatures and can no longer function. ⓘ *denatured proteins,* section 2.5C

Enzymes are so critical to life that just one faulty or missing enzyme can have dramatic effects. Lactose intolerance is one example. People whose intestinal cells do not secrete an enzyme called lactase cannot digest milk sugar (lactose). Fortunately, a product called Lactaid® can supply the missing enzyme. Phenylketonuria (PKU) is a much more serious disease. A PKU sufferer lacks an enzyme required to break down an amino acid called phenylalanine. When this amino acid accumulates in the bloodstream, it causes brain damage. People with PKU must avoid foods containing phenylalanine, including the artificial sweetener aspartame (NutraSweet). ⓘ *artificial sweeteners,* section 2.5

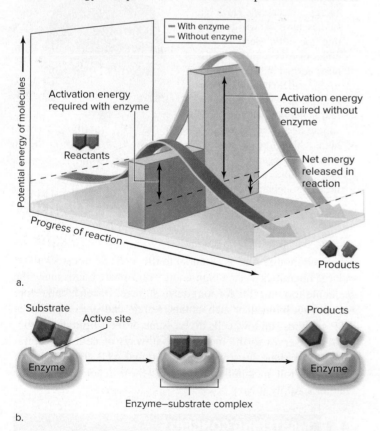

a.

b.

Figure 4.11 **How Enzymes Work.** (a) Enzymes lower the amount of energy required to start a reaction. The "walls" in this figure represent the activation energy for the same reaction, with and without an enzyme. (b) An enzyme's active site has a specific shape that binds to one or more substrates. After the reaction, the enzyme releases the products.

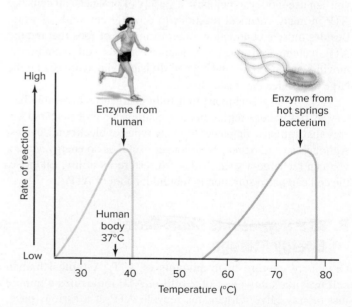

Figure 4.12 **Temperature Matters.** These graphs show how temperature affects the activity of enzymes from a human (*left*) and a bacterium that lives in hot springs (*right*). The microbes have heat-tolerant enzymes that function only at very high temperatures.

Enzymes also have household applications. Many detergents contain enzymes that break down food stains on clothing or dirty dishes. Raw pineapple contains an enzyme that breaks down protein. A gelatin dessert containing raw pineapple will fail to solidify because the fruit's enzymes destroy the gelatin protein. Some meat tenderizers contain the same enzyme, which breaks down muscle tissue and makes the meat easier to chew.

B. Enzymes Have Partners

Nonprotein "helpers" called **cofactors** are substances that must be present for an enzyme to catalyze a chemical reaction. Cofactors are often oxidized or reduced during the reaction, but, like enzymes, they are not consumed. Instead, they return to their original state when the reaction is complete.

Some cofactors are metals such as zinc, iron, and copper. Magnesium ions (Mg^{2+}), for example, help to stabilize many important enzymes. Other cofactors are organic molecules; an organic cofactor is called a coenzyme. The cell uses many water-soluble vitamins, including B_1, B_2, B_6, B_{12}, niacin, and folic acid, to produce coenzymes; vitamin C is a coenzyme itself. Diets lacking in vitamins can lead to reduced enzyme function and, eventually, serious illness or even death.

C. Cells Control Reaction Rates

The intricate network of metabolic pathways may seem chaotic, but in reality it is just the opposite. Cells precisely control the rates of their chemical reactions. If they did not, some vital compounds would always be in short supply, and others might accumulate to wasteful (or even toxic) levels.

One way to regulate a metabolic pathway is by **negative feedback** (also called feedback inhibition), in which a change triggers action that reverses the change. Figure 4.13 shows how negative feedback might slow the synthesis of a chemical

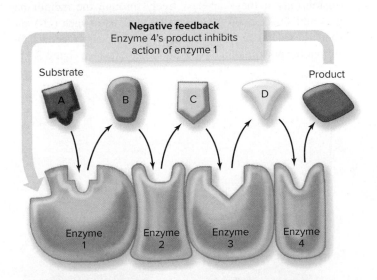

Figure 4.13 Negative Feedback. Some metabolic pathways require several enzyme-catalyzed steps. When the final product accumulates, it inhibits the activity of the first enzyme in the pathway, temporarily halting further production.

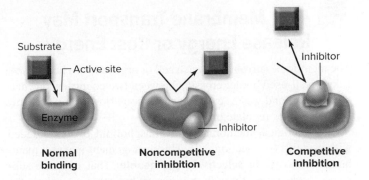

Figure 4.14 Enzyme Inhibitors. In noncompetitive inhibition, a substance binds to an enzyme at a location other than the active site, changing the enzyme's shape. A competitive inhibitor physically blocks an enzyme's active site.

product. As the product builds up, it inhibits the enzyme that catalyzes the initial reaction, and the entire pathway slows. Cells regulate amino acid production in this way. When an amino acid accumulates, it binds to an enzyme that acts early in the synthesis pathway. For a time, the synthesis of that amino acid stops. But when the level falls, the block on the enzyme lifts, and the cell can once again produce the amino acid.

Negative feedback works in two general ways to prevent too much of a substance from accumulating (figure 4.14). In **noncompetitive inhibition,** product molecules bind to the enzyme at a location other than the active site, in a way that alters the enzyme's shape so that it can no longer bind the substrate. (Figure 4.14 shows an example.) Alternatively, in **competitive inhibition,** the product of a reaction binds to the enzyme's active site, preventing it from binding substrate. It is "competitive" because the product competes with the substrate to occupy the active site.

Enzyme inhibitors have many practical applications. Aspirin relieves pain by binding to an enzyme that cells use to produce pain-related molecules called prostaglandins. Some antibiotics kill microorganisms—but not people—by inhibiting enzymes not present in our own cells. And the active ingredient in the herbicide Roundup® competitively inhibits an enzyme found in plant cells but not in animals.

The opposite of negative feedback is **positive feedback,** in which a process reinforces an existing condition. One example of a self-reinforcing pathway is blood clotting, which involves a series of chemical reactions initiated by damage to a blood vessel. Once a clot begins to form, the reactions accelerate, which further stimulates clotting, speeding up the reactions even more, and so on. Once the clot is large enough to stem the flow of blood, however, the pathway shuts down—an example of negative feedback. Positive feedback is much rarer than negative feedback in organisms.

4.4 MASTERING CONCEPTS

1. What do enzymes do in cells?
2. How does an enzyme lower a reaction's activation energy?
3. Distinguish between an enzyme and a coenzyme.
4. What are the roles of negative and positive feedback?
5. List three conditions that influence enzyme activity.

4.5 Membrane Transport May Release Energy or Cost Energy

The membrane surrounding each cell or organelle is a busy place. Like a well-used border crossing between two countries, raw materials enter and wastes exit in a continuous flow of traffic. How do membranes regulate this activity?

A biological membrane is a phospholipid bilayer studded with proteins (see section 3.3). This arrangement makes a membrane "choosy," or **selectively permeable.** That is, some substances pass freely through the bilayer, but others—such as the sugar from a digested energy bar—require help from proteins. This section describes the basic forms of traffic across the membrane; table 4.2 provides a summary.

Thanks to the regulation of membrane transport, the interior of a cell is chemically different from the outside. Concentrations of some dissolved substances (solutes) are higher inside the cell than outside, and others are lower. Likewise, the inside of each organelle in a eukaryotic cell may be chemically quite different from the solution in the rest of the cell.

The term *gradient* describes any such difference between two neighboring regions. In a **concentration gradient,** a solute is more concentrated in one region than in a neighboring region. For example, the images in table 4.2 all illustrate concentration gradients in which the solution on the right side of the membrane has a higher solute concentration than the solution on the left.

If a substance moves from an area where it is more concentrated to an area where it is less concentrated, it is said to be "moving down" or "following" its concentration gradient. As the solute moves, the gradient dissipates—that is, it disappears. Any concentration gradient will eventually dissipate *unless energy is expended to maintain it.* Why? Random molecular motion always increases the amount of disorder (entropy), and it costs energy to counter this tendency toward disorder. By the same token, however, an existing concentration gradient represents a form of stored potential energy. Cells therefore spend ATP to create some types of concentration differences, which they can later "cash in" to do work (see section 4.5B).

A. Passive Transport Does Not Require Energy Input

In **passive transport,** a substance moves across a membrane without the direct expenditure of energy. All forms of passive transport involve **diffusion,** the spontaneous movement of a substance from a region where it is more concentrated to a region where it is less concentrated. Because diffusion represents the dissipation of a chemical gradient—and the loss of potential energy—it does not require energy input.

For a familiar example of diffusion, picture what happens when you first place a tea bag in a cup of hot water: Near the tea bag, there are many more brown tea molecules than elsewhere in the cup (figure 4.15). Over time, however, the brownish color spreads to create a uniform brew.

How do the tea molecules "know" which way to diffuse? The answer is, of course, that atoms and molecules know nothing. Diffusion occurs because all substances have kinetic energy; that is, they are in constant, random motion. To simplify the tea example, suppose each molecule can move randomly along 1 of 10 possible paths (in reality, the number of possible directions is infinite). Assume further that only 1 path leads back to the tea bag. Since 9 of the 10 possibilities point away from the tea bag, the tea molecules tend to spread out; that is, they move down their concentration gradient.

If diffusion continues long enough, the gradient disappears. Diffusion *appears* to stop at that point, but the molecules do not stop moving. Instead, they continue to travel randomly back and forth at the same rate, so at equilibrium the concentration remains equal throughout the solution.

Simple Diffusion: No Proteins Required In a form of passive transport called **simple diffusion,** a substance moves down its concentration gradient without the use of a transport protein (see table 4.2). Substances may enter or leave cells by simple diffusion only if they can pass freely through the membrane. Lipids and small, nonpolar molecules such as oxygen (O_2) and carbon dioxide (CO_2), for example, diffuse easily across the hydrophobic portion of a biological membrane (see figure 3.11).

Figure 4.15 Diffusion in a Cup.
The solute particles leaving a tea bag can move in any direction, with only a few paths leading back to the source. Eventually, the solutes are distributed uniformly throughout the cup.

Solvent

Solute

If gradients dissipate without energy input, how can a cell use simple diffusion to acquire essential substances or get rid of toxic wastes? The answer is that the cell maintains the gradients, either by continually consuming the substances as they diffuse in or by producing more of the substances that diffuse out. For example, mitochondria consume O_2 as soon as it diffuses into the cell, maintaining the O_2 gradient that drives diffusion. Respiration also produces CO_2, which diffuses out because its concentration always remains higher in the cell than outside.

Osmosis: Diffusion of Water Across a Selectively Permeable Membrane

Two solutions of different concentrations may be separated by a selectively permeable membrane through which water, but not solutes, can pass. In that case, water will diffuse down its own gradient toward the side with the high solute concentration. **Osmosis** is this simple diffusion of water across a selectively permeable membrane (see table 4.2 and figure 4.16).

A human red blood cell demonstrates the effects of osmosis (figure 4.17). The cell's interior is normally **isotonic** to the surrounding blood plasma, which means that the plasma's solute concentration is the same as the inside of the cell (*iso-* means "equal," and *tonicity* is the ability of a solution to cause water movement). Water therefore moves into and out of the cell at equal rates. In a **hypotonic** environment, the solute concentration is lower than it is inside the cell (*hypo-* means "under," as in *hypodermic*). Water therefore moves by osmosis into a blood cell placed into hypotonic surroundings; since animal cells lack a cell wall, the membrane may even burst. Conversely, **hypertonic** surroundings have a higher concentration of solutes than the cell's cytoplasm (*hyper* means "over," as in *hyperactive*). In a hypertonic environment, a cell loses water, shrivels, and may die. (The movement of water out of cells also plays a role in some headaches; see the Burning Question in this section.)

Hypotonic and *hypertonic* are relative terms that can refer to the surrounding solution or to the solution inside the cell. The same solution might be hypertonic to one cell but hypotonic to another, depending on the solute concentrations inside the cells.

A plant's roots are often hypertonic to the soil, particularly after a heavy rain. Water rushes in, and the central vacuoles of the plant cells expand until the cell walls constrain their growth.

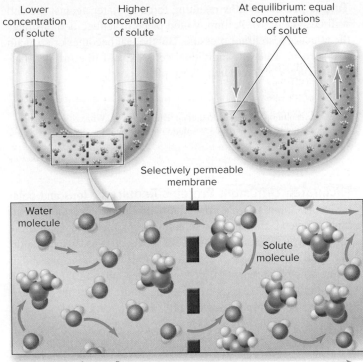

Figure 4.16 Osmosis. The selectively permeable membrane dividing this U-shaped tube permits water but not solutes to pass. Water diffuses from the left side (low solute concentration) toward the right side (high solute concentration). At equilibrium, water flow is equal in both directions, and the solute concentrations will be equal on both sides of the membrane.

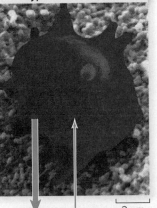

Figure 4.17 Osmosis and Red Blood Cells. (a) A human red blood cell is isotonic to the surrounding fluid. Water enters and leaves the cell at the same rate, and the cell maintains its shape. (b) When the salt concentration of the surrounding fluid decreases, water flows into the cell faster than it leaves. The cell swells and may even burst. (c) In salty surroundings, the cell loses water and shrinks.

Photos: (a–c): ©David M. Phillips/Science Source

Turgor pressure is the resulting force of water against the cell wall (figure 4.18). A limp, wilted piece of lettuce demonstrates the effect of lost turgor pressure. But the leaf becomes crisp again if placed in water, as individual cells expand like inflated balloons. Turgor pressure helps keep plants erect.

Figure It Out

A 0.9% salt solution is isotonic to human red blood cells. What will happen if you place a red blood cell in a 2.0% solution of salt water?

Answer: Water will leave the cell.

Facilitated Diffusion: Proteins Required

Ions and polar molecules cannot freely cross the hydrophobic layer of a membrane; instead, transport proteins form channels that help these solutes cross. **Facilitated diffusion** is a form of passive transport in which a membrane protein assists the movement of a polar solute down its concentration gradient (see table 4.2). Facilitated diffusion releases energy because the solute moves from where it is more concentrated to where it is less concentrated.

Glucose moves into red blood cells via facilitated diffusion. This sugar is too hydrophilic to pass freely across the membrane, but glucose transporter proteins form channels that allow it in. Respiration inside the red blood cells consumes the glucose and maintains the concentration gradient.

Membrane proteins can enhance osmosis, too. Although membranes are somewhat permeable to water, osmosis can be

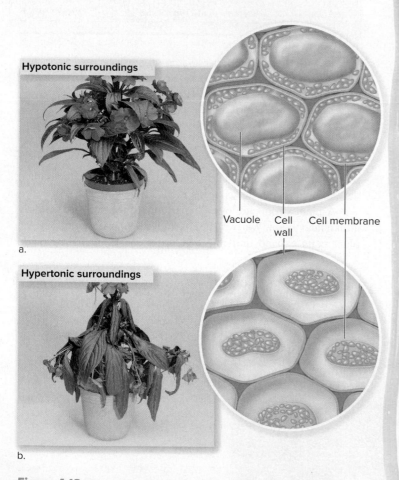

a.

b.

Figure 4.18 Osmosis and Plant Turgor Pressure. (a) The interior of a plant cell usually contains more concentrated solutes than its surroundings. Water enters the cell by osmosis, generating turgor pressure. (b) In a hypertonic environment, water leaves cells, so turgor pressure is low. The plant wilts.

Photos: (a, b): ©Nigel Cattlin/Science Source

TABLE 4.2 Movement Across Membranes: A Summary

Mechanism	Characteristics
Passive transport	Net movement is down concentration gradient; does not require energy input.
Simple diffusion	Substance moves across membrane without assistance from transport proteins.
Osmosis	Water diffuses across a selectively permeable membrane.
Facilitated diffusion	Substance moves across membrane with assistance from transport proteins.
Active transport	Net movement is against concentration gradient; requires transport protein and energy input, often from ATP.
Transport in vesicles	Vesicle carries large particles into or out of a cell; requires energy input.
Endocytosis	Membrane engulfs substance and draws it into cell.
Exocytosis	Vesicle fuses with cell membrane, releasing substances outside of cell.

slow. The cells of many organisms, including bacteria, plants, and animals, use membrane proteins called aquaporins to increase the rate of water flow. Kidney cells control the amount of water that enters urine by changing the number of aquaporins in their membranes.

B. Active Transport Requires Energy Input

Both simple diffusion and facilitated diffusion dissipate an existing concentration gradient. Often, however, a cell needs to do the opposite: create and maintain a concentration gradient. A plant's root cell, for example, may need to absorb nutrients from soil water that is much more dilute than the cell's interior. In **active transport,** a cell uses a transport protein to move a substance *against* its concentration gradient—from where it is less concentrated to where it is more concentrated (see table 4.2). Because a gradient represents a form of potential energy, the cell must expend energy to create it; this energy often comes from ATP.

Cells must contain high concentrations of potassium (K^+) and low concentrations of sodium (Na^+) to perform many functions. In animals, for example, sodium and potassium ion gradients are essential for nerve and muscle function (see chapters 26 and 29). One active transport system in the membranes of most animal cells is a protein called the **sodium–potassium pump,** which uses ATP as an energy source to expel three Na^+ for every two K^+ it admits (**figure 4.19**). Maintaining these ion gradients is costly: The million or more sodium–potassium pumps embedded in a cell's membrane use some 25% of the cell's ATP.

Concentration gradients are an important source of potential energy that cells can use to do work. For example, chapters 5 and 6 describe how cells establish concentration gradients of hydrogen ions (H^+) during photosynthesis and respiration. A chloroplast or mitochondrion can control how and when the H^+ gradient dissipates. As it does so, the organelle converts the potential energy stored in the gradient into another form of potential energy—that is, chemical energy in the bonds of ATP.

C. Endocytosis and Exocytosis Use Vesicles to Transport Substances

Most molecules dissolved in water are small, and they can cross cell membranes by simple diffusion, facilitated diffusion, or active transport. Large particles, however, must enter and leave cells with the help of a transport vesicle—a small sac that can pinch off of, or fuse with, a cell membrane.

In **endocytosis,** a cell membrane engulfs fluids and large molecules to bring them into the cell. When the cell membrane indents, a "bubble" of

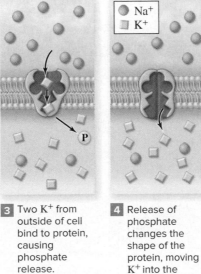

Outside of cell

Na⁺
K⁺

ATP ADP
P

P

P

Cytoplasm

1 ATP binds to transport protein along with three Na^+ from cytoplasm. ATP transfers phosphate to protein.

2 Phosphate changes the shape of the protein, moving Na^+ across the membrane.

3 Two K^+ from outside of cell bind to protein, causing phosphate release.

4 Release of phosphate changes the shape of the protein, moving K^+ into the cytoplasm.

Figure 4.19 Active Transport. The sodium–potassium pump is a protein embedded in the cell membrane. It uses energy released in ATP hydrolysis to move sodium ions (Na^+) out of the cell and potassium ions (K^+) into the cell. The process costs energy because both types of ions are moving from where they are less concentrated to where they are more concentrated.

membrane closes in on itself. The resulting vesicle traps the incoming substance (figure 4.20). The formation and movement of this vesicle along the cytoskeleton's "tracks" require energy.

The two main forms of endocytosis are pinocytosis and phagocytosis. In pinocytosis, the cell engulfs small amounts of fluids and dissolved substances. In **phagocytosis,** the cell captures and engulfs large particles, such as debris or even another cell (*phag-* means "eating"). The vesicle then fuses with a lysosome, where hydrolytic enzymes dismantle the cargo. ⓘ *lysosomes,* section 3.4B

When biologists first viewed endocytosis in white blood cells in the 1930s, they thought a cell would gulp in anything at its surface. They now recognize a more selective form of the process. In receptor-mediated endocytosis, a receptor protein on a cell's surface binds a biochemical; the cell membrane then indents, drawing the substance into the cell. Liver cells use receptor-mediated endocytosis to absorb cholesterol-toting proteins from the bloodstream.

Exocytosis, the opposite of endocytosis, uses vesicles to transport fluids and large particles out of cells (figure 4.21). Inside a cell, the Golgi apparatus produces vesicles filled with substances to be secreted. The vesicle moves to the cell membrane and joins with it, releasing the substance outside the membrane. For example, the tip of a neuron releases neurotransmitters by exocytosis; these chemicals then stimulate or inhibit neural impulses in a neighboring cell. The secretion of milk into milk ducts, depicted in figure 3.13, is another example. As in endocytosis, moving the transport vesicle requires energy.

4.5 MASTERING CONCEPTS

1. What is diffusion?
2. What types of substances diffuse freely across a biological membrane?
3. What would happen to a plant cell in a *hypertonic* environment?
4. Why does it cost energy for a cell to maintain a concentration gradient?
5. Distinguish among simple diffusion, facilitated diffusion, and active transport.
6. How do exocytosis and endocytosis use vesicles to transport materials across cell membranes?

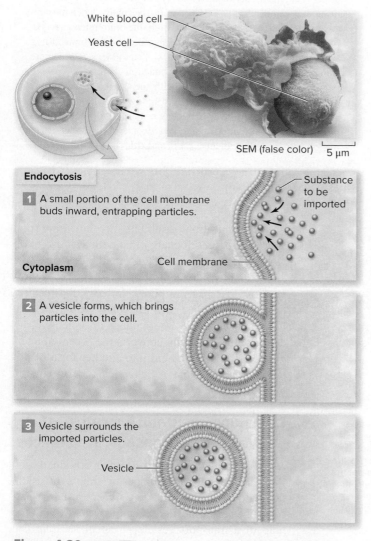

White blood cell

Yeast cell

SEM (false color)   5 µm

Endocytosis

1 A small portion of the cell membrane buds inward, entrapping particles.

Substance to be imported

Cytoplasm

Cell membrane

2 A vesicle forms, which brings particles into the cell.

3 Vesicle surrounds the imported particles.

Vesicle

Figure 4.20 **Endocytosis.** Large particles enter a cell by endocytosis. The inset (*top right*) shows a white blood cell engulfing a yeast cell by phagocytosis, a form of endocytosis.

Photo: ©Biology Media/Science Source

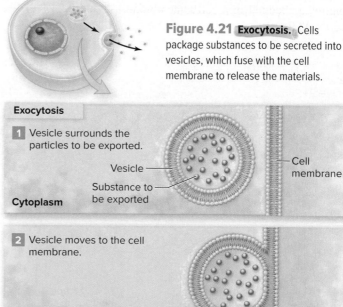

Figure 4.21 **Exocytosis.** Cells package substances to be secreted into vesicles, which fuse with the cell membrane to release the materials.

Exocytosis

1 Vesicle surrounds the particles to be exported.

Vesicle

Cell membrane

Substance to be exported

Cytoplasm

2 Vesicle moves to the cell membrane.

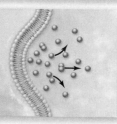

3 Vesicle merges with the membrane, releasing particles to the outside.

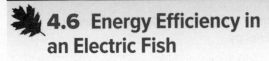

INVESTIGATING LIFE

4.6 Energy Efficiency in an Electric Fish

When you're alone, perhaps you enjoy soft music; at a party, however, you turn it up. Depending on the social situation, you control the volume.

Some types of fish, including the gold-lined knifefish, use a similar strategy. These animals generate an electric field that helps them communicate, hunt, and navigate in their native South American waters. The electric field comes from a specialized organ along the fish's flank and tail (**figure 4.22**). The electric organ consists of densely packed cells that generate nerve signals (action potentials) in unison. Each action potential begins with the movement of Na^+ into a cell, and it ends when K^+ ions exit. After the action potential is complete, the movement is reversed—an activity that requires the sodium–potassium pump and abundant ATP (see figure 4.19). Given that an active cell generates 75 or more action potentials per second, these pumps operate at a furious pace. ⓘ *action potential,* section 26.3

A knifefish keeps the electric field's "volume" low during times of inactivity, saving energy and lowering the chance of detection by a predator. An interaction with another fish, however, can trigger a boost in the electric field's strength within minutes. Researchers Michael Markham, Harold Zakon, and their colleagues at the University of Texas wanted to learn more about how an electric organ executes such rapid changes.

The researchers suspected that the cells of the electric organ might deploy ion channels strategically, moving them into and out of the cell membrane as needed. This strategy requires the participation of the endomembrane system, which shuttles ion channels and other proteins to the cell membrane in bubble-like vesicles (see figure 3.13). Exocytosis would deploy additional ion channels to the cell's membrane, and the intensity

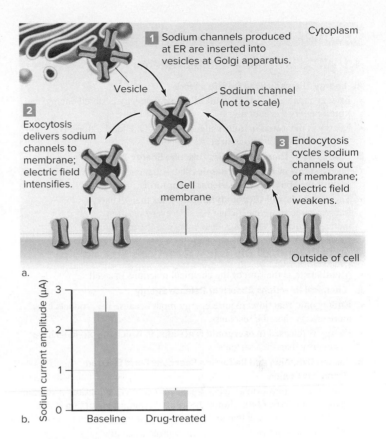

a.

b.

Figure 4.23 Recycling Center. (a) This model shows how a cell in an electric organ might use exocytosis and endocytosis to control the density of sodium channels in its membrane. (b) A drug that inhibits vesicle traffic prevented cells from adding sodium channels to its membrane, greatly reducing the sodium current. The error bars indicate a statistically significant difference (see appendix B).

of the electric field would rise. Endocytosis would have the opposite effect (**figure 4.23a**).

It is difficult to use a microscope to watch a cell's vesicle traffic as it happens, so the team used the next best thing to test their hypothesis: a drug that disrupts exocytosis. Sure enough, when they applied the drug to cells from an electric organ, the movement of Na^+ across the membrane slowed, nearly to a standstill (figure 4.23b).

Both endocytosis and exocytosis cost ATP, but maintaining a strong electric field is far more costly. The fish's strategy is therefore cheaper than leaving the organ set on "high" all the time. Overall, the fish benefits most by cranking up the power when needed and turning it down when it's time to chill.

Source: Markham, Michael R., M. Lynne McAnelly, Philip K. Stoddard, and Harold H. Zakon. 2009. Circadian and social cues regulate ion channel trafficking. *PLoS Biology,* vol. 7, issue 9; doi:10.1371/journal.pbio.1000203

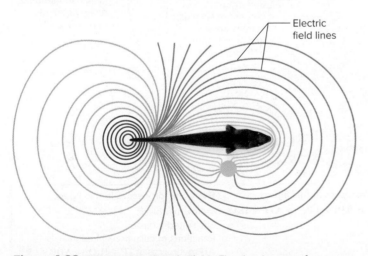

Figure 4.22 A Fish and Its Electric Field. The electric organ of a gold-lined knifefish extends along each side of its body. The fish uses its electric field to detect objects *(colored green)* and communicate with its neighbors.

4.6 MASTERING CONCEPTS

1. How does a knifefish adjust its electric field?
2. Why is changing electric field intensity adaptive?

CHAPTER SUMMARY

4.1 All Cells Capture and Use Energy

A. Energy Allows Cells to Do Life's Work
- **Energy** is the ability to do work. **Potential energy** is stored energy, and **kinetic energy** is action.
- Energy is measured in units called **calories.** One food Calorie is 1000 calories, or 1 **kilocalorie.**

B. The Laws of Thermodynamics Describe Energy Transfer
- The **first law of thermodynamics** states that energy cannot be created or destroyed but only converted to other forms.
- The **second law of thermodynamics** states that all energy transformations are inefficient because every reaction increases **entropy** and loses heat energy to the environment.

4.2 Networks of Chemical Reactions Sustain Life

- **Metabolism** is the sum of the chemical reactions in a cell.

A. Chemical Reactions Absorb or Release Energy
- **Endergonic reactions** require energy input because the products have more energy than the reactants.
- Energy is released in **exergonic reactions,** in which the products have less energy than the reactants.

B. Linked Oxidation and Reduction Reactions Form Electron Transport Chains
- Many energy transformations in organisms occur via **oxidation–reduction (redox) reactions. Oxidation** is the loss of electrons; **reduction** is the gain of electrons. Oxidation and reduction reactions occur simultaneously.
- In both photosynthesis and respiration, proteins shuttle electrons along **electron transport chains.**

4.3 ATP Is Cellular Energy Currency

A. Coupled Reactions Release and Store Energy in ATP
- **ATP** stores energy in its high-energy phosphate bonds. Cellular respiration generates ATP.
- Many energy transformations involve **coupled reactions,** in which the cell uses the energy released in ATP hydrolysis to drive another reaction.
- **Phosphorylation** is the transfer of a phosphate group from ATP to another molecule, causing the recipient to become energized or to change shape.

B. ATP Represents Short-Term Energy Storage
- ATP is too unstable for long-term storage. Instead, cells store energy as fats and carbohydrates.

4.4 Enzymes Speed Biochemical Reactions

A. Enzymes Bring Reactants Together
- **Enzymes** are organic molecules (usually proteins) that speed biochemical reactions by lowering the **activation energy.**
- Substrate molecules fit into the enzyme's **active site.**
- Enzymes have narrow ranges of conditions in which they function.

B. Enzymes Have Partners
- **Cofactors** are inorganic or organic substances that enzymes require to catalyze reactions. Like enzymes, cofactors are not consumed in the reaction.

C. Cells Control Reaction Rates
- A reaction product may temporarily shut down its own synthesis whenever its levels rise. This **negative feedback** may occur by **competitive inhibition** or **noncompetitive inhibition.**

- A reaction product may also stimulate its own production, an example of **positive feedback.**

4.5 Membrane Transport May Release Energy or Cost Energy

- Membranes have **selective permeability,** which means they admit only some substances.
- A **concentration gradient** is a difference in solute concentration between two neighboring regions, such as across a membrane. Gradients dissipate without energy input.

A. Passive Transport Does Not Require Energy Input
- All forms of **passive transport** involve **diffusion,** the dissipation of a chemical gradient by random molecular motion.
- In **simple diffusion,** a substance passes through a membrane along its concentration gradient without the aid of a transport protein.
- **Osmosis** is the simple diffusion of water across a selectively permeable membrane. Terms describing tonicity (**isotonic, hypotonic,** and **hypertonic**) predict whether cells will swell or shrink when the surroundings change. When plant cells lose too much water, the resulting loss of **turgor pressure** causes the plant to wilt.
- In **facilitated diffusion,** a membrane protein admits a substance along its concentration gradient without expending energy.

B. Active Transport Requires Energy Input
- In **active transport,** a carrier protein uses energy (ATP) to move a substance against its concentration gradient. For example, the **sodium–potassium pump** uses active transport to exchange sodium ions for potassium ions across an animal cell membrane.
- **Figure 4.24** shows how enzymes and ATP interact in a cell to generate energy for active transport and other activities that require energy input.

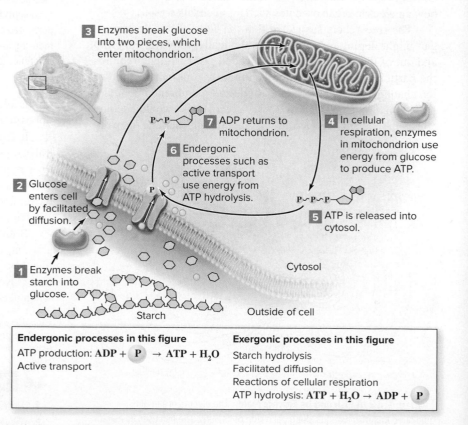

3 Enzymes break glucose into two pieces, which enter mitochondrion.

P~P **7** ADP returns to mitochondrion.

6 Endergonic processes such as active transport use energy from ATP hydrolysis.

4 In cellular respiration, enzymes in mitochondrion use energy from glucose to produce ATP.

2 Glucose enters cell by facilitated diffusion.

P~P~P **5** ATP is released into cytosol.

1 Enzymes break starch into glucose.

Starch

Cytosol

Outside of cell

Endergonic processes in this figure	Exergonic processes in this figure
ATP production: $ADP + P \rightarrow ATP + H_2O$	Starch hydrolysis
Active transport	Facilitated diffusion
	Reactions of cellular respiration
	ATP hydrolysis: $ATP + H_2O \rightarrow ADP + P$

Figure 4.24 Enzymes, Energy, and ATP.

C. Endocytosis and Exocytosis Use Vesicles to Transport Substances

- In **endocytosis,** a cell engulfs liquids or large particles. Pinocytosis brings in fluids; **phagocytosis** brings in solid particles.
- In **exocytosis,** vesicles inside the cell carry substances to the cell membrane, where they fuse with the membrane and release the cargo to the outside of the cell.

4.6 Investigating Life: Energy Efficiency in an Electric Fish

- A gold-lined knifefish can alter the strength of its electric field. Cells in the fish's electric organ use exocytosis and endocytosis to shuttle ion channel proteins into and out of the cell membrane.

MULTIPLE CHOICE QUESTIONS

1. Which of the following is the best example of potential energy in a cell?
   a. Cell division
   b. A molecule of glucose
   c. Movement of a flagellum
   d. Assembly of a cellulose fiber

2. Building proteins ____ energy; ATP hydrolysis ____ energy.
   a. releases; releases
   b. requires; requires
   c. releases; requires
   d. requires; releases

3. How does ATP participate in coupled reactions?
   a. Hydrolysis of ATP fuels endergonic reactions.
   b. Synthesis of ATP fuels endergonic reactions.
   c. Hydrolysis of ADP fuels exergonic reactions.
   d. Synthesis of ADP fuels exergonic reactions.

4. How does an enzyme affect the energy of a reaction?
   a. The activation energy is lowered.
   b. The net energy released is lowered.
   c. The energy of the reactants is raised.
   d. The energy of the products is raised.

5. Imagine that you place a drop of red food coloring in a glass of water. At first, the red molecules remain concentrated where you placed the drop, but the color eventually spreads throughout the water. This example illustrates
   a. hydrolysis.
   b. decreasing entropy.
   c. osmosis.
   d. diffusion.

6. An animal cell is at osmotic equilibrium with its environment. If you placed the cell into fresh water, what would happen?
   a. There would be no change.
   b. The cell would swell and (perhaps) burst.
   c. The cytoplasm would exert more turgor pressure on its cell wall.
   d. The cell would shrink.

7. How does ATP relate to membrane transport?
   a. The movement of a substance down its concentration gradient through transport proteins requires the hydrolysis of ATP.
   b. The higher the concentration of ATP in the cell, the more permeable the membrane is to water and small, nonpolar molecules.
   d. A cell uses the energy in ATP to transport substances against their concentration gradient.
   c. Digestion produces a high concentration of ATP outside the cell, and the ATP enters the cell via facilitated diffusion.

Answers to these questions are in appendix A.

WRITE IT OUT

1. Some people claim that life's high degree of organization defies the second law of thermodynamics. What makes this statement false?

2. List some examples of endergonic and exergonic reactions that have been introduced in previous chapters.

3. Why do electron transport chains release energy?

4. Provide an example of an appliance that uses coupled reactions.

5. Name at least four ways that a cell uses ATP.

6. Use what you know about enzymes to propose an explanation for why our bodies cannot digest the cellulose in dietary fiber. Also, why would a cell's fat-digesting enzymes not be able to digest an artificial fat such as Olestra (see chapter 2)?

7. Considering that enzymes are essential to all cells, including microbes, why might refrigeration and freezing help preserve food?

8. When a person eats a fatty diet, excess cholesterol accumulates in the bloodstream. Cells then temporarily stop producing cholesterol. What phenomenon described in the chapter does this control illustrate?

9. Diffusion is an efficient means of transport only over small distances. How does this relate to a cell's surface-area-to-volume ratio (see chapter 3)?

10. Liver cells are packed with glucose. If the concentration of glucose in a liver cell is higher than in the surrounding fluid, what mechanism could the cell use to import even more glucose? Why would only this mode of transport work?

11. List three ways the content in this chapter relates to an organism's ability to maintain homeostasis.

PULL IT TOGETHER

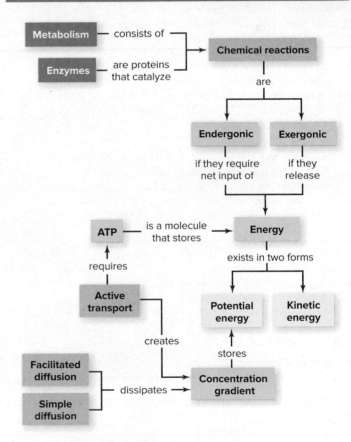

Figure 4.25 Pull It Together: The Energy of Life.

Refer to **figure 4.25** and the chapter content to answer the following questions.

1. Review the Survey the Landscape figure in the chapter introduction, and then add *life, cells,* and *respiration* to the Pull It Together concept map.

2. What are some examples of potential energy and kinetic energy?

3. Add the terms *substrate, active site,* and *activation energy* to this concept map.

5 Photosynthesis

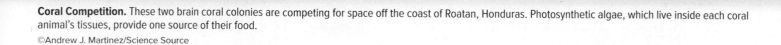

Coral Competition. These two brain coral colonies are competing for space off the coast of Roatan, Honduras. Photosynthetic algae, which live inside each coral animal's tissues, provide one source of their food.

©Andrew J. Martinez/Science Source

LEARN HOW TO LEARN
See What's Coming

Start by reviewing the Survey the Landscape figure at the start of each chapter to see how the material fits with the rest of the unit. Then check out the Learning Outline. Each heading is a complete sentence that summarizes the most important idea of the section. Read through these statements before you start each chapter. You can also flip to the end of the chapter before you start to read; the chapter summary and Pull It Together concept map can provide a preview of what's to come.

Is It Easier Being Green?

Food is expensive. It would be much cheaper and easier if we could feed ourselves using photosynthesis. Imagine the benefit of being photosynthetic: You could make your own food, free of charge, simply by sitting outside in the sun.

Of course, your body would have to have some new adaptations for photosynthesis to work. Your skin would have to be green, for starters. You might even have skin flaps that capture extra sunlight. You wouldn't eat, so you would need another way to acquire essential minerals; perhaps your feet would grow rootlike extensions that would absorb water and nutrients from soil.

Maybe photosynthetic cows, pigs, and chickens—or pets such as dogs and cats—would be a better idea. Feed-free animals would be a commercial and environmental triumph, costing less to own and generating less waste than the animals we raise now.

Fortunately or unfortunately, scientists will probably never be able to create photosynthetic people, chickens, or pooches. Mammals and birds move, breathe, pump blood, and maintain high body temperatures. All of this activity would likely require energy beyond what photosynthesis alone could supply.

Some animals, however, have adopted the "green" lifestyle by harboring live-in photosynthetic partners (see section 5.7). The closest to a true plant–animal hybrid is probably the sea slug *Elysia chlorotica,* a solar-powered mollusk with chloroplasts (photosynthetic organelles) in the cells lining its digestive tract. The chloroplasts come from algae in the slug's diet. As the animal grazes, it punctures the algal cells and discards everything but the chloroplasts, which migrate into the animal's cells. Light passes through the slug's skin and strikes the food-producing chloroplasts. Once its "solar panels" are in place, the animal may not eat again for months!

Perhaps the most famous animals to "farm" photosynthetic partners are corals. Inside the coral are single-celled protists called dinoflagellates, which use the sun's energy to feed the coral. In exchange, the animals provide a home for the protists. Sometimes, however, the partners break up. Corals under stress sometimes expel their dinoflagellates, or the protists may leave on their own. The reef then turns white. The coral animals eventually die, endangering the entire reef ecosystem. Pollution, disease, shading, excessively warm water, and ultraviolet radiation all trigger coral bleaching. Biologists predict that global climate change will only make this problem worse.

Corals and sea slugs are not the only animals whose lives depend on photosynthesis. Yours does, too, as you will learn in the next two chapters.

SURVEY THE LANDSCAPE
Science, Chemistry, and Cells

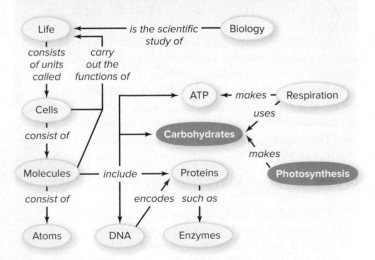

Photosynthetic cells use carbon dioxide, water, and sunlight to build carbohydrates. These molecules, which make up much of a plant's body, can also be used to store energy and generate ATP.

For more details, study the Pull It Together feature in the chapter summary.

5.1 Life Depends on Photosynthesis

It is spring. A seed germinates underground, its tender roots and pale yellow stem extending rapidly in a race against time. For now, the seedling's sole energy source is food stored in the seed itself. If its shoot does not reach light before its reserves run out, the seedling will die. But if it makes it, the shoot turns green and unfurls leaves that spread and catch the light. The seedling begins to feed itself, and an independent new life begins.

The plant is an **autotroph** ("self feeder"), meaning it uses inorganic substances such as water and carbon dioxide (CO_2) to produce organic compounds. The opposite of an autotroph is a **heterotroph,** which is an organism that obtains carbon by consuming preexisting organic molecules. You are a heterotroph, and so are all other animals, all fungi, and many other microorganisms.

Autotrophs underlie every ecosystem on Earth. It is not surprising, therefore, that if asked to designate the most important metabolic pathway, most biologists would not hesitate to cite **photosynthesis:** the process by which plants, algae, and some bacteria harness solar energy and convert it into chemical energy. With the exception of deep-ocean hydrothermal vent communities, all life on this planet ultimately depends on photosynthesis by autotrophs on land and in the water.

A. Photosynthesis Builds Carbohydrates Out of Carbon Dioxide and Water

Most plants are easy to grow (compared with animals, anyway) because their needs are simple. Give a plant water, essential elements in soil, carbon dioxide, light, and a favorable temperature, and it will produce food and oxygen that sustain its life. How can plants do so much with such simple raw materials?

In photosynthesis, pigment molecules in plant cells capture energy from the sun (**figure 5.1**). In a series of chemical reactions, that energy is then used to assemble CO_2 molecules into glucose ($C_6H_{12}O_6$) and other carbohydrates. The plant uses water in the process and releases oxygen gas (O_2) as a byproduct. The reactions of photosynthesis are summarized as follows:

$$6CO_2 + 6H_2O \xrightarrow{\text{light energy}} C_6H_{12}O_6 + 6O_2$$

Photosynthesis is an oxidation–reduction (redox) process. "Oxidation" means that electrons are removed from an atom or a molecule; "reduction" means electrons are added. As you will soon see, photosynthesis strips electrons from the oxygen atoms in H_2O (i.e., the oxygen atoms are oxidized). These electrons are eventually used to reduce the carbon in CO_2. Because oxygen atoms attract electrons more strongly than do carbon

atoms (as depicted in figure 2.5), moving electrons from oxygen to carbon requires energy input. The energy source for this endergonic reaction is, of course, sunlight. ⓘ *redox reactions,* section 4.2B

Photosynthesis provides not only food for the plant but also the energy, raw materials, and O_2 that support most heterotrophs. Animals, fungi, and other consumers eat the leaves, stems, roots, flowers, pollen, nectar, fruits, and seeds of the world's producers. Even the waste product of photosynthesis, O_2, is essential to much life on Earth.

Because humans live on land, we are most familiar with the contribution that plants make to Earth's terrestrial ecosystems. In fact, however, more than half of the world's photosynthesis occurs in the vast oceans, courtesy of countless algae and bacteria (**figure 5.2**). Several groups of bacteria are photosynthetic, some using pigments and metabolic pathways that are completely different from those in plants. For example, some photosynthetic bacteria do not use water as an electron source or generate oxygen gas. (This chapter focuses on photosynthesis as it occurs in plants and algae.)

B. Plants Use Carbohydrates in Many Ways

Several fates await the carbohydrates produced in photosynthesis. A plant's cells use about half of the sugar as fuel for their own cellular respiration, the metabolic pathway described in chapter 6. Roots, flowers, fruits, seeds, and other nonphotosynthetic plant parts could not grow without sugar shipments from green leaves and stems. Some sugars also participate indirectly in plant reproduction; sweet, sugary nectar and fruits attract animals that carry pollen and seeds (see chapter 24).

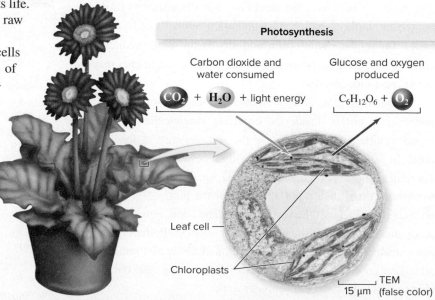

Photosynthesis

Carbon dioxide and water consumed

CO_2 + H_2O + light energy

Glucose and oxygen produced

$C_6H_{12}O_6$ + O_2

Leaf cell

Chloroplasts

TEM
15 μm   (false color)

Figure 5.1 Sugar from the Sun. In photosynthesis, a plant produces carbohydrates and O_2 from simple starting materials: CO_2, water, and sunlight.

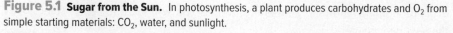

Electron micrograph by Wm. P. Wergin, courtesy of Eldon H. Newcomb, University of Wisconsin-Madison

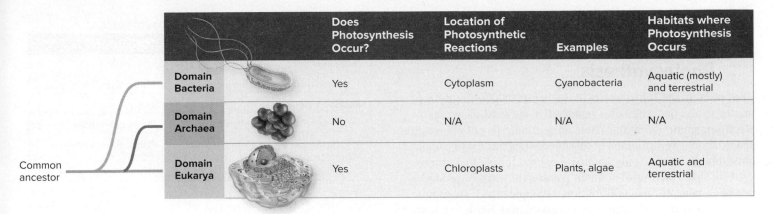

		Does Photosynthesis Occur?	Location of Photosynthetic Reactions	Examples	Habitats where Photosynthesis Occurs
Domain Bacteria		Yes	Cytoplasm	Cyanobacteria	Aquatic (mostly) and terrestrial
Domain Archaea		No	N/A	N/A	N/A
Domain Eukarya		Yes	Chloroplasts	Plants, algae	Aquatic and terrestrial

Figure 5.2 Autotroph Diversity. Photosynthesis produces carbohydrates in organisms belonging to domains Bacteria and Eukarya. Some archaeans can use light to produce ATP, but not to build carbohydrates.

A plant also uses glucose molecules as the building blocks of the cellulose wall that surrounds every plant cell. Wood is the remains of dead cells (see chapter 22), and it is mostly made of cellulose. The timber in the world's forests therefore stores enormous amounts of carbon. So do vast deposits of coal and other fossil fuels, which are the remains of plants and other organisms that lived long ago. Burning wood or fossil fuels releases this stored carbon into the atmosphere as CO_2. As the amount of CO_2 in the atmosphere has increased, Earth's average temperature has risen. Living forests help reduce climate change by locking carbon in wood. ⓘ *global climate change,* section 40.4B

In addition, plants often combine carbohydrates with other substances to manufacture additional compounds. Examples include amino acids and a host of economically important products such as rubber, medicines, and spices.

Finally, if a plant produces more glucose than it immediately needs, it may store the excess as starch. Carbohydrate-rich tubers and grains, such as potatoes, rice, corn, and wheat, are all energy-storing plant organs. Some plants, including sugarcane and sugar beets, store energy as sucrose instead. Table sugar comes from these crops, just as maple syrup comes from the sucrose-rich sap of a sugar maple tree. In addition, people use starch from corn kernels and sugar from sugarcane to produce biofuels such as ethanol (see chapter 19's Burning Question).

Figure It Out

Imagine that you plant a seedling weighing 1 kg in a pot containing 10 kg of soil. After the tree grows for 5 years, you uproot the tree, cut it up, and put the pieces in an oven to remove the water. When drying is complete, the tree weighs 50 kg. What is the main source of the atoms that make up the tree's dry weight?

Answer: The dried tree consists mainly of carbon, which was absorbed as CO_2.

C. The Evolution of Photosynthesis Changed Planet Earth

Most of today's organisms rely directly or indirectly on photosynthesis, so it may seem surprising that early life lacked the ability to capture sunlight. For the first billion or so years of life's history, however, all organisms were heterotrophs. As these early heterotrophs consumed carbon compounds from their surroundings, they released CO_2 into the environment. But these organisms could not use the carbon in CO_2, so they faced extinction as soon as they depleted the organic compounds in their habitats. ⓘ *first organic molecules,* section 15.1A

Some 3.5 billion years ago, however, some microbes developed a new talent: the ability to make their own food by photosynthesis. Little is known about the first photosynthetic pathway, but it is clear that the novel ability to convert light energy into chemical energy soon supported most other forms of life.

The evolution of photosynthesis radically altered Earth in other ways as well. Until about 2.4 billion years ago, the atmosphere contained little O_2. But thanks to photosynthesis by ancient cyanobacteria, O_2 gradually accumulated over the next billion or more years (see figure 15.6). The organisms that could use O_2 in respiration had an advantage: They could extract the most energy from food. Eventually, organisms using aerobic cellular respiration outcompeted most other life forms. With more energy available, life took on new shapes and sizes.

In addition, atmospheric O_2 reacted with free oxygen atoms to produce ozone (O_3). As ozone accumulated high in the atmosphere, it blocked harmful ultraviolet radiation from reaching the planet's surface, which prevented some genetic damage and allowed new varieties of life to arise. The ozone layer therefore also helps explain the explosion in the diversity of life that followed the evolution of photosynthesis.

5.1 MASTERING CONCEPTS

1. What is photosynthesis? Describe the reactants and products in words and in chemical symbols.
2. How is an autotroph different from a heterotroph?
3. Why is photosynthesis essential to life on Earth?
4. What happens to the sugars that plants produce?
5. How did the origin of photosynthesis alter Earth's atmosphere and the evolution of life?

5.2 Sunlight Is the Energy Source for Photosynthesis

Each minute, the sun converts more than 100,000 kilograms of matter to energy, releasing much of it outward as waves of electromagnetic radiation. After an 8-minute journey through space, about two billionths of this energy reaches Earth's upper atmosphere. Of this, only about 1% is used for photosynthesis, yet this tiny fraction of the sun's power ultimately produces enough sugar to fill 600 billion Olympic-sized swimming pools each year. Light may seem insubstantial, but it is a powerful force on Earth.

A. What Is Light?

Visible light is a small sliver of a much larger **electromagnetic spectrum,** the range of possible frequencies of radiation (**figure 5.3**). All electromagnetic radiation, including light, consists of **photons,** discrete packets of kinetic energy. A photon's **wavelength** is the distance it moves during a complete vibration. The shorter a photon's wavelength, the more energy it contains.

The sunlight that reaches Earth's surface consists of three main components: ultraviolet radiation, visible light, and infrared radiation. Of the three, ultraviolet radiation has the shortest wavelengths. Its high-energy photons damage DNA, causing sunburn and skin cancer. In the middle range is visible light, which provides the energy that powers photosynthesis; we perceive visible light of different wavelengths as distinct colors. Infrared radiation, with its longer wavelengths, contains too little energy per photon to be useful to organisms. Most of its energy is converted immediately to heat.

TABLE 5.1   Pigments of Photosynthesis

Pigment	Color(s)	Organisms
Major pigment		
Chlorophyll *a*	Blue-green	Plants, algae, cyanobacteria
Accessory pigments		
Chlorophyll *b*	Yellow-green	Plants, green algae
Carotenoids (carotenes and xanthophylls)	Red, orange, yellow	Plants, algae, bacteria

B. Photosynthetic Pigments Capture Light Energy

Plant cells contain several pigment molecules that capture light energy (see this chapter's Burning Question). The most abundant is **chlorophyll *a*,** a green photosynthetic pigment in plants, algae, and cyanobacteria. Photosynthetic organisms usually also have several types of **accessory pigments,** which are energy-capturing pigment molecules other than chlorophyll *a* (**table 5.1**). Chlorophyll *b* and carotenoids are accessory pigments in plants.

The photosynthetic pigments have distinct colors because they absorb only some wavelengths of visible light, while transmitting or reflecting others (**figure 5.4**). Chlorophylls *a* and *b* absorb red and blue wavelengths; they appear green because they reflect green light. Carotenoids, on the other hand, reflect longer wavelengths of light, so they appear red, orange, or yellow. These pigments also act as antioxidants that protect the plant from damage caused by free radicals. (Carrots, tomatoes,

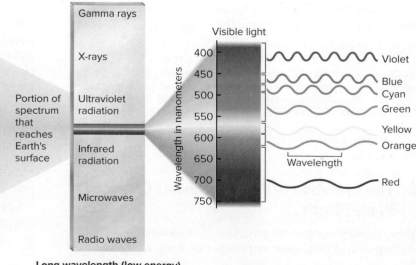

Figure 5.3 The Electromagnetic Spectrum. Sunlight reaching Earth consists of ultraviolet radiation, visible light, and infrared radiation, all of which are just a small part of a continuous spectrum of electromagnetic radiation. Photons with short wavelengths carry the most energy.

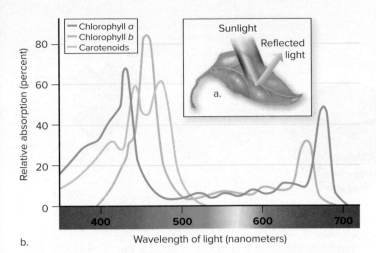

b.

Figure 5.4 **Everything but Green.** (a) Overall, a leaf reflects green and yellow wavelengths of light and absorbs the other wavelengths. (b) Each type of pigment absorbs some wavelengths of light and reflects others.

lobster shells, and the flesh of salmon all owe their distinctive colors to carotenoid pigments, which the animals must obtain from their diets.)

Only absorbed light is useful in photosynthesis. Accessory pigments absorb wavelengths that chlorophyll *a* cannot, so they extend the range of light wavelengths that a cell can harness. This is a little like the members of the same team on a quiz show, each contributing answers from a different area of expertise.

Figure It Out

If you could expose plants to just one wavelength of light at a time, would a wavelength of 300 nm, 450 nm, or 600 nm produce the highest photosynthetic rate?

Answer: 450 nm

C. Chloroplasts Are the Sites of Photosynthesis

In plants, leaves are the main organs of photosynthesis. Their broad, flat surfaces expose abundant surface area to sunlight. But light is just one requirement for photosynthesis. Water is also essential; roots absorb this vital ingredient, which moves up stems and into the leaves. And plants also exchange CO_2 and O_2 with the atmosphere. How do these gases get into and out of leaves?

The answer is that CO_2 and O_2 enter and exit a plant through **stomata** (singular: stoma), tiny openings in the epidermis of a leaf or stem (**figure 5.5a**). Stomata allow for gas exchange, but water evaporates through the same openings. When the plant loses too much water, pairs of specialized "guard cells" surrounding each stoma collapse against one another, closing the pores. Stomata therefore help balance the competing needs of gas exchange and water conservation.
ⓘ *leaf epidermis*, section 22.2B

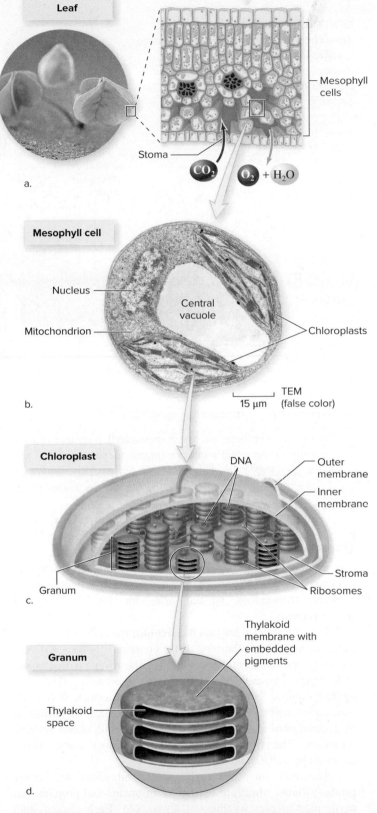

Figure 5.5 **Leaf and Chloroplast Anatomy.** (a) The tissue inside a leaf is called mesophyll. (b) Each mesophyll cell contains many chloroplasts. (c) A chloroplast contains light-harvesting pigments, embedded in (d) the stacks of thylakoid membranes that make up each granum.

Photos: (leaves): ©Steve Raymer/National Geographic Stock; (mesophyll): Electron micrograph by Wm. P. Wergin, courtesy of Eldon H. Newcomb, University of Wisconsin–Madison

Figure 5.6 Photosystem. This diagram shows one of the many photosystems embedded in a typical thylakoid membrane. Each photosystem consists of proteins (*purple*) and pigments (*green*).

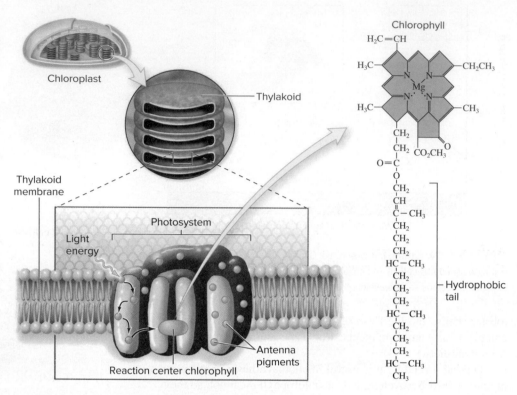

Most photosynthesis occurs in **mesophyll,** a collective term for the cells filling a leaf's interior (*meso-* means "middle," and *-phyll* means "leaf"). In many plants, at least part of the mesophyll has a "spongy" texture, reflecting the air spaces that maximize gas exchange within the leaf.

Leaf mesophyll cells contain abundant **chloroplasts,** the organelles of photosynthesis in plants and algae (see figure 5.5b, c). Most photosynthetic cells contain 40 to 200 chloroplasts, which add up to about 500,000 per square millimeter of leaf—an impressive array of solar energy collectors. Each chloroplast contains tremendous surface area for the reactions of photosynthesis.

Two membranes enclose the **stroma,** the chloroplast's fluid inner region. This gelatinous fluid contains ribosomes, DNA, and enzymes. (Be careful not to confuse the *stroma* with a *stoma,* or leaf pore.) Suspended in the stroma are between 10 and 100 **grana** (singular: granum), each composed of a stack of 10 to 20 pancake-shaped thylakoids (see figure 5.5d). Each **thylakoid,** in turn, consists of a membrane that is studded with photosynthetic pigments. The **thylakoid space** is the inner compartment enclosed by a thylakoid membrane.

Anchored in the thylakoid membranes are many **photosystems,** which are clusters of pigments and proteins that participate in photosynthesis (**figure 5.6**). Each photosystem includes about 300 chlorophyll *a* molecules and 50 accessory pigments. The photosystem's **reaction center** includes a special pair of chlorophyll *a* molecules that actually use the energy in photosynthetic reactions. The other pigments of the photosystem make up the light-harvesting complex that surrounds the reaction center. These additional pigments are called **antenna pigments** because they capture photon energy and funnel it to the reaction center. If the array of pigments in a photosystem is like a quiz show team, then the reaction center is analogous to the one member who announces the team's answer to the show's moderator.

Each photosystem has a few hundred chlorophyll molecules, so why does only the reaction center chlorophyll actually participate in the photosynthetic reactions? A single chlorophyll *a* molecule can absorb only a small amount of light energy. But because many pigment molecules are arranged close together, each antenna pigment can quickly pass its energy to the reaction center, freeing the antenna to absorb other photons as they strike. Thus, the photosystem's organization greatly enhances the efficiency of photosynthesis.

5.2 MASTERING CONCEPTS

1. What are the three main components of sunlight?
2. Describe the relationships among the chloroplast, stroma, grana, thylakoids, and photosystems.
3. How does it benefit a photosynthetic organism to have more than one type of pigment?
4. How does the reaction center chlorophyll interact with the antenna pigments in a photosystem?

5.3 Photosynthesis Occurs in Two Stages

Inside a chloroplast, photosynthesis occurs in two stages: the light reactions and the carbon reactions. **Figure 5.7** summarizes the entire process, and sections 5.4 and 5.5 describe each part in greater detail.

The **light reactions** convert solar energy to chemical energy. (You can think of the light reactions as the "photo-" part of photosynthesis.) In the chloroplast's thylakoid membranes, pigment molecules in two linked photosystems capture kinetic energy from photons and store it as potential energy in the chemical bonds of two molecules: ATP and NADPH.

Recall from chapter 4 that **ATP** is a nucleotide that stores potential energy in the covalent bonds between its phosphate groups. ATP forms when a phosphate group is added to ADP (see figure 4.9). The other energy-rich product of the light reactions, **NADPH,** is a coenzyme that carries pairs of energized electrons. In photosynthesis, these electrons come from one of the two reaction center chlorophyll molecules. Once the light reactions are under way, chlorophyll, in turn, replaces its "lost" electrons by splitting water molecules, yielding O_2 as a waste product. (i) *coenzymes, section 4.4B*

These two resources (energy and "loaded" electron carriers) set the stage for the second part of photosynthesis: the carbon reactions. In the carbon reactions, the chloroplast uses ATP, the high-energy electrons in NADPH, and CO_2 to produce sugar molecules. (These reactions are the "-synthesis" part of photosynthesis.) The ATP and NADPH come from the light reactions, and the CO_2 comes from the atmosphere. Once inside the leaf, CO_2 diffuses into a mesophyll cell and across the chloroplast membrane into the stroma, where the carbon reactions occur.

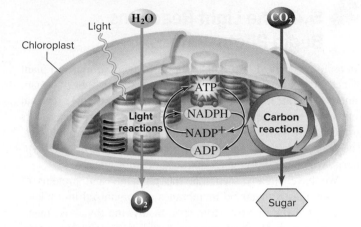

Figure 5.7 Overview of Photosynthesis. In the light reactions, pigment molecules capture light energy and transfer it to molecules of ATP and NADPH. The carbon reactions use this energy to build sugar molecules out of carbon dioxide.

Because the carbon reactions do not directly require light, they are sometimes called the "dark reactions" of photosynthesis. This term is misleading, however, because the carbon reactions occur mostly during the day, when the light reactions are producing ATP and NADPH. A more accurate alternative would be the "light-independent reactions."

5.3 MASTERING CONCEPTS

1. What happens in each of the two main stages of photosynthesis?
2. Explain the role of each of the products of the light reactions and the carbon reactions.

Burning Question

Why do leaves change colors in the fall?

Most leaves are green throughout a plant's growing season, although there are exceptions; some ornamental plants, for example, have yellow or purple foliage. The familiar green color comes from chlorophyll *a*, the most abundant pigment in photosynthetic plant parts. But the leaf also has other photosynthetic pigments. Carotenoids contribute brilliant yellow, orange, and red hues. Purple pigments, such as anthocyanins, are not photosynthetically active, but they do protect leaves from damage by ultraviolet radiation.

Carotenoids are less abundant than chlorophyll, so they usually remain invisible to the naked eye during the growing season. As winter approaches, however, deciduous plants prepare to shed their leaves. Anthocyanins accumulate while chlorophyll degrades, and the now "unmasked" accessory pigments reveal their colors for a short time as a spectacular autumn display. These pigments soon disappear as well, and the dead leaves turn brown and fall to the ground.

©Corbis RF

These carefully timed events help the plant conserve resources. After all, about 75% of a leaf's proteins occur in its chloroplasts. Rather than simply letting the first frost kill the leaves, the plant dismantles these proteins *before* the leaves die. The plant stores the valuable nitrogen and other nutrients from these molecules in living tissues that will survive the winter.

Spring brings a flush of fresh, green leaves. The energy to produce the foliage comes from glucose the plant produced during the last growing season and stored as starch. The new leaves make food throughout the spring and summer, so the tree can grow—both above the ground and below—and produce fruits and seeds. As the days grow shorter and cooler in autumn, the cycle will continue, and the colorful pigments will again participate in one of nature's great disappearing acts.

Submit your burning question to
Marielle.Hoefnagels@mheducation.com

5.4 The Light Reactions Begin Photosynthesis

An animal deprived of food can die of hunger; similarly, a plant placed in a dark closet literally starves. Without light, the plant cannot generate ATP or NADPH. And without these critical sources of energy and electrons, the plant cannot produce sugars to feed itself. Once its stored reserves are gone, the plant dies. The plant's life thus depends on the light reactions of photosynthesis, which occur in the membranes of chloroplasts.

We have already seen that the pigments and proteins of the chloroplast's thylakoid membranes are organized into photosystems (see figure 5.6). More specifically, the thylakoid membranes of algae and higher plants contain two types of photosystems, dubbed I and II. The two photosystems "specialize" in slightly different wavelengths of light. The reaction center chlorophyll of photosystem I, called P700, has a light absorption peak at 700 nm (see figure 5.4). Photosystem II's reaction center chlorophyll is called P680 because its long-wavelength peak is at 680 nm.

The entire sequence, from photosystem II through the enzyme that produces NADPH, is an electron transport chain. Recall from chapter 4 that an **electron transport chain** is a group of proteins that shuttle electrons from carrier to carrier, releasing energy with each step. As you will see, the photosynthetic electron transport chain provides both the energy required for ATP synthesis and the electrons required for the production of NADPH.

Figure 5.8 depicts the arrangement of the pigments and proteins in the thylakoid membrane. Refer to this illustration as you work through the rest of this section.

A. Light Striking Photosystem II Provides the Energy to Produce ATP

Photosynthesis begins in the cluster of pigment molecules of photosystem II. This may seem illogical, but the two photosystems were named as they were discovered. Photosystem II was discovered after photosystem I, but it functions first in the overall process.

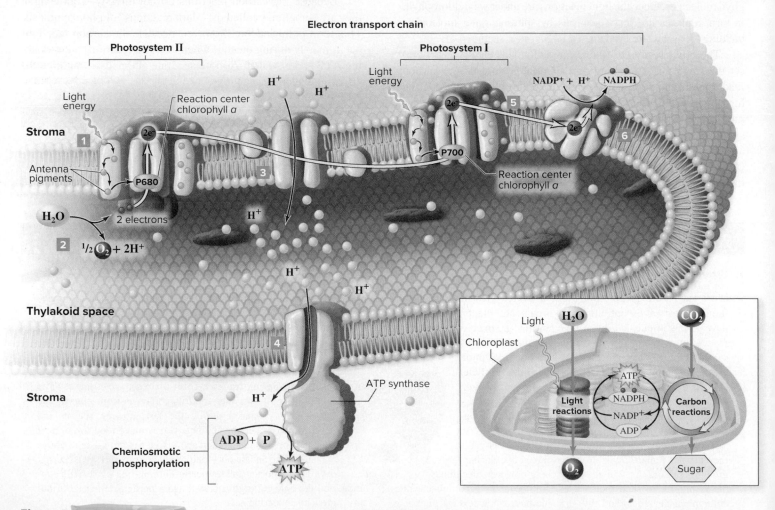

Figure 5.8 The Light Reactions. (*1*) Chlorophyll molecules in photosystem II transfer light energy to electrons. (*2*) Electrons are stripped from water molecules, releasing O_2. (*3*) The energized electrons pass to photosystem I. Each transfer releases energy that is used to pump protons (H^+) into the thylakoid space. (*4*) The resulting proton gradient is used to generate ATP. (*5*) In photosystem I, the electrons absorb more light energy and (*6*) are passed to $NADP^+$, creating the energy-rich NADPH. The inset (*lower right*) shows the light reactions in the context of the overall process of photosynthesis.

Apply It **Now**

Weed Killers

No plant can survive for very long in the dark. One low-tech way to kill an unwanted plant, therefore, is to deprive it of light. Gardeners who want to convert a lawn into a garden, for example, might kill the grass by covering it with layers of newspaper or cardboard for several weeks. The light reactions of photosynthesis cannot occur in the dark; the plants die.

©image100/Corbis RF

Another way to kill plants is to apply chemicals called herbicides. Some of these plant poisons also stop the light reactions (**figure 5.A**). For example, a weed killer called DCMU (also called diuron) blocks electron flow in photosystem II. Paraquat, noted for its use in destroying marijuana plants, diverts electrons from photosystem I. Either way, blocking electron flow prevents the production of ATP and NADPH. Without these critical products, photosynthesis cannot continue.

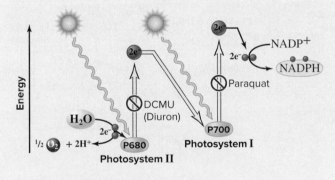

Figure 5.A Blocking the Light Reactions.

Pigment molecules in photosystem II absorb light and transfer the energy to a chlorophyll *a* reaction center, where it boosts two electrons to an orbital with a higher energy level. The "excited" electrons, now packed with potential energy, are ejected from the reaction center chlorophyll *a* molecule and begin their journey along the electron transport chain (figure 5.8, step 1). ⓘ *electron orbitals,* section 2.2A

How does the reaction center chlorophyll *a* molecule replace these two electrons? They come from water (H_2O), which donates two electrons when it splits into oxygen gas and two protons (H^+). Chlorophyll *a* picks up the electrons. The protons are released into the thylakoid space, and the O_2 is either used in the plant's respiration or released to the environment (step 2).

Meanwhile, the chloroplast uses the potential energy in the electrons to create a proton (H^+) gradient (step 3). As the electrons pass along the electron transport chain, the energy they lose drives the active transport of protons from the stroma into the thylakoid space. The resulting proton gradient across the thylakoid membrane represents a form of potential energy. ⓘ *active transport,* section 4.5B

ATP synthase is the enzyme complex that transforms the gradient's potential energy into chemical energy in the form of ATP (step 4). A channel in ATP synthase allows protons trapped inside the thylakoid space to return to the chloroplast's stroma. As the gradient dissipates, energy is released. The ATP synthase enzyme uses this energy to add phosphate to ADP, generating ATP.

This mechanism is similar to using a dam to produce electricity. Water accumulating behind a dam represents potential energy, like the proton gradient across a thylakoid membrane. To harness this potential energy, the dam's operators allow water to pour through a large pipe at the dam's base. The gushing water turns massive blades that spin an electric generator. Likewise, ATP synthase generates ATP as it allows accumulated protons to pass from the thylakoid space into the stroma.

The coupling of ATP formation to the release of energy from a proton gradient is called **chemiosmotic phosphorylation** because it is the addition of a phosphate to ADP (phosphorylation) using energy from the movement of protons across a membrane (chemiosmosis). As described in chapter 6, the same process also occurs in cellular respiration: An electron transport chain provides the energy to create a proton gradient, and ATP synthase uses the gradient's potential energy to produce ATP.

B. Electrons from Photosystem I Reduce NADP⁺ to NADPH

Photosystem I functions much as photosystem II does. Photon energy strikes energy-absorbing antenna pigment molecules, which pass the energy to the reaction center chlorophyll *a*. The reactive chlorophyll molecule boosts the electrons to a higher energy level and passes them along the chain (figure 5.8, step 5). The electrons from photosystem I are then replaced with electrons passing down from photosystem II.

At the end of the photosynthetic electron transport chain, the electrons energized at photosystem I reduce a molecule of NADP⁺ to NADPH (step 6). This NADPH is the electron carrier that will reduce carbon dioxide in the carbon reactions, while ATP will provide the energy. (The Apply It Now box explains how blocking the light reactions quickly leads to a plant's demise.)

5.4   MASTERING CONCEPTS

1. Describe how light striking photosystem II leads to the production of ATP.
2. What is water's role in the light reactions?
3. What happens after light strikes photosystem I?
4. How are the electrons from photosystem I replaced?

5.5 The Carbon Reactions Produce Carbohydrates

The carbon reactions, also called the Calvin cycle, occur in the chloroplast's stroma. The **Calvin cycle** is the metabolic pathway that uses NADPH and ATP to assemble CO_2 molecules into three-carbon carbohydrate molecules (**figure 5.9**). These products are eventually assembled into glucose and other sugars. (The pathway is named in honor of its discoverer, American biochemist Melvin Calvin.)

The first step of the Calvin cycle is **carbon fixation**—the initial incorporation of carbon from CO_2 into an organic compound. Specifically, CO_2 combines with **ribulose bisphosphate (RuBP),** a five-carbon sugar with two phosphate groups. **Rubisco** is the enzyme that catalyzes this first reaction. As an essential component of every plant, rubisco is one of the most abundant and important proteins on Earth.

The six-carbon product of the initial reaction immediately breaks down into two three-carbon molecules called phosphoglycerate (PGA). Further steps in the cycle convert PGA to phosphoglyceraldehyde (PGAL), which is the carbohydrate product that leaves the Calvin cycle. The cell can use PGAL to build larger carbohydrate molecules such as glucose and sucrose. Some of the PGAL, however, is rearranged to form additional RuBP, perpetuating the cycle.

ATP and NADPH produced in the light reactions provide the potential energy and electrons necessary to reduce CO_2. As long as ATP and NADPH are plentiful, the Calvin cycle continuously "fixes" the carbon from CO_2 into small organic molecules.

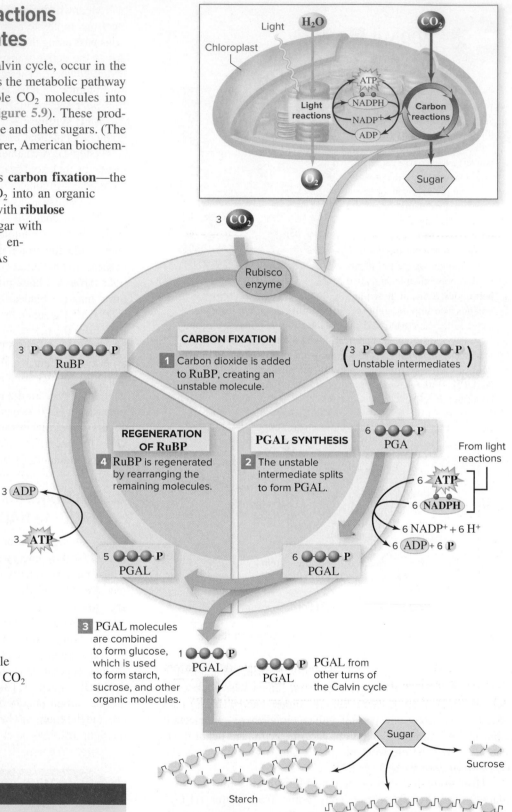

Figure 5.9 The Carbon Reactions. ATP and NADPH from the light reactions power the Calvin cycle, which assembles CO_2 molecules into carbohydrates.

5.5 MASTERING CONCEPTS

1. What is the relationship between the light reactions and the carbon reactions?

2. Use figure 5.9 to determine how many ATP and NADPH molecules are used to produce a six-carbon glucose molecule.

5.6 C₃, C₄, and CAM Plants Use Different Carbon Fixation Pathways

The Calvin cycle is also known as the **C₃ pathway** because a three-carbon molecule, PGA, is the first stable compound in the pathway. Although all plants use the Calvin cycle, C₃ plants use *only* this pathway to fix carbon from CO_2. About 95% of plant species are C₃, including cereals, peanuts, tobacco, spinach, sugar beets, soybeans, most trees, and some lawn grasses.

C₃ photosynthesis is obviously a successful adaptation, but it does have a weakness: inefficiency. Photosynthesis has a theoretical efficiency rate of about 30% in ideal conditions, but a plant's efficiency in nature is typically as low as 0.1% to 3%.

How do plants waste so much solar energy? One contributing factor is a metabolic pathway called **photorespiration,** a series of reactions that begin when the rubisco enzyme adds O_2 instead of CO_2 to RuBP. The product of this reaction does not enter the Calvin cycle. The plant therefore loses CO_2 that it fixed in previous turns of the cycle, wasting both ATP and NADPH.

A plant with open stomata minimizes its photorespiration rate. This is because CO_2 and O_2 compete for rubisco's active site; when stomata are open, CO_2 from the atmosphere enters the leaf, and O_2 produced in the light reactions diffuses out. But when the weather heats up, plants face a trade-off. If the stomata remain open too long, a plant may lose water, wilt, and die. If the plant instead closes its stomata, CO_2 runs low, and O_2 builds up in the leaves. Under those conditions, photorespiration becomes much more likely, and photosynthetic efficiency plummets.

In hot, dry climates, plants that minimize photorespiration have a significant competitive advantage. One way to improve efficiency is to ensure that rubisco always encounters high CO_2 concentrations. The C₄ and CAM pathways are two adaptations that do just that.

C₄ plants fix carbon twice, in different cells. A preliminary carbon fixation reaction called the C₄ pathway occurs in mesophyll cells. In the **C₄ pathway,** CO_2 combines with a three-carbon "ferry" molecule to form a four-carbon compound, oxaloacetate (hence the name C₄); rubisco does not participate in this reaction. The oxaloacetate is usually reduced to malate, another four-carbon molecule. Malate then moves via plasmodesmata into adjacent **bundle-sheath cells** that surround the leaf veins (**figure 5.10**). The CO_2 is liberated inside these cells, where a second carbon fixation reaction occurs—this time, rubisco fixes the carbon in the Calvin cycle. Meanwhile, at the cost of two ATP molecules, the three-carbon "ferry" returns to the mesophyll to pick up another CO_2.

C₄ plants owe their efficiency to the arrangement of cells in their leaves. Mesophyll cells are surrounded by the air spaces in the leaf, so their O_2 concentration is typically high. But O_2 exposure does not affect the CO_2-fixing enzyme in a C₄ plant's mesophyll cells; this enzyme does not bind O_2 at all, and it can function even if CO_2 concentrations are low. In contrast, the O_2-sensitive rubisco enzyme is confined to the thick-walled bundle-sheath cells, which are isolated from the leaf's air spaces. Malate

deliveries from the mesophyll cells mean that CO_2 is abundant in bundle-sheath cells. Moreover, the chloroplasts in bundle-sheath cells have adaptations that reduce the amount of O_2 they generate in the light reactions. Overall, the rubisco in a C₄ plant is much more likely to bind CO_2 than O_2, reducing photorespiration.

Thanks to their unique carbon fixation pathway, C₄ plants can close their stomata during the hottest times of day and still produce sugars, using the CO_2 trapped in the air spaces of the leaf. Since water loss occurs primarily through open stomata, C₄ plants require about half as much water as C₃ plants.

About 1% of plants use the C₄ pathway. All are flowering plants growing in hot, sunny environments, including crabgrass and crop plants such as sugarcane and corn. C₄ plants are less abundant, however, in cooler, moister habitats. In those environments, the ATP cost of ferrying each CO_2 from a mesophyll cell to a bundle-sheath cell apparently exceeds the benefits of reduced photorespiration.

Another energy- and water-saving strategy, called crassulacean acid metabolism (CAM), was first discovered in desert plants in the Crassulaceae family. Plants that use the **CAM pathway** also fix carbon twice, but they add a new twist: They open their stomata only at night, fix CO_2, then fix it again in the Calvin cycle during the day. Unlike in C₄ plants, both fixation reactions occur in the same cell.

At night, when the temperature drops and humidity rises, CO_2 diffuses in through the CAM plant's open stomata. Mesophyll cells incorporate the CO_2 into malate, which they store in large vacuoles. The stomata close during the heat of the day, but the stored malate moves from the vacuole to a chloroplast and releases its CO_2. The chloroplast then fixes the CO_2 in the Calvin cycle, which can occur only during the day (when ATP and NADPH are available from the light reactions). The

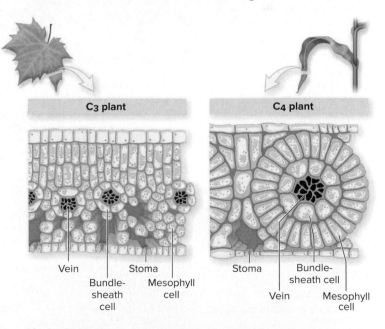

Figure 5.10 C₃ and C₄ Leaf Anatomy. In C₃ plants, the light reactions and the Calvin cycle occur in mesophyll cells. In C₄ plants, the light reactions occur in mesophyll, but the inner ring of bundle-sheath cells houses the Calvin cycle.

CAM pathway reduces photorespiration by generating high CO_2 concentrations inside chloroplasts.

About 3% to 4% of plant species, including pineapple and cacti, use the CAM pathway. All CAM plants are adapted to dry habitats. In cool environments, however, CAM plants cannot compete with C_3 plants. Their stomata are open only at night, so CAM plants have much less carbon available to their cells for growth and reproduction.

Figure 5.11 compares and contrasts C_3, C_4, and CAM plants.

5.6 | MASTERING CONCEPTS

1. Why is the Calvin cycle also called the C_3 pathway?
2. How does photorespiration counter photosynthesis?
3. What conditions maximize photorespiration?
4. Describe how a C_4 plant minimizes photorespiration.
5. How is the CAM pathway similar to C_4 metabolism, and how is it different?

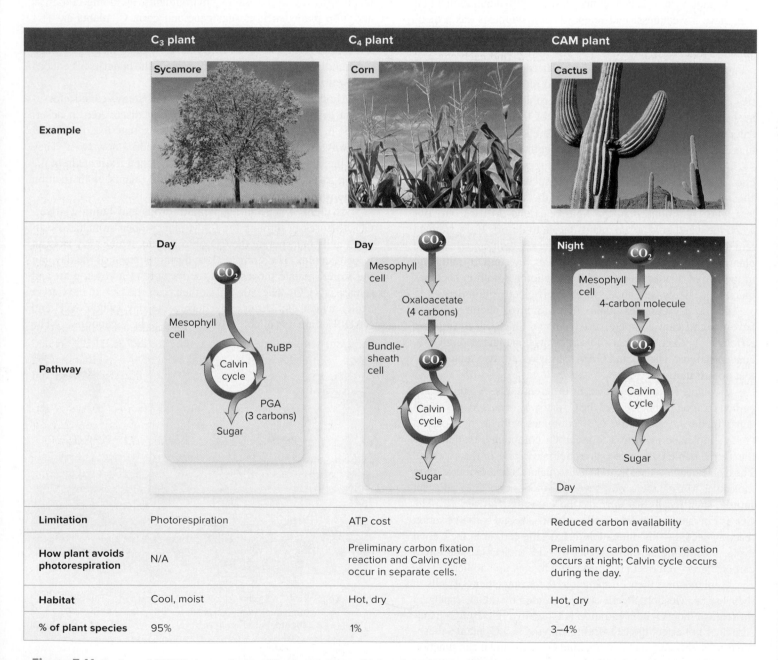

	C₃ plant	C₄ plant	CAM plant
Example	Sycamore	Corn	Cactus
Pathway	*(Day)*	*(Day)*	*(Night / Day)*
Limitation	Photorespiration	ATP cost	Reduced carbon availability
How plant avoids photorespiration	N/A	Preliminary carbon fixation reaction and Calvin cycle occur in separate cells.	Preliminary carbon fixation reaction occurs at night; Calvin cycle occurs during the day.
Habitat	Cool, moist	Hot, dry	Hot, dry
% of plant species	95%	1%	3–4%

Figure 5.11 C₃, C₄, and CAM Pathways Compared. Most plants use the C_3 pathway, which is vulnerable to photorespiration in hot, dry weather. The C_4 and CAM pathways are adaptations that minimize photorespiration.

Photos: (sycamore): ©Exactostock/SuperStock RF; (corn): ©National Geographic Image Collection/Alamy RF; (cactus): ©Digital Vision/Getty Images RF

INVESTIGATING LIFE

5.7 Solar-Powered Salamanders

This chapter's opening essay described two examples of solar-powered animals that live in the ocean: sea slugs and corals. But marine invertebrates are not the only animals with this unusual lifestyle. The eggs of spotted salamanders have live-in algae of their own.

Spotted salamanders are amphibians that live throughout the forests of North America. On rainy evenings in the spring, these animals mate in temporary ponds, where the females lay masses of fertilized eggs (figure 5.12). Each egg contains a tiny embryo and is surrounded by a thick jelly layer. Then, something unusual happens: Microscopic green algae somehow find each egg and enter the jelly layer. The algae reproduce and carry out photosynthesis in the protective confines of their new homes.

Biologists have known since 1986 that the green algae boost the O_2 concentration inside the salamander eggs, a real benefit to embryos that cannot yet breathe on their own. But do algae also feed the embryos a steady diet of sugars? Erin Graham, Robert Sanders, and two other researchers at Temple University wanted to learn more.

The team gathered algae-infected eggs from the wild and incubated them for nearly 2 hours in a solution containing CO_2 that was "tagged" with a radioactive isotope of carbon. They knew that only algae—not salamander eggs or embryos—would be able to use CO_2 directly in photosynthesis. After the incubation period, they rinsed off the excess solution and measured the amount of radioactive carbon in each egg and embryo. ⓘ *radioactive isotopes*, section 2.1C

The researchers reasoned that any radioactive carbon in a salamander embryo could come from one of two sources. The carbon might simply diffuse in from the solution, without any help from the algae. Alternatively, the algae might use the tagged carbon to produce sugars in photosynthesis, then

Figure 5.12 Spawning. A spotted salamander lays eggs in a pool of water.
©George Grall/National Geographic Creative

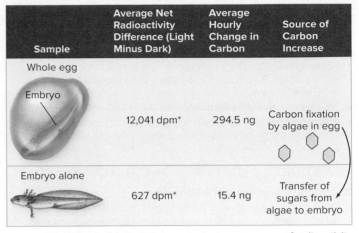

Sample	Average Net Radioactivity Difference (Light Minus Dark)	Average Hourly Change in Carbon	Source of Carbon Increase
Whole egg (Embryo)	12,041 dpm*	294.5 ng	Carbon fixation by algae in egg
Embryo alone	627 dpm*	15.4 ng	Transfer of sugars from algae to embryo

*dpm = disintegrations per minute, a measure of radioactivity

Figure 5.13 Thanks for the Snack. Using a radioactive isotope of carbon, researchers measured the amount of carbon transferred from egg-dwelling green algae to salamander embryos.

transfer some of the radioactive sugar to the embryos. ⓘ *diffusion*, section 4.5A

To differentiate between these two possibilities, the team incubated some eggs in the light, allowing both diffusion and photosynthesis to occur. A second set of eggs was incubated in total darkness. Photosynthesis is not possible in the dark, but diffusion continues. Subtracting the amount of radioactive carbon in dark-treated embryos from the amount in light-treated embryos should therefore reveal the effect of photosynthesis.

After the experiment was complete, radioactivity measurements revealed that eggs and embryos incubated in the light incorporated more radioactive carbon than did their dark-treated counterparts, a sure sign that the algae were sharing their carbon with their tiny hosts (figure 5.13). This sugar supplement can help a developing embryo survive.

It is not surprising that animals would take on photosynthetic partners, but what's in it for the algae? They probably benefit from the partnership as well. A developing embryo releases CO_2 in respiration (see chapter 6). Perhaps this extra shot of CO_2 makes photosynthesis more efficient for the algae, completing the exchange of materials between two allies from different kingdoms of life.

Source: Graham, Erin R., Scott A. Fay, Adam Davey, and Robert W. Sanders. 2013. Intracapsular algae provide fixed carbon to developing embryos of the salamander *Ambystoma maculatum. Journal of Experimental Biology*, vol. 216, pages 452–459.

5.7 MASTERING CONCEPTS

1. On average, what percentage of the 294.5 ng of carbon that the algae fix each hour is transferred to the embryo? Refer to figure 5.13.

2. Identify a standardized variable, an independent variable, and a dependent variable in the experiment.

CHAPTER SUMMARY

5.1 Life Depends on Photosynthesis

- **Autotrophs** produce their own organic compounds from inorganic starting materials such as CO_2 and water. **Heterotrophs** rely on organic molecules produced by other organisms.

A. Photosynthesis Builds Carbohydrates Out of Carbon Dioxide and Water

- **Photosynthesis** converts kinetic energy in light to potential energy in the covalent bonds of carbohydrates, according to the following chemical equation:

$$6CO_2 + 6H_2O \xrightarrow{\text{light energy}} C_6H_{12}O_6 + 6O_2$$

B. Plants Use Carbohydrates in Many Ways

- Plants use glucose and other sugars to grow, generate **ATP,** nourish nonphotosynthetic plant parts, and produce cellulose and many other biochemicals. Most store excess carbohydrates as starch or sucrose.

C. The Evolution of Photosynthesis Changed Planet Earth

- Before photosynthesis evolved, organisms were heterotrophs. The first autotrophs made new food sources available.
- Over billions of years, oxygen produced in photosynthesis changed Earth's climate and the history of life.

5.2 Sunlight Is the Energy Source for Photosynthesis

A. What Is Light?

- Visible light is a small part of the **electromagnetic spectrum.**
- **Photons** move in waves. The shorter the **wavelength,** the more kinetic energy per photon. Visible light occurs in a spectrum of colors representing different wavelengths.

B. Photosynthetic Pigments Capture Light Energy

- **Chlorophyll** *a* is the primary photosynthetic pigment in plants. **Accessory pigments** absorb wavelengths of light that chlorophyll *a* cannot absorb.

C. Chloroplasts Are the Sites of Photosynthesis

- Plants exchange gases with the environment through pores called **stomata.**
- Leaf **mesophyll** cells contain abundant **chloroplasts.**
- A chloroplast contains a gelatinous fluid called the **stroma.** The fluid surrounds the **grana,** which are stacks of pancake-shaped **thylakoid** membranes. Photosynthetic pigments are embedded in the thylakoid membranes, which enclose the **thylakoid space.**
- A **photosystem** consists of proteins, **antenna pigments,** and a **reaction center.**

5.3 Photosynthesis Occurs in Two Stages

- The **light reactions** of photosynthesis produce ATP and **NADPH;** these molecules provide energy and electrons for the sugar-producing **carbon reactions** (figure 5.14).

5.4 The Light Reactions Begin Photosynthesis

A. Light Striking Photosystem II Provides the Energy to Produce ATP

- Photosystem II captures light energy and sends electrons from reactive chlorophyll *a* along the **electron transport chain.**
- Electrons from chlorophyll are replaced with electrons from water. O_2 is the waste product.
- The energy released in the electron transport chain drives the active transport of protons (H^+) into the thylakoid space. The protons diffuse out through channels in **ATP synthase.** This movement powers the phosphorylation of ADP to ATP.
- The coupling of the proton gradient and ATP formation is called **chemiosmotic phosphorylation.**

B. Electrons from Photosystem I Reduce NADP+ to NADPH

- Light striking photosystem I re-energizes the electrons, which pass to an enzyme that uses them to reduce $NADP^+$. The product of this reaction is NADPH.

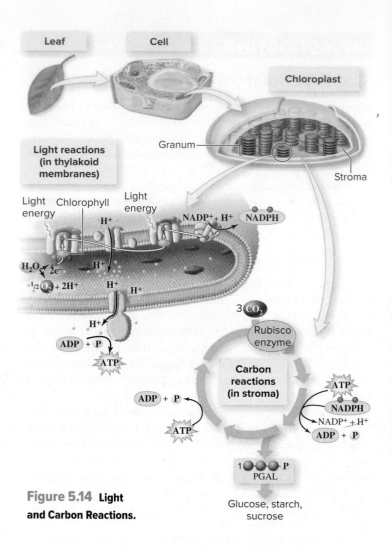

Figure 5.14 Light and Carbon Reactions.

5.5 The Carbon Reactions Produce Carbohydrates

- The carbon reactions use energy from ATP and electrons from NADPH in **carbon fixation** reactions that add CO_2 to organic compounds.
- In the **Calvin cycle, rubisco** catalyzes the reaction of CO_2 with **ribulose bisphosphate (RuBP)** to yield two molecules of PGA. These are converted to PGAL, the immediate product of photosynthesis. PGAL later becomes glucose and other carbohydrates.

5.6 C₃, C₄, and CAM Plants Use Different Carbon Fixation Pathways

- The Calvin cycle is also called the **C₃ pathway.** Most plant species are C₃ plants, which use only this pathway to fix carbon.
- **Photorespiration** wastes carbon and energy when rubisco reacts with O_2 instead of CO_2.
- The **C₄ pathway** reduces photorespiration by separating two carbon fixation reactions into different cells. In mesophyll cells, CO_2 is fixed as a four-carbon molecule, which moves to a **bundle-sheath cell** and liberates CO_2 to be fixed again in the Calvin cycle.
- In the **CAM pathway,** desert plants such as cacti open their stomata and take in CO_2 at night, storing the fixed carbon in vacuoles. During the day, they split off CO_2 and fix it in chloroplasts in the same cells.

5.7 Investigating Life: Solar-Powered Salamanders

- The eggs of spotted salamanders contain cells of green algae, which provide O_2 and carbon to the animal's embryo.

MULTIPLE CHOICE QUESTIONS

1. Where does the energy come from to drive photosynthesis?
   a. A chloroplast   c. The sun
   b. ATP   d. Glucose

2. Animals and other _____ rely on _____ that carry out photosynthesis.
   a. autotrophs; autotrophs   c. heterotrophs; autotrophs
   b. heterotrophs; heterotrophs d. autotrophs; heterotrophs

3. Photosynthesis is an example of an _____ chemical reaction because _____.
   a. exergonic; energy is released by the reaction center pigment
   b. endergonic; light energy is used to build chemical bonds
   c. exergonic; light energy is captured by pigment molecules
   d. endergonic; the reactions occur inside a cell

4. The evolution of photosynthesis resulted in
   a. an increase in the amount of O_2 in the atmosphere.
   b. the initial appearance of heterotrophs.
   c. global warming.
   d. an increase in the amount of CO_2 in the atmosphere.

5. Only high-energy light can penetrate the ocean and reach photosynthetic organisms in coral reefs. What color of light would you predict these organisms use?
   a. Red   c. Blue
   b. Yellow   d. Orange

6. The primary function of the light reactions is to _____, whereas the primary function of the carbon reactions is to _____.
   a. convert the sun's energy into chemical energy; store chemical energy
   b. use light energy to produce ATP; use chemical energy to produce ATP
   c. store light; use light energy to produce carbon
   d. transfer heat captured from light to electrons; use electrons to generate organic molecules

7. Photorespiration becomes more likely when
   a. CO_2 concentrations are high in leaf cells.
   b. stomata remain closed in C_3 plants.
   c. glucose concentrations are low in leaf cells.
   d. ATP binds to rubisco.

8. A plant that opens its stomata only at night is a
   a. C_2 plant.   c. C_4 plant.
   b. C_3 plant.   d. CAM plant.

Answers to these questions are in appendix A.

WRITE IT OUT

1. Imagine that multiple simultaneous volcanic eruptions send black ash into Earth's atmosphere, making photosynthesis impossible anywhere on Earth for many years. What would be the consequence to plants? To animals? To microbes?

2. Other stars in the galaxy emit light at different wavelengths than the sun. If photosynthesis evolved on a planet around one of these stars, how might it be different from and similar to photosynthesis on Earth?

3. Define these terms and arrange them from smallest to largest: *thylakoid membrane; photosystem; chloroplast; granum; reaction center.*

4. Would a plant grow better in a room painted blue or in a room painted green? Explain your answer.

5. Determine whether each of the following molecules is involved in the light reactions, the carbon reactions, or both and explain how: O_2, CO_2, carbohydrates, chlorophyll *a*, photons, NADPH, ATP, H_2O.

6. Of the many groups of photosynthetic bacteria, only cyanobacteria use chlorophyll *a*. How does this observation support the hypothesis that cyanobacteria gave rise to the chloroplasts of today's plants and algae?

7. In 1941, biologists exposed photosynthesizing cells to water containing a heavy oxygen isotope, designated ^{18}O. The "labeled" isotope appears in the O_2 gas released in photosynthesis, showing that the oxygen came from the water. Where would the ^{18}O have ended up if the researchers had used ^{18}O-labeled CO_2 instead of H_2O?

8. Over the past decades, the CO_2 concentration in the atmosphere has increased. (a) Predict the effect of increasing carbon dioxide concentrations on photorespiration. (b) Scientists suggest that increasing CO_2 concentrations are leading to higher average global temperatures. If temperatures are increasing, does this change your answer to part (a)?

9. How does photosynthesis help compensate for increasing atmospheric CO_2? Where does the CO_2 go? Does cutting down forests likely increase or decrease the rate of CO_2 accumulation in the atmosphere?

10. How is the CAM pathway adaptive in a desert habitat?

11. Explain how C_4 photosynthesis is based on a spatial arrangement of structures, whereas CAM photosynthesis is temporally based.

12. Explain why each of the following misconceptions about photosynthesis is false: (a) Only plants are autotrophs. (b) Plants do not need cellular respiration because they carry out photosynthesis. (c) Chlorophyll is the only plant pigment.

PULL IT TOGETHER

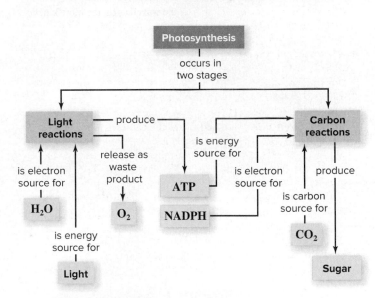

Figure 5.15 Pull It Together: Photosynthesis.

Refer to figure 5.15 and the chapter content to answer the following questions.

1. Review the Survey the Landscape figure in the chapter introduction, and then add *enzymes, cells, molecules,* and *respiration* to the Pull It Together concept map.

2. How would you incorporate the Calvin cycle, rubisco, C_3 plants, C_4 plants, and CAM plants into this concept map?

3. One possible connecting phrase in the concept map is "*Chlorophyll reflects _____ wavelengths of light.*" Fill in the blank and explain your answer.

4. Build another small concept map showing the relationships among the terms *chloroplast, stroma, grana, thylakoid, photosystem,* and *chlorophyll.*

5. Besides respiration, what happens to the sugar produced in photosynthesis?

6 Respiration and Fermentation

Huge Meal. This African rock python is consuming a Thomson's gazelle.

©Gunter Ziesler/Photoshot

LEARN HOW TO LEARN
Don't Skip the Figures

As you read the narrative in the text, pay attention to the figures; they are there to help you learn. Some figures summarize the narrative, making it easier for you to see the "big picture." Other illustrations show the parts of a structure or the steps in a process; still others summarize a technique or help you classify information. Also, remember that students use illustrations in different ways. Once you encounter a figure's callout, you may prefer to stop reading to absorb the entire figure, or you may switch back and forth between the narrative and the figure's parts. Being attentive to your preferences will help you to be more systematic as you study.

Eating for Life

The African rock python lay in wait for the lone gazelle. When the gazelle came close, the snake moved swiftly, entwining its 9-meter-long body snugly around the mammal. Each time the gazelle exhaled, the snake squeezed, shutting down the victim's heart and lungs in less than a minute.

Thanks to the adaptations of its digestive system, the snake can swallow and digest a meal over half its own size. The reptile begins by opening its jaws at an angle of 130 degrees (compared with 30 degrees for the most gluttonous human) and places its mouth over the gazelle's head, using strong muscles to gradually envelop and push along the carcass. Saliva coats the prey, easing its journey to the snake's stomach. After several hours, the huge meal arrives at the stomach, and the remainder of the digestive tract readies itself for several weeks of dismantling the gazelle. Hydrochloric acid (HCl) builds up in the snake's stomach, and the output of digestive enzymes in the intestines increases 60-fold.

As the gazelle passes through the snake's digestive system, it breaks into clumps of cells. These cells disintegrate, releasing proteins, carbohydrates, and lipids. After the snake digests these macromolecules, the component parts are small enough to enter the blood and move to the body's tissues. The animal's cells absorb these smaller nutrient molecules. Then, in cellular respiration, energy in the bonds of the food molecules is transferred to the high-energy phosphate bonds of ATP. Afterward, only a few chunks of hair and bone will remain to be eliminated.

Our own eating habits may not seem to have much in common with those of the African rock python. After all, we typically eat many small meals a day; our bodies are not adapted to digest enormous prey in a single gulp. Nevertheless, the fruits, vegetables, meats, eggs, dairy products, and other foods that we consume are doing for us precisely what the gazelle's body is doing for the snake: The food molecules break down into nutrients that the cells use to make ATP.

In humans, snakes, and every other organism, nearly all activities depend on energy stored in ATP. Yet no organism eats ATP directly. This chapter describes how cells convert what we do eat—glucose and other food molecules—into those little ATP molecules that nothing can live without.

SURVEY THE LANDSCAPE
Science, Chemistry, and Cells

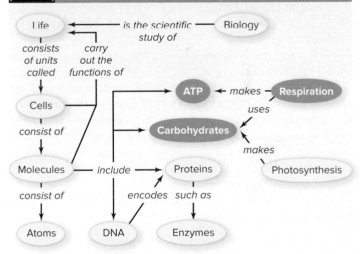

The reactions of aerobic respiration consume carbohydrates and oxygen gas. The overall function is to store energy in ATP, which powers cell activities.

For more details, study the Pull It Together feature in the chapter summary.

6.1 Cells Use Energy in Food to Make ATP

No cell can survive without **ATP**—adenosine triphosphate. Without this energy-toting molecule, you could not have developed from a fertilized egg into an adult. You could not breathe, chew, talk, circulate your blood, blink your eyes, walk, or listen to music. Without ATP, a plant could not take up soil nutrients, grow, or produce flowers, fruits, and seeds. A fungus could not produce mushrooms. A bacterial cell could not divide or move. Like a car without gasoline, a cell without ATP would simply die. (i) *ATP,* section 4.3

ATP is essential because it powers nearly every activity that requires energy input in the cell: synthesis of DNA, RNA, proteins, carbohydrates, and lipids; active transport across the membranes surrounding cells and organelles; separation of duplicated chromosomes during cell division; movement of cilia and flagella; muscle contraction; and many others. This constant need for ATP explains the need for a steady food supply. All organisms, from giant redwood trees to whales to bacteria, use the potential energy stored in food to make ATP.

Where does the food come from in the first place? Chapter 5 explains the answer: In most ecosystems, plants and other autotrophs use photosynthesis to make organic molecules such as glucose ($C_6H_{12}O_6$) out of carbon dioxide (CO_2) and water (H_2O). Light supplies the energy. The carbohydrates produced in photosynthesis feed not only autotrophs but also all of the animals, fungi, and microbes that share the ecosystem (see figure 4.2).

All cells need ATP, but they don't all produce it in the same way. The pathways that generate ATP from food fall into three categories. In aerobic cellular respiration, the main subject of this chapter, a cell uses oxygen gas (O_2) and glucose to generate ATP. Plants, animals, and most microbes, especially those in O_2-rich environments, use aerobic respiration. The other two pathways, anaerobic respiration and fermentation, generate ATP from glucose without using O_2. Section 6.8 describes these two processes, both of which are most common in microorganisms.

The overall equation for **aerobic respiration** is essentially the reverse of photosynthesis:

glucose + oxygen $\longrightarrow$ carbon dioxide + water + ATP

$$C_6H_{12}O_6 + 6O_2 \longrightarrow 6CO_2 + 6H_2O + 36ATP$$

This equation reveals that aerobic cellular respiration requires organisms to acquire O_2 and get rid of CO_2 (**figure 6.1**). These gases simply diffuse across the cell membranes of single-celled organisms, but more complex organisms have specialized organs of gas exchange such as gills or lungs. In humans and many other animals, O_2 from inhaled air diffuses into the bloodstream across the walls of microscopic air sacs in the lungs. The circulatory system carries the inhaled O_2 to cells, where gas exchange occurs. O_2 diffuses into the cell's mitochondria, the sites of respiration. Meanwhile, CO_2 diffuses out of the cells and into the bloodstream. After moving from the blood into the lungs, the CO_2 is exhaled.

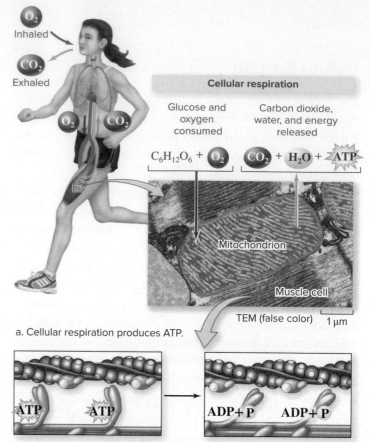

a. Cellular respiration produces ATP.

b. Muscle contraction consumes ATP.

Figure 6.1 Breathing and Cellular Respiration. (a) The athlete breathes in O_2, which is distributed to all cells. In mitochondria, the O_2 participates in the reactions of cellular respiration. CO_2, a metabolic waste, is exhaled. (b) Energy-rich ATP generated in cellular respiration is used in muscle contraction, among many other cellular activities.
Photo: (a): ©Thomas Deerinck, NCMIR/Science Source

Many people mistakenly believe that plants do not use cellular respiration because they are photosynthetic. In fact, plants use O_2 to respire about half of the glucose they produce. Why do plants have a reputation for producing O_2, if they also consume it? The reason is that plants incorporate much of the remaining glucose into cellulose, starch, and other stored organic molecules. Therefore, they absorb much more CO_2 in photosynthesis than they release in respiration, and they release much more O_2 than they consume.

The rest of this chapter describes how cells use the potential energy in food to generate ATP. Like photosynthesis, the journey entails several overlapping metabolic pathways and many different chemicals. But if we consider energy release in major stages, the logic emerges.

6.1 MASTERING CONCEPTS

1. Why do all organisms need ATP?
2. What are the three general ways to generate ATP from food, and which organisms use each pathway?
3. How do organisms get O_2 to their cells?
4. Why do plants carry out photosynthesis and respiration?

6.2 Cellular Respiration Includes Three Main Processes

The chemical reaction that generates ATP is straightforward: An enzyme tacks a phosphate group onto ADP, yielding ATP. As described in chapter 4, however, ATP synthesis requires an input of energy. The metabolic pathways of respiration harvest potential energy from food molecules and use it to make ATP. This section briefly introduces these pathways; later sections explain them in more detail.

Like photosynthesis, respiration is an oxidation–reduction reaction. The pathways of aerobic respiration oxidize (remove electrons from) glucose and reduce (add electrons to) O_2. Because of oxygen's strong attraction for electrons, this reaction is "easy," like riding a bike downhill. It therefore releases energy, which the cell traps in the bonds of ATP. ⓘ *redox reactions,* section 4.2B

This reaction does not happen all at once. If a cell released all the potential energy in glucose's chemical bonds in one uncontrolled step, the sudden release of heat would destroy the cell; in effect, it would act as a tiny bomb. Rather, the chemical bonds and atoms in glucose are rearranged one step at a time, releasing a tiny bit of energy with each transformation. According to the second law of thermodynamics, some of this energy is released as heat. But much of it is stored in the chemical bonds of ATP.

Biologists organize the intricate biochemical pathways of respiration into three main groups: glycolysis, the Krebs cycle, and electron transport (figure 6.2). In **glycolysis** (literally, "breaking sugar"), a six-carbon glucose molecule splits into two three-carbon molecules of **pyruvate.** This process harvests energy in two forms. First, some of the electrons from glucose are transferred to an electron carrier molecule called **NADH** (nicotinamide adenine dinucleotide). Second, glycolysis generates two molecules of ATP.

Additional reactions, including a "transition step" and the **Krebs cycle,** oxidize the pyruvate and release CO_2. Enzymes rearrange atoms and bonds in ways that transfer the pyruvate's potential energy and electrons to ATP, NADH, and another electron carrier molecule—**FADH$_2$** (flavin adenine dinucleotide).

By the time the Krebs cycle is complete, the carbon atoms that made up the glucose are gone—liberated as CO_2. The cell has generated a few molecules of ATP, but most of the potential energy from glucose now lingers in the high-energy electron carriers, NADH and FADH$_2$. The cell uses them to generate more ATP.

The **electron transport chain** transfers energy-rich electrons from NADH and FADH$_2$ through a series of membrane proteins. As electrons pass from carrier to carrier in the electron transport chain, the energy is used to create a gradient of hydrogen ions. (Recall that a hydrogen ion, H^+, is simply a hydrogen atom stripped of its electron, leaving just a proton; see figure 2.3.) The mitochondrion uses the potential energy stored in this proton gradient to generate ATP. An enzyme called

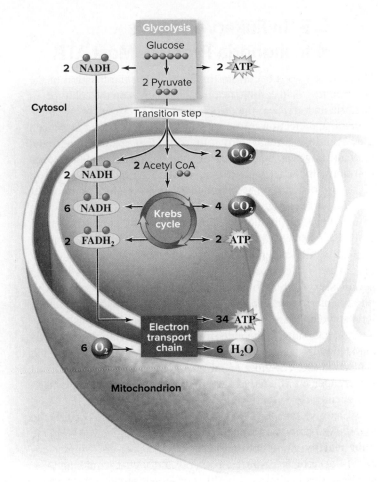

Figure 6.2 Overview of Aerobic Cellular Respiration. A six-carbon glucose molecule is broken down to six molecules of carbon dioxide in three main stages: glycolysis, the Krebs cycle, and the electron transport chain. Along the way, energy is harvested as ATP. Except for glycolysis, these reactions occur inside the mitochondria of eukaryotic cells.

ATP synthase forms a channel in the membrane, releasing the protons and using their potential energy to add phosphate to ADP. (As described in section 5.4, the same enzyme generates ATP in the light reactions of photosynthesis.) In the meantime, the "spent" electrons are transferred to O_2, generating water as a waste product.

A common misconception is that any ATP-generating pathway in a cell is considered "respiration." In fact, however, all forms of respiration, aerobic and anaerobic, use an electron transport chain. As you will see in section 6.8, fermentation is not respiration because it generates ATP from glycolysis only.

6.2 MASTERING CONCEPTS

1. Why do the reactions of respiration occur step-by-step instead of all at once?
2. What occurs in each of the three stages of cellular respiration?

6.3 In Eukaryotic Cells, Mitochondria Produce Most ATP

Glycolysis always occurs in the cytosol, but the location of the other pathways in aerobic respiration depends on the cell type. In bacteria and archaea, the enzymes of the Krebs cycle are in the cytosol, and electron transport proteins are embedded in the cell membrane. The eukaryotic cells of protists, plants, fungi, and animals, however, contain **mitochondria,** specialized organelles that house the other reactions of cellular respiration (**figure 6.3**).

A mitochondrion is bounded by two membranes: an outer membrane and a highly folded inner membrane. **Cristae** are folds of the inner membrane. The **intermembrane compartment** is the area between the two membranes, and the mitochondrial **matrix** is the fluid enclosed within the inner membrane.

In a eukaryotic cell, the two pyruvate molecules produced in glycolysis cross both of the mitochondrial membranes and move into the matrix. Here, enzymes cleave pyruvate and carry out the Krebs cycle. Then, $FADH_2$ and NADH from glycolysis and the Krebs cycle move to the inner mitochondrial membrane, which is studded with many copies of the electron transport proteins and ATP synthase. The inner membrane's cristae provide tremendous surface area on which the reactions of the electron transport chain can occur.

Electron transport chains and ATP synthase also occur in the thylakoid membranes of chloroplasts, which generate ATP in the light reactions of photosynthesis (see chapter 5). Similar enzymes operate in respiring bacteria and archaea, making ATP synthase one of the most highly conserved proteins over evolutionary time.

Mitochondria and chloroplasts share another similarity, too: Both types of organelles contain DNA and ribosomes. Mitochondrial DNA encodes ATP synthase and most of the proteins of the electron transport chain. Not surprisingly, a person with abnormal versions of these genes may be very ill or even die. The worst mitochondrial diseases affect the muscular and nervous systems. Muscle and nerve cells are especially energy-hungry; each one may contain as many as 10,000 mitochondria. When their mitochondria fail, these cells cannot carry out their functions.

6.3 MASTERING CONCEPTS

1. What are the parts of a mitochondrion?

2. Which respiratory reactions occur in each part of the mitochondrion?

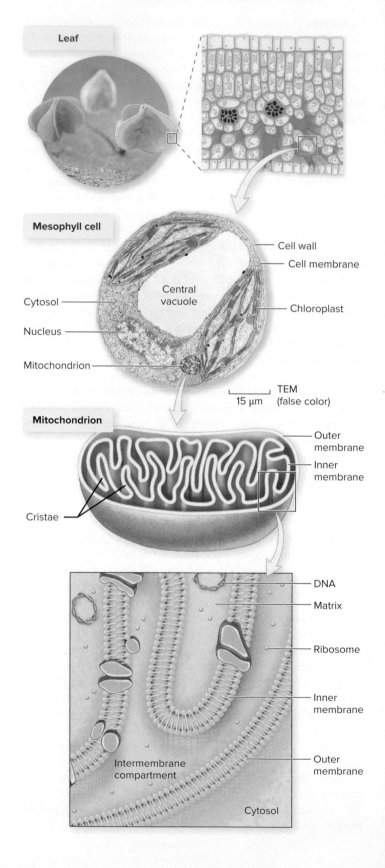

Figure 6.3 **Anatomy of a Mitochondrion.** Eukaryotic cells, such as the ones that make up leaves, contain mitochondria that provide most of the cell's ATP. Each mitochondrion includes two membranes. The inner membrane encloses fluid called the matrix, and the space between the inner and outer membranes is the intermembrane compartment.

Photos: (leaves): ©Steve Raymer/National Geographic Stock; (mesophyll): Electron micrograph by Wm. P. Wergin, courtesy of Eldon H. Newcomb, University of Wisconsin-Madison

6.4 Glycolysis Breaks Down Glucose to Pyruvate

Glycolysis is a more-or-less universal metabolic pathway that splits glucose into two three-carbon pyruvate molecules. The name of the pathway reflects its function: *glyco-* means "sugar," and *-lysis* means "to break."

The entire process of glycolysis requires 10 steps, all of which occur in the cell's cytosol (**figure 6.4**). None of the steps requires O_2, so cells can use glycolysis in both oxygen-rich and anaerobic environments.

The reactions of glycolysis are divided into two stages, the first of which is labeled "energy investment" in figure 6.4. The cell spends two molecules of ATP to activate glucose, redistributing energy in the molecule and splitting it in half. Then, in the "energy harvest" stage, the cell generates a return on its initial investment, producing two molecules of NADH plus four molecules of ATP. Overall, the net gain is two NADHs and two ATPs per molecule of glucose.

The ATP produced in glycolysis is formed by **substrate-level phosphorylation,** which means that an enzyme transfers a phosphate group directly from a high-energy "donor" molecule to ADP. Unlike chemiosmotic phosphorylation, which is described in section 6.5, this method of producing ATP does not require a proton gradient or the ATP synthase enzyme.

Glucose contains considerable bond energy, but cells recover only a small portion of it as ATP and NADH during glycolysis. Most of the potential energy of the original glucose molecule remains in the two pyruvate molecules. As you will see, the pathways of aerobic respiration extract much more of that energy.

6.4 MASTERING CONCEPTS

1. Overall, what happens in glycolysis?
2. How is substrate-level phosphorylation different from chemiosmotic phosphorylation?
3. What is the net gain of ATP and NADH for each glucose molecule undergoing glycolysis?

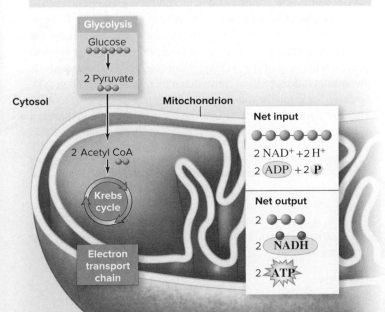

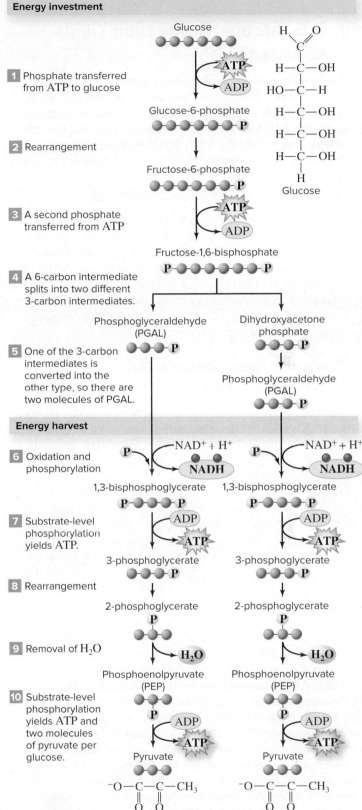

Figure 6.4 Glycolysis. Glucose splits into two pyruvate molecules, producing a net yield of two ATPs and two NADHs. The illustration of the cell shows an overview of glycolysis; the rest of the figure shows the entire 10-step process. (Each *gray sphere* represents a carbon atom.)

6.5 Aerobic Respiration Yields Abundant ATP

Overall, aerobic cellular respiration taps much of the potential energy remaining in the pyruvate molecules that emerge from the pathways of glycolysis. The Krebs cycle and electron transport chain are the key ATP-generating processes. This section explains what they do, and the Apply It Now box describes poisons that interfere with their work.

A. Pyruvate Is Oxidized to Acetyl CoA

After glycolysis, pyruvate moves into the mitochondrial matrix, but it is not directly used in the Krebs cycle. Instead, a preliminary "transition step" further oxidizes each pyruvate molecule (figure 6.5). First, a molecule of CO_2 is removed, and NAD^+ is reduced to NADH. The remaining two-carbon molecule, called an acetyl group, is transferred to a coenzyme to form acetyl coenzyme A (abbreviated acetyl CoA). **Acetyl CoA** is the compound that enters the Krebs cycle. ⓘ *coenzymes,* section 4.4B

B. The Krebs Cycle Produces ATP and Electron Carriers

The Krebs cycle completes the oxidation of each acetyl group, releasing CO_2 (figure 6.6). The cycle begins when acetyl CoA sheds the coenzyme and combines with a four-carbon molecule, oxaloacetate (step 1). The resulting six-carbon molecule is citrate; the Krebs cycle is therefore also called the citric acid cycle.

The remaining steps in the Krebs cycle rearrange and oxidize citrate through several intermediates. Along the way, two carbon atoms are released as CO_2 (steps 2 and 3). In addition, some of the transformations transfer electrons to NADH and $FADH_2$ (steps 2, 3, 5, and 6); others produce ATP by substrate-level phosphorylation (step 4). Eventually, the molecules in the Krebs cycle re-create the original acceptor molecule, oxaloacetate. The cycle can now repeat.

Since one glucose molecule yields two acetyl CoA molecules, the Krebs cycle turns twice for each glucose. Thus, the combined net output to this point (glycolysis, transition step, and the Krebs cycle) is 4 ATP molecules, 10 NADH molecules, and 2 $FADH_2$ molecules. All six carbon atoms are gone, released as CO_2.

Besides continuing the breakdown of glucose, the Krebs cycle also has another function not directly related to respiration. The cell uses intermediate compounds formed in the Krebs cycle to manufacture other organic molecules, such as amino acids or fats. Section 6.7 explains that the reverse process also occurs; amino acids and fats can enter the Krebs cycle to generate energy from food sources other than carbohydrates.

C. The Electron Transport Chain Drives ATP Formation

The products generated so far are CO_2, ATP, NADH, and $FADH_2$. The cell ejects the CO_2 as waste and uses ATP to fuel essential

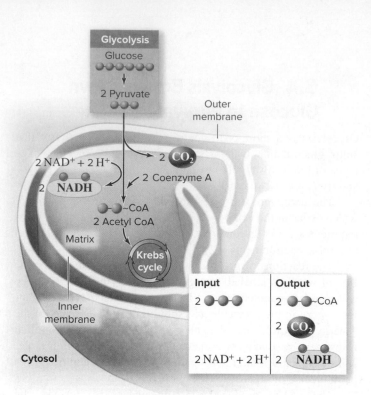

Figure 6.5 Transition Step. After pyruvate moves into a mitochondrion, it is oxidized to form an acetyl group, CO_2, and NADH. The acetyl group joins with coenzyme A to form acetyl CoA, the molecule that enters the Krebs cycle.

processes. But what becomes of the electron carriers (NADH and $FADH_2$)? They transfer their cargo to an electron transport chain in the inner mitochondrial membrane.

The electron transport chain harnesses the energy from these electrons in stages (figure 6.7). The first protein in the chain accepts electrons from NADH; $FADH_2$ donates its electrons to the second protein. The electrons then pass to the next protein in the chain, and the next, and so on. The final electron acceptor is O_2, which combines with H^+ to form water. Along the way, some of the proteins use energy from the electrons to pump H^+ from the matrix into the intermembrane compartment.

The electron transport chain therefore uses the energy in NADH and $FADH_2$ to establish a proton (H^+) gradient across the inner mitochondrial membrane. As explained in chapter 4, a gradient represents a form of potential energy. The mitochondrion harvests this energy as ATP in the final stage of cellular respiration, with the help of the ATP synthase enzyme. In **chemiosmotic phosphorylation,** protons move down their gradient through ATP synthase back into the matrix, and ADP is phosphorylated to ATP. The ATP synthase enzyme therefore captures the potential energy of the proton gradient and saves it in a form the cell can use: ATP.

6.5   MASTERING CONCEPTS

1. Pyruvate has three carbon atoms; an acetyl group has only two. What happens to the other carbon atom?
2. How does the Krebs cycle generate CO_2, ATP, NADH, and $FADH_2$?
3. How do electrons from NADH and $FADH_2$ power ATP formation?
4. What is the role of O_2 in the electron transport chain?

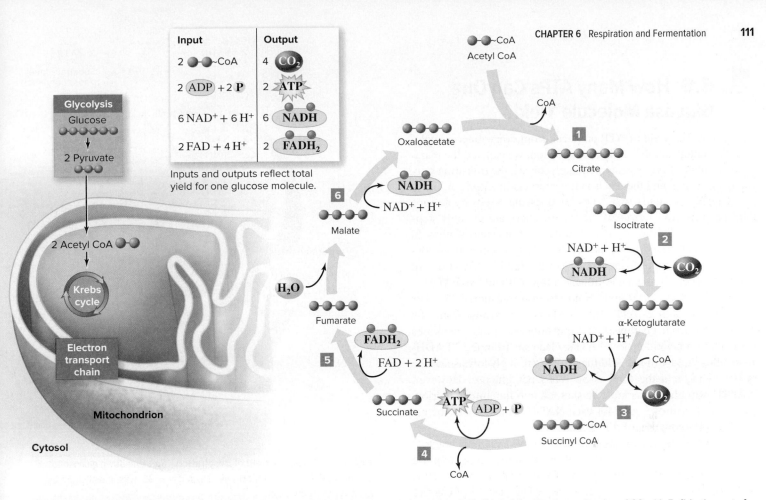

Figure 6.6 Krebs Cycle. In the mitochondrial matrix, (*1*) acetyl CoA enters the Krebs cycle and (*2, 3*) is oxidized to two molecules of CO_2. (*4, 5, 6*) In the rest of the Krebs cycle, potential energy is trapped as ATP, NADH, and $FADH_2$. The left half of the figure summarizes the inputs, outputs, and location of the Krebs cycle; the right half shows the entire cycle, step-by-step.

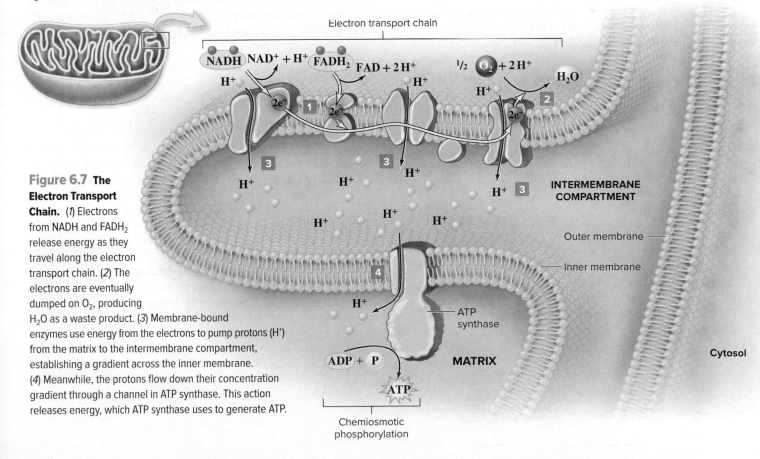

Figure 6.7 The Electron Transport Chain. (*1*) Electrons from NADH and $FADH_2$ release energy as they travel along the electron transport chain. (*2*) The electrons are eventually dumped on O_2, producing H_2O as a waste product. (*3*) Membrane-bound enzymes use energy from the electrons to pump protons (H^+) from the matrix to the intermembrane compartment, establishing a gradient across the inner membrane. (*4*) Meanwhile, the protons flow down their concentration gradient through a channel in ATP synthase. This action releases energy, which ATP synthase uses to generate ATP.

6.6 How Many ATPs Can One Glucose Molecule Yield?

To estimate the yield of ATP produced from every glucose molecule that enters aerobic cellular respiration, we can add the maximum number of ATPs generated in glycolysis, the transition step, the Krebs cycle, and the electron transport chain (figure 6.8).

Glycolysis yields two ATPs, as does the Krebs cycle (one ATP each from two turns of the cycle). These are the only steps that produce ATP directly by substrate-level phosphorylation. In addition, each glucose yields two NADH molecules from glycolysis and two more from acetyl CoA production. Two turns of the Krebs cycle yield an additional six NADHs and two FADH$_2$s.

In theory, the ATP yield from electron transport is 3 ATPs per NADH and 2 ATPs per FADH$_2$. Electrons from the 10 NADHs from glycolysis, the transition step, and the Krebs cycle therefore yield up to 30 ATPs; electrons from the 2 FADH$_2$ molecules yield 4 more. Add the 4 ATPs from glycolysis and the Krebs cycle, and the total is 38 ATPs per glucose. However, NADH from glycolysis must be shuttled into the mitochondrion, usually at a cost of 1 ATP for each NADH. This reduces the net theoretical production of ATPs to 36.

In reality, some protons leak across the inner mitochondrial membrane on their own, and the cell spends some energy to move pyruvate and ADP into the matrix. These "expenses" lower the actual ATP yield to about 30 per glucose. The number of calories stored in 30 ATPs is about 32% of the total calories stored in the glucose bonds; the rest of the potential energy in glucose is lost to the environment as heat. This may seem wasteful, but for a biological process, it is reasonably efficient. To put this energy yield into perspective, an automobile uses only about 20% to 25% of the energy contained in gasoline's chemical bonds; the rest is lost as heat.

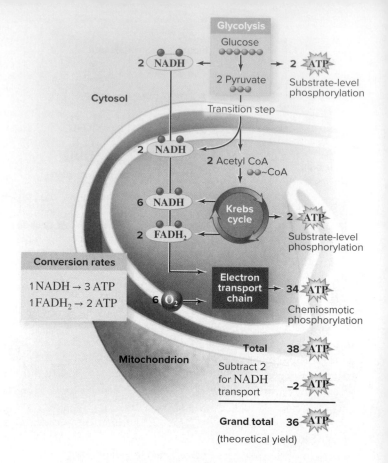

Figure 6.8 Energy Yield of Respiration. Breaking down glucose to carbon dioxide can theoretically yield as many as 36 ATPs, mostly from the electron transport chain.

6.6 MASTERING CONCEPTS

1. Explain how to arrive at the estimate that each glucose molecule theoretically yields 36 ATPs.
2. How does the actual ATP yield compare to the theoretical yield?

Apply It **Now**

Some Poisons Inhibit Respiration

Many toxic chemicals kill by blocking one or more reactions in respiration. Here are a few examples:

Krebs cycle inhibitor:
- Arsenic binds to part of a molecule needed for the formation of acetyl CoA. It therefore blocks the Krebs cycle.

Electron transport inhibitors:
- Some mercury compounds stop an oxidation–reduction reaction early in the electron transport chain. Mercury is used in some thermometers, in fluorescent lights, and in many industrial applications.
- Cyanide blocks the final transfer of electrons to O$_2$. When proteins in the electron transport chain have no place to "dump" their electrons, the process grinds to a halt. This deadly poison is used in mining and some other industries.

- Carbon monoxide (CO) blocks electron transport at the same point as cyanide. This colorless, odorless gas is a byproduct of incomplete fuel combustion. CO from unvented heaters, stoves, and fireplaces can accumulate to deadly levels in homes. Car exhaust and cigarette smoke are other sources of CO.

Chemiosmotic phosphorylation inhibitors:
- The insecticide 2,4-dinitrophenol (DNP) kills by making the inner mitochondrial membrane permeable to protons, blocking formation of the proton gradient necessary to drive ATP synthesis.
- Oligomycin blocks the phosphorylation of ADP by inhibiting the part of the ATP synthase enzyme that lets the protons through. Oligomycin is mostly used in laboratory studies of respiration.

©Corbis RF

🍁 6.7 Other Food Molecules Enter the Energy-Extracting Pathways

So far, we have focused on the complete oxidation of glucose. But food also includes starch, proteins, and lipids that contribute calories to the diet. These molecules also enter the energy pathways (**figure 6.9**).

The digestion of starch from potatoes, wheat, and other carbohydrate-rich food begins in the mouth and continues in the small intestine. Enzymes snip the long starch chains into individual glucose monomers, which generate ATP as described in this chapter. Another polysaccharide, glycogen, follows essentially the same path as starch. ⓘ *carbohydrates*, section 2.5B

Proteins are digested into monomers called amino acids. The cell does not typically use these amino acids to produce ATP; instead, most of them are incorporated into new proteins. When an organism depletes its immediate carbohydrate supplies, however, cells may use amino acids as an energy source. First, nitrogen is stripped from the amino acid and excreted, often as urea. The remainder of each molecule enters the energy pathways as pyruvate, acetyl CoA, or an intermediate of the Krebs cycle, depending on the amino acid. ⓘ *amino acids*, section 2.5C

Meanwhile, enzymes in the small intestine digest fat molecules from food into glycerol and three fatty acids, which enter the bloodstream and move into the body's cells. (This chapter's Burning Question describes a diet pill that blocks this process.)

Enzymes convert the glycerol to pyruvate, which then proceeds through the rest of cellular respiration as though it came directly from glucose. The fatty acids enter the mitochondria, where they are cut into many two-carbon pieces that become acetyl CoA. From here, the pathways continue as they would for glucose. ⓘ *lipids*, section 2.5E

Figure It Out

Suppose that each of a fat molecule's three fatty acid chains contains 16 carbon atoms. How many acetyl CoA molecules can a cell generate from this fat molecule?

Answer: 24 (8 acetyl CoA per chain × 3 chains)

Fats contain more calories per gram than any other food molecule; after all, a single fat molecule may yield dozens of two-carbon acetyl CoA groups for the Krebs cycle. Conversely, the body can also store excess energy from either carbohydrates or fat by doing the reverse: diverting acetyl CoA away from the Krebs cycle and using the two-carbon fragments to build fat molecules. These lipids are stored in fat tissue that the body can use for energy if food becomes scarce.

6.7 MASTERING CONCEPTS

1. At which points do digested polysaccharides, proteins, and fats enter the energy pathways?
2. How does the body store extra calories as fat?

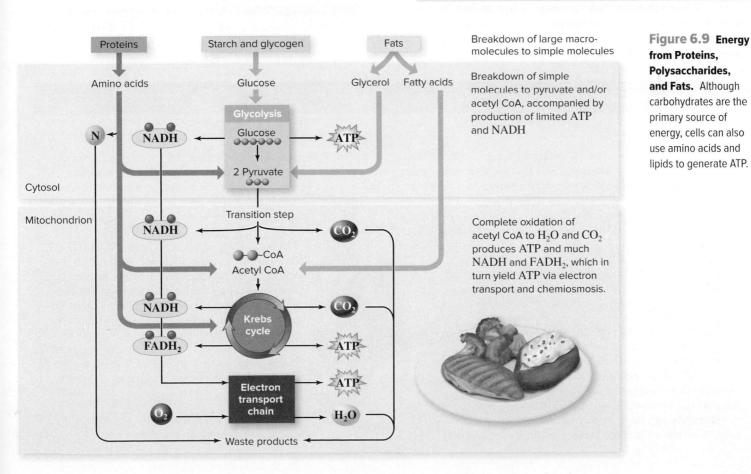

Breakdown of large macro-molecules to simple molecules

Breakdown of simple molecules to pyruvate and/or acetyl CoA, accompanied by production of limited ATP and NADH

Complete oxidation of acetyl CoA to H_2O and CO_2 produces ATP and much NADH and $FADH_2$, which in turn yield ATP via electron transport and chemiosmosis.

Figure 6.9 Energy from Proteins, Polysaccharides, and Fats. Although carbohydrates are the primary source of energy, cells can also use amino acids and lipids to generate ATP.

6.8 Some Energy Pathways Do Not Require Oxygen

Most of the known organisms on Earth, including humans, use aerobic cellular respiration. Nevertheless, life thrives without O_2 in waterlogged soils, deep puncture wounds, sewage treatment plants, and your own digestive tract, to name just a few places. In the absence of O_2, the microbes in these habitats generate ATP using anaerobic metabolic pathways. Two examples are anaerobic respiration and fermentation (figure 6.10).

A. Anaerobic Respiration Uses an Electron Acceptor Other Than O_2

Anaerobic respiration is essentially the same as aerobic respiration, except that an inorganic molecule other than O_2 is the electron acceptor at the end of the electron transport chain. Alternative electron acceptors include NO_3^- (nitrate), SO_4^{2-} (sulfate), and CO_2. The number of ATPs generated per molecule of glucose depends on the electron acceptor, but it is always lower than the ATP yield for aerobic respiration.

Many bacteria and archaea generate ATP by anaerobic respiration, and they play starring roles in nutrient cycles wherever O_2 is scarce. For example, in waterlogged, oxygen-poor soils, bacteria that use NO_3^- as an electron acceptor begin a chain reaction that

Figure 6.10 Alternative Metabolic Pathways. If O_2 is available, most organisms generate ATP in aerobic respiration. Two other pathways, anaerobic respiration and fermentation, can occur in the absence of O_2. Both alternatives yield fewer ATPs than does aerobic respiration.

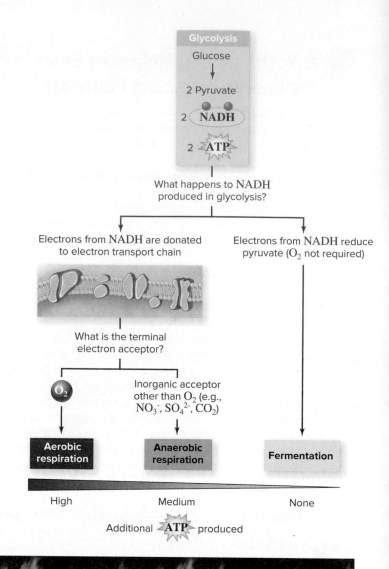

Burning Question

How do diet pills work?

Ads for diet pills are everywhere. Some are for weight-loss drugs that the U.S. Food and Drug Administration (FDA) has approved as safe and effective. Others are for dietary supplements that are not subject to FDA approval at all. These products, and their promises of effortless weight loss, may seem to be a dream come true. How do they work?

The FDA has approved three prescription weight-loss drugs. One is orlistat (Xenical); the over-the-counter drug Alli is a low-dose version of the same medicine. This drug interferes with lipase, the enzyme that digests fat in the small intestine. Undigested fat leaves the body in feces; orlistat therefore reduces calorie intake by reducing the body's absorption of high-energy fat molecules. The other two prescription weight-loss drugs are sibutramine (Meridia) and phentermine (Adipex-P). These medicines also reduce calorie intake, but in a different way: They suppress appetite.

All three prescription drugs can help a person lose weight but only if combined with exercise, a low-calorie diet, and behavior modification. Each also has side effects.

Dietary supplements greatly outnumber prescription weight-loss drugs. The FDA does not require the manufacturers of dietary supplements to show that the remedies are either safe or effective. Ads for "natural" supplements such as hoodia, green tea extract, and fucoxanthin make extraordinary promises of rapid weight loss, but the claims remain largely untested in scientific studies. The mechanism by which they work (if they work at all) usually remains unclear.

Unfortunately, some dietary supplements have serious side effects. Ephedra is one example. Before 2004, ephedra was marketed as a weight-loss aid and energy booster, but studies eventually linked it to fatal seizures, strokes, and heart attacks. The FDA therefore banned the sale of ephedra in the United States in 2004. An herb called bitter orange has taken its place in many "ephedra-free" weight-loss aids. But bitter orange has side effects that are similar to ephedra's, and its safety remains unknown.

Submit your burning question to
Marielle.Hoefnagels@mheducation.com

©Photodisc/Getty Images RF

ends with the production of nitrogen gas (N₂). This gas drifts into the atmosphere, leaving the soil less fertile for plant growth. Bacteria that live in wetlands may use SO_4^{2-}, producing smelly hydrogen sulfide (H_2S) as a byproduct. And archaea living inside the intestines of cattle use CO_2 as an electron acceptor, generating methane gas (CH_4). The methane, which the cattle emit as belches and flatulence, is one of the greenhouse gases implicated in global climate change. ⓘ *carbon cycle,* section 38.4B; *nitrogen cycle,* section 38.4C

B. Fermenters Acquire ATP Only from Glycolysis

Some microorganisms, including many inhabitants of your digestive tract, use fermentation. In these organisms, glycolysis still yields two ATPs, two NADHs, and two molecules of pyruvate per molecule of glucose. But the NADH does not donate its electrons to an electron transport chain, nor is the pyruvate further oxidized.

Instead, in **fermentation,** electrons from NADH reduce pyruvate. This process regenerates NAD⁺, which is essential for glycolysis to continue. But fermentation generates no additional ATP. This pathway is therefore far less efficient than respiration. Not surprisingly, fermentation is most common among microorganisms that live in sugar-rich environments where food is essentially unlimited.

Figure It Out

Compare the number of molecules of ATP generated from 100 glucose molecules undergoing aerobic respiration versus fermentation.

Answer: 3600 (theoretical yield) for aerobic respiration; 200 for fermentation

Some microorganisms make their entire living by fermentation. An example is *Entamoeba histolytica,* a protist that causes a form of dysentery in humans. Others, including the gut-dwelling bacterium *Escherichia coli,* use O₂ when it is available but switch to fermentation when it is not. Most multicellular organisms, however, require too much energy to rely on fermentation exclusively.

Of the many fermentation pathways that exist, one of the most familiar produces ethanol (an alcohol). In **alcoholic fermentation,** pyruvate is converted to ethanol and CO₂, while NADH is oxidized to produce NAD⁺ (**figure 6.11a**). Alcoholic fermentation produces wine from grapes, beer from barley, and cider from apples.

In **lactic acid fermentation,** a cell uses NADH to reduce pyruvate, but in this case, the products are NAD⁺ and lactic acid or its close relative, lactate (figure 6.11b). The bacterium *Lactobacillus,* for example, ferments the lactose in milk, producing the lactic acid that gives yogurt its sour taste. Bacteria can also ferment sugars in cabbage to produce the acids in sauerkraut.

Fermentation also occurs in human muscle cells. During vigorous exercise, muscles work so strenuously that they consume their available oxygen supply. In this "oxygen debt" condition, the muscle cells can acquire ATP only from glycolysis. The cells use lactic acid fermentation to generate NAD⁺ so that glycolysis can continue. Lactate concentrations therefore rise. After the workout, when the circulatory system catches up with the muscles' demand for O₂, liver cells convert lactate back to pyruvate. Mitochondria then process the pyruvate as usual. ⓘ *fermentation in muscle cells,* section 29.5

One common misconception about intense exercise is that lactic acid buildup causes a pH drop in muscle cells, provoking soreness a day or two later. Two lines of evidence, however, suggest that this idea is a myth. First, muscle cells produce lactate (not lactic acid); lactate does not change the cytoplasm's pH. Second, cells consume the lactate shortly after the workout ends, so it is unlikely to cause pain days later. Microscopic tears in muscle tissue are now thought to be the culprit responsible for delayed muscle soreness.

6.8 MASTERING CONCEPTS

1. What are some examples of alternative electron acceptors used in anaerobic respiration?
2. How many ATPs per glucose does fermentation produce?
3. What are two examples of fermentation pathways?

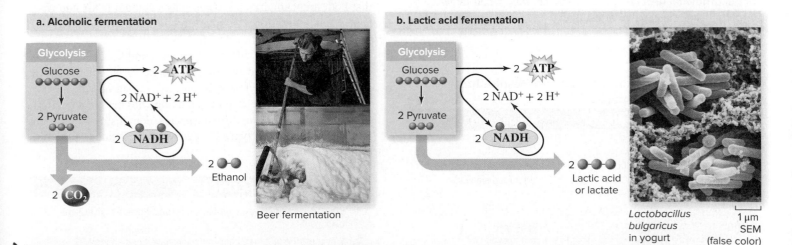

Figure 6.11 Fermentation. In fermentation, ATP comes only from glycolysis. (a) Yeasts produce ethanol and carbon dioxide by alcoholic fermentation. The man in the photograph is stirring a large vat of fermenting beer. (b) Lactic acid fermentation occurs in some bacteria and, occasionally, in mammalian muscle cells. The photograph shows *Lactobacillus* bacteria in yogurt. Photos: (a): ©Adam Woolfitt/Corbis; (b): ©Scimat/Science Source

6.9 Photosynthesis and Respiration Are Ancient Pathways

As you may have noticed, photosynthesis, glycolysis, and cellular respiration are intimately related (table 6.1 and figure 6.12). The carbohydrate product of photosynthesis—glucose—is the starting material for glycolysis. The O_2 released in photosynthesis becomes the final electron acceptor in aerobic respiration. CO_2 generated in respiration enters the carbon reactions in chloroplasts. Finally, photosynthesis splits water produced by aerobic respiration. Together, these energy reactions sustain life. How might they have arisen?

Glycolysis is probably the most ancient of the energy pathways because it occurs in virtually all cells. Glycolysis evolved when the atmosphere lacked or had very little O_2. These reactions enabled the earliest organisms to extract energy from simple organic compounds in the nonliving environment. Photosynthesis, in turn, may have evolved from glycolysis; some of the reactions of the Calvin cycle are the reverse of some of those of glycolysis. ⓘ Calvin cycle, section 5.5

The first photosynthetic organisms could not have been plants, because such complex organisms were not present on the early Earth. Rather, photosynthesis may have originated in an anaerobic cell that used hydrogen sulfide (H_2S) instead of water as an electron donor. These first photosynthetic microorganisms would have released sulfur, rather than O_2, into the environment. Eventually, changes in pigment molecules enabled some of these organisms to use water instead of H_2S as an electron source. Fossil evidence of cyanobacteria shows that oxygen-generating photosynthesis arose at least 3.5 billion years ago. Once this pathway started, the accumulation of O_2 in the primitive atmosphere made aerobic respiration possible and altered life on Earth forever (see section 5.1).

Later, in a process called endosymbiosis, a large "host" cell engulfed one of those ancient cyanobacteria and thereby transformed itself into a eukaryotic-like cell, complete with chloroplasts. Mitochondria evolved in a similar way, when larger cells engulfed bacteria capable of using O_2.

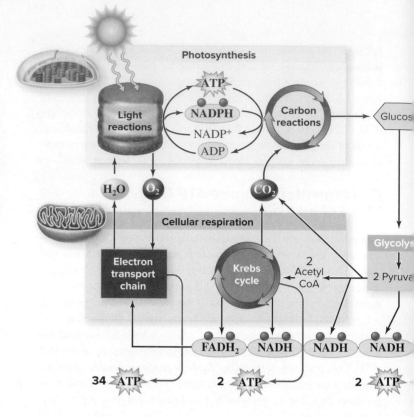

Figure 6.12 Connections Between Photosynthesis and Respiration. An overview of metabolism illustrates how biological energy reactions are interrelated.

The double-membrane structure of both chloroplasts and mitochondria is a consequence of endosymbiosis. The engulfed bacterium's cell membrane developed into each organelle's inner membrane, and the vesicle membrane remained as the outer membrane. Endosymbiosis therefore explains why the electron transport chain is in the bacterial cell membrane but in the inner mitochondrial membrane of a eukaryotic cell. The observation that both mitochondria and chloroplasts contain DNA and ribosomes lends additional support to the endosymbiosis theory. ⓘ endosymbiosis, section 15.2A

As time went on, different types of complex cells probably diverged, leading to the evolution of a great variety of eukaryotic organisms. Today, the interrelationships among photosynthesis, glycolysis, and aerobic respiration, along with the great similarities of these reactions in diverse species, demonstrate a unifying theme of biology: All types of organisms are related at the biochemical level.

6.9 MASTERING CONCEPTS

1. Which energy pathway is probably the most ancient? What is the evidence?
2. Why must the first metabolic pathways have been anaerobic?
3. What is the evidence that photosynthesis may have evolved from glycolysis?

TABLE **6.1** **Photosynthesis and Respiration Compared**

	Photosynthesis	Respiration
Food	Produced	Consumed
Energy	Stored as glucose and other sugars	Released from glucose and other food molecules
Light	Required	Not required
H_2O	Consumed	Released
CO_2	Consumed	Released
O_2	Released	Consumed

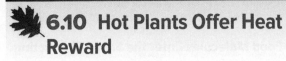

6.10 Hot Plants Offer Heat Reward

Think of an organism that feels warm. Did you think of yourself? A puppy? Your cat? Chances are you thought of a mammal or perhaps a bird, but certainly not a plant. Yet some plants, including *Philodendron,* do warm themselves (or at least their reproductive parts) to several degrees above ambient temperature (figure 6.13). How do they do it and, more important, what do they get out of it?

Philodendron flowers generate heat with a metabolic pathway involving the electron transport chain. As described in section 6.5, electrons from NADH and $FADH_2$ pass along a series of proteins embedded in the inner mitochondrial membrane. Along the way, the proteins pump H^+ into the space between the two mitochondrial membranes; ATP synthase uses the resulting proton gradient to generate ATP. The last protein in the electron transport chain dumps the electrons on O_2, yielding water as a waste product.

Plants and a few other types of organisms have another pathway, dubbed "alternative oxidase," that diverts electrons from the electron transport chain. NADH and $FADH_2$ still donate electrons to a protein in the chain, but the electrons next pass immediately to O_2 instead of traveling along the rest of the chain. The pathway therefore does not help the mitochondrion generate ATP. It does, however, generate more heat than conventional respiration—after all, potential energy that is not harvested as ATP must be released as heat.

So what does *Philodendron* gain by warming its flowers? One clue comes from the observation that the plant heats *just* its flowers and not its leaves, stem, or roots. Since flowers are reproductive parts, could the hot blooms somehow improve the plant's reproductive success?

In many plants, reproduction depends on animals that carry pollen from flower to flower. The plants may give away free meals of sweet nectar that lure pollinators such as insects, birds, and mammals. As the animal collects the offering, it brushes against the pollen-producing (male) flower parts. It then deposits the pollen on the female part of the next flower it visits.

Australian researcher Roger Seymour and his colleagues wondered whether heat from the flowers of *Philodendron solimoesense* helps the plant attract pollinators. They did a simple set of experiments to find out. First, they measured the temperature of *Philodendron* flowers. The central spike peaked at 40°C, about 15° above ambient

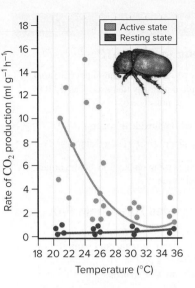

Figure 6.14 Energy Saver. Resting beetles respired at the same rate no matter what the temperature, but active beetles saved energy in warmer surroundings.

temperature, while the floral chamber was consistently a few degrees warmer than the surrounding air.

Next, the researchers turned their attention to beetles known to pollinate the flowers. The team used a device called a respirometer to measure the amount of CO_2 produced by active and resting beetles at a range of temperatures from 20°C to 35°C. Since respiration generates CO_2 as a waste product, a respirometer indirectly measures how much energy an organism uses. Resting beetles emitted approximately the same amount of CO_2 at all temperatures, but active ones (such as those that would visit flowers) produced only about one tenth as much CO_2 at 30°C as they did at 20°C (figure 6.14). This observation made sense, since warm flight muscles use energy more efficiently than cold ones.

Finally, the researchers used their data to calculate the "energy-saving factor" attributed to floral heat. They concluded that the beetles used 2.0 to 4.8 times more energy at ambient temperature than at the temperature of the warmed flower, depending on time of night. The beetles therefore save energy simply by loitering on or near the flowers, energy that they can use to find food or lure mates even as they pollinate the plant. The hot flowers—courtesy of the seemingly wasteful alternative oxidase pathway—therefore enhance the reproductive success of both *Philodendron* and the beetles.

Source: Seymour, Roger S., Craig R. White, and Marc Gibernau. November 20, 2003. Heat reward for insect pollinators. *Nature,* vol. 426, pages 243–244.

Figure 6.13 Hot Bloom. The central spike of this *Philodendron solimoesense* flower generates heat.
(both): ©Marc Gibernau, CNRS, Toulouse, France

6.10 MASTERING CONCEPTS

1. What hypothesis were the researchers testing, and what experiments did they design to help them test the hypothesis?

2. Suppose you hold one group of active beetles at 20°C and another group at 30°C. After several hours, you place each beetle in a device that measures how far the animal can fly at 20°C. Which group of beetles do you predict will fly farther?

CHAPTER SUMMARY

6.1 Cells Use Energy in Food to Make ATP

- Every cell requires **ATP** to power reactions that require energy input.
- **Aerobic respiration** is a biochemical pathway that produces ATP by extracting energy from glucose in the presence of oxygen:

$$glucose + oxygen \longrightarrow carbon\ dioxide + water + ATP$$

$$C_6H_{12}O_6 + 6O_2 \longrightarrow 6CO_2 + 6H_2O + 36ATP$$

- The cells of plants and animals use aerobic respiration. Some organisms use anaerobic respiration or fermentation.

6.2 Cellular Respiration Includes Three Main Processes

- **Glycolysis** is the first step in harvesting energy from glucose. In respiration, the **Krebs cycle** and an **electron transport chain** follow.
- The electron transport chain establishes a proton (H^+) gradient that powers ATP production by the enzyme **ATP synthase.**

6.3 In Eukaryotic Cells, Mitochondria Produce Most ATP

- The Krebs cycle and electron transport chain occur in the **mitochondria.**
- Each mitochondrion has two membranes enclosing a central **matrix.**
- Electron transport chain proteins are embedded in **cristae:** folds of the inner membrane. These proteins pump protons into the **intermembrane compartment.** ATP synthase also spans the inner membrane.

6.4 Glycolysis Breaks Down Glucose to Pyruvate

- In glycolysis, glucose is split into two molecules of **pyruvate.**
- During these reactions, electrons are added to NAD^+, forming **NADH.** Two ATPs are formed by **substrate-level phosphorylation** (figure 6.15).

6.5 Aerobic Respiration Yields Abundant ATP

A. Pyruvate Is Oxidized to Acetyl CoA
- Pyruvate moves into the mitochondrial matrix, where it is broken down into **acetyl CoA** and CO_2. This "transition step" also produces NADH.

B. The Krebs Cycle Produces ATP and Electron Carriers
- Acetyl CoA enters the Krebs cycle. This series of oxidation–reduction reactions occurs in the matrix and produces ATP, NADH, **FADH₂,** and CO_2. Substrate-level phosphorylation produces ATP in the Krebs cycle.

C. The Electron Transport Chain Drives ATP Formation
- Energy-rich electrons from NADH and FADH₂ fuel the electron transport chain. A series of proteins shuttle electrons and release their energy. O_2 accepts the electrons at the end of the chain, producing water.
- Proteins in the electron transport chain pump H^+ from the matrix into the intermembrane compartment. As protons diffuse back into the matrix through ATP synthase, their potential energy drives **chemiosmotic phosphorylation** of ADP to ATP.

6.6 How Many ATPs Can One Glucose Molecule Yield?

- In aerobic respiration, each glucose molecule theoretically yields 36 ATP molecules (figure 6.16). The actual yield is about 30 ATPs per glucose.

6.7 Other Food Molecules Enter the Energy-Extracting Pathways

- Polysaccharides are digested to glucose before undergoing cellular respiration. Amino acids enter the energy pathways as pyruvate, acetyl CoA, or an intermediate of the Krebs cycle. Fatty acids enter as acetyl CoA, and glycerol enters as pyruvate.

6.8 Some Energy Pathways Do Not Require Oxygen

A. Anaerobic Respiration Uses an Electron Acceptor Other Than O_2
- Nitrate or sulfate is the electron acceptor in **anaerobic respiration.**

B. Fermenters Acquire ATP Only from Glycolysis
- **Fermentation** pathways oxidize NADH to NAD^+, which is recycled to glycolysis. **Alcoholic fermentation** converts pyruvate to ethanol and carbon dioxide. **Lactic acid fermentation** reduces pyruvate to lactic acid or lactate.

6.9 Photosynthesis and Respiration Are Ancient Pathways

- Photosynthesis and respiration are interrelated, with common intermediates and some reactions that mirror those of other pathways.
- Eukaryotes may have arisen by endosymbiosis, in which cells engulfed bacteria that were forerunners to mitochondria and chloroplasts.

6.10 Investigating Life: Hot Plants Offer Heat Reward

- *Philodendron* plants use a modified respiratory pathway, creating a "heat reward" for their insect pollinators.

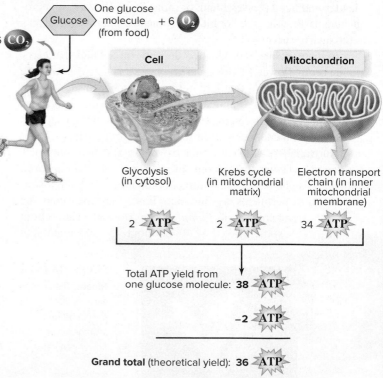

Figure 6.16 Theoretical ATP Yield for One Glucose Molecule.

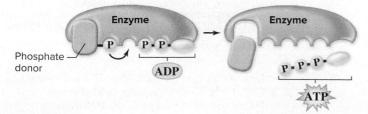

Figure 6.15 Substrate-Level Phosphorylation.

MULTIPLE CHOICE QUESTIONS

1. Which of the following best describes aerobic respiration?
   a. The production of ATP from glucose in the presence of oxygen
   b. The production of pyruvate in the absence of oxygen
   c. The production of pyruvate using energy from the sun
   d. The production of ATP from glucose in the absence of oxygen

2. Which stage in cellular respiration produces the most ATP?
   a. Glycolysis - 2
   b. Pyruvate oxidation
   c. Krebs cycle - 2
   d. Electron transport -32

3. What is the role of ATP synthase?
   a. It uses ATP to make glucose.
   b. It uses a proton gradient to make ATP.
   c. It uses ATP to make a proton gradient.
   d. It synthesizes ATP directly from glucose.

4. In a prokaryotic cell, glycolysis occurs in the _____; in a eukaryotic cell, glycolysis occurs in the _____.
   a. cytosol; cytosol
   b. cytosol; mitochondrial matrix
   c. cell membrane; cytosol
   d. cell membrane; mitochondrial matrix

5. If a substance causes holes to form in the inner mitochondrial membrane, which process would be affected first?
   a. The donation of electrons to O_2
   b. Glycolysis
   c. The production of ATP by ATP synthase
   d. The formation of acetyl CoA

6. Fats can be broken down into acetyl CoA for use in the Krebs cycle. Fats can also
   a. be built from excess acetyl CoA for energy storage.
   b. function as an electron carrier in the electron transport chain.
   c. be broken down directly into ATP.
   d. be broken down directly into NADH.

7. Why is it important to regenerate NAD^+ during fermentation?
   a. It helps maintain the reactions of glycolysis.
   b. So that it can transfer an electron to the electron transport chain
   c. To maintain the concentration of pyruvate in a cell
   d. To produce alcohol or lactic acid for the cell

8. What is endosymbiosis?
   a. A type of fermentation
   b. The transport of pyruvate into the matrix of the mitochondria
   c. An explanation for the origin of mitochondria
   d. The movement of electrons along the electron transport chain

Answers to these questions are in appendix A.

WRITE IT OUT

1. *Respiration* contains the Latin word root *spiro*, which means "to breathe." Why is the process described in this chapter called cellular respiration? What might your answer indicate about what scientists already knew when they first observed cellular respiration?

2. All steps of cellular respiration are closely connected. Describe the problems that would occur if glycolysis, the Krebs cycle, or the electron transport chain were not working.

3. How does aerobic respiration yield so much ATP from each glucose molecule, compared with glycolysis alone?

4. How might a mitochondrion's double membrane make cellular respiration more efficient than if it had a single membrane?

5. Health-food stores sell a product called "pyruvate plus," which supposedly boosts energy. Why is this product unnecessary? What would be a much less expensive substitute that would accomplish the same thing?

6. At what point does O_2 enter the energy pathways of aerobic respiration? What is the role of O_2? Why does respiration stop if a person cannot breathe?

7. Why would a cell die if it could not make ATP?

8. Describe the energy pathways that are available for cells living in the absence of O_2.

9. Some types of beer are bottled with yeast. These beers are not carbonated at bottling, but if you open them a few weeks later they will bubble. Explain the source of this carbonation.

10. Describe how aerobic respiration occurs in bacteria. How does this relate to how aerobic respiration occurs in mitochondria? Explain the relationship between bacteria and mitochondria.

11. Explain the fact that species as diverse as humans and yeasts use the same biochemical pathways to extract energy from nutrient molecules.

12. Compare the number of ATP molecules required to produce one glucose molecule in photosynthesis (see figure 5.9) with the number of ATP molecules generated per glucose in aerobic respiration (see figure 6.8). How do these numbers compare to the ATP yield from fermentation?

PULL IT TOGETHER

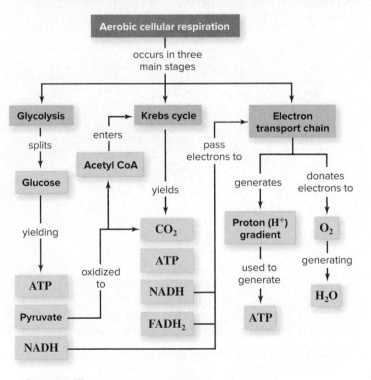

Figure 6.17 Pull It Together: Respiration and Fermentation.

Refer to figure 6.17 and the chapter content to answer the following questions.

1. Review the Survey the Landscape figure in the chapter introduction. Explain the connection between respiration and photosynthesis.

2. How many ATP, NADH, CO_2, $FADH_2$, and H_2O molecules are produced at each stage of respiration?

3. What do cells do with the ATP they generate in respiration?

4. Where would *fermentation, anaerobic respiration,* and *ATP synthase* fit into this concept map?

7 DNA Structure and Gene Function

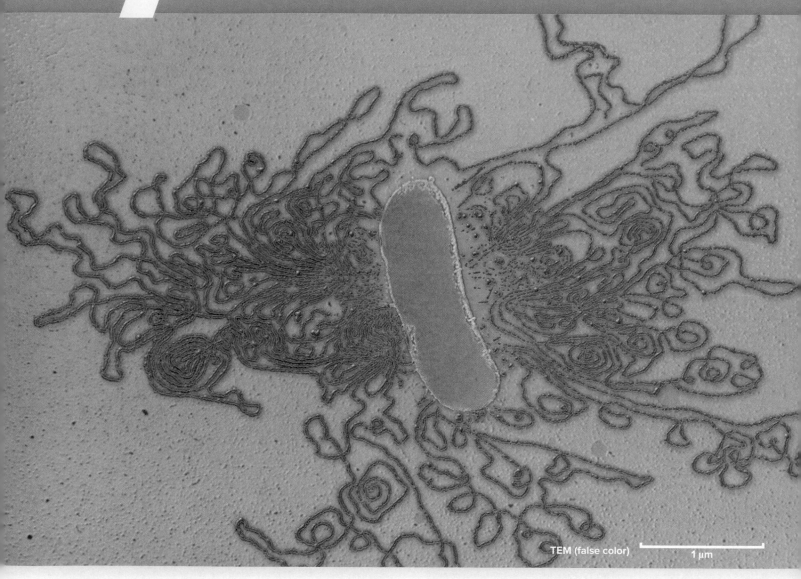

TEM (false color)

1 μm

Lots of DNA. Genetic material bursts from this bacterium, illustrating just how much DNA is packed into a single cell.

©Dr. Gopal Murti/Science Source

LEARN HOW TO LEARN
Pause at the Checkpoints

As you read, get out a piece of paper and see if you can answer the Figure It Out and Mastering Concepts questions. If not, you may want to study a bit more before you move on. Each section builds on the material that came before, and mastering one chunk at a time will make it much easier to learn whatever comes next.

UNIT 2

Our DNA Sequence Is Just the Beginning

All life requires DNA, the stringy substance spilling out of the bacterial cell shown here. Your own cells contain DNA, as do the cells of fruit flies, polar bears, and all other animals. Protists, fungi, and plants have DNA, too.

We know much more about DNA now than we did a generation ago. The technology needed to sequence DNA was in its infancy in the 1970s, but by the 1990s it had ignited a scientific revolution. Biologists began cranking out the first of hundreds of complete DNA sequences representing viruses and organisms from all three of life's domains. The 1990s also saw the dawn of the Human Genome Project. ⓘ *DNA sequencing,* section 11.2B

In a way, the Human Genome Project is old news. The completion of the DNA sequence in 2003, and the subsequent fanfare, might have led an outsider to conclude that science had learned all there was to know about human DNA. News stories suggested that parents would soon be able to screen their unborn children for every trait imaginable. Talk soon turned to "designer babies," whose genes would be artificially altered to boost health, attractiveness, intelligence, athletic ability, and other desirable characteristics.

The truth is that we still have a long way to go before we understand what all of that DNA actually does in our cells. Of course, we understand DNA's overall function: A gene's nucleotide sequence encodes a protein. But just knowing a gene's sequence does not provide instant insight into everything needed to make a human. By itself, a DNA sequence does not explain how the cell turns each gene on and off, the function of the protein, or what happens if the gene mutates. Nor does it explain the function of the huge swaths of DNA that do not code for protein.

We do know, however, that only a 0.1% difference separates any two individuals. Investigating these differences will likely answer such questions as why some people get cancer and others do not or why a medication helps some people but harms others. Even within an individual, cells express different combinations of genes. Understanding how the proteins produced in breast cancer cells differ from those in normal cells, for example, may reveal new targets for anticancer drugs (see the chapter 3 opening essay).

We begin this genetics unit with a look at the intimate relationship between DNA and proteins. Subsequent chapters describe how cells copy DNA just before they divide, how cell division leads to the fascinating study of inheritance, and how researchers find practical applications for knowledge about DNA.

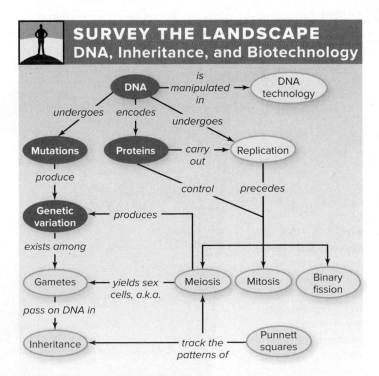

SURVEY THE LANDSCAPE
DNA, Inheritance, and Biotechnology

DNA is an information storage molecule; its main function is to carry the "recipes" for the proteins that carry out the cell's work. Mutations in DNA ultimately generate all genetic variation.

For more details, study the Pull It Together feature in the chapter summary.

7.1 Experiments Identified the Genetic Material

The nucleic acid DNA is one of the most familiar molecules, the subject matter of movies and headlines (figure 7.1). Criminal trials hinge on DNA evidence; the idea of human cloning raises questions about the role of DNA in determining who we are; and DNA-based discoveries are yielding new diagnostic tests, medical treatments, and vaccines.

More important than DNA's role in society is its role in life itself. DNA is a molecule with a remarkable function: It stores the information that each cell needs to produce proteins. These instructions make life possible. In fact, before a cell divides, it first makes an exact replica of its DNA. This process, described in chapter 8, copies all of the information that will enable the next generation of cells to live.

The discovery of DNA's role in life required many decades of research. By the early 1900s, biologists had recognized the connection between inheritance and protein. For example, English physician Archibald Garrod noted that people with inherited "inborn errors of metabolism" lacked certain enzymes. Other researchers linked abnormal or missing enzymes to unusual eye color in fruit flies and nutritional deficiencies in bread mold. But how were enzyme deficiencies and inheritance linked? Experiments in bacteria would answer the question.

A. Bacteria Can Transfer Genetic Information

In 1928, English microbiologist Frederick Griffith contributed the first step in identifying DNA as the genetic material (figure 7.2). Griffith studied two strains of a bacterium, *Streptococcus*

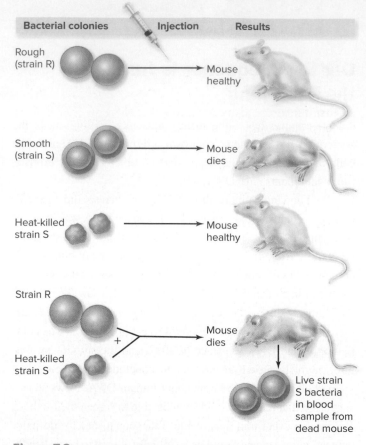

Figure 7.2 A Tale of Two Microbes. Griffith's experiments showed that a molecule in a lethal strain of bacteria (type S) could transform harmless type R bacteria into killers. Additional experiments showed that the molecule was DNA.

pneumoniae. Type R bacteria, named for their "rough" colonies, do not cause pneumonia when injected into mice. Type S ("smooth") bacteria, on the other hand, cause pneumonia. The polysaccharide capsule that encases type S bacteria is apparently necessary for infection (and makes their colonies appear smooth).

Griffith found that heat-killed type S bacteria did not cause pneumonia in mice. However, when he injected mice with a mixture of live type R bacteria plus heat-killed type S bacteria, neither of which could cause pneumonia alone, the mice died. Moreover, their bodies contained live type S bacteria encased in polysaccharide. Something in the heat-killed type S bacteria transformed the normally harmless type R strain into a killer.

How had the previously harmless bacteria acquired the ability to cause disease? In the 1940s, U.S. physicians Oswald Avery, Colin MacLeod, and Maclyn McCarty finally learned the identity of the "transforming principle." The researchers treated heat-killed type S bacteria with a protein-destroying enzyme and mixed them with live type R bacteria. The type R strain still changed into a killer; therefore, a protein was not transmitting the killing trait. But when they used a DNA-destroying enzyme instead, the type R bacteria remained harmless. The conclusion: DNA from type S cells altered the type R bacteria, enabling them to manufacture the smooth coat necessary to cause infection.

Figure 7.1 DNA—the Molecule in the Media. *Jurassic World* is a movie in which fictional scientists use DNA technology to produce genetically engineered dinosaurs. The dinosaur DNA came from blood found in ancient insects entombed in amber.
(poster): ©Universal Pictures/Photofest; (amber): ©Natural Visions/Alamy

B. Hershey and Chase Confirmed the Genetic Role of DNA

At first, biologists hesitated to accept DNA as the molecule of heredity. They knew more about proteins than about nucleic acids. They also thought that protein, with its 20 building blocks, could encode many more traits than DNA, which includes just four types of building blocks. In 1950, however, U.S. microbiologists Alfred Hershey and Martha Chase conclusively showed that DNA—not protein—is the genetic material.

Hershey and Chase used a very simple system. They infected the bacterium *Escherichia coli* with a bacteriophage, which is a virus that infects only bacteria. The virus consisted of a protein coat surrounding a DNA core. We now know that when the virus infects a bacterial cell, it injects its DNA, but the protein coat remains loosely attached to the outside of the bacterium. The viral DNA directs the bacterium to use its own energy and raw materials to manufacture more virus particles, which then burst from the cell. But much of this information was not available in 1950. In fact, Hershey and Chase wanted

to know which part of the virus controls its replication: the DNA or the protein coat. ⓘ *bacteriophages,* section 16.1A; *E. coli,* section 17.3

To answer the question, the researchers "labeled" two batches of viruses, one with radioactive sulfur that marked protein and the other with radioactive phosphorus that marked DNA. They used each type of labeled virus to infect a separate batch of bacteria (figure 7.3). Then they agitated each mixture in a blender, which removed the unattached viruses and empty protein coats from the surfaces of the bacteria. They poured the mixtures into test tubes and spun them at high speed. The infected bacteria settled to the bottom of each test tube because they were heavier than the liberated viral protein coats.

Hershey and Chase examined the bacteria and the fluid in each tube. In the test tube containing sulfur-labeled viral proteins, the bacteria were not radioactive, but the fluid portion of the material in the tube was. In the other tube, where the virus contained DNA marked with radioactive phosphorus, the infected bacteria were radioactive, but the fluid was not.

The "blender experiments" therefore showed that the part of the virus that could enter the bacteria and direct them to mass-produce viruses was the part with the phosphorus label—namely, the DNA. The genetic material, therefore, was DNA and not protein.

Figure 7.3 DNA's Role Confirmed. Hershey and Chase used radioactive isotopes to distinguish a bacteriophage's protein coat from its DNA. They showed that the virus transfers DNA (not protein) to the bacterium, and this viral DNA causes bacterial cells to produce viruses. The photo shows several bacteriophages.

Photo: ©Oliver Meckes/MPI—Tubingen/ Science Source

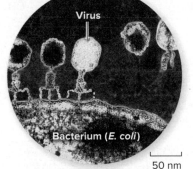

Virus

Bacterium (*E. coli*)

50 nm

TEM (false color)

7.1 MASTERING CONCEPTS

1. How did Griffith's research, coupled with the work of Avery and his colleagues, demonstrate that DNA, not protein, is the genetic material?

2. How did the Hershey–Chase "blender experiments" confirm Griffith's results?

Viral protein coat radioactively labeled (sulfur)

Protein coat

DNA

Virus

Bacterium

Virus

Viruses infect bacteria

Blended and spun at high speeds to separate bacteria from viral protein coats

Radioactive viral protein coats

Nonradioactive bacteria with viral DNA

Viral DNA radioactively labeled (phosphorus)

Protein coat

DNA

Virus

Bacterium

Virus

Viruses infect bacteria

Blended and spun at high speeds to separate bacteria from viral protein coats

Nonradioactive viral protein coats

Radioactive bacteria with viral DNA

7.2 DNA Is a Double Helix of Nucleotides

The early twentieth century also saw advances in the study of the structure of DNA. By 1929, biochemists had discovered the distinction between **RNA (ribonucleic acid)** and **DNA (deoxyribonucleic acid),** the two types of nucleic acid. They also had determined that nucleotides are the building blocks of nucleic acids. Finally, researchers knew that each nucleotide includes a sugar (ribose for RNA, deoxyribose for DNA), one of several nitrogen-containing bases, and one or more phosphorus-containing groups. But how were those nucleotides arranged?

In the early 1950s, biochemists raced to discover DNA's chemical structure. Two lines of evidence were considered critical. Austrian American biochemist Erwin Chargaff showed that the amount of the base guanine (G) in a DNA molecule always equals the amount of cytosine (C), and the amount of adenine (A) always equals the amount of thymine (T). English physicist Maurice Wilkins and chemist Rosalind Franklin bombarded DNA with X-rays, using a technique called X-ray diffraction to determine the three-dimensional shape of the molecule. The X-ray diffraction pattern revealed a regularly repeating structure of building blocks (figure 7.4a, b).

Early in 1953, chemist Linus Pauling proposed a triple-helix model of DNA; he was soon proved incorrect. But by April of 1953, U.S. biochemist James Watson and English physicist Francis Crick, working at the Cavendish laboratory in Cambridge in the United Kingdom, had solved the mystery. They used the "Chargaff rule" and Franklin's X-ray diffraction pattern to build a ball-and-stick model of the now-familiar DNA double helix. Watson, Crick, and Wilkins won the 1962 Nobel Prize in physiology or medicine for their discovery (figure 7.4c).

a. Rosalind Franklin

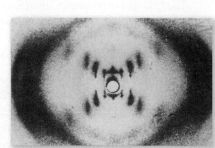

b. X–ray diffraction

c.

Figure 7.4 Discovery of DNA's Structure. (a) Rosalind Franklin produced (b) X-ray images of DNA that were crucial in the discovery of DNA's structure. (c) Maurice Wilkins, Francis Crick, and James Watson (*first, third, and fifth from the left*) shared the 1962 Nobel Prize in physiology or medicine for their discovery. Franklin had died in 1958, and by the rules of the award, she could not be included. (The other three men in the photo won Nobel Prizes in other disciplines.)

(a, b): ©Science Source; (c): ©Bettmann/Corbis

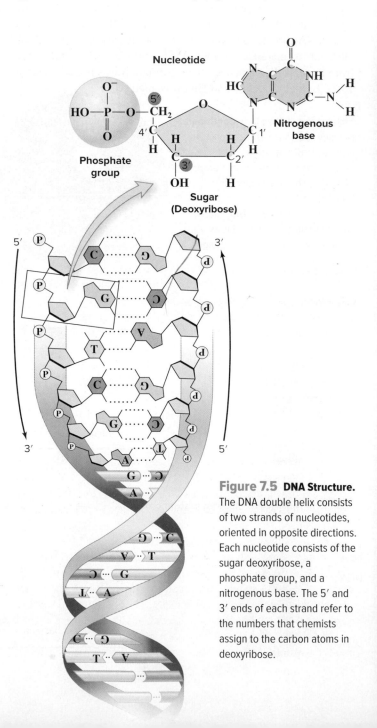

Figure 7.5 DNA Structure. The DNA double helix consists of two strands of nucleotides, oriented in opposite directions. Each nucleotide consists of the sugar deoxyribose, a phosphate group, and a nitrogenous base. The 5′ and 3′ ends of each strand refer to the numbers that chemists assign to the carbon atoms in deoxyribose.

Figure 7.6 Complementary Base Pairing.
(a) Adenine and guanine are purines; cytosine and thymine are pyrimidines. (b) Purines pair with complementary pyrimidines: cytosine with guanine, and adenine with thymine. Dotted lines represent hydrogen bonds; all other bonds are covalent.

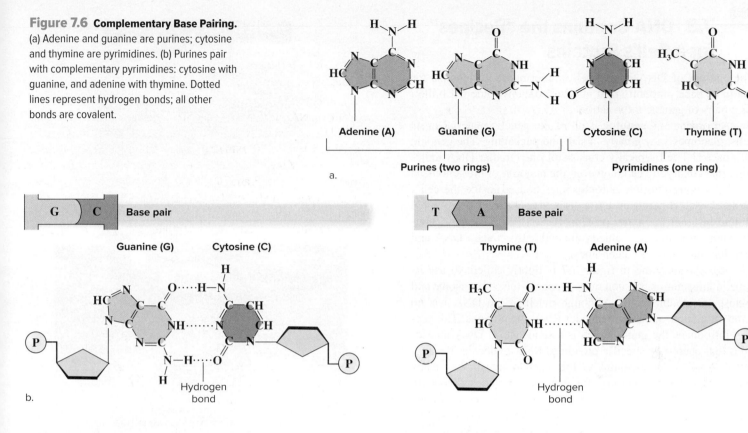

The DNA double helix resembles a twisted ladder (**figure 7.5**). The twin rails of the ladder, also called the sugar–phosphate "backbones," are alternating units of deoxyribose and phosphate joined with covalent bonds. ⓘ *covalent bonds,* section 2.2C

Although the two chains of the DNA double helix are parallel to each other, they are oriented in opposite directions, like the northbound and southbound lanes of a highway. This head-to-tail ("antiparallel") arrangement is apparent when the carbon atoms in deoxyribose are numbered. When the nucleotides are joined into a chain, opposite ends of the strand are designated **3 prime** (**3′**) and **5 prime** (**5′**). At the same end of the double helix, one chain therefore ends with a free (unbound) 3′ carbon, while the other chain ends with a free 5′ carbon.

The ladder's rungs are A–T and G–C base pairs joined by hydrogen bonds. These base pairs arise from the chemical structures of the nucleotides (**figure 7.6**). Adenine and guanine are purines, bases with a double ring structure. Cytosine and thymine are pyrimidines, which have a single ring. Each A–T pair is the same width as a C–G pair because each includes a purine and a pyrimidine.

The two strands of a DNA molecule are **complementary** to each other; that is, the sequence of each strand determines the sequence of the other. An A on one strand means a T on the opposite strand, and a G on one strand means a C on the other. The two strands are therefore somewhat like a photograph and its negative, since each is sufficient to define the other.

Why does A pair with T but not with C, even though both T and C have similar shapes? The explanation relates to the positions

Figure It Out

Write the complementary DNA sequence of the following:
  3′-ATCGGATCGCTACTG-5′

Answer: 5′-TAGCCTAGCGATGAC-3′

of the atoms in each nucleotide. Recall from section 2.2 that hydrogen bonds form between atoms carrying opposite partial charges (in this case, between H atoms with partial positive charges and oxygen or nitrogen atoms with partial negative charges). As you examine figure 7.6, note that cytosine and guanine can form three hydrogen bonds. On the other hand, adenine and thymine can form only two. This difference accounts for the specificity of the A–T and C–G base pairs. ⓘ *hydrogen bonds,* section 2.2D

You may also remember from section 2.2 that hydrogen bonds are weak compared to covalent bonds. While it is true that each hydrogen bond is weak, a DNA molecule consisting of millions of base pairs also has millions of hydrogen bonds. Collectively, these bonds are strong enough to hold the two strands together yet weak enough to pull apart when the cell needs to use its DNA.

7.2 MASTERING CONCEPTS

1. What are the components of DNA and its three-dimensional structure?
2. What evidence enabled Watson and Crick to decipher the structure of DNA?
3. Describe the 3′ and 5′ ends of a DNA strand.

7.3 DNA Contains the "Recipes" for a Cell's Proteins

The amount of DNA in any cell is enormous. In humans, for example, each pinpoint-sized nucleus contains some 6.4 billion base pairs of genetic information.

An organism's **genome** is all of the genetic material in its cells. Genomes vary greatly in size and packaging. The genome of a bacterial cell typically consists of one circular DNA molecule. In a eukaryotic cell, however, the majority of the genome is divided among multiple chromosomes housed inside the cell's nucleus; each **chromosome** is a discrete package of DNA coiled around histones and other proteins (**figure 7.7**). The mitochondria and chloroplasts of eukaryotic cells also contain DNA and therefore have their own genomes.

The chromosome in figure 7.7 is tightly coiled; to use its genetic information, the cell must "unpack" the chromosome and expose the double helix. Although much of the DNA has no known function, some of it encodes RNA and proteins. This section introduces the **gene**, which is a sequence of DNA nucleotides that encodes a specific protein or RNA molecule. Because many proteins are essential to life, each organism has many genes. The human genome, for example, includes 20,000 to 25,000 genes scattered on its 23 pairs of chromosomes. Likewise, a bacterial chromosome is also divided into multiple genes.

A. Protein Synthesis Requires Transcription and Translation

In the 1940s, biologists working with the fungus *Neurospora crassa* deduced that each gene somehow controls the production of one protein. In the following decade, Watson and Crick described this relationship between nucleic acids and proteins as a flow of information they called the "central dogma." ⓘ *Neurospora*, section 20.5

Figure 7.8 summarizes the process of protein production. First, in **transcription**, a cell "rewrites" a gene's DNA sequence to a complementary RNA molecule. Then, in **translation**, the information in RNA is used to assemble a different class of molecule: a protein (just as an interpreter translates one language into another).

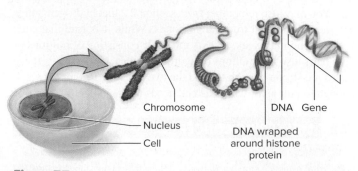

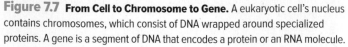

Figure 7.7 **From Cell to Chromosome to Gene.** A eukaryotic cell's nucleus contains chromosomes, which consist of DNA wrapped around specialized proteins. A gene is a segment of DNA that encodes a protein or an RNA molecule.

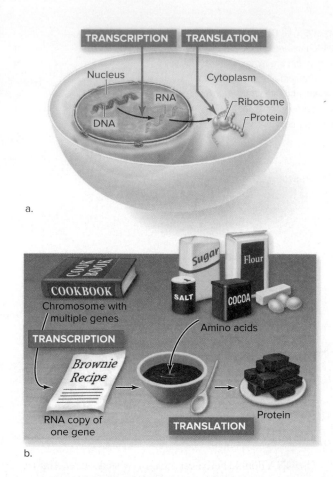

a.

b.

Figure 7.8 **DNA to RNA to Protein.** (a) The central dogma of biology states that information stored in DNA is copied to RNA (transcription), which is used to assemble proteins (translation). (b) DNA stores the information used to make proteins, just as a recipe stores the information needed to make brownies.

According to this model, a gene is therefore somewhat like a recipe in a cookbook. A recipe specifies the ingredients and instructions for assembling one dish, such as spaghetti sauce or brownies. Likewise, a protein-encoding gene contains the instructions for assembling a polypeptide, amino acid by amino acid (the polypeptide subsequently folds to become the finished protein). A cookbook that contains many recipes is analogous to a chromosome, which is an array of genes. A person's entire collection of cookbooks, then, is analogous to a genome.

To illustrate DNA's function with a concrete example, suppose a cell in a female mammal's breast is producing milk to feed an infant (see figure 3.13). One of the proteins in milk is albumin. The following steps summarize the production of albumin, starting with its genetic "recipe":

1. Inside the nucleus, an enzyme first transcribes the albumin gene's DNA sequence to a complementary sequence of RNA.

2. After some modification, the RNA emerges from the nucleus and binds to a ribosome.

3. At the ribosome, amino acids are assembled in a specific order to produce the albumin protein.

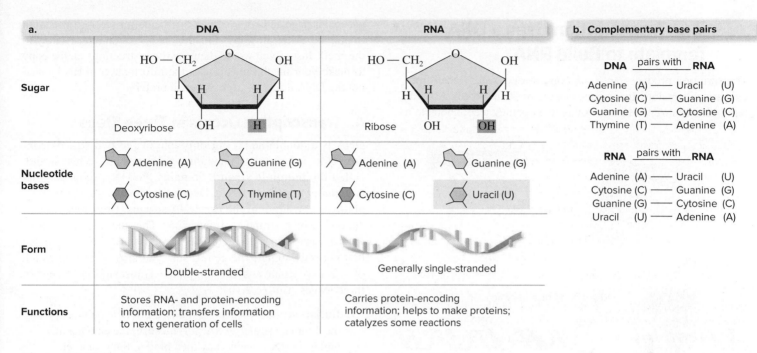

Figure 7.9 DNA and RNA Compared. (a) DNA and RNA differ in structure and function. (b) RNA contains uracil, not thymine. Like thymine, however, uracil pairs with adenine in complementary base pairs.

The amino acid sequence in albumin is dictated by the sequence of nucleotides in the RNA molecule. The RNA, in turn, was transcribed from DNA. In this way, DNA provides the recipe for albumin and every other protein in the cell.

B. RNA Is an Intermediary Between DNA and a Protein

RNA is a multifunctional nucleic acid that differs from DNA in several ways (figure 7.9). First, its nucleotides contain the sugar ribose instead of deoxyribose. Second, RNA has the nitrogenous base uracil, which behaves similarly to thymine; that is, uracil binds with adenine in complementary base pairs. Third, unlike DNA, RNA can be single-stranded (although it often folds into loops). Finally, RNA can catalyze chemical reactions, a role not known for DNA.

RNA is central to the flow of genetic information. Three types of RNA interact to synthesize proteins (table 7.1):

- **Messenger RNA (mRNA)** carries the information that specifies a protein. The mRNA is divided into genetic "code words" called codons; a codon is a group of three consecutive mRNA bases that corresponds to one amino acid.

- **Ribosomal RNA (rRNA)** combines with proteins to form a **ribosome,** the physical location where translation occurs. Some rRNAs help to correctly align the ribosome and mRNA, and others catalyze formation of the bonds between amino acids in the developing protein.

- **Transfer RNA (tRNA)** molecules are "connectors" that bind an mRNA codon at one end and a specific amino acid at the other. Their role is to carry each amino acid to the ribosome at the correct spot along the mRNA molecule.

The function of each type of RNA is further explained later in this chapter, beginning in section 7.4 with the first stage in protein production: transcription.

TABLE 7.1 Three Major Types of RNA

Molecule	Typical Number of Nucleotides	Function
mRNA	500–3000	Encodes amino acid sequence
rRNA	100–3000	Associates with proteins to form ribosomes, which structurally support and catalyze protein synthesis
tRNA	75–80	Physically links the message in mRNA to the amino acid sequence it encodes; binds mRNA codon on one end and an amino acid on the other

7.3 MASTERING CONCEPTS

1. What is the relationship between a gene and a protein?
2. How do transcription and translation use genetic information?
3. What are the three types of RNA, and how does each contribute to protein synthesis?

7.4 Transcription Uses a DNA Template to Build RNA

Transcription produces an RNA copy of one gene. Recall from our brownie analogy that a gene is a recipe for a protein. According to this analogy, transcription is like opening a cookbook to a particular page and copying just the recipe for the dish you want to prepare. After the copy is made, the book can return safely to the shelf. Just as you would then use the instructions on the copy to make your meal, the cell uses the information in RNA—and not the DNA directly—to make each protein.

A. Transcription Occurs in Three Steps

DNA is a double helix, but only one of the two strands contains the information encoding each protein. This strand, called the **template strand,** contains the DNA sequence that is actually copied to RNA. Depending on the gene, the template may be either of the two DNA strands. How does the cell "know" which strand to transcribe? The enzymes that carry out transcription recognize the **promoter,** a DNA sequence that not only signals the gene's start but also identifies which of the two strands is the template. Transcription occurs in three stages (**figure 7.10**):

1. **Initiation:** Enzymes unzip the DNA double helix, exposing the template strand. **RNA polymerase,** the enzyme that builds an RNA chain, may then bind to the promoter.

2. **Elongation:** RNA polymerase moves along the DNA template strand in a 3'-to-5' direction, adding nucleotides only to the 3' end of the growing RNA molecule.

3. **Termination:** A **terminator** sequence in DNA signals the end of the gene. Upon reaching the terminator, the RNA polymerase enzyme separates from the DNA template and releases the newly produced RNA. The DNA molecule then resumes its usual double-helix shape.

As the RNA molecule forms, it curls into a three-dimensional shape dictated by complementary base pairing within the molecule. The final shape determines whether the RNA functions as mRNA, tRNA, or rRNA.

The observation that the cell's DNA encodes all types of RNA—not just mRNA—has led to debate over the definition of the word *gene*. Originally, a gene was defined as any stretch of DNA that encodes one protein. More recently, however, the definition has expanded to include any DNA sequence that is transcribed to RNA. The phrase *gene expression* can therefore mean the production of either a functional RNA molecule or a protein.

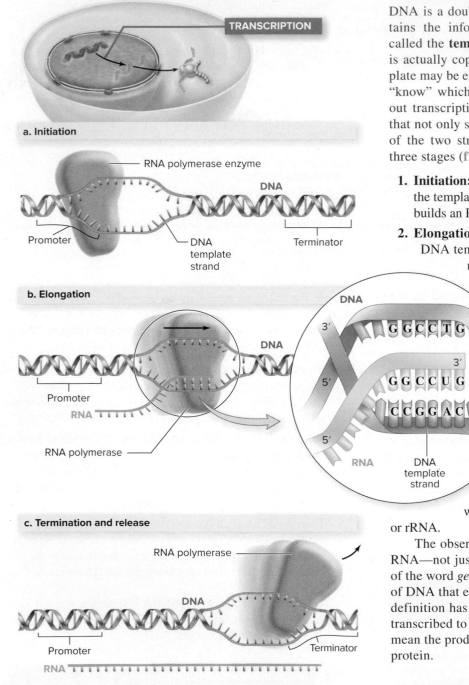

a. Initiation

RNA polymerase enzyme · DNA · Promoter · DNA template strand · Terminator

b. Elongation

DNA · DNA · Promoter · RNA · RNA polymerase · 3' GGCCTG · 5' GGCCUG 5' · CCGGAC 3' · DNA · RNA · DNA template strand

c. Termination and release

RNA polymerase · DNA · Promoter · Terminator · RNA

Figure 7.10 Transcription Builds mRNA. (a) During initiation, RNA polymerase binds to DNA. (b) During elongation, RNA polymerase adds nucleotides to the mRNA strand. (c) Termination occurs at the terminator sequence in DNA; RNA polymerase and mRNA detach from the DNA, which then returns to its double-helix form.

Figure It Out

Write the sequence of the mRNA molecule transcribed from the following DNA template sequence: 3'-TTACACTTGCAAC-5'

Answer: 5'-AAUGUGAACGUUG-3'

B. mRNA Is Altered in the Nucleus of Eukaryotic Cells

In bacteria and archaea, ribosomes may begin translating mRNA to a protein before transcription is even complete. In eukaryotic cells, however, the presence of the nuclear membrane prevents one mRNA from being simultaneously transcribed and translated. Moreover, in eukaryotes, mRNA is usually altered before it leaves the nucleus to be translated (**figure 7.11**).

5' Cap and Poly A Tail   After transcription, a short sequence of modified nucleotides, called a cap, is added to the 5′ end of the mRNA molecule. At the 3′ end, 100 to 200 adenines are added, forming a "poly A tail." Together, the cap and poly A tail enhance translation by helping ribosomes attach to the 5′ end of the mRNA molecule. The length of the poly A tail may also determine how long an mRNA lasts before being degraded.

Intron Removal   In archaea and in eukaryotic cells, only part of an mRNA molecule is translated into an amino acid sequence. Figure 7.11 shows that an mRNA molecule consists of alternating sequences called introns and exons. **Introns** are portions of the mRNA that are removed before translation. The word *intron* is short for *intragenic regions*, where *intra-* means "within" and

-genic refers to the gene. Small catalytic RNAs and proteins remove the introns from the mRNA.

The remaining portions, the **exons,** are spliced together to form the mature mRNA that leaves the nucleus to be translated. (A tip for remembering this is that *ex*ons are the portions of an mRNA molecule that are actually *ex*pressed or that *ex*it the nucleus.)

The amount of genetic material devoted to introns can be immense. The average exon is 100 to 300 nucleotides long, whereas the average intron is about 1000 nucleotides long. Some mature mRNA molecules consist of 70 or more spliced-together exons; the cell therefore simply discards much of the RNA produced in transcription. Although introns may seem wasteful, section 7.6 explains that this cutting and pasting is important in the regulation of gene expression.

7.4   MASTERING CONCEPTS

1. What happens during each stage of transcription?
2. Where in the cell does transcription occur?
3. What is the role of RNA polymerase in transcription?
4. What are the roles of the promoter and terminator sequences in transcription?
5. How is mRNA modified before it leaves the nucleus of a eukaryotic cell?

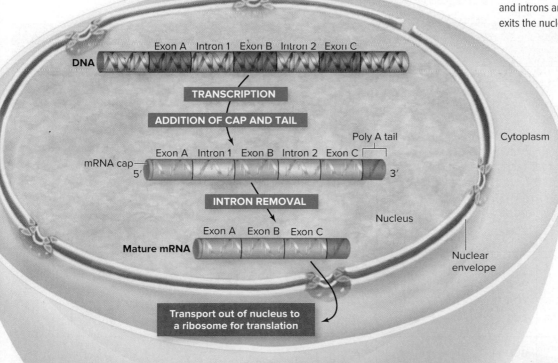

Figure 7.11 Processing mRNA. In eukaryotic cells, a nucleotide cap and poly A tail are added to mRNA, and introns are removed. Finally, the mature mRNA exits the nucleus.

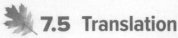

7.5 Translation Builds the Protein

Transcription copies the information encoded in a DNA base sequence into the complementary language of mRNA. Once transcription is complete and mRNA has left the nucleus, the cell is ready to translate the mRNA "message" into a sequence of amino acids. If mRNA is like a copy of a recipe, then translation is like preparing the dish.

A. The Genetic Code Links mRNA to Protein

The **genetic code** is the set of "rules" by which a cell uses the codons in mRNA to assemble amino acids into a protein (figure 7.12). Each codon is a group of three mRNA bases corresponding either to one amino acid or to a "stop" signal.

In the 1960s, however, researchers did not yet understand exactly how the genetic code worked. One early question was the number of RNA bases that specify each amino acid. Researchers reasoned that RNA contains only 4 different nucleotides, so a genetic code with a one-to-one correspondence of mRNA bases to amino acids could specify only 4 different amino acids—far fewer than the 20 amino acids that make up biological proteins. A code consisting of 2 bases per codon could specify only 16 different amino acids. A code with 3 bases per codon, however, yields 64 different combinations, more than enough to specify the 20 amino acids in life. Experiments later confirmed the triplet nature of the genetic code.

A second, and more difficult, problem was to determine which codons correspond to which amino acids. In the 1960s, researchers answered this question by synthesizing mRNA molecules in the laboratory. They added these synthetic mRNAs to test tubes containing all the ingredients needed for translation, extracted from *E. coli* cells. Analyzing the resulting polypeptides allowed scientists to finish deciphering the genetic code in less than a decade—a monumental task. Chemical analysis eventually showed that the genetic code also contains directions for starting and stopping translation. AUG is typically the first codon in mRNA, and the codons UGA, UAA, and UAG each signify "stop."

Nearly all species use the same mRNA codons to specify the same amino acids, although mitochondria and a handful of

Figure 7.12 The Genetic Code. In translation, mRNA codons are matched with amino acids as specified in the genetic code.

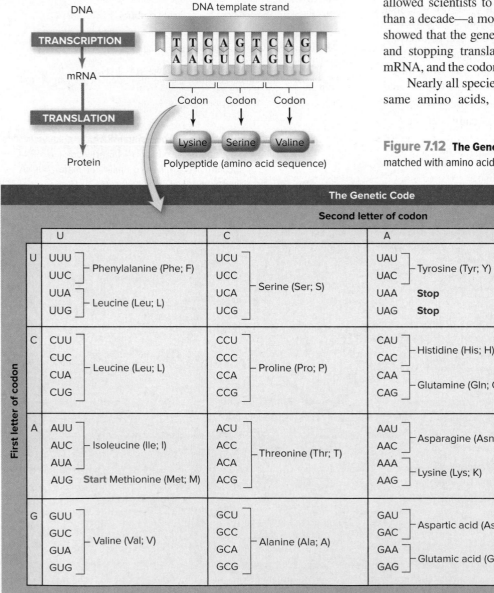

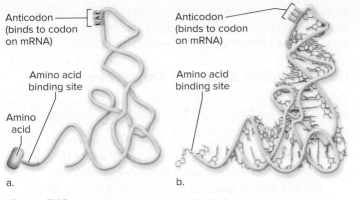

Figure 7.13 **Transfer RNA.** (a) A simplified tRNA molecule shows the anticodon at one end and the amino acid binding region at the opposite end. (b) Three-dimensional view of tRNA.

species use alternative codes that differ slightly from the code in figure 7.12. The best explanation for the (nearly) universal genetic code is that all life evolved from a common ancestor.

B. Translation Requires mRNA, tRNA, and Ribosomes

Translation—the actual construction of the protein—requires three main types of participants. We have already met the first type: mRNA, the molecule that contains the genetic information encoding a protein. As illustrated in figure 7.12, each three-base codon in mRNA specifies one amino acid.

The second type of participant is tRNA (figure 7.13). These "bilingual" molecules carry amino acids from the cytosol to the mRNA being translated. Each tRNA includes an **anticodon,** a three-base loop that is complementary to one mRNA codon. The other end of the tRNA molecule carries the amino acid

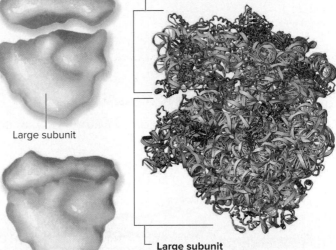

Figure 7.14 **The Ribosome.** A ribosome from a eukaryotic cell has two subunits containing a total of 82 proteins and four rRNA molecules.
Photo: ©Center for Molecular Biology of RNA, UC-Santa Cruz

corresponding to that codon. For example, a tRNA with the anticodon sequence AAG always picks up the amino acid phenylalanine, which is encoded by the codon UUC (see figure 7.12).

The remaining participant in translation is the ribosome. Each ribosome, which has one large and one small subunit, is built of rRNA and proteins (figure 7.14). Ribosomes are the sites of translation. That is, in the recipe analogy in figure 7.8, a ribosome is the "bowl" where the ingredients come together (and tRNA molecules

 Apply It **Now**

Some Poisons Disrupt Protein Production

We learned in chapter 6 that some poisons kill by interfering with respiration. Here we list a few poisons that inhibit protein synthesis; a cell that cannot make proteins quickly dies.

- **Amanatin:** This toxin occurs in the "death cap mushroom," *Amanita phalloides.* Amanatin inhibits RNA polymerase, making transcription impossible.
- **Diphtheria toxin:** Certain bacteria secrete a toxin that causes the respiratory illness diphtheria. This toxic compound inhibits an elongation factor, a protein that helps add amino acids to a polypeptide chain during translation.
- **Antibiotics:** Drugs that bind to bacterial ribosomes include clindamycin, chloramphenicol, tetracyclines, and gentamicin. When its ribosomes are disrupted, a cell cannot make proteins, and it dies.
- **Ricin:** Derived from seeds of the castor bean plant, ricin is a potent natural poison that consists of two parts. One part binds

to a cell, and the other enters the cell and inhibits protein synthesis by an unknown mechanism. Interestingly, the part of the molecule that enters the cell is apparently more toxic to cancer cells than to normal cells, making ricin a potential cancer treatment.

- **Trichothecenes:** Fungi in the genus *Fusarium* produce toxins called trichothecenes. During World War II, thousands of people died after eating bread made from moldy wheat, and many researchers believe trichothecenes were used as biological weapons during the Vietnam War. The mode of action is unclear, but the toxins seem to interfere somehow with ribosomes.

©Jacana/Science Source

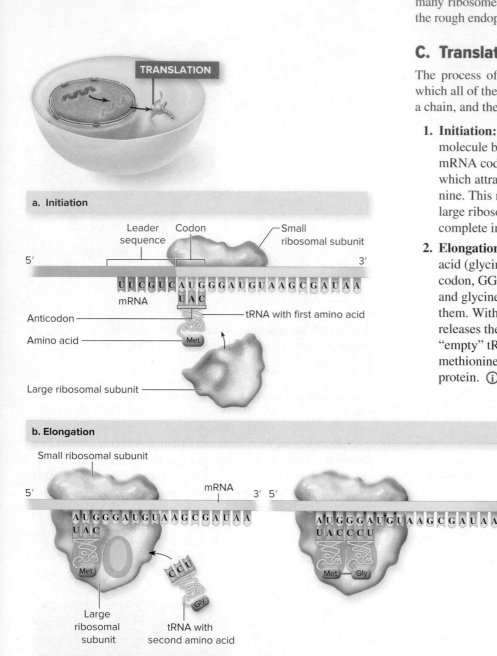

a. Initiation

Leader sequence — Codon — Small ribosomal subunit

5′

mRNA

Anticodon — tRNA with first amino acid

Amino acid — Met

Large ribosomal subunit

3′

UUCGUCAUGGGAUGUAAGCGAUAA
UAC

b. Elongation

Small ribosomal subunit

5′ — mRNA — 3′ 5′ — 3′ 5′ — 3′

AUGGGAUGUAAGCGAUAA
UAC

Met

Large ribosomal subunit

tRNA with second amino acid

CCU
Gly

AUGGGAUGUAAGCGAUAA
UACCCU

Met — Gly

AUGGGAUGUAAGCGAUAA
CCUACA

UAC   UUC

Met — Gly — Cys   Lys

c. Termination and release

Stop codon

5′ — 3′

AUGGGAUGUAAGCGAUAA
UUC   GCU

Met — Gly — Cys — Lys — Arg

Release factor protein

Polypeptide

are helpers that carry those ingredients to the bowl). Each cell has many ribosomes, which may be free in the cytosol or attached to the rough endoplasmic reticulum (see figure 3.15).

C. Translation Occurs in Three Steps

The process of translation is divided into three stages, during which all of the participants come together, link amino acids into a chain, and then dissociate again (**figure 7.15**).

1. **Initiation:** The leader sequence at the 5′ end of the mRNA molecule bonds with a small ribosomal subunit. The first mRNA codon to specify an amino acid is usually AUG, which attracts a tRNA that carries the amino acid methionine. This methionine signifies the start of a polypeptide. A large ribosomal subunit attaches to the small subunit to complete initiation.

2. **Elongation:** A tRNA molecule carrying the second amino acid (glycine in figure 7.15) then binds to the second codon, GGA in this case. The two amino acids, methionine and glycine, align, and a covalent bond forms between them. With that peptide bond in place, the ribosome releases the first tRNA. The cell may now reuse this "empty" tRNA; that is, the tRNA will pick up another methionine, which may be incorporated into another new protein. ⓘ *peptide bond,* section 2.5C

Figure 7.15 Translation Builds the Protein. Translation occurs in three stages. (a) Initiation brings together the ribosomal subunits, mRNA, and the tRNA carrying the first amino acid. (b) As elongation begins, a tRNA molecule bearing the second amino acid binds to the second codon. The first amino acid forms a covalent bond with the second amino acid. Additional tRNAs bring subsequent amino acids encoded in the mRNA. (c) Termination occurs when a release factor protein binds to the stop codon. All components of the translation machine are released, along with the completed polypeptide.

Next, the ribosome moves down the mRNA by one codon. A third tRNA enters, carrying its amino acid. This third amino acid aligns with the other two and forms a covalent bond to the second amino acid in the growing chain. The tRNA attached to glycine is released and recycled. With the help of proteins called elongation factors, the polypeptide grows one amino acid at a time, as tRNAs continue to deliver their cargo.

3. **Termination:** Elongation halts at a "stop" codon (UGA, UAG, or UAA). No tRNA molecules correspond to these stop codons. Instead, proteins called release factors bind to the stop codon, prompting the translation participants to separate from one another. The ribosome releases the last tRNA, the ribosomal subunits separate and are recycled, and the new polypeptide is released.

Figure It Out

If a DNA sequence is 3'-AAAGCAGTACTA-5', what is the corresponding amino acid sequence?

Answer: Phe-Arg-His-Asp

Protein synthesis can be very speedy. A cell in the human immune system, for example, can manufacture 2000 identical antibody proteins per second, helping the body respond quickly to infections. How can protein synthesis occur fast enough to meet all of a cell's needs? One way the cell maximizes efficiency is by producing multiple copies of each mRNA, giving the cell more than one genetic "recipe" for the ribosomes to read. In addition, dozens of ribosomes may simultaneously translate one mRNA molecule, following each other from the 5' end to the 3' end (**figure 7.16**). These ribosomes zip along the mRNA, incorporating some 15 amino acids per second. Thanks to this fast-moving "assembly line," a cell can make many copies of a protein from the same mRNA. ⓘ *antibodies,* section 34.3C

On the other hand, blocking protein synthesis can quickly kill a cell. This section's Apply It Now box describes a few poisons that make transcription or translation impossible.

D. Proteins Must Fold Correctly After Translation

The newly synthesized protein cannot do its job until it folds into its final shape (see figure 2.20). Some regions of the amino acid chain attract or repel other parts, contorting the polypeptide's overall shape. Enzymes catalyze the formation of chemical bonds, and "chaperone" proteins stabilize partially folded regions.

Proteins can fold incorrectly if the underlying DNA sequence is altered (see section 7.7), because the encoded protein may have the wrong sequence of amino acids. Serious illness may result. In some forms of cystic fibrosis, for example, a membrane transport protein does not fold correctly into its final form. The absence of the functional protein causes sticky mucus to accumulate in the lungs and other organs.

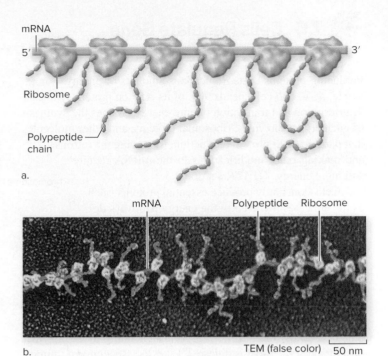

a.

b.
TEM (false color) ⊢ 50 nm ⊣

Figure 7.16 Efficient Translation. (a) Multiple ribosomes can simultaneously translate one mRNA. (b) This micrograph shows about 30 ribosomes producing proteins from the same mRNA.

(b): ©Dr. Elena Kiseleva/SPL/Science Source

Errors in protein folding can occur even if the underlying genetic sequence remains unchanged. Alzheimer disease, for example, is associated with a protein called amyloid, which folds improperly and then forms an abnormal mass in brain cells. Likewise, mad cow disease and similar conditions in sheep and humans are caused by abnormal clumps of misfolded proteins called prions in nerve cells. ⓘ *prions,* section 16.6B

In addition to folding, some proteins must be altered in other ways before they become functional. For example, insulin, which is 51 amino acids long, is initially translated as the 80-amino-acid polypeptide proinsulin. Enzymes cut proinsulin to form insulin. A different type of modification occurs when polypeptides join to form larger protein molecules. The oxygen-carrying blood protein hemoglobin, for example, consists of four polypeptide chains (two alpha and two beta) encoded by separate genes.

7.5 MASTERING CONCEPTS

1. How did researchers determine that the genetic code is a triplet and learn which codons specify which amino acids?
2. What happens in each stage of translation?
3. Where in the cell does translation occur?
4. How are polypeptides modified after translation?

7.6 Cells Regulate Gene Expression

Producing proteins costs tremendous amounts of energy. For example, an *E. coli* cell spends 90% of its ATP on protein synthesis. Transcription and translation require energy, as does the synthesis of nucleotides, enzymes, ribosomal proteins, and other molecules that participate in protein production. Removing the introns and making other modifications to the mRNA require still more energy. ⓘ *ATP,* section 4.3

Cells constantly produce essential proteins, such as the enzymes involved in the energy pathways described in chapters 5 and 6. Considering the high cost of making a protein, however, it makes sense that cells save energy by not producing unneeded proteins.

A related reason to regulate gene expression is that it gives cells flexibility to respond to changing conditions. For example, the python pictured at the beginning of chapter 6 ramps up its production of digestive enzymes shortly after it begins to swallow the gazelle. The genes encoding those enzymes turn off once the meal is digested. Likewise, specialized immune system cells churn out antibodies in response to an infection. Once the threat is gone, antibody production halts.

This section describes some of the mechanisms that regulate gene expression in cells.

A. Operons Are Groups of Bacterial Genes That Share One Promoter

Many bacteria, including *E. coli,* live in animal intestines; their food sources change from hour to hour. To maximize efficiency, the bacteria should produce enzymes that degrade only the foods that are actually available. For example, *E. coli* requires certain enzymes to absorb and degrade the sugar lactose. How does the bacterium "know" to produce these enzymes only when lactose is present?

In 1961, French biologists François Jacob and Jacques Monod discovered the answer. They showed that in *E. coli* and other bacteria, related genes are organized as operons. An **operon** is a group of related genes plus a promoter and an operator that control the transcription of the entire group at once. The promoter, as described earlier, is the site to which RNA polymerase attaches to begin transcription. The **operator** is a DNA sequence located between the promoter and the protein-encoding regions. If a protein called a **repressor** binds to the operator, it prevents the transcription of the genes.

Figure 7.17a shows one example: *E. coli*'s **lac operon,** which consists of the three genes that encode lactose-degrading proteins, plus a promoter and an operator. To understand how the *lac* operon works, first imagine an *E. coli* cell in an environment lacking lactose. Expressing the three genes would be a waste of energy. The repressor protein therefore binds to the operator, preventing RNA polymerase from transcribing the genes (figure 7.17b). The genes are effectively "off." But when lactose is present, the sugar attaches to the repressor, changing its shape so that it detaches from the DNA. RNA polymerase is now free to transcribe the genes (figure 7.17c). After translation, the resulting enzymes confer a (temporary) new trait: the ability to absorb and degrade the lactose.

Soon, geneticists discovered other groups of genes organized as operons. Some, like the *lac* operon, negatively control transcription by removing a block. Others produce factors that turn on transcription. As Jacob and Monod concluded in 1961, "The genome contains not only a series of blueprints, but a coordinated program of protein synthesis and means of controlling its execution."

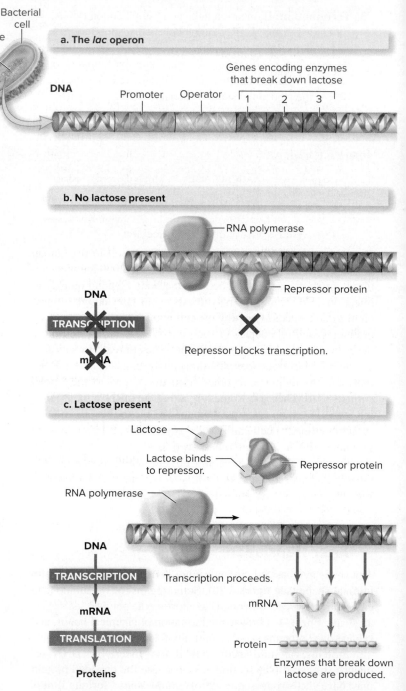

Figure 7.17 The *Lac* Operon. (a) An operon consists of a promoter, an operator, and a group of related genes. (b) In the absence of lactose, a repressor protein binds to the operator and prevents transcription. The enzymes are not produced. (c) If lactose is present, it binds to the repressor, which subsequently releases the operator. Transcription proceeds.

B. Eukaryotic Organisms Use Many Regulatory Mechanisms

Beyond energy savings, eukaryotic cells have an additional reason to regulate gene expression. Many eukaryotes are multicellular organisms consisting of multiple types of specialized cells. Humans, for example, have at least 200 different cell types. If all of the cells that make up your body contain the same set of genes, how does each cell acquire its unique function? The answer is that each type of cell expresses a different subset of its genes. A hair follicle cell, for example, produces a lot of keratin (the protein that makes up hair) but never makes hemoglobin. Conversely, a red blood cell produces a lot of hemoglobin but leaves the keratin gene turned off.

Cell specialization begins near the start of a multicellular organism's life. Early in an animal embryo's development, for example, protein signals "tell" cells whether they are at the head end of the body, the tail end, or somewhere in between. These signals, in turn, regulate the expression of unique combinations of genes that enable cells in each location to specialize. But gene regulation continues throughout life. For instance, in a flowering plant, genes that were silent early in life become active when external signals trigger flower formation.

To understand how and why multicellular organisms regulate gene expression, consider this analogy. A chef can use the same set of cookbooks to make breakfast, lunch, and dinner, just as your specialized hair follicle cells and red blood cells contain the same genetic information. The cook does not prepare all possible dishes from all cookbooks simultaneously; rather, he or she selects only the recipes required for each meal. By controlling gene expression, your cells accomplish a similar feat.

The rest of this section describes how eukaryotic cells control whether each gene is "on" or "off." **Figure 7.18** illustrates where each mechanism fits into the overall process of gene expression.

DNA Availability   A chromosome consists of DNA wrapped around proteins called histones (figure 7.18, part 1). Enzymes in a cell can modify stretches of DNA by adding and removing regulatory "tags" consisting of methyl groups ($-CH_3$). Tagged regions are folded in a way that makes the DNA unavailable; removing the tags means the genes may be expressed. Likewise, the histones can also be chemically modified in a way that either exposes DNA to be transcribed or tucks it away.

The study of these and other modifications is part of a field of research called **epigenetics.** The definition of this term is evolving, but biologists agree that it concerns changes in gene expression that do not involve changes to the DNA sequence itself. (The prefix *epi-*, which means "over" or "around," conveys this idea.)

Abnormalities in epigenetic mechanisms can mean that genes are activated or inactivated when they should not be, leading to birth defects, diabetes, and other serious conditions. Cancer, for example, is a family of illnesses in which cells divide out of control. Proteins provide the signals that normally

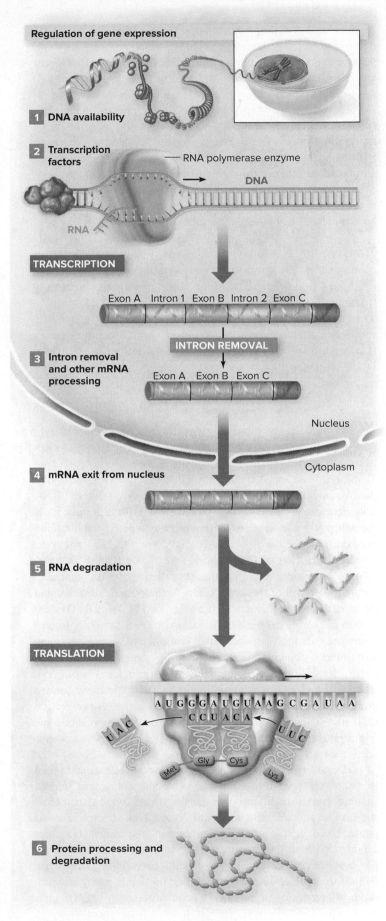

Figure 7.18 Regulating Gene Expression. Eukaryotic cells have many ways to control whether each gene is turned on or off.

regulate cell division. Since methyl groups and other tags help determine which genes are available to be expressed, faulty epigenetic markers can disrupt these fine-tuned signals. Cancer develops because the cells are unable to stop dividing. ⓘ *cancer,* section 8.6

The placement of epigenetic markers can persist in daughter cells after a cell divides. However, most of the tags are wiped away when sperm fertilizes egg, starting each new generation with a virtually clean slate. After conception, the tags on our chromosomes change throughout life in response to many environmental factors, including diet, exercise, stress, and exposure to toxins. These changes help explain why identical twins—who have matching DNA—are not exactly the same.

Transcription Factors   As we have already seen, DNA can be expressed only if it is uncoiled. Even then, RNA polymerase can bind to a promoter and initiate transcription of a gene only if transcription factors are also present (figure 7.18, part 2). **Transcription factors** are proteins that bind DNA at specific sequences that regulate transcription. A transcription factor may bind to a gene's promoter or to an **enhancer,** a regulatory DNA sequence that lies outside the promoter. An enhancer may be located near the gene (or even within it), but often they are thousands of base pairs away. ⓘ *enhancers and wing spots,* section 13.5

Figure 7.19 shows how groups of transcription factors prepare a promoter to receive RNA polymerase. The first transcription factor to bind is attracted to a part of the promoter called the TATA box. The TATA binding protein attracts other transcription factors, including proteins bound to an enhancer. Finally, RNA polymerase joins the complex, binding just in front of the start of the gene sequence. With RNA polymerase in place, transcription can begin.

Transcription factors respond to external stimuli that signal a gene to turn "on." In a process called signal transduction, a hormone or other molecule binds to the outside of a cell and triggers a series of chemical reactions inside the cell. The last step in the series can be the activation or deactivation of a transcription factor. Moreover, one stimulus may trigger many simultaneous changes in the cell. An example is the series of events that follows the fusion of egg and sperm (see chapter 35). The fertilized egg cell immediately begins its journey toward embryonic development, a process that requires the activation of many genes that were silent before the sperm reached the egg. ⓘ *signal transduction,* section 3.3; *steroid hormones,* section 28.2B

The ability to digest lactose provides an excellent example of the importance of transcription factors and enhancers in gene regulation. All infants produce lactase, the enzyme that digests the lactose in milk. But many adults are lactose-intolerant because their lactase-encoding gene remains turned off after infancy. Without the enzyme, lactose is indigestible. Some people, however, can continue to digest milk into adulthood. In these lactose-tolerant adults, an enhancer is modified in a way that promotes transcription of the lactase gene throughout life.

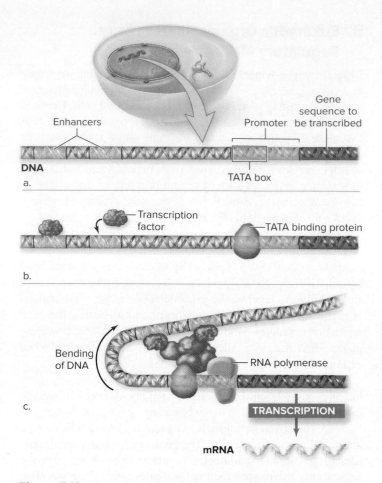

Figure 7.19 How Transcription Factors Work. RNA polymerase initiates transcription only if proteins called transcription factors are also present. (a) Enhancers and promoters are DNA sequences that regulate transcription. (b) Transcription factors, including TATA binding proteins, bind to enhancers and promoters. (c) DNA bends, bringing the transcription factors together with RNA polymerase.

Defects in transcription factors underlie some forms of cancer. As we have already seen, cells tightly control the production of proteins that regulate cell division. Transcription factors are star players in this process, so defective transcription factors can cause cells to divide out of control. In addition, some drugs interfere with transcription factors. The "abortion pill" RU486, for example, indirectly blocks transcription factors needed for the development of an embryo.

mRNA Processing   One gene can encode multiple proteins if different combinations of exons are included in the final mRNA. For example, if exon B were excluded in figure 7.18 (part 3), exons A and C would be spliced together. The mRNA would then encode a different protein. For mRNA molecules with dozens of exons, the number of alternative proteins is huge. Researchers know of a gene in fruit flies that can theoretically be spliced into more than 38,000 different configurations!

Burning Question

Is there a gay gene?

©Getty Images RF

Despite periodic headlines about newly discovered genes "for" homosexuality, the reality is a bit more complex.

Research linking human behavior to individual genes is extremely difficult for several reasons. First, the question of a "gay gene" is somewhat misleading. Genes encode RNA and proteins, not behaviors, so any relationship between DNA and sexual behavior must be indirect. Second, to establish a clear link to DNA, a researcher must be able to define and measure a behavior. This in itself is difficult, because people disagree about what it means to be homosexual. Third, multiple genes are likely to be involved. Fourth, an individual who possesses an allele associated with a trait will not necessarily express the allele; as illustrated in figure 7.18, many genes in each cell remain "off" at any given time. To complicate matters, combinations of environmental cues that are unique to each person also influence gene expression.

Despite these complications, research has yielded some evidence of a biological component to homosexuality, at least in males. For example, a male homosexual's identical twin is much more likely to also be homosexual than is a nonidentical twin, indicating a strong genetic contribution. In addition, the more older brothers a male has, the more likely he is to be homosexual. This "birth order" effect occurs only for siblings with the same biological mother; having older stepbrothers does not increase the chance that a male is homosexual. That means that events before birth, not social interactions with brothers, are responsible for the effect.

Other research has produced ambiguous results. Anatomical studies of cadavers have revealed differences in the size of a particular brain structure between heterosexual and homosexual men, but the relative contribution of genes and environment to this structure is unknown. One study linked homosexuality in males, but not in females, to part of the X chromosome; a subsequent study did not support this conclusion.

So is there a gay gene? The short answer is no, because there is unlikely to be a single gene that "causes" homosexuality. At the same time, research suggests a genetic contribution to sexual orientation. Sorting out the complex interactions between multiple genes and the environment, however, remains a formidable challenge.

Submit your burning question to
Marielle.Hoefnagels@mheducation.com

mRNA Exit from Nucleus For a protein to be produced, mRNA must leave the nucleus and attach to a ribosome (figure 7.18, part 4). If the mRNA fails to leave, the gene is silenced.

mRNA Degradation Not all mRNA molecules are equally stable. Some are rapidly destroyed, perhaps before they can be translated, whereas others persist long enough to be translated many times (figure 7.18, part 5).

Moreover, tiny RNA sequences called microRNAs can play a role in regulating gene expression. Each microRNA is only about 21 to 23 nucleotides long, and it does not encode a protein. Instead, a cell may produce a microRNA that is complementary to a coding mRNA. If the microRNA attaches to the mRNA, the resulting double-stranded RNA cannot be translated at a ribosome and is likely to be destroyed. Medical researchers are actively studying microRNAs; the ability to silence harmful genes may help treat illnesses ranging from cancer to influenza and HIV.

Protein Processing and Degradation Some proteins must be altered before they become functional (figure 7.18, part 6). Dozens of modifications are possible, including the addition of sugars or an alteration in the protein's structure. Producing insulin, for example, requires a precursor protein to be cut in two places. If these modifications fail to occur, the insulin protein cannot function.

In addition, to do its job, a protein must move from the ribosome to where the cell needs it. For example, a protein secreted in milk must be escorted to the Golgi apparatus and be packaged for export (see figure 3.13). A gene is effectively silenced if its product never moves to the correct destination.

Finally, like RNA, not all proteins are equally stable. Some are degraded shortly after they form, whereas others persist longer.

A human cell may express hundreds to thousands of genes at once. Unraveling the complex regulatory mechanisms that control the expression of each gene is an enormous challenge. Biologists now have the technology to begin navigating this regulatory maze. The work has just begun, but the payoff will be a much better understanding of cell biology, along with many new medical applications. The same research may also help scientists understand how external influences on gene expression contribute to complex traits, such as the one described in this section's Burning Question.

7.6 MASTERING CONCEPTS

1. What are some reasons that cells regulate gene expression?
2. How does a repressor protein help regulate the expression of a bacterial operon?
3. Explain how epigenetic modifications change the likelihood of transcription.
4. How do enhancers and transcription factors interact to regulate gene expression?

7.7 Mutations Change DNA Sequences

A **mutation** is any change in a cell's DNA sequence. The change may occur in a gene or in a regulatory region such as an enhancer. Many people think that mutations are always harmful, perhaps because some of them cause such dramatic changes (**figure 7.20**). Although some mutations do cause illness, they also provide the variation that makes life interesting (and makes evolution possible).

To continue the cookbook analogy introduced earlier, a mutation in a gene is similar to an error in a recipe. A small typographical error might be barely noticeable. A minor substitution of one ingredient for another might hurt (or improve) the flavor. But serious errors such as missing ingredients or truncated instructions are likely to ruin the dish.

A. Mutations Range from Silent to Devastating

A mutation may change one or a few base pairs or affect large portions of a chromosome. Some are detectable only by using DNA sequencing techniques, while others may be lethal. Table 7.2 illustrates some of the major types of mutations, using three-letter words to represent codons. The rest of this section describes them in detail. ⓘ *DNA sequencing,* section 11.2B

Substitution Mutations   A **substitution mutation** is the replacement of one DNA base with another. Such a mutation is "silent" if the mutated gene encodes the same protein as the original gene version. Mutations can be silent because more than one codon encodes most amino acids.

Often, however, a substitution mutation changes a base triplet so that it specifies a different amino acid. This change is called

TABLE 7.2   Types of Mutations

Type	Illustration
Original sentence	THE ONE BIG FLY HAD ONE RED EYE
Substitution (missense)	TH**Q** ONE BIG FLY HAD ONE RED EYE
Nonsense	THE ONE BIG
Insertion	THE ONE BIG **WET** FLY HAD ONE RED EYE
Insertion (frameshift)	THE ONE **Q**BI GFL YHA DON ERE DEY
Deletion	THE ONE BIG HAD ONE RED EYE
Expanding repeat	Generation 1: THE ONE BIG FLY HAD ONE RED EYE
	Generation 2: THE ONE BIG FLY **FLY FLY** HAD ONE RED EYE
	Generation 3: THE ONE BIG FLY FLY **FLY FLY FLY FLY** HAD ONE RED EYE

a missense mutation. The substituted amino acid may drastically alter the protein's shape, changing its function. Sickle cell disease results from this type of mutation in a gene encoding hemoglobin (**figure 7.21**).

In other cases, called nonsense mutations, a base triplet specifying an amino acid changes into one that encodes a "stop" codon. This shortens the protein product, which can profoundly influence the organism. At least one of the mutations that give rise to cystic fibrosis, for example, shortens a membrane protein from its normal 1480 amino acids to only 493. The faulty protein cannot function.

Figure It Out

Suppose that a substitution mutation replaces the first A in the following mRNA sequence with a U:

   5′-AAAGCAGUACUA-3′.

How many amino acids will be in the polypeptide chain?

Answer: Zero

Base Insertions and Deletions   An **insertion mutation** adds one or more nucleotides to a gene; a **deletion mutation** removes nucleotides. Either type of mutation may be a **frameshift mutation,** in which nucleotides are added or deleted by a number other than a multiple of three. Because triplets of DNA bases specify amino acids, such an addition or deletion disrupts the codon reading frame (**figure 7.22**). Frameshift mutations are therefore likely to alter the sequence of amino acids (missense) or cause premature stop codons (nonsense). Either way, a frameshift usually devastates a protein's function. Some mutations that cause cystic fibrosis, for example, reflect the addition or deletion of just one or two nucleotides.

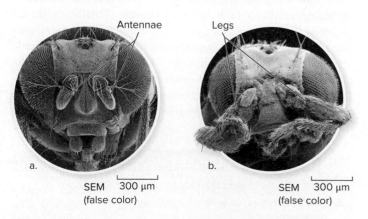

Antennae   Legs

a.   b.

SEM 300 μm   SEM 300 μm
(false color)   (false color)

Figure 7.20 Legs on the Head. Mutations in some genes can cause parts to form in the wrong places. (a) Normally, a fruit fly has two small antennae between its eyes. (b) This fly has legs growing where its antennae should be; it has a mutation that affects development.

(a): ©Andrew Syred/Science Source; (b): ©Eye of Science/Science Source

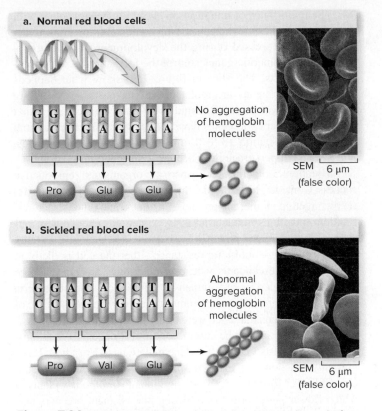

Figure 7.21 Sickle Cell Mutation. Sickle cell anemia usually results from a mutation in a hemoglobin gene. (a) Normal hemoglobin molecules do not aggregate, enabling the red blood cell to assume a rounded shape. (b) A substitution mutation causes hemoglobin molecules to clump (aggregate) into rods that deform the red blood cell.

Photos: (a): ©Micro Discovery/Corbis; (b): ©Dr. Gopal Murti/Science Source

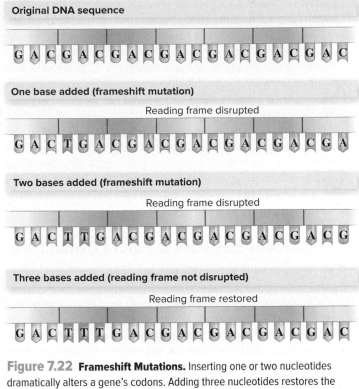

Figure 7.22 Frameshift Mutations. Inserting one or two nucleotides dramatically alters a gene's codons. Adding three nucleotides restores the reading frame, although the extra codon may still alter the encoded protein's shape.

Even if a small insertion or deletion does not shift the reading frame, the effect might still be significant if the change drastically alters the protein's shape. The most common mutation that causes severe cystic fibrosis deletes a single group of three nucleotides. The resulting protein lacks just one amino acid, but it cannot function.

Expanding Repeats In an **expanding repeat mutation,** the number of copies of a three- or four-nucleotide sequence increases over several generations. With each generation, the symptoms begin earlier or become more severe (or both). Expanding genes underlie several inherited disorders, including fragile X syndrome and Huntington disease. In Huntington disease, expanded repeats of GTC cause extra glutamines (an amino acid) to be incorporated into the gene's protein product. The abnormal protein forms fibrous clumps in the nuclei of some brain cells, which causes the symptoms of uncontrollable movements and personality changes.

Changes in Chromosome Structure Some mutations affect extensive regions of DNA (see figure 9.14). For example, a large part of a chromosome may be deleted or duplicated during meiosis. One region of a chromosome may also become inverted or fused with a different chromosome. Any of these events can bring together DNA segments that were not previously joined. When part of chromosome 9 breaks off and fuses with chromosome 22, for example, the resulting chromosome has a fused gene whose protein product causes a type of leukemia.

B. What Causes Mutations?

Some mutations occur spontaneously—that is, without outside causes. A spontaneous substitution mutation usually originates as a DNA replication error. Replication errors can also cause insertions and deletions, especially in genes with repeated base sequences, such as GCG CGC . . . . It is as if the molecules that guide and carry out replication become "confused" by short, repeated sequences, as a proofreader scanning a manuscript might miss the spelling errors in the words "happpiness" and "bananana." ⓘ *DNA replication,* section 8.2

The average rate of replication errors for most genes is about 1 in 100,000 bases, but it varies among organisms and among genes. The larger a gene, the more likely it is to mutate. In addition, the more frequently DNA replicates, the more it mutates. Bacteria accumulate mutations faster than cells of complex organisms, simply because their DNA replicates more often. Likewise, rapidly dividing skin cells tend to have more mutations than the nervous system's neurons, which divide slowly if at all.

Exposure to harmful chemicals or radiation may also damage DNA. A **mutagen** is any external agent that induces mutations. Examples include the ultraviolet radiation in sunlight, X-rays, radioactive fallout from atomic bomb tests and nuclear accidents, and chemicals in tobacco and in the environment (such as pollution in soil, air, or water). The more contact a person has with mutagens, the higher the risk for cancer. Coating skin with sunscreen, wearing a lead "bib" during dental X-rays, and avoiding tobacco all lower cancer risk by reducing exposure to mutagenic chemicals and radiation.

Movable DNA sequences are yet another source of mutations. A **transposable element,** or transposon for short, is a DNA sequence that can "jump" within the genome. A transposon can insert itself randomly into chromosomes. If it lands within a gene, the transposon can disrupt the gene's function; it can also leave a gap in the gene when it leaves.

C. Mutations May Pass to Future Generations

A **germline mutation** is a DNA sequence change that occurs in the cells that give rise to sperm and eggs. Germline mutations are heritable because the mutated DNA will be passed down in at least some of the sex cells that the organism produces. As a result, every cell of the organism's affected offspring will carry the mutation. Such mutations may run in families, or they can appear suddenly. For example, two healthy people of normal height may have a child with a form of dwarfism called achondroplasia. The child's achondroplasia arose from a new mutation that occurred by chance in the mother's or father's germ cell.

Most mutations, however, do not pass from generation to generation. A **somatic mutation** occurs in nonsex cells, such as those that make up the skin, intestinal tract, or lungs. All cells derived from the altered one will also carry the mutation, but the mutation does not pass to the organism's offspring. The children of a cigarette smoker with mutations that cause lung cancer, for example, do not inherit the parent's damaged genes.

D. Mutations Are Important

A mutation in a gene sometimes changes the structure of its encoded protein so much that the protein can no longer do its job. Inherited diseases, including cystic fibrosis and sickle cell anemia, stem from such DNA sequence changes. Some of the most harmful mutations affect the genes encoding the proteins that repair DNA. Additional mutations then rapidly accumulate in the cell's genetic material, which can kill the cell or lead to cancer.

Mutations are also extremely important because they produce genetic variability. They are the raw material of evolution because they generate new **alleles,** or variants of genes. Except for identical twins, everyone has a different combination of alleles for the 25,000 or so genes in the human genome. The same is true for any genetically variable group of organisms.

Some of these new alleles are "neutral" and have no effect on an organism's fitness. Your reproductive success, for example, does not ordinarily depend on your eye color or shoe size. As unit 3 explains, however, variation has important evolutionary consequences. In every species, individuals with some allele combinations reproduce more successfully than others. In natural selection, the environment "edits out" the less favorable allele combinations.

The importance of mutations in evolution became clear with the discovery of homeotic genes, which encode transcription factors that are expressed during the development of an embryo. Specifically, homeotic genes control the formation of the organism's body parts. The flies in figure 7.20 show what happens when homeotic genes are mutated. Having parts in the wrong places is usually harmful. But studies of many species reveal that mutations in homeotic genes have profoundly influenced animal evolution (see section 13.5). Limb modifications such as arms, wings, and flippers trace their origins to homeotic mutations.

Mutations sometimes enhance an organism's reproductive success. Consider, for example, the antibiotic drugs that kill bacteria by targeting membrane proteins, enzymes, and other structures. Random mutations in bacterial DNA sometimes encode new versions of these targeted proteins. Such a mutation may give a bacterium a new trait—antibiotic resistance—that the cell is likely to pass on to its descendants whenever antibiotics are present. The medical consequences are immense. Antibiotic-resistant bacteria have become more and more common, and many people now die of bacterial infections that once were easily treated with antibiotics.

Mutations can also be enormously useful in science and agriculture. Geneticists frequently induce mutations to learn how genes normally function. For example, biologists discovered how genes control flower formation by studying mutant *Arabidopsis* plants in which flower parts form in the wrong places. Plant breeders also induce mutations to develop new varieties of many crop species (**figure 7.23**). Some kinds of grapefruits, rice, cotton, oats, lettuce, begonias, and many other plants owe their existence to breeders who treated cells with radiation and then selected mutated individuals with interesting new traits. ⓘ *Arabidopsis,* section 19.5

7.7   MASTERING CONCEPTS

1. What is a mutation?
2. What are the types of mutations, and how does each alter the encoded protein?
3. What causes mutations?
4. What is the difference between a germline mutation and a somatic mutation?
5. How are mutations important?

Figure 7.23 Useful Mutants. (a) Rio Red grapefruits and several types of (b) rice and (c) cotton are among the many plant varieties that have been developed by using radiation to induce mutations.

(a): ©Erich Schlegel/Dallas Morning News/Corbis; (b): ©Pallava Bagla/Corbis; (c): ©Scott Olson/Getty Images

INVESTIGATING LIFE

7.8 Clues to the Origin of Language

As you chat with your friends and study for your classes, you may take language for granted. Although communication is not unique to humans, a complex spoken language does set us apart from other organisms. Every human society has language. Without it, people could not transmit information from one generation to the next, so culture could not develop. Its importance to human evolutionary history is therefore incomparable. But how and when did such a crucial adaptation arise?

One clue emerged in the early 1990s, when scientists described a family with a high incidence of an unusual language disorder. Affected family members had difficulty controlling the movements of their mouth and face, so they could not pronounce sounds properly. They also had lower intelligence compared with unaffected individuals, and they had trouble applying simple rules of grammar.

Researchers traced the language disorder to one mutation in a single gene on chromosome 7. Further research revealed that the gene belongs to the large *f*orkhead *b*ox family of genes, abbreviated *FOX*. All members of the *FOX* family encode transcription factors, proteins that bind to DNA and control gene expression. The "language gene" on chromosome 7, eventually named *FOXP2*, is not solely responsible for language acquisition. But the fact that the gene encodes a transcription factor explains how it can simultaneously affect both muscle control and the brain.

To learn more about the evolution of *FOXP2*, scientists Wolfgang Enard, Svante Pääbo, and colleagues at Germany's Max Planck Institute and at the University of Oxford compared the sequences of the 715 amino acids that make up the FOXP2 protein in humans, several other primates, and mice (**figure 7.24**). In the 70 million or so years since the mouse and primate lineages split, the FOXP2 protein has seldom changed. A mutation in the mouse *FOXP2* gene changed 1 amino acid; a different amino acid changed in orangutans. Yet, after humans split from chimpanzees—an event that occurred just 5 or 6 million years ago—the FOXP2 protein changed twice.

Initially, the new, human-specific *FOXP2* version would have been rare, as are all mutations. Today, however, nearly everyone has the same allele of *FOXP2*. The human-specific *FOXP2* allele evidently conferred such improved language skills that individuals with the allele consistently produced more offspring than those without it. That is, natural selection "fixed" the new, beneficial allele in the growing human population.

The research team used mathematical models to estimate that the human-specific mutations happened within the past 200,000 years. A subsequent study, however, revealed that Neandertal DNA contains the same two changes as those observed in modern humans. The mutations therefore must have occurred

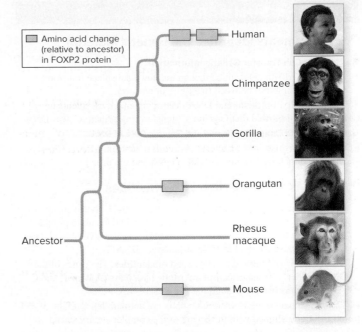

Figure 7.24 FOXP2 Protein Compared. This evolutionary tree shows how the 715 amino acids of the FOXP2 protein differ in mice and various primates. Each blue box represents a difference of 1 amino acid.

Photos: (human): ©Creatas/PictureQuest RF; (chimp): ©Darryl Estrine/Getty Images RF; (gorilla): ©Paul Souders/Corbis; (orangutan): ©Getty Images RF; (monkey): ©Judi Mowlem/Flickr/Getty Images RF; (mouse): ©imageBROKER/SuperStock RF

before modern humans and Neandertals split from their last common ancestor, some 300,000 to 400,000 years ago.

The study of *FOXP2* is important because it helps us understand a critical period in human history. The gene changed after humans diverged from chimpanzees, and then individuals with the new, advantageous allele had higher reproductive fitness than those with any other version. The new allele therefore quickly became fixed in the human population. Without those events, human communication and culture (including everything you chat about with your friends) might never have happened.

Sources: Enard, Wolfgang, Molly Przeworski, Simon E. Fisher, and five coauthors, including Svante Pääbo. August 22, 2002. Molecular evolution of *FOXP2*, a gene involved in speech and language. *Nature*, vol. 418, pages 869–872.

Krause, Johannes, Carles Lalueza-Fox, Ludovic Orlando, and 10 coauthors, including Svante Pääbo. November 6, 2007. The derived *FOXP2* variant of modern humans was shared with Neandertals. *Current Biology*, vol. 17, pages 1908–1912.

7.8 | MASTERING CONCEPTS

1. What question about the *FOXP2* gene were the researchers trying to answer?
2. What insights could scientists gain by intentionally mutating the *FOXP2* gene in a developing human? Would such an experiment be ethical?

CHAPTER SUMMARY

7.1 Experiments Identified the Genetic Material

A. Bacteria Can Transfer Genetic Information
- Frederick Griffith determined that an unknown substance transmits a disease-causing trait between two types of bacteria.
- With the help of protein- and DNA-destroying enzymes, scientists subsequently showed that Griffith's "transforming principle" was DNA.

B. Hershey and Chase Confirmed the Genetic Role of DNA
- Using viruses that infect bacteria, Alfred Hershey and Martha Chase confirmed that the genetic material is DNA and not protein.

7.2 DNA Is a Double Helix of Nucleotides

- Erwin Chargaff discovered that A and T, and G and C, occur in equal proportions in DNA. Maurice Wilkins and Rosalind Franklin provided X-ray diffraction data. James Watson and Francis Crick combined these clues to propose the double-helix structure of **DNA.**
- DNA is made of building blocks called **nucleotides.** The rungs of the DNA "ladder" consist of **complementary** base pairs (A with T, and C with G). Hydrogen bonds hold the two strands together.
- The two chains of the DNA double helix are antiparallel, with the **3′** end of one strand aligned with the **5′** end of the complementary strand.

7.3 DNA Contains the "Recipes" for a Cell's Proteins

- An organism's **genome** includes all of its genetic material. In eukaryotic cells, the genome is divided among multiple **chromosomes** (discrete packages of DNA and associated proteins).

A. Protein Synthesis Requires Transcription and Translation
- A **gene** is a sequence of DNA that is transcribed to **RNA,** typically encoding a protein. To produce a protein, a cell **transcribes** a gene's information to mRNA, which is **translated** into a sequence of amino acids (**figure 7.25** and **table 7.3**).

B. RNA Is an Intermediary Between DNA and a Protein
- Three types of RNA (**mRNA, rRNA,** and **tRNA**) participate in gene expression (**figure 7.26**).

TABLE 7.3 Miniglossary: Protein Synthesis

Term	Definition
Transcription	The first stage of protein synthesis, in which a cell builds an mRNA copy of DNA
Translation	The second stage of protein synthesis, in which a cell builds a protein using the information in mRNA
Template strand	The DNA strand that is transcribed
Codon	A three-nucleotide mRNA sequence that encodes one amino acid or a "stop translation" signal
Genetic code	The "dictionary" that relates each codon with an amino acid or a stop signal

7.4 Transcription Uses a DNA Template to Build RNA

A. Transcription Occurs in Three Steps
- Transcription begins when the **RNA polymerase** enzyme binds to a **promoter** sequence on the DNA **template strand.** RNA polymerase then builds an RNA molecule. Transcription ends when RNA polymerase reaches a **terminator** sequence in the DNA.

B. mRNA Is Altered in the Nucleus of Eukaryotic Cells
- After transcription, the cell adds a cap and a poly A tail to mRNA. **Introns** are cut out of RNA, and the remaining **exons** are spliced together. The finished mRNA molecule then leaves the nucleus.

7.5 Translation Builds the Protein

A. The Genetic Code Links mRNA to Protein
- Each group of three consecutive mRNA bases is a **codon** that specifies one amino acid (or signals translation to stop). The correspondence between codons and amino acids is the **genetic code.**

B. Translation Requires mRNA, tRNA, and Ribosomes
- mRNA carries a protein-encoding gene's information. rRNA associates with proteins to form **ribosomes,** which support and help catalyze protein synthesis.

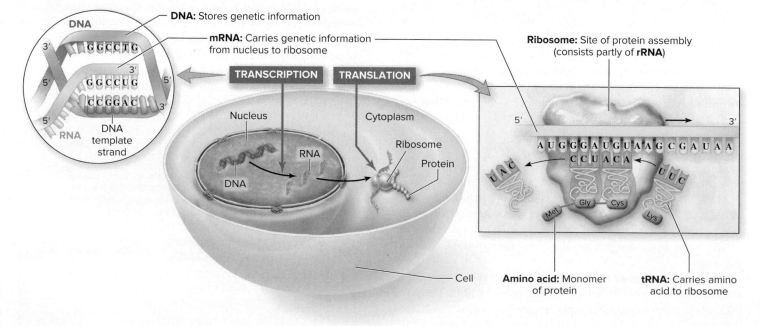

Figure 7.25 Protein Production: A Summary.

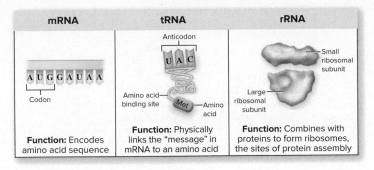

Figure 7.26 Three Types of RNA.

- Each type of tRNA has an end with an **anticodon** complementary to one mRNA codon; the other end of the tRNA carries the corresponding amino acid.

C. Translation Occurs in Three Steps

- The stages of translation are initiation, elongation, and termination.
- In initiation, mRNA joins with a small ribosomal subunit and a tRNA carrying an amino acid. A large ribosomal subunit then joins the small one.
- In the elongation stage, additional tRNA molecules carrying amino acids bind to subsequent mRNA codons. The ribosome moves along the mRNA as the chain grows.
- Termination occurs when the ribosome reaches a "stop" codon. The ribosome is released, and the new polypeptide breaks free.

D. Proteins Must Fold Correctly After Translation

- Chaperone proteins help fold the polypeptide, which may be shortened or combined with others to form the finished protein.

7.6 Cells Regulate Gene Expression

- Protein synthesis requires substantial energy input.

A. Operons Are Groups of Bacterial Genes That Share One Promoter

- In bacteria, **operons** coordinate the expression of grouped genes whose encoded proteins participate in the same metabolic pathway. *E. coli*'s *lac* **operon** is a well-studied example. Transcription does not occur if a **repressor** protein binds to the **operator** sequence of the DNA.

B. Eukaryotic Organisms Use Many Regulatory Mechanisms

- Cell specialization requires the regulation of gene expression.
- **Epigenetic** mechanisms—chemical modifications that alter the activity of regions of DNA—play important roles in regulating gene expression.
- In eukaryotic cells, proteins called **transcription factors** bind to promoters and **enhancers,** which are DNA sequences that regulate which genes a cell transcribes.
- Other regulatory mechanisms include alternative splicing; controls over mRNA stability and translation; and controls over protein folding and movement.

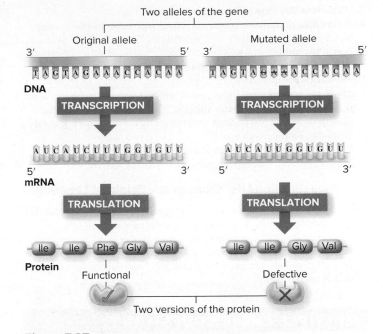

Figure 7.27 Mutations Generate New Alleles.

- Table 7.4 lists some of the most common regulatory mechanisms in the context of the steps in protein production.

7.7 Mutations Change DNA Sequences

- A **mutation** adds, deletes, or changes nucleotides in a DNA sequence (figure 7.27). The result is often an altered protein.

A. Mutations Range from Silent to Devastating

- A **substitution mutation** replaces one base with another. The resulting mRNA may encode the wrong amino acid or substitute a "stop" codon for an amino acid–coding codon. Substitution mutations can also be "silent."
- **Insertion mutations** and **deletion mutations** add or remove nucleotides. In a **frameshift mutation,** inserting or deleting nucleotides disrupts the reading frame and changes the sequence of the encoded protein. Insertions or deletions in groups of three nucleotides do not alter the reading frame.
- **Expanding repeat mutations** cause some inherited illnesses.

B. What Causes Mutations?

- A gene can mutate spontaneously, especially during DNA replication.
- **Mutagens** such as chemicals or radiation also induce some mutations.
- Problems in meiosis can cause mutations if portions of chromosomes are deleted, inverted, or moved.

TABLE **7.4** Regulated Points in Protein Production

Event	Description	Location in Eukaryotic Cell	Regulatory Mechanism(s)
Transcription	RNA polymerase enzyme uses DNA template to produce mRNA.	Nucleus	DNA availability (e.g., epigenetic markers); transcription factors
mRNA processing (eukaryotic cells)	Cap and tail are added; introns are removed.	Nucleus	Selection of introns to be removed; exit of mRNA from nucleus
Translation	Information in mRNA is used to assemble a protein.	Ribosome	mRNA degradation before translation
Protein modification	Protein is folded and may be shortened or combined with other polypeptides.	Anywhere in cell	Availability of proteins needed for processing; tagging for correct location in cell; protein degradation

C. Mutations May Pass to Future Generations

• A **germline mutation** originates in cells that give rise to gametes and therefore appears in every cell of an offspring that inherits the mutation. A **somatic mutation,** which occurs in nonsex cells, affects a subset of cells in the body but does not affect the offspring.

D. Mutations Are Important

• Mutations generate new alleles, which are the raw material for evolution.

• If a mutation occurs in a **homeotic** gene, an animal's body parts may form in the wrong places. Such mutations have revealed much about the genes that control development.

• Induced mutations help scientists deduce gene function and help plant breeders produce fruits and flowers with useful new traits.

7.8 Investigating Life: Clues to the Origin of Language

• A family with a language disorder led researchers to discover a gene that is apparently involved in the acquisition of language.

• Comparing the human version of the gene with that in other primates and in mice suggests that the gene apparently began evolving rapidly soon after modern humans arose. Eventually, one allele became fixed in the human population.

MULTIPLE CHOICE QUESTIONS

1. A nucleotide is composed of all of the following EXCEPT a
   a. sugar.
   c. sulfur-containing group.
   b. nitrogen-containing group.
   d. phosphorus-containing group.

2. A gene is a segment of DNA that encodes a
   a. protein or a functional RNA molecule.
   b. lipid or a protein.
   c. carbohydrate or a lipid.
   d. functional RNA molecule or a carbohydrate.

3. Transcription copies a _____ to a complementary _____ molecule.
   a. chromosome; DNA
   c. gene; RNA
   b. genome; RNA
   d. DNA sequence; ribosome

4. Choose the DNA sequence from which this mRNA sequence was transcribed: 5'-AUACGAUUA-3'.
   a. 3'-TATGCTAAT-5'
   c. 3'-UAUCGUAAU-5'
   b. 3'-UTUGCUTTU-5'
   d. 3'-CTCAGCTTC-5'

5. How many different three-codon sequences encode the amino acid sequence Phe-Val-Ala? *Hint:* Refer to figure 7.12.
   a. 1
   c. 8
   b. 4
   d. 32

6. What is the job of the tRNA during translation?
   a. It carries amino acids to the mRNA.
   b. It triggers the formation of a covalent bond between amino acids.
   c. It binds to the small ribosomal subunit.
   d. It triggers the termination of the protein.

7. How does the *lac* operon regulate lactose digestion in bacteria?
   a. The repressor protein becomes a lactose-digesting enzyme only when lactose is present.
   b. The repressor protein binds to the *lac* operon when lactose is present, blocking transcription.
   c. When lactose is present, it binds to the operator region of the *lac* operon, activating transcription of the repressor protein gene.
   d. The repressor protein falls off the *lac* operon when lactose is present, and lactose-digesting genes are expressed.

8. Certain portions of the mRNA transcribed from the tropomyosin gene can act as either introns or exons. As a result,
   a. one gene may encode many different possible proteins.
   b. each codon may encode many different amino acids.
   c. an amino acid may correspond to many different codons.
   d. a single protein may determine many different traits.

9. If adenine in the tenth nucleotide position of a gene mutates to cytosine, then the _____ amino acid in the protein encoded by this gene could change.
   a. first
   c. tenth
   b. fourth
   d. thirtieth

10. Are mutations harmful?
    a. Yes, because the DNA is damaged.
    b. No, because changes in the DNA result in better alleles.
    c. Yes, because mutated proteins don't function.
    d. It depends on how the mutation affects the protein's function.

Answers to these questions are in appendix A.

WRITE IT OUT

1. Explain how Griffith's experiment and Avery, MacLeod, and McCarty's experiment determined that DNA in bacteria transmits a trait that kills mice.

2. Explain Chargaff's observation that a DNA molecule contains equal amounts of A and T and equal amounts of G and C.

3. Write the complementary DNA sequence of each of the following base sequences:
   a. AGGCATACCTGAGTC
   b. GTTTAATGCCCTACA
   c. AACACTACCGATTCA

4. Put the following in order from smallest to largest: nucleotide, genome, nitrogenous base, gene, nucleus, cell, codon, chromosome.

5. What is the function of DNA?

6. List the three major types of RNA and their functions.

7. Some people compare DNA to a blueprint stored in the office of a construction company. Explain how this analogy would extend to transcription and translation.

8. List the sequences of the mRNA molecules transcribed from the following template DNA sequences:
   a. TGAACTACGGTACCATAC
   b. GCACTAAAGATC

9. How many codons are in each of the mRNA molecules that you wrote for question 8?

10. If a protein is 1259 amino acids long, what is the minimum size of the gene that encodes the protein? Why might the gene be longer than the minimum?

11. The amount of melanin in the skin is controlled by genes, yet melanin is not a protein. How can this be?

12. The roundworm *C. elegans* has 556 cells when it hatches. Each cell contains the entire genome but expresses only a subset of the genes. Therefore, the cells "specialize" in particular functions. List all of the ways that a roundworm cell might silence the unneeded genes.

13. Refer to the figure to answer these questions:
    a. Add labels for mRNA (including the 5′ and 3′ ends) and tRNA. In addition, draw in the RNA polymerase enzyme and the ribosomes, including arrows indicating the direction of movement for each.
    b. What are the next three amino acids to be added to polypeptide *b*?
    c. Fill in the nucleotides in the mRNA complementary to the template DNA strand.
    d. What is the sequence of the DNA complementary to the template strand (as much as can be determined from the figure)?
    e. Does this figure show the entire polypeptide that this gene encodes? How can you tell?
    f. What might happen to polypeptide *b* after its release from the ribosome?
    g. Does this figure depict a prokaryotic or a eukaryotic cell? How can you tell?

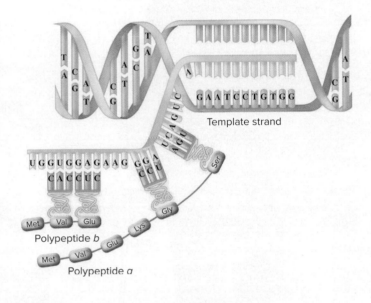

Template strand

Polypeptide *b*

Polypeptide *a*

14. A protein-encoding region of a gene has the following DNA sequence:
    T T T C A T C A G G A T G C A A C A

    Determine how each of the following mutations alters the amino acid sequence:
    a. Substitution of an A for the T in the first position
    b. Substitution of a G for the C in the 17th position
    c. Insertion of a T between the fourth and fifth DNA bases
    d. Insertion of a GTA between the 12th and 13th DNA bases
    e. Deletion of the first DNA nucleotide

15. Explain how a mutation in a protein-encoding gene, an enhancer, or a gene encoding a transcription factor can have the same effect on an organism.

16. Describe the mutation shown in figure 7.27 and explain how the mutation affects the amino acid sequence encoded by the gene.

17. Parkinson disease causes rigidity, tremors, and other motor symptoms. Only 2% of cases are inherited, and these tend to have an early onset of symptoms. Some inherited cases result from mutations in a gene that encodes the protein parkin, which has 12 exons. Indicate whether each of the following mutations in the *parkin* gene would result in a smaller protein, a larger protein, or no change in the size of the protein:
    a. Deletion of exon 3
    b. Deletion of six consecutive nucleotides in exon 1
    c. Duplication of exon 5
    d. Deletion of intron 2

18. In a disorder called gyrate atrophy, cells in the retina begin to degenerate in late adolescence, causing night blindness that progresses to total blindness. The cause is a mutation in the gene that encodes an enzyme, ornithine aminotransferase (OAT). Researchers sequenced the *OAT* gene for five patients, with the following results:

Patient	Mutation
A	A change in codon 209 of UAU to UAA
B	A change in codon 299 of UAC to UAG
C	A change in codon 426 of CGA to UGA
D	A two-nucleotide deletion at codons 64 and 65 that results in a UGA codon at position 79
E	Exon 6, including 1071 nucleotides, is entirely deleted

a. Which patient(s) have a frameshift mutation?
b. How many amino acids is patient E missing?
c. Which patient(s) will produce a shortened protein?

PULL IT TOGETHER

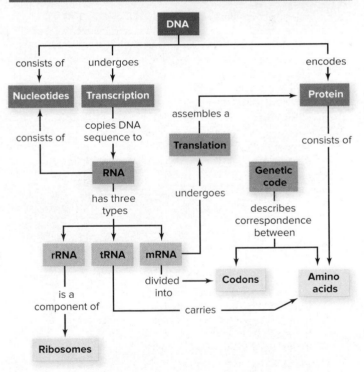

Figure 7.28 Pull It Together: DNA Structure and Gene Function.

Refer to **figure 7.28** and the chapter content to answer the following questions.

1. Why is protein production essential to cell function?

2. Where do promoters, terminators, stop codons, transcription factors, RNA polymerase, and enhancers fit into this concept map?

3. Use the concept map to explain how DNA nucleotides are related to amino acids.

4. Review the Survey the Landscape figure in the chapter introduction, and then explain why a mutation in DNA sometimes causes protein function to change.

8 DNA Replication, Binary Fission, and Mitosis

Big Man. One of the world's tallest men, Bao Xishun, stands at 2.36 meters (7 feet, 9 inches). Rapid cell division during adolescence explains his extreme height.
©Frederic J. Brown/AFP/Getty Images

LEARN HOW TO LEARN
Explain It, Right or Wrong

As you work through the multiple choice questions at the end of each chapter, make sure you can explain why each correct choice is right. You can also test your understanding by taking the time to explain why each of the other choices is wrong.

The Tallest and the Shortest

A quick glance at any crowd reveals that not everyone is the same height. In the United States, the average height is about 1.78 meters (5 feet, 10 inches) for men and about 1.62 m (5 feet, 4 inches) for women. But those statistics hide a great deal of variation; many people are much taller or shorter than average.

A person's adult height reflects cell division during childhood and adolescence, especially in the ends of the bones. During cell division, one cell becomes two. The next round of division yields four cells, and the following round produces eight. Each new cell grows as it produces proteins and takes in water and other nutrients. If the thousands of cells at each end of a bone are all actively dividing, then the bone tissue quickly expands, and the person gets taller. ⓘ *bone growth,* section 29.3B

Chemical signals regulate bone growth. The most important signal is human growth hormone, which is secreted by the pituitary gland. Human growth hormone "tells" bone cells to divide. The more growth hormone, the greater the rate of cell division. Small changes in the abundance of growth hormone may therefore profoundly affect development.

For example, gigantism results from an overproduction of growth hormone early in life. Over many years of bone growth, a small excess in growth hormone may produce bones that are much longer than normal. The tallest person known to have lived, Robert Wadlow, had gigantism (see figure 28.5). He never stopped growing, and he was 2.72 meters (nearly 9 feet) tall when he died.

At the other extreme, a growth hormone deficiency produces a type of dwarfism. In this disorder, bone cells do not receive enough signals to divide during childhood. While the brain develops normally, other body parts remain small. One person with this disorder is Chandra Bahadur Dangi of Nepal. At only about 0.55 m (1 foot, 9 inches) tall, he is the shortest known adult.

Dwarfism has other causes as well. The most common form of dwarfism, called achondroplasia, reflects a problem with the cells of the limb bones. Even though human growth hormone is present, the bone cells fail to divide, and the arms and legs remain abnormally short.

The natural variation in human height shines the spotlight on a larger message: Tissue growth and repair rely on cell division, the subject of this chapter. As you learn more about how cells divide, think back on this essay and remember that faulty signals can produce surprisingly large—or small—effects.

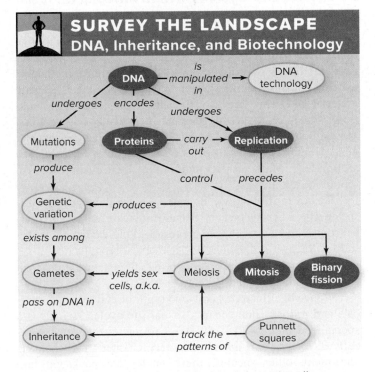

SURVEY THE LANDSCAPE
DNA, Inheritance, and Biotechnology

Prokaryotic cells divide by binary fission; eukaryotic cells use mitotic or meiotic division. Proteins regulate the entire process and copy all of the cell's DNA shortly before the split occurs.

For more details, study the Pull It Together feature in the chapter summary.

8.1 Cells Divide and Cells Die

Your cells are too small to see without a microscope, so it is hard to appreciate just how many you lose as you sleep, work, and play. Each minute, for example, you shed tens of thousands of dead skin cells. If you did not have a way to replace these lost cells, your body would literally wear away. Instead, cells in your deep skin layers divide and replace the ones you lose. Each new cell lives an average of about 35 days, so you will gradually replace your entire skin in the next month or so—without even noticing!

Cell division produces a continuous supply of replacement cells, both in your skin and everywhere else in your body. But cell division has other functions as well. No living organism can reproduce without cell division, and the growth and development of a multicellular organism also require the production of new cells.

This chapter explores the opposing but coordinated forces of cell division and cell death and considers what happens if either process goes wrong. We begin by exploring cell division's role in reproduction, growth, and development.

A. Sexual Life Cycles Include Mitosis, Meiosis, and Fertilization

Organisms must reproduce—generate other individuals like themselves—for a species to persist. For a single-celled organism, the most straightforward (and ancient) method is **asexual reproduction,** in which one cell replicates its genetic material and splits into two. Except for the occasional mutation, asexual reproduction generates genetically identical offspring. Bacteria and archaea, for example, reproduce asexually via a simple type of cell division called binary fission (see section 8.3). Many protists and multicellular eukaryotes also reproduce asexually.

Sexual reproduction, in contrast, is the production of offspring whose genetic makeup comes from two parents. Each parent contributes a sex cell, and the fusion of these cells signals the start of the next generation. Because sexual reproduction mixes up and recombines traits, the offspring are genetically different from each other and their parents.

Figure 8.1 illustrates how two types of cell division, meiosis and mitosis, interact in a sexual life cycle. **Meiosis** is a specialized process that gives rise to nuclei that are genetically different from one another (see chapter 9). In humans and many other species, these nuclei are packaged into **gametes:** sperm cells (produced by males) and egg cells (produced by females). The variation among gametes explains why siblings generally look different from one another (except for identical twins).

Fertilization is the union of the sperm and the egg cell, producing a zygote (the first cell of the new offspring). Immediately after fertilization, the other type of cell division—mitotic—takes over. **Mitosis** divides a eukaryotic cell's genetic information into two identical nuclei.

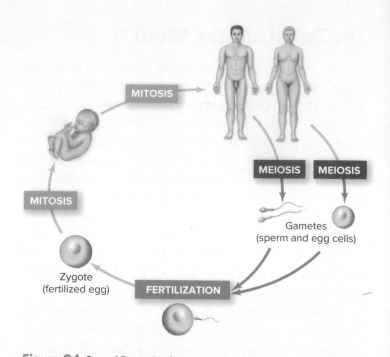

Figure 8.1 Sexual Reproduction. In the life cycle of humans and many other organisms, adults produce gametes by meiosis. Fertilization unites sperm and egg, forming a zygote. Mitotic cell division accounts for the growth of the new offspring.

Each of the trillions of cells in your body retains the genetic information that was present in the fertilized egg. Inspired by the astonishing precision with which this occurs, geneticist Herman J. Müller wrote in 1947:

> In a sense we contain ourselves, wrapped up within ourselves, trillions of times repeated.

This quotation eloquently expresses the powerful idea that every cell in the body results from countless rounds of cell division, each time forming two genetically identical cells from one.

Mitotic cell division explains how you repair damage after an injury, how you replace the cells that you lose every day, and how you grew from a single-celled zygote into an adult (**figure 8.2**). Likewise, mitotic cell division accounts for the growth and development of plants, mushrooms, and other multicellular eukaryotes and for asexual reproduction in protists and many other eukaryotes.

B. Cell Death Is Part of Life

The development of a multicellular organism requires more than just cell division. Cells also die in predictable ways, carving distinctive structures. **Apoptosis,** also called "programmed cell death," is a normal part of development. Like cell division, apoptosis is a precise, tightly regulated sequence of events (see section 8.7).

During early development, both cell division and apoptosis shape new structures. For example, the feet of both ducks and chickens start out as webbed paddles when the birds are embryos

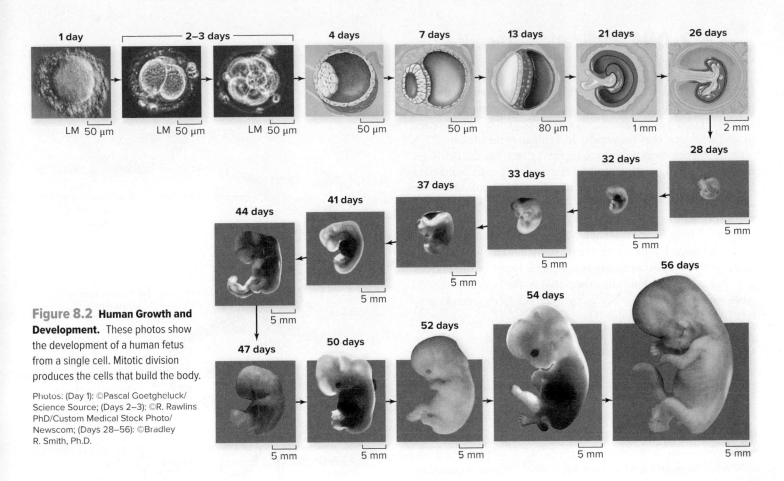

Figure 8.2 Human Growth and Development. These photos show the development of a human fetus from a single cell. Mitotic division produces the cells that build the body.

Photos: (Day 1): ©Pascal Goetgheluck/ Science Source; (Days 2–3): ©R. Rawlins PhD/Custom Medical Stock Photo/ Newscom; (Days 28–56): ©Bradley R. Smith, Ph.D.

Figure 8.3 Apoptosis Carves Toes. The feet of embryonic birds have webbing between the digits. (a) A duck's foot retains this webbing. (b) In a developing chicken foot, the toes take shape when the cells between the digits die.

Photos: (a): ©GK Hart/Vikki Hart/Getty Images RF; (b): ©Stockbyte RF

(figure 8.3). The webs of tissue remain in the duck's foot throughout life. In the chicken, however, individual toes form as cells between the digits die. Likewise, cells in the tail of a tadpole die as the young frog develops into an adult.

Throughout an animal's life, cell division and cell death are in balance, so tissue neither overgrows nor shrinks. Cell division compensates for the death of skin and blood cells, a little like adding new snow (cell division) to a snowman that is melting (apoptosis). Both cell division and apoptosis also help protect the organism. For example, cells divide to heal a scraped knee; apoptosis peels away sunburnt skin cells that might otherwise become cancerous.

Before learning more about mitosis and apoptosis, it is important to understand how the genetic material in a eukaryotic cell copies itself and condenses in preparation for cell division. Sections 8.2 and 8.4 explain these events.

8.1 MASTERING CONCEPTS

1. Explain the roles of mitotic cell division, meiosis, and fertilization in the human life cycle.
2. Why are both cell division and apoptosis necessary for the development of an organism?

8.2 DNA Replication Precedes Cell Division

Before any cell divides—by binary fission, mitotically, or meiotically—it must first duplicate its entire **genome,** which consists of all of the cell's genetic material. The genome may consist of one or more **chromosomes,** individual molecules of DNA with their associated proteins. Chapter 7 describes the cell's genome as a set of "cookbooks" (chromosomes), each containing "recipes" (genes) that encode proteins. In DNA replication, the cell copies all of this information, letter by letter. Without a full set of instructions, a new cell may die.

Recall from figure 7.5 that DNA is a double-stranded nucleic acid. Each strand of the double helix is composed of nucleotides. Hydrogen bonds between the nitrogenous bases of the nucleotides hold the two strands together. That is, the base adenine (A) pairs with thymine (T), whereas cytosine (C) forms complementary base pairs with guanine (G).

Figure It Out

How many cytosines are in the strand that is complementary to the following DNA sequence? 5′-TCAATACCGATTAT-3′

Answer: 1

When Watson and Crick reported DNA's chemical structure, they understood that they had uncovered the key to DNA replication. They envisioned DNA unwinding, exposing unpaired bases that would attract their complements, and neatly knitting two double helices from one. This route to replication, which turned out to be essentially correct, is called semiconservative, because each DNA double helix ends up with one complete strand from the original molecule (**figure 8.4**).

DNA does not, however, replicate by itself. Instead, an army of enzymes copies DNA just before a cell divides (**figure 8.5**). Enzymes called helicases unwind and "unzip" the DNA, while binding proteins prevent the two single strands from rejoining each other. Other enzymes then guide the assembly of new DNA strands. **DNA polymerase** is the enzyme that adds new DNA nucleotides that are complementary to the bases on each exposed strand. As the new DNA strands grow, hydrogen bonds form between the complementary bases. ⓘ *enzymes, section 4.4*

Curiously, DNA polymerase can add nucleotides only to an existing strand. A primase enzyme therefore must build a short, complementary piece of RNA, called an RNA primer, at the start of each DNA segment to be replicated. The RNA primer attracts the DNA polymerase enzyme. Once the new strand of DNA is in place, another enzyme removes each RNA primer. DNA polymerase then fills the gap with the correct DNA nucleotides. This enzyme cannot, however, join the existing strand with the last nucleotide to be placed in the gap. Enzymes called **ligases** form the covalent bonds that seal these nicks.

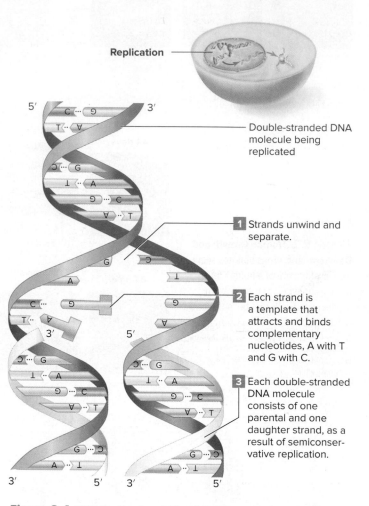

Figure 8.4 DNA Replication: A Simplified View. (*1*) DNA strands unwind and separate. (*2*) New nucleotides form complementary base pairs with each exposed strand. (*3*) The process ends with two identical double-stranded DNA molecules.

Furthermore, like the RNA polymerase enzyme described in section 7.4, DNA polymerase can add new nucleotides only to the exposed 3′ end—never the 5′ end—of a growing strand. Replication therefore proceeds continuously on only one new DNA strand, called the leading strand. On the other strand, called the lagging strand, replication occurs in short 5′ to 3′ pieces. These short pieces, each preceded by its own RNA primer, are called Okazaki fragments.

Replication enzymes work simultaneously at hundreds of points, called origins of replication, on each DNA molecule (**figure 8.6**). This arrangement is similar to the way that hurried office workers might split a lengthy report into short pieces and then divide the sections among many copy machines operating at the same time. Thanks to this division of labor, copying the billions of DNA nucleotides in a human cell takes only 8 to 10 hours.

DNA replication proteins

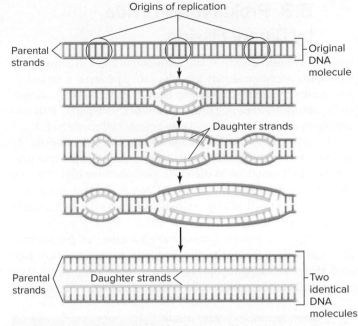

Helicase unwinds double helix.

Binding proteins stabilize each strand.

Primase adds short RNA primer to template strand.

DNA polymerase binds nucleotides to form new strands.

Ligase creates covalent bond between adjacent DNA segments.

Steps in DNA replication

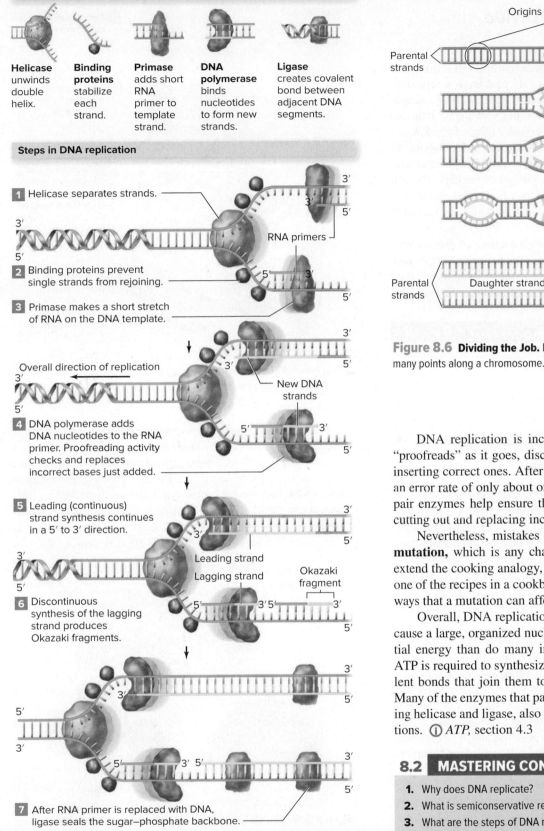

1 Helicase separates strands.

3′

5′

2 Binding proteins prevent single strands from rejoining.

3 Primase makes a short stretch of RNA on the DNA template.

RNA primers

3′
5′

Overall direction of replication

3′
5′

New DNA strands

4 DNA polymerase adds DNA nucleotides to the RNA primer. Proofreading activity checks and replaces incorrect bases just added.

5 Leading (continuous) strand synthesis continues in a 5′ to 3′ direction.

3′
5′

Leading strand

Lagging strand

Okazaki fragment

6 Discontinuous synthesis of the lagging strand produces Okazaki fragments.

3′

5′

7 After RNA primer is replaced with DNA, ligase seals the sugar–phosphate backbone.

Figure 8.5 DNA Replication in Detail. Many types of proteins participate in DNA replication. Although they are depicted separately for clarity, in reality they form a cluster that moves along the DNA molecule.

Origins of replication

Parental strands

Original DNA molecule

Daughter strands

Parental strands

Daughter strands

Two identical DNA molecules

Figure 8.6 Dividing the Job. DNA replication occurs simultaneously at many points along a chromosome.

DNA replication is incredibly accurate. DNA polymerase "proofreads" as it goes, discarding mismatched nucleotides and inserting correct ones. After proofreading, DNA polymerase has an error rate of only about one in a billion nucleotides. Other repair enzymes help ensure the accuracy of DNA replication by cutting out and replacing incorrect nucleotides.

Nevertheless, mistakes occasionally remain. The result is a **mutation,** which is any change in a cell's DNA sequence. To extend the cooking analogy, a mutation is similar to a mistake in one of the recipes in a cookbook. Section 7.7 describes the many ways that a mutation can affect the life of a cell.

Overall, DNA replication requires a great deal of energy because a large, organized nucleic acid contains much more potential energy than do many individual nucleotides. Energy from ATP is required to synthesize nucleotides and to create the covalent bonds that join them together in the new strands of DNA. Many of the enzymes that participate in DNA replication, including helicase and ligase, also require energy to catalyze their reactions. ⓘ *ATP,* section 4.3

8.2 MASTERING CONCEPTS

1. Why does DNA replicate?
2. What is semiconservative replication?
3. What are the steps of DNA replication?
4. Could DNA replication occur if primase were not present in a cell? Explain your answer.
5. Why do enzymes work at multiple origins of replication?

8.3 Prokaryotes Divide by Binary Fission

Like all organisms, bacteria and archaea transmit DNA from generation to generation as they reproduce. In prokaryotes, reproduction occurs by **binary fission,** an asexual process that replicates DNA and distributes it (along with other cell parts) into two daughter cells (**figure 8.7**). ⓘ *prokaryotic cells,* section 3.2

Each prokaryotic cell contains one circular chromosome. As the cell prepares to divide, its DNA replicates. The chromosome and its duplicate attach to the inner surface of the cell. The cell membrane grows between the two DNA molecules, separating them. Then the cell pinches in half to form two daughter cells from the original one.

In optimal conditions, some bacterial cells can divide every 20 minutes. Those few microbes that remain after you brush your teeth therefore easily repopulate your mouth as you sleep; their metabolic activities produce the foul-smelling "morning breath."

8.3   MASTERING CONCEPTS

1. Which cell types divide by binary fission?
2. What are the events of binary fission?

8.4 Chromosomes Condense Before Cell Division

The genetic information in a eukaryotic cell consists of multiple chromosomes inside a nucleus. Each species has a characteristic number of chromosomes. A mosquito's cell has 6 chromosomes; grasshoppers, rice plants, and pine trees all have 24; humans have 46; dogs and chickens have 78; carp have 104. In all of these species, the amount of information is immense. For example, if the DNA bases of all 46 human chromosomes were typed as A, C, T, and G, the several billion letters would fill 4000 books of 500 pages each.

With so much genetic information, a eukaryotic cell must balance two needs. On one hand, the cell must have access to the information in its DNA. On the other hand, a dividing cell must package its DNA into a portable form that can easily move into the two daughter cells (**figure 8.8**). DNA packing is therefore comparable to winding yarn into a compact ball. Just as a ball of yarn occupies less space and is more portable than a pile of loose yarn, condensed DNA is easier for the cell to manage than is an unwound chromosome.

To learn how cells maintain this balance, we must look closely at a chromosome's structure. Eukaryotic chromosomes consist of **chromatin,** which is a collective term for all of the cell's DNA and its associated proteins. These proteins include the many enzymes that help replicate the DNA and transcribe it to RNA (see chapter 7). Others serve as scaffolds around which DNA entwines, helping to pack the DNA efficiently inside the cell.

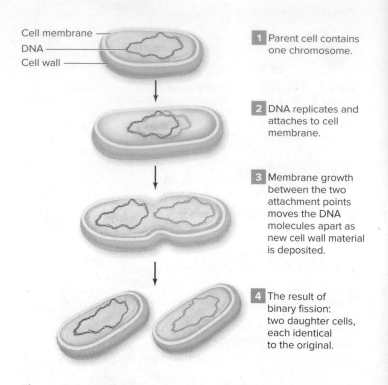

Figure 8.7 Binary Fission. A dividing prokaryotic cell replicates its DNA, grows, and then splits into two identical daughter cells.

Cell membrane
DNA
Cell wall

1 Parent cell contains one chromosome.

2 DNA replicates and attaches to cell membrane.

3 Membrane growth between the two attachment points moves the DNA molecules apart as new cell wall material is deposited.

4 The result of binary fission: two daughter cells, each identical to the original.

Chromatin is organized into **nucleosomes,** each consisting of a stretch of DNA wrapped around eight proteins (histones). A continuous thread of DNA connects nucleosomes like beads on a string (**figure 8.9**). When the cell is not dividing, chromatin is barely visible because the nucleosomes are loosely packed together. The cell can therefore access the information in the DNA to produce the proteins that it needs. DNA replication in preparation for cell division also requires that the cell's DNA be unwound.

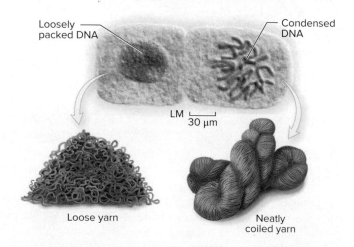

Loosely packed DNA
Condensed DNA
LM
30 μm
Loose yarn
Neatly coiled yarn

Figure 8.8 Two Views of DNA. Loosely packed DNA is available for replication and transcription. Before a cell divides, however, the DNA condenses into compact chromosomes. These two forms of DNA are comparable to loose and coiled yarn, respectively.

Photo: ©Clouds Hill Imaging Ltd./Corbis

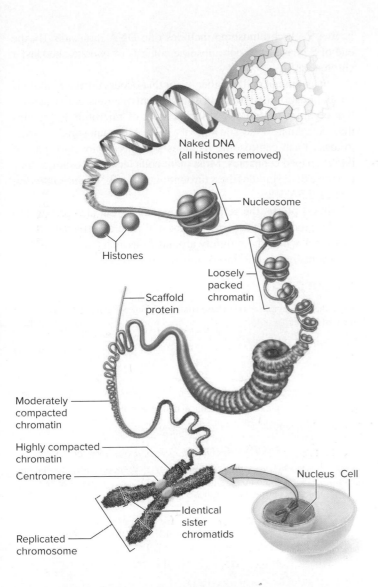

Figure 8.9 Parts of a Chromosome. DNA replicates just before a cell divides, and then the chromatin condenses into its familiar, compact form.

8.5 Mitotic Division Generates Exact Cell Copies

Suppose you scrape your leg while sliding into second base during a softball game. At first, the wound bleeds, but the blood soon clots and forms a scab. Underneath the dried crust, cells of the immune system clear away trapped dirt and dead cells. At the same time, undamaged skin cells bordering the wound begin to divide repeatedly, producing fresh, new daughter cells that eventually fill the damaged area.

Those actively dividing skin cells illustrate the **cell cycle,** which describes the events that occur in one complete round of cell division. Biologists divide the cell cycle into stages (**figure 8.10**). **Interphase** is the interval between successive cell divisions; protein synthesis, DNA replication, and many other events occur during interphase. Next is mitosis, during which the contents of the nucleus divide. In **cytokinesis,** the cell splits into daughter cells. After cytokinesis is complete, each daughter cell enters interphase, and the cell cycle begins anew.

Mitotic cell division occurs some 300 million times per minute in your body, replacing cells lost to abrasion or cell death. In each case, the products of cell division are two daughter cells, each receiving complete, identical genetic instructions plus the molecules and organelles they need for their own metabolism.

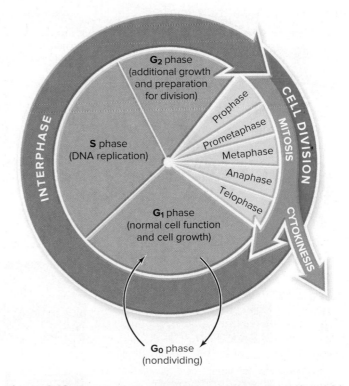

Figure 8.10 The Cell Cycle. Interphase includes gap phases (G_1 and G_2), when the cell grows and some organelles duplicate. Nondividing cells can leave G_1 and enter G_0 indefinitely. During the synthesis phase (S) of interphase, DNA replicates. Mitosis divides the replicated genetic material between two nuclei. Cytokinesis then splits the cytoplasm in half, producing two identical daughter cells.

The chromosome's appearance changes shortly after DNA replication. The nucleosomes gradually fold into progressively larger structures, making the DNA portable but unavailable for transcription. Once fully condensed, the chromosome has readily identifiable parts (see figure 8.9). Two **chromatids** make up the replicated chromosome. Because these paired chromatids have identical sequences, they are called "sister chromatids." The **centromere** is a small section of DNA and associated proteins that attaches the sister chromatids to each other. As a cell's genetic material divides, the centromere splits, and the sister chromatids move apart. At that point, each chromatid becomes an individual chromosome.

8.4 | MASTERING CONCEPTS

1. How are chromosomes and chromatin related?
2. Sketch and label the main parts of a duplicated chromosome.

A. DNA Is Copied During Interphase

Biologists once mistakenly described interphase as a time when the cell is at rest. The chromatin is unwound and therefore barely visible, so the cell appears inactive. However, interphase is actually a very active time. The cell produces proteins and carries out its functions, from photosynthesis to muscle contraction to insulin production to bone formation. DNA replication also occurs during this stage.

Interphase is divided into "gap" phases (designated G_1, G_0, and G_2), separated by a "synthesis" (S) phase. During **G_1 phase,** the cell grows, carries out its basic functions, and produces the new organelles and other components it will require if it divides. A cell in G_1 may enter a nondividing stage called G_0.

In **G_0 phase,** a cell continues to function, but it does not replicate its DNA or divide. At any given time, most cells in the human body are in G_0. Nerve cells in the brain are permanently in G_0, which explains both why the brain does not grow after it reaches its adult size and why brain damage is often irreparable.

During **S phase,** enzymes replicate the cell's genetic material and repair damaged DNA (see section 8.2). As S phase begins, each chromosome includes one DNA molecule. By the end of S phase, each chromosome consists of two attached sister chromatids.

In an animal cell, another event that occurs during S phase is the duplication of the centrosome. **Centrosomes** are structures that organize the mitotic **spindle,** a set of microtubule proteins that coordinates the movements of the chromosomes during mitosis. Each centrosome includes proteins enclosing a pair of barrel-shaped centrioles. Most plant cells lack centrosomes; they organize their spindle fibers throughout the cell. ⓘ *microtubules,* section 3.5A

In **G_2 phase,** the cell continues to grow but also prepares to divide, producing the proteins that will help coordinate mitosis. The DNA winds more tightly around its associated proteins, and this event signals the start of mitosis. Interphase has ended.

Figure It Out

A cell that has completed interphase contains ___ times as much DNA as a cell at the start of interphase.

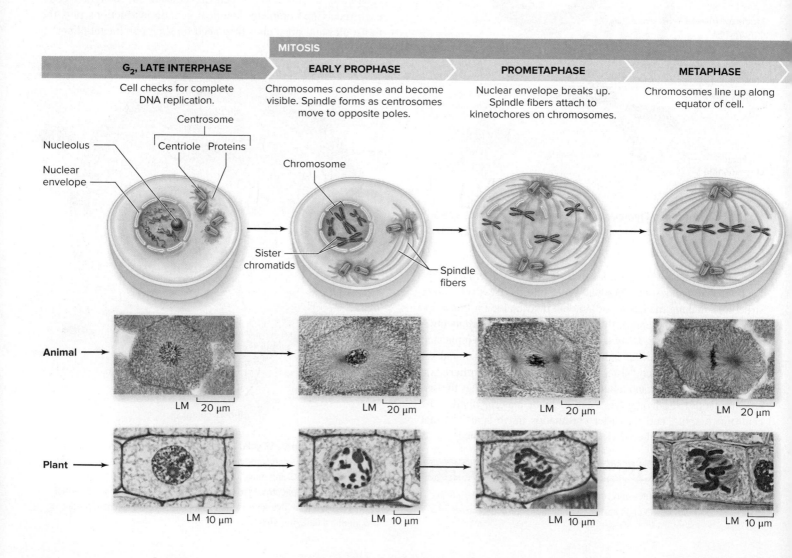

MITOSIS			
G_2, LATE INTERPHASE	**EARLY PROPHASE**	**PROMETAPHASE**	**METAPHASE**
Cell checks for complete DNA replication.	Chromosomes condense and become visible. Spindle forms as centrosomes move to opposite poles.	Nuclear envelope breaks up. Spindle fibers attach to kinetochores on chromosomes.	Chromosomes line up along equator of cell.

Nucleolus

Nuclear envelope

Centrosome

Centriole   Proteins

Chromosome

Sister chromatids

Spindle fibers

Animal →

LM 20 μm   LM 20 μm   LM 20 μm   LM 20 μm

Plant →

LM 10 μm   LM 10 μm   LM 10 μm   LM 10 μm

B. Chromosomes Divide During Mitosis

Overall, mitosis separates the genetic material that replicated during S phase. For the chromosomes to be evenly distributed, they must line up in a way that enables them to split equally into two sets that are then pulled to opposite poles of the cell.

Mitosis is a continuous process, but biologists divide it into stages for ease of understanding. Figure 8.11 summarizes the key events of mitosis; you may find it helpful to consult this figure as you read on.

During **prophase,** DNA coils very tightly, shortening and thickening the chromosomes (see figure 8.9). As the chromosomes condense, they become visible when stained and viewed under a microscope. For now, the chromosomes remain randomly arranged in the nucleus. Also during prophase, the two centrosomes migrate toward opposite poles of the cell, and the spindle begins to form. The nucleolus—the dark area in the nucleus—disappears.

Prometaphase occurs immediately after the formation of the spindle. The nuclear envelope breaks into small pieces, as does the surrounding endoplasmic reticulum. The spindle fibers are now free to reach the chromosomes. Meanwhile, proteins called **kinetochores** begin to assemble on each centromere; these proteins attach the chromosomes to the spindle.

As **metaphase** begins, the spindle lines up the chromosomes along the center, or equator, of the cell. This arrangement ensures that each cell will receive one copy of each chromosome.

In **anaphase,** the centromeres split, and some spindle fibers shorten as they pull the sister chromatids (now chromosomes) toward opposite poles of the cell. At the same time, other microtubules in the spindle lengthen in a way that moves the poles farther apart, stretching the dividing cell.

Figure 8.11 Stages of Mitosis. Mitotic cell division includes similar stages in all eukaryotes, including animals and plants. Notice that the cell entering mitosis has four chromosomes (*two red and two blue*), as does each of the two resulting daughter cells.

Photos: (all animal): ©Ed Reschke/Photolibrary/Getty Images; (all plant): ©Ed Reschke

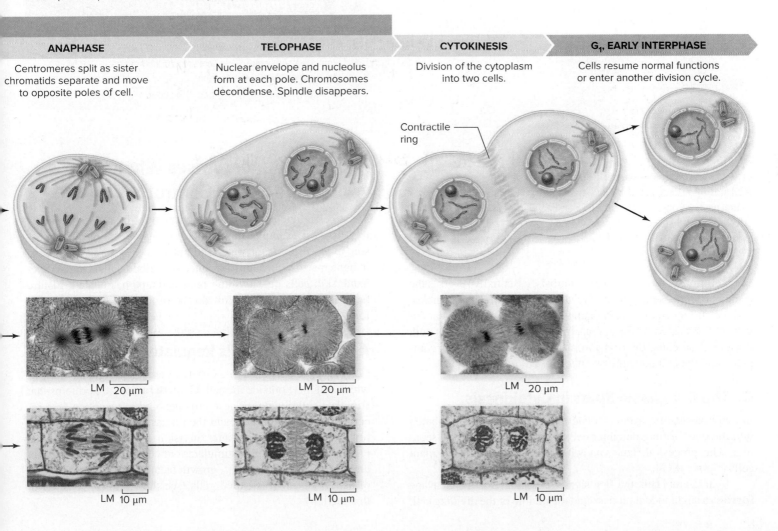

ANAPHASE	TELOPHASE	CYTOKINESIS	G₁, EARLY INTERPHASE
Centromeres split as sister chromatids separate and move to opposite poles of cell.	Nuclear envelope and nucleolus form at each pole. Chromosomes decondense. Spindle disappears.	Division of the cytoplasm into two cells.	Cells resume normal functions or enter another division cycle.

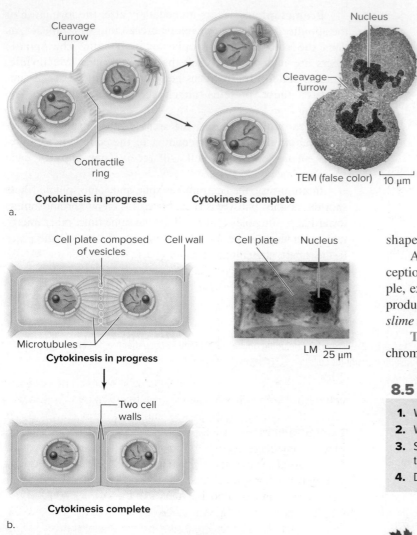

Figure 8.12 Cytokinesis. (a) In an animal cell, the first sign of cytokinesis is an indentation called a cleavage furrow, which is formed by a contractile ring consisting of actin and myosin proteins. (b) In plant cells, the cell plate is the first stage in the formation of a new cell wall.

Photos: (a): ©Dr. Gopal Murti/Science Source; (b): ©Ed Reschke/Photolibrary/Getty Images

Telophase, the final stage of mitosis, essentially reverses the events of prophase and prometaphase. The spindle disassembles, and the chromosomes begin to unwind. In addition, a nuclear envelope and nucleolus form at each end of the stretched-out cell. As telophase ends, the division of the genetic material is complete, and the cell contains two nuclei—but not for long.

C. The Cytoplasm Splits in Cytokinesis

In cytokinesis, the cytoplasm and the two nuclei are distributed into the two forming daughter cells, which then physically separate. The process differs somewhat between animal and plant cells (**figure 8.12**).

In an animal cell, the first sign of cytokinesis is the **cleavage furrow,** a slight indentation around the middle of the dividing cell.

A ring of actin and myosin proteins beneath the cell membrane contracts like a drawstring, separating the daughter cells.

Unlike animal cells, plant cells are surrounded by cell walls. A dividing plant cell must therefore construct a new wall that separates the two daughter cells. The first sign of cell wall construction is the **cell plate,** a structure that appears at the midline of the dividing plant cell. The cell plate grows and consolidates as vesicles from the Golgi apparatus deliver cellulose, other polysaccharides, and proteins. The resulting layer of cellulose fibers embedded in surrounding material makes a strong, rigid wall that gives a plant cell its shape. ⓘ *cell wall,* section 3.6B

Although cytokinesis typically follows mitosis, there are exceptions. Some types of green algae and slime molds, for example, exist as enormous cells containing thousands of nuclei, the products of many rounds of mitosis without cytokinesis. ⓘ *slime molds,* section 18.3A

Table 8.1 summarizes some of the vocabulary related to chromosomes and the cell cycle.

8.5	**MASTERING CONCEPTS**

1. What are the three main events of the cell cycle?

2. What happens during interphase?

3. Suppose a centromere does not split during anaphase. Describe the chromosomes in the daughter cells.

4. Distinguish between mitosis and cytokinesis.

🍁 8.6 Cancer Arises When Cells Divide Out of Control

Some cells divide more or less constantly. The cells at a plant's root tips, for example, may divide throughout the growing season, exploring soil for water and nutrients. Likewise, stem cells in bone marrow constantly produce new blood cells. On the other hand, skin cells quit dividing once a scrape has healed; mature brain cells rarely divide. How do any of these cells "know" what to do?

A. Chemical Signals Regulate Cell Division

Cells divide in response to a variety of chemical signals, many of which originate outside the cell. **Growth factors** are proteins that stimulate cell division. These proteins bind to receptors on a receiving cell's membrane, and then a cascade of chemical reactions inside the cell initiates division. At a wound site, for example, a growth factor stimulates our cells to produce new skin underneath a scab; in plants, growth factors induce the formation of abnormal growths called galls (see the Burning Question in this section).

In addition, several internal "checkpoints" ensure that a cell does not enter one stage of the cell cycle until the previous stage is complete. **Figure 8.13** illustrates a few cell cycle checkpoints:

- The G_1 checkpoint screens for damaged DNA. If the genetic material is damaged beyond repair, a protein called p53 promotes the expression of genes encoding DNA repair enzymes. Badly damaged DNA prompts p53 to trigger apoptosis, and the cell dies.

- Several S phase checkpoints ensure that DNA replication occurs properly. If the cell does not have enough nucleotides to complete replication or if a DNA molecule breaks, the cell cycle may pause or stop at this point.

- The G_2 checkpoint is the last one before the cell begins mitosis. If the cell does not contain two full sets of identical DNA or if the spindle-making machinery is not in place, the cell cycle may be delayed. Alternatively, the p53 protein may trigger apoptosis.

- The metaphase checkpoint ensures that all chromosomes are aligned and that the spindle fibers attach correctly to the chromosomes. If everything checks out, the cell proceeds to anaphase.

Two groups of internal signaling proteins guide a cell's progress through these checkpoints. The concentrations of proteins called cyclins fluctuate in predictable ways during each stage. For example, cyclin E peaks between the G_1 and S phases of interphase, whereas cyclin B is essentially absent at that time but has its highest concentration between G_2 and mitosis. Proteins that bind to each cyclin, in turn, translate these fluctuations into action by activating the transcription factors that stimulate entry into the next stage of the cell cycle. ⓘ *transcription factors,* section 7.6B

A cell that fails to pass a checkpoint correctly will not undergo the change in cyclin concentrations that allows it to progress to the next stage. These checkpoints are therefore somewhat like the guards that check passports at border crossings, denying entry to travelers without proper documentation.

Precise timing of the many chemical signals that regulate the cell cycle is essential. Too little cell division, and an injury may go unrepaired; too much, and an abnormal growth forms. Understanding these signals has helped reveal how diseases such as cancer arise.

B. Cancer Cells Break Through Cell Cycle Controls

What happens when the body loses control over cell division and cell death? Sometimes, a **tumor**—an abnormal mass of tissue—forms. Biologists classify tumors into two groups (**figure 8.14**). **Benign tumors** are usually slow-growing and harmless, unless they become large enough to disrupt nearby tissues or organs. A tough capsule surrounding the tumor prevents it from invading nearby tissues or spreading to other parts of the body. Warts and moles are examples of benign tumors of the skin.

In contrast, a **malignant tumor** invades adjacent tissue. Because it lacks a surrounding capsule, a malignant tumor is likely to **metastasize,** meaning that its cells can break away from the original mass and travel in the bloodstream or lymphatic system to colonize other areas of the body. **Cancer** is a class of diseases characterized by malignant cells.

Solid tumors of the breast, lung, skin, and other major organs are the most familiar forms of cancer. But cells in the blood-forming tissues of the bone marrow can also divide out of control. Leukemia is a group of cancers characterized by the excessive production of the wrong kinds of blood cells.

Whatever its form, cancer begins when a single cell breaks through its death and division controls. As the cell continues to divide, a tumor may develop, with each cancerous cell passing its loss of cell cycle control to its daughter cells. The growing tumor may crush vital organs, block the body's passageways, and divert nutrients from other body cells.

C. Cancer Cells Differ from Normal Cells in Many Ways

Cancer cells have multiple unique characteristics. First, a cancer cell looks different from a normal cell (**figure 8.15**). Its shape may be different, and it may lose some of the

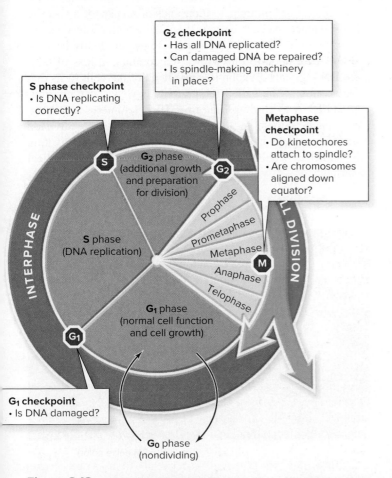

G₂ checkpoint
- Has all DNA replicated?
- Can damaged DNA be repaired?
- Is spindle-making machinery in place?

S phase checkpoint
- Is DNA replicating correctly?

Metaphase checkpoint
- Do kinetochores attach to spindle?
- Are chromosomes aligned down equator?

G₁ checkpoint
- Is DNA damaged?

INTERPHASE

S phase
(DNA replication)

G₂ phase
(additional growth and preparation for division)

Prophase
Prometaphase
Metaphase
Anaphase
Telophase

CELL DIVISION

G₁ phase
(normal cell function and cell growth)

G₀ phase
(nondividing)

Figure 8.13 Cell Cycle Control Checkpoints. These checkpoints ensure that a cell completes each stage of the cell cycle correctly before proceeding to the next.

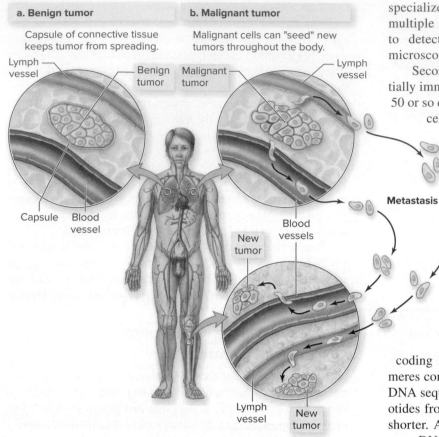

a. Benign tumor

Capsule of connective tissue keeps tumor from spreading.

Lymph vessel
Benign tumor
Capsule
Blood vessel

b. Malignant tumor

Malignant cells can "seed" new tumors throughout the body.

Malignant tumor
Lymph vessel
Metastasis
Blood vessels
New tumor
Lymph vessel
New tumor

Figure 8.14 **Benign and Malignant Tumors.** (a) A capsule of connective tissue prevents a benign tumor from invading adjacent tissues. (b) A malignant tumor lacks a capsule and therefore can spread throughout the body in blood and lymph.

specialized features of its parent cells. Some cancer cells have multiple nuclei. These visible differences allow pathologists to detect cancerous cells by examining tissue under a microscope.

Second, unlike normal cells, many cancer cells are essentially immortal, ignoring the "clock" that limits normal cells to 50 or so divisions. Given sufficient nutrients and space, cancer cells can divide uncontrollably and eternally. The cervical cancer cells of a woman named Henrietta Lacks vividly illustrate these characteristics. Shortly before Lacks died in 1951, researchers removed some of her cancer cells and began to grow them in a laboratory at Johns Hopkins University. Lacks's cells grew so well, dividing so often, that they quickly became a favorite of cell biologists seeking cells to culture that would divide indefinitely. Still used today, "HeLa" (for *H*enrietta *La*cks) cells replicate so vigorously that if just a few of them contaminate a culture of other cells, within days they completely take over.

The cellular "clock" resides in **telomeres,** the noncoding DNA at the tips of eukaryotic chromosomes. Telomeres consist of hundreds to thousands of repeats of a specific DNA sequence. At each cell division, the telomeres lose nucleotides from their ends, so the chromosomes gradually become shorter. After about 50 divisions, the cumulative loss of telomere DNA signals division to cease in a normal cell. Cells that produce an enzyme called **telomerase,** however, can continually add DNA to chromosome tips. Their telomeres stay long, which enables them to divide beyond the 50-or-so division limit. Cancer cells have high levels of telomerase; inactivating this enzyme could therefore have tremendous medical benefits.

Burning Question

What are the galls that form on plants?

Galls are abnormal growths that often form on plants. The growths may be smooth and perfectly round, as in the leaf gall shown here. They may also cause grotesque deformities on stems, flowers, roots, and other plant parts.

Many organisms cause plants to form galls, including fungi, bacteria, and even parasitic plants. The most common galls, however, are traced to a distinctive group of wasps. A female gall wasp lays an egg in the vein of a stem or leaf. When the egg hatches and develops into a larva, it secretes growth factors or other chemicals that stimulate the plant's cells to divide. The resulting gall does not usually hurt or help the tree, but it does form a protective shell that houses and feeds the young wasp until adulthood.

©Frank Greenaway/Dorling Kindersley/Getty Images

Submit your burning question to
Marielle.Hoefnagels@mheducation.com

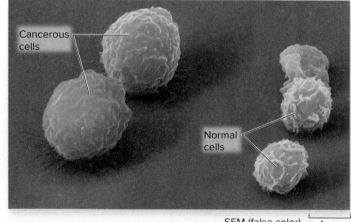

Cancerous cells

Normal cells

SEM (false color) 4 µm

Figure 8.15 **Cancer Cells Are Abnormal.** The two cancerous leukemia cells on the left are larger than the normal marrow cells on the right.
©Eye of Science/Science Source

TABLE **8.1** Miniglossary of Cell Division Terms

Term	Definition
Chromatin	Collective term for all of the DNA and associated proteins in a cell
Chromosome	A single, continuous molecule of DNA wrapped around protein. Eukaryotic cells contain multiple linear chromosomes, whereas bacterial cells typically have one circular chromosome.
Chromatid	One of two identical attached copies that make up a replicated chromosome
Centromere	A small part of a chromosome where sister chromatids attach to each other
Interphase	Stage of the cell cycle in which chromosomes replicate and the cell grows
G_1 phase	Gap stage of interphase in which the cell grows and carries out its functions
G_0 phase	Gap stage of interphase in which the cell functions but does not divide
G_2 phase	Gap stage of interphase in which the cell produces membrane components and spindle proteins
S phase	Synthesis stage of interphase when DNA replicates
Mitosis	Division of a cell's chromosomes into two identical nuclei
Prophase	Stage of mitosis when chromosomes condense and the spindle begins to form (*pro-* = before)
Prometaphase	Stage of mitosis when the nuclear membrane breaks up and spindle fibers attach to kinetochores
Metaphase	Stage of mitosis when chromosomes line up along the center of the cell (*meta-* = middle)
Anaphase	Stage of mitosis when the spindle pulls sister chromatids toward opposite poles of the cell
Telophase	Stage of mitosis when chromosomes arrive at opposite poles and nuclear envelopes form (*telo-* = end)
Cytokinesis	Distribution of cytoplasm to daughter cells following division of a cell's chromosomes
Cleavage furrow	Indentation in cell membrane of an animal cell undergoing cytokinesis
Cell plate	Material that forms the beginnings of the cell wall in a plant cell undergoing cytokinesis
Centrosome	Structure that organizes the microtubules that make up the spindle in animal cells
Spindle	Array of microtubule proteins that move chromosomes during mitosis
Kinetochore	Protein complex to which the spindle fibers attach on a chromosome's centromere

A third feature of cancer cells is their resistance to controls over cell division. Normal cells growing in culture exhibit **contact inhibition,** meaning that they stop dividing when they touch one another in a one-cell-thick layer. Normal cells also stop dividing once external growth factors are depleted, and they die (undergo apoptosis) when badly damaged. Cancer cells exhibit none of these features. They lack contact inhibition, so they tend to pile up in culture. They also may produce their own growth factors or be insensitive to "stop" signals.

Fourth, cancer cells send signals that stimulate a process called **angiogenesis,** the development of new blood vessels. The newly sprouted blood vessels boost tumor growth by delivering nutrients and removing wastes. Disrupting angiogenesis is a possible cancer-fighting strategy (see section 8.8).

Finally, we have already seen that cancer cells may invade nearby tissues and metastasize. Normal cells divide only when attached to a solid surface, a property called anchorage dependence. The observation that cancer cells lack anchorage dependence helps explain how metastasis occurs.

D. Cancer Treatments Remove or Kill Abnormal Cells

Medical professionals describe the spread of cancer cells as a series of stages. In one system used to classify colon cancer, for example, a stage I cancer has started invading tissue layers adjacent to the tumor's origin, but cancerous cells remain confined to the colon. At stage II, the tumor has spread to tissues around the colon but has not yet reached nearby lymph nodes. Stage III cancers have spread to organs and lymph nodes near the cancer's origin, and stage IV cancers have spread to distant sites. The names and criteria for each stage vary among cancers. In general, however, the lower the stage, the better the prospect for successful treatment.

Physicians use many techniques to estimate the stage of a patient's cancer. For example, X-rays, CT scans, MRIs, PET scans, ultrasound, and other imaging tests are noninvasive ways to detect and measure tumors inside the body. A physician can also use an endoscope to inspect the inside of some organs, such as the esophagus or intestines. The same tool can also collect a biopsy sample;

pathologists then use microscopes to search the tissue for suspicious cells. Blood tests can reveal more clues, including an abnormal number of white blood cells or a high level of a "tumor marker" such as prostate-specific antigen (PSA). Combining many such lines of evidence helps medical professionals diagnose cancer and determine the stage, which in turn helps guide treatment decisions.

Traditional cancer treatments include surgical tumor removal, drugs (chemotherapy), and radiation. Chemotherapy drugs, usually delivered intravenously, are intended to stop cancer cells anywhere in the body from dividing. Radiation therapy uses directed streams of energy from radioactive isotopes to kill tumor cells in limited areas. ⓘ *isotopes,* section 2.1C

Chemotherapy and radiation are relatively "blunt tools" that target rapidly dividing cells, whether cancerous or not. Examples of cells that divide frequently include those in the bone marrow, digestive tract, and hair follicles. The death of these cells accounts for the most notorious side effects of cancer treatment: fatigue, a weakened immune system, nausea, and hair loss. Fortunately, the healthy cells usually return after the treatment ends. Some patients, especially those who receive high doses of chemotherapy or radiation, also have bone marrow transplants to speed the replacement of healthy blood cells.

Basic research into the cell cycle has yielded new cancer treatments with fewer side effects. For example, drugs that target a cancer cell's unique molecules have been very successful in treating some forms of breast cancer and leukemia (see the opening essay for chapter 3). Drugs called angiogenesis inhibitors block a tumor's ability to recruit blood vessels, starving the cancer cells of their support system. In the future, cancer patients may receive gene therapy treatments that add functional genes to cells with faulty versions. ⓘ *gene therapy,* section 11.4D

The success of any cancer treatment depends on many factors, including the type of cancer and the stage in which it is detected. Surgery can cure cancers that have not spread. Once cancer metastasizes, however, it becomes difficult to locate and treat all of the tumors. Moreover, DNA replication errors introduce mutations in rapidly dividing cancer cells (see section 8.8). Treatments that shrank the original tumor may have no effect on this new, changed growth.

E. Genes and Environment Both Can Increase Cancer Risk

Proteins control both the cell cycle and apoptosis. Genes encode proteins, so genetic mutations (changes in DNA sequences) play a key role in causing cancer. So far, researchers know of hundreds of genetic mutations that contribute to cancer. But not all types of cancer trace to mutations; errors in the regulation of gene expression—epigenetics—also play a role. ⓘ *epigenetics,* section 7.6B

Two classes of cancer-related genes, oncogenes and tumor suppressor genes, are in a perpetual "tug of war" in determining whether cancer develops (**figure 8.16**). Genes called proto-oncogenes encode many types of cell cycle–related proteins, from the receptors that bind growth factors outside the cell to any of the participants in the series of reactions that trigger cell division. These proteins normally stimulate cell division (*onkos* is the Greek word for "mass" or "lump"). **Oncogenes** are mutated or overly active variants of proto-oncogenes; the encoded protein might be too active or expressed at too high a concentration. In that case, the cell cycle will be accelerated, and cancer may develop. Oncogenes cause some cancers of the cervix, bladder, and liver.

Recall from section 8.4 that our cells contain 23 pairs of chromosomes, with one member of each pair coming from each parent. Oncogenes are especially dangerous because only one of the two versions in a cell needs to be damaged for cancer to develop. The oncogene's abnormal protein is an "accelerator" that overrides the normal protein encoded by the proto-oncogene.

The other class of cancer-related genes, **tumor suppressor genes,** encode proteins that normally block cancer development; that is, they promote apoptosis or prevent cell division. Inactivating, deleting, or mutating these genes therefore eliminates crucial limits on cell division. Unlike oncogenes, usually both of a cell's versions of a tumor suppressor gene must be damaged for cancer

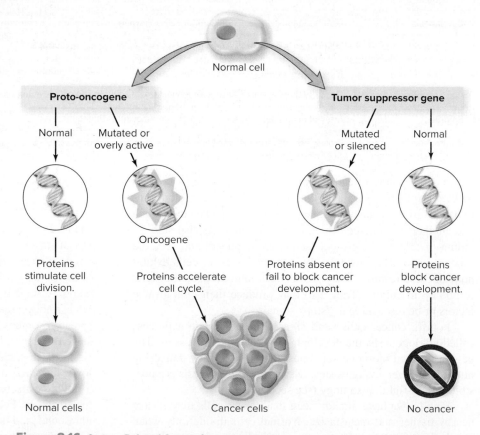

Figure 8.16 Cancer-Related Genes. Oncogenes and tumor suppressor genes both influence the cell cycle. When proto-oncogenes are mutated or overly active, they form oncogenes that accelerate cell division. Tumor suppressor genes encode proteins that normally inhibit cell division, but when the genes are mutated or silenced, cancer can develop.

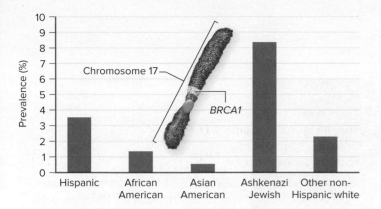

Figure 8.17 ***BRCA1* Mutations by Ethnic Group.** The likelihood that a woman with breast cancer has a mutation in *BRCA1* depends on her ethnicity. People of Ashkenazi Jewish descent are especially at risk.

to develop. That is, as long as one tumor suppressor gene is functioning, the cell continues to produce the protective proteins.

One example of a tumor suppressor gene is *p53,* which encodes a protein that participates in the cell cycle control checkpoints described earlier. Biologists suspect that mutations in *p53* cause about half of all human cancers.

A different tumor suppressor gene is associated with an unusually high risk for breast cancer. Ethnic groups vary widely in their risk of inheriting mutations in the gene, which is called *BRCA1* (**figure 8.17**). People of Ashkenazi Jewish descent have a much higher than average chance of carrying a particularly harmful *BRCA1* mutation. An evolutionary mechanism called the "founder effect" explains this observation. According to genetic studies, the mutation arose among Ashkenazi Jews many generations ago. Because Ashkenazi Jews historically avoided marrying outside their group, the mutation remains more frequent within this subgroup than in the human population at large. ⓘ *founder effect,* section 12.7B

The more oncogenes or mutated tumor suppressor genes in a person's cells, the higher the probability of cancer. Where do these mutations come from? Sometimes, a person inherits mutated DNA from one or both parents. The mutations may run in families, as in the case of many *BRCA1* mutations, or they may have arisen spontaneously in a parent's sperm- or egg-producing cells. Often, however, people acquire the cancer-causing mutations throughout their lifetimes.

Figure 8.18 depicts some choices a person can make to reduce cancer risk. Some of these strategies are straightforward. For example, UV radiation and many chemicals in tobacco are mutagens, which means they damage DNA. Reducing sun exposure and avoiding tobacco therefore directly reduce cancer risk (see the Apply It Now box). Likewise, condoms can help prevent infection with cancer-causing viruses that are sexually transmitted. ⓘ *mutagens,* section 7.7B

Other risk factors illustrated in figure 8.18 are less obvious. Obesity, for example, greatly increases the risk of death from cancers of the breast, cervix, uterus, and ovaries in women; obese men have an elevated risk of dying from prostate cancer. High-calorie foods that are rich in animal fats and low in fiber, coupled with a lack of exercise, contribute to high body weight. But scientists remain uncertain why obesity itself is a risk factor for cancer. Perhaps fat tissue secretes hormones that contribute to metastasis, or maybe obesity reduces immune system function. Research into the cancer–obesity connection is increasingly important as obesity rates continue to climb.

One thing is clear: An enormous variety of illnesses are grouped under the category of "cancer," and each is associated with a unique suite of risk factors. It therefore pays to be skeptical of claims that any one product can miraculously fight cancer. A healthy lifestyle remains the best way to reduce cancer risk.

8.6 MASTERING CONCEPTS

1. What prevents normal cells from dividing when they are not supposed to?
2. What happens at cell cycle checkpoints?
3. What is the difference between a benign and a malignant tumor?
4. How do cancer cells differ from normal cells?
5. List and describe the three most common cancer treatments.
6. What is the relationship between genetic mutations and cancer?
7. How does a person acquire the mutations associated with cancer?

To avoid or reduce the risk of cancer

Figure 8.18 **Cancer Risk.** Many aspects of a person's lifestyle influence the risk of cancer.

Eat a healthy diet, low in saturated fat and rich in fruits and vegetables.

Avoid obesity: get regular, vigorous exercise.

Stop using tobacco, or better yet, never start.

Avoid UV radiation from sunlight and tanning beds.

Use self tests and medical exams for early detection.

Use condoms to avoid exposure to viruses known to cause cancer.

Apply It **Now**

Detecting and Preventing Skin Cancer

Cancer has many forms, some inherited and others caused by radiation or harmful chemicals. Exposure to ultraviolet radiation from the sun or from tanning beds, for example, increases the risk of skin cancer because UV radiation damages DNA. If genetic mutations occur in genes encoding proteins that control the pace of cell division, cells may begin dividing out of control, forming a malignant tumor on the skin.

How might a person determine whether a mole, sore, or growth on the skin is cancerous? The abnormal skin may vary widely in appearance, and only a physician can tell for sure. Nevertheless, most skin cancers have a few features in common. "ABCD" is a shortcut for remembering these four characteristics:

©Gabriela Hasbun/Stone/Getty Images

- **Asymmetry:** One half of the area looks different from the other.
- **Borders:** The borders of the growth are irregular, not smooth.
- **Color:** The color varies within a patch of skin, from tan to dark brown to black. Other colors, including red, may also appear.

- **Diameter:** The diameter of a cancerous area is usually greater than 6 mm, which is about equal to the size of a pencil eraser.

Skin cancer is the most common form of cancer in the United States. Several types of skin cancer exist, including basal cell carcinoma, squamous cell carcinoma, and melanoma. Basal cell carcinoma is the most common, but melanoma is the deadliest because the cancerous cells quickly spread to other parts of the body.

Although anyone can get skin cancer, the highest risk occurs among people with light-colored skin, who have blue or green eyes, and who spend a lot of time in the sun. Avoiding exposure to UV radiation, both in the sun and in tanning beds, can help minimize this risk. Sunscreen is a must when outdoors. In addition, medical professionals recommend that people pay attention to changes in their skin. Carcinomas and melanomas are treatable if detected early.

8.7 Apoptosis Is Programmed Cell Death

Development relies on a balance between cell division and programmed cell death, or apoptosis (see figure 8.3). Apoptosis is different from necrosis, which is the "accidental" cell death that follows a cut or bruise. Whereas necrosis is sudden, traumatic, and disorderly, apoptosis results from a precisely coordinated series of events that dismantle a cell.

The process begins when a "death receptor" protein on a doomed cell's membrane receives a signal to die (figure 8.19). Within seconds, apoptosis-specific "executioner" proteins begin to cut apart the cell's proteins and destroy the cell. Immune system cells descend, and the cell is soon gone.

Apoptosis has two main functions in animals. First, apoptosis eliminates excess cells, carving out functional structures such as fingers, toes, nostrils, and ears as an animal grows. The second function of apoptosis is to weed out aging or defective cells that otherwise might harm the organism. A good example is the peeling skin that follows a sunburn. Sunlight contains UV radiation that can cause cancer by damaging the DNA in skin cells. Apoptosis helps protect against skin cancer by eliminating severely damaged cells, which die and simply peel away.

Plant cells die, too, but not in precisely the way that animal cells meet their programmed fate. Instead, plant cells are digested by enzymes in their own vacuoles; when the vacuole bursts, the cell dies. Plants also use a form of cell death to kill cells infected by fungi or bacteria, limiting the spread of the pathogen.

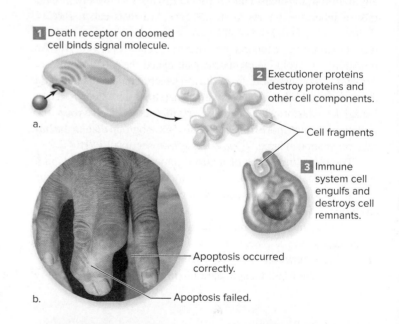

1 Death receptor on doomed cell binds signal molecule.

2 Executioner proteins destroy proteins and other cell components.

Cell fragments

a.

3 Immune system cell engulfs and destroys cell remnants.

Apoptosis occurred correctly.

b.   Apoptosis failed.

Figure 8.19 Death of a Cell. (a) Enzymes trigger apoptosis shortly after a cell's death receptor receives the signal to die. Afterwards, immune system cells mop up the debris. (b) Two fingers on this woman's hand appear fused together, thanks to a failure of apoptosis during development.
Photo: (b): ©Imaginechina/Corbis

8.7   MASTERING CONCEPTS

1. What events happen in a cell undergoing apoptosis?
2. Describe two functions of apoptosis.

8.8 Cutting Off a Tumor's Supply Lines in the War on Cancer

When Charles Darwin proposed natural selection as a mechanism of evolutionary change, he envisioned selective forces operating on tortoises, flowering plants, and other whole organisms. But the power of natural selection extends to a much smaller scale, including the individual cells that make up a tumor. The advance, retreat, resurgence, and death of these renegade cells command dramatic headlines in the war on cancer.

Our weapons against cancer include powerful chemotherapy drugs, but drug-resistant tumor cells are a significant barrier to successful treatment. Rapidly dividing tumor cells develop resistance to drugs because frequent cell division produces abundant opportunities for mutations. An alternative cancer-fighting strategy, therefore, might be to launch an indirect attack on a tumor's slower-growing support tissues instead.

Any tumor larger than 1 or 2 cubic millimeters needs a blood supply to carry nutrients, oxygen, and wastes. Blood travels throughout the body in vessels lined with endothelial tissue. For a blood vessel to grow, its endothelial cells must divide, which happens only rarely in adults. Cancer cells, however, secrete molecules that stimulate endothelial cells to divide and form new blood vessels. This sprouting of new "supply lines" is called angiogenesis.

Fortunately, some drugs can stop blood vessel growth. One such drug, called endostatin, keeps endothelial cells from dividing but does not kill resting endothelial cells or other cells in the body. It should therefore choke off a tumor's supply lines without toxic side effects. But does it really work? Cancer researchers Thomas Boehm, Judah Folkman, and their colleagues at the Dana Farber Cancer Center and Harvard Medical School set out to answer this question.

The researchers first induced cancer in mice by "seeding" each animal with one of three types of cancer cells. After tumors developed, the researchers injected the mice with endostatin. Injections continued for several days, until the tumors in endostatin-treated mice were barely detectable. When the tumors regrew, the researchers repeated the treatments.

The results were astounding. With each dose of endostatin, the tumors shriveled (**figure 8.20a**). Moreover, after two to six treatments with endostatin, the tumors never grew back, and the mice remained healthy. Standard chemotherapy drugs could temporarily shrink a tumor or delay its growth. But resistant cells would soon take over, and the tumor would regrow (figure 8.20b).

The results of subsequent clinical trials with human cancer patients, however, were mixed. The drug was ineffective in most patients, and its U.S. manufacturer stopped making it. The story is somewhat different in China, where the government has approved a modified version of endostatin. Research is ongoing; in many studies, the drug shows the most promise when combined with other forms of chemotherapy.

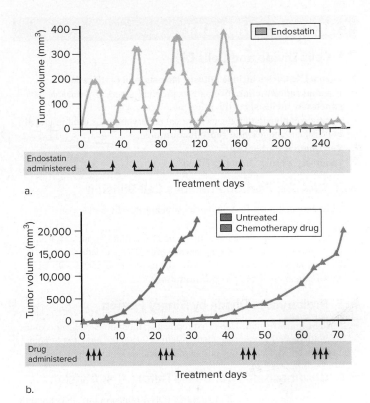

Figure 8.20 No Resistance. (a) In mice, endostatin repeatedly shrank tumors and eventually stopped them from regrowing. (b) A traditional chemotherapy drug delayed but did not prevent tumor growth.

What does endostatin have to do with evolution? The logic behind its use as an anticancer drug relies on natural selection. Because DNA may mutate every time it replicates, rapidly dividing cancer cells are genetically different from one another. A conventional chemotherapy drug may kill most cancer cells in a tumor, but a few have mutations that let them survive. These cells divide; over time, the entire tumor is resistant to the drug. Unlike other drugs, however, endostatin does not target the tumor; instead, it affects a blood vessel's endothelial cells. These cells rarely divide and therefore accumulate mutations slowly, reducing the chance that they will become resistant to endostatin.

This may seem comforting, but evolution will not stand still for our convenience. Understanding natural selection helps researchers know what to look for—and perhaps even launch new offensives in the war on cancer.

Source: Boehm, Thomas, Judah Folkman, Timothy Browder, and Michael S. O'Reilly. November 27, 1997. Antiangiogenic therapy of experimental cancer does not induce acquired drug resistance. *Nature*, vol. 390, pages 404–407.

8.8 MASTERING CONCEPTS

1. Why doesn't endostatin select for drug-resistant cancer cells, as other chemotherapy drugs do?

2. True or false: Cancer may increase the reproductive success of an individual cell but may decrease the reproductive success of a whole organism. Explain your answer.

CHAPTER SUMMARY

8.1 Cells Divide and Cells Die

A. Sexual Life Cycles Include Mitosis, Meiosis, and Fertilization

- In **sexual reproduction,** two parents produce genetically variable **gametes** by **meiosis. Fertilization** produces a zygote.
- **Mitotic** cell division produces identical eukaryotic cells used in growth, tissue repair, and asexual reproduction.

B. Cell Death Is Part of Life

- **Apoptosis** is programmed cell death.

8.2 DNA Replication Precedes Cell Division

- A dividing cell must first duplicate its **genome,** which may consist of one or more **chromosomes.**
- Helicase enzymes unwind and unzip the DNA; binding proteins keep the strands separate. **DNA polymerase** adds DNA nucleotides to an RNA primer. **Ligase** seals nicks after the primer is replaced with DNA.
- DNA replication errors produce **mutations.**

8.3 Prokaryotes Divide by Binary Fission

- During **binary fission,** DNA first replicates; then the two chromosomes attach to the cell membrane. Cell growth between the attachment points separates the chromosomes into two identical daughter cells.

8.4 Chromosomes Condense Before Cell Division

- A chromosome consists of **chromatin** (DNA plus protein). In eukaryotic cells, chromatin is organized into **nucleosomes.**
- A replicated chromosome consists of two identical sister **chromatids** attached at a section of DNA called a **centromere** (figure 8.21).

8.5 Mitotic Division Generates Exact Cell Copies

- The **cell cycle** (summarized in figure 8.22) is a sequence of events in which a cell is preparing to divide (**interphase**), dividing its DNA (**mitosis**), and dividing its cytoplasm (**cytokinesis**).

A. DNA Is Copied During Interphase

- Interphase includes gap periods, G_1 **phase** and G_2 **phase,** when the cell grows and produces molecules required for cell function and division. DNA replicates during the synthesis period (**S phase**). A cell that is not dividing is in G_0 **phase.**
- In animal cells, the **centrosome** duplicates during interphase. The two centrosomes organize the proteins that form the mitotic **spindle.**

B. Chromosomes Divide During Mitosis

- In **prophase,** the chromosomes condense and the spindle forms. In **prometaphase,** the nuclear envelope breaks up, and spindle fibers attach to **kinetochores.** In **metaphase,** replicated chromosomes line up along the cell's equator. In **anaphase,** the chromatids of each replicated chromosome separate. In **telophase,** the spindle breaks down, and nuclear envelopes form.

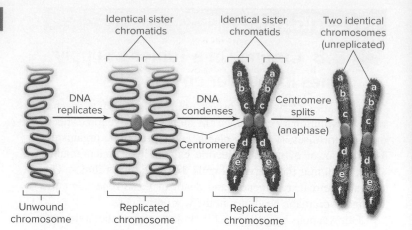

Figure 8.21 Chromosomes and Chromatids Compared.

C. The Cytoplasm Splits in Cytokinesis

- In animal cells, a contractile ring forms a **cleavage furrow,** dividing the cell in two. Plant cells divide as a **cell plate** forms at the midline of a dividing cell.

8.6 Cancer Arises When Cells Divide Out of Control

A. Chemical Signals Regulate Cell Division

- External signals called **growth factors** stimulate cell division.
- Cell division may pause or halt at multiple checkpoints during the cell cycle.

B. Cancer Cells Break Through Cell Cycle Controls

- **Tumors** can result from excess cell division or deficient apoptosis. A **benign tumor** does not spread, but a **malignant tumor** invades nearby tissues and **metastasizes** if it reaches the bloodstream or lymph.
- **Cancer** is a family of diseases characterized by malignant cells.

C. Cancer Cells Differ from Normal Cells in Many Ways

- A cancer cell divides uncontrollably.
- When **telomeres** become very short, division ceases. Cancer cells produce an enzyme called **telomerase,** which adds DNA to telomeres.
- Cancer cells may continue to divide even after growth factors are depleted.
- A cancer cell lacks **contact inhibition** and anchorage dependence, may not undergo apoptosis, and secretes chemicals that stimulate **angiogenesis** (the growth of new blood vessels).

D. Cancer Treatments Remove or Kill Abnormal Cells

- Surgery, chemotherapy, and radiation are common cancer treatments.

E. Genes and Environment Both Can Increase Cancer Risk

- **Oncogenes** speed cell division, and mutated **tumor suppressor genes** fail to stop excess cell division. Mutations in cancer-related genes may be inherited or acquired during a person's lifetime.

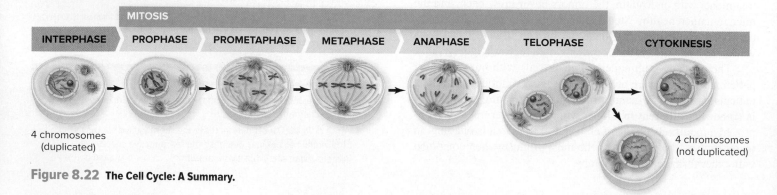

Figure 8.22 The Cell Cycle: A Summary.

8.7 Apoptosis Is Programmed Cell Death

- Apoptosis shapes structures and kills cells that could become cancerous.
- After a cell receives a signal to die, enzymes destroy the cell's components. Immune system cells dispose of the remains.

8.8 Investigating Life: Cutting Off a Tumor's Supply Lines in the War on Cancer

- Natural selection occurs inside tumors. As chemotherapy drugs eliminate susceptible cells, resistant ones survive and divide to regrow the tumor.
- Endostatin starves tumors by stopping the growth of blood vessels.

MULTIPLE CHOICE QUESTIONS

1. A DNA molecule is placed in a test tube containing fluorescently tagged nucleotides. DNA replication is induced. After replication,
   a. only one DNA molecule would have two fluorescent strands.
   b. both strands of each DNA molecule would be half-fluorescent.
   c. each DNA molecule would have one fluorescent strand.
   d. both DNA molecules would be completely fluorescent.

2. Which of the following best explains why binary fission can occur without a spindle like that found in mitotic cells?
   a. The cell is small, so there is less material to divide.
   b. There is only one chromosome, and it attaches to the membrane.
   c. The prokaryotic DNA does not need to replicate.
   d. Prokaryotic cells lack chromosomes, so the spindle is unnecessary.

3. Which stage of the cell cycle occurs immediately *after* the stage in which the chromosomes become visible?
   a. Anaphase
   b. Prophase
   c. Prometaphase
   d. Telophase

4. Imagine that gene *X* mutates and begins encoding a protein that accelerates the cell cycle. Before it mutated, gene *X* was most likely a(n)
   a. proto-oncogene.
   b. oncogene.
   c. tumor suppressor gene.
   d. apoptosis gene.

5. What is the role of executioner proteins in apoptosis?
   a. They kill the cell by destroying its proteins.
   b. They function as the "death receptors" on the surface of the cell.
   c. They are part of the immune response that eliminates the cells.
   d. They cause the cell to swell and burst.

Answers to these questions are in appendix A.

WRITE IT OUT

1. Explain how cell division and cell death work together to form a functional multicellular organism.

2. Write and explain an analogy for each of these DNA replication enzymes: helicase, binding proteins, ligase.

3. Tightly packed DNA cannot be used for protein synthesis. Why has evolution favored the histones and other proteins that help DNA fold into compact chromosomes?

4. Obtain a rubber band and twist it as many times as you can. What happens to the overall shape of the rubber band? How is this similar to what happens to chromosomes as a cell prepares to divide? How is it different?

5. Sketch and describe the events that occur when a bacterial cell divides.

6. If a cell somehow skipped G_1 of interphase during multiple cell cycles, how would the daughter cells change?

7. If you draw on your skin with a permanent marker, the markings will fade in a couple of days. What does this simple demonstration reveal about cell division in your skin? What can you infer about tattoos?

8. Describe what will happen to a cell if interphase happens but mitosis does not.

9. List the ways that binary fission is similar to and different from mitotic cell division.

10. Suppose you learn of a study in which ginger slowed tumor growth in mice for 30 days. What questions would you ask before deciding whether to recommend that a cancer-stricken relative eat more ginger?

11. In the early 1900s, scientists began to experiment with radiation as a cancer treatment. Many physicians who administered the treatment subsequently died of cancer. Why?

12. Why do chemotherapy and radiation sometimes kill hair follicle cells, while leaving many other body cells unaffected?

13. A protein called p53 promotes the expression of genes encoding DNA repair enzymes. Badly damaged DNA prompts p53 to trigger apoptosis, and the cell dies. Why might mutations in the gene encoding p53 be associated with a high risk for cancer?

PULL IT TOGETHER

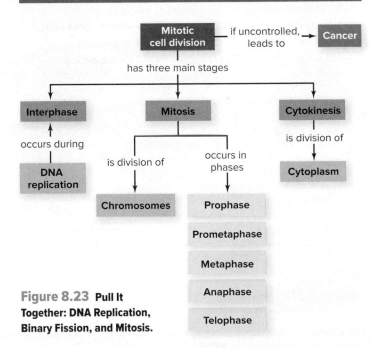

Figure 8.23 Pull It Together: DNA Replication, Binary Fission, and Mitosis.

Refer to figure 8.23 and the chapter content to answer the following questions.

1. Review the Survey the Landscape figure in the chapter introduction and then connect *DNA* and *proteins* to the Pull It Together concept map in at least two ways each.

2. Add *DNA polymerase, nucleotides,* and *complementary base pairing* to this concept map.

3. What is the relationship between mitotic cell division and apoptosis?

9 Sexual Reproduction and Meiosis

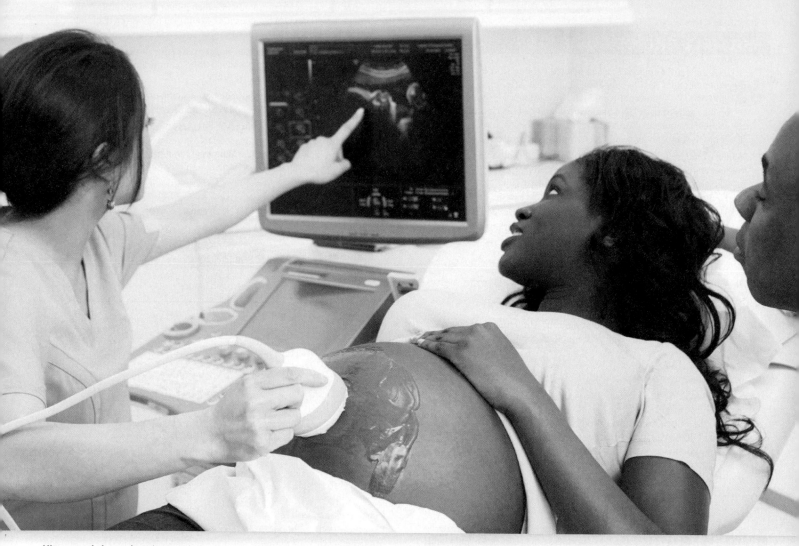

Ultrasound. A couple enjoys seeing an image of their unborn baby, which started developing months earlier. The sperm and egg cells that came together to make the child were the products of meiosis.

©monkeybusinessimages/Getty Images RF

LEARN HOW TO LEARN
Write It Out—Really!

Get out a pen and a piece of scratch paper, and answer the open-ended Write It Out questions at the end of each chapter. This tip applies even if the exams in your class are multiple choice. Putting pen to paper (as opposed to just saying the answer in your head) forces you to organize your thoughts and helps you discover the difference between what you know and what you only THINK you know.

Prenatal Diagnosis Highlights Ethical Dilemmas

Barbara is pregnant. Like many women, she periodically has her fetus examined by ultrasound. Barbara delights in seeing her unborn child, but her latest scan has revealed a possible abnormality. Her physician cannot be sure of the diagnosis without ordering a test of the fetus's chromosomes.

How is it possible to see chromosomes hidden inside the cells of a fetus, which is itself tucked into the mother's uterus? A technician begins by extracting a small amount of the fluid or tissue surrounding the developing fetus. Fetal cells in the fluid can then be used to prepare a photograph of the fetus's chromosomes.

The image may reveal several types of abnormalities, including extra chromosomes, missing chromosomes, or the movement of genetic material from one chromosome to another. If the physician detects a chromosomal abnormality, Barbara may consult a counselor who can advise her on how best to prepare for the birth of her baby. In the case of a severe abnormality, Barbara may decide to seek an abortion, ending the pregnancy. But this choice raises many difficult issues.

Prenatal diagnosis illustrates one of many intersections between morality and science. Few people would argue against Barbara's use of prenatal diagnosis to learn more about a possible illness. But what constitutes a "severe" abnormality? And what if a mother or family lacks the resources to care for a child with special needs?

In the future, as our knowledge of the human genome grows, prospective parents will be able to screen for an ever-expanding list of traits. The moral dilemmas will expand, too. Even though traits such as intelligence and height are the product of both genes and environment, wealthy people may be able to "stack the deck" by keeping only fetuses with the most promising allele combinations. If and when that becomes possible, should we be allowed to make these choices? These are difficult questions without scientific answers.

Science can, however, help us understand the origin of chromosomal abnormalities. Many of them trace to errors that occur during a specialized form of cell division called meiosis. In humans and many other organisms, meiosis plays a starring role in the production of sperm and egg cells, which lie at the heart of sexual reproduction. This chapter explains the chromosomal choreography of meiosis.

SURVEY THE LANDSCAPE
DNA, Inheritance, and Biotechnology

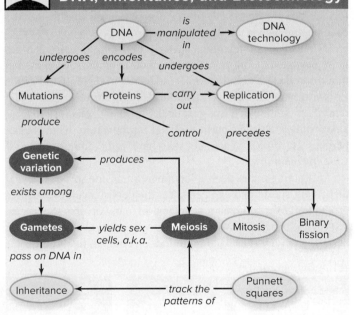

How can the same two parents produce offspring that look so different from one another? Meiosis explains the answer. This specialized process generates genetically variable nuclei that are packaged into gametes—the cells used in sexual reproduction.

For more details, study the Pull It Together feature in the chapter summary.

9.1 Why Sex?

Humans are so familiar with our way of reproducing that it can be hard to remember that there is any other way to make offspring. In fact, however, reproduction occurs in two main forms: asexual and sexual (**figure 9.1**). In **asexual reproduction,** an organism simply copies its DNA and splits the contents of one cell into two. Some genetic material may mutate during DNA replication, but the offspring are virtually identical. Examples of asexual organisms include bacteria, archaea, and single-celled eukaryotes such as the amoeba in figure 9.1a. Many plants, fungi, and other multicellular organisms also reproduce asexually.

Sexual reproduction, in contrast, requires two parents. The male parent contributes sperm cells, one of which fertilizes a female's egg cell to begin the next generation. Later in this chapter, you will learn that each time the male produces sperm, he scrambles the genetic information that he inherited from his own parents. A similar process occurs as the female produces eggs. The resulting variation among sex cells ensures that the offspring from two parents are genetically different from one another.

How did sexual reproduction evolve? Clues emerge from studies of reproduction and genetic exchange in diverse organisms. The earliest process that combines genes from two individuals appeared about 3.5 billion years ago. In **conjugation,** one bacterial cell uses an outgrowth called a sex pilus to transfer genetic material to another bacterium (see figure 17.9). This ancient form of bacterial gene transfer is still prevalent today. The unicellular eukaryote *Paramecium* uses a variation on this theme, exchanging nuclei via a bridge of cytoplasm.

Thanks to conjugation, bacteria and *Paramecium* can acquire new genetic information from their neighbors, even though they reproduce asexually. Unicellular green algae of the genus *Chlamydomonas,* however, exhibit a simple form of true sexual reproduction in which two genetically different cells fuse to form a new individual. The earliest sexual reproduction, which began perhaps 1.5 billion years ago, may have been similar to that of *Chlamydomonas.*

Attracting mates takes a lot of energy, as does producing and dispersing sperm and egg cells. Yet the persistence of sexual reproduction over billions of years and in many diverse species attests to its success. Why does such a costly method of reproducing persist, and why is asexual reproduction comparatively rare?

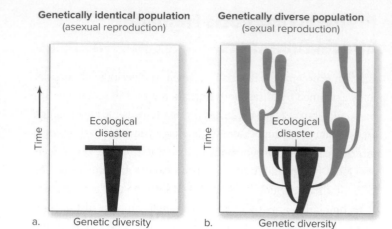

Figure 9.2 Why Sex? (a) In asexually reproducing organisms, the members of a population are usually very similar to one another; a single change in the environment can wipe out the population. (b) Sexual reproduction generates genetic variability, which boosts the chance that at least some members of the population (*blue*) will survive in a changing environment.

Although no one knows the full answer to this question, many studies point to the benefit of genetic diversity in a changing environment (**figure 9.2**). The mass production of identical offspring makes sense in habitats that never change, but conditions rarely remain constant in the real world. Temperatures rise and fall, prey species disappear, and new parasites emerge (see section 9.9). Genetic variability increases the chance that at least some individuals will have a combination of traits that allows them to survive and reproduce, even if some poorly suited individuals die. Asexual reproduction typically cannot create or maintain this genetic diversity, but sexual reproduction can.

9.1 MASTERING CONCEPTS

1. How do asexual and sexual reproduction differ?
2. How can asexually reproducing organisms acquire new genetic information?
3. Why does sexual reproduction persist even though it requires more energy than asexual reproduction?

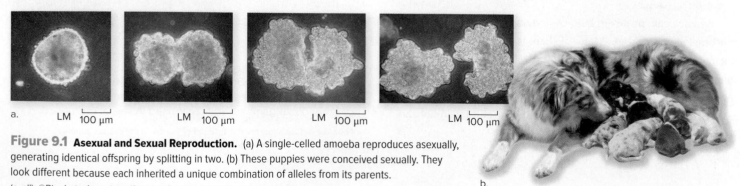

Figure 9.1 Asexual and Sexual Reproduction. (a) A single-celled amoeba reproduces asexually, generating identical offspring by splitting in two. (b) These puppies were conceived sexually. They look different because each inherited a unique combination of alleles from its parents.

(a, all): ©Biophoto Associates/Science Source; (b): ©Eric Isselee/123RF

9.2 Diploid Cells Contain Two Homologous Sets of Chromosomes

Before exploring sexual reproduction further, a quick look at a cell's chromosomes is in order. Recall from chapters 7 and 8 that a **chromosome** is a single molecule of DNA and its associated proteins.

A sexually reproducing organism consists mostly of **diploid cells** (abbreviated 2*n*), which contain two full sets of chromosomes; one set is inherited from each parent. Each diploid human cell, for example, contains 46 chromosomes (**figure 9.3**). The photo in figure 9.3 illustrates a **karyotype,** a size-ordered chart of all the chromosomes in a cell. Notice that the 46 chromosomes are arranged in 23 pairs; the mother and the father each contributed one member of each pair.

Of the 23 chromosome pairs in a human cell, 22 pairs consist of **autosomes**—chromosomes that are the same for both sexes. The remaining pair is made up of the two **sex chromosomes,** which determine whether an individual is female or male. Females have two X chromosomes, whereas males have one X and one Y chromosome.

The two members of most chromosome pairs are homologous to each other. In a **homologous pair,** the two chromosomes look alike and have the same sequence of genes. (The word *homologous* means "having the same basic structure.") The physical similarities between any two homologous chromosomes are evident in figure 9.3: They share the same size, centromere position, and pattern of light- and dark-staining bands. In addition, the two members of a homologous pair of chromosomes carry the same sequence of genes. For example, chromosome 21 includes 367 genes, always in the same order.

Homologous chromosomes, however, are not identical—after all, nobody has two identical parents! Instead, the two homologs differ in the combination of **alleles,** or versions, of the genes they carry (**figure 9.4**). As described in chapter 7, each allele of a gene encodes a different version of the same protein. A chromosome typically carries exactly one allele of each gene, so a person inherits one allele per gene from each parent. Depending on the parents' chromosomes, the two alleles may be identical or different. Overall, however, the members of each homologous pair of chromosomes are at least slightly different from each other.

Unlike the autosome pairs, the X and Y chromosomes are not homologous to each other. X is much larger than Y, and its genes are completely different. Nevertheless, in males, the sex chromosomes behave as homologous chromosomes during meiosis.

9.2 | MASTERING CONCEPTS

1. What are autosomes and sex chromosomes?

2. Review figure 7.8, which shows that a chromosome is like a cookbook. How could you extend this analogy to homologous chromosomes?

Figure 9.3 Human Karyotype. A diploid human cell contains 23 pairs of chromosomes, with one member of each pair inherited from each parent. Autosomes are numbered 1 through 22; X and Y are sex chromosomes. The insets show sex chromosomes for a female (XX) and a male (XY).

(all): ©CNRI/Science Source

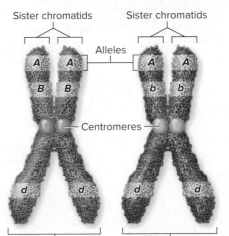

Figure 9.4 Homologous Pair. On these chromosomes, both alleles for gene *A* are the same, as are those for gene *D*. The two alleles for gene *B*, however, are different.

Photo: ©Andrew Syred/Science Source

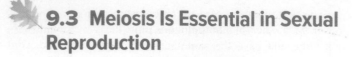

9.3 Meiosis Is Essential in Sexual Reproduction

Sexually reproducing species range from humans to ferns to the mold that grows on bread. This section describes some of the features that all sexual life cycles share.

A. Gametes Are Haploid Sex Cells

Sexual reproduction poses a practical problem: maintaining the correct chromosome number. We have already seen that most cells in the human body contain 46 chromosomes. If a baby arises from the union of a man's sperm and a woman's egg, then why does the child not have 92 chromosomes per cell (46 from each parent)? And shouldn't cells in the next generation have 184 chromosomes?

In fact, the chromosome number does not double with each generation. The explanation is that sperm

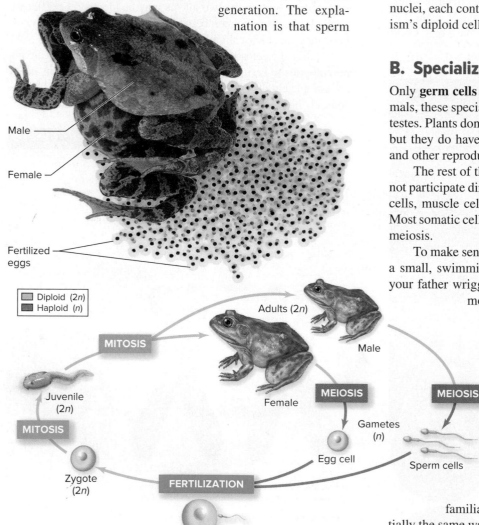

Figure 9.5 **Sexual Reproduction.** All sexual life cycles include meiosis and fertilization; mitotic cell division enables the organism to grow. The photo shows a male frog clasping his mate as he fertilizes her eggs.

Photo: ©Nature Picture Library/Britain On View/Getty Images

cells and egg cells are not diploid. Rather, they are **haploid cells** (abbreviated n); that is, they contain only one full set of genetic information instead of the two sets that characterize diploid cells.

These haploid cells, called **gametes,** are sex cells that combine to form a new offspring. **Fertilization** merges the gametes from two parents, creating a new cell: the diploid **zygote,** which is the first cell of the new organism (**figure 9.5**). The zygote has two full sets of chromosomes, one set from each parent. In most species, including plants and animals, the zygote begins dividing mitotically shortly after fertilization.

Thus, the life of a sexually reproducing, multicellular organism requires two ways to package DNA into new cells. **Mitosis,** described in chapter 8, divides a eukaryotic cell's chromosomes into two identical daughter nuclei. Mitotic cell division produces the cells needed for growth, development, and tissue repair. **Meiosis,** the subject of this chapter, forms genetically variable nuclei, each containing half as many chromosomes as the organism's diploid cells.

B. Specialized Germ Cells Undergo Meiosis

Only **germ cells** can undergo meiosis. In humans and other animals, these specialized diploid cells occur only in the ovaries and testes. Plants don't have the same reproductive organs as animals, but they do have specialized gamete-producing cells in flowers and other reproductive parts.

The rest of the body's diploid cells, called **somatic cells,** do not participate directly in reproduction. Leaf cells, root cells, skin cells, muscle cells, and neurons are examples of somatic cells. Most somatic cells can divide mitotically, but they do not undergo meiosis.

To make sense of this, consider your own life. It began when a small, swimming sperm cell carrying 23 chromosomes from your father wriggled toward your mother's comparatively enormous egg cell, also containing 23 chromosomes. You were conceived when the sperm fertilized the egg cell. At that moment, you were a one-celled zygote, with 46 chromosomes. That first cell then began dividing, generating identical copies of itself to form an embryo, then a fetus, an infant, a child, and eventually an adult (see figure 8.2). Once you reached reproductive maturity, diploid cells in your testes or ovaries produced haploid gametes of your own, perpetuating the cycle.

The human life cycle is, of course, most familiar to us, and many animals reproduce in essentially the same way. Gametes are the only haploid cells in our life cycle; all other cells are diploid. Sexual reproduction, however, can take many other forms as well. In some organisms, including plants, both the haploid and the diploid stages are multicellular. Section 9.8 describes the life cycle of a sexually reproducing plant in more detail.

C. Meiosis Halves the Chromosome Number and Scrambles Alleles

No matter the species, meiosis has two main outcomes. First, the resulting gametes contain half as many chromosomes as the rest of the body's cells. They therefore ensure that the chromosome number does not double with every generation. The second function of meiosis is to scramble genetic information, so that two parents can generate offspring that are genetically different from the parents and from one another. As described in section 9.1, genetic variability is one of the evolutionary advantages of sexual reproduction.

Figure It Out

A type of fish called a carp has gametes containing 52 chromosomes. How many chromosomes are in a carp's somatic cells?

Answer: 104

Although meiosis has unique functions, many of the events are similar to those of mitosis. As you work through the stages of meiosis, it may therefore help to think of what you already know about mitotic cell division. For example, a cell that is preparing to divide mitotically undergoes interphase, followed by the overlapping phases of mitosis and then cytokinesis (see figure 8.10).

Similarly, interphase occurs just before meiosis; the names of the phases of meiosis are similar to those in mitosis; and cytokinesis occurs after the genetic material is distributed.

Despite these similarities, meiosis has two unique outcomes, highlighted in figure 9.6. First, meiosis includes two divisions, which create four haploid cells from one specialized diploid cell. Second, meiosis shuffles genetic information, setting the stage for each haploid nucleus to receive a unique mixture of alleles.

Sections 9.4 and 9.5 explain in more detail how meiosis simultaneously halves the chromosome number and produces genetically variable nuclei. We then turn to problems that can occur in meiosis and describe how humans package haploid nuclei into individual sperm or egg cells.

9.3 | MASTERING CONCEPTS

1. How do haploid and diploid nuclei differ?
2. What are the roles of meiosis, gamete formation, and fertilization in a sexual life cycle?
3. What is a zygote?
4. What is the difference between somatic cells and germ cells?

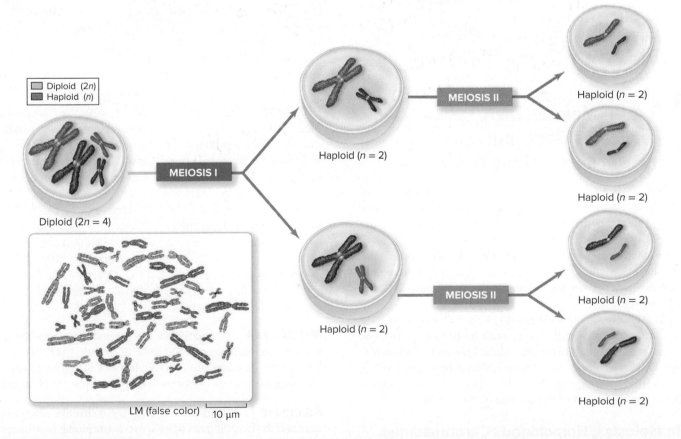

□ Diploid (2n)
■ Haploid (n)

Diploid (2n = 4)

MEIOSIS I

Haploid (n = 2)

MEIOSIS II

Haploid (n = 2)

Haploid (n = 2)

Haploid (n = 2)

MEIOSIS II

Haploid (n = 2)

Haploid (n = 2)

LM (false color) 10 μm

Figure 9.6 Summary of Meiosis. In meiosis, a diploid nucleus gives rise to four haploid nuclei with a mix of chromosomes from each parent. This illustration summarizes meiosis for a nucleus containing 4 chromosomes (2 homologous pairs). A diploid human cell, however, contains 46 chromosomes (23 homologous pairs, as shown in the inset).

Photo: ©James Cavallini/Science Source

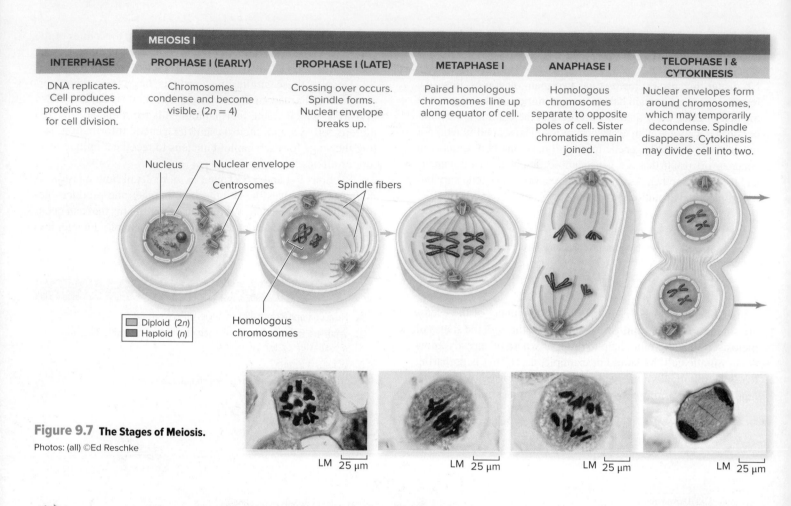

MEIOSIS I					
INTERPHASE	**PROPHASE I (EARLY)**	**PROPHASE I (LATE)**	**METAPHASE I**	**ANAPHASE I**	**TELOPHASE I & CYTOKINESIS**
DNA replicates. Cell produces proteins needed for cell division.	Chromosomes condense and become visible. (2n = 4)	Crossing over occurs. Spindle forms. Nuclear envelope breaks up.	Paired homologous chromosomes line up along equator of cell.	Homologous chromosomes separate to opposite poles of cell. Sister chromatids remain joined.	Nuclear envelopes form around chromosomes, which may temporarily decondense. Spindle disappears. Cytokinesis may divide cell into two.

Nucleus
Nuclear envelope
Centrosomes
Spindle fibers

☐ Diploid (2n)
■ Haploid (n)
Homologous chromosomes

Figure 9.7 **The Stages of Meiosis.**
Photos: (all) ©Ed Reschke

LM ⊢25 µm⊣   LM ⊢25 µm⊣   LM ⊢25 µm⊣   LM ⊢25 µm⊣

9.4 In Meiosis, DNA Replicates Once, but the Nucleus Divides Twice

Before meiosis occurs, a diploid cell first undergoes **interphase.** The cell grows during interphase and produces many proteins, including those required for DNA replication. After all of the cell's DNA is copied, each chromosome consists of two identical sister chromatids attached at a centromere. Finally, toward the end of interphase, the cell continues to grow and produce the enzymes and other proteins necessary to divide the cell. ⓘ *DNA replication,* section 8.2

The cell is now ready for meiosis to begin. During meiosis I, each chromosome physically aligns with its homolog. The homologous pairs split into two nuclei toward the end of meiosis I. Meiosis II then partitions the genetic material into four haploid nuclei. **Figure 9.7** diagrams the entire process; you may find it helpful to refer to it as you read the rest of this section.

A. In Meiosis I, Homologous Chromosomes Pair Up and Separate

Homologous pairs of chromosomes find each other and then split up during the first meiotic division.

Prophase I   During **prophase I** (that is, prophase of meiosis I), the replicated chromosomes condense. A **spindle** begins to form from microtubules assembled at the centrosomes, and spindle attachment points called **kinetochores** grow on each centromere. Meanwhile, the nuclear envelope breaks up; once it is gone, the spindle fibers can reach the chromosomes.

The events described so far resemble those of prophase of mitosis, but something unique happens during prophase I of meiosis: The homologous chromosomes line up next to one another. (Mules are sterile because their cells cannot complete this stage, as described in this chapter's Burning Question.) Section 9.5 describes how this arrangement allows for an allele-shuffling mechanism called crossing over.

Metaphase I   In **metaphase I,** the spindle arranges the paired homologs down the center "equator" of the cell. Each member of a homologous pair attaches to a spindle fiber stretching to one pole. The stage is therefore set for the homologous pairs to be separated.

Anaphase I, Telophase I, and Cytokinesis   Spindle fibers pull the homologous pairs apart in **anaphase I,** although the sister chromatids that make up each chromosome remain joined. The chromosomes complete their movement to opposite poles in **telophase I. Cytokinesis** typically occurs after telophase I, splitting the original cell into two. ⓘ *cytokinesis,* section 8.5C

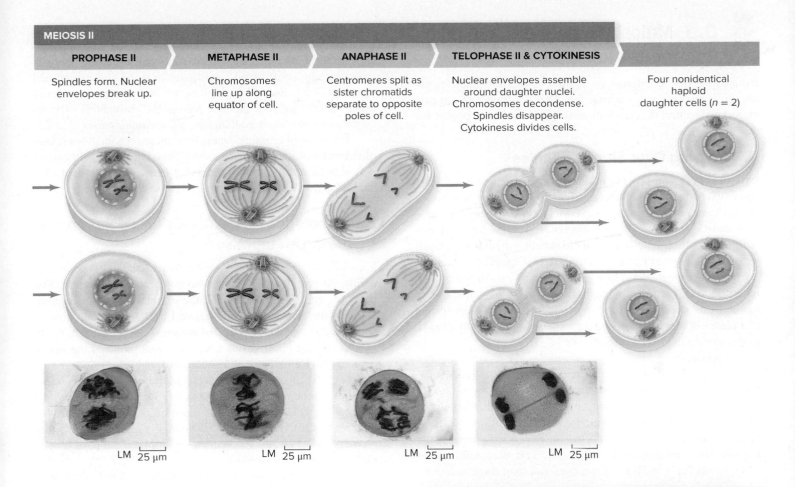

MEIOSIS II			
PROPHASE II	METAPHASE II	ANAPHASE II	TELOPHASE II & CYTOKINESIS
Spindles form. Nuclear envelopes break up.	Chromosomes line up along equator of cell.	Centromeres split as sister chromatids separate to opposite poles of cell.	Nuclear envelopes assemble around daughter nuclei. Chromosomes decondense. Spindles disappear. Cytokinesis divides cells.	Four nonidentical haploid daughter cells (n = 2)

LM ⊢25 μm⊣   LM ⊢25 μm⊣   LM ⊢25 μm⊣   LM ⊢25 μm⊣

B. Meiosis II Yields Four Haploid Nuclei

A second interphase precedes meiosis II in many species. During this time, the chromosomes unfold into very thin threads. The cell produces proteins, but the DNA does not replicate a second time.

Meiosis II strongly resembles mitosis. The process begins with **prophase II,** when the chromosomes again condense and become visible. In **metaphase II,** the spindle arranges the chromosomes down the center of each cell. In **anaphase II,** the centromeres split, and the separated sister chromatids move to opposite poles. In **telophase II,** nuclear envelopes form around the separated sets of chromosomes. Cytokinesis then separates the nuclei into individual cells. The overall result: One diploid cell has divided into four haploid cells.

Figure It Out

A cell that is entering prophase I contains __ times as much DNA as one daughter cell at the end of meiosis.

Answer: Four.

9.4   MASTERING CONCEPTS

1. What happens during interphase?
2. How do the events of meiosis I and meiosis II produce four haploid nuclei from one diploid nucleus?

Burning Question

If mules are sterile, then how are they produced?

©D. Normark/PhotoLink/ Getty Images RF

A mule is a human invention: It is the hybrid offspring of a male donkey and a female horse. Mules are desirable for some uses because they are surefooted; in addition, they are easier to maintain and more durable than horses. However, mules are usually sterile. Why?

A peek at the parents' chromosomes reveals the answer. Donkeys have 31 pairs of chromosomes, whereas horses have 32 pairs. When gametes from horse and donkey unite, the resulting hybrid zygote has 63 chromosomes (31 + 32). The zygote divides mitotically to yield the cells that make up the mule.

These hybrid cells cannot undergo meiosis for two reasons. First, they have an odd number of chromosomes, which disrupts meiosis because at least one chromosome lacks a homologous partner. Second, donkeys and horses have slightly different chromosome structures, so the hybrid's parental chromosomes cannot align properly during prophase I. The result is an inability to produce sperm and egg cells. The only way to produce more mules is to again mate horses with donkeys. ⓘ *hybrid infertility,* section 14.2B

Submit your burning question to Marielle.Hoefnagels@mheducation.com

9.5 Meiosis Generates Enormous Variability

By creating new combinations of alleles, meiosis generates astounding genetic variety among the offspring from just two parents. This section describes three mechanisms that account for this diversity.

A. Crossing Over Shuffles Alleles

Crossing over is a process in which two homologous chromosomes exchange genetic material (**figure 9.8**). During prophase I, the homologs align themselves precisely, gene by gene, in a process called synapsis. The chromosomes are attached at a few points along their lengths, called chiasmata (singular: chiasma), where the homologs exchange chromosomal material.

Consider what takes place in your own ovaries or testes. You inherited one member of each homologous pair from your mother; the other came from your father. Crossing over means that pieces of these homologous chromosomes physically change places during meiosis.

Suppose, for instance, that one chromosome carries the genes that dictate hair color, eye color, and finger length. Perhaps the version you inherited from your father has the alleles that specify blond hair, blue eyes, and short fingers. The homolog from your mother is different; its alleles dictate black hair, brown eyes, and long fingers. Now, suppose that crossing over occurs between the homologous chromosomes. Afterward, one chromatid might carry alleles for blond hair, brown eyes, and long fingers; another

would specify black hair, blue eyes, and short fingers. These two chromatids are termed "recombinant" because they combine alleles from your two parents. The two chromatids that did not form chiasmata, however, would remain unchanged and are termed "parental." Note that although all of the alleles in your ovaries or testes came from your parents, half of the chromatids—the recombinant ones—now contain new allele *combinations*.

The result of crossing over is four unique chromatids in place of two pairs of identical chromatids. Each chromatid will end up in a separate haploid cell. Thus, crossing over ensures that each haploid cell will be genetically different from the others.

B. Homologous Pairs Are Oriented Randomly During Metaphase I

Figure 9.9 reveals a second way that meiosis creates genetic variability. At metaphase I, pairs of homologous chromosomes line up at the cell's center. Examine the orientation of the chromosomes in the cell labeled "Alternative 1." All of the blue chromosomes are on top, whereas the red homologs are on the bottom. In anaphase I, the chromosomes separate, and the resulting nuclei contain either all blue or all red chromosomes.

The next time a cell in the same individual undergoes meiosis, the orientation of the chromosomes may be the same, or it may not be. The arrangement of chromosomes at metaphase I is random, and all four alternatives shown in figure 9.9 are equally probable. Most of the time, gametes will inherit a mix of genetic material from both parents.

The number of possible arrangements is related to the chromosome number. For two pairs of homologs, each resulting gamete may have any of four (2^2) unique chromosome configurations.

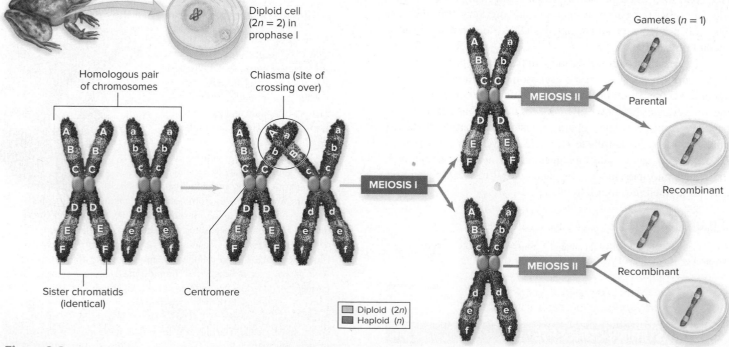

Figure 9.8 Crossing Over. In crossing over, portions of homologous chromosomes swap places. This process generates genetic diversity because each of the resulting chromatids has a unique combination of alleles. The capital and lowercase letters represent different alleles of six genes.

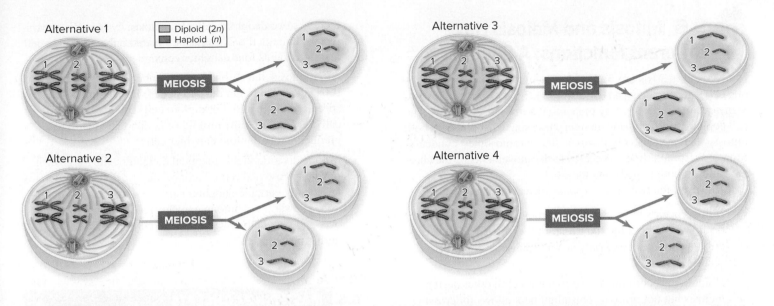

Figure 9.9 Random Orientation. A germ cell containing three homologous pairs of chromosomes can generate eight genetically different gametes. Note that this number does not include the effects of crossing over.

For three pairs, as shown in figure 9.9, eight (2^3) unique configurations can occur in the gametes. Extending this formula to humans, with 23 chromosome pairs, each gamete contains one of 8,388,608 (2^{23}) possible chromosome combinations—all equally likely.

C. Fertilization Multiplies the Diversity

We have already seen that every diploid cell undergoing meiosis is likely to produce haploid nuclei with different combinations of chromosomes. Furthermore, it takes two to reproduce. In one mating, any of a woman's 8,388,608 possible egg cells can combine with any of the 8,388,608 possible sperm cells of a partner. One couple could therefore theoretically create more than 70 trillion ($8,388,608^2$) genetically unique individuals! And this enormous number is an underestimate, because it does not take into account the additional variation from crossing over.

With so much potential variability, the chance of two parents producing genetically identical children seems exceedingly small. How do the parents of identical twins defy the odds? The answer is that identical twins result from just one fertilization event. The resulting zygote or embryo splits in half, creating separate, identical babies (**figure 9.10**). Identical twins are called *monozygotic* because they derive from one zygote. In contrast, nonidentical (fraternal) twins occur when two sperm cells fertilize two separate egg cells. The twins are therefore called *dizygotic*. See this chapter's Apply It Now box for more on multiple births.

9.5 | MASTERING CONCEPTS

1. How does crossing over shuffle alleles?
2. Explain how to arrive at the estimate that one human couple can produce over 70 trillion unique offspring.
3. How are identical twins different from fraternal twins?

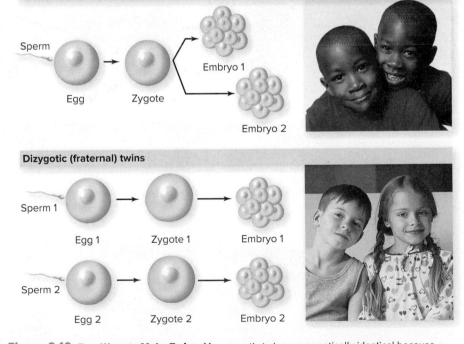

Figure 9.10 Two Ways to Make Twins. Monozygotic twins are genetically identical because they come from the same zygote. Dizygotic (fraternal) twins are no more alike than nontwin siblings because they start as two different zygotes.

Photos: (monozygotic twins): ©Barbara Penoyar/Getty Images RF; (dizygotic twins): ©Image Source Black/Getty Images RF

9.6 Mitosis and Meiosis Have Different Functions: A Summary

Mitosis and meiosis are both mechanisms that divide a eukaryotic cell's genetic material (**figure 9.11**). The two processes share many events, as revealed by the similar names of the stages. The cell copies its DNA during an interphase stage that precedes both mitosis and meiosis, after which the chromosomes condense. Moreover, spindle fibers orchestrate the movements of the chromosomes in both mitosis and meiosis.

However, the two processes also differ in many ways:

- Mitosis occurs in somatic cells throughout the body, and it occurs throughout the life cycle. In contrast, meiosis occurs only in germ cells and only at some stages of life (see section 9.8).

- Homologous chromosomes align with each other during meiosis but not mitosis. This alignment allows for crossing over, which also occurs only in meiosis.

- Following mitosis, cytokinesis occurs once for every DNA replication event. The product of mitotic division is

therefore two daughter cells. In meiosis, cytokinesis occurs twice, although the DNA has replicated only once. One cell therefore yields four daughter cells.

- After mitosis, the chromosome number in the two daughter cells is the same as in the parent cell. Depending on the species, either haploid or diploid cells can divide mitotically. In contrast, only diploid cells divide by meiosis, producing four haploid daughter cells.

- Mitotic division yields identical daughter cells for growth, repair, and asexual reproduction. Meiotic division generates genetically variable daughter cells used in sexual reproduction. The variation among gametes results from crossing over and the random orientation of chromosome pairs during metaphase I.

9.6 MASTERING CONCEPTS

1. In what ways are mitosis and meiosis similar?
2. In what ways are mitosis and meiosis different?

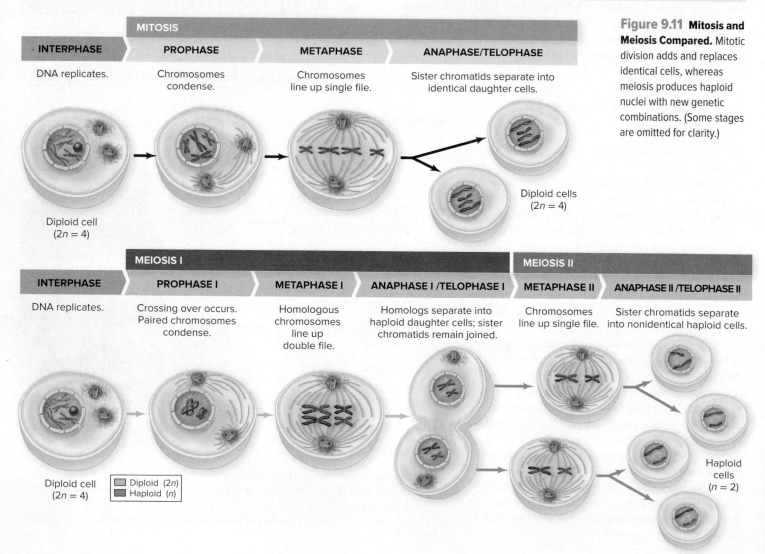

Figure 9.11 Mitosis and Meiosis Compared. Mitotic division adds and replaces identical cells, whereas meiosis produces haploid nuclei with new genetic combinations. (Some stages are omitted for clarity.)

9.7 Errors Sometimes Occur in Meiosis

We have already seen that DNA replication errors can cause mutations; these mistakes may occur during interphase preceding either mitosis or meiosis. Other types of errors can occur when a nucleus divides, producing daughter cells with extra or missing DNA. These errors may have especially serious consequences if they occur during gamete production. ⓘ *mutations,* section 7.7

A. Cells May Inherit Too Many or Too Few Chromosomes

An error in meiosis, such as the failure of the spindle to form properly, can produce a **polyploid cell** with one or more complete sets of extra chromosomes (*polyploid* means "many sets"). For example, if a sperm with the normal 23 chromosomes fertilizes an abnormal egg cell with two full sets (46), the resulting zygote will have three copies of each chromosome (69 total), a type of polyploidy called triploidy. Most human polyploids fail to live past the very early stages of development.

Polyploidy is an important force in plant evolution. In contrast to humans, about 30% of flowering plant species tolerate polyploidy well, and many crop plants are polyploids (see figure 14.10). The durum wheat in pasta, for example, is tetraploid (it has four sets of seven chromosomes), and the wheat in bread is hexaploid, with six sets of seven chromosomes.

An **aneuploid cell** typically has just one or a few extra or missing chromosomes. The cause of the abnormality is an error called **nondisjunction,** which occurs when chromosomes fail to separate at either anaphase I or anaphase II (**figure 9.12**). The problem traces to spindle fibers, which may not form properly or attach correctly to a chromosome. Regardless of how it happens, the result is a sperm or egg cell with two copies of a particular chromosome or none at all. When such a gamete fuses with another at fertilization, the resulting zygote has either 45 or 47 chromosomes instead of the normal 46.

Most embryos with incorrect chromosome numbers cease developing before birth. Extra genetic material, however, causes fewer problems than missing material. This is why most children with the wrong number of chromosomes have an extra one—a trisomy—rather than a missing one.

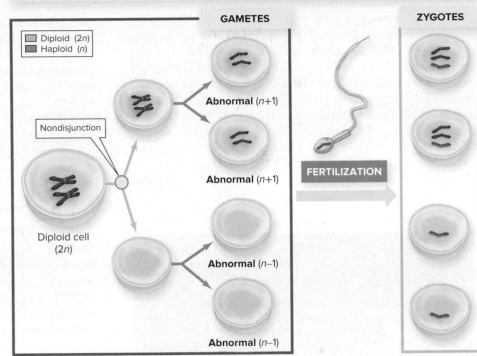

a. Nondisjunction in meiosis I

GAMETES
ZYGOTES

☐ Diploid (2*n*)
☐ Haploid (*n*)

Nondisjunction

Diploid cell (2*n*)

Abnormal (*n*+1)

Abnormal (*n*+1)

FERTILIZATION

Abnormal (*n*–1)

Abnormal (*n*–1)

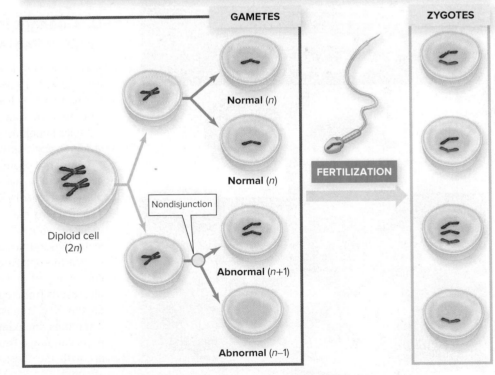

b. Nondisjunction in meiosis II

GAMETES
ZYGOTES

Diploid cell (2*n*)

Normal (*n*)

Normal (*n*)

Nondisjunction

FERTILIZATION

Abnormal (*n*+1)

Abnormal (*n*–1)

Figure 9.12 Nondisjunction. (a) A homologous pair of chromosomes fails to separate during meiosis I. The result is two nuclei with two copies of the chromosome and two nuclei that lack the chromosome. (b) Sister chromatids fail to separate during meiosis II. One nucleus has an extra chromosome, and one is missing the chromosome. The other two nuclei are unaffected. (All chromosomes other than the ones undergoing nondisjunction are omitted for clarity.)

The rest of this section describes some syndromes in humans resulting from too many or too few chromosomes.

Extra Autosomes: Trisomy 21, 18, or 13

A person with trisomy 21, the most common cause of Down syndrome, has three copies of chromosome 21 (figure 9.13). An affected person has distinctive facial features and a unique pattern of hand creases. Intelligence varies greatly; some children have profound mental impairment, whereas others learn well. Many affected children die before their first birthdays, often because of congenital heart defects. People with Down syndrome also have an above-average risk for leukemia and Alzheimer disease.

The probability of giving birth to a child with trisomy 21 increases dramatically as a woman ages. For women younger than 30, the chances of conceiving a child with the syndrome are 1 in 3000. For a woman of 48, the incidence jumps to 1 in 9. An increased likelihood of nondisjunction in older females apparently accounts for this age association, but no one knows for sure why older women might have problems completing meiosis.

Trisomy 21 is the most common autosomal trisomy, but that is only because the fetus is most likely to remain viable. Trisomies 18 and 13 are the next most common, but few infants with these genetic abnormalities survive. Trisomies undoubtedly occur with other chromosomes, but the embryos fail to develop at all.

Extra or Missing Sex Chromosomes: XXX, XXY, XYY, and XO

Nondisjunction can produce a gamete that contains two X or Y chromosomes instead of only one. Fertilization then produces a zygote with too many sex chromosomes: XXX, XXY, or XYY. A gamete may also lack a sex chromosome altogether. If one gamete contains an X chromosome and the other gamete has neither X nor Y, the resulting zygote is XO. Interestingly, medical researchers have never reported a person with one Y and no X chromosome. Table 9.1 summarizes some of the sex chromosome abnormalities.

B. Changes in Chromosome Structure May Be Harmful

Parts of a chromosome may be deleted, duplicated, inverted, or even moved to a new location (figure 9.14). These structural abnormalities may have many causes, ranging from radiation exposure to errors in crossing over. Because each chromosome includes hundreds or thousands of genes, even small changes in a chromosome's structure can affect an organism.

A chromosomal **deletion** results in the loss of one or more genes. Cri du chat syndrome (French for "cat's cry"), for example, is associated with deletion of several genes on chromosome 5. The illness is named for the odd cry of an affected child, similar to the mewing of a cat. The gene deletion also causes severe intellectual disabilities and developmental delays.

In the opposite situation, a **duplication** produces multiple copies of part of a chromosome. Fragile X syndrome, for example, results from repeated copies of a three-base sequence (CGG) on the X chromosome. The disorder can produce a range of symptoms, including intellectual disabilities. The number of repeats can range from fewer than 10 to more than 200. Individuals with the most copies of the repeat are the most severely affected.

The duplication of entire genes sometimes plays an important role in evolution (see figure 13.18). If one copy of the original gene continues to do its old job, then a mutation in a "spare" copy will not be harmful. Although these mutations often ruin the gene, they can also lead to new functions. As just one example, biologists have studied a gene that was originally required for

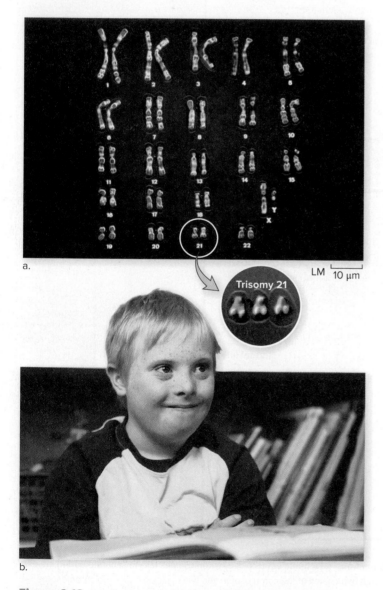

Figure 9.13   Trisomy 21. (a) A normal human karyotype reveals 46 chromosomes, in 23 pairs. (b) A child with three copies of chromosome 21 has Down syndrome.

(a, both): ©CNRI/Science Source; (b): ©George Doyle/Stockbyte/Getty Images RF

TABLE 9.1   Examples of Sex Chromosome Abnormalities

Chromosomes	Name of Condition	Approximate Incidence	Symptoms
XXX	Triplo-X	1 in 1500 females	Tall stature, menstrual irregularities, increased risk of giving birth to triplo-X daughters or XXY sons
XXY	Klinefelter, or XXY, syndrome	1 in 750 males	Variable, but often include sexual underdevelopment, long limbs, large hands and feet, development of breast tissue
XYY	Jacobs, or XYY, syndrome	1 in 1000 males	Often few noticeable symptoms; tall stature, acne, problems with speech and reading
XO	Turner syndrome	1 in 2000 females	Short stature, sexual underdevelopment, infertility

Figure 9.14 Structural Abnormalities. Portions of a chromosome can be (a) deleted, (b) duplicated, or (c) inverted. (d) In translocation, two nonhomologous chromosomes exchange parts. The micrograph shows a portion of chromosome 5 (larger pair) that has switched places with part of chromosome 14 (smaller pair).

Photo: ©Addenbrookes Hospital/Science Source

Normal   a. Deletion   b. Duplication   c. Inversion   d. Translocation (before)   Translocation (after)   LM (false color)   5 μm

the secretion of calcium in tooth enamel in vertebrates. Mutations in duplicate genes created new functions, including the production of calcium-rich breast milk.

In an **inversion,** part of a chromosome flips and reinserts, changing the gene sequence. Unless inversions break genes, they are usually less harmful than deletions, because all the genes are still present. Fertility problems can arise, however, if an adult has an inversion in one chromosome but its homolog is normal. During crossing over, the inverted chromosome and its noninverted partner may twist around each other in a way that generates chromosomes with deletions or duplications. Because the gametes will have extra or missing genes, the result may be a miscarriage or birth defects.

In a **translocation,** nonhomologous chromosomes exchange parts. Translocations often break genes, sometimes causing leukemia or other cancers. In about 95% of people with chronic myelogenous leukemia, for example, parts of chromosomes 9 and 22 switch places. The translocation creates a combined gene on chromosome 22; this gene, in turn, encodes a protein that speeds cell division and suppresses normal cell death (apoptosis). The

result is leukemia, a form of cancer in which blood cells divide out of control. ⓘ *apoptosis,* section 8.7

If no genes are broken in a translocation, then the person has the normal amount of genetic material; it is simply rearranged. Such a person is healthy but may have fertility problems. Some sperm or egg cells will receive one of the translocated chromosomes but not the other, causing a genetic imbalance—some genes are duplicated, and others are deleted. The consequences may be mild or severe, depending on which genes are disrupted.

9.7 MASTERING CONCEPTS

1. Draw a diagram to show how nondisjunction of all chromosomes during meiosis I in one parent could lead to polyploid offspring. (Use 2*n* = 6 for the starting cells; assume the other parent's gamete contributes the normal number of chromosomes.)

2. How can deletions, duplications, inversions, and translocations cause illness?

🍁 9.8 Haploid Nuclei Are Packaged into Gametes

The events of meiosis explain how a diploid cell produces four genetically different haploid nuclei. The same process occurs in both sexes, yet sperm and egg cells typically look very different from each other (figure 9.15). Usually, a sperm is lightweight and can swim; an egg cell is huge by comparison and packed with nutrients and organelles. How do males and females package those haploid nuclei into such different-looking gametes?

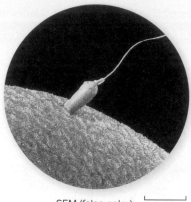

Figure 9.15 Human Gametes. Note the size difference between the sperm and the egg cell.
©Francis Leroy, Biocosmos/ Science Source

SEM (false color)   5 μm

A. In Humans, Gametes Form in Testes and Ovaries

The formation and specialization of sperm cells is called spermatogenesis (figure 9.16). Inside the testes, spermatogonia are diploid germ cells that divide mitotically to produce two kinds of cells: more spermatogonia and specialized cells called primary spermatocytes. During interphase, primary spermatocytes accumulate cytoplasm and replicate their DNA. The first meiotic division yields two equal-sized haploid cells called secondary spermatocytes.

Each secondary spermatocyte then completes its second meiotic division. The products are four equal-sized spermatids, each of which specializes into a mature, tadpole-shaped sperm cell. The entire process, from spermatogonium to sperm, takes about 74 days.

In comparison to a sperm cell, an egg cell is massive. The female produces these large cells by unequally packaging the cytoplasm from the two meiotic divisions. The egg cell gets most of the cytoplasm, and the other products of meiosis are tiny.

The formation of egg cells is called oogenesis (figure 9.17). It occurs in the ovaries and begins with a diploid germ cell, an oogonium. This cell can divide mitotically to produce more oogonia or a cell called a primary oocyte. In meiosis I, the primary oocyte divides into a small haploid cell with very little cytoplasm, called a polar body, and a much larger haploid cell called a secondary oocyte. In meiosis II, the secondary oocyte divides unequally to produce another polar body and the mature egg cell, or ovum, which contains a large amount of cytoplasm. The tiny polar bodies normally play no further role in reproduction.

Chapter 35 explores human reproduction and development in more detail.

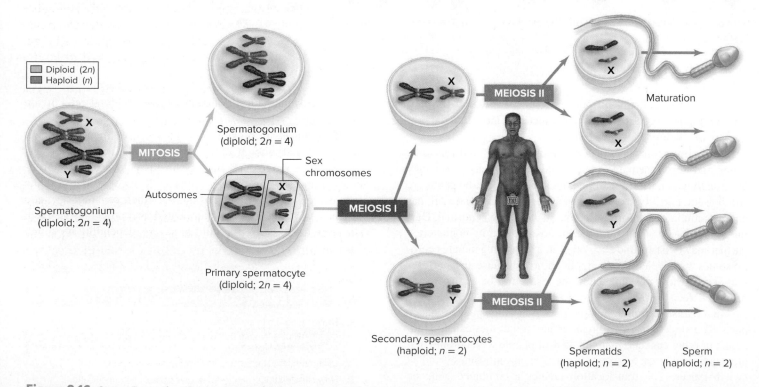

Figure 9.16 Sperm Formation (Spermatogenesis). In humans, diploid primary spermatocytes undergo meiosis, yielding four equal-sized, haploid sperm. Of the normal 23 pairs of chromosomes, only 1 pair of autosomes and 1 pair of sex chromosomes are shown.

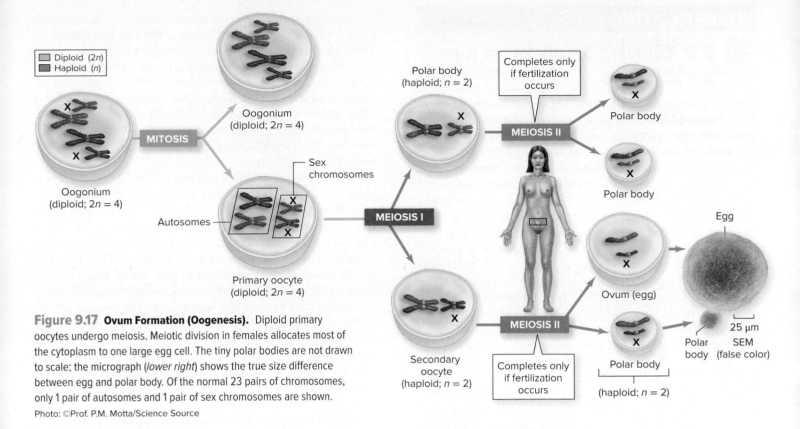

Figure 9.17 Ovum Formation (Oogenesis). Diploid primary oocytes undergo meiosis. Meiotic division in females allocates most of the cytoplasm to one large egg cell. The tiny polar bodies are not drawn to scale; the micrograph (*lower right*) shows the true size difference between egg and polar body. Of the normal 23 pairs of chromosomes, only 1 pair of autosomes and 1 pair of sex chromosomes are shown.
Photo: ©Prof. P.M. Motta/Science Source

B. In Plants, Gametophytes Produce Gametes

Plant life cycles include an **alternation of generations** between multicellular haploid and diploid individuals (**figure 9.18**). A diploid zygote divides mitotically and develops into a mature sporophyte. Germ cells in the diploid sporophyte undergo meiosis to produce haploid cells called spores. The spores germinate, dividing mitotically to produce a multicellular haploid plant called a gametophyte. The gametophyte, in turn, produces haploid sperm or egg cells by mitotic cell division. A sperm fertilizes an egg cell to form a zygote; the cycle begins anew.

In mosses and ferns, the gametophytes are small, green plants that are visible with the unaided eye. In flowering plants, however, the gametophyte is microscopic and relies on the sporophyte for nutrition. The egg-producing female gametophyte, for example, is buried deep within a flower.

Some plants produce swimming sperm cells. In mosses and ferns, the male gametes use flagella to swim in a film of water to the stationary egg cell. The sperm cells of conifers and flowering plants, however, do not swim. Instead, these plants produce pollen grains—male gametophytes—that travel in wind or on animals to reach female plant parts. Pollen germination delivers sperm cells directly to the stationary egg cell. Chapters 19 and 24 further describe plant reproduction.

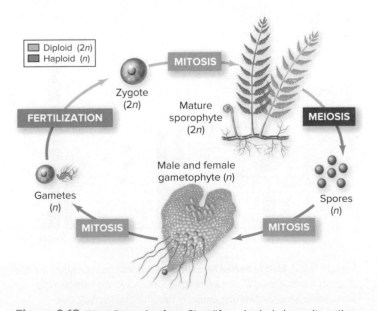

Figure 9.18 Plant Reproduction. Plant life cycles include an alternation of multicellular haploid and diploid generations.

9.8 MASTERING CONCEPTS

1. What are the stages of sperm development in humans?
2. What are the stages of egg cell development in humans?
3. How does gamete production in plants differ from that in animals?

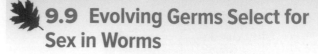

9.9 Evolving Germs Select for Sex in Worms

Sexual reproduction is a hassle. Why spend energy to attract mates when you could just make identical copies of yourself and ensure that your genome makes it to the next generation?

Some scientists suggest that parasites explain the evolution and persistence of sexual reproduction. Parasites often reproduce much faster than their hosts, so mutations frequently produce new variants. Perhaps sexually reproducing hosts, which generate new allele combinations in each generation, have the best chance to survive in an environment full of constantly changing parasites.

The predicted endless evolutionary battle between host and parasite gives this proposition its common name: the Red Queen hypothesis. In Lewis Carroll's book *Through the Looking Glass,* the Red Queen remarks to Alice, "It takes all the running you can do, to keep in the same place." Sexual reproduction may give a species just enough variation to hold its own against rapidly evolving parasites. Does any evidence support this idea?

A team of Indiana University researchers, led by Levi Morran and Curtis Lively, tested the predictions of the Red Queen hypothesis by studying disease-causing bacteria and their host, the microscopic roundworm *Caenorhabditis elegans* (**figure 9.19**). These worms are convenient laboratory workhorses. Hundreds of them can be reared, with or without bacteria, in a single petri dish. ⓘ *C. elegans,* section 21.7

Roundworms also have another trait that makes them ideal for research on reproduction: They can be hermaphrodites. Meiosis in hermaphrodites generates both male and female sex cells. A hermaphrodite worm can reproduce sexually with a male, or it can self-reproduce. Sexual reproduction between a male and a hermaphrodite generates variation, but when gametes from the same individual combine, the resulting offspring have genetic material identical to that of the parent. (See the Burning Question in section 35.5 to learn about human hermaphrodites.)

These two reproductive strategies allowed the researchers to study how a single population of *C. elegans* can shift toward or away from sexual reproduction when parasites are present. In one experiment, the research team exposed a population containing both sexually reproducing and self-reproducing *C. elegans* to three experimental conditions. One group of worms was allowed to reproduce in petri dishes with a "fixed" (unchanging) strain of disease-causing bacteria. A second group was reared with bacteria that had successfully infected and reproduced in hosts of the previous generation—in other words, these bacteria were evolving with the hosts. The remaining worms (the control group) were placed in petri dishes containing dead bacteria. In all three treatments, each generation of worm offspring was moved to fresh petri dishes (with live or dead bacteria) and allowed to reproduce.

The researchers documented the rate of sexual reproduction in each of the three *C. elegans* groups over 30 generations (**figure 9.20**). At the start of the experiment, sexual reproduction was rare in all three groups. But the rate of sexual reproduction increased within a few generations for both populations exposed to bacteria. Roundworms exposed to a fixed bacterial strain eventually shifted back toward self-reproduction, presumably because disease-resistance traits became more common. In contrast, the population exposed to evolving bacteria entered an evolutionary

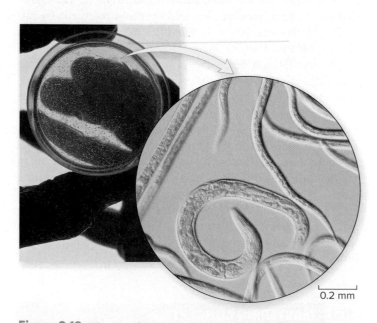

Figure 9.19 Microscopic Worms. The roundworm *C. elegans* replicates quickly in small petri dishes, making this species ideally suited for laboratory research.

(petri dish): ©James King-Holmes/Science Source; (roundworms): ©Sinclair Stammers/Science Source

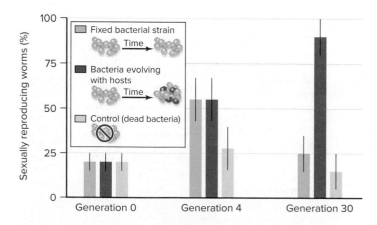

Figure 9.20 Shift to Sexual Reproduction. Researchers measured the rate of sexual reproduction in *C. elegans* populations exposed to a fixed bacterial strain, bacteria evolving with the worms, or dead bacteria (control). Over 30 generations of worms, the evolving bacteria selected for sexual reproduction in their hosts. (Error bars represent two standard errors; see appendix B.)

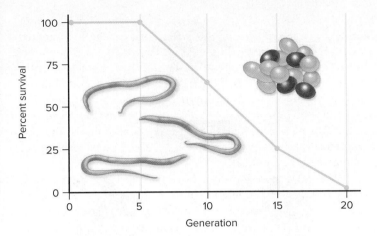

Figure 9.21 Hermaphrodite Extinction. When a completely hermaphroditic roundworm population is exposed to an evolving strain of disease-causing bacteria over multiple generations, the population eventually goes extinct.

race like the one the Red Queen hypothesis predicts. As each generation of bacteria got better at infecting familiar hosts, the worms became more likely to shuffle the genetic deck and produce a greater proportion of unfamiliar offspring.

In another experiment, a population consisting entirely of identical, hermaphroditic *C. elegans* was exposed to bacteria that evolved with their hosts throughout the experiment. The bacteria quickly became effective at infecting the roundworms. Meanwhile, each generation of identical offspring became more vulnerable to the evolving bacteria. After 20 generations, the roundworm population was extinct (figure 9.21). Evidently, without the ability to generate genetically unique offspring, the roundworms lost the evolutionary race against the bacteria.

These experiments reveal that parasites select for sexual reproduction in their hosts. In the evolutionary race between organisms and their parasites, sexual reproduction produces new allele combinations, boosting the chance that the host species will evade extinction in a rapidly changing pool of parasites. Sex costs time and energy, but the alternative—easy reproduction of a doomed allele combination—may be even costlier.

Source: Morran, Levi, Olivia Schmidt, Ian Gelarden, Raymond Parrish II, and Curtis Lively. July 8, 2011. Running with the Red Queen: Host-parasite coevolution selects for biparental sex. *Science*, vol. 333, pages 216–218.

9.9 MASTERING CONCEPTS

1. Why are the offspring of a male and a hermaphrodite more variable than the offspring of a self-reproducing hermaphrodite?

2. Under what conditions does evolution select for sexual reproduction in *C. elegans*?

Apply It **Now**

Multiple Births

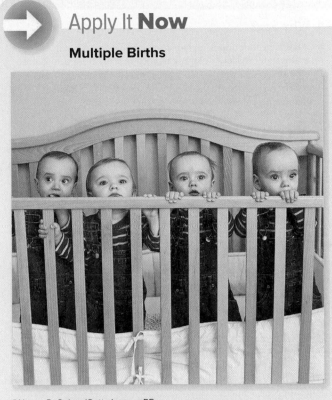

©Nancy R. Cohen/Getty Images RF

Triplets, quadruplets, and higher-order multiple births have become more common since the 1980s. How do they arise?

Triplets come about in several ways. The least common route is for a single embryo to split and develop into three genetically identical babies (monozygotic triplets). Alternatively, if three sperm fertilize three separate egg cells, the triplets will be fraternal (trizygotic). Most commonly, however, an embryo splits and forms two identical babies, and a separate embryo develops into an additional, nonidentical baby.

Higher-order multiples likewise usually include combinations of identical and fraternal siblings. Identical quadruplets are exceedingly rare, occurring perhaps once in 11 million deliveries. Monozygotic quintuplets are even more unusual, with only one set ever known to have been born.

Two trends account for the rising incidence of multiple births. First, older women are more likely than younger women to have multiple births, and childbearing among older women has become more common. Second, couples have increasingly sought treatment for infertility. Some fertility drugs stimulate a woman to release more than one egg cell. If sperm fertilize all of them, a multiple birth could result. Another infertility therapy is *in vitro* fertilization, in which sperm fertilize egg cells harvested from a woman's ovaries in the lab. One or more embryos judged most likely to result in a live birth are then implanted into the woman's uterus. Multiple births often result.

CHAPTER SUMMARY

9.1 Why Sex?

- **Asexual reproduction** is reproduction without sex. **Sexual reproduction** produces offspring by mixing traits from two parents.
- **Conjugation** is a form of gene transfer in some microorganisms.
- Asexual reproduction can be successful in a stable environment, but a changing environment selects for sexual reproduction.

9.2 Diploid Cells Contain Two Homologous Sets of Chromosomes

- **Diploid cells** have two full sets of **chromosomes,** one from each parent. A **karyotype** is a chart that displays all of the chromosomes from one cell.
- In humans, the **sex chromosomes** (X and Y) determine whether an individual is male or female. The 22 **homologous pairs** of **autosomes** do not determine sex.
- Homologous chromosomes share the same size, banding pattern, and centromere location, but they differ in the **alleles** they carry.

9.3 Meiosis Is Essential in Sexual Reproduction

A. Gametes Are Haploid Sex Cells
- **Meiosis** halves the genetic material to produce **haploid cells.** **Fertilization** occurs when **gametes** fuse, forming the diploid **zygote.** **Mitotic** cell division produces the body's cells during growth and development. **Figure 9.22** summarizes the events of a sexual life cycle.

B. Specialized Germ Cells Undergo Meiosis
- **Somatic cells** do not participate in reproduction, whereas diploid **germ cells** produce haploid sex cells.

C. Meiosis Halves the Chromosome Number and Scrambles Alleles
- The events of meiosis ensure that gametes are haploid and genetically variable (**figure 9.23**).

9.4 In Meiosis, DNA Replicates Once, but the Nucleus Divides Twice

- **Interphase** (including DNA replication) happens before meiosis.
- During meiosis, **spindle** fibers attached to **kinetochores** move the chromosomes.

A. In Meiosis I, Homologous Chromosomes Pair Up and Separate
- Homologous pairs of chromosomes align during **prophase I,** line up at the cell's center in **metaphase I,** then split apart during **anaphase I.** The chromosomes arrive at the poles in **telophase I,** and the cell often divides (**cytokinesis**).

B. Meiosis II Yields Four Haploid Nuclei
- The two products of meiosis I each divide again, yielding four haploid nuclei. The events of meiosis II include **prophase II, metaphase II, anaphase II,** and **telophase II.** Cytokinesis divides the cytoplasm.

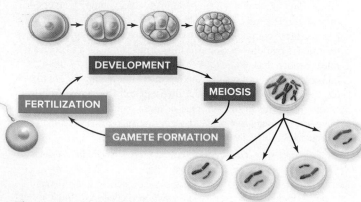

Figure 9.22 Sexual Life Cycle Events.

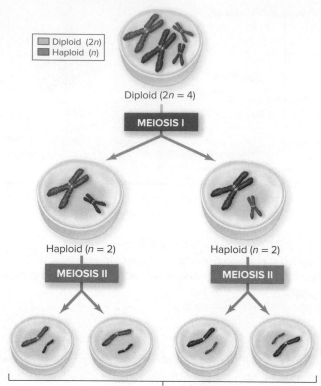

Figure 9.23 Summary of Meiosis.

9.5 Meiosis Generates Enormous Variability

A. Crossing Over Shuffles Alleles
- **Crossing over,** which occurs in prophase I, produces variability when portions of homologous chromosomes switch places.

B. Homologous Pairs Are Oriented Randomly During Metaphase I
- Every possible orientation of homologous pairs of chromosomes at metaphase I is equally likely. As a result, one person can produce over 8 million genetically different gametes.

C. Fertilization Multiplies the Diversity
- Because any sperm can fertilize any egg cell, a human couple can produce over 70 trillion genetically different offspring.
- Identical (monozygotic) twins arise when a zygote splits into two embryos. Fraternal (dizygotic) twins develop from separate zygotes.

9.6 Mitosis and Meiosis Have Different Functions: A Summary

- Mitotic division makes identical cell copies and occurs throughout life.
- Meiosis produces genetically different haploid cells. It occurs only in specialized cells and only during some parts of the life cycle.

9.7 Errors Sometimes Occur in Meiosis

A. Cells May Inherit Too Many or Too Few Chromosomes
- **Polyploid cells** have one or more extra sets of chromosomes.
- An **aneuploid cell** has one or a few extra or missing chromosomes.
- **Nondisjunction** is the failure of homologous chromosomes or sister chromatids to separate, and it causes gametes to have incorrect chromosome numbers. A sex chromosome aneuploidy is typically less severe than an autosomal aneuploidy.

B. Changes in Chromosome Structure May Be Harmful

- Chromosomal rearrangements can **delete** or **duplicate** genes. An **inversion** flips gene order; in a **translocation,** two nonhomologs exchange parts. Inversions and translocations may disrupt vital genes.

9.8 Haploid Nuclei Are Packaged into Gametes

A. In Humans, Gametes Form in Testes and Ovaries

- Spermatogenesis begins in the testes. Diploid germ cells undergo mitosis and then meiosis I and II before differentiating into four sperm cells.
- In oogenesis, diploid germ cells called oogonia divide mitotically and then meiotically, yielding a large egg cell and three small cells called polar bodies. Oogenesis occurs in the ovaries.

B. In Plants, Gametophytes Produce Gametes

- In plants, sexual reproduction involves an **alternation of generations** with multicellular haploid and diploid phases.

9.9 Investigating Life: Evolving Germs Select for Sex in Worms

- In lab experiments, populations of roundworms were more likely to reproduce sexually when exposed to disease-causing bacteria.

MULTIPLE CHOICE QUESTIONS

1. Compared to other forms of reproduction, the unique feature of sex is
   a. the ability of a cell to divide.
   b. the production of offspring.
   c. the ability to generate new genetic combinations.
   d. All of the above are correct.

2. Meiosis explains why
   a. you inherited half of your DNA from each of your parents.
   b. the sister chromatids in a chromosome are identical to each other.
   c. each of your somatic cells contains the same DNA.
   d. zygotes contain half as much DNA as somatic cells.

3. How many chromatids are in a human cell during metaphase I?
   a. 23   c. 92
   b. 46   d. 184

4. What event occurs soon after a cell completes meiosis I?
   a. Homologous chromosomes pair up.
   b. Homologous chromosomes move apart from each other.
   c. DNA replication occurs.
   d. Cytokinesis occurs.

5. Which of the following is *not* a mechanism that contributes to diversity among the offspring from two parents?
   a. Random fertilization
   b. Crossing over
   c. Cytokinesis
   d. Random chromosome orientation during metaphase I

6. Which of the following is true about the life cycle of a plant?
   a. Plants do not produce haploid cells.
   b. Spores are gametes.
   c. Gametes are formed by mitotic cell division.
   d. Meiosis produces a multicellular organism.

Answers to these questions are in appendix A.

WRITE IT OUT

1. Explain why evolution often selects traits that promote genetic diversity.
2. Describe a situation in which asexual reproduction might be more likely than sexual reproduction.
3. Sketch the relationships among mitosis, meiosis, and fertilization in a sexual life cycle.

4. What is the difference between haploid and diploid cells? Are your skin cells haploid or diploid? What about germ cells? Gametes?
5. How are the members of a homologous pair similar and different?
6. Where do the members of each pair of homologous chromosomes in a diploid cell come from?
7. Draw all possible metaphase I chromosomal arrangements for a cell with a diploid number of 8. How many unique gametes are possible for this species? Is this number an underestimate or an overestimate? Why?
8. In some animals, females can reproduce by themselves—that is, without males. In this process, called parthenogenesis, the young develop from unfertilized eggs. Use the Internet to find a species that uses parthenogenesis. How does parthenogenesis work in that species? What are the differences between meiosis in that species and the events of "typical" meiosis?
9. List examples of abnormalities in chromosome number and structure. Explain how each relates to an error in meiosis.
10. What is the relationship between nondisjunction and aneuploidy?
11. How does spermatogenesis differ from oogenesis, and how are the processes similar?
12. Provide examples to support or refute this statement: The products of meiosis are always haploid cells, whereas the products of mitotic division are always diploid cells.

PULL IT TOGETHER

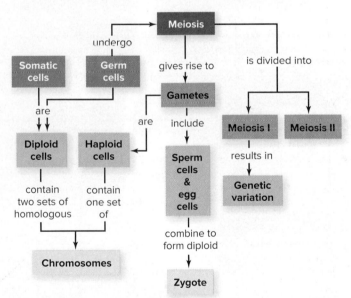

Figure 9.24 Pull It Together: Sexual Reproduction and Meiosis.

Refer to figure 9.24 and the chapter content to answer the following questions.

1. Review section 9.5 and the Survey the Landscape figure in the chapter introduction. What two processes in meiosis I generate genetic variation among gametes? What other process produces genetic variation?
2. Fit the following terms into this concept map: *chromatid, centromere, nondisjunction, fertilization,* and *mitosis.*
3. Create a separate concept map that includes these terms: *crossing over, synapsis, gamete, autosome,* and *homologous pair.* You may add other terms to the map as well.

Family Photo. Former University of Oklahoma basketball player Blake Griffin (*second from left*) walks with his brother and parents after a news conference announcing his intention to play in the professional leagues.
©Sue Ogrocki/AP Images

 LEARN HOW TO LEARN
Be a Good Problem Solver

This chapter is about the principles of inheritance, and you will find many sample problems within its pages; in addition, interactive problems are available online. Need help? The How to Solve a Genetics Problem guide at the end of this chapter shows a systematic, step-by-step approach to answering the most common types of questions. Keep using the guide until you feel comfortable solving any problem type.

From Mendel to Medical Genetics

Study the family portrait accompanying this introductory section. It is hard to resist scrutinizing the faces, searching for similarities and differences between parent and offspring. Our inherent interest in heredity is probably as old as humankind itself. But of all the people who have studied inheritance, one nineteenth-century investigator, Gregor Mendel, made the most lasting impression on what would become the science of genetics.

Mendel was born in 1822 and spent his early childhood in a small village in what is now the Czech Republic, where he learned how to tend fruit trees. After finishing school, Mendel became a priest at a monastery where he could teach and do research in natural science. The young man eagerly learned how to artificially pollinate crop plants to control their breeding. The monastery sent him to earn a college degree at the University of Vienna, where courses in the sciences and statistics fueled his interest in plant breeding. Mendel began to think about experiments to address a compelling question for plant breeders: Why do some traits disappear, only to reappear a generation later?

From 1857 to 1863, Mendel crossed and cataloged some 24,034 plants through several generations. He observed consistent ratios of traits in the offspring and deduced that the plants transmitted distinct units, or "elementen" (now called genes). Mendel described his work to the Brno Medical Society in 1865 and published it in the organization's journal the next year.

Interestingly, Charles Darwin was puzzling over natural selection and evolution at the same time that Mendel was tending his plants. No one knew it at the time, but each scientist was exploring genetic variation from a different point of view. Mendel focused on the fate of specific traits from generation to generation; Darwin studied larger-scale shifts in variation within populations. Thanks to another century of biological research, we now know that all variation traces to mutations in DNA. That insight ties together the ideas of Mendel, Darwin, and many other scientists. The so-called modern evolutionary synthesis integrates genetic variation, inheritance, and natural selection to explain evolutionary changes in populations. We take up this idea again in unit 3.

Biology has made great strides since Mendel and Darwin's time. Today, genetics and DNA are familiar to nearly everyone, and the entire set of genetic instructions to build a person—the human genome—has been deciphered. Even so, every family encounters the same principles of heredity that Mendel derived in his experiments with peas. Our look at genetics begins the traditional way, with Gregor Mendel, but we can now appreciate his genius in light of what we know about DNA.

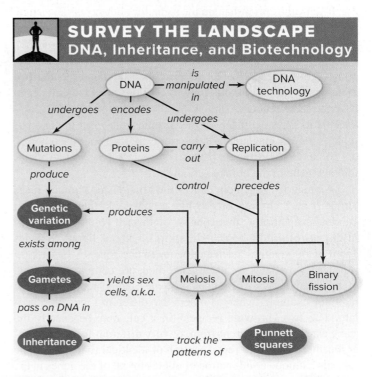

SURVEY THE LANDSCAPE
DNA, Inheritance, and Biotechnology

When gametes from two parents combine to produce offspring, what traits might the offspring inherit and express? Punnett squares help organize the possibilities.

For more details, study the Pull It Together feature in the chapter summary.

187

10.1 Chromosomes Are Packets of Genetic Information: A Review

A healthy young couple, both with family histories of cystic fibrosis, visits a genetic counselor before deciding whether to have children. The counselor suggests genetic tests, which reveal that both the man and the woman are carriers of cystic fibrosis. The counselor tells the couple that each of their future children has a 25% chance of inheriting this serious illness. How does the counselor arrive at that one-in-four chance? This chapter will explain the answer. ⓘ *genetic testing,* section 11.4C

First, however, it may be useful to review some concepts from earlier chapters. Chapter 7 explained that cells contain DNA, a molecule that encodes all of the information needed to sustain life. Human DNA includes about 25,000 genes. A **gene** is a portion of DNA whose sequence of nucleotides (A, C, G, and T) encodes a protein; the organism's proteins, in turn, help determine many of its characteristics. When a gene's nucleotide sequence mutates, the encoded protein—and the corresponding trait—may also change. Each gene can therefore exist as one or more **alleles,** or alternative forms, each arising from a different mutation.

The DNA in the nucleus of a eukaryotic cell is divided among multiple **chromosomes,** which are long strands of DNA associated with proteins. Recall that a **diploid cell** contains two sets of chromosomes, with one set inherited from each parent. The human genome consists of 46 chromosomes, arranged in 23 pairs (**figure 10.1a**). Of these, 22 pairs are **autosomes,** which are the chromosomes that are the same for both sexes. The single pair of **sex chromosomes** determines a person's sex: A female has two X chromosomes, whereas a male has one X and one Y.

With the exception of X and Y, the chromosome pairs are homologous (figure 10.1b). As described in chapter 9, the two members of a **homologous pair** of chromosomes look alike and have the same sequence of genes in the same positions. (A gene's *locus* is its physical place on the chromosome.) But the two homologs may or may not carry the same alleles. Since each homolog comes from a different parent, each person inherits two alleles for each gene in the human genome.

An analogy may help clarify the relationships among these terms. If each chromosome is like a cookbook, then the human genome is a "library" that consists of 46 such volumes, arranged in 23 pairs of similar books. The entire cookbook library includes about 25,000 recipes, each analogous to one gene.

The two alleles for each gene, then, are comparable to two of the many ways to prepare brownies; some recipes include nuts, for example, whereas others use different types of chocolate. The two "brownie recipes" in a cell may be exactly the same as, slightly different from, or very different from each other. Furthermore, with the exception of identical twins, everyone inherits a unique combination of alleles for all of the genes in the human genome.

Another important idea to review from chapter 9 is the role of meiosis and fertilization in a sexual life cycle (see figure 9.5). **Meiosis** is a specialized form of cell division that occurs in diploid germ cells and gives rise to **haploid cells,** each containing just one

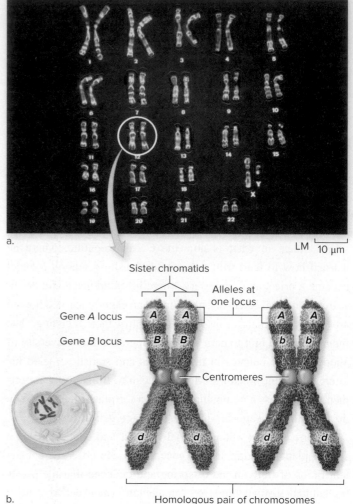

a.   LM |⎯10 μm⎯|

Figure 10.1 Homologous Chromosomes. (a) A human diploid cell contains 23 pairs of chromosomes. (b) Each chromosome has one allele for each gene. For the chromosome pair in this figure, both alleles for gene *A* are identical; the same is true for gene *D*. The chromosomes carry different alleles for gene *B*.

(a): ©CNRI/Science Source

set of chromosomes. In humans, these haploid cells are **gametes**—sperm or egg cells. **Fertilization** unites the gametes from two parents, producing the first cell of the next generation. Gametes are the cells that convey chromosomes from one generation to the next, so they play a critical part in the study of inheritance.

No one can examine a gamete and say for sure which allele it carries for every gene. As we shall see in this chapter, however, for some traits, we can use knowledge of a person's characteristics and family history to say that a gamete has a 100% chance, 50% chance, or 0% chance of carrying a specific allele. With this information for both parents, it is simple to calculate the probability that a child will inherit the allele.

10.1 | MASTERING CONCEPTS

1. Describe the relationships among chromosomes, DNA, genes, and alleles.

2. How do meiosis, fertilization, diploid cells, and haploid cells interact in a sexual life cycle?

10.2 Mendel's Experiments Uncovered Basic Laws of Inheritance

Gregor Mendel, the nineteenth-century researcher who discovered the basic principles of genetics (see the chapter opening essay), knew nothing about DNA, genes, chromosomes, or meiosis. But he nevertheless discovered how to calculate the probabilities of inheritance, at least for some traits. This section explains how he used careful observations of pea plants to draw his conclusions.

A. Why Peas?

As Mendel discovered, the pea plant is a good choice for studying heredity. Pea plants are easy to grow, develop quickly, and produce many offspring. Also, peas have many traits that appear in two easily distinguishable forms. For example, seeds may be round or wrinkled, yellow or green. Pods may be inflated or constricted. Stems may be tall or short.

Pea plants also have another advantage for studies of inheritance: It is easy to control which plants mate with which (**figure 10.2**). An investigator can take pollen from the male flower parts of one plant and apply it to the female part of the same plant (self-fertilization) or another plant (cross-fertilization). The resulting offspring are seeds that develop inside pods; each pea represents a genetically unique offspring, analogous to you and your siblings. Traits such as seed color or seed shape are evident right away; for other characteristics, such as plant height or flower color, the investigator must sow the seeds and observe each plant that develops.

B. Dominant Alleles Appear to Mask Recessive Alleles

Mendel's first experiments with peas dealt with single traits that have two expressions, such as yellow or green seed colors. He noted that some plants were always **true-breeding;** that is, self-fertilization always produced offspring resembling the parent plant for that trait. Plants derived from green seeds, for example, always produced green seeds when self-fertilized. But crosses

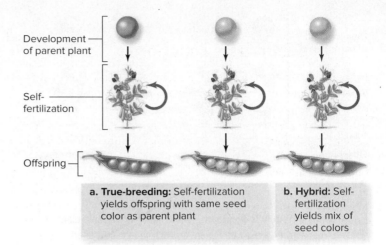

a. True-breeding: Self-fertilization yields offspring with same seed color as parent plant

b. Hybrid: Self-fertilization yields mix of seed colors

Figure 10.3 **True-Breeding and Hybrid Plants.** (a) Pea plants derived from green seeds were always true-breeding. Some plants grown from yellow seeds were true-breeding, but others (b) were hybrids.

involving plants grown from yellow seeds were more variable. Sometimes these plants were true-breeding, but others were **hybrids:** Their offspring were mixed, including both yellow and green peas (**figure 10.3**).

Cross-fertilization experiments yielded other intriguing results. For example, Mendel crossed plants derived from green seeds with plants grown from yellow seeds. Sometimes, the pods contained only yellow seeds; the green trait seemed to vanish, although it could reappear in the next generation. Other times, the pods contained green and yellow seeds (see **figure 10.4**). Mendel noticed a similar mode of inheritance when he studied other pea plant characteristics: One trait seemed to obscure the other. Mendel called the masking trait *dominant;* the trait being masked was called *recessive.* The yellow-seed trait, for example, is dominant over the green trait.

Although Mendel referred to *traits* as dominant or recessive, modern biologists reserve these terms for *alleles.* A **dominant allele** encodes a protein that exerts its effects whenever it is present; a **recessive allele** encodes a protein whose effect is masked if a dominant allele is also present. When a gene has only two

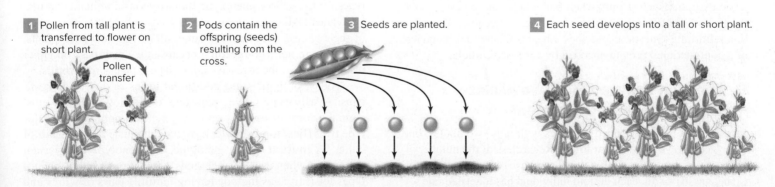

1 Pollen from tall plant is transferred to flower on short plant.

Pollen transfer

2 Pods contain the offspring (seeds) resulting from the cross.

3 Seeds are planted.

4 Each seed develops into a tall or short plant.

Figure 10.2 **Breeding Peas.** Gregor Mendel used this technique to set up carefully designed crosses of pea plants, so that he could observe the appearance of traits in the next generation.

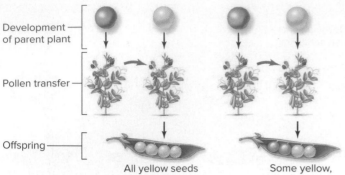

Figure 10.4 Yellow Is Dominant. When Mendel crossed a plant derived from a green seed with a plant grown from a yellow seed, the offspring could be all yellow, or they could be a mix of green and yellow peas. Crosses such as these led Mendel to conclude that yellow seed color is dominant over green.

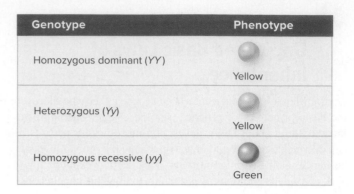

Figure 10.5 Genotypes and Phenotypes Compared. A pea's genotype for the "seed color" gene consists of the two alleles that the seed inherited from its parents. Its phenotype is its outward appearance: yellow or green.

alleles, it is common to symbolize the dominant allele with a capital letter (such as *Y* for yellow) and the recessive allele with the corresponding lowercase letter (*y* for green).

The "dominance" of an allele may seem to imply that it "dominates" in the population as a whole. The most common allele, however, is not always dominant. In humans, the allele that causes a form of dwarfism called achondroplasia is dominant, but it is very rare—as is the dominant allele that causes Huntington disease. Conversely, blue eyes are the norm in people of northern European origin, but the alleles that produce this eye color are recessive.

The term *dominant* may also conjure images of a bully that forces a weak, recessive allele into submission. After all, the recessive allele seems to hide when a dominant allele is present, emerging from its hiding place only if the dominant allele is absent. How does the recessive allele "know" what to do? In fact, alleles cannot hide, emerge, or know anything. A recessive allele remains a part of the cell's DNA, regardless of the presence of a dominant allele. It only seems to hide because it typically encodes a nonfunctional protein. If a dominant allele is also present, the organism usually has enough of the functional protein to maintain its normal appearance (although section 10.6 describes some exceptions). It is only when both alleles are recessive that the lack of the functional protein becomes noticeable. This section's Burning Question describes a health-related consequence of a nonfunctional protein encoded by a recessive allele.

C. For Each Gene, a Cell's Two Alleles May Be Identical or Different

Mendel chose traits encoded by genes with only two alleles, but some genes have hundreds of forms. Regardless of the number of possibilities, however, a diploid cell can have only two alleles for each gene. After all, each diploid individual has inherited one set of chromosomes from each parent, and each chromosome carries only one allele per gene.

For a given gene, a diploid cell's two alleles may be identical or different. The **genotype** expresses the genetic makeup of an individual, and it is written as a pair of letters representing the alleles (**figure 10.5**). An individual that is **homozygous** for a gene has two identical alleles, meaning that both parents contributed the same gene version. If both of the alleles are dominant, the individual's genotype is homozygous dominant (written as *YY,* for example). If both alleles are recessive, the individual is homozygous recessive (*yy*). An individual with a **heterozygous** genotype, on the other hand, has two different alleles for the gene (*Yy*). That is, the two parents each contributed different genetic information.

The organism's genotype is distinct from its **phenotype,** or observable characteristics (see figure 10.5). Seed color, flower color, and stem length are examples of pea plant phenotypes that Mendel studied. Your own phenotype includes not only your height, eye color, shoe size, number of fingers and toes, skin color, and hair texture but also other traits that are not readily visible, such as your blood type or the specific shape of your hemoglobin proteins (see figure 7.21). As described in section 10.9, most phenotypes result from a complex interaction between genes and environment. Mendel, however, chose traits controlled exclusively by genes.

Mendel's observation that some but not all yellow-seeded pea plants were true-breeding arises from the two possible genotypes for the yellow phenotype (homozygous dominant and heterozygous). All homozygous plants are true-breeding because all of their gametes contain the same allele. Heterozygous plants, however, are not true-breeding because they may pass on either the dominant or the recessive allele. These plants are hybrids.

Today, biologists use additional terms to describe organisms. A **wild-type** allele, genotype, or phenotype is the most common form or expression of a gene in a population. Wild-type fruit flies, for example, have two antennae and one pair of wings. A **mutant** allele, genotype, or phenotype is a variant that arises when a gene undergoes a mutation. Mutant phenotypes for fruit flies include having multiple pairs of wings and having legs instead of antennae growing out of the head (see figure 7.20).

D. Every Generation Has a Name

Mendel kept careful tallies of the offspring from countless crosses, which required a systematic accounting of multiple generations of plants. Biologists still use Mendel's system of standardized names to keep track of inheritance patterns.

The purebred **P generation** (for "parental") is the first set of individuals being mated; the **F₁ generation,** or first filial generation, is the offspring from the P generation (*filial* derives from the Latin word for "child"). The **F₂ generation** is the offspring of the F₁ plants, and so on. Although these terms are applicable only to lab crosses, they are analogous to human family relationships. If you

TABLE 10.1 Miniglossary of Genetic Terms

Term	Definition
Generations	
P	The parental generation
F₁	The first filial generation; offspring of P generation
F₂	The second filial generation; offspring of F₁ generation
Chromosomes and genes	
Chromosome	A continuous molecule of DNA plus associated proteins
Gene	A sequence of DNA that encodes a protein
Locus	The physical location of a gene on a chromosome
Allele	One of the alternative forms of a gene
Dominant and recessive	
Dominant allele	An allele that is expressed if present in the genotype
Recessive allele	An allele whose expression is masked by a dominant allele
Genotypes and phenotypes	
Genotype	An individual's allele combination for a particular gene
Homozygous	Possessing identical alleles of one gene
Heterozygous	Possessing different alleles of one gene
Phenotype	An observable characteristic
True-breeding	Homozygous; self-fertilization yields offspring identical to self for a given trait
Hybrid	Heterozygous; self-fertilization yields offspring with mixed genotypes and phenotypes
Wild-type	The most common allele, genotype, or phenotype in a population
Mutant	An allele, a genotype, or a phenotype resulting from a mutation in a gene

Burning Question

Why does diet soda have a warning label?

© Jill Braaten/McGraw-Hill Education

Foods containing the artificial sweetener aspartame carry a warning label that says "Contains phenylalanine." Since other sugar substitutes lack similar words of caution, aspartame must pose a unique threat. What is it?

A peek at aspartame's biochemistry reveals the answer. Aspartame contains an amino acid called phenylalanine. In most people, an enzyme converts phenylalanine into another amino acid. A mutated allele of the gene encoding this enzyme, however, results in the production of an abnormal, nonfunctional enzyme. People who have just one copy of this recessive allele are healthy because the cell has enough of the normal enzyme, thanks to the dominant allele. The recessive allele therefore seems to "vanish," just as in Mendel's pea plants.

Individuals who inherit two copies of the recessive allele, however, have a metabolic disorder called phenylketonuria (abbreviated PKU). These people cannot produce the normal enzyme. Phenylalanine accumulates to toxic levels, causing intellectual disability and other problems. Avoiding foods containing aspartame helps minimize the effects of PKU—hence the warning.

Submit your burning question to
Marielle.Hoefnagels@mheducation.com

consider your grandparents the P generation, your parents are the F₁ generation, and you and your siblings are the F₂ generation.

Table 10.1 summarizes the important terms encountered so far. The remainder of the chapter uses this basic vocabulary to integrate Mendel's findings with what biologists now know about genes, chromosomes, and reproduction.

10.2 MASTERING CONCEPTS

1. Why did Gregor Mendel choose pea plants as his experimental organism?
2. Distinguish between dominant and recessive; heterozygous and homozygous; phenotype and genotype; wild-type and mutant.
3. Create a flow chart with connections between the terms *P generation, F₁ generation,* and *F₂ generation.*

10.3 The Two Alleles of a Gene End Up in Different Gametes

Mendel used a systematic series of crosses to deduce the rules of inheritance, beginning with single genes.

A. The Simplest Punnett Squares Track the Inheritance of One Gene

Mendel began with a P generation consisting of true-breeding plants derived from yellow seeds (*YY*) and true-breeding green-seeded plants (*yy*). The F_1 offspring produced in this cross had yellow seeds (genotype *Yy*). The green trait therefore seemed to disappear in the F_1 generation.

Next, he used the F_1 plants to set up a **monohybrid cross:** a mating between two individuals that are both heterozygous for the same gene (figure 10.6). The resulting F_2 generation had both yellow and green phenotypes, in a ratio of 3:1; that is, for every three yellow seeds, he observed one green seed.

A **Punnett square** is a diagram that uses the genotypes of two parents to reveal which allele combinations their offspring may inherit. The Punnett square in figure 10.6, for example, shows how the green phenotype reappeared in the F_2 generation. Both parents are heterozygous (*Yy*) for the seed-color gene. Each therefore produces some gametes carrying the *Y* allele and some gametes carrying *y*. All three possible genotypes may therefore appear in the F_2 generation, in the ratio 1 *YY* : 2 *Yy* : 1 *yy*. The corresponding phenotypic ratio is three yellow seeds to one green seed, or 3:1. Mendel saw similar results for all seven traits that he studied (figure 10.7).

Yy
Female parent
Female gametes
Y
y
Yy
Male parent
Male gametes
Y
y
YY
Yy
Yy
yy
Genotypic ratio 1:2:1
(1 YY : 2 Yy : 1 yy)
Phenotypic ratio 3:1
(3 yellow : 1 green)

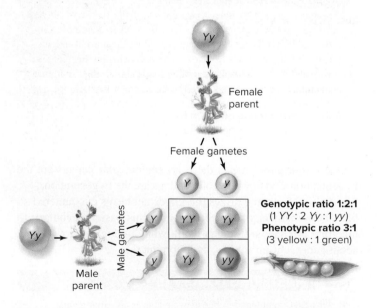

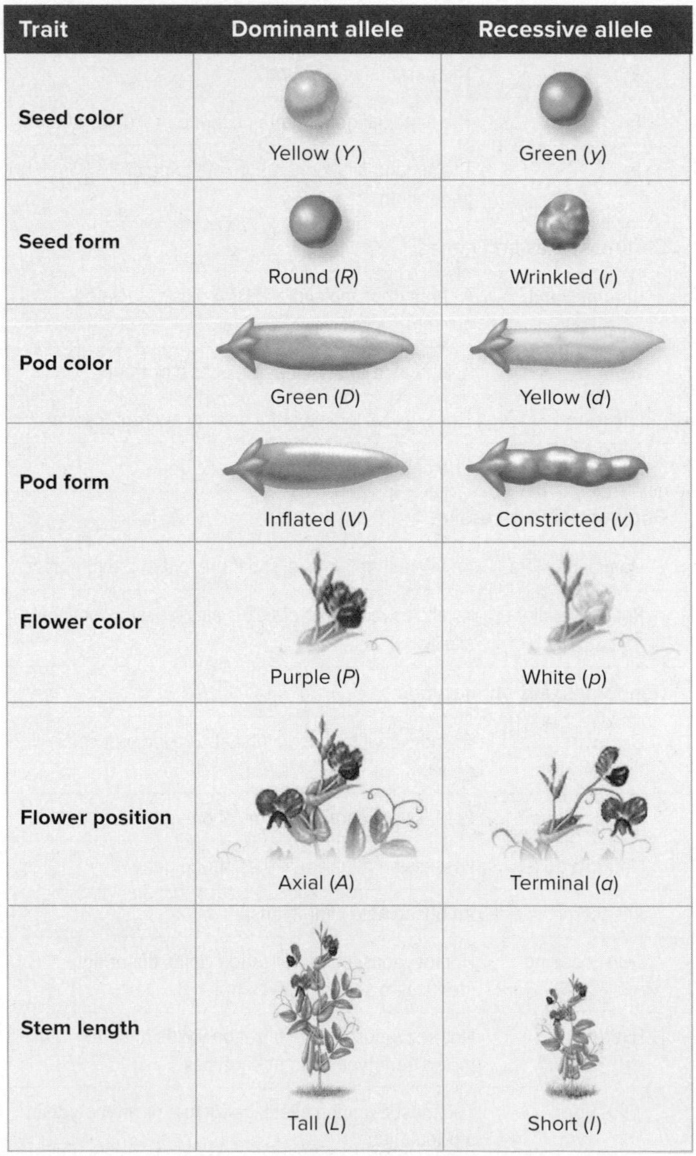

Figure 10.6 Punnett Square. This diagram depicts Mendel's monohybrid cross of two heterozygous plants grown from yellow seeds (*Yy*). The two possible types of female gametes are listed along the top of the square; the male gametes are listed on the left-hand side. The four compartments within the Punnett square contain the genotypes and phenotypes of all possible offspring.

Figure It Out

If Mendel mated a true-breeding tall plant with a heterozygous tall plant, what percentage of the offspring would also be tall?

Answer: 100%

Mendel could tally the plants with each phenotype, but he also needed to keep track of each genotype. He knew that green-seeded plants were always homozygous recessive (*yy*). But what was the genotype of each yellow seed, *YY* or *Yy*? He had no way

Trait
Dominant allele
Recessive allele
Seed color
Yellow (Y)
Green (y)
Seed form
Round (R)
Wrinkled (r)
Pod color
Green (D)
Yellow (d)
Pod form
Inflated (V)
Constricted (v)
Flower color
Purple (P)
White (p)
Flower position
Axial (A)
Terminal (a)
Stem length
Tall (L)
Short (l)

Figure 10.7 Pea Traits. Mendel's breeding studies helped him to deduce the inheritance patterns of these seven pea plant characteristics.

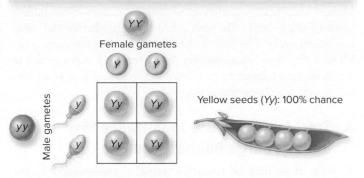

What is the genotype of a plant grown from a yellow seed?

YY   or   Yy   ?

Test cross results if plant is homozygous dominant (YY):

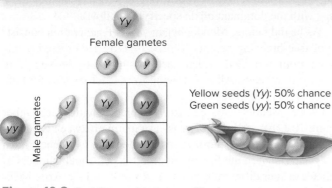

Yellow seeds (Yy): 100% chance

Test cross results if plant is heterozygous (Yy):

Yellow seeds (Yy): 50% chance
Green seeds (yy): 50% chance

Figure 10.8 Test Cross. A test cross with a homozygous recessive (yy) plant reveals whether a pea plant grown from a yellow seed is homozygous dominant (YY) or heterozygous (Yy).

to tell just by looking, so he set up breeding experiments called test crosses to distinguish between the two possibilities. A **test cross** is a mating between an individual of unknown genotype and a homozygous recessive individual (**figure 10.8**). If a yellow-seeded plant crossed with a *yy* plant produced only yellow seeds, Mendel knew the unknown genotype was *YY;* if the cross produced seeds of both colors, he knew it must be *Yy.*

B. Meiosis Explains Mendel's Law of Segregation

All of Mendel's breeding experiments and calculations added up to a brilliant description of basic genetic principles. Without any knowledge of chromosomes or genes, Mendel used his data to conclude that genes occur in alternative versions (which we now call alleles). He further determined that each individual inherits two alleles for each gene and that these alleles may be the same or different. Finally, he deduced his **law of segregation,** which states that the two alleles of each gene are packaged into separate gametes; that is, they "segregate," or move apart from each other, during gamete formation. (In science, a *law* is a statement about a phenomenon that is invariable, at least as far as anyone knows. Unlike a theory, a law does not necessarily explain the phenomenon.)

Mendel's law of segregation makes perfect sense in light of what we now know about reproduction. During meiosis I, homologous pairs of chromosomes separate and move to opposite poles of the cell. After a plant of genotype *Yy* undergoes meiosis, half of the gametes carry *Y* and half carry *y* (*red* chromosomes in **figure 10.9**). A *YY* plant (*blue*), on the other hand, can produce only *Y* gametes. When gametes from the two plants meet at fertilization, they combine at random. About 50% of the time, both gametes carry *Y;* the other 50% of the time, one contributes *Y* and the other, *y.*

Figure 10.9 Mendel's Law of Segregation. During meiosis, the four chromatids in a homologous pair of chromosomes segregate from one another and end up in separate gametes.

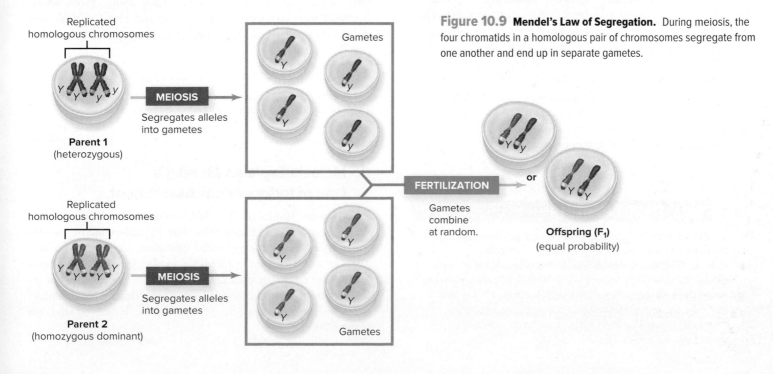

Replicated homologous chromosomes

MEIOSIS
Segregates alleles into gametes

Parent 1
(heterozygous)

Gametes

Replicated homologous chromosomes

MEIOSIS
Segregates alleles into gametes

Parent 2
(homozygous dominant)

Gametes

FERTILIZATION

Gametes combine at random.

or

Offspring (F₁)
(equal probability)

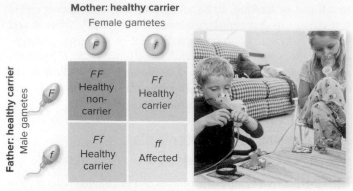

Mother: healthy carrier
Female gametes

Healthy noncarrier (*FF*): 25% chance
Healthy carrier (*Ff*): 50% chance
Affected (*ff*): 25% chance

Figure 10.10 Inheritance of Cystic Fibrosis. This Punnett square shows the possible results of a mating between two carriers of cystic fibrosis. The siblings in the photo have cystic fibrosis; the masks deliver medicine to treat their lung problems.

Photo: ©Paul Rodriguez/ZUMApress/Alamy

This principle of inheritance applies to all diploid species, including humans. Return for a moment to the couple and their genetic counselor introduced in section 10.1. Cystic fibrosis arises when a person has two recessive alleles for a particular gene on chromosome 7. Genetic testing revealed that the man and the woman are both carriers. In genetic terms, this means that although neither has the disease, both are heterozygous for the gene that causes cystic fibrosis. Just as in Mendel's monohybrid crosses, each of their children has a 25% chance of inheriting two recessive alleles (**figure 10.10**). Each child also has a 50% chance of being a carrier (heterozygous) and a 25% chance of inheriting two dominant alleles.

Note that all Punnett squares, including the one in figure 10.10, show the *probabilities* that apply to each offspring. That is, if the couple has four children, there will not necessarily be exactly one with genotype *FF,* two with *Ff,* and one with *ff.* Similarly, the chance of tossing a fair coin and seeing "heads" is 50%, but two tosses will not necessarily yield one head and one tail. If you toss the coin 1000 times, however, you will likely approach the expected 1:1 ratio of heads to tails. As Mendel discovered, pea plants are ideal for genetics studies in part because they produce many offspring in each generation.

10.3   MASTERING CONCEPTS

1. What is a monohybrid cross, and what are the genotypic and phenotypic ratios expected in the offspring of the cross?
2. How are Punnett squares helpful in following inheritance of single genes?
3. What is a test cross, and why is it useful?
4. How does the law of segregation reflect the events of meiosis?

10.4 Genes on Different Chromosomes Are Inherited Independently

Mendel's law of segregation arose from his studies of the inheritance of single traits. He next asked himself whether the same law would apply if he followed two characters at the same time. Mendel therefore began another set of breeding experiments in which he simultaneously examined the inheritance of two characteristics: pea shape and pea color.

A. Tracking Two-Gene Inheritance May Require Large Punnett Squares

A pea's shape may be round or wrinkled (determined by the *R* gene, with the dominant allele specifying round shape). At the same time, its color may be yellow or green (determined by the *Y* gene, with the dominant allele specifying yellow).

As he did before, Mendel began with a P generation consisting of true-breeding parents (**figure 10.11a**). He crossed plants grown from wrinkled, green seeds with plants derived from round, yellow seeds. All F_1 offspring were heterozygous for both genes (*Rr Yy*) and therefore had round, yellow seeds.

Next, Mendel crossed the F_1 plants with each other (figure 10.11b). A **dihybrid cross** is a mating between two individuals that are each heterozygous for the same two genes. Each *Rr Yy* individual in the F_1 generation produced equal numbers of gametes of four different types: *R Y, R y, r Y,* and *r y.* After Mendel completed the crosses, he found four phenotypes in the F_2 generation, reflecting all possible combinations of seed shape and color. The Punnett square predicts that the four phenotypes will occur in a ratio of 9:3:3:1. That is, 9 of 16 offspring should have round, yellow seeds; 3 should have round, green seeds; 3 should have wrinkled, yellow seeds; and just 1 should have wrinkled, green seeds. This prediction almost exactly matches Mendel's results.

Figure It Out

In a cross between an *Rr Yy* plant and an *rr yy* plant, what proportion of the offspring is homozygous recessive for both seed shape and seed color?

Answer: 25%

B. Meiosis Explains Mendel's Law of Independent Assortment

Based on the results of the dihybrid cross, Mendel proposed a second law of inheritance. The **law of independent assortment** states that during gamete formation, the alleles for one gene do not influence the alleles for another gene. That is, alleles *Y* and *y* are randomly packaged into gametes, independent of alleles *R* and *r.* With this second set of experiments, Mendel had again inferred a principle of inheritance based on meiosis (**figure 10.12**).

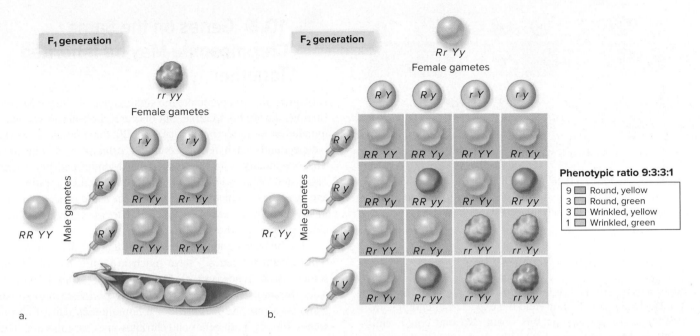

Figure 10.11 Generating a Dihybrid Cross. (a) In the parental generation, one parent is homozygous recessive for two genes; the other is homozygous dominant. The F₁ generation is therefore heterozygous for both genes. (b) A dihybrid cross is a mating between two plants from the F₁ generation. Phenotypes occur in a distinctive ratio in the resulting F₂ generation.

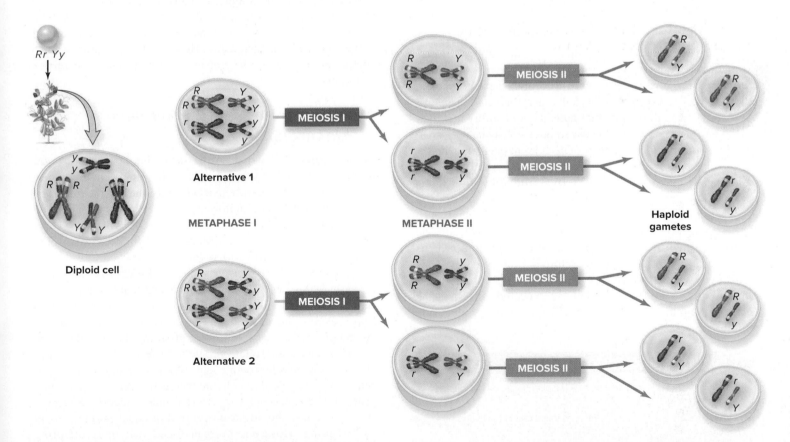

Figure 10.12 Mendel's Law of Independent Assortment. Homologous chromosome pairs are randomly oriented during metaphase I of meiosis. The exact allele combination in a gamete depends on which chromosomes happen to be packaged together. An individual of genotype *Rr Yy* therefore produces approximately equal numbers of four types of gametes: *R Y, r y, R y,* and *r Y.*

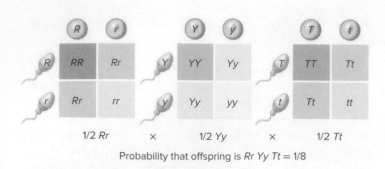

1/2 *Rr*   ×   1/2 *Yy*   ×   1/2 *Tt*

Probability that offspring is *Rr Yy Tt* = 1/8

Figure 10.13 The Product Rule. What is the chance that two parents that are heterozygous for three genes (*Rr Yy Tt*) will give rise to an offspring with that same genotype? To find out, multiply the individual probabilities for each gene.

Interestingly, Mendel found some trait combinations for which a dihybrid cross did not yield the expected phenotypic ratio. Mendel could not explain this result. No one could, until Thomas Hunt Morgan's work led to the chromosomal theory of inheritance. As you will see in section 10.5, the law of independent assortment does not apply to genes that are close together on the same chromosome.

C. The Product Rule Is a Useful Shortcut

Punnett squares become cumbersome when analyzing more than two genes. A Punnett square for three genes has 64 boxes; for four genes, 256 boxes. Fortunately, there is a shortcut that still relies on the rules of probability on which Punnett squares are based. According to the **product rule,** the probability that multiple independent events will occur simultaneously can be calculated by multiplying the chances of each event occurring alone.

As an example, the product rule can predict the chance of obtaining wrinkled, green seeds (*rr yy*) from dihybrid (*Rr Yy*) parents. The probability that two *Rr* plants will produce *rr* offspring is 25%, or ¼, and the chance of two *Yy* plants producing a *yy* individual is ¼. According to the product rule, the chance of dihybrid parents (*Rr Yy*) producing homozygous recessive (*rr yy*) offspring is therefore ¼ multiplied by ¼, or $^{1}/_{16}$. Now consult the 16-box Punnett square for Mendel's dihybrid cross (see figure 10.11). As expected, only 1 of the 16 boxes contains *rr yy*. **Figure 10.13** applies the product rule to three traits.

10.4   MASTERING CONCEPTS

1. What is a dihybrid cross, and what is the phenotypic ratio expected in the offspring of the cross?
2. How does the law of independent assortment reflect the events of meiosis?
3. How can the product rule be used to predict the results of crosses in which multiple genes are studied simultaneously?

🍁 10.5 Genes on the Same Chromosome May Be Inherited Together

Biologists did not appreciate the significance of Gregor Mendel's findings during his lifetime, but his careful observations laid the foundation for modern genetics. In 1900, three botanists working independently each rediscovered the principles of inheritance. They eventually found the paper that Mendel had published in 1866, and other scientists demonstrated Mendel's ratios again and again in several species. At about the same time, advances in microscopy were allowing scientists to observe and describe chromosomes for the first time.

It soon became apparent that what Mendel called "elementen" (later renamed "genes") have much in common with chromosomes. Both genes and chromosomes, for example, come in pairs. In addition, alleles of a gene are packaged into separate gametes, as are the members of a homologous pair of chromosomes. Finally, both genes and chromosomes are inherited in random combinations.

As biologists cataloged traits and the chromosomes that transmit them in several species, they realized that the number of traits far exceeds the number of chromosomes. Fruit flies, for example, have only four pairs of chromosomes, but they have dozens of different bristle patterns, body colors, eye colors, wing shapes, and other characteristics. How might a few chromosomes control so many traits? The answer: Each chromosome carries many genes.

A. Genes on the Same Chromosome Are Linked

Linked genes are carried on the same chromosome; they are therefore inherited together. Unlike genes on different chromosomes, they do not assort independently during meiosis. The seven traits that Mendel followed in his pea plants all happened to be transmitted on separate chromosomes. Had the same chromosome carried these genes, Mendel's dihybrid crosses would have generated markedly different results.

The inheritance pattern of linked genes was first noticed in the early 1900s, when Thomas Hunt Morgan at Columbia University studied the inheritance of pairs of traits in *Drosophila* fruit flies. In some of Morgan's experiments, two-gene crosses did not produce the proportions of offspring that Mendel's law of independent assortment predicts. **Figure 10.14**, for example, shows a cross between a male that is heterozygous for two genes and a female that is homozygous recessive for the same genes. If the two genes had assorted independently, four possible genotypes should have occurred in equal proportions in the offspring. However, Morgan found that most offspring were either heterozygous or homozygous recessive. Few offspring inherited genotypes differing from the parents. ⓘ *Drosophila,* section 21.7

Figure 10.14 Linked Genes. The Punnett square shows that if genes *A* and *B* are on separate chromosomes, the cross should yield approximately equal numbers of offspring of each genotype. However, the actual genotypic ratio is skewed toward two of the four offspring classes. This result reveals that genes *A* and *B* do not assort independently; that is, they are linked.

Morgan's data soon began to indicate four **linkage groups,** collections of genes that tended to be inherited together. Because the number of linkage groups was the same as the number of homologous pairs of chromosomes, scientists eventually realized that each linkage group was simply a set of genes transmitted on the same chromosome.

Nevertheless, the researchers did sometimes see offspring with trait combinations not seen in either parent. How could this occur? The answer turned out to involve yet another event in gamete formation: **crossing over,** an exchange of genetic material between homologous chromosomes during prophase I of meiosis (see figure 9.8). After crossing over, no two chromatids in a homologous pair of chromosomes are identical (**figure 10.15**).

Crossing over is a random process, and it might or might not occur between two linked genes. For any pair of genes on the same chromosome, then, most gametes receive a **parental chromatid,** which retains the allele combination from one parent. But when crossing over happens between two genes, some of the gametes receive a **recombinant chromatid** carrying a mix of maternal and paternal alleles. A recombinant offspring is one that inherits a recombinant chromatid.

B. Studies of Linked Genes Have Yielded Chromosome Maps

Morgan wondered why some crosses produced a higher proportion of recombinant offspring than others. Might the differences reflect the physical relationships between the genes on the chromosome? Alfred Sturtevant, Morgan's undergraduate assistant, explored this idea. In 1911, Sturtevant proposed that the farther apart two alleles are on the same chromosome, the more likely crossing over is to separate them—simply because more space separates the genes (**figure 10.16a**).

Sturtevant's idea became the basis for mapping genes on chromosomes. By determining the percentage of recombinant offspring, investigators can infer how far apart the genes are on one chromosome. Crossing over frequently separates alleles on opposite ends of the same chromosome, so recombinant offspring occur frequently. In contrast, a crossover would rarely separate alleles lying very close together on the chromosome, and the proportion of recombinant offspring would be small. Geneticists use this correlation between crossover frequency and the distance between genes to construct **linkage maps,** which are diagrams of gene order and spacing on chromosomes.

In 1913, Sturtevant published the first genetic linkage map, depicting the order of five genes on the X chromosome of the fruit fly (figure 10.16b). Researchers then rapidly mapped genes on all four fruit fly chromosomes. Linkage maps for human chromosomes followed over the next half century.

Figure It Out

Genes *A*, *B*, and *D* are on the same chromosome. The crossover frequency between *B* and *D* is 0.25, between *A* and *B* is 0.12, and between *D* and *A* is 0.13. Which two genes are farthest apart?

Answer: Genes *B* and *D* are farthest apart.

Figure It Out

Use the crossover frequencies in the previous question to create a linkage map of genes *A*, *B*, and *D*. What is the order of the genes on the chromosome?

Answer: *B–A–D*

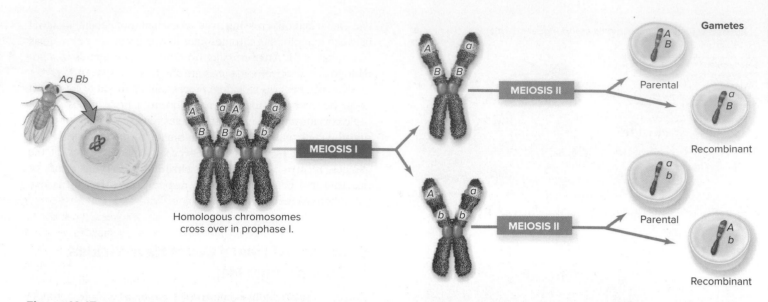

Figure 10.15 Crossing Over. Linkage between two alleles is interrupted if crossing over occurs at a point between the two genes. As a result, some gametes contain recombinant arrangements of the alleles.

At one time, phenotypes were the basis for linkage maps. Now, however, genetic marker technology can associate a known, detectable DNA sequence with a specific phenotype. The marker does not have to be part of the gene that controls the phenotype; the two must simply be located close enough together that the presence of the marker correlates strongly with the phenotype.

Genetic markers are useful tools for predicting the chance that a person will develop a particular inherited illness, even if the actual disease-causing gene is unknown. The first such use of genetic markers occurred in the 1980s. An extensive study of a small Venezuelan village with a high incidence of Huntington disease revealed that all family members with the disorder also carried a unique marker on chromosome 4. Healthy relatives did not have this sequence. Like a flag, the marker indicates the presence of the disease-causing allele to which it is closely linked. The actual

gene was not identified until the 1990s, but in the meantime, researchers could identify who was likely to carry the Huntington disease-causing allele long before the symptoms developed.

The linkage maps of the mid-twentieth century provided the rough drafts to which DNA sequence information was added later in the century. Today, entire genomes are routinely sequenced using powerful computers. ⓘ *DNA sequencing,* section 11.2B

10.5 | MASTERING CONCEPTS

1. How do patterns of inheritance differ for unlinked versus linked pairs of genes?
2. What is the difference between recombinant and parental chromatids, and how do they arise?
3. How do biologists use crossover frequencies to map genes on chromosomes?

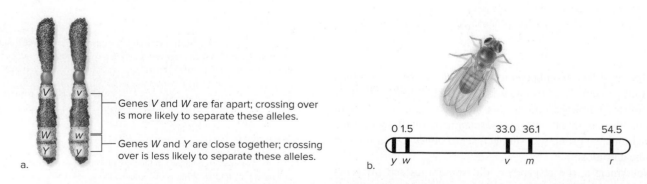

a. Genes *V* and *W* are far apart; crossing over is more likely to separate these alleles.

Genes *W* and *Y* are close together; crossing over is less likely to separate these alleles.

b. 0 1.5   33.0 36.1   54.5
    y w   v   m   r

Figure 10.16 Breaking Linkage. (a) Crossing over is more likely to separate the alleles of genes *V* and *W* (or *V* and *Y*) than to separate the alleles of genes *W* and *Y,* because there is more room for an exchange to occur. (b) A linkage map of a fruit fly chromosome, showing the locations of five genes. The numbers represent crossover frequencies relative to the leftmost gene, *y.*

10.6 Dominance Relationships Are Rarely Simple

Mendel's crosses yielded easily distinguishable offspring. A pea is either yellow or green, round or wrinkled; a plant is either tall or short. Often, however, offspring traits do not occur in the proportions that Punnett squares or probabilities predict. It may appear that Mendel's laws do not apply—but they do. The underlying genotypic ratios are there, but the nature of the phenotype, other genes, or the environment alters how traits appear. The rest of this chapter describes some situations that may produce phenotypic ratios other than those Mendel observed.

A. Incomplete Dominance and Codominance Add Phenotype Classes

For the traits Mendel studied, one allele is completely dominant and the other is completely recessive. The phenotype of a heterozygote is therefore identical to that of a homozygous dominant individual. For many genes, however, heterozygous offspring do not share the phenotype of either parent. As you will see, biologists apply unique notation to alleles for these traits. Designating the alleles with capital and lowercase letters does not work, because neither allele is necessarily dominant over the other.

When a gene shows **incomplete dominance,** the heterozygote has a phenotype that is intermediate between those of the two homozygotes. For example, a red-flowered snapdragon plant of genotype r_1r_1 crossed with a white-flowered r_2r_2 plant gives rise to an r_1r_2 plant with pink flowers (**figure 10.17**). The single copy of allele r_1 in the pink heterozygote directs less pigment production than the two copies in a red-flowered r_1r_1 plant. Although the red color seems to be "diluted" in the all-pink F_1 generation, the Punnett square shows that crossing two of these pink plants can yield F_2 offspring with red, white, or pink flowers.

In **codominance,** two different alleles are fully expressed in the phenotype. For example, the ABO blood typing system is important in determining whose blood a person can receive in a transfusion. A person's ABO blood type is determined by the I gene, which has three possible alleles: I^A, I^B, and i (**figure 10.18**). The I gene encodes an enzyme that inserts either an "A" or a "B" molecule onto the surfaces of red blood cells. Allele i is recessive, so a person with genotype ii produces neither molecule A nor molecule B

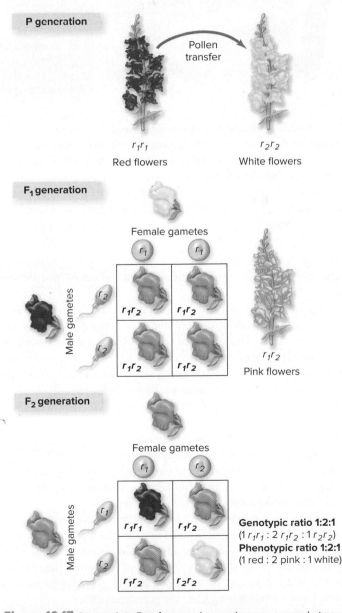

Genotypic ratio 1:2:1
($1\ r_1r_1 : 2\ r_1r_2 : 1\ r_2r_2$)
Phenotypic ratio 1:2:1
(1 red : 2 pink : 1 white)

Figure 10.17 Incomplete Dominance. In snapdragons, a cross between a plant with red flowers (r_1r_1) and a plant with white flowers (r_2r_2) produces heterozygous plants with pink flowers (r_1r_2). The red and white phenotypes reappear in the F_2 generation.

Genotypes	Phenotypes		
	Surface molecules	ABO blood type	
I^AI^A I^Ai	Only A	Type A	
I^BI^B I^Bi	Only B	Type B	
I^AI^B	Both A and B	Type AB	
ii	None	Type O	

Figure 10.18 Codominance. The I^A and I^B alleles of the I gene are codominant, meaning that both are fully expressed in a heterozygote. Allele i is recessive.

and therefore has type O blood. A person who produces only molecule A (genotype I^AI^A or I^Ai) has type A blood; likewise, someone with only molecule B (genotype I^BI^B or I^Bi) has type B blood. But genotype I^AI^B yields type AB blood. The I^A and I^B alleles are codominant because both are equally expressed when both are present. ⓘ *ABO blood groups,* section 30.2B

What is the difference between the recessive *i* allele and the codominant I^A and I^B alleles? The *i* allele encodes a nonfunctional protein, whereas alleles I^A and I^B code for functional proteins. People with genotype I^AI^B therefore have molecule A and molecule B on the surfaces of their red blood cells.

The ABO blood type example also illustrates what can happen when one gene has three or more possible alleles: The number of phenotypes increases. In this case, two codominant alleles (I^A and I^B) and one recessive allele (*i*) produce six genotypes and four phenotypes.

Figure It Out

A woman with type AB blood has children with a man who has type O blood. What is the probability that a child they conceive will have type B blood?

Answer: 50%; each child has an equal chance of inheriting I^Bi or I^Ai

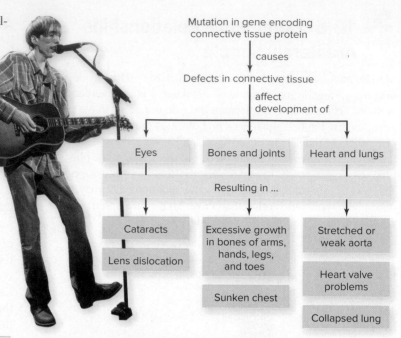

Figure 10.19 **Pleiotropy.** In Marfan syndrome, a single mutated gene encodes a defective connective tissue protein. The resulting effects on the body are widespread. Singer Bradford Cox was born with Marfan syndrome.
Photo: ©Roger Kisby/Getty Images

B. Some Inheritance Patterns Are Especially Difficult to Interpret

Some conditions are especially difficult to trace through families. For example, one gene may influence the phenotype in many ways; conversely, multiple genes may contribute to one phenotype. Although the basic rules of inheritance apply to each gene, the patterns of phenotypes that appear in grandparents, parents, and siblings may be hard to interpret.

Pleiotropy   In **pleiotropy,** a single gene has multiple effects on the phenotype (the word root *pleio-* comes from a Greek word meaning "very many"). Pleiotropy arises when one protein is important in different biochemical pathways or affects more than one body part or process. Although Mendelian rules of inheritance apply to pleiotropic genes, the conditions they cause can be difficult to trace through families because individuals with different subsets of symptoms may appear to have different illnesses.

A collection of disorders called Marfan syndrome offers one example of pleiotropy. In Marfan syndrome, a mutated gene on chromosome 15 encodes a defective version of a protein that normally occurs in connective tissue (**figure 10.19**). This tissue type is widespread throughout the body, and the normal protein is especially abundant in bones, lungs, the ligaments of the eye, and the aorta (a major blood vessel extending from the heart). The defective protein therefore affects many organ systems, with symptoms including long limbs, spindly fingers, a caved-in chest, a weakened aorta, lens dislocation, and other conditions. Abraham Lincoln may have had Marfan syndrome. Like achondroplasia and

Huntington disease, Marfan syndrome is an example of an uncommon disorder caused by a dominant allele. ⓘ *connective tissue,* section 25.2B

Cystic fibrosis offers another example of pleiotropy. A defective gene on chromosome 7 encodes misshapen membrane channel proteins. The abnormal protein blocks the movement of water into and out of cells. As a result, sticky mucus accumulates in the lungs, digestive tract, and sweat glands. The mucus, in turn, causes a variety of conditions, including increased susceptibility to lung infections, weight loss, fatigue, and salty sweat.

Protein Interactions   With thousands of genes active in a cell, it is not surprising that many proteins can interact to contribute to one phenotype. These interactions can complicate the analysis of inheritance patterns because mutations in different genes can produce similar or identical phenotypes.

Some interactions occur among multiple proteins involved in the same biochemical pathway. For example, blood clot formation requires 11 biochemical reactions. A different gene encodes each enzyme in the pathway, and clotting disorders may result from mutations in any of these genes. The phenotypes are the same (poor blood clotting), but the genotypes differ.

Another type of interaction is **epistasis,** which occurs when one gene's product affects the phenotype associated with another gene. As a familiar example of epistasis, consider male pattern baldness and the "widow's peak" hairline, both of which are

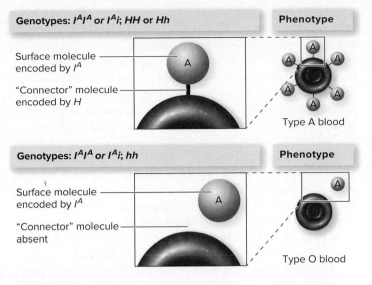

Figure 10.20 Epistasis. The *H* gene encodes a protein that normally attaches molecule A or B to the red blood cells of a person with allele I^A or I^B. If the genotype for the *H* gene is *hh*, then the molecule has nothing to attach to. The person's blood will test as type O, even though allele I^A or I^B is still present.

controlled by genes. The receding hairline associated with male pattern baldness hides the effects of the "widow's peak" allele.

Epistasis is also responsible for some ABO blood type inconsistencies between parents and their children. Rarely, a child whose blood type tests as O has parents whose genotypes indicate that type O offspring are not possible. Does this mean that the child does not belong with the parents? Not necessarily.

Most people with allele I^A or I^B have blood type A, B, or AB. However, a protein must physically link the A and B molecules to the blood cell's surface. If a gene called *H* is mutated, then that protein is nonfunctional, and A and B cannot attach (**figure 10.20**). A person with the extremely rare genotype *hh* therefore has blood that always tests as type O, even though he or she may have a genotype indicating type A, B, or AB blood. The interaction is epistatic because the *H* gene product affects the expression of the *I* gene product.

Moreover, a person with genotype *hh* can receive blood only from other *hh* individuals. The explanation is that the intact H protein in normal blood can induce an immune reaction in a person whose blood lacks that protein.

10.6 MASTERING CONCEPTS

1. How do incomplete dominance and codominance increase the number of phenotypes observed in a population?
2. Differentiate between pleiotropy and epistasis.
3. How can the same phenotype stem from many different genotypes?
4. Figures 10.18 and 10.20 show two ways that a person can have blood that tests as type O. Explain how each relates to specific alleles of genes *I* and *H*.

10.7 Sex-Linked Genes Have Unique Inheritance Patterns

Huntington disease, cystic fibrosis, and other diseases that are caused by genes on autosomes affect both sexes equally. A few conditions, including red–green color blindness and hemophilia, however, occur much more frequently in males than females. Phenotypes that affect one sex more than the other are **sex-linked**; that is, the alleles controlling them are on the X or Y chromosome.

A. X and Y Chromosomes Carry Sex-Linked Genes

In many species, including humans, the sexes have equal numbers of autosomes but differ in the types of sex chromosomes they have. Females have two X chromosomes, whereas males have one X and one much smaller Y chromosome (**figure 10.21**). People who seek to increase the odds of conceiving a boy or a girl can exploit this size difference between X and Y chromosomes, as described in this chapter's Apply It Now box.

The Y chromosome plays the largest role in human sex determination. All human embryos start with rudimentary female structures (see figure 35.17). An embryo having a working copy of a Y-chromosome gene called *SRY* (for *s*ex-*d*etermining *r*egion

Figure 10.21 Inheritance of Sex. In humans, each egg contains 23 chromosomes, 1 of which is a single X chromosome. A sperm cell's 23 chromosomes include either an X or a Y chromosome.

If a Y-bearing sperm cell fertilizes an egg, the baby will be a male (XY). If an X-bearing sperm cell fertilizes an egg, the baby will be a female (XX).

Photos: (both): ©Andrew Syred/ Science Source

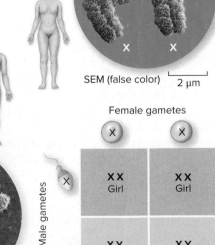

SEM (false color) ⌐ 2 µm

SEM (false color) ⌐ 2 µm

Female gametes

Male gametes

	X	X
X	**X X** Girl	**X X** Girl
Y	**X Y** Boy	**X Y** Boy

Girl (XX): 50% chance
Boy (XY): 50% chance

of the *Y*) develops into a male. *SRY* encodes a protein that switches on other genes that direct the undeveloped testes to secrete testosterone. Cascades of other gene activities promote the development of male sex organs while the embryonic female structures break down. (*SRY*'s effects illustrate pleiotropy; see section 10.6B.)

Despite this critical role, the Y chromosome carries fewer than 100 genes. Scientists therefore know of very few **Y-linked** disorders; most involve defects in sperm production. The human X chromosome, on the other hand, carries more than 1000 protein-encoding genes, most of which have nothing to do with sex determination. Most human sex-linked traits are therefore **X-linked:** They are controlled by genes on the X chromosome.

B. X-Linked Recessive Disorders Affect More Males Than Females

Thomas Hunt Morgan was the first to unravel the unusual inheritance patterns associated with genes on the X chromosome (**figure 10.22**). The eyes of fruit flies are normally red, but one day Morgan discovered a male with white eyes. To study the

inheritance of this odd phenotype, he created true-breeding lines of flies with each eye color. When he mated a parental generation of red-eyed females with white-eyed males, the F$_1$ flies were all red-eyed. The F$_2$ flies had a 3:1 ratio of red-eyed to white-eyed flies—but all the flies with white eyes were male. Morgan also did the reverse cross, mating a P generation of white-eyed females with red-eyed males. That time, all the males of the F$_1$ generation had white eyes, and all the females had red eyes.

Morgan reasoned that the recessive white-eye allele must be on the X chromosome. Because X and Y carry different genes, a male fly can never "mask" the white-eye allele on his X chromosome with a corresponding dominant allele. A female, however, has two X chromosomes. She will express the white-eye phenotype only if *both* of her X chromosomes carry the recessive eye color allele. X-linked inheritance therefore explained why white eyes appear more frequently in males than in females.

Recessive alleles cause most X-linked disorders in humans, although a few are associated with dominant alleles (**table 10.2**). As in fruit flies, a human female inherits an X chromosome from both parents; a male inherits his X chromosome from his mother

Figure 10.22 Fly Eye Color: An X-Linked Trait. (a) In a cross between a red-eyed female and a white-eyed male, white eyes reappear only in some males of the F$_2$ generation. (b) In the opposite cross, white eyes appear in all the F$_1$ males and in half of the F$_2$ males and females. If eye color were on an autosome rather than on the X chromosome, males and females would be equally likely to inherit each eye color.

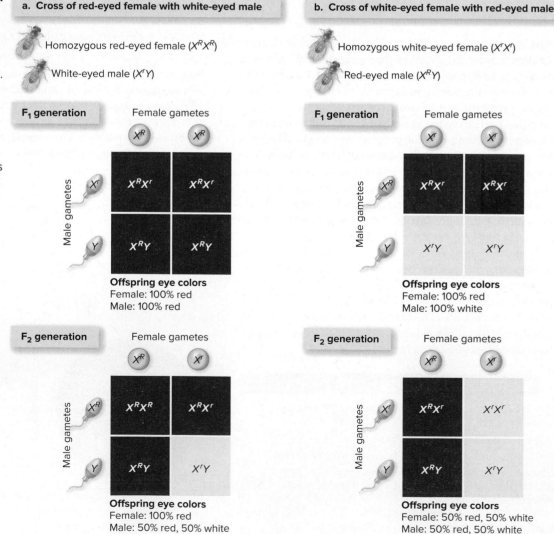

a. Cross of red-eyed female with white-eyed male

Homozygous red-eyed female (X^{R}X^R)

White-eyed male (X^{r}Y)

F$_1$ generation   Female gametes

X^R X^R

Male gametes: X^r, Y

X^{R}X^r X^{R}X^r

X^{R}Y   X^{R}Y

Offspring eye colors
Female: 100% red
Male: 100% red

F$_2$ generation   Female gametes

X^R X^r

Male gametes: X^R, Y

X^{R}X^R X^{R}X^r

X^{R}Y   X^{r}Y

Offspring eye colors
Female: 100% red
Male: 50% red, 50% white

b. Cross of white-eyed female with red-eyed male

Homozygous white-eyed female (X^{r}X^r)

Red-eyed male (X^{R}Y)

F$_1$ generation   Female gametes

X^r X^r

Male gametes: X^R, Y

X^{R}X^r X^{R}X^r

X^{r}Y   X^{r}Y

Offspring eye colors
Female: 100% red
Male: 100% white

F$_2$ generation   Female gametes

X^R X^r

Male gametes: X^r, Y

X^{R}X^r X^{r}X^r

X^{R}Y   X^{r}Y

Offspring eye colors
Female: 50% red, 50% white
Male: 50% red, 50% white

TABLE **10.2**   **Some X-Linked Disorders in Humans**

Disorder	Genetic Explanation	Characteristics
X-linked recessive inheritance		
Duchenne muscular dystrophy	Mutant allele of gene encoding dystrophin	Rapid muscle degeneration early in life
Fragile X syndrome	Unstable region of X chromosome has unusually high number of CCG repeats	Most common form of inherited intellectual disability
Hemophilia A	Mutant allele of gene encoding blood-clotting protein (factor VIII)	Uncontrolled bleeding, easy bruising
Red–green color blindness	Mutant alleles of genes encoding receptors for red or green (or both) wavelengths of light	Reduced ability to distinguish between red and green
Rett syndrome	Mutant allele encoding DNA-binding protein expressed in nerve cells	Multiple severe developmental problems; occurs almost exclusively in females; affected male fetuses rarely survive to birth
X-linked dominant inheritance		
Extra hairiness (congenital generalized hypertrichosis; some forms)	Mechanism unknown	Many more hair follicles than normal
Hypophosphatemic rickets (some forms)	Mutant allele of gene involved in phosphorus absorption	Defective bones caused by low blood phosphorus
Retinitis pigmentosa (some forms)	Mutant allele of cell-signaling protein; mechanism unknown	Partial blindness caused by defects in retina

(see figure 10.21). A female therefore exhibits an X-linked recessive disorder only if she inherits the recessive allele from both parents. A male, in contrast, expresses every allele on his X chromosome, whether dominant or recessive.

Figure 10.23 shows the inheritance of hemophilia A, a disorder with an X-linked recessive mode of inheritance. In hemophilia, a protein called a clotting factor is missing or defective. Blood therefore clots very slowly, and bleeding is excessive. The heterozygous female in the Punnett square does not exhibit symptoms because her dominant allele encodes a functional blood-clotting protein. When she has children with a normal male, however, each son has a 50% chance of being affected, and each daughter has a 50% chance of being a carrier.

Figure It Out

A woman who is heterozygous for the X-linked gene associated with color blindness marries a color-blind man. What is the probability that their first child will be a son who is color-blind?

Answer: 25%

C. X Inactivation Prevents "Double Dosing" of Proteins

As we have already seen, each diploid cell has two versions of every autosome, and the two alleles of each autosomal gene can be different (heterozygous) or the same (homozygous). But the sex chromosomes are different. Relative to males, female mammals have a "double dose" of every gene on the X chromosome. Cells balance this inequality by **X inactivation,** in which a cell shuts off all but one X chromosome in each cell. This process happens early in the embryonic development of a mammal.

X inactivation is directly observable. A turned-off X chromosome absorbs a stain much more readily than an active X chromosome does; the inactivated X forms a Barr body that is visible with a microscope. A normal male cell has no Barr bodies, because the single X chromosome remains active.

Which X chromosome becomes inactivated—the one inherited from the father or the one from the mother—is a random event. As a result, a female expresses the paternal X chromosome alleles in some cells and the maternal alleles in others. Moreover,

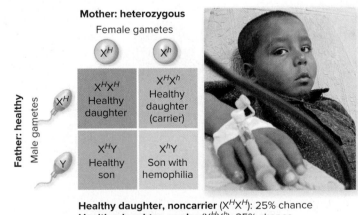

Healthy daughter, noncarrier ($X^H X^H$): 25% chance
Healthy daughter, carrier ($X^H X^h$): 25% chance
Healthy son ($X^H Y$): 25% chance
Affected son ($X^h Y$): 25% chance

Figure 10.23 Inheritance of Hemophilia. In a cross between a heterozygous female (a "carrier" of hemophilia A) and a healthy male, the chance of having a son with hemophilia is 25%. The boy in the photo is receiving treatment for hemophilia.

Photo: ©Waheed Khan/epa/Corbis

Figure 10.24 X Inactivation. In cats, the X chromosome carries a coat color gene with alleles for black or orange coloration. Calico cats are heterozygous for this gene; one of the two X chromosomes is inactivated in each colored patch. (A different gene accounts for the white background.) Can you explain why calico cats are almost always female?

Photo: ©Siede Preis/Getty Images RF

when a cell with an inactivated X chromosome divides mitotically, all of the daughter cells have the same X chromosome inactivated. Because the inactivation occurs early in development, females have patches of tissue that differ in their expression of X-linked alleles. **Figure 10.24** shows how inactivation of an X-linked coat color gene causes the distinctive orange and black fur patterns of calico and tortoiseshell cats, which are always female (except for rare XXY males).

X chromosome inactivation also explains another interesting observation: X-linked dominant disorders are typically less severe in females than in males. Thanks to X chromosome inactivation, a female who is heterozygous for an X-linked gene will

express a dominant disease-causing allele in only some of her cells. As a result, she experiences less severe symptoms than an affected male, who expresses the dominant allele in every cell.

An X-linked neurological disorder called Rett syndrome, for example, may be mild or severe in a girl, depending on how many of her cells express the Rett allele (**figure 10.25**). Male offspring who inherit the Rett allele typically die before birth. Nearly all people with Rett syndrome are therefore female.

10.7 MASTERING CONCEPTS

1. What determines a person's sex?
2. What is the role of the *SRY* gene in sex determination?
3. Why do males and females express recessive X-linked alleles differently?
4. Why does X inactivation occur in female mammals?

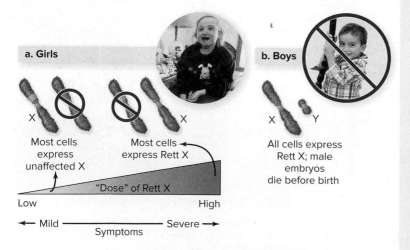

Figure 10.25 Rett Syndrome. In girls, symptoms of Rett syndrome depend on the proportion of cells expressing the normal allele (*yellow*) versus the Rett allele (*orange*). In males, every cell expresses the Rett allele, so affected male offspring typically die before birth.

Photos: (a): ©Andy Cross/The Denver Post via Getty Images; (b): ©Moof/Cultura/ Getty Images RF

Apply It **Now**

Choosing the Sex of Your Baby

Scientists have used the size difference between the X and Y chromosomes to develop technologies that may help people choose the sex of their babies. Some techniques use dyes to differentiate sperm carrying the X from those carrying the Y. Others use swimming speeds or centrifuges to sort the slower, heavier X-containing sperm from the faster, lighter sperm carrying the Y chromosome.

After Y-carrying sperm are separated from X-carrying sperm, the woman is inseminated with the desired fraction of the sperm. Alternatively, if a woman is using *in vitro* fertilization, the egg is fertilized in the laboratory with the sperm most likely to give a baby of the desired sex. None of the methods, however, works all the time; at best, the sperm samples are enriched in sperm that favor one sex over the other.

What about "natural" ways to boost the odds of having a boy or a girl? Folk wisdom includes advice on the mother's diet (including lists of "boy foods" and "girl foods"), the timing of intercourse, the sexual position, and even what objects to place under the pillow during intercourse. Some of these ideas may have a biological basis, and it may be possible to give a slight edge to X- or Y-containing sperm in the race to fertilize the egg.

But the evidence for all of these "do-it-yourself" techniques relies heavily on anecdotes. Testimonials like "It worked for me!" may seem compelling, but the odds of having a baby with the desired sex are near 50%, no matter what a person does. In reality, all such methods are difficult or impossible to test using controlled experiments, so it is hard to know for sure what (if anything) works.

🍁 10.8 Pedigrees Show Modes of Inheritance

Although Gregor Mendel did not study human genetics, our species nevertheless has some "Mendelian traits": those determined by single genes with alleles that are either dominant or recessive. For these traits, how do we know whether each phenotype is associated with a dominant or a recessive allele?

We saw in chapter 9 that karyotypes are useful for diagnosing chromosomal abnormalities. The same tool is useless in determining the inheritance of single-gene disorders, however, because different alleles look exactly alike under a microscope. To discover the inheritance pattern of an individual gene, researchers therefore track its incidence over multiple generations, much as Mendel did with his pea plants more than a century ago.

Genes on autosomes exhibit two modes of inheritance: autosomal dominant and autosomal recessive (table 10.3). An **autosomal dominant** disorder is expressed whether a person inherits one or two copies of the disease-causing allele. It therefore typically appears in every generation. Because the allele is dominant and located on an autosome, one or both of the affected individual's parents must also have the disorder (unless the disease-causing allele arose by a new mutation).

Inheriting an **autosomal recessive** disorder requires that a person receive the disease-causing allele from both parents. That is, each parent must either be homozygous recessive (and therefore have the disease) or be heterozygous. A person who is heterozygous is called a *carrier* because he or she is unaffected by the disorder but still has a 50% chance of passing the disease-causing allele to the next generation. If both parents are carriers, autosomal recessive conditions may seem to disappear in one generation, only to reappear in the next.

X-linked recessive conditions such as hemophilia produce unique inheritance patterns. Such disorders mostly affect males in a family; females are rarely affected, but many are heterozygous carriers for the disease-causing allele. Each child of a carrier has a 50% chance of inheriting the recessive allele from the mother. Any daughter who inherits the recessive allele from the mother will be a carrier—unless she also inherits the same allele from her father. If a son receives the recessive allele, however, he will have the condition; males cannot be symptomless carriers of X-linked traits.

TABLE 10.3 Some Autosomal Dominant and Autosomal Recessive Disorders in Humans

Disorder	Genetic Explanation	Characteristics
Autosomal dominant inheritance		
Achondroplasia	Mutant allele on chromosome 4 causes deficiency of receptor protein for growth factor.	Dwarfism with short limbs but normal-sized head and trunk
Familial hypercholesterolemia	Mutant allele on chromosome 2 encodes faulty cholesterol-binding protein.	High cholesterol, heart disease
Huntington disease	Mutant allele on chromosome 4 encodes protein that misfolds and forms clumps in brain cells.	Progressive, uncontrollable movements and personality changes; begins in middle age
Marfan syndrome	Mutant allele on chromosome 15 causes connective tissue disorder.	Long limbs, sunken chest, lens dislocation, spindly fingers, weakened aorta
Neurofibromatosis (type 1)	Mutant allele on chromosome 17 encodes faulty cell-signaling protein.	Brown skin marks (café-au-lait spots), benign tumors beneath skin
Polydactyly	Multiple genes on multiple chromosomes; mechanism is unknown	Extra fingers or toes or both
Autosomal recessive inheritance		
Albinism	Mutant allele on chromosome 11 encodes faulty protein required for pigment production.	Lack of pigmentation in skin, hair, and eyes
Cystic fibrosis	Mutant allele on chromosome 7 encodes faulty chloride channel protein.	Lung infections and congestion, infertility, poor fat digestion, poor weight gain, salty sweat
Phenylketonuria (PKU)	Mutant allele on chromosome 12 causes enzyme deficiency in pathway that breaks down the amino acid phenylalanine.	Intellectual disability caused by buildup of phenylalanine and related compounds
Tay-Sachs disease	Mutant allele on chromosome 15 causes deficiency of lysosome enzyme.	Nervous system degeneration caused by buildup of byproducts

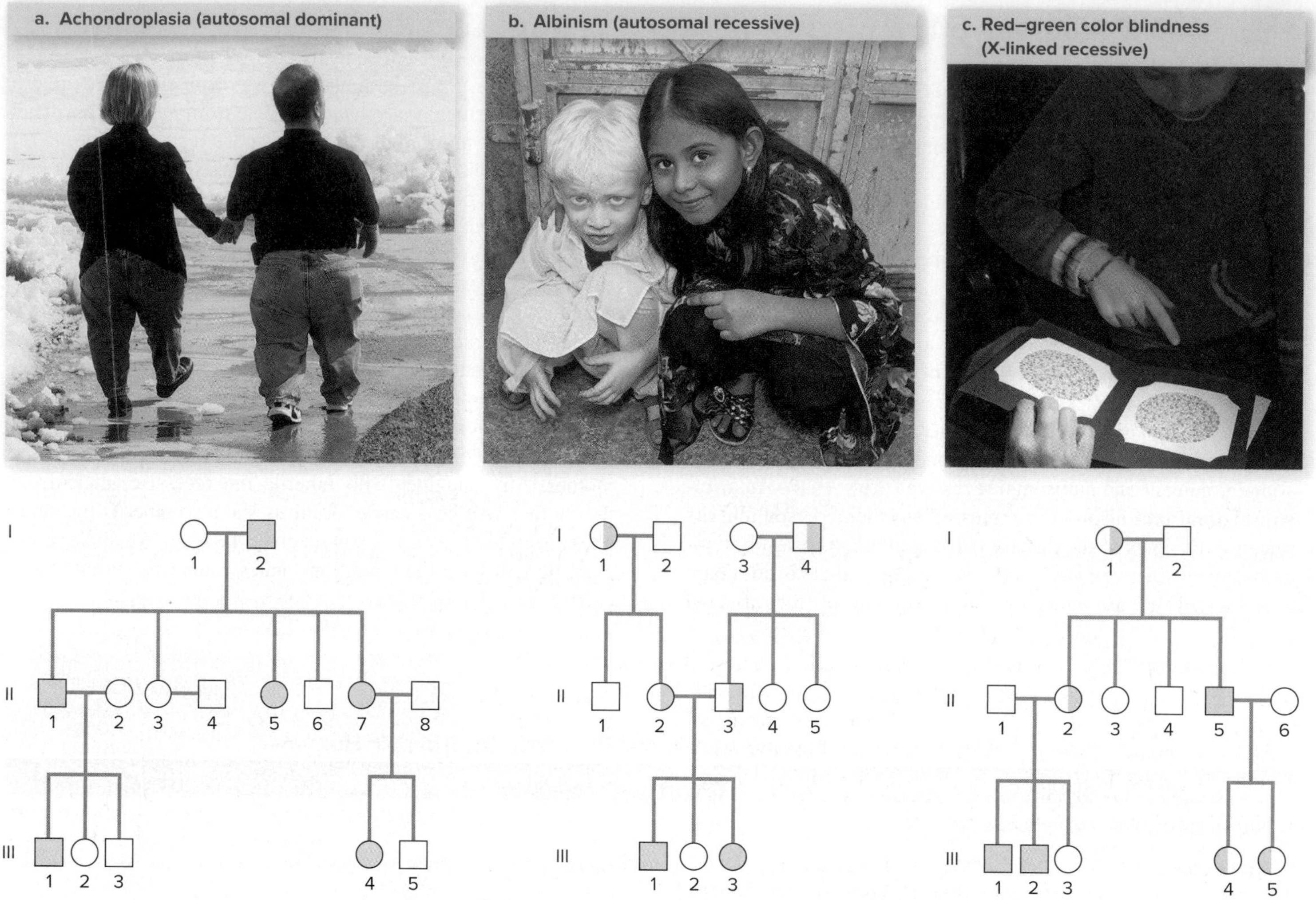

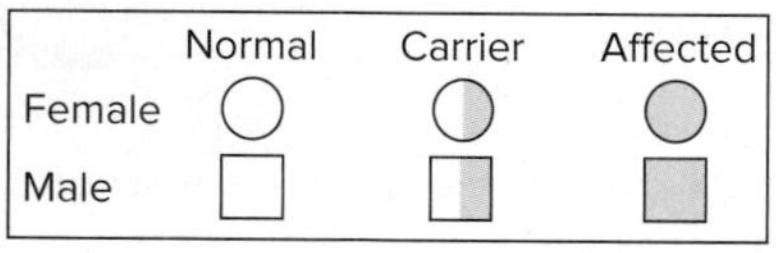

a. Achondroplasia (autosomal dominant)

b. Albinism (autosomal recessive)

c. Red–green color blindness (X-linked recessive)

Figure 10.26 Pedigrees. (a) A pedigree for an autosomal dominant trait; affected individuals are homozygous dominant or heterozygous. (b) A pedigree for an autosomal recessive trait; affected individuals are homozygous recessive. (c) A pedigree for an X-linked recessive trait. Males are affected if they inherit the allele on their single X chromosome; females are typically carriers.

Photos: (a): ©Rick Wilking/Reuters/Corbis; (b): ©Eric Lafforgue/Gamma-Rapho via Getty Images; (c): ©BSIP/Science Source

Pedigree charts depicting family relationships and phenotypes are useful for determining a disorder's mode of inheritance (figure 10.26). In a pedigree chart, squares indicate males, and circles denote females. Colored shapes indicate individuals with the disorder, and half-filled shapes represent known carriers. Horizontal lines connect parents. Siblings connect to their parents by vertical lines and to each other by an elevated horizontal line.

Figure 10.26 shows typical pedigrees for genes carried on autosomes and the X chromosome. In studying the differences among these three pedigrees, notice especially the patterns of the colored shapes. For example, affected individuals appear in every generation for autosomal dominant traits, whereas recessive conditions often skip generations. (Remember, however, that a recessive allele may remain hidden in carriers, even in a "skipped" generation.) Carriers appear only in pedigrees depicting the inheritance of recessive traits. Also, note that males are most commonly affected by X-linked recessive disorders.

Despite their utility, pedigrees can be difficult to construct and interpret for several reasons. People sometimes hesitate to supply information because they are embarrassed or want to protect their privacy. Adoption, children born out of wedlock, serial marriages, blended families, and assisted reproductive technologies all can complicate efforts to trace family ties. Moreover, as medical and family records are lost, many people simply lack the information to construct an accurate pedigree.

10.8   MASTERING CONCEPTS

1. How are pedigrees helpful in determining a disorder's mode of inheritance?

2. For each pedigree in figure 10.26, determine the genotype of individual #1 in the first row.

10.9 Most Traits Are Influenced by the Environment and Multiple Genes

Mendel's data were clear enough for him to infer principles of inheritance because he observed characteristics determined by single genes with two easily distinguished alleles. Moreover, the traits he selected are unaffected by environmental conditions. A genetic counselor can likewise be confident in telling two cystic fibrosis carriers that each of their children has a 25% chance of getting the disease.

But the counselor cannot calculate the probability that the child will be an alcoholic, have depression, be a genius, or wear size 9 shoes. The reason is that multiple genes and the environment control these and most other traits.

A. The Environment Can Alter the Phenotype

The environment can profoundly affect gene expression; that is, a gene may be active in one circumstance but inactive in another. As a simple example, temperature influences the quantity of pigment molecules in the fur of some animals. Siamese cats and Himalayan rabbits have light-colored bodies but dark ears, noses, paws, and tails, thanks to differences in gene expression between warm and cool body parts (**figure 10.27**). In crocodiles, the incubation temperature of the egg determines whether the baby will develop as a male or a female. Different combinations of genes are activated at each temperature, greatly altering the phenotype of the offspring. ⓘ *epigenetics,* section 7.6

In humans, fetal alcohol syndrome is an example of the effect of environment on phenotype: Prenatal exposure to alcohol can cause a developing baby to develop facial abnormalities or epilepsy. Likewise, personal circumstances ranging from hormone levels to childhood experiences to diet influence a person's susceptibility to depression, alcoholism, and type 2 diabetes.

These three diseases have a genetic component as well, but sorting out the relative contributions of nature and nurture is difficult. Studies of twins are often helpful, as are careful observations of everything from family composition to brain structures. The Burning Question in chapter 7, for example, explains some strategies that researchers use to explore the genetic connection to homosexuality. Discovering the exact mechanism by which genes interact with external stimuli is a very active area of research.

Even human diseases with simple, single-gene inheritance patterns can have an environmental component. Cystic fibrosis, for example, is a single-gene disorder. Because cystic fibrosis patients are so susceptible to infection, however, the course of the illness depends on which infectious agents a person encounters.

Figure 10.27 Temperature and Fur Color. Siamese cats have a mutation in a gene encoding an enzyme required for pigment production. The enzyme is active only at the relatively cool temperatures of the paws, ears, snout, and tail; these areas are darkly pigmented. At higher temperatures, the enzyme is inactive. Pigment production is therefore reduced where the skin is warmest. ©Carolyn A. McKeone/Science Source

Burning Question

Is obesity caused by genes or the environment?

©Getty Images RF

At first glance, the cause of obesity seems simple: If a person eats more calories than he or she expends, the body stores the excess calories as fat. As fat accumulates, body weight climbs. According to this view, a person's genes are irrelevant to his or her body weight. In reality, however, obesity reflects the combined action of genes and the environment.

Several genes are associated with obesity. One example is the gene that encodes leptin, a hormone that helps curb appetite. Individuals who inherit mutant alleles for this gene never feel full, leading to overeating and obesity.

The environment can also influence the expression of the genes that a person inherits. For example, scientists have found that mothers who ingest low amounts of carbohydrates—sugars and starches—give birth to children who are especially likely to become obese later in life. Evidence suggests that epigenetic modifications occur while a developing child is still in the womb, permanently altering gene expression patterns for life.

Fetuses presumably use the mother's diet to "prepare for" the environment that they will be born into. These changes occur before birth, giving a newborn the best chance for survival. But if a fetus prepared for a low-calorie life is born into an environment where food is actually plentiful, then obesity is likely.

Submit your burning question to
Marielle.Hoefnagels@mheducation.com

B. Polygenic Traits Depend on More Than One Gene

Unlike cystic fibrosis, most inherited traits are **polygenic,** meaning that the phenotype reflects the activities of more than one gene. Eye color is a familiar example of a polygenic trait. Multiple enzymes, encoded by multiple genes, influence the production and distribution of pigments in the eye's iris. Male pattern

a.

b.

Figure 10.28 Variation in Human Height. (a) These students from the University of Connecticut at Storrs lined up by height in 1920, demonstrating the characteristic bell-shaped distribution of polygenic traits. (b) A similar photo taken 57 years later reveals more tall students than the earlier lineup, illustrating the influence of improved health care and nutrition on height. (a, b): ©Peter Morenus/University of Connecticut

baldness is another example; multiple genes control the interactions between male sex hormones and hair follicle cells.

To complicate matters further, the environment also affects the expression of polygenic traits (see section 32.5 and the Burning Question in this section). For example, in plants, polygenic traits typically include flower color, the density of leaf pores (stomata), and crop yield. But genes alone do not determine these phenotypes. Soil pH can affect flower color, CO_2 concentration can influence the number of stomata per square centimeter, and nutrient and water availability helps determine crop production. ⓘ *stomata,* section 22.2B

When the frequencies of all the phenotypes associated with a polygenic trait are plotted on a graph, they form a characteristic bell-shaped curve. Human height, for example, ranges from very short to very tall, with most people somewhere in the middle (**figure 10.28**). Height is a product of genetics, childhood nutrition, and health care.

Figure 10.29 shows another example: the continuum of gene expression for skin color, a trait that is affected by genes and exposure to sunlight. Body weight and intelligence are other traits that are both polygenic and influenced by the environment.

10.9 MASTERING CONCEPTS

1. How can the environment affect a phenotype?
2. What is polygenic inheritance, and how is it different from codominance?

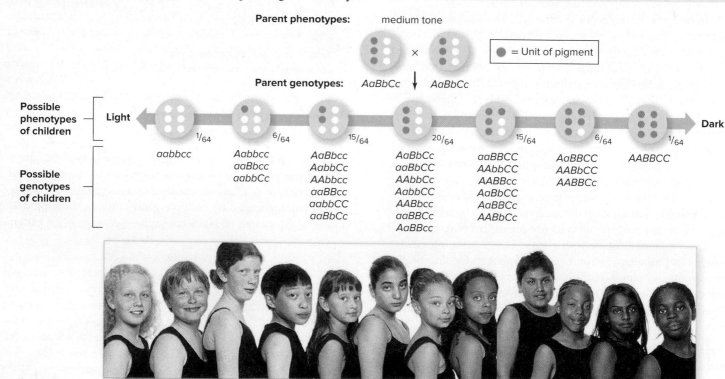

Figure 10.29 Variation in Human Skin Color. Multiple genes interact to determine the quantity of pigment in skin cells.
Photo: ©Sarah Leen/National Geographic Stock

INVESTIGATING LIFE

10.10 Heredity and the Hungry Hordes

Agriculture provides a steady food supply, but not just to humans; hungry insects and other animals also eat the plants we grow for food and fiber. Farmers continually seek new ways to kill these competitors, but each tactic selects for new adaptations in the insects. It is a long-standing, and seemingly unavoidable, evolutionary arms race.

The larvae of butterflies and moths are especially voracious. A good example is the pink bollworm (**figure 10.30**). The adults of this species are moths that lay eggs on cotton bolls. When the eggs hatch, the pink caterpillars tunnel into the boll and eat the seeds, damaging the cotton fibers.

One tool that keeps bollworms and other caterpillars at bay is a soil bacterium called *Bacillus thuringiensis,* abbreviated Bt. This microbe produces toxic proteins that bind to receptor molecules on the surface of the cells lining a bollworm's intestinal tract. The toxins effectively poke holes in the gut, leaving the animal vulnerable to infection and unable to digest food. Fortunately, Bt does not affect humans because we lack the receptor molecules to which the toxins bind.

Bt sprays are effective, but they cannot reach every insect. Some caterpillars, such as those that tunnel inside a cotton boll, escape death because they never eat the Bt sprayed on a plant's surface. In the 1990s, biotechnology solved this problem when scientists inserted the bacterial gene for a Bt toxin into plant cells. Every cell in these genetically modified "Bt plants" produces the toxin. Nibbling on any hidden or exposed plant part spells death to the caterpillar. ⓘ *transgenic organisms,* section 11.2A

Bt plants, however, pose a new problem: They kill most of the susceptible caterpillars, leaving the resistant individuals to produce the next generation. To combat this selective pressure, farmers growing Bt crops in the United States surround each field with a refuge—that is, a buffer strip planted with a conventional (non-Bt) variety of the same crop.

Figure 10.30 Hungry Caterpillar. Pink bollworms cost cotton producers tens of millions of dollars each year.
©Nigel Cattlin / Science Source

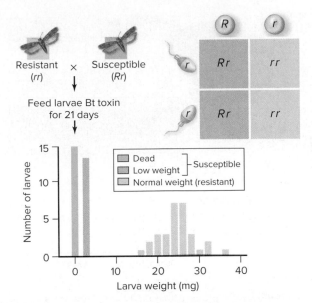

Figure 10.31 Recessive Resistance. When a heterozygous, susceptible insect is bred with a resistant mate, half the offspring should be resistant and half should be susceptible. Tests with Bt toxin show that this is indeed the case: Half of the larvae thrived in the presence of Bt toxin, while the susceptible ones died or were very small.

How does this strategy slow the evolution of Bt resistance? Biologists Shai Morin, Bruce Tabashnik, and their colleagues at the University of Arizona answered this question by studying several populations of pink bollworms. By comparing the genotypes of Bt-susceptible and Bt-resistant caterpillars, they discovered a gene with three recessive alleles conferring Bt resistance. Insects with two such resistance alleles are immune to the toxin, whereas those with one dominant allele are susceptible.

The buffer strategy decreases the rate of bollworm evolution because the Bt-resistance alleles are recessive (**figure 10.31**). A homozygous recessive, resistant moth emerging from the Bt crop has a good chance of encountering a susceptible mate from the buffer strip. All of their heterozygous offspring are susceptible; they will die if they eat the Bt plants. A ready pool of susceptible mates from the buffer strip should therefore keep the recessive allele rare.

So far, studies suggest that the refuge strategy works; Bt-resistant bollworms remain rare in cotton fields of the United States. But other countries might not have such stringent requirements, and a resistant variety of bollworms could quickly spread worldwide. Anticipating this problem, the companies that developed Bt cotton have added another weapon to the arms race. Instead of producing just one Bt toxin, the newest pest-resistant plants add a second Bt toxin with a different target receptor protein in the insect's gut (**figure 10.32a**). This strategy should greatly delay the emergence of resistant bollworms because random mutations would have to occur in two genes—encoding the two receptor proteins—to confer resistance to both toxins.

Tabashnik and his colleagues tested how well the two-toxin strategy worked. First, the team bred a lineage of bollworms that was resistant only to the new toxin. Then they placed 10 larvae of these bollworms on each of three types of bolls: non-Bt cotton,

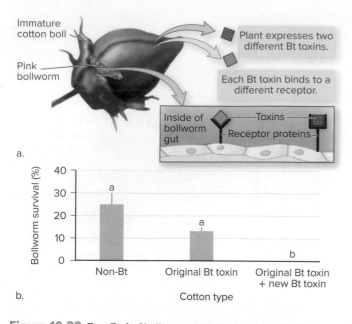

a.

b.

Figure 10.32 **Two-Toxin Challenge.** (a) Cotton plants producing two Bt toxins should slow the evolution of resistance in pink bollworms. (b) Bollworms resistant only to the newest Bt toxin were placed on cotton bolls. None survived on cotton containing two Bt toxins, suggesting the two-toxin strategy works. (Error bars represent standard errors; see appendix B.)

cotton producing only the original Bt toxin, and cotton producing both the original and the new Bt toxin. After 21 days, they counted the number of surviving larvae.

The results were encouraging (figure 10.32b): The bolls producing two Bt toxins killed every bollworm, but the other two cotton varieties did not. The researchers concluded that adding a second toxin is effective, even when the bollworms are already resistant to one toxin.

So far, the refuge strategy and new transgenic cotton varieties have put the brakes on the evolution of Bt resistance. The stakes are high. If resistance alleles become very common, Bt may become useless as a control measure. Growers may then switch back to broad-spectrum pesticides that kill many more types of insects—even beneficial ones. Careful monitoring will be crucial as we continue to wage war against the insects that compete for our crops.

Sources: Morin, Shai, Robert W. Biggs, Mark S. Sisterson, and 10 other authors (including Bruce E. Tabashnik). April 29, 2003. Three cadherin alleles associated with resistance to *Bacillus thuringiensis* in pink bollworm. *Proceedings of the National Academy of Sciences*, vol. 100, pages 5004–5009.

Tabashnik, Bruce E., and five other authors. July 21, 2009. Asymmetrical cross-resistance between *Bacillus thuringiensis* toxins Cry1Ac and Cry2Ab in pink bollworm. *Proceedings of the National Academy of Sciences*, vol. 106, pages 11889–11894.

10.10   MASTERING CONCEPTS

1. What do you predict will happen to the incidence of resistance alleles in pink bollworm populations if farmers choose not to plant the required refuge?

2. Suppose a variety of bollworm that is resistant to the original Bt toxin invades a cotton field planted with a two-toxin crop. Assuming that the farmer did not plant a non-Bt refuge, how might the invading bollworms become resistant to these new transgenic crops?

CHAPTER SUMMARY

10.1   Chromosomes Are Packets of Genetic Information: A Review

- **Genes** encode proteins; mutations create new **alleles** (figure 10.33a).
- A **chromosome** is a continuous molecule of DNA with associated proteins. In humans, a **diploid cell** contains 22 **homologous pairs** of **autosomes** and 1 pair of **sex chromosomes.**
- **Meiosis** gives rise to **haploid cells. Fertilization** unites haploid **gametes** and restores the diploid number.

10.2   Mendel's Experiments Uncovered Basic Laws of Inheritance

A.   Why Peas?
- Gregor Mendel studied inheritance in pea plants because they develop quickly, produce abundant offspring, and are easy to breed.

B.   Dominant Alleles Appear to Mask Recessive Alleles
- An individual is **true-breeding** for a trait if self-fertilization yields only offspring that resemble the parent (for that trait). A **hybrid** individual produces a mix of offspring when self-fertilized.
- A **dominant allele** is always expressed if it is present; a **recessive allele** is masked by a dominant allele.

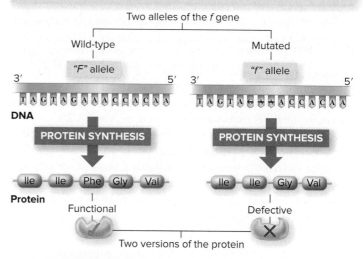

a. Mutations produce new alleles.

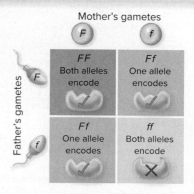

b. A Punnett square tracks the inheritance of these alleles among all possible offspring.

Figure 10.33 **From Mutation to Punnett Square.**

C. For Each Gene, a Cell's Two Alleles May Be Identical or Different

- An individual's **genotype** may be **heterozygous** (two different alleles for a gene) or **homozygous** (both alleles are the same).
- A **phenotype** is any observable characteristic of an organism.
- A **wild-type** allele is the most common in a population. A change in a gene is a **mutation** and may result in a **mutant** phenotype.

D. Every Generation Has a Name

- In genetic crosses, the purebred parental generation is designated **P**; the next generation is F_1; and the next is F_2.

10.3 The Two Alleles of a Gene End Up in Different Gametes

A. The Simplest Punnett Squares Track the Inheritance of One Gene

- A **monohybrid cross** is a mating between two individuals that are heterozygous for the same gene.
- **Punnett squares** are useful for predicting the allele combinations that the offspring of two parents might inherit (see figure 10.33b).
- A **test cross** is one way to reveal the unknown genotype of an individual with a dominant phenotype; the individual is crossed with a homozygous recessive mate.

B. Meiosis Explains Mendel's Law of Segregation

- Mendel's **law of segregation** states that the two alleles of the same gene separate into different gametes.

10.4 Genes on Different Chromosomes Are Inherited Independently

A. Tracking Two-Gene Inheritance May Require Large Punnett Squares

- A **dihybrid cross** is a mating between individuals that are heterozygous for two genes.

B. Meiosis Explains Mendel's Law of Independent Assortment

- According to Mendel's **law of independent assortment,** the inheritance of one gene does not affect the inheritance of another gene on a different chromosome. Independent assortment occurs because homologous chromosomes are oriented randomly during metaphase I of meiosis.

C. The Product Rule Is a Useful Shortcut

- The **product rule** is an alternative to Punnett squares for following the inheritance of two or more traits at a time.

10.5 Genes on the Same Chromosome May Be Inherited Together

A. Genes on the Same Chromosome Are Linked

- **Linked genes** are located on the same chromosome. **Linkage groups** are collections of genes that are often inherited together.
- The farther apart two genes are on a chromosome, the more likely **crossing over** is to separate their alleles. If crossing over occurs between two alleles, some offspring will inherit **recombinant chromatids;** otherwise, offspring will inherit **parental chromatids.**

B. Studies of Linked Genes Have Yielded Chromosome Maps

- Breeding studies reveal the crossover frequencies used to create **linkage maps**—diagrams that show the order of genes on a chromosome.

10.6 Dominance Relationships Are Rarely Simple

A. Incomplete Dominance and Codominance Add Phenotype Classes

- Heterozygotes for alleles with **incomplete dominance** have phenotypes intermediate between those of the two homozygotes. **Codominant** alleles are both expressed in a heterozygote.
- Figure 10.34 uses a lightbulb analogy to compare dominance relationships.

B. Some Inheritance Patterns Are Especially Difficult to Interpret

- A **pleiotropic** gene affects multiple phenotypes.
- When multiple proteins participate in a biochemical pathway, mutations in genes encoding any of the proteins can produce the same phenotype.
- In **epistasis,** one gene masks the effect of another.

10.7 Sex-Linked Genes Have Unique Inheritance Patterns

- Genes controlling **sex-linked** traits are located on the X or Y chromosome.

A. X and Y Chromosomes Carry Sex-Linked Genes

- In humans, the male has X and Y sex chromosomes, and the female has two X chromosomes. Scientists know of many more **X-linked** than **Y-linked** disorders.

B. X-Linked Recessive Disorders Affect More Males Than Females

- Females express an X-linked recessive disorder only if they inherit faulty alleles from both parents. Males express every allele on their single X chromosome.

C. X Inactivation Prevents "Double Dosing" of Proteins

- **X inactivation** randomly shuts off all but one X chromosome in each cell.

10.8 Pedigrees Show Modes of Inheritance

- **Pedigrees** trace phenotypes in families and reveal modes of inheritance.
- An **autosomal dominant** disorder can be inherited from one affected parent. An **autosomal recessive** disorder must be inherited from both parents. X-linked recessive disorders affect mostly males.

10.9 Most Traits Are Influenced by the Environment and Multiple Genes

A. The Environment Can Alter the Phenotype

- Most traits have environmental as well as genetic influences.

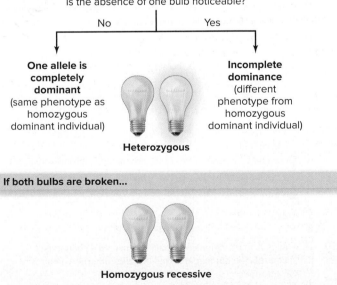

Figure 10.34 Dominance Relationships: An Analogy.

B. Polygenic Traits Depend on More Than One Gene

- A **polygenic trait** varies continuously in its expression.

10.10 Investigating Life: Heredity and the Hungry Hordes

- Researchers have developed a genetic test to monitor caterpillars for resistance to Bt, an insecticidal toxin produced in genetically modified cotton.

MULTIPLE CHOICE QUESTIONS

1. In the list of four terms below, which term is the *second* most inclusive?
   - a. Genome
   - b. Allele
   - c. Chromosome
   - d. Gene

2. According to Mendel, if an individual is *heterozygous* for a gene, the phenotype will correspond to that of
   - a. the recessive trait alone.
   - b. the dominant trait alone.
   - c. a blend of the dominant and recessive traits.
   - d. a wild-type trait.

3. If a plant that is homozygous dominant for a gene is crossed with a plant that is homozygous recessive for the same gene, what is the probability that the offspring will have the same genotype as one of the parents?
   - a. 100%
   - b. 50%
   - c. 25%
   - d. 0%

4. Each letter below represents an allele. Which of the following is an example of a dihybrid cross?
   - a. *R × R*
   - b. *Rr × Rr*
   - c. *Rr Yy × Rr Yy*
   - d. *RR yy × rr YY*

5. Which of the following is a possible gamete for an individual with the genotype *PP rr*?
   - a. *PP*
   - b. *P r*
   - c. *p r*
   - d. *rr*

6. Use the product rule to determine the chance of obtaining an offspring with the genotype *Rr Yy* from a dihybrid cross between parents with the genotype *Rr Yy*.
   - a. ½
   - b. ¼
   - c. ¼
   - d. ¹⁄₁₆

7. Refer to the linkage map in figure 10.16b. A crossover event is most likely to occur between which pair of genes?
   - a. *w* and *v*
   - b. *y* and *r*
   - c. *y* and *w*
   - d. *v* and *m*

8. How can epistasis decrease the number of phenotypes observed in a population?
   - a. One gene may mask the expression of another gene.
   - b. One allele may mask the expression of another allele of the same gene.
   - c. Genes may become silenced after exposure to certain environmental cues.
   - d. Genes may be inactivated if they are on an X chromosome.

9. Suppose a woman has a recessive X-linked disease. Her husband does not have the disease. What is the chance that their second child has the disease?
   - a. 100%
   - b. 50%
   - c. 25%
   - d. 0%

10. How does X inactivation contribute to a person's phenotype?
    - a. It controls the number and kind of genes inherited on an X chromosome.
    - b. It determines which X chromosome is expressed in a male.
    - c. It allows for the expression of either the maternal or the paternal X in different cells.
    - d. It enhances the expression of Y-linked genes in males.

Answers to these questions are in appendix A.

WRITE IT OUT

1. Select one gene mentioned in the chapter; then explain the link between an organism's genotype (for that gene) and its corresponding phenotype. Make sure to use the term *protein* in your answer.

2. List three genes (mentioned in this chapter or not) that do not affect physical appearance. Do these genes contribute to an organism's phenotype?

3. Some people compare a homologous pair of chromosomes to a pair of shoes. Explain the similarity. How would you extend the analogy to the sex chromosomes for females and for males?

4. How did Mendel use evidence from monohybrid and dihybrid crosses to deduce his laws of segregation and independent assortment? How do these laws relate to meiosis?

5. How does crossing over "unlink" genes?

6. Consider two genes that are near each other on a chromosome. After a germ cell undergoes meiosis, are the gametes likely or unlikely to contain a recombinant chromatid for these two genes? Explain.

7. Explain how each of the following produces phenotypic ratios other than those Mendel observed: incomplete dominance, codominance, pleiotropy, epistasis.

8. Many men with "Y chromosome infertility" are unable to produce sperm. Do you think this condition is typically inherited? Explain your answer.

9. Would you expect dominant X-linked illnesses to affect women as often as men? Explain your answer.

10. A family has an X-linked dominant form of congenital generalized hypertrichosis (excessive hairiness). Although the allele is dominant, males are more severely affected than females. Moreover, the women in the family often have asymmetrical, hairy patches on their bodies. How does X chromosome inactivation explain this observation?

11. X inactivation explains the large color patches in calico cat fur and the smaller patches in tortoiseshell cat fur. In which type of cat do you expect X inactivation occurs earlier in development? Why?

12. How do the pedigrees differ for autosomal dominant, autosomal recessive, and X-linked recessive conditions?

13. In the following pedigree, is the disorder's mode of inheritance autosomal dominant, autosomal recessive, or X-linked recessive? Explain your reasoning.

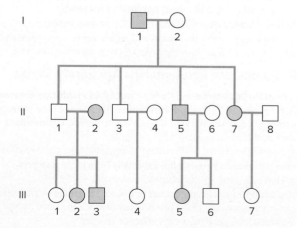

14. How do heart disease and cancer illustrate diseases that reflect both genetic and environmental influences?

15. Design an experiment using twins to determine the degree to which autism is genetic or environmental.

GENETICS PROBLEMS

See the How to Solve a Genetics Problem section for step-by-step guidance.

1. In rose bushes, red flowers (*FF* or *Ff*) are dominant to white flowers (*ff*). A true-breeding red rose is crossed with a white rose; two flowers of the F_1 generation are subsequently crossed. What will be the most common genotype of the F_2 generation?

2. In Mexican hairless dogs, a dominant allele confers hairlessness. However, inheriting two dominant alleles is lethal; the fetus dies before birth. Suppose a breeder mates two dogs that are heterozygous for the hair allele. Draw a Punnett square to predict the genotypic and phenotypic ratios of the puppies that are born.

3. A species of ornamental fish comes in two colors; red is dominant and gray is recessive. Emily owns a red fish, and she wants to know its genotype. Therefore, she mates her pet with a gray fish. If 50 of the 100 babies are red, what is the genotype of Emily's fish?

4. Two lizards have green skin and large dewlaps (genotype *Gg Dd*). (a) If 32 offspring are born, how many of the offspring are expected to be homozygous recessive for both genes? (b) What proportion of the offspring will have the dominant phenotype for both traits? (Assume that the traits assort independently.)

5. A fern with genotype *AA Bb Cc dd Ee* mates with another fern with genotype *aa Bb CC Dd ee*. What proportion of the offspring will be heterozygous for all genes? (Assume the genes assort independently.) *Hint:* Use the product rule.

6. Genes *Q*, *R*, and *S* are on the same chromosome. The crossover frequency between *S* and *Q* is 5%, the crossover frequency between *Q* and *R* is 30%, and the crossover frequency between *R* and *S* is 35%. Use this information to create a linkage map for the chromosome.

7. Three babies are born in the hospital on the same day. Baby X has type B blood; Baby Y has type AB blood; Baby Z has type O blood. Use the information in the following table to determine which baby belongs to which couple. (Assume that all individuals are homozygous dominant for the *H* gene.)

Couple	Father	Blood Type	Mother	Blood Type
1	Logan	B	Leslie	AB
2	Sam	A	Casey	A
3	Jordan	O	Taylor	B

8. In fraggles, males are genotype XY and females are XX. Silly, a male fraggle, has a rare X-linked recessive disorder that makes him walk backwards. He mates with Lilly, who is a carrier for the disorder. What proportion of their male offspring will walk backwards?

PULL IT TOGETHER

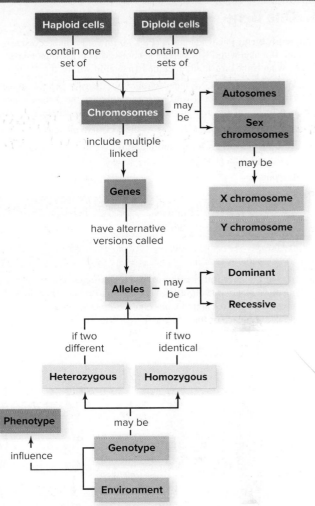

Figure 10.35 Pull It Together: Patterns of Inheritance.

Refer to **figure 10.35** and the chapter content to answer the following questions.

1. Compare the Survey the Landscape figure in the chapter introduction with the Pull It Together concept map. Explain the connections between alleles, genetic variation, gametes, Punnett squares, and inheritance.

2. Explain the effects of a mutation, using *allele, dominant, recessive, genotype,* and *phenotype* in your answer.

3. Add *meiosis, gametes, incomplete dominance, codominance, pleiotropy,* and *epistasis* to this concept map.

HOW TO SOLVE A GENETICS PROBLEM

A. One Gene

Sample problem: Phenylketonuria (PKU) is an autosomal recessive disorder. If a man with PKU marries a woman who is a symptomless carrier, what is the probability that their first child will be born with PKU?

1. **Write a key.** Pick ONE letter to represent the gene in your problem. Use the capital form of your letter to symbolize the dominant allele; use the lowercase letter to symbolize the recessive allele.

   Sample: The dominant allele is *K;* the recessive allele is *k.*

2. **Summarize the problem's information.** Make a table listing the phenotypes and genotypes of both parents.

   Sample:

	Male	**Female**
Phenotype	Has PKU	No PKU (carrier)
Genotype	*kk*	*Kk*

3. **Sketch the parental chromosomes and gametes.** Use the genotypes in your table to draw the alleles onto chromosomes. Then draw short arrows to show the homologous chromosomes moving into separate gametes for each parent.

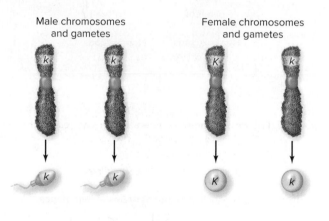

Male chromosomes and gametes

Female chromosomes and gametes

4. **Make a Punnett square.** Arrange the gametes you sketched in step 3 along the edges of the square, and fill in the genotypes of the offspring.

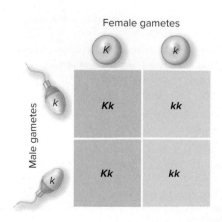

Female gametes

Male gametes

5. **Calculate the genotypic ratio.** Count the number of boxes that contain each offspring genotype.

   Sample: 2 *Kk* : 2 *kk*

6. **Calculate the phenotypic ratio.** Count the number of boxes that contain each offspring phenotype.

   Sample: 2 PKU carriers : 2 PKU sufferers

7. **Calculate the probability of each phenotype.** Divide each number in step 6 by the total number of boxes (4) and multiply by 100.

   Sample: 50% probability that a child will be a carrier; 50% probability that a child will have PKU

B. Two Genes (Punnett Square)

Sample problem: A student collects pollen (male sex cells) from a pea plant that is homozygous recessive for the genes controlling seed form and seed color. She brushes the pollen on the female flower parts of a plant that is heterozygous for both genes. What is the probability that an offspring plant has the same genotype and phenotype as the male parent? Assume the genes are not linked.

1. **Write a key.** Pick ONE letter to represent each of the genes in your problem. Use the capital form of your letter to symbolize the dominant allele; use the lowercase letter to symbolize the recessive allele.

   Sample: For seed form, the dominant allele (round) is *R;* the recessive allele (wrinkled) is *r;* for seed color, the dominant allele (yellow) is *Y;* the recessive allele (green) is *y.*

2. **Summarize the problem's information.** Make a table listing the phenotypes and genotypes of both parents.

   Sample:

	Male	**Female**
Phenotype	Wrinkled, green	Round, yellow
Genotype	*rr yy*	*Rr Yy*

3. **Sketch the parental chromosomes and gametes.** Use the genotypes in your table to draw the alleles onto two sets of chromosomes, one for each parent. The law of independent assortment means that you need to draw all possible configurations. So redraw the chromosomes, this time switching the order of the alleles in one pair. Then draw short arrows to show the chromosomes separating, and sketch the four possible gametes for each parent. (For homozygous parents, multiple gametes will have the same genotype.)

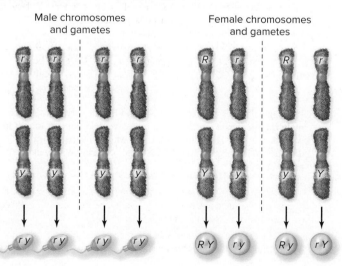

Male chromosomes and gametes

Female chromosomes and gametes

4. **Make a Punnett square.** Arrange the gametes you sketched in step 3 along the edges of the square, and fill in the genotypes of the offspring.

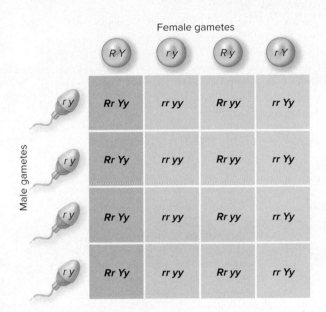

Female gametes

Male gametes

5. **Calculate the genotypic ratio.** Count the number of boxes that contain each offspring genotype.

   Sample: 4 *Rr Yy* : 4 *rr yy* : 4 *Rr yy* : 4 *rr Yy*

6. **Calculate the phenotypic ratio.** Count the number of boxes that correspond to each possible phenotype combination.

   Sample: 4 round, yellow : 4 wrinkled, green : 4 round, green : 4 wrinkled, yellow

7. **Calculate the probability of each phenotype.** Divide each number in step 6 by the total number of boxes (16) and multiply by 100.

   Sample: 25% probability that an offspring has the same genotype and phenotype as the male parent.

C. Two Genes (Product Rule)

The product rule is a simpler way to solve a multi-gene problem and eliminates the need for a large Punnett square. To use the product rule in the previous sample problem, first calculate the probability that the parents ($rr \times Rr$) produce an offspring with genotype *rr* (½, or 50%). Then calculate the chance that $yy \times Yy$ parents produce a *yy* offspring (½, or 50%). Multiply the two probabilities to calculate the probability that both events occur simultaneously: $\frac{1}{2} \times \frac{1}{2} = \frac{1}{4}$, or 25%. See section 10.4 for more on the product rule.

D. X-Linked Gene

Sample problem: Hemophilia is caused by an X-linked recessive allele. If a man who has hemophilia marries a healthy woman who is not a carrier, what is the chance that their child will have hemophilia?

1. **Write a key.** Pick ONE letter to represent the gene in your problem. Use the capital form of your letter to symbolize the dominant allele; use the lowercase letter to symbolize the recessive allele.

   Sample: The dominant allele is *H;* the recessive allele is *h.* Because these alleles are on the X chromosome, inheritance will differ between males and females. It is therefore best to designate the chromosomes and alleles together as X^H and X^h.

2. **Summarize the problem's information.** Make a table listing the phenotypes and genotypes of both parents.

   Sample:

	Male	**Female**
Phenotype	Has hemophilia	Healthy
Genotype	X^hY	X^HX^H

3. **Sketch the parental chromosomes and gametes.** Use the genotypes in your table to draw the alleles onto chromosomes. Then draw short arrows to show the chromosomes moving into separate gametes for each parent.

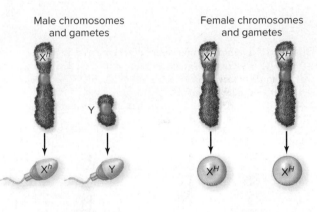

Male chromosomes and gametes

Female chromosomes and gametes

4. **Make a Punnett square.** Arrange the gametes you sketched in step 3 along the edges of the square; fill in the genotypes of the offspring.

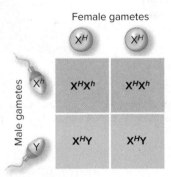

Female gametes

Male gametes

5. **Calculate the genotypic ratio.** Count the number of boxes that contain each offspring genotype.

   Sample: 2 X^HX^h : 2 X^HY

6. **Calculate the phenotypic ratio.** Count the number of boxes that correspond to each possible phenotype.

   Sample: 2 female carriers : 2 healthy males

7. **Calculate the probability of each phenotype.** Divide each number in step 6 by the total number of boxes (4) and multiply by 100.

   Sample: 50% probability that a child will be a female carrier; 50% probability that a child will be a healthy male. No child, male or female, will have hemophilia.

11 DNA Technology

Carbon Copies. Tabouli, left, plays with her sibling, Baba Ganoush. In an effort to publicize its pet gene banking and cloning services, a private company produced the kittens by cloning a 1-year-old cat.

©Ben Margot/AP Images

LEARN HOW TO LEARN
Vary Your Study Plan for Healthy Learning

Your study sessions may become stale if you do the same things over and over. After all, it is difficult to focus after watching countless animations or listening to hours of podcasts. Instead, try switching between strategies that are passive (watching and listening) and those that are active (drawing and writing). You might watch a video on Punnett squares and then try to draw one yourself. Or you could listen to a podcast about cloning and then write a paragraph describing the process in your own words. Keeping variety in your study plan will help you stay engaged.

The Clones Are Here

Imagine being able to produce a child, genetically identical to yourself, from a bit of skin or the root of a hair. Although humans cannot reproduce in this way, many organisms do the equivalent. They develop parts of themselves into genetically identical individuals—clones—which then detach and live independently.

Cloning, or asexual reproduction, has been a part of life since the first cell arose billions of years ago. Long before sex evolved, each individual simply reproduced by itself, without a partner to contribute half the offspring's genetic information.

In its simplest form, cloning consists of the division of a single cell. In bacteria, archaea, and other single-celled organisms, the cell's DNA replicates, and then the cell splits into two identical, individual organisms (see chapter 8).

Cloning is also a natural part of life for many species of plants, fungi, and animals. Hobbyists and commercial plant growers clone everything from fruit trees to African violets by cultivating cuttings from a parent plant's stems, leaves, and roots. Many fungi also are phenomenal breeders, asexually producing countless microscopic spores on bread, cheese, and every other imaginable food supply. Asexual reproduction is much less common in animals, but it does occur. For example, sponges and jellyfishes "bud" genetically identical clones that break away from the parent.

Mammals, however, normally reproduce only sexually. That is why biologists attracted worldwide attention in the 1990s by creating a lamb called Dolly, the first clone of an adult mammal (see section 11.3B). Since that time, many other mammals—but not humans—have been cloned.

As you will see, cloning is just one of many practical applications arising from knowledge of DNA's structure and function. Additional new uses for DNA technology may come from surprising sources. For example, studying how bacteria fight viral infections led to the discovery of CRISPR, one of the most promising new biotechnology tools. In bacteria, CRISPR is a segment of DNA that teams up with proteins to snip viral genes out of the genome. Modified CRISPR sequences now enable biologists to edit, activate, or inactivate any gene in any organism. The potential applications are endless. This chapter describes a sampling of many other uses of DNA technology.

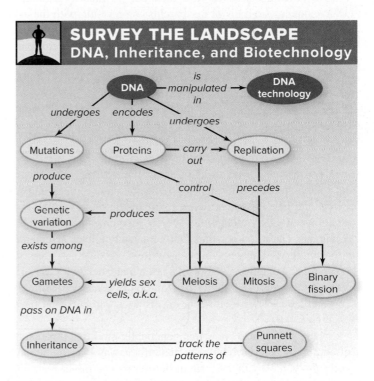

SURVEY THE LANDSCAPE
DNA, Inheritance, and Biotechnology

Biologists can manipulate DNA in many ways, from determining its nucleotide sequence to mixing genes from multiple species. As a result, we can now accomplish feats that were unimaginable a generation ago.

For more details, study the Pull It Together feature in the chapter summary.

11.1 DNA Technology Is Changing the World

The title of this section is not an exaggeration: DNA-based technologies have affected nearly every imaginable facet of society. **DNA technology** is a broad term that usually means the manipulation of genes for some practical purpose. This chapter describes some of the ways in which DNA technology has become a powerful tool in research, medicine, agriculture, criminal justice, and many other fields.

DNA technology became possible only after biologists learned the structure and function of DNA. Recall from chapter 7 that the DNA double helix is composed of nucleotides and that the function of DNA is to provide the "recipes" for the cell's proteins (figure 11.1). As described in chapter 8, enzymes copy these recipes as a cell prepares to divide, so that nearly every cell in a multicellular organism carries the same DNA sequence.

We have also learned that each person inherits a unique DNA sequence from his or her parents. As a result, with the exception of identical twins, each person is genetically different. Yet chapter 10 showed how we can trace the inheritance of particular alleles from child to parent to grandparent, and so on. Using the same logic, it is easy to see the power of DNA as a tool for tracing evolutionary history. Throughout the billions of years of life's history, descent with modification has produced countless unique species. Analyzing the differences in their DNA can reveal their relationships with one another.

DNA technology applies these facts (and many more) to open entirely new ways to learn about life's history, to prevent and relieve human suffering, to protect the environment, and to enforce the law. As you will see, many of the more familiar applications of DNA technology are in medicine. For example, technicians can test a person's DNA for many alleles associated with inherited illnesses, marking a huge advance in disease screening and diagnosis. Stem cells often make headlines as well, especially because the ability to manipulate gene expression in these cells may offer treatments for diseases that currently have no cure.

Many people also know that genetically modified organisms (often abbreviated GMOs) have made their way into the human food supply, mostly in the form of herbicide- and insect-resistant crop plants. Yet another familiar use of DNA technology is DNA profiling, which can help solve crimes and match parents with their offspring.

As helpful as DNA technology can be, the ability to manipulate DNA also carries both risks and ethical questions. This chapter describes not only some of the tools and applications but also some of the downsides of DNA technology.

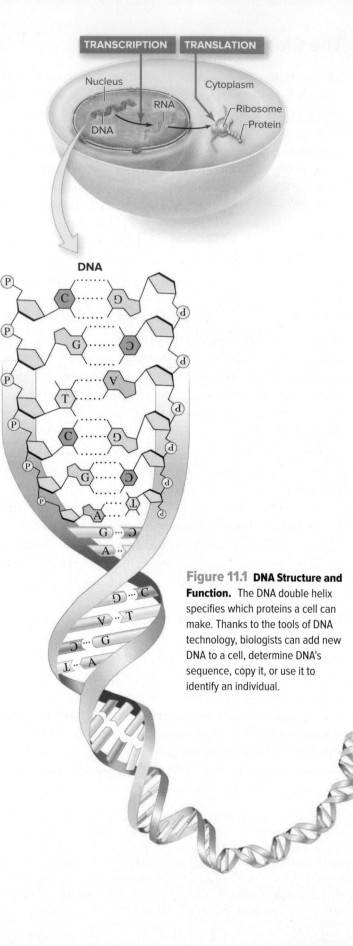

Figure 11.1 DNA Structure and Function. The DNA double helix specifies which proteins a cell can make. Thanks to the tools of DNA technology, biologists can add new DNA to a cell, determine DNA's sequence, copy it, or use it to identify an individual.

11.1 MASTERING CONCEPTS

1. What is DNA technology?

2. In what fields is DNA technology useful?

11.2 DNA Technology's Tools Apply to Individual Genes or Entire Genomes

Some applications of DNA technology require moving a gene from one cell to another; others require comparisons among multiple genomes. This section explores a few of the tools that biologists use to manipulate everything from short stretches of DNA to the entire genetic makeup of a cell.

A. Transgenic Organisms Contain DNA from Other Species

As we saw in chapter 7, virtually all species use the same genetic code. It therefore makes sense that one type of organism can express a gene from another species, even if the two are distantly related. Biologists take advantage of this fact by coaxing cells to take up **recombinant DNA,** which is genetic material that has been spliced together from multiple sources (typically different species). A **transgenic** organism is an individual that receives recombinant DNA.

Scientists first accomplished this feat of "genetic engineering" in *E. coli* in the 1970s, but many microbes, plants, and animals have since been genetically modified. When cells containing the recombinant DNA divide, all of their daughter cells also harbor the new genes. These transgenic organisms express their new genes just as they do their own, producing the desired protein along with all of the others that they normally make. (Note that new varieties of animals and plants may also come from selective breeding. These organisms, however, are not transgenic, as described in this section's Burning Question.) ⓘ *E. coli,* section 17.3

Transgenic Bacteria and Yeasts   How do scientists produce a transgenic organism? The first step is to obtain DNA from a source cell—usually a bacterium, a plant, or an animal (**figure 11.2**, step 1). The researcher may synthesize the DNA in the laboratory or extract it directly from the source cell. Extracting the DNA poses a problem, however, if the gene's source is a eukaryotic cell and the recipient will be a bacterium. Bacterial cells cannot remove introns from mRNA, so the DNA would encode a defective protein in bacteria.

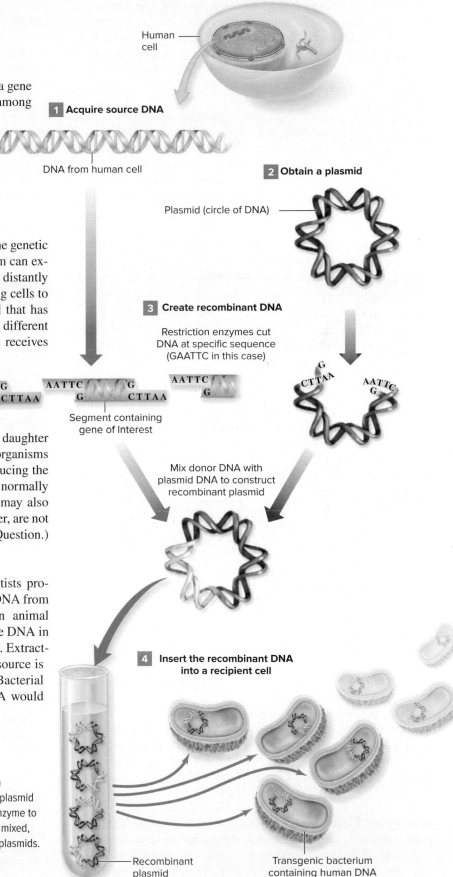

1 Acquire source DNA

DNA from human cell

Human cell

2 Obtain a plasmid

Plasmid (circle of DNA)

3 Create recombinant DNA

Restriction enzymes cut DNA at specific sequence (GAATTC in this case)

Segment containing gene of Interest

Mix donor DNA with plasmid DNA to construct recombinant plasmid

4 Insert the recombinant DNA into a recipient cell

Recombinant plasmid

Transgenic bacterium containing human DNA

Figure 11.2 Transgenic Bacteria. The first steps in creating a transgenic bacterium are to (*1*) isolate source DNA and (*2*) select a plasmid or other cloning vector. (*3*) Researchers use the same restriction enzyme to cut DNA from the donor cell and the plasmid. When the pieces are mixed, the "sticky ends" of the DNA fragments join, forming recombinant plasmids. (*4*) The plasmids are delivered into bacterial cells, which pass the recombinant DNA to their descendants as they reproduce.

Researchers therefore first isolate a mature mRNA molecule with the introns already removed. Then, they use an enzyme called **reverse transcriptase** to make a DNA copy of the mRNA. (As described in chapter 16, retroviruses such as HIV use this enzyme when they infect cells.) The resulting complementary DNA, or cDNA, encodes the eukaryotic protein but leaves out the introns.

Next, the researcher chooses a **cloning vector,** a self-replicating genetic structure that will carry the source DNA into the recipient cell. (In molecular biology, "cloning" means to make many identical copies of a DNA sequence.) A common type of cloning vector is a **plasmid,** which is a small circle of double-stranded DNA separate from the cell's chromosome (figure 11.2, step 2). Viruses are also used as vectors. They are altered so that they transport DNA but cannot cause disease.

The next step is to construct a recombinant plasmid (figure 11.2, step 3). To generate DNA fragments that can be spliced together, researchers use **restriction enzymes,** which are proteins that cut double-stranded DNA at a specific base sequence. Some restriction enzymes generate single-stranded ends that "stick" to each other by complementary base pairing. The natural function of restriction enzymes is to protect bacteria by cutting up DNA from infecting viruses. Biologists, however, use them to cut and paste segments of DNA from different sources. When plasmid and donor DNA is cut with the same restriction enzyme and the fragments are mixed, the single-stranded, sticky ends of some plasmids form base pairs with those of the donor DNA. Another enzyme, DNA ligase, seals the segments together.

Next, the researchers move the cloning vector with its recombinant DNA into a recipient cell (figure 11.2, step 4). Zapping a bacterial cell with electricity opens temporary holes that admit naked DNA. Alternatively, "gene guns" shoot DNA-coated pellets directly into cells. DNA can also be packaged inside a fatty bubble called a liposome, which fuses with the recipient cell's membrane, or it can be hitched to a virus that subsequently infects the recipient cell.

In the pharmaceutical industry, transgenic bacteria produce dozens of drugs, including human insulin to treat diabetes, blood-clotting factors to treat hemophilia, immune system biochemicals, and fertility hormones. Other genetically modified bacteria produce the amino acid phenylalanine, which is part of the artificial sweetener aspartame. Still others degrade petroleum, pesticides, and other soil pollutants. ⓘ *artificial sweeteners,* section 2.5

Single-celled fungi (yeasts) can also be genetically modified. For example, transgenic yeast cells produce a milk-curdling enzyme called chymosin used by many U.S. cheese producers. The baking, brewing, and wine industries, which rely on yeasts for fermentation, may increasingly use transgenic yeast cells in the future. ⓘ *fermentation,* section 6.8B

Transgenic Plants   One tool for introducing new genes into plant cells is a bacterium called *Agrobacterium tumefaciens* (**figure 11.3**). In nature, these bacteria enter the plant at a wound and inject a plasmid into the host's cells. The plasmid normally encodes proteins that stimulate the infected plant cells to divide rapidly, producing a tumorlike gall, where the bacteria live. (The name of the plasmid, Ti, stands for "tumor inducing.")

Scientists can replace some of the Ti plasmid's own genes with other DNA, such as a gene encoding a protein that confers herbicide resistance (figure 11.3, step 1). They allow the transgenic *Agrobacterium* to inject these recombinant plasmids into plant cells (step 2). All plants that grow from the infected cells should express the new herbicide-resistance gene (step 3). The farmer who plants the crop can therefore spray the field with herbicides, killing weeds without harming the genetically modified plants.

Biologists have used a similar technique to produce corn and cotton varieties that produce their own insecticides (see section 10.10). The insect-killing protein originated in a bacterium called *Bacillus thuringiensis,* abbreviated Bt. Any insect that nibbles on a plant expressing the Bt protein dies. These

Burning Question

Is selective breeding the same as genetic engineering?

USDA/Stephen Ausmus

Simply put, the answer to this question is no. Selective breeding, also called artificial selection, yields new varieties of plants and animals by selecting for or against traits that already occur in a population. For example, suppose that researchers want to develop a new variety of carrots lacking orange pigments. They would allow those rare plants with pale carrots to breed only among themselves. Over many generations, the result would be a line of white carrots. If, instead, only plants with darkly pigmented roots breed, the offspring might include purple or red carrots. Breeders used these selective breeding strategies to develop a rainbow of carrot colors.

Introducing new DNA—genetic engineering—is a totally different way to develop new plant and animal varieties. We have already seen, for instance, that Bt corn plants contain genes that were originally isolated from bacteria (see section 10.10). Likewise, some transgenic bacteria produce insulin and other human proteins, thanks to our ability to transfer DNA from one species to another.

A third technique for developing new varieties of plants and animals is random mutagenesis. Researchers use chemicals or radiation to induce genetic mutations in an organism's DNA. These mutations may cause damage, but they sometimes also bring about interesting new characteristics, such as sweeter fruits or higher yields (see figure 7.23). This technique falls somewhere in the middle of the spectrum between selective breeding and transgenic technology. It does not rely on preexisting mutations, as does selective breeding, but it is much less controlled than transgenic technology.

Submit your burning question to
Marielle.Hoefnagels@mheducation.com

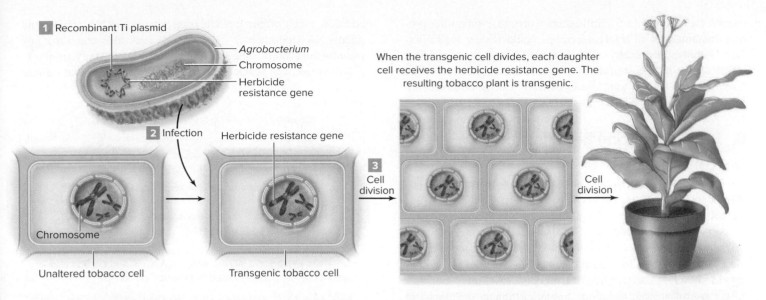

Figure 11.3 Transgenic Plant. (*1*) This genetically modified *Agrobacterium* cell contains a recombinant Ti plasmid encoding a gene that confers herbicide resistance. (*2*) The bacterium infects a tobacco plant cell, inserting the Ti plasmid into the plant cell's DNA. (*3*) The transgenic plant cells can be grown into tobacco plants that express the herbicide-resistance gene in every cell.

genetically modified Bt crops save farmers time and money because they greatly reduce the need for sprayed insecticides.

Besides tolerating herbicides or producing insecticides, transgenic crop plants may also resist viral infections, survive harsh environmental conditions, or contain nutrients that they otherwise wouldn't. Transgenic potato plants, for example, may someday be used to produce vaccines, and tobacco plants enhanced with bacterial enzymes may help degrade leftover explosives at contaminated military installations.

Transgenic Animals

So far, we have described the use of DNA technology to genetically modify bacteria, yeasts, and plants. Biologists use a different technique to produce transgenic mice and other animals. Typically, they pack recombinant DNA into viruses that can infect a gamete or fertilized egg. As the transgenic animal develops, it carries the foreign genes in every cell.

Transgenic animals have many applications. A transgenic mouse "model" for a human gene can reveal how a disease begins, enabling researchers to develop drugs that treat the disease in its early stages. Transgenic farm animals can secrete human proteins in their milk or semen, yielding abundant, pure supplies of otherwise rare substances that are useful as drugs. Transgenic salmon have been engineered for rapid growth, so the fish can reach a marketable size faster than their wild relatives. On a more whimsical note, a glow-in-the-dark zebra fish was the first genetically modified house pet (**figure 11.4**).

Ethical Issues

Although transgenic organisms have many practical uses, some people question whether their benefits outweigh their potential dangers. Some object to the "unnatural" practice of combining genes from organisms that would never breed in nature. Others fear that ecological disaster could result if genetically modified organisms spread their new genes to wild

Figure 11.4 Transgenic Animal. Glow-in-the-dark zebra fish have been genetically altered to produce a fluorescent protein.

(normal): ©Andrew Ilyasov/E+/ Getty Images RF; (altered): ©Edward Kinsman/Science Source

species (see section 11.5). Still others worry that unfamiliar protein combinations in transgenic crops could trigger food allergies. Finally, genetically modified seeds may be expensive, reflecting the high cost of developing and testing the plants. The farmers who stand to gain the most from transgenic plants are often unable to afford them.

B. DNA Sequencing Reveals the Order of Bases

Scientists often want to know the nucleotide sequences of genes, chromosomes, or entire genomes. Researchers can use DNA sequence information to predict protein sequences, as described in chapter 7, or they can compare DNA sequences among species to determine evolutionary relationships. How do investigators get the DNA sequence information they need?

Modern DNA sequencing instruments use a highly automated version of a basic technique Frederick Sanger developed in 1977 (**figure 11.5**). Step 1 in the figure shows the components of the reaction mixture. The DNA polymerase enzyme generates a series of DNA fragments that are complementary to the DNA being sequenced. The reaction mixture also contains short, single-stranded pieces of DNA called primers, which are required by DNA polymerase to begin replication. Also included are normal nucleotides, supplemented with low concentrations of specially modified "terminator" nucleotides tagged with fluorescent labels. Each time DNA polymerase incorporates one of these modified nucleotides instead of a normal one, the new DNA chain stops growing. (i) *DNA replication,* section 8.2

Step 2 in figure 11.5 shows the products of the replication reactions: a group of fragments that differ in length from one another by one end base. Once a collection of such pieces is generated, a technique called **electrophoresis** separates the fragments by size (step 3). The researcher can deduce the sequence by "reading" the fragments from smallest to largest (step 4). The data appear as a sequential readout of the wavelengths of the fluorescence from the labels.

The most famous application of DNA sequencing technology has been the Human Genome Project. This worldwide effort was aimed at sequencing all 3.2 billion base pairs of the human genome. The sequence, which was completed in 2003, revealed unexpected complexities. Although our genome includes approximately 25,000 protein-encoding genes, our cells can produce some 400,000 different proteins. Furthermore, only about 1.5% of the human genome sequence actually encodes protein.

How can so few genes specify so many proteins? Part of the answer lies in introns. By removing different combinations of introns from an mRNA molecule, a cell can produce several proteins from one gene—a departure from the old idea that each gene encodes exactly one protein. So far, no one understands exactly how a cell "decides" which introns to remove. (i) *introns,* section 7.4B

And what is the function of the 98.5% of our genome that does not encode proteins? Some of it consists of regulatory sequences, such as the enhancers that control gene expression. In addition, much of our DNA is transcribed to rRNA, tRNA, and microRNA (see chapter 7). Chromosomes also contain many **pseudogenes,** noncoding DNA sequences that are very similar to protein-encoding genes and that are transcribed but whose

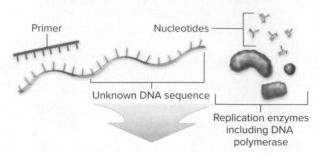

1 Each solution contains the unknown DNA sequence, replication enzymes, primers, normal nucleotides (**A**, **C**, **T**, and **G**), and a small amount of one type of labeled "terminator" nucleotide (A*, C*, T*, or G*).

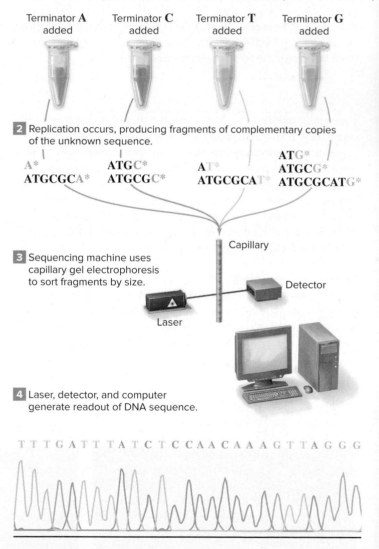

Figure 11.5 DNA Sequencing. The DNA polymerase enzyme makes complementary copies of an unknown DNA sequence. But the copies are terminated early, thanks to chemically modified "terminator" nucleotides in the reaction mix. Sorting the fragments by size reveals the sequence.

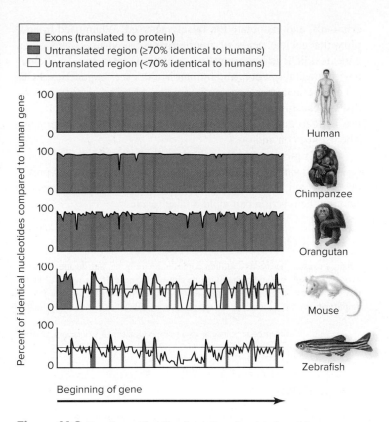

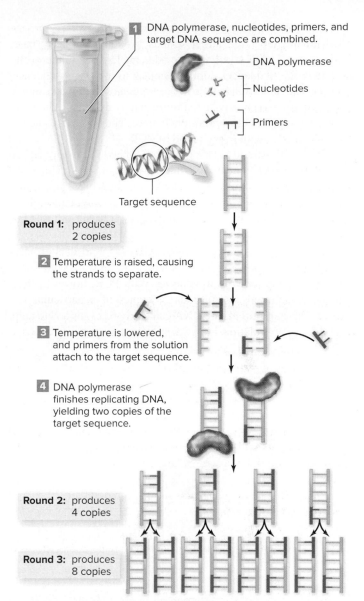

Figure 11.6 **One Gene, Five Species.** At each point along this gene fragment, researchers examined the 100 surrounding nucleotides and calculated the percent that are identical to the human version.

Figure 11.7 **Polymerase Chain Reaction.** A heat-stable DNA polymerase enzyme uses primers and plenty of nucleotides to produce millions of copies of a target DNA sequence.

mRNA is not translated into protein. Pseudogenes may be remnants of old genes that once functioned in our nonhuman ancestors; eventually, they mutated too far from the normal sequence to encode a working protein.

The human genome is also riddled with highly repetitive sequences that have no known function. The most abundant types of repeats are transposons (see section 7.7B), DNA sequences that can "jump" within the genome. Transposons make up about 45% of human DNA. The genome also contains many tandem repeats (or "satellite DNAs"). These sequences consist of one or more bases repeated many times without interruption, such as CACACA or ATTCGATTCG. The exact number of repeats varies from person to person. As described in section 11.2D, DNA profiling technology measures variation in these areas.

Researchers are comparing the human genome to the DNA sequences of dozens of other species, from bacteria and archaea to protists, fungi, plants, and other animals (**figure 11.6**). The similarities and differences have yielded insights into the sequences that unite all life and those that make each species unique.

C. PCR Replicates DNA in a Test Tube

The **polymerase chain reaction (PCR)** rapidly produces millions of copies of a selected DNA sequence in a test tube. Thanks to this extremely powerful and useful tool, trace amounts of DNA extracted from a single hair follicle or a few skin cells left at a crime scene can yield enough genetic material to reveal a person's unique DNA profile.

PCR borrows heavily from a cell's DNA copying machinery. As illustrated in step 1 of **figure 11.7**, a PCR reaction tube includes the target DNA sequence to be replicated, DNA polymerase enzymes, a supply of the four types of DNA nucleotides, and two types of short, laboratory-made primers that are complementary to opposite ends of the target sequence.

PCR occurs in an automated device called a thermal cycler, which controls key temperature changes. The reaction begins when heat separates the two strands of the target DNA (figure 11.7, step 2). Next, the temperature is lowered, and the short primers attach to the separated target strands by complementary base pairing (step 3). DNA polymerase adds nucleotides to the primers and builds sequences complementary to the target sequence (step 4). The new strands then act as templates in the next round of replication, which is initiated immediately by raising the temperature to separate the strands once more. The number of pieces of DNA doubles with every round of PCR.

The double-stranded DNA molecules become "unzipped" at about 95°C, which is nearly the boiling point of water. This temperature would denature most DNA polymerase enzymes, but PCR uses a heat-tolerant variety such as *Taq* polymerase. This enzyme is produced by *Thermus aquaticus,* a bacterium that inhabits hot springs.

Since its invention in the 1980s, PCR has found an enormous variety of applications. Forensic scientists often work with DNA samples that are too tiny to analyze. With PCR, however, they can make thousands or millions of copies of a particular sequence. Once amplified, the DNA can easily be examined to help establish family relationships, identify human remains, convict criminals, and exonerate the falsely accused. When used to amplify the nucleic acids of microorganisms, viruses, and other parasites, PCR is important in agriculture, veterinary medicine, environmental science, and human health care. In genetics, PCR is both a crucial basic research tool and a way to identify known disease-causing genes in a cell's genome. Evolutionary biologists use PCR to amplify DNA from long-dead plants and animals. The list goes on and on.

PCR's greatest weakness, ironically, is its extreme sensitivity. A blood sample contaminated by leftover DNA from a previous PCR run or by a stray eyelash dropped from the person running the reaction can yield a false result.

Figure It Out

Suppose a researcher needs a million copies of a viral gene. She decides to use PCR on a sample of fluid containing one copy of the gene. If one round of PCR takes 2 minutes, how long will it take the researcher to obtain her million-fold amplification?

Answer: Producing 1 million copies would take 20 rounds, or 40 minutes.

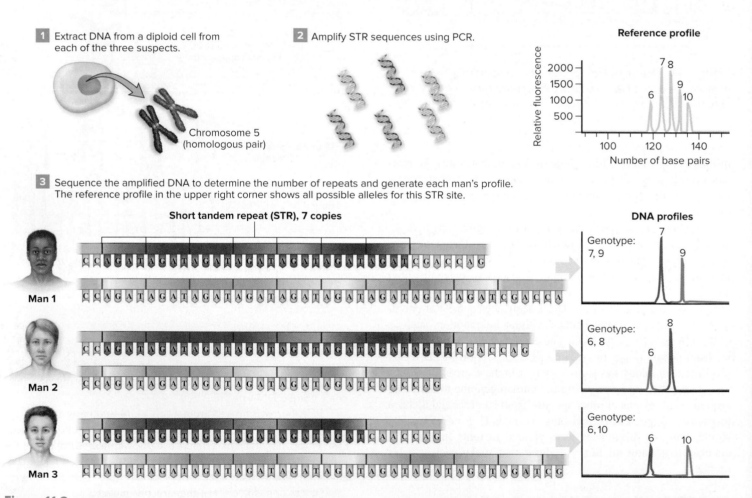

Figure 11.8 **DNA Profiling.** The human genome contains regions of short tandem repeats (STRs) that are genetically variable. DNA profiling techniques detect differences in the number of repeats at multiple STRs; this figure illustrates one STR site. (*1*) DNA extracted from cells of each man is (*2*) amplified using PCR. (*3*) Sequencing the DNA reveals each man's pattern at this STR site. Figure 11.9 shows complete profiles for all three men.

D. DNA Profiling Detects Genetic Differences

On average, each person's DNA sequence differs from that of a nonrelative by just 1 nucleotide out of 1000. Finding these small differences by sequencing and comparing entire genomes would be time-consuming, costly, and impractical. Instead, **DNA profiling** uses just the most variable parts of the genome to detect genetic differences between individuals.

The most common approach to DNA profiling is to examine **short tandem repeats (STRs)**, which are sequences of a few nucleotides that are repeated in noncoding regions of DNA. People within a population have different numbers of these repeats. **Figure 11.8** shows an STR site on chromosome 5 for three men. The first man has seven copies of the STR on the chromosome that he inherited from one of his parents; he has nine copies on his other chromosome. His genotype for this STR is therefore "7, 9." The other two men have different genotypes.

To generate a DNA profile, a technician extracts DNA from a person's cells (figure 11.8, step 1) and uses PCR to amplify the DNA at each of 13 STR sites, leaving the rest of the DNA alone (step 2). A fluorescent label is incorporated into the DNA at the STR sites during the PCR reaction. The technician can then use electrophoresis and a fluorescence imaging system to determine the number of repeats at each site (step 3).

Statistical analysis plays a large role in DNA profiling. For example, suppose that DNA extracted from a hair found on a murder victim's body matches DNA from a suspect at all 13 STR sites. What is the probability that the matching DNA patterns come from the same person—the suspect—rather than from two individuals who happen to share the same DNA sequences? To find out, investigators consult databases that compile the frequency of each STR variant in the population. The statistical analysis suggests that the probability that any two unrelated individuals have the same pattern at all 13 STR markers is 1 in 250 trillion.

Conversely, a suspect can use dissimilar DNA profiles as evidence of his or her innocence. Since 1989, DNA analysis of stored evidence has proved the innocence of more than 300 people serving time in prison for violent crimes they did not commit (figure 11.9).

In addition to STRs in nuclear DNA, analysis of mitochondrial DNA is also sometimes useful. Mitochondrial DNA is typically only about 16,500 base pairs long, far shorter than the billions of nucleotides in nuclear DNA. But because each cell contains multiple mitochondria, each of which contains many DNA molecules, mitochondria can often yield useful information even when nuclear DNA is badly degraded. Investigators extract mitochondrial DNA from hair follicles, bones, and teeth, then use PCR to amplify the variable regions for sequencing. (i) *mitochondria,* section 3.4C

Because everyone inherits mitochondria only from his or her mother, this technique cannot distinguish between siblings. It is very useful, however, for verifying the relationship between woman and child. For example, children who were kidnapped during infancy can be matched to their biological mothers or grandmothers. The study of human evolution also has benefited from mitochondrial DNA analysis, which has revealed the genetic relationships among subpopulations from around the world.

STR Locus		Genotypes			
Chromosome (Locus name)	Number of possible repeats	DNA from crime scene	Man 1 	Man 2 	Man 3 
2 (TPOX)	4–16	10, 11	9, 10	10, 11	8, 8
3 (D3S1358)	8–20	17, 19	19, 19	17, 19	15, 16
4 (FGA)	13–51	20, 24	22, 24	20, 24	21, 23
5 (D5S818)	**6–18**	**6, 8**	**7, 9**	**6, 8**	**6, 10**
5 (CSF1PO)	5–16	7, 10	10, 11	7, 10	6, 12
7 (D7S820)	5–16	13, 14	8, 12	13, 14	9, 10
8 (D8S1179)	7–20	11, 12	13, 14	11, 12	13, 13
11 (TH01)	3–14	7, 7	9, 12	7, 7	8, 9
12 (vWA)	10–25	15, 12	17, 19	15, 12	18, 19
13 (D13S317)	5–17	10, 10	14, 14	10, 10	9, 13
16 (D16S539)	4–16	9, 13	10, 11	9, 13	9, 9
18 (D18S51)	7–40	15, 21	20, 21	15, 21	14, 18
21 (D21S11)	12–41	30, 32	16, 18	30, 32	29, 30

a.

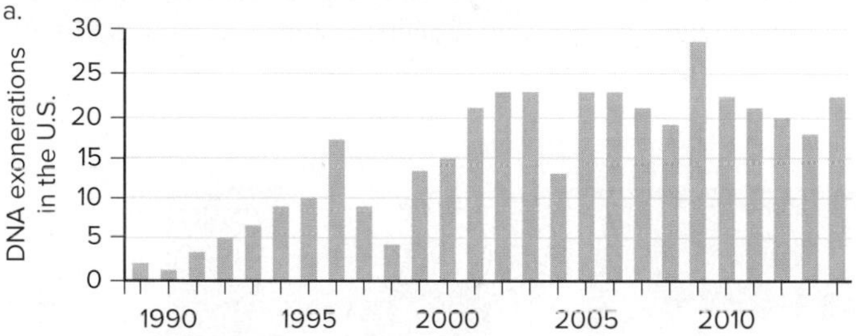

b.

Figure 11.9 **Guilt and Innocence.** (a) DNA evidence collected at a crime scene is compared to the DNA of three suspects. Man 2's profile matches the evidence from the crime scene. Note that the STR site highlighted in red is featured in figure 11.8. (b) In the United States, DNA evidence has exonerated hundreds of prisoners who were serving time for crimes they did not commit. (Data are from the National Registry of Exonerations.)

11.2 MASTERING CONCEPTS

1. What are some uses for transgenic organisms?
2. What are the steps in producing a transgenic organism?
3. How do researchers determine a sequence of DNA?
4. What is the function of the 98.5% of the human genome that does not encode protein?
5. How does PCR work, and why is it useful?
6. How are short tandem repeats used in DNA profiling?
7. Why do investigators sometimes analyze mitochondrial DNA instead of nuclear DNA?

11.3 Stem Cells and Cloning Add New Ways to Copy Cells and Organisms

The public debate over stem cells and cloning combines science, philosophy, religion, and politics in ways that few other modern issues do (figure 11.10). What is the biology behind the headlines?

A. Stem Cells Divide to Form Multiple Cell Types

A human develops from a single fertilized egg into an embryo and then a fetus—and eventually into an infant, a child, and an adult—thanks to mitotic cell division. As development continues, more and more cells become permanently specialized into muscle, skin, liver, brain, and other cell types. All contain the same DNA, but some genes become irreversibly "turned off" in specialized cells. Once committed to a fate, a mature cell rarely reverts to another type. ⓘ *regulation of gene expression,* section 7.6

Animal development therefore relies on stem cells. In general, a stem cell is any undifferentiated cell that can give rise to specialized cell types. When a stem cell divides mitotically to yield two daughter cells, one remains a stem cell, able to divide again. The other specializes.

a.

b.

Figure 11.10 **Stem Cell Controversy.** Debates over stem cells often pit (a) people such as actor Michael J. Fox who advocate the use of embryonic stem cells in medicine against (b) people with moral objections.

(a): ©Scott J. Ferrell/Congressional Quarterly/Getty Images; (b): ©Getty Images

Animals have two general categories of stem cells: embryonic and adult (figure 11.11). **Embryonic stem cells** give rise to all cell types in the body (including adult stem cells) and are therefore called "totipotent"; *toti-* comes from the Latin word for "entire." **Adult stem cells** are more differentiated and produce a limited subset of cell types. For example, stem cells in the skin replace cells lost through wear and tear, and stem cells in the bone marrow produce all of the cell types that make up blood. Adult stem cells are "pluripotent"; *pluri-* means "many" in Latin.

Stem cells are important in biological and medical research. With the correct combination of chemical signals, medical researchers should theoretically be able to coax stem cells to divide in the laboratory and produce blood cells, neurons, or any other cell type. Many people believe that stem cells hold special promise as treatments for neurological disorders such as Parkinson disease and spinal cord injuries, since neurons ordinarily do not divide to replace injured or diseased tissue. Stem cell therapies may also conquer blood cancers, eye problems, diabetes, heart disease, and many other illnesses that are currently incurable.

The practical benefits would extend beyond treating illness. Currently, pharmaceutical companies test new drugs primarily on whole organisms, such as mice and rats. The ability to test on just kidney or brain cells, for example, would allow researchers to better predict the likely side effects of a new drug. It might also reduce the need for laboratory animals.

Both embryonic and adult stem cells have advantages and disadvantages for medical use. Embryonic stem cells are extremely versatile, but a patient's immune system would probably reject tissues derived from another individual's cells. In addition, research on embryonic stem cells is controversial because of their origin. In fertility clinics, technicians fertilize eggs *in vitro,* and only a few of the resulting embryos are ever implanted into a woman's uterus. Researchers destroy some of the "spare" embryos at about 5 days old to harvest the stem cells. (The other embryos are either stored for possible later implantation or discarded.) Many people consider it unethical to use human embryos in research, even if those embryos would otherwise have been thrown away.

Biologists are also investigating adult stem cells in hair follicles, bone marrow, the lining of the small intestine, and other locations in the body. A patient's immune system would not reject tissues derived from his or her own adult stem cells. These stem cells are less abundant than embryonic stem cells, however, and they usually give rise to only some cell types.

New techniques may eliminate some of these drawbacks. Researchers have discovered how to induce adult cells to behave as embryonic stem cells. This technique could allow differentiated cells taken from an adult to be turned into stem cells, which could then be coaxed to develop into any other cell type. Time will tell how useful these "induced pluripotent stem cells" will be or whether they will match the medical potential of embryonic stem cells.

B. Cloning Produces Identical Copies of an Organism

Cloning simply means asexual reproduction. Although most plants, fungi, and animals reproduce sexually, at least some organisms in each kingdom also use asexual reproduction. Humans have exploited the power of cloning by producing countless identical copies of desirable species (see chapter 8).

Plants are especially easy to clone (**figure 11.12**). Commercial plant growers use a technique called tissue culture to clone carnivorous plants and many other species in petri dishes or small jars. But anyone can clone a favorite houseplant by placing a portion of a leaf or stem in water, allowing roots to develop, and transplanting the young plant into a pot of soil.

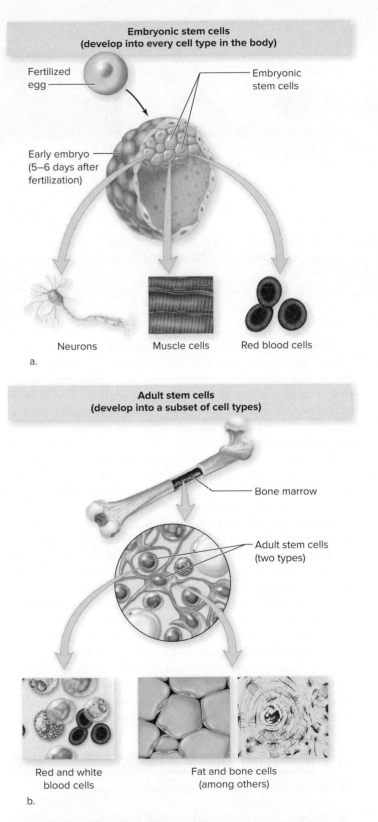

a.

b.

Figure 11.11 Stem Cells. (a) Human embryonic stem cells are derived from a ball of cells that forms several days after fertilization. These stem cells give rise to all of the body's cell types. (b) The adult body also contains stem cells, but they may not have the potential to develop into as many different cell types as do embryonic stem cells.

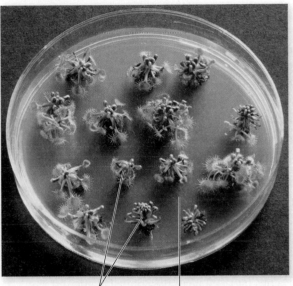

a.

b.

Figure 11.12 Cloning in a Dish. (a) When plant tissue is cultured with the correct combination of hormones and nutrients, it gives rise to genetically identical plantlets. This petri dish contains clusters of carnivorous sundew plantlets. (b) Plantlets are transferred to jars providing hormones, nutrients, and room to grow.

(a): ©Rosenfeld Images Ltd/Science Source; (b): © Philippe Psaila/Science Source

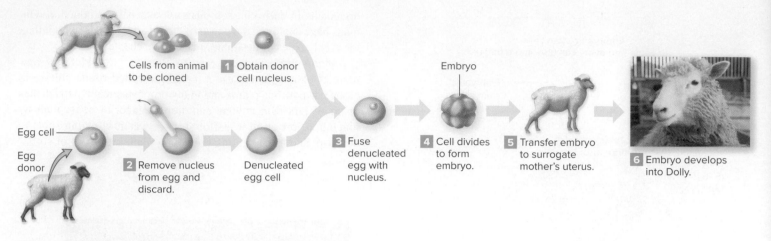

Figure 11.13 Making Dolly. The first steps in cloning an adult female sheep were to (*1*) obtain a nucleus from a cell of the ewe's udder and (*2*) remove the nucleus from an egg cell. (*3*) Placing the adult cell's nucleus into the egg yielded a new cell genetically identical to the DNA donor. (*4*) This cell developed into an embryo, which (*5*) was implanted into a surrogate mother sheep and (*6*) eventually was born as Dolly.

Photo: ©Paul Clements/AP Images

Unlike many other organisms, mammals do not naturally clone themselves. In 1996, however, researcher Ian Wilmut and his colleagues in Scotland used a new procedure to produce Dolly the sheep, the first clone of an adult mammal. The researchers used a cloning technique called **somatic cell nuclear transfer.** A *somatic cell* is any cell that makes up the animal's body, other than stem cells or gamete-forming cells. To clone an animal, a nucleus is moved from a DNA donor's somatic cell into an egg cell without a nucleus. This egg, now containing a full set of DNA, develops into a clone of the original donor.

Figure 11.13 illustrates the cloning technique in more detail. First, the researchers obtained the nucleus from a cell removed from a donor sheep's mammary gland (step 1). They then transferred this "donor" nucleus to a sheep's egg cell whose own nucleus had been removed (steps 2 and 3). The resulting cell divided mitotically to form an embryo (step 4), which the researchers implanted in a surrogate mother's uterus (step 5). The embryo then developed into a lamb, named Dolly (step 6).

Scientifically, this achievement was remarkable because it showed that the DNA from a differentiated somatic cell could "turn back the clock" and revert to an undifferentiated state. Mitotic cell division then produced every cell in Dolly's body.

Dolly appeared normal, and she gave birth to six healthy lambs via sexual reproduction. But she had arthritis in her hind legs, and she died of lung disease in 2003 at age 6. Her early death fueled speculation that clones inevitably have abnormally short lives. However, other sheep cloned from the same cell line as Dolly have had normal life spans.

Since Dolly's birth, researchers have used somatic cell nuclear transfer to clone other mammals as well, including dogs, cats, mice, cattle, goats, and a champion horse that had been castrated (and therefore could not reproduce). Although some of the clones have had shorter-than-average life spans, others have not; the debate over clone longevity therefore continues.

Cloning may even help rescue endangered species or recover extinct species, using preserved DNA implanted into eggs from closely related species. So far, however, researchers have struggled to implement this strategy. One problem is the low success rate for cloning animals in general. Compounding this issue is the fact that scientists know few key details about reproduction in most endangered animals.

Many people wonder whether humans can and should be cloned. Reproductive cloning could help infertile couples to have children. Scientists could also use cloned human embryos as a source of stem cells, which could be used to grow "customized" artificial organs that the patient's immune system would not reject. This application of cloning is called therapeutic cloning.

Despite the potential benefits, however, human cloning carries unresolved ethical questions. For example, most clones die early in development. Even the tiny percentage of clones that make it to birth often have abnormalities. This difficulty emphasizes the ethical issues surrounding human reproductive cloning. In addition, therapeutic cloning still requires the destruction of an embryo to harvest the stem cells. As we have already seen, many people question the practice of making human embryos only to destroy them. Finally, both reproductive and therapeutic cloning require unfertilized human eggs. The removal of eggs from a woman's ovaries is costly and poses medical risks.

11.3 MASTERING CONCEPTS

1. Describe the differences among embryonic, adult, and induced pluripotent stem cells.
2. What are the potential medical benefits of stem cells?
3. Summarize the steps scientists use to clone an adult mammal.
4. Why is the technique used to clone mammals called somatic cell nuclear transfer?

11.4 Many Medical Tests and Procedures Use DNA Technology

The list of human illnesses is long. Worldwide, the top causes of death include heart disease, stroke, cancer, and infection (see table 37.2); all of these diseases have environmental and genetic components. But some ailments, including hemophilia, Tay-Sachs disease, sickle cell disease, and dozens of others, are entirely caused by mutated alleles of single genes. This section describes how DNA technology can help prevent, detect, and treat genetic diseases. Although we use cystic fibrosis as an example, the same techniques are applicable (at least in theory) to any illness associated with a single gene.

A. DNA Probes Detect Specific Sequences

The ability to detect the alleles that cause cystic fibrosis and other genetic illnesses is crucial to the medical applications of DNA technology. At first glance, however, all DNA looks alike: a sequence of A, C, G, and T. With billions of nucleotides in a single cell, how can biologists search through an entire genome to find just the piece they need to "see"?

The answer is a **DNA probe,** a single-stranded sequence of nucleotides that is used to detect a complementary DNA sequence (**figure 11.14**). A typical probe is a short, synthetic strand of DNA that is labeled with either a radioactive isotope or a fluorescent tag. For example, a researcher can construct a probe that is complementary to part of an allele known to be associated with cystic fibrosis. A region of the DNA to be tested is separated into single strands and immobilized on a solid surface. If the DNA contains a nucleotide sequence that is complementary to the probe, then the probe binds to that region. The radioactivity or wavelength emitted by the probe reveals the presence of the cystic fibrosis allele.

B. Preimplantation Genetic Diagnosis Can Screen Embryos for Some Diseases

Imagine a young couple who want a child. Both of the prospective parents know they are symptomless carriers of cystic fibrosis. Can DNA technology help the couple ensure that their baby is free of the disease? Although no one can guarantee a cystic fibrosis–free baby, a technique called **preimplantation genetic diagnosis (PGD)** can screen embryos for the disease-causing allele and therefore greatly reduce the odds of having an affected child.

The process begins with *in vitro* fertilization (literally, fertilization "in glass"), in which the man's sperm fertilize several of the woman's eggs in a laboratory dish. The resulting zygotes develop into embryos, each consisting of eight genetically identical cells. A technician then selects an embryo for PGD. He or she removes one cell from the embryo (**figure 11.15**); the loss of this cell will not affect the embryo's subsequent development.

DNA extracted from that single cell undergoes PCR, amplifying the region of DNA where the cystic fibrosis gene is located. A DNA probe specific for one or more cystic fibrosis alleles can then determine whether the embryo's cells contain the disease-causing DNA sequences.

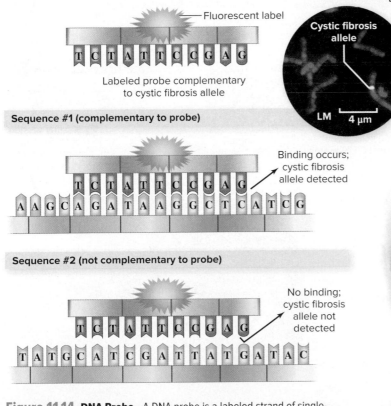

Figure 11.14 DNA Probe. A DNA probe is a labeled strand of single-stranded DNA that binds to a complementary target sequence. In this case, the probe reveals the presence of a cystic fibrosis allele.

Photo: ©Patrick Landmann/Science Source

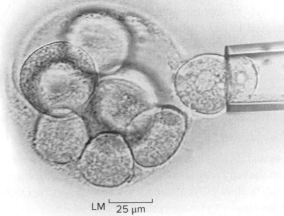

Figure 11.15 Preimplantation Genetic Diagnosis. A single cell from a human embryo is removed for testing.

©Rajau/Science Source

If the allele is detected, the embryo can be discarded, and others can be tested. Any embryo that lacks the disease-causing allele is a good candidate to be placed into the woman's body. If the embryo implants into the uterus and develops into a baby, the child is very likely to be born without cystic fibrosis.

There is a small chance, however, that the baby may be born with the disease despite PGD. Human error is one possible explanation. By amplifying DNA sequences that occur in just one or two copies from a single cell, PGD pushes PCR to its limits. As we have already seen, PCR is extremely sensitive to contamination; stray DNA that is accidentally amplified can lead to a false result. A second explanation relates to the fact that researchers have identified hundreds of mutations that can cause cystic fibrosis. PGD tests for the most common disease-causing alleles, but the baby may have inherited rare variants that the test cannot detect.

C. Genetic Testing Can Detect Existing Diseases

The same genetic tests used in PGD are also useful for testing fetuses, newborns, older children, and adults for disease-causing alleles. Instead of searching the DNA from an embryonic cell, however, the tests detect the alleles in DNA from cells taken from blood, saliva, or body tissues.

For example, newborns are routinely screened for a genetic disorder called phenylketonuria (PKU). Cells from unborn children can also be tested for disease-causing alleles; the parents can use the information to decide whether to terminate the pregnancy or to prepare for life with a special needs child.

Genetic testing has many applications in adults as well. People who suspect they may be heterozygous carriers of cystic fibrosis might choose to be tested for the disease-causing allele before deciding whether to have children. Likewise, a woman with a family history of breast cancer might be tested for damage to a gene called *BRCA1*, which is strongly associated with susceptibility to that disease. A positive test for the mutant allele might prompt the woman to have her breasts surgically removed to prevent the cancer from ever arising. And patients who already have breast cancer often have DNA from their tumors screened for genes encoding estrogen receptors. The results can indicate which treatments might be most promising.

D. Gene Therapy Uses DNA to Treat Disease

Cystic fibrosis and most other genetic illnesses currently have no cure, but **gene therapy** may someday provide new treatment options by adding healthy DNA to a person's cells. The new DNA supplements the function of a faulty gene (**figure 11.16**).

The gene therapy strategy illustrated in figure 11.16 shares some similarities with producing transgenic organisms (see section 11.2A) in that new DNA is introduced into existing cells. But the two techniques are also different in key ways. First, in gene therapy, the healthy gene introduced into a cell is from humans, not another species. Second, a typical transgenic organism can theoretically pass the foreign genes to the next generation, whereas a gene therapy patient would only receive new genes in the cell type that needs correction. Other cell types, including the germ-line cells that produce sperm and egg cells, would be left alone.

One variation on gene therapy is to use DNA to silence a gene whose activity is causing illness. For example, turning off cancer-associated genes might offer promising new cancer treatments. A technique called RNA interference can prevent the translation of mRNA from the cancer-related gene. In this strategy, researchers introduce DNA encoding mRNA that is complementary to the cancer gene's mRNA. If both mRNA molecules are produced at the same time, they bind to each other and form double-stranded RNA that cannot be translated to a protein. The cancer-related gene is therefore silenced.

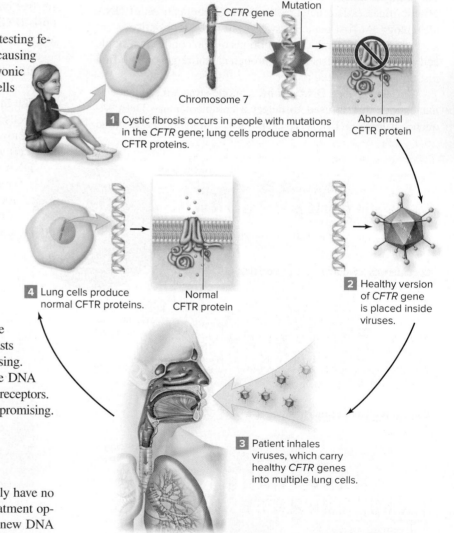

1 Cystic fibrosis occurs in people with mutations in the *CFTR* gene; lung cells produce abnormal CFTR proteins.

2 Healthy version of *CFTR* gene is placed inside viruses.

3 Patient inhales viruses, which carry healthy *CFTR* genes into multiple lung cells.

4 Lung cells produce normal CFTR proteins.

Figure 11.16 Gene Therapy. The overall goal of gene therapy is to supplement a faulty gene with a normal, healthy version. In this example, a genetically modified virus delivers a healthy *CFTR* gene to the lungs of a person with cystic fibrosis.

Gene therapy is challenging for several reasons. The new gene must be delivered directly to only those cells that express the faulty allele. Viruses may be ideal for carrying DNA into target cells, because they typically infect only a limited range of cells. But for gene therapy to be safe, the viruses must not trigger an immune reaction, and the new DNA must not induce mutations that cause cancer. In addition, the gene therapy patient must express the new genes long enough for his or her health to improve.

Gene therapy trials pose significant risks. In 1999, for example, 18-year-old Jesse Gelsinger received a massive infusion of viruses carrying a gene to correct an inborn error of metabolism. He died in days from an overwhelming immune system reaction. Gelsinger's death prompted a halt to several gene therapy studies and led to stricter rules for conducting experiments. Nevertheless, gene therapy research and clinical trials continue, with promising results for diseases including cystic fibrosis, sickle cell disease, and some forms of inherited blindness and immune disorders.

E. Medical Uses of DNA Technology Raise Many Ethical Issues

The use of DNA technology in medicine can prevent or reduce human suffering in many ways: by improving the chance of having healthy children, by detecting diseases early if they do occur, and by offering the prospect of new treatments for illnesses that currently have no cure.

But these techniques also present ethical dilemmas. A thorough treatment of ethics is beyond the scope of this book, but the rest of this section offers a small sampling of some questions that accompany the use of DNA technology in medicine.

In vitro fertilization and preimplantation genetic diagnosis, for example, are costly. Should these techniques be available only to the wealthy? And consider the diagnosis of a genetic disease in an unborn child. A woman who is pregnant with a fetus that carries a genetic abnormality may decide to end the pregnancy rather than carrying the child to term. Does the morality of her decision depend on the severity of the illness? In other words, should we reserve fetal screening for life-threatening illnesses, or is it morally permissible to use it for milder conditions as well? What about using genetic tests to select for or against embryos with traits that do not affect health at all, such as sex or eye color?

Genetic testing in older children and adults may also lead to sticky questions. For example, a genetic test that reveals a high risk for cancer may be beneficial if it leads to lifestyle changes that promote a longer, healthier life. On the other hand, genetic testing can detect alleles associated with diseases for which effective treatments are not yet available. The test results therefore may lead to depression or anxiety without improving the chance of treatment or a cure.

Gene therapy also comes with its share of dilemmas. This new form of treatment currently carries so many risks that its use is extremely limited. Once the technology is perfected,

however, how should it be used? Only for debilitating diseases, or for less serious conditions as well? Is it right to use the techniques of gene therapy to enhance a person's appearance or athletic performance, as described in this section's Apply It Now box? What about using DNA technology to alter the DNA in a person's germline, so that future generations contain the new gene? The answer to this question is not trivial; tinkering with germline DNA could affect our future evolution in unforeseen ways.

11.4 MASTERING CONCEPTS

1. Explain how and why a researcher might use a DNA probe.
2. Compare and contrast preimplantation genetic diagnosis and genetic testing.
3. What is gene therapy?
4. What are some examples of ethical questions raised by the medical use of DNA technology?

→ Apply It **Now**

Gene Doping

©Corbis RF

If we can use genes to cure diseases, it must also be possible to use DNA to make a healthy person even "better." For example, it should be possible to inject genes that make an athlete stronger, faster, or better able to withstand the physical stress of competition.

"Gene doping" is the use of DNA to enhance the function of a healthy person. The techniques would be essentially the same as those used in gene therapy: New genes would be introduced into existing cells, and the proteins encoded by those genes would change the cells' function. The difference is that rather than curing a disease, the goal of gene doping is to give an athlete a competitive edge. An introduced gene might induce the growth of extra muscle, for example. Alternatively, an endurance athlete might use gene doping to boost the production of erythropoietin (EPO), a protein that stimulates red blood cell formation.

For now, technical difficulties and risks have kept gene doping from developing into a practical option for athletes seeking an edge. As the process improves, however, gene doping may become common. That prospect has led many sports organizations to simultaneously ban the practice and seek improved detection methods.

INVESTIGATING LIFE

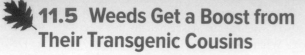

11.5 Weeds Get a Boost from Their Transgenic Cousins

Transgenic potatoes, corn, cotton, and soybeans have been wildly popular with farmers since they first became available in the 1990s. The most familiar transgenic plants either make their own insecticide (see section 10.10) or resist glyphosate, the active ingredient in the herbicide called Roundup (**figure 11.17**). Farmers who use glyphosate-resistant crops can spray their fields with the herbicide, killing weeds but leaving the crops unharmed.

Glyphosate kills most plants by inhibiting a key enzyme required for survival. Without that enzyme, biochemical pathways needed to produce certain amino acids and hormones are blocked. The plant dies. Some transgenic crops, however, contain a "booster" gene that enables the plant to produce 20 times more of this critical enzyme than normal plants. The extra dose of the enzyme allows the plant to survive glyphosate treatment, giving the crops a major advantage over their weedy competitors.

On the other hand, cultivated plants and their weedy (noncultivated) relatives may cross-pollinate. When they do, genes from transgenic plants could move into weed populations, leading to transgenic "superweeds" that farmers could not control with glyphosate. Such weeds would clearly have a fitness advantage in fields where farmers use glyphosate, but what about in other areas? A group of researchers, led by Bao-Rong Lu of Fudan University in Shanghai, wanted to learn more.

To make their transgenic rice, the team needed to cut and paste DNA from multiple sources (see figures 11.2 and 11.3). They isolated rice DNA and used restriction enzymes to cut out the gene encoding the critical enzyme. They combined that gene with a promoter from a different gene in corn plants. That step was important because plants constantly express whatever gene is attached to that promoter. The new promoter–gene combination was inserted into a plasmid and loaded into *Agrobacterium,* which then infected rice cells to produce the transgenic plants. The "always on" booster gene worked like a charm; the transgenic rice plants produced much more of the enzyme than did the original plants.

The researchers next transferred pollen from their transgenic rice to weedy rice, producing two types of hybrid offspring: weedy rice expressing the booster gene and weedy rice not expressing the gene (the control group). They then grew both types of hybrid rice in an herbicide-free environment—conditions the weeds would encounter outside of a farm—and measured the ability of the plants to reproduce. For example, in a laboratory study they monitored the germination of hundreds of seeds from both types of rice. In experimental field plots, they counted the number of seeds produced by dozens of plants.

Their results were clear: Even in the absence of herbicides, weeds with the booster gene had a striking reproductive advantage. They had a higher seed germination rate and produced more seeds than did weeds without the beneficial gene (**figure 11.18**). Plant growth also increased in weeds with the booster gene.

Figure 11.17 Plant-Killing Chemicals. In high concentrations, the herbicide Roundup is toxic to humans when ingested, inhaled, or applied to skin. This farmer is therefore wearing protective clothing while mixing chemicals.

USDA Natural Resources Conservation Service/Tim McCabe

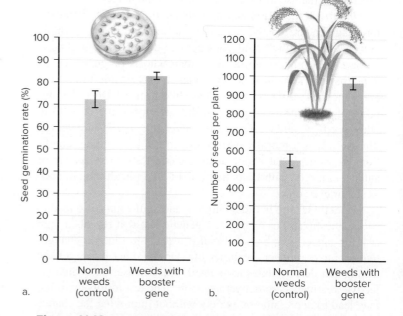

Figure 11.18 Getting Weedier. (a) Weedy rice with the glyphosate booster gene had a higher seed germination rate and (b) produced more seeds than similar plants without the gene. Error bars represent standard errors; see appendix B.

The implications of these findings are unknown, but they could be serious. In the short term, glyphosate-resistant "super-weeds" mean farmers have to use more herbicides (or different ones). Looking into the future, climate change is driving researchers to develop new transgenic crops that resist drought and heat. If these new genes move to other plants, then weeds are likely to become even "weedier." The new plant varieties, which might be difficult to control, could prove to be formidable competitors for the crops that sustain our food supply.

Source: Wang, Wei, and nine coauthors. 2014. A novel 5-enolpyruvoylshikimate-3-phosphate (EPSP) synthase transgene for glyphosate resistance stimulates growth and fecundity in weedy rice (*Oryza sativa*) without herbicide. *New Phytologist,* vol. 202, pages 679–688.

11.5 | MASTERING CONCEPTS

1. How and why did scientists develop the weedy rice expressing the "booster" gene?

2. Why did the researchers conclude that weeds inheriting the "booster" gene had a reproductive advantage?

CHAPTER SUMMARY

11.1 DNA Technology Is Changing the World

- Many disciplines benefit from **DNA technology,** the practical application of knowledge about DNA. **Figure 11.19** summarizes some of the tools and techniques of DNA technology.

11.2 DNA Technology's Tools Apply to Individual Genes or Entire Genomes

A. Transgenic Organisms Contain DNA from Other Species
- **Transgenic** organisms are important in industry, research, and agriculture.
- **Restriction enzymes** and **plasmids** are tools that help researchers construct **recombinant DNA** and introduce it to recipient cells.

B. DNA Sequencing Reveals the Order of Bases
- In DNA sequencing, DNA polymerase incorporates modified nucleotides into a copy of DNA, generating DNA fragments of various lengths. Using **electrophoresis** to sort the fragments by size reveals the DNA sequence.
- Only 1.5% of the 3.2 billion base pairs of the human genome encode protein. The remaining 98.5% of the human genome encodes rRNA, tRNA, microRNA, regulatory sequences, pseudogenes, transposons, and other repeats.

C. PCR Replicates DNA in a Test Tube
- In the **polymerase chain reaction (PCR),** DNA separates into two strands, and DNA polymerase adds complementary nucleotides to each strand. Repeated cycles of heating and cooling allow for rapid amplification of the target DNA sequence.
- PCR finds many applications in research, forensics, medicine, agriculture, and other fields.

D. DNA Profiling Detects Genetic Differences
- Individuals vary genetically in single bases and **short tandem repeats (STRs). DNA profiling** detects these differences.
- Investigators can use known frequencies of alleles in the population to calculate the probability that two DNA samples match purely by chance.
- Analysis of mitochondrial DNA can verify maternal relationships.

Technology or Tool	Definition
Restriction enzyme	Protein that cuts double-stranded DNA at a specific base sequence
Recombinant DNA	Genetic material that has been cut with restriction enzymes and spliced with DNA from other organisms
Transgenic organism	An individual with recombinant DNA 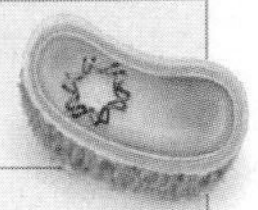
DNA sequencing	Determines the nucleotide sequence of DNA fragments
PCR (polymerase chain reaction)	Amplifies DNA in a test tube using the cell's replication machinery
DNA profiling	Uses DNA sequencing and PCR to detect genetic differences among individuals
Stem cells 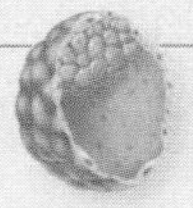	Cells found in embryos and some adult tissues that can give rise to other cell types
Cloning	Makes an identical copy of an organism 
Somatic cell nuclear transfer	A type of cloning that combines a nucleus taken from one individual's body cell with a denucleated egg cell from another individual to produce the first cell of a new organism
DNA probe	A single-stranded sequence of DNA, labeled with a radioactive isotope or fluorescent tag, used to detect the presence of a known sequence of nucleotides
Preimplantation genetic diagnosis	Uses PCR and DNA probes to detect genetic diseases in embryos that might later be implanted in a woman's uterus
Genetic testing	Uses PCR and DNA probes to detect genetic diseases in fetuses, newborns, children, and adults
Gene therapy 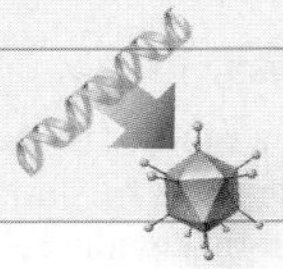	Employs viruses to insert healthy genes into cells

Figure 11.19 Miniglossary of DNA Technology.

11.3 Stem Cells and Cloning Add New Ways to Copy Cells and Organisms

A. Stem Cells Divide to Form Multiple Cell Types
- **Embryonic stem cells** give rise to all cells in the body; **adult stem cells** produce only a limited subset of cell types.
- Induced pluripotent stem cells are adult cells that are converted to stem cells. They may eliminate some ethical issues associated with embryonic stem cells.

B. Cloning Produces Identical Copies of an Organism
- Researchers use a technique called **somatic cell nuclear transfer** to clone adult mammals.
- Human reproductive and therapeutic cloning have potential medical applications, but they also involve ethical dilemmas.

11.4 Many Medical Tests and Procedures Use DNA Technology

A. DNA Probes Detect Specific Sequences
- A **DNA probe** is a single-stranded fragment of DNA that is labeled. The probe binds to any complementary DNA, revealing its location.

B. Preimplantation Genetic Diagnosis Can Screen Embryos for Some Diseases
- In **preimplantation genetic diagnosis (PGD),** a human embryo can be tested for a variety of diseases before being implanted into a woman's uterus.

C. Genetic Testing Can Detect Existing Diseases
- With the help of DNA probes, genetic material extracted from cells of a fetus, a child, or an adult can be tested for the presence of disease-causing alleles.

D. Gene Therapy Uses DNA to Treat Disease
- **Gene therapy** places a functional gene into cells that are expressing a faulty gene.

E. Medical Uses of DNA Technology Raise Many Ethical Issues
- Because of its risks, high expense, and potential to alter human life, DNA technology raises a number of ethical questions.

11.5 Investigating Life: Weeds Get a Boost from Their Transgenic Cousins

- Weedy species of rice with a "booster" gene conferring herbicide resistance produced more seeds and had a higher seed germination rate than did their counterparts without the gene.

MULTIPLE CHOICE QUESTIONS

1. If a restriction enzyme cuts between the G and the A whenever it encounters the sequence GAATTC, how many fragments will be produced when the enzyme is digested with DNA with the following sequence? TGAGAATTCAACTGAATTCAAATTCGAATTCTTAGC
   a. Two   c. Four
   b. Three   d. Five

2. Which of the following is *not* a reason that scientists make transgenic organisms?
   a. To increase the global diversity of organisms
   b. To produce human drugs
   c. To increase the human food supply
   d. To help plants tolerate harsh environments

3. Suppose an investigator at the scene of a murder finds a pair of gloves. Back at the lab, she discovers small fragments of skin and hair on the gloves. What fast test might she do with this evidence to help solve the case?
   a. Remove the nucleus of a skin cell and place it in an egg cell
   b. Extract DNA and determine the genotypes at multiple STR sites
   c. Extract DNA and sequence the genome
   d. Use DNA probes to search for specific gene sequences

4. What is an induced pluripotent stem cell?
   a. A cell from which the nucleus has been removed
   b. A cell extracted from an early embryo
   c. A specially treated somatic cell that can develop into any cell type
   d. A specially treated embryonic stem cell that develops into one specialized cell type

5. A DNA probe with sequence TCAGGCTTCAG would bind most strongly to which of the following DNA fragments?
   a. AGTCCGAAGTC   c. GACTTCGGACT
   b. TCAGGCTTCAG   d. UGAGGCUUGAG

6. Preimplantation genetic diagnosis would be least useful in detecting a _____ disease-causing allele.
   a. dominant
   b. recessive
   c. common
   d. rare

7. What is the role of a virus in gene therapy?
   a. It causes the disease that gene therapy is aiming to cure.
   b. It carries the healthy DNA into the patient's cells.
   c. It carries the faulty DNA out of the patient's cells.
   d. It reveals which cells carry the DNA causing the disease.

Answers to these questions are in appendix A.

WRITE IT OUT

1. What techniques might researchers use to create transgenic bacteria that produce human growth hormone (a drug used to treat extremely short stature)?

2. Transgenic crops often require fewer herbicides and insecticides than conventional crops. In that respect, they could be considered environmentally friendly. Use the Internet to research the question of why some environmental groups oppose transgenic technology.

3. Explain how the ingredients in a PCR reaction tube replicate DNA.

4. Compare and contrast the use of the DNA polymerase enzyme in DNA sequencing and PCR.

5. Why are entire genomes not used for DNA profiling?

6. In a 2013 investigation, researchers discovered that meat packaged as "100% beef" in Europe actually contained traces of horse meat and pork. What DNA technology techniques might the researchers have used to uncover the truth about the origin of the meat?

7. Unneeded genes in an adult animal cell are permanently inactivated, making it impossible for most specialized cells to turn into any other cell type. How does this arrangement save energy inside a cell? Why does the ability to clone an adult mammal depend on techniques for reactivating these "dormant" genes?

8. Scientists are interested in cloning an extinct animal called the gastric brooding frog. This strange frog swallows its eggs and broods its young within its stomach. So far, scientists have successfully used cloning to make an embryo of the frog, but they have yet to raise one to maturity. What steps might the scientists have used to clone this extinct species? Why was it important for scientists to determine before the experiments that the great barred frog is a close relative to the gastric brooding frog?

9. Describe gene therapy, and explain the ethical issues that gene therapy presents.

10. If a cell's genome is analogous to a cookbook and a gene is analogous to a recipe, what is an analogy for preimplantation genetic diagnosis? For gene testing? For gene therapy?

PULL IT TOGETHER

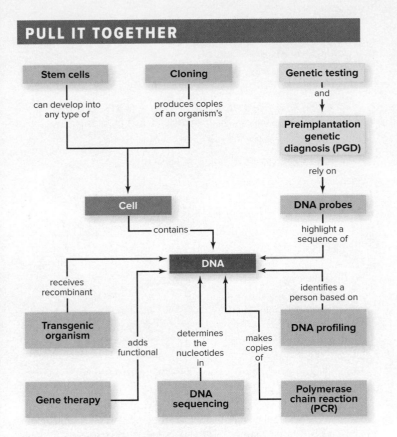

Figure 11.20 Pull It Together: DNA Technology.

Refer to figure 11.20 and the chapter content to answer the following questions.

1. Review the Survey the Landscape figure in the chapter introduction. Given DNA's role in the cell, why do the basic tools of DNA technology summarized in figure 11.20 have applications in such diverse fields of study?

2. How does PCR relate to DNA profiling and preimplantation genetic diagnosis?

3. Add the terms *restriction enzyme, plasmid, virus, DNA polymerase,* and *short tandem repeat* to this concept map.

12 The Forces of Evolutionary Change

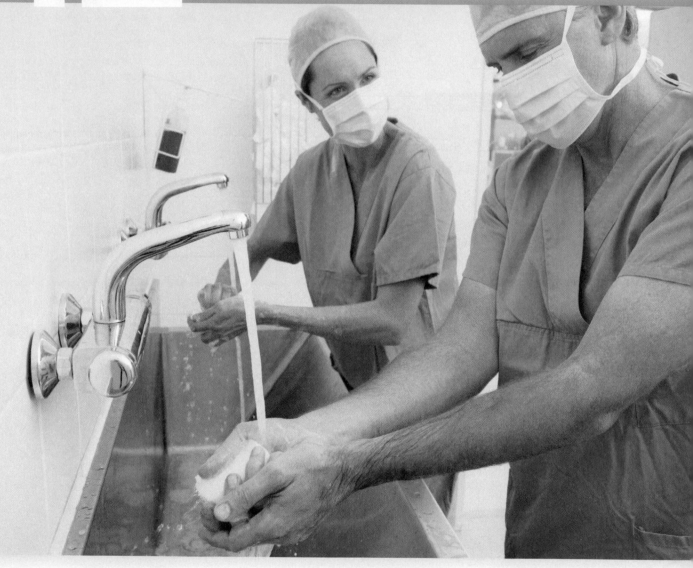

Washing Off Bacteria. Hospitals are breeding grounds for antibiotic-resistant bacteria, which have become less susceptible to drugs in recent decades. Maintaining a bacteria-free environment is especially important during surgery, when a patient's skin is broken and cannot act as a barrier to infection.
©Chris Ryan/Getty Images RF

LEARN HOW TO LEARN
Practice Your Recall

Here's an old-fashioned study tip that still works. When you finish reading a passage, close the book and write what you remember—in your own words. In this chapter, for example, you will learn about several forces of evolutionary change. After you read about them, can you list and describe them without peeking at your book? Try it and find out!

UNIT 3

The Unending War with Bacteria

Do you owe your life to antibiotics? Even if you have never had a serious infection, chances are that one or more of your ancestors did, and antibiotics may have saved their lives. Had it not been for these "wonder drugs," you may never have been born.

Many antibiotics are naturally occurring chemicals. Soil fungi and bacteria secrete these compounds into their surroundings, giving them an edge against microbial competitors. Biologists discovered antibiotics in the early 1900s, but it took decades for chemists to figure out how to mass-produce them. Once that occurred, antibiotics revolutionized medical care in the twentieth century and enabled people to survive many once-deadly bacterial infections.

Unfortunately, the miracle of antibiotics is under threat, and many infections that once were easily treated are reemerging as killers. Ironically, the overuse and misuse of antibiotics are partly responsible for the problem. Physicians sometimes prescribe antibiotics for viral infections, even though the drugs kill only bacteria; moreover, many patients fail to take the drugs as directed. Agricultural practices contribute to the problem as well. Producers of cattle, chickens, and other animals use antibiotics to treat and prevent disease, even adding small amounts to the animals' feed to promote growth. Farmers also spray antibiotics on fruit- and vegetable-producing plants to treat bacterial infections.

By saturating the environment with antibiotics, we have profoundly affected the evolution of bacteria. Microbes that can defeat the drugs are becoming increasingly common, and the explanation is simple. Antibiotics kill susceptible bacteria and leave the resistant ones alone. The survivors multiply, producing a new generation of antibiotic-resistant bacteria. This is an example of natural selection in action, and the public health consequences are both widespread and severe. ⓘ *antibiotics*, section 17.2

Antibiotic-resistant bacteria appeared just 4 years after these drugs entered medical practice in the late 1940s, and researchers responded by discovering new drugs. But the microbes kept pace. Today, two of the biggest threats are *Clostridium difficile* (CDIFF) and carbapenem-resistant Enterobacteriaceae (CRE), both of which are most common in hospital patients. The most dangerous of these so-called superbugs are resistant to all (or nearly all) antibiotics. Researchers fear that the discovery of new drugs will not keep up with the global proliferation of resistant strains.

Natural selection is just one mechanism of evolution, a process that occurs in every species and is obvious in many ways. This chapter explains how evolution occurs.

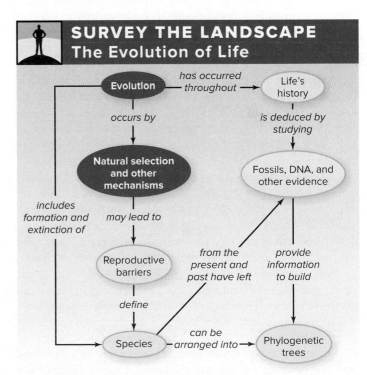

SURVEY THE LANDSCAPE
The Evolution of Life

Evolution occurs in many ways. The most familiar mechanism is natural selection, but genetic changes from one generation to the next also happen by mutation, genetic drift, nonrandom mating, and migration.

For more details, study the Pull It Together feature in the chapter summary.

12.1 Evolution Acts on Populations

Scientific reasoning has profoundly changed thinking about the origin of species. Just 250 years ago, no one knew Earth's age. A century later, scientists learned that Earth is millions of years old (or older), but many believed that a creator made all species in their present form. Today's scientists, relying on a wide range of evidence, accept evolution as the explanation for life's diversity.

But what *is* evolution? A simple definition of **evolution** is descent with modification. "Descent" implies inheritance; "modification" refers to changes in traits from generation to generation. For example, we see evolution at work in the lions, tigers, and leopards that descended from one ancestral cat species.

Evolution has another, more specific, definition as well. Recall from chapter 7 that a gene is a DNA sequence that encodes a protein; in part, an organism's proteins determine its traits. Moreover, each gene can have multiple versions, or alleles. We have also seen that a **population** consists of interbreeding members of the same species (see figure 1.2). Biologists say that evolution occurs in a population when some alleles become more common, and others less common, from one generation to the next. A more precise definition of evolution, then, is genetic change in a population over multiple generations.

According to this definition, evolution is detectable by examining a population's **gene pool**—its entire collection of genes and their alleles. Evolution is a change in **allele frequencies;** an allele's frequency is calculated as the number of copies of that allele, divided by the total number of alleles in the population. Suppose, for example, that a gene has 2 possible alleles, *A* and *a*. In a population of 100 diploid individuals, the gene has 200 alleles. If 160 of those alleles are *a*, then the frequency of *a* is 160/200, or 0.8. In the next generation, *a* may become either more or less common. Because an individual's alleles do not change, *evolution occurs in populations, not in individuals.*

The allele frequencies for each gene determine the characteristics of a population (figure 12.1). Many people in Sweden, for example, have alleles conferring blond hair and blue eyes; a population of Asians would contain many more alleles specifying darker hair and eyes. If Swedes migrate to Asia and interbreed with the locals (or vice versa), regional allele frequencies change.

Some people use the term *microevolution* to refer to the small, generation-by-generation changes occurring in every population or species. This chapter describes the most common ways that such changes occur. Over long periods, these same processes give rise to what is sometimes called *macroevolution,* which includes the appearance of new species (see chapter 14). Evolution does not, however, explain how life began. Chapter 15 describes what science can tell us about life's ancient origin.

12.1 MASTERING CONCEPTS

1. What are two ways to define evolution?
2. Why can't evolution act on individuals?

Figure 12.1 Same Genes, Different Alleles. Human populations originating in different regions of the world have unique allele frequencies. Blond hair and blue eyes are typical of people from northern European countries, whereas people originating on the Asian continent usually have darker coloration.

(woman): ©Stockdisc/PunchStock RF; (man): ©Red Chopsticks/Getty Images RF

12.2 Evolutionary Thought Has Evolved for Centuries

Although Charles Darwin typically receives credit for developing the theory of evolution, people began pondering life's diversity well before his birth. This section offers a brief glimpse into the history of evolutionary thought.

A. Many Explanations Have Been Proposed for Life's Diversity

People have tried to explain the diversity of life for a very long time (figure 12.2). In ancient Greece, Aristotle recognized that all organisms are related in a hierarchy of simple to complex forms, but he believed that all members of a species were created identical to one another. This idea influenced scientific thinking for nearly 2000 years.

Several other ideas were also considered fundamental principles of science well into the 1800s. Among them was the concept of a "special creation," the sudden appearance of organisms on Earth. People believed that this creative event was planned and purposeful, that species were fixed and unchangeable, and that Earth was relatively young.

What About Fossils?   Scientists struggled to reconcile these beliefs with compelling evidence that species could, in fact, change. Fossils, which had been discovered at least as early as 500 BCE, were at first thought to be oddly shaped crystals or faulty attempts at life that arose spontaneously in rocks. But by the mid-1700s, the increasingly obvious connection between living organisms and fossils argued against these ideas.

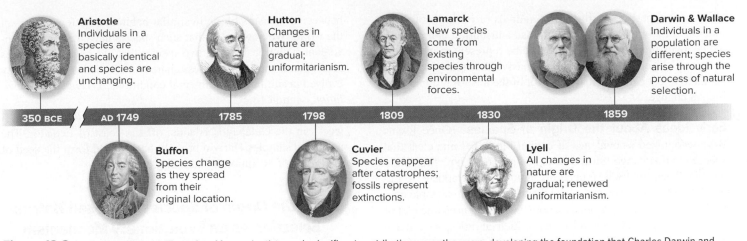

Aristotle Individuals in a species are basically identical and species are unchanging.

Hutton Changes in nature are gradual; uniformitarianism.

Lamarck New species come from existing species through environmental forces.

Darwin & Wallace Individuals in a population are different; species arise through the process of natural selection.

350 BCE   AD 1749   1785   1798   1809   1830   1859

Buffon Species change as they spread from their original location.

Cuvier Species reappear after catastrophes; fossils represent extinctions.

Lyell All changes in nature are gradual; renewed uniformitarianism.

Figure 12.2 Early Evolutionary Thought. Many scientists made significant contributions over the years, developing the foundation that Charles Darwin and Alfred Russel Wallace used to describe natural selection as the mechanism for evolution.

Photos: (Aristotle): ©Science Source; (Buffon, Hutton): ©Getty Images; (Cuvier): ©George Bernard/Science Source; (Lamarck): ©Bettmann/Corbis; (Lyell): ©Corbis; (Darwin): ©Richard Milner Archives; (Wallace): ©Hulton Archive/Getty Images

Scientists used religious stories to explain the existence of fossils without denying the role of a creator. Yet some of the fossils depicted organisms not seen before. Because people believed that species created by God could not become extinct, these fossils presented a paradox. The conflict between ideology and observation widened as geologists discovered that different rock layers revealed different groups of fossilized organisms, many of them now extinct.

New Ideas from Geology

In 1749, French naturalist Georges-Louis Buffon (1707–1788) became one of the first to openly suggest that closely related species arose from a common ancestor and were changing—a radical idea at the time. By moving the discussion into the public arena, he made possible a new consideration of evolution and its causes from a scientific point of view. Still, no one had proposed how species might change.

Meanwhile, in the 1700s and 1800s, much of the study of nature focused on geology. In 1785, physician James Hutton (1726–1797) proposed the theory of **uniformitarianism,** which suggested that the processes of erosion and sedimentation that act in modern times have also occurred in the past, producing profound changes in Earth over time. On the other side, Georges Cuvier (1769–1832) was convinced of **catastrophism,** the theory that a series of brief, violent, global upheavals such as enormous floods, volcanic eruptions, and earthquakes were responsible for most geological formations.

Cuvier also used his knowledge of anatomy to identify fossils and to describe the similarities among organisms. He was the first to recognize the **principle of superposition**—the idea that lower layers of rock (and the fossils they contain) are older than those above them (figure 12.3). Although he had to accept that some species must have become extinct, he refused to believe that they were not originally formed through creation. He argued that catastrophes would destroy most of the organisms in an area, but then new life would arrive from surrounding areas.

Geologist Charles Lyell (1797–1875) renewed the argument for uniformitarianism in 1830. He suggested that natural processes are slow and steady, and that Earth is much older than

Newer rock layers

Older rock layers

Figure 12.3 Earth's History Revealed in Rocks. The Grand Canyon's colorful layers of sedimentary rock originated as sand, mud, and gravel that were deposited in ancient seas. Rock layers sometimes contain fossil evidence of organisms that lived (and died) when the layer was formed. Layers near the bottom are older than those on top.

©Terry Moore/Stocktrek Images/Getty Images RF

6000 years—perhaps millions or hundreds of millions of years old. One obvious conclusion from his contribution is that gradual changes in some organisms could be represented in successive fossil layers. Lyell was so persuasive that many scientists began to reject catastrophism in favor of the idea of gradual geological change.

Early Ideas About the Origin of Species   Once fossils were recognized as evidence of extinct life, it became clear that species could change. Still, no one had proposed how this might happen. Then, in 1809, French taxonomist Jean Baptiste de Lamarck (1744–1829) proposed the first scientifically testable evolutionary theory. He reasoned that organisms that used one part of their body repeatedly would increase their abilities, very much like weight lifters developing strong arms. Conversely, disuse would weaken an organ until it disappeared. Lamarck surmised (incorrectly) that these changes would pass to future generations.

With these new theories and ideas, people were beginning to accept the concept of evolution but did not yet understand how it could result in the formation of new species. Ultimately, Charles Darwin solved this puzzle.

B. Charles Darwin's Voyage Provided a Wealth of Evidence

Charles Darwin (1809–1882) was the grandson of Erasmus Darwin, a noted physician and poet who had anticipated evolutionary theory by writing in 1796 that all animals arose from a single "living filament."

Young Darwin attended Cambridge University in England and, at the urging of his family, completed studies to enter the clergy. Meanwhile, he also followed his own interests. He joined geological field trips and met several eminent geology professors.

Eventually, Darwin was offered a position as a collector and captain's companion aboard the HMS *Beagle*. Before the ship set sail for its 5-year voyage in 1831 (figure 12.4), the botany professor who had arranged Darwin's position gave the young man the first volume of Lyell's *Principles of Geology*. Darwin picked up the second and third volumes in South America. By the time he had finished reading these works, Darwin was an avid proponent of uniformitarianism.

He recorded his observations as the ship journeyed around the coast of South America. He noted forces that uplifted new land, such as earthquakes and volcanoes, and the constant erosion that wore it down. He marveled at forest plant fossils interspersed with sea sediments and at shell fossils in a mountain cave. Darwin tried to reconstruct the past from contemporary observations and wondered how each fossil had arrived where he found it.

He was particularly aware of similarities and differences among organisms. If there had been a single special creation, then why was one sort of animal or plant created to live on a mountaintop in one part of the world, yet another type was on mountains elsewhere? Even more puzzling was the resemblance

between organisms living in similar habitats in different parts of the world. We now know that such species have undergone convergent evolution. That is, two species that live on opposite sides of the planet may nevertheless share characteristics because they evolved in similar environmental conditions. (i) *convergent evolution,* section 13.4C

In the fourth year of the voyage, the HMS *Beagle* spent 5 weeks in the Galápagos Islands, off the coast of Ecuador. The notes and samples Darwin brought back would form the seed of his theory of evolution by natural selection.

C. *On the Origin of Species* Proposed Natural Selection as an Evolutionary Mechanism

Toward the end of the voyage, Darwin began to assimilate all he had seen and recorded. Pondering the great variety of organisms in South America and their relationships to fossils and geology, he began to think that these were clues to how new species originate.

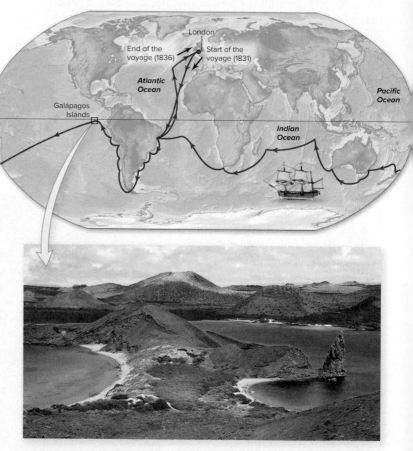

Figure 12.4   The Voyage of the *Beagle*. Darwin observed life and geology throughout the world during the journey of the HMS *Beagle*. Many of Darwin's ideas about natural selection and evolution had their origins in the observations he made on the Galápagos Islands.

Photo: ©David Zurick RF

Descent with Modification

Darwin returned to England in 1836, and by 1837 he had begun assembling his notes in earnest. In March 1837, Darwin consulted ornithologist (bird expert) John Gould about the finches and other birds that the *Beagle* brought back from the Galápagos Islands. Gould could tell from bill structures that some of the finches ate small seeds, whereas others ate large seeds, fruits, or insects. In all, he described 13 distinct types of finch, each different from the birds on the mainland yet sharing some features.

Darwin thought that all of the finch species on the Galápagos had probably descended from a single ancestral type of finch that had flown to the islands and, finding a relatively unoccupied new habitat, flourished. Over the next few million years, the finch population gradually branched in several directions. Different groups ate insects, fruits, and seeds of different sizes, depending on the resources each island offered. Darwin also noted changes in other species, including Galápagos tortoises. He coined the phrase "descent with modification" to describe gradual changes from an ancestral type.

Malthus's Ideas on Populations

In September 1838, Darwin read a work that helped him further understand the diversity of finches on the Galápagos Islands. Economist and theologian Thomas Malthus's *Essay on the Principle of Population*, written 40 years earlier, stated that food availability, disease, and war limit the size of a human population. Wouldn't other organisms face similar limitations? If so, then individuals that could not obtain essential resources would die.

The insight Malthus provided was that individual members of a population were not all the same, as Aristotle had taught. Instead, individuals better able to obtain resources were more likely to survive and reproduce. This would explain the observation that more individuals are produced in a generation than survive; they do not all obtain enough vital resources to live. Over time, environmental challenges would eliminate the more poorly equipped variants, and gradually the population would change.

The Concept of Natural Selection

Darwin used the term *natural selection* to describe "this preservation of favourable variations and the rejection of injurious variations." Biologists later modified the definition to add modern genetics terminology. We now say that **natural selection** occurs when individuals with certain genotypes—those that are best suited to the environment—have greater reproductive success than other individuals.

Darwin got the idea of natural selection from thinking about artificial selection (also called selective breeding). In **artificial selection,** a human chooses one or a few desired traits, such as milk production or leaf size, and then allows only the individuals that best express those qualities to reproduce (**figure 12.5**). Artificial selection is responsible not only for agriculturally important varieties of animals and plants but also for the many breeds of domesticated cats and dogs (see this chapter's Apply It Now box). Darwin himself raised pigeons and developed several new breeds by artificial selection.

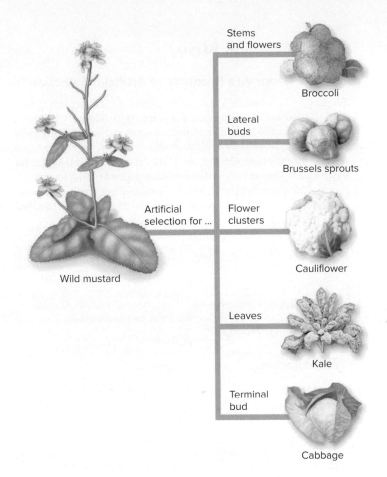

Figure 12.5 **Artificial Selection.** By selecting for different traits, plant breeders used one type of wild mustard to create all five of these vegetable varieties.

How did natural selection apply to the diversity of finch species on the Galápagos? Long ago, some finches flew from the mainland to one island. Eventually, that island population outgrew the supply of small seeds, and birds that could eat nothing else starved. But finches that could eat other things, perhaps because of an inherited difference in bill structure, survived and reproduced. Since their food was plentiful, these once-unusual birds gradually came to make up more of the population.

Darwin further realized that he could extend this idea to multiple islands, each of which had a slightly different habitat and therefore selected for different varieties of finches. A new species might have arisen when a population adapted to so many new conditions that its members could no longer breed with the original group (see chapter 14). In a similar way, new species have evolved throughout the history of life as populations have adapted to different resources. All species are therefore ultimately united by common ancestry.

Apply It **Now**

Dogs Are Products of Artificial Selection

People have been breeding dogs for thousands of years, beginning with domesticated wolves. Dog fanciers now recognize hundreds of breeds, each the product of artificial selection for a different trait that originally occurred as natural genetic variation. Bloodhounds, for example, are selected for their keen sense of smell. Border collies herd livestock (or anything else that moves), and the sleek greyhound is bred for speed.

Behind the carefully bred traits, however, lurk small gene pools and extensive inbreeding, which may harm the health of purebred show animals (table 12.A). Dog breeders can select for desired characteristics, but they can't always avoid the hereditary health problems associated with each breed.

Two examples appear in figure 12.A. Pugs have broad heads and short snouts, which makes them gasp and snort for air when they become excited. Pugs' compressed bodies also may have misaligned vertebrae, sometimes leading to paralysis. At the other extreme, dachshunds have extended torsos. These dogs often suffer from painful degeneration of the discs between the vertebrae.

TABLE **12.A** Purebred Plights

Breed	Health Problem(s)
Cocker spaniel	Nervousness, ear infections, hernias, kidney problems
Collie	Blindness, bald spots, seizures
Dalmatian	Deafness
German shepherd	Hip dysplasia
Golden retriever	Lymphatic cancer, muscular dystrophy, skin allergies, hip dysplasia, absence of one testicle
Great Dane	Heart failure, bone cancer
Labrador retriever	Dwarfism, blindness
Shar-pei	Skin disorders

Pug   Dachshund

Figure 12.A Artificial Selection in Dogs. Whereas the small-statured pug was bred as a lap dog, the short legs and bold demeanor of dachshunds make them excellent hunters of badgers and other den-dwelling animals. These traits originally occurred as natural genetic variation in their wolf ancestors.
(pug): ©Stockbyte/Getty Images RF; (dachshund): ©Punchstock/BananaStock RF

Publication of *On the Origin of Species*   Darwin continued to work on his ideas until 1858, when he received a manuscript from British naturalist Alfred Russel Wallace (1823–1913). Wallace had observed the diverse animals of South America and southeast Asia, and his manuscript independently proposed that natural selection was the driving force of evolution.

Both Darwin's and Wallace's papers were presented at the Linnaean Society meeting later that year. In 1859, Darwin finally published the 490-page *On the Origin of Species by Means of Natural Selection, or Preservation of Favoured Races in the Struggle for Life*. It would form the underpinning of modern life science.

Table 12.1 summarizes Darwin's main arguments in support of natural selection. He began with three observations. First, individuals in a species are different from one another, and at least some of this variation is heritable. Second, essential resources such as food and space are limited in every habitat. Third, in every population, more offspring are born than can survive. These observations led Darwin to infer that organisms engage in a "struggle for existence"—that is, they must compete for scarce resources. He also inferred that those individuals with the most adaptive traits would be most likely to "win" the competition, reproduce, and pass those favorable traits to the next generation. Darwin's final inference was that over many generations, natural selection could change a population or even give rise to new species.

TABLE **12.1** The Logic of Natural Selection: A Summary

Observations of nature
1. **Genetic variation:** Within a species, no two individuals (except identical siblings) are exactly alike. Some of this variation is heritable.
2. **Limited resources:** Every habitat contains limited supplies of the resources required for survival.
3. **Overproduction of offspring:** More individuals are born than survive to reproduce.
Inferences from observations
1. **Struggle for existence:** Individuals compete for the limited resources that enable them to survive.
2. **Unequal reproductive success (natural selection):** The inherited characteristics of some individuals make them more likely to obtain resources, survive, and reproduce.
3. **Descent with modification:** Over many generations, a population's characteristics can change by natural selection, even giving rise to new species.

Some members of the scientific community happily embraced Darwin's efforts. Upon reading *On the Origin of Species,* his friend Thomas Henry Huxley remarked, "How stupid of me not to have thought of that." Others, however, were less appreciative. People in some religious denominations perceived a clash with their beliefs that all life arose from separate special creations, that species did not change, and that nature was harmonious and purposeful. Perhaps most disturbing to many people was the idea that humans were just one more species competing for resources.

D. Evolutionary Theory Continues to Expand

Although Charles Darwin's arguments were fundamentally sound, he could not explain all that he saw. For instance, he did not understand the source of variation within a population, nor did he know how heritable traits were passed from generation to generation. Ironically, Austrian monk Gregor Mendel was solving the puzzle of inheritance at the same time that Darwin was pondering natural selection (see the opening essay for chapter 10). Mendel's work, however, remained obscure until after Darwin's death.

Since Darwin's time, scientists have learned much more about genes, chromosomes, and the origin and inheritance of genetic variation (**figure 12.6**). In the 1930s, scientists finally recognized the connection between natural selection and genetics. They unified these ideas into the **modern evolutionary synthesis,** which suggests that genetic mutations create heritable variation and that this variation is the raw material upon which natural selection acts.

After the discovery of DNA's structure in the 1950s, the picture became even clearer. We now know that mutations are changes in DNA sequence (see chapter 7) and that mutations occur at random in all organisms. Sexual reproduction amplifies this variability by shuffling and reshuffling parental alleles to produce genetically different offspring (see chapter 9).

Today, overwhelming evidence supports the theory of evolution by natural selection; chapter 13 describes some of the data in detail. Contemporary biologists therefore accept evolution as the best explanation for the fact that diverse organisms use the same genetic code, the same chemical reactions to extract energy from nutrients, and many of the same (or very similar) enzymes and other proteins. Descent from a common ancestor explains both this great unity of life and the spectacular diversity of organisms today. Coupled with a wide variety of changing habitats and enormous amounts of time, the result of natural selection is a planet packed with millions of variations of the same underlying biochemical theme.

12.2	**MASTERING CONCEPTS**

1. How does the history of evolutionary thought illustrate the process of science?
2. What did Darwin observe that led him to develop his ideas about the origin of species?
3. How might artificial selection and natural selection produce the same result? Which process would be faster? Why?
4. What is the modern evolutionary synthesis?

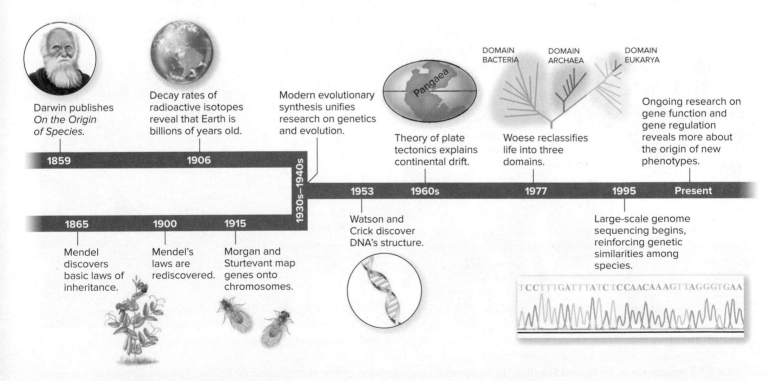

Figure 12.6 Evolutionary Theory Since Darwin. Charles Darwin and Gregor Mendel laid a foundation for evolutionary theory, but thousands of scientists since that time have added to our understanding of evolution.

12.3 Natural Selection Molds Evolution

Natural selection is the most famous, and often the most important, mechanism of evolution. This section explains the basic requirements for natural selection to occur. The rest of this chapter describes natural selection and several other forces of evolutionary change in more detail.

A. Adaptations Enhance Reproductive Success

As Darwin knew, organisms of the same species are different from one another, and every population produces more individuals than resources can support (**figure 12.7**). Some members of any population will not survive to reproduce. A struggle for existence is therefore inevitable.

The variation inherent in each species means that some individuals in each population are better than others at obtaining nutrients and water, avoiding predators, tolerating temperature changes, attracting mates, or reproducing. The heritable traits conferring these advantages are **adaptations**—features that provide a selective advantage because they improve an organism's ability to survive and reproduce.

The word *adaptation* can be confusing because it has multiple meanings. For example, a student might say, "I have adapted well to college life," but short-term changes in an individual do not constitute evolution. Adaptations in the evolutionary sense include only those structures, behaviors, or physiological processes that are heritable and that contribute to reproductive success.

In any population, individuals with the best adaptations are most likely to reproduce and pass their advantage to their offspring. Because of this "differential reproductive success," a population changes over time, with the best available adaptations to the existing environment becoming more common with each generation (**figure 12.8**).

Natural selection requires preexisting genetic diversity. Ultimately, this diversity arises largely by chance. Nevertheless, it is important to realize that natural selection itself is not a random process. Instead, it selectively eliminates most of the individuals that are least able to compete for resources or cope with the prevailing environment.

One additional important note is that not all variation within a population is subject to natural selection. Some features are "selectively neutral," meaning that they neither increase nor decrease reproductive success. As a simple example, the shape of a person's earlobes has little to do with how many offspring he or she produces. Earlobe shape differences therefore represent neutral variation in the human population. Other types of neutral variation do not change an individual's characteristics at all. For example, a mutation may occur in a stretch of DNA that does not encode a protein. Even if a mutation occurs in a coding region of DNA, the protein's amino acid sequence may not change, thanks to redundancy in the genetic code. Natural selection neither favors nor selects against these "silent" mutations.

B. Natural Selection Eliminates Poorly Adapted Phenotypes

Recall from chapter 10 that an individual's phenotype is its observable properties, most of which arise from a combination of

a. Genetic variation

b. Overproduction of offspring

Figure 12.7 Requirements for Natural Selection. (a) This group of people illustrates genetic variation within the human population. (b) Dandelions produce many offspring, but few survive.

(a): ©Punchstock/Stockbyte RF; (b): ©Angelo Cavalli/Getty Images RF

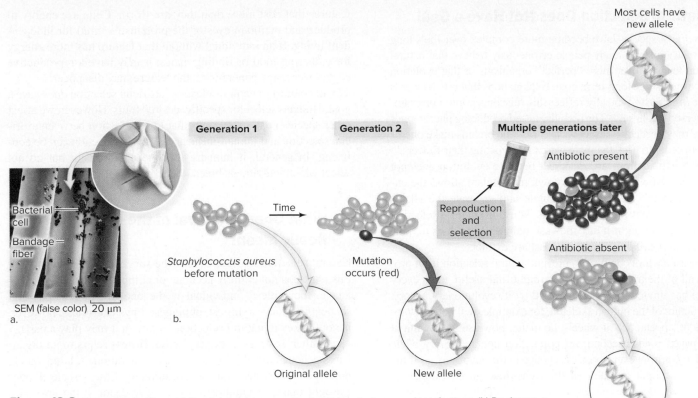

Figure 12.8 Natural Selection. (a) *Staphylococcus aureus* is a bacterium that causes skin infections. (b) By chance, a cell undergoes a random genetic mutation. The population is then exposed to an antibiotic. The drug kills most of the unmutated cells, but the mutated cell is unaffected and can reproduce. After many generations of exposure to the antibiotic, the mutation is common.

Photo: (a): ©Paul Gunning/Science Source

environmental influences and the action of multiple genes. By "weeding out" individuals with poorly adapted phenotypes, natural selection indirectly changes allele frequencies in the population. The scientific literature contains countless examples, some of which are described in the Investigating Life sections of this book. For example, disease-causing bacteria select for sexual reproduction in nematode worms (see section 9.9); genetically modified cotton plants producing the Bt toxin eliminate Bt-susceptible moths from the population (see section 10.10); and exposure to the sun's ultraviolet radiation has selected against light skin pigmentation in our own species (see section 25.6).

Environmental conditions constantly change, so evolution never really stops. After all, the phenotype that is "best" depends entirely on the time and place; a trait that is adaptive in one set of circumstances may become a liability in another. Some orchids, for example, produce flowers that are pollinated by only one or a few species of wasp (**figure 12.9**). The orchids release chemicals that mimic a female wasp's pheromones. When a male wasp visits the flower, seeking a mate, pollen from the orchid sticks to its body. The insect later deposits the pollen on another orchid. As long as wasps are present, this exclusive relationship benefits the plants; they do not waste energy by attracting animals that are unlikely to visit another flower of the same species. But if the wasps went extinct, the orchids could no longer reproduce. In that case, having an exclusive relationship with one or a few pollinator species would doom the orchids.

Figure 12.9 Evolving Together. The intimate relationship between this orchid and its pollinator (a wasp) is efficient for the plant. But if the insect goes extinct, the orchid pays the price, too.

©Tim Gainey/Alamy

C. Natural Selection Does Not Have a Goal

Because most species have become more complex over life's long evolutionary history, many people erroneously believe that natural selection leads to ever more "perfect" organisms or that evolution works toward some long-term goal. Explanations that use the words *need* or *in order to* typically reflect this misconception. For example, a person might say, "The orchids started producing pheromones because they needed to attract wasps" or "The orchids make pheromones in order to trick the male wasps into visiting their flowers."

Both explanations are incorrect because evolution does not have a goal. How could it? No known mechanism allows the environment to tell DNA how to mutate and generate the alleles needed to confront future conditions. Nor does natural selection strive for perfection; if it did, the vast majority of species in life's history would still exist. Instead, most are extinct.

Several factors combine to prevent natural selection from producing all of the traits that a species might find useful. First, every genome has limited potential, imposed by its evolutionary history. The structure of the human skeleton, for example, will not allow for the sudden appearance of wheels, no matter how useful they might be on paved roads. Second, no population contains every allele needed to confront every possible change in the environment. If the right alleles aren't available at the right time, an environmental change may wipe out a species (**figure 12.10**). Third, disasters such as floods and volcanic eruptions can indiscriminately eliminate the best allele combinations, simply by chance (see section 12.6). And finally, natural selection cannot act on some harmful genetic traits, such as diseases that appear only after reproductive age.

"Need-based" evolution is among the most common misconceptions about biology; another is that traits automatically disappear when they are no longer needed. Blind cave fish, for example, lack eyes. Likewise, the protist that causes malaria can survive only inside a living host, having lost many features that allow independent living. Note, however, that "absence of need" is not a selective force. Instead, the environment selects against features that cost more than they are worth. Cells use energy to produce and maintain eyes or the proteins required for independent living. If an individual without that feature has more energy for gathering food or finding mates, it may have a reproductive edge. Over many generations, the feature may disappear.

In contrast to natural selection, artificial selection *does* have a goal: Humans select for specific, desired traits. However, we affect other species in so many ways that the distinction between artificial selection and human-influenced natural selection can be confusing. In general, if humans alter the environment but do not select which individuals breed, the term *natural selection* applies.

D. What Does "Survival of the Fittest" Really Mean?

Natural selection is often called "the survival of the fittest," but this phrase is not entirely accurate or complete. In everyday language, the "fittest" individual is the one in the best physical shape: the strongest, fastest, or biggest. Physical fitness, however, is not the key to natural selection (although it may play a part).

Rather, in an evolutionary sense, **fitness** refers to an organism's genetic contribution to the next generation. A large, quick, burly elk scores zero on the evolutionary fitness scale if poor eyesight makes it vulnerable to an early death in the jaws of a wolf. On the other hand, a mayfly that dies in the act of producing thousands of offspring is highly fit.

Because successful reproduction is the only way for an organism to perpetuate its genes, fitness depends on the ability to survive just long enough to reproduce. Plants that germinate, grow, flower, produce seeds, and die within just a few weeks may have fitness equal to a redwood that lives for centuries (**figure 12.11**).

These examples illustrate an important point: *By itself, survival is not enough.* Paradoxically, natural selection promotes any trait that increases fitness, even if the trait virtually guarantees an individual's death. For example, a male praying mantis may not resist if the

a.   2 cm   b.

Figure 12.10 **Extinction.** (a) Sea scorpions once thrived worldwide but became extinct some 250 million years ago. (b) These petrified trees in the Arizona desert are from a family of trees that is now extinct in the northern hemisphere, thanks to continental drift.

(a): ©Francois Gohier/Science Source; (b): ©B.A.E. Inc./Alamy RF

Burning Question

Is there such a thing as a "pinnacle of evolution"?

One common illustration of human evolution depicts a slouching chimpanzee that transforms by several stages into a caveman and then into an upright modern human. The misleading implications are that chimps evolved into humans and that the process was goal-oriented and directional—in other words, that humans are the "pinnacle of evolution."

But viewing evolutionary processes as a ladder, with humans at the top and every other species evolving toward that "goal," is a mistake. Instead, evolution has produced a tree of life, with successful organisms at every branch tip. We occupy one of those tips.

We might therefore reframe the question to ask whether any species could colonize every habitat on Earth. Would humans fit the bill? No. We can survive in a great variety of locations, but we can make only brief excursions into water and onto the highest mountaintops. What about other species? Again, the answer is no. To take an extreme example, a trout's adaptations to a cold mountain stream are worthless in parched desert sands. Even ubiquitous pests such as cheat grass, dandelions, cockroaches, and rats can't withstand conditions that are too hot, too cold, too dry, or too wet. The variety of habitats on Earth—ocean, freshwater, tundra, prairie, desert, forest—is too great for one species to be able to live everywhere.

©Sean Thompson/Photodisc/Getty Images RF

Submit your burning question to
Marielle.Hoefnagels@mheducation.com

female begins to eat his head during copulation. The male's passive behavior is adaptive because the extra food the female obtains in this way will enhance the chance of survival for their young. Likewise, section 35.7 describes a male spider that somersaults his abdomen into the jaws of his mate during copulation. Unlike the mantis, his body is too small to offer the female a nutritional benefit. Instead, the male's suicidal behavior prolongs copulation, so that he sires the most possible offspring. As in the case of the mantis, the male spider does not survive—but his alleles will.

Fitness includes not only the total number of offspring produced but also the proportion that reach reproductive age. Some organisms, including humans, have few offspring but invest large amounts of energy in each one. Insects and many other species produce thousands of young but invest minimally in each. The optimal balance between "quality" and "quantity" may vary greatly, even among individuals within a population. Section 37.4 further describes this evolutionary trade-off.

Many adaptations contribute to an organism's overall fitness. Being able to overcome poor weather conditions, combat parasites and other disease-causing organisms, evade predators, and compete for resources enhances an organism's chances of reaching reproductive age. At that point, the ability to attract mates (or pollinators, in the case of many flowering plants) affects the number of offspring an organism produces.

Some people wonder whether humans (or any other type of organism with an unusually high ability to outcompete other species) could be considered a "pinnacle of evolution." This chapter's Burning Question discusses this possibility.

a.   b.

Figure 12.11 Fitness Is Reproductive Success. One key to fitness is living long enough to reproduce. (a) For a redwood, that may take centuries. (b) In an annual plant, it may take just a few weeks.

(a): ©Comstock/Jupiter Images RF; (b): ©Corbis RF

12.3 | MASTERING CONCEPTS

1. What is an adaptation, and how do adaptations become more common within a population?
2. What is the role of genetic variation in natural selection?
3. How can natural selection favor different phenotypes at different times?
4. Why doesn't natural selection produce perfectly adapted organisms?
5. What is evolutionary fitness?

12.4 Evolution Is Inevitable in Real Populations

Shifting allele frequencies in populations are the small steps of change that collectively drive evolution. Given the large number of genes in any organism and the many factors that can alter allele frequencies (including but not limited to natural selection), evolution is not only possible but unavoidable. This section explains why.

A. At Hardy–Weinberg Equilibrium, Allele Frequencies Do Not Change

The study of population genetics relies on the intimate relationship between allele frequencies and **genotype frequencies.** Each genotype's frequency is the number of individuals with that genotype, divided by the total size of the population. For example, if 64 of the 100 individuals in a population are homozygous recessive, then the frequency of that genotype is 64/100, or 0.64.

Hardy–Weinberg equilibrium is the highly unlikely situation in which allele frequencies and genotype frequencies do not change from one generation to the next. It occurs only in populations that meet the following assumptions: (1) natural selection does not occur; (2) mutations do not occur, so no new alleles arise; (3) the population is infinitely large, or at least large enough to eliminate random changes in allele frequencies; (4) individuals mate at random; and (5) individuals do not migrate into or out of the population.

Hardy–Weinberg equilibrium is named after mathematician Godfrey H. Hardy and physician Wilhelm Weinberg. They independently developed two simple equations that represent the relationship between allele frequencies and genotype frequencies (**figure 12.12**). To understand their logic, begin by assuming that a gene has only two possible alleles, with frequencies p and q. The first equation represents the frequencies of both alleles in the population:

$$p + q = 1$$

The two frequencies add up to 1 because the two alleles represent all the possibilities in the population. For example, in the ferrets illustrated in figure 12.12, the frequency of the dark fur allele (D) is 0.6; the frequency of the alternative allele d, which confers tan fur, is 0.4. (Tally the D and d alleles in the picture of the ferrets to verify these numbers.)

At Hardy–Weinberg equilibrium, we can use allele frequencies to calculate genotype frequencies, according to the second equation in figure 12.12:

$$p^2 + 2pq + q^2 = 1$$

In this equation, the proportion of the population with genotype DD equals p^2 (0.36 for our ferrets) and the proportion with genotype dd equals q^2 (0.16). To calculate the frequency of the heterozygous class, multiply pq by 2 (0.48). Since the

Assumptions of the Hardy–Weinberg model

1. Natural selection does not occur.
2. Mutations do not occur.
3. The population is infinitely large.
4. Individuals mate at random.
5. Individuals do not migrate into or out of the population.

Allele frequencies: $p + q = 1$

Definition/equation	Example
p = frequency of dominant allele	p = frequency of D (dark fur) = 0.6
q = frequency of recessive allele	q = frequency of d (tan fur) = 0.4
$p + q = 1$	$0.6 + 0.4 = 1$

Genotype frequencies: $p^2 + 2pq + q^2 = 1$

Definition/equation	Example
p^2 = frequency of DD genotype	$0.6 \times 0.6 = 0.36$
$2pq$ = frequency of Dd genotype	$2 \times 0.6 \times 0.4 = 0.48$
q^2 = frequency of dd genotype	$0.4 \times 0.4 = 0.16$
$p^2 + 2pq + q^2 = 1$	$(0.6)^2 + (2 \times 0.6 \times 0.4) + (0.4)^2 = 1$

Reproduction (random mating)

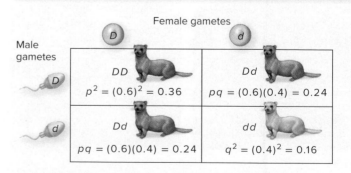

Figure 12.12 Hardy–Weinberg Equilibrium. At Hardy–Weinberg equilibrium, allele frequencies remain constant from one generation to the next; evolution does not occur.

homozygotes and the heterozygotes account for all possible genotypes, the sum of their frequencies must add up to 1.

Figure It Out

In a species of ladybug, one gene controls whether the beetle has spots or not, with the allele conferring spots being dominant. Suppose you find a swarm consisting of 1000 ladybugs; 250 of them lack spots. What are the frequencies of the dominant and recessive alleles?

Answer: $q^2 = 0.25$, so $q = 0.5$. Therefore, $p = 0.5$ as well.

Figure It Out

A population of 100 sea stars is in Hardy–Weinberg equilibrium. The trait for long arms is completely dominant to the trait for short arms. In this population, 40% of the alleles for this trait are dominant, and 60% are recessive. What is the frequency of the heterozygous genotype in this population?

Answer: $2 \times 0.4 \times 0.6 = 0.48$

The ferrets mate at random, as illustrated in the Punnett square. In the next generation, allele and genotype frequencies remain the same. Evolution therefore has not occurred. Conversely, a change in allele and genotype frequencies in the next generation would have indicated that the population had evolved.

Besides providing a framework for determining whether evolution has occurred, these two equations are useful because they allow us to infer characteristics of a population based on limited information. One application is the use of known allele frequencies to estimate genotype frequencies in a population. DNA profiling, for example, relies on population databases that contain the known frequencies of each allele at 13 sites in the human genome. Forensic analysts can use this information to calculate the probability that two people share the same genotype across all 13 sites. ⓘ *DNA profiling,* section 11.2D

Usually, however, we do not know the exact frequency of every allele in a population. In that case, we can use the Hardy–Weinberg equations to estimate allele frequencies based on the known frequency of one genotype. But how do we get that information? Recall from chapter 10 that a distinctive phenotype is often associated with a homozygous recessive genotype. As a result, determining q^2 for some genes is relatively easy. Knowing q, in turn, makes it possible to calculate p. The values of p and q can then be plugged into the second equation to estimate the frequencies of homozygous dominant and heterozygous genotypes. For an example based on a genetic disease called cystic fibrosis, see the Figure It Out question.

Figure It Out

Assume that 1 in 3000 Caucasian babies in the United States is born with cystic fibrosis, a disease caused by a recessive allele. The value of q^2 is therefore $1/3000 = 0.0003$; q is the square root of 0.0003, or .018. Use this information to estimate the frequency of heterozygotes (symptomless carriers) in the American Caucasian population.

Answer: If $q = 0.018$, then $p = 0.982$; the frequency of heterozygotes is $2 \times 0.982 \times 0.018 = 0.035$, or 3.5%.

B. In Reality, Allele Frequencies Always Change

A population at Hardy–Weinberg equilibrium does not evolve. A real population, however, violates the assumptions of Hardy–Weinberg equilibrium when any of the following occurs:

- some phenotypes are better adapted to the environment than others (natural selection);
- mutations introduce new alleles;
- allele frequencies change due to chance (genetic drift);
- individuals remain in closed groups, mating among themselves rather than with the larger population (nonrandom mating);
- individuals migrate among populations.

All of these events are common. If you think about the human population, for example, we can sometimes counter natural selection by medically correcting phenotypes that would otherwise prevent some of us from having children. But natural selection still acts to reduce the frequency of alleles that cause deadly childhood illnesses. After all, children who inherit these illnesses will not survive long enough to pass the alleles on.

Moreover, genetic mutations can and do occur. Some mutations happen randomly when DNA replicates. External agents, including harmful chemicals and radiation, can also cause mutations.

Finally, genetic drift, nonrandom mating, and migration all happen in real populations. The large size of the human population usually minimizes the chance of random changes in allele frequencies, but they occasionally occur (see section 12.7B). Humans are deliberate in our choices of mates and we may choose to live across the world from our parents; allele frequencies change as we move and mix.

These forces act on populations of other species as well, so allele frequencies always change over multiple generations. In other words, evolution is inevitable. Even though its assumptions do not apply to real populations, the concept of Hardy–Weinberg equilibrium does serve as a basis of comparison (a "null hypothesis") to reveal when evolution is occurring. Additional studies can then reveal which mechanism of evolution is acting on the population. Sections 12.5 through 12.7 describe these mechanisms of evolution in more detail.

12.4 MASTERING CONCEPTS

1. What are the five conditions required for Hardy–Weinberg equilibrium?
2. Why is the concept of Hardy–Weinberg equilibrium important?
3. Explain the components and meaning of the equation $p^2 + 2pq + q^2 = 1$.
4. Why doesn't Hardy–Weinberg equilibrium occur in real populations?

12.5 Natural Selection Can Shape Populations in Many Ways

Of all the mechanisms by which a population can evolve, natural selection is probably the most important. Natural selection changes the genetic makeup of a population by favoring the alleles that contribute to reproductive success and selecting against those that do not.

Natural selection, however, does not eliminate alleles directly. Instead, individuals with the "best" phenotypes are most likely to pass their alleles to the next generation; those with poorly suited phenotypes are less likely to survive long enough to reproduce. Three modes of natural selection—directional, disruptive, and stabilizing—are distinguished by their effects on the phenotypes in a population (figure 12.13).

In **directional selection,** one extreme phenotype is fittest, and the environment selects against the others. A change in tree trunk color from light to dark, for example, may select for dark-winged moths and against light-winged individuals. The rise of antibiotic resistance among bacteria, described in this chapter's opening essay, also reflects directional selection, as does the increase in pesticide-resistant insects (see section 10.10). The fittest phenotype may initially be rare, but its frequency increases over multiple generations as the environment changes—for example, after exposure to the antibiotic or insecticide.

In **disruptive selection** (sometimes called diversifying selection), two or more extreme phenotypes are fitter than the intermediate phenotype. Consider, for example, a population of marine snails that live among brown rocks encrusted with white barnacles. The white snails near the barnacles are camouflaged, and the dark brown ones on the bare rocks likewise blend in. The snails that are neither white nor dark brown are most often seen and eaten by predatory shorebirds.

In a third form of natural selection, called **stabilizing selection** (or normalizing selection), extreme phenotypes are less fit than the optimal intermediate phenotype. Human birth weight illustrates this tendency to stabilize. Very small or very large newborns are less likely to survive than babies of intermediate weight. By eliminating all but the individuals with the optimal phenotype, stabilizing selection tends to reduce the variation in a population. It is therefore most common in stable, unchanging environments.

These three models of natural selection might seem to suggest that, for each trait, only one or a few beneficial alleles ought to persist in the population. The harmful alleles should gradually become less common until they disappear, while the others become "fixed" in the population.

For some genes, however, natural selection maintains a **balanced polymorphism,** in which multiple alleles of a gene persist indefinitely in the population at more or less constant frequencies. (*Polymorphism* means "multiple forms.") In a balanced

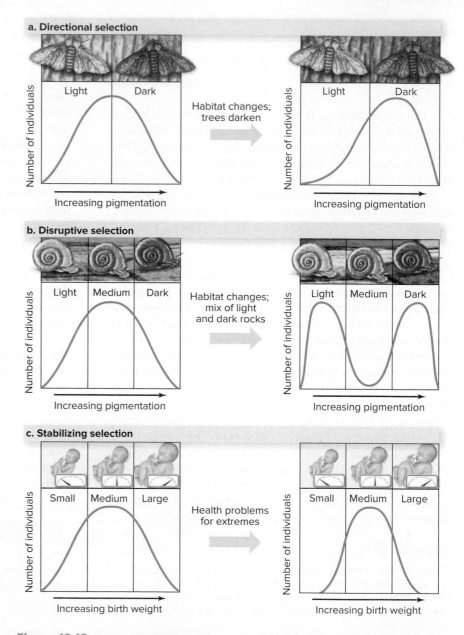

Figure 12.13 Types of Natural Selection. (a) Directional selection results from selection against one extreme phenotype. (b) In disruptive selection, two extreme phenotypes each confer a selective advantage over the intermediate phenotype. (c) Stabilizing selection maintains an intermediate expression of a trait by selecting against extreme variants.

polymorphism, even harmful alleles may remain in the population. This situation seems contrary to natural selection; how can it occur?

One circumstance that can maintain a balanced polymorphism is a **heterozygote advantage,** which occurs when an individual with two different alleles for a gene (a heterozygote) has greater fitness than those whose two alleles are identical (homozygotes). Heterozygotes can maintain a harmful recessive allele in a population, even if homozygous recessive individuals have greatly reduced fitness.

The best documented example of heterozygote advantage in humans is sickle cell disease. The disease-causing allele encodes an abnormal form of hemoglobin. The abnormal hemoglobin proteins do not fold properly; instead, they form chains that bend a red blood cell into a characteristic sickle shape (see figure 7.21). In a person who is homozygous recessive for the sickle cell allele, all the red blood cells are affected. Symptoms include anemia, joint pain, a swollen spleen, and frequent, severe infections; the person may not live long enough to reproduce. On the other hand, a person who is heterozygous for the sickle cell allele is only mildly affected. Some of his or her red blood cells may take abnormal shapes, but the resulting mild anemia is not usually harmful.

Still, if the sickle cell allele causes such severe problems in homozygous recessive individuals, why hasn't natural selection eliminated it from the population? The answer is that heterozygotes also have a reproductive edge over people who are homozygous for the *normal* hemoglobin allele. Specifically, heterozygotes are resistant to a severe infectious disease, malaria. When a mosquito carrying a protist called *Plasmodium* feeds on a human with normal hemoglobin—a homozygous dominant person—the parasite enters the red blood cells. Eventually, infected blood cells burst, and the parasite travels throughout the body. The resulting bouts of fever, chills, fatigue, and nausea are severe; if damaged blood cells block blood vessels, the patient may die of organ failure. But the sickled red blood cells of an infected carrier—a heterozygote—halt the parasite's spread. ⓘ *malaria,* section 18.4D

Not surprisingly, the sickle cell allele's frequency is highest in parts of the world where malaria is most common (**figure 12.14**). In these areas, the heterozygotes remain healthiest; they are resistant to malaria, but they are not seriously ill from sickle cell disease. They therefore have more children (on average) than do people who are homozygous for either allele. Unfortunately, two carriers have a 25% chance of producing a child who is homozygous recessive. These children pay the evolutionary price for the genetic protection against malaria.

Cystic fibrosis provides another example of heterozygote advantage. This illness originates with a mutation in the gene encoding a membrane protein called CFTR; the mutated allele encodes a faulty (nonfunctional) version of the protein. People born with two mutated *CFTR* alleles have cystic fibrosis and are unlikely to reproduce. Heterozygotes, however, have an advantage in areas where an infectious disease called cholera is common. The cholera toxin overstimulates normal CFTR

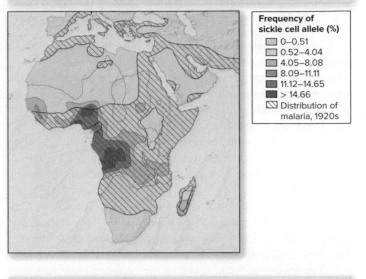

a. Distribution of sickle cell allele and malaria

Frequency of sickle cell allele (%)
- ☐ 0–0.51
- ☐ 0.52–4.04
- ☐ 4.05–8.08
- ☐ 8.09–11.11
- ☐ 11.12–14.65
- ■ > 14.66
- ▨ Distribution of malaria, 1920s

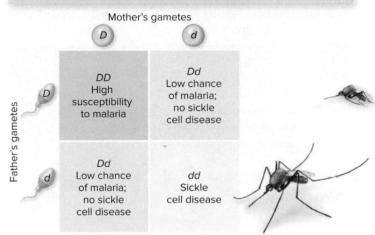

b. Cross between two carriers of sickle cell disease

Mother's gametes

	D	d
D	DD High susceptibility to malaria	Dd Low chance of malaria; no sickle cell disease
d	Dd Low chance of malaria; no sickle cell disease	dd Sickle cell disease

Father's gametes

Figure 12.14 **Heterozygote Advantage.** (a) The sickle cell allele is most common where malaria is prevalent. (b) In regions where malaria thrives, heterozygotes for the sickle cell allele are most likely to survive long enough to reproduce. However, two heterozygotes have a 25% chance of producing a child with sickle cell anemia.

proteins in the intestines, causing watery diarrhea that can lead to dehydration and death. Carriers of the recessive allele are unlikely to die of dehydration from cholera because half of their CFTR proteins are abnormal. Heterozygotes therefore have a fitness advantage over both types of homozygotes, allowing the harmful allele to remain in the population.

12.5 | MASTERING CONCEPTS

1. Distinguish among directional, disruptive, and stabilizing selection.
2. How can natural selection maintain harmful alleles in a population?

12.6 Sexual Selection Directly Influences Reproductive Success

In many vertebrate species, the sexes look alike. The difference between a male and a female house cat, for example, is not immediately obvious. In some species, however, natural selection can maintain a **sexual dimorphism,** which is a difference in appearance between males and females. One sex may be much larger or more colorful than the other, or one sex may have distinctive structures such as horns or antlers.

Some of these sexually dimorphic features may seem to violate natural selection. For example, female cardinals are brown and inconspicuous, but their male counterparts have vivid red feathers that make the birds much more visible to predators. Similarly, the extravagant tail of a peacock is brightly colored and makes flying difficult. How can natural selection allow for traits that apparently reduce survival?

The answer is that a special form of natural selection is at work. **Sexual selection** is a type of natural selection resulting from variation in the ability to obtain mates, and it occurs in two forms. In **intrasexual selection,** the members of one sex compete among themselves for access to the opposite sex (**figure 12.15**). Male bighorn sheep, for example, use their horns to battle for the right to mate with multiple females. The strongest rams are therefore the most likely to pass on their alleles.

In **intersexual selection,** the members of one sex choose their mates from among multiple individuals of the opposite sex (**figure 12.16**). If female birds of paradise prefer brightly colored males, then showy plumage directly increases a male's chance of reproducing. Because the brightest males get the most opportunities to mate, alleles that confer those feathers are common in the population.

Why do males usually show the greatest effects of sexual selection? In most (but not all) vertebrate species, females spend more time and energy on rearing offspring than do males. Because of this high investment in reproduction, females tend to be selective about their mates. Males are typically less choosy and must compete for access to females.

The evolutionary origin of the males' elaborate ornaments remains an open question. One possibility is that long tail feathers and bright colors are costly to produce and maintain; they are therefore indirect advertisements of good health or disease resistance. Likewise, the ability to win fights with competing males could also be an indicator of good genes. A female who chooses a high-quality male will increase not only his fitness but also her own.

12.6 | MASTERING CONCEPTS

1. How does sexual selection promote traits that decrease survival?
2. What is the difference between intrasexual selection and intersexual selection?

a.

b.

Figure 12.15 Intrasexual Selection. (a) Two bighorn rams butt heads in Montana. (b) Male ox beetles fight during the mating season.

(a): ©mlharing/Getty Images RF; (b): ©James H. Robinson/Science Source

a.   b.

Figure 12.16 Intersexual Selection. (a) Male weaver birds build nests; females select their mates based on nest quality. (b) The male bird of paradise displays bright plumes that attract females.

(a): ©James Warwick/Stone/Getty Images; (b): ©Michael S. Yamashita/Corbis

12.7 Evolution Occurs in Several Additional Ways

Natural selection is responsible for adaptations that enhance survival and reproduction, but it is not the only mechanism of evolution. This section describes four more ways that a population can evolve: mutation, genetic drift, nonrandom mating, and gene flow. All occur frequently, and each can, by itself, disrupt Hardy–Weinberg equilibrium. The changes in allele frequencies that constitute evolution therefore occur nearly all the time.

A. Mutation Fuels Evolution

A change in an organism's DNA sequence introduces a new allele to a population. The new trait may be harmful, neutral, or beneficial, depending on how the mutation affects the sequence of the encoded protein. ⓘ *mutations,* section 7.7

Mutations are the raw material for evolution because genes contribute to phenotypes, and natural selection acts on phenotypes. For example, random mutations in bacterial DNA may change the shapes of key proteins in the cell's ribosomes or cell wall. Some of these mutations may mean a new phenotype—resistance to an antibiotic, for example. If exposure to antibiotics selects for that phenotype, the mutations will pass to the next generation.

As we saw in section 12.3C, one common misconception about evolution is that a mutation produces a novel adaptation precisely when a population "needs" it to confront a new environmental challenge. For example, many people mistakenly believe that antibiotics *create* resistance—that is, that resistance arises in bacteria *in response* to exposure to the drugs.

In reality, genes do not "know" when to mutate; the chance that a mutation will occur is independent of whether a new phenotype would benefit the organism. The only way antibiotic resistance arises is if some bacteria happen to have a mutation that confers resistance *before* exposure to the drug. The drug creates a situation in which these variants can flourish. That trait will then become more common within the population by natural selection. If no bacteria start out resistant, the drug kills the entire population.

Because bacterial populations are often enormous, however, it is likely that at least a few individuals carry such a mutation.

The rate at which mutations occur varies, both among different genes and within a gene. The average rate is around one DNA sequence change per 10^9 base pairs. At first, this number may seem too low to pose a significant force in evolution. Each genome, however, has an enormous number of base pairs, and a large number of cell divisions occur throughout life. Extensive cell division means ample opportunities for mutation.

A mutation affects evolution only if subsequent generations inherit it. In asexually reproducing organisms such as bacteria, each mutated cell gives rise to mutant offspring (if the mutation does not prevent reproduction). In a multicellular organism, however, a mutation can pass to the next generation only if it arises in a germ cell (i.e., one that will give rise to gametes; see chapter 9). For example, a cigarette smoker with lung cancer will not pass any smoking-induced mutations to her children, because her egg cells will not contain the altered DNA.

B. Genetic Drift Occurs by Chance

Genetic drift (sometimes called sampling drift) is a change in allele frequencies that occurs purely by chance. Unlike mutation, which increases diversity, genetic drift tends to eliminate alleles from a population.

All forms of genetic drift are rooted in sampling error, which occurs when a sample does not match the larger group from which it is drawn (**figure 12.17**). Such sampling errors are most likely to affect small populations. Suppose, for example, that one allele of a gene occurs at a very low frequency in a population. If, by chance, none of the individuals carrying the rare allele happens to reproduce, that variant will disappear from the population. Even if some do reproduce, the allele still might not pass to the next generation. After all, the events of meiosis ensure that each allele has only a 50% chance of passing to each offspring. A rare allele can therefore vanish from a population—not because it reduces fitness but simply by chance.

Genetic drift is inevitable; that is, allele frequencies for any gene will fluctuate at random in *every* generation.

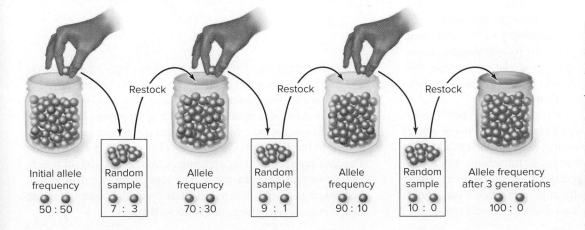

Initial allele frequency 50 : 50   Random sample 7 : 3   Restock   Allele frequency 70 : 30   Random sample 9 : 1   Restock   Allele frequency 90 : 10   Random sample 10 : 0   Restock   Allele frequency after 3 generations 100 : 0

Figure 12.17 Sampling Error. Ten marbles are drawn at random from the first jar. Even though 50% of the marbles in the jar are blue, the random sample contains only 30% blue marbles. In the next "generation," 30% of the marbles are blue, but the random sample contains just one blue marble. The third sample contains no blue marbles and therefore eliminates the blue "allele" from the population, purely by chance.

Hardy–Weinberg equilibrium requires populations to be very large (approaching infinity) to minimize the effects of these random changes in allele frequencies. In reality, of course, many populations are small, so genetic drift can be an important evolutionary force.

Random chance can eliminate rare alleles, but genetic drift can also operate in other ways. As described below, sampling errors (such as those illustrated in figure 12.17) can occur when a small population separates from a larger one or when a large population is reduced to a very small size.

The Founder Effect   One cause of genetic drift is the **founder effect,** which occurs when a small group of individuals leaves its home population and establishes a new, isolated settlement. The small group's random "allele sample" may not represent the allele frequencies of the original population. Some traits that were rare in the original population may therefore be more frequent in the new population. Likewise, other traits will be less common or may even disappear.

The Amish people of Pennsylvania provide a famous example of the founder effect. About 200 followers of the Amish denomination immigrated to North America from Switzerland in the 1700s. One couple, who immigrated in 1744, happened to carry the recessive allele associated with Ellis–van Creveld syndrome (**figure 12.18**). This allele is extremely rare in the population at large. Intermarriage among the Amish, however, has kept the disease's incidence high in this subgroup more than two centuries after the immigrants arrived.

On a much broader scale, the history of human migration out of Africa has been a series of founder effects (see figure 15.27). A subset of people migrated from Africa to Asia about 40,000 to 70,000 years ago, and from there a smaller subset later migrated to the Americas. Meanwhile, a different subset migrated from Africa to Europe. Each migration divided the genetic variation of the previous population. As populations became isolated in new environments, phenotypes diverged. These "serial founder effects" therefore explain the unique allele combinations among native occupants of each region (see figure 12.1).

The Bottleneck Effect   Genetic drift also may result from a population **bottleneck,** which occurs when a population's size drops rapidly over a short period. The bottleneck randomly eliminates many alleles that were present in the larger ancestral population (**figure 12.19**). Even if the few remaining individuals mate and restore the population's numbers, the loss of genetic diversity is permanent. Note that the loss of alleles is random; the bottleneck effect is therefore different from natural selection, which weeds out alleles that reduce fitness.

Cheetahs are currently undergoing a population bottleneck. Until 10,000 years ago, these cats were common in many areas. Today, just two isolated populations live in South and East Africa, numbering only a few thousand animals. Inbreeding has made the South African cheetahs so genetically alike that even unrelated animals can accept skin grafts from each other. Researchers attribute this uniformity to at least two bottlenecks: one that

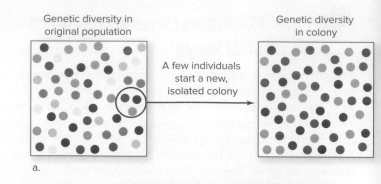

Genetic diversity in original population

A few individuals start a new, isolated colony

Genetic diversity in colony

a.

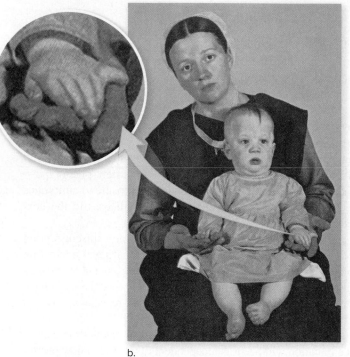

b.

Figure 12.18   The Founder Effect.   (a) A few individuals leave their original population and begin a new colony. The new population has an altered allele frequency—and less genetic diversity—compared to the original population. (b) Ellis–van Creveld syndrome is characterized by dwarfism, extra fingers, and other symptoms. This recessive disorder occurs in 7% of the Old Order Amish population of Lancaster County, Pennsylvania, but is extremely rare elsewhere.

Photos: (b, both): Courtesy of Dr. Victor A. McKusick/Johns Hopkins Hospital

occurred at the end of the most recent ice age, when habitats changed drastically, and another when humans slaughtered many cheetahs during the 1800s.

North American species have undergone severe bottlenecks as well. Bison, for example, nearly went extinct because of overhunting in the 1800s. And habitat loss has caused the population of greater prairie chickens to plummet from about 100 million in 1900 to several hundred today. The loss of genetic diversity in cheetahs, bison, and prairie chickens invites disaster: A single change in the environment might doom them all.

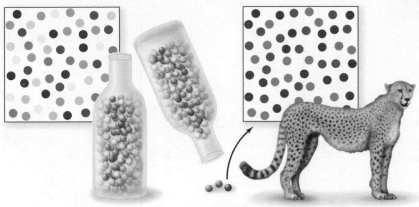

Original cheetah population contains 25 different alleles of a particular gene.

Cheetah population is drastically reduced.

Repopulation occurs. Only three different alleles remain.

Figure 12.19 **The Bottleneck Effect.** A drastic decline in a population's size—a population bottleneck—eliminates many alleles at random. Even if the survivors rebuild the population, their genetic diversity is greatly reduced compared to the original population.

C. Nonrandom Mating Concentrates Alleles Locally

In a population with completely random mating, every individual has an equal chance of mating with any other member of the population. Wind-pollinated plants illustrate random mating.

In most species, however, mating is not random. As we have already seen, many animals exhibit some preference in mate choice, including sexual selection (figure 12.20). Many other factors also influence mating, including geographical restrictions and physical access to the opposite sex.

Figure 12.20 **Nonrandom Mating.** If mating among toads were random, all individuals would have an equal chance of reproducing. Instead, a male mates only if his song attracts a willing female. This male toad inflates his throat pouch and generates a call.
©Creatas/PunchStock RF

The practice of artificial selection is another way to reduce random mating. Humans select those animals or plants that have a desired trait; they then allow only those "superior" individuals to mate. The result is a wide variety of subpopulations (such as different breeds of dogs) that humans maintain by selective breeding.

D. Gene Flow Moves Alleles Between Populations

Under Hardy–Weinberg equilibrium, no alleles ever leave or enter a population. In reality, **gene flow** moves alleles from one population to another. Migration is one common way that gene flow occurs (figure 12.21). The migrating brown rabbit in the figure will add new alleles to the population of black rabbits, increasing the local genetic diversity in its new home. But gene flow does not require the movement of entire individuals. Wind can carry a plant's pollen for miles, spreading one individual's alleles to a new population.

Over time, sustained migration has reduced the genetic differences between human populations. For example, isolated European populations once had unique allele frequencies for many genetic diseases. Geographical barriers, such as mountain ranges and large bodies of water, historically restricted migration and kept the gene pools separate. Highways, trains, and airplanes, however, have eliminated physical barriers to migration. Eventually, gene flow should make these regional differences disappear.

12.7	**MASTERING CONCEPTS**

1. How do mutations affect an organism's phenotype?
2. South China tigers have two color patterns (orange/black and blue/gray), which do not affect an individual's reproductive success. Over many decades, the tiger population has drastically declined, and all blue/gray individuals have disappeared. What evolutionary process eliminated this color pattern?
3. Compare and contrast the founder effect and a population bottleneck.
4. How do nonrandom mating and gene flow result in evolutionary change?

Figure 12.21 **Gene Flow.** A migrating animal can transfer its alleles to another population.

INVESTIGATING LIFE

12.8 Size Matters in Fishing Frenzy

Studying the mechanisms of evolution helps us to understand life's history, but it also has practical consequences. A good example of natural selection is unfolding in fisheries worldwide. The selective force stems from a surprisingly mundane source—fishing regulations—but it affects everything from restaurant menus to the future of the ocean ecosystem.

The past several decades have seen devastating declines in the numbers of large predatory fishes such as swordfish, marlin, and sharks, as well as smaller animals, including tuna, cod, and flounder. From a biological point of view, the reason for the fisheries decline is simple: The animals' death rate exceeds their reproductive rate. Industrial-scale fishing is the culprit. Since the 1950s, fishing fleets have employed larger ships and improved technologies in pursuit of their prey.

Fishing regulations usually allow the harvest of only those fish that exceed some minimum size. This strategy is logical, because the smallest fish are most likely to be juveniles. Protecting the youngsters should permit the population to recover from the harvest of adult fish. But these regulations may also have evolutionary side effects. If humans harvest large individuals, fish that are small at maturity are the most likely to survive long enough to reproduce. Large fish may become more scarce over many generations. The same policy should also select for slow-growing fish, since they would be last to exceed the minimum allowed size.

Fish ecologists David Conover and Stephan Munch of Stony Brook University in New York tested these predictions by studying small coastal fish called Atlantic silversides (*Menidia menidia*). Conover and Munch randomly divided a large, captive population of Atlantic silversides into six tanks, each containing about 1100 juvenile fish. After about 6 months, the researchers assigned two tanks to each of the following treatments:

- **Large-harvested:** Remove the largest 90% of the fish (about 1000 fish) from two of the tanks, leaving the smallest 10% to reproduce.

- **Small-harvested:** Remove the smallest 90% of the fish.

- **Random-harvested (control):** Remove 90% of the fish, without size bias.

After the harvests, the 100 or so survivors in each tank reproduced, and their descendants were reared in identical conditions until it was again time to harvest 90% of each population. The researchers repeated the treatments over four generations of fish.

Predictably, the average weight of the removed fish was initially highest for the large-harvested population. Over four generations of size-biased fish removal, however, the small-harvested treatment favored both large size and rapid growth; the opposite was true for the large-harvested treatment (**figure 12.22**). The three treatments imposed different selective forces on the populations.

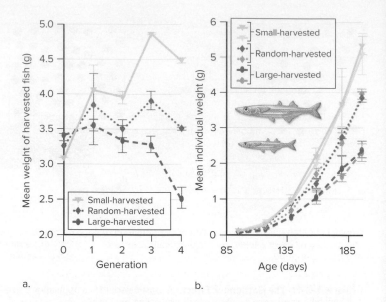

a.   b.

Figure 12.22 Fish Harvests. (a) Over four generations, the average fish in the large-harvested population weighed much less than those in the small-harvested population. (b) Harvesting the smallest individuals selected for fast-growing fish.

The results in figure 12.22 represent only the first 4 years of a decade-long experiment. The fifth generation of fish was treated like the first 4, but in years 6 through 10, Conover and his colleagues tested whether the evolutionary changes they observed could be reversed. For 5 generations they harvested 90% of the fish at random from all six tanks. The complete results suggested that the size of the fish in the large-harvested population might return to normal after about 12 generations.

Conover and Munch's experiment is more than a straightforward demonstration of natural selection in action; it also has economic and ecological applications. Revised regulations that protect the smallest and the largest fish would spare the juveniles that are critical to a species' future reproduction while also selecting for fast-growing fish. Imposing a maximum size limit would also have ecological benefits, restoring the feeding patterns and other "ecosystem services" of the largest fish. Moreover, the results of the second half of the experiment suggest that fish populations might rebound in coming decades if stocks are protected. This wide range of implications beautifully illustrates the powerful ideas that spring from understanding one fundamental idea: natural selection.

Sources: Conover, David O., and Stephan B. Munch. 2002. Sustaining fisheries yields over evolutionary time scales. *Science,* vol. 297, pages 94–96.

Conover, David O., Stephan B. Munch, and Stephen A. Arnott. 2009. Reversal of evolutionary downsizing caused by selective harvest of large fish. *Proceedings of the Royal Society B,* vol. 276, pages 2015–2020.

12.8 MASTERING CONCEPTS

1. What hypothesis did Conover and Munch test?
2. Draw how the graph in figure 12.22a might look once data from the second half of the experiment are added.

CHAPTER SUMMARY

12.1 Evolution Acts on Populations

- Biological **evolution** is descent with modification. One way to detect evolution is to look for a shift in the **gene pool** of a **population; allele frequencies** change from one generation to the next when evolution occurs.
- A small-scale genetic change within a species is called microevolution. Over the long term, microevolutionary changes also explain macroevolutionary events such as the emergence of new species.

12.2 Evolutionary Thought Has Evolved for Centuries

A. Many Explanations Have Been Proposed for Life's Diversity
- Geology laid the groundwork for evolutionary thought. Some people explained the distribution of rock strata with the idea of **catastrophism** (a series of upheavals). The more gradual **uniformitarianism** (continual remolding of Earth's surface) became widely accepted.
- The **principle of superposition** states that lower rock strata are older than those above, suggesting an evolutionary sequence for fossils within them.
- Lamarck proposed a testable mechanism of evolution, but it was mistakenly based on use and disuse of traits acquired during an organism's life.

B. Charles Darwin's Voyage Provided a Wealth of Evidence
- During the voyage of the HMS *Beagle,* Darwin observed the distribution of organisms in diverse habitats and their relationships to geological formations. After much thought and consideration of input from other scientists, he developed his theory of the origin of species by means of **natural selection.**

C. *On the Origin of Species* Proposed Natural Selection as an Evolutionary Mechanism
- Natural selection is based on multiple observations: Individuals vary for inherited traits; many more offspring are born than survive; and life is a struggle to acquire limited resources. The environment eliminates poorly adapted individuals, so only those with the best adaptations reproduce.
- **Artificial selection** is based on similar requirements, except that a human breeder allows only certain individuals to reproduce.
- *On the Origin of Species* offered abundant evidence for descent with modification.

D. Evolutionary Theory Continues to Expand
- The **modern evolutionary synthesis** unifies ideas about DNA, mutations, inheritance, and natural selection.

12.3 Natural Selection Molds Evolution

- Natural selection is one mechanism of microevolution.

A. Adaptations Enhance Reproductive Success
- Individuals with the best **adaptations** to the current environment are most likely to leave offspring, and therefore their alleles become more common in the population over time.
- Natural selection requires variation, which arises ultimately from random mutations.

B. Natural Selection Eliminates Poorly Adapted Phenotypes
- Natural selection weeds out some phenotypes, causing changes in allele frequencies over multiple generations.

C. Natural Selection Does Not Have a Goal
- Natural selection does not work toward an objective, nor can it achieve perfectly adapted organisms.

D. What Does "Survival of the Fittest" Really Mean?
- Organisms with the highest evolutionary **fitness** are the ones that have the greatest reproductive success. Many traits contribute to an organism's fitness.

12.4 Evolution Is Inevitable in Real Populations

- Calculations of allele frequencies and **genotype frequencies** allow biologists to detect whether evolution has occurred.

A. At Hardy–Weinberg Equilibrium, Allele Frequencies Do Not Change
- We can calculate the frequencies of genotypes and phenotypes in a population by inserting allele frequencies into this equation: $p^2 + 2pq + q^2 = 1$.
- If a population meets all assumptions of **Hardy–Weinberg equilibrium,** evolution does not occur because allele frequencies do not change from generation to generation.

B. In Reality, Allele Frequencies Always Change
- The conditions for Hardy–Weinberg equilibrium do not occur together in natural populations, suggesting that allele frequencies always change from one generation to the next. Figure 12.23 summarizes the main mechanisms of evolution.

12.5 Natural Selection Can Shape Populations in Many Ways

- In **directional selection,** one extreme phenotype becomes more prevalent in a population.
- In **disruptive selection,** two or more extreme phenotypes survive at the expense of intermediate forms.
- In **stabilizing selection,** an intermediate phenotype has an advantage over individuals with extreme phenotypes.
- In **balanced polymorphism,** natural selection indefinitely maintains more than one allele for a gene.
- Harmful recessive alleles may remain in a population because of a **heterozygote advantage,** in which carriers have a reproductive advantage over homozygotes.

12.6 Sexual Selection Directly Influences Reproductive Success

- **Sexual dimorphisms** differentiate the sexes. They result from **sexual selection,** a form of natural selection in which inherited traits—even those that seem nonadaptive—make an individual more likely to mate.
- **Intrasexual selection** is competition that does not involve a choice by the opposite sex; **intersexual selection** reflects mate choice by members of the opposite sex.

12.7 Evolution Occurs in Several Additional Ways

A. Mutation Fuels Evolution
- Mutation alters allele frequencies by changing one allele into another, sometimes providing new phenotypes for natural selection to act on. Many mutations do not pass to the next generation.

B. Genetic Drift Occurs by Chance
- In **genetic drift,** allele frequencies change purely by chance events, especially in small populations. The **founder effect** and population **bottlenecks** are forms of genetic drift.

C. Nonrandom Mating Concentrates Alleles Locally
- Nonrandom mating causes some alleles to concentrate in subpopulations.

D. Gene Flow Moves Alleles Between Populations
- Allele movement between populations, as by migration, is **gene flow.**

12.8 Investigating Life: Size Matters in Fishing Frenzy

- Fishing regulations that spare only the smallest fish in a population select for small, slow-growing individuals. Studies of Atlantic silversides suggest that protecting the largest fish as well would increase fishery productivity in the long run.

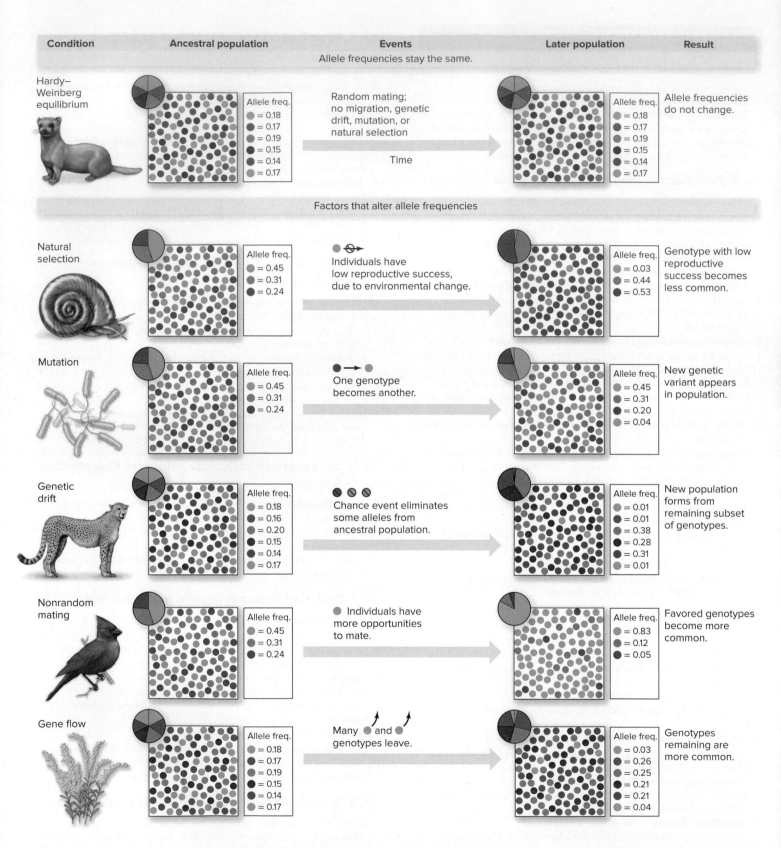

Condition	Ancestral population	Events	Later population	Result
		Allele frequencies stay the same.		
Hardy–Weinberg equilibrium	Allele freq. = 0.18, = 0.17, = 0.19, = 0.15, = 0.14, = 0.17	Random mating; no migration, genetic drift, mutation, or natural selection	Time	Allele freq. = 0.18, = 0.17, = 0.19, = 0.15, = 0.14, = 0.17	Allele frequencies do not change.
		Factors that alter allele frequencies		
Natural selection	Allele freq. = 0.45, = 0.31, = 0.24	Individuals have low reproductive success, due to environmental change.	Allele freq. = 0.03, = 0.44, = 0.53	Genotype with low reproductive success becomes less common.
Mutation	Allele freq. = 0.45, = 0.31, = 0.24	One genotype becomes another.	Allele freq. = 0.45, = 0.31, = 0.20, = 0.04	New genetic variant appears in population.
Genetic drift	Allele freq. = 0.18, = 0.16, = 0.20, = 0.15, = 0.14, = 0.17	Chance event eliminates some alleles from ancestral population.	Allele freq. = 0.01, = 0.01, = 0.38, = 0.28, = 0.31, = 0.01	New population forms from remaining subset of genotypes.
Nonrandom mating	Allele freq. = 0.45, = 0.31, = 0.24	Individuals have more opportunities to mate.	Allele freq. = 0.83, = 0.12, = 0.05	Favored genotypes become more common.
Gene flow	Allele freq. = 0.18, = 0.17, = 0.19, = 0.15, = 0.14, = 0.17	Many and genotypes leave.	Allele freq. = 0.03, = 0.26, = 0.25, = 0.21, = 0.21, = 0.04	Genotypes remaining are more common.

Figure 12.23 Mechanisms of Evolution: A Summary. At Hardy–Weinberg equilibrium, allele frequencies remain unchanged. Each mechanism of evolution changes allele frequencies. In these images, different-colored dots represent the alleles for a particular gene. The left image of each pair shows the allele frequencies in the ancestral population; the right image shows the frequencies after evolutionary change has occurred.

MULTIPLE CHOICE QUESTIONS

1. What does "descent with modification" mean?
   a. Populations that change quickly are likely to become extinct.
   b. Inherited traits change from generation to generation.
   c. Tracing an evolutionary tree from top to bottom reveals changes in species diversity.
   d. Parents change their own features before passing them on to offspring.

2. In population genetics, evolution describes how _____ change from one generation to the next.
   a. individuals
   b. allele frequencies
   c. phenotype frequencies
   d. communities

3. What is the most accurate way to explain the relationship between antibiotics and antibiotic-resistant bacteria?
   a. Exposing bacteria to antibiotics causes DNA mutations that confer antibiotic resistance.
   b. A population of bacteria evolves antibiotic resistance alleles in order to survive exposure to antibiotics.
   c. Bacteria with alleles that confer antibiotic resistance are most likely to survive when exposed to antibiotics.
   d. Antibiotics bind to alleles that confer antibiotic susceptibility, causing changes that make the gene pool more resistant.

4. After an environmental change, foxes with shorter legs than average are most likely to survive and reproduce. What type of selection will act on this population in the coming generations?
   a. Directional selection
   b. Stabilizing selection
   c. Disruptive selection
   d. Normalizing selection

5. Assume that the dominant allele for the D gene occurs at a frequency of 0.8 in a population. What percentage of the population is heterozygous for the D gene, based on Hardy–Weinberg equilibrium?
   a. 80%
   b. 64%
   c. 32%
   d. 20%

6. Darwin observed that different types of organisms live on either side of a geographical barrier. Such barriers prevent
   a. gene flow.
   b. genetic drift.
   c. sexual selection.
   d. mutation.

Answers to these questions are in appendix A.

WRITE IT OUT

1. List and describe five mechanisms of evolution.

2. How did James Hutton, Georges Cuvier, Georges-Louis Buffon, Jean Baptiste de Lamarck, Charles Lyell, and Thomas Malthus influence Charles Darwin's thinking?

3. Explain how understanding evolution is important to medicine, agriculture, and maintaining the diversity of organisms on Earth.

4. How does variation arise in an asexually reproducing population? A sexually reproducing population?

5. Write a paragraph that describes the connections among the following terms: *gene, nucleotide, allele, phenotype, population, genetic variation, natural selection,* and *evolution.*

6. Explain how harmful recessive alleles can persist in populations, even though they prevent homozygous individuals from reproducing.

7. Fraggles are mythical, mouselike creatures that live underground beneath a large vegetable garden. Of the 100 Fraggles in this population, 84 have green fur, and 16 have gray fur. A dominant allele F confers green fur, and a recessive allele f confers gray fur. Assuming Hardy–Weinberg equilibrium is operating, answer the following questions. (a) What is the frequency of the gray allele f? (b) What is the frequency of the green allele F? (c) How many Fraggles are heterozygotes (Ff)? (d) How many Fraggles are homozygous recessive (ff)? (e) How many Fraggles are homozygous dominant (FF)?

8. Which mechanisms of evolution cause random changes in allele frequencies? Which cause nonrandom changes?

9. Jellyfish Lake, located on the Pacific island of Palau, is home to millions of jellyfish. Many years ago, sea levels dropped and the jellyfish were trapped in the basin. The lake contains no predators, and the jellyfish's sting has weakened. Jellyfish Lake is now a popular tourist attraction where snorkelers can swim among the jellyfish. Explain how Jellyfish Lake is evidence for evolution.

10. Suppose that gene A does not affect fitness in a small population of earthworms contained in a glass enclosure. For 20 worm generations, a pair of curious scientists study the frequencies of the A gene's two alleles (A and a). In generation 1, 75% of the alleles are A. By generation 5, the a allele has become more common than A. Then, by generation 20, the A allele is most common again. What is the most likely explanation for this oscillation in allele frequencies of the A gene?

PULL IT TOGETHER

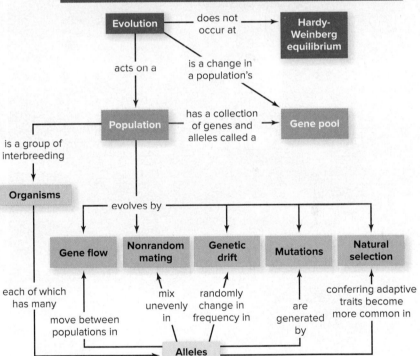

Figure 12.24 Pull It Together: The Forces of Evolutionary Change.

Refer to figure 12.24 and the chapter content to answer the following questions.

1. Review the Survey the Landscape figure in the chapter introduction. When has evolution occurred in life's history? How do scientists know that evolution has occurred in the past?

2. Describe a situation in which each of the five mechanisms of evolution shown in the concept map would occur.

3. Add the terms *genotype, phenotype, allele frequencies, founder effect, bottleneck effect,* and *sexual selection* to this concept map.

13 Evidence of Evolution

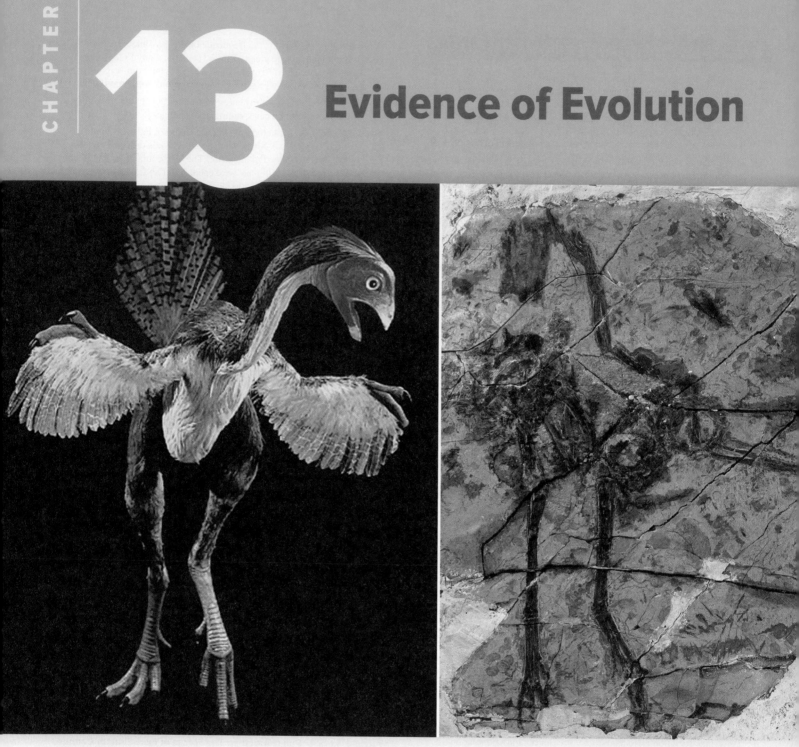

Feathery Dinosaur. The theropod dinosaur *Protarchaeopteryx* lived about 145 million years ago (MYA) in China. It had downy feathers on its body and tail but also larger, barbed feathers at the end of the tail and perhaps elsewhere. The model (*left*) shows what the animal might have looked like, based on evidence from fossils (*right*).

(both): ©O. Louis Mazzatenta/National Geographic Creative Stock

LEARN HOW TO LEARN
Make a Chart

One way to organize the information in a chapter is to make a summary chart or matrix. The chart's contents will depend on the chapter. For this chapter, for example, you might write the following headings along the top of a piece of paper: "Type of Evidence," "Definition," "How It Works," "What It Tells Us," and "Example." Then you would list the lines of evidence for evolution along the left edge of the chart. Start filling in your chart, using the book at first to find the information you need. Later, you should be able to re-create your chart from memory.

Are Birds Dinosaurs?

Like a puzzle with many pieces missing, the history of life has many possible interpretations. But as more information surfaces, the pieces fit together, and the story becomes clearer. Investigating the relationship between birds and dinosaurs is one such story of life.

The idea that birds are closely related to small, meat-eating dinosaurs called theropods began with the discovery of a 150-million-year-old fossil of a theropod with feathers in Bavaria. It was 1860, the year after Darwin published *On the Origin of Species.* The animal, named *Archaeopteryx lithographica,* was the size of a blue jay, with a mix of features seen in birds and nonavian reptiles: wings and feathers, a toothed jaw, and a long, bony tail.

In 1870, Thomas Henry Huxley reported that he had identified 35 features that only ostriches and theropod skeletons shared, which he interpreted as evidence of a close relationship. Others dismissed Huxley's ideas. At the time, dinosaurs were not known to have flown, and theropods were thought to lack bones needed for flight.

In the 1960s, however, Yale University paleontologist John Ostrom strengthened the bird–dinosaur link with an exhaustive comparison of bones discovered in the 1920s. Since then, the similarities between dinosaurs and birds have added up. Many of the structures that make flight possible were present in dinosaurs that lived millions of years before birds. For example, some theropods had hollow bones, an upright stance in which they stood on their toes, a horizontal back, long arms, and a short tail. Rare finds of dinosaur nests reveal behavioral and reproductive similarities to birds.

Decades later, fossils of three types of feathered dinosaurs were unearthed. Researchers discovered *Sinosauropteryx* in China in 1996. This turkey-sized animal had a fringe of downlike structures along its neck, back, and flanks. Then, in 1998, a bonanza fossil find introduced *Protarchaeopteryx* and *Caudipteryx,* both even more feathery than *Sinosauropteryx.*

Still another piece of the puzzle fell into place in 2007, when scientists revealed amino acid sequences of collagen protein molecules extracted from *Tyrannosaurus rex* fossils. Of all the vertebrate species studied, the *T. rex* collagen sequence was more similar to that of a chicken than it was to other reptiles, including alligators and lizards.

The surprising bird–dinosaur link illustrates the changeable nature of science and reinforces the importance of evidence. Every chapter of this textbook contains a section, titled Investigating Life, that explains how biologists test hypotheses about evolution. This chapter summarizes the main types of evidence they use.

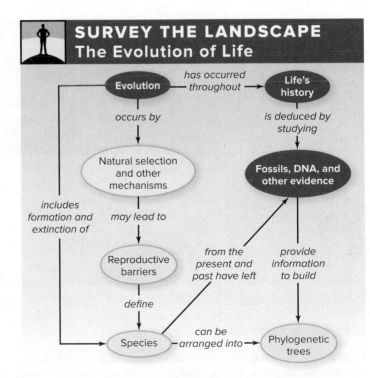

SURVEY THE LANDSCAPE
The Evolution of Life

Life has a rich evolutionary history. Traces of evolution are in fossils, geography, anatomy, patterns of embryonic development, and the sequences of life's molecules.

For more details, study the Pull It Together feature in the chapter summary.

261

13.1 Clues to Evolution Lie in the Earth, Body Structures, and Molecules

The millions of species alive today did not just pop into existence all at once; they are the result of continuing evolutionary change that started with organisms living billions of years in the past. Many types of clues enable us to hypothesize about how modern species evolved from extinct ancestors and to understand the relationships among organisms that live today (see this chapter's Burning Question).

Comparisons among living organisms provided early evidence of the evolutionary relationships among species. Abundant additional data came from **paleontology,** the study of fossil remains or other clues to past life (**figure 13.1**). The discovery of many new types of fossils in the early 1800s created the climate that allowed Charles Darwin to make his tremendous breakthrough (see chapter 12). As people recognized that fossils must represent snapshots from the history of life, scientists developed theories to explain that history. The geographical locations of fossils and modern species provided additional clues.

In Darwin's time, some scientists suspected that Earth was hundreds of millions of years old, but no one knew the exact age. We now know that Earth's history is about 4.6 billion years long, a duration that most people find nearly unimaginable. Scientists describe the events along life's long evolutionary path in the context of the **geologic timescale,** which divides Earth's history into a series of eons and eras defined by major geological or biological events such as mass extinctions (**figure 13.2**). ⓘ *mass extinctions,* section 14.5B

Fossils and biogeographical studies provided the original evidence for evolution, revealing when species most likely diverged from common ancestors in the context of other events happening on Earth. Comparisons of embryonic development and anatomical structures provided additional supporting data. An entirely new type of evidence emerged in the 1960s and 1970s, when scientists began analyzing the sequences of DNA, proteins, and other biological molecules. Since then, the explosion of molecular data has revealed in unprecedented detail how species are related to one another.

Chapter 12 explained how natural selection and other processes drive evolutionary changes, both in the past and today. This chapter examines the different approaches to studying evolution in both living and extinct species. Chapter 14 explains how new species form and become extinct. Chapter 15 continues on this theme by offering a brief history of life on Earth.

13.1 MASTERING CONCEPTS

1. What is the geologic timescale?
2. What types of information provide the clues that scientists use in investigating evolutionary relationships?

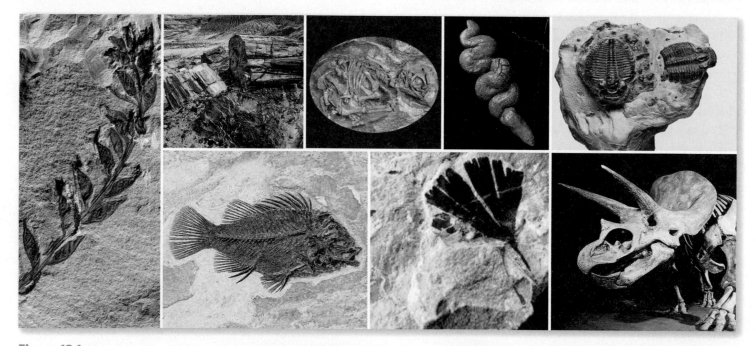

Figure 13.1 Diverse Fossils. Plants and animals have left a rich fossil record. Clockwise from left: *Archaefructus,* an early flowering plant that lived in China more than 100 million years ago; petrified wood; a 190-million-year-old dinosaur egg; fossilized feces of a turtle from the Miocene epoch; trilobites, which were arthropods that became extinct some 250 million years ago; a skull of *Triceratops,* a dinosaur that lived in North America until 65 million years ago; a *Ginkgo* leaf; an exceptionally well-preserved fish.

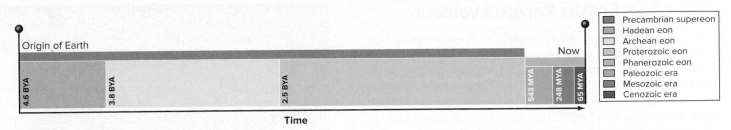

a.

Eon	Era	Period	Epoch	MYA	Important events
			Recent		Human civilization
		Quaternary		0.01	
			Pleistocene		*Homo sapiens*, large mammals; ice ages
				1.8	
		Neogene	Pliocene		Early humans; modern whales
				5.3	
	Cenozoic era		Miocene		First great apes; other mammals continue to diversify; modern birds; expansion of grasslands
	"Age of Mammals"			23.8	
			Oligocene		Elephants, horses; grasses
		Paleogene		33.7	
			Eocene		Mammals and flowering plants continue to diversify; first whales
				54.8	
			Paleocene		First primates; mammals, birds, and pollinating insects diversify
				— 65	
Phanerozoic eon	Mesozoic era "Age of Reptiles"	Cretaceous			Widespread dinosaurs until extinction at end of Cretaceous; flowering plants diversify; present-day continents form
				144	
		Jurassic			First birds; cycads and ferns abundant; giant reptiles on land and in water; first flowering plants
				206	
		Triassic			First dinosaurs; first mammals; therapsids and thecodonts; forests of conifers and cycads
				248	
		Permian			First conifers; fewer amphibians, more reptiles; cotylosaurs and pelycosaurs; Pangaea supercontinent forms
				290	
	Paleozoic era "Age of Amphibians"	Carboniferous			First reptiles; ferns abundant; amphibians diversify; first winged insects
				354	
	"Age of Fishes"	Devonian			First bony fishes, corals, crinoids; first amphibians; first seed plants; arthropods diversify
				417	
		Silurian			First vascular plants and terrestrial invertebrates; first fish with jaws
				443	
		Ordovician			Algae, invertebrates, jawless fishes; first land plants
				490	
		Cambrian			"Explosion" of sponges, worms, jellyfish, "small shelly fossils"; ancestors of all modern animals appear; trilobites
				543	
Precambrian supereon	Proterozoic eon				O₂ from photosynthesis accumulates in atmosphere; first eukaryotes; first multicellular organisms; Ediacaran organisms
				2500	
	Archean eon				Life starts; first bacteria and archaea
				3800	
	Hadean eon				Earth forms
				4600	

b.

Figure 13.2 The Geologic Timescale. (a) Scientists divide Earth's 4.6-billion-year history into four eons. The three earliest eons are combined into the Precambrian supereon, which lasted more than 4 billion years. The most recent eon, the Phanerozoic, started 543 MYA (million years ago). (b) Fossil evidence paints a detailed portrait of life's history during the Phanerozoic, which includes three eras (Paleozoic, Mesozoic, and Cenozoic). Red lines indicate the five largest mass extinction events of the Phanerozoic eon.

13.2 Fossils Record Evolution

A **fossil** is any evidence of an organism from more than 10,000 years ago (the end of the Pleistocene epoch). Fossils come in all sizes, documenting the evolutionary history of everything from microorganisms to dinosaurs to humans. These remains, the oldest of which formed more than 3 billion years ago, give us our only direct evidence of organisms that preceded human history. They occur all over the world and represent all major groups of organisms, revealing much about the geological past. For example, the abundant remains of extinct marine animals called ammonites in Oklahoma indicate that a vast, shallow ocean once submerged what is now the central United States (**figure 13.3**).

Fossils do more than simply provide a collection of ancient remains of plants, animals, and microbes. They also allow researchers to test predictions about evolution. Section 13.7 describes an excellent example: the discovery of *Tiktaalik*, an extinct animal with characteristics of both fishes and amphibians. Based on many lines of evidence showing the close relationship between these two groups, biologists had long predicted the existence of such an animal. *Tiktaalik* finally provided direct fossil evidence of the connection.

A. Fossils Form in Many Ways

Evidence of past life comes in many forms (**figure 13.4**). Often, a fossil forms when an organism dies, becomes buried in sediments, and then is chemically altered. For example, coal, oil, and natural gas (also called fossil fuels) are the decomposed remains of plants and other organisms preserved by compression. Alternatively, minerals can replace the organic matter left by a

Figure 13.3 Big Change. Ammonite fossils such as these are common in land-locked Oklahoma, indicating that what is now the central United States was once covered by an ocean.
©Jean-Claude Carton/Photoshot

decaying organism, which literally turns to stone. Petrified wood forms in this way, as did the ammonite fossils in figure 13.3.

Impression fossils form when an organism presses against soft sediment, which then hardens, leaving an outline or imprint of the body. An impression fossil may also be evidence of an animal's movements, such as footprints. If the imprint later fills with mud that hardens into rock, the resulting cast is a rocky replica of the ancient organism. Teeth, bones, arthropod exoskeletons, and tree trunks often leave casts.

Burning Question

Is evolution really testable?

Some people believe that evolution cannot be tested because it happened in the past. Although no experiment can re-create the conditions that led to today's diversity of life, evolution is testable. In fact, its validity has been verified repeatedly over the past 150 years.

All scientific theories, including evolution, not only explain existing data but also predict future observations. Some other explanations for life's diversity, including intelligent design, are unscientific because they do not offer testable predictions; that is, no future observation can disprove the idea that an intelligent creator designed life on Earth. One strength of evolutionary theory, in contrast, is its ability to predict future discoveries.

For example, if vertebrate life started in the water and then moved onto land, an evolutionary biologist might predict that fishes should appear in the fossil record before reptiles or mammals. A newly found fossil that contradicted the prediction would require investigators to form a new hypothesis. On the other hand, fossils that confirmed the prediction would lend additional weight to the theory's validity.

Evaluating common descent is somewhat similar to solving a crime. The perpetrator may leave footprints, tire tracks, fingerprints,

and DNA at a crime scene. Detectives would develop hypotheses about possible suspects by piecing together the physical and biological evidence. Innocent suspects can exonerate themselves by providing additional information, such as a verified alibi, that contradicts the hypothesis. Ultimately, the best explanation is the one that is consistent with all available evidence.

A mountain of evidence supports the idea of common descent. Extinct organisms have left traces of their existence, both as fossils and as the genetic legacy that all current organisms have inherited. The distribution of life on Earth offers other clues, as does the study of everything from anatomical structures to protein sequences. Laboratory experiments and field observations of natural populations likewise suggest likely mechanisms for evolutionary change and have even documented the emergence of new species. No other scientifically testable hypothesis explains and unifies all of these observations as well as common descent.

Submit your burning question to
Marielle.Hoefnagels@mheducation.com

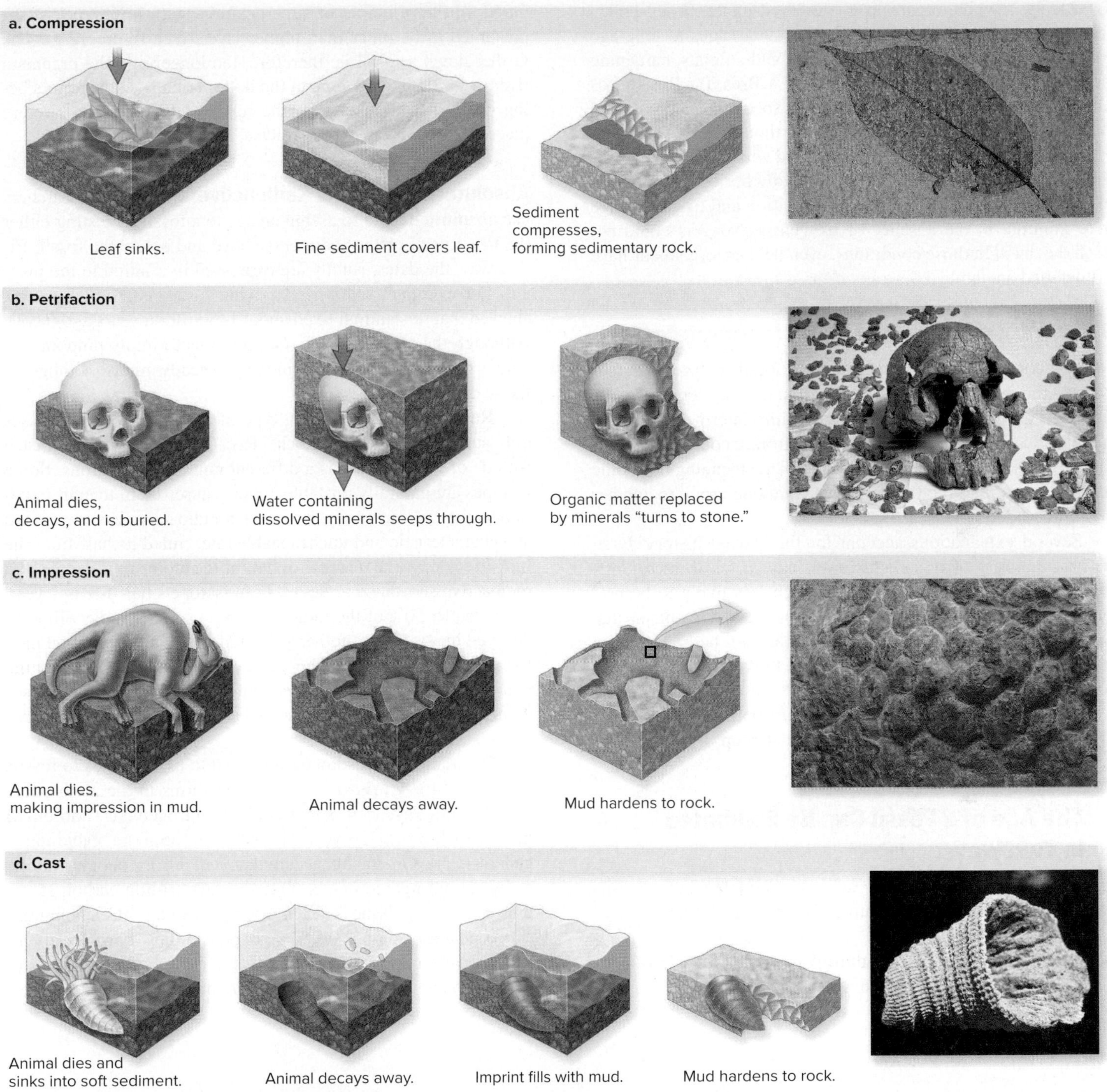

a. Compression

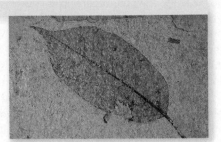

Leaf sinks.

Fine sediment covers leaf.

Sediment compresses, forming sedimentary rock.

b. Petrifaction

Animal dies, decays, and is buried.

Water containing dissolved minerals seeps through.

Organic matter replaced by minerals "turns to stone."

c. Impression

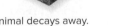

Animal dies, making impression in mud.

Animal decays away.

Mud hardens to rock.

d. Cast

Animal dies and sinks into soft sediment.

Animal decays away.

Imprint fills with mud.

Mud hardens to rock.

e. Intact preservation

Oozing sap traps an insect.

Figure 13.4 How Fossils Form. (a) A compression fossil of a leaf preserves part of the plant. (b) Most fossils of human ancestors consist of mineralized bones and teeth, usually found in fragments. (c) An impression fossil reveals an imprint of anatomical details, such as the scaly skin of a dinosaur. (d) This horn coral is a cast. Once-living material dissolved and was replaced by mud that hardened into rock. (e) Fossils can also be preserved intact in tree resin, which hardens to form amber.

Photos: (a): ©Wong Hock Weng/123RF; (b): ©John Reader/Science Source; (c): ©The Natural History Museum/Alamy; (d): ©Robert Gossington/Photoshot; (e): ©Natural Visions/Alamy

Rarely, a whole organism is preserved intact. Fossil pollen grains in waterlogged lake sediments provide clues to long-ago climates. Sticky tree resin entraps plants and animals, hardening them in translucent amber tombs. The La Brea Tar Pits in Los Angeles have preserved more than 660 species of animals and plants that became stuck in the gooey tar thousands of years ago.

The most striking fossils have formed when sudden catastrophes, such as mud slides and floods, rapidly buried organisms in an oxygen-poor environment. Decomposition and tissue damage were minimal in the absence of oxygen; scavengers could not reach the dead. In those conditions, even delicate organisms have left detailed anatomical portraits (figure 13.5).

B. The Fossil Record Is Often Incomplete

Researchers sometimes collect groups of fossils that reveal, step-by-step, the evolution of one species into another. For example, biologists have found many fossils revealing intermediate stages in the evolution of whales and dolphins from land animals. Usually, however, the fossil record is incomplete, meaning that some of the features marking the transition from one group to another are not recorded in fossils.

Several explanations account for this partial history. First, the vast majority of organisms never leave a fossil trace. Soft-bodied organisms, for example, are much less likely to be preserved than are those with teeth, bones, or shells. Organisms that decompose or are eaten after death, rather than being buried in sediments, are also unlikely to fossilize. Second, erosion or the movements of Earth's continental plates have destroyed many fossils that did form. Third, scientists are unlikely to ever discover the many fossils that must be buried deep in the Earth or submerged under water.

C. The Age of a Fossil Can Be Estimated in Two Ways

Scientists use two general approaches to estimate when a fossilized organism lived: relative dating and absolute dating.

Relative Dating   Relative dating places a fossil into a sequence of events without assigning it a specific age. It is usually based on the principle of superposition, with lower rock strata presumed to be older than higher layers (see figure 12.3). The farther down a fossil is, therefore, the longer ago the organism lived—a little like a memo at the bottom of a stack of papers being older than a sheet near the top. Relative dating therefore places fossils in order from "oldest" to "most recent."

Absolute Dating and Radioactive Decay   Researchers use **absolute dating** to assign an age to a fossil by testing either the fossil itself or the sediments above and below the fossil. Either way, the dates usually are expressed in relation to the present. For example, scientists studying fossilized *Archaeopteryx* showed that this animal lived about 145 million years ago (MYA). Although the term *absolute dating* seems to imply pinpoint accuracy, absolute dating techniques typically return a range of likely dates.

Radiometric dating is a type of absolute dating that uses radioactive isotopes as a "clock." Recall from chapter 2 that each isotope of an element has a different number of neutrons. Some isotopes are naturally unstable, which causes them to emit radiation as they radioactively decay. Each radioactive isotope decays at a characteristic and unchangeable rate, called its half-life. The **half-life** is the time it takes for half of the atoms in a sample of a radioactive substance to decay. If an isotope's half-life is 1 year, for example, 50% of the radioactive atoms in a sample will have decayed in a year. In another year, half of the remaining radioactive atoms will decay, leaving 25%, and so on. If we measure the amount of a radioactive isotope in a sample, we can use the isotope's known half-life to deduce when the fossil formed. ⓘ *isotopes,* section 2.1C

One radioactive isotope often used to assign dates to fossils is carbon-14 (^{14}C; figure 13.6), which forms in the atmosphere when cosmic rays from space bombard nitrogen gas. Carbon-14 has a half-life of 5730 years; it decays to the more stable nitrogen-14 (^{14}N). Organisms accumulate ^{14}C during photosynthesis or by eating organic matter. One in every trillion carbon atoms present in living tissue is ^{14}C; most of the rest are ^{12}C, a nonradioactive (stable) isotope. When an organism dies, however, its intake of carbon, including ^{14}C, stops. As the body's ^{14}C decays without being replenished, the ratio of ^{14}C to ^{12}C decreases.

Figure 13.5 **Doomed Dinosaur Embryos.**
(a) Thousands of dinosaur eggs containing about-to-hatch babies were buried in mud in present-day Argentina about 89 MYA. The newborns would have been only 40 centimeters long but would have grown to an adult length of over 13 meters. (b) Preserved embryonic teeth were about 1 mm long. (c) The skin from some of the embryos was remarkably well preserved.
(a): ©Tim Boyle/Getty Images; (b, c): ©Dr. Luis M. Chiappe

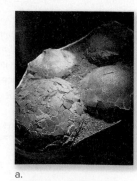
a.

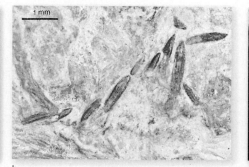
b.

c.

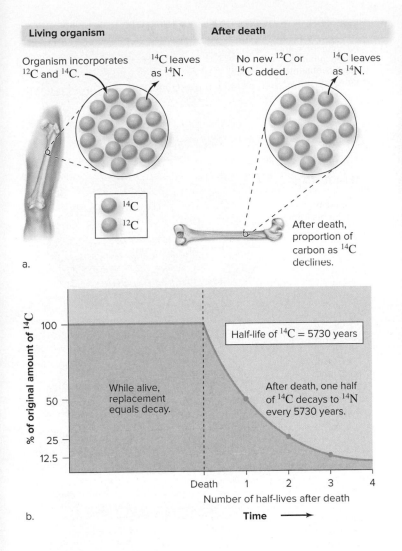

a.

b.

ç.

Figure 13.6 **Carbon-14 Dating.** (a) Living organisms accumulate radioactive carbon-14 (^{14}C) by photosynthesis or eating other organisms. During life, ^{14}C is replaced as fast as it decays to nitrogen-14 (^{14}N). After death, no new ^{14}C enters the body, so the proportion of ^{14}C to ^{12}C declines. (b) During one half-life, 50% of the remaining radioactive atoms in a sample decay. (c) Measuring the proportion of ^{14}C to ^{12}C allows scientists to determine how long ago a fossilized organism—such as this woolly mammoth—died.

(c): ©Ethan Miller/Getty Images

This ratio is then used to determine when death occurred, up to about 40,000 years ago.

For example, radioactive carbon dating determined the age of fossils of vultures that once lived in the Grand Canyon. The birds' remains have about one fourth the ^{14}C-to-^{12}C ratio of a living organism. Therefore, about two half-lives, or about 11,460 years, passed since the animals died. It took 5730 years for half of the ^{14}C to decay, and another 5730 years for half of what was left to decay to ^{14}N.

Another widely used radioactive isotope, potassium-40 (^{40}K), decays to argon-40 (^{40}Ar) with a half-life of 1.3 billion years. It is valuable in dating very old rocks containing traces of both isotopes. Chemical analysis can detect the accumulation of ^{40}Ar in amounts corresponding to fossils that are about 300,000 years old or older.

One limitation of ^{14}C and potassium–argon dating is that they leave a gap, resulting from the different half-lives of the radioactive isotopes. To cover the missing years, researchers use isotopes with intermediate half-lives or turn to other techniques, such as tree-ring comparisons. ⓘ *tree rings*, section 22.4D

Figure It Out

Kennewick Man is a human whose remains were found in 1996 in Washington State. Radiometric dating of a bone fragment suggests that he lived about 9300 years ago. At the time scientists dated the bone, about what percentage of the original amount of ^{14}C remained in his bones?

Answer: Approximately 30%

13.2   **MASTERING CONCEPTS**

1. What are some of the ways that fossils form?
2. Why will the fossil record always be incomplete?
3. Distinguish between relative and absolute dating of fossils.
4. How does radiometric dating work?

13.3 Biogeography Considers Species' Geographical Locations

Geographical barriers such as mountains and oceans greatly influence the origin of species (see chapter 14). It is therefore not surprising that the studies of geography and biology overlap in one field, **biogeography,** the study of the distribution of species across the planet.

A. The Theory of Plate Tectonics Explains Earth's Shifting Continents

Earth's geological history has been extremely eventful. Fossils tell the story of ancient seafloors rising all the way to Earth's "ceiling": the Himalayan Mountains. Littering the Kali Gandaki River in the mountains of Nepal are countless fossilized ammonites, large mollusks similar to the chambered nautilus. How did fossils of marine animals end up more than 3600 meters above sea level?

The answer is that Earth's continents are in motion, an idea called continental drift (**figure 13.7**). According to the theory of **plate tectonics,** Earth's surface consists of several rigid layers, called tectonic plates, that move in response to forces acting deep within the planet. In some places where plates come together, one plate dives beneath another, forming a deep trench. In other areas, mountain ranges form as the plates become wrinkled and distorted. Long ago, the plate that carries the Indian subcontinent moved slowly north and collided with the Eurasian plate. The mighty Himalayas—once an ancient seafloor—rose at the boundary, lifting the marine fossils toward the sky.

Meanwhile, at areas where plates move apart, molten rock seeps to Earth's surface and forms new plate material at the seafloor. As a result, oceans now separate continents that were once joined. This slow-motion dance of the continental plates has dramatically affected life's history as oceans shifted, land bridges formed and disappeared, and mountain ranges emerged.

It may seem hard to imagine that Earth's continents have not always been located where they are now. But a wealth of evidence, including the distribution of some key fossils, indicates that the continents were once united (**figure 13.8**). Deep-sea probes that measure seafloor spreading, along with the locations of the world's earthquake-prone and volcanic "hot spots," reveal that the continents continue to move today.

B. Species Distributions Reveal Evolutionary Events

Biogeographical studies have shed light on past evolutionary events. The rest of this section describes three examples.

The Case of the Missing Marsupials   Marsupials are pouched mammals such as kangaroos, koalas, and sugar gliders. The newborns are tiny, hairless, blind, and helpless. As soon as

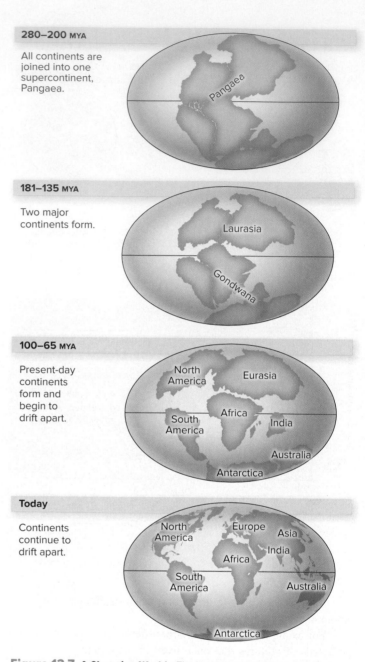

280–200 MYA

All continents are joined into one supercontinent, Pangaea.

Pangaea

181–135 MYA

Two major continents form.

Laurasia

Gondwana

100–65 MYA

Present-day continents form and begin to drift apart.

North America   Eurasia

Africa

South America   India

Australia

Antarctica

Today

Continents continue to drift apart.

North America   Europe   Asia

India

Africa

South America

Australia

Antarctica

Figure 13.7 A Changing World. The locations of Earth's continents have changed with time, due to shifting tectonic plates.

they are born, they crawl along the mother's fur to tiny, milk-secreting nipples inside her pouch.

Marsupials were once the most widespread land mammals on Earth. By about 110 MYA, however, a second group, the placental mammals, had evolved. The young of placental mammals develop within the female's body, nourished in the uterus by the placenta. Baby placental mammals are born more fully developed than are marsupials, giving them a better chance of survival after birth. Because of this reproductive advantage, placental mammals soon displaced marsupials on most continents, including North America. ⓘ *mammals,* section 21.16

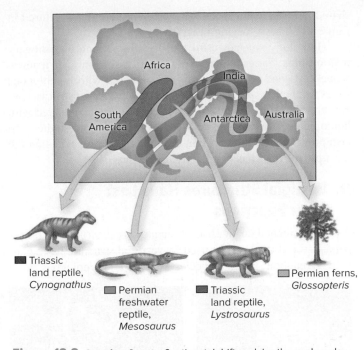

Figure 13.8 Growing Apart. Continental drift explains the modern-day distributions of fossils representing life in the southern hemisphere at the time of Pangaea.

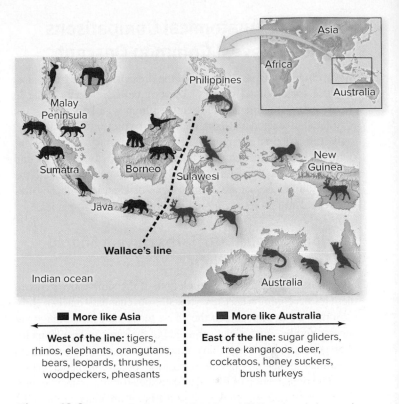

Figure 13.9 Wallace's Line. As Alfred Russel Wallace traveled around the Malay Archipelago, he noticed distinct patterns of animal life on either side of an imaginary boundary, which eventually came to be called Wallace's line.

Nevertheless, fossil evidence suggests that marsupials were diverse and abundant in South America until about 1 or 2 MYA, long after their counterparts on most other continents had disappeared. The reason is that water separated South America from North America until about 3 MYA. But sediments eroding from both continents eventually created a new land bridge, the isthmus of Panama, which permitted migration between North and South America. The resulting invasion of placental mammals spelled extinction for most South American marsupials.

Australia's marsupials remained isolated from competition with placental mammals for much longer. Until about 140 MYA, Antarctica and Australia were part of Gondwana (see figure 13.7). Then, about 60 or 70 MYA, Australia separated from Antarctica and began drifting toward Eurasia. The isolation from the other continents meant marsupials remained free of competition from placentals. In fact, Australia remains unique in that most of its native mammals are still marsupials.

Wallace's Line Biogeography figured prominently in the early history of evolutionary thought. Alfred Russel Wallace, the British naturalist who independently discovered natural selection along with Charles Darwin, had noticed unique assemblages of birds and mammals on either side of an imaginary line in the Malay Archipelago (figure 13.9). The explanation for what came to be called "Wallace's line" turned out to be a deep-water trench that separated the islands, even as sea levels rose and fell over tens of millions of years. The watery barrier prevented the migration of most species, so evolution produced a unique variety of organisms on each side of Wallace's line.

Island Biogeography A smaller-scale application of biogeography is the study of species on island chains. Hawaii, the Galápagos, and other island groups house many unique species that appear nowhere else on Earth. Some of these organisms are downright bizarre; see, for example, the photo that accompanies the introduction to chapter 14.

Although an island's inhabitants may have unusual features, they are typically closely related to species on the nearest mainland. The explanation is that the only organisms that can colonize a newly formed, isolated island are those that can swim, fly, or raft from an inhabited location. Such colonization events may be extremely rare. Those organisms that do reach the island encounter conditions that are different from those on the mainland. With limited migration between the island and the ancestral populations, the relocated organisms have evolved and diversified into multiple new species, a process called adaptive radiation. Just as they did for the marsupials and the species near Wallace's line, geographical barriers have influenced the course of evolution on island chains as well. ⓘ *adaptive radiation, section 14.4B*

13.3 MASTERING CONCEPTS

1. How have the positions of Earth's continents changed over the past 200 million years?
2. How does biogeography provide evidence for evolution?

13.4 Anatomical Comparisons May Reveal Common Descent

Many clues to the past come from the present. As unit 1 explains, all life is made of cells, and eukaryotic cells are very similar in the structure and function of their membranes and organelles. On a molecular scale, cells share many similarities in their enzymes, signaling proteins, and metabolic pathways. Unit 2 describes another set of common features: the relationship between DNA and proteins, and the mechanisms of inheritance. We now turn to the whole-body scale, where comparisons of anatomy and physiology reveal still more commonalities among modern species.

A. Homologous Structures Have a Shared Evolutionary Origin

Two structures are termed **homologous** if the similarities between them reflect common ancestry. (Recall from chapter 9 that two chromosomes are homologous if they have the same genes, though not necessarily the same alleles, arranged in the same order.) Homologous genes, chromosomes, anatomical structures, or other features are similar in their configuration, position, or developmental path.

The organization of the vertebrate skeleton illustrates homology. All vertebrate skeletons support the body, are made of the same materials, and consist of many of the same parts. Amphibians, birds and other reptiles, and mammals typically have four limbs, and the numbers and positions of the bones that make up those appendages are strikingly similar (**figure 13.10**). The simplest explanation is that modern vertebrates descended from a common ancestor that originated this skeletal organization. Each group gradually modified the skeleton as species adapted to different environments.

Note that homologous structures share a common evolutionary origin, but they may not have the same function. The middle ear bones of mammals, for example, originated as bones that supported the jaws of primitive fishes, and they still exist as such in some vertebrates. These bones are homologous and reveal our

shared ancestry with fishes. Likewise, the forelimbs pictured in figure 13.10 have varying functions.

Homology is a powerful tool for discovering evolutionary relationships. For example, as described in section 13.6, a newly sequenced gene or genome can be compared with homologous genes from other species to infer how closely related any two species are. Similarly, fossilized structures are often compared with homologous parts in known species. Comparative studies can also provide clues to the origin of human features, including the hiccups (see this chapter's Apply It Now box).

B. Vestigial Structures Have Lost Their Functions

As environmental changes select against some structures, others persist even if they are not used. A **vestigial** structure has no apparent function in one species, yet it is homologous to a

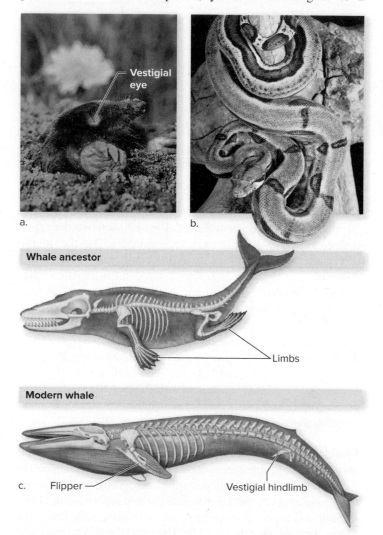

Figure 13.11 **Vestigial Structures.** (a) A mole's tiny vestigial eyes are covered with skin and fur. (b) Boa constrictors and some other snakes have vestigial hindlimbs. (c) Whales descended from mammals with four limbs; some modern whales retain a vestigial pelvis and hindlimbs.

Photos: (a): ©Giel/Getty Images; (b): ©Pascal Goetgheluck/Science Source

Figure 13.10 **Homologous Limbs.** Although their forelimbs have different functions, all of these vertebrates have skeletons that are similarly organized and composed of the same type of tissue.

functional organ in another. (Darwin compared vestigial structures to silent letters in a word, such as the "g" in *night;* they are not pronounced, but they offer clues to the word's origin.)

Figure 13.11 shows three examples of animals with vestigial structures: a mole, a boa constrictor, and a whale. A mole has vestigial eyes (and greatly reduced ears); this animal spends most of its life underground. Boa constrictors and pythons lack external limbs, as do all snakes, but their skeletons reveal the bones of tiny hindlimbs. (See sections 13.5 and 13.7 for more on snake evolution.) Whales also have vestigial hindlimbs, retained from vertebrate ancestors that used legs to walk on land. Their aquatic habitat, meanwhile, selected for forelimbs modified into flippers.

Humans have several vestigial organs. The tiny muscles that make hairs stand on end helped our furry ancestors conserve heat or show aggression; in us, they apparently serve only as the basis of goose bumps. Human embryos have tails, which usually disintegrate long before birth; in other vertebrates, tails persist into adulthood. Above our ears, a trio of muscles (which most of us can't use) help other mammals move their ears in a way that improves hearing. Each vestigial structure links us to other animals that still use these features.

Vestigial structures abound among other organisms, too. They include mitochondria and chloroplasts in some protists; male flower parts and chloroplasts in some plants; and everything from shells to wings in some animals.

C. Convergent Evolution Produces Superficial Similarities

Some body parts that appear superficially similar in structure and function are not homologous. Rather, they are **analogous,** meaning that the structures evolved independently. Flight, for example, evolved independently in birds and in insects. The bird's wing is a modification of vertebrate limb bones, whereas the insect's wing is an outgrowth of the exoskeleton that covers its body (figure 13.12). The wings have the same function—flight—and enhance fitness in the face of similar environmental challenges. The differences in structure, however, indicate they do not have a common developmental pathway. (In evolutionary biology, *homoplasy* is the technical term for a similarity between structures that evolved independently.)

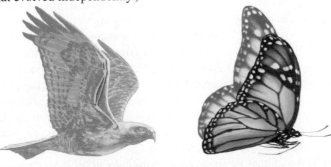

Figure 13.12 Analogous Structures. The wings of birds and butterflies both function in flight, but they are analogous structures because they are not made of the same materials, nor are they organized in the same way. They are not inherited from a recent common ancestor.

a. Cave animals

b. Desert plants

Figure 13.13 Convergent Evolution. (a) The blind cave salamander (*left*) and the cave crayfish (*right*), both from Florida, lack both eyes and pigment. (b) Cardon cacti from the desert in Mexico (*left*) have adaptations similar to those of *Euphorbia* plants from the Namib Desert in Africa (*right*).

(a, both): ©Danté Fenolio/Science Source; (b, Cardon): ©DLILLC/Corbis RF; (b, Euphorbia): ©Natphotos/Getty Images RF

Analogous structures are often the product of **convergent evolution,** which produces similar adaptations in organisms that do not share the same evolutionary lineage. The loss of pigmentation and eyes in cave animals provides a compelling example of convergent evolution, as does the similar appearance of unrelated desert plants from Mexico and Africa (figure 13.13).

Likewise, the similarities between sharks and dolphins illustrate the power of selective forces in shaping organisms. A shark is a fish, whereas a dolphin is a mammal that evolved from terrestrial ancestors. The two animals are not closely related; their last common ancestor lived hundreds of millions of years ago. Nevertheless, their marine habitat and predatory lifestyle have selected for many shared adaptations, including the streamlined body and the shape and locations of the fins or flippers.

13.4 MASTERING CONCEPTS

1. What can homologous structures reveal about evolution?
2. What is a vestigial structure? What are some examples of vestigial structures in humans and other animals?
3. What is convergent evolution?

13.5 Embryonic Development Patterns Provide Evolutionary Clues

Because related organisms share many physical traits, they must also share the processes that produce those traits. Developmental biologists study how the adult body takes shape from its single-celled beginning. Careful comparisons of developing body parts can be enlightening. As just one example, figure 13.14 shows how skulls that look alike as fetuses can develop in different ways, depending on how each part grows in proportion to the others.

Developmental biologists have also photographed embryos and fetuses of a variety of vertebrate species (figure 13.15). The images reveal homologies in the overall structure of the embryonic bodies and in specific features such as the tail.

More recently, the discovery of genes that contribute to development has spawned the field of evolutionary developmental biology (or "evo-devo" for short). Recall from chapter 7 that a

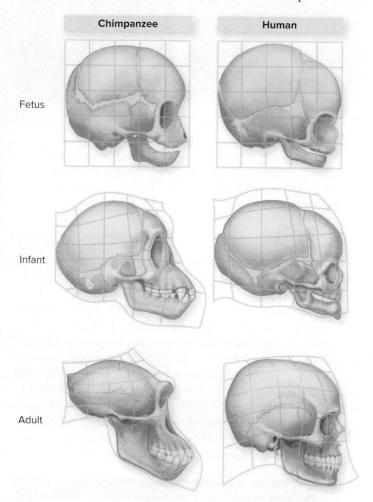

Figure 13.14 **Same Parts, Different Proportions.** The fetuses of humans and chimpanzees have very similar skulls, but the two species follow different developmental pathways. The grids superimposed on the images show the relative growth of each skull part over time. In an adult human, the brain is larger and the jaw is smaller than in chimpanzees.

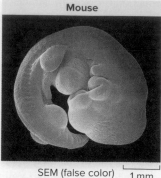

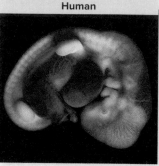

Figure 13.15 Embryo Resemblances. Vertebrate embryos appear alike early in development. As development continues, parts grow at different rates in different species, and the embryos begin to look less similar.

(chick): ©Oxford Scientific/Getty Images; (mouse): ©Steve Gschmeissner/Science Source; (human): ©Science Source

gene is a region of DNA that encodes a protein. Some genes encode proteins that dictate how an organism will develop. One goal of evo-devo research is to identify these genes and determine how mutations can give rise to new body forms.

A basic question in developmental biology, for example, is how a clump of identical cells transforms into a body with a distinct head, tail, segments, and limbs. One way to learn more about this process is to study genes encoding proteins that regulate development. Mutations in these genes can produce dramatic new phenotypes—see, for example, the fruit fly with legs growing out of its head in figure 7.20.

Homeotic is a general term describing any gene that, when mutated, leads to organisms with structures in abnormal or unusual places. Many homeotic genes encode proteins that regulate the expression of other genes. Homeotic genes occur in all animal phyla studied to date, as well as in plants and fungi, and they provide important clues about development.

Consider, for example, a pair of homeotic genes that influence limb formation in vertebrates (figure 13.16). In the chick embryo in figure 13.16a, the gene labeled A prompts wing development, whereas the gene labeled B stimulates formation of the legs. No limbs occur where both genes are expressed, such as in the midsection of the body. Now compare the chick with the pattern of gene expression in the python (figure 13.16b). The same two genes are expressed along most of the snake's body; as a result, the animal never develops forelimbs at all. (Another gene prevents the development of the python's vestigial hindlimbs.)

Researchers have also discovered that new phenotypes can come from mutations in DNA that does not encode proteins. As described in chapter 7, eukaryotic cells require proteins called

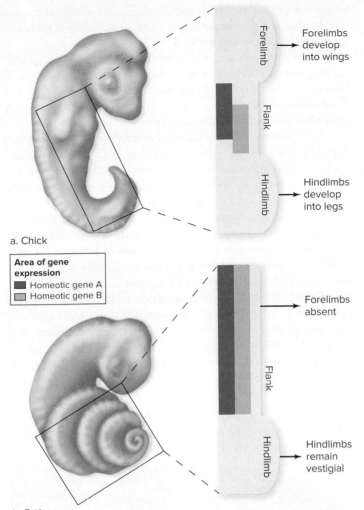

a. Chick

Area of gene expression
- ■ Homeotic gene A
- ▨ Homeotic gene B

b. Python

Figure 13.16 **Homeotic Genes.** (a) Two homeotic genes are expressed unequally along the length of a chicken embryo. Where both genes are expressed, no limbs form. (b) The same pair of homeotic genes prevents the development of forelimbs in a python embryo.

transcription factors for gene expression. The transcription factors bind to areas of DNA called enhancers, signaling a gene to turn on. **Figure 13.17** shows that mutations in these enhancers can cause new features such as wing spots to develop in fruit flies. Perhaps more interesting is the observation that changes in other enhancers can cause the loss of a trait (in this case, pigmentation), even though the gene encoding the pigment remains intact.

Evo-devo studies can even help biologists peer hundreds of millions of years into the past. For example, sponges—the simplest animals—contain the same developmental genes as do much more complex animals. If this genetic "toolbox" was already in place at the dawn of animal evolution, it may explain how diverse phyla arose so quickly during the Cambrian period (see section 21.17).

These examples only scratch the surface of the types of information about evolution that biologists can learn by studying development. The relatively new evo-devo field is sure to yield many more insights into evolution in the future.

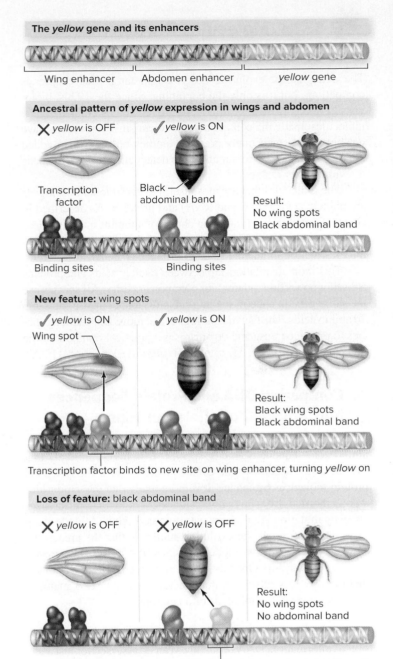

Figure 13.17 **Gene Regulation.** In fruit flies, black pigment is deposited wherever the gene called *yellow* is turned on. Different combinations of transcription factors bound to enhancers determine whether *yellow* is activated in a fly's wings and abdomen. These markings are important in the courtship rituals of many fruit fly species.

13.5 MASTERING CONCEPTS

1. How does the study of embryonic development reveal clues to a shared evolutionary history?
2. Why are evolutionary biologists interested in how genes influence development?

13.6 Molecules Reveal Relatedness

The evidence for evolution described so far in this chapter is compelling, but it is only the beginning. Since the 1970s, biologists have compiled a wealth of additional data by comparing the molecules inside the cells of diverse organisms. The results have not only confirmed many previous studies but have also added unprecedented detail to our ability to detect and measure the pace of evolutionary change.

The molecules that are most useful to evolutionary biologists are nucleic acids (DNA and RNA) and proteins. As described in chapter 2, nucleic acids are long chains of subunits called nucleotides, whereas proteins are composed of amino acids. Cells use the information in nucleic acids to produce proteins.

This intimate relationship between nucleic acids and proteins is itself a powerful argument for common ancestry. All species use the same genetic code in making proteins, and cells use the same 20 amino acids. The fact that biologists can move DNA among species to produce transgenic organisms is a practical reminder of the universal genetic code. ⓘ *transgenic organisms,* section 11.2A

A. Comparing DNA and Protein Sequences May Reveal Close Relationships

To study molecular evolution, biologists compare nucleotide and amino acid sequences among species. It is highly unlikely that two unrelated species would evolve precisely the same DNA and protein sequences by chance. It is more likely that the similarities were inherited from a common ancestor and that differences arose by mutation after the species diverged. Thus, an underlying assumption of molecular evolution studies is that the greater the similarity between two modern species, the closer their evolutionary relationship. One advantage of molecular comparisons is that they are less subjective than deciding whether two anatomical structures are homologous or analogous.

DNA   The ability to rapidly sequence DNA has led to an explosion of information. DNA differences can be assessed for just a few bases, for one gene, for families of genes with related structures or functions, or for whole genomes. Biologists routinely locate a gene in one organism, then scan huge databases to study homologous genes in other species. ⓘ *DNA sequencing,* section 11.2B

Most of a cell's genetic material is in its nucleus, but the cell's numerous mitochondria also contain DNA. Mitochondrial DNA (mtDNA) presents unique possibilities. Each cell contains many mitochondria, each containing many DNA molecules. Some of those mtDNA copies often remain intact in extinct organisms and museum specimens, even if the nuclear DNA is badly degraded. It is therefore sometimes easier to trace long-term evolutionary events with mtDNA than with nuclear DNA. ⓘ *mitochondria,* section 3.4C

For example, many studies show that people from Africa have the most diverse mtDNA sequences. Because mutations take time to accumulate, these results suggest that Africans have existed longer than other modern populations. The idea that early humans originated in Africa and then migrated to the other continents is called the single origin ("out of Africa") hypothesis. ⓘ *human evolution,* section 15.4

The Y chromosome, which passes only from father to son, has also been useful in tracking human migration. Because the Y chromosome does not exchange much genetic material with the X chromosome during meiosis, it accumulates changes much more slowly than do the other chromosomes.

Biologists have also sequenced noncoding DNA, including transposons and pseudogenes (see section 11.2B). Neither transposons nor pseudogenes have any known function, yet their sequences are similar in closely related species. The best explanation for these observations is common descent.

To make a large-scale tree that incorporates all life, researchers study DNA sequences that evolve slowly and are common to all organisms. The genes encoding ribosomal RNA fit the bill. Carl Woese used these genes to deduce the existence of the three domains (Archaea, Bacteria, and Eukarya).

The recent explosion of genetic information has also helped explain how evolution works. We now know that cells may add new functions by acquiring DNA from other organisms (see figure 17.9) and by duplicating genes (figure 13.18). And we have already seen that studying gene expression can reveal differences among closely related species (see figure 13.17 and section 15.5). The list of additional applications is endless, ranging from studies pinpointing the origin of diseases (see

Figure 13.18 Gene Duplication. Chimpanzees consume little starch compared to humans. This dietary difference apparently selected for additional copies of the gene encoding amylase, the starch-digesting enzyme.
Photos: (chimp): ©FLPA/Alamy Stock Photo; (girl): ©Brand X Pictures/Alamy RF

Cytochrome c Evolution	
Organism	Number of amino acid differences from humans
Chimpanzee	0
Rhesus monkey	1
Rabbit	9
Cow	10
Pigeon	12
Bullfrog	18
Fruit fly	25
Yeast	40

Figure 13.19
Cytochrome c Comparison. The more recent the shared ancestor with humans, the fewer the differences in the amino acid sequence for the respiratory protein cytochrome c.

section 16.7) to monitoring the evolution of pesticide-resistant insects (see section 10.10).

Proteins   Like DNA, homologous protein sequences also often support fossil and anatomical evidence of evolutionary relationships. One study, for instance, found 7 of 20 proteins to be identical in humans and chimps, our closest relatives. Many other proteins have only minor sequence differences from one species to another. The keratin of sheep's wool, for example, is virtually identical to that of human hair. The similarity reflects the shared evolutionary history of all mammals.

To study even broader groups of organisms, biologists use proteins that are present in all species. One example is cytochrome c, which is part of the electron transport chain in respiration (see section 6.5C). **Figure 13.19** shows that the more closely related two species are, the more alike is their cytochrome c.

B. Molecular Clocks Help Assign Dates to Evolutionary Events

A biological molecule can act as a "clock." A **molecular clock** uses DNA sequences to estimate the time when the organisms diverged from a common ancestor. In the example shown in **figure 13.20**, biologists used the known mutation rate for a gene, plus the number of DNA sequence differences between species 1 and species 2, to estimate that the two species last shared an ancestor about 50 million years ago.

DNA sequences helped estimate the time since humans and chimpanzees diverged. Many human and chimp genes differ in 4% to 6% of their nucleotides, and substitutions occur at an estimated rate of 1% per 1 million years. Therefore, the two species diverged about 4 million to 6 million years ago.

Researchers can use either nuclear DNA or mtDNA as a molecular clock. Mitochondrial DNA is especially valuable in tracking recent evolutionary events because its molecular clock "ticks" 5 to 10 times faster than the nuclear DNA clock.

Molecular clock studies, however, are not quite as straightforward as glancing at a wristwatch. DNA replication errors occur in different regions of a chromosome at different rates. In addition, if too much time has passed since two species diverged, the same site may have undergone multiple changes, which would be impossible to detect. Researchers must account for these limitations when interpreting a molecular clock.

13.6   MASTERING CONCEPTS

1. How does analysis of DNA and proteins support other evidence for evolution?
2. What is an advantage of using mtDNA instead of nuclear DNA in tracing evolution?
3. How can molecular clocks help determine when two species diverged from a common ancestor?

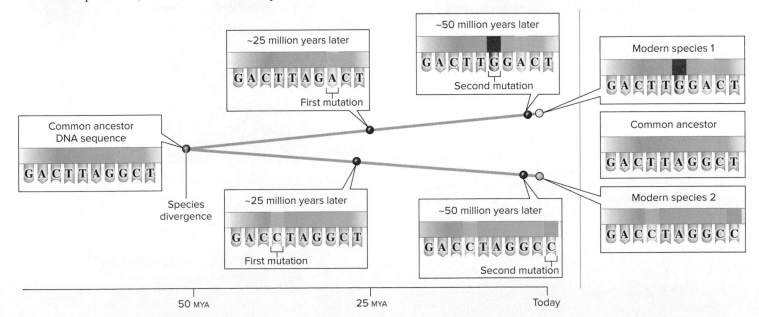

Figure 13.20 Molecular Clock. If a sequence of DNA accumulates mutations at a regular rate, then the number of sequence changes can act as a clock that tells how much time has passed since two species last shared a common ancestor. In this example, mutations occur about once every 25 million years.

INVESTIGATING LIFE

13.7 Limbs Gained and Limbs Lost

Every now and then, spectacular fossils grab headlines worldwide. That is exactly what happened with news of fossils that shed light on two important questions in vertebrate evolutionary biology: How did terrestrial vertebrates gain four limbs as they evolved from their fish ancestors, and how did snakes lose their limbs?

Early amphibians crawled onto land some 375 MYA, and the descendants of those early colonists are today's amphibians, reptiles, and mammals. Fossils discovered over the past half-century, including *Acanthostega* and *Ichthyostega,* have clarified the fish–amphibian transition (see figure 15.15). Still, some details of this fascinating event remain poorly understood.

Scientists Edward B. (Ted) Daeschler of Philadelphia's Academy of Natural Sciences, Neil Shubin of the University of Chicago, and Harvard University's Farish Jenkins added new insights when they published back-to-back papers in the journal *Nature.* The articles described fossils of an extinct animal, *Tiktaalik roseae,* that the researchers had unearthed in Arctic Canada (figure 13.21). *Tiktaalik* either crawled or paddled in shallow tropical streams about 380 MYA, during the late Devonian period. (Although today's Canada is anything but tropical, the entire North American continent was near the equator during the Devonian.)

The animal had an uncanny mix of characteristics from fish and terrestrial vertebrates. Like a fish, *Tiktaalik* had scales and gills. Like a land animal, it had lungs, and its ribs were robust enough to support its body. It could also move its head independently of its shoulders, something that a fish cannot do. But the appendages got the most attention. *Tiktaalik* had movable wrist bones that were sturdy enough to support the animal in shallow water or on short excursions to land. Although the bones were clearly limblike, the "limbs" were fringed with fins, not toes.

The *Tiktaalik* fossils caught the world's eye because they were extraordinarily complete and exquisitely preserved. Scientifically, *Tiktaalik* is important for two reasons. First, it adds to our knowledge about the evolution of terrestrial vertebrates.

Apply It **Now**

An Evolutionary View of the Hiccups

It happens to everyone: We eat or drink too fast, or we laugh too hard, and then the hiccups begin. A hiccup is an involuntary muscle spasm that causes a person to inhale sharply. At the same time, a flap of tissue called the epiglottis blocks the airway to the lungs, producing the classic "hic" sound. This quick intake of air doesn't prevent or solve any known problem. So why do we get the hiccups at all?

Hiccups result from irritation of one or both of the phrenic nerves that trigger contraction of the diaphragm, which controls breathing. The great distance between the origin of these nerves (in the neck) and the diaphragm (below the lungs) offers many opportunities for irritation and, thus, hiccups. A more practical arrangement would be for the nerves that control breathing to emerge from the spinal cord nearer the diaphragm. But we inherited these nerves from our fishy ancestors, whose gills are near where the breathing-control nerves emerge (figure 13.A).

A second clue to the origin of hiccups comes from a close examination of young amphibians. Tadpoles have both lungs and gills, so they can gulp air or extract oxygen from water. In the latter case, they pump mouthfuls of water across their gills; at the same time, the glottis closes to keep water out of the lungs. The tadpole's breathing action almost exactly matches what happens in a human hiccup.

Taken together, these two lines of anatomical evidence suggest that the hiccup is an accident of evolution—a remnant of our shared evolutionary history with the vertebrates that came before us.

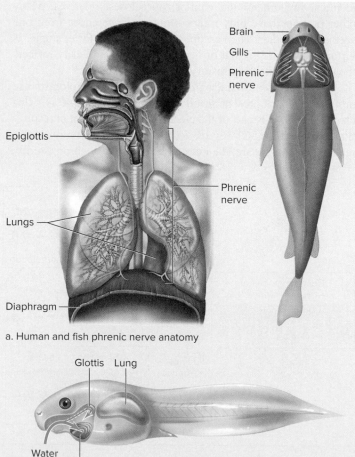

a. Human and fish phrenic nerve anatomy

b. Tadpole pumping water through gills (glottis blocks lungs)

Figure 13.A **Hiccup Origins.** (a) Humans have long phrenic nerves compared to those of fish. (b) When a tadpole breathes with its gills, its glottis closes. This action resembles a human hiccup.

Figure 13.21 Fossil "Fishapod." This photo shows a portion of the *Tiktaalik* fossil.

©Ted Daeschler/VIREO/Academy of Natural Sciences

Second, it highlights the predictive power of evolutionary biology. The researchers did not simply stumble on *Tiktaalik* by accident. Instead, they were looking for a fossil representing the fish–amphibian transition, based on previous knowledge of how, when, and where vertebrates moved onto land hundreds of millions of years ago. Finding *Tiktaalik* confirmed the prediction.

Fossils have answered the question of how terrestrial vertebrates got their limbs, but until recently, the same was not true for another important issue in vertebrate evolution: How did snakes *lose* their limbs? Molecular and anatomical information, including vestigial legs in some snakes (see figure 13.11), clearly indicated that snakes evolved from lizards. Yet the precise four-legged ancestor remained elusive.

Scientists proposed two competing hypotheses to explain the origin of snakes. Noting that snakes resemble two existing groups of burrowing lizards, some scientists suggested that snakes evolved on land. Opposing scientists, citing skull and jaw similarities, contend that snakes descend from mosasaurs, marine reptiles that thrived during the Cretaceous period.

Argentinian paleontologist Sebastián Apesteguía, along with Brazilian colleague Hussam Zaher, added a critical clue to the debate over snake origins when they reported finding three fossilized snakes in the Patagonia region of Argentina (figure 13.22). The snakes, which they named *Najash rionegrina,* lived during the Upper Cretaceous period, about 90 MYA.

Najash is different from other ancient snakes for at least two reasons. First, it is the first snake ever found to have not only functional legs and a pelvis but also a sacrum—a bone connecting the pelvis to the spine. The sacrum is important because lizards and other terrestrial vertebrates have the same bone. *Najash* is therefore more primitive than any snake ever found, including fossils of marine snakes (see figure 13.22b). Second, both the fossil's features and the rock where it was found suggest that it was terrestrial. Taken together, these two pieces of evidence seem to settle the matter: Snakes originated on land.

Spectacular fossils such as *Tiktaalik* and *Najash* spark a flurry of excitement that obscures the countless hours of tedious, labor-intensive work needed to interpret fossils. Researchers scrutinize the scales, limbs, skull, jaw, teeth, vertebrae, and other parts of each new find to glean every possible piece of information about the animal and how it lived. In this way, fossils contribute immeasurably to our understanding of life's long history.

Sources: Apesteguía, Sebastián, and Hussam Zaher. April 20, 2006. A Cretaceous terrestrial snake with robust hindlimbs and a sacrum. *Nature,* vol. 440, pages 1037–1040.

Daeschler, Edward B., Neil H. Shubin, and Farish A. Jenkins, Jr. April 6, 2006. A Devonian tetrapod-like fish and the evolution of the tetrapod body plan. *Nature,* vol. 440, pages 757–763.

Shubin, Neil H., Edward B. Daeschler, and Farish A. Jenkins, Jr. April 6, 2006. A pectoral fin of *Tiktaalik roseae* and the origin of the tetrapod limb. *Nature,* vol. 440, pages 764–771.

13.7 MASTERING CONCEPTS

1. How might the ability to crawl on land for short periods have enhanced the reproductive fitness of *Tiktaalik*?
2. How might the loss of hindlimbs enhance the reproductive fitness of a burrowing animal such as *Najash*?

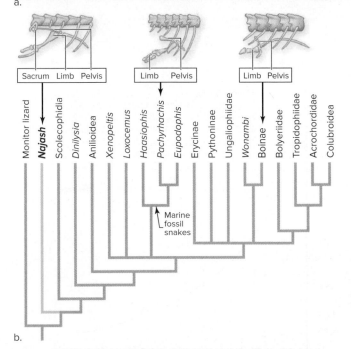

a.

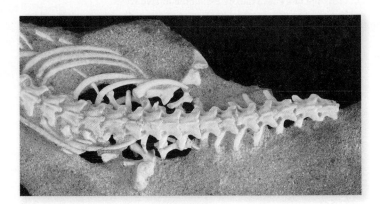

b.

Figure 13.22 *Najash,* **the Fossil Snake.** (a) *Najash* had hindlimbs, a pelvis, and a sacrum. (b) *Najash* was one of the most primitive snakes, whereas all three types of extinct marine snakes with legs arose later. This evidence suggests that snakes evolved on land and later colonized water.

(a): ©Dr. Hussam Zaher

CHAPTER SUMMARY

13.1 Clues to Evolution Lie in the Earth, Body Structures, and Molecules

- The **geologic timescale** divides Earth's history into segments defined by major events such as mass extinctions.
- Evidence for evolutionary relationships comes from **paleontology** (the study of past life), from biogeography, and from comparison of the physical and biochemical characteristics of species.

13.2 Fossils Record Evolution

- **Fossils** are the remains of ancient organisms.

A. Fossils Form in Many Ways
- Fossils may form when mineral replaces tissue gradually, or they may consist of footprints or feces. Rarely, organisms are preserved whole.

B. The Fossil Record Is Often Incomplete
- Many organisms that lived in the past did not leave any fossil evidence.

C. The Age of a Fossil Can Be Estimated in Two Ways
- The position of a fossil in the context of others provides a **relative date.**
- **Radiometric dating** uses radioactive isotopes to estimate the **absolute date** when an organism lived (**figure 13.23**). The length of an isotope's **half-life** determines whether it is useful for ancient or recent objects.

13.3 Biogeography Considers Species' Geographical Locations

- **Biogeography** is the study of the distribution of species on Earth.

A. The Theory of Plate Tectonics Explains Earth's Shifting Continents
- The **plate tectonics** theory indicates that forces deep inside Earth have moved the continents, creating and eliminating geographical barriers.

B. Species Distributions Reveal Evolutionary Events
- Biogeography provides insight into large- and small-scale evolutionary events.

13.4 Anatomical Comparisons May Reveal Common Descent

A. Homologous Structures Have a Shared Evolutionary Origin
- **Homologous** anatomical structures and molecules have similarities that indicate they were inherited from a shared ancestor.

B. Vestigial Structures Have Lost Their Functions
- **Vestigial** structures have no function in an organism but are homologous to functioning structures in related species.

C. Convergent Evolution Produces Superficial Similarities
- **Analogous** structures are similar in function but do not reflect shared ancestry. **Convergent evolution** can produce analogous structures.
- **Figure 13.24** illustrates the evolutionary origins of homologous, analogous, and vestigial structures.

13.5 Embryonic Development Patterns Provide Evolutionary Clues

- Evolutionary developmental biology combines the study of development with the study of DNA sequences. Many genes, including **homeotic** genes, influence the development of an organism's body parts; mutations in homeotic genes therefore may lead to new phenotypes.

13.6 Molecules Reveal Relatedness

A. Comparing DNA and Protein Sequences May Reveal Close Relationships
- Similarities in molecular sequences are unlikely to occur by chance; descent from a shared ancestor is more likely.
- DNA sequence comparisons provide an indication of the relationships among species, as can the amino acid sequences of proteins.

B. Molecular Clocks Help Assign Dates to Evolutionary Events
- A **molecular clock** compares DNA sequences to estimate the time when two species diverged from a common ancestor.

13.7 Investigating Life: Limbs Gained and Limbs Lost

- Fossils of an extinct animal named *Tiktaalik roseae* reflect the transition between fishes and amphibians, which occurred about 375 MYA.
- Another fossil, *Najash rionegrina*, has helped researchers understand the origin of snakes. The features of *Najash* suggest that snakes arose on land from burrowing ancestors.

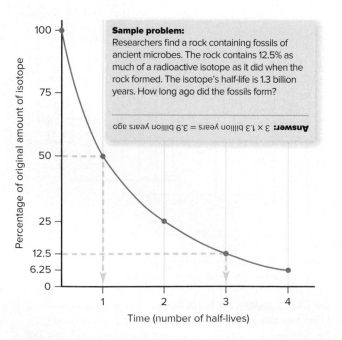

Sample problem:
Researchers find a rock containing fossils of ancient microbes. The rock contains 12.5% as much of a radioactive isotope as it did when the rock formed. The isotope's half-life is 1.3 billion years. How long ago did the fossils form?

Answer: 3 × 1.3 billion years = 3.9 billion years ago

Figure 13.23 Radiometric Dating.

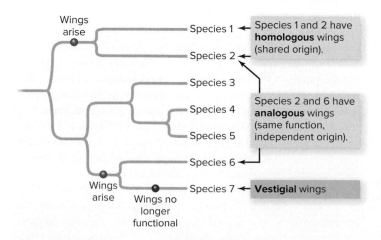

Figure 13.24 Homologous, Analogous, and Vestigial Structures Compared.

MULTIPLE CHOICE QUESTIONS

1. You discover that a 24,000-year-old fossil has one fourth the concentration of a radioactive isotope compared to a living organism. What is the half-life of this isotope?
   a. 3000 years
   b. 6000 years
   c. 8000 years
   d. 12,000 years

2. In fossils found in deeper layers of the Earth, how does the ^{14}C compare to that of fossils found in the upper layers?
   a. Deeper fossils have more ^{14}C.
   b. Deeper fossils have less ^{14}C.
   c. All fossils have equal amounts of ^{14}C.
   d. The layers have different types of ^{14}C.

3. The study of biogeography is most concerned with the
   a. correct placement of species on the evolutionary tree.
   b. precise rock layer in which a fossil is found.
   c. current and past distribution of species on Earth.
   d. predicted locations of future extinction "hot spots."

4. Ground beetles (*Carabus solieri*) have useless hind wings. In related species, the hind wings function in flight. What term describes ground beetles' hind wings?
   a. homeotic
   b. vestigial
   c. analogous
   d. convergent

5. Scorpions occupy every continent except Antarctica, and all scorpions fluoresce under ultraviolet light. What do these observations most likely suggest about the origin of scorpion fluorescence?
   a. The common ancestor of scorpions fluoresced.
   b. Scorpion fluorescence evolved independently on each continent.
   c. Scorpion fluorescence is a vestigial characteristic.
   d. Scorpion fluorescence evolved recently.

6. How does the activity of a homeotic gene relate to evolution?
   a. Homeotic genes serve as markers for convergent evolution.
   b. Organisms with similar homeotic genes have the same vestigial structures.
   c. Mutations in homeotic genes can produce new body plans.
   d. The presence of homeotic genes helps to identify a fossil as that of an animal.

7. Which of the following would be most useful for comparing ALL known groups of organisms?
   a. DNA encoding ribosomal RNA
   b. DNA encoding the keratin protein
   c. Mitochondrial DNA
   d. Y chromosome DNA

Answers to these questions are in appendix A.

WRITE IT OUT

1. Explain the significance of the geologic timescale in the context of evolution.

2. What types of information are used to hypothesize how species are related to one another by descent from a shared ancestor? Give an example of how multiple types of evidence can support one another.

3. Describe six types of fossils and how they form. What present environmental conditions might preserve today's organisms to form the fossils of the future?

4. The bubonic plague swept through western Europe in 1348. Suppose researchers use ^{14}C dating on the skeletons of suspected plague victims, and they discover that about 50% of the original amount of ^{14}C remains in the bones. Are the bones the remains of plague victims?

5. Index fossils represent organisms that were widespread but lived during relatively short periods of time. How might index fossils be useful in relative dating?

6. What hypothesis did Alfred Russel Wallace make about the unique birds and mammals on either side of an imaginary line in the Malay Archipelago? How did the eventual explanation for Wallace's line demonstrate the predictive power of evolution?

7. Suppose that collaborating research teams found fossils of the same extinct species in eastern South America and western Africa. What can the researchers conclude about the age of these fossils without using absolute or relative dating techniques?

8. Why is it important for evolutionary biologists to be able to distinguish between homologous and analogous anatomical structures?

9. Suppose that plants in the San Francisco Bay area and in southern Chile have the same seed-dispersal method. Scientists determine that the evolutionary divergence of these plants happened long before this seed-dispersal method arose in each plant. What term relates the seed-dispersal method of the San Francisco Bay plant to the seed-dispersal method of the southern Chile plant? Explain your answer.

10. Explain why a researcher might combine evidence from fossils and from the anatomy of existing organisms when creating an evolutionary tree.

11. Many species look similar as embryos. What causes them to appear different as adults? Why does the study of development give insights into evolutionary relationships?

12. How do biologists use sequences of proteins and genes to infer evolutionary relationships?

13. Some genes are more alike between human and chimp than other genes are from person to person. Does this mean that chimps are humans or that humans with different alleles are different species? What other explanation fits the facts?

PULL IT TOGETHER

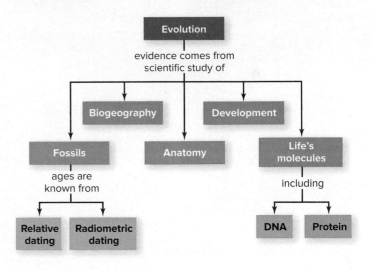

Figure 13.25 Pull It Together: Evidence of Evolution.

Refer to figure 13.25 and the chapter content to answer the following questions.

1. Review the Survey the Landscape figure in the chapter introduction. What diagrams do scientists use to visualize evolutionary relationships? Add this term to the concept map.

2. Write a phrase to connect *fossils* and *biogeography* and a separate phrase to connect *development* and *DNA*.

3. Add the following terms to this concept map: *homologous structures, vestigial structures, homeotic genes,* and *molecular clock*.

14 Speciation and Extinction

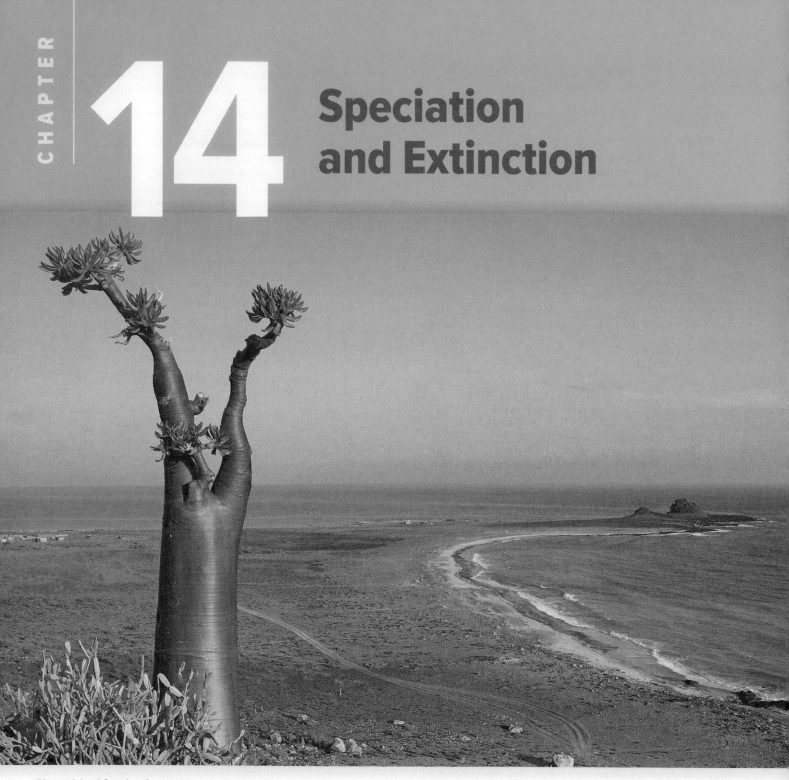

Bizarre Island Species. Socotra, an isolated island in the Indian Ocean, is home to many strange organisms found nowhere else on Earth. In this plant, called the desert rose, the dry climate has selected for a thick trunk that stores water.

©Morandi Bruno/Hemis/Corbis

LEARN HOW TO LEARN
Don't Neglect the Boxes

You may be tempted to skip the chapter opening essays and boxed readings because they're not "required." Read them anyway. The contents should help you remember and visualize the material you are trying to learn. And who knows? You may even find them interesting.

Islands Provide Windows on Speciation and Extinction

Over billions of years, many new species have appeared on Earth in a process called speciation. Yet most species that have ever lived are now extinct. The opposing, ongoing processes of speciation and extinction have defined life's history.

Islands provide ideal opportunities to study speciation. Their small land areas house populations that are relatively easy to monitor. In addition, few organisms can travel vast distances across oceans to reach isolated islands. The descendants of those that have done so have diversified and exploited multiple habitats. Islands located far from a mainland therefore often have groups of closely related species found nowhere else.

Many examples illustrate the spectacular diversity that islands may host. For example, about 800 species of *Drosophila* flies and their close relatives inhabit the Hawaiian Islands. Twenty-eight species of plants called silverswords thrive in every imaginable Hawaiian island habitat, yet all apparently descended from one ancestor that colonized Hawaii long ago. The same islands are also home to birds called honeycreepers, each with a bill adapted to a different food source.

At the same time, island species are especially vulnerable to extinction. Many factors conspire against them. A single hurricane, fire, or flood may destroy a small island population, as can the extinction of a prey species or the introduction of a new predator. Even random fluctuations in birth and death rates can doom a small population.

A volcanic island called Mauritius, in the Indian Ocean, illustrates how humans increase the extinction risk that island populations face. Until the sixteenth century, Mauritius teemed with tall forests, colorful birds, scurrying insects, and basking reptiles. One inhabitant was the large, flightless dodo bird, described by one scientist as "a magnificently overweight pigeon." When European sailors arrived in the 1500s, however, the men ate dodo meat; their pet monkeys and their pigs ate dodo eggs. Rats and mice swam ashore from ships and attacked native insects and reptiles. The sailors' Indian myna birds inhabited nests of the native echo parakeet, while imported plants crowded the seedlings of native trees. By the mid-1600s, only 11 of the original 33 species of birds remained. The dodo was exterminated by 1681, the first of many recorded extinctions caused by human activities.

Both speciation and extinction have likely been a part of evolution since life began. This chapter explains how these processes occur.

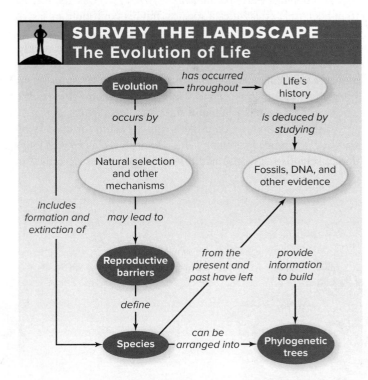

SURVEY THE LANDSCAPE
The Evolution of Life

Evolutionary processes sometimes lead to reproductive barriers within a population, and the isolated groups may become unique species. Meanwhile, existing species may go extinct. Phylogenetic trees track these large-scale evolutionary changes.

For more details, study the Pull It Together feature in the chapter summary.

14.1 What Is a Species?

Throughout the history of life, the types of organisms have changed: New species have appeared, and existing ones have gone extinct. The term **macroevolution** describes these large, complex changes in life's panorama. Macroevolutionary events tend to span very long periods, whereas the microevolutionary processes described in chapter 12 happen so rapidly that we can sometimes observe them over just a few years (see the Burning Question in this section). Nevertheless, the many small changes that accumulate in a population by microevolution eventually lead to large-scale macroevolution.

Evolution has produced an obvious diversity of life. A bacterium, for example, is clearly distinct from a tree or a bird (figure 14.1). At the same time, some organisms are more closely related than others, as the two cats in figure 14.2 illustrate. To make sense of these observations, biologists recognize the importance of grouping similar individuals into **species**—that is, distinct types of organisms. This task requires agreement on what the word *species* means. Perhaps surprisingly, the definition has changed over time and is still the topic of vigorous debate among biologists.

A. Linnaeus Devised the Binomial Naming System

Swedish botanist Carolus Linnaeus (1707–1778) was not the first to ponder what constitutes a species, but his contributions last to this day. Linnaeus defined species as "all examples of creatures that were alike in minute detail of body structure." Importantly, he was the first investigator to give every species a two-word name. Each name combines the broader classification *genus* (plural: genera) with a second word that designates the species. The scientific name for humans, for example, is *Homo sapiens*.

Linnaeus also devised a hierarchical system for classifying species. He grouped similar genera into orders, classes, and kingdoms; scientists now use additional categories, as described in

Figure 14.2 Similar Species. These two species of felines—a leopard and a bobcat—have many characteristics in common.
(leopard): ©Photodisc RF; (bobcat): ©Holly Kuchera/Shutterstock RF

section 14.6. Linnaeus's classifications organized the great diversity of life and helped scientists communicate with one another.

His system did not, however, consider the role of evolutionary relationships. Linnaeus thought that each species was created separately and could not change. Therefore, species could not appear or disappear, nor were they related to one another.

Charles Darwin (1809–1882) finally connected species diversity to evolution. He predicted that classifications would come to resemble genealogies, or extended "family trees." As the theory of evolution by natural selection became widely accepted in the nineteenth and twentieth centuries, scientists no longer viewed classifications merely as ways to organize life. They considered them to be hypotheses about life's evolutionary history.

B. Species Can Be Defined Based on the Potential to Interbreed

In the 1940s, Harvard biologist Ernst Mayr amended the work of Linnaeus and Darwin by considering reproduction and genetics. Mayr defined a **biological species** as a population, or group of populations, whose members can interbreed and produce fertile

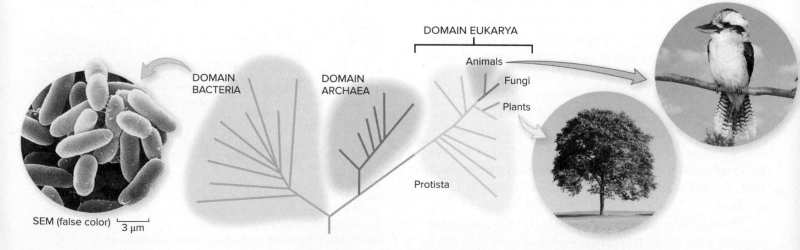

Figure 14.1 Distinctive Species. Bacteria, a tree, and a bird are about as different as three types of organisms can be.
Photos: (bacteria): ©S. Lowry/University Ulster/Getty Images; (tree): ©Jorg Greuel/Getty Images RF; (bird): ©IT Stock/PunchStock RF

Burning Question

Can people watch evolution happen?

Evolution is ongoing in every species, so it is not surprising that biologists have documented many instances of evolution "in action." Medicine provides the most familiar context. One example is the discovery that HIV evolved from a virus that occurs in chimpanzees (see section 16.7); another is the rise of antibiotic resistance among the bacteria that cause staph infections and tuberculosis (see the opening essay for chapter 12).

USDA

The use of pesticides has provided plentiful examples as well. Many people know that populations of DDT-resistant insects skyrocketed shortly after people began using DDT to kill mosquito larvae. Likewise, section 10.10 describes the selection for moth larvae that are resistant to Bt, an insecticidal protein.

The formation of new species is observable as well. Scientists have witnessed the birth of new plant species in genus *Tragopogon,* a group that was introduced to western North America decades ago. In addition, mosquito populations that have been isolated for more than 100 years in the tunnels of the London Underground can no longer breed with their aboveground counterparts—a sure sign that a new species has formed.

Submit your burning question to
Marielle.Hoefnagels@mheducation.com

Figure 14.3 How Many Species? Linnaeus would have categorized these butterflies based on their physical appearance. Mayr's biological species definition, however, provides an objective rule for determining whether each group really is a separate species.
©IT Stock Free/Alamy RF

offspring. **Speciation,** the formation of new species, occurs when some members of a population can no longer successfully interbreed with the rest of the group.

How might this happen? A new species can form if a population somehow becomes divided. Recall from chapter 12 that a population's **gene pool** is its entire collection of genes and their alleles. An intact, interbreeding population shares a common gene pool. After a population splits in two, however, microevolutionary changes such as mutations, natural selection, and genetic drift can lead to divergence between the groups. With the accumulation of enough differences in their separate gene pools, the two groups can no longer produce fertile offspring even if they come into contact once again. In this way, microevolution becomes macroevolution.

The biological species definition does not rely on physical appearance, so it is much less subjective than Linnaeus's observations. Under the system of Linnaeus, it would be impossible to determine whether two similar-looking butterflies belong to different species (**figure 14.3**). Using Mayr's definition, however, we can say that they belong to one species if the two groups can produce fertile offspring together.

Nevertheless, Mayr's species definition raises several difficulties. First, the biological species concept cannot apply to asexually reproducing organisms, such as bacteria, archaea, and many fungi and protists. Second, it is impossible to apply the biological

species definition to extinct organisms known only from fossils. Third, some types of organisms have the *potential* to interbreed in captivity, but they do not do so in nature. Fourth, reproductive isolation is not always absolute. Closely related species of plants, for example, may occasionally produce fertile offspring together, even though their gene pools mostly remain separate.

As a result, the biological species concept does not provide a perfect way to determine the "boundaries" of each species. DNA sequence analysis has helped to fill in some of these gaps. Biologists working with bacteria and archaea, for example, use a stretch of DNA that encodes ribosomal RNA to define species. If the DNA sequences of two specimens are more than 97% identical, they are considered to be the same species. These genetic sequences, however, still present some ambiguity because they cannot reveal whether genetically similar organisms currently share a gene pool. ⓘ *DNA sequencing,* section 11.2B

Despite these difficulties, reproductive isolation is the most common criterion used to define species. The rest of this chapter therefore uses the biological species concept to describe how speciation occurs.

14.1 MASTERING CONCEPTS

1. What is the relationship between macroevolution and microevolution?
2. How does the biological species concept differ from Linnaeus's definition?
3. What are some of the challenges in defining species?

14.2 Reproductive Barriers Cause Species to Diverge

In keeping with Mayr's biological species concept, a new species forms when one portion of a population can no longer breed and produce fertile offspring with the rest of the population. That is, the separate groups no longer share a gene pool, and each begins to follow its own, independent evolutionary path.

One portion of a population can become reproductively isolated in many ways, because successful reproduction requires so many complex events. Any interruption in courtship, fertilization, embryo formation, or offspring development can be a reproductive barrier.

Biologists divide the many mechanisms of reproductive isolation into two broad groups: prezygotic and postzygotic. Prezygotic reproductive barriers prevent the formation of a zygote, or fertilized egg; postzygotic barriers reduce the fitness of a hybrid offspring. (A **hybrid,** in this context, is the offspring of individuals from two different species.) Figure 14.4 summarizes the reproductive barriers; the rest of this section describes them in detail.

Barrier	Description	Example	Illustration
PREZYGOTIC REPRODUCTIVE ISOLATION			
Habitat isolation	Different environments	Ladybugs feed on different plants.	
Temporal isolation	Active or fertile at different times	Field crickets mature at different rates.	
Behavioral isolation	Different courtship activities	Frog mating calls differ.	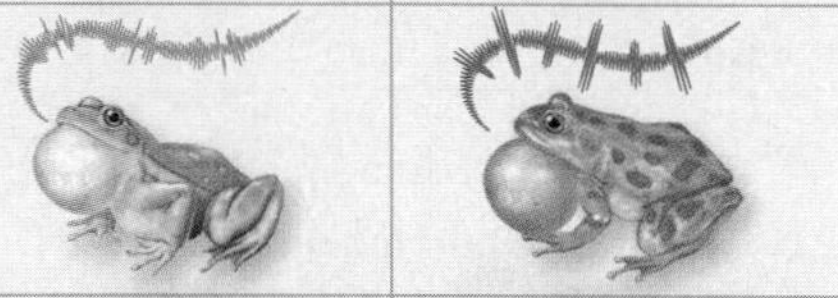
Mechanical isolation	Mating organs or pollinators incompatible	Sage species use different pollinators.	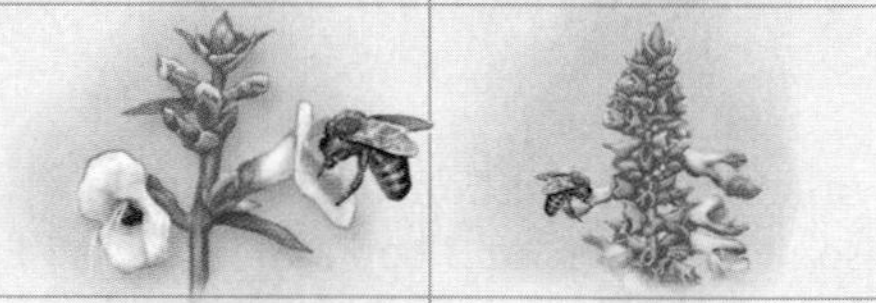
Gametic isolation	Gametes cannot unite.	Sea urchin gametes are incompatible.	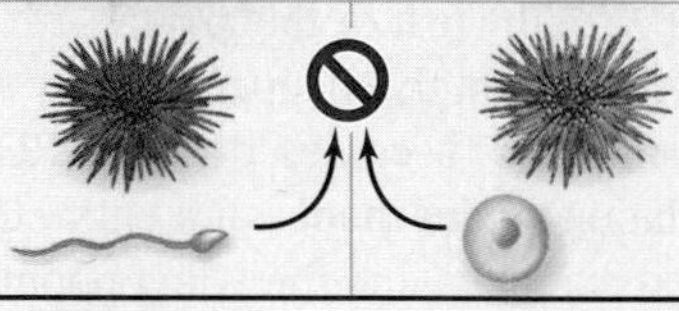
POSTZYGOTIC REPRODUCTIVE ISOLATION			
Hybrid inviability	Hybrid offspring fail to reach maturity.	Hybrid eucalyptus seeds and seedlings are not viable.	
Hybrid infertility (sterility)	Hybrid offspring unable to reproduce	Lion-tiger cross (liger) is infertile.	
Hybrid breakdown	Second-generation hybrid offspring have reduced fitness.	Offspring of hybrid mosquitoes have abnormal genitalia.	

Figure 14.4 **Reproductive Barriers.** Prezygotic barriers prevent gametes from combining to produce a hybrid offspring. Postzygotic barriers keep species separate by selecting against hybrid offspring.

A. Prezygotic Barriers Prevent Fertilization

Mechanisms of **prezygotic reproductive isolation** affect the ability of two species to combine gametes and form a zygote. These reproductive barriers include the following:

- **Ecological (or habitat) isolation:** A difference in habitat preference can separate two populations in the same geographical area. For example, one species of ladybird beetle eats one type of plant, while a closely related species eats a different plant. The two species never occur on the same host plant, although they interbreed freely in the laboratory. The different habitat preferences are the reproductive barrier that keeps the gene pools of the two species separate.

- **Temporal isolation:** Two species that share a habitat will not mate if they are active at different times of day or reach reproductive maturity at different times of year. Among field crickets in Virginia, for example, one species takes much longer to develop than another, so the adults of the two species may not meet until late in the season, if at all. Similarly, the different emergence schedules for 13- and 17-year periodical cicadas suggest that temporal isolation prevents interbreeding between these two species.

- **Behavioral isolation:** Behavioral differences may prevent two closely related species from mating. The males of two species of tree frogs, for instance, use distinct calls to attract mates. Female frogs choose males of their own species based on the unique calls. Likewise, sexual selection in many birds is based on intricate mating dances. Any variation in the ritual from one group to another could prevent them from being attracted to one another. ⓘ *sexual selection,* section 12.6

- **Mechanical isolation:** In many animal species, male and female parts fit together almost like a key in a lock. Any change in the shape of the gamete-delivering or -receiving structures may prevent groups from interbreeding. In plants, males and females do not copulate, but mechanical barriers still apply. For example, although two species of sage plant in California can interbreed, in practice they rarely do because they use different pollinators. One species has flowers that attract large bees, whereas the other accommodates small to medium bees. The different pollinators effectively isolate the gene pools of the two plant species. ⓘ *pollination,* section 24.2C

- **Gametic isolation:** If a sperm cannot fertilize an egg cell, then no reproduction will occur. For example, many marine organisms, such as sea urchins, simply release sperm and egg cells into the water. These gametes display unique surface molecules that enable an egg to recognize sperm of the same species. In the absence of a "match," fertilization will not occur, and the gene pools will remain separate.

B. Postzygotic Barriers Prevent the Development of Fertile Offspring

Individuals of two different species may produce a hybrid zygote. Even then, **postzygotic reproductive isolation** may keep the species separate by selecting against the hybrid offspring, effectively preventing genetic exchange between the populations. Collectively, these postzygotic barriers are sometimes called hybrid incompatibility.

Postzygotic reproductive barriers include the following:

- **Hybrid inviability:** A hybrid embryo may die before reaching reproductive maturity, typically because the genes of its parents are incompatible. For example, two species of eucalyptus trees coexist in California forests, but hybrid offspring are rare. Either the hybrid seeds fail to germinate or the seedlings die soon after sprouting. Since the hybrid offspring cannot reproduce, the gene pools of the parent species remain isolated from one another.

- **Hybrid infertility (sterility):** Some hybrids are infertile. A familiar example is the mule, a hybrid offspring of a female horse and a male donkey. Mules are infertile because a horse's egg has one more chromosome than a donkey's sperm cell. Meiosis does not occur in the mule's germ cells, so the animal cannot produce gametes. Similarly, a liger is the hybrid offspring of a male lion and a female tiger. Like mules, ligers are usually sterile. Although mules and ligers contain the mixed genes of two species, they cannot pass their genes back into either parent population. Therefore, genetic exchange between parent populations is not sustained. ⓘ *mules,* section 9.4

- **Hybrid breakdown:** Some species produce hybrid offspring that are fertile. When the hybrids reproduce, however, their offspring may have abnormalities that reduce their fitness. Some second-generation hybrid offspring of the mosquito species *Aedes aegypti* and *Aedes mascarensis*, for example, have abnormal genitalia that make mating difficult. The strong selective pressure against hybrid offspring limits gene flow between the two mosquito species.

Successful hybridization is rare in animals, but it frequently occurs in plants. One of the problems with introducing nonnative species of plants into a region is the potential production of hybrids that displace the native plants. On the other hand, some food crops are the result of hybridization. A fruit called a tangelo, for example, is the hybrid offspring of a tangerine and a pomelo (grapefruit); a plumcot is a cross between a plum and an apricot.

14.2 MASTERING CONCEPTS

1. How do reproductive barriers lead to speciation?
2. Write a real or fictitious example (other than those listed in figure 14.4) of each type of reproductive barrier.

14.3 Spatial Patterns Define Three Types of Speciation

Reproductive barriers keep related species apart, but how do these barriers arise in the first place? More specifically, how can two populations of the same species evolve along different pathways, eventually yielding two species?

The most obvious way is to physically separate the populations so that they do not exchange genes. Evolution would then act independently in the two populations. Eventually, the genetic differences between the populations would give rise to one or more reproductive barriers.

Yet speciation can also occur when two populations have physical contact with each other. They may divide into new species even though they inhabit neighboring regions or even share a habitat. Biologists recognize these different circumstances by dividing the geographical setting of speciation into three categories: allopatric, parapatric, and sympatric (**figure 14.5**).

A. Allopatric Speciation Reflects a Geographical Barrier

In **allopatric speciation,** a new species forms when a geographical barrier physically separates a population into two groups that cannot interbreed (*allo-* means "other," and *patria* means "fatherland"). The barrier may be a river, desert, glacier, mountain range, large body of water, dam, farm, or city. Rising sea levels may also trap populations on isolated islands.

If the separate parts of a population cannot contact each other, gene flow between them stops. Meanwhile, the forces of microevolution continue to alter allele frequencies in each group. Mutations and genetic drift both occur at random, while natural selection favors different allele combinations in each habitat. Eventually, the result of these ongoing microevolutionary changes may be one or more reproductive barriers. And when the descendants of the original two populations can no longer interbreed, one species has branched into two.

Isolated springs offer ideal opportunities for allopatric speciation. The Devil's Hole pupfish, which inhabits a warm spring near Death Valley, California, provides one example (**figure 14.6**). The spring was isolated from other bodies of water about 50,000 years ago, preventing genetic exchange between the fish trapped in the spring and those in the original population. Generation after generation, different alleles have accumulated in each pupfish population. Without reproduction between the pupfish groups, changes in one population were not shared with the others; each population took its own evolutionary path. Since the time that the spring became isolated, the gene pools have shifted enough that a Devil's Hole pupfish cannot mate with fish from nearby springs. It has become a distinct species.

In addition to isolated springs, island archipelagos also offer opportunities for allopatric speciation. For example, 11 subspecies of tortoise occupy the Galápagos Islands. Researchers have compared the DNA collected from 161 individuals and used the sequences to reconstruct many of the events that occurred after tortoises first colonized the islands a couple of million years ago (**figure 14.7**).

According to the DNA analysis, a few newcomers from the South American mainland first colonized either San Cristóbal or Española (also called Hood Island) a couple of million years ago. Tortoises soon floated on ocean currents to nearby islands, where they encountered new habitats that selected for different adaptations, especially in shell shape. Dry islands with sparse vegetation selected for notched shells that enable the tortoises to reach for higher food sources. The Hood Island tortoise in figure 14.7 illustrates this characteristic "saddleback" shape. On islands with lush, low-growing vegetation, the tortoises have domed shells; the tortoise from Santa Cruz Island is an example.

Although many of the subspecies look noticeably different from one another, the tortoises can interbreed. They are not yet separate species, but the genetic similarities among tortoises on each island suggest that gene flow from island to island was historically rare. The Galápagos tortoises illustrate an ongoing process of allopatric speciation.

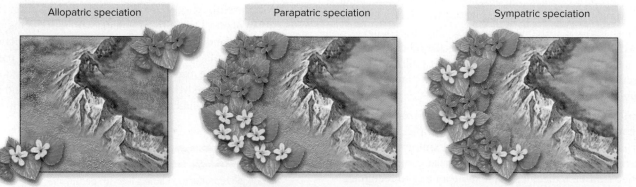

| Allopatric speciation | Parapatric speciation | Sympatric speciation |
| No contact between populations | Populations share a border area | Continuous contact between populations |

Figure 14.5 Speciation and Geography. The three modes of speciation are distinguished based on whether populations are separated by a physical barrier or mingle within a shared border area.

Figure 14.6 Allopatric Speciation in Pupfish. The pupfish species that lives in Devil's Hole cannot breed with pupfish from nearby springs.

Photos: (cave): ©Robyn Beck/AFP/Getty Images; (both fish): ©Stone Nature Photography/Alamy

Separation by a geographical barrier has been considered the most common mechanism of speciation, because the evidence for it is the most abundant and obvious. The diversification of plant and animal species on island archipelagos such as Hawaii and the Galápagos provides many striking examples. So do the world's fishes: Of the 29,000 or so known species of fishes, 36% live in freshwater habitats, although these places account for only 1% of Earth's surface. Compared with the vast, interconnected oceans, the countless springs, lakes, ponds, streams, and rivers provide diverse habitats and ample barriers to genetic exchange.

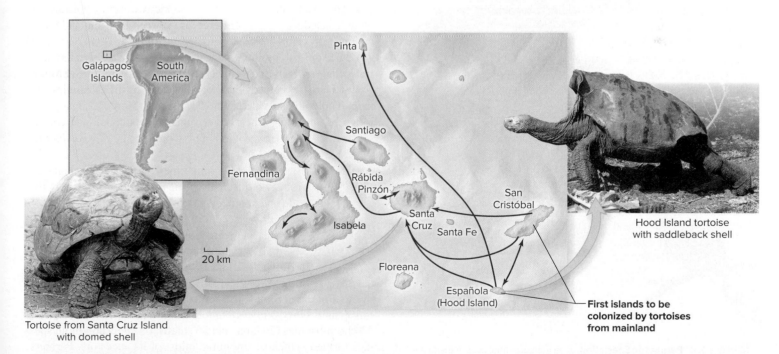

Figure 14.7 Allopatric Speciation in Tortoises. Descendants of the first tortoises to arrive on the Galápagos have colonized most of the islands, evolving into many subspecies. Unique habitats have selected for different sets of adaptations in the tortoises.

Photos: (domed): ©Millard H. Sharp/Science Source; (saddleback): ©Nancy Nehring/Getty Images

B. Parapatric Speciation Occurs in Neighboring Regions

In **parapatric speciation,** part of a population enters a new habitat bordering the range of the parent species (*para-* means "alongside"). Most individuals mate within their own populations, although gene flow may still occur among individuals that venture into the shared border zone.

Despite this limited gene flow, parapatric speciation can occur if the two habitats are different enough to drive disruptive selection. As described in section 12.5, disruptive selection occurs when individuals with intermediate forms have lower fitness than those at either extreme. Parapatric speciation might begin if one phenotype conferred high fitness in one habitat but a different phenotype was adaptive in the neighboring area. As allele frequencies changed independently in each habitat, the two subpopulations would mingle less and less. Over time, the two populations would become genetically (and reproductively) isolated.

Consider the little greenbul, a small, green bird that lives in the tropical rain forest of Cameroon, West Africa (**figure 14.8**). The birds also inhabit the isolated patches of forest that characterize the transitional areas (called ecotones) between rain forest and grassland. In one study, researchers captured birds from six tropical rain forest sites and six ecotone sites. The birds in the ecotone patches had greater weight, deeper bills, and longer legs and wings than their rain forest counterparts.

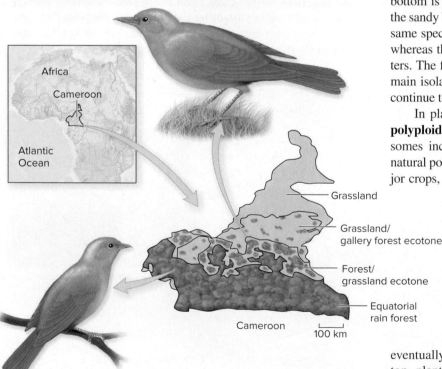

Figure 14.8 Parapatric Speciation. The little greenbul lives in the rain forest and in ecotones, which are patches of forest in the border zone with grasslands. Birds from the ecotone patches are larger than their rain forest counterparts.

What forces might have selected for these differences? The researchers speculated that longer wings increase fitness in the ecotone by improving flight, since the little greenbuls are vulnerable to predation in the open grassland areas separating the patches of forest. Presumably, the foods available in the ecotone and rain forest habitats account for the difference in bill depth.

The little greenbuls from the ecotones can still mate with those from the rain forest, so speciation has not yet occurred. Nevertheless, the researchers concluded that the forces of natural selection are greater than the gene flow between the two populations, gradually taking the groups farther apart. We are likely seeing speciation in action.

C. Sympatric Speciation Occurs in a Shared Habitat

In **sympatric speciation,** a new species arises while living in the same physical area as its parent species (*sym-* means "together"). Among evolutionary biologists, the idea of sympatric speciation can be controversial. After all, how can a new species arise in the midst of an existing population?

Often, sympatric speciation reflects the fact that a habitat that appears uniform actually consists of many microenvironments. Fishes called cichlids, for example, have diversified within African lakes. **Figure 14.9** shows two cichlids in Cameroon's tiny Lake Ejagham. This 18-meter-deep lake has distinct ecological zones. Its bottom is muddy near the center, whereas leaves and twigs cover the sandy bottom near the shore. The two types of fish belong to the same species, but the larger ones consume insects near the shore, whereas the smaller ones eat tiny, floating prey in the deeper waters. The fish breed where they eat, so the two forms typically remain isolated. As genetic differences between the subpopulations continue to accumulate, sympatric speciation may occur.

In plants, a common mechanism of sympatric speciation is **polyploidy,** which occurs when the number of sets of chromosomes increases. Nearly half of all flowering plant species are natural polyploids, as are about 95% of ferns. Moreover, many major crops, including wheat, corn, sugarcane, potatoes, and coffee, are derived from polyploid plants. ⓘ *polyploidy, section 9.7*

Polyploidy sometimes arises when gametes from two different species fuse. Cotton plants provide an example (**figure 14.10**). An Old World species of wild cotton has 26 large chromosomes, whereas one from Central and South America has 26 small chromosomes. The two species interbred, forming a diploid hybrid with 26 chromosomes (13 large and 13 small). This hybrid was sterile. But eventually the chromosome number doubled. The resulting cotton plant, *Gossypium hirsutum,* is a fertile polyploid with 52 chromosomes (26 large and 26 small); this new polyploid species formed sympatrically in the midst of its ancestors. Farmers around the world cultivate this species to harvest cotton for cloth.

Polyploidy can also occur when meiosis fails. A diploid individual may occasionally produce diploid sex cells. Self-fertilization

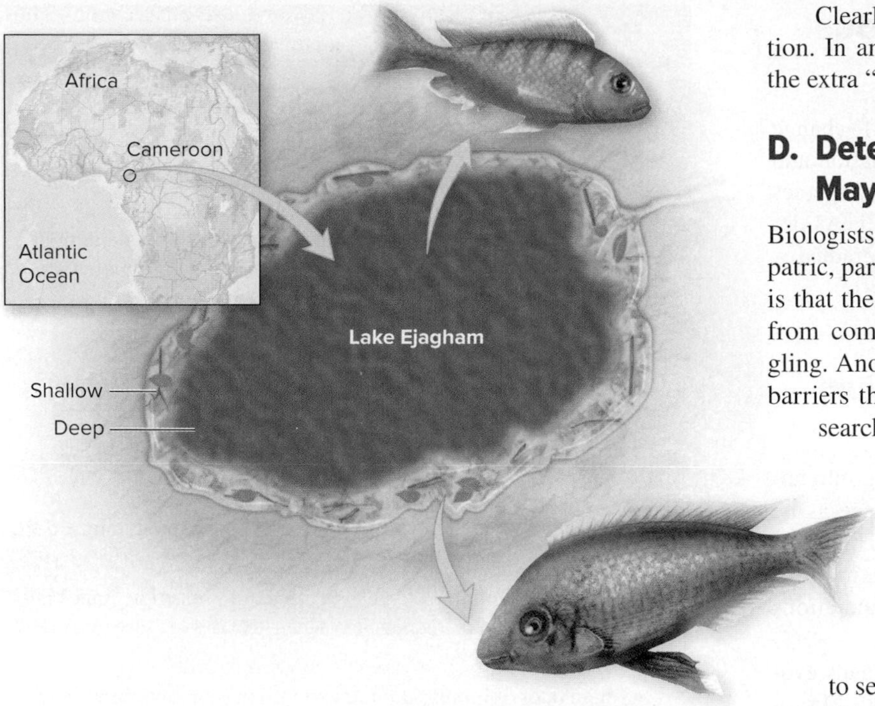

Clearly, polyploidy is extremely important in plant evolution. In animals, however, polyploidy is rare, possibly because the extra "dose" of chromosomes is usually fatal.

D. Determining the Type of Speciation May Be Difficult

Biologists sometimes debate whether a speciation event is allopatric, parapatric, or sympatric. One reason for the disagreement is that the definitions represent three points along a continuum, from complete reproductive isolation to continuous intermingling. Another difficulty is that we may not be able to detect the barriers that are important to other species. For example, a researcher might perceive a patch of forest to be uniform and conclude that speciation events occurring there are sympatric. But to a tiny insect, the distance between the forest floor and the treetops may represent an insurmountable barrier. In that case, speciation would be considered allopatric.

The problem of perspective also leads to debate over the size of the geographical barrier needed to separate two populations, which depends on the distance over which a species can spread its gametes. A plant with wind-blown pollen or a fungus producing lightweight spores encounters few barriers to gene flow. Pollen and spores can travel thousands of miles in the upper atmosphere. On the other hand, a desert pupfish cannot migrate out of its aquatic habitat, so the isolation of its pool instantly creates a geographical barrier. The same circumstance would not deter gene flow in species that walk or fly between pools.

Figure 14.9 Sympatric Speciation. Cichlids in the deepest waters of Lake Ejagham have smaller bodies than do shallow-water fish.

will produce tetraploid offspring (with four sets of chromosomes). This organism represents an "instant species," because it is reproductively isolated from its diploid ancestors. After all, when the tetraploid plant's gametes (each containing two sets of chromosomes) combine with haploid gametes from the parent species, each offspring inherits three sets of chromosomes. This hybrid is infertile because it cannot complete meiosis: Three chromosome sets cannot form the required homologous pairs.

14.3	**MASTERING CONCEPTS**

1. Distinguish among allopatric, parapatric, and sympatric speciation, and provide examples of each.
2. How can polyploidy contribute to sympatric speciation?
3. Why is it sometimes difficult to determine whether speciation is allopatric, parapatric, or sympatric?

Gametes

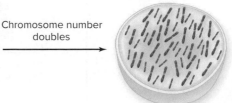

Old World cotton
n=13
(2*n*=26)

FERTILIZATION

South and Central American cotton
n=13
(2*n*=26)

Sterile hybrid (diploid)
2*n*=26

Chromosome number doubles

Cultivated American cotton (polyploid)
4*n*=52

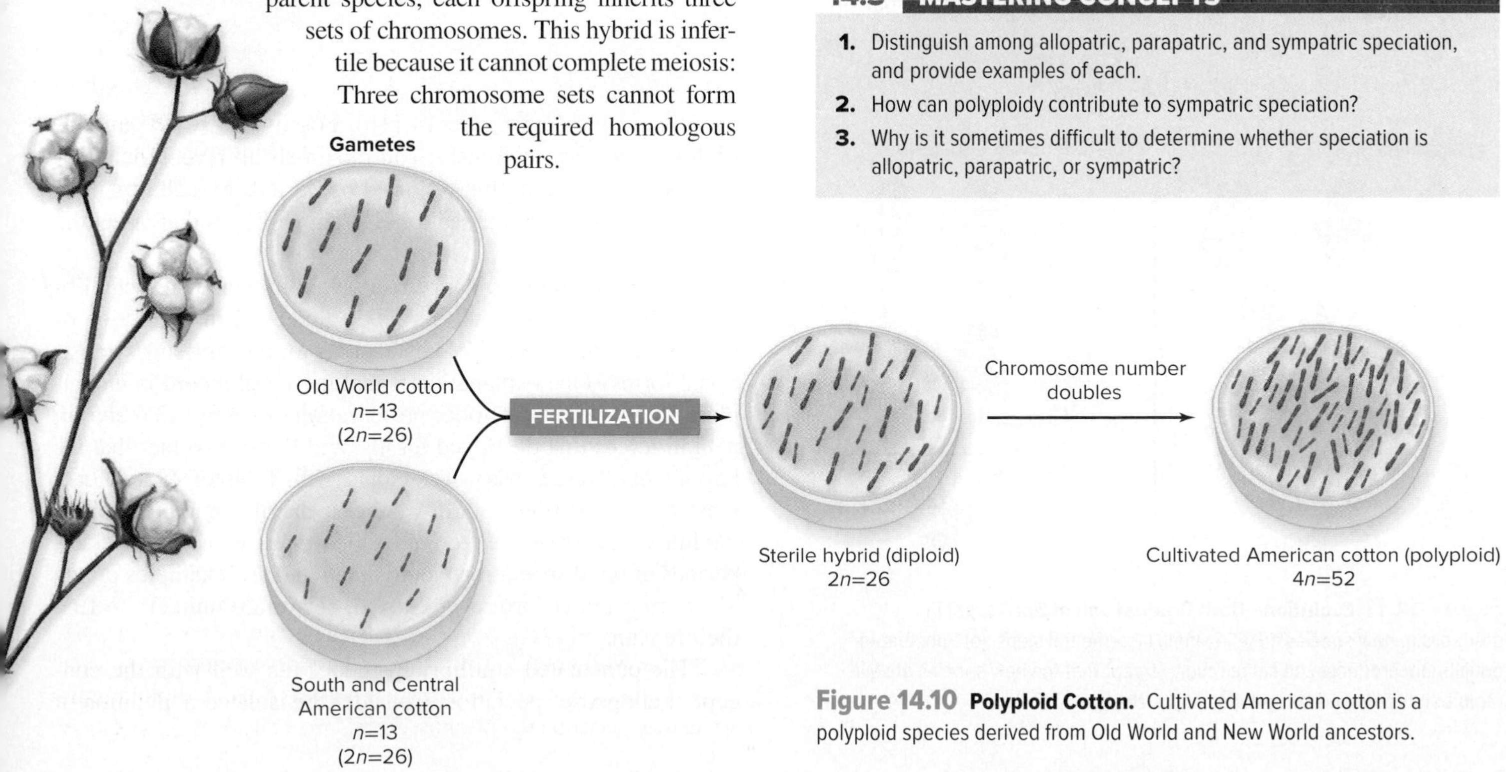

Figure 14.10 Polyploid Cotton. Cultivated American cotton is a polyploid species derived from Old World and New World ancestors.

14.4 Speciation May Be Gradual or May Occur in Bursts

On a global level, Earth has seen times of relatively little change in the living landscape but also periods of rapid speciation and times of mass extinctions. Even within a short time, one species can evolve rapidly, while another hardly changes at all. As described in this section, speciation can happen quickly, gradually, or at any rate in between.

A. Gradualism and Punctuated Equilibrium Are Two Models of Speciation

Darwin envisioned one species gradually transforming into another through a series of intermediate stages. The pace as he saw it was slow, although not necessarily constant. This idea, which became known as **gradualism,** held that evolution proceeds in small, incremental changes over many generations (**figure 14.11a**).

If the gradualism model is correct, then slow and steady evolutionary change should be evident in the fossil record. Microscopic protists such as foraminiferans and diatoms, for example, have evolved gradually. Vast populations of these asexually reproducing organisms span the oceans, leaving a rich fossil record in sediments. Since isolated populations rarely form in this uniform environment, it is unsurprising that speciation has been gradual.

Much of the fossil record, however, supports a different pattern. In 1972, paleontologists Stephen Jay Gould and Niles Eldredge coined the term **punctuated equilibrium** to describe relatively brief bursts of rapid evolution interrupting long periods

Burning Question

Why does evolution occur rapidly in some species but slowly in others?

Even though the same forces of microevolution operate on all populations, species evolve at different rates. Think of all the different ways that allele frequencies can change, including mutation, genetic drift, natural selection, and migration (see chapter 12). Several conditions favor a rapid pace of evolution:

- high mutation rates (mutations create new alleles)
- rapid reproduction (the faster cells divide, the more opportunities for mutation)
- small population size (small populations are most susceptible to genetic drift)
- sexual reproduction (meiosis shuffles genes, producing ample genetic variation)
- rapid environmental change (natural selection quickly changes allele frequencies when conditions are highly variable)
- high rates of immigration or emigration (migrants carry new alleles into populations)

Biologists have much to learn about evolution rates. A reptile called a tuatara offers one example of an unsolved mystery. DNA sequences from living and fossilized tuataras suggest they have the highest known molecular evolution rate. Yet the tuataras themselves have changed little. The solution to the mystery awaits further research.

**Submit your burning question to
Marielle.Hoefnagels@mheducation.com**

of little change (see figure 14.11b). Fossils of diverse animals such as bryozoans, mollusks, and mammals all reveal many examples of rapid evolution followed by periods of stability. (The Burning Question in this section lists some factors that affect the pace of evolution.)

Evolution seems to happen suddenly in punctuated equilibrium, with few transitional fossils documenting the evolution of one species into another. What accounts for the "missing" transitional forms? One explanation is that the fossil record is incomplete, for many reasons: poor preservation of biological material, natural forces that destroyed fossils, and the simple fact that we haven't discovered every fossil on Earth. Chapter 13 explores these reasons in more detail. Another is that the predicted "missing links" may have been too rare to leave many fossils. After all, periods of rapid speciation would mean that few examples of any given transitional form ever existed, so we are unlikely to find their remains.

The punctuated equilibrium model fits well with the concept of allopatric speciation. Consider the isolated population of

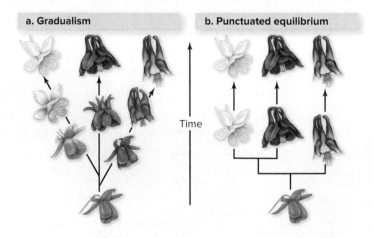

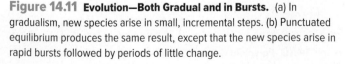

Figure 14.11 Evolution—Both Gradual and in Bursts. (a) In gradualism, new species arise in small, incremental steps. (b) Punctuated equilibrium produces the same result, except that the new species arise in rapid bursts followed by periods of little change.

Upper trunk/canopy

Short limbs, very large toe pads, long tail, typically green (but many can change color)

Cuba—*Anolis porcatus*
Hispaniola—*A. chlorocyanus*
Jamaica—*A. grahami*
Puerto Rico—*A. evermanni*

A. evermanni

Midtrunk

Intermediate limb length, short tail, gray color

Cuba—*Anolis loysiana*
Hispaniola—*A. distichus*
Jamaica—none found
Puerto Rico—none found

A. loysiana

Lower trunk/ground

Stocky body, long hindlimbs, brown color

Cuba—*Anolis sagrei*
Hispaniola—*A. cybotes*
Jamaica—*A. lineatopus*
Puerto Rico—*A. cristatellus*

A. cristatellus

Grass/bush

Slender body, very long tail, long limbs, brown color

Cuba—*Anolis alutaceus*
Hispaniola—*A. olssoni*
Jamaica—none found
Puerto Rico—*A. pulchellus*

A. pulchellus

Figure 14.12 Adaptive Radiation. More than 150 anole lizard species have evolved on four Caribbean islands. Each island has a unique collection of species, yet the adaptations to each habitat are similar—a striking example of convergent evolution (see section 13.4C).

B. Bursts of Speciation Occur During Adaptive Radiation

Speciation can happen in rapid bursts during an **adaptive radiation,** in which a single species gives rise to multiple specialized forms in a relatively short time. (In this context, the term *radiation* means spreading outward from a central source.)

In adaptive radiation, speciation typically occurs in response to the availability of new resources. For example, a few individuals might colonize a new, isolated habitat such as a mountaintop or an island. Multiple food sources, such as plants with different-sized seeds, would simultaneously select for different adaptations. Over time, multiple species would develop.

Adaptive radiation is especially common in island groups. One stunning example occurred in the Caribbean islands of Cuba, Hispaniola, Jamaica, and Puerto Rico (**figure 14.12**). Adaptive radiation occurred separately on each island, where at least 150 species of small lizards called anoles have adapted in very similar ways to different parts of their habitats. Other examples of adaptive radiation occur on Hawaii (as described in this chapter's opening essay), the Galápagos, and the Malay Archipelago, where Alfred Russel Wallace collected a multitude of beetles and butterflies.

As another path to adaptive radiation, some members of a population inherit a key adaptation that gives them an advantage. The first flowering plants, for example, appeared around 144 million years ago (in the late Jurassic period). Their descendants diversified rapidly during the Cretaceous, and all of today's major lineages were already in place 100 million years ago. Hundreds of thousands of flowering plant species now inhabit Earth. The new adaptation—the flower—apparently unleashed an entirely new set of options for reproduction, prompting rapid diversification. ⓘ *flowering plants,* section 19.5

A third type of adaptive radiation occurs when some members of a population inherit adaptations that enable them to survive a major environmental change. After the poorly suited

desert pupfish that became genetically distinct from its ancestral population over 50,000 years (see figure 14.6). If the climate changed and the spring containing the new species rejoined its "old" spring, the fossil record might show that a new fish species suddenly appeared with its ancestors—after all, 50,000 years is a blink of an eye in geologic time. Afterward, unless the environment changed again, no new selective forces would drive the formation of new species. Thus, a period of stability would ensue.

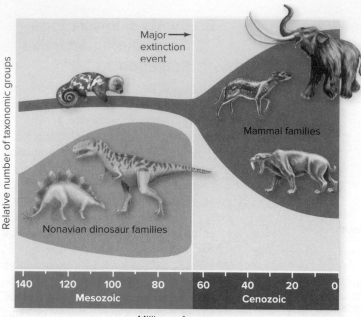

Figure 14.13 Speciation Following a Mass Extinction. Many ecological niches were vacated when most dinosaurs became extinct about 65 million years ago. Mammals flourished in the aftermath.

organisms perish, the survivors diversify as they exploit the new resources in the changed environment. Mammals, for example, underwent an enormous adaptive radiation when dinosaur extinctions opened up many new habitats (**figure 14.13**).

14.4 MASTERING CONCEPTS

1. Describe the theories of gradualism and punctuated equilibrium. How can the fossil record support both?
2. What are three ways that adaptive radiation can occur?

14.5 Extinction Marks the End of the Line

A species goes **extinct** when all of its members have died. If speciation is the birth of a species, extinction represents its death.

A. Many Factors Can Combine to Put a Species at Risk

Extinctions have many causes. Habitat loss, new predators, and new diseases all can wipe out a species. Extinction may also be a matter of bad luck: Sometimes no individual of a species survives a catastrophe such as a volcanic eruption or an asteroid impact.

No matter what the external trigger of an extinction, the root cause is always the same: Species die out if evolution fails to meet the pace of environmental change. Any species will eventually vanish if its gene pool does not contain the "right" alleles necessary for individuals to produce fertile offspring and sustain the population.

When faced with shifting conditions, what is the chance that a species will become extinct? The answer depends in part on the pace and scale of environmental change. Species such as elephants, which reach reproductive maturity at 30 years, are much less likely to survive a sudden change than are mice, which produce three generations a year. And if the environment changes on a massive scale—the climate changes from warm and wet to cool and dry, for example—extinctions are inevitable. Research suggests that plants rarely thrive in a habitat that is very different from that of their ancestors (see section 39.6).

In addition, the smaller the initial size of a population, the less likely it is to endure a major challenge. Small populations experience fewer genetic mutations, which are the ultimate source of new adaptations (see section 12.7). Low genetic diversity within a population, in turn, poses two problems. First, the population may contain too few beneficial alleles to withstand a new challenge. An emerging disease, for example, may wipe out most individuals, leaving too few survivors to maintain the population's size. Second, inbreeding in a small population tends to bring together lethal recessive alleles, which can weaken the organisms and make them less able to survive and reproduce.

B. Extinction Rates Have Varied over Time

Biologists distinguish between two different types of extinction events. The **background extinction rate** results from the steady, gradual loss of species due to normal evolutionary processes. Paleontologists have used the fossil record to calculate that most species exist from 1 million to 10 million years before becoming extinct. Thus, the background rate is roughly 0.1 to 1.0 extinctions per year per million species. Most extinctions overall have occurred as part of this more-or-less constant background rate.

Earth has also witnessed several periods of **mass extinctions,** when a great number of species disappeared in a relatively short time. The geologic timescale in figure 13.2 shows five major mass extinction events over the past 500 million years (red lines indicate mass extinctions). These events have had a great influence on Earth's history because they have periodically opened vast new habitats for adaptive radiation to occur.

Paleontologists study clues in Earth's sediments to understand the catastrophic events that contribute to mass extinctions. Two theories have emerged to explain these events, although several processes have probably contributed to mass extinctions.

The first explanation, called the **impact theory,** suggests that meteorites or comets have crashed to Earth, sending dust, soot, and other debris into the sky, blocking sunlight and setting into motion a deadly chain reaction. Without sunlight, plants died. The animals that ate plants, and the animals that ate those animals, then perished. Evidence for the impact theory of the extinction at the end of the Cretaceous period includes centimeter-thick layers of earth that are rich in iridium, an element rare on Earth but common in meteorites (**figure 14.14**).

Apply It **Now**

Recent Species Extinctions

Species extinctions have occurred throughout life's long history. They continue today, often accelerated by human activities (see chapter 40). Overharvesting contributes to species extinctions, as does habitat loss to agriculture, urbanization, damming, and pollution. Introduced plants and animals can deplete native species by competing with or preying on them.

The International Union for Conservation of Nature keeps track of recent extinctions of animal and plant species (figure 14.A). Of the hundreds of documented extinctions over the past 500 years, most have occurred on islands. This chapter's opening essay explores some of the reasons that island species are especially vulnerable.

Table 14.A lists a few species of vertebrate animals that have disappeared during the past few centuries. This list is far from complete; many more species of animals and plants have become extinct during the same time. Countless others are threatened or endangered, meaning that they are at risk for extinction. In fact, some conservationists predict that up to 3 in 10 species will become extinct in the next half-century, exceeding the background extinction rate by a factor of more than 1000.

Many people are trying to slow or even reverse today's species extinctions. Ultimately, this effort will be essential to human survival: We need other organisms to provide our food, medicine, energy, shelter, clean air, clean water, and many other vital services.

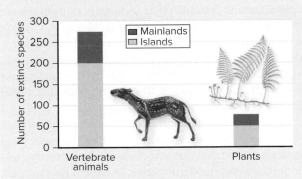

Figure 14.A **Recent Extinctions.** Island species of vertebrates and plants are most vulnerable to extinction.

TABLE **14.A**	Recent Vertebrate Extinctions	
Name	**Cause of Extinction**	**Former Location**
Fishes		
Chinese paddlefish (*Psephurus gladius*)	Habitat destruction	China
Las Vegas dace (*Rhinichthys deaconi*)	Habitat destruction	North America
Amphibians		
Palestinian painted frog (*Discoglossus nigriventer*)	Habitat destruction	Israel
Southern day frog (*Taudactylus diurnus*)	Undetermined	Australia
Monteverde golden toad (*Bufo periglenes*)	Habitat destruction, disease	Costa Rica
Reptiles		
Yunnan box turtle (*Cuora yunnanensis*)	Habitat destruction, overharvesting	China
Martinique lizard (*Leiocephalus herminieri*)	Undetermined	Martinique
Birds		
Dodo (*Raphus cucullatus*)	Habitat destruction, overharvesting	Mauritius
Moa (*Megalapteryx diderius*)	Overharvesting	New Zealand
Laysan honeycreeper (*Himatione sanguinea*)	Habitat destruction	Hawaii
Black mamo (*Drepanis funerea*)	Habitat destruction, introduced predators	Hawaii
Passenger pigeon (*Ectopistes migratorius*)	Overharvesting	North America
Great auk (*Alca impennis*)	Overharvesting	North Atlantic
Mammals		
Quagga (*Equus quagga quagga*)	Overharvesting	South Africa
Steller's sea cow (*Hydrodamalis gigas*)	Overharvesting	Bering Sea
Bali tiger (*Panthera tigris balica*)	Habitat destruction, overharvesting	Indonesia
Javan tiger (*Panthera tigris sondaica*)	Habitat destruction, overharvesting	Indonesia
Caspian tiger (*Panthera tigris virgata*)	Habitat destruction, overharvesting	Central Asia
Yangtze River dolphin (*Lipotes vexillifer*)	Habitat destruction, overharvesting	China

Figure 14.14 Impact Theory Evidence. This distinctive layer of rock marks the end of the Cretaceous period, about 65 million years ago.

Photo: ©Francois Gohier/Science Source

A second theory is that movements of Earth's crust may explain some mass extinctions. The crust, or uppermost layer of the planet's surface, is divided into many pieces, called tectonic plates. During Earth's history, movement of tectonic plates caused continents to drift apart, then come back together. These continental wanderings have profoundly affected life, both on land and in water. Organisms that had thrived in their old habitats had new competitors. Climates changed as continents moved toward or away from the poles, and colliding continents caused shallow coastal areas packed with life to disappear. Mountain ranges grew, destroying some habitats and creating new ones. ⓘ *plate tectonics,* section 13.3A

Many biologists warn that we are in the midst of a sixth mass extinction—this one caused by human actions. Ecologists estimate that the extinction rate is now about 20 to 200 extinctions per million species per year. Habitat loss and habitat fragmentation, pollution, introduced species, and overharvesting combine to imperil many species (see chapter 40). This chapter's Apply It Now box lists a few of the many vertebrate species that have recently become extinct, but the problem extends throughout all kingdoms of life. The loss of so many species is likely to severely disrupt the ecosystems we rely on.

14.5 | MASTERING CONCEPTS

1. What factors can cause or hasten extinction?
2. Distinguish between background extinction and mass extinctions.
3. How have humans influenced extinctions?

14.6 Biological Classification Systems Are Based on Common Descent

Darwin proposed that evolution occurs in a branched fashion, with each species giving rise to other species as populations occupy and adapt to new habitats. As described in chapter 13, ample evidence has shown him to be correct.

The goal of modern classification systems is to reflect this shared evolutionary history. **Systematics,** the study of classification, therefore incorporates two interrelated specialties: taxonomy and phylogenetics. **Taxonomy** is the science of describing, naming, and classifying species; **phylogenetics** is the study of evolutionary relationships among species. This section describes how biologists apply the evidence for evolution to the monumental task of organizing life's diversity into groups.

A. The Taxonomic Hierarchy Organizes Species into Groups

Carolus Linnaeus, the biologist introduced at the start of this chapter, made a lasting contribution to systematics. He devised a way to organize life into a hierarchical classification scheme that assigned a consistent, scientific name to each type of organism.

Linnaeus's idea is the basis of the taxonomic hierarchy used today. Biologists organize life into nested groups of taxonomic levels, based on similarities (**figure 14.15**). The three domains—Archaea, Bacteria, and Eukarya—are the most inclusive levels. Each domain is divided into kingdoms, which in turn are divided into phyla, then classes, orders, families, genera, and species. A **taxon** (plural: taxa) is a group at any rank; that is, domain Eukarya is a taxon, as is the order Liliales and the species *Aloe vera.* Some disciplines also use additional ranks, such as superfamilies and subspecies.

The more features two organisms have in common, the more taxonomic levels they share (**figure 14.16**). A human, a squid, and a fly are all members of the animal kingdom, but their many differences place them in separate phyla. A human, rat, and pig are more closely related—all belong to the same kingdom, phylum, and class (Mammalia). A human, an orangutan, and a chimpanzee are even more closely related, sharing the same kingdom, phylum, class, order, and family (order Primates, family Hominidae). As humans, our full classification is Eukarya-Animalia-Chordata-Mammalia-Primates-Hominidae-*Homo*-*Homo sapiens.*

This hierarchy is useful, but it has a flaw: The ranks are not meaningful in an evolutionary context. That is, the eight main levels might give the impression that evolution took exactly eight "leaps" to produce each modern species. In reality, the ranks are arbitrary, leading to frequent disputes about whether collections of species should be lumped into one taxon or divided among many. Despite its imperfections, the system remains in use because biologists have not agreed on a better approach.

B. A Cladistics Approach Is Based on Shared Derived Traits

Biologists illustrate life's diversity in the form of phylogenetic (evolutionary) trees, also called phylogenies, which depict relationships based on descent from shared ancestors. Multiple lines of evidence are used to construct phylogenetic trees. Anatomical features of fossils and existing organisms are useful, as are behaviors, physiological adaptations, and molecular sequences.

Taxonomic group	*Aloe vera* plant found in:	Number of species
Domain	Eukarya	Several million
Kingdom	Plantae	~375,000
Phylum	Anthophyta	~235,000
Class	Liliopsida	~65,000
Order	Liliales	~1200
Family	Asphodelaceae	785
Genus	*Aloe*	500
Species	*Aloe vera*	1

Figure 14.15 Taxonomic Hierarchy. Life is divided into domains, then kingdoms, then numerous smaller categories. This diagram shows the complete classification for the plant *Aloe vera*.

a.   b.

Figure 14.16 More Similarities, More Shared Levels. (a) A squid and a fly are both eukaryotes classified in the animal kingdom, but they do not share other taxonomic levels. (b) A chimpanzee and an orangutan share enough similarities to be classified in the same family.

Photos: (a, squid): ©Frank & Joyce Burek/Getty Images RF; (a, fly): ©Kimberly Hosey/Getty Images RF; (b, chimp): ©Alexander Rieber/EyeEm/Getty Images RF; (b, orangutan): ©MedioImages/SuperStock RF

In the past, systematists constructed phylogenetic tree diagrams by comparing as many characteristics as possible among species. Those organisms with the most characteristics in common would be neighbors on the tree's branches. Basing a tree entirely on similarities, however, can be misleading. As just one example, many types of cave animals are eyeless and lack pigments (see figure 13.13a). But these resemblances do not mean that all cave animals are closely related to one another; instead, they are the product of convergent evolution. If the goal is to group related organisms together, then attending only to similarities might lead to an incorrect classification. ⓘ *convergent evolution,* section 13.4C

A cladistics approach solves this problem. Widely adopted beginning in the 1990s, **cladistics** is a phylogenetic system that defines groups by distinguishing between ancestral and derived characters. **Ancestral characters** are inherited attributes that resemble those of the ancestor of a group; an organism with **derived characters** has features that are different from those found in the group's ancestor (**figure 14.17**).

In making a diagram such as figure 14.17, how do researchers know which characters are ancestral and which are derived? They choose an **outgroup** consisting of comparator organisms that are not part of the group being studied. For example, in a cladistic analysis of mammals that give birth to live young, an appropriate outgroup might be monotremes. Features that are present in all mammals, such as mammary glands and hair, are assumed to be ancestral. For placental mammals, derived features include the placenta and other characteristics that do not appear in monotremes or marsupials.

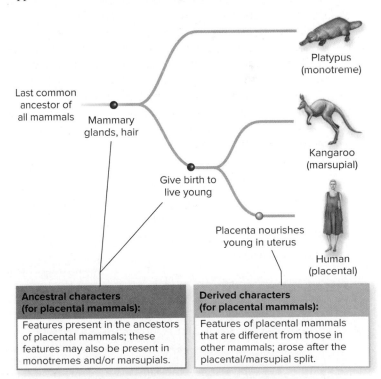

Ancestral characters (for placental mammals):	Derived characters (for placental mammals):
Features present in the ancestors of placental mammals; these features may also be present in monotremes and/or marsupials.	Features of placental mammals that are different from those in other mammals; arose after the placental/marsupial split.

Figure 14.17 Ancestral and Derived Characters. The placenta is a derived character that was not present in the common ancestor that placental mammals share with marsupials and monotremes.

C. Cladograms Depict Hypothesized Evolutionary Relationships

The result of a cladistics analysis is a **cladogram,** a treelike diagram built using shared derived characters (figure 14.18). The basic "unit" of a cladogram is a **clade,** also called a **monophyletic** group, which is a set of organisms consisting of a common ancestor and all of its descendants. A clade is therefore a group of species united by a single evolutionary pathway. For example, birds form a clade because they all descended from the same group of reptiles. A clade may contain any number of species, as long as all of its members share an ancestor that organisms outside the clade do not share.

Figure It Out

How many clades are represented in figure 14.18?

All cladograms have features in common. The tips of the branches represent taxa. Existing species, such as birds and turtles, are at the tips of longer branches in figure 14.18; the nonavian dinosaurs are extinct and therefore occupy a shorter branch. Each node in a cladogram indicates where two groups arose from a common ancestor. A branching pattern of lines therefore represents populations that diverged genetically, splitting off to form a new species. The branching pattern also implies the passage of time, as indicated by the arrow at the bottom of figure 14.18.

The emphasis in a cladogram is not physical similarities but rather historical relationships. To emphasize this point, imagine a lizard, a crocodile, and a chicken. Which resembles the lizard more closely: the crocodile or the chicken? Clearly, the most *similar* animals are the lizard and the crocodile. But these resemblances are only superficial. The shared derived characters tell a more complete story of evolutionary history. Based on the evidence, crocodiles are more closely related to birds than they are to lizards.

The evolutionary relationships in a cladogram are depicted as nested hierarchies, with small clades contained entirely within larger ones. You can therefore think of a cladogram as a tree with a trunk, main branches, smaller branches, and twigs. According to this analogy, a clade is the piece that would "fall off" if you made a single cut to any part of the tree. The larger the branch you cut, the more inclusive the clade. Moreover, each clade can rotate on its branch without changing the meaning of the cladogram (figure 14.19). As a result, many equivalent cladograms can tell the same evolutionary story.

A common mistake in interpreting cladograms is to incorrectly assume that a taxon must be closely related to both groups that appear next to it on the tree. In figure 14.18, for example, mammals are adjacent to both turtles and amphibians. Does this mean that rabbits are as closely related to frogs as they are to turtles? To find out, look at the amount of time that has passed since mammals last shared a common ancestor with each group. Because the common ancestor of mammals and turtles existed more recently, these groups are more closely related than are mammals and amphibians.

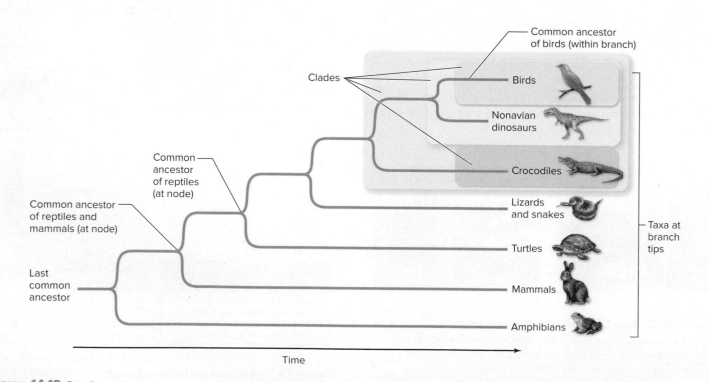

Figure 14.18 Reading a Cladogram. Each clade consists of a common ancestor and all of its descendants. The more recently any two groups share a common ancestor, the more closely related they are.

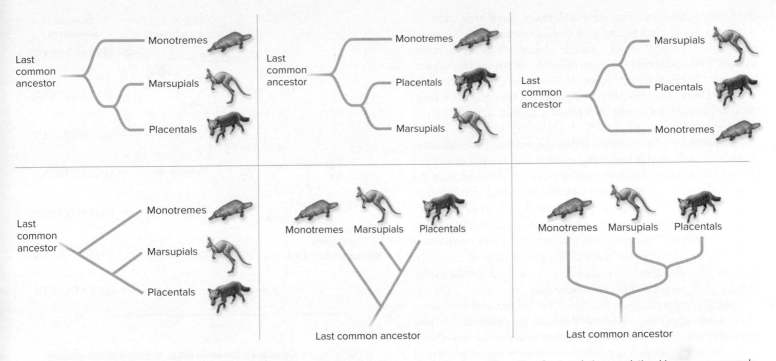

Figure 14.19 Different Cladograms, Same Relationships. All of these cladograms depict the same information about evolutionary relationships among mammals.

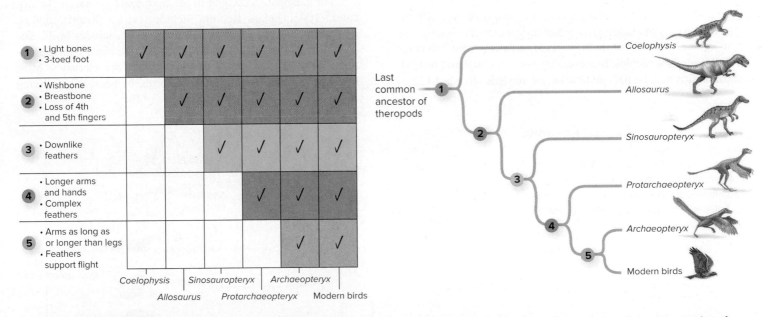

Figure 14.20 Constructing a Cladogram. To build a cladogram, tally up the derived characters that species share. Pairs that share the greatest number of derived characters are hypothesized to be most closely related.

Source: Cladogram data based on Kevin Padian, June 25, 1998, When is a bird not a bird? *Nature,* vol. 393, page 729.

Figure 14.20 demonstrates how to construct a cladogram using the derived characters that were important in the evolution of birds. The first step is to select the traits to be studied in each species, such as the presence of feathers and lightweight bones. Next, the researcher collects the data and makes a chart showing which species have which traits. Then, in the tree, species sharing the most derived characters occupy the branches farthest from the root. The resulting cladogram shows the hypothesized relationship between modern birds and their close relatives, the nonavian dinosaurs.

The cladogram in figure 14.20 is based on the physical features of birds and extinct dinosaurs. Molecular sequences are also extremely useful because they add many more characters on which to build cladograms and deduce evolutionary relationships. **Figure 14.21** shows a simple cladogram based on mutations in a 10-nucleotide sequence of DNA. (In reality, cladograms are typically based on far longer DNA sequences.) Scientists can analyze DNA not only from living species but also from long-dead organisms preserved in museums, amber, and permafrost (see section 19.6).

A problem with cladistics is that the analysis becomes enormously complicated when many species and derived characters are included. Mathematically, many trees can accommodate the same data set; for instance, just 10 taxa can be arranged into millions of possible trees. Sorting through all the possibilities requires tremendous computing power. How do researchers settle on the "best" tree? One strategy is to select the most **parsimonious** tree, which is the one that requires the fewest steps to construct. The most parsimonious tree therefore invokes the fewest evolutionary changes needed to explain the data.

All phylogenetic trees are based on limited and sometimes ambiguous information. They are therefore not peeks into the past but rather tools that researchers can use to construct hypotheses about the relationships among different types of organisms. These investigators can then add other approaches to test the hypotheses.

D. Many Traditional Groups Are Not Clades

Contemporary scientists using a cladistics approach typically assign names only to clades; groups that reflect incomplete clades or that combine portions of multiple clades are not named. Many familiar groups of species, however, are not monophyletic. Instead, these groups may be paraphyletic or polyphyletic (**figure 14.22**).

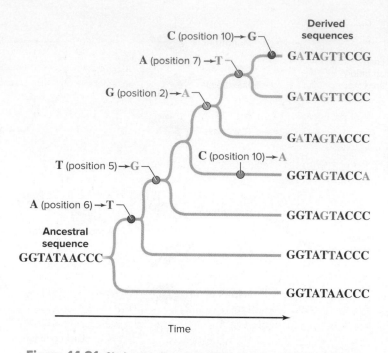

Figure 14.21　Cladogram Based on DNA. Mutations in DNA produce unique sequence variations, characters that can be used to build cladograms.

For example, according to the traditional Linnaean classification system, class Reptilia includes turtles, lizards, snakes, crocodiles, and the extinct dinosaurs, but it excludes birds. The cladogram, however, places birds in the same clade with the reptiles based on their many shared derived characteristics. Most biologists therefore now consider birds to be reptiles—living

Figure 14.22　Paraphyletic and Polyphyletic Groups. The Linnaean class Reptilia is paraphyletic; it excludes birds and therefore does not form a clade. A group containing only endothermic animals is polyphyletic because it excludes the most recent common ancestor of these animals.

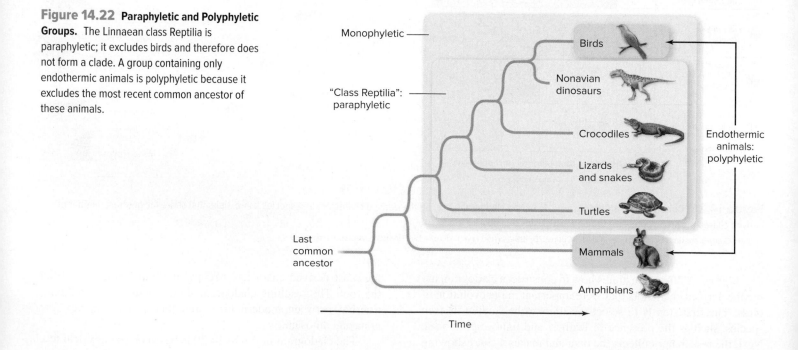

dinosaurs, actually—so they make a distinction between birds and nonavian ("nonbird") reptiles.

The traditional class Reptilia is considered **paraphyletic** because it contains a common ancestor and some, but not all, of its descendants. The kingdom Protista is also paraphyletic (figure 14.23). Protists include mostly single-celled eukaryotes that do not fit into any of the three eukaryotic kingdoms (plants, fungi, and animals). Yet all three of these groups share a common eukaryotic ancestor with the protists. Biologists are currently struggling to divide kingdom Protista into monophyletic groups, an immense task.

Polyphyletic groups exclude the most recent common ancestor shared by all members of the group. For example, a group consisting of endothermic (formerly called "warm-blooded") animals includes only birds and mammals. This group is polyphyletic because it excludes the most recent common ancestor of birds and mammals, which was an ectotherm (formerly called "cold-blooded"). Likewise, the term *algae* reflects a polyphyletic grouping of many unrelated species of aquatic organisms that carry out photosynthesis. In general, polyphyletic groups reflect characteristics—such as ectothermy or photosynthesis—that have evolved independently in multiple species.

Systematists try to avoid paraphyletic and polyphyletic groups, which do not reflect a shared evolutionary history. Nevertheless, many such group names remain in everyday usage.

Table 14.1 summarizes some of the terminology used in systematics.

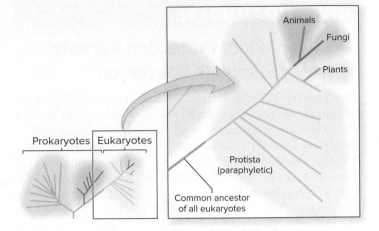

Figure 14.23 Paraphyletic Protists. Although domains Bacteria, Archaea, and Eukarya are all monophyletic, the group of eukaryotes called protists is paraphyletic because it excludes plants, fungi, and animals.

14.6 MASTERING CONCEPTS

1. Describe the taxonomic hierarchy.
2. What is the advantage of a cladistics approach over a more traditional approach to phylogeny?
3. Distinguish between ancestral and derived characters.
4. What sorts of evidence do biologists use in a cladistics analysis?
5. How is a cladogram constructed?
6. How do paraphyletic and polyphyletic groups differ from monophyletic groups?

TABLE **14.1** Miniglossary of Systematics Terms

Term	Definition
Ancestral characters	Features present in the common ancestor of a clade
Clade	Group of organisms consisting of a common ancestor and all of its descendants; a monophyletic group
Cladistics	Phylogenetic system that groups organisms by characters that best indicate shared ancestry
Cladogram	Phylogenetic tree built on shared derived characters
Derived characters	Features of an organism that are different from those found in a clade's ancestors
Monophyletic group	Group of organisms consisting of a common ancestor and all of its descendants; a clade
Outgroup	Comparator group outside the group being studied; useful for identifying ancestral traits
Paraphyletic group	Group of organisms consisting of a common ancestor and some, but not all, of its descendants
Phylogenetic tree	Diagram depicting hypothesized evolutionary relationships
Polyphyletic group	Group of organisms that excludes the most recent common ancestor
Systematics	The combined study of taxonomy and evolutionary relationships among organisms

INVESTIGATING LIFE

14.7 Plant Protection Rackets May Stimulate Speciation

Most people are familiar with the nectar in flowers, the sex organs of most plants. An insect or a bird seeking nectar deep within a flower may brush against the flower's pollen, then unwittingly carry those sex cells to the next plant it visits. Many plants, however, also store nectar in other places, such as in the leaves or stems (figure 14.24). These structures, collectively called extrafloral nectaries, have evolved independently in more than 100 plant families.

ⓘ *pollination*, section 24.2C

Extrafloral nectaries do not directly participate in reproduction, yet their abundance suggests that they contribute to plant fitness. Researchers hypothesize that a "protection racket" analogy applies to these interactions. In a protection racket, a business pays money to a criminal group in exchange for defense against harm from third parties. Similar deals abound in the natural world. Thousands of species of ferns and flowering plants "pay" sugary nectar to animals (often ants or wasps), which then fight off leaf-gobbling herbivores that would otherwise consume the plant's leaves. In essence, the plant boosts its reproductive success by enlisting the enemies of its enemies.

Do these arrangements have macroevolutionary implications? Perhaps plants that recruit insect defenders can exploit areas that were unavailable to their unprotected ancestors. These new

Figure 14.24 Extrafloral Nectary. An ant drinks from an extrafloral nectary on a plant's leaf in the rain forest of Ecuador.
©Dr. Morley Read/Stockbyte/Getty Images RF

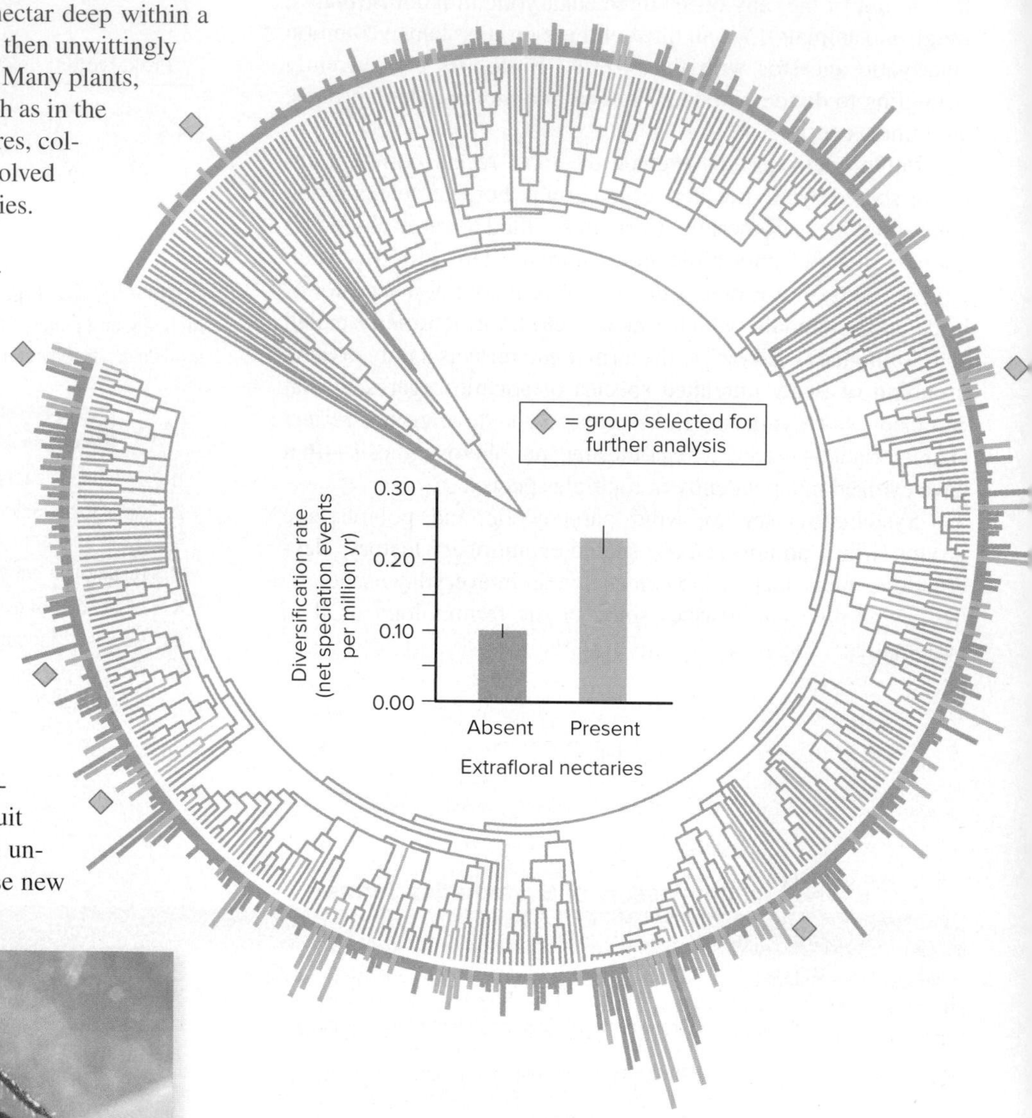

Figure 14.25 Tree of Plants. The common ancestor of plants is at the center of circular evolutionary tree. The height of each outer bar represents the relative size of plant family; color indicates the presence of extrafloral nectaries. The inner graph co the diversification rates of families with and without extrafloral nectaries.

Source: Modified from Weber, Marjorie G., and Anurag A. Agrawal. 2014. Defense mutualisms enhance plant diversification. *Proceedings of the National Academy of Sciences*, vol. 111, pages 16442–16447.

◇ = group selected for further analysis

niches would likely have unique selective pressures, driving evolutionary changes that could lead to speciation. If so, then plant lineages with extrafloral nectaries should have higher speciation rates than lineages without these adaptations.

To test this hypothesis, researchers Marjorie Weber and Anurag Agrawal of Cornell University looked to the scientific literature. First, they used fossils and published data to construct a detailed evolutionary tree depicting most families in the plant kingdom. Next, they marked all families containing at least one species with extrafloral nectaries.

Then the researchers calculated the diversification rate of each family, based on the family's age and the number of species it contains. A relatively young plant family with many species would have a high diversification rate; conversely, an old family with just a few species would have a low rate. Finally, Weber and Agrawal compared the average diversification rate for all families with extrafloral nectaries with the average for their counterparts lacking these sugary rewards.

Figure 14.25 summarizes the result of their analysis. On average, plant lineages with members containing extrafloral nectaries have diversified much more quickly than lineages without them. These data support the researchers' hypothesis.

However, Weber and Agrawal identified a possible limitation of their approach. They categorized a plant family as having extrafloral nectaries even if only one species actually has the adaptation. Since the largest families have the most members, they are most likely to have at least one species with extrafloral nectaries, simply by chance. Therefore, the "extrafloral nectaries present" category is biased toward large families and a correspondingly high diversification rate.

To compensate for this limitation, Weber and Agrawal analyzed in detail the evolutionary trees of six distantly related groups (indicated with orange diamonds in figure 14.25). Within each group, some members have extrafloral nectaries and some do not. They tested whether the emergence of extrafloral nectaries *within* a plant group caused the diversification rate to climb. This follow-up analysis supported the previous results for four of the six groups (table 14.2).

With these results in mind, the researchers wondered how extrafloral nectaries might contribute to plant diversification. The emergence of nectaries in a lineage could somehow have a *direct* effect, immediately increasing the speciation rate or decreasing the extinction rate. Or they could *indirectly* influence diversification by opening up new habitats—and therefore evolutionary pressures—to plants. In that case, the researchers expected to see a delay between the emergence of extrafloral nectaries and a boost in the diversification rate.

Weber and Agrawal used the same six plant groups to answer their question, mapping when extrafloral nectaries arose in each lineage and then analyzing when the diversification rate accelerated. The researchers consistently found that increases in the diversification rate occurred long after extrafloral nectaries emerged. This delay suggests that extrafloral nectaries alone do not cause speciation or decrease extinctions. Rather, having insect defenders may allow plants to occupy new habitats, which select for the trait changes that may eventually lead to speciation events.

The plant kingdom contains more than 250,000 species, most of them flowering plants. This huge number is all the more remarkable because flowering plants began diversifying rapidly only in the past 110 million or so years. During that time, plants forged many ecological interactions with animals, but few are as intriguing as the "protection racket" explored in this study. Amazingly, this mutually beneficial interaction—a criminal act in human society—seems to have spawned the evolution of many new species throughout the plant kingdom.

TABLE **14.2** Extrafloral Nectaries and Diversification Rates

Plant Group	Difference in Diversification Rates Between Groups with and Without Extrafloral Nectaries	Statistically Significant Difference?
Genus *Pleopeltis*	7x higher in groups with extrafloral nectaries	Yes
Genus *Viburnum*	4x higher in groups with extrafloral nectaries	Yes
Genus *Senna*	3x higher in groups with extrafloral nectaries	Yes
Genus *Byttneria*	Approximately equal	No
Genus *Turnera*	6x higher in groups with extrafloral nectaries	Yes
Tribe Polygoneae*	2x higher in groups with extrafloral nectaries	No

*Tribe Polygoneae is a clade consisting of multiple genera within family Polygonaceae.

Source: Weber, Marjorie G., and Anurag A. Agrawal. 2014. Defense mutualisms enhance plant diversification. *Proceedings of the National Academy of Sciences*, vol. 111, pages 16442–16447.

14.7 MASTERING CONCEPTS

1. In figure 14.25, the six plant families marked with symbols each have members with extrafloral nectaries. Do these six families form a monophyletic group, a paraphyletic group, or a polyphyletic group? Are extrafloral nectaries an ancestral character or a derived character in each of the six plant families? Explain your answers.

2. What do the researchers speculate is the connection between extrafloral nectaries and rapid speciation in plants?

CHAPTER SUMMARY

14.1 What Is a Species?

- **Macroevolution** refers to large-scale changes in life's diversity, including both extinction and the appearance of new **species.**

A. Linnaeus Devised the Binomial Naming System

- Linnaeus's species designations and classifications helped scientists communicate. Darwin added evolutionary meaning.

B. Species Can Be Defined Based on the Potential to Interbreed

- Mayr added the requirement for reproductive isolation to define **biological species.**
- **Speciation** is the formation of a new species, which occurs when a population's **gene pool** is divided and each part takes its own evolutionary course.

14.2 Reproductive Barriers Cause Species to Diverge

A. Prezygotic Barriers Prevent Fertilization

- **Prezygotic reproductive isolation** occurs before or during fertilization. It includes obstacles to mating such as space, time, and behavior; mechanical mismatches between male and female; and molecular mismatches between gametes.

B. Postzygotic Barriers Prevent the Development of Fertile Offspring

- **Postzygotic reproductive isolation** results in **hybrid** offspring that die early in development, are infertile, or produce a second generation of offspring with abnormalities.
- Figure 14.26 summarizes the main types of reproductive barriers.

14.3 Spatial Patterns Define Three Types of Speciation

A. Allopatric Speciation Reflects a Geographical Barrier

- **Allopatric speciation** occurs when a geographical barrier separates a population. The two populations then diverge genetically to the point that their members can no longer produce fertile offspring together.

B. Parapatric Speciation Occurs in Neighboring Regions

- **Parapatric speciation** occurs when two populations live in neighboring areas but share a border zone. Genetic divergence between the two groups exceeds gene flow, driving speciation.

C. Sympatric Speciation Occurs in a Shared Habitat

- **Sympatric speciation** enables populations that occupy the same area to diverge. **Polyploidy** (one or more extra chromosome sets) may create the reproductive barrier that triggers sympatric speciation.

D. Determining the Type of Speciation May Be Difficult

- The distinction between allopatric, parapatric, and sympatric speciation is not always straightforward, partly because it is difficult to define the size of a geographical barrier.

14.4 Speciation May Be Gradual or May Occur in Bursts

A. Gradualism and Punctuated Equilibrium Are Two Models of Speciation

- Evolutionary change occurs at many rates, from slow and steady **gradualism** to the periodic bursts that characterize **punctuated equilibrium.**

B. Bursts of Speciation Occur During Adaptive Radiation

- In **adaptive radiation,** an ancestral species rapidly branches into several new species.

14.5 Extinction Marks the End of the Line

- **Extinction** is the disappearance of a species.

A. Many Factors Can Combine to Put a Species at Risk

- Rapid environmental change, a slow reproductive rate, and low genetic diversity make species vulnerable to extinction.

B. Extinction Rates Have Varied over Time

- The **background extinction rate** reflects steady, ongoing losses of species.
- Historically, **mass extinctions** have resulted from global changes such as continental drift. The **impact theory** suggests that a meteorite or comet can also cause mass extinctions, such as the one at the end of the Cretaceous period.
- Human activities are increasing the extinction rate.

14.6 Biological Classification Systems Are Based on Common Descent

- **Systematics** includes **taxonomy** (classification) and **phylogenetics** (species relationships).

A. The Taxonomic Hierarchy Organizes Species into Groups

- Biologists use a taxonomic hierarchy to classify life's diversity, with **taxa** ranging from domain to species.

B. A Cladistics Approach Is Based on Shared Derived Traits

- **Cladistics** defines groups using **ancestral** and **derived characters.** An **outgroup** helps researchers detect ancestral characters.

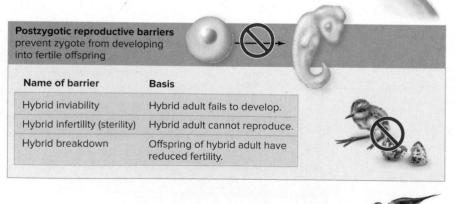

Barriers that maintain reproductive isolation between related species

Prezygotic reproductive barriers
prevent formation of zygote

Name of barrier	Basis	
Habitat isolation	Space	No mating
Temporal isolation	Time	
Behavioral isolation	Mating rituals	
Mechanical isolation	Reproductive organs	Mating but no fertilization
Gametic isolation	Chemical signals on gametes	

Fertilization occurs if no prezygotic barriers are present; zygote forms

Postzygotic reproductive barriers
prevent zygote from developing into fertile offspring

Name of barrier	Basis
Hybrid inviability	Hybrid adult fails to develop.
Hybrid infertility (sterility)	Hybrid adult cannot reproduce.
Hybrid breakdown	Offspring of hybrid adult have reduced fertility.

A viable, fertile offspring forms only if no reproductive barriers are present

Figure 14.26 **Reproductive Barriers: A Summary.**

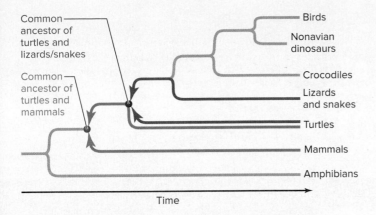

Figure 14.27 Finding Common Ancestors in a Cladogram.

C. Cladograms Depict Hypothesized Evolutionary Relationships

- A **cladogram** shows evolutionary relationships as a branching hierarchy with nodes representing common ancestors. A **clade,** or **monophyletic** group, is an ancestor plus all of its descendants.
- Figure 14.27 shows how to use a cladogram to find the common ancestor shared by two groups.
- The most **parsimonious** cladogram is the simplest tree that fits the data.

D. Many Traditional Groups Are Not Clades

- **Paraphyletic** groups exclude some descendants of a common ancestor; **polyphyletic** groups exclude the common ancestor of its members.

14.7 Investigating Life: Plant Protection Rackets May Stimulate Speciation

- Many plants produce sugary nectar on their leaves or stems, attracting animals that defend the plants against herbivores. Researchers have concluded that the presence of extrafloral nectaries may boost the speciation rate.

MULTIPLE CHOICE QUESTIONS

1. The biological species concept defines species based on
   a. external appearance.
   b. the number of adaptations to the same habitat.
   c. ability to interbreed.
   d. DNA and protein sequences.

2. A mule is the offspring of a male donkey and a female horse. Mules are unable to produce offspring. What reproductive barrier separates horses and donkeys?
   a. Mechanical isolation   c. Hybrid inviability
   b. Gametic isolation   d. Hybrid infertility

3. A mountain range separates a population of gorillas. After many generations, the gorillas on one side of the mountains cannot produce viable, fertile offspring with their counterparts on the other side. What has happened?
   a. Sympatric speciation   c. Parapatric speciation
   b. Allopatric speciation   d. Adaptive radiation

4. Why might speciation occur at an unusually rapid pace?
   a. A new phenotype enables organisms to exploit the environment in new ways.
   b. The environment changes relatively rapidly, selecting for new phenotypes.
   c. A dominant group of organisms goes extinct, paving the way for the evolution of new species.
   d. All of the above are possibilities.

5. Flying animals have diverse evolutionary histories. They therefore form
   a. a monophyletic group.   c. a polyphyletic group.
   b. a paraphyletic group.   d. an outgroup.

Answers to these questions are in appendix A.

WRITE IT OUT

1. What type of reproductive barrier applies to each of these scenarios?
   a. Water buffalo and cattle can mate, but the embryos die early in development.
   b. Scientists try to mate two species of dragonfly that inhabit the same pond at the same time of day. However, females never allow males of the other species to mate with them.
   c. One species of reed warbler is active in the upper parts of the canopy, while another species of reed warbler is active in the lower canopy. Both species are active during the day.
   d. Scientists mate two parrots from different populations to see if speciation has occurred. The parrots mate over and over again, but the male's sperm never fertilizes the female's egg.

2. How does natural selection predict a gradualistic mode of evolution? Does the presence of fossils that are consistent with punctuated equilibrium mean that natural selection does not occur?

3. Why do species become extinct? Choose a species that has recently become extinct, and describe some possible evolutionary consequences to other species that interacted with that species before its extinction.

4. Which of the groups in figure 1.9 represent clades? Which groups do not represent clades? Explain your answers.

5. On figure 14.20, circle a monophyletic group, a paraphyletic group, and a polyphyletic group. Describe the qualities that define how each group is classified.

PULL IT TOGETHER

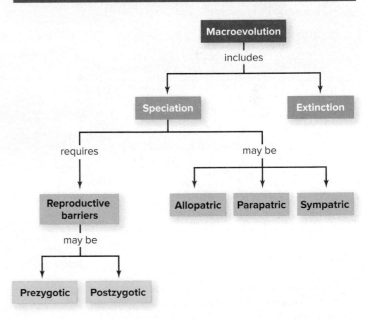

Figure 14.28 Pull It Together: Speciation and Extinction.

Refer to figure 14.28 and the chapter content to answer the following questions.

1. Review the Survey the Landscape figure in the chapter introduction and then add *phylogenetic trees, species, fossils, DNA,* and *anatomical structures* to figure 14.28.

2. Add *fertilization* and *offspring* to this concept map.

3. Draw pictures of allopatric, parapatric, and sympatric speciation.

4. Add *gradualism* and *punctuated equilibrium* to this concept map.

15

The Origin and History of Life

Bombardment. Meteoroids enter Earth's atmosphere every day, as they have done for eons. Some scientists hypothesize that ancient objects from space carried the organic molecules needed for life to begin on Earth.

©Joe Tucciarone/Science Source

LEARN HOW TO LEARN
Write Your Own Test Questions

Have you ever tried putting yourself in your instructor's place by writing your own multiple choice test questions? It's a great way to pull the pieces of a chapter together. The easiest questions to write are based on definitions and vocabulary, but those will not always be the most useful. Try to think of questions that integrate multiple ideas or that apply the concepts in a chapter. Write 10 questions, and then let a classmate answer them. You'll probably both learn something new.

Life from Space

In a 1908 book entitled *Worlds in the Making,* Swedish chemist Svante Arrehenius suggested that life came to Earth from the cosmos. He later broadened the idea, calling it "panspermia" and proposing that life-carrying spores (or the organic chemicals needed for life) arrived on interstellar dust, comets, asteroids, and meteorites.

What is the evidence for panspermia? Modern proponents point to several intriguing clues:

- Some microorganisms can survive under extreme conditions, which would be necessary to endure the high radiation and cold temperatures of space during a journey that could take millions of years. A bacterium called *Deinococcus radiodurans,* for example, tolerates a thousand times the radiation level that a person can; it even lives in nuclear reactors! Microbes also live in pockets of water within the ice of Antarctic lakes, surroundings not unlike the icy insides of a comet. In addition, researchers have revived some bacteria after a dormancy lasting millions of years.

- Meteorites that have fallen to Earth from space sometimes contain organic compounds such as amino acids.

- The surface of one of Jupiter's moons, Europa, is covered with a thick layer of ice. The frozen surface may conceal an ocean of liquid water, a prerequisite for life. ⓘ *essential water,* section 2.3

- Canyons, shorelines, and other physical features of Mars leave little doubt that liquid water once flowed on the red planet. The Martian climate is now too cold for liquid water, but vast deposits of ice—and the possibility of life—remain below the planet's surface. In 2004, an orbiting spacecraft called *Mars Express* found high concentrations of water vapor and methane in the Martian atmosphere, suggesting that methane-producing microbes might live in liquid water beneath Mars's surface. The Mars lander *Phoenix* "tasted" Martian soil and verified the presence of subsurface ice in 2008. A second lander, *Curiosity,* arrived in 2012; one of the mission's objectives is to learn whether Mars ever supported life.

Panspermia is not widely accepted, in part because it sidesteps the question of life's ultimate origin in the universe. Instead, most scientists accept that life probably arose from simple chemical substances on Earth. This process and its astounding aftermath are the subjects of this chapter.

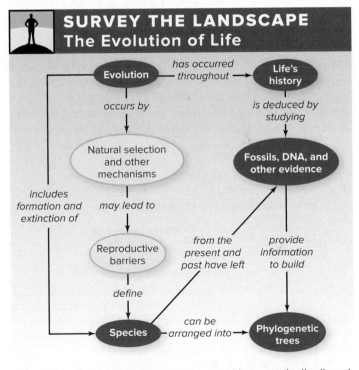

SURVEY THE LANDSCAPE
The Evolution of Life

After life started billions of years ago, organisms gradually diversified into a spectacular array of bacteria, archaea, protists, plants, fungi, and animals. This chapter's brief tour of life's history includes only the highlights.

For more details, study the Pull It Together feature in the chapter summary.

305

15.1 Life's Origin Remains Mysterious

Reconstructing life's start is like reading all the chapters of a novel except the first. A reader can get some idea of the events and setting of the opening chapter from clues throughout the novel. Similarly, scattered clues from life through the ages reflect events that may have led to the origin of life.

Scientists describe the origin and history of life in the context of the **geologic timescale,** which divides time into eons, eras, periods, and epochs defined by major geological or biological events. **Figure 15.1** shows a simplified geologic timescale; see figure 13.2 for a complete version.

The study of life's origin begins with astronomy and geology. Earth and the solar system's other planets formed about 4.6 BYA (billion years ago) as solid matter condensed out of a vast expanse of dust and gas swirling around the early sun. The red-hot ball that became Earth cooled enough to form a crust by about 4.2 to 4.1 BYA, when the surface temperature ranged from 500°C to 1000°C, and atmospheric pressure was 10 times what it is now.

The geological evidence paints a chaotic picture of this Hadean eon, including volcanic eruptions, earthquakes, and ultraviolet radiation. Analysis of craters on other objects in the solar system suggests that comets, meteorites, and possibly asteroids bombarded Earth's surface during its first 500 million to 600 million years (**figure 15.2**). These impacts repeatedly boiled off the seas and vaporized rocks to carve the features of the fledgling world.

Still, organic molecules could probably aggregate and interact in protected pockets of the environment. So harsh and unsettled were conditions on the early Earth that organized groups of chemicals may have formed many times and at many places, only to be torn apart by heat, debris from space, or radiation. We can't know.

Figure 15.2 Early Earth. Intense bombardment by meteorites marked Earth's first 500 million to 600 million years.

At some point, however, an entity arose that could survive, thrive, reproduce, and diversify. The clues from geology and paleontology suggest that early in the Precambrian supereon—sometime between 4.2 and 3.85 BYA—simple cells (or their precursors) arose.

This section describes some of the major steps in the chemical evolution that eventually led to the first cell; **figure 15.3** summarizes one possible version of the process. (Today, however, new life is unlikely to originate from nonliving matter; the Burning Question in this section explains why.)

A. The First Organic Molecules May Have Formed in a Chemical "Soup"

Early Earth was different from today's planet, both geologically and chemically. The atmosphere today contains gases such as nitrogen (N_2), oxygen (O_2), carbon dioxide (CO_2), and water vapor (H_2O). What might it have been like 4 BYA?

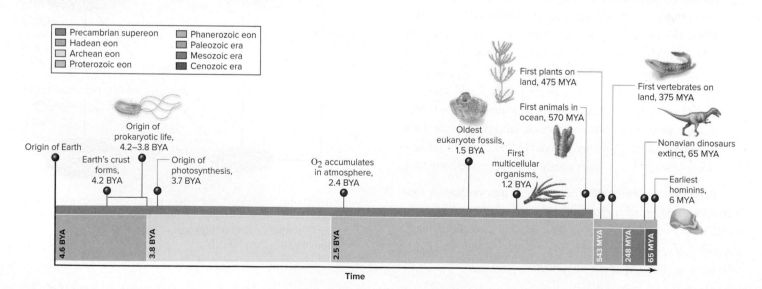

Figure 15.1 Highlights in Life's History. In this simplified geologic timescale, the size of each eon and era is proportional to its length in years. (BYA = billion years ago; MYA = million years ago)

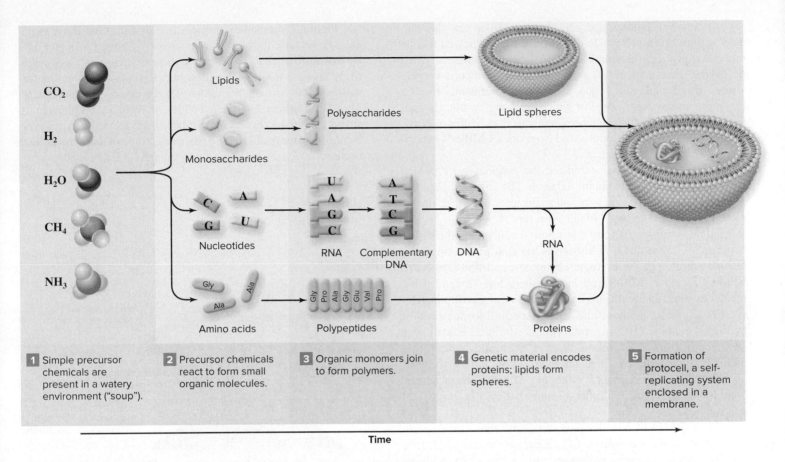

Figure 15.3 Pathway to a Cell. The steps leading to the origin of life on Earth may have started with the formation of organic molecules from simple precursors. However it originated, the first cell would have contained self-replicating molecules enclosed in a phospholipid bilayer membrane.

Russian chemist Alex I. Oparin hypothesized in his 1938 book, *The Origin of Life,* that a hydrogen-rich, or reducing, atmosphere was necessary for organic molecules to form on Earth. Oparin thought that this long-ago atmosphere included methane (CH_4), ammonia (NH_3), water, and hydrogen (H_2), similar to the atmospheres of the outer planets today. These simple chemicals appear in step 1 of figure 15.3. Note that O_2 was not present in the atmosphere at the time life originated. In the absence of O_2, Oparin suggested, the chemical reactions that form amino acids and nucleotides could have occurred (see figure 15.3, step 2). ⓘ *organic molecules,* section 2.5

Miller's Experiment   In 1953, Stanley Miller, a graduate student in chemistry at the University of Chicago, and his mentor, Harold Urey, decided to test whether Oparin's atmosphere could indeed give rise to organic molecules. Miller built a sterile glass enclosure to contain Oparin's four gases, through which he passed electrical discharges to simulate lightning (**figure 15.4**). He condensed the gases in a narrow tube; from there, the liquid passed into a flask. Boiling the fluid caused gases to evaporate back into the synthetic "atmosphere," completing the loop.

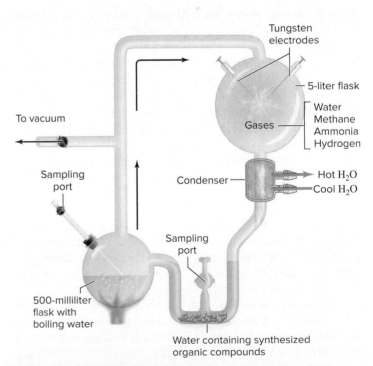

Figure 15.4 The Miller Experiment. When Stanley Miller passed an electrical spark through heated gases, the mixture generated amino acids and other organic molecules.

After a few failures and adjustments, Miller saw the condensed liquid turn yellowish. Chemical analysis showed that he had made glycine, the simplest amino acid in organisms. When he let the brew cook a full week, the solution turned varying shades of red, pink, and yellow-brown; he subsequently found a few more amino acids, some of which are found in life. (In fact, the solution contained even more chemicals than Miller realized; a 2008 reanalysis of material saved from one of Miller's experiments revealed 22 amino acids.)

A prestigious journal published the original work, which Urey gallantly refused to put his name on. The 25-year-old Miller made headlines in newspapers and magazines reporting (incorrectly) that he had created "life in a test tube."

Life is far more than just a few amino acids, but "the Miller experiment" went down in history as the first **prebiotic simulation,** an attempt to re-create chemical conditions on Earth before life arose. Miller and many others later extended his results by altering conditions or using different starting materials. For example, methane and ammonia could form clouds of hydrogen cyanide (HCN), which produced amino acids in the presence of ultraviolet light and water. Prebiotic "soups" that included phosphates yielded nucleotides, including the biological energy molecule ATP. Other experiments produced carbohydrates and phospholipids similar to those in biological membranes.

The experiment has survived criticisms that Earth's early atmosphere actually contained CO_2, a gas not present in Miller's original setup. Organic molecules still form, even with an adjusted gas mixture.

Hydrothermal Vents as a Model

More recent prebiotic simulations mimic deep-sea hydrothermal vents (see chapter 39's introduction). Here, in a zone where hot water meets cold water, chemical mixtures could have encountered a rich brew of minerals spewed from Earth's interior. One laboratory version combines mineral-rich lava with seawater containing dissolved CO_2; under high temperature and pressure, simple organic compounds form. In another vent model, nitrogen compounds and water mix with an iron-containing mineral under high temperature and pressure. The iron catalyzes reactions that produce ammonia—one of the components of the original Miller experiment.

The Possible Role of Clays

Once the organic building blocks (monomers) of macromolecules were present, they had to have linked into chains (polymers). This process, depicted in step 3 of figure 15.3, may have happened on hot clays or other minerals that provided ample, dry surfaces.

Clays may have played an important role in early organic chemistry for at least three reasons. First, the sheetlike minerals in clays form templates on which chemical building blocks could have linked to build larger molecules. Second, iron pyrite and some other minerals in clay can release electrons, providing energy to form chemical bonds. Third, these minerals may also have acted as catalysts to speed chemical reactions.

Prebiotic simulations demonstrate that the first RNA molecules could have formed on clay surfaces (**figure 15.5**). Not only do the positive charges on clay's surface attract and hold negatively charged RNA nucleotides, but clays also promote the formation of the covalent bonds that link the nucleotides into chains. They even attract other nucleotides to form a complementary strand. About 4 BYA, clays might have been fringed with an ever-increasing variety of growing polymers. Some of these might have become the macromolecules that would eventually build cells.

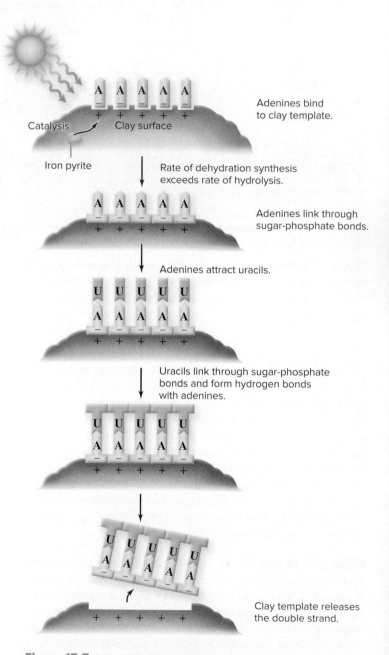

Adenines bind to clay template.

Rate of dehydration synthesis exceeds rate of hydrolysis.

Adenines link through sugar-phosphate bonds.

Adenines attract uracils.

Uracils link through sugar-phosphate bonds and form hydrogen bonds with adenines.

Clay template releases the double strand.

Figure 15.5 A Possible Role for Clay. Chains of nucleotides may have formed on clay templates. In this hypothesized scenario, iron pyrite ("fool's gold") was the catalyst for polymer formation, and sunlight provided the energy.

B. Some Investigators Suggest an "RNA World"

Life requires an informational molecule. That molecule may have been RNA, or something like it, because RNA is a versatile molecule. It stores genetic information and uses it to manufacture proteins. RNA can also catalyze chemical reactions and duplicate on its own. As Stanley Miller summed it up, "The origin of life is the origin of evolution, which requires replication, mutation, and selection. Replication is the hard part. Once a genetic material could replicate, life would have just taken off."

Perhaps pieces of RNA on clay surfaces formed, accumulated, grew longer, became more complex in sequence, and changed as replication errors led to mutations. Some members of this accumulating community of molecules would have been more stable than others, leading to an early form of natural selection. The term **"RNA world"** has come to describe how self-replicating RNA may have been the first independent precursor to life on Earth.

At some point, RNA might have begun encoding proteins, just short chains of amino acids at first. An RNA molecule may eventually have grown long enough to encode the enzyme reverse transcriptase, which copies RNA to DNA. With DNA, the chemical blueprints of life found a much more stable home. Protein enzymes eventually took over some of the functions of catalytic RNAs. Step 4 in figure 15.3 shows this stage in life's origin.

C. Membranes Enclosed the Molecules

Meanwhile, lipids would have been entering the picture. Under the right temperature and pH conditions, and with the necessary precursors, phospholipids could have formed membranelike structures, some of which left evidence in ancient sediments. Laboratory experiments show that pieces of membrane can indeed grow on structural supports and break free, forming a bubble called a liposome. ⓘ *phospholipids,* section 3.3

Perhaps an ancient liposome enclosed a collection of nucleic acids and proteins to form a cell-like assemblage, or protocell (see figure 15.3, step 5). Carl Woese, who described the domain Archaea, gave the term **progenotes** to these hypothetical, ancient aggregates of RNA, DNA, proteins, and lipids. These assemblages were precursors of cells but they were not nearly as complex.

The capacity of nucleic acids to mutate may have enabled progenotes to become increasingly self-sufficient, giving rise eventually to the reaction pathways of metabolism. No one knows what these first metabolic pathways were or how they eventually led to respiration, photosynthesis, and thousands of other chemical reactions that support life today. Despite intriguing similarities between glycolysis and some photosynthetic reactions (see section 6.9), these early stages of metabolism may not have left enough evidence for us ever to understand their origins.

Burning Question

Does new life spring from inorganic molecules now, as it did in the past?

It is intriguing to think of the possibility that new life could be forming from nonliving matter now, just as it did long ago in Earth's history. Although theoretically possible, scientists have never seen life emerging from a collection of simple chemicals. Such a finding would be a major blow to the cell theory, which says that cells come only from preexisting cells (see section 3.1B).

The emergence of new life from simple molecules seems improbable today. One reason is that when Earth was young, no life existed, so the first simple cells encountered no competition. Now, however, life thrives nearly everywhere on Earth (see chapter 17's Burning Question). Perhaps new life *is* forming, but before it has a chance to become established, a hungry microbe gobbles it up. Such an event would be extremely difficult to detect.

So does the ancient chemical origin for life on Earth violate the cell theory? The answer is no, because the cell theory applies to today's circumstances. The chemical and physical environment on the young Earth was nothing like that of today's world. Scientists have never observed the formation of life from nonliving matter—but that does not mean that it did not happen in the distant past.

Submit your burning question to
Marielle.Hoefnagels@mheducation.com

D. Early Life Changed Earth Forever

Unfortunately, direct evidence of the first life is likely gone because most of Earth's initial crust has been destroyed. Erosion tears rocks and minerals into particles, only to be built up again into sediments, heated and compressed. Seafloor is dragged into Earth's interior at deep-sea trenches, where it is melted and recycled. The oldest rocks that remain today, from an area of Greenland called the Isua formation, date to about 3.85 BYA. They house the oldest hints of life: quartz crystals containing organic deposits rich in the carbon isotopes found in organisms. ⓘ *plate tectonics,* section 13.3A

Whatever they were, the first organisms were simpler than any cell known today. Several types of early cells probably prevailed for millions of years, competing for resources and sharing genetic material. Eventually, however, a type of cell arose that was the last shared ancestor of all life on Earth today.

The first cells lived in the absence of O_2 and probably used organic molecules as a source of both carbon and energy. Another source of carbon, however, was the CO_2 in the atmosphere.

Photosynthetic bacteria and archaea eventually evolved that could use light for energy and atmospheric CO_2 as a carbon source (see chapter 5). These microbes no longer relied on organic compounds in their surroundings for food.

Photosynthesis probably originated in aquatic bacteria that used hydrogen sulfide (H_2S) instead of water as an electron donor. These first photosynthetic microorganisms would have released sulfur, rather than O_2, into the environment. Eventually, changes in pigment molecules enabled some of the microorganisms to use H_2O instead of H_2S as an electron donor. Cells using this new form of photosynthesis released O_2 as a waste product. The O_2 would have bubbled out of the water and been released into the air, changing the composition of Earth's atmosphere (**figure 15.6a**).

Some of the oldest known fossils are from 3.7-billion-year-old rocks in Australia and South Africa. The fossils strongly resemble large formations of cyanobacteria called stromatolites. These ancient cyanobacteria, along with many others, would have consumed CO_2 and released O_2 over hundreds of millions of years. Extensive iron deposits dating to about 2 billion years ago provide evidence of the changing atmosphere. As O_2 from photosynthesis built up in the oceans, iron that was previously dissolved in seawater reacted with the O_2 and sank to the bottom of the sea, producing distinctive layers of iron-rich sediments.

The evolution of photosynthesis forever altered life on Earth. Photosynthetic organisms formed the base of new food chains. In addition, natural selection began to favor aerobic organisms that could use O_2 in metabolism, while anaerobic species would persist in pockets of the environment away from O_2. Ozone (O_3) also formed from O_2 high in the atmosphere, blocking the sun's damaging ultraviolet radiation. The overall result was an explosion of new life that eventually gave rise to today's microbes, plants, fungi, and animals (figure 15.6b).

Although most people associate photosynthesis only with plants, microbes living in water or along shorelines were the only photosynthetic organisms for most of life's history. Plants did not even colonize dry land until about 475 million years ago. Even today, about half of the O_2 generated in photosynthesis comes from aquatic organisms, mostly living in the oceans.

15.1 MASTERING CONCEPTS

1. How were conditions on Earth before life began different from current conditions?
2. What can we learn from simulations of early Earth?
3. Why is RNA likely to have been pivotal in life's beginnings?
4. How did early life change Earth?

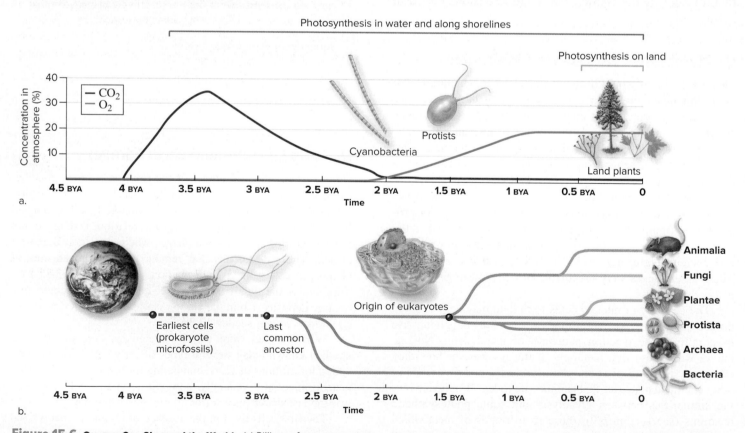

Figure 15.6 Oxygen Gas Changed the World. (a) Billions of years ago, photosynthetic microbes began to pump O_2 into the water and atmosphere. (b) As O_2 accumulated, life's diversity exploded.

15.2 Complex Cells and Multicellularity Arose over a Billion Years Ago

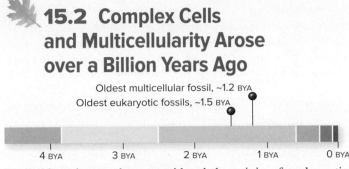

Oldest multicellular fossil, ~1.2 BYA
Oldest eukaryotic fossils, ~1.5 BYA

4 BYA   3 BYA   2 BYA   1 BYA   0 BYA

Until this point, we have considered the origin of prokaryotic cells. Fossil evidence shows that eukaryotic cells emerged during the Proterozoic era, at least 1.9 to 1.4 BYA. Australian fossils consisting of organic residue 1.69 billion years old are chemically similar to eukaryotic membrane components and may have come from a very early unicellular eukaryote.

Recall from chapter 3 that prokaryotic cells are structurally simple compared with compartmentalized eukaryotic cells. We may never know the origin of the nuclear envelope, endoplasmic reticulum, Golgi apparatus, and other membranes within the eukaryotic cell. The membranes of these organelles consist of phospholipids and proteins, as does the cell's outer membrane. Perhaps the outer membrane of an ancient cell repeatedly folded in on itself, eventually pinching off inside the cell to form a complex internal network of organelles (**figure 15.7**).

Unfortunately, that hypothesis is difficult or impossible to test, so we can only speculate about that aspect of eukaryotic cell evolution. Some details, however, are becoming clear. For example, the endosymbiont theory may explain the origin of two types of membrane-bounded organelles.

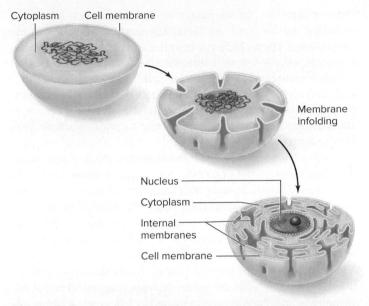

Figure 15.7 Membrane Infolding. A highly folded cell membrane may have formed an internal membrane network as a possible step in the origin of eukaryotic cells.

A. Endosymbiosis Explains the Origin of Mitochondria and Chloroplasts

The **endosymbiont theory** proposes that mitochondria and chloroplasts originated as free-living bacteria that began living inside other cells (**figure 15.8**). The term *endosymbiont* derives from *endo-*, meaning "inside," and *symbiont*, meaning

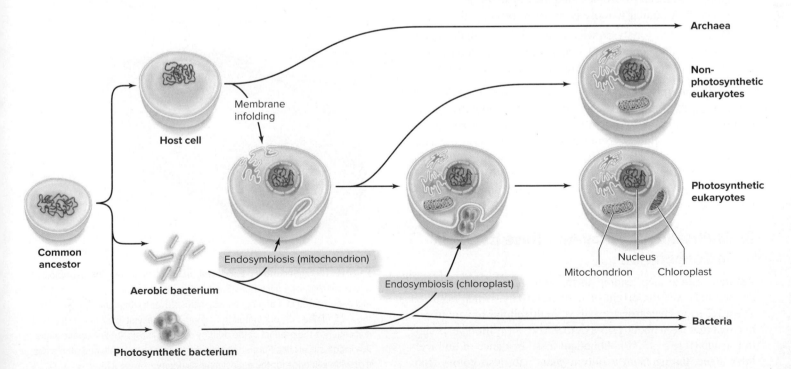

Figure 15.8 The Endosymbiont Theory. Mitochondria and chloroplasts may have originated from a union of bacterial cells with ancient host cells.

"to live together." In the first endosymbiotic event, a host cell engulfed one or more bacteria that could carry out aerobic respiration. These bacteria eventually developed into mitochondria. (We know that mitochondria must have come first because virtually all eukaryotes have these organelles.) In a later endosymbiosis, some descendants of this early eukaryote took in photosynthetic bacteria, which became chloroplasts in the lineages that eventually gave rise to photosynthetic protists and plants.

After the ancient endosymbiosis events, many genes moved from the DNA of the organelles to the nuclei of the host cells. These genetic changes made the captured microorganisms unable to live on their own outside their hosts, as described in the Burning Question in this section. Over time, according to the endosymbiont theory, they came to depend on one another for survival.

Biologist Lynn Margulis proposed this theory in the late 1960s. Since that time, the evidence supporting the idea that mitochondria and chloroplasts originated as independent organisms has mounted. The evidence includes

- similarities in size, shape, and membrane structure between the organelles and some types of bacteria;

- the double membrane surrounding mitochondria and chloroplasts, a presumed relic of the original engulfing event;

- the observation that mitochondria and chloroplasts are not assembled in cells but instead divide, as do bacterial cells;

- the similarity between the photosynthetic pigments in chloroplasts and those in cyanobacteria;

- the observation that mitochondria and chloroplasts contain DNA, RNA, and ribosomes (and the ribosomes are structurally similar to those in bacterial cells); and

- DNA sequence analysis, which shows a close relationship between mitochondria and aerobic bacteria (proteobacteria), as well as between chloroplasts and cyanobacteria.

Endosymbiosis has been a potent force in eukaryote evolution. In fact, the chloroplasts of some types of photosynthetic protists apparently derive from a secondary endosymbiosis—that is, from a eukaryotic cell engulfing a eukaryotic red or green alga (**figure 15.9**; see also figure 18.2). In these species, three or four membranes surround the chloroplasts; some of their cells even retain remnants of the engulfed cell's nucleus.

B. Multicellularity May Also Have Its Origin in Cooperation

Another critical step leading to the evolution of plants, fungi, and animals was the origin of multicellularity, which occurred about 1.2 BYA. The earliest fossils of multicellular life are from a red alga that lived about 1.25 BYA to 950 MYA (million years ago) in Canada (**figure 15.10**). Abundant fossil evidence of multicellular algae, dating from a billion years ago, also comes from eastern Russia.

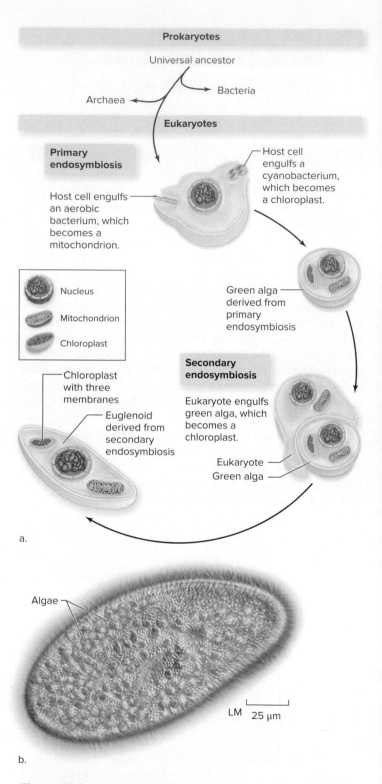

Figure 15.9 **Secondary Endosymbiosis.** (a) In primary endosymbiosis, a host cell engulfs a bacterium. Some chloroplasts, including those in green algae, originated in a primary endosymbiosis event. Other chloroplasts, like those of euglenoids, originated by secondary endosymbiosis, in which a host cell engulfed the product of the primary endosymbiosis. (b) Eukaryotic algae live inside this protist, *Paramecium bursaria*. This partnership and others like it provide evidence for the endosymbiosis theory.

(b): ©Michael Abbey/Science Source

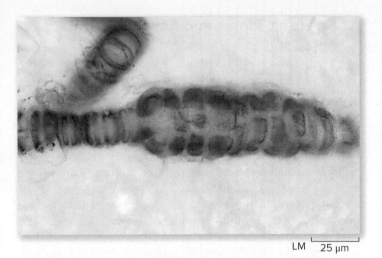

Figure 15.10 Early Multicellularity. This 1.2-billion-year-old fossil of a red alga offers some of the oldest evidence of multicellularity.
©Dr. N.J. Butterfield

No one knows how eukaryotes came to adopt a multicellular lifestyle. The fossil record is essentially silent on the transition, mostly because the first multicellular organisms lacked hard parts that fossilize readily. We do know, however, that multicellularity arose independently in multiple lineages. After all, genetic evidence clearly suggests that plants, fungi, and animals arose from different lineages of multicellular protists.

We also know that some multicellular organisms consist of cells that bear an uncanny resemblance to one-celled protists; **figure 15.11** shows two examples. How might the transition to multicellularity have occurred? Perhaps many individual cells came together, joined, and took on specialized tasks to form a multicellular organism. The life cycle of modern-day protists

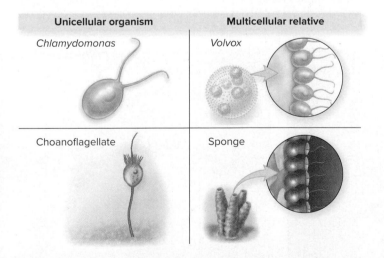

Figure 15.11 From One to Many. A single-celled green alga called *Chlamydomonas* shares many similarities with its close relative, the many-celled *Volvox*. Likewise, a protist called a choanoflagellate resembles a collar cell on the inner surface of a sponge.

Burning Question

Why can't mitochondria and chloroplasts survive on their own?

The idea that mitochondria and chloroplasts are derived from free-living bacteria is fascinating, and it raises the intriguing question of whether these organelles could be extracted from living cells and survive on their own.

The brief answer to this question is that they cannot. The endosymbiosis event that eventually gave rise to mitochondria occurred more than a billion years ago, and since that time the organelles have lost most of the genes present in their bacterial ancestors.

Gene numbers tell the story. A typical chloroplast genome encodes about 100 genes; in humans and other mammals, the mitochondrial genome encodes a mere 37. These numbers are tiny compared to the several thousand genes in a bacterial cell. Among the genes that mitochondria and chloroplasts have lost are those that encode defensive chemicals, many transport proteins, and other molecules that are essential to free-living bacteria.

Mitochondria and chloroplasts did, however, retain DNA encoding proteins that are essential to their present-day functions: aerobic cellular respiration for mitochondria, and photosynthesis for chloroplasts. They also have their own ribosomes, so they can produce these proteins without help from the rest of the cell. However, their limited set of genes is nowhere near enough to enable mitochondria and chloroplasts to regain their independence.

Submit your burning question to
Marielle.Hoefnagels@mheducation.com

called cellular slime molds illustrates this possibility (see figure 18.13). Alternatively, a single-celled organism may have divided, and the daughter cells may have remained stuck together rather than separating. After many rounds of cell division, these cells may have begun expressing different subsets of their DNA. The result would have been a multicellular organism with specialized cells—similar to the way in which modern animals and plants develop from a single fertilized egg cell.

Whichever way it happened, the origin of multicellularity ushered in the possibility of specialized cells, which allowed for new features such as attachment to a surface or an upright orientation. The resulting explosion in the variety of body sizes and forms introduced new evolutionary possibilities and opened new habitats for other organisms. The rest of this chapter describes the diversification of multicellular life.

15.2 MASTERING CONCEPTS

1. How might the endoplasmic reticulum, nuclear envelope, and other internal membranes have arisen?
2. What is the evidence that mitochondria and chloroplasts descend from simpler cells engulfed long ago?
3. What are two ways that multicellular organisms may have originated?

15.3 Life's Diversity Exploded in the Past 500 Million Years

It would take many thousands of pages to capture all of the events that passed from the rise of the first cells to life today—if we even knew them. This section highlights a few key events in the history of multicellular, eukaryotic life.

As you read through this section, keep in mind what you have already learned about evolution, speciation, and extinction. Species have come and gone in response to major environmental changes such as rising and falling temperatures, the increasing concentration of O_2 in the atmosphere, the shifting continents, and the advance and retreat of glaciers. Short-term catastrophes such as floods, volcanic eruptions, and meteorite impacts have also taken their toll on some species and created new opportunities for others. ⓘ *mass extinctions,* section 14.5B

A. The Strange Ediacarans Flourished Late in the Precambrian

The "Precambrian" is an informal name for the eventful 4 billion years that preceded the Cambrian period of the Paleozoic era. During the Precambrian supereon, life originated, reproduced, and diversified into many forms. In addition, photosynthesis evolved, O_2 accumulated in Earth's atmosphere, and eukaryotes arose, as did the first multicellular algae and animals.

Perhaps the most famous Precambrian residents were the mysterious Ediacaran organisms, which left no known modern descendants (figure 15.12). One example, *Dickinsonia,* could reach 1 meter in diameter but was less than 3 millimeters thick. Biologists have interpreted these fossils as everything from worms to ferns to fungi. The Ediacarans vanished from the fossil record about 544 MYA. (In 2004, geologists named the last portion of the Precambrian, from about 600 to 543 MYA, the Ediacaran period in honor of these strange marine creatures.)

B. Paleozoic Plants and Animals Emerged onto Land

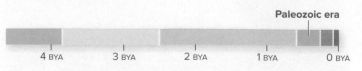

Cambrian Period (543 to 490 MYA)
Fossils of all major phyla of animals appeared within a few million years of one another in the Cambrian seas, a spectacular period of diversification sometimes called the "Cambrian explosion" (see section 21.17). During this time, remnants of the Ediacaran world coexisted with abundant red and green algae, sponges, jellyfishes, and worms. Most notable were the earliest known organisms with hard parts, such as insectlike trilobites, nautiloids, scorpion-like eurypterids, and brachiopods, which resembled clams. Many of these invertebrates left remnants identified only as "small, shelly fossils." The early Cambrian seas were also home to diverse wormlike, armored animals, some of which would die out.

The Burgess Shale from the Canadian province of British Columbia preserves a glimpse of life from this time. A mid-Cambrian mud slide buried enormous numbers of organisms, including animals with skeletons and soft-bodied invertebrates not seen elsewhere. The Burgess Shale animals were abundant, diverse, and preserved in exquisite detail (figure 15.13).

Ordovician Period (490 to 443 MYA)
During the Ordovician period, the seas continued to support huge communities of algae and invertebrates such as sponges, corals, snails, clams, and cephalopods. The first vertebrates to leave fossil evidence, jawless fishes called ostracoderms, appeared at this time. Fossilized spores indicate that life had ventured onto land, in the form of primitive plants that may have resembled modern liverworts (see section 19.2).

Figure 15.12

Ediacarans Were . . . Different.
(a) *Dickinsonia* had segments and two different ends. But just what it was remains unclear.
(b) No one knows what type of animal *Spriggina* was, either.

Photos: (a): ©De Agostini Picture Library/Getty Images; (b): ©The Natural History Museum/The Image Works

a.   1 cm   b.   0.5 cm   Stalked or motile?

Figure 15.13 Cambrian Life. *Marrella* is one of many strange animals whose fossils have been discovered in the Burgess Shale.
©O. Louis Mazzatenta/National Geographic Creative Stock

Figure 15.14 Early Land Plant. *Cooksonia*, which lived during the Silurian, is the oldest known vascular plant.
©The Natural History Museum/The Image Works

The Ordovician period ended with a mass extinction that killed huge numbers of marine invertebrates. Apparently the supercontinent of Gondwana drifted toward the South Pole, causing temperatures on the landmass to fall. Sea levels dropped as glaciers accumulated, destroying shoreline habitats.

Silurian Period (443 to 417 MYA)

The first plants with specialized water- and mineral-conducting tissues, the vascular plants, evolved during the Silurian (**figure 15.14**). These plants were larger than their ancestors, so they provided additional food and shelter for animals. The first terrestrial animals to leave fossils resembled scorpions, which may have preyed upon other small animals exploring the land. Fungi likely colonized land at the same time.

Aquatic life also continued to change. Fishes with jaws arose, as did the first freshwater fishes, but their jawless counterparts were still widespread during the Silurian. The oceans also contained abundant corals, trilobites, and mollusks.

Devonian Period (417 to 354 MYA)

The Devonian period was the "Age of Fishes." The seas continued to support more life than did the land. The now prevalent invertebrates were joined by fishes with skeletons of cartilage or bone. Corals and animals called crinoids that resembled flowers were abundant.

The fresh waters of the Devonian were home to the lobe-finned fishes, including *Eusthenopteron* (**figure 15.15a**). These animals had fleshy, powerful fins that they could use like feet, and they could obtain O_2 through both gills and primitive lunglike structures. Toward the end of the Devonian period, about 375 MYA, the first amphibians appeared (figure 15.15b and c). *Acanthostega* had a fin on its tail like a fish and used its powerful tail to move underwater, but it also had hips, paddlelike legs, and toes. Preserved footprints indicate that the animal could venture briefly onto land. A contemporary of *Acanthostega,* called *Ichthyostega,* had more powerful legs and a rib cage strong enough to support the animal's weight on land, yet it had a skull shape and finned tail reminiscent of fish ancestors.

a. *Eusthenopteron*   b. *Acanthostega*   c. *Ichthyostega*

Figure 15.15 The Vertebrate Transition to Land. (a) This lobe-finned fish, *Eusthenopteron*, had fleshy fins with bones that closely resemble those of a terrestrial vertebrate's limbs. (b) *Acanthostega* stayed mostly in the water but had legs and other adaptations that permitted it to spend short periods on land. (c) *Ichthyostega* could spend longer periods on land because its rib cage was stronger.

(a): ©Jan Sovak/Stocktrek Images, Inc./Alamy Stock Photo; (c): ©DEA Picture Library/Getty Images

Many fossils indicate that by this time, plants were diversifying to ferns, horsetails, and seed plants. Scorpions, millipedes, and other invertebrates lived on the land.

A mass extinction of marine life marks the end of the Devonian period. Warm-water invertebrates and jawless fishes were hit especially hard, and jawed fishes called placoderms became extinct. Life on land, however, was largely spared. The cause of this mass extinction remains unknown.

Carboniferous Period (354 to 290 MYA)

Amphibians flourished from about 350 to 300 MYA, giving this period the name the "Age of Amphibians." These animals spent time on land, but they had to return to the water to wet their skins and lay eggs. Although today's amphibians are small, some of their ancient relatives were huge—up to 9 meters long!

During the Carboniferous, some amphibians arose that coated their eggs with a hard shell. These animals branched from the other amphibians, eventually giving rise to reptiles. The first vertebrates capable of living totally on land, the primitive reptiles, appeared about 300 MYA.

Carboniferous swamps included ferns and early seed plants, some of which towered to 40 meters (figure 15.16). The air was alive with the sounds of grasshoppers, crickets, and giant dragonflies with 75-centimeter wingspans. Land snails and other invertebrates flourished in the sediments. By the end of the period, many of the plants had died, buried beneath the swamps to form coal beds during the coming millennia. In fact, the term *Carboniferous* means "coal-bearing"; see this chapter's Apply It Now box for more on coal.

Meanwhile, in the oceans, the bony fishes and sharks were beginning to resemble modern forms, and protists called foraminiferans were abundant. Bryozoans and brachiopods were plentiful, but trilobites were becoming less common.

Permian Period (290 to 248 MYA)

During the Permian, seed plants called gymnosperms became more prominent. Reptiles were also becoming more prevalent. The reptile introduced a new adaptation, the amniote egg, in which an embryo could develop completely on dry land (see figure 21.29). Amniote eggs persist today in reptiles, birds, and a few mammals.

The Permian period foreshadowed the dawn of the dinosaur age. Cotylosaurs were early Permian reptiles that gave rise to the dinosaurs and all other reptiles. They coexisted with their immediate descendants, the pelycosaurs, or sailed lizards.

The Permian period ended with what paleontologists call "the mother of mass extinctions." It affected marine life the most, wiping out more than 90% of species in shallow areas of the sea. On the land, many types of insects, amphibians, and reptiles disappeared, paving the way for the age of dinosaurs.

Paleontologists hypothesize that the Permian extinctions were partly the result of a drop in sea level, which dried out

a.

b.

Figure 15.16 Carboniferous Forest Life. (a) About 300 MYA, lush forests dominated swampy landscapes. In this illustration, the trees near the shore are related to today's seedless vascular plants and seed-bearing gymnosperms. A giant millipede, early amphibian, and flying insect are also visible. (b) These forests were eventually preserved in massive coal beds containing the remains of Carboniferous plants, such as this fossilized fern frond.

(a): ©Richard Bizley/Science Source; (b): ©Kevin Schafer/Corbis

coastline communities. Carbon dioxide from oxidation of organic molecules accumulated in the atmosphere, raising global temperature and depleting the sea's dissolved O_2. The loss of O_2, in turn, may have favored bacteria that produce toxic hydrogen sulfide (H_2S) as a waste. A long series of volcanic eruptions, beginning 255 MYA and lasting a few million years, further altered global climate. Finally, sea level rose again, drowning coastline communities.

C. Reptiles and Flowering Plants Thrived During the Mesozoic Era

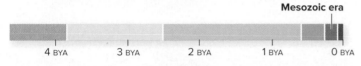

Triassic Period (248 to 206 MYA)   During the Triassic period, the first archosaurs flourished, ushering in the "Age of Reptiles" (as the Mesozoic era is sometimes called). These first archosaurs were the ancestors of the now-extinct dinosaurs and of modern birds and crocodiles. Early archosaurs shared the forest of cycads, ginkgos, and conifers with other animals called therapsids, which were the ancestors of mammals.

At the close of the Triassic period, yet another mass extinction affected life in the oceans and on land. Many marine animals were wiped out, as were many reptiles and amphibians on land. As a result, much larger animals began to infiltrate a wide range of habitats. These new, well-adapted animals were the dinosaurs, and they would dominate for the next 120 million years.

Jurassic Period (206 to 144 MYA)   By the Jurassic period, giant reptiles were everywhere (**figure 15.17**). Ichthyosaurs, plesiosaurs, and giant marine crocodiles swam in the seas alongside sharks and rays, feasting on fish, squid, and ammonites. Apatosaurs and stegosaurs roamed the land. Carnivores, such as allosaurs, preyed on the herbivores. Pterosaurs glided through the air, as did *Protoarchaeopteryx* and then *Archaeopteryx*—the first birds (see chapter 13's opening essay).

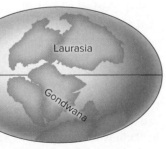

At the same time, the first flowering plants (angiosperms) appeared on land. The forests, however, still consisted largely of tall ferns and conifers, ginkgos, club mosses, and horsetails. The first frogs and the first true mammals, which were no larger than rats, appeared as well.

Figure 15.17 **Marine Reptile.** Plesiosaurs were enormous carnivorous reptiles that swam in the Jurassic seas. ©Mike Kemp/In Pictures/Corbis

Cretaceous Period (144 to 65 MYA)   The Cretaceous period was a time of great biological change. By around 100 MYA, flowering plants had spread in spectacular diversity; many modern insects arose at about the same time. Marine reptiles hunted mollusks and fish, and birds and pterosaurs roamed the skies. Duck-billed maiasaurs traveled in groups of thousands in what is now Montana. Huge herds of apatosaurs migrated from the plains of Alberta to the Arctic, northern Europe, and Asia, which

were joined as one continent at the time. Near the end of the Cretaceous, *Tyrannosaurus* roamed in what is now western North America, and *Triceratops* was so widespread that some paleontologists call it the "cockroach of the Cretaceous."

The reign of the giant reptiles ended about 65 MYA, with the extinction of ichthyosaurs, plesiosaurs, mosasaurs, pterosaurs, and nonavian dinosaurs. Ammonites vanished, too, as did many types of foraminiferans, sea urchins, and bony fishes. In all, nearly 75% of species perished. The mass extinction opened up habitats for many species that survived, including flowering plants, mollusks, amphibians, some smaller reptiles, birds, and mammals. Many of these groups, including the mammals that gave rise to our own species, subsequently flourished.

The mass extinction that ended the Mesozoic era coincides with an asteroid impact near the Yucatán peninsula. The asteroid, which was about 10 kilometers in diameter, left a debris- and clay-filled crater offshore and a huge semicircle of sinkholes onshore (figure 14.14 shows the distinctive iridium layer, an important piece of evidence for the impact theory). Biologists estimate that photosynthesis was almost nonexistent for 3 years, as debris that was thrown into the sky circulated in the atmosphere, blocking sunlight. Plankton, which provide microscopic food for many larger marine dwellers, died as well, causing a devastating chain reaction.

Figure It Out

Suppose that a 100-meter track represents Earth's 4.6-billion-year history. How close to the end of the track would you mark the Permian and Cretaceous extinctions?

Answer: 5.4 m (Permian) and 1.4 m (Cretaceous)

D. Mammals Diversified During the Cenozoic Era

Cenozoic era

4 BYA	3 BYA	2 BYA	1 BYA	0 BYA

Tertiary Period (65 to 1.8 MYA) The Cenozoic era, sometimes called the "Age of Mammals," began with the Tertiary period. The Tertiary was a time of great adaptive radiation for mammals, according to the fossil record (see figure 14.13). Within just 1.6 million years, 15 of the 18 modern orders of placental mammals arose. ⓘ *adaptive radiation,* section 14.4B

At the start of the Tertiary period, diverse hoofed mammals grazed the grassy Americas. Many may have been marsupials (pouched mammals) or egg-laying monotremes, ancestors of the platypus. Then placental mammals appeared, and fossil evidence indicates that they rapidly dominated the mammals.

Geology and the resulting climate changes molded the comings and goings of species throughout the Cenozoic. The era began with the formation of new mountains and coastlines as tectonic plates shifted. The wet warmth of the Paleocene epoch, which opened up many habitats for mammals, continued into the Eocene, providing widespread forests and woodlands. Grasslands began to replace the forests by the end of the Eocene, when the temperature and humidity dropped. Extinctions of some mammals paralleled the changing plant populations as the forests diminished, but grazing mammals thrived throughout the remaining three epochs of the Tertiary period.

Although the Cenozoic is best known for its spectacular assemblage of mammals, other types of animals diversified as well. One especially fruitful source of fossils from the Eocene epoch is the Green River formation in Wyoming (**figure 15.18**). Here, countless animals and plants were trapped in lake sediments that subsequently dried up and became exposed. Fossil collectors have scooped up millions of exquisitely preserved fish fossils similar to the one in figure 15.18a. In addition, the limestone of the Green River formation contains a rich fossil collection representing crocodiles, snakes, birds, mammals, snails, and insects. One particularly important example is the bat skeleton in figure 15.18b; this specimen is the oldest known fossilized bat. Leaves of ferns, sycamores, cattails, and other flowering plants were buried in the lake sediments as well. Amazingly, the Green River fossils not only cover a continuous 6-million-year-long period but also reveal how life responded to the changing climate during the Eocene.

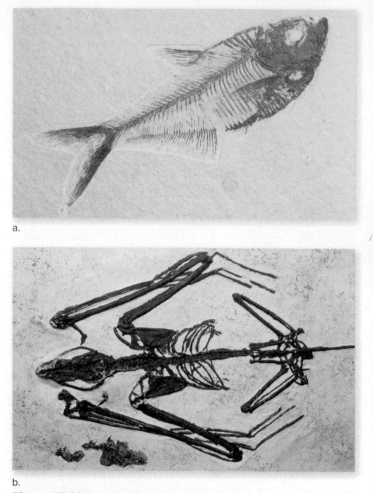

a.

b.

Figure 15.18 Eocene Fossils. Wyoming's Green River formation has preserved a showcase of Eocene life from about 50 million years ago.
(a) Freshwater fish are especially well represented in the fossil beds.
(b) This skeleton is the oldest known fossil of a bat.
(a): ©Jason Edwards/Getty Images RF; (b): ©Kevin Schafer/Corbis

Quaternary Period (1.8 MYA to Present) The Pleistocene epoch accounts for all but the last 10,000 years of the Quaternary period. During several Pleistocene ice ages, huge glaciers covered about 30% of Earth's surface and then withdrew again.

Many organisms of the time were similar to those that are familiar now, including flowering plants, insects, birds, and mammals. Some Pleistocene species, however, are extinct (**figure 15.19**). The camels and horses that were native to North America are gone (today's wild horses are descended from domesticated European horses). And the woolly mammoths, mastodons, saber-toothed cats, giant ground sloths, and other large mammals that once roamed North America, Asia, and Europe are known only from their fossils and the occasional DNA fragment (see section 19.6).

Apply It **Now**

Coal's Costs

The relationship between ancient plants and today's electronic gadgetry may seem remote, but it is not. Gadgets require electricity, and nearly half of the electricity in the United States comes from coal-fired power plants. Each lump of coal, in turn, is mined from the long-dead remains of the Carboniferous forests (**figure 15.A**).

The ferns and other plants in these forests produced their own food by photosynthesis. That is, they used water, energy from the sun, and CO_2 from the atmosphere to build the sugars and complex molecules that made up their bodies. Many of these leaves and stems did not decompose when the plants died; instead, they were buried under sediments. Heat and pressure eventually transformed the plant matter into coal. When we burn the coal, we release the potential energy and CO_2 that ancient plants trapped in photosynthesis hundreds of millions of years ago.

Burning coal to generate electricity is relatively inexpensive, compared to other energy sources. Coal powered the Industrial Revolution and continues to be economically important today. This fuel does not, however, come without costs.

First, coal mining has environmental and health consequences. Extracting coal from underground deposits can destroy soil, plants, and wildlife habitat. Water quality can suffer as exposed mountainsides erode away. And although coal mining in developed countries is safer than in times past, miners still risk injury and death in mine collapses and explosions.

Second, burning coal releases CO_2 into the atmosphere. This heat-trapping gas plays a large role in global climate change. Coal combustion also releases potentially harmful heavy metals such as mercury, which can accumulate in body tissues. Moreover, nitrogen- and sulfur-

Figure 15.A Coal Mining. Two miners extract coal deep within a mine shaft.
©Digital Vision/PunchStock RF

rich gases from coal-fired power plants contribute to acid deposition. Chapter 40 describes these environmental issues in more detail.

Feeding the human population's voracious demand for energy requires enormous amounts of coal and other fossil fuels. As you consider the costs of coal, keep in mind that no energy source is free, not even renewable ones such as sunlight and wind; all require equipment, roads, power lines, and maintenance. The best way to avoid these costs is to reduce the overall demand for energy.

Figure 15.19 American Mastodon. These enormous, elephant-like mammals lived in North America from about 3.7 million years ago until they went extinct about 10,000 years ago.
©Julie Dermansky/Science Source

The Pleistocene epoch was also eventful for our own branch of the mammal family tree, as multiple species of *Homo* came and went. Our species, *Homo sapiens* ("the wise human"), probably first appeared about 200,000 years ago in Africa and had migrated throughout most of the world by about 10,000 years ago. Other species of *Homo*, including Neandertals, vanished during the Pleistocene, leaving *Homo sapiens* as the sole human species.

The roots of our family tree, however, extend much deeper into history. We pick up the story of human evolution in more detail in section 15.4.

15.3 | MASTERING CONCEPTS

1. When did the Ediacarans live, and what were they like?
2. What types of organisms flourished in the Cambrian?
3. How did Paleozoic life diversify during the Ordovician, Silurian, Devonian, Carboniferous, and Permian periods?
4. How did the Paleozoic era end?
5. Which organisms came and went during the Mesozoic era?
6. Which new organisms arose during the Cenozoic era?

15.4 Fossils and DNA Tell the Human Evolution Story

In many ways, humans are Earth's dominant species. True, we are not the most numerous—more microbes occupy one person's intestinal tract than there are people on Earth. But in the short time of human existence, we have colonized most continents, altered Earth's surface, eliminated many species, and changed many others to fit our needs. Where did we come from?

A. Humans Are Primates

If you watch the monkeys or apes in a zoo for a few minutes, it is almost impossible to ignore how similar they seem to humans. Young ones scramble about and play. They sniff and handle food. Babies cling to their mothers. Adults gather in small groups or sit quietly, snoozing or staring into space.

It is no surprise that we see ourselves reflected in the behaviors of monkeys and apes. All **primates**—including monkeys, apes, and humans—share a suite of physical characteristics (**figure 15.20**). First, primates have grasping hands with opposable thumbs that can bend inward to touch the pads of the fingers. Some primates also have grasping feet with opposable big toes. Second, a primate's fingers and toes have flat nails instead of claws. Third, eyes set in the front of the skull give primates binocular vision with overlapping fields of sight that produce excellent depth perception. Fourth, the primate brain is large by comparison with body size.

Compared with many other groups of mammals, primate anatomy is unusually versatile. For instance, bat wings are useful for flight but not much else; likewise, horse hooves are best for fast running. In contrast, primates have multipurpose fingers and toes that are useful not only for locomotion but also for grasping and manipulating small objects. Primate limbs are similarly versatile.

The Primate Lineage The primate lineage, which originated some 60 MYA, contains three main groups: prosimians, monkeys, and hominoids (**figure 15.21**). **Prosimian** is an informal umbrella term for lemurs, aye-ayes, lorises, tarsiers, and bush babies. Monkeys are divided into two main groups: Old World monkeys (native to Africa and Asia) and New World monkeys (native to South and Central America). The **hominoids** are apes, including humans.

The apes are further divided into two groups. One contains the gibbons, or "lesser apes." The other, the **hominids,** contains all of the "great apes": orangutans, gorillas, chimpanzees (including bonobos), and humans. **Hominins** are humans and their extinct ancestors on the human branch of the primate evolutionary tree. **Table 15.1** summarizes the groups of primates.

While studying figure 15.21, keep in mind two important concepts. First, humans are not descended from other groups of modern apes. Instead, all living humans and chimpanzees share a common ancestor and diverged from that ancestor about 6 MYA. Second, gibbons, orangutans, gorillas, and chimpanzees are not "less evolved" than humans. All living species are on an equal evolutionary footing, although some may belong to older lineages.

By examining the physical characteristics of skeletons, paleontologists have learned much about the evolutionary relationships among primates (**figure 15.22**). Most human fossils consist of bones and teeth. Comparing these remains with existing primates reveals surprisingly detailed information about locomotion and diet. For this reason, knowledge of primate skeletal anatomy is essential to interpreting human fossils.

Dietary Adaptations in Primates Some of the most important skeletal characteristics, including the size and shape of the teeth, are related to diet (**figure 15.23**). Upper and lower

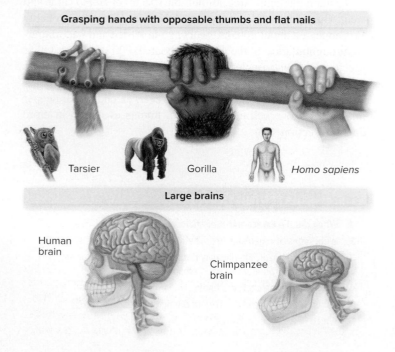

Grasping hands with opposable thumbs and flat nails

Tarsier Gorilla *Homo sapiens*

Large brains

Human brain

Chimpanzee brain

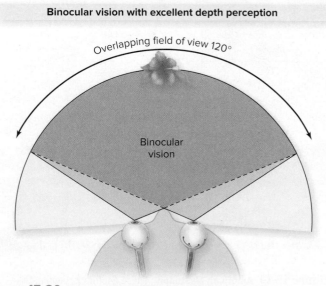

Binocular vision with excellent depth perception

Overlapping field of view 120°

Binocular vision

Figure 15.20 Primate Characteristics. Primates share several characteristics, including opposable thumbs, flattened nails, binocular vision, and large brains relative to body size.

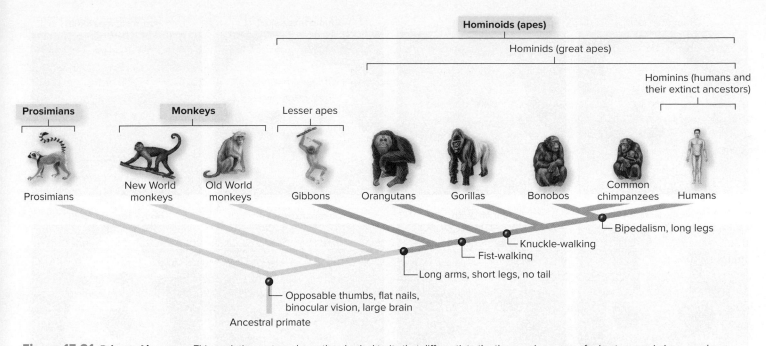

Figure 15.21 Primate Lineages. This evolutionary tree shows the physical traits that differentiate the three main groups of primates: prosimians, monkeys, and hominoids (apes). Molecular data support this hypothesis of the evolutionary relationships among primates.

TABLE **15.1** **Miniglossary of Human Evolution Terms**

Term	Definition
Primate	A mammal with grasping hands, opposable thumbs, flat nails, and binocular vision; includes prosimians, monkeys, and apes
Hominid	A great ape; includes orangutans, gorillas, chimpanzees, and humans
Hominin	Any hominid in genus *Ardipithecus, Australopithecus, Paranthropus,* or *Homo*
Human	A member of genus *Homo,* of which the only living representative is *Homo sapiens*

molar teeth have ridges that fit together, much as the teeth of gears intermesh. Food caught between these surfaces is ground, crushed, and mashed. The size of these teeth is an adaptation that reflects the toughness of the diet.

As you examine the skulls in figure 15.23, notice the differences in the sagittal crests. This bony ridge runs lengthwise along the top of the skull and is an attachment point for muscles. A prominent sagittal crest indicates particularly strong jaws, another clue to an animal's diet.

Other important features in primate skulls include the size of the jaw bones, the prominence of the ridge of bone above the eye, the degree to which the jaw protrudes, and the shape of the curve of the tooth row. All of these characteristics allow paleoanthropologists, the scientists who study human fossils, to assign each new discovery to a species.

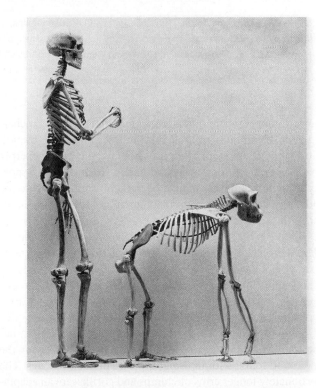

Figure 15.22 Clues from Bones. The skeleton on the left, from a human, shares many similarities with the gorilla skeleton on the right.
©Tom McHugh/Science Source

Primate Locomotion Additional important characteristics in hominoid skeletons are adaptations related to locomotion. Brachiation is swinging from one arm to the other while the body dangles below. Many apes move through the treetops in this way;

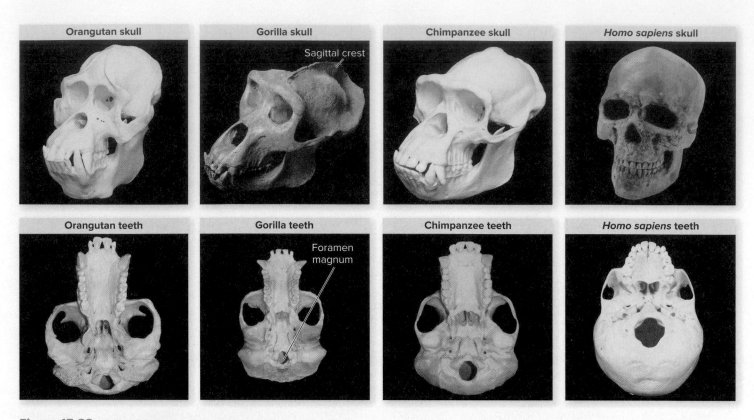

Figure 15.23 Skulls and Teeth. The skulls and teeth of an orangutan, a gorilla, a chimpanzee, and a human reveal details about posture, diet, and jaw strength. Biologists compare fossils of extinct species to bones of existing primates to learn how our ancestors lived. The top row of photos shows side views; the bottom row shows skulls from underneath, revealing the hole through which the spinal cord enters the brain.

(orangutan & chimp skull, orangutan, gorilla & chimp teeth): Skulls Unlimited International, Inc ©David Liebman Pink Guppy; (gorilla skull): ©Pascal Goetgheluck/Science Source; (*Homo sapiens* skull): ©Ingram Publishing/Alamy RF; (*Homo sapiens* teeth): ©McGraw-Hill Education/Christine Eckel

in contrast, monkeys run on all fours along the tops of branches. Orangutans spend most of their lives in trees and move by brachiation when they are in treetops. Gibbons, the most superbly acrobatic apes, have long arms and hands. The size and opposability of their thumbs are reduced, but their arms connect to the shoulders by ball-and-socket joints that allow free movement of the arms in 360 degrees. In addition, long collarbones act as braces and keep the shoulders from collapsing toward the chest.

Heavier-bodied chimpanzees and gorillas don't brachiate as much as gibbons and orangutans, but like humans, they can do so. Humans seldom brachiate, with the exception of small, light-bodied children playing on schoolyard "monkey bars." Adult human arms are typically too weak to support the heavy torso and legs.

Chimpanzees and gorillas move by knuckle-walking, a behavioral modification that allows an animal to run rapidly on the ground on all fours, with its weight resting on the knuckles. The proportionately longer arms of chimps and gorillas are an adaptation to knuckle-walking.

One important feature distinguishes humans from the other great apes: bipedalism, or the ability to walk upright on two legs. Adaptations to bipedalism include relatively short arms and longer, stronger leg bones. Foot bones form firm supports for walking, with the big toe fixed in place and not opposable. The bowl-shaped pelvis supports most of the weight of the body, and the vertebrae of the lower back are robust enough to bear some body weight.

Bipedalism is also reflected in the bones of the head. The foramen magnum is the large hole in the skull where the spinal cord leaves the brain (see figure 15.23). In modern humans, this hole is tucked beneath the skull. In gorillas and chimps, the foramen magnum is located somewhat closer to the rear of the skull; in animals that run on all fours, such as horses and dogs, the hole is at the back of the skull.

B. Molecular Evidence Documents Primate Relationships

Fossil evidence and anatomical similarities were once the only lines of evidence that paleoanthropologists could analyze in tracing the course of human evolution. Around 1960, however, scientists began to use molecular sequences to investigate relationships among primates. Studies of blood proteins and DNA presented a new picture of primate evolution, as it became clear that humans are a species of great ape. One of the astounding findings of these molecular studies was that the DNA sequences of humans and chimpanzees are nearly identical (see section 15.5). The evolutionary tree in figure 15.21 takes into account both the anatomical characteristics and the molecular data.

Other research has further eroded the distinctions between humans and other great apes. Previously, humans had been placed in a separate group, supposedly characterized by upright walking, tool-making, and language. Then, in the 1970s, chimpanzees and wild gorillas were observed to use tools, and captive great apes learned to use sign language to communicate with their trainers. The only characteristic that now remains unique to humans is bipedal locomotion.

C. Human Evolution Is Partially Recorded in Fossils

Even though DNA and proteins provide overwhelming evidence of the relationships between living primates, these molecules deteriorate with time. Scientists therefore cannot usually use molecular data to establish relationships of prehistoric hominins. For this, we must turn to studies of fossilized remains.

To interpret fossils from our family tree, paleoanthropologists compare details of the ancient skeletal features with modern primates and try to reconstruct as much information as they can. So far, hominin fossils have fallen into four groups: *Ardipithecus, Australopithecus, Paranthropus,* and *Homo* (figure 15.24).

Discovered in Ethiopia in 1994, *Ardipithecus ramidus* (or "Ardi" for short) dates to about 4.4 MYA and is the oldest representative of the hominin lineage. Fossils reveal several clues about Ardi's life. The teeth indicate that Ardi was an omnivore. The pelvis supported both upright walking and powerful climbing, as did the feet, which had opposable big toes. The flexible hands had long, grasping fingers with which Ardi could have carried objects while walking upright.

The next oldest fossils in the hominin lineage have been assigned to the genus *Australopithecus,* meaning "Southern apeman." The downward position of the foramen magnum indicates that these small apes walked upright. *Australopithecus afarensis* (including the famous "Lucy" fossil in figure 15.25) and *A. africanus* are members of this group, which dates from about 4 to 2.5 MYA.

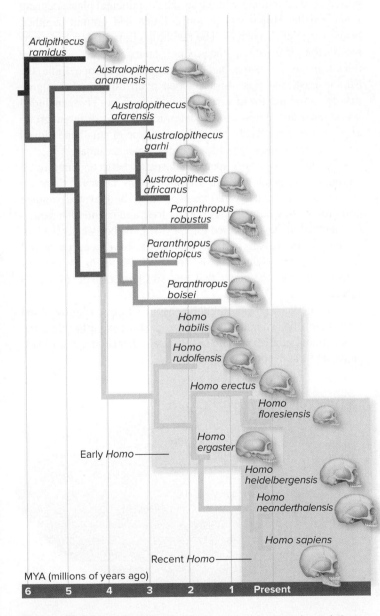

Figure 15.24 Family Tree. Fossilized remains place our close relatives into four main groups (*Ardipithecus, Australopithecus, Paranthropus,* and *Homo*). Skull sizes are approximately to scale and range from about 400 cm³ for *A. afarensis* to about 1450 cm³ for modern humans. The evolutionary relationships in this tree are hypothesized.

Figure 15.25 Old Bones. This reconstruction of the *Australopithecus afarensis* specimen called "Lucy" was based on her 3.2-million-year-old bones.
©Philippe Plailly & Atelier Daynes/Science Source

Paranthropus, which dates from about 3 to 1.5 MYA, seems to be an evolutionary dead end that descended from *Australopithecus* but gave rise to no other group. The name literally means "beside humans." These primates had extremely large teeth, protruding jaws, and skulls with a sagittal crest. All of these specializations probably relate to the large jaw muscles needed to crush tough plants or crack nuts.

All members of genus *Homo* are considered humans, and fossils in this group are associated with stone tools. *Homo* species tend to have larger bodies and larger brains than do australopiths. *Homo habilis, H. ergaster,* and *H. erectus* are among the cluster of extinct species belonging to the group called "early *Homo*." These species lived from about 2.5 MYA to about 500,000 years ago. New fossil discoveries have called this portion of the family tree into question; some fossils currently classified as unique species may, in fact, represent normal variation within *H. erectus*.

Regardless of how this issue is settled, early *Homo* species eventually gave rise to "recent *Homo*." Recent species of *Homo* have smaller teeth, lighter and less protruding jaws, larger braincases, and lighter brow ridges. Their fossils are associated with evidence of culture (**figure 15.26**). Recent *Homo* species include *H. heidelbergensis, H. neanderthalensis, H. floresiensis,* and *H. sapiens*. The only human species alive today is *Homo sapiens*.

One interesting trend in human evolution has been an adaptive radiation of species, followed by extinctions. Fossil evidence shows that about 1.8 MYA, as many as five species of humans lived together in Africa. About 200,000 years ago, three species of recent *Homo* coexisted in Europe. Today, however, all except *Homo sapiens* are extinct.

What happened to the other *Homo* species? No one knows, but many anthropologists wonder whether *H. sapiens* contributed to their extinction. Scientists have speculated that Neandertals interbred with *H. sapiens*, effectively causing the Neandertals to disappear from the fossil record about 30,000 years ago. Analyses of DNA from Neandertals and *H. sapiens* have supported that hypothesis, although the question remains unsettled.

D. Environmental Changes Have Spurred Human Evolution

What provoked the first hominins to abandon brachiation in favor of bipedal, upright walking? What allowed the large brains that are characteristic of recent *Homo* to develop? To find the answers, we have to consider a related question: Where did hominins evolve?

Charles Darwin was one of the first to speculate that humans evolved in Africa. About 12 MYA, tectonic movements caused a period of great mountain building. The continental plates beneath India and the Himalayan region collided and ground together, heaving up the Himalayas. The resulting climatic shift had enormous ecological consequences. Cooler temperatures reduced the thick tropical forests that had covered much of Europe, India, the Middle East, and East Africa. Open plains appeared, bringing new opportunities for species that could live there. These included less competition, a new assortment of foods, and a different group of predators. Experts speculate that one type of small ape moved out of the trees and began life on the African savannas.

Perhaps at first this species alternated between running on all fours and bipedal walking. On open plains, however, there are advantages to bipedal walking, especially the elevated vantage point for sensing danger and spotting food and friends. This environment would have selected for apes with the best skeletal adaptations to bipedalism, and the trait would have been preserved and honed in the plains. Bipedalism also freed the hands to carry objects and use the tools that are so characteristic of *Homo* species.

No one knows what might have spurred the evolution of the large brain that characterizes humans. Some experts relate the development of a large brain to tool use; others connect it to language and life in social groups.

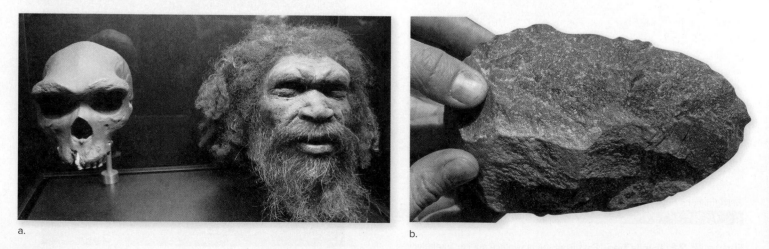

a.

b.

Figure 15.26 *Homo heidelbergensis.* (a) A skull replica (*left*) is paired with a model showing what *H. heidelbergensis* might have looked like (*right*). (b) This hand axe, nicknamed "Excalibur," was discovered with a pit of 350,000-year-old *H. heidelbergensis* bones in Spain.

(a): ©Patrick Kovarik/AFP/Getty Images; (b): ©Javier Trueba/Science Source

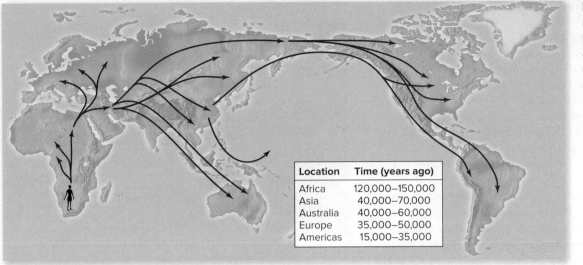

Figure 15.27 Out of Africa. Researchers used mitochondrial DNA sequences to deduce approximately when humans originally settled each continent after migrating out of Africa.

Location	Time (years ago)
Africa	120,000–150,000
Asia	40,000–70,000
Australia	40,000–60,000
Europe	35,000–50,000
Americas	15,000–35,000

E. Migration and Culture Have Changed *Homo sapiens*

DNA sampled from people around the world has revealed a compelling portrait of human migration out of Africa (figure 15.27). Asia, Australia, and Europe all were colonized at least 40,000 years ago, but it took somewhat longer for humans to reach the Americas.

As humans spread throughout the world, new habitats selected for different adaptations. Near the equator, for example, sunlight is much more intense than at higher latitudes. One component of sunlight is ultraviolet (UV) radiation, which is both harmful and beneficial. On the one hand, UV radiation damages DNA and causes skin cancer. On the other hand, some UV wavelengths help the skin produce vitamin D, which is essential to bone development and overall health.

These two counteracting selective pressures help explain why skin pigmentation is strongly correlated with the amount of ultraviolet radiation striking the Earth (see section 25.6). One pigment that contributes to skin color, melanin, blocks UV radiation. Intense UV radiation selects for alleles that confer abundant melanin. People whose ancestry is near the equator, such as in Africa and Australia, therefore tend to have very dark brown skin. In northern Europe and other areas with weak sunlight, however, heavily pigmented skin would block so much UV light that indigenous people would suffer from vitamin D deficiency. These conditions are correlated with a high frequency of alleles conferring pale, pinkish skin.

No matter where people roamed, one byproduct of the large human brain was **culture:** the knowledge, beliefs, and behaviors that we transmit from generation to generation. Among the earliest signs of culture is cave art from about 14,000 years ago, which indicates that our ancestors had developed fine hand coordination and could use symbols.

By 10,000 years ago, depending on which native plants and animals were available for early farmers to domesticate, agriculture began to replace a hunter–gatherer lifestyle in many places. Agriculture meant increased food production, which profoundly changed societies. Freed from the necessity of producing their own food, specialized groups of political leaders, soldiers, weapon-makers, religious leaders, scientists, engineers, artists, writers, and many other types of workers arose. These new occupations meant improved transportation and communication, better technologies, and the ability to explore the world for new lands and new resources.

Undoubtedly, humans are a special species. We can modify the environment much more than, for example, a slime mold or an earthworm can. We can also alter natural selection, in our own species and in others. Finally, culture allows each generation to build on information accumulated in the past. The knowledge, beliefs, and behaviors that shape each culture are constantly modified within a person's lifetime, in stark contrast to the millions of years required for biological evolution. Our species is therefore extremely responsive to short-term changes.

Despite our unique set of features, however, we are descended from ancestors with which we share many characteristics. It is intriguing to think about where the human species is headed, which species will vanish, and how life will continue to diversify in the next 500 million years.

15.4 MASTERING CONCEPTS

1. Name and describe the three groups of contemporary primates. To which group do humans belong?
2. What can skeletal anatomy and DNA sequences in existing primates tell us about the study of human evolution?
3. What are the four groups of species in the hominin family tree, and which still exist today?
4. Which conditions may have contributed to the evolution of humans?

INVESTIGATING LIFE

15.5 What Makes Us Human?

Perhaps no scientific issue is more tantalizing and entangled with philosophy than the question of what makes us human. One way to look for answers is to study the similarities and differences between humans and chimpanzees, our closest living relatives (**figure 15.28**). A key piece of this puzzle has been the ability to compare the 3 billion or so DNA nucleotides that make up the human and chimpanzee genomes. ⓘ *DNA sequencing, section 11.2B*

Long before the human and chimpanzee genome sequencing projects began, the startling genetic similarities among humans, chimps, and other great apes were well known. Their chromosomes, for example, are extremely similar (**figure 15.29**). Chimpanzees, gorillas, and orangutans, however, have one more pair of chromosomes than do humans. This difference in number traces to chromosome 2. As you can see in the figure, human chromosome 2 reflects an ancient fusion of two smaller chromosomes.

The complete DNA sequences for both species reveal exactly how much we have in common: Overall, at least 96% of our nucleotides are identical. However, scientists must still scrutinize both genomes to identify the 25,000 or so coding regions (the sequences that specify proteins). They must also sequence the genomes of additional individuals to locate the variable regions, and they must learn which alleles confer which traits. The result will be an unprecedented view of human biology and evolution.

Here is a sampling of questions we may soon be able to answer:

- Which genes define humans? With the chimp genome complete, scientists can now search for regions that differ from those in our closest relatives. These are the sequences that define humans.

- What accounts for our uniquely human features? Considering how genetically similar humans and chimps are, we have strikingly different phenotypes. Are human and chimp proteins essentially the same but expressed at different times? Or have human-specific mutations caused some previously functional genes to stop working (see section 26.7)?

- Why do humans and chimps have different diseases? A surprising number of diseases affect humans but not chimps, including Alzheimer disease and carcinomas (a type of cancer). Similarly, HIV progresses to AIDS in humans but not in chimps. At least some of these differences will certainly lie in our genes, and their discovery may yield new disease cures.

More recently, biologists have sequenced the genomes of orangutans and gorillas. The DNA sequences of these apes will reveal the answers to many questions that were previously impossible to approach. For example, researchers once suspected that a gene involved in hearing was one key to the human acquisition of

Figure 15.28 Close Relative. The genomes of humans and chimpanzees are nearly identical. What are the differences that make us human?
©Karen Huntt/Corbis

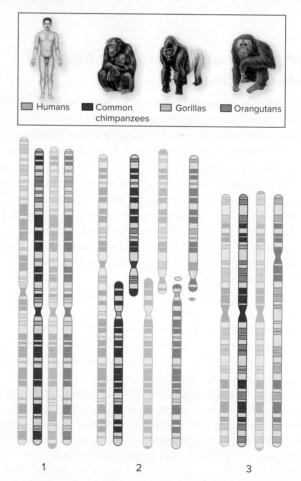

☐ Humans ■ Common chimpanzees ☐ Gorillas ■ Orangutans

1   2   3

Figure 15.29 Similar Chromosomes. Striking similarities in chromosome size and banding patterns are evident in this diagram showing chromosomes 1, 2, and 3 for humans, chimpanzees, gorillas, and orangutans. The diagram also shows that our own chromosome 2 is the product of a human-specific chromosomal fusion event.

Source: Yunis, J.J., and O. Prakash. "The Origin of Man: A Chromosomal Pictorial Legacy," *Science*, 1982 Mar: 215(4539): 1527.

complex language. The latest genome sequence, however, revealed that gorillas—which do not have complex language—have the same version of the gene as humans; other genes must therefore be responsible. The gorilla and orangutan genome sequences are sure to help researchers discover many additional details of our evolutionary history.

Whatever the source of our humanity, it is revealed partly in our ability to make reasoned decisions and in our compassion for others. Chimpanzees are endangered in the wild, where they succumb to habitat loss, hunting, the pet trade, and biomedical research. The Chimpanzee Sequencing and Analysis Consortium's paper advocates the protection of chimpanzees in the wild, and it ends with this statement: "We hope that elaborating how few differences separate our species will broaden recognition of our duty to these extraordinary primates that stand as our siblings in the family of life."

Sources: Chimpanzee Sequencing and Analysis Consortium. September 1, 2005. Initial sequence of the chimpanzee genome and comparison with the human genome. *Nature*, vol. 437, pages 69–87.

Locke, Devin P., and 100 coauthors. January 27, 2011. Comparative and demographic analysis of orang-utan genomes. *Nature*, vol. 469, pages 529–533.

Olson, Maynard V., and Ajit Varki. January 2003. Sequencing the chimpanzee genome: Insights into human evolution and disease. *Nature Reviews Genetics*, vol. 4, pages 20–28.

Scally, Aylwyn, and 70 coauthors. March 8, 2012. Insights into hominid evolution from the gorilla genome sequence. *Nature,* vol. 483, pages 169–175.

15.5 | MASTERING CONCEPTS

1. What information can researchers gain by comparing the human and chimpanzee genome sequences?
2. How do the genome sequences of orangutans and gorillas help scientists further resolve the evolutionary history of humans?

CHAPTER SUMMARY

15.1 Life's Origin Remains Mysterious

- The solar system formed about 4.6 BYA, and life first left evidence on Earth by about 3.7 BYA. The **geologic timescale** describes these and many other events in life's history.

A. The First Organic Molecules May Have Formed in a Chemical "Soup"

- **Prebiotic simulations** combine simple inorganic chemicals to form life's organic building blocks, including amino acids and nucleotides.
- These monomers may have linked together to form polymers on hot clay or mineral surfaces.

B. Some Investigators Suggest an "RNA World"

- The **RNA world** theory proposes that RNA preceded formation of the first cells. Proteins provided enzymes and structural features. Reverse transcriptase could have copied RNA's information into DNA.

C. Membranes Enclosed the Molecules

- Phospholipid sheets that formed bubbles around proteins and nucleic acids may have formed cell precursors, or **progenotes.**

D. Early Life Changed Earth Forever

- Early organisms permanently changed the physical and chemical conditions in which life continued to evolve.

15.2 Complex Cells and Multicellularity Arose over a Billion Years Ago

- The internal membranes of eukaryotic cells may have formed when the outer membrane folded in on itself repeatedly.

A. Endosymbiosis Explains the Origin of Mitochondria and Chloroplasts

- The **endosymbiont theory** proposes that chloroplasts and mitochondria originated as free-living bacteria that were engulfed by larger cells.

B. Multicellularity May Also Have Its Origin in Cooperation

- The evolution of multicellularity, which occurred about 1.2 BYA, is poorly understood.
- Figure 15.30 summarizes the steps leading from the origin of life to today's diversity or prokaryotic and eukaryotic organisms.

15.3 Life's Diversity Exploded in the Past 500 Million Years

A. The Strange Ediacarans Flourished Late in the Precambrian

- The Ediacarans were soft, flat organisms that were completely unlike modern species. They lived during the late Precambrian and early Cambrian periods.

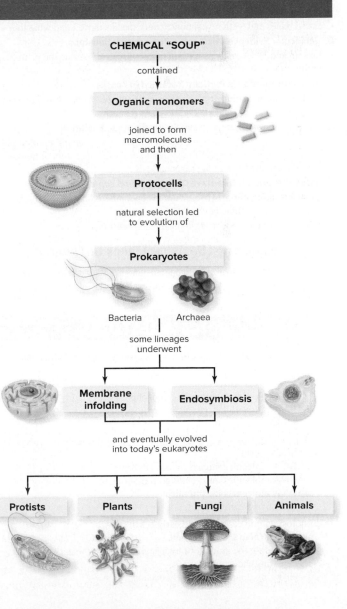

Figure 15.30 Origin and Diversification of Life: A Summary.

B. Paleozoic Plants and Animals Emerged onto Land

- The Cambrian explosion introduced many species, notably those with hard parts. Amphibian-like animals ventured onto land about 360 MYA, followed by reptiles, birds, and mammals. Invertebrates, ferns, and forests flourished.

C. Reptiles and Flowering Plants Thrived During the Mesozoic Era

- Dinosaurs prevailed throughout the Mesozoic era, when forests were largely cycads, ginkgos, and conifers. In the middle of the era, flowering plants became prevalent. When the nonavian dinosaurs died out 65 MYA, resources opened up for mammals.

D. Mammals Diversified During the Cenozoic Era

- Mammals diversified during the Tertiary period. Humans arose during the Pleistocene epoch of the Quaternary period. Repeated ice ages and the extinction of many large mammals also occurred during the Pleistocene.

15.4 Fossils and DNA Tell the Human Evolution Story

A. Humans Are Primates

- **Primates** have grasping hands, opposable thumbs, binocular vision, large brains, and flat nails. The three groups of primates are **prosimians,** monkeys, and **hominoids** (apes).
- **Hominids** are the "great apes," whereas **hominins** are humans and their extinct ancestors on the human branch of the primate evolutionary tree.
- Fossil bones and teeth reveal how extinct species moved and what they ate.

B. Molecular Evidence Documents Primate Relationships

- Protein and DNA analysis has altered how scientists draw the primate family tree.

C. Human Evolution Is Partially Recorded in Fossils

- Early hominins included *Ardipithecus, Australopithecus, Paranthropus,* and *Homo.*

D. Environmental Changes Have Spurred Human Evolution

- Millions of years ago, new mountain ranges arose, causing climate shifts. Savannas replaced tropical forests, and apes—the ancestors of humans—moved from the trees to the savanna.

E. Migration and Culture Have Changed *Homo sapiens*

- After migrating out of Africa, humans encountered new habitats that selected for new allele combinations.
- Humans owe our success to language and **culture.**

15.5 Investigating Life: What Makes Us Human?

- Comparing the chimpanzee and human genomes will lead to unprecedented insights into the genes, phenotypes, and diseases that are unique to humans.

MULTIPLE CHOICE QUESTIONS

1. Place the following events in Earth's history in order. Which happened *second*?
   a. Eukaryotes arose.
   b. O_2 gas accumulated in the atmosphere.
   c. First multicellular organisms arose.
   d. Prokaryotes arose.

2. Which of the following must be true for natural selection to occur in an "RNA world"?
   a. RNA molecules must turn into DNA molecules.
   b. RNA molecules must undergo mutations.
   c. RNA molecules must replicate.
   d. Both b and c are correct.

3. Photosynthetic cells affected early Earth by
   a. adding O_2 to the atmosphere.
   b. increasing the amount of hydrogen sulfide in the early oceans.
   c. depleting the ozone layer.
   d. changing the pH of the early oceans.

4. Why is multicellularity adaptive?
   a. Multicellular organisms reproduce more quickly than unicellular organisms.
   b. Cells work together, each specializing in specific functions.
   c. Multicellular organisms are motile, whereas unicellular organisms are not.
   d. All of the above are correct.

5. What characteristic of mitochondria and chloroplasts makes them different from other organelles and suggests they were once independent organisms?
   a. They have a specialized function in the cell.
   b. They are surrounded by a membrane.
   c. They contain enzymes.
   d. They have their own DNA and ribosomes.

6. Which of the following represents the correct order of appearance, from earliest to most recent?
   a. Fishes, reptiles, Ediacarans, flowering plants
   b. Ediacarans, fishes, reptiles, flowering plants
   c. Ediacarans, flowering plants, fishes, reptiles
   d. Flowering plants, reptiles, fishes, Ediacarans

7. Why was the Mesozoic era extinction significant to the history of mammals?
   a. As photosynthesis slowed, meat-eating mammals prevailed.
   b. Fewer plankton species meant more open ocean ecosystems for large, aquatic mammals.
   c. Dinosaur extinctions opened up new habitats to mammals.
   d. Only the smartest primates knew how to survive the extinction.

8. Primates share all of the following characteristics EXCEPT
   a. opposable thumbs.
   b. excellent depth perception.
   c. bipedalism.
   d. flat fingernails.

9. Humans evolved from
   a. monkeys.
   b. chimpanzees.
   c. gorillas.
   d. an ancestor shared with chimpanzees.

10. DNA evidence suggests that modern humans
    a. share a single origin.
    b. arose independently in several isolated populations.
    c. have not evolved in the last 150,000 years.
    d. are evolving from chimpanzees today.

Answers to these questions are in appendix A.

WRITE IT OUT

1. Explain how the origin of self-replicating molecules was critical to life's origin.

2. List three ways that studying the history of life helps us understand life's current diversity, and predict how diversity might change in the future.

3. Review the structures of nucleic acids and proteins in chapter 2. What chemical elements had to have been in primordial "soup" to generate these organic molecules?

4. List a logical sequence of events that starts with an early prokaryote and ends with a modern multicellular eukaryote.

5. The amoeba *Pelomyxa palustris* is a single-celled eukaryote with no mitochondria, but it contains symbiotic bacteria that can live in the presence of O_2. How does this observation support the endosymbiont theory?

6. The antibiotic streptomycin kills bacterial cells but not eukaryotic cells; diphtheria toxin kills eukaryotic cells but not bacteria. Which of these two substances do you predict would kill mitochondria and chloroplasts? Explain your answer.

7. List the major events of the Precambrian supereon and of the Paleozoic, Mesozoic, and Cenozoic eras.

8. How has the emergence of new species changed Earth's history?

9. Distinguish among the terms *primate, hominid, hominin,* and *Homo.*

10. Explain how opposable thumbs, large brains, and binocular vision are adaptive to primates.

11. What can scientists learn by comparing the fossilized skeletons of extinct primates with the bones of modern species?

12. How do you predict a scientist would respond to a question about whether humans "evolved from monkeys"?

13. Use the Internet to learn about *National Geographic*'s Genographic Project. What are the main objectives and components of the project, and how are researchers using the information they gather to learn more about human evolution?

14. In what ways has culture been an important factor in human evolution?

15. Compare today's changing culture to biological evolution by means of natural selection. How are they similar? How are they different? Can you think of examples of knowledge, beliefs, and behaviors that were appropriate in one set of conditions but that humans selected against as conditions changed?

16. The video game "Spore" invites players to design creatures and guide them through "five stages of evolution." Search the Internet for information about "Spore"; describe how evolution in this game is similar to, and different from, the evolution of life on Earth.

PULL IT TOGETHER

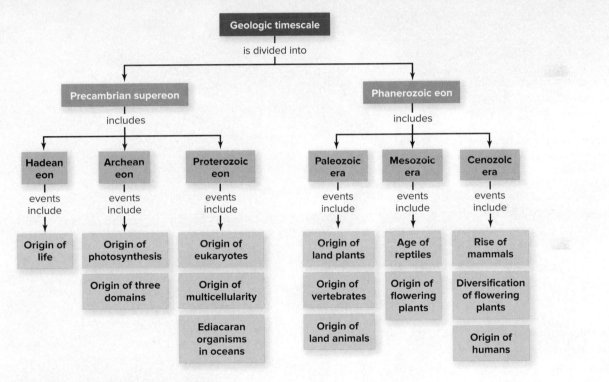

Figure 15.31 Pull It Together: The Origin and History of Life.

Refer to figure 15.31 and the chapter content to answer the following questions.

1. Review the Survey the Landscape figure in the chapter introduction and add *life's history, DNA evidence,* and *fossil evidence* to the Pull It Together concept map.

2. Arrange the 14 major events at the bottom of the concept map in chronological order; indicate when the major extinction events occurred. What was the implication of each extinction?

3. Create an additional concept map that depicts the evolution of humans.

16 Viruses

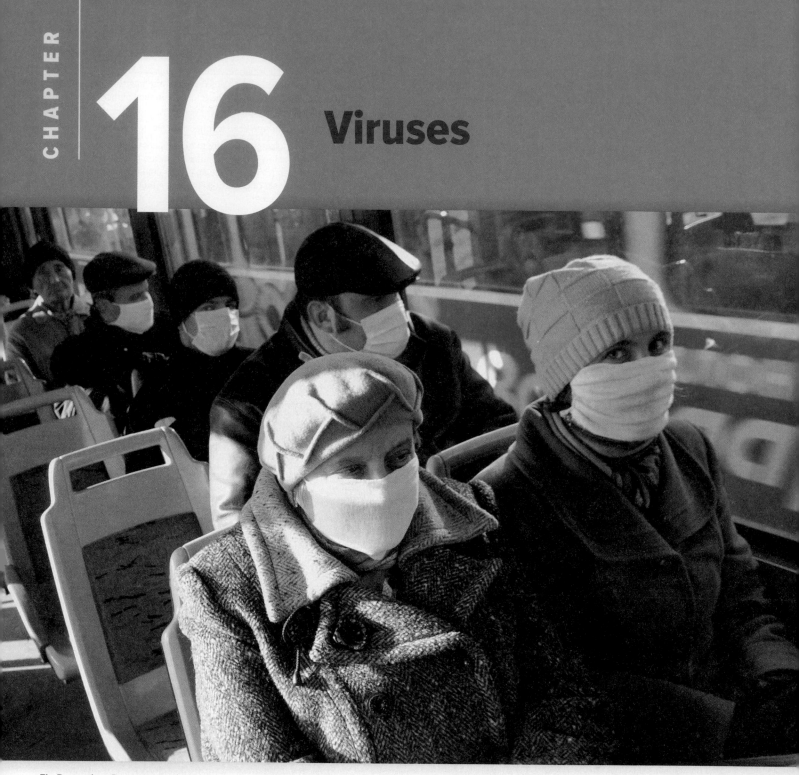

Flu Protection. Protective masks help prevent the spread of influenza among passengers on this tram in western Ukraine.
©Yuriy Dyachyshyn/AFP/Getty Images

LEARN HOW TO LEARN
Take the Best Possible Notes

Some students take notes only on what they consider "important" during a lecture. Others write down words but not diagrams, or they write what's on the board but not what the instructor is saying. All of these strategies risk losing vital information and connections between ideas that could help in later learning. Instead, write down as much as you can during lecture, including sketches of the diagrams and notes on what the instructor is telling you about the main ideas. It will be much easier to study later if you have a complete picture of what happened in every class.

UNIT 4

Viruses Lurk on the Farm and in the Jungle

Most viruses infect only a few, closely related species. But some—including those that cause influenza and Ebola hemorrhagic fever—jump among more distant relatives.

Influenza, known as "flu" for short, moves among pigs (swine flu), birds (avian flu), and humans. The cells that line a pig's throat carry receptors that bind to both the avian and the human versions of flu viruses. In the pig's cells, these viruses exchange segments of their genomes. The new viral varieties sometimes cause serious epidemics; after all, most humans lack immunity against viruses that normally infect birds or pigs.

The worst flu pandemic in history was the "Spanish flu," which killed tens of millions of people between 1918 and 1920—more than the number that perished in World Wars I and II, the Korean War, and the Vietnam War combined. No one knows whether this flu virus originated in chickens, pigs, or other animals.

Less severe flu pandemics also occurred in 1957, 1968, 1977, and 2009. Because flu viruses evolve rapidly, epidemiologists watch carefully for new variants that are both deadly and easily transmitted among humans. A new virus with both qualities might trigger an outbreak rivaling the disastrous 1918 flu pandemic.

As horrifying as it was, the Spanish flu killed only about 2% of infected people; the 2009 flu pandemic had a fatality rate of just 0.03%. At the other end of the spectrum, the Ebola virus epidemic that started in 2013 had a fatality rate as high as 70%. It took nearly 2 years to bring the disaster under control; nearly all of the 30,000 affected people lived in the West African countries of Liberia, Sierra Leone, and Guinea. Before the epidemic, only 1716 Ebola cases had ever been known to occur.

Once an outbreak begins in human populations, infection with Ebola virus causes sore throat, headache, vomiting, diarrhea, and bleeding. It spreads to new victims in blood and other body fluids. Between outbreaks, however, little is known about Ebola. For example, no one knows where the virus "hides" in nature. Bats are considered a likely candidate, but dogs, pigs, chimpanzees, gorillas, and other animals can also become infected with Ebola. Nor does anyone know how or why the virus periodically jumps from wild animals to humans.

Viruses infect every type of organism, not just humans and other animals. This chapter explains what viruses are, how they replicate and cause disease, and why they have proven so hard to defeat.

LEARNING OUTLINE

16.1 Viruses Are Genes Wrapped in a Protein Coat

16.2 Viral Replication Occurs in Five Stages

16.3 Viruses May Kill Bacteria Immediately or Their DNA May "Hide" in the Cell

16.4 Illnesses Caused by Animal Viruses May Be Mild or Severe

16.5 Viruses Cause Diseases in Plants

16.6 Viroids and Prions Are Other Noncellular Infectious Agents

16.7 Investigating Life: Scientific Detectives Follow HIV's Trail

SURVEY THE LANDSCAPE
The Diversity of Life

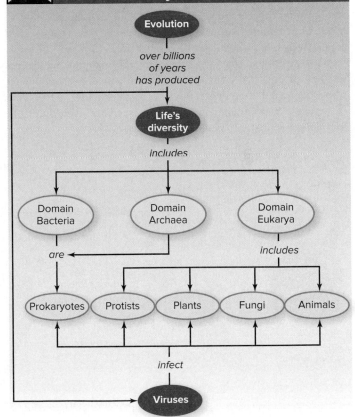

This first chapter in the diversity unit covers viruses, which are noncellular particles that straddle the boundary between life and nonlife. These infectious agents are hijackers that turn living cells into virus-making factories.

For more details, study the Pull It Together feature in the chapter summary.

16.1 Viruses Are Genes Wrapped in a Protein Coat

Smallpox, influenza, the common cold, rabies, polio, chickenpox, warts, AIDS, Zika fever—this diverse list includes illnesses that range from merely inconvenient to deadly. All have one thing in common: They are infectious diseases caused by viruses.

Many people mistakenly lump viruses and bacteria together as "germs." Viruses, however, are not bacteria. In fact, they are not even cells. A **virus** is a small, infectious agent that is simply genetic information enclosed in a protein coat. The 2000 or so known species of viruses therefore straddle the boundary between the chemical and the biological.

A. Viruses Are Smaller and Simpler Than Cells

A typical virus is much smaller than a cell (**figure 16.1**). At about 10 μm (microns) in diameter, an average human cell is perhaps one-tenth the diameter of a human hair. A bacterium is about one-tenth again as small, at about 1 μm (1000 nm) long. The average virus, with a diameter of about 80 nm, is more than 12 times smaller than a bacterium. However, not all viruses are quite so tiny: The largest known example, an ocean virus called pandoravirus, has a diameter of about 1 μm.

Viruses are simple structures that lack many of the characteristics of cells. A virus does not have a nucleus, organelles, ribosomes, a cell membrane, or even cytoplasm. Only a few types of viruses contain enzymes. All viruses, however, have the following two features in common:

- **Genetic information.** All viruses contain genetic material (either DNA or RNA) that carries the "recipes" for their proteins (see chapter 7). The major criterion for classifying viruses is whether the genetic material is DNA or RNA. Either type of nucleic acid may be single- or double-stranded. ⓘ *nucleic acids,* section 2.5D

- **Protein coat.** The **capsid,** or protein coat, surrounds the genetic material. The capsid's shape determines a virus's overall form, which is another characteristic used in classification (**figure 16.2**). Many viruses are spherical or icosahedral (a 20-faced shape built of triangular sections). Others are rod-shaped, oval, or filamentous.

Some viruses have other features as well. For example, some have an **envelope,** a lipid- and protein-rich outer layer derived from parts of the host cell's membrane. Despite this origin, the viral envelope does not share the cell membrane's function. That is, the envelope does not control what enters and leaves the virus. Instead, proteins embedded in the envelope help the virus invade a new host cell. ⓘ *cell membrane,* section 3.3

The presence or absence of an envelope is another criterion for virus classification. One example of an enveloped virus is the human immunodeficiency virus (HIV), which causes acquired immunodeficiency syndrome (AIDS). The influenza and Ebola viruses also have an envelope; the virus that causes polio does not.

Despite having relatively few components, a virus's overall structure can be quite intricate and complex. For example, **bacteriophages** (sometimes simply called "phages") are viruses that infect bacteria. Some phages have parts that resemble tails, legs, and spikes; they look like the spacecrafts once used to land on the moon (see figure 16.2b).

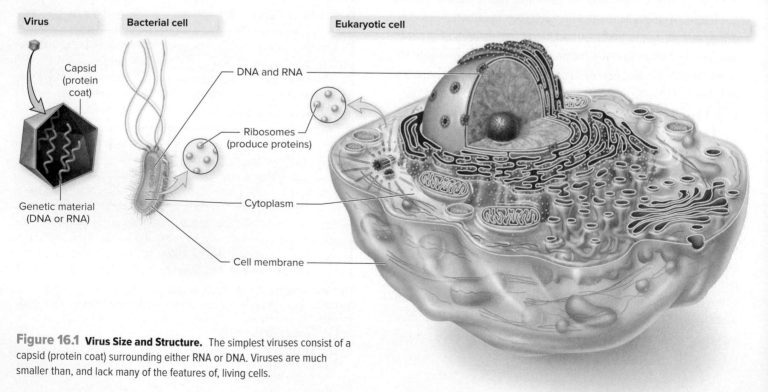

Figure 16.1 Virus Size and Structure. The simplest viruses consist of a capsid (protein coat) surrounding either RNA or DNA. Viruses are much smaller than, and lack many of the features of, living cells.

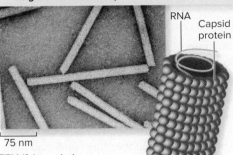

a. Tobacco mosaic virus (filamentous; single-stranded RNA)

RNA

Capsid protein

75 nm

TEM (false color)

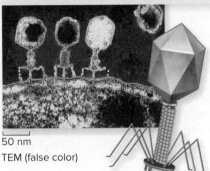

b. T-even bacteriophage (spaceship; double-stranded DNA)

50 nm

TEM (false color)

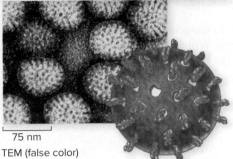

c. Rotavirus (spherical; double-stranded RNA)

75 nm

TEM (false color)

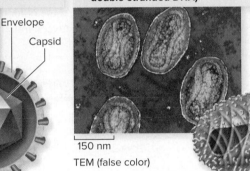

d. Herpesvirus (icosahedral, enveloped; double-stranded DNA)

Envelope

Capsid

200 nm

TEM (false color)

e. Poxvirus (oval, enveloped; double-stranded DNA)

150 nm

TEM (false color)

Figure 16.2 Virus Variety. Each type of virus has a characteristic structure, visible only with an electron microscope. (a) The capsid of the tobacco mosaic virus (TMV) is long and filamentous. (b) T-even viruses look like tiny spaceships. (c) Rotavirus causes severe diarrhea in young children. (d) Herpesviruses cause cold sores and rashes. (e) Poxviruses cause smallpox.

Photos: (a): ©G. Wanner/ScienceFoto/Getty Images; (b): ©Eye of Science/Science Source; (c): CDC/Dr. Erskine Palmer & Byron Skinner; (d): ©G. Murti/Science Source; (e): ©James Cavallini/BSIP SA/Alamy

B. A Virus's Host Range Consists of the Organisms It Infects

Virtually all species of animals, fungi, plants, protists, and bacteria get viral infections. The **host range** of a virus is the types of organisms or cells that it can infect. A virus can enter only a cell that has a specific target attachment molecule, or receptor, on its surface. The reason a bacteriophage cannot attack human cells is that our cell surfaces lack the correct target molecules.

Some target molecules occur on a very small subset of cells in an organism, whereas others are more widespread. HIV, for example, infects only certain types of human immune system cells. The rabies virus, on the other hand, can infect the muscle and nerve cells of any mammal, including humans, skunks, foxes, raccoons, bats, and dogs. All of these animals have the target molecule that the rabies virus uses to recognize a potential host.

The **reservoir** of a virus is the host that acts as a continual source of the viral infection for other host species. For many viruses that infect humans, the reservoir is a host animal that may or may not show symptoms of infection. Examples of reservoirs for viruses that cause diseases in humans include wild birds (avian influenza and West Nile encephalitis), rodents (hantavirus pulmonary syndrome), mosquitoes (yellow fever), and raccoons (rabies).

C. Are Viruses Alive?

Most biologists do not consider a virus to be alive because it does not metabolize, respond to stimuli, or reproduce on its own. Instead of dividing as cells do, viruses are assembled from pieces that are manufactured inside infected host cells. ⓘ *what is life?* section 1.1

Nevertheless, viruses do have some features in common with life, including genetic material. Both DNA and RNA can mutate, which means that viruses evolve just as life does. Each time a host cell produces a virus, random mutations occur. The genetic variability among the new viruses is subject to natural selection. That is, some variants are better than others at infecting host cells, replicating, and passing on their genes to a new crop of viruses.

Although viruses evolve, no one knows how they originated; their extreme diversity suggests that they do not share a single common ancestor. Viruses are therefore not included in the taxonomic hierarchy. Instead, scientists group viruses based on the type of nucleic acid, the structure of the virus, how it replicates, and the type of disease it causes. ⓘ *taxonomic hierarchy,* section 14.6A

16.1 MASTERING CONCEPTS

1. How are viruses similar to and different from cells?
2. What determines a virus's host range?
3. How do viruses evolve?

16.2 Viral Replication Occurs in Five Stages

The production of new viruses is very different from cell division. When a cell divides, it doubles all of its components and splits in two. Virus production, on the other hand, more closely resembles the assembly of cars in a factory.

A cell infected with one virus may produce hundreds of new viral particles. Whatever the host species or cell type, the same basic processes occur during a viral infection (figure 16.3):

1. **Attachment:** A virus attaches to a host cell by adhering to a receptor molecule on the cell's surface. Generally, the virus can attach only to a cell within which it can reproduce. HIV cannot infect skin cells, for example, because its receptors occur only on certain white blood cells.

2. **Penetration:** The viral genetic material can enter the cell in several ways. Animal cells engulf virus particles and bring them into the cytoplasm via endocytosis. Viruses that infect plants often enter their host cells by hitching a ride on the mouthparts of herbivorous insects. Many bacteriophages inject their genetic material through a hole in the cell wall, somewhat like a syringe. ⓘ *endocytosis,* section 4.5C

3. **Synthesis:** The host cell produces multiple copies of the viral genome; mutations during this stage are the raw material for viral evolution. In addition, the information

encoded in viral DNA is used to produce the virus's proteins. The host cell provides all of the resources required for the production of new viruses: ATP, ribosomes, nucleotides, tRNA, amino acids, and enzymes.

4. **Assembly:** The subunits of the capsid join, and then genetic information is packed inside.

5. **Release:** Once the virus particles are assembled, they are ready to leave the cell. Some bacteriophages induce production of an enzyme that breaks down the host's cell wall, killing the cell as it releases the viruses. Enveloped viruses such as HIV and herpesviruses, on the other hand, bud from the host cell by exocytosis. The cell may die as they carry off segments of the cell membrane. ⓘ *exocytosis,* section 4.5C

The amount of time between initial infection and cell death varies. Bacteriophages need as little as half an hour to infect a cell and replicate. At the other extreme, for some animal viruses, years may elapse between initial attachment and the final burst of viral particles.

16.2 | MASTERING CONCEPTS

1. Describe the five steps in viral replication.
2. What is the source of energy and raw materials for the synthesis of viruses in a host cell?

Figure 16.3 Viral Replication. These five basic steps of viral replication apply to any virus, whether the host cell is prokaryotic or eukaryotic.

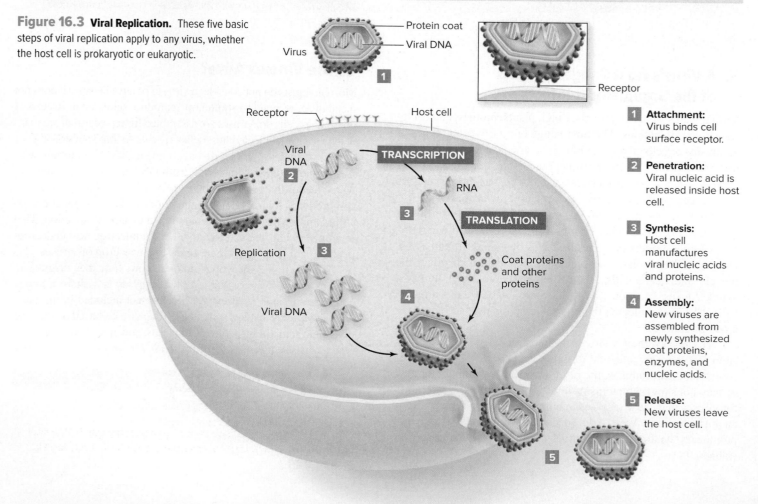

16.3 Viruses May Kill Bacteria Immediately or Their DNA May "Hide" in the Cell

Following attachment to the host cell and penetration of the viral genetic material, viruses may or may not immediately cause cell death. Bacteriophages, the viruses that infect bacteria, can do either. The two viral replication strategies in bacteriophages are called lytic and lysogenic infections.

In a **lytic infection,** a virus enters a bacterium, immediately replicates, and causes the host cell to burst (lyse) as it releases a flood of new viruses (**figure 16.4a**). The newly released viruses infect other bacteria, repeating the process until all of the cells are dead.

Some researchers have investigated the possibility of using lytic bacteriophages to treat bacterial infections in people. This strategy, called "phage therapy," would have two main advantages over antibiotics (drugs that kill bacteria).

First, unlike drugs, viruses evolve along with their bacterial hosts, and they keep killing until all host cells are dead. Bacterial populations are therefore unlikely to acquire resistance to the phages. Second, each bacteriophage targets only one or a few types of bacteria, so the treatment is precisely tailored to the infection. Paradoxically, phage therapy's main weakness is related to this strength. Medical personnel must first identify the exact type of bacteria causing infection before beginning phage therapy. This delay could be deadly.

In a **lysogenic infection,** the genetic material of a virus is replicated along with the bacterial chromosome, but the cell lives and reproduces as usual (figure 16.4b). At some point, however, the virus reverts to a lytic cycle, releasing new viruses and killing the cell.

Many lysogenic viruses use enzymes to cut the host cell DNA and join their own DNA with the host's. A **prophage** is the DNA of a lysogenic bacteriophage that is inserted into the host chromosome. Other lysogenic viruses maintain their DNA apart from the chromosome. Either way, however, when the cell divides, the viral genes replicate, too. The virus gets a "free ride," infecting new cells without actually triggering viral production.

During a lysogenic stage, the viral DNA does not damage the host cell. Only a few viral proteins are produced, most functioning as a "switch" that determines whether the virus should become lytic. At some signal, such as stress from DNA damage or cell starvation, these viral proteins trigger a lytic infection cycle. The cell dies and releases new viruses, which infect other cells. The next generation of viruses may enter a lytic or lysogenic replication cycle, depending on the condition of the host cells.

Much of what biologists know about viral replication comes from research on bacteriophages. The Focus on Model Organisms box describes some of what scientists have learned from studies of one such bacteriophage, called lambda.

16.3   MASTERING CONCEPTS

1. What is a lytic infection?
2. How is a lysogenic infection similar to and different from a lytic infection?

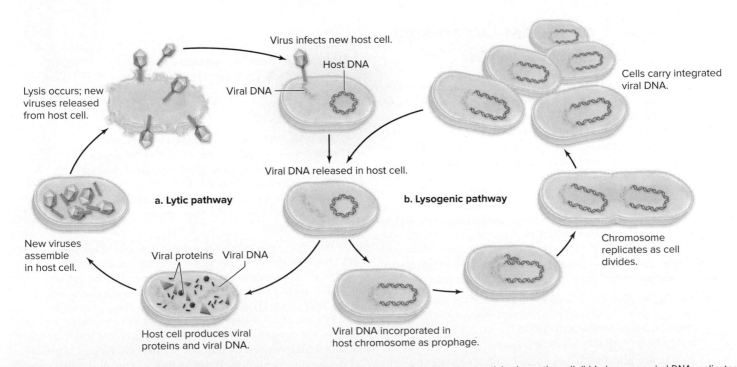

Figure 16.4 Lysis and Lysogeny. (a) In the lytic pathway, the host cell bursts (lyses) when new virus particles leave the cell. (b) In lysogeny, viral DNA replicates along with the cell, but new viruses are not produced. An environmental change, such as stress in the host cell, may trigger a lysogenic virus to switch to the lytic pathway.

16.4 Illnesses Caused by Animal Viruses May Be Mild or Severe

A person can acquire a viral infection by inhaling the respiratory droplets of an ill person or by ingesting food or water contaminated with viruses. Some viruses, such as HIV and hepatitis B, enter a person's bloodstream via a blood transfusion, sexual contact, or the use of contaminated needles. The infected person often becomes ill (although some infections are symptomless).

A. Symptoms Result from Cell Death and the Immune Response

Once a viral infection is established, the death of infected cells produces a wide range of symptoms that reflect the types of host cells destroyed. If enough cells die, the disease may be severe.

To illustrate how symptoms develop, consider the flu. Influenza viruses infect cells lining the human airway. As cells produce and release new viruses, the infection spreads rapidly in the lungs, trachea, throat, and nose. The dead and damaged cells in the airway cause the respiratory symptoms of influenza, including cough and sore throat.

What about the other symptoms, such as fever and body aches? These classic signs of influenza do not directly involve the respiratory system, where the infected cells are located. Instead, these whole-body symptoms result from the immune system's response to the viral infection. Cells of the immune system release signaling molecules called cytokines, which in turn induce fever; a high body temperature speeds other immune responses.

Cytokines also trigger inflammation, which causes body ache and fatigue. These (and other) immune reactions usually defeat the influenza virus, but in the meantime, they can also make the host rather miserable! Chapter 33 describes the human immune system in more detail.

B. Some Animal Viruses Linger for Years

In a **latent infection,** viral genetic information inside an animal cell lies dormant; even as the host cell divides, new viruses are not produced. However, the virus may be reactivated later. A latent virus in an animal cell is therefore similar to a lysogenic phage in a bacterial cell.

Many people harbor latent infections of the herpes simplex virus type I, which causes cold sores on the lips. After initial infection, the viral DNA remains in skin cells indefinitely. When a cell becomes stressed or damaged, new viruses are assembled and leave the cell to infect other cells. Cold sores, which reflect the localized death of virus-infected skin cells, periodically recur near the site of the original infection.

Some latent viruses persist by signaling their host cells to divide continuously, possibly leading to cancer. A latent infection by some strains of human papillomavirus can cause cervical cancer (see the Burning Question box). Epstein–Barr virus is another example. More than 80% of the human population carries this virus, which infects B cells of the immune system. A person who is initially exposed to the virus may develop mononucleosis (often called "mono"). The virus later maintains a latent infection in B cells. In a few people, especially those

F O CUS on Model Organisms

Bacteriophage Lambda

Although viruses are not organisms, they have nevertheless contributed enormously to the scientific understanding of life. This box focuses on a virus that has never made you sick because it does not infect human cells (figure 16.A). Rather, it kills *Escherichia coli* bacteria that live in the intestines of humans and other mammals. Bacteriophage lambda (Greek letter λ) injects its genetic material into its host, which subsequently turns into a virus-making factory. In so doing, phage lambda has revealed many processes fundamental to all life.

Phage lambda's double-stranded DNA genome of 50,000 base pairs contains all the information it takes to make more phages, and its 60 or so genes must turn on and off in proper sequence for the new viruses to form properly. Biologists learn the functions of viral proteins by studying viral mutants—phages with missing proteins. These studies have revealed that phage lambda's proteins fall into three general groups:

- **Capsid proteins:** Phage lambda's protein coat consists of a nearly spherical head (enclosing the viral DNA) and a tube-shaped tail with proteins that bind to *E. coli*'s surface. DNA enters the host cell through the tail.

- **Regulatory proteins:** Some of phage lambda's proteins bind to viral DNA and either promote or prevent transcription of particular genes. Other regulatory proteins stop the virus from entering the lysogenic pathway or prevent the replication of other viruses that may have also infected the cell.

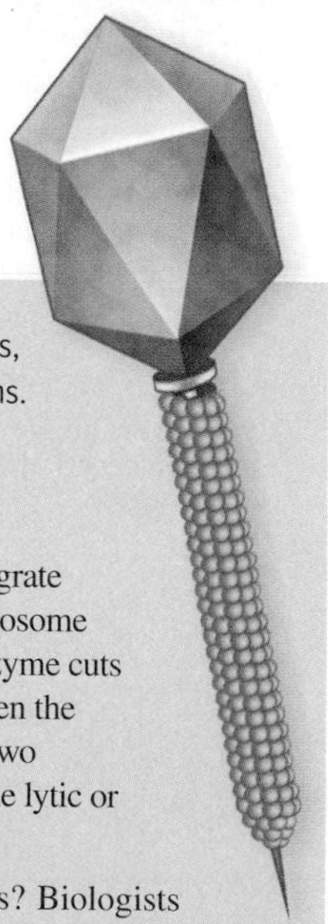

Figure 16.A Bacteriophage Lambda. This virus, deadly to the bacterium *E. coli,* is harmless to humans.

- **Enzymes:** A protein called integrase helps integrate phage lambda's DNA into the host cell's chromosome when it enters the lysogenic phase. Another enzyme cuts the viral DNA back out of the chromosome when the lytic phase begins. The balance between these two enzymes determines whether the virus enters the lytic or lysogenic phase.

Why study gene regulation in bacteriophages? Biologists can learn a lot about complex systems by studying the simplest models, since similar mechanisms of gene repression and activation also work in our own cells. When those mechanisms fail, cancer may result. ⓘ *regulation of gene expression,* section 7.6

Phage lambda has taught scientists about not only gene regulation but also recombination and protein folding, both of which are fundamental life processes. In addition, this virus has become something of a laboratory workhorse. Phage lambda is unusual because it can complete its replication cycle even if a large amount of foreign DNA is inserted into its genome. This discovery has made phage lambda an extremely important tool for ferrying recombinant DNA into *E. coli.* ⓘ *protein folding,* section 2.5C; *transgenic organisms,* section 11.2A

with weakened immune systems, the virus eventually causes a form of cancer called Burkitt lymphoma. ⓘ *cancer,* section 8.6

HIV is another virus that can remain latent inside a human cell (**figure 16.5**). HIV belongs to a family of viruses called retroviruses, all of which have an RNA genome. The virus attaches to and penetrates a helper T cell, an essential type of white blood cell that helps coordinate the immune system. The virus's reverse transcriptase enzyme transcribes the viral RNA to DNA. The DNA then inserts itself into the host cell's DNA. Cells with active infections express the viral genes, producing and releasing many new HIV particles. These viruses go on to infect other T cells, so the number of helper T cells gradually declines. Eventually, the loss of T cells leaves the body defenseless against infections or cancer. AIDS is the result. ⓘ *helper T cells,* section 34.3A

In some of the body's T cells, however, HIV's genetic information remains dormant, forming a latent reservoir of infected but inactive cells. These latent infections are especially hard to treat with anti-HIV drugs, most of which work only in cells that are actively producing viruses (see the Apply It Now box for more on anti-HIV drugs). Finding ways to coax the virus out of latency would greatly improve the efficacy of anti-HIV drugs; this is one of many active areas of HIV research.

C. Drugs and Vaccines Help Fight Viral Infections

Halting a viral infection is a challenge, in part because viruses invade living cells. Some antiviral drugs interfere with enzymes or other proteins that are unique to viruses, but overall, researchers have developed few medicines that inhibit viruses without killing infected host cells. Many viral diseases therefore remain incurable.

Antiviral drug development is complicated by the genetic variability of many viruses. Consider the common cold. Many different cold viruses exist, and their genomes mutate rapidly. As a result, a different virus strain is responsible every time you get the sniffles. Developing drugs that work against all of these variations has so far proved impossible.

Vaccination remains our most potent weapon against many viral diseases. A **vaccine** "teaches" the immune system to recognize one or more molecular components of a virus without actually exposing the person to the disease. Some vaccines confer immunity for years, whereas others must be repeated annually. The influenza vaccine is an example of the latter. Flu viruses mutate rapidly, so this year's vaccine is likely to be ineffective against next year's strains. ⓘ *vaccines,* section 34.4

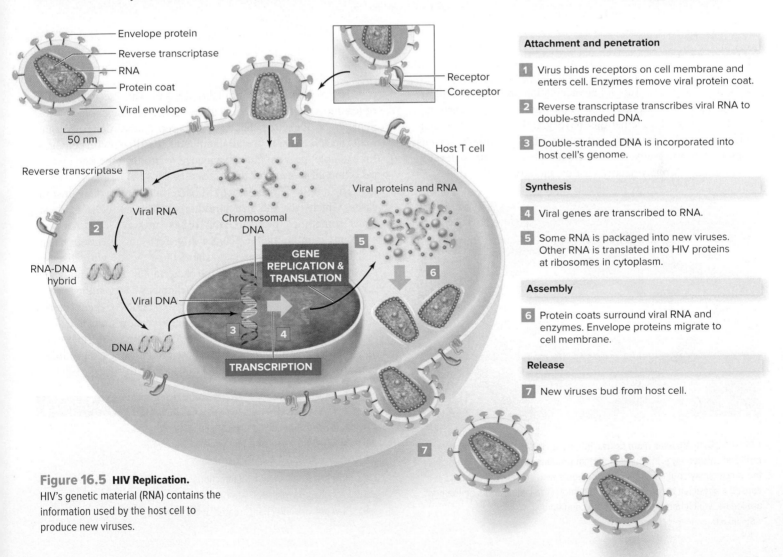

Attachment and penetration

1 Virus binds receptors on cell membrane and enters cell. Enzymes remove viral protein coat.

2 Reverse transcriptase transcribes viral RNA to double-stranded DNA.

3 Double-stranded DNA is incorporated into host cell's genome.

Synthesis

4 Viral genes are transcribed to RNA.

5 Some RNA is packaged into new viruses. Other RNA is translated into HIV proteins at ribosomes in cytoplasm.

Assembly

6 Protein coats surround viral RNA and enzymes. Envelope proteins migrate to cell membrane.

Release

7 New viruses bud from host cell.

Figure 16.5 HIV Replication.
HIV's genetic material (RNA) contains the information used by the host cell to produce new viruses.

Labels in figure: Envelope protein, Reverse transcriptase, RNA, Protein coat, Viral envelope, 50 nm, Reverse transcriptase, Viral RNA, RNA-DNA hybrid, DNA, Chromosomal DNA, Viral DNA, GENE REPLICATION & TRANSLATION, TRANSCRIPTION, Receptor, Coreceptor, Host T cell, Viral proteins and RNA

Burning Question

Can a person get cancer by having sex?

Cancer usually has nothing to do with sex, but there are exceptions. Most cancers of the cervix, anus, and throat, for example, are caused by a sexually transmitted virus called human papillomavirus (HPV). The virus can spread through vaginal, anal, and oral sex.

More than 200 strains of this virus exist, some causing harmless ailments such as warts on the soles of the feet. Thirty or so strains infect the genitals and are passed among male and female sexual partners. Usually, the infection goes unnoticed. But some strains of HPV cause genital warts; two of these strains are considered high-risk because of their strong association with cancer.

What are the best ways to prevent HPV infection? Abstaining from sex, limiting the number of sexual partners, and using a condom are all common-sense strategies. A vaccine can also help by protecting against four strains of HPV. Two of these strains are associated

©Christopher Kerrigan/
McGraw-Hill Education

with 70% of cervical cancer cases; the other two cause 90% of genital warts.

Health professionals recommend vaccination at age 11 or 12. Vaccinating girls before they become sexually active makes sense because the vaccine does not work in people who have already been exposed to the virus. Since the vaccine was introduced, HPV infection among teenage girls has dropped by over 50%. Experts also recommend vaccinating 11- to 12-year-old boys, partly to prevent the transmission of HPV to females and partly because males are also at risk of contracting HPV-associated cancers.

Submit your burning question to
Marielle.Hoefnagels@mheducation.com

Thanks to successful global vaccination programs, smallpox has vanished from human populations, and polio is nearly defeated. Childhood vaccinations have greatly reduced the incidence of measles, mumps, and many other potentially serious illnesses. Unfortunately, researchers have been unable to develop vaccines against many deadly viruses, including HIV.

To produce viruses in a laboratory for vaccine manufacture or testing, scientists must inoculate host cells. The choice of host depends on the virus; bacteria, cultured animal cells, live animals, and live plants are all candidates. Fertilized chicken eggs, for example, provide the cells needed to produce the raw materials of the influenza vaccine (**figure 16.6**).

The antibiotic drugs that kill bacteria never work against viruses, which lack the structures targeted by antibiotics. Physicians sometimes prescribe the drugs anyway. This needless exposure to antibiotics selects for drug-resistance genes in harmless bacteria, which can later share those genes with disease-causing microbes. Drug-resistant bacteria are an enormous and growing public health problem (see chapter 12's opening essay).

Nevertheless, a viral infection can sometimes trigger bacterial growth. For example, patients sometimes develop sinus infections as a complication of influenza or the common cold. Physicians may prescribe antibiotics to treat these secondary bacterial infections, but the drugs will not affect the underlying virus.

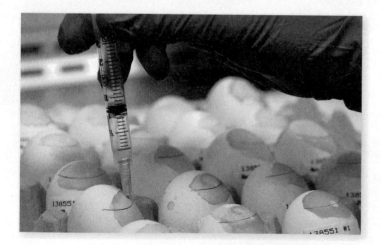

Figure 16.6 Viruses from Eggs. Influenza viruses replicate inside fertilized chicken eggs in a laboratory that produces the flu vaccine. Because the vaccine may contain egg residues, people with egg allergies should consult a physician before receiving a flu shot. (The yellow spots on the eggs are iodine, which helps maintain sterile conditions.)
©Stephen D. Cannerelli/The Image Works

16.4   MASTERING CONCEPTS

1. How can a person acquire a viral infection?
2. How do symptoms of a viral infection develop?
3. What is a latent animal virus?
4. How are some latent viral infections linked to cancer?
5. Describe how HIV replicates in host cells.
6. How are viral infections treated and prevented?

Apply It **Now**

Anti-HIV Drugs

HIV spreads by contact with infected blood, semen, vaginal secretions, and breast milk. As the virus replicates, it kills T cells in the immune system. An infected person therefore becomes increasingly vulnerable to cancer and infectious diseases. Fortunately, decades of painstaking research on HIV at the molecular level has enabled biologists to develop medications that slow the replication of this deadly virus. Here are some ways that these drugs work:

- **Keep viruses out of uninfected host cells.** Entry inhibitors block the receptors that HIV uses to recognize and enter a host cell. The drug enfuvirtide (Fuzeon) is an example.

- **Inhibit replication of viral genetic information.** Azidothymidine (AZT) and a few other drugs stop reverse transcriptase from making a DNA copy of HIV's RNA genome.

- **Inhibit viral DNA integration into host DNA.** Integrase inhibitors keep viral DNA from inserting into the host cell's chromosome.

- **Inhibit assembly of new viruses inside infected host cells.** Protease inhibitors prevent viral enzymes called proteases from cleaving the proteins that make up HIV's protein coat. The protein coat cannot form properly, so fewer mature viruses leave infected cells, and the rate of infection slows.

Researchers are also beginning to use gene therapy to treat HIV infections, genetically altering T cells to omit a coreceptor that HIV requires for entry. Without the coreceptor, HIV cannot bind (figure 16.B). An extension of this strategy is to delete the gene encoding the coreceptor from stem cells that produce immune cells, so that the patient's own body can generate receptor-free T cells. Gene therapy, however, is not yet a mainstream treatment for HIV. ⓘ *gene therapy,* section 11.4D

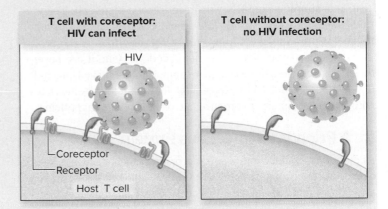

Figure 16.B **Role of Coreceptors in HIV Infection.**

16.5 Viruses Cause Diseases in Plants

Like all organisms, plants can have viral infections. The first virus ever discovered was tobacco mosaic virus (see figure 16.2a), which affects not only tobacco but also tomatoes, peppers, and more than 120 other plant species.

To infect a plant cell, a virus must penetrate waxy outer leaf layers and thick cell walls. Most viral infections spread when plant-eating insects such as leafhoppers and aphids move virus-infested fluid from plant to plant on their mouthparts. ⓘ *plant cell wall,* section 3.6B

Once inside a plant, viruses multiply at the initial site of infection. The killed plant cells often appear as small dead spots on the leaves. Over time, the viruses spread from cell to cell through plasmodesmata (bridges of cytoplasm between plant cells). They can also move throughout a plant by entering the vascular tissues that distribute sap. Depending on the location and extent of the viral infection, symptoms may include blotchy, mottled leaves or abnormal growth. A few symptoms, such as the streaking of some flower petals, appear beautiful to us (figure 16.7).

Although plants do not have the same forms of immunity as do animals, they can fight off viral infections. For example, virus-infected cells may "commit suicide" before the infection has a chance to spread to neighboring cells. Alternatively, a plant cell may destroy the mRNA transcribed from viral genes. Since the viral mRNA is never translated into proteins, this defense prevents the assembly of new viruses.

a.   b.

Figure 16.7 **Sick Plants.** (a) Cucumber mosaic virus causes a characteristic mottling (spotting) of squash leaves. (b) A virus has also caused the streaking on the petals of these tulips.

(a): ©Nigel Cattlin/Science Source; (b): ©Photodisc/Getty Images/Getty Images RF

16.5 **MASTERING CONCEPTS**

1. How do viruses enter plant cells and spread within a plant?
2. What are some symptoms of a viral infection in plants?

16.6 Viroids and Prions Are Other Noncellular Infectious Agents

The idea that something as simple as a virus can cause devastating illness may seem amazing, yet some infectious agents are even simpler than viruses.

A. A Viroid Is an Infectious RNA Molecule

A **viroid** is a highly wound circle of RNA that lacks a protein coat; it is simply naked RNA that can infect a plant cell. The viroid coils tightly to form a double-stranded RNA. This configuration helps prevent degradation by host cell enzymes.

Although viroid RNA does not encode protein, it can nevertheless cause severe disease in many important crop plants, including potatoes (figure 16.8). Apparently, the viroid's RNA interferes with the plant's ability to produce one or more essential proteins.

B. A Prion Is an Infectious Protein

Another type of infectious agent is a **prion,** which stands for "proteinaceous infectious particle." A prion protein (PrP for short) is a normal membrane protein that can exist in multiple three-dimensional shapes, at least one of which is abnormal and can cause disease (figure 16.9). Upon contact with an abnormal form of PrP, a normal prion protein switches to the abnormal PrP configuration. The change triggers another round of protein refolding, and so on. As a result of this chain reaction, masses of abnormal prion proteins accumulate inside cells. ⓘ *protein folding,* section 2.5C

The misshapen prion proteins cause brain cells to die. The brain eventually becomes riddled with holes, like a sponge. Prion

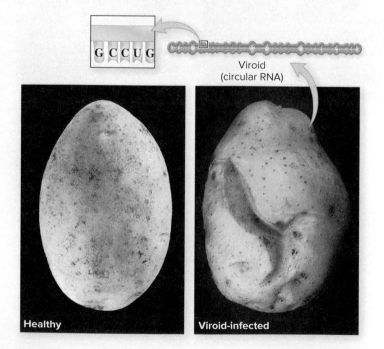

Figure 16.8 Viroid Disease. Potatoes affected by the potato spindle tuber viroid often have prominent cracks.

Photos: (healthy): ©SerAlexVi/Getty Images RF; (infected): ©Nigel Cattlin/Alamy

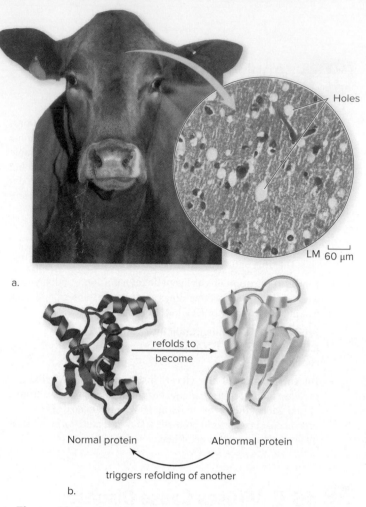

a.

b.

Figure 16.9 Prion Disease. (a) A brain affected by bovine spongiform encephalopathy ("mad cow disease") is riddled with holes caused by prions. (b) A normal prion protein refolds into an abnormal shape, which can trigger refolding of more proteins in a chain reaction.

Photos: (a, cow): ©Pixtal/age fotostock RF; (a, tissue): ©Ralph Eagle Jr./Science Source

diseases are therefore called spongiform encephalopathies. "Mad cow disease" is one example. A cow may acquire this disease after ingesting infected cattle. Affected cattle grow fearful, then aggressive, and then lose weight and die.

Prions cause a few human diseases, all of them fatal (but rare). Kuru is associated with cannibalism. Creutzfeldt–Jakob disease (CJD) may occur spontaneously, be acquired in medical procedures using tainted materials, or be inherited as a gene that encodes the abnormal prion protein. Another heritable prion disease causes insomnia, dementia, and death.

Prions are extremely hardy; heat, radiation, and chemical treatments that destroy bacteria and viruses have no effect on prions. Luckily, commonsense precautions that keep brains and spinal cords out of food and medical products can prevent the transmission of prion diseases.

16.6   MASTERING CONCEPTS

1. How are viroids and prions different from viruses?
2. How do viroids and prions cause disease?
3. What is the best way to avoid prion diseases?

INVESTIGATING LIFE

16.7 Scientific Detectives Follow HIV's Trail

Anyone who watches crime dramas on TV knows that detectives collect many types of evidence when trying to match a suspect to a crime. Do footprints found at the scene match the suspect's shoes? Do skin cells found under the victim's fingernails contain DNA matching the suspect's? Was the suspect in the victim's neighborhood at the time of the crime?

Epidemiologists use similar tactics when they test hypotheses about the evolution of viruses. For example, scientists have suggested that the ancestor of HIV is a virus called simian immunodeficiency virus (SIV). Many strains of SIV exist. They typically cause symptomless infections in monkeys and apes, although some can cause an AIDS-like disease in chimpanzees.

Researchers have tested hypotheses about the origin of HIV by asking five independent questions: (1) Do the genomes of HIV and SIV consist of the same type of genetic material and have the same order, number, and types of genes? (2) Do the viral genes (and their encoded proteins) share similar sequences? (3) Is SIV common enough in its natural host that it has a chance of spreading to other hosts, including humans? (4) Do SIV and HIV occur in the same geographic region? (5) Is there a plausible transmission pathway for SIV to spread from its original host to humans?

The two major types of HIV, called HIV-1 and HIV-2, apparently arose independently. Scientists had learned by the early 1990s that HIV-2 originated from SIV in sooty mangabey monkeys. But evidence for the origin of HIV-1 pointed to a different source: SIVcpz (for chimpanzee). In the late 1990s, a team led by University of Alabama researcher Beatrice Hahn reported genetic similarities between HIV-1 and SIVcpz.

Over the following decade, researchers gathered more evidence to piece together the evolutionary history of HIV and SIV. Much of this work focused on the central African country of Cameroon, where HIV is thought to have originated. A breakthrough came with the discovery that antibodies against SIV could be detected in feces, allowing researchers to study viral infection rates in large numbers of wild primates for the first time. Martine Peeters of France's University of Montpellier and a team of collaborators analyzed 725 ape droppings from 15 sites in Cameroon. If a sample contained anti-SIV antibodies, the group extracted primate DNA and viral RNA to reveal the type of ape and to learn more about the SIV strain.

Amino acid sequences of viral proteins confirmed the close relationship between strains of SIV and HIV-1 groups M, N, and O (figure 16.10). HIV-1 group M causes an estimated 98% of infections in humans; it is closely related to a strain of SIV found in a central chimpanzee from southeast Cameroon. As many as 34% of chimpanzees in this region are infected with SIV. HIV-1 group N is closely related to a different strain of SIV. Surprisingly, HIV-1 group O was most closely related to a strain of SIV found only in gorilla droppings. No one yet knows whether the chimp virus jumped independently to gorillas and humans or whether it moved first to gorillas and from there to humans.

These studies have fulfilled the epidemiological criteria to demonstrate the origin of HIV-1. Clearly, SIV and HIV-1 share genetic similarities (criteria 1 and 2). Some strains of SIV are common, with as many as a third of chimpanzees in a population harboring infection (criterion 3). Moreover, the habitat of central chimpanzees overlaps with the region of Africa where HIV-1 occurs (criterion 4), and humans who hunt chimps for meat provide a likely transmission route (criterion 5). Other research indicates that this event likely occurred around the 1920s.

Questions remain about HIV-1's origin from SIV. For example, why do HIV-1 groups N and O remain rare, whereas group M is causing the global AIDS epidemic? Can we use HIV-1 and HIV-2 to learn about the potential for other viruses to emerge as human pathogens? The scientific detectives continue to search for the answers.

Sources: Gao, Feng, Elizabeth Bailes, David L. Robertson, and nine coauthors, including Beatrice Hahn. 1999. Origin of HIV-1 in the chimpanzee *Pan troglodytes troglodytes*. *Nature*, vol. 397, pages 436–441.

van Heuverswyn, Fran, and 17 coauthors, including Martine Peeters. 2007. Genetic diversity and phylogeographic clustering of SIVcpz*Ptt* in wild chimpanzees in Cameroon. *Virology*, vol. 368, pages 155–171.

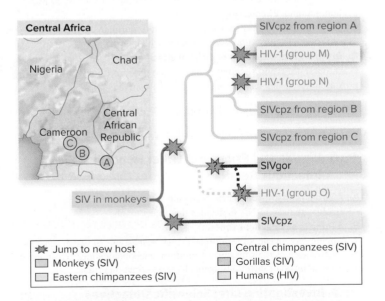

Figure 16.10 HIV-1 from SIV. Viral protein sequences reveal that the central chimpanzee is the ultimate source of all known HIV-1 groups. SIVcpz has jumped from chimps to humans or gorillas multiple times. Dotted lines indicate unknown evolutionary relationships. HIV-1 group M, which causes most HIV infections, is highlighted.

16.7 MASTERING CONCEPTS

1. How did researchers gather data about SIV infection in wild primates?
2. What conclusions did the researchers make based on the relationships in figure 16.10?

CHAPTER SUMMARY

16.1 Viruses Are Genes Wrapped in a Protein Coat

A. Viruses Are Smaller and Simpler Than Cells

- A **virus** is a nucleic acid (DNA or RNA) in a **capsid**. An **envelope** derived from the host cell's membrane surrounds some viruses. Many viruses, including some **bacteriophages,** have relatively complex structures.
- A virus must infect a living cell to reproduce.
- Table 16.1 compares cells and viruses.

B. A Virus's Host Range Consists of the Organisms It Infects

- The species that a virus infects constitute its **host range.** The **reservoir** of a virus is the long-term host that acts as a source of viral infection to other hosts.

C. Are Viruses Alive?

- Viruses are intracellular parasites that most biologists do not consider to be alive.
- Many viral genomes mutate rapidly.
- Scientists classify viruses based on the type of genetic material, the shape of the capsid, the presence or absence of an envelope, the replication strategy, and the type of disease.

16.2 Viral Replication Occurs in Five Stages

- After a virus infects a cell, its host manufactures many copies of the viral proteins and nucleic acids, then assembles these components into new viruses.
- The five stages of viral replication within a host cell are attachment, penetration, synthesis, assembly, and release (figure 16.11).

16.3 Viruses May Kill Bacteria Immediately or Their DNA May "Hide" in the Cell

- In a **lytic infection,** new viruses are immediately synthesized, assembled, and released.
- In a **lysogenic infection,** the virus's nucleic acid replicates along with that of a dividing cell without causing symptoms. The viral DNA may integrate as a **prophage** into the host chromosome.

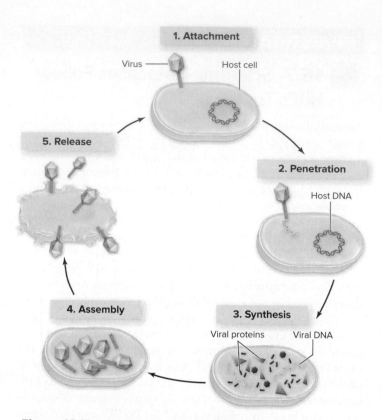

Figure 16.11 **Viral Replication: A Summary.**

16.4 Illnesses Caused by Animal Viruses May Be Mild or Severe

A. Symptoms Result from Cell Death and the Immune Response

- The effects of a virus depend on the cell types it infects. Viruses cause disease by killing infected cells and by stimulating immune responses.

B. Some Animal Viruses Linger for Years

- HIV and some other viruses remain **latent,** or dormant, inside animal cells. Some latent viruses are associated with cancer.

C. Drugs and Vaccines Help Fight Viral Infections

- Antiviral drugs and **vaccines** combat some viral infections.
- Antibiotics, drugs that kill bacteria, are ineffective against viruses.

16.5 Viruses Cause Diseases in Plants

- Viruses infect plant cells, then spread via plasmodesmata.

16.6 Viroids and Prions Are Other Noncellular Infectious Agents

A. A Viroid Is an Infectious RNA Molecule

- **Viroids** are naked RNA molecules that infect plants.

B. A Prion Is an Infectious Protein

- A **prion** protein can take multiple shapes, at least one of which can cause diseases such as transmissible spongiform encephalopathies. Treatments that destroy other infectious agents have no effect on prions.

16.7 Investigating Life: Scientific Detectives Follow HIV's Trail

- Scientists have tested hypotheses about HIV's origin by collecting multiple types of information.
- The subgroups of HIV-1 evolved independently from simian immunodeficiency viruses (SIV) infecting chimpanzees.

TABLE **16.1**   Cells and Viruses Compared

Feature	Cells	Viruses
Size	Typically 1–100 μm	Typically ~80 nm
Genetic material	DNA	DNA or RNA
Protein coat (capsid)	No	Yes
Cell membrane	Yes	No
Viral envelope	No	Some viruses
Nucleus/membrane-bounded organelles	Eukaryotes only	No
Cytoplasm	Yes	No
Ribosomes	Yes	No
Enzymes	Yes	Some viruses
Metabolism	Yes	No
Independent replication	Yes	No

MULTIPLE CHOICE QUESTIONS

1. Which of the following is NOT a feature associated with viruses?
   a. Cytoplasm
   b. Genetic information
   c. Protein coat
   d. Envelope

2. Which of the following is the largest?
   a. HIV
   b. RNA molecule
   c. *E. coli* cell
   d. Human T cell

3. Which of the following characteristics of life does a virus have?
   a. Ribosomes
   b. Evolution
   c. Homeostasis
   d. Growth

4. At which stage in viral replication does the genetic information enter the host cell?
   a. Penetration
   b. Synthesis
   c. Assembly
   d. Release

5. Although some viruses are complete after the assembly stage, others do not complete replication until they acquire _____ during the release stage.
   a. DNA
   b. RNA
   c. proteins
   d. the envelope

6. Which type of infection is most similar to a lysogenic infection in bacteria?
   a. A lytic infection in bacteria
   b. An influenza infection in humans
   c. A latent infection in animals
   d. A viroid infection in plants

7. Which enzyme copies HIV's genetic material, forming DNA?
   a. DNA polymerase from the host cell
   b. Reverse transcriptase from the virus
   c. RNA polymerase from the host cell
   d. Coat protein from the virus

8. The severity of the symptoms associated with a viral infection is related to
   a. the response of the immune system.
   b. the type of genetic material in the virus.
   c. the number and types of cells that become infected.
   d. both a and c.

9. What is a prion?
   a. A highly wound circle of RNA
   b. A virus that has not yet acquired its envelope
   c. A protein that can alter the shape of a second protein
   d. The protein associated with a latent virus

Answers to these questions are in appendix A.

WRITE IT OUT

1. Describe the basic parts of a virus and how each contributes to viral replication.

2. Your biology lab instructor gives you a petri dish of agar covered with visible colonies. Your lab partner says the colonies are viruses, but you disagree. How do you know the colonies are bacteria?

3. Why is it inaccurate to refer to the "growth" of viruses?

4. Why are lytic viruses better suited as agents of "phage therapy" than are lysogenic viruses? How would you test whether such a treatment would be effective? Would you be willing to take a "viral antibiotic"?

5. Rhinoviruses replicate in the mucus-producing cells in a person's nose, throat, and lungs, causing the common cold. Papillomaviruses, which infect skin cells, cause growths called warts. HIV infects T cells and causes AIDS. How do these three types of viruses "know" which human cells to infect?

6. As described in this chapter's Burning Question, human papillomavirus (HPV) infects cells of the skin and genitals, but it has never been shown to infect T cells. If a researcher put HPV and T cells in the same petri dish, which step of HPV replication would fail? Explain your answer.

7. Refer to figure 16.B; then explain why a mutation in a gene encoding a T cell coreceptor might be beneficial.

8. Search the Internet for information about the injectable flu vaccine (a "flu shot"). Why is the flu shot administered annually, when many other vaccines last for years? Is it possible for a flu shot to cause influenza?

9. Why do antibiotics kill bacteria but leave viruses unharmed?

10. Kuru is a prion disease associated with cannibalism in humans. Using your knowledge about prions, why might eating other members of the same species make prion diseases more likely to occur? What can you infer from the fact that some humans have acquired prions after eating beef from a prion-infected cow?

11. Use the Internet to find three examples of viruses that infect humans but are not mentioned in the chapter. Describe the symptoms associated with each infection.

12. How is a biological virus similar to and different from a computer virus?

PULL IT TOGETHER

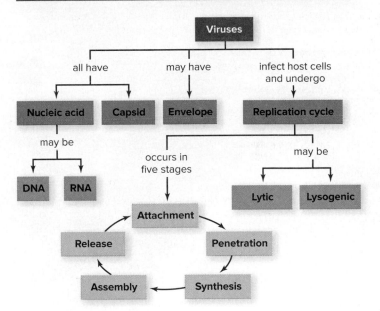

Figure 16.12 Pull It Together: Viruses.

Refer to **figure 16.12** and the chapter content to answer the following questions.

1. Review the Survey the Landscape figure in the chapter introduction to recall what types of organisms viruses infect. Then, add *bacteriophages, animal viruses,* and *plant viruses* to the Pull It Together concept map.

2. How is a virus similar to and different from a bacterium, a viroid, and a prion?

3. Add *host range, latent,* and *vaccines* to the concept map.

Bacteria and Archaea

Bacteria

Bacterial secretions

SEM (fals

Refreshing. A routine scrubbing in the morning removes the bacterial biofilms (inset) that accumulate on our teeth as we sleep.
©Drazen Lovric/E+/Getty Images RF; (inset): ©Steve Gschmeissner/SPL/Getty Images RF

LEARN HOW TO LEARN
Skipping Class?

Attending lectures is important, but you may need to skip class once in a while. How will you find out what you missed? If your instructor does not provide complete lecture notes, you may be able to copy them from a friend. Whenever you borrow someone else's notes, it's a good idea to compare them with the assigned reading to make sure the notes are complete and accurate. You might also want to check with the instructor if you have lingering questions about what you missed.

"Mob Mentality" on a Microscopic Scale

Bacteria are one-celled organisms, so it is tempting to think of them as rugged individualists. Instead, however, they often build complex communities in which cells communicate, protect one another, and even form structures with specialized functions. These organized aggregations of bacterial cells are called biofilms.

A biofilm forms when bacteria settle and reproduce on a solid surface. Once the cells reach a critical density, they express genes that trigger the secretion of a sticky slime made of polysaccharides. As cells continue to divide, they pile up into mushroom-shaped structures that maximize the biofilm's surface area. Occasionally, cells released from the biofilm colonize new habitats, starting the process anew.

Microbiologists are still learning about the cues that trigger biofilm formation. For example, bacteria use signaling molecules to detect the density of cells around them. These "quorum-sensing" signals allow the cells in a biofilm to coordinate their activities, somewhat like a multicellular organism. Shortly after the biofilm forms, the bacteria turn off genes controlling the production of a flagellum; life in a biofilm does not require the ability to swim. Meanwhile, the cells produce new proteins that enable them to attach to a solid surface.

Interest in biofilms extends far beyond idle curiosity about bacterial life. These microbial mats degrade sewage in wastewater treatment plants, help mine copper ore, and coat the surfaces of plants' roots. Much of the attention that biofilms receive, however, is medical. Persistent biofilms can form in catheters, on teeth, and on the mucous membranes of lungs and sinuses. Bacteria in biofilms are much more resistant to immune defenses and antibiotic treatment than are individual cells.

As microbiologists learn more about biofilm formation, they may be able to develop new treatments to combat medically important bacteria. For example, it may be possible to disrupt biofilm formation by silencing quorum-sensing signals. Conversely, learning how to trigger biofilm formation in beneficial soil bacteria may enhance efforts to clean up toxic wastes.

The more scientists learn about prokaryotes—bacteria and archaea—the more we can appreciate the amazing capabilities of Earth's simplest organisms. This chapter offers a taste of the microbial world, beginning with a tour of prokaryotic cells. A sampling of their diversity comes next, and the chapter ends by explaining why microbes are essential to all other life. As you will see, microbes are much more than just "germs."

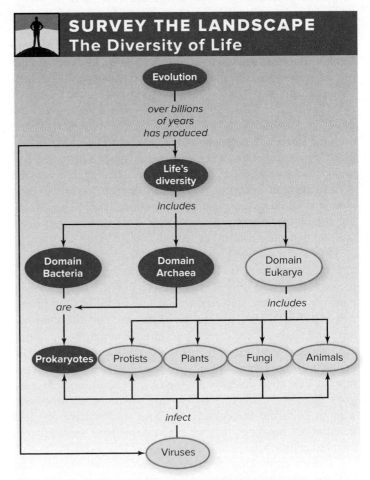

SURVEY THE LANDSCAPE
The Diversity of Life

Domains Bacteria and Archaea contain prokaryotic organisms. These microscopic cells are essential to all other organisms, playing critical roles at scales from individual health to the functioning of the entire biosphere.

For more details, study the Pull It Together feature in the chapter summary.

17.1 Prokaryotes Are a Biological Success Story

The microscopic world of life may be invisible to the naked eye, but its importance is immense. Chapter 16 described viruses, tiny infectious particles that straddle the line between life and nonlife. The remaining chapters in this unit focus squarely on life, beginning with the prokaryotes.

A **prokaryote** is a single-celled organism that lacks a nucleus and membrane-bounded organelles. DNA sequences and other lines of evidence suggest the existence of two prokaryotic domains: **Bacteria** and **Archaea** (figure 17.1).

Microbiologists have probably discovered just a tiny fraction of prokaryotic life on Earth. Soil, water, and even the human body teem with microbes that have yet to be named and described. The total number of species in both domains may be anywhere between 100,000 and 10,000,000; no one knows.

Although we have much to learn of their diversity, it is clear that prokaryotic cells have had a huge influence on Earth's natural history. The earliest known fossils closely resemble today's bacteria, suggesting that the first cells were prokaryotic. Ancient photosynthetic microbes also contributed oxygen gas (O_2) to Earth's atmosphere, creating a protective ozone layer and paving the way for aerobic respiration. Along the road of evolution, bacteria probably gave rise to the chloroplasts and mitochondria of eukaryotic cells. ⓘ *aerobic respiration,* section 6.1; *endosymbiosis,* section 15.2A

The reign of the prokaryotes continues today. Virtually no place on Earth is free of bacteria and archaea. Their cells live within rocks and ice; high in the atmosphere; far below the ocean's surface; in thermal vents, nuclear reactors, hot springs, animal intestines, and plant roots; and practically everywhere else. Many species prefer hot, cold, acidic, alkaline, or salty habitats that humans consider "extreme."

Most of what we know about the biology of bacteria and archaea comes from studies in the laboratory. However, many prokaryotes thrive only if they receive just the right types of nutrients, and they may require a specific combination of salt and water availability, oxygen concentration, light, pH, temperature, and so on. These complex requirements may be unique to each species. Consequently, microbiologists can only simulate the natural environment of relatively few species, which do not represent the total diversity of all prokaryotes. Microbial ecologists therefore extract DNA directly from the environment, without culturing the cells first. These studies have revealed much about microbes that we could never have learned from cultured cells alone.

Before embarking on a tour of prokaryote biology and ecology, it is worth noting that the term *prokaryote* has become somewhat controversial among microbiologists. One reason is that the word falsely implies a close evolutionary relationship between bacteria and archaea, despite strong evidence that archaea are actually more closely related to eukaryotes. A related reason is that prokaryotes do not form a single, complete clade;

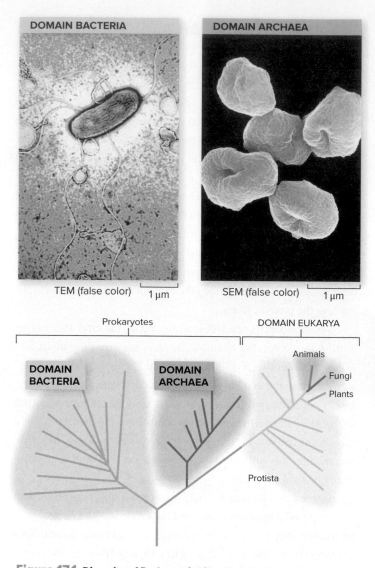

Figure 17.1 Diversity of Prokaryotic Life. Domains Bacteria and Archaea form two of the three main branches of life.
Photos: (bacteria): ©Kwangshin Kim/Science Source; (archaea): ©Eye of Science/ Science Source

that is, they are a paraphyletic group. The tree in figure 17.1 shows that the prokaryotes exclude some of the descendants of the common ancestor that bacteria and archaea share. Nevertheless, many biologists continue to use the term as a handy shortcut for describing all cells that lack nuclei. ⓘ *paraphyletic groups,* section 14.6D

17.1 MASTERING CONCEPTS

1. What are two domains that contain prokaryotes?
2. List several ways that prokaryotes have influenced evolution.
3. In what habitats do bacteria and archaea live?
4. Why are most species of prokaryotes little understood?

17.2 Prokaryote Classification Traditionally Relies on Cell Structure and Metabolism

At about 1 to 10 μm long, a typical prokaryotic cell is 10 to 100 times smaller than most eukaryotic cells (see figure 3.2). Bacteria and archaea also lack the membranous organelles that characterize eukaryotic cells (see figure 3.31). How do microbiologists classify the diversity of life within these two domains, given the tiny cell sizes and scarcity of distinctive internal structures?

The answer to that question has evolved over time. For hundreds of years, biologists classified microbes based on close scrutiny of their cells and metabolism. This section describes some of the most important features; figure 17.2 provides a partial list of the similarities and differences between bacteria and archaea.

A. Microscopes Reveal Cell Structures

Viewing cells with a microscope is an essential step in identifying bacteria and archaea. Light microscopes (and sometimes electron microscopes) reveal the internal and external features unique to each species. Figure 17.3 illustrates a typical bacterial cell; in reading through this section, remember that a given cell may have some or all of the structures pictured.

Internal Cell Structures   Like the cells of other organisms, all bacteria and archaea are bounded by a cell membrane that encloses cytoplasm, DNA, and ribosomes. A prokaryotic cell's DNA typically consists of one circular chromosome. The **nucleoid** is the region where this DNA is located, along with some RNA and a few proteins. Unlike the nucleus of a eukaryotic cell, a membranous envelope does not surround the nucleoid.

The cells of many bacteria and archaea also contain one or more **plasmids,** circles of DNA apart from the chromosome. The genes on a plasmid may encode the proteins necessary to copy the plasmid and transfer it to another cell. Other genes may provide the ability to resist a drug or toxin, cause disease, or alter the

Similarities between Bacteria and Archaea
• Prokaryotic cells (no nucleus or other membrane-bounded organelles)
• Size ~1-10 μm
• Circular chromosome
• Predominantly unicellular
• Some can fix nitrogen or grow at temperatures above 80°C

Features Unique to Bacteria	Features Unique to Archaea
• Cell wall typically composed of peptidoglycan	• Cell wall composed of molecules other than peptidoglycan
• Membrane based on fatty acids	• Membrane based on nonfatty acid lipids
• Some use chlorophyll in photosynthesis	• Do not use chlorophyll
• Cannot generate methane	• Some generate methane
• Sensitive to streptomycin	• Insensitive to streptomycin
• Genes do not contain introns	• Genes may contain introns

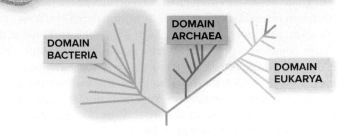

DOMAIN BACTERIA

DOMAIN ARCHAEA

DOMAIN EUKARYA

Figure 17.2 **Bacteria and Archaea Compared.**

cell's metabolism. Recombinant DNA technology uses plasmids to ferry genes from one kind of cell to another. ⓘ *transgenic organisms,* section 11.2

Ribosomes are structures that use the information in RNA to assemble proteins, a process described in chapter 7. Bacterial, archaean, and eukaryotic ribosomes all make proteins in essentially the same way, but they are structurally different from one another. Some antibiotics, such as streptomycin, kill bacteria without harming eukaryotic host cells by exploiting this difference. This chapter's Apply It Now box describes more examples of how antibiotics work.

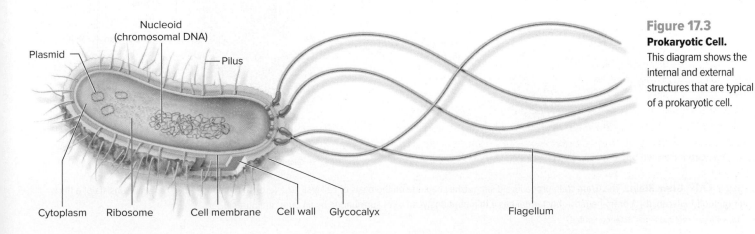

Nucleoid (chromosomal DNA)

Plasmid

Pilus

Cytoplasm   Ribosome   Cell membrane   Cell wall   Glycocalyx

Flagellum

Figure 17.3
Prokaryotic Cell.
This diagram shows the internal and external structures that are typical of a prokaryotic cell.

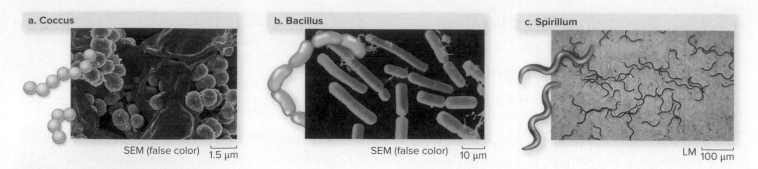

a. Coccus

SEM (false color) ⊢ 1.5 μm

b. Bacillus

SEM (false color) ⊢ 10 μm

c. Spirillum

LM ⊢ 100 μm

Figure 17.4 Cell Shapes. (a) Cocci are spherical. (b) A bacillus is rod-shaped. (c) A spirillum is spiral-shaped.

Photos: (a): National Institute of Allergy and Infectious Diseases (NIAID)/NIH/USHHS; (b): ©Scimat/Science Source; (c): ©Ed Reschke/Photolibrary/Getty Images

External Cell Structures

The **cell wall** is a rigid barrier that surrounds the cells of most bacteria and archaea. In most species of bacteria, the cell wall contains **peptidoglycan,** a complex polysaccharide that does not occur in the cell walls of archaea. The antibiotic penicillin inhibits the reproduction of bacteria (but not archaea) by interfering with the final steps in peptidoglycan synthesis.

The wall gives the cell its shape (**figure 17.4**). Three of the most common forms are **coccus** (spherical), **bacillus** (rod-shaped), and **spirillum** (spiral or corkscrew-shaped). In addition, the arrangement of the cells in pairs, clusters (*staphylo-*), or chains (*strepto-*) is sometimes important in classification. The disease-causing bacterium *Staphylococcus,* for example, forms grapelike clusters of spherical cells.

The **Gram stain** reaction distinguishes between two types of cell walls (**figure 17.5**). After a multistep staining procedure, gram-positive cells appear purple, whereas gram-negative bacteria stain pink.

Structural differences between the cell walls account for the distinctive colors. The walls of gram-positive bacteria are made primarily of a thick layer of peptidoglycan. Gram-negative cell walls consist of a thin inner layer of peptidoglycan plus a protective outer membrane of lipid, polysaccharide, and protein. This outer membrane causes the toxic effects of many medically important gram-negative bacteria, such as *Salmonella.* Parts of the outer layer trigger a strong immune response, including fever and inflammation, which help the body eliminate the bacteria (see chapter 33).

Medical technicians often use Gram staining as a first step in identifying bacteria that cause infections. The distinction is important because gram-positive and gram-negative bacteria are susceptible to different antibiotic drugs.

Many prokaryotic cells have other distinctive structures outside the cell wall (**figure 17.6**). A **glycocalyx,** also sometimes called a capsule or slime layer, is a sticky layer of proteins or polysaccharides that may surround the cell wall. The glycocalyx has many functions, including attachment to surfaces, resistance to drying, and protection from immune system cells. It also plays a role in the formation of biofilms, as described in this chapter's opening essay.

Some cells have **pili** (singular: pilus), which are short, hairlike projections made of protein (figure 17.6b). Attachment pili enable cells to adhere to objects. The bacterium that causes

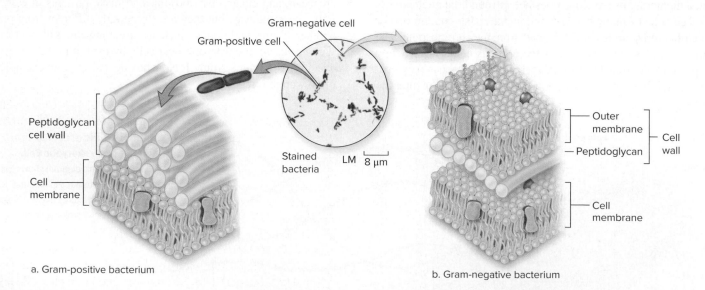

Gram-negative cell

Gram-positive cell

Peptidoglycan cell wall

Cell membrane

Stained bacteria

LM ⊢ 8 μm

Outer membrane
Peptidoglycan
} Cell wall

Cell membrane

a. Gram-positive bacterium

b. Gram-negative bacterium

Figure 17.5 Gram Stain. The Gram stain procedure distinguishes bacteria on the basis of cell wall structure. (a) A gram-positive cell wall consists of a thick layer of peptidoglycan. (b) A gram-negative bacterium has a thin peptidoglycan layer surrounded by an outer membrane.

Photo: ©Biophoto Associates/Science Source

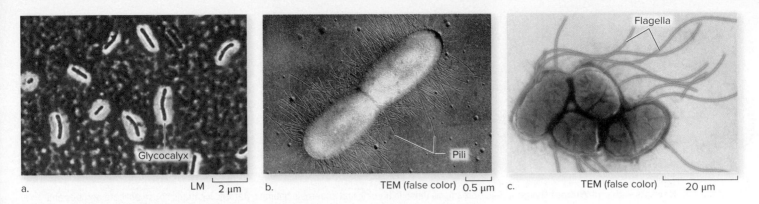

a.   LM 2 μm   b.   TEM (false color) 0.5 μm   c.   TEM (false color) 20 μm

Figure 17.6 External Structures of Prokaryotic Cells. (a) A glycocalyx is a layer of sticky secretions that enables a bacterium to adhere to a surface. (b) Pili attach cells to objects, surfaces, and other cells. This is *E. coli.* (c) The numerous flagella on each of these bacteria enable the cells to move.
(a): ©Michael Abbey/Science Source; (b): ©CNRI/Science Source; (c): ©Kwangshin Kim/Science Source

cholera, for example, uses pili to attach to a human's intestinal wall. Other projections, called sex pili, aid in the transfer of DNA from cell to cell, as described in section 17.2D.

Not all prokaryotes can move, but many can. In a response called **taxis,** cells move toward or away from an external stimulus such as food, toxins, oxygen, or light. Cells that can move have a **flagellum,** which is a whiplike extension that rotates like a propeller. The bacteria in figure 17.6c have many flagella; other cells have one or a few. (Some eukaryotic cells also have flagella, but they are not homologous to those on bacterial or archaean cells.)

Endospores   Some types of gram-positive bacteria produce **endospores,** which are dormant, thick-walled structures that can survive harsh conditions (figure 17.7). The endospore wall surrounds DNA and a small amount of cytoplasm. An endospore can withstand boiling, drying, ultraviolet radiation, and disinfectants. Once environmental conditions improve, the endospore germinates and develops into a normal cell. Some researchers have reported reviving endospores preserved millions of years ago in amber.

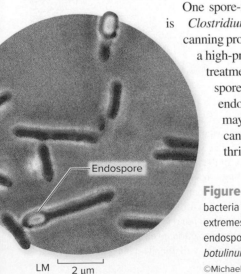

One spore-forming soil bacterium is *Clostridium botulinum.* Food-canning processes typically include a high-pressure heat sterilization treatment to destroy endospores of this species. If any endospores survive, they may germinate inside the can, producing cells that thrive in the absence of

Figure 17.7 Endospores. Some bacteria survive environmental extremes by forming thick-walled endospores. This is *Clostridium botulinum.*
©Michael Abbey/Science Source

LM   2 μm

oxygen. The cells produce a toxin that causes botulism, a severe (and sometimes deadly) form of food poisoning. Green beans, corn, and other vegetables that are improperly home-canned are the most frequent sources of food-borne botulism.

Another spore-forming bacterium is *Bacillus anthracis.* This organism, ordinarily found in soil, can cause a deadly disease called anthrax when inhaled. Cultures of *B. anthracis* can be dried to induce endospore formation, then ground into a fine powder that remains infectious for decades. This property makes anthrax a potential biological weapon.

B. Metabolic Pathways May Be Useful in Classification

Over billions of years, bacteria and archaea have developed a tremendous variety of chemical reactions that allow them to metabolize everything from organic matter to metal. One way to group microorganisms is to examine some of these key metabolic pathways.

The methods by which organisms acquire carbon and energy form one basis for classification. **Autotrophs,** for example, are "self-feeders"; they assemble their own organic molecules using inorganic carbon sources such as carbon dioxide (CO_2). Plants and algae are the most familiar autotrophs. **Heterotrophs,** on the other hand, are "other-feeders," acquiring carbon by consuming organic molecules produced by other organisms. *Escherichia coli,* a notorious intestinal bacterium, is a heterotroph. The organism's energy sources are also important. **Phototrophs** derive energy from the sun; **chemotrophs** oxidize inorganic or organic chemicals.

By combining these terms, a biologist can describe how a microbe fits into the environment. Plants and cyanobacteria, for example, are photoautotrophs; they use sunlight (*photo-*) for energy and CO_2 (*auto-*) for carbon, as described in chapter 5. Many disease-causing bacteria are chemoheterotrophs because they use organic molecules from their hosts as sources of both carbon and energy. Animals and fungi are also chemoheterotrophs.

Body surface
(aerobic)

Digestive
tract
(anaerobic)

Waterlogged
soil (anaerobic)

Surfaces of
soil and plant
(aerobic)

Well-drained
soil (aerobic)

Lake

Surface
water
(aerobic)

Deep water
(anaerobic)

Sediments
(anaerobic)

Figure 17.8 Habitats with and Without Oxygen. Microorganisms thrive in aerobic and anaerobic habitats. Aerobic locations house obligate aerobes and facultative anaerobes; anaerobic habitats are home to obligate and facultative anaerobes.

In addition, oxygen requirements are often important in classification. **Obligate aerobes** require O_2 for generating ATP in cellular respiration (see chapter 6). For **obligate anaerobes,** O_2 is toxic, and they live in habitats that lack it. *Clostridium tetani,* the bacterium that causes tetanus, is one example. **Facultative anaerobes,** which include the intestinal microbes *E. coli* and *Salmonella,* can live either with or without O_2. Obligate and facultative anaerobes use fermentation or anaerobic respiration to generate ATP in the absence of O_2 (see section 6.8).

Figure 17.8 shows habitats with varying O_2 availability. O_2-rich areas include a mouse's skin, a plant's leaves, and the surface of a lake. Anaerobic habitats include the animal's digestive tract and the lake's sediments. Note that soil may contain abundant O_2 or be anaerobic, depending on whether it is waterlogged or well drained.

Apply It **Now**

Antibiotics and Other Germ Killers

When a person develops a bacterial infection, a physician may prescribe antibiotics. These drugs typically inhibit structures and functions present in bacteria but not in host cells. Some mechanisms of action include

- **Inhibiting cell wall synthesis:** Penicillin is an antibiotic that interferes with cell wall formation. A bacterium that cannot make a rigid cell wall will burst and die.

- **Disrupting cell membranes:** All life depends on an intact membrane that regulates what enters and leaves a cell. Polymyxin antibiotics exploit differences between bacterial and eukaryotic cell membranes. (i) *membranes,* section 3.3

- **Inhibiting transcription:** Gene expression requires RNA synthesis. Rifamycin antibiotics prevent RNA synthesis in bacteria by binding to a bacterial form of RNA polymerase. (i) *transcription,* section 7.4

- **Inhibiting protein assembly:** The antibiotics streptomycin, chloramphenicol, and erythromycin bind to different parts of bacterial ribosomes, but all have the same effect—they kill bacteria without killing us. (i) *ribosomes,* section 7.5B

- **Inhibiting metabolic enzymes:** Theoretically, antibiotics could block any bacterial metabolic pathway that does not occur in host cells. Sulfanilamide, for example, mimics the substrate of a bacterial enzyme that participates in an essential chain of chemical reactions. (i) *enzymes,* section 4.4

Unfortunately, the misuse of antibiotics in medicine and agriculture has selected for antibiotic-resistant bacteria. How does this occur? Bacterial DNA accumulates mutations each time a cell divides; cells may also acquire new genes, as described in section 17.2D. Therefore,

©Rick Gomez/Corbis RF

new bacterial strains that are resistant to one or more classes of antibiotics occasionally emerge. Each time we use an antibiotic drug, we kill the susceptible strains and select for the resistant ones. The result is an ever-increasing incidence of antibiotic-resistant bacteria (see section 17.5).

Ordinary people can help prevent antibiotic-resistant bacteria. Patients should not demand antibiotics for viral infections, such as the flu or the common cold. And if a physician does prescribe antibiotics, it is important to finish the entire prescription, even if you feel better. Otherwise, only the weakest bacteria will die, leaving the strongest to reproduce and spread.

C. Molecular Data Reveal Evolutionary Relationships

Traditionally, the classification of prokaryotes has relied on easy-to-observe characteristics such as cell shape. This method of organizing prokaryotes, however, groups together organisms that are only distantly related to one another.

Molecular data such as DNA sequences are harder to obtain and analyze, but they have triggered a revolution in microbial taxonomy and evolutionary biology. Once microbiologists began to analyze the DNA sequences encoding ribosomal RNA (rRNA), they realigned all organisms into three domains (see figure 17.1). The prokaryotes fell into two domains—Archaea and Bacteria—rather than one kingdom, as previously thought. Moreover, studies of DNA extracted from soil, water, and other habitats have revealed many previously unknown species of microbes.

DNA sequence data have brought scientists closer to a classification system that reflects evolutionary relationships. Still, phylogenetic trees depicting their diversity remain too preliminary to include in this chapter.

D. Horizontal Gene Transfer Complicates Classification

Like all organisms, bacteria and archaea transmit DNA from generation to generation as they reproduce, a process sometimes called **vertical gene transfer.** In prokaryotes, reproduction occurs by binary fission, an asexual process that replicates DNA and distributes it and other cell parts into two daughter cells (see figure 8.7). Genetic diversity typically arises from mutations in a cell's DNA.

However, bacteria and archaea can also acquire new genetic material from **horizontal gene transfer:** A cell receives DNA from another cell that is not its ancestor. Horizontal gene transfer occurs in three ways (**figure 17.9**). First, a dying cell may release its genetic material as it bursts; **transformation** occurs when other cells absorb stray bits of its DNA. Second, viruses sometimes mistakenly package host cell DNA along with their own. In **transduction,** a virus transfers this combined DNA to a bacterial cell. Third, in **conjugation,** one cell receives DNA via direct contact with another cell. A **sex pilus** is the appendage through which DNA passes from donor to recipient.

Horizontal gene transfer has profound implications in fields as diverse as origin-of-life research, medicine, systematics, and biotechnology. Both transduction and conjugation, for example, move antibiotic-resistance genes among bacteria, a serious and growing public health problem (see chapter 12's opening essay). At the same time, however, biologists take advantage of horizontal gene transfer when they create new types of genetically modified bacteria. ⓘ *transgenic bacteria,* section 11.2A

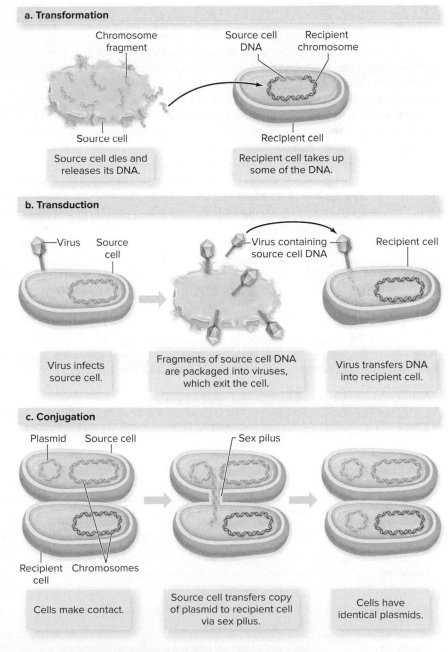

a. Transformation

Chromosome fragment — Source cell — Source cell DNA — Recipient chromosome — Recipient cell

Source cell dies and releases its DNA. | Recipient cell takes up some of the DNA.

b. Transduction

Virus — Source cell — Virus containing source cell DNA — Recipient cell

Virus infects source cell. | Fragments of source cell DNA are packaged into viruses, which exit the cell. | Virus transfers DNA into recipient cell.

c. Conjugation

Plasmid — Source cell — Sex pilus — Recipient cell — Chromosomes

Cells make contact. | Source cell transfers copy of plasmid to recipient cell via sex pilus. | Cells have identical plasmids.

Figure 17.9 Horizontal Gene Transfer. Three forms of horizontal gene transfer include (a) transformation, (b) transduction, and (c) conjugation.

17.3 Prokaryotes Include Two Domains with Enormous Diversity

For many decades, the tendency to lump together all prokaryotic organisms hid much of the diversity in the microbial world. We now know of so many species of bacteria and archaea in so many habitats that it would take many books to describe them all—and many more species remain undiscovered. This section contains a small sampling of this extraordinary diversity; the Burning Question box explores some of the few locations that are microbe-free.

A. Domain Bacteria Includes Many Familiar Groups

Scientists have identified 23 phyla within domain Bacteria, but the evolutionary relationships among them remain unclear. Table 17.1 lists a few of the main groups, and figure 17.10 illustrates two examples.

Phylum Proteobacteria is a group of gram-negative bacteria that exemplify the overall diversity within the domain. Some proteobacteria, including the purple sulfur bacteria, carry out photosynthesis. Others play important roles in nitrogen or sulfur cycling, whereas still others form a medically important group that includes enteric bacteria and vibrios. *Helicobacter*, the bacterium that causes ulcers in humans, is a proteobacterium, as are the intestinal bacteria *E. coli* and *Salmonella*. (*E. coli* is the subject of this chapter's Focus on Model Organisms.)

Cyanobacteria is a second phylum in domain Bacteria. Billions of years ago, these autotrophs were the first to produce O_2 as a byproduct of photosynthesis. They also gave rise to the chloroplasts inside the cells of land plants and green algae.

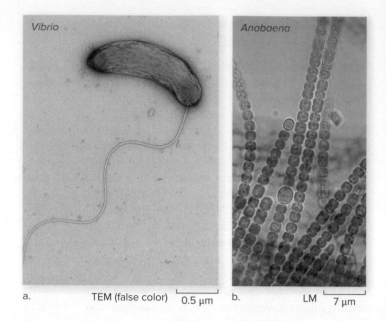

a.   TEM (false color)   0.5 μm   b.   LM   7 μm

Figure 17.10 Examples of Bacteria. (a) *Vibrio cholerae*, a proteobacterium. (b) Filaments of *Anabaena*, a cyanobacterium.

(a): ©Dr Gopal Murti/Science Source; (b): ©John Walsh/Science Source

Cyanobacteria remain important in ecosystems: They make up some of the photosynthetic plankton at the base of aquatic food chains, and they form symbiotic relationships with fungi on land (see section 20.7). In nutrient-polluted water, however, huge populations of cyanobacteria form unsightly mats and release toxic chemicals that can harm people and aquatic animals. ⓘ *eutrophication*, section 38.4E

TABLE **17.1**   Selected Phyla in Domain Bacteria

Phylum	Features	Example(s)
Proteobacteria	Largest group of gram-negative bacteria	
Purple sulfur bacteria	Bacterial photosynthesis using H_2S (not H_2O) as electron donor	*Chromatium vinosum*
Enteric bacteria	Rod-shaped, facultative anaerobes in animal intestinal tracts	*Escherichia coli, Salmonella* species (cause gastrointestinal disease)
Vibrios	Comma-shaped, facultative anaerobes common in aquatic environments	*Vibrio cholerae* (causes cholera)
Cyanobacteria	Photosynthetic; some fix nitrogen; free-living or symbiotic with plants, fungi (in lichens), or protists	*Nostoc, Anabaena*
Spirochaetes	Spiral-shaped; some pathogens of animals	*Borrelia burgdorferi* (causes Lyme disease), *Treponema pallidum* (causes syphilis)
Firmicutes	Gram-positive bacteria with a low proportion of G+C in their DNA; aerobic or anaerobic; rods or cocci	*Bacillus anthracis* (causes anthrax), *Clostridium tetani* (causes tetanus), *Staphylococcus, Streptococcus*
Actinobacteria	Filamentous gram-positive bacteria with a high proportion of G+C in their DNA	*Streptomyces*
Chlamydiae	Grow only inside host cells; cell walls lack peptidoglycan	*Chlamydia*

Are there areas on Earth where no life exists?

Sand, bare rock, and polar ice may seem devoid of life, but they are not. Scientists using microscopes and molecular tools have discovered microbes living in the hottest, coldest, wettest, driest, saltiest, highest, most radioactive, and most pressurized places on the planet—including places where no other organism can survive.

There are a few places, however, that humans keep artificially microbe-free for the sake of our own health. For example, people in many professions use autoclaves, radiation, and filters to sterilize everything from surgical tools to medicines and bandages to processed foods. Artificial sterilization kills microbes that could otherwise cause infections, food poisoning, or other illnesses.

Our own bodies are home to many, many microbes, both inside and out, yet we maintain many germ-free fluids and tissues; they include the sinuses, muscles, brain and spinal cord, ovaries and testes, blood, urine in kidneys and the bladder, and semen before it enters the urethra. These areas are among the few places where microbes do not ordinarily live; if a bacterial infection does occur, the resulting illness can be deadly.

Submit your burning question to
Marielle.Hoefnagels@mheducation.com

SEM
(false color)   1 μm

Figure 17.11 Extremophiles. Archaea such as *Sulfolobus* thrive in boiling mud pools. This is Krafla caldera in Iceland.
©Ralph Eagle Jr./Science Source; (inset): ©Eye of Science/Science Source

A third phylum, Spirochaetes, contains some medically important bacteria. These spiral-shaped organisms include *Borrelia burgdorferi,* a bacterium that can cause Lyme disease when transmitted to humans in a tick's bite. *Treponema pallidum* causes the sexually transmitted disease syphilis.

Phylum Firmicutes contains gram-positive bacteria with a unique genetic signature: They have a low proportion of guanine and cytosine (G+C) in their DNA. Some of the firmicutes form endospores; examples include *Bacillus anthracis* (the cause of anthrax) and *Clostridium tetani* (the cause of tetanus). Others are *Staphylococcus* and *Streptococcus,* both of which can cause lethal infections throughout the body (see section 17.5).

The actinobacteria form another phylum of gram-positive bacteria. These filamentous, soil-dwelling microbes are also medically important: They are the source of infection-fighting antibiotics, including streptomycin.

Bacteria in phylum Chlamydiae are unusual in that they cannot generate ATP on their own, so they are obligate parasites—that is, they must live inside a host cell. When they infect cells lining the human genital tract, they cause a sexually transmitted disease called chlamydia. The cell walls of these bacteria lack peptidoglycan, so penicillin cannot kill them. Fortunately, a chlamydia infection can be treated with certain other antibiotics.

B. Many, but Not All, Archaea Are "Extremophiles"

Archaea are often collectively described as "extremophiles" because scientists originally found them in places that were extremely hot, acidic, or salty (figure 17.11). At first, the organisms were informally grouped by habitat. The thermophiles, for example, live in hot springs and hydrothermal vents, whereas the halophiles prefer salt concentrations of up to 30%. Acidophiles tolerate a pH as low as 1.0—low enough to dissolve metal! As more archaea are discovered in moderate environments such as soil or the open ocean, however, formal classification is becoming more important.

Microbiologists now tentatively divide domain Archaea into three phyla. One phylum, Euryarchaeota, contains methanogens that live in stagnant waters and the anaerobic intestines of many animals, generating large quantities of methane gas. The same phylum also includes halophilic archaea that use light energy to produce ATP in very salty habitats such as seawater, evaporating ponds, and salt flats.

A second phylum, Crenarchaeota, includes species that thrive in acidic hot springs or at hydrothermal vents on the ocean floor. The same phylum also contains a wide variety of soil and water microorganisms with moderate temperature requirements.

Other thermophiles are classified into a third phylum, Korarchaeota, known mostly from genes extracted from their habitats. They seem to be most closely related to the Crenarchaeota.

The importance of archaea in ecosystems is slowly becoming clearer as scientists decipher more about their roles in global carbon, nitrogen, and sulfur cycles. Many live in ocean waters and sediments, a hard-to-explore habitat in which the role of archaea is especially poorly understood. Their immense numbers, however, suggest that archaea are critical players in ocean ecology.

17.3 MASTERING CONCEPTS

1. In what ways are bacteria and archaea similar and different?
2. What are some examples of phyla within domain Bacteria?
3. What are the three phyla in domain Archaea?

17.4 Bacteria and Archaea Are Essential to All Life

Many people think of microbes as harmful "germs" that cause disease. Indeed, most of the familiar examples of bacteria listed in the previous section are pathogens. Although some bacteria do make people sick, most microbes do not harm us at all. This section describes some of the ways that bacteria and archaea affect our lives.

A. Microbes Form Vital Links in Ecosystems

Although it may seem hard to believe that one-celled organisms can be essential, the truth is that all other species would die without bacteria and archaea. For example, microbes play crucial roles in the global carbon cycle. They decompose organic matter in soil and water, releasing CO_2. Other microorganisms absorb CO_2 in photosynthesis. And all kinds of microbes, both heterotrophs and autotrophs, are eaten by countless organisms in every imaginable habitat. Chapter 39 explains these community interactions in more detail. ⓘ *carbon cycle,* section 38.4B

Another essential process is **nitrogen fixation,** the chemical reactions in which prokaryotes convert atmospheric nitrogen gas (N_2) into forms that plants and other organisms can absorb, such as ammonium (NH_4^+). Nitrogen is a component of protein, DNA, and many other organic molecules. The only organisms that can use N_2 directly by fixing nitrogen are a few species of bacteria and archaea. Ultimately, most of Earth's nitrogen would be locked in the atmosphere if not for nitrogen fixers. The nitrogen cycle—and therefore all life—would eventually cease without these crucial microbes. ⓘ *nitrogen cycle,* section 38.4C

Some nitrogen-fixing bacteria live in soil or water. Others, such as those in the genus *Rhizobium,* induce the formation of nodules in the roots of clover and other host plants in the legume family (**figure 17.12**). Inside the nodules, *Rhizobium* cells share the nitrogen that they fix with their hosts; in exchange, the bacteria receive nutrients and protection.

B. Bacteria and Archaea Live in and on Us

No matter how hard you scrub, it is impossible to escape the fact that you are a habitat for microorganisms. A menagerie of microbes lives on the skin and in the mouth, large intestine, urogenital tract, and upper respiratory tract (**figure 17.13**). These microscopic companions are beneficial because they crowd out disease-causing bacteria, "train" the immune system to ignore harmless molecules, and produce certain vitamins (among other benefits).

Most people never notice these invisible residents unless something disrupts their personal microbial community. Suppose, for example, that your cat scratches your leg and the wound becomes infected. If you take antibiotics to fight the infection, the drug will probably also kill off some of the normal microbes in your body. As they die, harmful ones can take their place. The resulting microbial imbalance in the intestines or genital tract causes unpleasant side effects such as diarrhea or a vaginal yeast infection. These problems subside in time as the normal microbes divide and restore their populations. (Read more about illnesses connected to microbial imbalances in chapter 32's opening essay.)

Although most bacteria in and on the human body are harmless, some cause disease. (So far, no archaea are linked to human illnesses.) To cause an infection, bacteria must first enter the body. Animal bites transmit some bacteria, as can sexual activity;

F⊙CUS on Model Organisms

Escherichia coli

Of all the model organisms profiled in this unit, *Escherichia coli,* commonly called *E. coli,* may be the best understood (**figure 17.A**). Dr. Theodor Escherich (1857–1911) discovered this normal resident of the human intestinal tract in 1885, and it quickly became a popular lab organism. Biologists have studied every aspect of this bacterium for over a century, and its contributions to biology are enormous. The following are a few highlights:

- **DNA is genetic material:** In 1950, Hershey and Chase used virus-infected *E. coli* cells to demonstrate that DNA, not protein, is the genetic material. ⓘ *Hershey and Chase,* section 7.1B

- **Genetic exchange:** In the 1940s and 1950s, biologists studying *E. coli* discovered conjugation and transduction (see section 17.2D). Both phenomena have important implications for bacterial evolution and transgenic technology.

- **DNA replication:** In the late 1950s, Matthew Meselson and Franklin Stahl used *E. coli* in their famous experiments that demonstrated how DNA is copied. ⓘ *DNA replication,* section 8.2

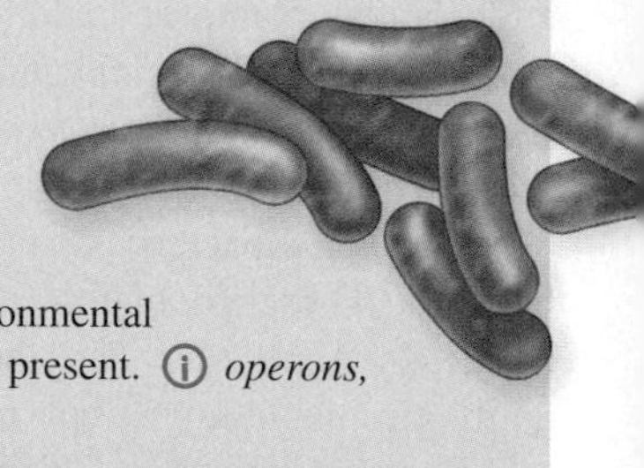

Figure 17.A *Escherichia coli.*

- **Gene regulation:** The 1961 description of the *lac* operon in *E. coli* revealed how cells can turn genes on or off, depending on environmental conditions such as the type of sugar present. ⓘ *operons,* section 7.6A

- **Transgenic technology:** In 1973, *E. coli* became the first organism to receive a gene from another species. The researchers used the newly discovered restriction enzyme *Eco*R1, derived from *E. coli,* to produce the recombinant plasmids. Since then, *E. coli* cells have received countless genes from many other species. ⓘ *transgenic organisms,* section 11.2

- **Gene function:** The *E. coli* genome has been sequenced, and many scientists are working to describe the functions of hundreds of *E. coli* genes. Many genes in other organisms, including humans, will no doubt have similar DNA sequences and functions. Such studies of homologous genes yield new insights into disease and cell function.

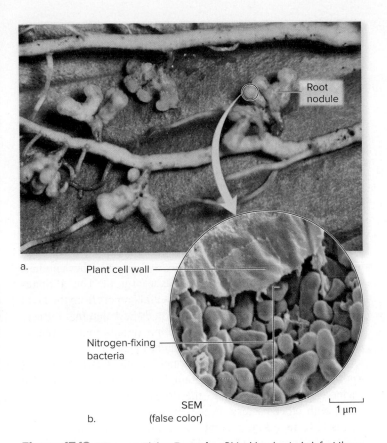

a.

Plant cell wall

Nitrogen-fixing bacteria

SEM
(false color)

b.

1 μm

Figure 17.12 **Nitrogen-Fixing Bacteria.** *Rhizobium* bacteria infect these sweet clover roots, producing root nodules, where nitrogen fixation occurs. The inset shows a portion of a root nodule, revealing the bacteria inside one root cell.

(a): ©Biophoto Associates/Science Source; (b): ©CMSP/Newscom

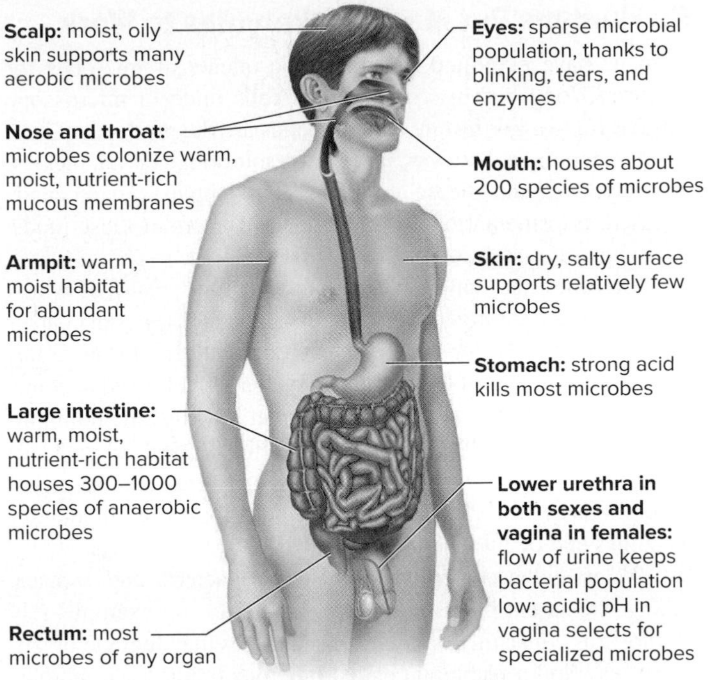

Scalp: moist, oily skin supports many aerobic microbes

Nose and throat: microbes colonize warm, moist, nutrient-rich mucous membranes

Armpit: warm, moist habitat for abundant microbes

Large intestine: warm, moist, nutrient-rich habitat houses 300–1000 species of anaerobic microbes

Rectum: most microbes of any organ

Eyes: sparse microbial population, thanks to blinking, tears, and enzymes

Mouth: houses about 200 species of microbes

Skin: dry, salty surface supports relatively few microbes

Stomach: strong acid kills most microbes

Lower urethra in both sexes and vagina in females: flow of urine keeps bacterial population low; acidic pH in vagina selects for specialized microbes

Figure 17.13 **A Human Habitat.** Many parts of the body house thriving populations of microorganisms. The blood, nervous system, urinary bladder, and some other areas normally remain sterile.

section 35.4 describes sexually transmitted diseases in more detail. A person can also inhale air containing respiratory droplets from a sick coworker or ingest bacteria in contaminated food or water. Bacteria can also enter the body through open wounds.

Once inside the host, pili or slime capsules attach the pathogens to host cells. As the invaders multiply, disease symptoms may develop. Some symptoms result from damage caused by the bacteria themselves. The cells may produce enzymes that break down host tissues, for example, or they may release toxins that harm the host's circulatory, digestive, or nervous system. These toxins may help the pathogens invade the host, acquire nutrients, escape the immune system, or spread to new hosts. ⓘ *enzymes,* section 4.4

Microbiologists divide bacterial toxins into two categories: exotoxins and endotoxins. Exotoxins are toxic proteins that diffuse out of a bacterial cell. *Staphylococcus aureus* and *Clostridium botulinum* are two examples of bacteria that produce exotoxins; *S. aureus* causes toxic shock syndrome and infections of the skin and sinuses, and *C. botulinum* causes botulism.

Rather than diffusing out of a bacterial cell, endotoxins form a component of the cell itself—specifically, the outer membrane of a gram-negative cell wall. Consider, for example, *E. coli,* a normal inhabitant of animal intestines. Sometimes, cattle droppings containing *E. coli* contaminate water, milk, raw fruits and vegetables, hamburger, and other foods. Most cells of *E. coli* are harmless, but one particularly nasty variety multiplies inside the body. Its endotoxin can cause belly pain, bloody diarrhea, and in some cases life-threatening kidney failure. Outbreaks of foodborne *E. coli* have led to widely publicized recalls of everything from raw spinach to ground beef to unpasteurized apple juice.

Raw eggs and other foods contaminated with animal feces also may contain *Salmonella*, a close relative of *E. coli. Salmonella, Staphylococcus aureus,* and *Bacillus cereus* are all examples of microbes that thrive in foods that have been improperly refrigerated or inadequately cooked. The toxins they produce in the food—not infection with the bacteria themselves—produce the vomiting and diarrhea associated with food poisoning.

Besides the effects of toxins, a bacterial infection may also trigger an immune reaction. As a result, the classic signs and symptoms of a bacterial infection develop: fever, swollen lymph nodes, pain, and nausea, among others (see chapter 33). Eventually, strong immune defenses usually defeat a bacterial infection.

But the arms race extends to the bacterial side as well. Natural selection favors bacteria that can evade the immune system by hiding inside host cells or forming protective biofilms (see the chapter opening essay). Most pathogens also spread efficiently to new hosts. Bacteria exit the body in many ways: in respiratory droplets, feces, vaginal discharge, or semen, for example. Bloodfeeding animals such as ticks and mosquitoes also transmit some pathogenic bacteria.

C. Humans Put Many Prokaryotes to Work

Humans have exploited the metabolic talents of microbes for centuries, long before we could see cells under a microscope (**figure 17.14**). For instance, many foods are the products of bacterial metabolism. Vinegar, sauerkraut, sourdough bread, pickles, olives, yogurt, and cheese are just a few examples; organic acids released in fermentation produce the tart flavors of these foods. ⓘ *fermentation,* section 6.8B

Microbes also have industrial applications, many of which are related to the burgeoning field of biotechnology (see chapter 11). Vats of fermenting bacteria can produce enormous quantities of vitamin B_{12} and useful chemicals such as ethanol and acetone. Transgenic bacteria mass-produce human proteins, including insulin and blood-clotting factors. In addition, heat-tolerant enzymes isolated from bacteria can degrade proteins and fats in hot water, boosting the cleaning power of detergents used in laundry machines and dishwashers.

Water and waste treatment also use bacteria and archaea. Sewage treatment plants in most communities, for example, rely on biofilms consisting of countless microbes that degrade organic wastes. And a technique called bioremediation uses microorganisms to metabolize and detoxify pollutants such as petroleum and mercury. Given the right conditions, microbes can consume huge quantities of organic wastes. For example, oil- and gas-munching bacteria consumed much of the oil spilled in the Gulf of Mexico after a disastrous well blowout in 2010.

17.4   MASTERING CONCEPTS

1. In what ways are bacteria and archaea essential to eukaryotic life?
2. How are the microbes that colonize your body beneficial?
3. What adaptations enable pathogenic bacteria to enter the body and cause disease?
4. What are some practical uses of bacteria and archaea?

17.5 Bacterial Evolution Goes "Hog Wild" on the Farm

Although infectious diseases were once the leading cause of human death, antibiotics had made many bacteria-caused diseases manageable by the mid-1900s. Since that time, bacteria have become resistant not only to the original penicillin but also to the many manufactured antibiotics that followed it. Now antibiotic-resistant bacteria are common, creating new obstacles for physicians treating infectious disease.

Medical settings contribute to the rise of antibiotic-resistant bacteria, but so do farms. Antibiotics promote rapid growth when added to the food of cattle, chickens, swine, and other animals. Approximately 80% of the antibiotics used in the United States are fed to farm animals, according to a 2009 report from the Food and Drug Administration. A similar study revealed that farmers in China use over 100 million kilograms of antibiotics a year—about four times more than American farmers.

The unrestrained use of antibiotics on animal farms comes at a cost. The animals' manure contains not only antibiotics but also bacteria that are resistant to the drugs. These microbes swap genes with other bacteria in the environment (see figure 17.9). Farms have therefore become "breeding grounds" for bacteria resistant to antibiotics (**figure 17.15**).

Researchers Yong-Guan Zhu and James Tiedje, along with colleagues in China and the United States, wanted to know more about the consequences of antibiotic use in Chinese pig farms. Instead of monitoring the number of resistant bacteria, as previous researchers had done, the team looked for resistance genes. This strategy allowed the scientists to detect resistance in bacteria that are difficult to grow in the lab.

Zhu and colleagues first asked if the use of antibiotics in animal feed increased the diversity of resistance genes in pig manure. The

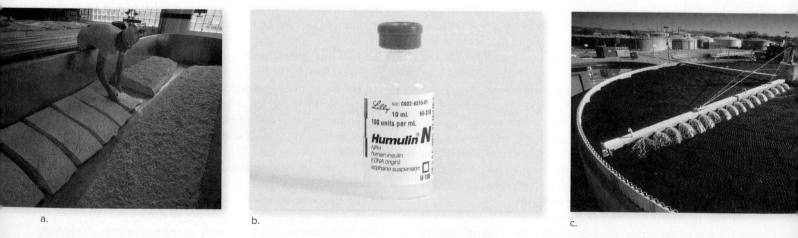

a.   b.   c.

Figure 17.14   Bacteria at Work. (a) Bacteria of genus *Lactococcus* manufacture cheddar cheese from fermenting milk. (b) Transgenic bacteria produce many drugs, including human insulin. (c) Raw sewage is sprayed on a trickling filter at a municipal wastewater treatment plant. Bacterial biofilms on the filter degrade the organic matter in the sewage.

pigs than in control manure. In compost, 44 genes met this standard; in soil, 17 did. For some genes, the difference between antibiotic-treated samples and control samples was immense: One gene showed up 28,000 times in treated manure for every time it was detected in control samples!

These results have serious implications, and not just for farm workers. Antibiotic-resistant bacteria from farms can easily spread to the general public. Since animal manure is composted and spread over fields, crops may become contaminated with antibiotic-resistant bacteria. Meat from treated livestock may also harbor resistance genes. When we eat the crops or the meat, bacteria in the human gut may take up the resistance genes via horizontal gene transfer. Insects that land on the manure may carry the resistant microbes to nearby communities. Bacteria might also leach from the field and enter streams and groundwater.

Increased regulation of antibiotics will likely slow the accumulation of resistance genes. In Denmark, the use of several antibiotics has been banned or voluntarily stopped on livestock farms, and the number of resistant bacteria in pigs has fallen sharply. In the United States, the Food and Drug Administration has called for animal drug manufacturers to voluntarily amend the approved uses of antibiotics, with the long-term goal of reducing their use in animal feed. Consumer awareness may hasten change. Many meat labels now indicate if the animals were raised without antibiotics. As demand for these meats grows, farmers will have an economic incentive to find alternatives to the drugs.

The need for change is urgent. According to some estimates, antibiotics may soon become useless if current practices continue. Evolution never stops, but a thorough understanding of natural selection and bacteria can help us slow the rise of antibiotic resistance.

Source: Zhu, Yong-Guan, and seven coauthors, including James M. Tiedje. 2013. Diverse and abundant antibiotic resistance genes in Chinese swine farms. *Proceedings of the National Academy of Sciences*, vol. 110, pages 3435–3440.

Figure 17.15 Pig Farm. Antibiotic-resistant bacteria thrive at pig farms. The animals are crowded into pens with manure-covered floors, and they receive copious antibiotics.
©Dario Sabljak/Alamy

research team took four samples of manure from each of three Chinese farms where antibiotics are used. They also collected samples from pigs that had never been exposed to antibiotics. They extracted DNA from each sample and amplified it using polymerase chain reaction (PCR) primers targeting 244 known antibiotic-resistance genes. At the end of the reactions, the scientists tallied the number of resistance genes that each sample contained (figure 17.16a). The manure from antibiotic-treated animals had many more resistance genes than did the control manure. ⓘ *PCR*, section 11.2C

But pig manure does not remain in the barn for long; farmers often compost pig manure and then spread it on their croplands. Do the genes persist under those conditions? To find out, the researchers collected samples from compost piles and from the soil in nearby fields. Control samples came from a forest that is isolated from synthetic antibiotic input. Again, DNA analysis of these samples revealed that compost and soil from farms using antibiotics had more diverse resistance genes than did forest soil (figure 17.16b).

The researchers also measured the abundance of each antibiotic-resistance gene in every sample. In manure, 56 genes were significantly more common in manure from antibiotic-treated

17.5 MASTERING CONCEPTS

1. What did the research team use as control samples in their experiment? Why was each control appropriate for comparison with antibiotic-treated samples?

2. How did the samples collected from farms that use antibiotics differ from control samples?

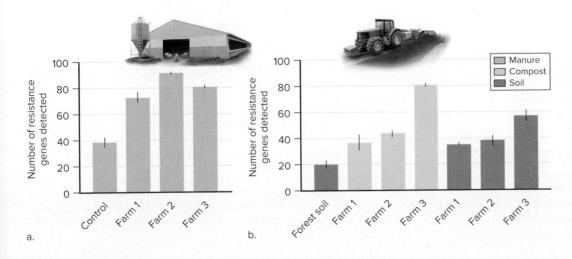

a.

b.

Figure 17.16 Antibiotic Resistance, on and off the Farm. (a) The diversity of antibiotic-resistance genes was significantly higher in the manure of antibiotic-fed pigs than in that of untreated animals. (b) Pig compost and compost-treated soil had significantly more diverse resistance genes than did untreated forest soil (control). Error bars represent the standard error of the mean (see appendix B).

CHAPTER SUMMARY

17.1 Prokaryotes Are a Biological Success Story

- **Prokaryotes** (domains **Bacteria** and **Archaea**) were the first organisms.
- Bacteria and archaea are very abundant and diverse, and they occupy a great variety of habitats.

17.2 Prokaryote Classification Traditionally Relies on Cell Structure and Metabolism

A. Microscopes Reveal Cell Structures

- Like other organisms, prokaryotic cells contain DNA and **ribosomes**, and they are surrounded by a cell membrane.
- The chromosome is located in an area called the **nucleoid**. **Plasmids** are circles of DNA apart from the chromosome.
- A **cell wall** surrounds most prokaryotic cells and confers the cell's shape: a spherical **coccus**, rod-shaped **bacillus**, or spiral-shaped **spirillum**.
- **Gram staining** reveals differences in cell wall architecture. Gram-positive bacteria have a thick **peptidoglycan** layer; gram-negative bacteria have a cell wall consisting of an outer membrane surrounding a thin peptidoglycan layer.
- A **glycocalyx** outside the cell wall provides attachment to surfaces or protection from host immune system cells.
- **Pili** are projections that allow cells to adhere to surfaces or transfer DNA to other cells.
- **Flagella** rotate to allow a motile cell to move toward or away from a stimulus, a process called **taxis.**
- Some bacteria survive harsh conditions by forming **endospores.**

B. Metabolic Pathways May Be Useful in Classification

- Bacteria and archaea are metabolically diverse. **Autotrophs** acquire carbon from inorganic sources, and **heterotrophs** obtain carbon from other organisms. A **phototroph** derives energy from the sun, and a **chemotroph** acquires energy from organic or inorganic chemicals.
- **Obligate aerobes** require oxygen, **facultative anaerobes** can live whether or not oxygen is present, and **obligate anaerobes** cannot function in the presence of oxygen.

C. Molecular Data Reveal Evolutionary Relationships

- Microbiologists are using molecular data to reconsider the traditional classification of bacteria and archaea.

D. Horizontal Gene Transfer Complicates Classification

- Binary fission (cell division) is a form of **vertical gene transfer,** in which DNA is transferred to the next generation.
- Prokaryotes can also acquire new DNA directly from the environment (**transformation**), a virus (**transduction**), or another cell via a **sex pilus** (**conjugation**). All three are routes of **horizontal gene transfer.**

17.3 Prokaryotes Include Two Domains with Enormous Diversity

A. Domain Bacteria Includes Many Familiar Groups

- Proteobacteria, cyanobacteria, spirochetes, firmicutes, actinobacteria, and chlamydias are a few examples of diversity within domain Bacteria.

B. Many, but Not All, Archaea Are "Extremophiles"

- Domain Archaea contains methanogens, halophiles, thermophiles, and many organisms that thrive in moderate conditions.

17.4 Bacteria and Archaea Are Essential to All Life

- Bacteria and archaea are nearly everywhere (figure 17.17).

A. Microbes Form Vital Links in Ecosystems

- All life depends on the bacteria and archaea that contribute gases to the atmosphere, recycle organic matter, and **fix nitrogen.**

Figure 17.17 Microbes in a Selection of Habitats.

B. Bacteria and Archaea Live in and on Us

- Most prokaryotes are beneficial or harmless, but pathogenic bacteria adhere to host cells and colonize tissues. Disease symptoms often result from the immune system's reaction to the infection.
- Toxins produced by bacteria, plus the ability to spread to new hosts, also contribute to disease.

C. Humans Put Many Prokaryotes to Work

- Bacteria and archaea are used in the manufacture of many foods, drugs, and other chemicals; in sewage treatment; and in bioremediation.

17.5 Investigating Life: Bacterial Evolution Goes "Hog Wild" on the Farm

- On Chinese pig farms, the routine use of antibiotics is associated with increased abundance and diversity of antibiotic-resistance genes in manure, compost, and soil.

MULTIPLE CHOICE QUESTIONS

1. A prokaryotic cell is one that
   a. lacks DNA.
   b. has membrane-bounded organelles.
   c. lacks a nucleus.
   d. lacks a plasma membrane.

2. Which of these is a distinguishing characteristic between the domains Bacteria and Archaea?
   a. Their size
   b. The chemical composition of the cell wall and cell membrane
   c. Their ability to grow at high temperatures
   d. The shape of their mitochondria

3. What feature distinguishes the cell walls of gram-positive and gram-negative bacteria?
   a. Peptidoglycan in the cell wall of gram-positive bacteria
   b. Peptidoglycan in the cell wall of gram-negative bacteria
   c. A second lipid bilayer in the cell wall of gram-positive bacteria
   d. A second lipid bilayer in the cell wall of gram-negative bacteria

4. What type of organism may use inorganic chemicals for both energy and a carbon source?
   a. A photoautotroph
   b. A photoheterotroph
   c. A chemoautotroph
   d. A chemoheterotroph

5. The primary advantage of conjugation is that it
   a. is the first step in the formation of a biofilm.
   b. allows prokaryotes to inject toxins into host cells.
   c. shuffles genes—including some that enhance survival—among cells.
   d. provides a mechanism by which bacteria avoid being infected with viruses.

6. Bacteria produce two kinds of toxins: endotoxins and exotoxins. Whereas endotoxins are _____, exotoxins are _____.
   a. part of the bacterium; chemicals that diffuse out of the bacterium
   b. injected into the host cell; secreted onto the surface of the host cell
   c. chemicals that target the host's digestive tract; chemicals that target the host's outer covering (e.g., skin)
   d. released while the bacterium is alive; released when the bacterium dies

7. How do scientists induce prokaryotes to produce human proteins?
   a. They insert human genes into bacterial genomes.
   b. They cross bacterial strains until the proteins arise at random.
   c. They inject bacteria into human muscles.
   d. All of the above are correct.

8. Which of the following processes occurs ONLY in prokaryotes?
   a. Nitrogen fixation
   b. Photosynthesis
   c. Asexual reproduction
   d. All of the above are correct.

Answers to these questions are in appendix A.

WRITE IT OUT

1. Explain why the antibiotics penicillin and polymyxin are not effective against archaea. (Review this chapter's Apply It Now box.)

2. Why do some microbiologists disagree with classifying bacteria and archaea as "prokaryotes"?

3. Give five examples that illustrate how bacteria and archaea are important to other types of organisms.

4. If you were developing a new "broad-spectrum" antibiotic to kill a wide variety of bacteria, which cell structures and pathways would you target? Which of those targets also occur in eukaryotic cells, and why is that important? How would your strategy change if you were designing a new "narrow-spectrum" antibiotic active against only a few types of bacteria?

5. Describe your own metabolic classification: Are you a photoautotroph, photoheterotroph, chemoautotroph, or chemoheterotroph? Are you an obligate aerobe, an obligate anaerobe, or a facultative anaerobe? Are you a nitrogen fixer?

6. A prokaryote with which type of metabolism would be especially challenging to isolate and culture? Explain your answer. (*Hint:* Refer to section 17.2B.)

7. Ernst Mayr defined a biological species as a population whose members can exchange genetic material during reproduction (see chapter 14). How does horizontal gene transfer complicate this definition of species?

8. Why did the discovery of archaea generate interest in searching for cells on other planets?

9. In an article in *Nature* magazine, Sean Nee wrote about the diversity of species on Earth: "Earth's real biodiversity is invisible, whether we like it or not." What does that statement mean?

10. Ecosystems rely on nitrogen-fixing bacteria, which convert atmospheric nitrogen (N_2) into ammonium (NH_4^+). Into what types of organic molecules do plants incorporate the nitrogen in ammonia?

11. Stomach ulcers, once thought to be entirely a product of spicy food or high stress, are now known to be caused by bacteria (*Helicobacter pylori*). How has ulcer treatment changed because of this new knowledge?

12. Use the Internet to learn about sexually transmitted diseases (STDs). Which of the most common STDs are caused by bacteria? Choose one STD caused by bacteria to study in more depth; describe how the pathogen spreads and its symptoms, treatment, and prevention.

13. Probiotics are dietary supplements consisting of live bacteria that are normally found in the human digestive tract. Some people claim that consuming probiotics promotes digestive health. Design an experiment that would help you determine (a) whether the bacteria in a probiotic supplement survive the trip from the mouth, through the stomach and small intestine, and into the large intestine and (b) whether probiotics actually do promote digestive health.

14. Botox is a toxin produced by the bacterium *Clostridium botulinum*. When ingested with tainted food, Botox can kill by paralyzing muscles needed for breathing and heartbeat. Physicians inject small quantities of diluted Botox into facial muscles to paralyze them and reduce the appearance of wrinkles. Some people have expressed concern about "Botox parties" at hair salons and other nonmedical settings. What are the risks of getting injections in such a setting?

15. After treatment with antibiotics, your belly aches. Why might taking antibiotics cause digestive problems? How might the problem fix itself?

16. *Mycobacterium tuberculosis* causes most cases of tuberculosis. Strains of this bacterium that are resistant to all known antibiotics have become increasingly common. Explain how this change occurred; use the terms *mutation, DNA,* and *natural selection* in your answer.

PULL IT TOGETHER

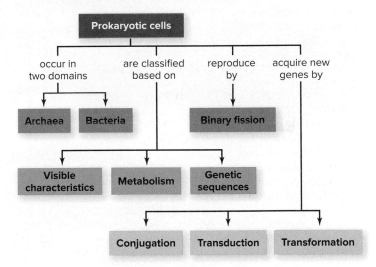

Figure 17.18 Pull It Together: Bacteria and Archaea.

Refer to figure 17.18 and the chapter content to answer the following questions.

1. Review the Survey the Landscape figure in the chapter introduction, and connect *viruses* and *evolution* to the Pull It Together concept map in at least two ways each.

2. Add *autotrophs, heterotrophs, phototrophs,* and *chemotrophs* to this concept map.

3. Where do obligate aerobes, obligate anaerobes, and facultative anaerobes fit on this map?

4. Where could you place humans on this concept map?

5. Create a new concept map that includes the internal and external parts of a prokaryotic cell.

6. What are the cell features and metabolic criteria by which biologists classify microbes?

Pfiesteria Fish Lesions. Toxins produced by the protist *Pfiesteria* cause gaping sores on fish; the inset shows one *Pfiesteria* cell.
©Richard Ellis/Alamy; (inset): Courtesy of Burkholder Laboratory

LEARN HOW TO LEARN
What's the Point of Rewriting Your Notes?

Your notes are your record of what happened in class, so why should you rewrite them after the lecture is over? One answer is that the abbreviations and shorthand that make perfect sense while you take notes will become increasingly mysterious as time goes by. Rewriting the information in complete sentences not only reinforces learning but also makes your notes much easier to study before an exam.

Science Lessons from a Killer Cell

Of the many thousands of species of protists, the vast majority are harmless to animals. One exception, however, is *Pfiesteria*, which lives in water off the east coast of the United States. *Pfiesteria* belongs to a group of protists called dinoflagellates. Typically, *Pfiesteria* is nontoxic, feeding on algae and bacteria in coastal waters. But a few of the stages in its life cycle can produce extremely potent toxins that kill fish and accumulate in shellfish, making them poisonous to humans.

Pfiesteria produces at least two toxins. One is a powerful neurotoxin, and the other causes disintegration of the skin of a fish. Although the toxins break down rapidly, the infected fish often develop open sores. Many fish die. In humans, *Pfiesteria* toxins can produce rashes, open sores, fatigue, erratic heartbeat, breathing difficulty, personality changes, and extreme memory loss.

The on-again, off-again nature of the toxins has made them very difficult for scientists to study. Controversy erupted after one team of scientists suggested that a species called *Pfiesteria shumwayae* did not produce toxins. In making their case, the researchers used several experimental strategies, none of which yielded evidence of toxins.

Later, however, other researchers refuted these results, saying that the first group raised the *Pfiesteria* cells under conditions that suppressed toxin production (the cells must grow with live fish to produce toxins). This second group used the same form of *P. shumwayae* as the first group, but they also included two control forms: one known to make toxins and the other known to be nontoxic. The results indicated that, under some conditions, *P. shumwayae* did indeed produce toxins that harmed fish.

These conflicting studies illustrate two important features of science. First, the conditions under which scientists conduct experiments can greatly affect their conclusions. Second, communication is vital. When the first group published their results, they described their *Pfiesteria* forms and experimental techniques in detail. The second group used this information to identify conditions known to suppress toxin production. The careful reporting of materials and methods enables scientists to evaluate one another's work.

This chapter's content illustrates a third feature of science: adaptability to new information. Protists were once considered a single kingdom, but we now know that the protists are much too diverse to fit neatly into one category. This chapter presents a sampling of that diversity.

LEARNING OUTLINE

18.1 Protists Lie at the Crossroads Between Simple and Complex Organisms

18.2 Algae Are Photosynthetic Protists

18.3 Some Heterotrophic Protists Resemble Fungi

18.4 Protozoa Are Diverse Heterotrophic Protists

18.5 Protist Classification Is Changing Rapidly

18.6 Investigating Life: Shining a Spotlight on Danger

SURVEY THE LANDSCAPE
The Diversity of Life

The protists include a vast array of organisms that have little in common with one another. The one feature they share is that they are eukaryotes that are not classified as plants, fungi, or animals.

For more details, study the Pull It Together feature in the chapter summary.

18.1 Protists Lie at the Crossroads Between Simple and Complex Organisms

Our tour of life's diversity began with viruses, infectious particles that lie on the border between the living and the nonliving. Chapter 17 then described the two domains of prokaryotes, Bacteria and Archaea. Early prokaryotes played pivotal roles in the evolution of life, inventing countless metabolic pathways and releasing oxygen into Earth's atmosphere.

Nearly 2 billion years ago, the prokaryotes gave rise to a new, more complex cell type: the eukaryote. Unlike a prokaryote, a **eukaryotic cell** has a nucleus and other membrane-bounded organelles, such as mitochondria and chloroplasts. Figures 3.8 and 3.9 clearly show the compartmentalization and division of labor typical of eukaryotic cells, and section 3.2 offers a more complete description of the distinctions among the cell types. We now embark on a tour of the protists, the simplest eukaryotes.

A. What Is a Protist?

Until recently, biologists recognized four eukaryotic kingdoms: Protista, Plantae, Fungi, and Animalia. The plants, fungi, and animals are distinguished based on their characteristics. Kingdom Protista, in contrast, was defined by *exclusion*. An organism was designated a **protist** if it was a eukaryote that did not fit the description of a plant, a fungus, or an animal. Kingdom Protista was, in effect, a convenient but artificial "none of the above" category (figure 18.1). Not surprisingly, the nearly 100,000 named species of protists are extremely diverse, displaying great variety in size, nutrition, locomotion, reproduction, and cell surfaces.

Biologists have traditionally grouped the protists based on the organisms they resemble: the plantlike algae, funguslike slime molds, and animal-like protozoa. Modern systematists, however, group organisms based on evolutionary relationships. DNA sequences provide the most objective measure of relatedness. Based on these new molecular data, the former kingdom Protista has shattered into dozens of groups whose relationships to one another remain uncertain. ⓘ *systematics,* section 14.6

Because the classification of protists is in transition and many of the new groupings are not universally accepted, this chapter uses the traditional approach to classification. Section 18.5, however, revisits modern trends in protist taxonomy.

B. Protists Are Important in Many Ways

The metabolic diversity among protists means they have an astonishingly wide variety of functions and roles in human life. In ecosystems, the autotrophic ("self-feeding") algae carry out photosynthesis, producing much of the O_2 in Earth's atmosphere and supporting food webs in oceans, lakes, rivers, and ponds. In addition, lichens are unique "dual organisms" consisting of algae living among the threads of fungi. Their activities help build soil from bare rock. ⓘ *food webs,* section 38.3A; *lichens,* section 20.7

Medically important protists include parasites that infect plant and animal hosts (including humans), at a cost of billions of dollars, incalculable suffering, and millions of deaths annually. As you will see, insects transmit some of these organisms; others enter new hosts in contaminated food or water. Symptoms of infection by the latter group may include diarrhea, an adaptation that allows the parasite to move from an infected host to water.

Protists also have found their way into diverse industrial applications. Some algae, for example, help make chocolate smooth and creamy, whereas others help make paints reflective. Still other species have left distinctive fossils that point the way to petroleum reserves. Conversely, biologists may one day be able to reduce our reliance on petroleum by harnessing protist photosynthesis and reproduction in "bioreactors" that use sunlight and CO_2 to produce oil and other biofuels (see chapter 19's Burning Question).

C. Protists Have a Lengthy Evolutionary History

Evolutionary biologists are especially interested in protists. One reason is that the cells of today's protists may retain clues to important milestones in eukaryote history. The **endosymbiont theory,** for example, suggests that early eukaryotes originally

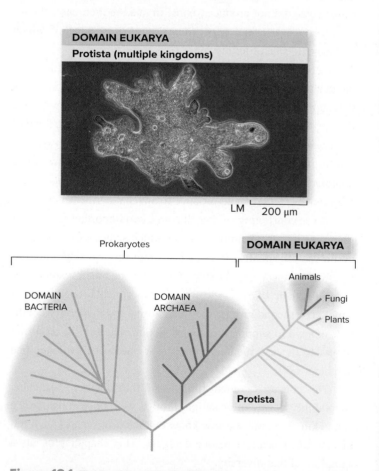

Figure 18.1 The Protists. "Kingdom" Protista consists of many lineages, each of which may eventually be considered its own kingdom.

Photo: ©Melba Photo Agency/PunchStock RF

obtained mitochondria and chloroplasts by engulfing bacterial cells (figure 18.2). Nearly all eukaryotes have mitochondria, so the aerobic bacteria that became these organelles were probably engulfed first. ⓘ *endosymbiosis*, section 15.2A

Chloroplasts have an especially colorful evolutionary history (see figure 18.2). Biologists studying chloroplast DNA, membrane structure, and photosynthetic pigments suggest that the chloroplasts of red algae, green algae, and land plants all arose from cyanobacteria engulfed by some ancient eukaryotic cell. The chloroplasts in these species are surrounded by two membranes. In other photosynthetic protists, three or more membranes surround each chloroplast. The evidence suggests that these organelles originated long ago when red algae or green algae were themselves engulfed by other cells—an event called secondary endosymbiosis. We are just beginning to unravel the events surrounding the origins of all the different types of chloroplasts (see section 18.5).

Another important milestone in eukaryote history is the origin of multicellularity. Biologists do not know how unicellular eukaryotes adopted a multicellular lifestyle. We do know, however, that multicellularity arose independently in multiple lineages. After all, genetic evidence clearly suggests that plants, fungi, and animals arose from different lineages of unicellular

protists. Additional clues may come from studies of colonies in which individual protists interact as they move and obtain food. ⓘ *multicellularity*, section 15.2B

A second reason that protists are important to studies of evolutionary biology is that these simple eukaryotes shed light on the evolutionary history of plants, fungi, and animals. Genetic and cellular similarities, for example, illustrate the strong evolutionary connection between green algae and plants. Likewise, DNA evidence suggests that heterotrophic protists called choanoflagellates are the closest existing relatives to sponges, the simplest animals (see figure 21.1). ⓘ *sponges*, section 21.2

Clearly, depicting the diversity of the protists in just one chapter is a challenge. In exploring these organisms, remember that the examples in each section represent just a small sampling of some of the most intriguing members of this uniquely variable group.

18.1 MASTERING CONCEPTS

1. What features define the protists?
2. Describe examples of how protists are important.
3. Why are evolutionary biologists interested in protists?

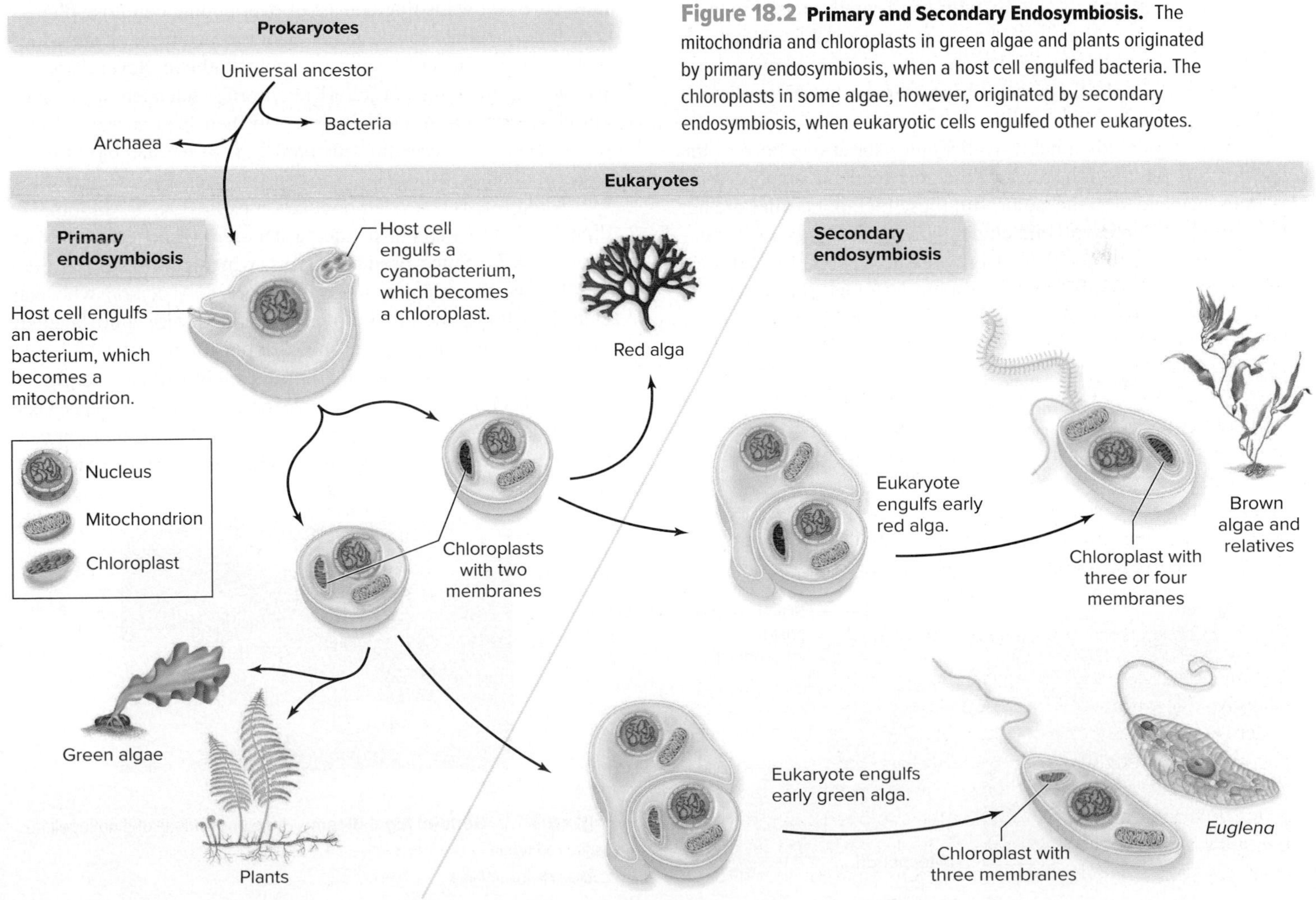

Figure 18.2 Primary and Secondary Endosymbiosis. The mitochondria and chloroplasts in green algae and plants originated by primary endosymbiosis, when a host cell engulfed bacteria. The chloroplasts in some algae, however, originated by secondary endosymbiosis, when eukaryotic cells engulfed other eukaryotes.

18.2 Algae Are Photosynthetic Protists

Most people probably think of algae as pond scum, but **algae** is a general term that refers to any photosynthetic protists that live in water. Although the cyanobacteria were traditionally called "blue-green algae," most biologists now reserve the term *algae* for eukaryotes.

The cells of algae contain chloroplasts that house a rainbow of yellow, gold, brown, red, or green photosynthetic pigments. These organelles use light energy and CO_2 to produce carbohydrates and other organic molecules that support freshwater and marine food webs. They also release O_2 as a waste product.

Algae may be single-celled, colonial, filamentous, or multicellular. Some of the more complex species produce differentiated tissues. Although the body forms may resemble those of plants, algae are considered protists because they lack the distinctive reproductive structures that define plants. This section describes the major types of algae.

A. Euglenoids Are Heterotrophs and Autotrophs

The **euglenoids** are unicellular protists with elongated cells (figure 18.3). Most have a long, whiplike flagellum used in locomotion and a short flagellum that does not extend from the cell. Supporting the cell membrane is a pellicle, a protective layer made of rigid or elastic protein strips. An eyespot helps the cell orient toward light.

Most euglenoids inhabit fresh water. About one third of the species are photosynthetic, and the rest feed on organic compounds suspended in the water. But these metabolic roles are not always fixed. Photosynthetic euglenoids such as *Euglena,* for example, may occasionally feed on organic matter. In darkness, their cells become entirely heterotrophic, although photosynthesis resumes once light returns.

Figure 18.3 *Euglena.* The pond-dwelling *Euglena* has a flagellum and chloroplasts, but it can also ingest food particles.
Photo: ©blickwinkel/Alamy

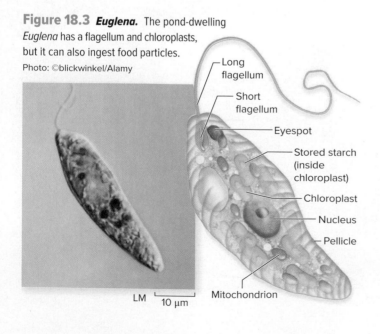

- Long flagellum
- Short flagellum
- Eyespot
- Stored starch (inside chloroplast)
- Chloroplast
- Nucleus
- Pellicle
- Mitochondrion

LM 10 μm

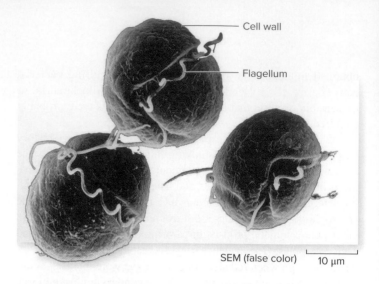

- Cell wall
- Flagellum

SEM (false color) 10 μm

Figure 18.4 **Dinoflagellates.** Note the two flagella and the cellulose plates that make up the cell walls of these dinoflagellates.
©David M. Phillips/Science Source

B. Dinoflagellates Are "Whirling Cells"

The marine protists known as **dinoflagellates** have two flagella of different lengths (figure 18.4). One of the flagella propels the cell with a whirling motion (the Greek *dinein* means "to whirl"); the other mainly acts as a rudder. In addition, many dinoflagellates have cell walls that consist of overlapping cellulose plates.

Dinoflagellates are a major component of the ocean's food webs. About half the species are photosynthetic. Several species live within the tissues of jellyfishes, corals, sea anemones, or giant clams, providing carbohydrates to their host animals. Many other species are predators or parasites. Some are bioluminescent, producing flashing lights in coastal waters (see section 18.6).

A red tide is a sudden population explosion, or "bloom," of dinoflagellates that turn the water red, orange, or brown (figure 18.5). Some produce toxins, which can make red tides deadly (see this chapter's opening essay). A person who eats clams, scallops, oysters, or mussels tainted with dinoflagellate toxins may develop paralytic shellfish poisoning.

A red tide is one type of harmful algal bloom: an overgrowth of algae that release toxins or harm ecosystems in other ways. Usually, these blooms occur in response to a boost in the nutrient content of the water, as described in this chapter's Burning Question.

Figure 18.5 **Harmful Algal Bloom.** Huge populations of dinoflagellates cause red tides.
©Julian Nieman/Alamy

Burning Question

Why and how do algae form?

Figure 18.A **Algae on a Pond Surface.**
©Michael Marten/Science Source

Algae are common aquatic protists, but they are often inconspicuous. Sometimes, however, their populations grow so large that they seem to take over; ponds and poorly maintained swimming pools can turn bright green with algae (figure 18.A). This population explosion occurs when nutrients and sunlight are abundant.

Algal blooms are normal in some ecosystems, such as in many ponds. A bloom where water is normally clear, however, usually indicates that nutrients from sewage, fertilizer, or animal waste are polluting the waterway. The use of lawn fertilizers is a common cause of algal blooms in ponds in residential settings.

Algae may seem harmless, but algal blooms induced by nutrient pollution can devastate aquatic life. After the algae die, bacteria and other microbes decompose their bodies. Respiration by these decomposers depletes oxygen in the water, causing fishes and other aquatic species to suffocate. (i) *eutrophication*, section 38.4E

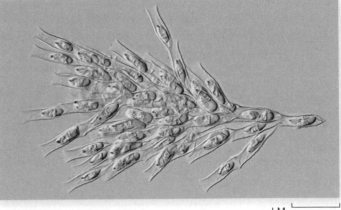

Figure 18.6 **Golden Algae.** *Dinobryon* is a colonial golden alga.
©Frank Fox/Science Source

Like *Euglena,* golden algae can act as autotrophs or heterotrophs. In some aquatic ecosystems, photosynthesis by golden algae provides a significant source of food for zooplankton. When light or nutrient supplies dwindle, however, many golden algae can consume bacteria or other protists.

Some golden algae can be harmful. In nutrient-rich streams and lakes, blooms of golden algae can release toxins that kill fish (although humans are apparently unaffected).

Diatoms **Diatoms** are unicellular algae with ornate, two-part silica cell walls that fit together like a shoebox and its lid (figure 18.7). These protists occupy just about every moist habitat on Earth, from streams to fish tanks to damp soil. Their populations are sensitive to water pH, salinity, and other environmental conditions. Biologists therefore periodically sample freshwater diatom populations as indicators of environmental quality.

Figure 18.7 **Diatoms.** The "glass houses" (silica cell walls) of these photosynthetic protists exhibit a dazzling variety of forms.
(multiple diatoms): ©Jan Hinsch/SPL/Getty Images; (single diatom): ©Steve Gschmeissner/Science Source

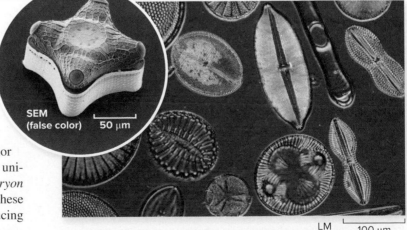

C. Golden Algae, Diatoms, and Brown Algae Contain Yellowish Pigments

Three groups of algae contain a yellowish photosynthetic pigment called fucoxanthin in addition to chlorophylls *a* and *c*. This accessory pigment gives these organisms a golden, olive green, or brown hue.

Golden Algae The **golden algae** are named for their color (figure 18.6). Their cells usually have two flagella. Most are unicellular, but filamentous and colonial forms also exist. *Dinobryon* is an example of a freshwater genus of golden algae; in these protists, individual vase-shaped cells stack end to end, producing branched or unbranched chains.

Although diatoms occur nearly everywhere, most live in oceans. Their tiny photosynthetic cells can reach huge densities, removing CO_2 from the atmosphere and providing food for zooplankton. Over millions of years, the glassy shells of diatoms have accumulated as thick deposits on the ocean floor. The abrasive shells mined from these deposits are used in swimming pool filters, polishes, toothpaste, and many other products. Diatoms also impart the reflective quality of paints used in roadway signs and license plates.

Few microorganisms fossilize as well as diatoms, and biologists know of some 35,000 extinct species. In some undisturbed lake sediments, diatoms have left complete records of the evolution of entirely new species.

Brown Algae   The **brown algae** are the largest and most complex protists. Although they are multicellular, they resemble the golden algae in their pigments and their reproductive cells, each of which bears two flagella.

Brown algae live in marine habitats all over the world. The Sargasso Sea in the northern Atlantic Ocean, for example, is named after floating masses of brown algae called *Sargassum*. The kelps, which are the largest of the brown algae, produce enormous underwater forests that provide food and habitat for many animals (figure 18.8). Some kelps exceed 30 meters in length.

Humans consume several species of kelp. *Laminaria digitata,* for example, is an ingredient in many Asian dishes. Algin, a chemical extracted from the cell walls of brown algae, is used as an emulsifying, thickening, and stabilizing agent in products including ice cream, candies, chocolate, salad dressings, toothpaste, cosmetics, polishes, latex paint, and paper.

Figure 18.9 **A Red Alga.** The blades of this alga, called dulse, grow to 50 cm. People on the northern Atlantic coast consider it a healthy snack.
©Andrew J. Martinez/Science Source

D. Red Algae Can Live in Deep Water

Most **red algae** are relatively large (figure 18.9), although some are microscopic. These marine organisms are somewhat similar to green algae in that they store carbohydrates as a modified form of starch, have cell walls containing cellulose, and produce chlorophyll *a*.

Red algae, however, can live in water exceeding 200 meters in depth, thanks to reddish and bluish photosynthetic pigments that absorb wavelengths of light that chlorophyll *a* cannot capture. All light becomes dimmer with increasing depth, but the wavelengths do not dissipate equally. The pigments in red algae can use some of the wavelengths that persist in deep water.

Humans use red algae in many ways. Agar, for example, is a polysaccharide in the cell walls of some species. This jellylike substance is used as a culture medium for microorganisms in petri dishes; agar is sometimes also used as a gel in canned meats and as a thickener in ice cream and yogurt. Another useful product is carrageenan, a polysaccharide that emulsifies fats in chocolate bars and stabilizes paints, cosmetics, and creamy foods. A red alga called nori is used for wrapping sushi.

Figure It Out

Consult figure 5.4 to see the light absorption spectrum for chlorophyll *a*. Predict the most likely range of wavelengths absorbed by the photosynthetic pigments unique to red algae.

Answer: Between 500 and 650 nm

E. Green Algae Are the Closest Relatives of Land Plants

The **green algae** are the protists that share the most similarities with plants. They use chlorophyll *a* and *b* as photosynthetic pigments, use starch as a storage carbohydrate, and have cell walls containing cellulose. Like plants, many green algae also have life cycles that feature an **alternation of generations,** in which a multicellular haploid (gametophyte) phase is followed by a diploid (sporophyte) phase (figure 18.10).

Blade (nutrient absorption and photosynthesis)

Bladder (buoyancy)

Holdfast (anchorage)

Figure 18.8   **Giant Kelp.** These brown algae form huge underwater forests near coastlines. Each individual may be dozens of meters long.
Photo: ©Ralph A. Clevenger/Corbis

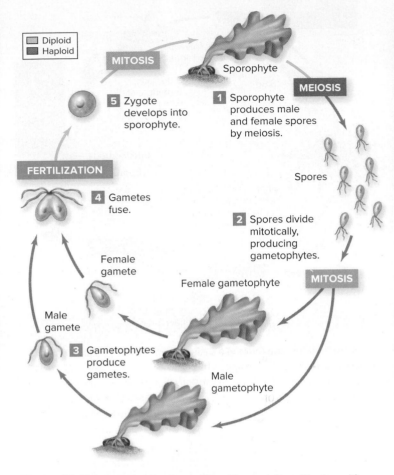

Figure 18.10 **Life Cycle of a Green Alga.** The sea lettuce *Ulva* has a life cycle that features an alternation of haploid and diploid generations.

The habitats and body forms of green algae are diverse (figure 18.11). Most live in fresh water or in moist habitats on land, although some live in symbiotic relationships with fungi, forming lichens. The size ranges from the smallest eukaryote (*Micromonas*), only 1 µm in diameter, to sea lettuce (*Ulva*), exceeding 1 meter in length. Green algae may be unicellular, filamentous, colonial, or multicellular. The multicellular species may have rootlike and stemlike structures, but these structures are far less specialized than the true roots, stems, and leaves of plants.

One well-studied green alga is *Chlamydomonas*, a unicellular organism that reproduces asexually and sexually. Scientists study these algae to learn about the evolution of sex, how an individual's sex is determined, and how cells of opposite sexes recognize each other. A classroom favorite is the colonial green alga *Volvox*. Hundreds to thousands of *Volvox* cells form hollow balls; the cells move their flagella to move the sphere. New colonies remain within the parental ball of cells until they burst free.

Other green algae include the geometrically shaped desmids (*Micrasterias*), mermaid's wineglass (*Acetabularia*), the ribbonlike *Spirogyra*, and the tubular *Codium*. One species, *Chlorella*, is being considered as a food and oxygen source on prolonged space flights; this species can multiply very quickly, as evidenced by the rapid "greening" of a poorly maintained aquarium or swimming pool.

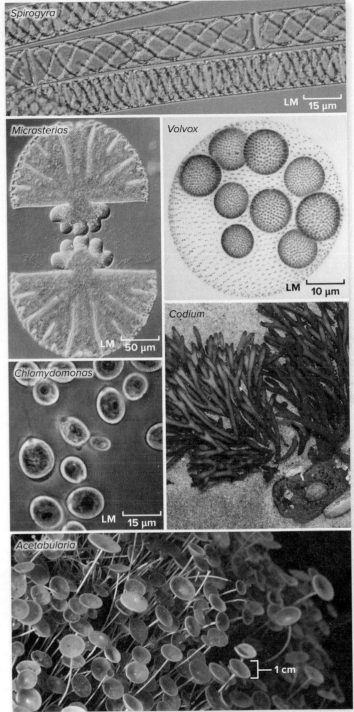

Figure 18.11 **Gallery of Green Algae.** Green algae have a variety of forms, from solitary microscopic cells to complex multicellular bodies.

(*Spirogyra*): ©Nuridsany et Perennou/Science Source; (*Micrasterias*): ©M. I. Walker/Science Source; (*Volvox*): ©Frank Fox/Science Source (*Chlamydomonas*): ©Biophoto Associates/Science Source; (*Codium*): ©Darlyne A. Murawski/National Geographic/Getty Images; (*Acetabularia*): ©incamerastock/Alamy

18.2 MASTERING CONCEPTS

1. What mode of nutrition do the algae use?
2. Describe several criteria for classifying the algae.
3. List and describe the characteristics of the major groups of algae.

18.3 Some Heterotrophic Protists Resemble Fungi

Slime molds and water molds are protists that resemble fungi in some ways: They are heterotrophic, and some produce filamentous feeding structures. They also commonly occur alongside fungi in many habitats. Nevertheless, DNA sequences clearly indicate that they are not closely related to fungi.

A. Slime Molds Are Unicellular and Multicellular

The slime molds are informally divided into two groups whose relationship to each other remains unclear. Both types live in damp habitats such as forest floors. In addition, each type of organism exists as single, amoeboid cells and as large masses that behave as one multicellular organism. The major difference between the two types is reflected in their names: plasmodial and cellular slime molds.

The feeding stage of a **plasmodial slime mold** consists of a plasmodium, which is a huge cell: a mass of thousands of diploid nuclei enclosed by one cell membrane. This structure gives these organisms their other common name, the "acellular" slime molds. The plasmodium may be a conspicuous, slimy, bright yellow or orange network up to 25 cm in diameter (**figure 18.12**). It migrates along the forest floor, engulfing bacteria and other microorganisms on leaves, debris, and rotting logs.

In times of drought or starvation, the plasmodium halts and forms fruiting bodies, which produce thick-walled reproductive cells called spores. When favorable conditions return, the spores germinate and form new cells that resume feeding. Two of these cells may fuse, forming a diploid zygote nucleus that divides repeatedly by mitosis. The resulting plasmodium contains multiple nuclei.

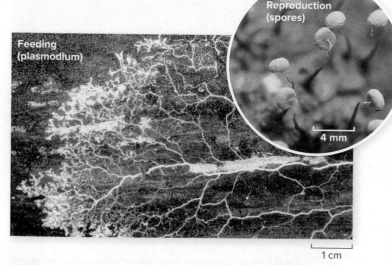

Figure 18.12   Plasmodial Slime Mold. *Physarum* oozes across a decaying log. The inset shows the spore-producing structures of a slime mold.

(plasmodium): ©Steven P. Lynch/McGraw-Hill Education; (spores): ©Ray Simons/ Science Source

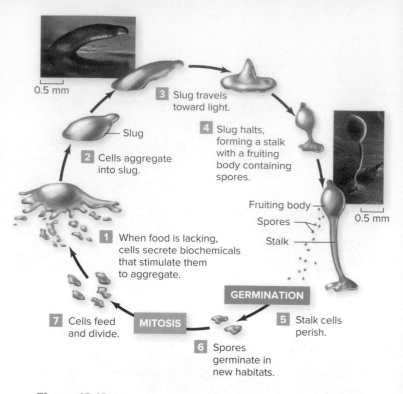

Figure 18.13   Life Cycle of a Cellular Slime Mold. (*1*) Starvation stimulates cells to (*2*) aggregate into a multicellular "slug," which (*3*) crawls to a new habitat and (*4*) forms a fruiting body that releases spores. (*5*) Stalk cells die, but (*6*) spores develop into (*7*) amoeboid cells that consume bacteria. Only asexual reproduction is shown; all cells shown are haploid.

Photos: (slug): ©Carolina Biological Supply Company/Phototake; (fruiting body): ©David Scharf/Science Source

Scientists use plasmodial slime molds to study cell division and the movements of cytoplasm inside a cell.

In contrast to the plasmodial slime molds, individual cells of a **cellular slime mold** retain their membranes throughout the life cycle. The cells exist as haploid feeding amoebae, engulfing bacteria in fresh water, moist soil, and decaying vegetation.

When food becomes scarce, the amoebae secrete chemical attractants that stimulate the neighboring cells to aggregate into a sluglike structure (**figure 18.13**). The "slug" moves toward light, stops, and forms a stalk topped by a fruiting body that produces haploid spores. The cells of the stalk perish, but the spores survive; wind, water, or animals carry them to new habitats. The spores then germinate to form haploid amoebae, and the cycle begins anew. Sexual reproduction also occurs in some conditions. *Dictyostelium discoideum* is a cellular slime mold used in many scientific studies (see this chapter's Focus on Model Organisms).

B. Water Molds Are Decomposers and Parasites

The **water molds,** or oomycetes, are decomposers or parasites of plants and animals in moist environments (**figure 18.14**). The filaments of water molds secrete digestive enzymes into their surroundings and absorb the nutrients. In addition, like fungi called chytrids, water molds produce swimming spores that aid their dispersal in water and wet soil. Despite the similarities, however, water molds are unlike fungi in many ways. The

a.   b.

Figure 18.14 **Water Molds.** (a) *Phytophthora infestans* is a water mold that causes late blight of potatoes and was responsible for the Irish potato famine in the mid-1840s. (b) *Saprolegnia* infects a tetra.

(a): ©W.E. Fry, Plant Pathology, Cornell University; (b): ©sdbower/iStock/Getty Images RF

filaments of water molds are diploid, for example, whereas most fungal filaments are haploid. Also, fungi have cell walls containing chitin, but water mold cell walls contain cellulose.

The best-known water molds are those that ruin crops, causing such diseases as downy mildew of grapes and lettuce. In the 1870s, downy mildew of grapes nearly destroyed the French wine industry. The water mold *Phytophthora infestans*, which means "plant destroyer," causes late blight of potato. This disease caused the devastating Irish potato famine from 1845 to 1847, during which more than a million people starved and millions more emigrated from Ireland. The Irish potato famine followed several rainy seasons, which fostered the rapid spread of the plant disease. A newly

discovered relative of *P. infestans,* called *Phytophthora ramorum,* causes a tree disease called sudden oak death. Another well-known water mold is *Saprolegnia,* a protist that forms cottony masses on fishes and other aquatic organisms that are weakened or dead.

18.3 MASTERING CONCEPTS

1. What mode of nutrition do the slime molds and water molds use?
2. Compare and contrast the plasmodial and cellular slime molds.
3. What has been the role of water molds in the environment and history?

F🔍CUS on Model Organisms

Cellular Slime Mold: *Dictyostelium discoideum*

Slime molds have such an unappealing name that it may seem hard to imagine why anyone would study them. But *Dictyostelium discoideum* is an unusual organism, one that straddles the boundary between the unicellular and the multicellular. As illustrated in figure 18.13, its life cycle includes individual amoeba-like cells, a multicellular migrating "slug," and a spore-producing structure.

Dictyostelium discoideum is useful as a model because, like other model organisms, it is easy to grow in the laboratory and has a short generation time. In addition, its cells are readily accessible to microscopy and genetic studies. As a result, *D. discoideum* (affectionately called "Dicty" by its researchers) remains a fascinating organism. Researchers have made several discoveries in this species:

- **Cell movement:** A "Dicty" cell eats by producing extensions that engulf and absorb food particles by phagocytosis. Scientists have discovered that this movement is possible because proteins such as actin and myosin move rapidly within the cell. These same proteins produce muscle movement in animals. ⓘ *muscle movement*, section 29.4
- **Cytokinesis:** Researchers observing cell division in *D. discoideum* have discovered that the protein myosin is also

required for cytokinesis (the physical division of one cell into two). ⓘ *cytokinesis*, section 8.5C

- **Chemotaxis:** Starving Dicty cells move toward one another and form a multicellular "slug." This movement toward a chemical stimulus, called chemotaxis, requires membrane proteins that not only detect the signals from other Dicty cells but also transmit the information to the inside of the cell. Similar signal transduction systems occur in many organisms. ⓘ *cell membrane*, section 3.3
- **Cell differentiation:** When individual Dicty cells come together, chemical signals presumably determine which cells will become stalk cells (and die) and which will become spore cells (and survive). Such research may help answer questions about the origin of multicellularity. ⓘ *multicellularity*, section 15.2B

18.4 Protozoa Are Diverse Heterotrophic Protists

Finding a list of characteristics that unite the diverse **protozoa** is difficult. Most are unicellular, and the vast majority are heterotrophs, but several autotrophic species exist. They move by flagella, cilia, or pseudopodia. Some are free-living, and others are obligate parasites. Most are asexual, but sexual reproduction occurs in many species.

This section describes four groups of distantly related protozoa that are defined by locomotion and morphology. New molecular techniques are redefining the protozoa, but until the newer system of classification is better defined and more widely accepted, these four groups remain practical for general biology, education, and medicine.

A. Several Flagellated Protozoa Cause Disease

The **flagellated protozoa** are unicellular organisms with one or more flagella (**figure 18.15**). Most are free-living in fresh water, the ocean, and soil. The euglenoids and dinoflagellates, groups already described with the algae, are flagellates. This section turns to a few of the heterotrophic species.

One example of a flagellated protozoan is *Trichonympha*, a protist that lives in the intestines of termites. The cells of *Trichonympha*, in turn, harbor bacteria that digest cellulose. This bacterium-within-protist living organization enables termites to "digest" wood. Exposing termites to high oxygen or high temperature kills the symbionts. The insects soon die, with guts full of undigested wood.

Some parasitic flagellated protozoa cause disease in humans. For example, *Trichomonas vaginalis* resides in the urogenital tracts of both men and women. It is sexually transmitted and causes a form of vaginitis in females. *Giardia intestinalis* (also known as *Giardia lamblia*) causes "hiker's diarrhea," or giardiasis. People ingest the cysts of the organism in contaminated water. As *Giardia* cells divide in the small intestine, they impair the host's ability to absorb nutrients, resulting in diarrhea and cramping.

Another group of disease-causing flagellates are the **trypanosomes,** whip-shaped parasites that invade the bloodstream and brain. Insects transmit trypanosomes to humans. Tsetse flies, for example, carry *Trypanosoma brucei*, the organism that causes African sleeping sickness. In South and Central America, kissing bugs transmit *Trypanosoma cruzi* to humans from rodents, armadillos, and dogs. The resulting illness, called Chagas disease, kills 45,000 people annually. The sand fly transmits a related parasite, *Leishmania*.

B. Amoeboid Protozoa Produce Pseudopodia

The **amoeboid protozoa** produce cytoplasmic extensions called pseudopodia (Latin, meaning "false feet"), which are important in locomotion and food capture via phagocytosis. The most studied species is *Amoeba proteus,* a common freshwater microbe that engulfs bacteria, algae, and other protists in its pseudopodia (**figure 18.16**). The human digestive tract may be invaded by another species, *Entamoeba histolytica,* which can cause fever and severe diarrhea. ⓘ *phagocytosis,* section 4.5C

The **foraminiferans,** or forams, are an ancient group of mostly marine amoeboid protozoa. They have complex, brilliantly colored tests (shells) made primarily of calcium carbonate (**figure 18.17a**). Their populations are immense: About one third of the ocean floor is made of the shells of the marine foram *Globigerina.* England's White Cliffs of Dover, among other limestone and chalk deposits, are made largely of the tests of forams and other marine organisms that have been lifted out of the water by geological forces. Paleontologists studying extinct forams have learned which species correlate with oil and gas deposits. The shells are also useful in dating rock strata.

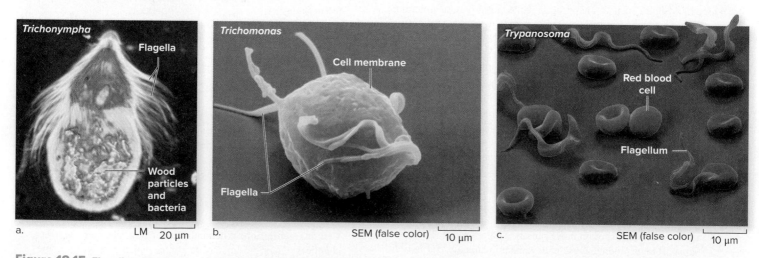

Figure 18.15 Flagellated Protozoa. (a) *Trichonympha* lives inside the intestines of termites. Note the fringe of flagella. (b) *Trichomonas vaginalis* causes the sexually transmitted disease trichomoniasis. These organisms have multiple flagella. (c) A blood smear from a patient with African sleeping sickness reveals *Trypanosoma brucei* (*purple*) among the blood cells (*red*). Each trypanosome features a single flagellum.

(a): ©Eric V. Grave/Science Source; (b, c): ©Eye of Science/Science Source

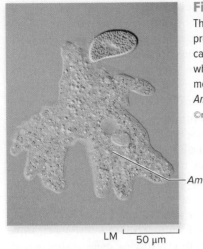

Figure 18.16 Amoeba.
The cells of amoeboid organisms produce temporary projections called pseudopodia ("false feet"), which enable the organism to move or take in food. This *Amoeba* is consuming a ciliate.
©micro_photo/iStock/Getty Images RF

Amoeba

LM   50 µm

The **radiolarians** are among the oldest protozoa. They are planktonic organisms with intricate tests made of silica (figure 18.17b); pseudopodia extend through holes in the shells. "Radiolarian ooze" is sediment consisting of large numbers of their tests. On the ocean floor, radiolarian ooze can be as thick as 4000 meters.

C. Ciliates Are Common Protozoa with Complex Cells

The **ciliates** are complex, mostly unicellular protists characterized by abundant hairlike cilia (**figure 18.18**). The cilia have multiple functions. Waves of moving cilia propel the organism through the water. Cilia also sweep bacteria, algae, and other ciliates into the cell's oral groove. ⓘ *cilia,* section 3.5B

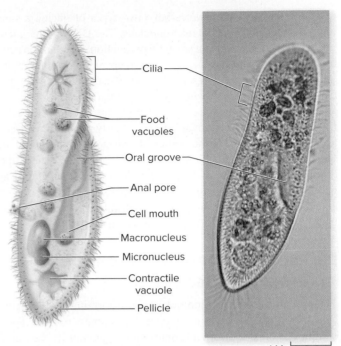

Cilia
Food vacuoles
Oral groove
Anal pore
Cell mouth
Macronucleus
Micronucleus
Contractile vacuole
Pellicle

LM   25 µm

Figure 18.18 Ciliate. Numerous hairlike cilia on the cell's exterior give ciliates their name. This is *Paramecium.*
Photo: ©Nancy Nehring/Getty Images RF

Ciliate cells have other distinctive features as well. A food vacuole surrounds and transports each captured meal inside the cell, and a permanent anal pore releases the wastes. In freshwater habitats, water may enter the cell by osmosis. An organelle called a contractile vacuole helps maintain water balance by pumping the excess fluid out of the cell. ⓘ *osmosis,* section 4.5A

Test (shell)
LM   200 µm
Pseudopodia
a.

b.   LM   500 µm

Figure 18.17 Foraminiferans and Radiolarians. (a) Thin threads of cytoplasm extend from the shell of this foram. The White Cliffs of Dover are composed largely of foram tests. (b) These intricate silica shells are the remains of radiolarians.
(a, cliffs): ©Dave Carr/Flickr/Getty Images RF; (a, foram): ©Peter Parks/Newscom; (b): ©Eric V. Grave/Science Source

In addition, many ciliates have two types of nuclei, a small micronucleus and a larger macronucleus. The DNA in the micronucleus is passed on during sexual reproduction, whereas the genes in the macronucleus have metabolic and developmental functions.

The habitats of ciliates are diverse. Most are free-living, motile cells such as *Paramecium* and its predator, *Didinium*. Several other ciliate species, such as *Stentor*, are sessile or attached forms living on a variety of substrates. Nearly one third of ciliates are symbiotic, living in the bodies of crustaceans, mollusks, and vertebrates. Some inhabit the stomachs of cattle, where they house bacteria that break down the cellulose in grass. Others are parasites. *Ichthyophthirius multifilis*, for example, is familiar to aquarium owners as the cause of a freshwater fish disease called "ich." Symptoms include white spots on the skin and gills.

D. Apicomplexans Include Nonmotile Animal Parasites

The **apicomplexans** are nonmotile, spore-forming, internal parasites of animals. The name *apicomplexa* comes from the apical complex, a cluster of microtubules and organelles at one end of the cell. This structure, visible only with an electron microscope, apparently helps the parasite attach to and invade host cells.

Apicomplexans include several organisms that cause illness. This chapter's Apply It Now box describes *Cryptosporidium*, a genus containing several species that cause waterborne disease. Another example is *Toxoplasma gondii*, a protist that infects cats and other mammals. A person who handles feces from infected cats can accidentally ingest *Toxoplasma* cysts. The resulting infection may remain symptomless or develop into an illness called toxoplasmosis, especially in people with weakened immune systems. In the most severe cases, the parasite can damage the brain and eyes. The infection can also pass to a fetus, which is why pregnant women should avoid cat litter boxes.

Malaria is another example of a human illness caused by an apicomplexan. Four species of *Plasmodium* cause mosquito-borne malaria. (*Plasmodium*, a genus of apicomplexans, is not to be confused with the plasmodium produced by some slime molds.) The life cycle is complex, involving many stages in multiple hosts and including both asexual and sexual reproduction (**figure 18.19**).

A cycle of malaria begins when an infected mosquito of any of 60 *Anopheles* species feeds on human blood. The insect's saliva transmits small haploid cells called sporozoites to the human host (figure 18.19, step 1). The sporozoites travel in the bloodstream (step 2) and enter the liver cells, where they multiply

Apply It **Now**

Don't Drink That Water

The sign says "Please shower before entering pool." What is the purpose of that request? Isn't showering a wasteful prelude to taking a refreshing plunge?

The truth is that the preswim shower is an important public health measure. Washing thoroughly with soap and hot water helps eliminate harmful microbes before they have a chance to contaminate the pool's water.

Cryptosporidium, or "crypto," is one example of a contagious microorganism that spreads easily in water. This apicomplexan protist lives in the intestinal tracts of infected humans, entering the body through the mouth and exiting in feces. It produces tough-walled cysts that can survive for days in the chlorinated water of public pools, water parks, splash pads, and other places where people gather to play in the water.

Even tiny amounts of feces, invisible to the unaided eye, can contaminate water with *Cryptosporidium* cysts, triggering an outbreak. Swimmers who accidentally swallow the tainted water typically become ill within a week, as cells released from the cysts invade the lining of the digestive tract. Symptoms include diarrhea, cramps, fever, vomiting, and dehydration.

Crypto outbreaks periodically occur in communities throughout the United States. One notable episode occurred in Milwaukee, Wisconsin, in 1993. A malfunctioning water treatment plant distributed contaminated water to hundreds of thousands of households, sickening about 25% of Milwaukee's population and killing 100 people.

Cryptosporidium cysts

LM 5 μm

Scattered small outbreaks also have occurred in pools and other recreational water venues, where the victims are often young children who ingest the water. Simple preventive measures include keeping children with diarrhea out of the water, washing hands thoroughly after using the toilet or changing diapers, and—as the sign says—showering before entering the pool.

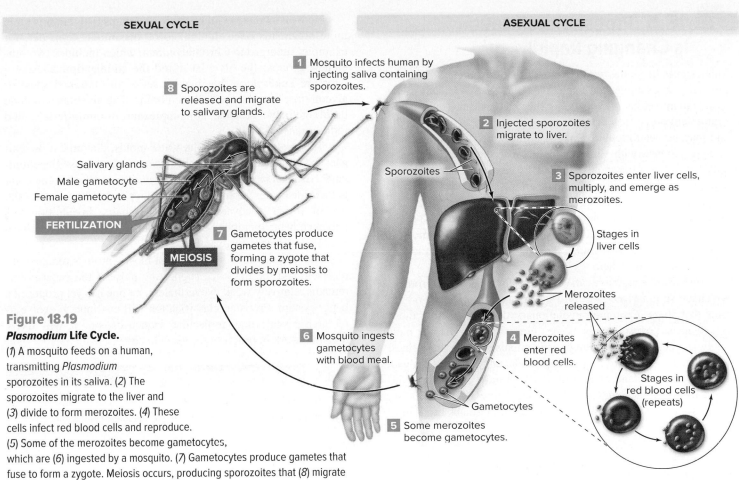

SEXUAL CYCLE

ASEXUAL CYCLE

1 Mosquito infects human by injecting saliva containing sporozoites.

8 Sporozoites are released and migrate to salivary glands.

Salivary glands
Male gametocyte
Female gametocyte

FERTILIZATION

MEIOSIS

7 Gametocytes produce gametes that fuse, forming a zygote that divides by meiosis to form sporozoites.

6 Mosquito ingests gametocytes with blood meal.

2 Injected sporozoites migrate to liver.

Sporozoites

3 Sporozoites enter liver cells, multiply, and emerge as merozoites.

Stages in liver cells

Merozoites released

4 Merozoites enter red blood cells.

Stages in red blood cells (repeats)

Gametocytes

5 Some merozoites become gametocytes.

Figure 18.19

***Plasmodium* Life Cycle.**

(*1*) A mosquito feeds on a human, transmitting *Plasmodium* sporozoites in its saliva. (*2*) The sporozoites migrate to the liver and (*3*) divide to form merozoites. (*4*) These cells infect red blood cells and reproduce. (*5*) Some of the merozoites become gametocytes, which are (*6*) ingested by a mosquito. (*7*) Gametocytes produce gametes that fuse to form a zygote. Meiosis occurs, producing sporozoites that (*8*) migrate to the mosquito's salivary glands.

rapidly (step 3). Eventually, the cells emerge as merozoites, which continue the infection within the human host. Some merozoites reproduce in red blood cells; every 48 to 72 hours, they burst out and infect other red blood cells (step 4). This release is often synchronized throughout the victim's body, causing the characteristic recurrent chills and fever of malaria.

Other merozoites become specialized as male and female sexual forms called gametocytes (step 5). When mosquitoes ingest the gametocytes from an infected person's blood (step 6), the cells produce gametes that unite in the insect's stomach (step 7). After several additional steps, sporozoites form. These move to the mosquito's salivary glands (step 8), ready to enter a new host when the insect seeks its next blood meal.

Around 214 million people a year suffer from malaria, with nearly 90% of them in sub-Saharan Africa. Globally, about 438,000 people die each year from this disease. Even survivors may retain dormant forms of the parasite, becoming ill again months or years after the initial infection.

Despite decades of research, malaria continues to be the world's most significant infectious disease. No effective vaccine exists, and nations troubled by poverty and civil unrest struggle to distribute drugs to prevent or treat malaria. Moreover, *Plasmodium* continues to develop resistance to drugs that were

effective in the past. Malaria prevention efforts therefore focus on repelling and killing mosquitoes (even as the use of insecticides has selected for resistance in mosquitoes). The careful use of antimalarial drugs, alone and in combination, helps prevent and treat malaria and reduces selection for resistant parasites.

Not everyone is equally susceptible to malaria. People with one copy of the recessive sickle cell allele are much less likely to contract malaria than are people with two dominant alleles. In areas of the world where malaria is endemic, human populations have a relatively high incidence of the sickle cell allele. In malaria-free areas, the sickle cell allele is much rarer. This pattern illustrates the selective force that malaria exerts on the human population. ⓘ *sickle cell mutation,* section 7.7A; *balanced polymorphism,* section 12.5

18.4 MASTERING CONCEPTS

1. What mode of nutrition do protozoa use?
2. What are the characteristics of each major group of protozoa?
3. List three diseases caused by flagellated protozoa.
4. Compare and contrast amoebae, foraminiferans, and radiolarians.
5. How do ciliates move and eat?
6. What are the distinguishing characteristics of apicomplexans?

18.5 Protist Classification Is Changing Rapidly

This chapter illustrates some of the difficulties in classifying the protists. For example, the euglenoids and dinoflagellates could easily fall into either of two groups: the algae (because they photosynthesize) or the flagellated protozoa (because they have flagella and may be heterotrophic). Likewise, the water molds are traditionally grouped with slime molds because they share a habitat with fungi, but water molds are actually closely related to brown algae. Clearly, the traditional scheme groups unrelated organisms.

New research based on genetic sequences is helping to assign each species into a lineage with its closest relatives. Nevertheless, biologists have not yet firmly established the number of taxonomic groups, their ranks, their names, or the evolutionary relationships among them.

One scheme organizes all eukaryotes into six "supergroups," which are summarized in table 18.1. A supergroup unites organisms that share a common evolutionary lineage, whether those organisms are microorganisms (i.e., most of the protists) or multicellular (i.e., the plants, fungi, and animals). Along these lines, note that the supergroup Archaeplastida unites the red algae, green algae, and land plants, all of which have chloroplasts derived from primary endosymbiosis (see figure 18.2). Likewise, supergroup Opisthokonta includes fungi, animals, and their common ancestor.

Note also that several of the supergroups listed in this table unite organisms once thought to be dissimilar. Consider, for example, supergroup Chromalveolata, which includes two other new groups: the alveolates and the stramenopiles. Alveolates are eukaryotes that have a series of flattened sacs, or alveoli, just beneath the cell membrane. The alveolates include dinoflagellates, ciliates, apicomplexans, foraminiferans, and radiolarians.

The stramenopiles are the water molds, diatoms, brown algae, and golden algae. The word *stramenopile* means "flagellum-hair" (*stramen* = straw or flagellum; *pilos* = hair). At some point in their life cycles, stramenopiles produce cells with two flagella, one of which is covered with tubular hairs. In addition, the photosynthetic members of this group produce the yellowish accessory pigment fucoxanthin.

The placement of many—if not most—protists remains unresolved, and the evidence supporting most of the supergroups remains relatively weak. Nevertheless, no one has yet proposed a better system. Protistan classification will continue to evolve as research reveals new molecular sequences, but it will likely remain a work in progress for years to come.

18.5 MASTERING CONCEPTS

1. How have molecular sequences changed protist classification?
2. What features unite some of the major lineages of eukaryotes?

TABLE **18.1** Proposed Eukaryotic "Supergroups": A Summary

Supergroup	Distinguishing Features	Examples
Archaeplastida (a)	Photosynthetic eukaryotes with chloroplasts derived from primary endosymbiosis	Red algae, green algae, land plants
Opisthokonta (b)	Motile cells with one flagellum	Choanoflagellates, animals, fungi
Chromalveolata (c)	Chloroplasts (if present) are derived from secondary endosymbiosis	
Alveolates	Flattened sacs (alveoli) beneath cell membrane	Dinoflagellates, apicomplexans, ciliates (including *Paramecium*)
Stramenopiles	Motile cells with two flagella, one of which has tubular hairs; fucoxanthin is an accessory pigment in photosynthetic forms	Water molds, diatoms, brown algae, golden algae
Amoebozoa (d)	Amoeboid movement via pseudopodia; feed by phagocytosis; slime molds form spores	*Amoeba*, many slime molds (including *Physarum* and *Dictyostelium*)
Excavata (e)	Unicellular, flagellated protists; may lack mitochondria; photosynthetic or parasitic; chloroplasts (when present) are derived from secondary endosymbiosis	*Trichomonas, Trichonympha, Giardia, Euglena,* trypanosomes
Rhizaria (f)	Amoeboid movement; many produce shells	Radiolarians, foraminiferans

Red alga — a.
Choanoflagellate — b.
Paramecium — c.
Brown alga — d. *Amoeba*
Giardia — e.
Euglena
Foraminiferan — f.

INVESTIGATING LIFE

18.6 Shining a Spotlight on Danger

As waves break after sunset near La Jolla, California, blue-green light explodes from millions of algae (figure 18.20). It is a beautiful sight that is not limited to this southern California beach—bioluminescent protists are common throughout the world's oceans. But their glow, however awe-inspiring, did not evolve for the curious eyes of humans. Why has evolution selected some algae to release light when they are disturbed?

Bioluminescence, or the production of light by an organism, has been studied in dinoflagellates for several decades. In that time, many research groups have tested hypotheses about how and why these algae emit light when agitated. For example, some proposed that the light startles small herbivores called copepods, which graze on dinoflagellates at night. But that does not explain why copepods might avoid the light, which is harmless. After all, natural selection should favor bold grazers that take a meal whether it is flashing or not.

Researchers Mark Abrahams and Linda Townsend of the Pacific Biological Station in Canada hoped to shed light on the evolution of protist bioluminescence. They hypothesized that emitting light is adaptive to dinoflagellates because it attracts the predators of the copepods (figure 18.21a). If so, then copepods grazing on bioluminescent dinoflagellates should face a higher risk of predation than do copepods given only non-bioluminescent dinoflagellates to eat.

Abrahams and Townsend collected a dinoflagellate species called *Gonyaulax polyedra*, which is native to the Pacific coast of North America. The researchers conducted their experiment in 20 large jars in the laboratory. Each jar contained 500 copepods and one stickleback fish, a copepod predator. Half of the jars received dinoflagellates that would bioluminesce when copepods approached. The other 10 jars served as controls—each contained dinoflagellates that would not emit light. The experimenters then darkened the room, allowing copepods to graze on dinoflagellates and sticklebacks to prey on copepods. After 3.5 hours, they counted the remaining copepods.

Figure 18.20 Blue Glow of Algae. A man and his son admire the neon blue light emitted by algae inhabiting coastal waters near La Jolla.
©North County Times/Newscom

The results aligned with the predictions: Stickleback fish eat more copepods when bioluminescent dinoflagellates are present (figure 18.21b). This outcome suggests that a protist can avoid being eaten by increasing the threat of predation on its grazers; the copepod benefits more from fleeing than from continuing to graze. Since light-emitting protists are less likely to be a copepod's dinner, natural selection maintains bioluminescence in dinoflagellates as well.

If you ever visit La Jolla, drop a tiny pebble in the water. You'll see sparkles of light from the protists you disturbed—a beacon to predators that a grazer may be nearby. And if you look closely, you might also see a fish looking for a copepod dinner.

Source: Abrahams, Mark V., and Linda D. Townsend. 1993. Bioluminescence in dinoflagellates: A test of the burglar alarm hypothesis. *Ecology,* vol. 74, pages 258–260.

18.6 MASTERING CONCEPTS

1. Use the food chain in figure 18.21a to explain why bioluminescence is adaptive to dinoflagellates.
2. What selects for a copepod's light-avoidance response?

Figure 18.21 Dangerous Light. (a) A simplified experimental food chain. (b) Copepods sharing a jar with light-emitting dinoflagellates were more likely to be eaten by stickleback fish than were copepods in a jar with dinoflagellates that did not produce light.

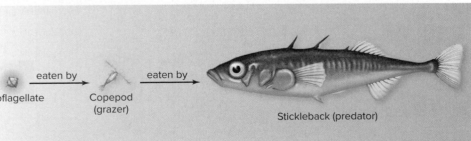

Dinoflagellate → eaten by → Copepod (grazer) → eaten by → Stickleback (predator)

a.

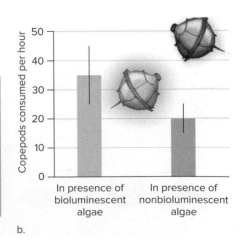

Copepods consumed per hour

In presence of bioluminescent algae | In presence of nonbioluminescent algae

b.

CHAPTER SUMMARY

18.1 Protists Lie at the Crossroads Between Simple and Complex Organisms

A. What Is a Protist?
- **Protists** are **eukaryotes** that are not plants, fungi, or animals. The protists are diverse in ecology, motility, and other traits (**table 18.2**).
- Classification of protists is changing as molecular data are considered.

B. Protists Are Important in Many Ways
- Photosynthetic protists support food webs and release oxygen, whereas parasitic protists cause disease. Some protists have industrial uses.

C. Protists Have a Lengthy Evolutionary History
- According to the **endosymbiont theory,** mitochondria and chloroplasts originated as prokaryotes that were engulfed by other cells. Subsequent secondary endosymbiosis events explain differences in chloroplasts among the photosynthetic protists.
- Some protists may provide clues to the origins of multicellularity and to the ancestors of the plant, fungi, and animal lineages.

18.2 Algae Are Photosynthetic Protists

- The photosynthetic protists are known as **algae.**

A. Euglenoids Are Heterotrophs and Autotrophs
- **Euglenoids** are unicellular, elongated flagellates.

B. Dinoflagellates Are "Whirling Cells"
- The **dinoflagellates** have two different-sized flagella at right angles that generate a whirling movement. Dinoflagellates cause red tides.

C. Golden Algae, Diatoms, and Brown Algae Contain Yellowish Pigments
- The **golden algae** are photosynthetic but can consume other microorganisms when light or nutrient supplies decline.
- **Diatoms** are microscopic phytoplankton with intricate silica shells.
- **Brown algae** are large, multicellular seaweeds such as kelp.

D. Red Algae Can Live in Deep Water
- **Red algae** contain pigments that expand their photosynthetic range.

E. Green Algae Are the Closest Relatives of Land Plants
- **Green algae** store carbohydrates as starch and use the same pigments as plants. Many have **alternation of generations.** The green algae have diverse body forms, ranging from microscopic to multicellular.

18.3 Some Heterotrophic Protists Resemble Fungi

A. Slime Molds Are Unicellular and Multicellular
- **Plasmodial slime molds** form plasmodia, oozing networks containing thousands of diploid nuclei. A plasmodium feeds by engulfing other cells.
- In **cellular slime molds,** individual cells retain their separate cell membranes throughout the life cycle.

B. Water Molds Are Decomposers and Parasites
- **Water molds** are filamentous heterotrophs that live in moist or wet environments. Some are pathogens and others are decomposers.

18.4 Protozoa Are Diverse Heterotrophic Protists

- Most **protozoa** are heterotrophs, and most have motile cells.

A. Several Flagellated Protozoa Cause Disease
- Several species of **flagellated protozoa,** such as *Giardia, Trichomonas,* the **trypanosomes,** and *Leishmania,* cause disease in humans.

TABLE 18.2   The Protists: A Summary

Group	Autotrophic (Photosynthetic)?	Heterotrophic?	Parasitic?	Motile?	Supergroup
Euglenoids (a)	Some	Yes	Some	Yes	Excavata
Dinoflagellates (b)	Some	Some	Some	Yes	Chromalveolata
Golden algae (c)	Yes	Yes	No	Yes	Chromalveolata
Diatoms (d)	Yes	No	No	No	Chromalveolata
Brown algae (e)	Yes	No	No	Gametes only	Chromalveolata
Red algae (f)	Yes	No	No	No	Archaeplastida
Green algae (g)	Yes	No	No	Some gametes	Archaeplastida
Slime molds (h)	No	Yes	No	Yes	Amoebozoa, Excavata, and Rhizaria
Water molds (i)	No	Yes	Some	Spores only	Chromalveolata
Flagellated protozoa (j)	Some	Most	Many	Yes	Excavata
Amoeboid protozoa (k)	No	Yes	Some	Some	Rhizaria, Amoebozoa, and Chromalveolata
Ciliates (l)	No	Yes	Some	Some	Chromalveolata
Apicomplexans (m)	No	Yes	Yes	No	Chromalveolata

a.   b.   c.   d.   e.   f.   g.   h.   i.   j.   k.   l.   m.

B. Amoeboid Protozoa Produce Pseudopodia
- **Amoeboid protozoa** move by means of "false feet," or pseudopodia. This group includes amoebae, **foraminiferans,** and **radiolarians.**

C. Ciliates Are Common Protozoa with Complex Cells
- **Ciliates** have complex cells with cilia, vacuoles, and two types of nuclei.

D. Apicomplexans Include Nonmotile Animal Parasites
- **Apicomplexans** are obligate parasites characterized by an apical complex of organelles. These protists cause malaria and toxoplasmosis.

18.5 Protist Classification Is Changing Rapidly
- The traditional means of classifying protists is giving way to newer eukaryote "supergroups" that reflect shared ancestry.
- The relationships among most lineages of protists remain unclear.

18.6 Investigating Life: Shining a Spotlight on Danger
- Light produced by bioluminescent dinoflagellates may attract stickleback fish, which eat the copepods that would otherwise graze on the dinoflagellates.

MULTIPLE CHOICE QUESTIONS

1. Which of the following is NOT a characteristic of all protists?
   a. Unicellular
   b. Cells containing membrane-bounded organelles
   c. Cells containing a nucleus
   d. Eukaryotic

2. Suppose you are studying a protist under a microscope. If it is a known species of protist, then it could be
   a. a parasite.
   b. a decomposer.
   c. photosynthetic.
   d. All of the above are correct.

3. Some protist lineages arose from secondary endosymbiosis. How many membranes would surround the chloroplasts of these organisms?
   a. 0   c. 2
   b. 1   d. 3 or more

4. Which protist group is mismatched with its description?
   a. Algae: photosynthetic autotrophs that live in water
   b. Protozoa: decomposers that form threadlike filaments in damp soil
   c. Water molds: funguslike protists that have swimming spores
   d. Plasmodial slime molds: masses of cells that behave as multicellular organisms

5. Why are DNA sequences useful in the classification of protists?
   a. Because only protists have DNA
   b. Because genetic sequences have confirmed the traditional categories of protists
   c. Because DNA reveals evolutionary relationships, even among organisms that look different
   d. All of the above are correct.

Answers to these questions are in appendix A.

WRITE IT OUT

1. Explain why evolutionary biologists are interested in choanoflagellates, green algae, and organisms with mitochondria whose genomes resemble those of bacteria.

2. The amoeba *Pelomyxa palustris* is a single-celled eukaryote with no mitochondria, but it contains symbiotic bacteria that can live in the presence of oxygen. How does this observation support the endosymbiont theory?

3. Describe the relationship between nutrient pollution and harmful algal blooms. Why might harmful algal blooms be more frequent in summer? What steps could coastal communities take to prevent nutrient pollution?

4. Explain why the fossil record for diatoms is much more complete than that of other protists, such as amoebae and slime molds.

5. How is it adaptive for a red alga to have pigments other than chlorophyll?

6. How are kelp similar to trees? How are they different?

7. Natural selection favors stalk formation in cellular slime molds even though the cells of the stalk die. Explain this observation.

8. Why might overwatering your plants make them more susceptible to infection by some kinds of heterotrophic protists?

9. Give three examples of protists for which the classifications have recently changed. In each case, what was the justification for the old category, and what is the justification for the change?

10. Suppose someone hands you a microscope and a single-celled organism. Create a flow chart that you could use to identify the specimen.

PULL IT TOGETHER

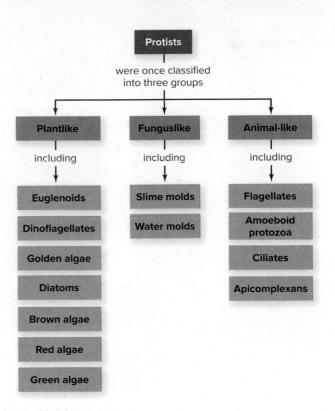

Figure 18.22 Pull It Together: Protists.

Refer to figure 18.22 and the chapter content to answer the following questions.

1. Review the Survey the Landscape figure in the chapter introduction, and then add *evolution, domain Eukarya, plants, fungi,* and *animals* to the Pull It Together concept map.

2. Name at least one unique characteristic of each protist group in figure 18.22.

3. Molecular data have changed protist classification. Use a red pen to circle the protist groups classified as stramenopiles; use a blue pen to circle the groups classified as alveolates.

Natural Biofuel. *Sphagnum,* the source of peat moss, grows in bogs. This man is cutting peat that will be used as fuel.

©G. R. "Dick" Roberts/Natural Sciences Image Library

LEARN HOW TO LEARN
Take Notes on Your Reading

Many classes have reading assignments. Taking notes as you read should help you not only retain information but also identify what you don't understand. Before you take notes, skim through the assigned pages once; otherwise, you may have trouble distinguishing between main points and minor details. Then read them again. This time, pause after each section and write the most important ideas in your own words. What if you can't remember the material or don't understand it well enough to summarize the passage? Read it again, and if that doesn't work, ask for help with whatever isn't clear.

Peat Moss, Pot Scrubbers, and Drugs: The Many Uses of Plants

Most people know why plants are essential to animal life: Vegetation provides food and habitat, and photosynthesis produces oxygen gas. Yet plants serve us in many unexpected ways as well. Here are some interesting examples from the four main plant lineages.

- **Peat moss:** Gardeners and houseplant lovers recognize peat moss as a major ingredient in potting mixes. Peat comes from partially decomposed sphagnum moss harvested from enormous bogs. The dried moss is unusually spongy, absorbing 20 times its weight in water. When mixed with soil, peat slowly releases water to plant roots. People also burn peat as cooking fuel or to generate electricity.

- **Horsetail:** *Equisetum,* or the horsetail, is a seedless plant related to ferns. Some horsetails are called "scouring rushes" because their stems and leaves contain abrasive silica particles. Native Americans used horsetails to polish bows and arrows, and early colonists and pioneers used them to scrub pots and pans. Some people believe that horsetails, taken as a dietary supplement, can strengthen fingernails and prevent osteoporosis.

- **Pacific yew:** *Taxus brevifolia,* the Pacific yew, is a conifer that contains the compound paclitaxel in its bark. Paclitaxel has anticancer properties, particularly for treating breast cancer. But Pacific yews are slow-growing trees, and harvesting them for their bark would mean their extinction. Fortunately, paclitaxel is now synthesized in the laboratory and marketed under the trade name Taxol.

- **Cotton:** You probably own many garments made of cotton, a cloth that comes from flowering plants in genus *Gossypium.* Cotton seeds develop in a dense web of cellulose fibers, which textile manufacturers spin into threads that make up T-shirts, blue jeans, underwear, towels, sheets, and many other cloth products. The seeds themselves also produce cooking oil. More than 90% of the U.S. cotton crop is genetically modified to produce its own insecticides (see section 10.10), to resist herbicides, or both.

Peat moss, horsetail, the Pacific yew, and cotton are just four of the many plants that humans use, and they represent only a tiny percentage of the diverse kingdom Plantae. This chapter highlights some of the history and diversity of these essential organisms.

LEARNING OUTLINE

19.1 Plants Have Changed the World

19.2 Bryophytes Are the Simplest Plants

19.3 Seedless Vascular Plants Have Xylem and Phloem but No Seeds

19.4 Gymnosperms Are "Naked Seed" Plants

19.5 Angiosperms Produce Seeds in Fruits

19.6 Investigating Life: Genetic Messages from Ancient Ecosystems

SURVEY THE LANDSCAPE
The Diversity of Life

Evolution

over billions of years has produced

Life's diversity

includes

Domain Bacteria   Domain Archaea   Domain Eukarya

are   *includes*

Prokaryotes Protists **Plants** Fungi Animals

infect

Viruses

Nearly everywhere on land, plants produce the carbohydrates and oxygen that sustain entire food webs. The ancestors of plants lived in water, but natural selection gradually selected for adaptations that enabled them to dominate terrestrial ecosystems.

For more details, study the Pull It Together feature in the chapter summary.

19.1 Plants Have Changed the World

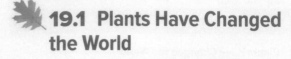

If you glance at your surroundings in almost any outdoor setting, the plants are the first things you see. Grasses, trees, shrubs, ferns, or mosses exist nearly everywhere, at least on land (figure 19.1). Members of kingdom **Plantae** dominate habitats from moist bogs to parched deserts. They are so familiar that it is difficult to imagine a world without plants.

Plants are autotrophs: Like the cyanobacteria and algae described in chapters 17 and 18, they use sunlight as an energy source to assemble CO_2 and H_2O into sugars (chapter 5 describes the reactions of photosynthesis). The sugars, in turn, provide the energy and raw materials that maintain and build a plant's body.

Moreover, photosynthesis releases oxygen gas, O_2, as a waste product. The evolution of photosynthesis billions of years ago set into motion a complex series of changes that would profoundly affect both the nonliving and the living worlds. The explosion of photosynthetic activity altered the atmosphere, gradually lowering

Figure 19.1 Plants Galore. Nearly every ecosystem on land is dominated by plants.

(snow): ©Design Pics/Carson Ganci/Getty Images RF; (prairie): ©Tetra Images/Tetra Images/Corbis RF; (forest): ©Ted Mead/Getty Images RF

CO_2 levels and raising O_2 content (see figure 15.6). Animals and other organisms that use aerobic respiration need this gas; in addition, O_2 in the atmosphere helps form the ozone layer that protects life from the sun's harmful ultraviolet radiation.

Hundreds of millions of years ago, plants emerged from water and transformed the terrestrial landscape. As plants gradually expanded from the water's edge to the world's driest habitats, they formed the bases of intricate food webs, providing diverse habitats for many types of animals, fungi, and microbes.

Of course, plants remain essential to life today. Herbivores consume living leaves, stems, roots, seeds, and fruits. Dead leaves accumulating on the soil surface feed countless soil microorganisms, insects, and worms. When washed into streams and rivers, this leaf litter supports a spectacular assortment of fishes and other aquatic animals.

From a human perspective, farms and forests provide the foods we eat, the paper we read, the lumber we use to build our homes, many of the clothes we wear, and some of the fuel we burn. (This chapter's Burning Question introduces biofuels derived from plants.) The list goes on and on. It is amazing to think that plants do so much with such modest raw materials: sunlight, water, minerals, and CO_2.

A. Green Algae Are the Closest Relatives of Plants

All of the plants listed in table 19.1, from mosses to maple trees, are multicellular organisms with eukaryotic cells. With the exception of a few parasitic species, plants are autotrophs. A careful reading of section 18.2, however, will reveal that some algae have the same combination of traits. Which of the many lineages of algae gave rise to plants?

The answer is that green algae apparently share the most recent common ancestor with plants. About 475 million years ago, or perhaps earlier, one group of green algae related to today's

TABLE 19.1 Phyla of Plants

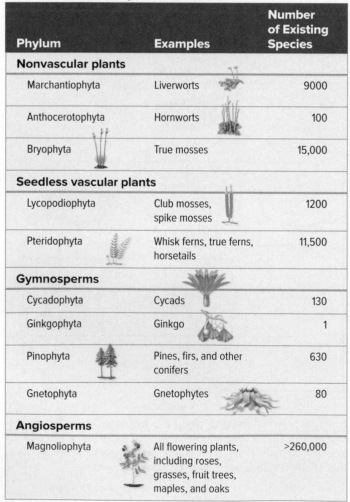

Phylum	Examples	Number of Existing Species
Nonvascular plants		
Marchantiophyta	Liverworts	9000
Anthocerotophyta	Hornworts	100
Bryophyta	True mosses	15,000
Seedless vascular plants		
Lycopodiophyta	Club mosses, spike mosses	1200
Pteridophyta	Whisk ferns, true ferns, horsetails	11,500
Gymnosperms		
Cycadophyta	Cycads	130
Ginkgophyta	Ginkgo	1
Pinophyta	Pines, firs, and other conifers	630
Gnetophyta	Gnetophytes	80
Angiosperms		
Magnoliophyta	All flowering plants, including roses, grasses, fruit trees, maples, and oaks	>260,000

Figure 19.2 Charophyte.
This green alga, called *Chara,* may resemble the ancestors of land plants.

©Paulo de Oliveira/Newscom

LM 2 mm

algae contain the same photosynthetic pigments. In addition, like green algae, plants have cellulose-rich cell walls and use starch as a nutrient reserve. Similar DNA sequences offer additional evidence of a close relationship. ⓘ *green algae,* section 18.2E; *polysaccharides,* section 2.5B

Nevertheless, the body forms of algae are quite different from those of plants, in part because water presents selective forces that are far different from those in the terrestrial landscape. Consider the aquatic habitat. Light, water, minerals, and dissolved gases surround the whole body of a submerged green alga, and the buoyancy of water provides physical support. In sexual reproduction, an alga simply releases swimming gametes into the water.

On land, water and minerals are in the soil, and only the aboveground part of the plant is exposed to light. Air provides much less physical support than does water, and it dries out the plant's stem and leaves. Furthermore, the dispersal of gametes for sexual reproduction becomes more complicated on dry land.

These conditions have selected for unique adaptations in the body forms and reproductive strategies of plants. As described in the next section, biologists use some of these features to organize land plants into four main groups (**figure 19.3**): the bryophytes, seedless vascular plants, gymnosperms, and angiosperms.

charophytes likely gave rise to plants (**figure 19.2**). Evidence for this evolutionary connection includes chemical and structural similarities. For example, the chloroplasts of plants and green

Figure 19.3 Plant Diversity.
Biologists classify plants according to the presence or absence of vascular tissue, seeds, flowers, and fruits.

Photo: USDA/Keith Weller

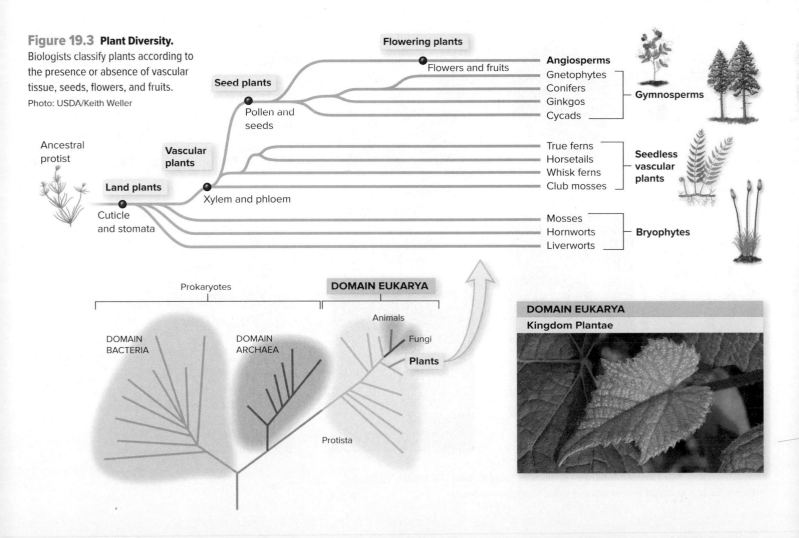

B. Plants Are Adapted to Life on Land

Figure 19.4 illustrates many of the adaptations that enable plants to produce food, grow upright, retain moisture, survive, and reproduce on land. Refer to this figure often as you read this section.

Obtaining Resources   To carry out photosynthesis, plants need light, CO_2, water, and minerals. Aboveground stems support leaves, which capture sunlight and CO_2. Below the ground surface, highly branched root systems not only absorb water and minerals but also anchor the plant in the soil.

A plant that dries out will not survive. After all, water is a critical ingredient in photosynthesis. Water also exerts turgor pressure on cell walls, enabling the plant to stay upright and grow. One water-conserving adaptation is the **cuticle,** a waxy coating that minimizes water loss from the aerial parts of a plant. Dry habitats such as deserts select for extra-thick cuticles; plants in moist habitats typically have thin cuticles. ⓘ *turgor pressure,* section 4.5A

The waxy cuticle is impermeable not only to water but also to gases such as CO_2 and O_2. Plants exchange these gases with the atmosphere through **stomata,** which are pores in the epidermis of stems and leaves. Two guard cells surround each stoma and control whether the pore is open or closed. Water also escapes from the plant's tissues through open stomata. Plants close their stomata in dry weather, minimizing water loss. ⓘ *plant epidermis,* section 22.2B

Internal Transportation and Support   The division of labor in a plant poses a problem: Roots need the food produced at the leaves, whereas leaves and stems need water and minerals from soil. In bryophytes, cell-to-cell diffusion meets these needs. Other plants have **vascular tissue,** a collection of tubes that transport sugar, water, and minerals throughout the plant.

The two types of vascular tissue are xylem and phloem. **Xylem** conducts water and dissolved minerals from the roots to the leaves. **Phloem** transports sugars produced in photosynthesis to the roots and other nongreen parts of the plant. This internal transportation system has supported the evolution of specialized roots, stems, and leaves, many of which have adaptations that enable plants to exploit extremely dry habitats.

In addition, xylem is rich in **lignin,** a complex polymer that strengthens cell walls. The additional support from lignin means that vascular plants can grow tall and form branches, important adaptations in the intense competition for sunlight.

Reproduction   Plants have a life cycle called **alternation of generations,** in which a multicellular diploid stage alternates with a multicellular haploid stage (**figure 19.5**). In the **sporophyte** (diploid) generation, some cells undergo meiosis and produce haploid spores; these **spores** divide mitotically to form the gametophyte. The haploid **gametophyte,** in turn, produces **gametes** by mitotic cell division; these sex cells fuse at

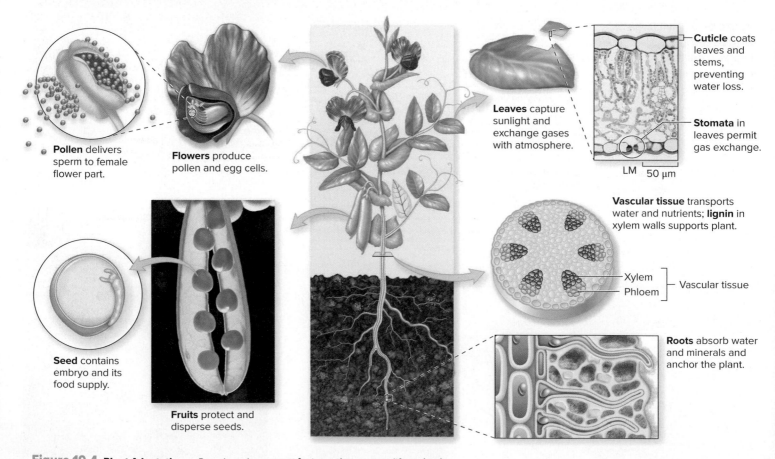

Pollen delivers sperm to female flower part.

Flowers produce pollen and egg cells.

Seed contains embryo and its food supply.

Fruits protect and disperse seeds.

Leaves capture sunlight and exchange gases with atmosphere.

Cuticle coats leaves and stems, preventing water loss.

Stomata in leaves permit gas exchange.

LM   50 μm

Vascular tissue transports water and nutrients; **lignin** in xylem walls supports plant.

Xylem
Phloem — Vascular tissue

Roots absorb water and minerals and anchor the plant.

Figure 19.4 Plant Adaptations. Pea plants have many features that support life on land.

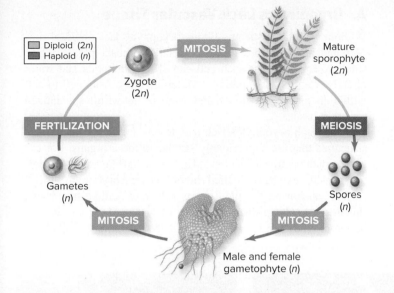

Figure 19.5 Alternation of Generations. Plants have multicellular haploid (gametophyte) and diploid (sporophyte) generations.

fertilization. The sporophyte develops from the resulting zygote, starting the cycle anew.

A prominent trend among land plants is a change in the relative sizes and independence of the gametophyte and sporophyte generations (figure 19.6). In a moss, for example, the gametophyte is the most prominent generation, and the sporophyte depends on it for nutrition. In more complex plants, the sporophyte is photosynthetic and much larger than the gametophyte.

Plant reproduction has other variations as well. The sperm cells of mosses and ferns swim in a film of water to reach an egg, limiting the distance over which gametes can spread. Gymnosperms and angiosperms can reproduce over far greater distances, thanks to pollen. **Pollen** consists of the male gametophytes of seed plants; each pollen grain produces sperm. In **pollination,** wind or animals deliver pollen to female plant parts, eliminating the need for free water in sexual reproduction.

Gymnosperms and angiosperms also share another reproductive adaptation: seeds. A **seed** is a dormant plant embryo packaged with a food supply; a tough outer coat keeps the seed's interior from drying out. The food supply sustains the young plant between the time the seed germinates and when the seedling begins photosynthesis.

The origin of pollen and seeds was a significant event in the evolution of plants. The gametes and spores of mosses and ferns—the seedless plants—take little energy to produce, but they are short-lived and tend to remain close to the mother plant. Gymnosperms and angiosperms, in contrast, use pollen and seeds to disperse over great distances, even in dry conditions. Moreover, seeds can remain dormant for years, germinating when conditions are favorable. These adaptations give gymnosperms and angiosperms a competitive edge in many habitats.

Two additional reproductive adaptations occur only in the angiosperms: flowers and fruits (see figure 19.4). **Flowers** are reproductive structures that produce pollen and egg cells. After fertilization, parts of the flower develop into a **fruit** that contains the seeds. Flowers and fruits help angiosperms protect and disperse both their pollen and their offspring. These adaptations are spectacularly successful; angiosperms far outnumber all other plants, both in numbers and species diversity.

Biologists classify plants based on the presence or absence of transport tissues, seeds, flowers, and fruits. The next four sections of this chapter describe the diversity of plants; unit 5 delves more deeply into the anatomy and physiology of angiosperms.

19.1   MASTERING CONCEPTS

1. What is the evidence that plants are closely related to green algae?
2. How have plants changed the landscape, and how are they vital to life today?
3. How does vascular tissue adapt plants to land?
4. Describe the reproductive adaptations of plants.

	Bryophytes	Seedless Vascular Plants	Gymnosperms	Angiosperms
Gametophyte (haploid generation)		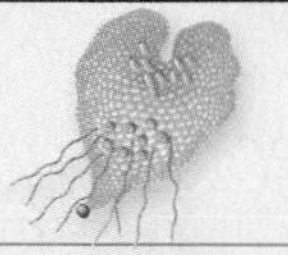	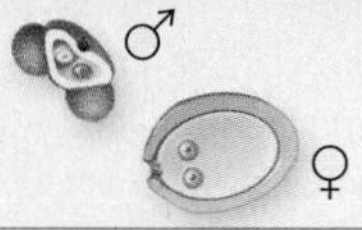	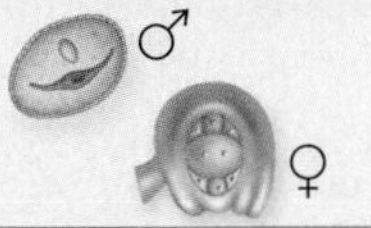
Size relative to sporophyte?	Varies	Small	Microscopic	Microscopic
Depends on sporophyte for nutrition?	No	No	Yes	Yes
Sporophyte (diploid generation)				
Size relative to gametophyte?	Varies	Large	Large	Large
Depends on gametophyte for nutrition?	Yes	No	No	No

Figure 19.6 Changes in the Generations. As plants became more complex, the gametophyte generation was reduced to just a few cells that depend on the sporophyte for nutrition.

19.2 Bryophytes Are the Simplest Plants

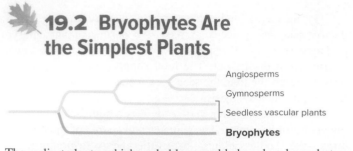

Angiosperms
Gymnosperms
Seedless vascular plants
Bryophytes

The earliest plants, which probably resembled modern bryophytes, emerged onto land during the Ordovician period some 475 million years ago. All **bryophytes** are seedless plants that lack vascular tissue, but evidence suggests that they do not form a single clade (a group that includes one common ancestor and all of its descendants). This section describes them together, however, because they share some important features. ⓘ *clades,* section 14.6C

A. Bryophytes Lack Vascular Tissue

Without vascular tissue and lignin, bryophytes lack physical support and are therefore typically small, compact plants. Their small size means that each cell can absorb minerals and water directly from its surroundings. Materials move from cell to cell within the plant by diffusion and osmosis, not within specialized transport tissues.

Although bryophytes lack true leaves and roots, many have structures that are superficially similar to these organs. For example, photosynthesis occurs at flattened, leaflike areas. In addition, hairlike extensions called rhizoids cover a bryophyte's lower surface, anchoring the plant to its substrate. Unlike true roots, rhizoids cannot tap distant sources of water. Many species are therefore restricted to moist, shady habitats that are unlikely to dry out.

a. Liverwort

b. Hornwort

c. Moss

Figure 19.7 A Gallery of Bryophytes. (a) Umbrella-shaped structures produce sperm or egg cells in these liverworts. (b) Tapered, hornlike sporophytes characterize the hornworts. (c) Moss sporophytes are topped with spore-bearing capsules. (a): ©Dr. Jeremy Burgess/Science Source; (b, c): ©Steven P. Lynch RF

Burning Question

What are biofuels?

Petroleum and other fossil fuels consist of organisms buried millions of years ago; burning these fuels releases their ancient carbon into the atmosphere as CO_2. Biofuels are plant-based substitutes for fossil fuels. They have attracted attention recently, in part because they can help decrease reliance on foreign oil. The biofuel crops also carry out photosynthesis, temporarily removing CO_2 from the atmosphere and helping to reduce global climate change. ⓘ *global climate change,* section 40.4

Two types of biofuels are biodiesel and ethanol. Currently, most biodiesel comes from oil extracted from crushed soybeans or canola seeds. To avoid driving up the price of these food crops, researchers are looking for economical, nonfood sources of biodiesel. Examples include everything from green algae to the seeds of a plant called the jatropha tree. This long-lived tree uses little water and tolerates poor soil, so it does not compete with food plants for rich farmland.

Ethanol, the other main biofuel, is a gasoline substitute. Corn kernels are the main source of ethanol in the United States. Starch extracted from the corn kernel is enzymatically digested to sugar, which is fermented into ethanol. Sugarcane, the main source of ethanol in Brazil, is a more economical alternative in the tropics. Its tissues are high in sugar, not starch, so

biofuel manufacturers can omit the costly enzymes from the ethanol production process. ⓘ *fermentation,* section 6.8B

Researchers are also searching for nonfood sources of sugar to use in ethanol production. The inedible stems of corn or of prairie grasses such as switchgrass would be ideal; bacterial and fungal enzymes easily break the cellulose in the plant cell walls into simple sugars. One problem, however, is that the stems also contain lignin, a complex molecule that interferes with cellulose extraction. So far, the heat and acid treatment needed to eliminate the lignin is too costly and inefficient to make cellulose-derived ethanol economical.

Burning biodiesel or ethanol returns the CO_2 absorbed in photosynthesis to the atmosphere, but it is important to realize that biofuels are not exactly "carbon-neutral." Most biofuel crops require fertilizers and pesticides, both of which come from fossil fuels—and cause additional environmental problems of their own.

Submit your burning question to
Marielle.Hoefnagels@mheducation.com

©Creatas/PunchStock RF

Biologists classify the 24,000 or so species of bryophytes into three phyla (**figure 19.7**):

- **Liverworts** (phylum Marchantiophyta) have a variety of gametophyte forms, from flat to upright and leafy. The liverworts may be the bryophytes most closely related to ancestral land plants.

- **Hornworts** (phylum Anthocerotophyta) are the smallest group of bryophytes, with only about 100 species. They are named for their sporophytes, which are shaped like tapered horns.

- **Mosses** (phylum Bryophyta) are the closest living relatives to the vascular plants. The gametophytes resemble short "stems" with many "leaves."

Bryophytes play important roles in ecosystems. For example, mosses can survive on bare rock or in a very thin layer of soil. As their tissues die, they contribute organic matter, helping build soil that larger plants subsequently colonize.

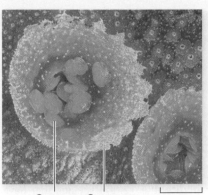

Figure 19.8 Asexual Reproduction. The gametophytes of liverworts can produce gemmae. Raindrops splash these fragments from the cups; each gemma develops into an identical new plant.
©M. I. Walker/Science Source

Gemma   Gemma cup   1.5 mm

B. Bryophytes Have a Conspicuous Gametophyte

Bryophytes reproduce asexually or sexually. Asexual reproduction in liverworts occurs at gemmae, small pieces of tissue that detach and grow into new plants (**figure 19.8**). Mosses also produce gemmae.

The sexual life cycle of a moss is illustrated in **figure 19.9**. The sporophyte is a stalk attached to the gametophyte. At the tip of the stalk, specialized cells inside a sporangium undergo meiosis and produce haploid spores. After the spores are released, they germinate, giving rise to new haploid gametophytes. Gametes form by mitosis in separate sperm- and egg-producing structures on the gametophyte. Sperm swim to the egg cell in a film of water that coats the plants. Sexual reproduction therefore requires water, another factor that limits these plants to moist areas. The sporophyte generation begins at fertilization, with the formation of the diploid zygote. This cell divides mitotically, producing the sporophyte's stalk.

Windblown spores offer the main means of dispersal for bryophytes. Thanks to their tough walls, which protect against drying out, bryophyte spores are preserved in the fossil record. In fact, the earliest evidence of plants on land is a fossilized bryophyte spore.

19.2 | MASTERING CONCEPTS

1. Describe the three main groups of bryophytes.
2. Name two reasons mosses usually live in moist, shady habitats.

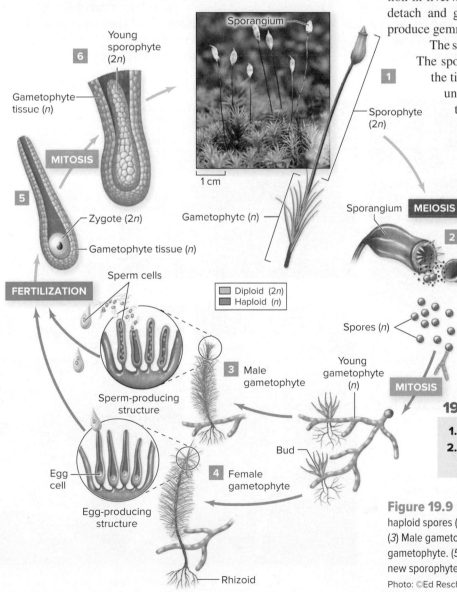

Figure 19.9 Life Cycle of a Moss. (*1*) In the sporophyte, meiosis yields haploid spores (*2*) that develop into the male and female gametophytes. (*3*) Male gametophytes produce sperm that swim to (*4*) egg cells in a female gametophyte. (*5*) Gametes join and form a zygote, which (*6*) develops into a new sporophyte.

Photo: ©Ed Reschke/Photolibrary/Getty Images

19.3 Seedless Vascular Plants Have Xylem and Phloem but No Seeds

Angiosperms

Gymnosperms

Seedless vascular plants

Bryophytes

Most nonvascular plants are small and easily overlooked. Not so for the **seedless vascular plants,** the 12,000 species that have xylem and phloem but do not produce seeds. These plants have much larger representatives than their bryophyte counterparts.

A. Seedless Vascular Plants Include Ferns and Their Close Relatives

The earliest species of seedless vascular plants are extinct, but fossil evidence suggests that they originated in the middle of the Silurian period, about 425 million years ago. The club mosses (not to be confused with the true mosses, which are bryophytes) are descendants of these early vascular plants. The other seedless vascular plants, including the ferns, first appeared about 50 million years later.

As described in section 19.1, vascular tissue enabled plants to grow much larger than nonvascular plants, both in height and in girth. This increase in size was adaptive because taller plants have the edge over their shorter neighbors in the competition for sunlight. Larger plants also triggered evolutionary changes in other organisms by providing new habitats and more diverse food sources for arthropods, vertebrates, and other land animals.

Unlike the bryophytes, the seedless vascular plants typically have true roots, stems, and leaves. In many species, the leaves and roots arise from underground stems called rhizomes. Rhizomes sometimes also store carbohydrates that provide energy for the growth of new leaves and roots.

The seedless vascular plants include two phyla divided into four main lineages (figure 19.10):

- **Club mosses** (phylum Lycopodiophyta) are small plants in genus *Lycopodium*. These plants have simple leaves that resemble scales or needles. The name reflects their club-shaped reproductive structures. Their close relatives are the spike mosses (*Selaginella*). Collectively, club mosses and spike mosses are sometimes called lycopods.

a.

b.

c.

d.

e.

Figure 19.10 A Gallery of Seedless Vascular Plants. (a) The club moss *Lycopodium* produces upright stems. (b) The spike moss, *Selaginella,* has scalelike foliage. (c) Yellowish spore-producing structures are visible on this whisk fern, *Psilotum*. (d) *Equisetum* is a horsetail. (e) This narrow beech fern has the fronds typical of a true fern.

(a): ©ImageBROKER/Alamy RF; (b): ©Howard Rice/Garden Picture Library/Getty Images; (c): ©Biosphoto/Superstock; (d): ©Ed Reschke/Photolibrary/Getty Images; (e): ©Rod Planck/Science Source

- **Whisk ferns** (phylum Pteridophyta) are simple plants that have rhizomes but not roots. Most species have no obvious leaves. Their name comes from the highly branched stems of *Psilotum*, which resemble whisk brooms.

- **Horsetails** (phylum Pteridophyta) grow along streams or at the borders of forests. The only living genus of horsetails, *Equisetum*, includes plants with branched rhizomes that give rise to green aerial stems bearing spores at their tips. Horsetails are also called scouring rushes because their stems and leaves contain abrasive silica particles (see the chapter opening essay).

- **True ferns** (phylum Pteridophyta) make up the largest group of seedless vascular plants, with about 11,000 species. The fronds, or leaves, of ferns are their most obvious feature; some species are popular as ornamental plants. Ferns were especially widespread and abundant during the Carboniferous period, when their huge fronds dominated warm, moist forests. Their remains form most coal deposits (see the Apply It Now box in section 15.3).

Most seedless vascular plants live on land, where their roots and rhizomes help stabilize soil and prevent erosion. Some species of ferns and horsetails are especially adept at colonizing disturbed soils such as road cuts. But not all species are terrestrial. The tiny fern *Azolla* lives in water, where its leaves house cyanobacteria that fix nitrogen. In Asia, farmers cultivate *Azolla* within rice paddies to help fertilize their crops. ⓘ *nitrogen fixation*, section 23.1C

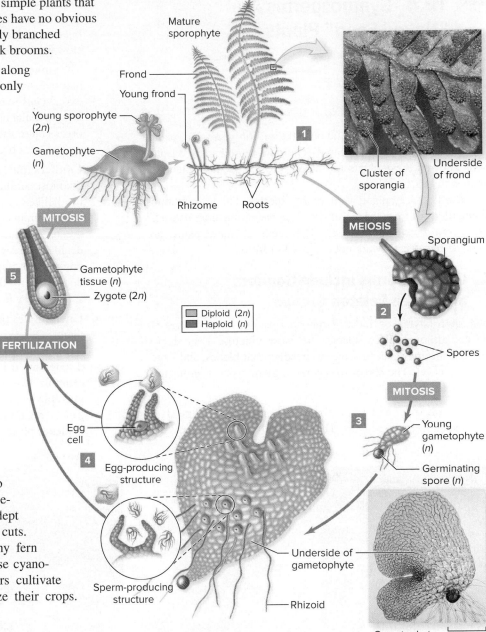

B. Seedless Vascular Plants Have a Conspicuous Sporophyte and Swimming Sperm

Figure 19.11 illustrates the life cycle of a fern. The sporophyte produces haploid spores by meiosis in collections of sporangia on the underside of each frond. Once shed, the spores germinate and develop into tiny, heart-shaped gametophytes that produce gametes by mitotic cell division. The swimming sperm require a film of water to reach the egg cell. The gametes fuse, forming a zygote. This diploid cell divides mitotically and forms the sporophyte, which quickly dwarfs the gametophyte.

Many seedless vascular plants live in shady, moist habitats. Like bryophytes, these plants produce swimming sperm and therefore cannot reproduce sexually in the absence of water.

Figure 19.11 Life Cycle of a True Fern. (*1*) Sporangia on the sporophyte's fronds house cells that (*2*) produce spores by meiosis. (*3*) A haploid spore develops into a gametophyte, which (*4*) produces egg cells and sperm cells. Sperm swim in a film of water to reach eggs. (*5*) These gametes join and produce a zygote, which develops into the sporophyte.

Photos: (spores): ©Ed Reschke/Photolibrary/Getty Images; (gametophyte): ©Les Hickok and Thomas Warne, C-Fern

19.3   MASTERING CONCEPTS

1. Describe the four groups of seedless vascular plants.
2. How do seedless vascular plants reproduce?
3. How are seedless vascular plants similar to and different from bryophytes?

19.4 Gymnosperms Are "Naked Seed" Plants

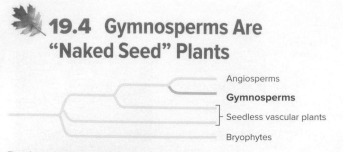

- Angiosperms
- **Gymnosperms**
- Seedless vascular plants
- Bryophytes

During the Permian period, about 300 million years ago, plants with pollen and seeds appeared. The new reproductive adaptations allowed these plants to outcompete the seedless plants in many habitats.

The first seed plants were gymnosperms. The term **gymnosperm** derives from the Greek words *gymnos,* meaning "naked," and *sperma,* meaning "seed." The seeds of these plants are "naked" because they are not enclosed in fruits.

A. Gymnosperms Include Conifers and Three Related Groups

The sporophytes of most gymnosperms are woody trees or shrubs, although a few species are more vinelike. Leaf shapes range from tiny reduced scales to needles, flat blades, and large, fernlike leaves. The 800 or so species of gymnosperms group into four phyla (**figure 19.12**):

- **Cycads** (phylum Cycadophyta) live primarily in tropical and subtropical regions. They have palmlike leaves, and they produce large cones. Cycads dominated Mesozoic era landscapes. Today, cycads are popular ornamental plants, but many species are near extinction in the wild because of slow growth, low reproductive rates, and shrinking habitats.

- The **ginkgo** (phylum Ginkgophyta), also called the maidenhair tree, has distinctive, fan-shaped leaves. Only one species exists: *Ginkgo biloba*. It no longer grows wild in nature, but it is a popular cultivated tree. Ginkgos have male and female organs on separate plants; landscapers avoid planting female ginkgo trees because the fleshy seeds produce a foul odor.

- **Conifers** (phylum Pinophyta) such as pine trees are by far the most familiar gymnosperms. These plants often have needlelike or scalelike leaves, and they produce egg cells and pollen in cones. Conifers are commonly called "evergreens" because most retain their leaves all year, unlike deciduous trees. This term is somewhat misleading, however, because conifers do shed their needles. They just do it a few needles at a time, turning over their entire needle supply every few years.

- **Gnetophytes** (phylum Gnetophyta) include some odd plants. One example is *Welwitschia,* a desert plant with a single pair of large, strap-shaped leaves that persist throughout its life. The *Ephedra* in figure 19.12d is another example. Botanists struggle with the classification of gnetophytes. Some of the details of their life history suggest a close relationship with flowering plants, but molecular evidence places these plants with the conifers.

a.   Cycad   b.   Ginkgo   c.   Conifer   d.   Gnetophyte

Figure 19.12 **A Gallery of Gymnosperms.** (a) Cycads are ancient seed plants with cones that form within a crown of large leaves. A seed cone is shown here. (b) The leaves of *Ginkgo biloba* turn yellow in the fall. The lower photo shows the fleshy seed. (c) This pinyon pine is an example of a conifer. The seed cone has woody scales. (d) *Ephedra* is a gnetophyte with cones that resemble tiny flowers.

B. Conifers Produce Pollen and Seeds in Cones

Pine trees illustrate the gymnosperm life cycle (figure 19.13). The mature sporophyte produces **cones,** the organs that bear the reproductive structures. Each female cone scale bears two ovules on its upper surface; the **ovules** produce the female reproductive cells (and eventually develop into seeds). Inside each ovule, a sporangium undergoes meiosis and produces four haploid megaspores, only one of which develops into a female gametophyte. Over many months, the female gametophyte gives rise to two to six egg cells. At the same time, male cones bear sporangia on thin, delicate scales. Through meiosis, these sporangia produce microspores, which eventually become wind-blown pollen grains (male gametophytes). Pollination occurs when pollen grains settle between the scales of female cones and adhere to a sticky secretion.

The pollen grain germinates, giving rise to a pollen tube that grows through the ovule toward the egg cell. Two haploid sperm nuclei develop inside the pollen tube; one sperm nucleus fertilizes the haploid egg cell, and the other disintegrates. The resulting zygote is the first cell of the sporophyte. The whole process is so slow that fertilization occurs about 15 months after pollination.

Within the ovule, the haploid tissue of the female gametophyte nourishes the developing diploid embryo, which soon becomes dormant. Meanwhile, a tough, protective seed coat develops. Eventually, the seed is shed and dispersed by wind or animals. If conditions are favorable, the seed germinates, giving rise to a new tree that can begin the cycle again.

19.4 | MASTERING CONCEPTS

1. What are the characteristics of gymnosperms?
2. What are the four groups of gymnosperms?
3. What is the role of cones in conifer reproduction?
4. What happens during and after pollination in gymnosperms?

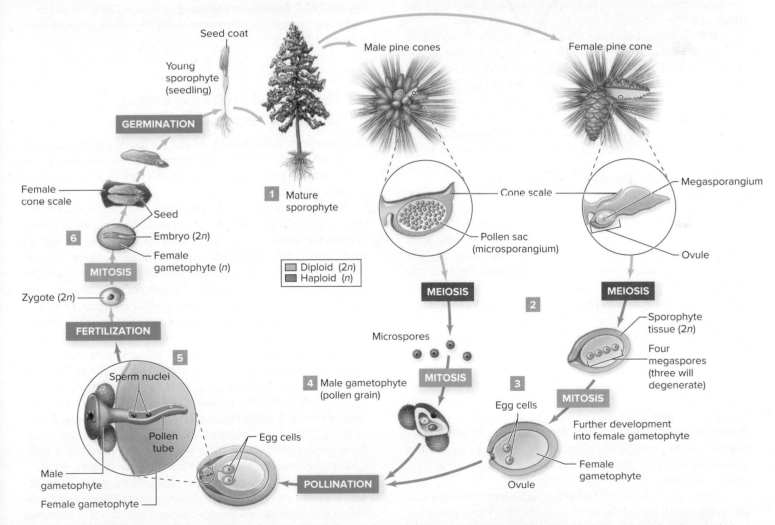

Figure 19.13 Life Cycle of a Pine. (1) The mature sporophyte produces male and female cones. (2) Cells in the male and female cone scales undergo meiosis, producing spores that develop into haploid gametophytes consisting of just a few cells each. (3) On the female cones, each scale has two ovules (only one is shown), each of which yields an egg-producing gametophyte. (4) The male cones produce pollen, the male gametophytes. (5) A pollen grain delivers a sperm nucleus to an egg cell via a pollen tube. The fertilized egg (zygote) will become the embryo. (6) The embryo is packaged inside a seed, which will eventually germinate and yield a pine seedling.

19.5 Angiosperms Produce Seeds in Fruits

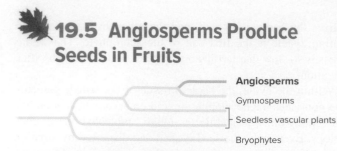

The bryophytes, seedless vascular plants, and gymnosperms make up less than 5% of all modern plant species. The other 95% are **angiosperms,** or flowering plants (phylum Magnoliophyta). Examples include apple trees, corn, roses, petunias, lilies, grasses, and many other familiar plants, including those we grow for our own food.

Flowers and fruits are reproductive structures that are unique to the angiosperms. The plant's flowers produce pollen and egg cells. After pollination and fertilization, flowers develop into fruits that enclose the plant's seeds. The name *angiosperm* is a tribute to this unique life cycle; the prefix *angio-* derives from the Greek word for "vessel." A fruit, then, is a "seed vessel."

Multiple lines of evidence place the origin of angiosperms in the Jurassic period, at least 144 million years ago. By 100 million years ago, all of today's major lineages of angiosperms were in place. Biologists have long puzzled about the sudden appearance and rapid diversification of the flowering plants. The adaptive radiation of the angiosperms coincided with the diversification of vertebrates and arthropods on land, but no one has established a definite cause-and-effect link. (i) *adaptive radiation,* section 14.4B

A. Most Angiosperms Are Eudicots or Monocots

Biologists are still working to sort out the evolutionary relationships among the angiosperms; figure 19.14 shows one hypothesis. The two largest clades, the eudicots and the monocots, together account for 97% of all flowering plants. The **eudicots** have two cotyledons (the first leaf structures to arise in the embryo), and their pollen grains feature three or more pores. About 175,000 species exist, representing about two thirds of all angiosperms. The diverse eudicots include roses, daisies, sunflowers, oaks, tomatoes, beans, and many others. *Arabidopsis thaliana*, the subject of this chapter's Focus on Model Organisms, is a eudicot.

Most of the other angiosperms are **monocots,** which are named for their single cotyledon; in addition, their pollen grains have just one pore. (Monocots and eudicots also differ by other characteristics; see chapter 22.) Examples of the 70,000 species of monocots are orchids, lilies, grasses, bananas, and ginger. The grasses include lawn plants, sugarcane, and grains such as rice, wheat, and corn.

Finally, about 3% of the flowering plants form a paraphyletic group informally called the basal angiosperms. These plants belong to lineages that diverged from ancestral plants before eudicots and monocots evolved (see figure 19.14). Examples of basal angiosperms are magnolias, nutmeg, avocados, black pepper, water lilies, and star anise. (i) *paraphyletic group,* section 14.6D

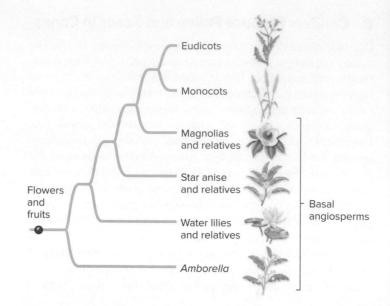

Figure 19.14 Angiosperm Phylogeny. This phylogenetic tree depicts the proposed evolutionary relationships among the eudicots, monocots, and basal angiosperms.

B. Flowers and Fruits Are Unique to the Angiosperm Life Cycle

The angiosperm life cycle is similar to that of gymnosperms in some ways (figure 19.15). For example, the sporophyte is the only conspicuous generation, and both types of plants produce pollen and seeds.

Yet the angiosperm and gymnosperm life cycles differ in important ways. Most obviously, the reproductive organs in angiosperms are flowers, not cones. Another difference is that an angiosperm's ovules develop into seeds inside the flower's ovary. The ovary develops into the fruit, which helps to protect and disperse the seeds. By comparison, a gymnosperm's seeds are produced "naked" on the female cone's scales (see figure 19.13).

Pollination triggers one other unique feature of the angiosperm life cycle. In double fertilization, two sperm nuclei enter the female gametophyte. One sperm nucleus fertilizes the egg, producing the zygote that will develop into the embryo. The other sperm nucleus fertilizes a pair of nuclei in the female gametophyte's central cell. The resulting triploid nucleus develops into the endosperm, a tissue that supplies nutrients to the germinating seedling. The embryo and endosperm, together with a seed coat, make up the seed; one or more seeds develop inside each fruit. Chapter 24 considers angiosperm reproduction in more detail.

The endosperm supplies nutrients until the young plant begins to produce its own food in photosynthesis. Endosperm tissue therefore often contains energy-rich starch or oils. For example, the endosperm of wheat and other grains is starchy; bakers grind these seeds into flour to make bread and other baked goods. Coconuts and castor seeds are two sources of useful oils derived from endosperm.

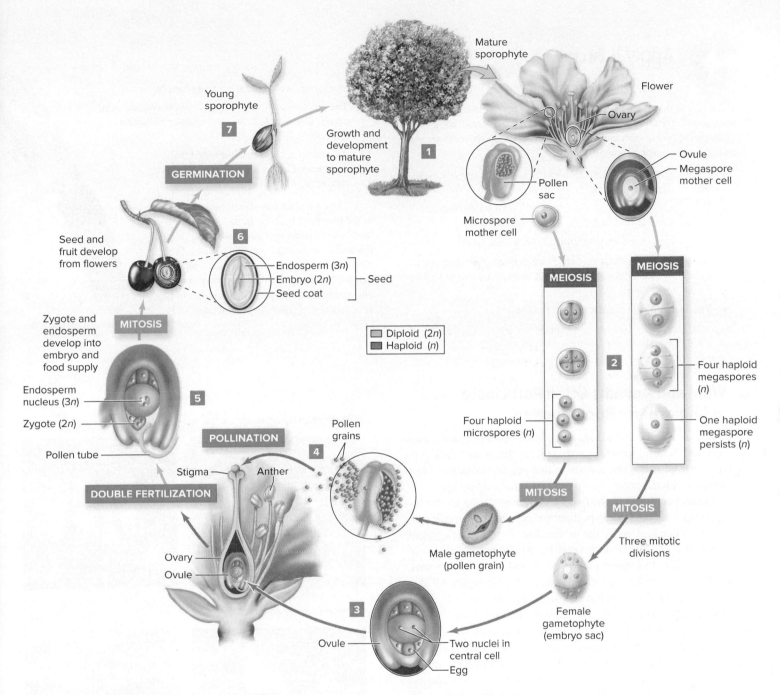

Figure 19.15 Life Cycle of an Angiosperm. (1) The mature sporophyte produces flowers. (2) Cells in a flower's pollen sac and ovary undergo meiosis, producing spores that develop into haploid male and female gametophytes. (3) Inside each ovule, the female gametophyte includes one egg cell and two nuclei in a central cell. (4) The pollen sac produces pollen, the male gametophytes. (5) A pollen grain delivers two sperm nuclei. One fertilizes the egg and the other fertilizes the nuclei in the central cell, forming a triploid cell that develops into the endosperm. (6) Each ovule develops into a seed; the fruit develops from the ovary wall. (7) Seed germination reveals the young sporophyte.

Endosperm also contains proteins. Gluten, for example, is a mixture of proteins produced in the endosperm of wheat and other grains. People with gluten sensitivity must take care to avoid foods containing these grains. Reactions to gluten vary widely. In people with moderate sensitivity, eating gluten causes bloating and abdominal discomfort. If gluten sensitivity is severe (celiac disease), gluten triggers inflammation in the small intestine. The result may be painful digestive problems,

vitamin deficiencies, and many other symptoms. People with a gluten allergy may have a rapid whole-body reaction to ingested gluten, leading to hives, swelling, headache, or even anaphylactic shock. Fortunately, gluten-free foods are increasingly easy to find.

On the other hand, people with corn allergies may struggle to find foods lacking corn. As explained in this chapter's Apply It Now box, corn is everywhere.

Apply It **Now**

Corn, Corn, Everywhere

Even if you don't eat the kernels off the cob, you likely have much more corn in your diet than you imagine. Besides fresh corn, frozen corn, canned corn, corn meal, corn oil, and corn starch, these are some of the many unexpected fates of corn:

- Baking powder and confectioners ("powdered") sugar, which often contain corn starch
- Corn syrup, a liquid sugar derived from corn starch, that sweetens soft drinks, ketchup, and many other foods
- Vanilla extract, which is often made with corn syrup
- Dextrin and maltodextrins, which are derived from corn starch; these polysaccharides thicken syrups and add texture to low-fat foods (among other uses)
- The sugars dextrose (glucose) and fructose

- Margarine, which often contains corn oil
- The grain alcohol in bourbon whiskey; ethanol from corn is also a biofuel (see this chapter's Burning Question)
- Animal feed; chickens, hogs, and cattle on commercial farms are fed corn, effectively converting the nutrients in corn to meat

The corn plant has also played a major role in the history of biology. In the 1940s, Barbara McClintock discovered transposons, or "jumping genes," in Indian corn. She won the 1983 Nobel Prize in physiology or medicine for this work, which has led to new ways to mutate genes and, in turn, new insights into gene function. ⓘ *transposons*, section 7.7B

©C Squared Studios/Getty Images RF

C. Wind and Animals Often Participate in Angiosperm Reproduction

Pollination and seed dispersal are at the heart of angiosperm sexual reproduction. Natural selection favors plants that scatter their pollen and seeds far and wide, a strategy that simultaneously reduces competition with the parent and promotes genetic diversity.

Wind and animals are the two main transportation modes for pollen (**figure 19.16**). Wind pollination is relatively inefficient, since most pollen carried by the breeze does not reach a flower of the same species. Plants pollinated by wind produce abundant pollen in flowers that are typically plain and easily overlooked. Grasses, along with the maples and the cattails in figure 19.16a, are common examples of wind-pollinated plants.

Many angiosperms rely on animal "couriers" that unwittingly carry pollen from flower to flower. Because animals typically visit only a limited number of plant species, this strategy is much more efficient than is wind pollination. However, arthropods and vertebrates do not pollinate plants as an act of charity; they usually visit flowers in search of food. Animal pollination therefore has a different cost: The plants invest in large petals, bright colors, alluring scents, and sweet rewards such as nectar that attract animals. The chamomile and banana flowers in figure 19.16b illustrate some of these adaptations.

The relationship between angiosperms and their pollinators is sometimes so tight that one cannot reproduce or survive without the other. This situation can lead to **coevolution,** in which a genetic change in one species selects for subsequent change in another species. For example, an alteration in the shape or color of a flower can select for new adaptations in its pollinator (see section 14.7). The reverse can also occur: A change in the curve of a hummingbird's beak, for example, can select for corresponding modifications in the flower that it pollinates.

a. Pollination by wind

b. Pollination by animals

Figure 19.16 A Gallery of Angiosperms. (a) Wind-pollinated plants include (*left*) boxelder maple trees and (*right*) cattails. The brown cylinders on the cattail plant are elongated clusters of tiny brown flowers. (b) Animal-pollinated plants include (*left*) German chamomile and (*right*) banana trees. Bats pollinate banana flowers in the wild, but commercial varieties bear fruit without pollination.

Photos: (a, maple tree): ©Steven P. Lynch/McGraw-Hill Education; (a, cattails): ©Hans Reinhard/Okapia/Science Source; (b, chamomile): ©McGraw-Hill Education; (b, banana tree): ©Igor Prahin/Flickr Open/Getty Images RF

F⊙CUS on Model Organisms

Mouse-Ear Cress: *Arabidopsis thaliana*

The choice of a "star" model organism for plants is simple: *Arabidopsis thaliana* easily beats out all others (figure 19.A). This tiny angiosperm, a mustard relative, has become a staple of plant biology laboratories because of its small size, easy cultivation, and prolific reproduction. In addition, biologists can use a bacterium called *Agrobacterium tumefaciens* to carry new genes into *Arabidopsis* cells.

The amount of attention devoted to this plant may seem extravagant, considering its insignificance as a commercial plant. Because angiosperms are closely related to one another, however, discoveries in *Arabidopsis* will likely apply directly to economically important crop plants. For example, genetic and molecular studies of *Arabidopsis* can help researchers identify genes that enable plants to grow on poor soil. Understanding these genes may help plant scientists develop improved varieties of food crops such as rice, barley, wheat, and corn.

Moreover, many genes in *Arabidopsis* have counterparts in humans and other organisms, so research on this plant can have far-reaching applications. For example, scientists have discovered that *Arabidopsis* has at least 139 genes that correspond to human disease genes. Studies of *Arabidopsis* can therefore help us understand illnesses from colitis to Alzheimer disease to arthritis.

The following list includes a few of the important discoveries resulting from work on *Arabidopsis*:

- **Control of gene expression:** Each cell type in a multicellular organism turns on a different combination of genes. One way that cells regulate gene expression is to attach methyl groups to unneeded DNA; another is to produce small pieces of RNA that bind to genes that have already been transcribed. Researchers have studied both processes in *Arabidopsis*, in part because problems in gene expression cause some types of cancer in humans. ⓘ *regulation of gene expression*, section 7.6B

- **Genome duplication:** The *Arabidopsis* genome sequencing project was completed in 2000. Analysis of the DNA suggests that the entire genome has duplicated two or three times, fueling speculation that all plants are polyploids. This finding may help shed light on the evolutionary history of plants. ⓘ *polyploidy*, section 9.7A

- **Disease resistance:** Some plants construct a sort of "fire break" around the spot where a bacterium or fungus has entered. Small areas of surrounding plant tissue die, and this zone of dead cells prevents the invader from spreading throughout the plant. Study of this response in *Arabidopsis* has led to new insights into how genes regulate apoptosis (programmed cell death); this research may one day help plant breeders improve disease resistance in other crops as well. ⓘ *apoptosis*, section 8.7

- **Response to the environment:** A plant cannot avoid extremes of temperature, light availability, and salinity by moving to a better location. Instead, it must adjust its physiology. *Arabidopsis* has genes that control its response when the weather turns cold, a finding that could help crop plants survive freezing.

- **Hormones:** Ethylene is a gas that helps control fruit ripening and plant senescence (aging). Mutant *Arabidopsis* plants that do not respond to ethylene have helped researchers find ethylene receptor proteins. Researchers have also discovered that the ethylene response requires copper, which the plant transports using a protein similar to one that transports copper in humans. (When faulty in humans, this protein causes Menkes disease.) ⓘ *ethylene*, section 24.4B

- **Circadian rhythms:** Circadian processes occur in 24-hour cycles. In *Arabidopsis*, for example, proteins encoded by clock genes ensure that the expression of genes needed for photosynthesis peaks at around noon. The same genes are repressed at night. Pigments called phytochromes "reset" the clock each day. ⓘ *phytochrome*, section 24.5B

- **Flowering:** Angiosperms delay flowering until they reach reproductive maturity. How do they "know" when the time comes, and how do they build flower parts in the right places? *Arabidopsis* research has revealed genes that control the timing of flowering, the differentiation of cells that give rise to flowers, the development of individual flower parts, and the development of ovules inside the flower. For example, some genetic mutations induce flowers to develop into shoots; others promote early flowering.

Figure 19.A
Arabidopsis thaliana.

Like pollination, seed dispersal also usually involves wind or animals. Some fruits, such as those of dandelions and maples, have "parachutes" or "wings" that promote wind dispersal. Others, however, spread only with the help of animals. Some have burrs that cling to animal fur. Others are sweet and fleshy, attracting animals that eat the fruits and later spit out or discard the seeds in feces. ⓘ *seed dispersal*, section 24.2G

19.5 | MASTERING CONCEPTS

1. What are the two largest clades of angiosperms?
2. In what ways are the life cycles of angiosperms similar to and different from those of gymnosperms?
3. What is the relationship between flowers and fruits?
4. Describe two ways animals participate in angiosperm reproduction and dispersal.

INVESTIGATING LIFE

19.6 Genetic Messages from Ancient Ecosystems

Psychics claim to be able to communicate with the spirits of the dead. Although scientists do not assert the same spiritual connection, they can bring back the genetic remnants of species that lived and died long ago.

When an organism dies, its DNA usually degrades rapidly. But in some special cases, the genetic material remains intact indefinitely. Freezing is one way to preserve DNA. An ideal source of diverse ancient DNA is a landscape that once teemed with life but has since become permanently frozen.

One such example is the land bridge, called Beringia, that once connected present-day northeastern Siberia to Alaska. Long ago, giant mammals such as mammoth, bear, bison, and large cats roamed the grassy Beringian landscape. But climates shift, and much of Beringia is now permanently frozen land in Siberia, Alaska, and the Yukon. Can DNA preserved in Siberian soil reveal which plants supported the ecosystem hundreds of thousands of years ago?

Danish researchers Eske Willerslev, Anders Hansen, and their colleagues drove metal cylinders deep into permafrost and removed long, thin rods (called "cores") of ice, soil, and organic material (figure 19.17). The deepest holes yield the oldest deposits, because new sediments accumulate over old ones. Radiometric dating, pollen analysis, and other techniques helped them estimate the age of each layer of material in the cores. ⓘ *radiometric dating,* section 13.2C

Willerslev and colleagues tried to extract DNA from eight sediment samples ranging in age from modern to about 2 million years old. Wherever DNA was present, they searched for two sets of genes. One target was part of the gene encoding the protein rubisco, which is essential for photosynthesis (see chapter 5). The presence of this gene indicates plant material. The other targets were fragments of genes found only in the mitochondria of vertebrate animals such as mammals.

Sediments from 300,000 to 400,000 years old contained plant DNA, including a tremendous diversity of mosses, herbs (grasses and other nonwoody plants), shrubs, and trees. On the other hand, the oldest mitochondrial DNA from vertebrates was only about 20,000 to 30,000 years old. The presence of mitochondrial genes from extinct mammoth and bison suggests that the sediment DNA was authentic and not simply a modern contaminant.

These gene fragments can help scientists reconstruct ancient ecosystems. The chloroplast DNA, for example, reveals that herbs dominated the Beringian landscape 300,000 or so years ago but lost ground to shrubs over time (figure 19.18). The most dramatic decline of grasses occurred in the past 10,000 years, a time that coincided with the extinction of the mammoth and bison. Did one event cause the other? What role did increasing human populations play? These questions remain unanswered for now.

Willerslev's team hopes their work will inspire others to extract DNA from ancient sediments around the world. The resulting patchwork of gene fragments, pieced together, will reveal valuable information about the changes that shaped ancient ecosystems—without the help of a psychic.

Source: Willerslev, E., A. J. Hansen, J. Binladen, et al. May 2, 2003. Diverse plant and animal genetic records from Holocene and Pleistocene sediments. *Science,* vol. 300, pages 791–795.

19.6 MASTERING CONCEPTS

1. Why do researchers collect DNA from permafrost?
2. What are some alternative hypotheses for why the researchers failed to recover any DNA from sediments that were more than 400,000 years old? How would you test your hypotheses?

Figure 19.17 Core Sample. Cores of frozen sediment taken from permafrost may contain ancient DNA.

Photo: Courtesy of K. Schaefer

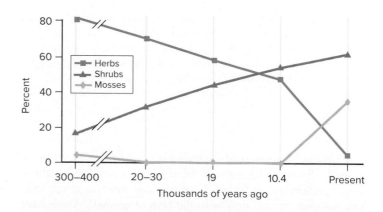

Figure 19.18 Changing Community. Chloroplast DNA isolated from sediment samples reveals changes in the Siberian plant community over the past 400,000 or so years.

CHAPTER SUMMARY

19.1 Plants Have Changed the World

- Members of kingdom **Plantae** provide food and habitat for other organisms, remove CO_2 from the atmosphere, and produce O_2. Humans rely on plants for food, lumber, clothing, paper, and many other resources.

A. Green Algae Are the Closest Relatives of Plants

- The ancestor of land plants may have resembled green algae called **charophytes.** Plants emerged onto land about 475 million years ago.
- Like many green algae, plants are multicellular, eukaryotic autotrophs that have cellulose cell walls and use starch as a carbohydrate reserve. Green algae and plants also use the same photosynthetic pigments. Unlike green algae, most plants live on land.

B. Plants Are Adapted to Life on Land

- Adaptations that enable plants to obtain and conserve resources include roots, leaves, a waterproof **cuticle,** and **stomata.**
- **Vascular tissue** is a transportation system inside many plants. **Xylem** transports water and minerals; **phloem** carries sugars. **Lignin** strengthens xylem cell walls, providing physical support.
- **Figure 19.19** illustrates the **alternation of generations** in the plant life cycle. The diploid **sporophyte** produces haploid **spores** by meiosis; the haploid **gametophyte** produces haploid **gametes** by mitosis. Fertilization restores the diploid number.
- In the simplest plants, the gametophyte generation is most prominent; in more complex plants, the sporophyte dominates (see figure 19.19).
- In gymnosperms and angiosperms, reproductive adaptations include **pollen** and **seeds. Pollination** delivers sperm to egg. The resulting

zygote develops into an embryo, which is packaged along with a food supply into a seed. Angiosperms also produce **flowers** and **fruits.**
- Plants are classified by the presence or absence of vascular tissue, seeds, flowers, and fruits.

19.2 Bryophytes Are the Simplest Plants

A. Bryophytes Lack Vascular Tissue

- **Bryophytes** are small plants lacking vascular tissue, leaves, roots, and stems.
- The three groups of bryophytes are **liverworts, hornworts,** and **mosses.**

B. Bryophytes Have a Conspicuous Gametophyte

- In bryophytes, the gametophyte stage is dominant. Many bryophytes reproduce asexually by fragmentation of the gametophyte. Sperm require water to swim to egg cells.

19.3 Seedless Vascular Plants Have Xylem and Phloem but No Seeds

A. Seedless Vascular Plants Include Ferns and Their Close Relatives

- **Seedless vascular plants** have vascular tissue but lack seeds. This group includes **club mosses, whisk ferns, horsetails,** and **true ferns.**

B. Seedless Vascular Plants Have a Conspicuous Sporophyte and Swimming Sperm

- The diploid sporophyte generation is the most obvious stage of a fern life cycle, but the haploid gametophyte forms a tiny separate plant.
- In the sexual life cycle of ferns, collections of sporangia appear on the undersides of fronds. Meiosis occurs in the sporangia and yields haploid spores, which germinate in soil and develop into gametophytes. The gametophytes produce egg cells and swimming sperm.

19.4 Gymnosperms Are "Naked Seed" Plants

A. Gymnosperms Include Conifers and Three Related Groups

- **Gymnosperms** are vascular plants with seeds that are not enclosed in fruits. The four groups of gymnosperms are **cycads, ginkgos, conifers,** and **gnetophytes.**

B. Conifers Produce Pollen and Seeds in Cones

- In pines (a type of conifer), **cones** house the reproductive structures. Male cones release pollen, and female cones produce egg cells inside **ovules.** Pollen germination yields a pollen tube through which a sperm nucleus travels to the egg cell. After fertilization, the resulting embryo remains dormant in a seed until germination.
- In conifers, the sperm do not require water to swim to the egg cell. Instead, most gymnosperms rely on wind to spread pollen.

19.5 Angiosperms Produce Seeds in Fruits

A. Most Angiosperms Are Eudicots or Monocots

- **Angiosperms** are vascular plants that produce flowers and fruits. The two largest clades of angiosperms are **eudicots** and **monocots.**

B. Flowers and Fruits Are Unique to the Angiosperm Life Cycle

- Flowers produce pollen and egg cells. After pollination, the flower parts develop into the fruit, which protects the seeds.

C. Wind and Animals Often Participate in Angiosperm Reproduction

- Most angiosperms are wind- or animal-pollinated. Corresponding adaptations in animals and plants illustrate **coevolution.**
- Angiosperm fruits aid in seed dispersal, usually by wind or animals.
- **Figure 19.20** summarizes plant diversity.

19.6 Investigating Life: Genetic Messages from Ancient Ecosystems

- Researchers have extracted DNA from plants and animals buried long ago in frozen sediments. The DNA evidence reveals changes in the landscape over hundreds of thousands of years.

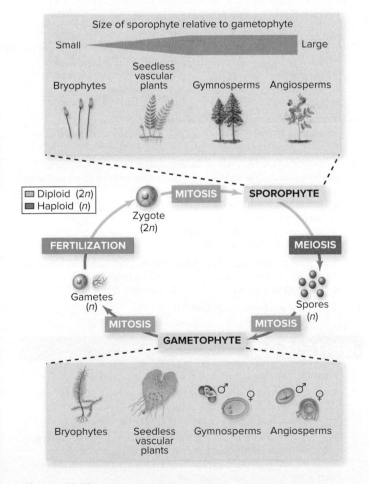

Figure 19.19 Alternation of Generations: A Summary.

Figure 19.20 Plant Diversity: A Summary.

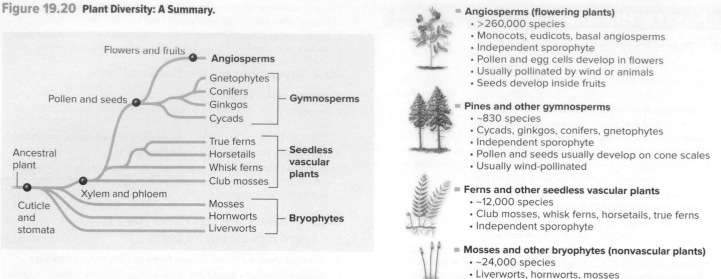

■ **Angiosperms (flowering plants)**
• >260,000 species
• Monocots, eudicots, basal angiosperms
• Independent sporophyte
• Pollen and egg cells develop in flowers
• Usually pollinated by wind or animals
• Seeds develop inside fruits

■ **Pines and other gymnosperms**
• ~830 species
• Cycads, ginkgos, conifers, gnetophytes
• Independent sporophyte
• Pollen and seeds usually develop on cone scales
• Usually wind-pollinated

■ **Ferns and other seedless vascular plants**
• ~12,000 species
• Club mosses, whisk ferns, horsetails, true ferns
• Independent sporophyte

■ **Mosses and other bryophytes (nonvascular plants)**
• ~24,000 species
• Liverworts, hornworts, mosses
• Sporophyte depends on gametophyte for nutrition

Group	Swimming Sperm	Vascular Tissue	Pollen	Seeds	Flowers	Fruits
Bryophytes	Yes	No	No	No	No	No
Seedless vascular plants	Yes	Yes	No	No	No	No
Gymnosperms	No	Yes	Yes	Yes	No	No
Angiosperms	No	Yes	Yes	Yes	Yes	Yes

MULTIPLE CHOICE QUESTIONS

1. Which of the following is NOT a similarity between land plants and green algae?
   a. Photosynthesis
   b. Starch as a storage form of energy
   c. Cellulose cell walls
   d. The presence of a cuticle and stomata

2. In the alternation of generations in plants, the gametophyte is _____ and produces gametes by _____.
   a. haploid; mitosis
   b. diploid; mitosis
   c. haploid; meiosis
   d. diploid; meiosis

3. What conditions did plants face when they moved to land?
   a. Air provides less physical support than water.
   b. Drying out became more likely.
   c. They were adapted to gamete dispersal in water.
   d. All of the above are correct.

4. Which of the following is present in all land plants?
   a. Wind-blown sperm
   b. A seed coat that prevents embryos from drying out
   c. Vessels that transport water and nutrients throughout the plant
   d. None of the above are present in all land plants.

5. When moss spores are released from a sporangium, they develop into
   a. a sperm cell.
   b. a gametophyte.
   c. a sporophyte.
   d. a gemma.

6. Which adaptation to land is present in the most plant species?
   a. Pollen
   b. Seeds
   c. Phloem
   d. Fruits

7. Which type of plant may have a gametophyte that is large relative to its sporophyte?
   a. Bryophyte
   b. Gymnosperm
   c. Angiosperm
   d. True fern

8. Reproduction in a pine tree is associated with
   a. male and female cones.
   b. wind-blown pollen.
   c. the formation of pollen tubes.
   d. All of the above are correct.

9. In comparing the life cycle of an angiosperm to that of a human, pollination is analogous to _____ and the embryo is analogous to the _____.
   a. childbirth; growth of the child
   b. sexual intercourse; baby
   c. production of egg cells; mother
   d. production of sperm; uterus

10. What plant group is correctly matched with an adaptation of its members?
    a. Cycads: vascular tissue
    b. Liverworts: true leaves
    c. Whisk ferns: seeds
    d. Conifers: fruits

Answers to these questions are in appendix A.

WRITE IT OUT

1. What characteristics do all land plants have in common?

2. Analyze the alternation of generations common to all plants. If you analyzed the DNA of all of the gametes produced by one gametophyte, would you see variation among the gametes?

3. How are terrestrial habitats different from aquatic habitats? List the adaptations that enable land plants to obtain resources, transport materials, and reproduce; explain how each adaptation contributes to a plant's reproductive success on land.

4. List the characteristics that distinguish the four major groups of plants; then provide an example of a plant within each group.

5. Give at least two explanations for the observation that bryophytes are much smaller than most vascular plants. How can increased height be adaptive? In what circumstances is small size adaptive?

6. A fern plant can produce as many as 50 million spores a year. (a) How are these spores similar to and different from seeds? (b) In a fern population that is neither shrinking nor growing, approximately what proportion of these spores is likely to survive long enough to reproduce? (c) What factors might determine whether an individual spore successfully produces a new fern plant?

7. How do the adaptations of gymnosperms and angiosperms enable them to live in drier habitats than bryophytes and seedless vascular plants?

8. Your friend John is admiring what he calls "little flowers" on a moss. How would you correct his statement? In what way might those structures be similar to flowers?

9. Describe how the petals, ovary, and ovule of flowers participate in reproduction. What happens to each part after fertilization?

10 The immature fruit of the opium poppy produces many chemicals that affect animal nervous systems. In what way might these chemicals benefit the plant?

11. Scientists have studied plant populations that have moved into areas with low herbivory compared to the plant's previous habitat. After many generations in the new habitat, the plants produce fewer defensive chemicals and grow larger than plants in the ancestral population. How does natural selection explain these observations? (*Hint:* Producing defensive chemicals uses energy.)

12. The spurge-laurel is a species of shrub. It produces berries that, if consumed, will cause internal bleeding and death in humans. On the other hand, some birds can eat the berries and remain unharmed. Why might it be advantageous for the plant to prevent mammals but not birds from eating its fruits?

13. In a sentence or two, either support or refute the following statement: The pollen grains of angiosperms are homologous to the spores of bryophytes.

14. An angiosperm called "ghost plant" does not produce chlorophyll or other photosynthetic pigments and is almost completely white. Based only on this information, what assumptions can you make about ghost plants?

15. Compare and contrast the life cycles of the four groups of plants. How does each group represent a variation on the common theme of alternation of generations?

16. A slight change in a plant species' flower structure might favor a different pollinator. How might such a change in flower structure lead to a new plant species? How does coevolution between flowering plants and animal pollinators help explain the huge diversity of angiosperms?

17. Suppose you and a friend are hiking and you see an unfamiliar plant. What observations would you make in trying to determine which type of plant it is?

18. Human activities and natural phenomena can drive plant species to extinction. The U.S. Department of Agriculture Natural Resources Conservation Service maintains lists of threatened and endangered plant species. What are some examples of threatened or endangered species in your area? What are the most important threats to those species? What are the potential consequences of a plant species extinction? What steps should we take to save threatened and endangered plants?

19. What are the pros and cons of pursuing biofuels as alternatives to fossil fuels? In your opinion, do the pros outweigh the cons, or vice versa? Justify your answer.

PULL IT TOGETHER

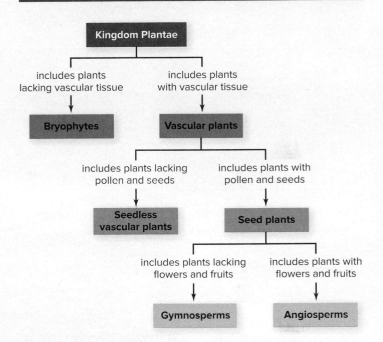

Figure 19.21 Pull It Together: Plants.

Refer to **figure 19.21** and the chapter content to answer the following questions.

1. Review the kingdoms within domain Eukarya, using the Survey the Landscape figure in the chapter introduction. To which group of eukaryotes are plants most closely related? Use a connecting phrase to add that group to the Pull It Together concept map.

2. Circle each plant group that produces spores.

3. How do bryophytes and seedless vascular plants reproduce if they lack pollen and seeds?

4. What is the relationship between pollen and seeds?

20 Fungi

Mushrooms in the Morning. Parasol mushrooms sometimes seem to come from nowhere, appearing overnight after a rain. But what you see is not the whole organism; most of the fungus remains underground.

©Frans Lemmens/Corbis

LEARN HOW TO LEARN
Use All Your Resources

Whether you are using a print book or an ebook, Connect and other websites offer a wealth of online quizzes, animations, and other resources that can help you learn biology. As you study, pay attention to which concepts are sinking in and which are more difficult to understand. Take regular breaks from reading and try a digital learning resource associated with your book. For example, practice questions can help you learn the material and test your understanding. Also, check for animations that take you through complex processes one step at a time. Sometimes the motion of an animation can help you understand what's happening more easily than studying a static image.

The Mushroom Mystique

For many people, the word *fungi* conjures images of mushrooms, along with emotions that range from delight to revulsion. Enthusiasts know that mushrooms can be delectable, yet fairy tales associate toadstools with dank, lonely forests teeming with witches and evil spirits.

Mushroom imagery is common in everyday language. A phrase such as "a mushrooming business venture," for example, evokes the amazing growth rates of fungi. When warm weather follows a rain, mushrooms seem to pop up overnight on lawns and around trees. This extraordinarily rapid growth adds to their mystique.

How can mushrooms appear so suddenly? Like an apple on a tree, a mushroom is a reproductive structure attached to a much larger organism. Most of the fungal body, however, is underground. Fungi produce filaments that feed by penetrating their food, absorbing nutrients as they go. Sometimes two threads unite, and the combined organism prepares to reproduce. When the temperature and moisture are just right, the cells of tiny "pre-mushrooms" absorb water, expanding rapidly. Mushrooms erupt out of the ground and shed reproductive cells called spores. Afterward, the mushrooms wilt, vanishing as quickly as they appeared.

Picking and eating wild mushrooms can be extremely dangerous. One of the deadliest mushrooms is the death angel (*Amanita virosa*). The poisons can be deceptive. Initial symptoms of nausea and diarrhea are followed by a lag of a day or two, during which the mushroom's toxins quietly destroy the liver, kidney, and other organs. By the time symptoms return, the damage is irreversible, and an emergency organ transplant is the only hope for survival. Unfortunately, there is no simple way to tell poisonous species from safe ones.

Commercial mushroom growers cultivate many edible fungi. The familiar white "button mushroom" grows on a compost of horse manure and straw. Shiitakes live on hardwood logs or compressed sawdust; oyster mushrooms prefer wheat or rice straw. Hobbyists can purchase kits to grow these species at home.

Mushrooms represent only a small subset of kingdom Fungi. Diverse fungi spoil food and cause Dutch elm disease, chestnut blight, athlete's foot, diaper rash, ringworm, and yeast infections. On the other hand, they are essential decomposers in ecosystems, and they help humans manufacture cheese, bread, alcoholic beverages, soy sauce, antibiotics, dyes, and many other useful substances. Moreover, research on yeasts and other simple fungi has helped biologists learn about genetics and basic cell processes. This chapter opens the door to this fascinating group.

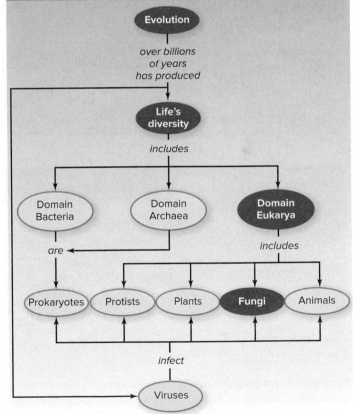

SURVEY THE LANDSCAPE
The Diversity of Life

Fungi may not be glamorous, but they are the unsung heroes of eukaryotic life. Although some are parasites, many others use their unique metabolic capabilities to play ecological roles that no other organisms can fill.

For more details, study the Pull It Together feature in the chapter summary.

20.1 Fungi Are Essential Decomposers

The members of kingdom **Fungi** live nearly everywhere—in soil, in and on plants and animals, in water, even in animal dung (figure 20.1). They range in size from the microscopic fungi infecting the cells of protists to massive fungi extending enormous

a.   SEM (false color) ⊢—⊣ 5 µm   b.

c.   d.

e.

Figure 20.1 A Gallery of Fungi. (a) These microscopic yeast cells are reproducing by budding. (b) Powdery mildew fungi live on leaf surfaces and produce microscopic reproductive structures (inset). (c) This spider was killed by a parasitic fungus, which has produced eerie white spikes. (d) A cluster of "honey mushrooms" erupts from this decaying tree. (e) A man harvests mushrooms from a commercial crop.

(a): ©Scimat/Science Source; (b): ©Steven P. Lynch/McGraw-Hill Education;
(b, micrograph): ©Scenics & Science/Alamy; (c): ©Morley Read/Alamy;
(d): ©WSL Handout/Reuters/Corbis; (e): ©Ed Young/Corbis

distances. For example, a single underground fungus extends over nearly 9 million square meters in an Oregon forest.

Mycologists (biologists who study fungi) have identified about 80,000 species in five phyla, but 1.5 million or so species are thought to exist (table 20.1). These organisms have a rather unsavory reputation, mainly because most people know that fungi can cause disease and turn foods moldy. Overall, however, this reputation is undeserved. Fungi benefit humans directly in many ways: Some are edible, others aid in food and beverage production, and still others are useful in biological research. In addition, fungi are vitally important in ecosystems. Many fungi secrete digestive enzymes that break down dead plants and animals, releasing inorganic nutrients and recycling them to plants. These decomposers are, in a sense, the garbage processors of the planet. Other fungi help plants absorb minerals or fight disease.

A. Fungi Are Eukaryotic Heterotrophs That Digest Food Externally

The evolutionary history of fungi remains unclear. The hypothetical common ancestor of all fungi is an aquatic, single-celled, flagellated protist resembling contemporary fungi called chytrids. The identity of the closest living relative to the fungi, however, remains controversial.

Biologists do know, however, that fungi are more closely related to animals than to plants. This finding may surprise those who notice the superficial similarities between plants and fungi. Unlike plants, however, fungi cannot carry out photosynthesis. Moreover, fungi share many chemical and metabolic features with animals.

Fungi have a unique combination of characteristics:

- The cells of fungi are eukaryotic, as are the cells of protists, plants, and animals.

- Fungi are heterotrophs, as are animals, but these two groups acquire food in different ways. Animals ingest their food and digest it internally; fungi secrete enzymes that break down organic matter outside their bodies. The fungus then absorbs the nutrients.

- Fungal cell walls are composed primarily of the modified carbohydrate chitin. This tough, flexible molecule also forms the exoskeletons of some animals. ⓘ *carbohydrates*, section 2.5B

TABLE 20.1 Phyla of Fungi

Phylum	Examples	Number of Existing Species
Chytridiomycota	Parasite of frog skin	1000
Zygomycota	Black bread mold	1000
Glomeromycota	Arbuscular mycorrhizal fungi	200
Ascomycota	Morels, truffles	More than 50,000
Basidiomycota	Mushrooms, puffballs	30,000

Burning Question

Why does food get moldy?

Fungal spores are everywhere. They germinate and grow into colonies on any surface that provides enough food, oxygen, and moisture. Fresh bread, cheese, fruits, and vegetables are all perfect for fungal growth. Often, a fruit remains mold-free until its protective skin is punctured, but fungi quickly take over once they gain access to the moist interior.

Perhaps this would be a better question: Why *don't* some foods get moldy? Humans have devised many ways to keep perishable foods fresh. Refrigeration dramatically slows the rate of fungal growth. Salt and sugar, in sufficiently high concentrations, also retard mold growth by limiting the fungus's ability to take up water by osmosis. Dried foods are

©Emily Keegin/fStop/
Getty Images RF

preserved in the same way. Cooking and pickling prevent spoilage by damaging microbial enzymes. ⓘ *osmosis,* section 4.5A

One additional method is to add chemical preservatives to foods. Organic acids such as sodium benzoate are common food additives that inhibit mold growth by disrupting fungal cell membranes. Many processed foods are so laden with preservatives that their shelf lives extend for years, a remarkable accomplishment in a world full of hungry microbes (see the Burning Question in section 2.5). ⓘ *membranes,* section 3.3

Submit your burning question to
Marielle.Hoefnagels@mheducation.com

- The storage carbohydrate of fungi is glycogen, the same as for animals (see figure 2.18).

- Most fungi are multicellular, but **yeasts** are unicellular (see figure 20.1a). The distinction between yeasts and multicellular fungi, however, is not absolute; some species can switch between uni- and multicellular phases.

- Fungi have unique reproductive cycles. Some fungi remain haploid throughout most of the life cycle. Others have a **dikaryotic** stage, which forms when cells of two different individuals unite but the nuclei from the two parents remain separate (**figure 20.2**). Only fungi have dikaryotic cells. The two nuclei eventually fuse, producing the diploid zygote; in most fungi, the zygote is the only diploid cell. The zygote immediately undergoes meiosis and yields haploid nuclei,

which then divide mitotically as the organism grows. (Chapters 8 and 9 describe mitosis and meiosis in detail.)

The body of a fungus is much more extensive than just a mushroom or the visible fuzz on a moldy piece of food (**figure 20.3**). Instead, fungi usually consist of an enormous number of **hyphae** (singular: hypha), which are microscopic, threadlike filaments. Hyphae branch rapidly within a food source, growing and absorbing nutrients at their tips. A **mycelium** is a mass of aggregated hyphae that may form visible strands in soil or decaying wood.

While the feeding hyphae remain hidden in the food, the reproductive structures emerge at the surface. Most fungi produce

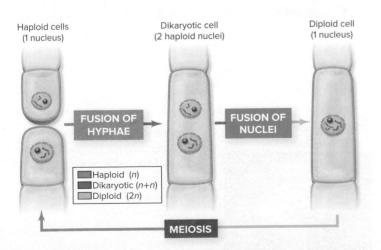

Haploid cells
(1 nucleus)

Dikaryotic cell
(2 haploid nuclei)

Diploid cell
(1 nucleus)

FUSION OF HYPHAE

FUSION OF NUCLEI

- Haploid (*n*)
- Dikaryotic (*n+n*)
- Diploid (2*n*)

MEIOSIS

Figure 20.2 Haploid, Dikaryotic, or Diploid? In most fungi, dikaryotic cells form when two haploid cells fuse but their nuclei remain separate. Fusion of the nuclei produces the diploid zygote.

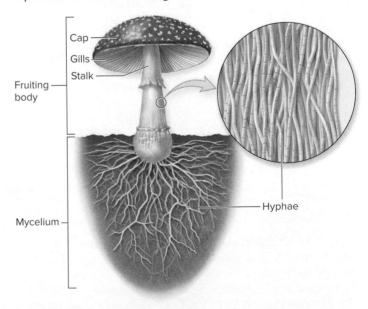

Cap

Gills

Stalk

Fruiting body

Mycelium

Hyphae

Figure 20.3 The Fungal Body. A mushroom arises from hyphae penetrating the fungus's food source. The mushroom itself is composed of hyphae that are tightly aligned to form a solid structure.

Conidia

Hypha

SEM (false color)   50 µm

Figure 20.4 **Conidia.** This fungus, a common mold, is producing abundant asexual spores.

©Steve Gschmeissner/Science Source

abundant **spores,** which are microscopic reproductive cells (see this chapter's Burning Question). Spores that land on a suitable habitat can germinate and give rise to hyphae, starting a new colony.

Spores can be asexually or sexually produced. **Conidia** are asexual spores (figure 20.4); the greenish or black powder on moldy food consists entirely of conidia. The production of sexual spores can be considerably more complex. In most fungal species, hyphae aggregate to form a **fruiting body,** a specialized sexual spore-producing organ such as a mushroom, puffball, or truffle. In the mushroom illustrated in figure 20.3, for example, the stalk supports a cap with numerous gills, each of which produces many spores.

B. Fungal Classification Is Traditionally Based on Reproductive Structures

Mycologists classify fungi into five phyla based on the presence and types of sexual structures (figure 20.5). The chytridiomycetes, or chytrids, produce gametes and asexual spores with flagella. Zygomycetes produce thick-walled sexual zygospores. Glomeromycetes do not reproduce sexually at all; instead, they have large, distinctive asexual spores. Ascomycetes produce sexual spores in characteristic sacs, and basidiomycetes release sexual spores from club-shaped structures.

Biologists are still studying the evolutionary relationships among the chytrids, zygomycetes, and glomeromycetes. The ascomycetes and basidiomycetes, however, are clearly sister groups, as figure 20.5 indicates. These are the most complex fungi. Their fruiting bodies, which include morels, truffles, mushrooms, and puffballs, are usually visible with the unaided eye.

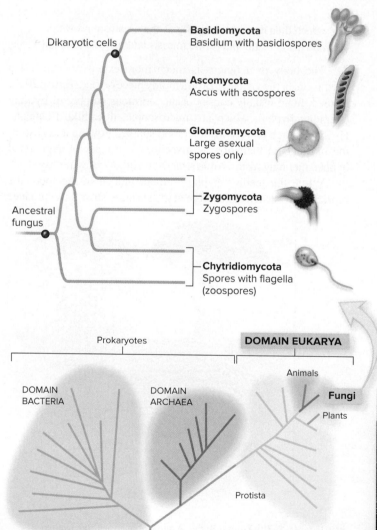

Dikaryotic cells

Basidiomycota
Basidium with basidiospores

Ascomycota
Ascus with ascospores

Glomeromycota
Large asexual spores only

Zygomycota
Zygospores

Ancestral fungus

Chytridiomycota
Spores with flagella (zoospores)

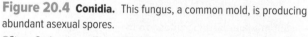

Prokaryotes

DOMAIN EUKARYA

DOMAIN BACTERIA

DOMAIN ARCHAEA

Animals

Fungi

Plants

Protista

20.1   MASTERING CONCEPTS

1. In what ways are fungi important in ecosystems?
2. What characteristics define the fungi?
3. What evidence suggests that fungi are more closely related to animals than to plants?
4. Describe the major structures of a fungus.

DOMAIN EUKARYA
Kingdom Fungi

Figure 20.5 **Fungal Diversity.**
The fungal kingdom contains five phyla, distinguished mainly on the basis of spore type.

Photo: ©Corbis RF

20.2 Chytridiomycetes Produce Swimming Spores

- Basidiomycetes
- Ascomycetes
- Glomeromycetes
- Zygomycetes
- **Chytridiomycetes**

The 1000 or so species of **chytridiomycetes** (phylum Chytridiomycota) may provide a glimpse of what the earliest fungi were like. Their body forms vary from single cells to slender hyphae. Chytrids are the only fungi to produce swimming cells, typically with a single flagellum. Some of these motile cells act as gametes; others are asexually produced, in which case they are termed **zoospores** (figure 20.6). Most chytrid life cycles are poorly understood.

These microscopic fungi are powerful decomposers, secreting enzymes that degrade cellulose, chitin, and keratin. One ecosystem where resident chytrids are particularly valuable is a ruminant's digestive tract. There, anaerobic chytrids start digesting the cellulose in a cow's grassy meal, paving the way for bacteria to continue the process.

Chytrids also contribute to the ongoing worldwide decline in amphibian populations. Some chytrids feed on keratin in a frog's skin, impairing the animal's ability to breathe through its body surface (figure 20.7). The fungus spreads to new hosts by releasing zoospores into the water. ⓘ *amphibians*, section 21.14

The chytrids and zygomycetes are sometimes called the basal fungi because of their simple structures and position close to the base of the fungal family tree. Each group was once considered to be monophyletic, but molecular data have argued against that hypothesis. Instead, each phylum apparently contains representatives from multiple lineages, and many fungi once considered to be chytrids have since been reclassified as zygomycetes.

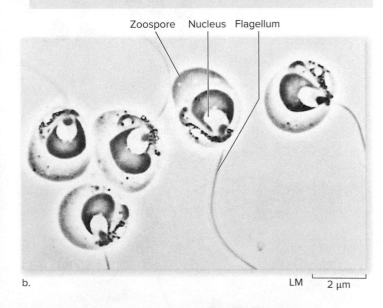

SEM ⌐ 15 μm

Figure 20.7 Chytrid Infection. In these infected frog skin cells, the arrow shows a tube through which a chytrid's zoospores leave the host. The inset shows a cricket frog, a species that is often infected with the chytrid.
(chytrid): ©Lee Berger, James Cook University; (frog): ©Dr. Janalee Caldwell

20.2 MASTERING CONCEPTS

1. How are flagellated cells adaptive in moist environments?
2. What do chytrids eat?

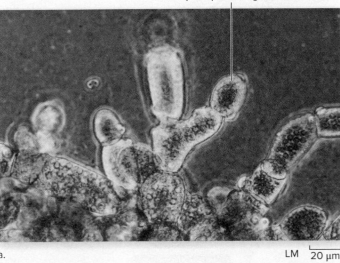

Zoospore-producing structure

a.   LM ⌐ 20 μm

Zoospore   Nucleus   Flagellum

b.   LM ⌐ 2 μm

Figure 20.6 Chytrids. (a) The rounded sacs at the tips of *Allomyces* hyphae produce flagellated zoospores. (b) Zoospores from *Blastocladiella*.

20.3 Zygomycetes Are Fast-Growing and Prolific

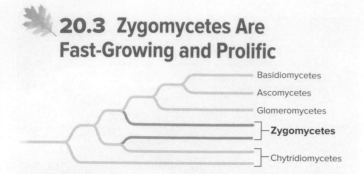

- Basidiomycetes
- Ascomycetes
- Glomeromycetes
- **Zygomycetes**
- Chytridiomycetes

The **zygomycetes** (phylum Zygomycota) account for only about 1% of identified fungi, but the 1000 or so species include some familiar organisms. The black mold *Rhizopus stolonifer,* which grows on bread, fruits, and vegetables, is a zygomycete. In addition to forming black fuzz on refrigerated leftovers, zygomycetes occur on decaying plant and animal matter in soil. Some are parasites of insects, and others must pass through the digestive system of an herbivore to complete their life cycles.

The zygomycetes are known for their spectacular growth rates, but they take their name from their mode of sexual reproduction: The prefix *zygo-* is derived from a Greek word that implies a pairing, or the joining of two parts. As illustrated in **figure 20.8**, this "pairing" occurs when two hyphae fuse. Then their haploid nuclei merge into a new structure, a diploid **zygospore** protected within a distinctive spiny, dark wall (the zygosporangium). After the merger, the hyphae appear as two vacated areas that hug the zygosporangium.

The diploid zygospore nucleus undergoes meiosis, and a haploid hypha emerges. The hypha immediately produces a spore sac, which breaks open and releases numerous haploid spores; as the products of meiosis, they are genetically variable. Each spore then gives rise to a hypha that grows into a haploid mycelium.

Zygomycetes typically produce few zygospores but many conidia. These asexual spores are the products of mitosis, so the spores produced by each hypha are genetically identical. The abundant haploid conidia spread easily to new habitats; **figure 20.9** shows two of the more unusual dispersal mechanisms. Spore germination produces a hypha, which quickly grows and branches as it digests its food. Within days, the new mycelium sprouts its own spore sacs, and the cycle begins anew.

20.3 MASTERING CONCEPTS

1. Where do zygomycetes occur?
2. How do zygomycetes reproduce asexually and sexually?
3. How does the zygospore fit into the zygomycete life cycle?

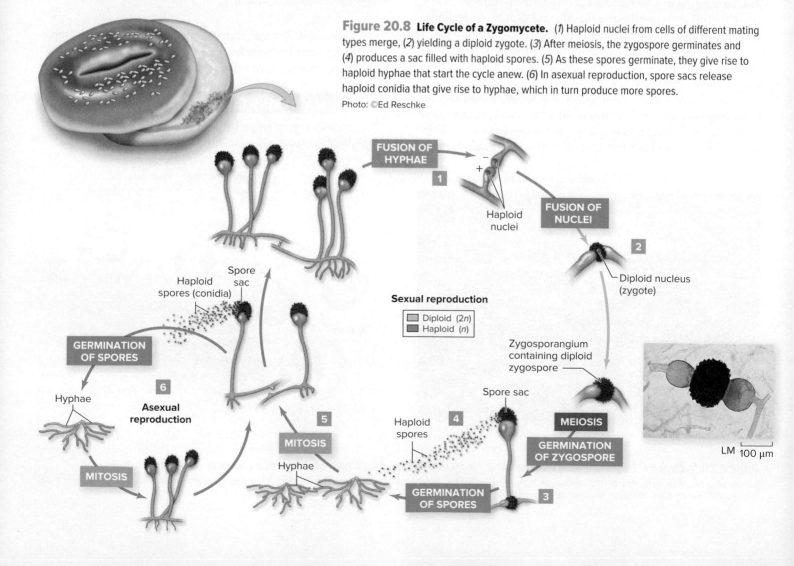

Figure 20.8 Life Cycle of a Zygomycete. (*1*) Haploid nuclei from cells of different mating types merge, (*2*) yielding a diploid zygote. (*3*) After meiosis, the zygospore germinates and (*4*) produces a sac filled with haploid spores. (*5*) As these spores germinate, they give rise to haploid hyphae that start the cycle anew. (*6*) In asexual reproduction, spore sacs release haploid conidia that give rise to hyphae, which in turn produce more spores.
Photo: ©Ed Reschke

FUSION OF HYPHAE
1
+
−
Haploid nuclei

FUSION OF NUCLEI
2
Diploid nucleus (zygote)

Sexual reproduction
Diploid (2*n*)
Haploid (*n*)

Zygosporangium containing diploid zygospore

Spore sac

MEIOSIS

GERMINATION OF ZYGOSPORE

LM 100 μm

Haploid spores

4

3

GERMINATION OF SPORES

Haploid spores

Hyphae

MITOSIS

5

Spore sac

Haploid spores (conidia)

GERMINATION OF SPORES

Hyphae

6

Asexual reproduction

MITOSIS

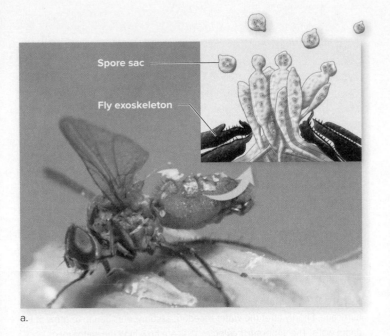

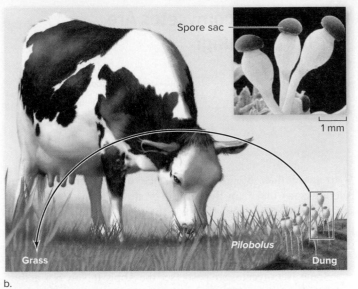

Figure 20.9 Zygomycete Spore Dispersal. (a) Spores of a zygomycete produce hyphae that penetrate a fly's exoskeleton. The fly is soon dead. When the fungus emerges, its spores (*inset*) infect additional flies. (b) *Pilobolus* stalks on dung are topped with black sacs (*inset*) that explosively release their spores onto nearby grass. Cattle eat the grass and spores, which later germinate in the animal's dung.
Photos: (a): ©Dwight Kuhn; (b): ©Andrew Syred/Science Source

20.4 Glomeromycetes Colonize Living Plant Roots

The **glomeromycetes** (phylum Glomeromycota) form the smallest group of fungi, with only about 200 known species. Their place in kingdom Fungi is still uncertain. These organisms were once classified as zygomycetes, but molecular evidence suggests they form their own clade.

Glomeromycetes have some unusual features. First, they live only in association with living plant roots; they cannot grow in laboratory petri dishes. Their partnership with plants is called a **mycorrhiza** (literally, "fungus-root"). In a mycorrhiza, specialized fungal hyphae colonize plant roots in a way that benefits both partners. The hyphae absorb water and minerals from the soil and share these resources with the plant; in return, the plant transfers carbohydrates to the fungus.

Glomeromycetes form a type of mycorrhiza in which the exchange of materials occurs at structures called arbuscules. The fungus produces the highly branched arbuscules inside the root cells (**figure 20.10**). Ascomycetes and basidiomycetes form a different type of mycorrhiza; section 20.7 describes these partnerships in more detail.

A second notable feature of the glomeromycetes is their unusually large asexual spores (see figure 20.10); some of the

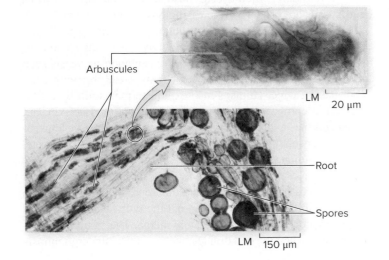

Figure 20.10 Mycorrhizal Root. The arbuscules and large spores of a glomeromycete fill this root. A stain colors the fungal structures blue, making them easily visible with the microscope.
(both): ©Joseph B. Morton

largest are visible with the unaided eye. At the same time, the glomeromycetes are not known to produce sexual spores. Because biologists classify most fungi based on their sexual structures, the absence of a sexual life cycle is partly responsible for the difficulty in placing these fungi in the phylogenetic tree.

20.4 MASTERING CONCEPTS

1. What are the distinctive features of glomeromycetes?
2. Describe arbuscular mycorrhizae.

20.5 Ascomycetes Are the Sac Fungi

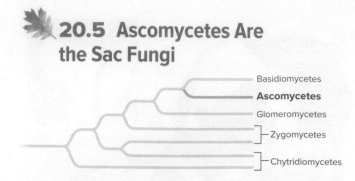

- Basidiomycetes
- **Ascomycetes**
- Glomeromycetes
- Zygomycetes
- Chytridiomycetes

With more than 50,000 species, the **ascomycetes** (phylum Ascomycota) make up the largest group of fungi. Their lifestyles vary widely. Some colonize living insects, others decompose organic matter, and still others form partnerships with photosynthetic organisms (see section 20.7). A few are carnivores (figure 20.11).

Some ascomycetes are pests (see the Apply It Now box at the end of this section). They cause most fungal diseases of plants, including Dutch elm disease and chestnut blight. A few species cause skin infections such as athlete's foot and other human diseases. Many of the common molds that ravage flood-damaged homes are ascomycetes (figure 20.12).

Nevertheless, ascomycetes also benefit humans. Truffles and morels are prized for their delicious flavors (figure 20.13). The common mold *Penicillium* is famous for secreting penicillin, the first antibiotic discovered. *Penicillium* species also lend their sharp flavors to Roquefort cheese. Fermentation by yeasts such as *Saccharomyces* is essential for baking bread, brewing beer, and making wine. Humans also exploit fermentation by a filamentous ascomycete to produce soy sauce, sake, and rice vinegar. Cyclosporine, a drug that suppresses the immune systems of organ transplant recipients, comes from an ascomycete. ⓘ *fermentation*, section 6.8B

Ascomycetes can produce enormous numbers of asexual and sexual spores (figure 20.14). In sexual reproduction, the hyphae of compatible mating types fuse, but the individual nuclei from the two parents do not immediately merge. The resulting cell is dikaryotic; when the cell divides, the two nuclei undergo mitosis separately.

The fruiting body consists of tightly woven dikaryotic hyphae, often in a cuplike or bottlelike shape. Eventually, the two nuclei in certain dikaryotic cells fuse, forming a diploid zygote.

Figure 20.12 Mold Everywhere. This home was flooded during a hurricane. Shortly after the water receded, ascomycetes began to grow in the damp walls. The uppermost colonies indicate how high the water rose.
©Julie Dermansky/Science Source

The zygote immediately undergoes meiosis and produces four haploid nuclei, each of which usually divides once by mitosis. The result is eight haploid **ascospores,** so named because they form in a saclike **ascus** (plural: asci). After dispersal by wind, water, or animals, ascospore germination yields a new haploid individual. (The Focus on Model Organisms box explains how the asci and metabolism of the red bread mold, *Neurospora crassa,* have contributed to biological research.)

20.5 MASTERING CONCEPTS

1. List examples of ascomycetes that are important to humans.
2. Describe asexual and sexual reproduction in an ascomycete.
3. How do dikaryotic cells fit into the ascomycete life cycle?
4. How does an ascus come to contain eight ascospores?

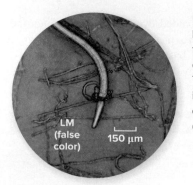

Figure 20.11 A Carnivorous Fungus. Threadlike loops of the fungus *Drechslerella anchonia* constrict around a nematode worm. The fungus will thread its hyphae into its prey, releasing digestive enzymes and eating it from within.
©Photo Researchers/Science Source

a.   b.

Figure 20.13 Edible Ascomycetes. (a) Truffles are familiar ascomycetes; they produce asci on internal folds of tissue. (b) Each "pit" in a morel's cap produces thousands of asci. This morel is about 10 cm tall.
(a): ©DEA/G. Cozzi/Getty Images; (b): ©Robert Marien/Corbis RF

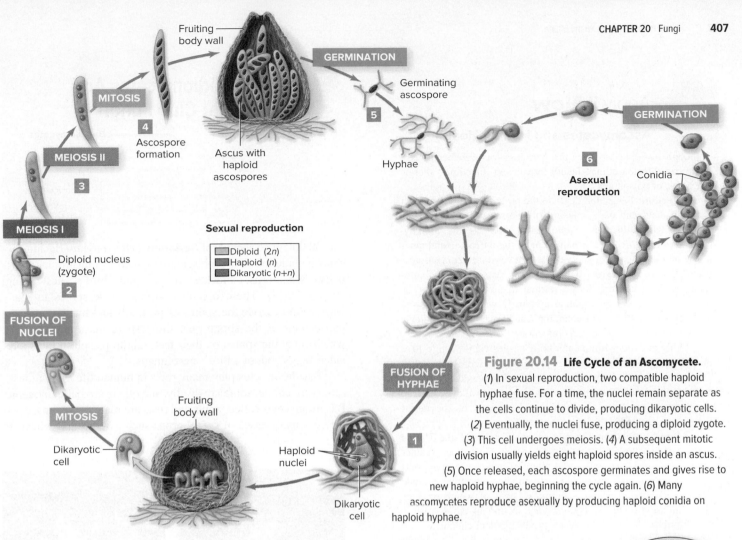

Figure 20.14 **Life Cycle of an Ascomycete.**
(*1*) In sexual reproduction, two compatible haploid hyphae fuse. For a time, the nuclei remain separate as the cells continue to divide, producing dikaryotic cells. (*2*) Eventually, the nuclei fuse, producing a diploid zygote. (*3*) This cell undergoes meiosis. (*4*) A subsequent mitotic division usually yields eight haploid spores inside an ascus. (*5*) Once released, each ascospore germinates and gives rise to new haploid hyphae, beginning the cycle again. (*6*) Many ascomycetes reproduce asexually by producing haploid conidia on haploid hyphae.

F🔍CUS on Model Organisms

Red Bread Mold: *Neurospora crassa*

Figure 20.A *Neurospora crassa.*
This photo shows the asci inside a fruiting body. Note that each ascus contains exactly eight spores.
©Steven P. Lynch/McGraw-Hill Education

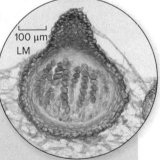

Why would biologists study *Neurospora crassa* (figure 20.A), a filamentous ascomycete commonly known as red bread mold? First, *N. crassa* can tell us about other ecologically, industrially, and medically important fungi. Second, *N. crassa* has the traits that all model organisms share: It is small and easy to grow, and it has a rapid reproductive cycle. Third, this fungus has special advantages for tracing inheritance patterns. Because its hyphae are haploid, *N. crassa* expresses every recessive or mutated allele in its DNA (in diploid organisms, dominant alleles can mask the presence of their recessive counterparts).

A full accounting of *Neurospora*'s contributions to basic biology could fill a book. The following list, however, details some ways this fungus has enhanced our understanding of life:

- **Crossing over:** In the 1920s and 1930s, *Neurospora* provided the first clear evidence of crossing over, which occurs during prophase I of meiosis. *Neurospora*'s narrow, tube-shaped asci are especially well suited to such studies because they retain the products of meiosis in the original order in which the cells divided (see figure 20.14). ⓘ *crossing over,* section 9.5A

- **One-gene/one-enzyme hypothesis:** In the 1940s, George Beadle, Edward Tatum, and their colleagues isolated *Neurospora*

mutants with unusual nutritional requirements. Their observation that each mutant had a single enzyme deficiency provided convincing evidence for the idea that one gene encodes each protein in a cell, a topic described in detail in chapter 7.

- **Circadian rhythms:** *Neurospora* cultures produce conidia at roughly 24-hour intervals, a phenomenon called a circadian rhythm (*circa,* "about"; *dies,* "day"). Mutants that produce conidia constantly or not at all have yielded insights into regulation of the so-called clock-controlled genes.

- **Gene regulation:** Some genes in a cell are "silent" at any given time. *Neurospora* has helped reveal some of the ways cells regulate gene expression. For example, cells can "tag" DNA with chemical groups that prevent transcription, or they may destroy RNA after transcription. ⓘ *regulation of gene expression,* section 7.6

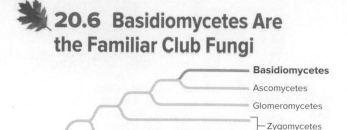

20.6 Basidiomycetes Are the Familiar Club Fungi

The 30,000 or so species of **basidiomycetes** (phylum Basidiomycota) include familiar representatives such as mushrooms, puffballs, stinkhorns, bracket fungi, and bird's nest fungi (**figure 20.15**). These organisms spread their spores in many ways. Wind carries the spores of puffballs and mushrooms; the putrid odor of the stinkhorn's slimy spore mass attracts flies, which carry the spores on their feet; raindrops splash the spore-laden "eggs" out of a bird's nest fungus.

Basidiomycetes play many roles in human life. Some mushrooms are edible, some are deadly, and others are hallucinogenic. Basidiomycetes called smuts and rusts are plant pathogens, causing serious diseases of cereal crops such as corn and wheat. In

Figure 20.15 **A Gallery of Basidiomycetes.** (a) Puffballs. (b) Stinkhorns. (c) Turkey tail bracket fungi. (d) Bird's nest fungi.

(a): ©RF Company/Alamy RF; (b): ©George McCarthy/Corbis; (c): ©imageBROKER/Alamy RF; (d): ©Steven P. Lynch/McGraw-Hill Education

Apply It Now

Ascomycetes and Human Health

Although most ascomycetes are harmless (or even beneficial), some can threaten human health by causing infections, allergic reactions, or poisonings.

Infection: Fungi that degrade the protein keratin can infect our skin, hair, and nails, causing ringworm, athlete's foot, and other irritating diseases. Some fungi, such as the ascomycete *Candida albicans,* normally inhabit the mucous membranes of the mouth, intestines, and vagina. In a yeast infection, populations of these fungi grow out of control. Yeast infections commonly occur when the body's normal microbial community is disrupted or when the immune system is weakened. Wearing loose-fitting clothing and changing out of wet clothes quickly helps reduce the risk of a yeast infection.

Other infectious fungi enter body tissues via the lungs. People who inhale dust containing spores of an ascomycete called *Coccidioides immitis,* for example, may develop a deadly disease called valley fever. The simplest way to prevent valley fever is to avoid breathing dusty air in high-risk regions of the desert Southwest, especially during the summer.

Allergic reactions: People in many parts of the United States receive "mold spore counts" along with their weather reports, and residents with mold allergies may avoid going outside when the counts are high. Common allergenic fungi include species of *Alternaria* and *Aspergillus,* both of which produce abundant asexual spores. ⓘ *allergies,* section 34.5C

Toxicity: Some ascomycetes produce potent chemicals, and people sometimes report headaches and memory loss after living or working in moldy buildings. *Stachybotrys atra,* the toxic "black mold," commonly takes the blame, despite scant proof that fungal toxins cause these symptoms.

The fungus that causes a plant disease called ergot (**figure 20.B**) has pronounced effects on the human nervous system. The drug lysergic acid diethylamide (LSD) comes from a chemical found in ergots. People who eat bread made from ergot-contaminated grain can therefore develop convulsions, gangrene, and psychotic delusions. Some of the women in Salem, Massachusetts, who were executed as witches in the late 1690s may have been suffering from ergotism, although their uncontrollable movements were attributed to demons.

Ergotism is uncommon today because of simple preventive measures. Plowing fields deeply and rotating crops annually minimizes ergot infection. In addition, when ergot-infested grain is placed in a bath of salty water, the healthy grains sink. Meanwhile, the ergots float to the surface and can be skimmed off.

Figure 20.B **Ergot.** This black ergot is a mass of hyphae.
©Carmen Rieb/Shutterstock RF

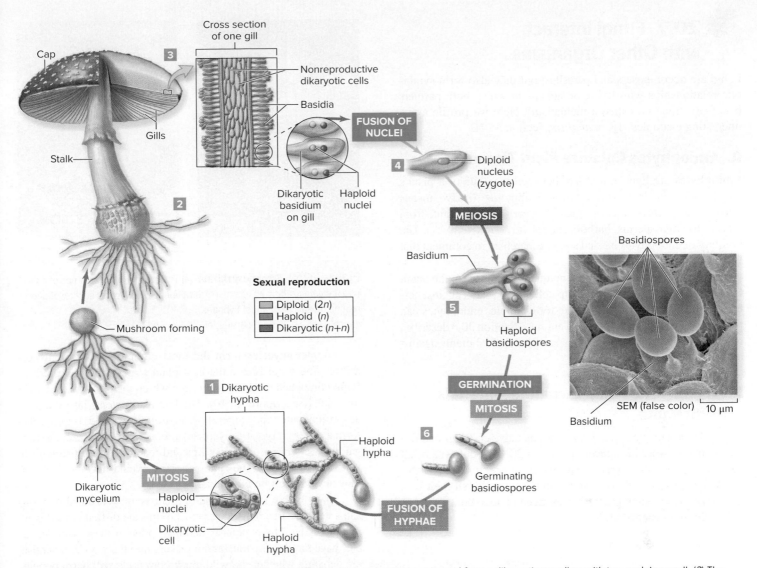

Figure 20.16 Life Cycle of a Basidiomycete. (*1*) Hyphae of compatible mating types unite and form a dikaryotic mycelium with two nuclei per cell. (*2*) The dikaryotic fungus grows and forms a mushroom. (*3*) Basidia form on gills on the underside of the mushroom cap. (*4*) The two nuclei in the basidium fuse, creating a diploid zygote that undergoes meiosis. (*5*) The resulting haploid nuclei migrate into four basidiospores, which the mature mushroom sheds. (*6*) The spores germinate, and new haploid hyphae grow.

Photo: ©Cultura RM Exclusive/Rolf Ritter/Getty Images

forests, wood-decaying fungi return carbon locked in trees to Earth's atmosphere. Unfortunately, their talent for degrading the cellulose and lignin in fallen logs also makes them serious pests in another context: They cause dry rot in wooden wall studs and other building materials.

Basidiomycetes can reproduce asexually via conidia, but the sexual portion of the life cycle is usually the most prominent. As **figure 20.16** shows, the fusion of two haploid hyphae creates a dikaryotic mycelium. This mycelium typically grows unseen within its food source. When environmental conditions are favorable, however, one or more mushrooms emerge. Lining each mushroom's gills are numerous dikaryotic, club-shaped cells called **basidia** (singular: basidium). Inside each basidium, the haploid nuclei fuse, giving rise to a diploid zygote. The zygote immediately undergoes meiosis, yielding four haploid nuclei.

Each nucleus migrates into a **basidiospore,** which germinates after dispersal. The cycle begins anew.

Sometimes a circle of mushrooms emerges from the ground all at once. The growth pattern of the underground mycelium explains this phenomenon. Hyphae extend outward in all directions from a colony's center. Mushrooms poke up at the margins of the mycelium, creating the "fairy rings" of folklore.

20.6 MASTERING CONCEPTS

1. What are some familiar basidiomycetes?
2. How are basidiomycetes different from ascomycetes?
3. Describe the sexual life cycle of basidiomycetes.
4. How does a "fairy ring" form?

20.7 Fungi Interact with Other Organisms

Fungi are decomposers and parasites, but they also form symbiotic relationships with living organisms in which both partners benefit (a situation called a mutualism). Here we profile a few interesting examples. ⓘ *mutualism,* section 38.1B

A. Endophytes Colonize Plant Tissues

Endophytes are hyphae that live between the cells of a plant's tissues without triggering disease symptoms (*endo-* means "inside," and *-phyte* means "plant"). Every known plant, from mosses to angiosperms, harbors endophytes (figure 20.17). The ubiquity of endophytes has led some researchers to comment that "all plants are part fungi."

Some of these internal fungi neither help nor harm the plant. Others secrete substances that help defend the plants against herbivores. In grasses such as fescue, for example, endophytes can produce toxins that sicken grazing animals. Section 20.8 describes research showing that endophytes can also defend plants against other fungi.

B. Mycorrhizal Fungi Exchange Materials with Roots

Mycorrhizae are specialized associations between fungi and living roots that were introduced in section 20.4. About 80% of all land plants, including grasses, shrubs, and trees, form mycorrhizae. Some plants, such as orchids, cannot live without their fungal associates. Other plant species depend less on the fungi, especially in nutrient-rich soils.

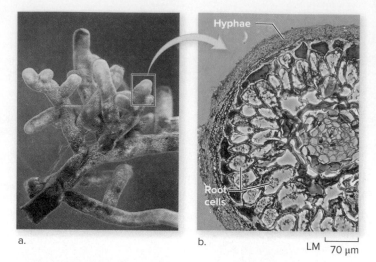

Figure 20.18 Ectomycorrhizae. (a) The creamy white root tips of a pine tree are colonized by an ectomycorrhizal fungus. (b) A cross section of one root tip reveals a sheath of hyphae.

(a): ©R Henrik Nilsson; (b): ©Biology Pics/Science Source

Glomeromycetes form the most common types of mycorrhizae. The fungi pierce the host plant's root cells and produce highly branched arbuscules through which the partners exchange materials (see figure 20.10). Hyphae also extend into the soil, absorbing water and minerals. Basidiomycetes and ascomycetes form a different type of mycorrhiza. In ectomycorrhizae, the fungal hyphae wrap around root tips and reach into the surrounding soil (figure 20.18). The filaments may extend into the root, but they never penetrate individual cells.

Many edible fungi depend on their ectomycorrhizal relationships with live tree roots; these mushrooms are difficult to cultivate commercially. The popularity and high price of these wild delicacies have lured many mushroom pickers into the woods. Scientists are debating whether the wild mushroom trade will harm populations of fungi—and the trees that rely on them—in the long term.

Although both endophytes and mycorrhizal fungi occupy roots, these fungus–root interactions typically differ in at least two ways. First, the hyphae of mycorrhizal fungi produce specialized structures such as arbuscules that exchange materials with the host plant. Those of endophytes do not. Second, many mycorrhizal fungi cannot survive on their own; endophytes are often less dependent on their plant hosts.

C. Some Ants Cultivate Fungi

The leaf-cutter ants of Central and South America and the southern United States cultivate a basidiomycete in special underground chambers. Into each chamber, the ants transport a paste made from saliva and the disks that they cut from green leaves (figure 20.19). The fungi grow on the paste. Adult and larval ants eat the hyphae so quickly that mushrooms never get a chance to form.

The ants and their fungal gardens constitute a mutualistic partnership; both the ants and the fungi have a steady food supply. Both benefit. But how do the ants keep out competitors or pathogens? Biologists once thought that the ants simply ate all interlopers, but then they discovered a third partner: bacteria coating parts of the

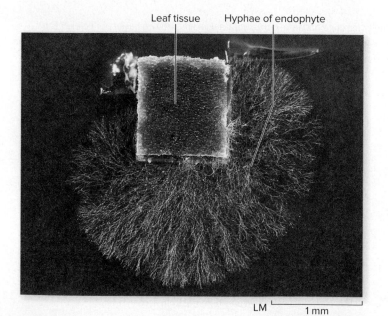

Leaf tissue   Hyphae of endophyte

LM ⌐ 1 mm

Figure 20.17 Endophytes. Hyphae of an endophytic fungus grow out of a tiny piece of leaf tissue.
©Dr. Elizabeth Arnold

Figure 20.19 Ants at Work. Leaf-cutter ants use the vegetation they harvest to farm fungi—their food source.
©Gail Shumway/Photographer's Choice/Getty Images

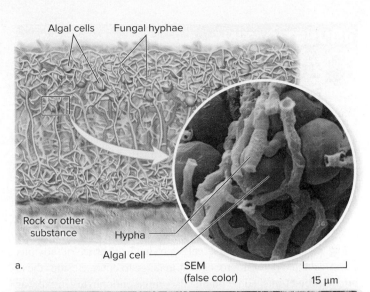

a. SEM (false color) 15 μm

ants' exoskeletons. These microbes secrete a potent toxin, which kills an ascomycete that attacks the cultivated fungus.

D. Lichens Are Dual Organisms

A **lichen** forms when a fungus, either an ascomycete or a basidiomycete, harbors green algae or cyanobacteria among its hyphae (figure 20.20). The photosynthetic partner contributes carbohydrates; the fungus absorbs water and essential minerals.

Lichens are sometimes called "dual organisms" because the two species—the fungus and its photosynthetic partner—appear to be one individual when viewed with the unaided eye. The body forms can vary widely. Many lichens are colorful, flattened crusts, but others form upright structures that resemble mosses or miniature shrubs. Still others are long, scraggly growths that dangle from tree branches.

Lichens are important ecologically because they secrete acids that break down rock, a first step in the soil-building process. Many also harbor nitrogen-fixing cyanobacteria, which make usable nitrogen available to plants. In addition, some animals, including caribou, eat lichens. ⓘ *nitrogen fixation,* section 23.1C

Just about any stable surface, from tree bark to boulders to soil, can support lichens. They live everywhere from the driest deserts to the wettest tropical rain forests to the frozen Arctic. Lichens survive dehydration by suspending their metabolism, only to revive when moisture returns. One type of habitat, however, is hostile to lichens: polluted areas. Lichens absorb toxins but cannot excrete them. Toxin buildup hampers photosynthesis, and the lichen dies. Disappearance of native lichens is a sign that pollution is disturbing the environment; scientists therefore use lichens to monitor air quality.

20.7 MASTERING CONCEPTS

1. What are endophytes?
2. Which types of fungi form arbuscular mycorrhizae and ectomycorrhizae?
3. How do ants, plants, fungi, and bacteria interact in leaf-cutter ant colonies?
4. How do fungi and autotrophs interact in a lichen?
5. How do scientists use lichens to monitor pollution?

c. SEM (false color) 2 mm d.

Figure 20.20 Anatomy of a Lichen. (a) A cross section of a lichen reveals fungal hyphae wrapping tightly around their photosynthetic "partner" cells. (b) This rock is home to an assortment of colorful lichens. (c) The tiny, stalked cups of this trumpet lichen are fruiting bodies. (d) Orange lichens colonize a tree branch.

Photos: (a): ©Eye of Science/Science Source; (b): ©William H. Mullins/Science Source; (c): ©Eye of Science/Science Source; (d): ©Steven P. Lynch/McGraw-Hill Education

INVESTIGATING LIFE

20.8 The Battle for Position in Cacao Tree Leaves

Chocolate has many friends; its delectable taste and healthful antioxidants make it a favorite food. But chocolate also has its detractors, beyond low-fat diet enthusiasts. Most of its enemies are microbial pathogens of the cacao tree (*Theobroma cacao;* figure 20.21), the source of cocoa. Protists and fungi frequently attack the tree's foliage, branches, and seed pods.

All organisms, including plants, defend themselves against disease. Waxy coverings on leaves and stems prevent the entry of many pathogens, and plant cells produce an array of noxious chemicals that deter microbes that manage to enter. Besides these innate protections, some plants also enlist endophytes in the war against predators and pathogens.

The cacao tree acquires its resident fungi from the environment. A new leaf is endophyte-free, but fungal spores soon arrive in wind or rain. Hyphae penetrate the waxy cuticle of the young leaf and set up shop within the plant's tissues, absorbing water and nutrients that leak out of the leaf veins. The endophytes do not trigger the plant's defenses, raising an interesting question: What does the tree gain from allowing plant-eating fungi to live inside its tissues?

University of Arizona biologist A. Elizabeth Arnold, along with a research team at the Smithsonian Tropical Research Institute in Panama, wondered whether the endophytes help protect cacao trees from disease. The researchers needed endophyte-free plants to test their hypothesis, so they planted seeds from cacao plants in a greenhouse and protected them from fungal spores that might blow in from outside. Once the trees were 100 days old, the team ensured that the plants were endophyte-free.

Figure 20.21 **The Cacao Tree.** The pods of *Theobroma cacao* are the source of chocolate.
©Robert van der Hilst/Corbis

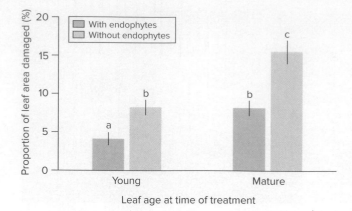

Figure 20.22 **Helpful Partners.** Endophytic fungi protected young and mature cacao leaves from damage by the pathogen *Phytophthora*. (Letters on bars indicate statistical significance; see appendix B.)

Next, they sprayed spores from several species of endophyte fungi on some of the leaves, so each tree had some treated leaves and some that remained free of the fungi. Once the fungi were thriving in the sprayed leaves, the researchers were ready to test their hypothesis. They inoculated both endophyte-treated and control leaves with spores of a protist called *Phytophthora,* which causes black pod disease of cacao.

Fifteen days later, Arnold and her team counted the number of dead leaves on each tree and measured the *Phytophthora*-damaged area on surviving leaves. These combined measures showed that plant parts without endophytes lost twice as much leaf area as those with the resident fungi (figure 20.22).

No one is sure what the endophytes gain from the relationship, other than the nutrients they absorb from inside the leaf. One hypothesis is that their biggest gain comes after their home leaf ages, dies, and falls to the ground. Perhaps the endophytes colonize a living leaf to get "first dibs" on the dead tissue.

Win–win relationships, such as the one between the cacao tree and its resident fungi, are common in life. Scientists are trying to develop some of these cooperative organisms into biological control agents that might prevent disease in cacao plants without the use of harmful chemicals. Our endophyte allies are already living quietly in the stems and leaves of our crop plants. Perhaps one day farmers will enlist these friendly fungi in a more aggressive fight to preserve our chocolate fix.

Source: Arnold, A. Elizabeth, Luis Carlos Mejia, Damond Kyllo, et al. December 23, 2003. Fungal endophytes limit pathogen damage in a tropical tree. *Proceedings of the National Academy of Sciences,* U.S.A., vol. 100, pages 15649–15654.

20.8 MASTERING CONCEPTS

1. According to figure 20.22, were young leaves with endophytes significantly less damaged than those without endophytes? Explain how you know.

2. How would you design an experiment to determine whether one species of endophyte or some combination of multiple species protects leaves against pathogens?

CHAPTER SUMMARY

20.1 Fungi Are Essential Decomposers

- **Fungi** are widespread and profoundly affect ecosystems by feeding on living and dead organic material.

A. Fungi Are Eukaryotic Heterotrophs That Digest Food Externally

- Fungi are more closely related to animals than to plants.
- Fungi are heterotrophs that produce chitin cell walls and glycogen. Fungi have both asexual and sexual reproduction, and some have unique **dikaryotic** cells with two genetically different nuclei.
- A fungal body includes a **mycelium** built of threads called **hyphae,** which may form a **fruiting body.** Some fungi occur both as filamentous hyphae and as unicellular **yeasts.**
- Fungi reproduce using asexual and sexual **spores** (figure 20.23). **Conidia** are asexually produced spores.

B. Fungal Classification Is Traditionally Based on Reproductive Structures

- The five main groups of fungi are chytridiomycetes, zygomycetes, glomeromycetes, ascomycetes, and basidiomycetes.
- Molecular evidence indicates that ascomycetes and basidiomycetes are sister groups.
- Figure 20.24 summarizes the diversity of the fungi.

20.2 Chytridiomycetes Produce Swimming Spores

- **Chytridiomycetes** are microscopic fungi that produce flagellated cells; if the cells are asexually produced, they are termed **zoospores.**
- Chytrids decompose major biological carbohydrates such as cellulose, chitin, and keratin.

20.3 Zygomycetes Are Fast-Growing and Prolific

- **Zygomycetes** reproduce rapidly via asexually produced, haploid, thin-walled spores.
- Sexual reproduction occurs when hyphae of different mating types fuse their nuclei into a **zygospore.** The zygospore undergoes meiosis, then generates a spore sac. Spore germination yields haploid hyphae, continuing the cycle.

20.4 Glomeromycetes Colonize Living Plant Roots

- **Glomeromycetes** form **mycorrhizae,** which are mutually beneficial associations between fungi and roots. Mycorrhizal plants have extra surface area for absorbing nutrients, and the fungus acquires carbohydrates from its plant host.
- Glomeromycetes have no known sexual stage.

20.5 Ascomycetes Are the Sac Fungi

- **Ascomycetes** include important plant pathogens, but they also have many uses in industry.
- Hyphae are haploid for most of the life cycle. Asexual reproduction typically occurs via conidia.
- In sexual reproduction, haploid hyphae join, producing a brief dikaryotic stage. After the nuclei fuse, meiosis yields haploid **ascospores** in a saclike **ascus.**

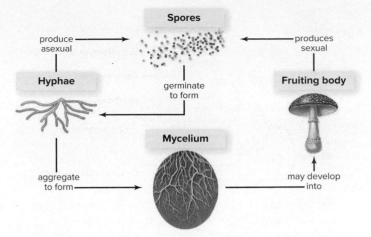

Figure 20.23 **Fungal Reproduction: A Summary.**

20.6 Basidiomycetes Are the Familiar Club Fungi

- **Basidiomycetes** include mushrooms and other familiar fungi. Wood decay fungi are important in ecosystems but also rot building materials. Some are plant pathogens.
- Basidiomycetes are dikaryotic for most of the life cycle. Asexual reproduction occurs by budding, by fragmentation, or with asexual spores.
- Sexual reproduction occurs when haploid nuclei in **basidia** along the gills of mushrooms fuse. The resulting zygote then undergoes meiosis, producing haploid **basidiospores.** Shortly after spore germination, hyphae fuse, regenerating the dikaryotic state.

20.7 Fungi Interact with Other Organisms

A. Endophytes Colonize Plant Tissues

- **Endophytes** occupy the spaces between living plant cells without triggering disease symptoms.

B. Mycorrhizal Fungi Exchange Materials with Roots

- Glomeromycetes form arbuscular mycorrhizae, whereas ascomycetes and basidiomycetes form ectomycorrhizae.

C. Some Ants Cultivate Fungi

- Leaf-cutter ants cultivate basidiomycetes on a nutritious paste made from leaf disks, and bacteria on the ants kill a parasitic ascomycete.

D. Lichens Are Dual Organisms

- A **lichen** is a compound organism that consists of a fungus in intimate association with cyanobacteria or green algae.
- Lichens are useful air pollution indicators.

20.8 Investigating Life: The Battle for Position in Cacao Tree Leaves

- Experimental evidence indicates that endophytes protect cacao tree leaves from *Phytophthora* pathogens.

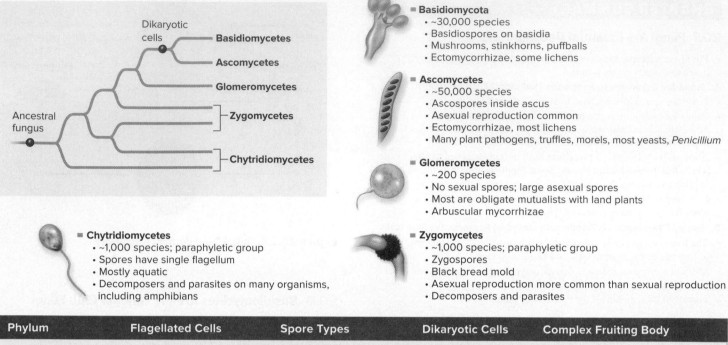

Basidiomycota
- ~30,000 species
- Basidiospores on basidia
- Mushrooms, stinkhorns, puffballs
- Ectomycorrhizae, some lichens

Ascomycetes
- ~50,000 species
- Ascospores inside ascus
- Asexual reproduction common
- Ectomycorrhizae, most lichens
- Many plant pathogens, truffles, morels, most yeasts, *Penicillium*

Glomeromycetes
- ~200 species
- No sexual spores; large asexual spores
- Most are obligate mutualists with land plants
- Arbuscular mycorrhizae

Zygomycetes
- ~1,000 species; paraphyletic group
- Zygospores
- Black bread mold
- Asexual reproduction more common than sexual reproduction
- Decomposers and parasites

Chytridiomycetes
- ~1,000 species; paraphyletic group
- Spores have single flagellum
- Mostly aquatic
- Decomposers and parasites on many organisms, including amphibians

Phylum	Flagellated Cells	Spore Types	Dikaryotic Cells	Complex Fruiting Body
Chytridiomycota	Yes	Zoospores	No	Absent
Zygomycota	No	Zygospores; conidia	No	Absent
Glomeromycota	No	Large asexual spores only	No	Absent
Ascomycota	No	Ascospores in ascus; conidia	Yes	Present
Basidiomycota	No	Basidiospores on basidium; conidia	Yes	Present

Figure 20.24 Fungal Diversity: A Summary.

MULTIPLE CHOICE QUESTIONS

1. The ancestral fungus shown in the evolutionary tree illustrated in figure 20.24 was probably most similar to what type of organism?
   a. A bacterium
   b. A plant
   c. A basidiomycete
   d. A chytrid

2. In addition to the cell wall of a fungus, where would you find the carbohydrate chitin?
   a. The exoskeleton of some animals
   b. The plant cell wall
   c. The outer membrane of a bacterium
   d. The nuclear membrane of a protist

3. Fungi are traditionally classified based on their
   a. habitat.
   b. dikaryotic cells.
   c. spore type.
   d. method of acquiring energy.

4. Which phylum of fungi is not known to have a sexual stage in its life cycle?
   a. Chytridiomycetes
   b. Glomeromycetes
   c. Basidiomycetes
   d. Zygomycetes

5. Fungi are considered _____ because they get their carbon and energy from _____.
   a. autotrophs; photosynthesis
   b. autotrophs; organic matter
   c. heterotrophs; organic matter
   d. heterotrophs; photosynthesis

6. A dikaryotic cell develops into a zygote when
   a. mitosis divides the two nuclei into separate cells, which then combine at fertilization.
   b. mitosis occurs without cytokinesis.
   c. the two haploid nuclei within the cell fuse.
   d. meiosis occurs once and cytokinesis occurs twice.

7. The fruiting body of a basidiomycete is often a mushroom, which
   a. is the product of asexual reproduction.
   b. produces haploid spores.
   c. is composed mostly of dikaryotic cells.
   d. Both b and c are correct.

8. The feeding hyphae of a basidiomycete are mostly
   a. haploid.
   b. diploid.
   c. dikaryotic.
   d. acellular.

9. What is a lichen?
   a. A type of photosynthetic fungus
   b. A combination of a fungus and an alga or a cyanobacterium
   c. A combination of two phyla of fungi
   d. A combination of a fungus and a root

10. How does a mycorrhizal basidiomycete differ from a free-living basidiomycete?
    a. It gains its nutrients from a living plant.
    b. It has no sexual stage in its life cycle.
    c. It is not edible.
    d. All of the above are correct.

Answers to these questions are in appendix A.

WRITE IT OUT

1. Give examples of fungi that are important economically, ecologically, and as food for humans.

2. Fungi and animals are both heterotrophs. What does this mean? How do fungi and animals differ in how they obtain food?

3. Review figure 20.5. Are fungi more closely related to animals or to plants? What characteristics do fungi share with plants? What characteristics do fungi share with animals?

4. In figure 20.5, which groups of fungi are monophyletic? What term can you use to describe the other groups?

5. What characteristics distinguish each phylum of fungi?

6. Many fungi produce chemicals that inhibit bacterial growth. Why might the genes encoding these chemicals be adaptive to fungi?

7. Review figure 19.5, which shows the alternation of generations in plants. Compare and contrast the life cycles of zygomycetes, ascomycetes, and basidiomycetes with the basic plant life cycle.

8. Sketch each of the following structures: mycelium; ascus with ascospores; zoospore; zygospore; basidium with basidiospores.

9. Create a table with three columns. Write the following terms in the first column: mycelium, zygospore, basidium, conidium, and hyphae. Define the terms in the second column, and classify the terms as diploid, dikaryotic, or haploid in the third column.

10. Does the absence of fuzzy mold on bread mean that it is free of fungi? Why or why not?

11. Describe the difference between sexual and asexual reproduction in zygomycetes. Why might asexual reproduction be more common than sexual reproduction?

12. Other than shared DNA sequences, what characteristics place ascomycetes and basidiomycetes together as sister groups?

13. Each ascus within an ascomycete fruiting body contains eight cells. Are these cells haploid, diploid, or dikaryotic? Is every cell within an ascus genetically unique? What is the fate of each cell?

14. List some places where you might find ascomycetes.

15. Compare and contrast endophytes, mycorrhizae, and lichens.

16. Use the Internet to find examples of chytrids, zygomycetes, ascomycetes, and basidiomycetes that cause diseases in plants or animals. How does each fungus infect a host and spread to new hosts? What can humans do to fight each disease? What are the costs and benefits of doing so?

17. Some endophytes produce compounds that fight bacteria, making them potential new sources of antibiotics. Describe an experimental method that would allow you to screen endophytes for antibacterial activity.

18. Describe how experiments might show that
   a. chytrids are killing amphibians.
   b. fungi benefit more from a lichen relationship than do algae.
   c. bacteria help leaf-cutting ants cultivate one fungus while killing another.
   d. overharvesting of mycorrhizal basidiomycetes harms forest health.
   e. mold on a building's walls can cause illness in the building's occupants.

19. Why might it be challenging to produce drugs that treat fungal infections without harming human cells?

20. White nose syndrome is an illness that weakens and kills bats roosting in caves. Affected animals have a white fungus around their muzzle and on their wings, but no one knows if the fungus causes the illness or simply infects bats that are weakened by some other disease. Search the Internet for the latest information about white nose syndrome; then propose a testable hypothesis to explain how the fungus might have entered bat populations. What information would you need to determine whether the fungus actually causes the disease? How can people help prevent the spread of the fungus from cave to cave? If the disease continues to spread, how might the widespread death of bats affect ecosystems?

21. Create a graph showing the number of species in each group of fungi. Which group is least diverse? Based on the habitats and reproductive structures of these fungi, propose an explanation for their relatively low diversity.

PULL IT TOGETHER

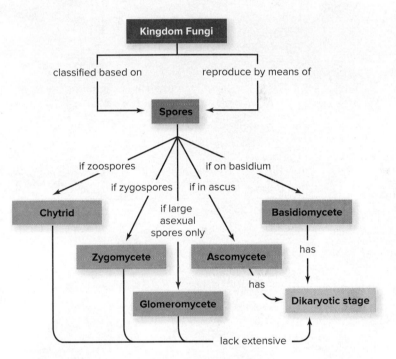

Figure 20.25 Pull It Together: Fungi.

Refer to **figure 20.25** and the chapter content to answer the following questions.

1. Review the Survey the Landscape figure in the chapter introduction, and add *eukaryotes, animals, plants, protists,* and *prokaryotes* to the Pull It Together concept map. Then connect each of the five groups of fungi to another group of organisms to indicate interactions between the two groups.

2. Add at least one characteristic of each group of fungi to the concept map.

3. Connect *hyphae* and *mycelium* to the concept map.

21 Animals

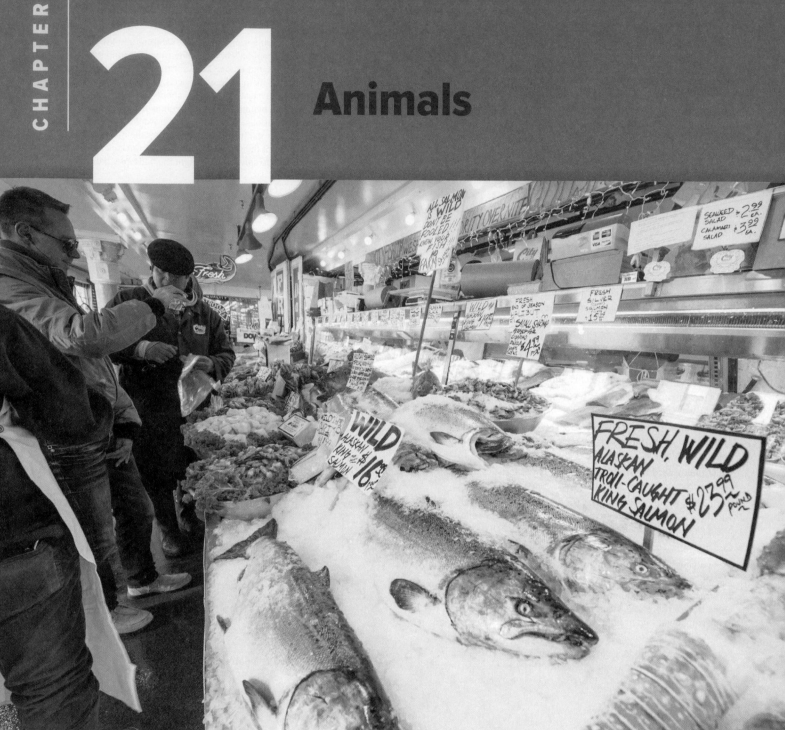

Fresh Seafood. Many markets sell fresh fish, such as these salmon. Store shelves also hold a wide variety of processed foods containing animal products.
©Michael DeFreitas/robertharding/Getty Images

LEARN HOW TO LEARN
Think While You Search the Internet

Some assignments may require you to use the Internet. But the Internet is full of misinformation, so you must evaluate every site you visit. Collaborative sites such as Wikipedia may be unreliable because anyone can change any article. For other sites, ask the following questions: Are you looking at someone's personal page? Is there an educational, governmental, nonprofit, or commercial sponsor? Is the author reputable? Does the page contain opinions or facts? Are facts backed up with documentation? Answering these questions will help ensure that the sites you use are credible.

Animal Products in Processed Foods

Most people consume many types of animals. In American grocery stores, the most common animal meats come from chordates (tuna, chickens, turkeys, pigs, and cattle), arthropods (crabs and lobsters), and mollusks (clams and oysters). We also consume many other animal products, including milk, eggs, and honey.

Vegans know to avoid these foods, but they should also carefully check labels for hidden ingredients derived from animals. For example, gelatin is an animal protein used as a jelling agent in yogurt, ice cream, and candies. Glycerin, or glycerol, also typically comes from animals; it occurs in candy, marshmallows, and some dairy products. Food manufacturers sometimes use insect products, too. Carmine, a bright red insect pigment, is often used as food coloring. Shellac is a waxy insect extract used as a glaze on some candies.

An ingredient may be called a "natural flavoring" if its function is to add flavor, not nutrients; it may come from meat, milk, plants, or other sources. Therefore, vegans might consider avoiding foods that contain natural flavoring—or perhaps all processed foods.

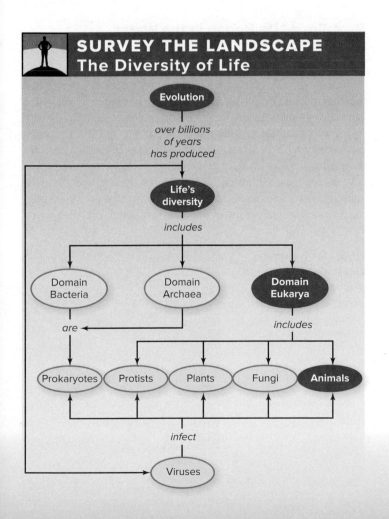

SURVEY THE LANDSCAPE
The Diversity of Life

Evolution

over billions of years has produced

Life's diversity

includes

Domain Bacteria — Domain Archaea — **Domain Eukarya**

are — *includes*

Prokaryotes — Protists — Plants — Fungi — **Animals**

infect

Viruses

Animal life began in the oceans, with a diverse array of simple body forms. Muscular and nervous systems gradually developed, expanding the range of lifestyles. Once plants invaded land, they provided new sources of food and shelter that sparked the evolution of diverse terrestrial animals.

For more details, study the Pull It Together feature in the chapter summary.

21.1 Animals Live Nearly Everywhere

Think of any animal. There's a good chance that the example that popped into your head was a mammal such as a dog, cat, horse, or cow. Although it makes sense that we think first of our most familiar companions, the 5800 species of mammals represent only a tiny subset of organisms in kingdom **Animalia.**

Biologists have described about 1,300,000 animal species, distributed among 37 phyla. This chapter considers 9 of the largest phyla in detail (**table 21.1**); the Burning Question box in section 21.8 briefly describes 3 others. The vast majority of animals are **invertebrates** (animals without backbones). The phylum Arthropoda, for example, includes more than 1 million species of invertebrates such as insects, crustaceans, and spiders. Only about 57,000 known species are **vertebrates** (animals with backbones) such as mammals and birds.

The diversity of animals is astonishing. Animals live in us, on us, and around us. They are extremely diverse in size, habitat, body form, and intelligence. Whales are immense; roundworms can be microscopic. Bighorn sheep scale mountaintops; crabs scuttle on the deep-ocean floor. Earthworms are squishy; clams surround themselves in heavy armor. Sponges are witless; humans, chimps, and dolphins are clever. This chapter explores some of this amazing variety; unit 6 explores animal anatomy and physiology in more detail.

A. What Is an Animal?

Animals are diverse, but their shared evolutionary history means that all animal phyla have some features in common. First, they are multicellular organisms with eukaryotic cells lacking cell walls. (i) *animal cell,* section 3.2C

Second, all animals are heterotrophs, obtaining both carbon and energy from organic compounds produced by other organisms. Most animals ingest their food, break it down in a digestive tract, absorb the nutrients, and eliminate the indigestible wastes.

Third, animal development is unlike that of any other type of organism. After fertilization, the diploid zygote (the first cell of the new organism) divides rapidly. The early animal embryo begins as a solid ball of cells that quickly hollows out to form a **blastula,** a sphere of cells surrounding a fluid-filled cavity. (Embryonic stem cells, the source of so much controversy, come from embryos at this stage of development.) No other organisms go through a blastula stage of development. (i) *stem cells,* section 11.3A

Fourth, animal cells secrete and bind to a nonliving substance called the extracellular matrix. This complex mixture of proteins and other substances enables some cells to move, others to assemble into sheets, and yet others to embed in supportive surroundings, such as bone or shell. (Section 21.17 explains how key genes encoding extracellular matrix proteins have helped shed light on the Cambrian explosion.) (i) *extracellular matrix,* section 25.2

B. Animal Life Began in the Water

The animal you thought of at the start of this section probably lives on land, since terrestrial animals are the ones we see most often. Nevertheless, only 10 of the 37 known phyla include species that live on land, and no phylum contains only terrestrial animals. All of today's animals clearly have their origins in aquatic ancestors.

The first animals, which arose about 570 million years ago (MYA), may have been related to aquatic protists called choanoflagellates (**figure 21.1**), organisms that strongly resemble collar cells in sponges. Although no one knows exactly what the first animal looked like, the Ediacaran organisms that thrived during the Precambrian left some of the oldest animal fossils ever found

TABLE **21.1**   Nine Phyla of Animals

Phylum	Examples	Number of Existing Species
Porifera	Sponges	5000
Cnidaria	Hydras, jellyfishes, corals, sea anemones	11,000
Platyhelminthes (flatworms)	*Planaria*, tapeworms, flukes	25,000
Mollusca	Bivalves, chitons, snails, slugs, squids, octopuses	112,000
Annelida	Earthworms, leeches, polychaetes	15,000
Nematoda (roundworms)	Pinworms, hookworms, *C. elegans*	80,000
Arthropoda	Horseshoe crabs, spiders, scorpions, crustaceans, insects	More than 1,000,000
Echinodermata	Sea stars, sea urchins, sand dollars	7000
Chordata	Tunicates, lancelets, fishes, amphibians, reptiles, mammals	60,000

Figure 21.1 Animal Ancestor? The ancestor to all animals may have resembled an aquatic protist (a choanoflagellate).

Solitary   Colonial

(see figure 15.12). Animal life diversified spectacularly during the Cambrian period, which ended about 490 MYA. Most of today's phyla of animals, including sponges, jellyfishes, arthropods, mollusks, and many types of worms, originated in the Cambrian seas. Their fossils are exceptionally abundant in an area of British Columbia called the Burgess Shale (see figure 21.43). ⓘ *Cambrian explosion,* section 15.3B

Aquatic animals were already diverse by the time plants and fungi colonized the land about 475 MYA. Arthropods, vertebrates, and other animals soon followed, diversifying as they adapted to new food sources and habitats.

As you read through this chapter, keep in mind the importance of habitat as a selective force shaping animal adaptations. Animal life started in the oceans, which provide buoyancy and generally stay within a certain range of temperature, salt concentration, oxygen concentration, and other environmental variables. Fresh water also provides buoyancy but contains less salt, may vary more in temperature, and may disappear seasonally or in the case of a drought. Terrestrial habitats nearly always have abundant oxygen, but air does not provide physical support as water does, and animals are at risk of drying out. While you study, note how each set of challenges selects for unique combinations of adaptations contributing to support, locomotion, reproduction, digestion, circulation, gas exchange, and waste disposal.

C. Animal Features Reflect Shared Ancestry

Figure 21.2 compiles the nine animal phyla described in this chapter into a phylogenetic tree. As you will see, the members of each phylum share similarities because they evolved from a common ancestor with those features. Moreover, the phyla are themselves grouped based on shared features of their appearance, physiology, embryonic development, and DNA. As you read this section, refer back to the tree to recall the positions of the branching points.

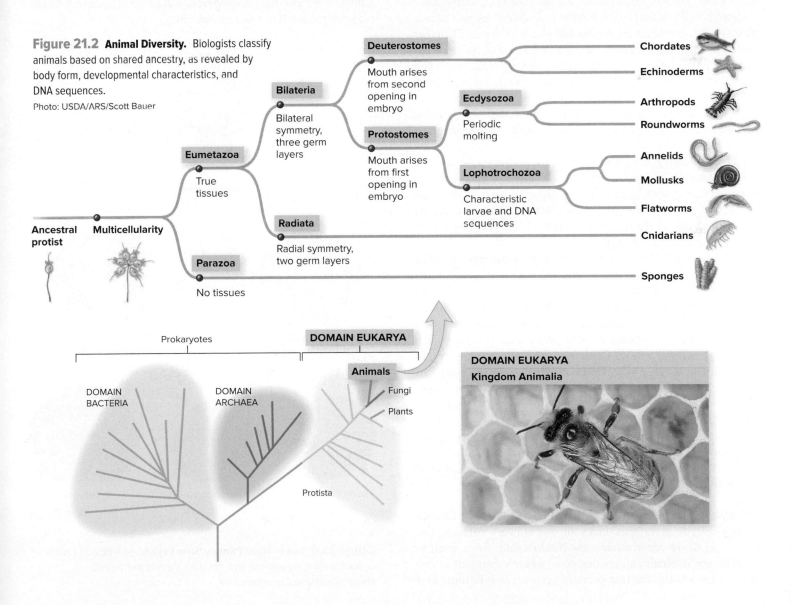

Figure 21.2 Animal Diversity. Biologists classify animals based on shared ancestry, as revealed by body form, developmental characteristics, and DNA sequences.

Photo: USDA/ARS/Scott Bauer

Figure 21.3 Types of Symmetry. (a) Many sponges are asymmetrical. (b) A hydra has radial symmetry. (c) A crayfish has bilateral symmetry. Animals with bilateral symmetry have a front (anterior) and a rear (posterior) end, as well as a dorsal (back or top) and a ventral (bottom or belly) side.

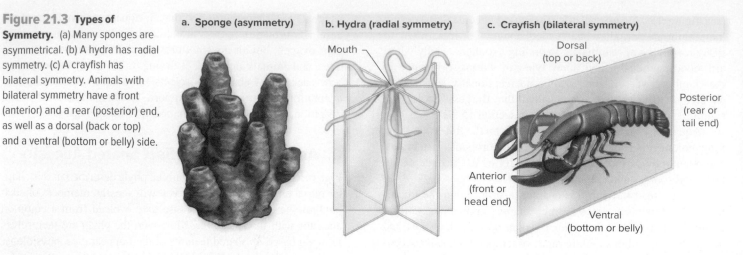

a. Sponge (asymmetry)

b. Hydra (radial symmetry)

Mouth

c. Crayfish (bilateral symmetry)

Dorsal (top or back)

Posterior (rear or tail end)

Anterior (front or head end)

Ventral (bottom or belly)

Cell and Tissue Organization

The first major branching point separates animals based on whether their bodies contain true tissues. The simplest animals, the **parazoans** (sponges), have several specialized cell types, but the cells do not interact to provide specific functions as they would in a true tissue. The other clade contains **eumetazoans,** animals with tissues. In most eumetazoans, multiple tissue types interact to form organs such as a heart, brain, or kidney. The organs, in turn, work together to circulate and distribute blood, dispose of wastes, and carry out other functions. These interactions boost efficiency, which explains in part why eumetazoans are typically more active than sponges.

Body Symmetry and Cephalization

Body symmetry is another major criterion (**figure 21.3**). Many sponges are asymmetrical; that is, they lack symmetry. Other sponges, hydras, jellyfishes, adult sea stars, and their close relatives have **radial symmetry,** a body form in which multiple similar parts are arranged around a central axis.

Most animals, however, have **bilateral symmetry,** in which only one plane can divide the animal into mirror images. Bilaterally symmetrical animals such as crayfish have head (anterior) and tail (posterior) ends, and they typically move through their environment head first. Bilateral symmetry is correlated with **cephalization,** the tendency to concentrate sensory organs and a brain at an animal's head. Compared with radial symmetry, in which simple sensory cells are distributed across the body surface, bilateral symmetry is typically accompanied by more complex sense organs and a greater ability to evaluate and respond to environmental stimuli.

Bilateral symmetry is also associated with an elongated body form, often with paired appendages and organs on either side of the body. Among other benefits, these structures opened new options for animal locomotion.

Embryonic Development: Two or Three Germ Layers

Early embryos give other clues to evolutionary relationships (**figure 21.4**). In eumetazoans, the blastula folds in on itself to generate the **gastrula,** a cup-shaped structure composed of two or three layers of tissue (the primary germ layers). **Ectoderm** is the outer tissue layer, and **endoderm** is the inner layer. Jellyfishes and their relatives have only those two layers. All other eumetazoans also have **mesoderm,** a third germ layer that forms between the ectoderm and endoderm.

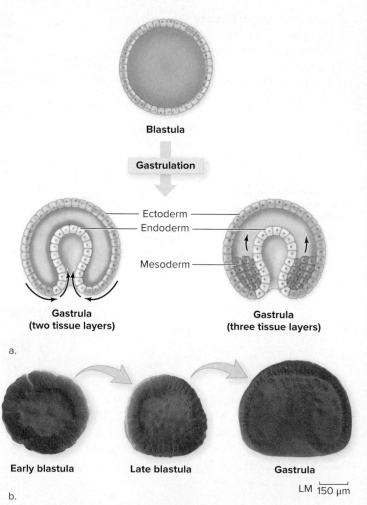

Blastula

Gastrulation

Ectoderm
Endoderm

Mesoderm

Gastrula (two tissue layers)

Gastrula (three tissue layers)

a.

Early blastula

Late blastula

Gastrula

LM 150 µm

b.

Figure 21.4 Two or Three Primary Germ Layers. (a) A blastula folds in on itself, forming the gastrula. (b) A sea star's blastula and gastrula.

Photos: (all): ©Herve Conge/Phototake

These germ layers eventually give rise to all of the body's tissues and organs. Ectoderm develops into the skin and nervous system, whereas endoderm becomes the digestive tract and the organs derived from it. Mesoderm gives rise to the muscles, the circulatory system, and many other specialized structures. Muscles, in turn, not only enhance locomotion but also power the movement of food through the digestive tract, boosting the speed and efficiency with which animals extract nutrients from food. Overall, animals with three germ layers have much greater variety in body forms and functions than do animals with two germ layers.

Embryonic Development: Protostomes and Deuterostomes

After an embryo has folded into a gastrula with three layers (see figure 21.4), the inner cell layer fuses with the opposite side of the embryo, forming a tube with two openings. This cylinder of endoderm will develop into the animal's digestive tract, with one opening becoming the mouth and the other becoming the anus.

But which end is which? The answer determines whether the animal belongs to the protostome or deuterostome clade (see figure 21.2). In most **protostomes,** the gastrula's first indentation develops into the mouth, and the anus develops from the second opening. (*Protostome* means "mouth first.") In **deuterostomes,** the first indentation becomes the anus, and the mouth develops from the second opening. (*Deuterostome* means "mouth second.") We now know that some animals classified as protostomes do not conform to the "mouth first" pattern. Nevertheless, DNA sequences support their close relationship to other animals in the protostome clade.

As you can see from figure 21.2, protostomes are further divided into two main groups: ecdysozoans and lophotrochozoans. These groups are largely defined by their DNA sequences rather than by a combination of easily observable characteristics. Ecdysozoans, however, do share a visible feature (molting).

D. Biologists Also Consider Additional Characteristics

Other adaptations have also been important milestones in animal evolution. This section describes some other characteristics that you will encounter as you learn about the animal phyla in this chapter.

Body Cavity (Coelom)

A bilaterally symmetrical animal may or may not have a coelom (**figure 21.5**). The **coelom** (pronounced SEA-loam) is a fluid-filled body cavity that forms completely within the mesoderm. Animals that have a coelom include earthworms, snails, insects, sea stars, and chordates. In contrast, roundworms have a **pseudocoelom** ("false coelom"), a cavity that is lined partly with mesoderm and partly with endoderm. Flatworms lack a coelom, although evidence suggests their ancestors may have had body cavities.

The coelom's chief advantage is flexibility. As internal organs such as the heart, lungs, liver, and intestines develop, they push into the coelom. The fluid of the coelom cushions the organs, protects them, and enables them to shift as the animal bends and moves.

In many animals, the coelom or pseudocoelom serves as a hydrostatic skeleton that provides support and movement. In a **hydrostatic skeleton,** muscles push against a constrained fluid. An earthworm, for example, burrows through soil by alternately contracting and relaxing muscles surrounding its coelom. Note that jellyfishes, flatworms, and other invertebrates also have hydrostatic skeletons, even though they lack a coelom or pseudocoelom. Instead, their muscles push against fluid in the digestive tract or between body cells.

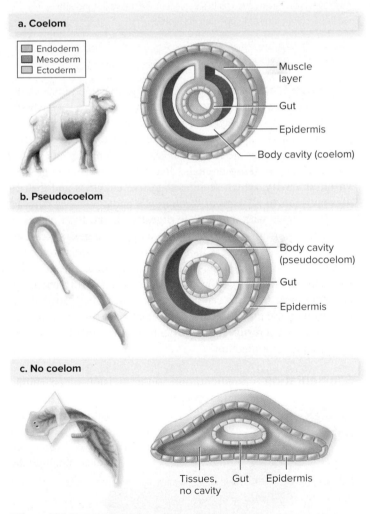

Figure 21.5 Body Cavities. (a) A sheep has a coelom, (b) a roundworm has a pseudocoelom, and (c) a flatworm lacks a coelom. Note that these drawings are abstractions. In the sheep, for example, the internal organs grow into the coelom, greatly distorting the cavity's shape.

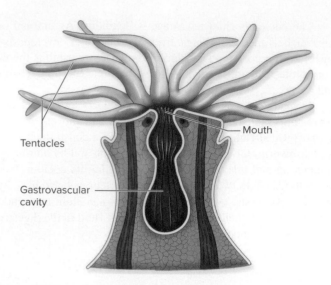

Figure 21.6 Incomplete Digestive Tract. An animal with a gastrovascular cavity, such as this sea anemone, takes in food and ejects wastes through the same opening (its mouth).

Digestive Tract   A sponge lacks a digestive tract; instead, the animal has pores through which water enters and leaves the body. In other animals, the digestive tract may be incomplete or complete. Cnidarians and flatworms have an **incomplete digestive tract,** in which the mouth both takes in food and ejects wastes (**figure 21.6**). In these animals, digestion occurs in the **gastrovascular cavity,** which secretes digestive enzymes and distributes nutrients to all parts of the animal's body.

In animals with a **complete digestive tract,** food passes in one direction from mouth to anus. A complete digestive tract allows the animal to process its food stepwise in specialized compartments. For example, cells near the mouth can secrete digestive enzymes into the tract, "downstream" cells can absorb nutrients, and those near the anus can help eject wastes. These specialized regions increase the efficiency with which nutrients are extracted from food. As a result, more nutrients are available for hunting, defense, and reproduction.

Segmentation   **Segmentation** is the division of an animal body into repeated parts (**figure 21.7**). In centipedes, millipedes, and earthworms, the segments are clearly visible. Insects and vertebrates also have segmented bodies, although the

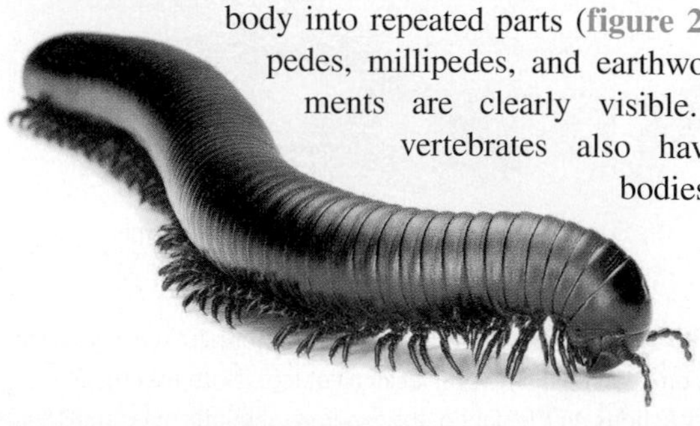

Figure 21.7 Segmentation. This millipede illustrates segmentation—the division of the body into repeated parts.
©Don Farrall/Getty Images RF

subdivisions may be less obvious. Segmentation adds to the body's flexibility, and it enormously increases the potential for the development of specialized body parts. Activating different combinations of genes in each segment can create regions with unique functions. Antennae can form on an insect's head, for example, while wings or legs sprout from other segments.

Reproduction and Development   Most animals reproduce sexually, and the development of the resulting embryo follows either of two paths (**figure 21.8**). Animals that undergo **direct development** have no larval stage; at hatching or birth, they already resemble adults. A newborn elephant or newly hatched cricket, for example, looks like a smaller version of the adult.

In contrast, an animal with **indirect development** may spend part of its life as a **larva,** which is an immature stage that does not resemble the adult. The larvae eventually undergo **metamorphosis,** in which their bodies change greatly as they mature into adults. Larvae often live in different habitats and eat different foods than the adults, an adaptation that may help reduce competition between the generations. Insects (such as butterflies) and amphibians (such as frogs) are the most familiar examples of animals that undergo indirect development.

21.1   MASTERING CONCEPTS

1. What characteristics do all animals share?
2. When and in what habitat did animals likely originate?
3. What features were used to build the animal phylogenetic tree?
4. What are the events of early embryonic development in an animal?
5. What are the two main types of digestive tracts?
6. What advantages does segmentation confer?
7. Differentiate between direct and indirect development.

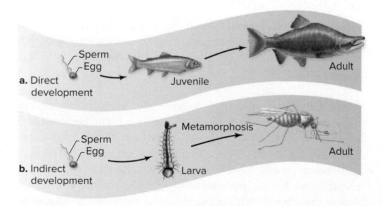

Figure 21.8 Development. (a) The young of animals that undergo direct development resemble the adult. (b) In indirect development, metamorphosis transforms a larva into a very different-looking adult.

🍁 21.2 Sponges Are Simple Animals That Lack Differentiated Tissues

The **sponges** belong to phylum Porifera, which means "pore-bearers"—an apt description of these simple animals (figure 21.9). Sponges have all of the animal-defining features described in section 21.1B, including multicellularity. Unlike in other animals, a sponge's cells do not interact to form tissues. Their structural simplicity means that sponges bear little resemblance to the rest of the animal kingdom.

All sponges are aquatic. Most are marine, although some live in fresh water. In addition, sponges are generally considered sessile, meaning they remain anchored to their substrate. Together, their watery habitat and sessile lifestyle have selected for a unique combination of features.

A sponge's body is either radially symmetrical or asymmetrical. It is also hollow, and its body wall is riddled with pores. These specialized channels allow water to pass through the body wall, into a main chamber, and out a hole at the top. This arrangement enables sponges to acquire food and oxygen, dispose of wastes, and reproduce.

To understand how sponges carry out these functions without moving, examine the body wall in figure 21.9. Several types of cells are embedded in a jellylike matrix (the mesohyl). Lining the inner surface of the body wall is a layer of "collar cells," which strongly resemble choanoflagellates. (These protists, which are close relatives of animals, are illustrated in figure 21.1.) As the flagella on the collar cells wave, water moves into the sponge through the pores. This water current carries not only oxygen but also microscopic particles of organic matter—the sponge's food.

Collar cells trap and partially digest the food particles and pass them to another cell type, the amoebocyte. These cells help digest food and distribute nutrients to other cells. They also secrete skeletal components, including protein fibers and spicules (sharp slivers of silica or calcium carbonate; see figure 21.9). The spicules, along with an arsenal of toxic chemicals, help these sessile animals deter predators.

The porous body wall not only participates in feeding but also produces gametes. Sponges are hermaphrodites, which means the same individual makes both sperm and egg cells. The sperm are released into the water, but the animal retains its eggs. Meanwhile, sperm from nearby sponges enter its body through the pores. After fertilization, the zygote develops into a blastula, which is released and drifts briefly before settling into a new habitat. Some sponges also reproduce asexually by budding or fragmentation.

Some people use natural sponges in bathing. Also, the chemicals that protect sponges from predators may yield useful anti-cancer and antimicrobial drugs. Collecting sponges, however, can harm ecosystems.

Key features

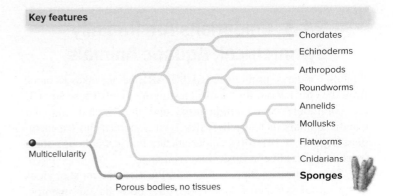

Chordates
Echinoderms
Arthropods
Roundworms
Annelids
Mollusks
Flatworms
Cnidarians
Sponges

Multicellularity

Porous bodies, no tissues

Diversity

Anatomy

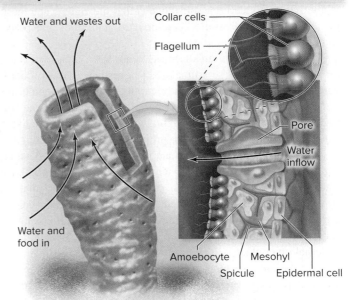

Water and wastes out
Collar cells
Flagellum
Pore
Water inflow
Water and food in
Amoebocyte   Mesohyl
Spicule   Epidermal cell

Figure 21.9 Sponges (Phylum Porifera).

Photos: (diversity green): ©Getty Images RF; (diversity pink): ©Laurence F Tapper/YAY Micro/age fotostock RF

21.2 MASTERING CONCEPTS

1. What characteristics distinguish the sponges? (*Hint:* See figure 21.9.)
2. How is a sponge's body adapted to its aquatic habitat and sessile life?
3. Explain how the arrangement of cells in a sponge is adaptive to its feeding strategy.
4. How do sponges reproduce sexually and asexually?

21.3 Cnidarians Are Radially Symmetrical, Aquatic Animals

Phylum Cnidaria (pronounced nigh-DARE-ee-ah) takes its name from the Greek word for "nettle," a stinging plant. These animals are commonly called **cnidarians,** and they include familiar aquatic animals such as jellyfishes, hydras, corals, and sea anemones (**figure 21.10**). Most cnidarians are marine, although some (such as hydra) live in fresh water.

The cnidarians share several similarities, the most obvious being radial symmetry. One end of the body has an opening, the mouth, which is surrounded by a ring of tentacles. (The mouth is clearly visible on the sea anemones in figure 21.10.) In a sessile cnidarian, called a **polyp,** a stalk holds the tentacles upward. Hydras, corals, and sea anemones are examples of polyps. In a **medusa,** tentacles dangle downward from a free-swimming bell. A jellyfish illustrates the medusa body form. In both body forms, the mouth leads to the dead-end gastrovascular cavity.

The cross sections in figure 21.10 reveal another characteristic unique to the cnidarians. The body wall is composed of just two thin layers: an outer epidermis derived from ectoderm and an inner layer derived from endoderm. Between these two layers is a jellylike, noncellular substance called mesoglea. The mesoglea acts as a hydrostatic skeleton that maintains the body's shape in water.

Although a cnidarian's tissues do not form organs such as a brain or muscles, these animals can nevertheless make coordinated movements as they swim or capture prey. In the epidermis, groups of linked neurons called nerve nets coordinate the contraction of specialized cells. In this way, for example, a jellyfish can force water out of its bell to propel itself through water.

One additional feature unique to the cnidarians is the stinging cells embedded in the epidermis (see the inset in figure 21.10). These cells, called **cnidocytes,** contain tiny harpoons that can sting the animal's prey or a predator. Anyone who has accidentally stepped on a jellyfish on the beach knows that the sting can cause skin irritation or cramps; a few species have toxins that can be lethal on contact.

Cnidarians use their stinging tentacles to sense, grab, and paralyze passing prey, which can range from tiny water fleas to small fish. The tentacles stuff the meal into the mouth, and cells lining the digestive cavity secrete enzymes that digest the food. After absorbing the nutrients, the animal ejects indigestible matter through the mouth.

Thanks to their digestive cavity, cnidarians can handle much larger prey than can the filter-feeding sponges. The resulting boost in feeding efficiency means cnidarians can acquire much more energy for movement and reproduction.

The simple cnidarian body also is adapted to gas exchange in water. The thin body wall has a high surface area, and all body parts are in contact with the water. This arrangement means that each cell can use simple diffusion to acquire O_2 from the environment and to dispose of CO_2 and other metabolic wastes.

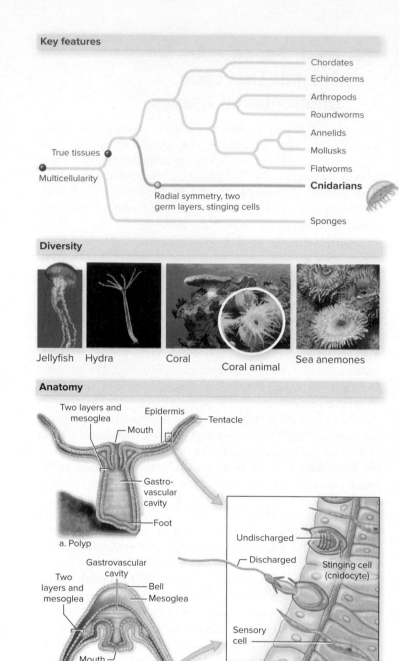

Figure 21.10 **Cnidarians (Phylum Cnidaria).**

Photos: (jellyfish): ©Kevin Schafer/Alamy RF; (hydra): ©Ted Kinsman/Science Source; (coral): ©Comstock Images/PictureQuest RF; (coral animal): ©Leslie Newman & Andrew Flowers/Science Source; (anemone): ©Russell Illig/Getty Images RF

Cnidarians can be very prolific breeders, and reproduction may be sexual and asexual (**figure 21.11**). In the moon jelly *Aurelia,* for example, male and female medusae release gametes into the water, where fertilization occurs. The resulting larvae attach to a surface and develop into the polyp form. The polyp

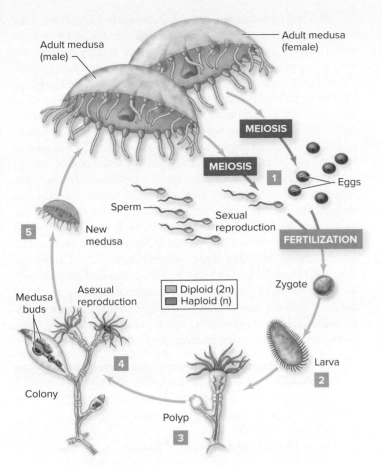

Figure 21.11 Cnidarian Reproduction. The life cycle of *Aurelia* includes an alternation of generations between medusa and polyp stages. (*1*) Medusae release sperm and egg cells. After fertilization, (*2*) a larva develops, attaches to a surface, and becomes (*3*) a polyp. The polyp reproduces asexually to form (*4*) a colony, from which (*5*) new medusae are released.

then buds asexually, generating a colony of additional polyps and eventually medusae. In some areas, huge swarms of jellyfish are becoming increasingly common, presenting a nuisance for tourist destinations and fisheries.

Among the cnidarians in figure 21.10, the coral animals are unusual in that they secrete calcium carbonate exoskeletons. Over many generations, the exoskeletons secreted by countless coral animals have built magnificent coral reefs. These complex structures house many commercially important species of fishes and other animals, and they protect coastlines from erosion. ⓘ *coral reefs,* section 39.5A

21.3 MASTERING CONCEPTS

1. What features do all cnidarians share? (*Hint:* See figure 21.10.)
2. Compare and contrast a polyp and a medusa.
3. How do cnidarians feed, move, and reproduce?
4. In what ways are cnidarians important?

21.4 Flatworms Have Bilateral Symmetry and Incomplete Digestive Tracts

Phylum Platyhelminthes includes the **flatworms.** (*Platy* means "flat"; *helminth* means "worm.") Some of these animals are surprisingly beautiful, whereas others look downright scary (**figure 21.12**). The three main classes of flatworms are the **free-living flatworms** (such as marine flatworms and freshwater planarians) and the parasitic **flukes** and **tapeworms.**

The phylogenetic tree in figure 21.12 shows that flatworms are part of a large clade characterized by bilateral symmetry and three germ layers. As explained in section 21.1C, animals in all seven phyla in this clade therefore have the potential to develop more complex bodies and more varied responses to the environment than do the sponges and cnidarians.

Flatworms, annelids, and mollusks occupy the protostome ("mouth first") branch of the evolutionary tree (see section 21.1C). Within the protostome clade, the most obvious feature that defines the flatworms is the flattened body shape that gives this phylum its name. A related, but less conspicuous, feature is the absence of a coelom. In many other animals, the fluid-filled

Key features

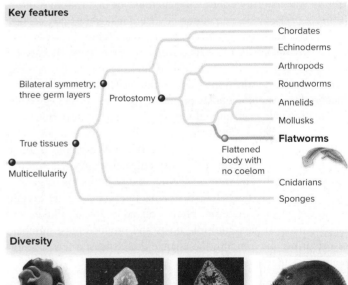

Diversity

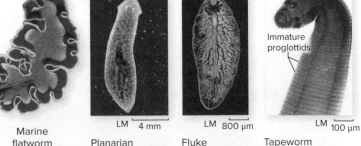

Figure 21.12 Flatworms (Phylum Platyhelminthes).

Photos: (flatworm): ©Leslie Newman & Andrew Flowers/Science Source; (planarian): ©NHPA/M. I. Walker RF; (fluke): ©Volker Steger/Science Source; (tapeworm): ©Biophoto Associates/Science Source

coelom helps circulate oxygen throughout the body. Thanks to the flatworm's shape, however, each of the animal's cells is near enough to the body surface to exchange gases directly with its environment. In fact, the flat body enables even the largest of flatworms to meet its metabolic needs without a specialized circulatory or respiratory system.

All flatworms share these characteristics, but each lineage within this phylum also has unique traits. **Figure 21.13**, for example, shows some adaptations of free-living flatworms. These animals usually are predators or scavengers in aquatic food webs. The mouth opens into a muscular, tubelike pharynx at the body's midpoint (figure 21.13a). This structure delivers food to the incomplete digestive tract. The highly branched gut ensures that every body cell is near enough to the digestive tract to acquire nutrients by simple diffusion. Undigested food is ejected from the body through the pharynx and mouth.

The flatworm nervous system can sense stimuli and coordinate movements. The planarian in figure 21.13b, for example, has a ladderlike arrangement of nerve cords running the length of its body. The head end features a simple brain and sensory structures that detect touch, chemicals, and light intensity. Eyespots cannot form an image but can detect light, helping the animal find shelter by avoiding bright light. The worm can then creep or swim toward food or shelter, either by gliding on cilia along a film of mucus or by contracting its muscles in a rolling motion. Loosely packed cells surrounded by fluid inside the body act as a hydrostatic skeleton. ⓘ *invertebrate eyes*, section 27.4A

Many flatworms reproduce asexually. Free-living species, for example, may simply pinch in half and regenerate the missing parts. Sexual reproduction is also common. These animals are hermaphrodites, simultaneously producing both sperm and egg cells. In a mating pair, each animal fertilizes the eggs of its partner.

The two other flatworm groups, the flukes and the tapeworms, are adapted to a parasitic lifestyle. For example, a tough outer layer protects against the host's digestive enzymes and immune system. In addition, investing energy into some organs—such as eyes—is not adaptive for an internal parasite; certain organ systems are therefore often reduced or absent. Flukes and tapeworms also have multiple hosts and produce huge numbers of offspring, maximizing the chance that at least a few will encounter a suitable host.

Figure 21.14 illustrates the life cycle of the blood fluke. Larval worms enter the human body through the skin and mature in blood vessels surrounding the intestines or urinary bladder. As they develop, the parasites feed on blood and other host tissues. Adult flukes mate, and the fertilized eggs migrate into the host's bladder or gut. From there, they leave the host's body in urine or feces and hatch into larvae upon reaching water. These larvae infect their intermediate host, a snail, where the flukes continue to mature. The resulting larvae—now at a different developmental stage—emerge from the snail and infect a new human host.

Tapeworms have a different set of adaptations. A tapeworm lacks a mouth and digestive system. Instead, hooks or suckers on its head attach to the host's intestines, and the worm absorbs the host's already-digested food through its body wall. Most of the tapeworm body consists of repeated organs called proglottids, which contain fertilized eggs; proglottids break off from the worm and leave the host in feces. When an intermediate host such as a pig swallows proglottids in contaminated water, the eggs hatch. The larvae can migrate to the host's muscles, which may be consumed by a human host; this is how people acquire tapeworm infections by eating undercooked fish, beef, or pork.

21.4 MASTERING CONCEPTS

1. What features do all flatworms share? (*Hint:* See figure 21.12.)
2. How does the body shape of a flatworm enhance gas exchange with the environment?
3. List and describe the three classes of flatworms.

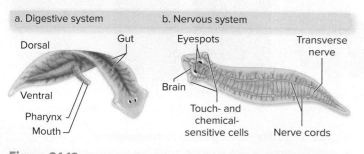

Figure 21.13 Flatworm Anatomy. (a) The pharynx is a muscular tube that opens into the incomplete digestive tract. (b) A brain and ladderlike nerve cords make up the nervous system.

a. Digestive system — Dorsal, Gut, Ventral, Pharynx, Mouth

b. Nervous system — Eyespots, Transverse nerve, Brain, Touch- and chemical-sensitive cells, Nerve cords

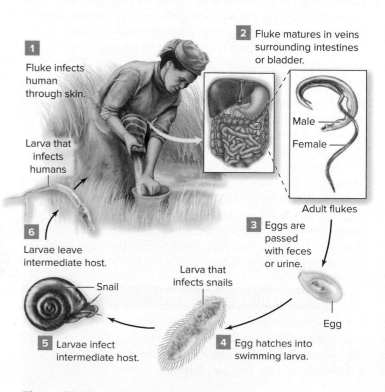

1. Fluke infects human through skin.
2. Fluke matures in veins surrounding intestines or bladder.
Larva that infects humans
Male
Female
Adult flukes
3. Eggs are passed with feces or urine.
Egg
4. Egg hatches into swimming larva.
Larva that infects snails
5. Larvae infect intermediate host.
Snail
6. Larvae leave intermediate host.

Figure 21.14 Life Cycle of a Blood Fluke.

21.5 Mollusks Are Soft, Unsegmented Animals

Mollusks form the second-largest phylum (after arthropods), and they include many familiar animals on land, in fresh water, and in the ocean (figure 21.15). The word *mollusk* comes from the Latin word for "soft," reflecting the fleshy bodies in this phylum.

The four largest classes of mollusks are the chitons, bivalves, gastropods, and cephalopods. **Chitons** are marine animals with eight flat shells that overlap like shingles. **Bivalves,** such as oysters, clams, scallops, and mussels, have two-part, hinged shells. **Gastropods** ("stomach-foot") are snails, slugs, sea slugs (nudibranchs), and limpets. Their name comes from the broad, flat foot on which they crawl. The **cephalopods** include marine animals such as octopuses, squids, and nautiluses. *Cephalopod* means "head-foot," a reference to the arms connected to the head of the animal. Cephalopods include the largest known invertebrate—the colossal squid, which may be up to 14 meters long.

Other than the fact that their soft bodies are unsegmented, the mollusks do not look much alike. Yet they share several features that reveal their close evolutionary relationship. One is the **mantle,** a fold of tissue that secretes a calcium carbonate shell in most species. The hard shells of chitons, bivalves, and snails protect against many predators, but they are heavy and limit the exposed surface area available for gas exchange. In the cephalopods, the shell is internal or absent.

The mantle is folded into a space called the mantle cavity, which is exposed to the environment. The mantle cavity plays an important role in both gas exchange and excretion. In terrestrial snails and slugs, the lung is derived from the mantle cavity. In aquatic mollusks, a tubelike siphon pulls water into the mantle cavity, which houses the gills that exchange O_2 and CO_2 with the environment. Meanwhile, an excretory organ empties wastes into the mantle cavity to be flushed away.

In some mollusks, the mantle cavity participates in locomotion and filter feeding. Cephalopods, for example, are the speediest invertebrates; they move by "jet propulsion" as they shoot water out of the mantle cavity's siphon. Clams use their two siphons in a completely different way. One siphon draws water into the mantle cavity; the clam absorbs oxygen and food particles, and the water leaves the body through the other siphon.

Another characteristic common to all mollusks is a muscular **foot,** which typically provides locomotion. All mollusks have a hydrostatic skeleton; the animal moves when muscles act on the constrained fluid of the coelom. The speed and type of locomotion, however, vary among the mollusks. On land, snails and slugs glide on a trail of mucus. Bivalves use the muscular foot to burrow into sediments. In cephalopods, the foot is modified into multiple arms or tentacles that participate in locomotion but also grasp prey.

Besides the mantle and foot, the third major region of a mollusk's body is the **visceral mass,** which contains the digestive, circulatory, excretory, and reproductive organs. Consider,

Key features

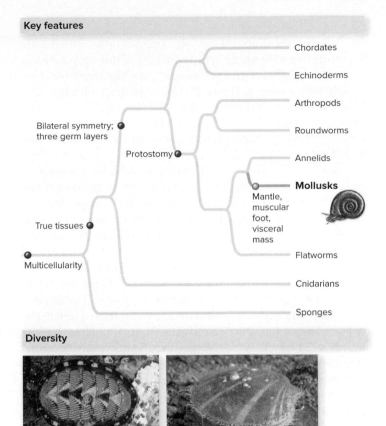

Diversity

Chiton

Scallop (bivalve)

Snail (gastropod)

Slug (gastropod)

Octopus (cephalopod)

Squid (cephalopod)

Anatomy

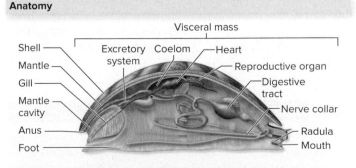

Figure 21.15 **Mollusks (Phylum Mollusca).**

Photos: (chiton): ©Kjell B. Sandved/Science Source; (bivalve): ©Andrew J. Martinez/Science Source; (snail): ©Digital Vision RF; (slug): ©Steven P. Lynch RF; (octopus): ©Rich Carey/Shutterstock RF; (squid): ©Comstock Images/PictureQuest RF

for example, the digestive system. The mollusks, along with the remaining animals we will encounter in this chapter, have a complete digestive tract (you can trace the digestive tract from mouth to anus in figure 21.15). Note that cnidarians and flatworms have a digestive tract with just one opening, the mouth. As described in section 21.1D, the main advantage of a two-opening digestive tract is improved nutrient extraction from food.

The mouth cavity of many mollusks contains a **radula,** a tonguelike strap with teeth made of chitin (a tough polysaccharide). Chitons, snails, and slugs use the radula to scrape algae off rocks or tear vegetation apart. But other mollusks use different feeding strategies. For example, cephalopods capture fast-moving prey, and bivalves are filter feeders.

Natural selection has also modified the molluskan nervous system in many ways. The tentacles on a snail's head, for example, contain sensory cells that detect odors and eyespots that detect light. The nervous system in these animals is relatively simple. An octopus's nervous system is much more complex, with a large brain, highly developed eyes, and an excellent sense of touch. These organs work together to coordinate the animal's camouflage, reproducing the color and texture of its surroundings. Moreover, in lab studies, octopuses have displayed remarkable problem-solving abilities.

With all the variation we have seen so far, it is not surprising that mollusks have diverse reproductive strategies. Many bivalves shed gametes into the water, where external fertilization occurs. Gastropods and cephalopods, on the other hand, mate and fertilize eggs internally. In many marine mollusks, a ciliated, pear-shaped larva settles to the seafloor and develops into an adult (indirect development). In cephalopods and snails, however, the hatchlings resemble adults (direct development). Snail growth is especially unusual. The body twists inside its shell as it develops, so that in the adult, the anus empties digestive wastes near the head.

Mollusks have diverse effects on human life, health, and environmental quality. We harvest pearls from oysters, many people collect the shells of bivalves, and we eat clams, mussels, oysters, snails, squids, and octopuses. Bivalves can become poisonous, however, if they accumulate pollutants or toxins produced by protists called dinoflagellates. On land, snails are garden pests; in water, some snails host parasitic worms (see figure 21.14), and the venom of a cone snail can kill a human. Invasive zebra mussels have disrupted aquatic ecosystems in the central United States. ⓘ *dinoflagellates,* section 18.2B; *invasive species,* section 40.5A

21.5 | MASTERING CONCEPTS

1. What features do all mollusks share? (*Hint:* See figure 21.15.)
2. What are the four largest classes of mollusks, and where do they live?
3. How do mollusks feed, move, reproduce, and protect themselves?
4. Propose possible disadvantages to having a shell.

21.6 Annelids Are Segmented Worms

Earthworms and other segmented worms are **annelids.** The name of the phylum, Annelida, derives from the Latin word *annulus* ("little ring") and is a reference to the segmented bodies of these animals (figure 21.16).

Biologists recognize two main classes of annelids. One class contains the **earthworms** and **leeches,** the most familiar types of annelids. These segmented worms have a distinctive feature in common: a saddlelike thickening near the head end (this region is especially obvious in an earthworm). This structure secretes a protective "cocoon" for the fertilized eggs when the animal reproduces.

Although earthworms and leeches are closely related, they live in very different habitats, which have selected for unique adaptations in each group. Earthworms ingest soil, digest the organic matter, and eliminate the indigestible particles as castings. Each segment of an earthworm's body sports a few bristles that provide traction as the animal burrows through soil.

Most leeches, on the other hand, live in fresh water. The leech body consists of segments lacking bristles. Some leeches suck the blood of vertebrates (including humans), but most eat small animals such as arthropods, snails, or other annelids. Each end of the flattened, darkly pigmented leech has a sucker with which it attaches itself to a surface or to its prey.

The other class of annelids, the **polychaetes,** contains the marine segmented worms. Most polychaetes have pairs of fleshy, paddlelike appendages that they use in locomotion. The name *polychaete* comes from the many bristles (*chaetae*) embedded in each of these appendages.

Annelids lack a specialized respiratory system. Instead, earthworms and leeches exchange gases by diffusion across the body wall. (Some polychaetes have feathery gills.) Gas exchange can occur only across a moist surface. Leeches and polychaetes, which live in water, are always wet, but earthworms are vulnerable. They rapidly dry out and die if removed from moist soil.

The segmented worms do, however, have other organ systems. The unbranched digestive tract extends along the length of the animal, as do two main blood vessels. Multiple aortic arches pump blood from one vessel to the other. Within each segment, small channels connect the main vessels, completing the loop.

The nervous system includes a simple "brain," a mass of nerve cells at the head end. These cells connect to a ventral nerve cord, with lateral nerves branching into each segment. Together, the nerves stimulate contraction of the circular and longitudinal muscles in the body wall. These muscles push against the coelom as the worm crawls, burrows, or swims to find food or avoid predators.

Some organs are repeated in each segment. For example, each segment contains a pair of excretory organs. These structures draw in fluid from the coelom, return some ions and other substances to the blood, and discharge the waste-laden fluid outside the body through a pore.

Key features

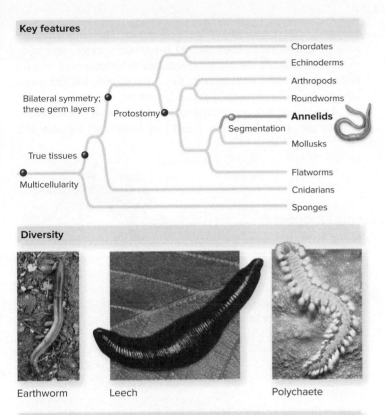

- Chordates
- Echinoderms
- Arthropods
- Roundworms
- **Annelids**
- Mollusks
- Flatworms
- Cnidarians
- Sponges

Bilateral symmetry; three germ layers
Protostomy
Segmentation
True tissues
Multicellularity

Diversity

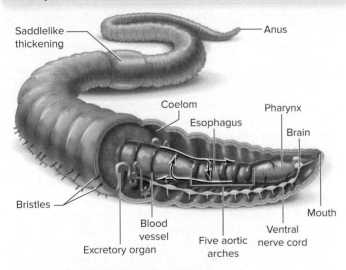

Earthworm   Leech   Polychaete

Anatomy

Saddlelike thickening
Anus
Coelom
Pharynx
Esophagus
Brain
Bristles
Blood vessel
Mouth
Excretory organ
Five aortic arches
Ventral nerve cord

Figure 21.16 Annelids (Phylum Annelida).

Photos: (earthworm) ©David Chapman/Alamy; (leech): ©Edward Kinsman/Science Source; (polychaete): ©L. Newman & A. Flowers/Science Source

Leeches and earthworms are hermaphrodites, so each individual has the reproductive organs of both sexes. Two individuals copulate, each discharging sperm for the other's eggs. After fertilization, the animal secretes fluid that forms a protective "cocoon" into which it lays its eggs. The baby worms hatch a couple of weeks later.

Humans benefit from annelids in many ways. Earthworms aerate and fertilize soil, and worm farms raise many types of worms for sale as fishing bait or as soil conditioners (see this section's

Apply It **Now**

Start Your Own Worm Farm

You may have heard of cow manure being used as a fertilizer, but what about worm castings? These pinhead-sized droppings are rich in the water-soluble nutrients that plants need. You can buy plant fertilizers made of worm castings, but if you produce food scraps in your kitchen, you can easily build a small-scale worm farm of your own.

To start the farm, acquire two stackable plastic tubs, one nesting inside the another. Drill holes in the bottom of the upper tub, which will contain the worms. The lower tub will collect the waste fluids that drain through the holes; like the castings, this liquid is an excellent fertilizer.

Purchase live worms locally or online. Common garden earthworms are not a good choice; they prefer to burrow deep into soil and are therefore poorly suited for life in a shallow container. Instead, red wigglers (*Eisenia fetida*) work best because they eat a lot and reproduce rapidly. But they are introduced from Europe and (like any nonnative species) may damage local ecosystems, so they should not be released from the worm farm.

Create bedding by shredding black-and-white newsprint or computer paper. Toss in a couple of handfuls of garden soil and some crushed, cooked egg shells. After you moisten the mixture, add your worms and some food. Worms can eat vegetable scraps, most fruits, stale bread, leftover pasta, coffee grounds, tea leaves, paper, and cardboard. Avoid meat and dairy products, which will produce unpleasant odors. In addition, onions and garlic repel worms, and citrus fruits cause the bins to become too acidic. Keep your farm in a dark, quiet location at room temperature.

When the bedding and food have decayed, it is time to harvest the castings. Drill holes into a third tub and prepare a fresh bedding mixture. Place the new bin on top of the one containing the worms. If the bottom of the tub is resting lightly on the old bedding, the worms will crawl through the holes into the new tub. After about 2 weeks, the lower tub will contain nothing but castings, which you can add to your plants.

Apply It Now box). In medicine, a blood-thinning chemical from leeches can stimulate circulation in surgically reattached digits and ears, and physicians sometimes apply leeches to remove excess blood that accumulates after damage to the nervous system.

21.6   MASTERING CONCEPTS

1. What defining feature arose in the annelid lineage that distinguishes annelids from mollusks? Explain why natural selection might have favored this characteristic. (*Hint:* See figure 21.16.)
2. Compare the digestive tract of an earthworm with that of a flatworm. How is each worm's digestive system adaptive?
3. What features distinguish the two classes of annelids, and where do members of each group live?
4. How do annelids feed, exchange gases, and move?
5. In what ways are annelids important?

21.7 Nematodes Are Unsegmented, Cylindrical Worms

Most **roundworms** (phylum Nematoda) are barely visible to the unaided eye, but they are extremely abundant in every habitat—on land, in fresh water, and in oceans from the tropics to the poles. A small scoop of mud can yield millions of tiny nematodes, representing hundreds of species. Overall, biologists have described more than 80,000 nematode species, and hundreds of thousands (or even millions) more species may remain undescribed. Most roundworms are microscopic, but the giant intestinal roundworm (*Ascaris*) can reach 40 cm long.

The nematodes are just one of three groups of worms in the protostome clade. You may recall that flatworms are flat and

FOCUS on Model Organisms

Caenorhabditis elegans and *Drosophila melanogaster*

Two invertebrates, a roundworm and an arthropod, share the spotlight in this box. Both have the characteristics common to all model organisms: small size, easy cultivation in the lab, and rapid life cycles. Each has provided crucial insights into life's workings.

The nematode: *Caenorhabditis elegans*

This roundworm, a soil inhabitant, is arguably the best understood of all animals. This was the first animal to have its genome sequenced (in 1998), revealing about 18,000 genes. A small sampling of the contributions derived from research on *C. elegans* includes the following:

- **Animal development:** An adult *C. elegans* consists of only about 1000 cells. Because the worm is transparent, biologists can watch each organ form, cell by cell, as the animal develops from a zygote into an adult. Eventually, researchers hope to understand every gene's contribution to the development of this worm.

- **Apoptosis:** Programmed cell death, or apoptosis, is the planned "suicide" of cells as a normal part of development. Researchers already know which cells die at each stage. Learning about genes that promote apoptosis may help researchers to better understand cancer, a family of diseases in which cell division is unregulated. ⓘ *cancer,* section 8.6; *apoptosis,* section 8.7

- **Muscle function:** The first *C. elegans* gene to be cloned, *unc-54*, revealed the amino acid sequence of one part of myosin, a protein required for muscle contraction. ⓘ *myosin,* section 29.4A

- **Drug development:** Nematodes provide a good forum for preliminary testing of new pharmaceutical drugs. For example, researchers might identify a *C. elegans* mutant lacking a functional insulin gene, then test new diabetes drugs for the ability to replace the function of the missing gene. ⓘ *diabetes,* section 28.4D

- **Aging:** Worms with mutations in some genes have life spans that are twice as long as normal. Insights into aging in *C. elegans* may eventually help increase the human life span.

- **Origin of sex:** *C. elegans* is a hermaphrodite, so the same individuals produce both sperm and eggs. They can also reproduce asexually. Section 9.9 describes how hermaphroditic nematodes have helped biologists understand the evolution of sexual reproduction.

The fruit fly: *Drosophila melanogaster*

Drosophila melanogaster is only about 3 mm long, but like *C. elegans,* it is a giant in the biology lab. The flies are easy to rear in plugged jars containing rotting fruit or a mix of water, yeast, sugar, cornmeal, and agar. The fruit fly's genome sequence was completed in 1999; many of its 13,600 genes have counterparts in humans. But these relatively recent findings belie *Drosophila*'s century-long history as a model organism. This list includes some of the most important research areas:

- **Heredity:** In the early 1900s, Thomas Hunt Morgan and his colleagues used *Drosophila* to show that chromosomes carry the information of heredity. Studies on mutant flies with different-colored eyes led to the discovery of sex-linked traits. Morgan's group also demonstrated that genes located on the same chromosome are often inherited together. In the process, they discovered crossing over. ⓘ *crossing over,* section 9.5A; *linked genes,* section 10.5; *sex linkage,* section 10.7

- **Human disease:** The similarity of some *Drosophila* genes to those in the human genome has led to important insights into muscular dystrophy, cancer, and many other diseases. For example, researchers have studied the fly version of the human *p53* gene, which induces damaged cells to commit suicide (apoptosis). When that gene is faulty, the cell may continue to divide uncontrollably. The result: cancer.

- **Animal development:** Homeotic genes are "master switch" genes that regulate the overall development of the body, including segmentation and wing placement. Researchers discovered these genes in mutant flies with dramatic abnormalities, such as legs growing in place of antennae on the fly's head (see figure 7.20). Later, researchers discovered comparable genes in many organisms, including mice, leading to new insights into mammalian development (see the Focus on Model Organisms box in section 21.16).

- **Circadian rhythms:** The expression of some genes in bacteria, plants, fungi, and animals cycles throughout a 24-hour day. How do the rhythmically expressed genes "know" what time it is? In *Drosophila,* clock genes called *period* and *timeless* encode proteins that turn off their own expression, much as a thermostat turns off a heater when the temperature is too high. This "master clock" controls the animal's other daily cycles of hormone secretion and behavior.

unsegmented, and they lack a coelom; earthworms (annelids) are cylindrical and segmented, and they have a coelom. Nematodes, in contrast, have cylindrical, unsegmented bodies that have a pseudocoelom (**figure 21.17**). The name of the phylum reflects this long, thin shape; the Greek word *nema* means "thread."

A nematode's pseudocoelom has multiple functions. For example, fluid in the pseudocoelom distributes nutrients, O_2, and CO_2 throughout the body; these animals lack specialized circulatory or respiratory organs. Specialized cells maintain salt balance and eliminate nitrogenous wastes from the pseudocoelom via an excretory pore. A second function of the pseudocoelom is to act as a hydrostatic skeleton. Nematodes are limited to back-and-forth, thrashing motions because only longitudinal (lengthwise) muscles act on the pseudocoelom. As a result, a nematode can neither crawl nor lift its body above its substrate.

Although nematodes may superficially resemble other worms, the evolutionary tree in figure 21.17 shows that arthropods

are actually their closest relatives. Molting provides one clue to this shared evolutionary heritage. That is, both nematodes and arthropods shed and replace their tough external covering (called a cuticle) several times during development.

The nematode nervous system includes a simple brain connected to two nerve cords that run the length of the body. Although the thick cuticle is a barrier to many sensory stimuli, bristles and other structures on the body surface enable the worms to detect touch and chemicals. For example, soil-dwelling nematodes may remain dormant until they detect distinctive molecules secreted by a suitable food source.

Free-living nematodes live in soil or the sediments of freshwater or marine ecosystems. Many of them have the remarkable ability to survive extreme heat, cold, or drying by entering a state of suspended animation; life resumes when favorable conditions return. The worms eat bacteria, protists, fungi, plants, insect larvae, or decomposing organic matter, playing essential roles in nutrient cycling. (Biologists also use the free-living nematode *Caenorhabditis elegans* in scientific research; see the Focus on Model Organisms box in this section.)

Nematodes can also parasitize plants. The worms use their spearlike mouthparts to pierce the cells of roots or shoots. They then suck out the contents, reducing the yields of important crops such as cotton and soybeans. Nematodes may spread harmful viruses on their mouthparts as they move from plant to plant.

But the most familiar nematodes infect humans and other animals. Examples include intestinal parasites such as pinworms, hookworms, and *Ascaris*. Other parasitic roundworms, including *Trichinella*, live in the muscle tissue of humans and pigs and are transmitted by eating undercooked pork. Still other nematodes cause elephantiasis or African river blindness; in both cases, insects transmit the worms. Moreover, heartworms are nematodes that infect the hearts, lungs, and blood vessels of dogs and cats.

Like the flukes and tapeworms described in section 21.4, roundworms that infect animals have several adaptations to the parasitic lifestyle. For example, the worm may attach to host tissues with hooks or suckers, then suck blood or consume digested food in the host's intestines. The thick cuticle protects against the host's digestive enzymes and immune defenses. The worms may even secrete chemicals that suppress the host's immune system.

Parasitic worms also produce huge numbers of fertilized eggs, which subsequently leave the host's body (often in feces). The tough eggs can survive drying and exposure to damaging chemicals, making them very hard to kill. The eggs hatch into larvae that infect new hosts by ingestion or through the skin.

Key features

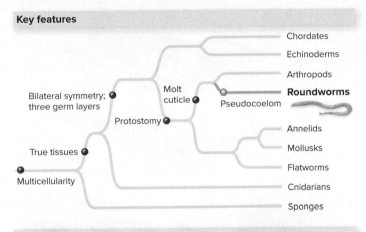

Diversity

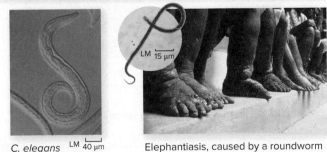

C. elegans   LM 40 μm

Elephantiasis, caused by a roundworm

Anatomy

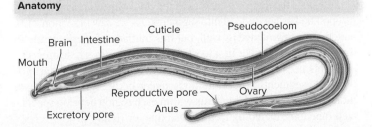

Figure 21.17 **Nematodes (Phylum Nematoda).**

Photos: (*C. elegans*): ©Sinclair Stammers/Science Source; (*Wuchereria*): CDC/Dr. Mae Melvin; (elephantiasis) ©R. Umesh Chandran, TDR, WHO/Science Source

21.7 MASTERING CONCEPTS

1. What features do all roundworms share? (*Hint:* See figure 21.17.)
2. What are some examples of roundworms?
3. What evidence places roundworms in a clade with arthropods?
4. How do nematodes feed, excrete metabolic wastes, move, and reproduce?
5. In what ways are roundworms important?

21.8 Arthropods Have Exoskeletons and Jointed Appendages

If diversity and sheer numbers are the measure of biological success, then phylum Arthropoda certainly is the most successful group of animals. They thrive on land, in air, in the oceans, and in fresh water. More than 1 million species of **arthropods** have been recorded, and biologists speculate that this number could double.

Arthropods intersect with human society in about every way imaginable. Mosquitoes, flies, fleas, and ticks transmit infectious diseases as they consume human blood. Bees and scorpions sting, termites chew wood in our homes, and many insects destroy crops. Yet entire industries rely on arthropods—consider beeswax, honey, silk, and delicacies such as shrimp, crabs, and lobsters. Insects pollinate many plants, and spiders eat crop pests. On a much smaller scale, dust mites eat the flakes of skin that we shed as we move about our homes, while follicle mites inhabit our pores (see the Apply It Now box in this section).

A. Arthropods Have Complex Organ Systems

Arthropods are the fifth (and final) group of protostomes described in this chapter (**figure 21.18**). These animals share a branch of the evolutionary tree with nematodes; evidence for this close relationship includes the observation that both types of animals periodically molt as they grow (see section 21.7).

Arthropoda means "jointed foot," a reference to the most distinctive feature of this phylum: their jointed appendages. These appendages include not only feet and legs but also antennae, copulatory organs, ornaments, weapons, and mouthparts. The many modifications to their mouthparts mean that arthropods eat almost everything imaginable, including dead organic matter, plant parts, and other animals.

In addition, all arthropods have an **exoskeleton,** a rigid outer covering that protects and supports the body. The arthropod exoskeleton is made mostly of chitin, protein, and (sometimes) calcium salts. This tough material protects the animal, keeps it from drying out on land, gives the animal's body its shape, and enables it to move. Thin, flexible areas create the movable joints between body segments and within appendages. Internal muscles attach to the exoskeleton on each side of the joints; contracting these muscles generates precise, forceful movements as the animal crawls, jumps, swims, or flies. This arrangement enables ants to carry items that are much heavier than their own bodies; likewise, thanks to the exoskeleton and associated muscles, grasshoppers and fleas can jump great distances.

Although the exoskeleton is versatile and lightweight, it has a drawback: To grow, an animal must molt and secrete a bigger one, leaving the animal vulnerable while its new exoskeleton is still soft. However, the exoskeleton is not the only arthropod defense. Many arthropods can bite, sting, pinch, make noises, or emit foul odors or toxins that deter predators. Some have excellent camouflage that enables them to blend into their surroundings. Others have defensive behaviors; they may jump, run, roll into a ball, dig

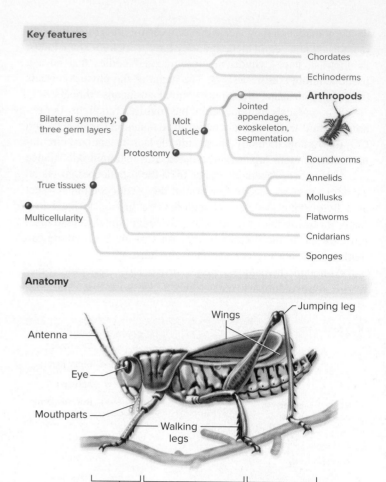

Figure 21.18 **Arthropods (Phylum Arthropoda).**

into soil, or fly away when threatened. Some moths display wings with dramatic eyespots that startle or confuse predators.

Besides the distinctive exoskeleton and jointed appendages, arthropod bodies are divided into segments. As we saw in section 21.6, annelids are segmented as well, but this trait arose independently in the two groups. Unlike in a leech or an earthworm, an arthropod's segments usually do not all function alike. Instead, in many arthropods, the segments group into three major body regions: the head, thorax, and abdomen. Segments within each region develop specialized functions, including feeding, walking, or flying.

Thanks to a nervous system with a brain and ventral nerve cords, many arthropods are active, fast, and sensitive to their environment. Consider, for example, the speed with which a fly can avoid a swatter. Arthropod eyes, bristles, antennae, and other sensory structures can detect light, sound, touch, vibrations, air currents, and chemical signals. All of these clues help arthropods find food, identify mates, and escape predation.

Arthropods have open circulatory systems (see figure 30.1). A tubelike heart propels the circulating fluid freely in the space surrounding most of the animal's organs (**figure 21.19**). An open circulatory system is generally associated with slow-moving animals such as snails and clams. Yet some arthropods are among the most active animals known. How can this be?

Part of the answer relates to the arthropod respiratory system. In most land arthropods, the body wall is perforated with holes (spiracles) that open into a series of branching tubes called tracheae, transporting O_2 and CO_2 directly to and from tissues (see figure 31.2). Spiders and scorpions have stacked folds of tissue called book lungs, which have a large surface area for gas exchange. Aquatic arthropods have extensively branched gills.

Most arthropods have separate sexes. In aquatic arthropods, both external and internal fertilization occurs. But on land, gametes released in external fertilization would dry out; natural selection therefore has selected for internal fertilization in terrestrial arthropods. The male commonly produces a waterproof packet of sperm. The female takes the sperm packet into her body, often after a

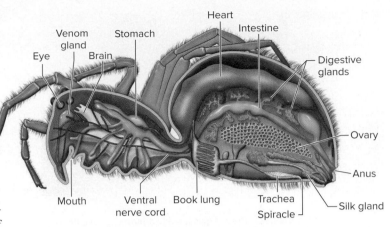

Figure 21.19 Arthropod Organs. The internal anatomy of a spider reveals complex organ systems.

courtship ritual that may involve cannibalism (see section 35.7). In most species, the female then lays the fertilized eggs, although mites and scorpions bear live young. Ants and bees tend their young, but in most other arthropods, parental care is minimal.

Burning Question

Are there really only nine kinds of animals?

The 9 phyla described in this chapter represent a diverse cross section of the 37 known animal phyla. But it would be a mistake to conclude that these 9 groups contain the only important or interesting animals. Here are 3 additional invertebrate phyla, each representing microscopic animals with unusual features (**figure 21.A**):

Phylum Placozoa ("flat animals"): This phylum contains just one named species, *Trichoplax adhaerens*. An adult consists of a few thousand cells that are differentiated into just four cell types. The transparent, asymmetrical body resembles a microscopic sandwich, with an upper surface, a lower surface, and a connecting layer in between. Cilia enable the animal to glide or flow along a solid surface. Biologists first discovered placozoans living in aquaria, and the phylogenetic position of these strange animals has been debated ever since. *Trichoplax* has the smallest genome of any animal, and its body plan is even simpler than that of a sponge. Yet genetic sequences suggest the placozoans are eumetazoans.

Phylum Rotifera ("wheel bearer"): Like placozoans, the rotifers have tiny, transparent, unsegmented bodies. But the 2000 or so species of rotifers are considerably more complex, with bilateral symmetry and a complete digestive tract. These animals are named for the wheel-like tufts of cilia that sweep particles of decomposing organic matter into the mouth. Tiny, hard "jaws" then grind the food. Rotifers inhabit fresh water and moist soil, and they can survive drying for years by suspending their metabolism. As with the placozoans, the phylogenetic position of rotifers remains controversial. The cuticle and pseudocoelom suggest a close relationship to nematodes, but molecular data place rotifers in a clade with annelids, mollusks, and flatworms.

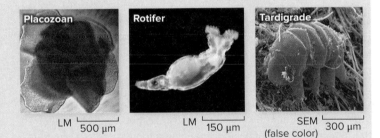

Figure 21.A A Placozoan, a Rotifer, and a Tardigrade.
(placozoan): ©Ana Signorovitch/Yale; (rotifer): ©John Walsh/Science Source; (tardigrade): ©Andrew Syred/Science Source

Phylum Tardigrada ("slow walker"): These charismatic little animals are commonly called "water bears," owing to their aquatic habitat and overall shape. Their segmented bodies have eight legs, each ending in a claw. Covering the body is a cuticle of chitin, and the animal molts as it grows. These features place the tardigrades in a clade with arthropods and nematodes. More than 1000 species have been collected from habitats all over the world, from the poles to the tropics. Like rotifers, tardigrades can enter suspended animation, a state in which they can survive extreme cold or heat, desiccation, the vacuum of space, high pressure, or radiation doses that would kill a human.

Submit your burning question to
Marielle.Hoefnagels@mheducation.com

Figure 21.20 Trilobites. These extinct marine arthropods have three distinct regions along the length of the body: a long central lobe plus flanking right and left lobes.

©Francois Gohier/Science Source

B. Arthropods Are the Most Diverse Animals

Phylum Arthropoda is divided into five subphyla. One contains the 17,000 species of extinct **trilobites** (figure 21.20). Biologists infer the evolutionary relationships among the other four subphyla partly from mouthpart shape (see figure 21.18). Spiders, scorpions, and other **chelicerates** have grasping, clawlike mouthparts called chelicerae. The three subphyla of **mandibulates** have chewing, jawlike mouthparts termed mandibles.

Chelicerates: Spiders and Their Relatives
Most chelicerates have two major body regions: an abdomen and a fused head and thorax (figure 21.21). They also have four or more pairs of walking legs, but they lack antennae.

The two most familiar groups of chelicerates are horseshoe crabs and arachnids. **Horseshoe crabs** are primitive-looking animals whose name refers to the hard, horseshoe-shaped exoskeleton, which covers a wide abdomen and a long tailpiece. The four

Apply It **Now**

Your Tiny Companions

Even when you think you are alone, you aren't; your body may host a diverse assortment of arthropods. Head lice and body lice are biting insects that cause skin irritation. Ticks latch onto the skin and suck your blood, sometimes transmitting the bacteria that cause Lyme disease. The tiny larvae of chigger mites produce saliva that digests small areas of skin tissue, causing intense itching.

Lice, ticks, and chiggers are hard to ignore, but you may never notice one inconspicuous companion: the follicle mite, *Demodex* (figure 21.B). This arachnid, which is less than half a millimeter long, lives in hair follicles and nearby oil glands, where it eats skin secretions and dead skin cells. *Demodex* mites are by no means rare. Nearly everyone has them, and each follicle may house up to 25 of the tiny animals. (If you would like to see your own follicle mites, carefully remove an eyebrow hair or eyelash and examine it with a compound microscope.) Luckily, the infestation is typically symptomless, although occasionally the mites may cause a rash.

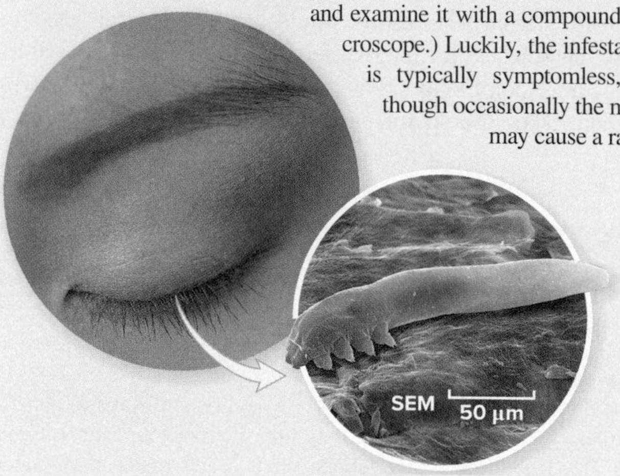

Figure 21.B Follicle Mites. Tiny *Demodex* mites live in skin pores and hair follicles.

(brow): ©Ingram Publishing RF; (mite): ©Andrew Syred/Science Source

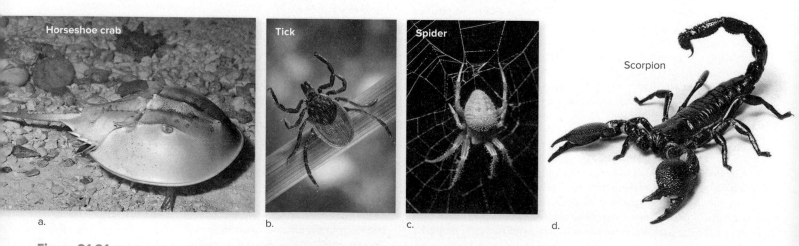

Figure 21.21 Chelicerate Arthropods. Chelicerates include (a) horseshoe crabs, (b) ticks, (c) spiders, and (d) scorpions.

(a): ©Nature's Images/Science Source; (b): CDC/James Gathany; William Nicholson; (c): ©Corbis RF; (d): ©Digital Vision/Punchstock RF

species of horseshoe crabs are not true crabs, which are crustaceans. Humans have found an unusual way to exploit the horseshoe crab's blood, which contains a unique immune system compound that binds to bacteria. Technicians routinely use this compound to test medical supplies for bacterial contamination.

The more than 100,000 species of **arachnids** are eight-legged arthropods, including mites and ticks; spiders; harvestmen ("daddy longlegs"); and scorpions. Spiders make "silk" and use it to produce webs, tunnels, egg cases, and spiderling nurseries. Some spiderlings use silk driftlines to float to a new habitat.

Mandibulates: Millipedes and Centipedes

About 13,000 species of **millipedes** and **centipedes** make up a group of terrestrial arthropods called myriapods (**figure 21.22a**). In these animals, the head features mandibles and one pair of antennae. The rest of the body is divided into repeating subunits, each with one pair (centipedes) or two pairs (millipedes) of appendages. Millipedes eat decaying plants and are generally harmless to humans, but centipedes are predators, and their venomous bite can be painful.

Mandibulates: Crustaceans

The **crustaceans** form a group of about 52,000 species, including crabs, shrimp, and lobsters. Smaller aquatic crustaceans include brine shrimp, water fleas (*Daphnia*), copepods, barnacles, and krill. Isopods, commonly known as pill bugs or "roly-polies," are the only terrestrial crustaceans; all other crustaceans live in water. Their bodies are extremely variable, but all have two pairs of antennae.

Mandibulates: Insects

Scientists know of well over 1 million species of **insects,** with many more awaiting formal description. All of these animals have mandibles; one pair of antennae; a body divided into a head, a thorax, and an abdomen; six legs; and (usually) two pairs of wings.

The ancestors of today's insects colonized land shortly after plants, about 475 MYA, and they diversified rapidly in one of life's great adaptive radiations. Why did this group diversify into so many species? Biologists point to several possible explanations. For example, mutations in homeotic genes can modify the body segments of insects into seemingly unlimited variations; some biologists compare the insect body plan to the versatility of a Swiss army knife. ⓘ *homeotic genes,* section 7.7D

Wings may also partly account for insect success. Although flight later evolved independently in birds and in bats, insects were the first animals to fly. They use their wings to disperse to new habitats, escape predators, court mates, and find food that other animals cannot reach. Many of today's flowering plants evolved in conjunction with flying insects, trading nectar for rapid, efficient pollination services. ⓘ *pollination,* section 24.2C

Insect reproductive strategies may also have played a role. Insects have high reproductive rates, and their eggs can survive in dry habitats. In some insect species, such as crickets, the offspring change only gradually from molt to molt. Most insect life cycles, however, include a metamorphosis, in which a larva (such as a caterpillar or maggot) undergoes a transformation into an adult (such as a butterfly or housefly). As described in section 21.1D, indirect development helps reduce competition for food between adults and their young.

Whatever the explanation for their success, the variety of insect species alive today almost defies description (see figure 21.22c). Familiar examples include silverfish, mayflies, dragonflies, roaches, crickets, grasshoppers, lice, cicadas, aphids, beetles, ants, wasps, bees, butterflies, moths, fleas, and flies. (The fruit fly, *Drosophila melanogaster,* is profiled in the Focus on Model Organisms box in section 21.7.) Insects range in size from wingless soil-dwellers less than 1 mm long to fist-sized

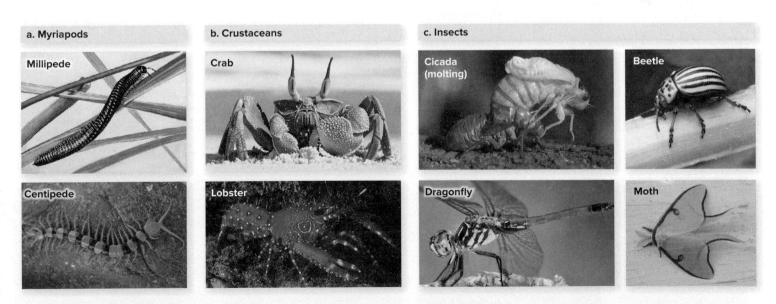

Figure 21.22 Mandibulate Arthropods. (a) Myriapods include millipedes, which have two pairs of legs per segment, and centipedes, which have one pair per segment. (b) Crustaceans include crabs and lobsters. (c) Insects include cicadas, dragonflies, beetles, and moths.

Photos: (a, millipede): ©De Agostini Picture Library/Getty Images; (a, centipede): ©Matthijs Kuijpers/Alamy RF; (b, crab): ©Pete Atkinson/Photographer's Choice/Getty Images RF; (b, lobster): ©Photoshot Holdings Ltd/Alamy; (c, cicada): ©Rob Crandall/Shutterstock RF; (c, dragonfly): ©Thomas Shahan/Flickr/Getty Images; (c, beetle): USDA/Scott Bauer; (c, moth): ©Steven P. Lynch RF

beetles, foot-long walking sticks, and flying insects with foot-wide wingspans. Some extinct dragonflies were even larger—one had a wingspan of about 75 cm! Most of these species are terrestrial, but some insects live or reproduce in fresh water. The ocean, high altitudes, and extremely cold habitats are about the only places that are nearly devoid of insects.

21.8 MASTERING CONCEPTS

1. Approximately how many arthropods have scientists identified? What subphylum of arthropods has the richest diversity of species?

2. What features do all arthropods share? (*Hint:* See figure 21.18.)

3. How do arthropods use their jointed appendages?

4. Describe how arthropods feed, respire, sense their environment, move, reproduce, and defend themselves.

5. What are the advantages and disadvantages of an exoskeleton?

6. In what ways are arthropods important?

7. How are chelicerates different from mandibulates?

8. Give an example of an animal in each subphylum of arthropods. Describe one characteristic unique to each group.

21.9 Echinoderm Adults Have Five-Part, Radial Symmetry

The **echinoderms** (phylum Echinodermata) include some of the most colorful and distinctive sea animals (figure 21.23). This phylum includes sea urchins, sea stars, sea cucumbers, and sand dollars. Many of these animals are familiar to people who visit tide pools and beaches.

It is hard to imagine animals less similar to ourselves than sea stars, yet chordates and echinoderms share a branch on the evolutionary tree. What is the evidence for this close relationship? Early development patterns provide part of the answer: Echinoderms and chordates are both deuterostomes. That is, during the gastrula stage of development, the first indentation becomes the anus (see figure 21.4). Molecular evidence, such as ribosomal RNA sequences, confirms the close relationship between echinoderms and chordates.

The evolutionary tree in figure 21.23 also holds another surprise. Even though echinoderms belong to the large clade containing animals with bilateral symmetry, adult sea stars and other echinoderms have radially symmetrical bodies divided into five parts. This body form is most obvious in the sea stars and brittle stars, which usually have five arms. Although other echinoderms lack arms, close inspection reveals the five-part symmetry. Radial symmetry is typically associated with a sessile lifestyle, as we saw in the sponges and cnidarians. However, most of today's echinoderms are motile. Biologists continue to debate how and why radial symmetry evolved in this group.

Regardless of the answer, a close look at a developing sea star reveals that echinoderms are related to the other bilateral animals. Echinoderms usually reproduce sexually, with male and female gametes from separate individuals combining in the sea. The larvae start out with bilateral symmetry (see figure 21.23), but a radical metamorphosis soon occurs. Through a complex

Key features

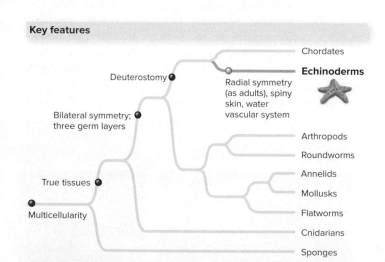

Figure 21.23 Echinoderms (Phylum Echinodermata).

Diversity

Sea urchin

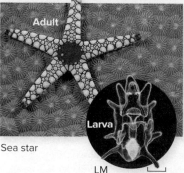

Adult

Larva

LM   500 μm

Sea star

Sand dollar

Sea cucumber

series of events, the left side of a larva develops into the underside of the young animal, while the right side becomes the upper surface. Once the transformation is complete, the animal's body betrays few clues to its bilateral beginnings.

Echinoderms have other unique characteristics as well. For example, the name *echinoderm* literally means "spiny skin," a feature clearly visible in the sea urchins. The spines are connected to the animal's internal skeletal plates (in sea urchins and sand dollars, these plates fuse into a protective shell). The spines have multiple functions. Defense is the most obvious one, but some sea urchins can use their spines to "walk" along the seafloor, and some echinoderms have spines modified for feeding or deflecting waves.

Besides spines, echinoderm skin also offers other defenses. Small pincers embedded in the skin may deter predators. The epidermis may also contain glands that secrete toxins, along with pigments that provide camouflage or warn predators to stay away.

The bodies of echinoderms are unusual in other ways, too. They have a unique, collagen-rich tissue that can rapidly interchange between soft and hard, allowing the animal to squeeze into tight spaces or, alternatively, stiffen to aid in feeding or to defend against predation. They can also regenerate arms lost to predators; in some species, even a small part of an arm can grow into an entire new animal!

Another unique feature of echinoderms is the **water vascular system,** a series of enclosed, water-filled canals that end in hollow tube feet (**figure 21.24**). Coordinated muscle contractions extend and retract each foot, bending it from side to side or creating a suction-cup effect when applied to a hard surface. The wavelike pumping of water into and out of the tube feet allows echinoderms to glide slowly while maintaining a firm grip on the substrate, a clear advantage in a wave-pounded environment.

Tube feet have many other functions besides locomotion. For example, sensory cells in tube feet can detect light and chemicals. Moreover, tube feet can also help echinoderms acquire food. Consider a sea star, which is a predator that eats mussels and other bivalves. The sea star attaches its tube feet to a mussel's shell and steadily pulls until the prey's muscles tire and the shell opens. Part of the sea star's stomach comes out through its mouth and enters the mussel's shell, secreting digestive enzymes into the mussel. When the stomach retracts, it carries with it the partially digested food. Digestive glands along the sea star's arms secrete additional digestive enzymes and absorb nutrients. Undigested food is ejected through the anus on the dorsal surface.

Echinoderms lack complex circulatory, respiratory, and excretory systems, but the versatile water vascular system fulfills many of the same functions. The animal's internal canals can exchange water with the ocean via a specialized pore. As a result, the thin-walled tube feet can function as gills, exchanging gases between ocean water and the internal fluid. Some metabolic wastes can also diffuse out of the tube feet and into the ocean.

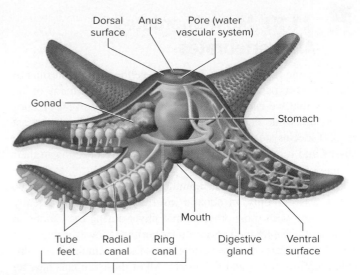

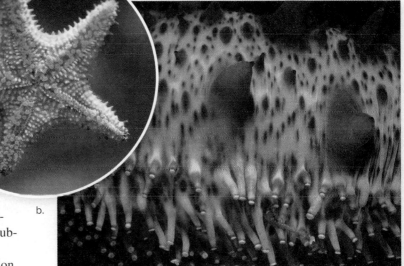

Figure 21.24 **Sea Star Anatomy.** (a) A sea star's internal anatomy includes a complete digestive tract, reproductive organs, and a unique water vascular system. (b) The ventral surface of a sea star reveals the mouth and tube feet. (c) A close-up view of a sea cucumber's tube feet.

Photos: (b): ©McGraw-Hill Education; (c): ©Jeff Rotman/Alamy

21.9 MASTERING CONCEPTS

1. What features do all echinoderms share? (*Hint:* See figure 21.23.)
2. What is a water vascular system, and why is it adaptive?
3. What are some examples of echinoderms?
4. How do echinoderms eat, respire, excrete metabolic wastes, sense their environment, move, reproduce, and defend themselves?

21.10 Most Chordates Are Vertebrates

Many people find phylum Chordata to be the most interesting of all, at least in part because it contains humans and many of the animals that we eat, keep as pets, and enjoy observing in zoos and in the wild. The **chordates** are a diverse group of at least 60,000 species (table 21.2). From the tiniest tadpole to fearsome sharks and lumbering elephants, chordates are dazzling in their variety of forms.

No one knows what the common ancestor of chordates was, but it was certainly an aquatic invertebrate, and it apparently arose along with most other animal phyla during the Cambrian explosion. Despite their invertebrate heritage, however, the most familiar chordates are the vertebrates—fishes, amphibians, non-avian reptiles, birds, and mammals. All of these animals have an internal skeleton (endoskeleton) that includes a protective, flexible, segmented backbone.

A. Four Key Features Distinguish Chordates

The common ancestor of chordates had several key features; this section concentrates on four of the most important ones. Every chordate has inherited these features and expresses each one at some point during its life (figure 21.25):

1. **Notochord:** The notochord is a flexible rod that extends along the length of a chordate's back. In most vertebrates, the notochord does not persist into adulthood but rather is replaced by the backbone that surrounds the spinal cord.

2. **Dorsal, hollow nerve cord:** The dorsal, hollow nerve cord is parallel to the notochord. In many chordates, the nerve cord develops into the spinal cord and enlarges at the head end, forming a brain. ⓘ *central nervous system*, section 26.6

3. **Pharyngeal slits (or pouches):** In most chordate embryos, slits or pouches form in the pharynx, the muscular tube that begins at the back of the mouth. These structures have multiple functions. Invertebrate chordates feed by straining food particles out of water that passes through the slits. In vertebrates, the pouches develop into gills, the middle ear cavity, or other structures.

4. **Postanal tail:** A muscular tail extends past the anus in all chordate embryos. In humans, chimpanzees, and gorillas, the body absorbs most of the tail before birth; only the tailbone remains as a vestige. (Rarely, a human baby is born with a tail, which is removed in minor surgery.) In fishes, salamanders, lizards, cats, and many other species, adults retain the tail.

TABLE **21.2** Major Taxonomic Groups in Phylum Chordata

Group	Examples	Approximate Number of Species
Tunicates (subphylum Urochordata)	Sea squirt	3000
Lancelets (subphylum Cephalochordata)	Amphioxus	30
Hagfishes and lampreys (superclass Agnatha)	Slime hag, sea lamprey	70 (hagfishes) 38 (lampreys)
Cartilaginous and bony fishes (superclass Osteichthyes)	Shark, salmon, lungfish, coelacanth	30,000
Amphibians (class Amphibia)	Frog, salamander, caecilian	6000
Reptiles (class Reptilia and class Aves)	Turtle, lizard, snake, tuatara, crocodile, chicken, ostrich	8000 to 10,000 (nonavian reptiles) 9000 to 10,000 (birds)
Mammals (class Mammalia)	Platypus, kangaroo, dog, whale, human	5800

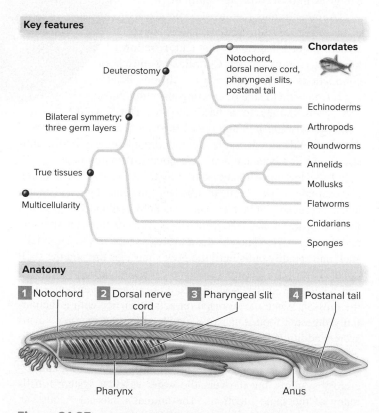

Figure 21.25 Chordates (Phylum Chordata). A simple animal called a lancelet illustrates four characteristics present in all chordates.

B. Many Features Reveal Evolutionary Relationships Among Chordates

Figure 21.26 depicts the evolutionary relationships within phylum Chordata. If you have previously studied the chordates, you may be surprised at the absence of a separate branch for birds in the phylogenetic tree. At one time, the nonavian reptiles such as snakes, lizards, turtles, and crocodiles were considered to form a clade separate from the birds. We now know, from molecular, fossil, and anatomical evidence, that birds are a type of reptile. Figure 21.26 reflects this new understanding.

This section explains the features that determine the branches on the chordate evolutionary tree, and the rest of this chapter considers each chordate group separately.

Cranium Most chordates have a **cranium,** a bony or cartilage-rich case that surrounds and protects the brain (figure 21.27). Hagfishes and vertebrates form two clades of **craniates,** animals that have a cranium.

Vertebrae **Vertebrae** are a series of small bone or cartilage structures that make up the vertebral column, or backbone (see figure 21.27). Vertebrae protect the spinal cord and provide attachment points for muscles, giving the animal a greater range of movement. (Vertebrates are chordates that have a backbone.)

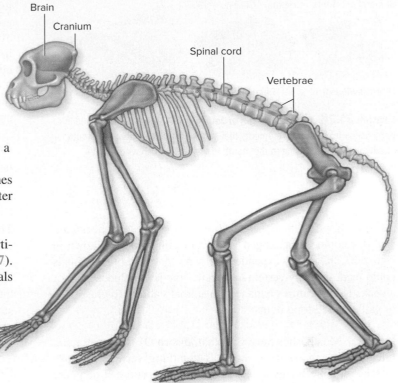

Figure 21.27 Skull and Backbone. A cranium and vertebrae are two characteristics that are common to most chordates.

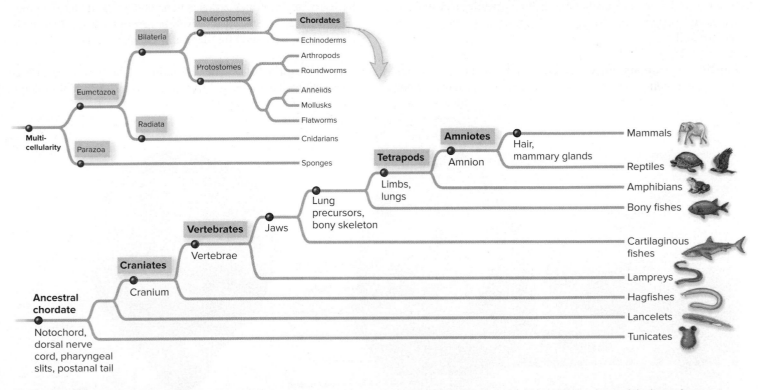

Figure 21.26 Chordate Diversity. Chordates include several groups of animals, including the invertebrate tunicates and lancelets and the better-known fishes, amphibians, reptiles (including birds), and mammals.

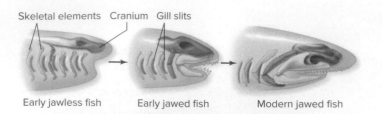

Figure 21.28 Possible Origin of Jaws. Jaws may have developed in early fishes from skeletal elements that supported gill slits near the mouth. The two elements closest to the mouth of the jawless fish were lost in animals with jaws.

Jaws

Jaws are the bones that frame the entrance to the mouth. The development of hinged jaws from gill supports, shown in figure 21.28, greatly expanded the ways that vertebrate animals could feed. In many vertebrate species, the jaw includes teeth or a beak. These features enhance the animal's ability to grasp prey or gather small food items.

Lungs

Most fishes have gills that absorb O_2 from water and release CO_2. In contrast, most air-breathing vertebrates have internal saclike **lungs** as the organs of respiration. Lungs are homologous to the swim bladders of bony fishes. Both structures apparently arose from simple outgrowths of the esophagus. These sacs, which allowed fishes to gulp air in shallow water, developed into air-breathing lungs in the ancestors of terrestrial vertebrates. (Section 31.1 shows gills and lungs in more detail.)

Limbs

Tetrapods are vertebrates with two pairs of limbs that enable the animals to walk on land (*tetrapod* means "four legs"). Amphibians, reptiles (including birds), and mammals are all tetrapods. Some animals classified as tetrapods, however, have fewer than four limbs. Snakes, for example, lack limbs entirely. The limbs of whales, dolphins, and sea lions are either modified into flippers or too small to project from the body. Anatomical and molecular evidence, however, clearly links all of these animals to tetrapod ancestors (see figure 13.11 and section 13.7).

Amnion

The jellylike eggs of fishes and amphibians do not have a shell; they must remain in water, or the embryos inside will die. In contrast, reptiles and mammals can breed in arid habitats. The egg of a reptile (including a bird) has a leathery or hard outer layer, so the embryo does not dry out and die on land. This **amniotic egg** contains several membranes (the amnion, chorion, and allantois) that cushion the embryo, provide for gas exchange, and store metabolic wastes (figure 21.29). Meanwhile, the yolk nourishes the developing embryo. The internal membranes of the amniotic egg are homologous to the protective structures that surround a developing fetus in the uterus of a female mammal. The **amniote** clade, which consists of reptiles and mammals, reflects this shared evolutionary history.

Body Coverings

Fish scales are bony, and amphibians have naked, unscaly skin. In the three groups of amniotes, the body coverings all are composed of the same protein—keratin (figure 21.30). Nonavian reptiles such as snakes and crocodiles have dry, tough scales all over their bodies. Birds have similar scales on their legs; feathers cover the rest of the body. Mammals have hair (often called fur, depending on the animal).

Thermoregulation

The regulation of body temperature is an additional characteristic that is important in animal biology.

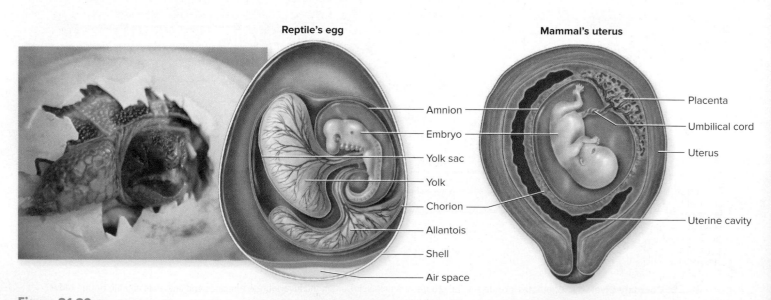

Figure 21.29 The Amnion. The amnion is a sac that encloses the developing embryo of a reptile or mammal. In an amniotic egg, the embryo is encased in a hard, protective shell, and it is supported internally by three membranes—the amnion, allantois, and chorion. Placental mammals also enclose embryos in an amnion.
Photo: ©Creatas/PunchStock RF

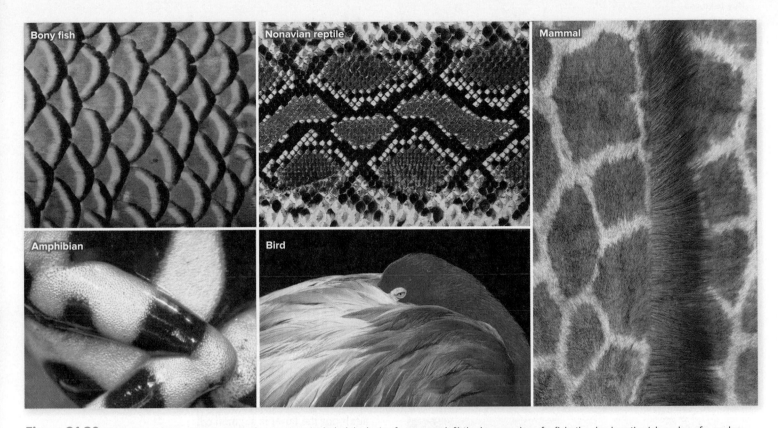

Figure 21.30 Body Coverings. Vertebrate body coverings include (*clockwise from upper left*) the bony scales of a fish; the dry, keratin-rich scales of a snake (a reptile); the fur of a mammal; the feathers of a bird; and the naked, unscaly skin of an amphibian.

(fish): ©Jeffrey L. Rotman/Photonica/Getty Images; (reptile, mammal): ©Siede Preis/Getty Images RF; (amphibian): ©Danita Delimont/Gallo Images/Getty Images; (bird): ©Corbis RF

Thermoregulation strategies vary widely, but in general, biologists often classify animals as ectotherms or endotherms. The body temperature of an **ectotherm** tends to fluctuate with the environment; these animals lack internal mechanisms that keep their temperature within a narrow range. Invertebrates, fishes, most amphibians, and most nonavian reptiles are ectotherms. Many behaviors, such as basking in the sun or burrowing into the ground, help an ectotherm adjust its temperature.

Endotherms maintain their body temperature mostly by using heat generated from their own metabolism. Birds and mammals are endotherms, as are a few other types of animals. Endothermy requires an enormous amount of energy, which explains why birds and mammals must eat so much more food than ectotherms of the same size. Section 33.6 explores the origin of fur and feathers, which help endothermic animals conserve body heat.

Heart Chambers The structure of the heart marks another set of evolutionary milestones in vertebrates (see figure 30.2). Fishes have a two-chambered heart, with one atrium and one ventricle. In amphibians and most nonavian reptiles, the heart has two atria and one ventricle. Four chambers (two atria and two ventricles) make up the hearts of crocodiles, birds, and mammals. The more heart chambers, the better the separation of oxygen-rich blood from oxygen-poor blood, and the greater the efficiency with which blood delivers oxygen—a necessity in an

energy-hungry endothermic animal. ⓘ *vertebrate circulatory systems,* section 30.1A

In the first half of this chapter, the emphasis was on large-scale developmental patterns, which highlight the dramatic evolutionary transitions between the simplest animals and the most complex. The rest of this chapter takes a slightly different approach. Chordates share a similar embryonic body plan and have complex organ systems. We therefore focus here on how unique selective forces in each group's habitat have shaped its anatomy and physiology.

21.10 MASTERING CONCEPTS

1. What are four key defining characteristics of chordates? (*Hint:* See figure 21.25.)
2. Which chordates are craniates, and which of those are also vertebrates?
3. How did the origin of jaws, lungs, limbs, and the amnion affect the course of vertebrate evolution?
4. How do the body coverings of fishes, amphibians, nonavian reptiles, birds, and mammals differ?
5. Differentiate between an ectotherm and an endotherm.
6. How does the number of heart chambers affect the efficiency of oxygen delivery to body tissues?

Key features

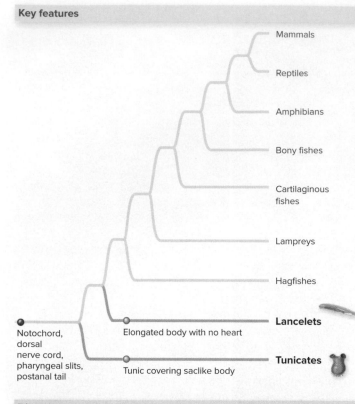

Mammals

Reptiles

Amphibians

Bony fishes

Cartilaginous fishes

Lampreys

Hagfishes

Lancelets

Tunicates

Notochord, dorsal nerve cord, pharyngeal slits, postanal tail

Elongated body with no heart

Tunic covering saclike body

Diversity

Tunicates

Lancelet

Tunicate anatomy

Exit siphon   Intake siphon

Water and wastes out

Water and food in

Anus

Pharynx

Genital duct

Pharyngeal slits

Intestine

Gonads (ovary and testis)

Stomach

Tunic

Heart

Lancelet anatomy

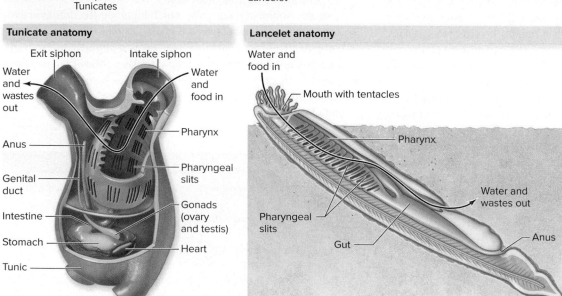

Water and food in

Mouth with tentacles

Pharynx

Water and wastes out

Pharyngeal slits

Gut

Anus

Figure 21.31 Tunicates and Lancelets.

Photos: (blue tunicate): ©Nancy Sefton/Science Source; (orange tunicates): ©Janna Nichols; (lancelet): ©Natural Visions/Alamy

21.11 Tunicates and Lancelets Are Invertebrate Chordates

Tunicates and lancelets form two subphyla of invertebrate chordates; their bodies have neither a cranium nor vertebrae (figure 21.31). These animals would probably attract little attention if not for one fact: They are the modern organisms that most resemble the ancestral chordates. Biologists continue to debate which group is more closely related to vertebrates.

The 3000 species of **tunicates** take their name from the tunic, a protective, flexible body covering that the epidermis secretes. The best-studied tunicates, the ascidians, are marine animals that resemble a bag with two siphons. Cilia pull water in through one siphon. As the water moves across slits in the pharynx, oxygen diffuses into nearby blood vessels, and carbon dioxide diffuses out. Mucus covering the slits traps suspended food particles, and the water exits through the other siphon. These animals are also called sea squirts because they can forcibly eject water from their siphons if disturbed.

The free-swimming tunicate larva resembles a tadpole, and it has all four chordate characteristics. Once it settles headfirst onto a solid surface, however, the tail and notochord disappear, and the nerve cord shrinks to nearly nothing. The adults, which are usually sessile, retain only the pharyngeal slits. Neither the adult nor the larva is segmented.

Some invasive tunicate species are a major nuisance in coastal areas. Huge colonies of rapidly reproducing sea squirts coat dock pilings and boats, and they smother shellfish that are economically and environmentally important.

Lancelets (also called amphioxus) resemble small, eyeless fishes with translucent bodies. They live in shallow seas, with the tail buried in sediment and the mouth extending into the water. Cilia and mucus secreted onto the pharyngeal slits trap and move food particles into the digestive tract. The water passes out of the pharynx through the slits, leaving the body through a separate pore. Blood distributes nutrients to body cells, but gas exchange occurs directly across the skin. To reproduce, males and females release gametes into the water at the same time. The larvae, like the adults, resemble tiny fish.

Biologists have identified some 30 species of lancelets. These animals clearly display all four major chordate characteristics, as well as inklings of the organ systems that appear in the vertebrates.

Like vertebrates, lancelets have segmented blocks of muscles. Furthermore, the lancelet nervous system consists of a nerve cord with a slight swelling at the head end, plus sensory receptors on the body. The simple lancelet brain appears to share some of the same divisions as the more elaborate vertebrate brain. However, these animals lack the sophisticated sensory organs, complex brains, and mobility of vertebrates.

21.11 MASTERING CONCEPTS

1. Compare and contrast the features of tunicates and lancelets. (*Hint:* See figure 21.31.)

2. How do tunicates use their siphons in feeding and gas exchange?

3. What is the relationship among tunicates, lancelets, and vertebrate chordates?

Figure 21.32 Hagfishes and Lampreys.
Photos: (hagfish mouth): ©Steven Senne/AP Images; (hagfish): ©Mark Conlin/Alamy; (lamprey): ©David Hosking/Alamy; (lamprey mouth): ©Russ Kinne/Science Source

Key features

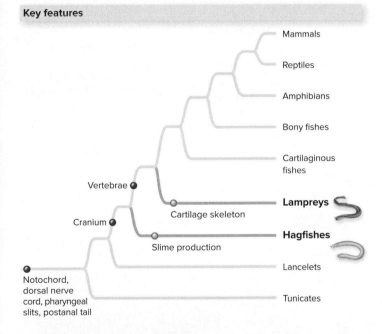

Diversity

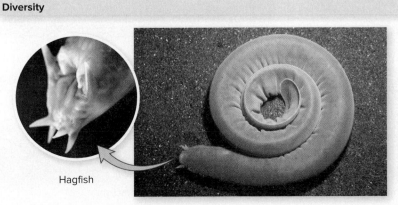

Hagfish

21.12 Hagfishes and Lampreys Are Craniates Lacking Jaws

Hagfishes and lampreys are two groups of chordates that share several similarities (figure 21.32): They have long, slender bodies with gills and specialized sense organs clustered near the head end, and their mouths lack jaws. Scientists are still debating the relationship between hagfishes and lampreys; in this section we discuss them together as jawless, fishlike animals.

The common ancestor of the craniates may have resembled a **hagfish.** In a hagfish, cartilage makes up the cranium and supports the tail. But because vertebrae do not surround the nerve cord, hagfishes are not vertebrates.

The 70 or so species of hagfishes live in cold ocean waters, eating marine invertebrates such as shrimp and worms or using their raspy tongues to scavenge the soft tissues of dead or near-dead animals. Their eyes are poorly developed, but touch- and chemical-sensitive tentacles near the mouth help locate food.

These animals have some unusual abilities. They can slide their flexible bodies in and out of knots to pull on food, escape predation, or clean themselves. Hagfishes are also called "slime hags," in recognition of the glands that release copious amounts of a sticky, white slime when the animal is disturbed.

Lampreys are the simplest organisms to have cartilage around the nerve cord, so they are vertebrates. About 38 species exist in salty or fresh waters around the world. They spend most of their lives as larvae, straining food from the water column. The adults eat small invertebrates, although some species use their suckers to consume the blood of fish. Over the past century, sea lampreys have ventured beyond their natural Lake Ontario range into the other Great Lakes, where they have been largely responsible for the decline in populations of lake trout and whitefish. ⓘ *invasive species,* section 40.5A

21.12 MASTERING CONCEPTS

1. Compare the characteristics of hagfishes and lampreys. (*Hint:* See figure 21.32.)

2. How do hagfishes eat and defend themselves?

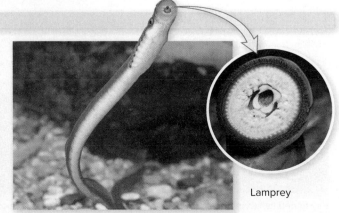

Lamprey

21.13 Fishes Are Aquatic Vertebrates with Jaws, Gills, and Fins

Fishes are the most diverse and abundant of the vertebrates, with more than 30,000 known species that vary greatly in size, shape, and color (**figure 21.33**). They occupy nearly all types of water, from fresh to salty, from clear to murky, and from frigid to warm, although they cannot tolerate hot springs.

Fishes play important roles in their aquatic habitats. They graze on algae, scavenge dead organic matter, or prey on other animals, eating everything from mosquito larvae and other small invertebrates to one another. Tuna and many other fish species are also an important source of dietary protein for people (and their pets) on every continent. Angling for trout, bass, salmon, and other fishes remains a popular sport. Fishes also inspire a wide range of emotions, from an intense fear of sharks to the peace and tranquility that come from watching tropical fish in a home aquarium.

Fishes originated some 500 MYA from an unknown ancestor with jaws, gills, and paired fins. Millions of years later, they diversified into two main clades: the cartilaginous fishes and the bony fishes.

A. Cartilaginous Fishes Include Sharks, Skates, and Rays

The **cartilaginous fishes,** the most ancient clade of fishes, include about 800 species of sharks, skates, and rays. As the name implies, their skeletons are made of cartilage.

Sharks are the most notorious of the cartilaginous fishes. Although some sharks feed on plankton, the carnivorous species are famous for their ability to detect blood in the water. Extending along each side of a fish is the **lateral line,** a sense organ that detects vibration in nearby water and helps the animal to find prey and escape predation.

Many people believe that a shark will die if it stops swimming. In fact, this is true only for some species of sharks and other cartilaginous fishes, which must swim to maintain a constant flow of water through the mouth and over the gills. Other cartilaginous fishes, however, can pump water over their gills while at rest.

B. Bony Fishes Include Two Main Lineages

The **bony fishes** form a clade that includes 96% of existing fish species. They have skeletons of bony tissue reinforced with mineral deposits of calcium phosphate (**figure 21.34**). Like sharks,

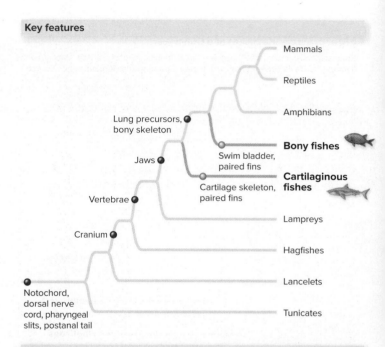

Key features

- Mammals
- Reptiles
- Amphibians
- Lung precursors, bony skeleton
- **Bony fishes**
- Jaws — Swim bladder, paired fins
- **Cartilaginous fishes**
- Vertebrae — Cartilage skeleton, paired fins
- Cranium
- Lampreys
- Hagfishes
- Lancelets
- Notochord, dorsal nerve cord, pharyngeal slits, postanal tail
- Tunicates

Diversity

Figure 21.33 The Fishes.

Photos: (ray): ©MedioImages / SuperStock RF; (shark): ©Michele Westmorland/Getty Images RF; (snappers): ©Corbis RF; (lungfish): ©Peter E. Smith/Natural Sciences Images Library; (coelacanth): ©Peter Scoones/Planet Earth Pictures/Getty Images

Stingray (cartilaginous fish)

Shark (cartilaginous fish)

Ray-finned fish (bony fish)

Lungfish (bony fish)

Coelacanth (bony fish)

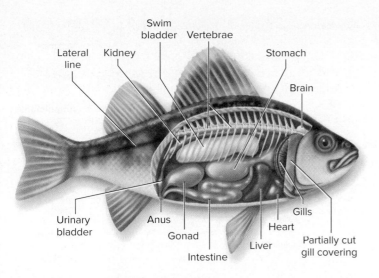

Figure 21.34 Anatomy of a Bony Fish. This illustration shows some of a fish's adaptations to life in water, including fins, a swim bladder, and gills.

bony fishes have a lateral line system. Unlike cartilaginous fishes, however, the bony fishes have a hinged gill covering that can direct water over the gills, eliminating the need for constant swimming. In addition, most bony fishes have a swim bladder. By expanding or contracting the volume of its swim bladder, a bony fish can adjust its buoyancy.

Bony fishes are divided into two classes, the ray-finned and lobe-finned fishes.

Ray-Finned Fishes   The **ray-finned fishes** include nearly all familiar fishes: eels, minnows, catfish, trout, tuna, salmon, and many others. Their name comes from their fan-shaped fins, which consist of slender, bony spines (the "rays") supporting thin, flexible webs of skin.

The extraordinary diversity and abundance of the ray-finned fishes reflect their superb adaptations to a watery world. In particular, the ray-finned fishes are notable for the wide-ranging modifications of their jaws. Thanks to a large number of movable bones in the skull, fish jaws are extremely versatile. While some fish consume algae or small animals from the water column, others have specialized jaws reflecting limited diets, such as the scales, fins, eggs, or eyes of other fish.

Lobe-Finned Fishes   The 10 or so existing species of **lobe-finned fishes** are the bony fishes most closely related to the tetrapods, based on the anatomical structure of their fleshy paired fins consisting of bone and muscle. This group includes the lungfishes and the coelacanths. **Lungfishes** have lungs that are homologous to those of tetrapods. During droughts, a lungfish burrows into the mud beneath stagnant water, gulping air and temporarily slowing its metabolism. **Coelacanths** are called "living fossils"; they originated during the Devonian and remain the oldest existing lineage of vertebrates with jaws.

C. Fishes Changed the Course of Vertebrate Evolution

Several features arose in fishes that would have profound effects on vertebrate evolution. A segmented backbone, with its multiple muscle attachment points, expanded the range of motion. Jaws opened new feeding opportunities, which in turn selected for a more complex brain that could develop a hunting strategy or plan an escape route.

Two of the adaptations that enabled vertebrates to thrive on land originated in fishes: lungs and limbs. Lungs developed in a few species of fishes, and the air-breathing descendants of these animals eventually colonized the land. No fish has true limbs, but some fishes have pectoral fins with stronger bones and more flesh than the delicate, swimming fins of other fishes. These robust fins may have enabled the ancestors of tetrapods to move along the sediments of their shallow water homes. Whatever their original selective advantage, pectoral and pelvic fins eventually evolved into the limbs that define tetrapods (**figure 21.35**).

21.13 MASTERING CONCEPTS

1. What features do all fishes share? What distinguishes cartilaginous fishes from bony fishes? (*Hint:* See figure 21.33.)
2. What are the major types of cartilaginous and bony fishes?
3. Describe the major vertebrate adaptations that arose in fishes and were adaptive on land.

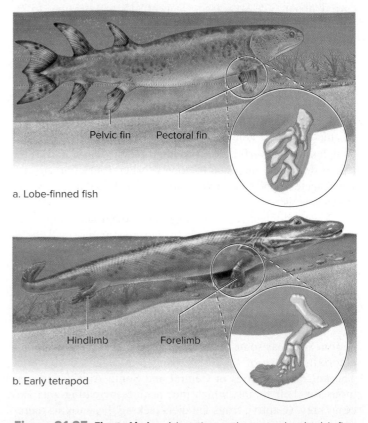

a. Lobe-finned fish

Pelvic fin   Pectoral fin

Hindlimb   Forelimb

b. Early tetrapod

Figure 21.35 Fins to Limbs. Adaptations to the pectoral and pelvic fins in some fish lineages led to the four limbs characteristic of the tetrapods.

21.14 Amphibians Lead a Double Life on Land and in Water

The word **amphibian** is Greek for "double-life," referring to the ability of these tetrapod vertebrates to live in fresh water and on land (**figure 21.36**). Amphibians are important in ecosystems, controlling both algae and populations of insects that transmit human disease. The chorus of mating calls from frogs, ranging from squeaks to grunts and croaks, adds ambience to springtime evenings in many areas. Scientists are also studying toxins in amphibian skin as possible painkilling drugs.

Many biologists are concerned about dramatic declines in amphibian populations worldwide. The destruction of wetlands and forests devastates amphibians' breeding areas, and infection with a chytrid fungus has killed many of them outright (see section 20.2). Collecting animals for the pet trade also takes a toll, as described in this section's Apply It Now box.

In addition, amphibians have porous skin that makes them especially vulnerable to pollution. They are therefore useful as indicators of environmental quality.

A. Amphibians Were the First Tetrapods

It is easy to envision how gulping lungfishes might have foreshadowed the lungs of amphibians or how the fleshy fins of lobe-finned fishes might have become legs. New fossil finds continue to fill in the pieces of the fish–amphibian transition, which occurred about 375 MYA (see figure 15.15 and section 13.7). ⓘ *Paleozoic life,* section 15.3B

Life on land offered amphibian ancestors space, shelter, food, and plentiful oxygen, compared with the crowded aquatic habitat. But the land also presented new challenges. The animals faced wider swings in temperature, and delicate gills collapsed without the buoyancy of water. The new habitat therefore selected for new adaptations (**figure 21.37**). Lungs improved, and circulatory systems (including a three-chambered heart) grew more complex and powerful. The skeleton became denser and better able to withstand the force of gravity. Natural selection also favored acute hearing and sight, with tear glands and eyelids keeping eyes moist.

Yet amphibians retain a strong link to the water. Amphibian eggs, which lack protective shells and membranes, will die if they dry out. Also, the larvae respire through external gills, which require water. Although adults typically have lungs, these organs are not very efficient in amphibians. The thin skin provides an additional gas exchange surface and must therefore remain moist.

An amphibian's skin may also help the animal avoid predation. Whereas some are camouflaged, others display vibrant colors that warn of toxins secreted from glands in the skin. The poison dart frogs of Central and South America are famous for their toxins, which they acquire by eating ants and other prey. (Captive frogs fed diets lacking these toxins therefore eventually become nonpoisonous, although their bright colors remain.)

Key features

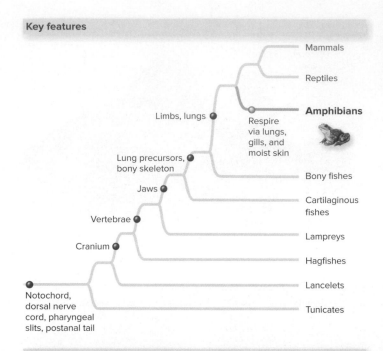

Diversity

Frog   Caecilian

Salamander

Figure 21.36 The Amphibians.

Photos: (frog): ©Creatas/PunchStock RF; (caecilian): ©E.D. Brodie Jr.; (salamander): ©Suzanne L. Collins & Joseph T. Collins/Science Source

B. Amphibians Include Three Main Lineages

The amphibians are grouped into three orders: frogs, salamanders and newts, and caecilians.

Frogs   Most amphibian species are **frogs,** a group that includes the smooth-skinned "true frogs" and the warty-skinned toads. Adults have large mouths, and they are "neckless"; that is, their heads are fused to their trunks. Their bodies lack tails and scales.

Apply It **Now**

Wild-Caught Pets

Wild animals make fascinating and unique pets, but owning one is not usually a good idea. This box summarizes four arguments against owning animals caught in the wild.

- **Population pressures:** Many vertebrate species are already endangered by habitat loss and pollution. Collecting for the pet trade only adds to the threat of extinction. For example, huge numbers of North American box turtles are captured each year. These animals have low reproductive rates, so it is difficult for the wild populations to recover. The problem is not limited to turtles; the pet trade puts pressure on many animal species, from primates to fishes and frogs (figure 21.C). Trapping these animals threatens not only their populations but also the ecosystems where they normally live. To make matters worse, many wild-caught animals die in transit.

- **Threats to native species:** Virtually every animal in the pet industry is an introduced species, including birds, ferrets, gerbils, sugar gliders, snakes, lizards, amphibians, fishes, and invertebrates. If an exotic pet escapes or if its owner releases it, the animal may prey on, compete with, or spread disease to native organisms. Domesticated Asian goldfish, for example, are

a nuisance species in the Pacific Northwest.

- **Physical dangers:** Large animals such as crocodiles and large cats represent a physical threat to their owners, becoming more dangerous as they grow.

- **Disease:** Wild-caught animals can carry disease. In 2003, monkeypox spread among people who kept prairie dogs as pets. The prairie dogs caught the disease from African rodents imported into the United States. Wild birds can also carry diseases that can infect native birds, poultry, and people. Likewise, hamsters, turtles, and iguanas can carry *Salmonella*.

Figure 21.C **Frogs Collected for the Pet Trade.**
©saravuth-photohut/iStock/Getty Images RF

In most species, the female frog lays her eggs directly in the water as a male clasps her back and releases sperm. The fertilized eggs hatch into legless, aquatic tadpoles. Most tadpoles feed on algae. As they mature, tadpoles typically undergo a dramatic change in body form—a metamorphosis. They develop legs and lungs, lose the tail, and acquire carnivorous tastes.

A frog's toxic skin is a potent defense, but frogs also have behavioral adaptations that help them escape predation. For example, frogs can use their powerful legs to jump away. They can also play dead or inflate their mouths so that a predator cannot swallow them.

Salamanders and Newts Salamanders and newts have tails and four legs, so they resemble lizards. Unlike lizards, however, their skin lacks scales, their digits lack claws, and they always live near water. In most salamanders, the male deposits a sperm packet near the female; she takes the packet into her body and subsequently lays the fertilized eggs in water, moist soil, or another damp location. Free-swimming larvae with a finlike tail hatch from the eggs. Both adults and young are carnivores, eating arthropods, worms, snails, fish, and other salamanders. In some groups of salamanders, the adults have larval features. The North American mudpuppy, for example, swims on pond bottoms and retains external gills.

Caecilians The **caecilians** are unusual amphibians in that their bodies lack limbs; as a result, they resemble giant earthworms. Most species burrow under the soil in tropical forests, but a few inhabit shallow freshwater ponds. In reproduction, they use internal fertilization; a male uses a sex organ to deliver sperm inside a female's body. Caecilians are carnivores, eating insects and worms.

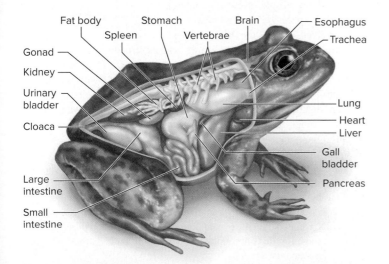

Figure 21.37 **Anatomy of a Frog.** Although amphibians have adaptations to life on land, their small lungs, thin skin, and reproductive requirements tie them to moist or wet habitats.

21.14 **MASTERING CONCEPTS**

1. What are the features of a typical amphibian? (*Hint:* See figure 21.36.)
2. What role does water play in amphibian gas exchange and reproduction?
3. What features distinguish the three orders of amphibians?

21.15 Reptiles Were the First Vertebrates to Thrive on Dry Land

The changeability of scientific knowledge is evident in any modern discussion of **reptiles** and **birds.** The word *reptile* traditionally referred only to snakes, lizards, crocodiles, and other amniotes with dry, scaly skin. Birds had feathery body coverings and were considered a separate lineage. But that point of view has changed. We now know that birds form one of several clades of reptiles (**figure 21.38**). As a consequence, modern use of the term *reptile* includes both the nonavian reptiles and the birds.

Reptiles evolved from amphibians about 310 to 320 MYA. They dominated animal life during the Mesozoic era, until their decline beginning 65 MYA. Although many reptile species survived to the present day, many others are known only from fossils. The extinct groups include the marine ichthyosaurs and plesiosaurs, the flying pterosaurs, and the terrestrial dinosaurs. Of these, the dinosaurs especially capture the imagination (see the Burning Question in this section). ⓘ *Mesozoic life,* section 15.3C

Unlike their amphibian ancestors, most reptiles have a suite of adaptations that enable them to live and reproduce on dry land (**figure 21.39**). Tough, keratin-rich scales reduce water loss from the skin, and the kidneys excrete only small amounts of water. Internal fertilization and amniotic eggs mean that reptiles do not require moist habitats to reproduce (see figure 21.29). Finally, reptiles have greater lung capacity and more efficient circulation than their aquatic ancestors.

A. Nonavian Reptiles Include Four Main Groups

The 8000 or so species of nonavian reptiles include turtles and tortoises; lizards and snakes; tuataras; and crocodilians (**figure 21.40**). Like fishes and amphibians, the nonavian reptiles are typically ectothermic.

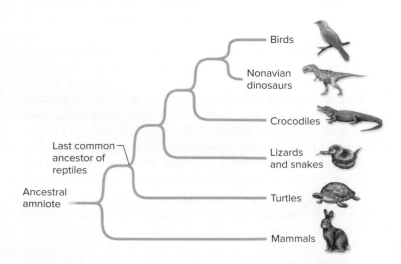

Figure 21.38 The Amniotes. Evidence suggests that birds form one of several clades of reptiles.

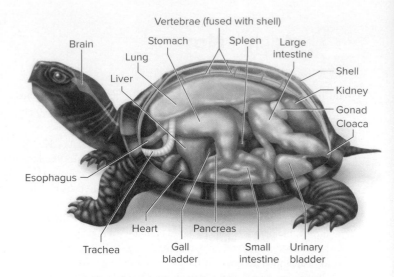

Figure 21.39 Anatomy of a Tortoise. The skin, lungs, heart, and reproductive system of a tortoise adapt this animal to life on land.

Burning Question

Did humans and dinosaurs ever coexist?

To answer this question, it is important to first understand what dinosaurs were—and what they were not. As the term is traditionally used, *dinosaurs* were terrestrial reptiles that lived during the Mesozoic era, between 245 and 65 MYA. They ranged in stature from chicken-sized to truly gargantuan—the largest could have peeked into the sixth story of a modern-day building.

Many people mistakenly believe that all reptiles (or even all mammals) that lived during the Mesozoic era were dinosaurs. In reality, lizards, snakes, crocodiles, and other terrestrial reptiles lived alongside the dinosaurs. The Mesozoic also saw marine reptiles such as plesiosaurs and flying reptiles such as pterodactyls, but these animals belonged to clades that were distinct from the dinosaurs. Nor did all dinosaurs live at the same time. As some species appeared, others went extinct throughout the Mesozoic.

©Andersen Ross/ Getty Images RF

Although several comic strips, cartoons, and feature films suggest the contrary, humans never coexisted with *T. rex, Apatosaurus,* or any other nonavian dinosaur. These dinosaurs were gone by the time our own mammalian lineage—the primates—was just getting started. Tens of millions of years later, humans finally roamed the Earth and found the fossils that prove that these huge reptiles once existed, leaving only birds as their modern descendants.

Submit your burning question to
Marielle.Hoefnagels@mheducation.com

Key features

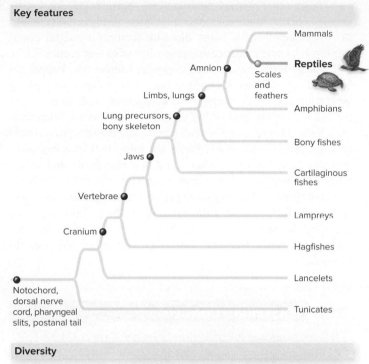

Diversity

Turtle   Lizard

Snake   Alligator   Bird

Figure 21.40 The Reptiles.

Photos: (turtle): ©Ed Reschke/Peter Arnold/Getty Images; (lizard): ©Creatas/ PunchStock RF; (snake): ©Dorling Kindersley/Getty Images; (alligator): ©LaDora Sims//Flickr/Getty Images RF; (bird): ©Image Source RF

Movies depict snakes, alligators, and crocodiles as terrifying killers, but most reptiles are inconspicuous animals that cannot harm people. Yet they are an important link in ecosystems, controlling populations of rodents and insects while providing food for owls and other predatory birds. In addition, some people keep snakes, lizards, or turtles as pets. In some parts of the world, reptiles have an even greater economic effect; the skins of farm-raised snakes and crocodiles are the raw material for boots, belts, and wallets, and some restaurants serve alligator meat.

Turtles and Tortoises   With a reputation for being "slow and steady," the aquatic **turtles** and their terrestrial relatives (tortoises) may not seem to have a recipe for success. Yet they have persisted since the Triassic period. Nowadays, habitat destruction threatens many turtles and tortoises, as does the pet trade.

This group's trademark feature is its shell, made of bony plates and a covering derived from the animal's epidermis. The shell's plates are fused to the animal's vertebrae and ribs, so it forms an integral part of the skeleton.

The largest member of this group is the giant leatherback sea turtle. These migratory animals are endangered worldwide because they are hunted for food, become trapped in fishing nets, swallow garbage, and ingest toxic chemicals. The destruction of nesting areas also takes its toll.

Lizards and Snakes   Almost 95% of nonavian reptile species are **snakes** or **lizards.** Snakes probably evolved from burrowing lizards, so these animals share many similarities. The most obvious difference, however, is that lizards usually have legs and snakes do not. Also, snakes never have external ear openings or movable eyelids, and most lizards lack the distinctive forked tongue that characterizes snakes.

Several adaptations help lizards and snakes feed and avoid becoming food. An iguana might detach its tail as it escapes a predator; a basilisk lizard can scamper across water. Camouflage and the ability to hold perfectly still enable snakes to surprise their prey, then subdue it by injecting venom or wrapping coils around the victim and squeezing the life out of it. Unique jaw adaptations then enable a snake to open its jaw so far that it can swallow prey much larger than its own body (see the opening figure for chapter 6).

Tuataras   **Tuataras** are reptiles that closely resemble lizards. The order containing tuataras once contained many species, all but two of which are now extinct. A captive breeding program may help restore populations of these endangered animals, which are native only to a few islands near New Zealand.

Crocodilians   The **crocodilians** (crocodiles, alligators, and their relatives) are carnivores that live in or near water. Their horizontally held heads have eyes on top and nostrils at the end of the elongated snout. Heavy scales cover their bodies; four legs project from the sides. Unlike most other nonavian reptiles, they have a four-chambered heart.

These reptiles look somewhat primitive, and indeed they are ancient animals. They arose some 230 MYA, about the same time that the first dinosaurs roamed the land. Yet they have acute senses and complex behaviors, comparable to those of birds. For example, crocodilians lay eggs in nests, which the adults guard. The adults also care for the hatchlings, which stay with their mothers while they are young and call to the adults when they are in danger. Adults mark their territories by smacking their heads or jaws on the water surface. Like many birds and mammals, they even have dominance hierarchies.

B. Birds Are Warm, Feathered Reptiles

The behavioral similarities between crocodilians and birds are unsurprising in light of evolutionary history. Birds, dinosaurs, and crocodilians all belonged to a reptilian group called archosaurs, of which only the birds and crocodilians survive today.

Chickens and crocodiles look quite different, so what evidence suggests that birds really are reptiles? As described in chapters 12 and 13, *Archaeopteryx* and fossils of feathered reptiles provide important clues to the evolutionary history of birds, as do the skeletal similarities between birds and nonavian reptiles. The bird's amniotic egg, with its hard, calcium-rich shell, is another clue, as are the keratin-rich scales on a bird's legs and the four-chambered heart of birds and crocodilians.

Of course, the 9000 or so species of birds have unique features that set them apart from other reptiles. Most birds can fly, thanks to anatomical adaptations including wings and a tapered body with a streamlined profile (**figure 21.41**). Their lightweight bones are hollow, with internal struts that add support. The powerful heart and unique lungs supply the oxygen that supports the high metabolic demands of flight muscles (see figures 30.2 and 31.5).

Birds are the only modern animals that have **feathers,** keratin-rich epidermal structures that provide insulation and enable birds to fly. Feathers are also important in mating behavior, as anyone who has watched a peacock show off his plumage can attest. In fact, evidence of glossy, iridescent feathers in the fossils of winged dinosaurs suggests that the first feathers may have been adaptive not in flight but in sexual signaling.

Unlike other reptiles, birds are endothermic. Paleontologists debate whether dinosaurs, the ancestors of birds, were endotherms or ectotherms. Many dinosaur features argue for endothermy, but some evidence suggests otherwise (see section 33.6).

Today, birds are a part of everyday human life. People eat hundreds of millions of chickens and turkeys every year, along with countless chicken eggs. We keep caged birds as pets, and we use feathers in everything from hats to blankets. Songbirds enrich the lives of many birdwatchers, and hunters pursue wild turkeys, doves, and ducks. Birds are important in ecosystems as well. Some pollinate plants and disperse fruits and seeds, whereas others eat rodents, insects, and other vermin. But birds can also be pests. Starlings and pigeons are a nuisance in cities, fouling buildings and sidewalks with their droppings and speeding the rusting of bridges. Moreover, ducks and other domesticated birds transmit bird flu and other diseases to humans.

21.15   MASTERING CONCEPTS

1. What features do all reptiles share? (*Hint:* See figure 21.40.)
2. How do scales and amniotic eggs adapt reptiles to land?
3. Describe each of the major groups of nonavian reptiles.
4. What characteristics suggest that birds are reptiles?
5. What are the functions of feathers?
6. What adaptations enable birds to fly?

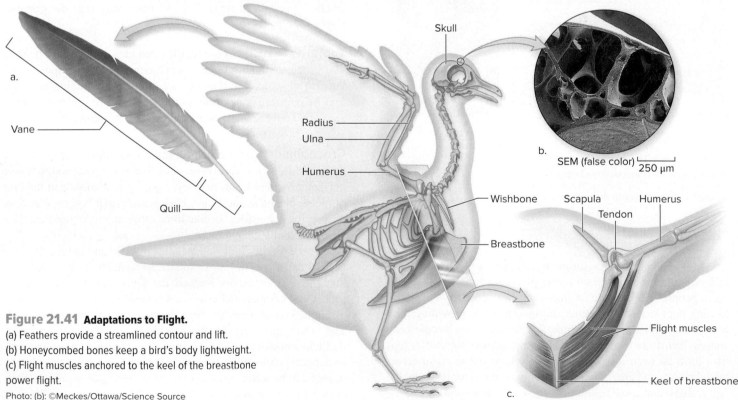

Figure 21.41 Adaptations to Flight.
(a) Feathers provide a streamlined contour and lift.
(b) Honeycombed bones keep a bird's body lightweight.
(c) Flight muscles anchored to the keel of the breastbone power flight.

Photo: (b): ©Meckes/Ottawa/Science Source

21.16 Mammals Are Warm, Furry Milk-Drinkers

Mammals are by far the most familiar vertebrates, not only because we *are* them but also because we surround ourselves with them. We keep dogs, cats, rabbits, gerbils, hamsters, and ferrets as pets. Farmers raise many types of mammals for their meat or milk, including cows, pigs, and goats. Lambs and adult sheep provide meat and wool, and leather from cows makes up everything from upholstery to shoes. Horses, oxen, mules, and dogs are important work animals. Many people enjoy hunting deer for food and sport, and trappers kill mink, fox, beaver, and other mammals for their fur.

Mammals also play important roles in ecosystems. Coyotes, wolves, and foxes keep populations of herbivorous deer, rodents, rabbits, and other mammals in check. Some bats eat insects, and others pollinate plants. On the other hand, pests such as rats, mice, and skunks thrive alongside human populations. Some transmit diseases, including hantavirus and rabies.

A. Mammals Share a Common Ancestor with Reptiles

The 5800 or so species of mammals trace their ancestry to the late Triassic period (about 200 MYA). DNA similarities and the existence of egg-laying mammals suggest that mammals and reptiles share a common ancestor (**figure 21.42**). Fossil evidence, particularly details of skull structure, also provides clues to their origin.

Sometime after mammals and reptiles diverged, the traits that are unique to the mammalian clade arose. For example, the common ancestor of the mammals had **mammary glands,** structures that secrete milk in the female. (The word *mammal* derives from the Latin *mammae* for "breast.") Infant mammals are nourished by their mother's milk. In addition, the skin is keratin-rich and waterproof, and it produces hair—another distinctive mammalian feature. Hair is composed of keratin and helps conserve body heat. (Even whales and dolphins have hair at birth, but they lose it as they mature into their streamlined shapes.) Depending on the species, mammals may also produce other structures out of keratin, including horns, hooves, claws, and nails.

Because mammary glands and hair do not leave fossils, ancient mammals are distinguished from reptilian fossils in other ways. A mammal has three middle ear bones, compared with the reptilian one or two, and a mammal's lower jaw consists of one bone, compared with the reptile's several. Mammalian teeth are distinctive, too, with four types: molars, premolars, canines, and incisors. Reptile teeth are much more uniform in size and shape.

Mammals also share a few other features. First, like birds, mammals have a four-chambered heart, which evolved independently in the two groups. Second, the outer layer of the mammalian brain is very well developed, enabling mammals to learn, remember, plan, and purposefully respond to stimuli. Third, only mammals have a dome-shaped, muscular diaphragm, which draws air into the lungs.

Key features

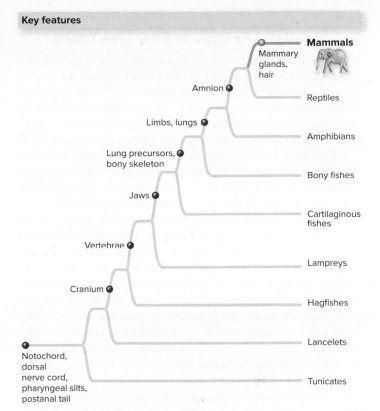

Diversity
Monotremes

Platypus   Echidna

Marsupials

Kangaroo   Opossum

Placental mammals

Human   Dolphin   Bat

Figure 21.42 The Mammals.

Photos: (platypus): ©JohnCarnemolla/Getty Images RF; (echidna): ©Paul Hobson/ Nature Picture Library; (kangaroo): ©Anan Kaewkhammul/123RF; (opossum): ©Frank Lukasseck/Photographer's Choice/Getty Images; (human): ©JGI/Blend Images LLC RF; (dolphin): ©imageBROKER/Alamy RF; (bat): ©Grzegorz Gust/Alamy RF

B. Mammals Lay Eggs or Bear Live Young

Most mammals were small until after the mass extinction that occurred 65 MYA. The loss of so many reptiles paved the way for the rapid diversification (adaptive radiation) of large mammals. At the same time, flowering plants became increasingly prominent, providing new types of food and habitats for mammals. ⓘ *adaptive radiation,* section 14.4B; *Cenozoic life,* section 15.3D

Biologists divide mammals into two subclasses, one containing egg-laying monotremes and the other containing live-bearing mammals. The latter is subdivided into two clades, the marsupials and the placental mammals (see figure 21.42).

The **monotremes** are mammals that lay eggs, such as the duck-billed platypus and the echidna. The name *monotreme* comes from the distinctive anatomy of these animals: The urinary, digestive, and reproductive tracts share a single (*mono*) opening to the outside of the body. Reptiles have a similar anatomy. The amniotic eggs in this group reveal another link to reptiles. When a helpless young monotreme hatches, it crawls along its mother's fur until it reaches milk-secreting pores in the skin. After a few months of suckling, the offspring leaves the safety of its mother's burrow and begins hunting its own food.

The other subclass contains the live-bearing marsupials and placental mammals. **Marsupials,** such as kangaroos and opossums, give birth to tiny, immature young about 4 to 5 weeks after conception. In many marsupials, the babies crawl from the mother's vagina to a **marsupium,** or pouch, where they suckle milk and continue developing. Some species, however, have poorly developed pouches, and the young drink from exposed nipples.

The **placental mammals,** also called eutherians, are the most diverse group. In these species, the young develop inside the female's uterus, where a **placenta** connects the maternal and fetal circulatory systems (see figure 21.29). The placenta nourishes and removes wastes from the developing offspring. (This section's Focus on Model Organisms describes the mouse, a placental mammal.)

Placental mammals diversified and displaced most marsupials early in the Cenozoic era, about 65 MYA. Australian marsupials, however, continued to diversify long after marsupials on other continents had gone extinct; section 13.3 describes how continental drift explains this biogeographical pattern.

The two largest groups of placental mammals are rodents and bats, but the clade also includes carnivores (dogs and cats), hoofed mammals, elephants, and many other familiar animals. Some groups reinvaded the water, including manatees, otters, seals, beavers, hippos, whales, and dolphins. Humans are placental mammals as well. We belong to an order, Primates, that arose some 60 MYA. Section 15.4 describes the evolution of primates.

21.16   MASTERING CONCEPTS

1. Which characteristics do all mammals share? (*Hint:* See figure 21.42.)
2. What evidence suggests that mammals share a common ancestor with reptiles?
3. How do monotremes, marsupials, and placental mammals differ in how they reproduce?

FOCUS on Model Organisms

The Mouse: *Mus musculus*

The history of the mouse, *Mus musculus,* as the stereotypical lab animal dates back to the early 1900s. Researchers discovered that the mouse's small size made it an excellent research animal. Also, mice are famous for their prolific breeding. They reach sexual maturity at the age of about 4 weeks, are sexually receptive every few days, and give birth to litters of 1 to 10 pups after a gestation of only about 3 weeks. Over a life span of 1.5 to 3 years, a single pair of mice can produce hundreds of offspring.

Researchers have benefited from biotechnology in their studies of mice. Transgenic mice have been available since the 1980s (see chapter 7). These mice are modified in countless ways, including altered susceptibility to human diseases. Another biotechnology, cloning, was applied to mice in 1998, making possible the production of genetically identical animals ideal for testing new disease treatments. ⓘ *cloning,* section 11.3B

The mouse genome sequence was completed in 2002, revealing about 30,000 genes divided among 20 chromosomes. Not surprisingly, most mouse genes have counterparts in the human genome. This genetic similarity has made possible some of the following ways in which *Mus musculus* has contributed to biological research:

- **Immune function:** In the 1930s, the discovery that mice reject transplants from all but their very close relatives led to the discovery of the major histocompatibility complex (MHC). Since that time, biologists have discovered an array of genes related to immune function (see chapter 34).
- **Human disease:** Mice have been used to study human disease since the 1930s, when researchers discovered that mice could contract yellow fever. Vaccines were subsequently tested in mice. The availability of transgenic mice has opened new possibilities for research on the cause and treatment of human disease, including muscular dystrophy, Alzheimer disease, obesity, Parkinson disease, cancer, and HIV/AIDS. ⓘ *cancer,* section 8.6; *HIV,* section 16.4B
- **X chromosome inactivation:** In the 1960s, biologist Mary Lyon proposed that in female mammals, one of the two X chromosomes is inactivated early in embryonic development. This phenomenon, which is now often illustrated using calico cats, was first proved in mice with mottled coats. ⓘ *X inactivation,* section 10.7C
- **Stem cells:** These undifferentiated cells, which can be derived from embryos or adults, can specialize into many other cell types. Mouse stem cell research has shown great promise in treating spinal cord injuries and many other ailments. ⓘ *stem cells,* section 11.3A

INVESTIGATING LIFE

21.17 Sponges Fill Holes in Animal Evolution

The Cambrian explosion fascinates biologists. After all, the major phyla of animals appeared during an eventful 25-million-year interval of the Cambrian period, which lasted from 543 to 490 MYA. Some of the best Cambrian fossils are preserved in exquisite detail at the Burgess Shale, a site in the Canadian Rockies (figure 21.43). These remains, representing animals from sponges to early chordates, provide a fascinating window into the past. But how did evolution occur so quickly during the Cambrian explosion? DNA evidence from today's animals may help answer that key question.

A research group led by Nicole King at the University of California, Berkeley, used sponges to "soak up" new information about animal diversification. Sponges have simpler body plans than do eumetazoans, such as arthropods and mammals. Nonetheless, the researchers hypothesized that sponges contain some of the same critical genes that help guide the development of more complex animal bodies.

To test their hypothesis, the group isolated genetic material from marine sponges called *Oscarella carmela*. They then used databases to search for genes known to guide body plan development in eumetazoans. Comparing the gene sets revealed that diverse animal groups contain many of the 68 developmental genes found in *O. carmela* sponges (figure 21.44). Moreover, the sponges contained nearly all major types of developmental genes found in eumetazoans.

The data suggest that many of the genes encoding complex animal body plans probably originated in the parazoan ancestors of both modern sponges and eumetazoans. This finding makes the rapid diversification of eumetazoans during the Cambrian less surprising. The genetic toolbox was already in place. When

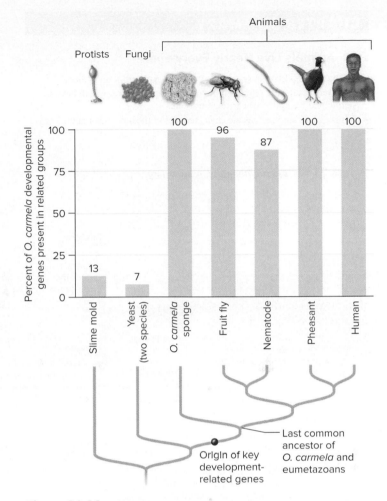

Figure 21.44 Developmental Genes Across Taxa. Diverse animal groups share many of the developmental genes found in sponges, whereas protists and fungi do not.

a favorable environment arose, these genes began to diversify, leading to an explosion of new animal forms.

Since fossils cannot reveal how DNA changed in the earliest ancestors of animals, the appearance of new animal groups during the Cambrian explosion may simply represent the most visible step in a largely hidden evolutionary process. This study clearly illustrates why scientists analyze several lines of evidence before making decisive conclusions. As scientists continue to challenge one another's findings, future studies will surely lead to additional revisions to the history of animal phyla.

Source: Nichols, Scott A., William Dirks, John S. Pearse, and Nicole King. August 15, 2006. Early evolution of animal cell signaling and adhesion genes. *Proceedings of the National Academy of Science,* vol. 103, pages 12451–12456.

Figure 21.43 Burgess Shale Fossil. Animals with complex bodies arose during the Cambrian explosion. This trilobite fossil is one of many relics from the period of rapid animal diversification that began about 543 MYA.
©Michael Melford/National Geographic Creative

21.17 MASTERING CONCEPTS

1. How did the researchers use the DNA of living sponges to draw conclusions about events occurring hundreds of millions of years ago?

2. If the researchers had analyzed the developmental genes of *Arabidopsis* plants, would you have expected the similarity to sponge developmental genes to be low or high? Explain your answer.

CHAPTER SUMMARY

21.1 Animals Live Nearly Everywhere

- Of the 1,300,000 or so species in kingdom **Animalia,** most are **invertebrates;** only one phylum contains **vertebrates,** which have a segmented backbone of cartilage or bone.
- Figure 21.45 summarizes the characteristics of the nine major animal phyla.

A. What Is an Animal?
- Animals are multicellular, eukaryotic heterotrophs whose cells secrete extracellular matrix but do not have cell walls. Most digest their food internally.
- The **blastula** is a stage in embryonic development that is unique to animals (figure 21.46).

B. Animal Life Began in the Water
- The immediate ancestor of animals probably resembled a protist called a choanoflagellate.
- The earliest fossil evidence of animals is from about 570 MYA. These first animals lived in water, and existing animal diversity strongly reflects this aquatic heritage.

Figure 21.45 Animal Diversity: A Summary.

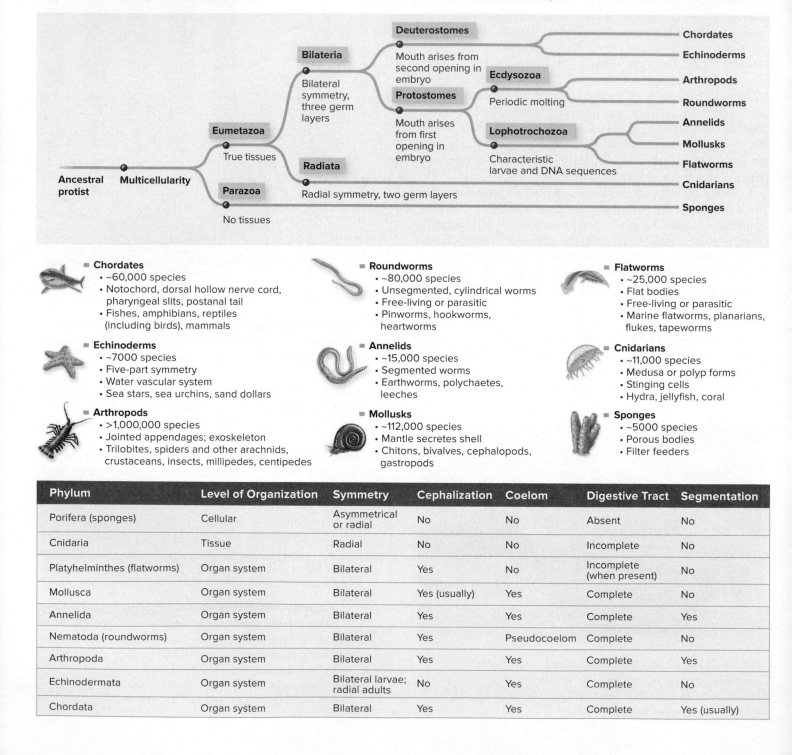

- **Chordates**
  - ~60,000 species
  - Notochord, dorsal hollow nerve cord, pharyngeal slits, postanal tail
  - Fishes, amphibians, reptiles (including birds), mammals

- **Echinoderms**
  - ~7000 species
  - Five-part symmetry
  - Water vascular system
  - Sea stars, sea urchins, sand dollars

- **Arthropods**
  - >1,000,000 species
  - Jointed appendages; exoskeleton
  - Trilobites, spiders and other arachnids, crustaceans, insects, millipedes, centipedes

- **Roundworms**
  - ~80,000 species
  - Unsegmented, cylindrical worms
  - Free-living or parasitic
  - Pinworms, hookworms, heartworms

- **Annelids**
  - ~15,000 species
  - Segmented worms
  - Earthworms, polychaetes, leeches

- **Mollusks**
  - ~112,000 species
  - Mantle secretes shell
  - Chitons, bivalves, cephalopods, gastropods

- **Flatworms**
  - ~25,000 species
  - Flat bodies
  - Free-living or parasitic
  - Marine flatworms, planarians, flukes, tapeworms

- **Cnidarians**
  - ~11,000 species
  - Medusa or polyp forms
  - Stinging cells
  - Hydra, jellyfish, coral

- **Sponges**
  - ~5000 species
  - Porous bodies
  - Filter feeders

Phylum	Level of Organization	Symmetry	Cephalization	Coelom	Digestive Tract	Segmentation
Porifera (sponges)	Cellular	Asymmetrical or radial	No	No	Absent	No
Cnidaria	Tissue	Radial	No	No	Incomplete	No
Platyhelminthes (flatworms)	Organ system	Bilateral	Yes	No	Incomplete (when present)	No
Mollusca	Organ system	Bilateral	Yes (usually)	Yes	Complete	No
Annelida	Organ system	Bilateral	Yes	Yes	Complete	Yes
Nematoda (roundworms)	Organ system	Bilateral	Yes	Pseudocoelom	Complete	No
Arthropoda	Organ system	Bilateral	Yes	Yes	Complete	Yes
Echinodermata	Organ system	Bilateral larvae; radial adults	No	Yes	Complete	No
Chordata	Organ system	Bilateral	Yes	Yes	Complete	Yes (usually)

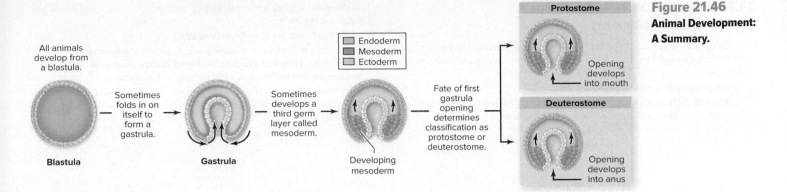

Figure 21.46
Animal Development: A Summary.

All animals develop from a blastula.

Blastula

Sometimes folds in on itself to form a gastrula.

Gastrula

Sometimes develops a third germ layer called mesoderm.

Developing mesoderm

Fate of first gastrula opening determines classification as protostome or deuterostome.

Endoderm
Mesoderm
Ectoderm

Protostome — Opening develops into mouth

Deuterostome — Opening develops into anus

C. Animal Features Reflect Shared Ancestry

- Animal bodies exhibit varying levels of organization. **Parazoans** are animals without tissues; **eumetazoans** have tissues that are typically arranged into organs and organ systems.
- In most phyla, animal bodies have **radial symmetry** or **bilateral symmetry. Cephalization** is correlated with bilateral symmetry.
- An animal zygote divides mitotically to form a blastula and then usually a **gastrula.** In some animals, the gastrula has two tissue layers (**ectoderm** and **endoderm**). In others, a third layer (**mesoderm**) forms between the other two.
- Bilaterally symmetrical animals are **protostomes** if the gastrula's first indentation forms into the mouth. In **deuterostomes,** the first indentation develops into the anus.

D. Biologists Also Consider Additional Characteristics

- Some types of animals have a body cavity (**coelom** or **pseudocoelom**) that can act as a **hydrostatic skeleton.**
- An animal may have an **incomplete digestive tract** (**gastrovascular cavity**) or a **complete digestive tract.**
- **Segmentation** improves flexibility and increases the potential for specialized body parts.
- Most animals reproduce sexually. The resulting embryo may undergo **direct development** or **indirect development,** in which **metamorphosis** transforms a **larva** into an adult.

21.2 Sponges Are Simple Animals That Lack Differentiated Tissues

- **Sponges** are aquatic, sessile animals that are either asymmetrical or radially symmetrical. Their porous bodies filter small particles out of water.
- Although they lack tissues, sponges have specialized cell types, including collar cells and amoebocytes.
- A sponge's skeleton consists of spicules or organic fibers (or both).
- Sponges may bud asexually and reproduce sexually. After fertilization, the offspring develops into a blastula but not a gastrula.

21.3 Cnidarians Are Radially Symmetrical, Aquatic Animals

- **Cnidarians** are mostly marine animals. Examples include corals, hydras, and jellyfishes.
- The cnidarian body form is radially symmetrical and may be a sessile **polyp** or a swimming **medusa.**
- The body wall consists of jellylike material constrained between two tissue layers. Cnidarians move by contracting muscle cells that act on this hydrostatic skeleton.
- Cnidarians capture prey with tentacles and stinging **cnidocytes.** They digest food in a gastrovascular cavity.
- Reproduction is sexual or asexual.

21.4 Flatworms Have Bilateral Symmetry and Incomplete Digestive Tracts

- **Flatworms** are unsegmented protostomes that lack a coelom. This phylum includes **free-living flatworms** such as planarians. **Flukes** and **tapeworms** are parasitic.
- The flat body shape allows individual cells to exchange gases with their environment.
- Free-living flatworms and flukes have an incomplete digestive tract; tapeworms absorb the host's food directly through the body wall. Flukes and tapeworms have many additional adaptations to the parasitic lifestyle.

21.5 Mollusks Are Soft, Unsegmented Animals

- **Mollusks** are unsegmented protostomes with a complete digestive tract. The main groups of mollusks are **chitons, bivalves, gastropods,** and **cephalopods.**
- The mollusk body includes a **mantle,** a muscular **foot,** and a **visceral mass.**
- The mantle secretes a shell in most mollusks. Natural selection has shaped the mantle and foot into many variations.
- Mollusks are filter feeders, herbivores, or predators. Many have a tonguelike **radula.**
- Cephalopods have complex sensory and nervous systems.
- Sexes are usually separate in the mollusks; fertilization may be internal or external.

21.6 Annelids Are Segmented Worms

- **Annelids** are segmented protostomes. The group includes **earthworms, leeches,** and **polychaetes.**
- These animals lack a specialized respiratory system, but they have a complete digestive tract and a closed circulatory system.
- Excretory structures, blood vessels, nerves, and muscles are repeated in each segment. The coelom acts as a hydrostatic skeleton.
- Leeches and earthworms are hermaphrodites that lay eggs in a protective "cocoon."

21.7 Nematodes Are Unsegmented, Cylindrical Worms

- **Roundworms** are unsegmented worms with a pseudocoelom. They molt periodically, a feature they share with arthropods.
- The pseudocoelom aids in circulation and acts as a hydrostatic skeleton.
- Nematodes include parasitic and free-living species in soil and aquatic sediments.

21.8 Arthropods Have Exoskeletons and Jointed Appendages

- **Arthropods** are segmented animals with jointed appendages and a chitin-rich **exoskeleton.** Like nematodes, arthropods molt periodically as they grow.

A. Arthropods Have Complex Organ Systems

- Arthropods include over a million species that live on land and on water. Depending on the habitat, natural selection has sculpted a great variety of defenses, feeding and reproductive strategies, respiratory systems, and nervous systems.

B. Arthropods Are the Most Diverse Animals

- The extinct **trilobites** were arthropods, as are the **chelicerates** (**horseshoe crabs** and **arachnids**). **Mandibulate** arthropods include **centipedes** and **millipedes, crustaceans,** and **insects.**

- Insects are by far the most diverse arthropods. Segmented bodies, flight, and metamorphosis all may account for today's enormous number of insect species.

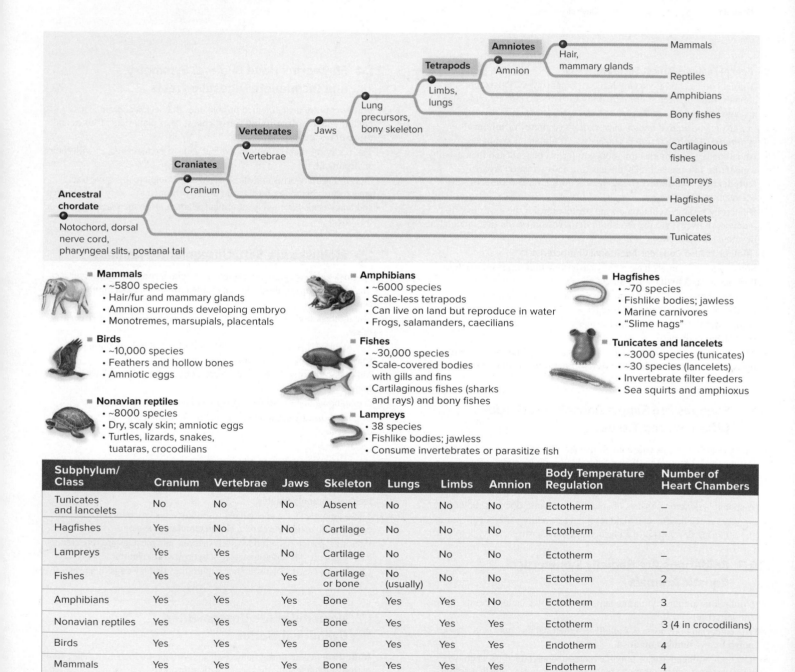

Mammals
- ~5800 species
- Hair/fur and mammary glands
- Amnion surrounds developing embryo
- Monotremes, marsupials, placentals

Birds
- ~10,000 species
- Feathers and hollow bones
- Amniotic eggs

Nonavian reptiles
- ~8000 species
- Dry, scaly skin; amniotic eggs
- Turtles, lizards, snakes, tuataras, crocodilians

Amphibians
- ~6000 species
- Scale-less tetrapods
- Can live on land but reproduce in water
- Frogs, salamanders, caecilians

Fishes
- ~30,000 species
- Scale-covered bodies with gills and fins
- Cartilaginous fishes (sharks and rays) and bony fishes

Lampreys
- 38 species
- Fishlike bodies; jawless
- Consume invertebrates or parasitize fish

Hagfishes
- ~70 species
- Fishlike bodies; jawless
- Marine carnivores
- "Slime hags"

Tunicates and lancelets
- ~3000 species (tunicates)
- ~30 species (lancelets)
- Invertebrate filter feeders
- Sea squirts and amphioxus

Subphylum/ Class	Cranium	Vertebrae	Jaws	Skeleton	Lungs	Limbs	Amnion	Body Temperature Regulation	Number of Heart Chambers
Tunicates and lancelets	No	No	No	Absent	No	No	No	Ectotherm	–
Hagfishes	Yes	No	No	Cartilage	No	No	No	Ectotherm	–
Lampreys	Yes	Yes	No	Cartilage	No	No	No	Ectotherm	–
Fishes	Yes	Yes	Yes	Cartilage or bone	No (usually)	No	No	Ectotherm	2
Amphibians	Yes	Yes	Yes	Bone	Yes	Yes	No	Ectotherm	3
Nonavian reptiles	Yes	Yes	Yes	Bone	Yes	Yes	Yes	Ectotherm	3 (4 in crocodilians)
Birds	Yes	Yes	Yes	Bone	Yes	Yes	Yes	Endotherm	4
Mammals	Yes	Yes	Yes	Bone	Yes	Yes	Yes	Endotherm	4

Figure 21.47   Chordate Diversity: A Summary.

21.9 Echinoderm Adults Have Five-Part, Radial Symmetry

- **Echinoderms** are spiny-skinned marine animals. They are deuterostomes, as are the chordates.
- Adult echinoderms have radial symmetry. The bilaterally symmetrical larvae look very different from the adults.
- The **water vascular system** enables echinoderms to move, sense their environment, acquire food, exchange gases, and get rid of metabolic wastes.
- Most echinoderms reproduce sexually.

21.10 Most Chordates Are Vertebrates

A. Four Key Features Distinguish Chordates
- **Chordates** share many characteristics, including a **notochord**; a **dorsal, hollow nerve cord; pharyngeal slits** or pouches in the pharynx; and a **postanal tail.**
- Figure 21.47 summarizes diversity within the chordates.

B. Many Features Reveal Evolutionary Relationships Among Chordates
- The **craniates** (hagfishes and vertebrates) have a bony or cartilage-rich **cranium** protecting the brain. Tunicates and lancelets lack a cranium.
- Other features that distinguish chordates from one another include **vertebrae, jaws, lungs,** and the presence of limbs in **tetrapods.**
- Reptiles lay **amniotic eggs,** in which the amnion and other membranes protect the developing embryo. A mammal's amnion is homologous to that of the amniotic egg; mammals and reptiles form a clade called the **amniotes.**
- Each group of chordates has a characteristic type of body covering.
- **Ectotherms** lack internal temperature-control mechanisms, whereas **endotherms** use heat from metabolism to maintain body temperature. Endotherms require much more energy than do ectotherms.
- Fishes have two heart chambers, amphibians and most nonavian reptiles have three, and birds and mammals have four.
- Figure 21.48 summarizes vertebrate adaptations.

Key Vertebrate Adaptations		
Adaptation	**Adaptive significance**	**Animals with adaptation**
Vertebrae 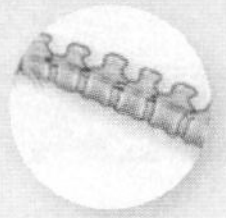	Expand range of motion	Lampreys, fishes, amphibians, reptiles, mammals
Jaws 	Increase feeding versatility	Fishes, amphibians, reptiles, mammals
Lungs 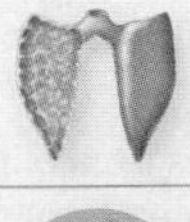	Enable animal to breathe air	Bony fishes (a few species), amphibians, reptiles, mammals
Limbs 	Allow for locomotion on land	Amphibians, reptiles, mammals
Amnion 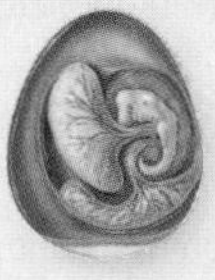	Enables reproduction away from water	Reptiles, mammals

Figure 21.48 Vertebrate Adaptations: A Summary.

21.11 Tunicates and Lancelets Are Invertebrate Chordates

- **Tunicates** obtain food and oxygen with a siphon system. A tunicate larva has all four chordate characteristics, but adults retain only the pharyngeal slits.
- **Lancelets** resemble eyeless fishes; adults have all four chordate characteristics.

21.12 Hagfishes and Lampreys Are Craniates Lacking Jaws

- **Hagfishes** have a cranium of cartilage, but they are not vertebrates. They secrete slime and lack jaws.
- **Lampreys** are vertebrates that resemble fishes, but they are jawless.

21.13 Fishes Are Aquatic Vertebrates with Jaws, Gills, and Fins

- **Fishes** are abundant and diverse vertebrates.

A. Cartilaginous Fishes Include Sharks, Skates, and Rays
- The **cartilaginous fishes** are the most ancient lineage of fishes. They include sharks, skates, and rays, all of which have a skeleton of cartilage.
- Sharks detect vibrations from prey with a **lateral line** system.

B. Bony Fishes Include Two Main Lineages
- The **bony fishes** account for 96% of existing fish species. Bony fishes have lateral line systems and swim bladders, which enable them to control their buoyancy.
- The two groups of bony fishes are **ray-finned fishes** and **lobe-finned fishes,** which are further subdivided into **lungfishes** and **coelacanths.**

C. Fishes Changed the Course of Vertebrate Evolution
- Adaptations in fishes that allowed vertebrates to move onto land include lungs and fleshy, paired fins that were later modified as limbs.

21.14 Amphibians Lead a Double Life on Land and in Water

A. Amphibians Were the First Tetrapods
- **Amphibians** breed in water and must keep their skin moist to breathe. Adaptations to life on land include a sturdy skeleton, lungs, limbs, and a three-chambered heart.

B. Amphibians Include Three Main Lineages
- Amphibians include **frogs, salamanders** and newts, and **caecilians.**

21.15 Reptiles Were the First Vertebrates to Thrive on Dry Land

- **Reptiles** (including **birds**) have efficient excretory, respiratory, and circulatory systems. Internal fertilization and amniotic eggs permit reproduction on dry land.

A. Nonavian Reptiles Include Four Main Groups
- Nonavian reptiles include **turtles** and tortoises, **lizards** and **snakes, tuataras,** and **crocodilians.**

B. Birds Are Warm, Feathered Reptiles
- Birds retain scales and egg-laying from reptilian ancestors. Like crocodilians, birds have a four-chambered heart.
- Honeycombed bones, streamlined bodies, and **feathers** are adaptations that enable flight.
- Like mammals but unlike the other reptiles, birds are endothermic.

21.16 Mammals Are Warm, Furry Milk-Drinkers

A. Mammals Share a Common Ancestor with Reptiles

- **Mammals** have fur, secrete milk from **mammary glands,** and have distinctive teeth and highly developed brains. The four-chambered mammalian heart evolved independently from that of crocodilians and birds.

B. Mammals Lay Eggs or Bear Live Young

- **Monotremes** are mammals that hatch from an amniotic egg. The young of **marsupial** mammals are born after a short pregnancy and often develop inside the mother's **marsupium. Placental mammals** have longer pregnancies; the young are nourished by the **placenta** in the mother's uterus.

21.17 Investigating Life: Sponges Fill Holes in Animal Evolution

- Researchers have discovered that key genes required for the development of complex animals were already present in sponges, the simplest animals. These genes may have set the stage for the Cambrian explosion.

MULTIPLE CHOICE QUESTIONS

1. Following gastrulation, the cells that have folded inward develop into
   a. endoderm.
   b. mesoderm.
   c. ectoderm.
   d. All of the above are correct.

2. Which of the following groups includes all of the others?
   a. Protostomes
   b. Echinoderms
   c. Eumetazoans
   d. Ecdysozoans

3. How is the body structure of an annelid different from that of an arthropod?
   a. Annelids lack jointed appendages.
   b. Annelids have a complete digestive tract.
   c. Annelids have cephalization.
   d. Annelids have bilateral symmetry.

4. Which of the following applies to a squid?
   a. Cephalization
   b. True coelom
   c. Complete digestive tract
   d. All of the above are correct.

5. What is a key characteristic of all arthropods?
   a. Six legs
   b. Pseudocoelom
   c. Mandibles
   d. Exoskeleton

6. Refer to figure 21.45. What group of eumetazoans has radially symmetrical adults and only two embryonic germ layers?
   a. Chordates
   b. Cnidarians
   c. Echinoderms
   d. Sponges

7. What shared characteristic supports the surprisingly close relationship between echinoderms and chordates?
   a. Deuterostomy
   b. Bilateral symmetry
   c. Multicellularity
   d. Presence of a notochord

8. Arthropods molt because
   a. they undergo indirect development.
   b. the exoskeleton prevents the organism from growing.
   c. they are protostomes.
   d. they have an open circulatory system.

9. Into what structures do pharyngeal slits (or pouches) develop?
   a. Food-straining organs
   b. Gills
   c. The middle ear
   d. All of the above are correct.

10. Since a tunicate is considered to be a chordate, it must have a(n)
    a. cranium.   c. amniotic egg.
    b. notochord.   d. lung.

11. Lobe-finned fishes are important because they
    a. were the first vertebrates.
    b. were the earliest animals.
    c. are closely related to tetrapods.
    d. lack jaws.

12. To which of the following is a salamander most closely related?
    a. A snail   c. A shark
    b. A beetle   d. A catfish

13. How do reptiles and mammals differ from amphibians?
    a. Only reptiles and mammals are amniotes.
    b. Only reptiles and mammals are tetrapods.
    c. Only reptiles and mammals have lungs.
    d. All of the above are correct.

14. What feature(s) do birds and crocodilians share?
    a. Four-chambered hearts
    b. Egg-laying
    c. Scales
    d. All of the above are correct.

15. Since a whale is a mammal, it must
    a. have scales.
    b. have gills.
    c. produce milk.
    d. All of the above are correct.

Answers to these questions are in appendix A.

WRITE IT OUT

1. Compare the nine major animal phyla in the order in which the chapter presents them, listing the features for each group.

2. Suppose you watch a video showing the development of an unknown animal. What clues can the developmental pattern give you about how this organism is classified? Creating a flow chart might be useful.

3. List the criteria used to distinguish (a) animals from other organisms; (b) vertebrates from invertebrates; (c) protostomes from deuterostomes; (d) ectotherms from endotherms.

4. Distinguish between (a) radial and bilateral symmetry; (b) blastula and gastrula; (c) direct and indirect development; (d) complete and incomplete digestive tracts; (e) coelom and pseudocoelom.

5. Analyze the evolutionary tree in figure 21.2, and then write an argument supporting or refuting this statement: Annelids are more closely related to flatworms than to roundworms.

6. Like sponges, plants are sessile organisms, but plants and sponges have dissimilar appearances. What important difference between sponges and plants explains this observation?

7. Using the evolutionary trees in this chapter, compare cnidarians to sponges and to the clade containing flatworms, mollusks, and annelids.

8. Compare the defining features of flatworms to those of other protostomes and deuterostomes. What features are similar among these groups? What features are different?

9. Compare and contrast the roundworm body structure with those of a flatworm and an annelid.

10. Make a chart showing the characteristics of each subphylum of arthropods. Suppose you could examine a fossilized arthropod; use your chart to describe how you would assign the fossil to a subphylum.

11. List the features that determine the echinoderms' position on the evolutionary tree.

12. Create lists of animal phyla that (a) are cephalized, (b) have an incomplete digestive tract, (c) have segmented bodies, and (d) have a coelom.

13. How do tunicates and lancelets differ from fishes and tetrapods?

14. Draw from memory a phylogenetic tree that traces the evolutionary history of tetrapods. Include the features that mark each branching point in your tree.

15. List the evidence that biologists use to classify earthworms, caecilians, and snakes in different clades despite the superficial similarities among these animals.

16. List five adaptations that enable (a) fishes to live in water; (b) amphibians to live on land; (c) snakes to live in the desert; (d) birds to fly.

17. How are a fish's and a bird's skeletons similar in structure and function?

18. Fishes are adapted to life in water, and tetrapods to life on land. Cite two criteria for assessing which group has been more successful.

19. Summarize the evidence for the idea that birds are reptiles. How does the changing placement of birds in the vertebrate family tree illustrate the scientific process? Why does this type of research matter?

20. If you found a fossil and were not sure whether it was from a reptile or a mammal, how could you tell the difference?

21. Early tunicates and lancelets did not leave fossil evidence because their bodies lacked hard parts. The hair and mammary glands that distinguish mammals are not hard either, yet mammals have a rich fossil record. Explain this difference.

22. If the success of an animal group is determined by the number of species it contains, how successful are mammals compared to other chordate classes? How successful are mammals compared to insects?

23. How are fishes, amphibians, nonavian reptiles, birds, and mammals important to humans? How are they important in ecosystems?

24. Give three examples of interactions between animals classified in different phyla.

25. Explain how a sessile or slow-moving lifestyle, such as that of sponges, sea cucumbers, and tunicates, might select for bright colors and an arsenal of toxic chemicals.

26. Search the Internet to answer this question: Other than *Caenorhabditis elegans*, *Drosophila melanogaster*, and *Mus musculus*, what are some examples of invertebrate or vertebrate animals that have contributed to scientists' knowledge of general biology and animal biology? What other genomes of animals have scientists sequenced, and what are some resulting discoveries?

27. Invasive animal species are disrupting ecosystems around the world. Search the Internet for a list of invasive animal species. Which phyla are represented in the list? What harm do invasive species do? How important is it to try to eradicate invasive species?

PULL IT TOGETHER

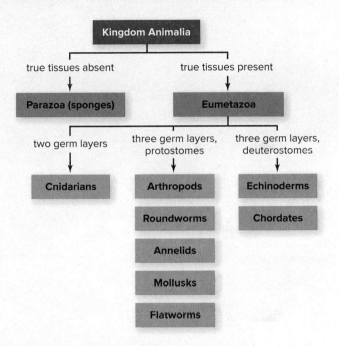

Figure 21.49 Pull It Together: Animals.

Refer to figure 21.49 and the chapter content to answer the following questions.

1. Review the Survey the Landscape figure in the chapter introduction. Describe the position of kingdom Animalia in the tree of life.

2. Write connecting phrases to separate arthropods and roundworms from annelids, mollusks, and flatworms.

3. Draw a concept map that summarizes the chordates, including both invertebrates and vertebrates.

22

Plant Form and Function

Cash Crop. Women harvest coca in Peru; the leaves are the raw material for producing cocaine.

©Gustavo Gilabert/Corbis SABA

LEARN HOW TO LEARN
Bite-Sized Pieces

Many students think they need to read a whole chapter in one sitting. Instead, try working through one topic at a time. Read just one section of the chapter, compare it to your class notes, and test yourself on what you have learned. Think of each chapter as a meal: You eat a sandwich one bite at a time, so why not tackle biology the same way?

UNIT 5

Candy, Herbs, and Drugs: Plants Are Chemical Factories

Food, beverages, paper, lumber, textiles, oil, rope—most people know that these products come from plants. The cells of roots, stems, leaves, flowers, and fruits produce less conspicuous resources as well: a vast array of potent chemicals that people use every day.

Consider, for example, the refreshing flavor of peppermint. You may associate peppermint with candy, but the flavor originally comes from a plant. Adorning the leaves of peppermint plants are microscopic hairs that produce and store chemicals called terpenes. Menthol, the most abundant terpene in peppermint, produces a cooling sensation when eaten or inhaled. Spearmint, sage, basil, lavender, rosemary, thyme, and oregano are other aromatic herbs in the mint family; they produce different terpenes, each with its own distinctive scent.

Like menthol, hashish is also a leaf hair extract. Hashish is purified resin from the leaves and flowers of *Cannabis sativa*. A chemical called tetrahydrocannabinol (THC) produces the narcotic effect of hashish and marijuana. THC is part of a class of chemical compounds called phenolics.

Alkaloids are another class of powerful plant-derived compounds. Most people are familiar with alkaloids; capsaicin, for example, makes chili peppers taste "hot." Other examples include caffeine (from coffee beans), nicotine (from tobacco leaves), morphine and codeine (from opium poppy flowers), and cocaine (from coca leaves). Alkaloids can also have medicinal value. The antimalarial medicine quinine, for example, comes from the bark of Cinchona trees, and vincristine is an antileukemia drug from the leaves of a periwinkle, *Catharanthus roseus*.

Terpenes, phenolics, alkaloids, and many other potent chemicals are called secondary metabolites because they do not participate directly in photosynthesis, respiration, reproduction, or other essential processes. So why do plant cells invest energy to make them?

The answer lies in natural selection. Many secondary metabolites are bad-tasting or toxic to disease-causing organisms and herbivores. Quite simply, plants that defend themselves with these weapons retain more leaf area than those that don't. The more leaf area, the greater the potential for photosynthesis and, ultimately, the more resources available to produce offspring.

A chemical arsenal that repels or poisons hungry animals is just one defense against predation. As you'll see in this chapter, a plant's growth pattern and structural defenses play important roles as well.

SURVEY THE LANDSCAPE
Plant Life

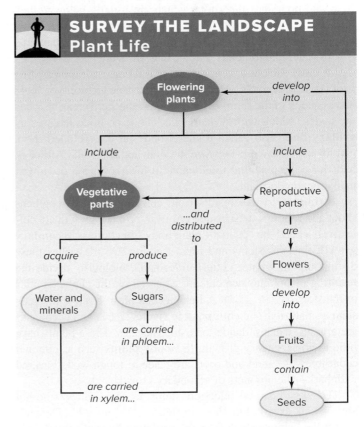

Like all other organisms, plants are made of cells. This chapter explores the cell types that make up a plant's tissues, which in turn compose the familiar organs of the vegetative plant body: stems, leaves, and roots.

For more details, study the Pull It Together feature in the chapter summary.

22.1 Vegetative Plant Parts Include Stems, Leaves, and Roots

Imagine a rose bush that produces a bounty of sweet-smelling flowers. Biologists would divide your plant into two sets of parts. The gorgeous flowers, which will eventually give rise to fruits called rose hips, are the reproductive parts. Chapter 24 describes flowers and fruits in detail. The rest of the plant consists of **vegetative** parts, which do not participate directly in reproduction.

This chapter and the next focus on the **anatomy** (form) and **physiology** (function) of a plant's vegetative organs—its stems, leaves, and roots. As you will see, a plant's **organs** are functional units that are each composed of multiple interacting tissues. A **tissue,** in turn, is a group of cells that interact to provide a specific function. This section begins our tour of the plant body with an overview of its major organs.

A plant's vegetative organs work together (**figure 22.1**). The **shoot** is the aboveground part of the plant; a **terminal bud** contains undeveloped tissue at the shoot tip. The shoot's **stem** supports the **leaves,** which produce carbohydrates such as sucrose by photosynthesis. A large portion of this sugar moves down the stem and nourishes the **roots,** which are usually belowground. Root cells depend on the shoots to provide energy for their metabolism. At the same time, however, roots anchor the plant and absorb water and minerals that move to the stem and leaves.

A close look at a stem reveals that it consists of alternating nodes and internodes. A **node** is a point at which one or more leaves attach to the stem. **Internodes** are the stem areas between the nodes. Along with at least one leaf, each node features an **axillary bud** (also called a lateral bud), an undeveloped shoot that forms in the angle between the stem and leaf stalk. Although buds have the potential to elongate to form a branch or flower, many remain dormant.

Primary growth results from cell division that lengthens shoot and root tips; tissues derived from primary growth are collectively called the "primary plant body." In contrast, **secondary growth** thickens stems and roots (section 22.4 describes these distinctions in greater detail). Informally, biologists divide the resulting plants into two categories based on the characteristics of the stem (**figure 22.2**). A **herbaceous plant** has a soft, green stem at maturity. The chili plant in figure 22.1 is herbaceous, as are grasses, daisies, dandelions, and radishes. These plants have little to no secondary growth. In **woody plants** such as elm and cedar trees, the stem and roots are made of tough wood covered with bark—the products of secondary growth.

Thanks to natural selection, stems, leaves, and roots do not always look exactly like those in figures 22.1 and 22.2. Depending on the habitat, plants may contend with hungry animals, extreme drought, continuous flooding, or frozen winters. These selective forces have sculpted the vegetative plant body into a tremendous diversity of forms (**figure 22.3**).

Stems, for example, often have specialized functions (see figure 22.3a). The stems of climbing plants may form tendrils that coil around objects. Tendrils allow a plant to climb from

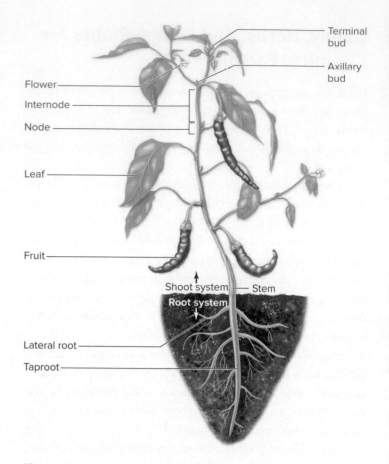

Figure 22.1 Parts of a Flowering Plant. A plant consists of a shoot system and a root system. Stems, leaves, and roots are vegetative organs; flowers and fruits are reproductive structures.

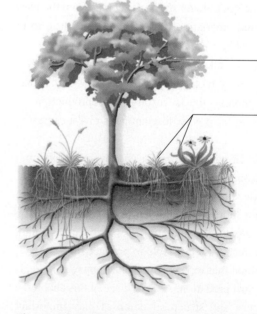

Figure 22.2 Herbaceous and Woody Plants. Herbaceous plants, such as grasses and daisies, have soft, green stems. Trees are woody plants; they have stems and roots made of wood.

a. Specialized stems

Climbing
(grape tendrils)

Underground nutrient storage
(iris rhizomes)

Water storage
(cactus)

Defense
(honey locust thorns)

b. Specialized leaves

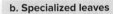

Nutrient storage
(onion)

Pollinator attraction
(poinsettia)

Carnivory
(Venus flytrap)

Asexual reproduction
(kalanchöe)

c. Specialized roots

Nutrient storage
(carrot)

Oxygen absorption
(mangrove trees)

Photosynthesis
(orchid aerial roots)

Support
(prop roots of screw pine)

Figure 22.3 Specialized Stems, Leaves, and Roots.

Photos: (a, vine): ©Franz Krenn/Science Source; (a, iris): ©Dwight Kuhn; (a, cactus): ©G.C. Kelly/Science Source; (a, thorns): ©Kenneth W. Fink/Science Source; (b, onion): ©YAY Media AS/Alamy RF; (b, poinsettia): ©Design Pics/Don Hammond RF; (b, flytrap): ©Win Initiative/Getty Images RF; (b, kalanchöe): ©Steven P. Lynch RF; (c, carrots): ©Huw Jones/ Photolibrary/Getty Images; (c, mangrove): ©Tim Laman/Getty Images RF; (c, orchid): ©Settawut Visedbubpha/123R; (c, screw pine): ©Steven P. Lynch/McGraw-Hill Education

the forest floor toward the canopy, maximizing its exposure to sunlight. Rhizomes are thickened, underground stems that produce shoots and roots; figure 22.3a shows the rhizomes of an iris plant. Stolons are horizontal stems that sprout from an existing stem and grow aboveground, forming roots and new shoots at their nodes. A strawberry plant's "runners" are stolons. Tubers, such as potatoes, are swollen regions of rhizomes or stolons that store starch. The succulent, fleshy stems of cacti stockpile a different resource: water. Still other stems are protective; some types of thorns, such as those on honey locust trees, are modified branches that may help prevent herbivores from grazing. (Note that thorns are distinct from spines; the sharp cactus spines in figure 22.3a are actually modified leaves, not stems.)

TABLE 22.1   Many Parts Are Edible

Plant Part*	Edible Example(s)
Vegetative	
Axillary (lateral) bud	Brussels sprout
Terminal bud	Cabbage
Bark	Cinnamon
Bulb	Onion, garlic
Leaf blade	Basil, parsley, spinach, chive, kale
Leaf petiole	Celery, rhubarb
Rhizome	Ginger "root"
Root	Carrot, parsnip, beet, sweet potato
Seedling	Alfalfa sprout, bean sprout
Stem	Asparagus, kohlrabi, sugarcane
Tuber	Potato
Reproductive	
Flower bud	Broccoli, cauliflower, artichoke
Style and stigma (flower parts)	Saffron
Seeds and fruits	Tomato, string bean, olive, pumpkin, bell pepper, peppercorn, rice, peanut, nutmeg

*Why no wood? Because it's mostly empty cell walls made of cellulose and lignin, neither of which we can digest.

Leaves typically carry out photosynthesis, but they can have many other functions, too (see figure 22.3b). Onion bulbs, for example, are collections of the fleshy bases of leaves that store nutrients; the plant uses the food reserves in its bulb when resuming growth after winter or a drought. The brightly colored leaves of a poinsettia attract pollinators to the plant's inconspicuous flowers. In a few carnivorous plant species, the leaves attract, capture, and digest prey, as described in chapter 23. (Carnivorous plants obtain nitrogen from the prey they catch; the leaves produce their own sugars by photosynthesis.) And in kalanchöe, the leaves produce tiny, identical plantlets, each of which may fall to the ground and take root. Kalanchöe is nicknamed "the maternity plant" in honor of its reproductive output.

Roots may also have specialized functions, including storage, gas exchange, support, and even photosynthesis (figure 22.3c). Beet and carrot roots stockpile starch, and desert plant roots may store water. In oxygen-poor habitats such as swamps, the specialized roots of some trees grow up into the air, allowing oxygen to diffuse in. Orchids produce aerial roots that carry out photosynthesis. Thick buttress roots at the base of a tree provide support, as do prop roots that arise from a corn plant's stem.

Burning Question

What's the difference between fruits and vegetables?

Have you ever heard the rumor that tomatoes are really fruits? If so, you may also have wondered about a larger question: the difference between fruits and vegetables.

Botanically speaking, tomatoes are fruits because they contain seeds. Any seed-bearing structure produced by a flowering plant is a fruit. Apples, cherries, oranges, peaches, and raspberries are fruits. So are foods that we think of as vegetables, such as zucchinis, green beans, bell peppers, and pumpkins. But the term *vegetable* also includes plant parts that don't contain seeds, such as spinach leaves, carrot roots, and Brussels sprouts.

Unlike fruits, vegetables do not have a botanical definition—just a culinary one. That is, foods that we consider vegetables may come from any part of a plant, but they are typically used in salty or savory dishes, not sweet ones.

Submit your burning question to
Marielle.Hoefnagels@mheducation.com

©Getty Images RF

Prop roots are examples of **adventitious roots,** which arise from any plant part other than the roots. Many but not all plants can form adventitious roots. People who cultivate plants often exploit this ability to clone plants from cuttings. For example, if you cut the stem of *Philodendron* (a common houseplant) just below a node and place the cutting in water, adventitious roots may begin to grow within a few days. In some plant species, adventitious roots make up a large part of the root system; the iris in figure 22.3a is one example.

Humans have used artificial selection to modify stems, leaves, roots, flowers, and fruits to suit our own needs. Table 22.1 shows a variety of plants that farmers cultivate for their edible parts. Most people divide these foods into two categories: fruits and vegetables. This chapter's Burning Question explains more about these overlapping terms.

22.1 | MASTERING CONCEPTS

1. How do stems, leaves, and roots support one another?
2. What is the relationship between the node and the internode of a stem?
3. Give examples of stems, leaves, and roots with specialized functions.
4. Categorize each structure in figure 22.1 as vegetative or reproductive.

22.2 Plant Cells Build Tissues

A cactus, an elm tree, and a dandelion have distinctly different stems, leaves, roots, and growth patterns. They may seem to have little in common, but a closer look reveals that all consist of the same types of cells and tissues. Before examining these building blocks, it may be helpful to review the structure of a plant cell in figure 3.9 and of the cell wall in figure 3.28. Note that in many plant cells, the wall has two layers: a thin, flexible primary one and a thick, rigid secondary one that forms after the cell is fully grown.

A. Plants Have Several Cell Types

Plants consist of several cell types. This section lists the most common ones.

Ground Tissue Cells   **Ground tissue,** which has a wide variety of functions, makes up the majority of the primary plant body. It consists of three main cell types: parenchyma, collenchyma, and sclerenchyma (**figure 22.4**).

Parenchyma cells are the most abundant cells in the primary plant body. They are alive at maturity, have thin primary cell walls, and retain the ability to divide, which enables them to differentiate in response to injury or a changing environment. It is parenchyma cells, for example, that divide to produce the adventitious roots that emerge from a houseplant cutting placed in water. Parenchyma cells are structurally unspecialized but carry out vital functions, including photosynthesis, respiration, gas exchange, secretion, wound repair, and the storage of starch and other materials.

Collenchyma cells are elongated living cells with unevenly thickened primary walls that can stretch as the cells grow. These cells provide elastic support without interfering with the growth of young stems or expanding leaves. Collenchyma strands often form near the angular ridges on the stems of some plants, such as those in the mint family. Collenchyma is perhaps most familiar, however, as the tough, flexible "strings" in celery stalks.

Sclerenchyma cells provide inelastic support to parts of a plant that are no longer growing. These cells, which are dead at maturity, have thick, rigid secondary cell walls that occupy most of the cell's volume. The secondary walls typically contain **lignin,** a tough, complex molecule that adds great strength to the cell walls.

Cell Type	Description	Alive at Maturity	Functions	
Parenchyma	• Most abundant cell type in primary plant body • Thin primary cell walls • Unspecialized • Can divide at maturity	Yes	Make up most nonwoody tissues; carry out photosynthesis, respiration, gas exchange, secretion, wound repair, and storage	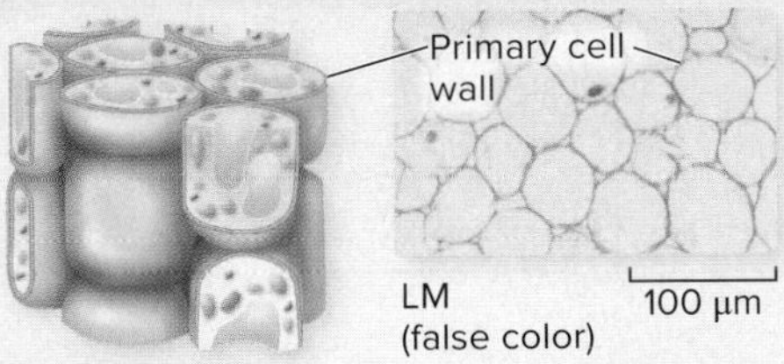 Primary cell wall LM (false color) 100 μm
Collenchyma	• Elongated cells • Unevenly thickened primary cell walls	Yes	Elastic support for growing stems and leaves	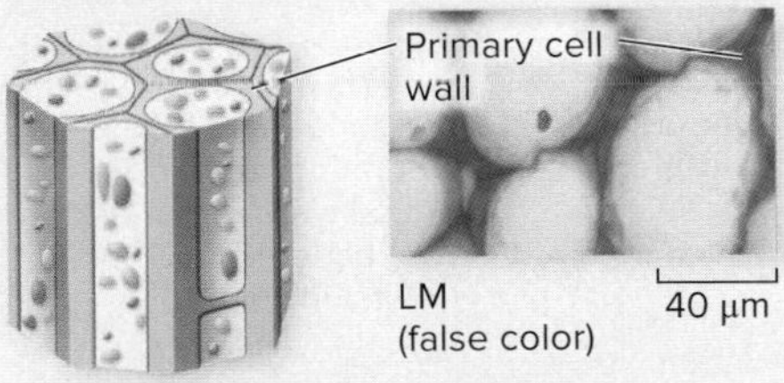 Primary cell wall LM (false color) 40 μm
Sclerenchyma: Fiber	• Long, slender cells • Thick secondary cell walls high in lignin	No	Inelastic support for nongrowing plant parts	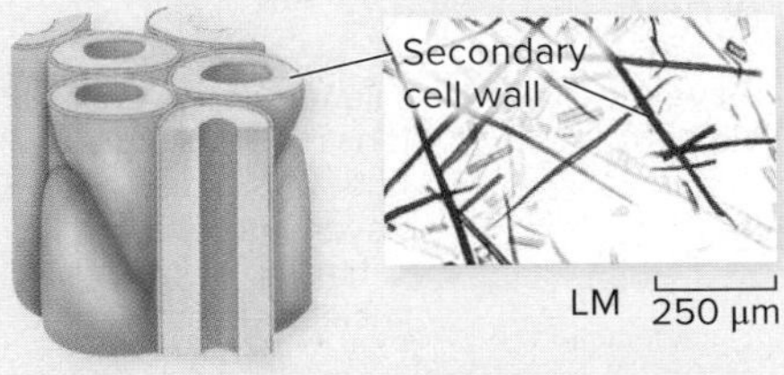 Secondary cell wall LM 250 μm
Sclerenchyma: Sclereid	• Variable shapes, generally not elongated • Thick secondary cell walls high in lignin	No	Inelastic support for nongrowing plant parts	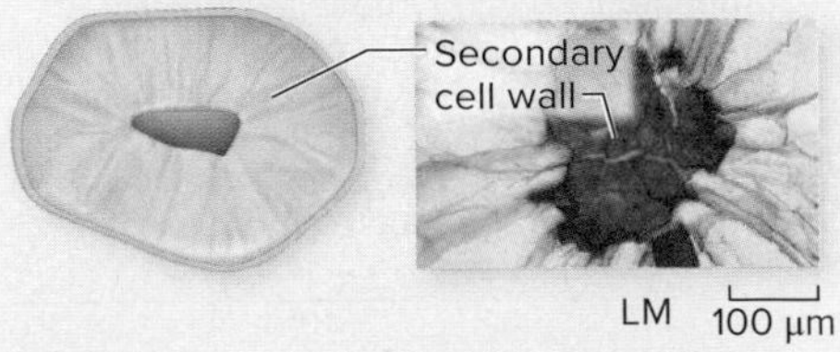 Secondary cell wall LM 100 μm

Figure 22.4 Ground Tissue Cell Types: A Summary. Parenchyma, collenchyma, and sclerenchyma cells compose the majority of the primary plant body.

Photos: (parenchyma): ©Malcolm Park microimages/Alamy; (collenchyma): ©Biophoto Associates/Science Source; (fibers): ©Steven P. Lynch/McGraw-Hill Education; (sclereid): ©Garry DeLong/Oxford Scientific/Getty Images

Two types of sclerenchyma are fibers and sclereids. **Fibers** are elongated cells that usually occur in strands. Linen, for example, comes from the soft fibers of the stems of flax plants. Sisal, which comes from the leaves of *Agave sisalana,* is a hard fiber used in coarse fabrics and rope. The other sclerenchyma cells, **sclereids,** occur in many shapes, although they are generally shorter than fibers. Small groups of sclereids create a pear's gritty texture. Sclereids also form hard layers in nutshells, apple cores, and the pits of cherries and plums.

Conducting Cells in Xylem and Phloem

Vascular tissues transport water, minerals, carbohydrates, hormones, and other dissolved compounds throughout the plant. Two types of vascular tissue are xylem and phloem; **figure 22.5** summarizes the most important cells in each.

Xylem is a tissue that transports water, dissolved minerals, and hormones from the roots to all parts of the plant. The water-conducting cells of xylem are elongated and have thick, lignin-rich secondary walls. They are dead at maturity, which means that no cytoplasm blocks water flow.

The two kinds of water-conducting cells in xylem are tracheids and vessel elements. **Tracheids** are long, narrow cells that overlap at their tapered ends. Water moves from tracheid to tracheid through pits, which are thin areas in the cell wall. Thanks to their small diameters and overlapping end walls, however, water moves slowly in tracheids.

All vascular plants have tracheids, but only angiosperms also have vessel elements. **Vessel elements** are short, wide, barrel-shaped conducting cells that stack end to end, forming long, continuous tubes. Their side walls have pits, but their end walls are either perforated or absent. Water moves much faster in vessels than in tracheids, both because of their greater diameter and because water can pass easily through each vessel element's end wall. On the other hand, the narrower tracheids are less vulnerable to air bubble formation, which may stop the flow of water in xylem (see section 23.2C).

Phloem tissue transports dissolved organic compounds, primarily sugars produced in photosynthesis. Phloem sap also contains proteins, RNA, hormones, ions, and sometimes viruses.

Cell Type	Description	Alive at Maturity	Functions	
XYLEM				
Tracheid	• Narrow diameter • Overlapping walls • Pits on all walls	No	Conduct water and minerals through pits	
Vessel element	• Wide diameter • Aligned end to end, forming vessel • End walls perforated or absent • Pits on side walls	No	Conduct water and minerals through pits and perforated end walls	
PHLOEM				
Sieve tube element	• Aligned end to end, forming sieve tube • End walls have sieve plates • Little cytoplasm; nucleus absent	Yes	Conduct dissolved sucrose and other organic compounds through sieve plates	
Companion cell	• Organelles typical of plant cells • High metabolic activity • Plasmodesmata connect cytoplasm with sieve tube element	Yes	Transfer materials into and out of sieve tube elements	

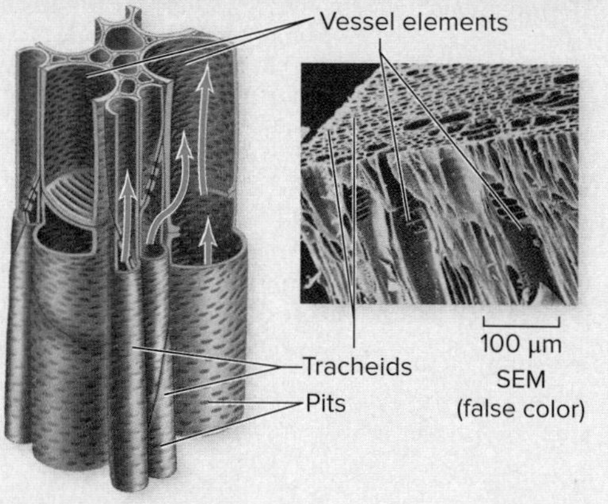

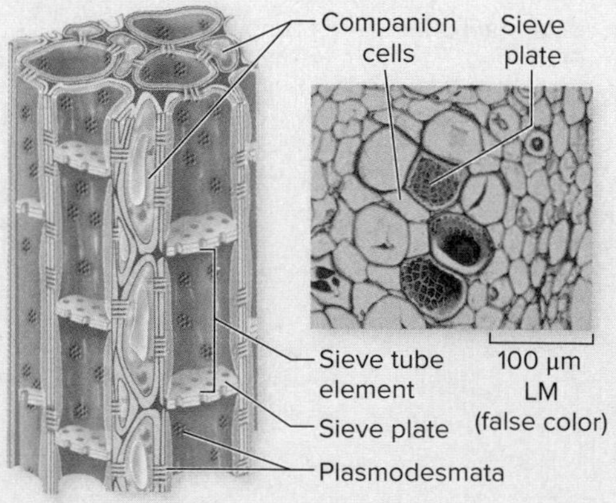

Figure 22.5 Vascular Tissue Cell Types: A Summary. Xylem and phloem are vascular tissues that transport materials within a plant. This figure summarizes the main cell types in each type of vascular tissue.

Photos: (both): ©Biophoto Associates/Science Source

The main conducting cells in phloem are **sieve tube elements,** which align end to end to make a single functional unit called a sieve tube. Most sieve tube elements have **sieve plates,** where modified plasmodesmata form large pores at the ends of the cells. Materials pass from one sieve tube element to the next through these openings. Sieve tube elements are alive at maturity, but they have little cytoplasm and lack nuclei entirely. This arrangement presumably eases the flow of phloem sap.

Adjacent to each sieve tube element is at least one **companion cell,** a specialized parenchyma cell that transfers carbohydrates into and out of the sieve tube elements. Companion cells are connected to sieve tube elements by strands of cytoplasm that pass through plasmodesmata. Unlike sieve tube elements, companion cells retain all of their organelles; sieve tube elements rely on these "partners" for energy and proteins.

B. Plant Cells Form Three Main Tissue Systems

The cells that make up a plant form three main tissue systems: ground tissue, dermal tissue, and vascular tissue (**figure 22.6** and **table 22.2**). For comparison, animals have four types of tissues (see chapter 25). Each tissue type derives its properties from a unique combination of specialized cells. Together, these cells carry out all of the plant's functions.

Ground Tissue   Ground tissue often fills the spaces between more specialized cell types inside roots, stems, leaves, fruits, and seeds. For example, the pulp of an apple, the photosynthetic area inside a leaf, and the starch-containing cells of a potato all consist

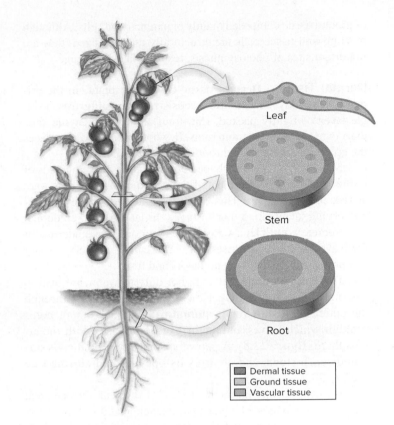

Figure 22.6 Three Tissue Types. Dermal, ground, and vascular tissues make up the leaves, stem, and roots of a plant.

TABLE **22.2**   **Plant Tissue and Cell Types: A Summary**

Tissue System	Location	Cell Types	Functions
Ground	Forms bulk of interior of stems, leaves, and roots, including cortex, pith, and leaf mesophyll	Parenchyma	Photosynthesis, storage, respiration
		Collenchyma	Elastic support
		Sclerenchyma (fibers and sclereids)	Inelastic support
Dermal	Surface of roots and stems	Epidermal parenchyma	Protects primary plant body, controls gas exchange in stems and leaves, absorbs water and minerals in roots
		Cork cells in periderm	Protects woody roots and stems
Vascular: xylem	Vascular bundles in stems, veins in leaves, vascular bundles or vascular cylinder in roots	Tracheids, vessel elements	Conduct water and minerals in stems, leaves, and roots
		Parenchyma	Storage
		Sclerenchyma (fibers and sclereids)	Inelastic support
Vascular: phloem	Vascular bundles in stems, veins in leaves, vascular bundles or vascular cylinder in roots	Sieve tube elements	Transport of organic molecules
		Companion cells (specialized parenchyma)	Transfer of carbohydrates to/from sieve tube elements
		Sclerenchyma (fibers and sclereids)	Inelastic support

of ground tissue composed mainly of parenchyma cells. Although most ground tissue cells are structurally unspecialized, they are important sites of photosynthesis, respiration, and storage.

Dermal Tissue   **Dermal tissue** covers the plant. In the primary plant body, dermal tissue consists of the **epidermis,** a single layer of tightly packed, flat, transparent parenchyma cells that cover leaves, stems, and roots. In woody plants, bark replaces the epidermis in stems and roots (section 22.4D).

Natural selection has shaped dermal tissue over hundreds of millions of years. When plants first moved onto land, they were at risk for drying out. This condition selected for new water-conserving adaptations. For instance, in land plants, epidermal cells secrete a **cuticle,** a waxy layer that coats the epidermis of the leaves and stem (figure 22.7). The cuticle conserves water and protects the plant from predators and fungi.

The cuticle is impermeable not only to water but also to gases such as O_2 and CO_2. How do these gases pass through the cuticle? The answer is **stomata** (singular: stoma), pores through which leaves and stems exchange gases with the atmosphere (figure 22.8). A pair of specialized **guard cells** surrounds each stoma and controls its opening and closing (see section 23.2).

Stomata occupy about 1% to 2% of the leaf surface area, allowing for a balance between gas exchange and water conservation. Open stomata let CO_2 diffuse into a leaf for photosynthesis but also allow water to diffuse out; the pores typically close when conditions are too dry. Moreover, the arrangement of stomata on a leaf's surface can also minimize water loss. In many species, for example, stomata are most abundant on the

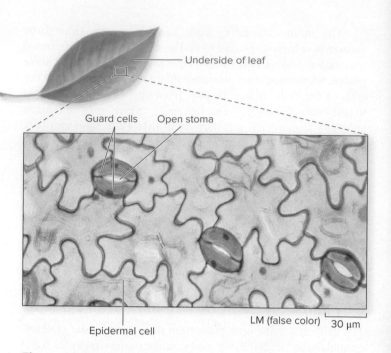

Figure 22.8 Stomata. Plants exchange gases with the atmosphere through open stomata; guard cells surround each stoma.

Photo: ©Steven P. Lynch/McGraw-Hill Education

shaded undersides of horizontal leaves; in pine needles, stomata are recessed in grooves.

Vascular Tissue   Vascular tissues—xylem and phloem—form a continuous distribution system embedded in the ground tissue of shoots and roots. In stems and leaves, a **vascular bundle** is a strand of tissue containing xylem and phloem, often together with collenchyma tissue or sclerenchyma fibers. The tough fibers protect against animals that might otherwise tap into the phloem's rich sugar supply.

The transition from water to land selected for vascular tissues, which accommodate the division of labor between roots and shoots. Roots absorb water and minerals; shoots produce food. Xylem and phloem shuttle these materials throughout the plant's body. (Chapter 23 further explores how water, nutrients, sugars, and other resources travel in vascular tissue.)

Vascular tissue also has another function: support. Lignin strengthens the walls of xylem cells and sclerenchyma fibers. As described in chapter 19, this physical support enables vascular plants to tower over their nonvascular counterparts, an important adaptation in the intense competition for sunlight.

Table 22.2 summarizes the locations, cell types, and functions of the tissue systems in plants.

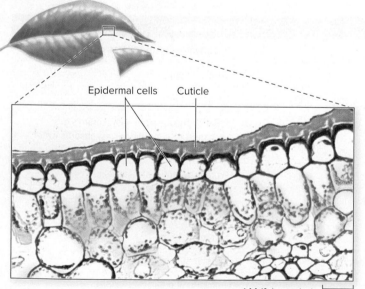

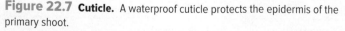

Figure 22.7 Cuticle. A waterproof cuticle protects the epidermis of the primary shoot.

©Steven P. Lynch/McGraw-Hill Education

22.2   MASTERING CONCEPTS

1. Where in the plant does ground tissue occur?
2. Compare and contrast the structure and function of tracheids, vessel elements, and sieve tube elements.
3. How does the structure of dermal tissue contribute to its functions?

22.3 Tissues Build Stems, Leaves, and Roots

The tissues described in section 22.2 make up the stems, leaves, and roots of vascular plants. We now return to these organs to examine their structures more closely. Note that flowering plants (angiosperms) differ in the internal anatomy of their main vegetative organs. As described in chapter 19, two groups—eudicots and monocots—account for 97% of all angiosperms. Eudicots, the largest group, include everything from chili peppers to elm trees. Monocots include orchids, grasses, corn, rice, wheat, and bamboo. The rest of this chapter describes the structural similarities and differences between these two groups of plants.

A. Stems Support Leaves

Ground tissue occupies most of the volume of the stem of a herbaceous plant. The ground tissue, which consists mostly of parenchyma cells, stores water and starch. The cells are often loosely packed, allowing for gas exchange between the stem interior and the atmosphere.

Vascular bundles are embedded in the stem's ground tissue. The vascular bundles, which typically have phloem to the outside and xylem toward the inside, are arranged differently in monocots and eudicots (**figure 22.9**). In most monocot stems, vascular bundles are scattered throughout the ground tissue. Most eudicot stems, in contrast, have a single ring of vascular bundles. Ground tissue occupies most of the rest of the eudicot stem: The **cortex** fills the area between the epidermis and vascular tissue, and a soft, spongy **pith** occupies the center. Among other functions, the cortex and pith store starch and water.

B. Leaves Are the Primary Organs of Photosynthesis

Most leaves have two main parts: The stalklike **petiole** attaches to the stem and supports the broad, flat **blade.** The leaf blade maximizes the surface area available to capture light. For example,

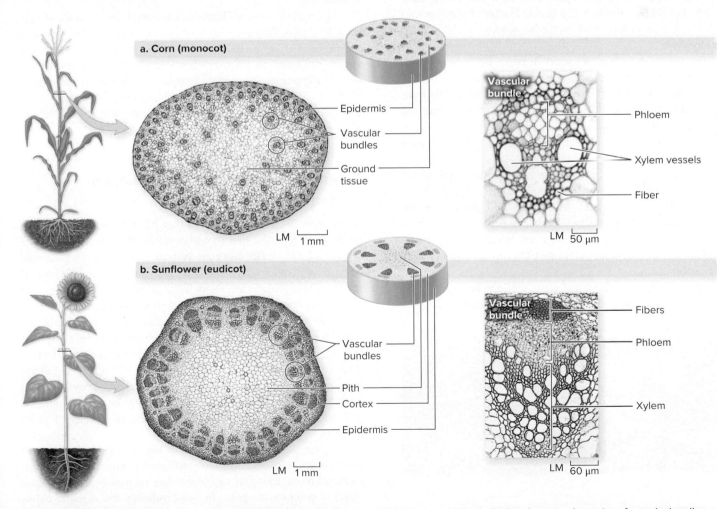

a. Corn (monocot)

Epidermis
Vascular bundles
Ground tissue
LM | 1 mm

Vascular bundle
Phloem
Xylem vessels
Fiber
LM | 50 µm

b. Sunflower (eudicot)

Vascular bundles
Pith
Cortex
Epidermis
LM | 1 mm

Vascular bundle
Fibers
Phloem
Xylem
LM | 60 µm

Figure 22.9 **Stem Anatomy.** (a) In a monocot stem, vascular bundles are scattered in ground tissue. (b) A eudicot stem has a ring of vascular bundles surrounding a central pith.

Figure 22.10 Leaf Types. A simple leaf has an undivided blade, whereas compound leaves consist of multiple leaflets. An axillary bud defines the base of each leaf.

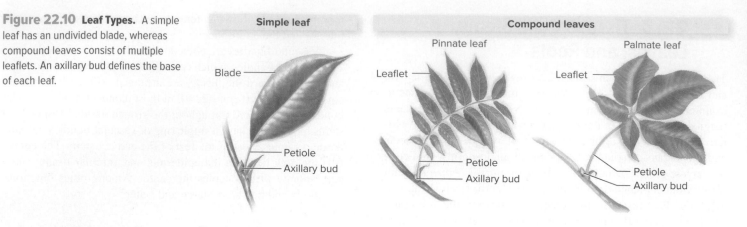

Simple leaf

Blade

Petiole

Axillary bud

Compound leaves

Pinnate leaf

Leaflet

Petiole

Axillary bud

Palmate leaf

Leaflet

Petiole

Axillary bud

a large maple tree has about 100,000 leaves, with a total surface area that would cover six basketball courts (about 2500 square meters).

Biologists categorize leaves according to their basic forms (figure 22.10). A **simple leaf** has an undivided blade. **Compound leaves** are divided into leaflets, typically either paired along a central line (pinnate) or all attached to one point at the top of the petiole, like fingers on a hand (palmate). How can you tell the difference between a simple leaf and one leaflet of a compound leaf? A leaf has an axillary bud at its base, whereas an individual leaflet does not.

Veins are vascular bundles inside leaves, and they are often a leaf's most prominent external feature. Networks of veins occur in two main patterns (figure 22.11). Many monocots have parallel veins, with several major longitudinal veins connected by smaller minor veins. Most eudicots have netted veins, with minor veins branching in all directions from large, prominent midveins.

The ground tissue inside a leaf is called **mesophyll,** and it is composed mostly of parenchyma cells (figure 22.12). Most mesophyll cells have abundant chloroplasts and produce sugars by photosynthesis. Often, large air spaces separate the mesophyll cells. When the leaf's stomata are open, photosynthetically active mesophyll cells exchange CO_2 and O_2 directly with the atmosphere.

The monocot and eudicot leaf cross sections in figure 22.12 look different from each other. For example, the corn leaf cross

section has only one type of mesophyll cells, whereas the eudicot leaf has two types: Long, column-shaped mesophyll cells maximize light absorption along the upper leaf surface, and irregularly shaped, loosely packed mesophyll cells are below. In addition, stomata are often present on both surfaces of a monocot leaf but are typically most abundant on the lower surface of a eudicot leaf.

Materials exchanged between mesophyll cells and a leaf vein must first pass through a layer of bundle sheath cells. Xylem in the tiniest leaf veins delivers water and minerals. Meanwhile, sugars produced in photosynthesis move from the mesophyll cells to the phloem's companion cells and then to the sieve tube elements. The sugars, along with other organic compounds, travel within the phloem to the roots and other nonphotosynthetic plant parts (see figure 23.14).

C. Roots Absorb Water and Minerals and Anchor the Plant

Roots grow in two main patterns that differ based on the fate of the primary root, the first root to develop after a seed germinates (figure 22.13). A **fibrous root system** consists of a widespread network of slender adventitious roots that arise from the base of the stem and replace the short-lived primary root. Grasses and other monocots usually have fibrous root systems. Because they are relatively shallow, these roots rapidly absorb minerals and water near the soil surface and prevent erosion. In a **taproot system,** on the other hand, the primary root enlarges to form a thick root that persists throughout the life of the plant. Lateral branches emerge from this main root. Taproots grow fast and deep, maximizing support and enabling a plant to use minerals and water deep in the soil. Most eudicots develop taproot systems.

Figure 22.13 also reveals that the vascular cylinders of monocot and eudicot roots have different arrangements. In most monocot roots, a ring of vascular tissue surrounds a central core (pith) of parenchyma cells. In most eudicots, the vascular cylinder consists of a solid core of xylem, with ridges that project toward the root's exterior. Phloem strands are generally located between the "arms" of the xylem core.

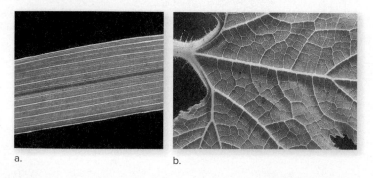

a.

b.

Figure 22.11 Leaf Veins. (a) Monocot leaves typically have prominent parallel veins. (b) Eudicot leaves have a netlike pattern of veins.

(both): ©Dwight Kuhn

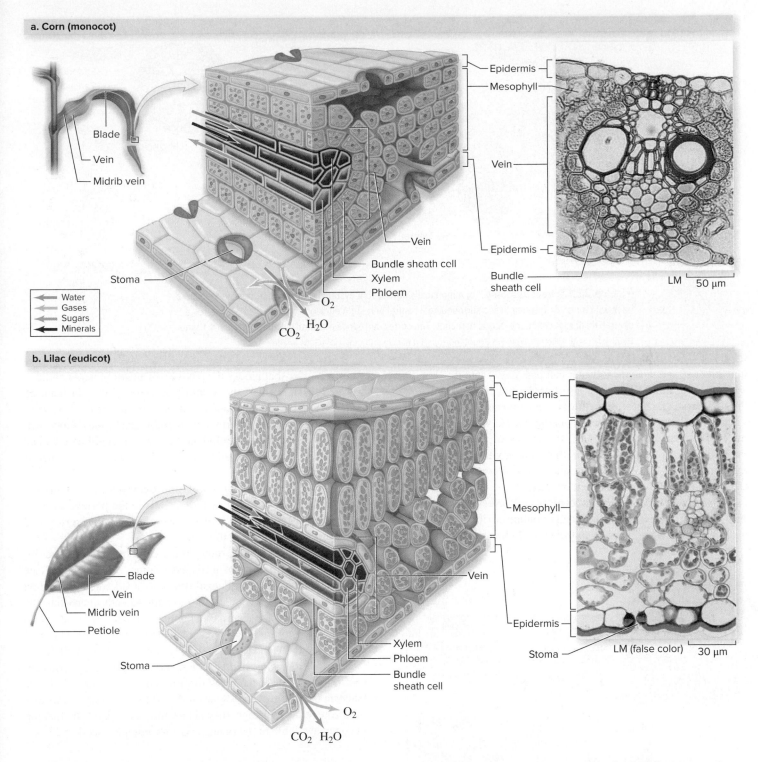

Figure 22.12 Leaf Anatomy. Leaf mesophyll consists of cells that carry out photosynthesis in (a) monocots and (b) eudicots. Leaf veins deliver water and minerals, and they carry off the products of photosynthesis.

Photos: (a): ©Steven P. Lynch/McGraw-Hill Education; (b): ©M. I. Walker/Science Source

In both fibrous and taproot systems, countless root tips explore the soil for water and nutrients. The **root cap** protects the growing tip from abrasion. Root cap cells, which slough off and are continually replaced, secrete a slimy substance that lubricates the root as it pushes through the soil. The cells of the root cap also play a role in sensing gravity (see section 24.6).

The epidermis surrounds the entire root except the root cap. In contrast to the stem and leaf, the root epidermal cells have a very

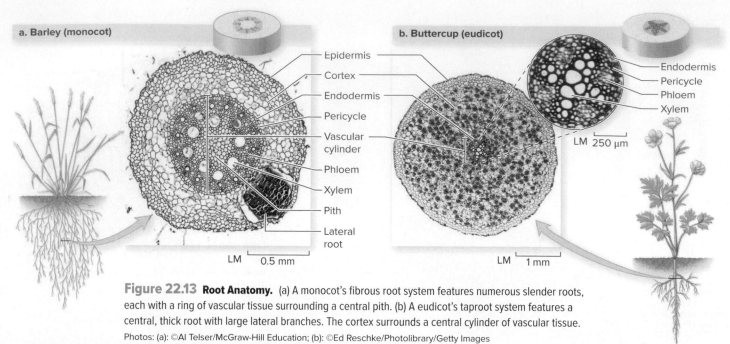

a. Barley (monocot)

- Epidermis
- Cortex
- Endodermis
- Pericycle
- Vascular cylinder
- Phloem
- Xylem
- Pith
- Lateral root

LM 0.5 mm

b. Buttercup (eudicot)

- Endodermis
- Pericycle
- Phloem
- Xylem

LM 250 μm

LM 1 mm

Figure 22.13 Root Anatomy. (a) A monocot's fibrous root system features numerous slender roots, each with a ring of vascular tissue surrounding a central pith. (b) A eudicot's taproot system features a central, thick root with large lateral branches. The cortex surrounds a central cylinder of vascular tissue.

Photos: (a): ©Al Telser/McGraw-Hill Education; (b): ©Ed Reschke/Photolibrary/Getty Images

thin cuticle or none at all, so water can pass easily into the root. In addition, **root hairs** are extensions of epidermal cells, maximizing the surface area for absorption of water and minerals (**figure 22.14**).

Just internal to the epidermis is the cortex, which makes up most of the primary root's bulk. It consists of loosely packed, interconnected parenchyma cells that may store starch or other materials. The spaces between the cells allow for both gas exchange and the free movement of water.

The **endodermis** is the innermost cell layer of the cortex. The walls of its tightly packed cells contain a ribbon of waxy, waterproof material. These waxy deposits form the **Casparian strip,** a barrier that blocks the passive diffusion of water and dissolved substances into the xylem (see figure 23.6). Instead, all materials entering the root's vascular tissue must first pass through the selectively permeable membranes of the endodermal cells. The endodermis therefore acts as a filter, enabling the plant to exclude toxins and regulate the entry of water and minerals into the xylem.

Internal to the endodermis is the **pericycle,** the outermost layer of the root's vascular cylinder. Cells in the pericycle divide to produce lateral roots, which push through the cortex and epidermis before entering the soil.

In most plant species, roots do not explore the soil alone. Instead, they form mycorrhizal associations with fungi (see figures 20.10 and 20.18). Fungal filaments inside the root feed on the plant's sugars and extend into the soil, greatly increasing the root's surface area for absorption. Fungi and plants are ancient partners; hundreds of millions of years ago, these two groups of organisms moved onto land together. ⓘ *mycorrhizae,* section 20.7B

In addition, some plant species, such as peas and beans, form root nodules in association with nitrogen-fixing bacteria. These bacteria use some of the sugar produced by their host plants as a food source. In exchange, they act as built-in fertilizer by making nitrogen available to the plants. ⓘ *root nodules,* section 23.1C

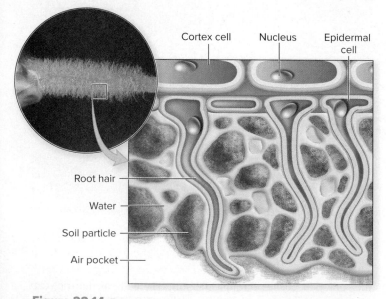

- Cortex cell
- Nucleus
- Epidermal cell
- Root hair
- Water
- Soil particle
- Air pocket

Figure 22.14 Root Hairs. These epidermal cell outgrowths greatly increase the surface area of this radish seedling's root.

Photo: ©Dr. Jeremy Burgess/Science Source

22.3 MASTERING CONCEPTS

1. Name the cell layers in the stem of a monocot and a eudicot, moving from the epidermis to the innermost tissues.
2. List the parts of a simple and a compound leaf.
3. Describe the anatomy of a typical leaf.
4. Corn is a monocot and sunflower is a eudicot. Make a chart that compares the stems, leaves, and roots of these plants.

22.4 Plants Have Flexible Growth Patterns, Thanks to Meristems

Consider the plight of a green plant. Rooted in place, it seems vulnerable and defenseless against drought, flood, wind, fire, and hungry herbivores. Nor can it escape changes in temperature, moisture, sunlight, and soil nutrient availability, all of which may fluctuate throughout a plant's life. Yet plants dominate nearly every habitat on land. How do they do it? Some plant defenses come from their arsenal of secondary metabolites, as described at the start of this chapter. But in many ways, plants owe their success to modular growth.

A. Plants Grow by Adding New Modules

To understand modular growth, imagine a landowner who plans to build a motel. Money is tight at first, so she starts with just a few rooms. As her business grows, however, she adds more units to the basic plan. Plant growth is similar. Shoots become larger by adding units ("modules") consisting of repeated nodes and internodes; roots can branch repeatedly as they explore the soil.

Some plants, such as dandelions, have **determinate growth,** meaning that they stop growing after they reach their mature size. Most plants, however, can grow indefinitely by adding module after module—think of ivy climbing a building. Such **indeterminate growth** can continue as long as environmental conditions allow it.

Modular growth enhances a plant's ability to respond to the environment. For example, plants produce the most root tips in pockets of soil with the richest nutrients. Likewise, a shrub growing in partial shade can add new branches where it receives the most sunlight, while the shaded limbs remain unchanged. The motel owner mentioned earlier might use a similar strategy. That is, if clients will pay more for an ocean view than for a city view, she would be wise to build additional rooms overlooking the sea.

In addition, modular growth means that the loss of a branch or root does not harm a plant as much as, say, the loss of a leg affects a cat. Neighboring branches can add modules to compensate for a broken tree limb, but the cat's body cannot regenerate a leg. Modular growth is one key feature that distinguishes plants from animals.

B. Plant Growth Occurs at Meristems

All of a plant's new cells come from **meristems,** regions that undergo active mitotic cell division (see chapter 8). Meristems are patches of "immortality" that allow a plant to grow, replace damaged parts, and respond to environmental change. The cells of the meristem are analogous to an animal's stem cells (see chapter 11), which can divide indefinitely and can become many tissue types.

Three main types of meristems occur in plants (table 22.3). **Apical meristems** are small patches of actively dividing cells near the tips of roots and shoots; dormant apical meristems

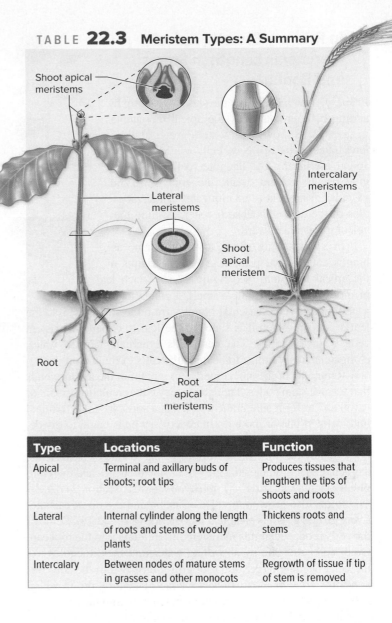

TABLE 22.3 Meristem Types: A Summary

Type	Locations	Function
Apical	Terminal and axillary buds of shoots; root tips	Produces tissues that lengthen the tips of shoots and roots
Lateral	Internal cylinder along the length of roots and stems of woody plants	Thickens roots and stems
Intercalary	Between nodes of mature stems in grasses and other monocots	Regrowth of tissue if tip of stem is removed

also reside in a plant's axillary buds. The function of an apical meristem is to add length to a growing root or shoot. When the cells in the meristem divide, they give rise to new cells that differentiate into all of the tissue types described in section 22.3.

Woody plants also have **lateral meristems,** which produce cells that thicken a stem or root. A lateral meristem is usually an internal cylinder of cells extending along most of the length of the plant. When the cells divide, they typically produce tissues to both the inside and the outside of the meristem.

In some monocots, **intercalary meristems** occur at the base of a leaf blade or between the nodes of a mature stem, usually at the base of an internode. (*Intercalary* means "inserted between," referring to the position of these meristems.) Grasses, for example, tolerate grazing and mowing because they have intercalary meristems whose cells divide to regrow a leaf from its base when the tip is munched off.

C. In Primary Growth, Apical Meristems Lengthen Stems and Roots

Primary growth lengthens the shoot or root tip by adding cells produced by the apical meristems. **Figure 22.15** shows how a stem grows and differentiates at its tip. New cells originate at the apical meristem. The daughter cells eventually give rise to ground tissue, the epidermis, and vascular tissue. The stem elongates as the vacuoles of the new cells absorb water, pushing the apical meristem upward.

In a developing shoot, leaves originate as bumps called "leaf primordia" on the flanks of the apical meristem (see figure 22.15). Each leaf primordium develops into a small, dormant leaf bud at a stem node, where vascular bundles from the stem pass into the leaf's petiole. When it is time for the leaf to expand, meristem cells begin dividing. The new cells eventually absorb water, elongate, and specialize to become the cells of the mature leaf.

Remnants of the apical meristem remain in the axillary buds that form at stem nodes. These buds may either remain dormant or "awaken" to form side branches. When a shoot loses its terminal bud, cells in one or more dormant axillary buds begin to divide. The result is a bushy growth form. Gardeners exploit this phenomenon by pinching off the tips of young tomato or basil stems, a practice that promotes the growth of side branches and therefore greatly increases yields. ⓘ *apical dominance,* section 24.4A

Roots also grow at their tips. Just behind the tip of each actively growing root is an apical meristem (**figure 22.16**). Some of the cells produced at this meristem differentiate into the root cap. Other cells elongate by absorbing water into their vacuoles. As the cells become larger, the root grows farther into the soil. Beyond this zone of elongation is a zone of maturation, in which cells complete their differentiation and mature into the functional ground, dermal, and vascular tissues that make up the root.

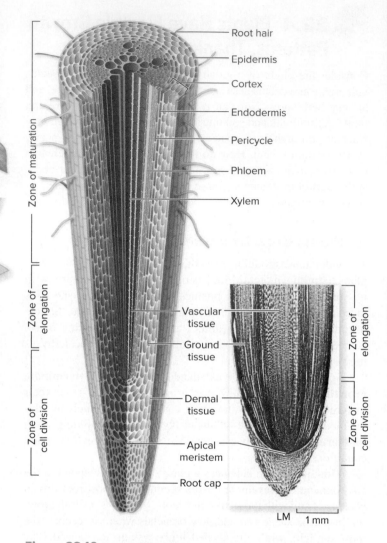

Figure 22.16 Root Apical Meristem. Roots become longer when cells in the apical meristem divide; the new cells develop into the root cap, ground tissue, vascular tissue, and epidermis.

Photo: ©Steven P. Lynch/McGraw-Hill Education

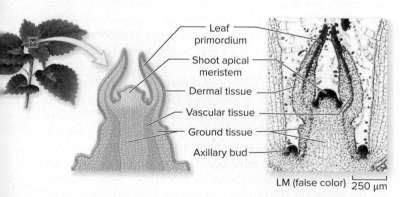

Figure 22.15 Shoot Apical Meristem. Cell division in the apical meristem adds length to the shoot tip; as the new cells differentiate, they give rise to the specialized tissues that make up the mature shoot.

Photo: ©Steven P. Lynch/McGraw-Hill Education

D. In Secondary Growth, Lateral Meristems Thicken Stems and Roots

In many habitats, plants compete for sunlight. The tallest plants reach the most light, so selection for height has been a powerful force in the evolutionary history of plants. But primary tissue is not strong enough to support a very tall plant. The increasing competition for light therefore selected for additional support, in the form of secondary growth that increases the girth of stems and roots in woody plants.

The Origin of Wood and Bark Wood and bark are tissues that arise from secondary growth originating at two types of lateral meristems: vascular cambium and cork cambium (**figure 22.17**). These meristems occur in gymnosperms (conifers and their relatives) and many eudicots, but not in monocots.

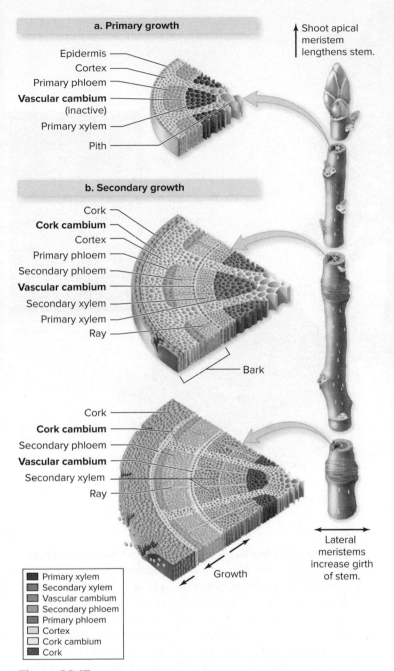

a. Primary growth

Epidermis
Cortex
Primary phloem
Vascular cambium (inactive)
Primary xylem
Pith

Shoot apical meristem lengthens stem.

b. Secondary growth

Cork
Cork cambium
Cortex
Primary phloem
Secondary phloem
Vascular cambium
Secondary xylem
Primary xylem
Ray

Bark

Cork
Cork cambium
Secondary phloem
Vascular cambium
Secondary xylem
Ray

Growth

Lateral meristems increase girth of stem.

■ Primary xylem
■ Secondary xylem
■ Vascular cambium
■ Secondary phloem
■ Primary phloem
□ Cortex
□ Cork cambium
■ Cork

Figure 22.17 Secondary Growth. (a) In a primary shoot, the microscopic vascular cambium has not yet started producing secondary vascular tissue. (b) The two lower diagrams show the vascular cambium producing secondary xylem toward the inside of the stem and secondary phloem toward the outside. Cork cambium, meanwhile, produces cork cells to the outside and parenchyma to the inside (not shown).

The **vascular cambium** is an internal cylinder of meristem tissue that produces most of the diameter of a woody root or stem. This lateral meristem forms a thin layer between the primary xylem and phloem (see figure 22.17a). When a cell in the vascular cambium divides, it produces two daughter cells. One of the two cells remains a meristem cell. If the other cell matures to the inside of the cambium, it becomes secondary xylem; if it matures to the outside, it becomes secondary phloem.

The vascular cambium also produces **rays,** bands of parenchyma cells that extend from the center of the stem or root like spokes on a bicycle wheel. Rays transport water and nutrients laterally within secondary xylem and phloem.

Overall, the vascular cambium produces much more secondary xylem than secondary phloem, so the stem or root acquires most of its girth from growth internal to the cambium. A stem or root's **wood** consists of secondary xylem, and it can accumulate to massive proportions. For example, a giant sequoia tree in California is 100 meters tall and more than 7 meters in diameter. Secondary phloem occupies much less volume. This tissue forms the live, innermost layer of **bark,** a collective term for all tissues to the outside of the vascular cambium. (Some trees have bark with unique adaptations that boost fire resistance; see this chapter's Apply It Now box.)

The **cork cambium** is a lateral meristem that gives rise to cork cells to the outside (see figure 22.17b) and, in some plants, parenchyma cells to the inside. Cork consists of layers of densely packed, waxy cells on the surfaces of mature stems and roots. The cells are dead at maturity and form waterproof, insulating layers that protect the plant. The cork used in wine bottles comes from oak trees that grow in the Mediterranean region. Every 10 years, harvesters remove much of the cork cambium and the thick cork layer, which grows back. Cork is also important in the history of biology; in 1665, Robert Hooke became the first person to see cells when he used a primitive microscope to gaze at cork.

The nonliving cork cells, the living parenchyma cells (if present), and the cork cambium that produces both layers collectively make up the **periderm,** the protective dermal tissue that covers a woody stem or root. Whereas secondary phloem is the innermost layer of bark, periderm makes up the outer portion.

A close comparison of the top and bottom cross sections in figure 22.17 reveals that secondary growth eventually destroys the stem or root's primary phloem, cortex, and epidermis. However, the secondary phloem takes over responsibility for phloem sap transport, and the periderm replaces the primary epidermis.

Wood Is Durable and Useful

Few plant products are as versatile or economically important as wood. Lumber forms the internal frame that supports many buildings. Firewood provides heat and cooking fuel. Throughout history, humans have fashioned wood into paper, furniture, pencils, cabinets, boats, baseball bats, serving bowls, roofing shingles, jewelry, picture frames, and countless other items.

Wood varies in hardness. Angiosperms such as oak, maple, and ash are often called hardwood trees, whereas conifers such as pine, spruce, and fir are called softwood trees. Angiosperm wood contains both tracheids and vessels, whereas wood from softwood trees consists mainly of tracheids. Eudicot wood is usually stronger and denser than wood from conifers. The soft, light wood from the balsa tree, an angiosperm, is a notable exception.

Figure 22.18 **Anatomy of a Woody Stem.** (a) Wood is secondary xylem, and bark is all the tissue outside the vascular cambium. (b) At the center of the stem, the darker-colored heartwood is nonfunctional secondary xylem. (c) Wood that forms in the spring has larger cells than wood that forms in the summer, thanks to differences in soil moisture. This size difference is visible as tree rings.

Photos: (b): ©Siede Preis/Getty Images RF; (c): ©Herve Conge/Phototake

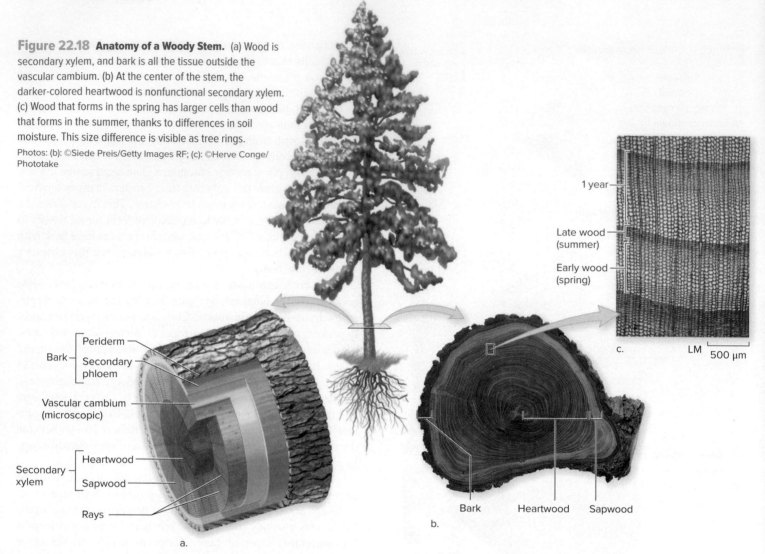

Figure 22.18a illustrates the internal anatomy of a tree trunk, including the bark, vascular cambium, and secondary xylem. Nearly all of the trunk consists of secondary xylem, or wood. The **heartwood,** or innermost wood, is darker than the **sapwood,** or outer portion. This color difference arises as the tree ages. The lighter-colored sapwood, located nearest the vascular cambium, transports water and dissolved minerals. Meanwhile, as the years pass, the oldest secondary xylem at the center of the trunk—the heartwood—gradually becomes unable to conduct water. As its function declines, dark-colored chemicals accumulate in the heartwood.

Another feature of a trunk's cross section is tree rings. In temperate climates, cells in the vascular cambium are dormant in winter, but they divide to produce wood during the spring and summer. During the moist days of spring, the vascular cambium produces large, water-conducting cells. During the drier days of summer, new wood has smaller cells. The contrast between the summer wood of one year and the spring wood of the next highlights each annual tree ring (figure 22.18b and c).

Fortunately, it is not necessary to cut down a tree to see its growth rings; a slender core extracted from the trunk reveals the pattern without harming the tree. By counting the rings, a forester can estimate a tree's age. Growth rings also provide clues about climate and significant events throughout a tree's life. A thick ring indicates plentiful rainfall and good growing conditions. Narrow tree rings may reflect stress from herbivory, disease, or fierce competition for light or water. A fire leaves behind a charred "burn scar."

Researchers can also combine tree ring data from multiple trees to look into ancient history. For example, the oldest known living tree is a bristlecone pine growing in the White Mountains of California. It is more than 4760 years old. By aligning distinctive patterns in its early tree rings with the same patterns in older, dead trees from the same area, researchers have deduced rainfall data going back 8200 years!

22.4 MASTERING CONCEPTS

1. What is the difference between determinate and indeterminate growth?
2. What are the locations and functions of meristems?
3. What are the two lateral meristems in a woody stem or root, and which tissues does each meristem produce?
4. What are the functions of wood and bark?
5. How do softwoods differ from hardwoods?
6. Explain the origin of tree rings.

INVESTIGATING LIFE

22.5 An Army of Tiny Watchdogs

For many people, a dog is both a friend and a protector, warning of burglars and other intruders. In exchange, we provide our canine companions with shelter and food. Likewise, some plants welcome ants and other invertebrates with open arms—or, more precisely, with open stems.

In the tropics, hundreds of tree species provide ants with room and board. Some of the trees' stems have unique anatomical features such as hollow areas, called domatia, where ants live. Moreover, their young leaves secrete a sugary nectar or other food that their guests eat.

Supplying nectar to ants costs energy, which a plant could otherwise put toward its own reproduction. How can natural selection allow plants to extend such hospitality to ants? Biologist Laurence Gaume, an expert on plant–insect interactions at France's University of Montpellier, wanted to know the answer to that question. Working with colleagues from the Indian Institute of Science and the University of Montpellier, Gaume conducted a series of field experiments in India to find out.

The research team chose a tree called *Humboldtia brunonis* to learn more about ant–plant interactions (**figure 22.19**). In

H. brunonis, the domatia are swollen, hollowed-out stem internodes up to 10 centimeters long and 1 centimeter wide. A hole near the node allows ants and other small invertebrates to enter and leave each stem cavity.

In most tree species, either all of the individuals produce domatia or none do. But *H. brunonis* is unusual: Within the same population, some individuals have the stem cavities and others

Figure 22.19 Home Sweet Home. Some types of ants make their homes in hollow stems called domatia.

Photo: ©Dr. Morley Read/Science Source

Apply It **Now**

Torching the Competition

Fast-moving wildfires are common in certain parts of the world. Some residents of California, the Mediterranean coast, and southern Australia must be constantly ready to evacuate during fire season. Trees and shrubs, on the other hand, cannot move out of a fire's path. Instead, plants native to these areas—such as giant sequoias and lodgepole pines—typically tolerate flames. Their cones may even release seeds after the fire subsides. (i) *shrublands*, section 39.3D

But a few plants, including eucalyptus trees, actually *intensify* fires. The leaves of eucalypts contain highly flammable oils even after they have fallen to the ground. These oils add fuel to nearby fires. The flames race up the trunk and reach the forest canopy. The trees explode with fire, sending embers for miles. Intensifying a fire means that many neighboring plants—would-be competitors of the eucalyptus trees—will die.

Unlike many of their neighbors, the eucalyptus trees quickly regrow in the torched aftermath. How do they do it? The secrets to their success are epicormic buds, remnants of the apical meristem that remain buried deep within their bark. Plant hormones, such as auxins and cytokinins, typically suppress their growth. But fire releases the epicormic buds from dormancy, and new branches emerge from the trunk (**figure 22.A**). With up to 300 epicormic buds per meter, eucalyptus trees have high potential for regrowth. Moreover, even if the entire trunk burns to the ground, new shoots emerge from buds

within large, woody tubers near the soil surface. Like the mythical phoenix, the eucalyptus tree emerges from the ashes. (i) *plant hormones*, section 24.4

Figure 22.A Regrowth After Fire. Shoots emerge from a torched eucalyptus tree after a fire in western Australia.

©Gerhard Saueracker/Getty Images

do not. This unusual property made *H. brunonis* ideal for an experiment comparing trees with and without domatia.

In the first experiment, Gaume's research team counted the number of fruits on 104 *H. brunonis* trees as a measure of reproductive success. They noted the height of each tree and whether or not domatia were present, and they identified the ants exploring each tree. After entering the data into a statistical analysis program, they found that the trees with domatia tended to produce the most fruits. The presence of one ant species, called the tramp ant (*Technomyrmex albipes*), was also correlated with fruit production.

A correlation, however, is not the same thing as a causal relationship. One possible interpretation is that the domatia house tramp ants, which somehow promote fruit production. But an alternative explanation is that trees in the best locations have access to the most resources and therefore can "afford" both domatia and high fruit production. In that case, the apparent link between domatia and reproductive success would be a mere coincidence.

Based on studies of other "ant plants," the researchers hypothesized that the tramp ant indirectly promoted the tree's reproductive success by chasing off insects that would otherwise eat the tree's leaves. They therefore predicted that excluding ants from leaves should increase the amount of herbivory. To set up their experiment, the team randomly selected 20 trees, 5 of which were patrolled exclusively by the tramp ant; other ant species patrolled the other 15 trees. On each tree, the researchers selected two young compound leaves, each consisting of four leaflets. They dabbed glue at the base of one of the two leaves. Ants would be free to chase herbivores off the control leaves but would get stuck at the base of a glue-treated leaf.

After 10 days, the researchers scored each leaflet as either intact or damaged. As illustrated in figure 22.20, trees patrolled by the tramp ant suffered less herbivory overall than trees harboring other ant species. In addition, leaves from which the tramp ant was excluded had more damaged leaflets.

In a follow-up experiment, the team observed how tramp ants reacted upon discovering hungry caterpillars on the leaves of "their" trees. The ants took an average of 62 seconds to discover a caterpillar, and the most common reaction was to bite and evict the intruder. Clearly, the ants took an active role in deterring herbivores, presumably to reduce competition for the nectar secreted by the tender young leaves.

Overall, Gaume and the rest of the research team concluded that domatia improve the reproductive success of *H. brunonis* indirectly. Tramp ants that occupy the domatia protect the trees from herbivores. The resulting increased leaf area translates into more photosynthesis; more food could, in turn, promote reproductive output.

That finding stimulated another question: Why do *H. brunonis* trees sometimes lack domatia? No one knows, but the researchers speculate that the production of domatia is a heritable trait and that the hollow stems may have a cost as well as a benefit. For example, stems with domatia may break easily or attract birds that damage the trees while exploring the domatia in search of ant prey. Perhaps variation in the production of domatia is an example of a balanced polymorphism, such as the one that maintains the cystic fibrosis and sickle cell anemia traits in the human population. (i) *balanced polymorphism, section 12.5*

This study helped researchers understand the evolutionary forces that have created ant–plant partnerships. Natural selection has shaped the anatomy of the stem and every other part of the plant body. In this case, the unique hollowed-out stems create a home for ants—an army of tiny watchdogs that protect the plant by chasing off intruders.

Source: Gaume, Laurence, Merry Zacharias, Vladimir Grosbois, and Renee M. Borges. 2005. The fitness consequences of bearing domatia and having the right ant partner: Experiments with protective and non-protective ants in a semi-myrmecophyte. *Oecologia*, vol. 145, pp. 76–86.

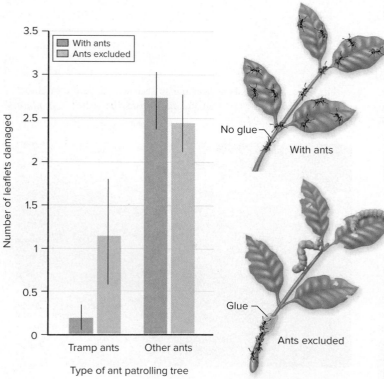

Figure 22.20 Ants on Patrol. When tramp ants were excluded, leaves suffered extra damage from herbivores. Excluding other types of ants did not have an effect. (Error bars represent standard errors; see appendix B.)

22.5   MASTERING CONCEPTS

1. What is the overall question the researchers were asking in this study? Summarize the data they used to arrive at their conclusion.

2. Propose an explanation for the observation that the trees secrete nectar only on young leaves and not on tougher, more mature leaves.

CHAPTER SUMMARY

22.1 Vegetative Plant Parts Include Stems, Leaves, and Roots

- **Anatomy** is the study of an organism's form, and **physiology** is the study of its function. A plant's body consists of **organs** composed of **tissues** with specific functions.
- The **vegetative** plant body includes a **shoot** and roots that depend on each other.
- The **stem** is the central axis of a shoot, which ends in a **terminal bud.** A stem consists of **nodes,** where **leaves** attach, and **internodes** between leaves, where the stem elongates. An **axillary bud** is located at each node.
- Cell division that lengthens root and shoot tips provides **primary growth. Herbaceous plants** typically have soft, green stems; **woody plants** have stems and roots strengthened with wood, the product of **secondary growth.**
- Leaves are the main sites of photosynthesis.
- Roots absorb water and dissolved minerals. **Adventitious roots** arise from stems or leaves.
- Environmental conditions select for modified stems, leaves, and roots.

22.2 Plant Cells Build Tissues

A. Plants Have Several Cell Types

- **Parenchyma** cells are alive at maturity and have thin cell walls. They are relatively unspecialized and often function in metabolism or storage.
- **Collenchyma** cells are also alive. Their thick primary cell walls provide elastic support to growing shoots.
- **Sclerenchyma** cells, including long **fibers** and shorter **sclereids,** are dead at maturity. Their thick secondary cell walls contain **lignin,** supporting plant parts that are no longer growing.
- Water-conducting cells in **xylem** include long, narrow, less specialized **tracheids** and more specialized, barrel-shaped **vessel elements.** Water moves through pits in tracheids and through the end walls of vessel elements. Both cell types have thick, lignin-rich cell walls and are dead when functioning.
- Sugar-conducting cells in **phloem** include **sieve tube elements.** Pores cluster at **sieve plates,** allowing nutrient transport between adjacent cells via strands of cytoplasm. **Companion cells** help transfer carbohydrates into sieve tubes.

B. Plant Cells Form Three Main Tissue Systems

- Most of the primary plant body consists of **ground tissue,** parenchyma cells that fill the space between dermal and vascular tissues.
- **Dermal tissue** includes the **epidermis,** a single cell layer covering the plant. The epidermis secretes a waxy **cuticle** that coats the shoot. Gas exchange in the shoot occurs through **stomata** bounded by **guard cells.**
- **Vascular tissue** is conducting tissue. Xylem transports water, dissolved minerals, and hormones from roots upward. Phloem transports dissolved carbohydrates, hormones, and other substances throughout a plant. Xylem and phloem occur together with other tissues to form **vascular bundles.**
- **Figure 22.21** summarizes the tissue types in a leaf.

22.3 Tissues Build Stems, Leaves, and Roots

A. Stems Support Leaves

- Vascular bundles are scattered in the ground tissue of monocot stems but form a ring of bundles in eudicot stems. Between a eudicot stem's epidermis and vascular tissue lies the **cortex,** made of ground tissue. **Pith** is ground tissue in the center of a stem.

B. Leaves Are the Primary Organs of Photosynthesis

- A stalklike **petiole** supports each leaf **blade.** A **simple leaf** has one undivided blade, and a **compound leaf** forms leaflets. **Veins** are vascular bundles in leaves; they may be in either netted or parallel formation.
- Leaves arise at the flanks of the apical meristem. The cells that make up a leaf's epidermis are tightly packed, transparent, and mostly nonphotosynthetic. Leaf ground tissue includes **mesophyll** cells that carry out photosynthesis. Stomata enable gas exchange.

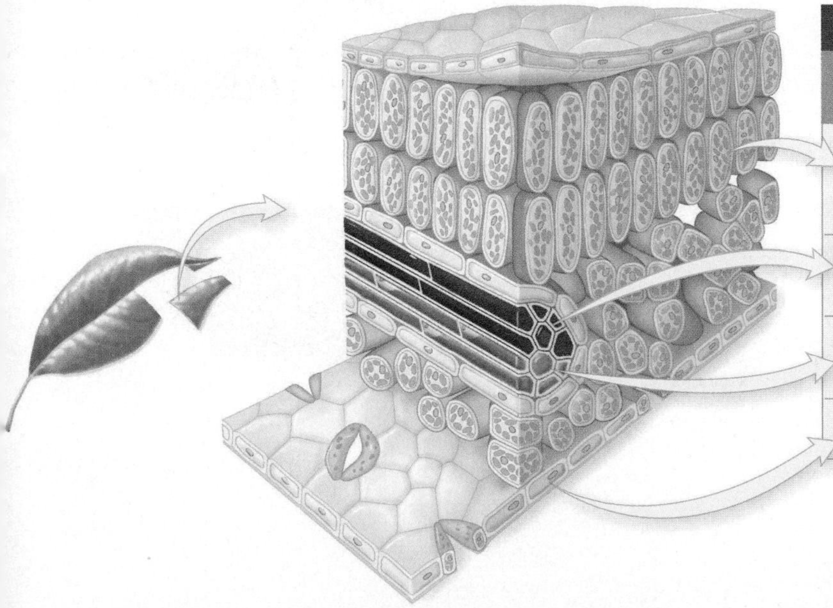

Multiple Tissue Types Interact in a Leaf		
Tissue type	Example(s) of cell types in leaf	Function(s) in leaf
Ground	Parenchyma cells in mesophyll	Carries out photosynthesis
Vascular: Xylem	Tracheids and vessel elements	Transports water and minerals to mesophyll
Vascular: Phloem	Sieve tube elements and companion cells	Transports sugars from mesophyll to rest of plant
Dermal	Epidermal parenchyma, including guard cells	Protects leaf; controls gas exchange

Figure 22.21 Plant Tissue Types: A Summary.

C. Roots Absorb Water and Minerals and Anchor the Plant

- **Fibrous root systems** consist of shallow, branched, relatively fine roots, whereas **taproot systems** have a large, persistent major root with small branches emerging from it.

- A **root cap** protects the tip of a growing root, and **root hairs** behind the cap provide tremendous surface area. The root cortex consists of storage parenchyma and **endodermis.** The waxy **Casparian strip** ensures that the solution entering the xylem first passes through the living cells of the endodermis. The **pericycle** gives rise to branch roots. In monocots, pith fills the center of the root.

- Mycorrhizal fungi help roots take up soil nutrients; nitrogen-fixing bacteria in root nodules provide some plants with nitrogen.

- **Figure 22.22** summarizes the structural differences between monocots and eudicots.

22.4 Plants Have Flexible Growth Patterns, Thanks to Meristems

A. Plants Grow by Adding New Modules

- Plants with **determinate growth** stop growing when they reach their mature size; plants with **indeterminate growth** grow indefinitely.

B. Plant Growth Occurs at Meristems

- **Meristems** are localized collections of cells that retain the ability to divide throughout the life of the plant. **Apical meristems** at the plant's root and shoot tips provide primary growth. **Lateral meristems** are cylinders of cells at the periphery of a stem or root that add girth, or secondary growth. In some plants, **intercalary meristems** occur between nodes.

C. In Primary Growth, Apical Meristems Lengthen Stems and Roots

- Cells in the shoot apical meristem divide to add length; they give rise to the dermal tissue, vascular tissue, and ground tissue in stems and leaves.

- The root apical meristem produces cells that add length as they differentiate into epidermis, cortex, and vascular tissues.

D. In Secondary Growth, Lateral Meristems Thicken Stems and Roots

- The **vascular cambium** increases the girth of the stem or root. It produces **wood** (secondary xylem), secondary phloem, and **rays.**

- The **cork cambium** produces the **periderm,** which makes up the majority of a woody plant's **bark.**

- **Heartwood** is the central, nonfunctional wood in a tree. The light-colored **sapwood** transports water and minerals within a tree.

- Tree rings form in response to seasonal differences in the sizes of conducting cells in wood.

22.5 Investigating Life: An Army of Tiny Watchdogs

- Many species of tropical plants produce hollowed-out stem internodes called domatia, which often house ants. The plants also produce nectar, which is a food source for the ants.

- Some types of ants protect their host plants by chasing off herbivores.

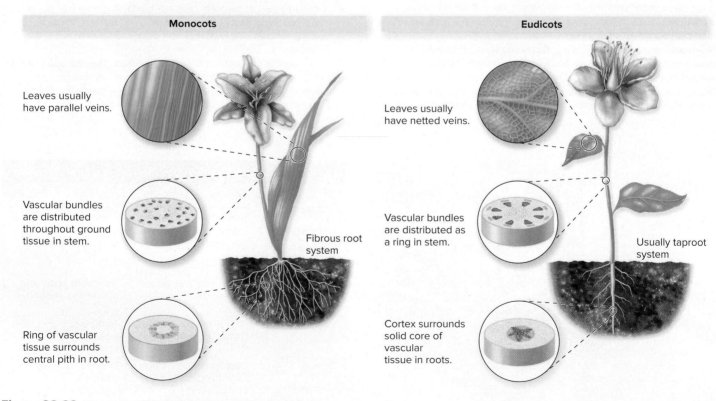

Leaves usually have parallel veins.

Vascular bundles are distributed throughout ground tissue in stem.

Ring of vascular tissue surrounds central pith in root.

Fibrous root system

Leaves usually have netted veins.

Vascular bundles are distributed as a ring in stem.

Cortex surrounds solid core of vascular tissue in roots.

Usually taproot system

Figure 22.22 Monocot and Eudicot Anatomy: A Summary.

MULTIPLE CHOICE QUESTIONS

1. Which of the following is NOT a vegetative organ in a plant?
   a. Stem
   b. Leaf
   c. Flower
   d. Root

2. What is an example of a specialized plant stem that you can buy in the produce section of the grocery store?
   a. Yellow onion
   b. Potato
   c. Carrot
   d. Spinach

3. Which of the following is a supportive cell type with a thick secondary cell wall?
   a. Collenchyma cell
   b. Sclerenchyma cell
   c. Guard cell
   d. Vascular cambium cell

4. Where do tracheids occur in stems?
   a. Ground tissue
   b. Mesophyll
   c. Cortex
   d. Vascular bundles

5. What plant tissue has cells that are dead at maturity?
   a. Xylem
   b. Phloem
   c. Ground tissue
   d. Dermal tissue

6. If you spray herbicide onto a weed, which barrier might prevent the chemical from entering the leaves?
   a. The stomata
   b. The Casparian strip
   c. The cuticle
   d. The pith

7. The ability of a sunflower plant to become taller is directly due to its
   a. apical meristem.
   b. lateral meristems.
   c. mesophyll cells.
   d. petiole.

8. What structures provide abundant surface area for water and mineral absorption?
   a. Leaves
   b. Meristems
   c. Veins
   d. Root hairs

9. Ground tissue occupies the
   a. root cortex.
   b. leaf mesophyll.
   c. stem pith.
   d. All of the above are correct.

10. Which tissue type occupies most of the volume of a woody stem? *Hint:* See figure 22.18.
    a. Secondary xylem
    b. Secondary phloem
    c. Cork cambium
    d. Vascular cambium

Answers to these questions are in appendix A.

WRITE IT OUT

1. List the main vegetative organs of a plant, and explain how each relies on the others.

2. Thorns, spines, and tendrils are so highly modified that it can be difficult to tell whether they derive from leaves or stems. How could a biologist use his or her knowledge of plant anatomy to determine their origin?

3. Imagine you are conducting an experiment on plant growth. You buy two seedlings, plant them in identical soil, and water them the same amount each week. Each week you also measure the plants' heights. For the first 2 months, the plants grow at the same rate. Then, one plant continues growing while the growth of the other plant slows and eventually stops. Propose a possible cause for this observation.

4. Many biology labs use slides of root tips to demonstrate the stages of mitosis. Why is this a better choice than using a slide of a mature leaf?

5. Write nonbiological analogies for vessel elements, phloem, stomata, and lignin.

6. Propose why it might be more adaptive for a tree to produce thousands of small leaves rather than one huge leaf.

7. Suppose you find a flowering plant that has leaves with netted veins. Using only this information, what inferences can you make about the internal anatomy of the plant?

8. List the structures that help a plant maintain water homeostasis.

9. Explain how conditions in the terrestrial environment selected for each of the following adaptations: cuticle, stomata, vascular tissue, roots, stems, and leaves.

10. Draw a diagram that includes the following labels: leaf, petiole, axillary bud, node, internode.

11. The Cork Forest Conservation Alliance is a nonprofit environmental organization dedicated to preserving the cork forests of the Mediterranean. Visit its website, and then explain to a friend why buying wine closed with natural cork is more environmentally friendly than buying wine closed with a screw cap.

12. Suppose you drive a metal spike from the outermost bark layer to the center of a tree's trunk. Which tissues does your spike encounter as it moves through the stem? What type of meristem produced each type?

13. List a function of each organ, tissue, and cell type described in this chapter, and then list at least one feature that facilitates that function.

PULL IT TOGETHER

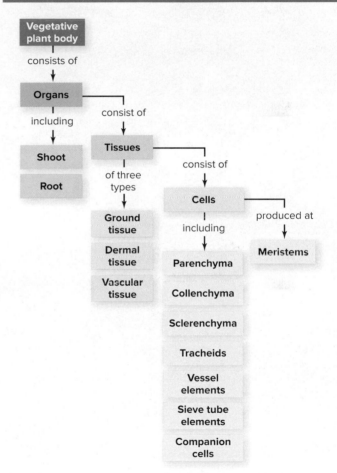

Figure 22.23 Pull It Together: Plant Form and Function.

Refer to figure 22.23 and the chapter content to answer the following questions.

1. Review the Survey the Landscape figure in the chapter introduction, and then add *flowering plants, reproductive parts,* and *flowers* to the Pull It Together concept map. Make a connection between *flowers* and at least one vegetative part.

2. Add the terms *cortex, pith, mesophyll, endodermis,* and *Casparian strip* to this concept map.

3. Connect *monocots* and *eudicots* to *vascular tissue.* Your connecting phrases should indicate a difference between the two plant groups.

23

Plant Nutrition and Transport

Carnivorous Plant. A sundew's leaves bear dozens of tentacles, each of which secretes sticky droplets that attract the plant's prey.
©Matauw/iStock/Getty Images RF

LEARN HOW TO LEARN
A Quick Once-Over

Unless your instructor requires you to read your textbook in detail before class, try a quick preview. Read the chapter outline to identify the main ideas; then look at the figures and the key terms in the narrative. Previewing a chapter should help you follow the lecture, because you will already know the main ideas. In addition, note taking will be easier if you recognize new vocabulary words from your quick once-over. Return to your book for an in-depth reading after class to help nail down the details.

Carnivorous Plants

Little Shop of Horrors tells the story of a plant that grows to enormous size and develops a taste for humans. Fortunately, such bloodthirsty plants exist only in science fiction—right?

Although it is true that no plants eat people, some do consume small animals, especially insects and other arthropods. The following examples illustrate just a few of the world's 450 or so species of carnivorous plants:

- The Venus flytrap has hinged leaves that fold in half lengthwise, like a clamshell (see figure 22.3b). The upper surface of each leaf bears highly sensitive trigger hairs. When an insect wanders onto a leaf and bends the trigger hairs, the two halves snap shut. The leaf's epidermis secretes enzymes that digest the captured animal. These plants are endangered in the wild because of habitat loss, and collecting them is illegal. (Most Venus flytraps sold commercially are propagated asexually in greenhouses.)

- The sundew plant has paddle-shaped leaves with tiny, nectar-covered hairs that lure insects. Once the insect lands on the plant and starts dining on its sweet, sticky meal, the surrounding leaf hairs begin to move. Gradually they fold inward, entrapping the helpless visitor and forcing it down toward the leaf's center. Here, powerful digestive enzymes dismantle the insect's body and release its nutrients. Afterward, all that remains of the guest are a few indigestible bits.

- The pitcher plant produces nectar along the rim of a slippery, tube-shaped leaf. Rainwater collects at the bottom of the trap. When a hapless insect falls in and drowns, an army of bacteria, protists, rotifers, mosquito larvae, and other invertebrates goes to work, decomposing the victim. The plant absorbs the nutrients through the thin cuticle lining the pitcher tube. (Section 23.5 examines pitcher plant nutrition in more detail.)

It may seem strange that a plant would consume insects. Plants are, after all, photosynthetic, so they can make their own food. The answer to this puzzle is that carnivorous plants use insects as a source of nitrogen or phosphorus, not energy. Carnivorous plants are most abundant in boggy, acidic soils containing few nutrients. The leafy traps of carnivorous plants are adaptations that extract minerals from animals instead of from the soil. As this chapter describes, however, most plants have less exotic ways of acquiring nutrients.

SURVEY THE LANDSCAPE
Plant Life

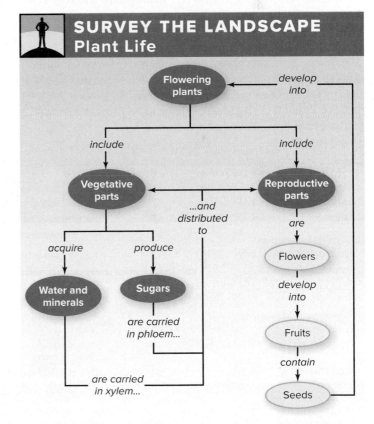

The typical plant body is divided into underground roots and aboveground shoots, each acquiring resources from its part of the plant's habitat. A transportation network composed of xylem and phloem "pipes" supports this division of labor.

For more details, study the Pull It Together feature in the chapter summary.

23.1 Soil and Air Provide Water and Nutrients

To stay healthy, a person needs water and the right dietary mix of fat, protein, carbohydrates, vitamins, and minerals (see section 32.4). Plants have similar needs, but they do not acquire these raw materials by eating and drinking. Instead, they are autotrophs that use photosynthesis to assemble organic molecules from elements absorbed from their surroundings.

This section describes the sources of the elements that plants require. The rest of the chapter focuses on how plants absorb and transport water, minerals, and organic substances such as sugar within their tissues.

A. Plants Require 16 Essential Elements

Like every organism, a plant requires certain **essential nutrients,** which are chemicals that are vital for metabolism, growth, and reproduction. Biologists have identified at least 16 elements essential to all plants (**figure 23.1**). Nine are **macronutrients,** meaning that they are needed in fairly large amounts. The macronutrients are carbon (C), oxygen (O), hydrogen (H), nitrogen (N), potassium (K), calcium (Ca), magnesium (Mg), phosphorus (P), and sulfur (S). The others are **micronutrients,** which are required in much smaller amounts.

Among the essential elements, C, H, and O are by far the most abundant, together accounting for about 96% of the dry weight of a plant. The six other macronutrients account for another 3.5%. Of these, N, P, and K are the most common ingredients in commercial fertilizers (see this chapter's Apply It Now box).

Gardeners and farmers use fertilizers to prevent or treat nutrient deficiencies such as those shown in **figure 23.2**. At the other end of the nutritional spectrum are plants that accumulate unusually high concentrations of some elements, such as zinc or nickel. This "hyperaccumulation" of heavy metals is an adaptation that reduces herbivory.

B. Soils Have Distinct Layers

Plants can grow in containers of water with dissolved nutrients, a technique called hydroponics. Most roots, however, extract water and mineral nutrients from **soil,** a complex mixture of rock particles, organic matter, air, and water. Soil is also home to bacteria, archaea, fungi, protists, and animals. The microbes decay organic matter, releasing inorganic nutrients that plant roots absorb.

Most people give little thought to the ground under their feet, but soil is critical to ecosystem function. Plants growing in soil provide food, habitat, and oxygen to humans and other animals. In addition, rain and melting snow seep into soil, which releases the moisture slowly, preventing flooding. Wastewater is also purified as it trickles through soil.

Soil formation begins when physical and chemical processes cause rocks to "weather," or disintegrate into small particles. Sandy soils have coarse particles; silty soils have finer ones. Clay particles are the smallest. Soil scientists classify a soil's texture based on the relative amounts of sand, silt, and clay. Soil texture

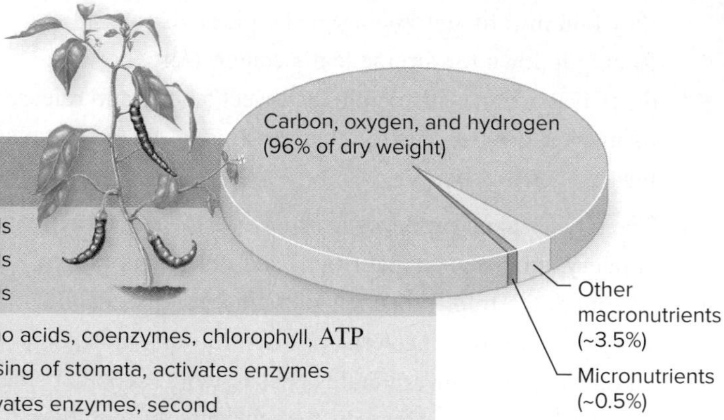

Macronutrients	Form Taken Up by Plants	Percent Dry Weight	Selected Functions
Carbon (C)	CO_2	45	Part of organic compounds
Oxygen (O)	H_2O, O_2, CO_2	45	Part of organic compounds
Hydrogen (H)	H_2O	6	Part of organic compounds
Nitrogen (N)	NO_3^-, NH_4^+	1.5	Part of nucleic acids, amino acids, coenzymes, chlorophyll, ATP
Potassium (K)	K^+	1.0	Controls opening and closing of stomata, activates enzymes
Calcium (Ca)	Ca^{2+}	0.5	Cell wall component, activates enzymes, second messenger in signal transduction, maintains membranes
Magnesium (Mg)	Mg^{2+}	0.2	Part of chlorophyll, activates enzymes, participates in protein synthesis
Phosphorus (P)	$H_2PO_4^-$, HPO_4^{2-}	0.2	Part of nucleic acids, sugar phosphates, ATP, coenzymes, phospholipids
Sulfur (S)	SO_4^{2-}	0.1	Part of cysteine and methionine (amino acids), coenzyme A

Micronutrients	Form Taken Up by Plants	Percent Dry Weight	Selected Functions
Chlorine (Cl)	Cl^-	0.01	Water balance
Iron (Fe)	Fe^{3+}, Fe^{2+}	0.01	Chlorophyll synthesis, cofactor for enzymes, part of electron carriers
Boron (B)	BO_3^-, $B_4O_7^{2-}$	0.002	Growth of pollen tubes, sugar transport, regulates certain enzymes
Zinc (Zn)	Zn^{2+}	0.002	Hormone synthesis, activates enzymes, stabilizes ribosomes
Manganese (Mn)	Mn^{2+}	0.005	Activates enzymes, electron transfer, photosynthesis
Copper (Cu)	Cu^{2+}	0.0006	Part of plastid pigments, lignin synthesis, activates enzymes
Molybdenum (Mo)	MoO_4^{2-}	0.00001	Nitrate reduction

Carbon, oxygen, and hydrogen (96% of dry weight)

Other macronutrients (~3.5%)

Micronutrients (~0.5%)

Figure 23.1

Essential Nutrients for Plants. The nine most abundant elements in plants are macronutrients; the seven micronutrients occur in much lower concentrations.

a.   b.

Figure 23.2 Nutrient Deficiencies. (a) Phosphorus deficiency causes plants to develop purplish leaves. (b) Iron deficiency causes yellowed leaves, but the veins remain green.

(both): ©1991 Regents University of California Statewide IPM Project

is important to plant growth. The finer the particles, the more surface area per unit of soil volume and the higher the soil's capacity to store water. A soil with too much sand therefore dries out quickly, whereas a clay soil easily becomes waterlogged. Most plants grow best in a more balanced soil.

Litter

Topsoil
(A horizon)

Minerals
and clay
(B horizon)

Partially
weathered
rock
(C horizon)

Figure 23.3 Soil Horizons. Soil consists of layers called horizons. Above the rich topsoil (the A horizon) is a litter layer. Water passing through the topsoil deposits clay and minerals in the B horizon, below which is partially weathered rock, the C horizon.

Photo: ©Stramyk/iStock/Getty Images RF

The disintegrating rock particles interact with water, plants, and other organisms. Over time, distinct soil layers develop (figure 23.3). Decomposing leaves and stems, called litter, lie on the soil's surface. Microbes decay the litter and release most of the carbon to the atmosphere as carbon dioxide (CO_2). Some carbon, however, remains in the upper layer of soil as **humus,** a chemically complex, hard-to-digest, spongy organic substance. Most humus is in the upper soil layer, the **topsoil,** also called the A horizon. Topsoil typically supplies most of a plant's water and nutrients, and it may be anywhere from a few centimeters to several meters thick. Below topsoil is the B horizon, which has less organic matter, although roots extend to this depth. Still lower is the C horizon, which consists mostly of partially weathered pieces of rocks and minerals. Below the C horizon is bedrock.

Plant roots normally stabilize the topsoil and prevent **erosion,** the "wearing away" of soil by water and wind. Construction, overgrazing, farming, logging, road building, and many other human activities remove plants and therefore promote erosion. Bare soil washes away, choking streams and lakes with sediment. Few plants grow in the absence of topsoil, so ecosystem recovery is slow in badly eroded soils.

Soil properties vary around the world; a rich prairie soil, for example, is very different from a desert soil or Arctic permafrost. Why? Part of the answer is that the mineral composition of the underlying bedrock varies. In addition, climate alters how minerals weather. Hot temperatures speed weathering, and rainfall leaches water-soluble nutrients through the soil. Climate also influences the plants and other organisms that shape the soil's properties.

C. Leaves and Roots Absorb Essential Elements

Plants obtain their three most abundant elements (C, H, and O) from water and the atmosphere. Water (H_2O) enters the plant through the roots, as described in section 23.2. Carbon and oxygen atoms come from the atmosphere in the form of CO_2 gas, which diffuses into the leaf or stem through pores called stomata. ⓘ *diffusion,* section 4.5A

As roots absorb water, they also take up all of the other mineral elements listed in figure 23.1. These nutrients dissolve in the soil's water when rock particles disintegrate or when microbes decompose dead organisms.

Plants often have help from soil organisms in obtaining water and nutrients. Recall from chapter 20 that the roots of most land plants are colonized with mycorrhizal fungi. In exchange for sugars produced in photosynthesis, the fungal filaments explore the soil and absorb water and minerals that the roots could not otherwise reach. In particular, phosphorus is poorly soluble in water and does not move easily to roots. Mycorrhizae therefore especially boost phosphorus absorption. ⓘ *mycorrhizae,* section 20.7B

The source of nitrogen also deserves special mention. As described in chapter 2, nitrogen atoms occur in proteins, nucleic acids, and chlorophyll. Nitrogen gas (N_2) makes up 78% of the

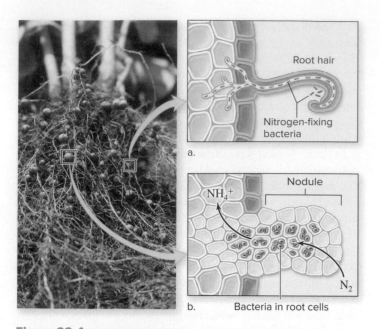

a.

b.   Bacteria in root cells

Figure 23.4 **Root Nodules.** (a) Nitrogen-fixing bacteria enter through root hairs and trigger nodule development. (b) In a mature nodule, the bacteria live symbiotically within the root's cells. The plant provides the energy the bacteria need to fix nitrogen.
Photo: ©Kelly Marken/Shutterstock RF

atmosphere, but a strong triple covalent bond holds the two nitrogen atoms together. Plants cannot break this bond and therefore cannot use N_2 directly. Instead, roots must take up nitrogen from soil in the form of nitrate (NO_3-) or ammonium (NH_4+). These ions can be scarce, so nitrogen availability often limits plant growth.

Fortunately for plants (and ultimately animals), several types of bacteria use **nitrogen-fixing** enzymes to convert N_2 to NH_4^+. Many nitrogen-fixing bacteria are free-living; others live in **nodules,** small swellings on the roots of some types of plants (figures 23.4 and 17.12). The bacteria consume sugars that the host plant produces by photosynthesis. The plant, in turn, incorporates the nitrogen atoms from NH_4^+ into its own tissues. When the plant dies, decomposers make the nitrogen available to other organisms. The first step—nitrogen fixation—is therefore the key to the entire nitrogen cycle, bringing otherwise inaccessible nitrogen to plants, microbes, and all other life (see figure 38.21 for a detailed look at the nitrogen cycle).

The most famous nitrogen-fixing bacteria, in genus *Rhizobium,* stimulate nodule formation on the roots of plants called legumes (clover, beans, peas, peanuts, soybeans, and alfalfa). A farming technique called crop rotation alternates legume crops with nitrogen-hungry plants such as corn or cotton. The legume replenishes the soil's nitrogen, reducing the need for fertilizer.

A few plants acquire nutrients from more unusual sources. For example, carnivorous plants, described in the chapter opening essay, often live in waterlogged, acidic soils where nutrients are scarce. They obtain nitrogen and phosphorus from their prey.

23.1   MASTERING CONCEPTS

1. Which macro- and micronutrients do all plants require?
2. How do plants acquire C, H, O, N, and P?
3. How would planting nodule-producing legumes be useful in crop rotations?

 Apply It **Now**

Boost Plant Growth with Fertilizer

Farmers and gardeners often amend soil with commercial synthetic fertilizers or with nutrient-rich organic matter, such as manure or compost. Plant growth surges if the added material provides a nutrient that was previously scarce.

Commercial fertilizer labels prominently display three numbers that indicate the content of nitrogen, phosphorus, and potassium (figure 23.A). These are the three elements that are most commonly deficient in soils. For example, a product marked 6-12-6 contains 6% elemental nitrogen (N), 12% phosphate (P_2O_5), and 6% potash (K_2O) by weight; the rest is filler. The label also lists other macro- and micronutrients in the fertilizer.

Nitrogen promotes leaf growth, whereas potassium and phosphorus encourage flowering and fruit development. Lawn fertilizer sold at the hardware store might therefore have an N-P-K ratio of 29-3-4. A fertilizer for vegetables would be more balanced at 10-20-20. Both types of synthetic fertilizers deliver an efficient, highly concentrated shot of nutrients to the soil. By comparison, household compost and cow manure are balanced but not very concentrated, with an N-P-K ratio of about 0.5-0.5-0.5. These nutrients are released slowly and in low doses.

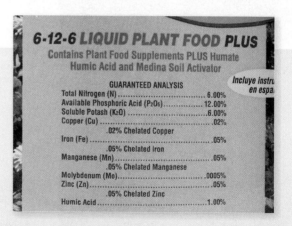

Figure 23.A **Read the Label.** Commercial fertilizer labels show the concentrations of each nutrient in the product.
©David Tietz/Editorial Image, LLC

Overuse of synthetic fertilizers can dehydrate plants and pollute waterways; these risks are lower with compost and manure. Moreover, soil organic matter not only boosts aeration but also retains moisture. Many environmentally conscious farmers therefore avoid synthetic fertilizers in favor of organic alternatives.

23.2 Water and Minerals Are Pulled Up to Leaves in Xylem

As you learned in chapter 22, xylem and phloem together form a continuous system of **vascular tissue,** a transportation network that connects the plant's roots, stems, leaves, flowers, and fruits (**figure 23.5**). This section describes how **xylem** transports **xylem sap,** a dilute solution consisting mostly of water and dissolved minerals absorbed from soil. Note that understanding xylem transport requires some background knowledge about water and osmosis. It might therefore be useful to review sections 2.3A and 4.5A before reading this section.

A. Water Evaporates from Leaves in Transpiration

The amount of water that passes through plants each day is staggering. The corn plants occupying an area the size of a football field, for example, might use nearly 20,000 liters in one summer day, and a typical tree might consume 265 liters daily. Throughout a growing season, a plant uses 200 to 1000 liters of water to produce just 1 kilogram of tissue.

Why so much water? A plant's cells are mostly water, which is a medium for most of its metabolic reactions. Water participates in some of those reactions, including hydrolysis and photosynthesis. Furthermore, the watery cytoplasm exerts turgor pressure on the cell wall, which enables plant cells to elongate and helps the entire plant to stay upright. Finally, the surfaces of leaf mesophyll cells must remain moist for CO_2 to diffuse inside. These uses, however, add up to only a small fraction of the water that a plant's roots pull in. The rest simply evaporates. (i) *hydrolysis,* section 2.5A; *turgor pressure,* section 4.5A

The easiest way to visualize how plants acquire and transport water is to begin at the end. Plants lose water through transpiration, the evaporation of water from a leaf. Heat from the sun causes water in the cell walls to evaporate into the spaces between the leaf's cells. This evaporation helps cool the leaf, but it also establishes a gradient. That is, the concentration of water molecules inside the leaf is higher than in the air surrounding the leaf. Water vapor therefore diffuses from inside the leaf to the outside air. Most transpiration occurs through pores called stomata, but some water escapes through the cuticle that coats the epidermis.

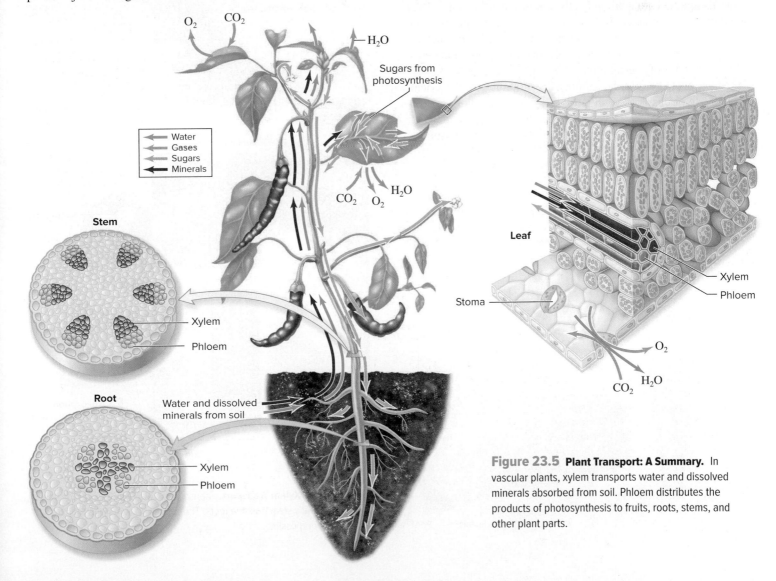

Figure 23.5 Plant Transport: A Summary. In vascular plants, xylem transports water and dissolved minerals absorbed from soil. Phloem distributes the products of photosynthesis to fruits, roots, stems, and other plant parts.

The bigger the concentration gradient between the leaf interior and the surrounding air, the faster the transpiration rate. Low humidity, wind, and high temperatures therefore all cause the transpiration rate to soar. As described in section 23.2D, however, the plant's stomata close when air is too hot or too dry. When the stomata close, transpiration slows—and so does photosynthesis. But because plants store extra food as starch, a temporary drop in sugar production causes far less harm than does a rapid loss of water.

B. Water and Dissolved Minerals Enter at the Roots

Water and dissolved minerals can travel through the root's epidermis and cortex in two ways (figure 23.6). In the extracellular pathway, the solution moves in the spaces between cells or along the cell walls. Alternatively, water can move from cell to cell via plasmodesmata in the intracellular pathway. (i) *plasmodesmata,* section 3.6B

The solution flows through the outer portion of the root until it reaches the endodermis, the innermost layer of the cortex. At that point, a waxy barrier called the **Casparian strip** forces the materials that had gone around cells to now enter the cells of the endodermis. Water enters by osmosis, because the concentration of solutes in cells is generally higher than in the soil. But not all solutes can enter the endodermal cells, thanks to membrane proteins that admit only certain ions. (i) *osmosis,* section 4.5A

Materials that cross the endodermis continue into the xylem and enter the transpiration stream. The water eventually returns to the atmosphere through open stomata in the leaves and stem; the dissolved nutrients are incorporated into the plant's tissues.

C. Xylem Transport Relies on Cohesion

The functional cells of the xylem—tracheids and vessel elements—are dead at maturity (see figure 22.5). Metabolic activity in xylem cells therefore cannot drive water transport in a plant. So how does water in the xylem get from roots to the mesophyll in leaves?

The **cohesion–tension theory** explains how xylem sap moves within a plant (figure 23.7). As its name implies, the cohesion–tension theory hinges on the cohesive properties of water—the tendency for water molecules to form hydrogen bonds and "cling" together. As water molecules evaporate from the leaf, additional water diffuses out of leaf veins and into the mesophyll (figure 23.7, step 1). Water molecules leaving the vein attract molecules adjacent to them, pulling them under tension toward the vein ending. Each water molecule tugs on the one behind it (step 2). Finally, additional water enters roots from the soil through root hairs (step 3). These specialized cells contribute enormous surface area for water and mineral absorption. (i) *cohesion,* section 2.3A

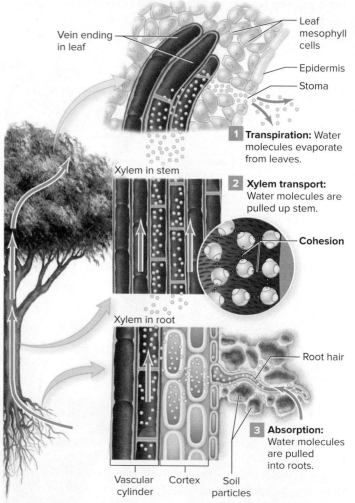

Figure 23.7 Xylem Transport. Transpiration of water from leaves pulls water up a plant's stem from the roots. The cohesiveness of water makes xylem transport possible.

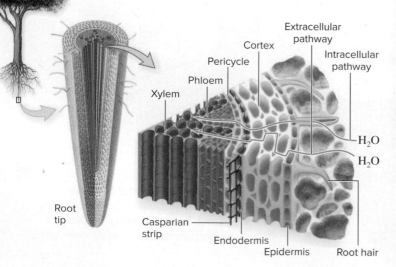

Figure 23.6 Two Routes into Roots. Water and dissolved minerals can travel along two routes through a root's epidermis and cortex: in the spaces between cells or through living cytoplasm. The waxy Casparian strip ensures that all incoming material passes through the living cells of the endodermis to reach the xylem.

As long as sufficient moisture is available in the soil, the cohesion between water molecules is enough to move continuous, narrow columns of xylem sap upward against the force of gravity. This mechanism exploits the physical properties of water and requires no energy input from the plant.

What if the soil is too dry to replace water lost in transpiration? The tension on the xylem sap may become so great that the water column stretches and breaks in one or more vessels and tracheids. This is a serious problem, because water molecules below an air bubble can no longer be pulled upward by the molecules evaporating from the leaf. Water movement therefore stops. The plant's cells quickly lose turgor, and the leaves wilt. This interruption in xylem flow explains why cut flowers droop when their stems are removed from water.

Most plants face a trade-off between efficient water movement and the risk of air bubble formation; xylem anatomy reflects this balance. Recall from chapter 22 that tracheids occur in all vascular plants, but only angiosperms have vessels. When water is plentiful, a continuous vessel conducts xylem sap more efficiently than do the narrow tracheids with their overlapping end walls. But when water is scarce, an air bubble can disable an entire vessel, which may extend for several meters. Tracheids, on the other hand, function individually. An air bubble that forms in a tracheid therefore disrupts the function of just that one cell.

The observation that air bubbles can interrupt xylem function provides evidence supporting the cohesion–tension theory of water movement. If water were being pushed from below, air bubbles in the xylem stream would not pose a problem.

D. The Cuticle and Stomata Help Conserve Water

The **cuticle** is an important water-saving adaptation in land plants. This waxy layer covers the epidermis of the leaves and primary stem (see figure 22.7). Impermeable to water and gases, the cuticle prevents the plant's tissues from drying out.

In addition, **stomata** are pores that permeate the cuticle and permit the leaf to exchange gases with the atmosphere. As long as the soil contains enough moisture to replace water lost in transpiration, the plant will flourish. But if the soil dries out, the plant will wilt—unless it closes its stomata and temporarily shuts down the transpiration stream.

A pair of **guard cells** determines whether each stoma is open or closed (**figure 23.8**). Note that the inner wall of each guard cell—the wall facing the pore—is thick and elastic; the outer walls are much thinner. When water is plentiful, membrane proteins in the guard cells import potassium ions (K^+) from adjacent cells in the epidermis, a process that costs ATP. The concentration of K^+ therefore becomes greater inside than outside the guard cells. Water follows by osmosis. The incoming water causes the thin outer walls of the guard cells to balloon outward, which forces the thick inner walls to curve outward as well. The stoma opens.

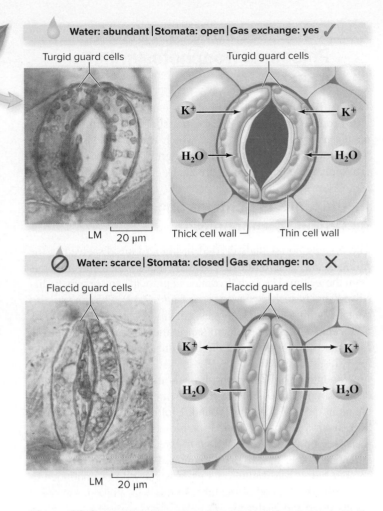

Figure 23.8 Guard Cell Function. When water is abundant, guard cells use energy to import potassium ions (K^+) from adjacent cells. Water follows by osmosis; the guard cells swell, opening the stoma. When water is scarce, hormonal signals stimulate K^+ to leave the guard cells. Water follows, and the stoma closes as the flaccid guard cells collapse.

Photos: (both): ©Ray Simons/Science Source

During drought stress, on the other hand, a plant hormone called abscisic acid binds to the guard cell membranes. The hormone indirectly triggers the loss of K^+ from the guard cells. As water exits the guard cells by osmosis, the cells lose turgor and collapse against each other. The pore between them closes, and the transpiration stream stops. ⓘ *abscisic acid,* section 24.4B

A plant that closes its stomata, however, also cuts off its supply of CO_2 for use in photosynthesis. Most plants therefore conserve water by closing their stomata after dark, when photosynthesis cannot occur anyway. The C_4 and CAM pathways of photosynthesis also save water. ⓘ *C_4 and CAM plants,* section 5.6

23.2 MASTERING CONCEPTS

1. What are the components of xylem sap?
2. Trace the path of water and dissolved minerals from soil, into the root's xylem, and up to the leaves.
3. How do the cuticle and stomata help plants conserve water?

23.3 Sugars Are Pushed in Phloem to Nonphotosynthetic Cells

With sufficient light, water, and nutrients, a photosynthetic cell will produce sugars that can be transported in **phloem** to the plant's nonphotosynthetic cells, which cannot produce food on their own. The major transport structures of phloem are the microscopic sieve tubes (see figure 22.5).

A. Phloem Sap Contains Sugars and Other Organic Compounds

The organic compounds carried in phloem are dissolved in the **phloem sap,** a solution that also includes water and minerals from the xylem. The carbohydrates in phloem sap are mostly dissolved sugars such as sucrose (see figure 2.18). Phloem sap also contains amino acids, hormones, enzymes, and messenger RNA molecules. (Although phloem sap is the most common vehicle for sugar transport, it is not the only one, as this chapter's Burning Question explains.)

Studying phloem composition is difficult. An investigator cannot simply cut a stem and squeeze out the phloem sap because the plant quickly plugs the wound. Biologists receive help, however, from a surprising source: aphids (**figure 23.9**). These insects feed on phloem sap without triggering the wound response. Scientists found that if they amputated the feeding tube of an aphid while it was dining on a plant, sugary drops continued to flow from the cut end of the tube. In this way, phloem sap could be harvested and analyzed.

B. The Pressure Flow Theory Explains Phloem Function

The explanation of phloem transport is called the **pressure flow theory,** which suggests that phloem sap moves under positive pressure from "sources" to "sinks." A **source** is any plant part that produces or releases sugars; a **sink** is any plant part that receives these sugars. Sinks, which are typically nonphotosynthetic, include flowers, fruits, shoot apical meristems, roots, and

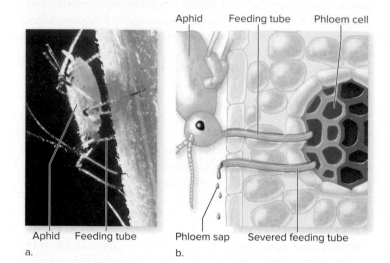

Aphid   Feeding tube   Phloem cell

Aphid   Feeding tube
a.

Phloem sap   Severed feeding tube
b.

Figure 23.9   Phloem Feeders. (a) An aphid is an insect with a feeding tube that penetrates a plant's sugar-rich phloem. (b) If the feeding tube is severed, phloem sap flows out of the tube under pressure. Researchers collect this fluid to study the sap's composition.

(a): ©Oxford Scientific/Getty Images

---

storage organs. If these cells do not receive enough sugar to generate the ATP they require, the plant may die or fail to reproduce. ⓘ *meristems,* section 22.4

Figure 23.10 illustrates the pressure flow theory. Inside a leaf or other sugar source, companion cells load sucrose into sieve tube elements by active transport (step 1). Because sucrose becomes so much more concentrated in the sieve tubes than in the adjacent xylem, water moves by osmosis out of the xylem and into the phloem sap (step 2). The resulting increased turgor pressure drives phloem sap through the sieve tubes (step 3). ⓘ *active transport,* section 4.5B

At a root, flower, fruit, or other sink, cells take up the sucrose (and other compounds in the phloem sap) through facilitated diffusion or active transport (step 4). As the sucrose is unloaded from the sieve tubes, the concentration of solutes in the phloem sap declines. Water therefore moves by osmosis from the sieve tube to the surrounding tissue, which is often xylem (step 5). Movement of water out of the sieve tube relieves the pressure, so the phloem sap in the sieve tube continues to flow toward the sink.

A given organ may act as either a sink or a source. For example, a developing potato tuber is a sink, storing the plant's sugars in the form of starch. Later, when the plant uses those stored

reserves to fuel the growth of new tissues, the same tuber becomes a source. The starch in the tuber breaks down into simple sugars, which are loaded into phloem sap for transport to other plant parts.

A similar mobilization occurs each spring when a deciduous tree produces new stems and leaves, using carbohydrates stored in roots. Later in the growing season, leaves approach their mature size and produce sugar of their own. The leaves are then sources, and the roots are again sinks.

Evidence for the pressure flow theory includes the observation that phloem sap moves through aphids under positive pressure, a little like toothpaste being squeezed from a tube. Microscopes that produce images of living tissue "in action" have also confirmed some details of phloem function.

23.3 MASTERING CONCEPTS

1. What are the components of phloem sap?
2. Explain the pressure flow theory of phloem transport.
3. Distinguish between a source and a sink. How can the same plant part act as both a source and a sink?

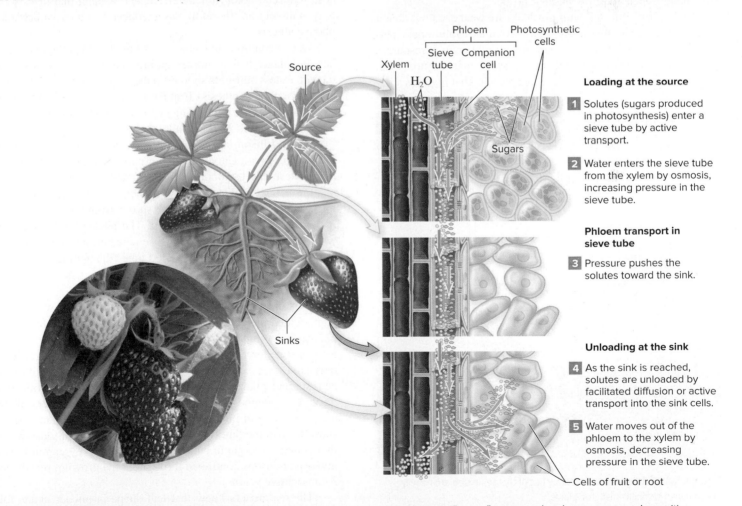

Loading at the source

1 Solutes (sugars produced in photosynthesis) enter a sieve tube by active transport.

2 Water enters the sieve tube from the xylem by osmosis, increasing pressure in the sieve tube.

Phloem transport in sieve tube

3 Pressure pushes the solutes toward the sink.

Unloading at the sink

4 As the sink is reached, solutes are unloaded by facilitated diffusion or active transport into the sink cells.

5 Water moves out of the phloem to the xylem by osmosis, decreasing pressure in the sieve tube.

Figure 23.10 Phloem Transport. According to the pressure flow theory, sugar produced in green "source" organs such as leaves moves under positive pressure to roots, fruits, and other nonphotosynthetic "sinks."

Photo: ©Ingram Publishing RF

23.4 Parasitic Plants Tap into Another Plant's Vascular Tissue

Of the hundreds of thousands of plant species, most are self-sufficient. They produce their own food by photosynthesis, and they absorb their own nutrients and water from soil (often with the help of mycorrhizal fungi). Some plant species, however, are parasites that exploit the hard-won resources of other plants.

Parasitic plants acquire water, minerals, and food by tapping into the vascular tissues of their hosts. The story begins with the parasite's seeds, which are carried to the host by birds or released explosively from seed pods. Either way, a seed germinates, and the seedling secretes an adhesive that sticks the young plant to its host. The seedling's root pushes through the host's epidermis and connects the parasite's vascular tissues to those of the host.

Dodder is one example of a parasitic plant. Its leafless stems, which lack chlorophyll, drape over host plants like cooked spaghetti. The most common parasites, however, are the many species of mistletoe. These dark green shrubs live in the branches of host trees throughout the United States (**figure 23.11**). Mistletoe produces some of its own food by photosynthesis, but other parasitic plants depend entirely on their hosts.

Most plants infected with mistletoe are weakened but do not die, an observation that makes sense from an evolutionary perspective. The most successful parasites extract enough resources to survive and reproduce—but not so much that the host dies. After all, a dead plant is of no use to a parasite that requires a living host.

23.4 | MASTERING CONCEPTS

1. Explain why the only effective treatment for mistletoe is to prune off infected branches.
2. Briefly describe how a parasitic plant infects a host.

Mistletoe plants

Figure 23.11 Mistletoe. The branches of this apple tree are heavily infested with parasitic mistletoe plants.
©Mark Boulton/Alamy

INVESTIGATING LIFE

23.5 The Hidden Cost of Traps

Carnivorous plants seem to have it made: They live in boggy habitats with plentiful water, their invertebrate prey provide nitrogen and phosphorus, and they can use energy from sunlight to produce their own food.

But if the carnivorous lifestyle is so easy, why don't all plants eat bugs? After all, invertebrates are so abundant that natural selection should favor plants that exploit animal sources of nutrients. Yet rather than taking over the world, most carnivorous plants are limited to sunny, nutrient-poor wetlands.

Biologists hypothesize that these plants cannot compete in most other habitats because carnivory involves an evolutionary trade-off. To catch an animal, a plant must construct a trap. These structures vary from a sticky surface (as in sundews) to a fluid-filled chamber (as in pitcher plants) to a clamshell-shaped snap trap (as in Venus flytraps) to a hollow bladder with a door that closes after an insect enters (as in bladderworts). Although the plant uses the nutrients extracted from the prey, the trap has a cost: reduced photosynthetic efficiency. A plant that invests energy in insect traps therefore has a reduced leaf area available for photosynthesis.

One prediction that emerges from this hypothetical cost–benefit balance is that supplementing a carnivorous plant's habitat with excess nutrients should tip the balance toward leaves that maximize photosynthesis. That is, if a carnivorous plant has sufficient nutrients, then investing in photosynthetic leaf area should benefit the plant more than building new traps.

Aaron Ellison, a biologist at Harvard Forest in Massachusetts, worked with University of Vermont biologist Nicholas Gotelli to test this prediction. They studied the northern pitcher plant, *Sarracenia purpurea* (**figure 23.12**). The prey-capturing pitchers are modified leaves that sprout from the underground rhizomes of this carnivorous plant. The pitcher's tube develops when the edges of a folded leaf fuse together; a flaplike keel reinforces the seam. Once the sealed tube fills with rainwater, invertebrates can fall in, drown, and decay. The pitchers absorb the nutrients through a thin cuticle.

Ellison and Gotelli set up a field experiment in which they manipulated the concentrations of nutrients available to pitcher plants. Ninety adult plants were randomly assigned to one of nine treatments: two controls (distilled water or micronutrients only) and seven solutions containing different amounts of nitrogen and phosphorus (**figure 23.13a**). Every 2 weeks from June through September, the researchers poured 5 milliliters of the assigned solution into every open pitcher of each plant. Afterward, they immediately plugged each pitcher with glass wool; this cottony, inorganic substance prevented prey capture. The nutrient additions continued throughout the growing seasons of 3 consecutive years.

The researchers knew that leaf shape in pitcher plants follows a continuum, with larger keels generally associated with

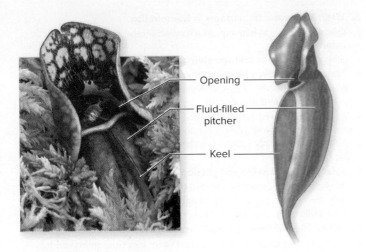

Figure 23.12 **Anatomy of a Pitcher.** Insects enter the fluid-filled pitcher through the mouthlike opening. The keel is the flap of tissue running the length of the pitcher.

Photo: ©Ed Reschke/Photolibrary/Getty Images

greater photosynthetic efficiency. A normal pitcher has a narrow keel and a wide pitcher opening, maximizing the potential for trapping prey. On the other hand, a leaf optimized for photosynthesis has a wide keel and a small pitcher opening.

Every September, the researchers measured the shapes of all of the leaves on each study plant; figure 23.13 summarizes the results. The control plants, with low nutrient availability, had prominent pitchers with wide openings and small keels. Supple-

mental phosphorus did not affect leaf shape, but added nitrogen did. The high-nitrogen treatments were associated with the largest keels and the smallest pitcher openings.

But do leaves with large keels really maximize photosynthesis at the expense of prey capture? To find out, Ellison and Gotelli measured the photosynthetic rate of the largest leaf on each plant. The data supported the prediction: The wider the keel, the greater the rate at which a leaf absorbed CO_2.

Overall, the study showed that a shortage of nitrogen favored prey capture at the expense of photosynthetic leaf area. Nutrient manipulation, coupled with careful leaf measurements, has therefore helped shed light on an interesting twist on plant evolution. Carnivory apparently represents a significant trade-off; in most terrestrial habitats, the traps would simply cost more energy than they were worth.

Source: Ellison, Aaron M., and Nicholas J. Gotelli. 2002. Nitrogen availability alters the expression of carnivory in the northern pitcher plant, *Sarracenia purpurea. Proceedings of the National Academy of Sciences,* vol. 99, no. 7, pp. 4409–4412.

23.5 MASTERING CONCEPTS

1. Describe the hypothesis and experimental design in Ellison and Gotelli's study.
2. Explain the purpose of the two control solutions.
3. Predict how the graph in figure 23.13 would look if nitrogen did not affect leaf structure.

a. Treatment descriptions

Treatment	Nutrient solution
■ Control (dH₂O)	Distilled water only; no added nutrients
■ Control (micros)	Micronutrients only; no added N or P
● High P	0.25 mg PO₄-P/liter + micronutrients
● Low P	0.025 mg PO₄-P/liter + micronutrients
▲ Low N	0.1 mg NH₄-N/liter + micronutrients
♦ Low N + High P	0.1 mg NH₄-N/liter + 0.25 mg PO₄-P/ liter + micronutrients
♦ High N + High P	1.0 mg NH₄-N/liter + 0.25 mg PO₄-P/liter + micronutrients
♦ High N + Low P	0.1 mg NH₄-N/liter + 0.025 mg PO₄-P/liter + micronutrients
▲ High N	1.0 mg NH₄-N/liter + micronutrients

b. Pitcher descriptions

Size of keel	Size of pitcher opening
Small	Large
Small	Large
Fairly small	Large
Fairly small	Large
Medium	Medium
Medium	Small
Fairly large	Small
Large	Small
Large	Small

c. Graph of results

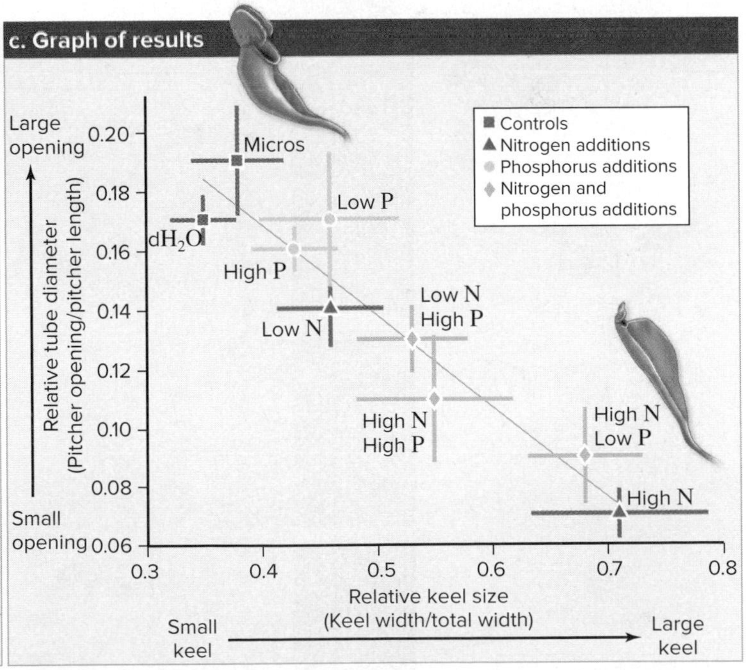

Legend:
■ Controls
▲ Nitrogen additions
● Phosphorus additions
♦ Nitrogen and phosphorus additions

Y-axis: Relative tube diameter (Pitcher opening/pitcher length), Large opening 0.20 to Small opening 0.06

X-axis: Relative keel size (Keel width/total width), 0.3 to 0.8; Small keel to Large keel

Data points: Micros, dH₂O, Low P, High P, Low N, Low N High P, High N High P, High N Low P, High N

Figure 23.13 **Nitrogen Matters.** (a) Control, nitrogen, and phosphorus treatments used in the study. (b, c) Pitcher plants that received supplemental nitrogen had larger keels and smaller pitchers than plants that received little nitrogen. The addition of phosphorus did not have the same effect. In the graph, each point represents a treatment mean; horizontal and vertical bars represent 95% confidence intervals.

CHAPTER SUMMARY

23.1 Soil and Air Provide Water and Nutrients

A. Plants Require 16 Essential Elements
- Like all organisms, plants require water and **essential nutrients.**
- In all plants, the nine essential **macronutrients** are C, H, O, P, K, N, S, Ca, and Mg. The seven **micronutrients** are Cl, Fe, B, Mn, Zn, Cu, and Mo.

B. Soils Have Distinct Layers
- **Soil** consists of rock and mineral particles mixed with air, water, and decaying organic molecules. Soil layers are called horizons.
- **Topsoil** contains **humus,** which is a reservoir for both water and nutrients. **Erosion** destroys topsoil and hinders plant growth.

C. Leaves and Roots Absorb Essential Elements
- Plants obtain CO_2 and O_2 from the atmosphere, and they acquire hydrogen and oxygen atoms from H_2O in soil. The other elements also come from soil.
- Mycorrhizal fungi add to the root's surface area for absorbing water and nutrients, especially phosphorus. Several types of **nitrogen-fixing** bacteria live in root **nodules,** converting atmospheric nitrogen into forms that plants can use.

23.2 Water and Minerals Are Pulled Up to Leaves in Xylem

- **Vascular tissue,** which consists of xylem and phloem, transports materials within plants (**figure 23.14**).

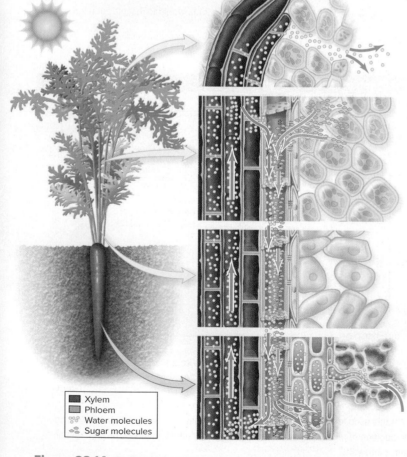

Figure 23.14 Vascular Tissue Interactions: A Summary.

A. Water Evaporates from Leaves in Transpiration
- **Xylem** transports **xylem sap,** which consists mostly of water and dissolved minerals.
- Leaves lose water by **transpiration** through open stomata.

B. Water and Dissolved Minerals Enter at the Roots
- Water enters roots by osmosis because the solute concentration in the soil is less than that of root cells.
- Water and dissolved minerals move through the root's epidermis and cortex, either between cells or through cell interiors. The endodermis, with its impermeable **Casparian strip,** controls which substances enter the xylem.

C. Xylem Transport Relies on Cohesion
- According to the **cohesion–tension theory,** water molecules evaporating from leaves are replaced by those pulled up from below, a consequence of the cohesive properties of water.

D. The Cuticle and Stomata Help Conserve Water
- The waxy, waterproof **cuticle** prevents water loss from stems and leaves.
- Plants can open and close **stomata** in response to water availability. When **guard cells** bordering each stoma absorb water, the pore opens; when the guard cells lose water and collapse, the stoma closes.

23.3 Sugars Are Pushed in Phloem to Nonphotosynthetic Cells

A. Phloem Sap Contains Sugars and Other Organic Compounds
- **Phloem** transports **phloem sap,** which includes sugars, hormones, viruses, and other substances, plus water and minerals from xylem.

B. The Pressure Flow Theory Explains Phloem Function
- According to the **pressure flow theory,** phloem sap flows under positive pressure through sieve tubes from a **source** to a **sink.**
- Sources are photosynthetic or sugar-storing parts that load carbohydrates into phloem. Water follows by osmosis. The increase in pressure drives phloem sap to nonphotosynthetic sinks such as roots, flowers, and fruits.

23.4 Parasitic Plants Tap into Another Plant's Vascular Tissue

- Mistletoe, dodder, and other parasitic plants absorb water, minerals, and sugar from a host plant's xylem and phloem.

23.5 Investigating Life: The Hidden Cost of Traps

- Carnivorous plants produce leaves that specialize in prey capture or photosynthesis. Experiments show that nutrient additions tip the balance in favor of photosynthetic leaves.

MULTIPLE CHOICE QUESTIONS

1. What is the difference between a macronutrient and a micronutrient?
   a. The size of the atoms
   b. The amount required by the plant
   c. The use by multicellular organisms (macroorganisms) versus microorganisms
   d. The source of the nutrient

2. Through which of the following structures might a water molecule flow as it moves through a plant?
   a. Sieve tubes
   b. Tracheids
   c. Vessel elements
   d. All of the above are correct.

3. How does water move through xylem?
   a. Water moves into the plant at leaves, and gravity pulls it down to the roots.
   b. Water moves into the plant at leaves, and phloem helps push it to the roots.
   c. Water moves into the plant at roots, and phloem helps push it to the leaves.
   d. Water moves into the plant at roots, and cohesion pulls it to the leaves.

4. What process creates the pressure that drives phloem toward sinks?
   a. Water moving by osmosis into the sieve tubes at sources
   b. Active transport of water into companion cells
   c. Water moving by osmosis out of sieve tubes at sinks
   d. Both a and c are correct.

5. The main function of stomata is to allow for
   a. transpiration.   c. the release of excess heat.
   b. the absorption of sunlight.   d. Both a and b are correct.

6. Where do the simple sugars in phloem sap originate in most plants?
   a. Root hairs, which absorb decaying matter in the soil
   b. Photosynthetic cells in leaves
   c. Mutualistic fungi
   d. Storage cells within fruits

7. Which of the following is typically NOT a sink?
   a. A root   c. A flower
   b. A mature leaf   d. A meristem

8. In contrast with a typical plant, a parasitic plant
   a. consumes insects as a source of nitrogen.
   b. carries out photosynthesis.
   c. extracts nutrients from another plant.
   d. has adaptations that promote survival and reproduction.

Answers to these questions are in appendix A.

WRITE IT OUT

1. Explain the relationship between transpiration and how plants obtain nitrogen and phosphorus.

2. How might transgenic technology (see chapter 11) be used to endow plants with the ability to fix nitrogen without the aid of bacteria? In what ways would this new feature change agriculture?

3. Draw a diagram of a plant's roots growing into the soil. Label the soil layers and describe the composition of each.

4. Why might warm temperatures and heavy precipitation in rain forests result in topsoils with low nutrient concentrations?

5. Which nutrients are absorbed from soil by roots, and which are absorbed from the atmosphere by leaves?

6. When Chris mows the grass, she faces a choice between discarding the clippings or leaving them on the lawn. How would each choice influence the nutrient content of the soil? Explain your answer.

7. If left in the same pot for multiple years, a houseplant may become "root-bound," meaning that the roots grow in circles along the inner surface of the pot. Why do root-bound plants eventually show signs of nutrient deficiency?

8. Explain the role of hydrogen bonding in xylem transport.

9. Suppose you use a rubber band to secure a clear plastic bag around a few leaves on a live plant. What do you think will happen?

10. How can phloem transport occur either with or against gravity?

11. Basil is common in vegetable gardens. Many gardeners grow this plant for its leaves, which provide flavor in sauces and other dishes. Leaf production is higher when young flowers are pruned off the plant before they have a chance to develop. Explain this observation.

12. Explain how plant transport might change in the winter after a deciduous tree has lost its leaves.

13. Suppose that a scientist exposes a leaf to CO_2 labeled with radioactive carbon-14, which is incorporated into organic compounds in photosynthetic cells. At various times after exposure, the scientist can determine the location of the radioactive carbon in the plant. In what tissues would you expect to find the radioactive material immediately after exposure to the labeled carbon? What about during transport? When transport is complete, will the radioactive material be in plant parts above the leaf, below the leaf, or both?

14. In the early 1600s, Jean Baptista van Helmont investigated how plants acquire new mass as they grow. He weighed a willow shoot and planted it into soil that he had also weighed. After 5 years of adding nothing but water to the plant, he found that the soil had lost only a little weight, while the plant had grown from 2 kg to about 76 kg. He therefore concluded, incorrectly, that water was the sole source of the added plant material. What other source did he fail to consider in his experiment?

15. Write nonbiological analogies for the Casparian strip and for pressure flow theory.

PULL IT TOGETHER

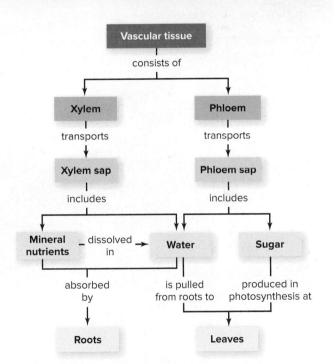

Figure 23.15 Pull It Together: Plant Nutrition and Transport.

Refer to figure 23.15 and the chapter content to answer the following questions.

1. Review the Survey the Landscape figure in the chapter introduction, and then add *flowering plants, vegetative parts, reproductive parts,* and *flowers* to the Pull It Together concept map. Connect *vegetative parts* and *reproductive parts* to the concept map in at least two ways each.

2. Add the terms *soil, source, sink, pressure flow,* and *transpiration* to the concept map.

3. Write a phrase connecting *water* to *sugar.*

Partners. This beekeeper is tending to her beehives. The honeybees occupying the hives aid in the reproduction of many types of flowering plants, including raspberries and other crops that humans depend on for food.

(bee keeper): ©Liu Jin/AFP/Getty Images; (bee): ©Stephen Dalton/Science Source

LEARN HOW TO LEARN
Know Yourself

Setting aside time to study is one important ingredient for academic success; another is paying attention to your work habits throughout the day and night. Are you most alert in the morning, afternoon, or evening? Block off time to study during periods when you are at your best. Your study time will be much more productive if you are not fighting to stay awake.

Imperiled Pollinators

Nearly all animals depend on plants for food, habitat, and oxygen. Yet many flowering plant species, including many of the crops we raise for food, depend on animals as well. Their flowers produce nectar, a sweet substance that lures insects, birds, and small mammals. As the animals explore the blooms for food, they unwittingly transfer pollen from one flower to another. As we will see in this chapter, pollination is a critical link in the life cycle of the flowering plant.

Bees are proficient pollinators. European honeybees, which were introduced into the United States in 1621, visit many flowering plant species. In fact, honeybees gather pollen and nectar from so many plants that they have displaced native bees.

Unfortunately, bee populations worldwide are plummeting. In 2006, beekeepers began to report a mysterious ailment called colony collapse disorder. The bees in affected hives disappear for no known reason. The insects have weakened defenses and harbor unusually high populations of harmful bacteria and fungi, but so far no one knows exactly why the bees are dying.

Some possible culprits, however, have been ruled out. For example, researchers have concluded that neither radiation from cell phone towers nor insecticide-producing genes from genetically modified plants are responsible for the die-off.

So what is causing colony collapse disorder? The problem is especially difficult to study because a combination of conditions may be to blame. The use of insecticides in agriculture has killed many bees and stressed many more. Moths and mites threaten bees, too. Caterpillars of the greater wax moth tunnel through honeycombs and eat beeswax, pollen, and cocoons. Tracheal mites colonize a bee's respiratory system and consume the host's blood; an external mite sucks the blood of young bees and adults. Infectious diseases also take a toll. A bacterial disease called American foulbrood kills bee larvae, and single-celled parasites called microsporidia damage the bee's digestive tract and increase susceptibility to other diseases. Researchers hypothesize that the combination of these stresses has made bees vulnerable to a particular virus, but no one knows for sure.

Biologists predict that falling bee populations will soon lead to declining crops and the extinctions of many native plant species. What can ordinary people do to help? Entomologists suggest boosting populations of wild bees, such as bumblebees, to offset the loss of honeybees. Strategies include avoiding the use of pesticides, planting clusters of diverse native plants in yards, providing nesting habitat for bees, and surrounding agricultural fields with native grasses, herbs, and trees.

SURVEY THE LANDSCAPE
Plant Life

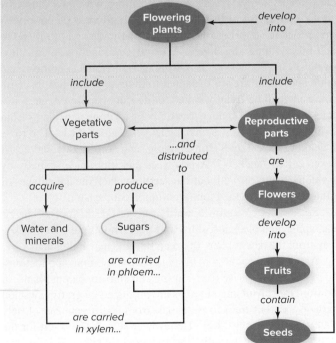

Have you ever wondered how an apple's seeds get inside the fruit? The answer traces to sexual reproduction in flowering plants. After pollination, the seeds develop inside the apple flower, while surrounding flower parts develop into the fruit.

For more details, study the Pull It Together feature in the chapter summary.

24.1 Angiosperms Reproduce Asexually and Sexually

Flowering plants (angiosperms) dominate many terrestrial landscapes. From grasslands to deciduous forests, and from garden plots to large-scale agriculture, angiosperms have been extremely successful. These plants first evolved only about 144 million years ago, and their subsequent rapid diversification has been one of the most extraordinary examples of adaptive radiation in the history of life. ⓘ *adaptive radiation,* section 14.4B

Angiosperms owe their success to three adaptations. First, pollen enables sperm to fertilize an egg in the absence of free water. In contrast, the sperm cells of mosses and ferns must swim to the egg, so these plants can reproduce sexually only in moist habitats. Second, the seed protects the embryo during dormancy and nourishes the developing seedling. Third, flowers not only promote pollination but also develop into fruits that help disperse the seeds far from the parent plant.

Most angiosperms reproduce sexually, although some species also reproduce asexually. This section describes asexual reproduction and briefly introduces sexual reproduction, a topic described in detail in section 24.2. (For comparison, chapter 19 describes the life cycles of the mosses, ferns, and conifers.)

A. Asexual Reproduction Yields Clones

Many plants reproduce asexually, forming new individuals by mitotic cell division. In **asexual reproduction,** a parent organism produces offspring that are genetically identical to it and to each other—they are clones. Asexual reproduction, also called vegetative reproduction, is advantageous when conditions are stable and plants are well adapted to their surroundings, because the clones will be equally suited to the same environment.

Plants often reproduce asexually by forming new plants from portions of their roots, stems, or leaves (**figure 24.1**). The roots of quaking aspen trees, for example, can sprout identical aerial shoots called "suckers." Aspen trees can reproduce sexually by producing seeds as well, but the seedlings lack the advantages of being connected to the clone's common root system.

Suckers can also grow upward from buds on the roots of cherry, pear, apple, and black locust trees. If these shoots are cut or broken away from the parent plant, they can become new individuals.

Asexual reproduction has many commercial applications. Most houseplants arise from cuttings taken from parent plants. In addition, most fruit and nut trees are produced by grafting a scion (part of a parent tree) to rootstock from a different but closely related plant (**figure 24.2**). The grower selects the scion for the quality of its fruit. Usually, the rootstock is either disease- and pest-resistant or especially well adapted to dry or salty soil. Grafting the scion to the rootstock gives the grower the advantages of both.

Scientists can use tissue culture to produce huge numbers of identical plants in a laboratory (see figure 11.12). The first step is to place pieces of plant tissue in a dish containing nutrients and hormones. After a few days, the plant cells lose their specialized

a.

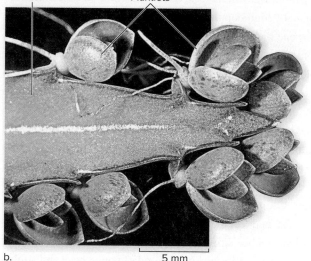

Leaf   Plantlets

b.   5 mm

Figure 24.1 Asexual Reproduction. (a) Quaking aspen trees are clones connected by a common root system. Suckers are shoots from the clone, whereas seedlings are the product of sexual reproduction. (b) The leaf of this kalanchöe plant is producing genetically identical plantlets.
Photos: (a, b): ©Steven P. Lynch RF

characteristics and form a white lump called a callus. The cells of the callus divide for a few weeks. The lump is then transferred to a dish containing a new mix of hormones, prompting some cells to develop into tiny, genetically identical plantlets with shoots and roots. Each plantlet can then be separated from the others and grown into an individual plant.

Callus growth is unique to plants. The human equivalent would be a cultured skin cell multiplying into a blob of unspecialized tissue and then sprouting tiny humans!

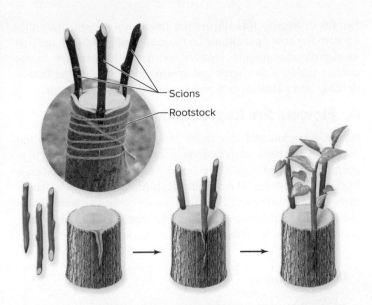

Figure 24.2 Grafting a Fruit Tree. Commercial fruit trees are often propagated by grafting. This example shows a bark graft, in which the scions are inserted under the rootstock's bark and secured with twine.

Photo: ©Dorling Kindersley/Getty Images

B. Sexual Reproduction Generates Variability

Asexually generated plants have predictable characteristics because they are essentially identical to their parents (except for mutations). In contrast, **sexual reproduction** yields genetically unique offspring with a mix of traits derived from two parents. Sexual reproduction is adaptive in habitats where selective pressures frequently change. For example, in laboratory experiments, asexually reproducing worms exposed to an evolving assortment of disease-causing bacteria died out within a few generations. Sexually reproducing animals fared better (see section 9.9). The same is true for plants: Producing variable offspring improves reproductive success in an uncertain world. ⓘ *why sex?* section 9.1

The major groups of multicellular organisms (fungi, animals, and plants) all have the same basic pattern of sexual reproduction. Meiosis produces the cells that begin the haploid generation; fertilization unites the gametes to begin the diploid generation. But organisms vary in the proportion of the life cycle spent as haploid or diploid cells. For example, sperm and egg cells are usually the only haploid animal cells; the rest of an animal's cells are diploid.

In contrast, the basic plant life cycle has an **alternation of generations,** with multicellular diploid and haploid stages (see figure 19.5). The **sporophyte,** or diploid generation, produces haploid **spores** by meiosis. A spore divides mitotically to produce a multicellular haploid **gametophyte,** which undergoes mitosis to generate haploid **gametes** (egg cells or sperm). In **fertilization,** these gametes fuse to form a diploid **zygote.** The zygote develops into an embryo as its cells divide mitotically. With additional growth, the embryo becomes a mature sporophyte, and the cycle begins anew.

Figure 24.3 applies this life cycle to angiosperms. Notice that flowering plants produce two types of spores. Microspores give rise

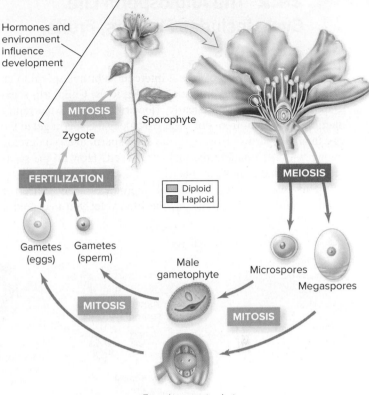

Figure 24.3 Alternation of Generations in Angiosperms. In angiosperms, the sporophyte produces flowers. Inside the flower, diploid cells undergo meiosis to produce haploid microspores and megaspores. These cells, in turn, develop into the gametophytes that produce sperm and egg cells by mitosis. Fertilization yields the zygote, the first cell of the next sporophyte generation.

to male gametophytes, and the larger megaspores produce female gametophytes. The prefix *mega-* may suggest huge size, but that is only in comparison with the size of microspores. In flowering plants, the spores, gametophytes, and gametes are all microscopic; the entire haploid gametophyte generation consists of only a few cells. The greatly reduced gametophyte is one of the many features that distinguish angiosperms from other plants (see section 19.1).

The rest of this chapter begins with a look at sexual reproduction in flowering plants. As you will see, the plant packages its offspring inside a seed. When the seed germinates, the young plant faces a host of challenges. The second half of this chapter explores the hormonal signals and environmental factors that influence the development of the angiosperm throughout its life.

24.1 MASTERING CONCEPTS

1. What adaptations contribute to the reproductive success of angiosperms?
2. What are some examples of asexual reproduction in plants?
3. When are sexual and asexual reproduction each adaptive?
4. Describe alternation of generations in the plant life cycle.

24.2 The Angiosperm Life Cycle Includes Flowers, Fruits, and Seeds

When humans reproduce, sexual intercourse brings sperm to an egg cell, and the embryo develops into a fetus. Childbirth separates woman from baby. Clearly, flowering plant sexual reproduction is different from our own. How does the sperm get to the egg in angiosperms? How does the angiosperm embryo develop (along with surrounding tissues) into a seed? How do the seeds separate from the "mother" plant?

This section explains sexual reproduction in flowering plants. **Figure 24.4** summarizes the life cycle; you may find it

helpful to refer to this illustration frequently as you study this section. For now, concentrate on the relationship between the two most prominent players: flowers and fruits. **Flowers** are reproductive organs where eggs and sperm unite. Parts of the flower develop into a **fruit,** which protects and disperses the seeds.

A. Flowers Are Reproductive Organs

After a seed germinates (see figure 24.4, step 1), the plant grows and develops to maturity (step 2). Flowers then begin to form. The trigger for flower production is a protein called florigen, which leaf cells load into the phloem. The protein travels in sieve tubes to the shoot tips, where it stimulates the expression of genes that transform a shoot apical meristem into a floral meristem. The meristem cells divide to produce the flower. ⓘ *meristems,* section 22.4

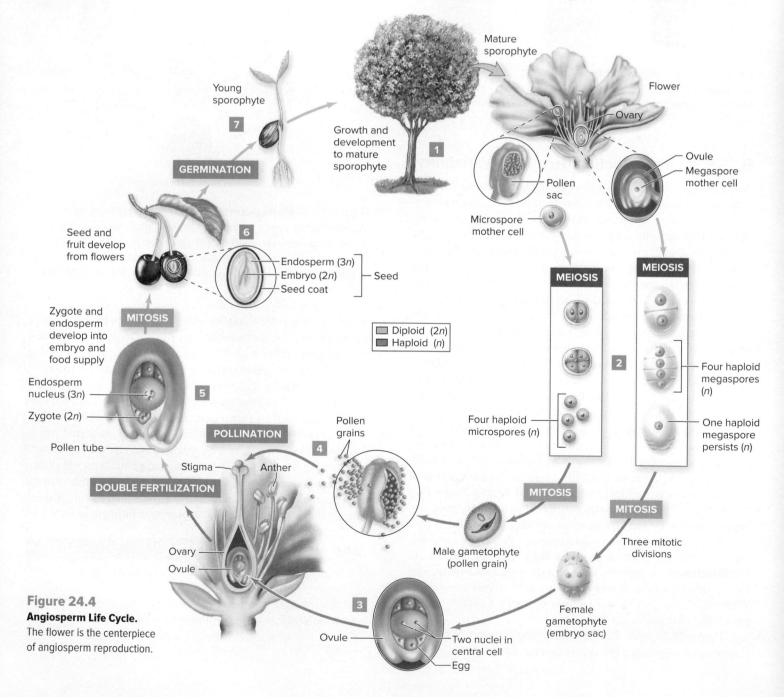

Figure 24.4

Angiosperm Life Cycle.
The flower is the centerpiece of angiosperm reproduction.

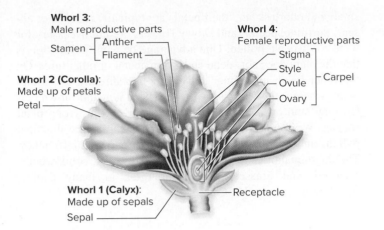

Whorl 3:
Male reproductive parts

Stamen — Anther
          Filament

Whorl 2 (Corolla):
Made up of petals

Petal

Whorl 4:
Female reproductive parts

Stigma
Style
Ovule   Carpel
Ovary

Whorl 1 (Calyx):
Made up of sepals

Sepal

Receptacle

Figure 24.5 Parts of a Flower. A cross section of a complete flower reveals four whorls: sepals, petals, stamens, and one or more carpels.

Figure 24.5 shows the anatomy of a typical flower—in this case, a cherry blossom. A part of the floral stalk called the **receptacle** is the attachment point for four types of structures, all of which are modified leaves composed mostly of parenchyma cells (see chapter 22). The outermost whorl (ring of parts) is the calyx. It consists of **sepals,** leaflike structures that enclose and protect the inner floral parts. Next is the corolla, which is a whorl of **petals.** The calyx and corolla do not play a direct role in sexual reproduction. In many flowers, however, colorful petals attract pollinators.

The two innermost whorls of a flower are essential for sexual reproduction. The male flower parts consist of **stamens,** which are filaments that bear pollen-producing bodies called **anthers** at their tips. The female part, at the center of a flower, is composed of one or more **carpels.** The base of each carpel is called an **ovary,** and it encloses one or more egg-bearing **ovules.** The upper part of each carpel is a stalklike **style** that bears a structure called a **stigma** at its tip. Stigmas receive pollen.

The flower in figure 24.5 is "complete" because it contains all four whorls, including both male and female parts. In some species, however, the sexes are separate. That is, each flower has either male or female parts, but not both. An individual plant may have both types of single-sex flowers, or the plant may produce just one flower type. A holly plant, for example, is either male or female; only the female plants produce the distinctive red berries.

Recall from chapters 19 and 22 that the two largest clades of angiosperms are monocots and eudicots. Flower structure is one feature that distinguishes the two groups. Most monocots, such as lilies and tulips, have petals, stamens, and other flower parts in multiples of three. Most eudicots, on the other hand, have flower parts in multiples of four or five. Buttercups and geraniums are examples of eudicots with five prominent petals on each flower.

B. The Pollen Grain and Embryo Sac Are Gametophytes

Once the flowers have formed, the next step is to produce the microscopic male and female gametophytes (see figure 24.4, steps 3 through 5). Inside the anther's **pollen sacs,** diploid cells

divide by meiosis to produce four haploid **microspores.** Each microspore then divides mitotically and produces a two-celled, thick-walled structure called a **pollen** grain, which is the young male gametophyte. One of the haploid cells inside the pollen grain divides by mitosis to form two sperm nuclei.

Meanwhile, meiosis also occurs in the female flower parts. The ovary may contain one or more ovules, each containing a diploid cell that divides by meiosis to produce four haploid **megaspores.** In many species, three of these cells quickly disintegrate, leaving one large megaspore. The megaspore undergoes three mitotic divisions to form the **embryo sac,** which is a female gametophyte. At first, the embryo sac contains eight nuclei. Cell walls soon form, dividing the female gametophyte into seven cells. One of these cells is the egg. In addition, a large, central cell contains two nuclei; as we will see, both the egg and the central cell's two nuclei participate in fertilization.

C. Pollination Brings Pollen to the Stigma

Eventually, the pollen sac opens and releases millions of pollen grains. The next step is **pollination:** the transfer of pollen from an anther to a receptive stigma (**figure 24.6**).

a.   b.

c.

d.   e.   f.

Figure 24.6 Pollination. Animal pollinators include (a) hummingbirds, (b) butterflies, and (c) bats. (d, e) Some flowers attract insects with markings visible only in UV light. (f) Birch trees are wind-pollinated.

(a): ©Corbis RF; (b): ©MedioImages/Getty Images RF; (c): ©Merlin D. Tuttle/Bat Conservation International/Science Source; (d, e): ©Leonard Lessin/Science Source; (f): ©Dr. Jeremy Burgess/Science Source

Flower color, shape, and odor attract animal pollinators. For example, hummingbirds are attracted to red, tubular flowers. Beetles visit dull-colored flowers with spicy scents, whereas blue or yellow sweet-smelling blooms attract bees. Bee-pollinated flowers often have markings that are visible only at ultraviolet wavelengths of light, which bees can perceive. Butterflies prefer red or purple flowers with wide landing pads. Moths and bats pollinate white or yellow, heavily scented flowers, which are easy to locate at night.

Many animals benefit from their association with plants: They obtain food in the form of pollen or sugary nectar, seek shelter among the petals, or use the flower for a mating ground (see section 6.10). Some plants, however, lure pollinators with a false reward. Skunk cabbage and the "carrion" flowers of South African *Stapelia* plants emit foul odors that attract flies. Flies that land on the flowers gain nothing, but the plant benefits if the insect unwittingly carries pollen grains to another flower of the same species.

Sometimes the flower and its animal partner have matching parts. For example, some hummingbirds have long, curved bills that fit precisely into the tubular flowers from which they sip nectar. The connection between plant and pollinator may be so strong that the partners directly influence each other's evolution. In **coevolution,** a genetic change in one species selects for subsequent change in the genome of another species. Coevolution is likely to be occurring when a plant has an exclusive relationship with just one pollinator species.

About 10% of angiosperms (and most gymnosperms) use wind, not animals, to carry pollen. Wind-pollinated flowers are small and odorless, and their petals are typically reduced or absent; perfume, nectar, and showy flowers are not necessary for wind to disperse pollen. One advantage of wind pollination is that the plant does not spend energy on nectar or other lures. On the other hand, an animal delivers pollen directly to another plant, whereas the wind is "wasteful." That is, wind-blown pollen may land on the ground, on water, or on the wrong plant species. Wind-pollinated plants therefore manufacture abundant pollen, an adaptation that compensates for this inefficiency. The large quantities of pollen produced by oaks, cottonwoods, ragweed, and grasses provoke allergies in many people. ⓘ *allergies,* section 34.5C

D. Double Fertilization Yields Zygote and Endosperm

After a pollen grain lands on a stigma of the correct species, a pollen tube emerges (**figure 24.7**, step 1). The pollen grain's two haploid sperm nuclei enter the pollen tube as it grows through the tissue of the style toward the ovary (figure 24.7, step 2). When the pollen tube reaches an ovule, it discharges its two sperm nuclei into the embryo sac.

Then, in **double fertilization,** the sperm nuclei fertilize the egg and the central cell's two nuclei (figure 24.7, step 3; see also figure 24.4, step 5). That is, one sperm nucleus fuses with the haploid egg nucleus and forms a diploid zygote, which will develop into the embryo. The second sperm nucleus fuses with the haploid nuclei in the central cell. The resulting triploid nucleus

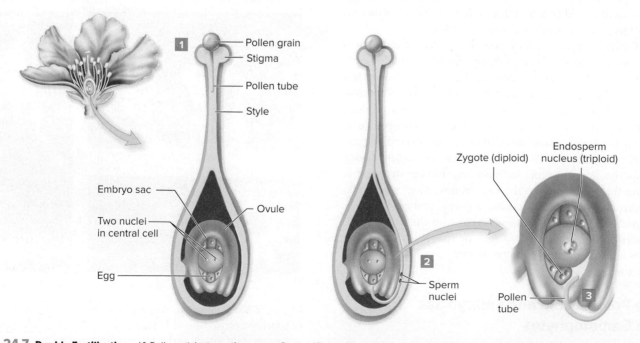

Figure 24.7 Double Fertilization. (*1*) Pollen sticks to a stigma on a flower. (*2*) A pollen tube grows toward the ovule and transports two sperm nuclei. (*3*) One sperm nucleus fertilizes the egg to form a diploid zygote, and the other fertilizes the two nuclei in the central cell to yield the triploid endosperm.

divides to form a tissue called **endosperm,** which is composed of parenchyma cells that store food for the developing embryo. Familiar endosperms are the "milk" and "meat" of a coconut and the starchy part of a rice grain. The starchy endosperm of corn is an important source of not only food but also ethanol, a biofuel (see the Burning Question in chapter 19).

Double fertilization reduces the energetic cost of reproduction. In gymnosperms, which lack this adaptation, the female gametophyte stockpiles food for the embryo in advance of fertilization. The investment is wasted if no zygote ever forms. In contrast, double fertilization saves energy because the angiosperm produces food for the embryo only if a sperm nucleus actually fertilizes the egg.

E. A Seed Is an Embryo and Its Food Supply Inside a Seed Coat

Immediately after fertilization, the ovule contains an embryo sac with a diploid zygote and a triploid endosperm. The ovule eventually develops into a **seed:** a plant embryo together with its stored food, surrounded by a seed coat (see figure 24.4, step 6). Where do these parts come from?

The zygote divides to form the embryo (**figure 24.8**). Among the first features of the developing embryo are the **cotyledons,** or seed leaves. (The cotyledons are called seed leaves

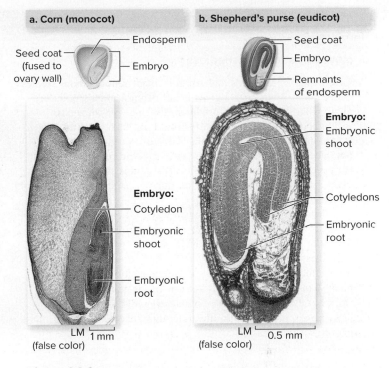

Figure 24.9 Mature Seeds. (a) Corn is a monocot. A grain of corn is a fruit containing one seed, including a starchy endosperm and an embryo with one cotyledon. (b) In a eudicot, the two cotyledons may absorb much of the endosperm as the seed develops.

Photos: (a): ©Steven P. Lynch/McGraw-Hill Education; (b): ©Steven P. Lynch RF

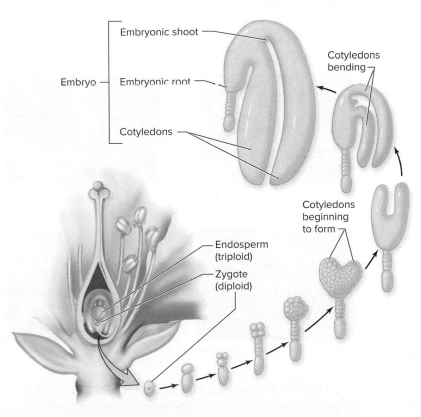

Figure 24.8 From Zygote to Embryo. As a seed develops, the zygote divides repeatedly to form a tiny embryonic plant.

because in many species, they emerge from the soil with the seedling and carry out photosynthesis for a short time. But they are not true leaves.) Soon, the shoot and root apical meristems form at opposite ends of the embryo.

What about the embryo's food supply? Inside the developing seed, the endosperm cells divide more rapidly than the zygote and thus form a large, multicellular mass. This endosperm supplies nutrients to the developing embryo, although the timing depends on the type of plant (**figure 24.9**). In monocots, the mature seed retains the endosperm, and the cotyledon transfers stored nutrients from the endosperm to the rest of the embryo during germination. In many eudicots, the paired cotyledons absorb the endosperm during seed development.

As the embryo and endosperm develop, the seed coat begins to form. In many species, the **seed coat** is a tough, sclereid-rich outer layer that protects both the embryo and its food supply from damage and hungry animals. In other plants, including corn, the thin seed coat fuses with the ovary wall as the seed matures.

Burning Question

How can a fruit be seedless?

Given the role of seeds and fruits in plant reproduction, seedless fruits might seem puzzling. After all, from a plant's point of view, what's the point of producing a fruit without seeds inside? Natural selection clearly would not favor such a trait in the wild! But the crop plants that people grow do not necessarily live by the same rules as wild plants. Because humans often consider seeds a nuisance, plant breeders have selected for many seedless varieties.

Seedless fruits can form in two ways. Often, the fruit develops in the absence of fertilization. Seedless oranges and watermelons are two examples. Alternatively, if fertilization does occur, the embryos may die during development. Tiny, immature seeds remain inside the fruit, but they do not interfere with eating. Seedless grapes and bananas illustrate this second path to seedlessness.

©Deborah Jaffe/Getty Images RF

Since seedless fruits lack seeds, how do farmers grow more of them? Most are produced asexually by grafting or taking cuttings. But surprisingly, seedless watermelons do come from seeds! To make "seedless watermelon seeds," plant breeders first cross a regular diploid watermelon plant with a tetraploid plant (with four sets of chromosomes). Farmers then sow the triploid seeds arising from this union. The new triploid plants produce flowers, which the grower pollinates with diploid pollen. Pollination stimulates fruit production, but like a mule, the triploid plant is sterile (see the Burning Question in chapter 9). The "seeds" that do develop inside seedless watermelons are actually empty, soft hulls that are easy to chew and swallow.

Submit your burning question to
Marielle.Hoefnagels@mheducation.com

At some point in seed development, hormonal signals "tell" cells in the embryo and endosperm to stop dividing, and the seed gradually loses moisture and enters dormancy. The ripe, mature seeds are firm, dry, and ready for dispersal. Depending on the species, the dormant period can last for days, weeks, months, years, decades, or even centuries.

Seed dormancy is a crucial adaptation because it ensures that seeds have time to disperse away from the parent plant before germinating. Moreover, dormancy enables seeds to postpone development if the environment is unfavorable, such as during a drought or frost. Favorable conditions trigger embryo growth to resume when young plants are more likely to survive.

Producing seeds is costly: The parent plant uses precious sugars, lipids, and other organic molecules to produce both the embryo and its stored food supply. The seed continues to consume its parent's resources until it enters dormancy. From that time until the young seedling begins carrying out photosynthesis on its own, however, the only energy available to the embryo is the fuel stored inside the seed.

F. The Fruit Develops from the Ovary

The rest of the flower also changes as the seeds develop (see figure 24.4, step 6). When a pollen tube begins growing, the stigma produces ethylene, a hormone that stimulates the stamens and petals to wither and drop; these flower parts are no longer needed. Developing seeds also produce another hormone, auxin, which triggers fruit formation.

In many angiosperms, the ovary grows rapidly to form the fruit, which may contain one or more seeds. (This chapter's Burning Question explores how seedless fruits develop.) In some species, additional plant parts also contribute to fruit development. The pulp of an apple, for example, derives from a cup-shaped region of the receptacle. The apple's core is derived from the carpel walls, which enclose the seeds. **Figure 24.10** shows how the parts of an apple flower give rise to the fruit.

Fruits come in many forms (**table 24.1**). A simple fruit develops from a flower with one carpel. The carpel may have one seed, as in a cherry, or many seeds, as in a tomato. An aggregate fruit develops from one flower with many carpels. Strawberries

Pollination occurs.   Petals are shed.

Fruit protects and disperses seeds.   Ovary and receptacle swell as seeds develop.

Figure 24.10 Development of a Fruit. After pollination and fertilization, the apple tree's flower begins to develop into a fruit. The fleshy part of an apple develops from the receptacle, which enlarges along with the ovary wall as the fruit develops.

Photos: (all): ©Brent Seabrook

TABLE **24.1** Types of Fruits: A Summary

Fruit Type	Characteristics	Example(s)
Simple ©Ingram Publishing/ Alamy RF	Derived from one flower with one carpel	Olive, cherry, peach, plum, coconut, grape, tomato, pepper, eggplant, apple, pear
Aggregate ©Corbis RF	Derived from one flower with many separate carpels	Blackberry, strawberry, raspberry, magnolia
Multiple ©Ingram Publishing/ Alamy RF	Derived from tightly clustered flowers whose ovaries fuse as the fruit develops	Pineapple, fig

and raspberries are examples of aggregate fruits. A multiple fruit develops from clusters of flowers that fuse into a single fruit as they mature. Pineapples and figs are multiple fruits.

A few fruits develop underground. After the yellow flowers of peanut plants are fertilized, the petals wither, and the young fruit buries itself in the soil. Three to 5 months later, farmers dig up the plants to harvest the mature fruits. Each fruit consists of a fibrous shell enclosing one to three peanuts—the seeds.

G. Fruits Protect and Disperse Seeds

Fruits have two main functions, one of which is the protection of the seeds. Many developing fruits contain chemicals that animals find distasteful, so the immature seeds inside remain safe from herbivory. Once the seeds are mature, however, the fruit changes: It becomes soft, sweet, and appealing to animals.

These changes relate to the second function of fruits: to promote seed dispersal by animals, wind, and water (**figure 24.11**). Regardless of the transportation mode, the seeds are often deposited far from the parent plant, promoting reproductive success by minimizing competition between parent and offspring.

Many animals disperse fruits and seeds. Colored berries attract birds and other animals that carry the ingested seeds to new locations, only to release them in their droppings. Birds and mammals spread seeds when spiny fruits attach to their feathers or fur. Squirrels and other nut-hoarding animals also disperse seeds. Although they later eat many of the seeds they hide, they also forget some of their cache locations. The uneaten seeds may germinate.

Wind and water can also distribute seeds. Wind-dispersed fruits, such as those of dandelions and maples, have wings or other structures that catch air currents. Water-dispersed fruits include gourds and coconuts, which may drift across the ocean before colonizing distant lands.

24.2 MASTERING CONCEPTS

1. Redraw figure 24.5 and write the function of each part of the flower.
2. How does pollen move from one flower to another, and why is this process essential for sexual reproduction?
3. Describe the difference between pollination and seed dispersal.
4. What are the components of a seed?
5. What are the two main functions of fruits?

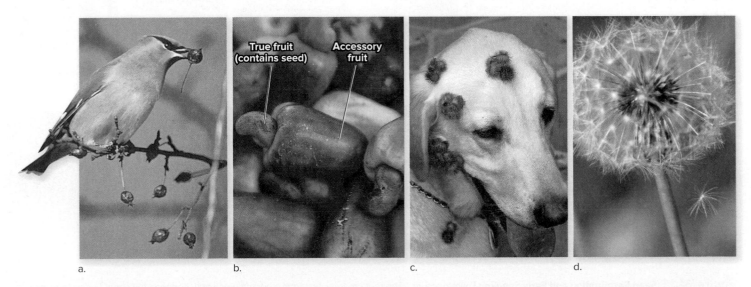

a. b. c. d.

Figure 24.11 **Seed Dispersal.** (a) This cedar waxwing helps disperse the seeds of winterberry, a type of holly. (b) The toxic, seed-containing true fruit of the cashew tree is attached to an accessory fruit that attracts animals. (c) Hooks on the surface of the burdock fruit easily attach to the fur of a passing animal. (d) Dandelion fruits have fluff that enables them to float on a breeze.

(a): ©Rod Planck/Science Source; (b): ©Barry Barker/McGraw-Hill Education; (c) ©Scott Camazine/Science Source; (d): ©Ingram Publishing/SuperStock RF

In the figure: "True fruit (contains seed)" and "Accessory fruit"

24.3 Plant Growth Begins with Seed Germination

Germination is the resumption of growth and development after a period of seed dormancy. It usually requires water, oxygen, and a favorable temperature. First, the seed absorbs water. The incoming water swells the seed, rupturing the seed coat and exposing the plant embryo to oxygen. Water also may cause the embryo to release hormones that stimulate the production of starch-digesting enzymes. The stored starch in the seed breaks down to sugars, providing energy for the now-growing embryo.

Growth and development continue after the seed coat ruptures (figure 24.12). Rapidly dividing cells in apical meristems add length to both shoots and roots (see section 22.4). The new cells differentiate into the ground tissue, vascular tissue, and dermal tissue that make up the plant body. ⓘ *plant tissue types,* section 22.2

The seedling soon begins to take on its mature form. Young roots grow downward in response to gravity, anchoring the plant in the soil and absorbing water and minerals. The shoot produces leaves as it grows upward toward the light. Initially, the energy source for the seedling's growth is stored food inside the seed. By the time the seedling has depleted these reserves, the new green leaves should begin producing food by photosynthesis. But if a seed is buried too deep in the soil, the young plant will never emerge—in effect, it will starve to death before reaching the light.

The size of a plant's seeds therefore reflects an evolutionary trade-off. Large, heavy seeds contain ample nutrient reserves to fuel seedling growth but may not travel far. Small seeds, on the other hand, store limited nutrients but tend to disperse far and wide. Interestingly, the crops that humans cultivate typically have larger seeds than do their wild ancestors. We gather and plant the seeds ourselves, removing the selection pressure favoring small seed size.

Depending on the species, a plant may keep growing for weeks, months, years, decades, or even centuries. When the plant reaches reproductive maturity, it, too, will develop flowers, seeds, and fruits, continuing the life cycle.

24.3 MASTERING CONCEPTS

1. Why must seeds absorb water before germinating?
2. What are the events of early seedling development?
3. How does natural selection influence seed size?

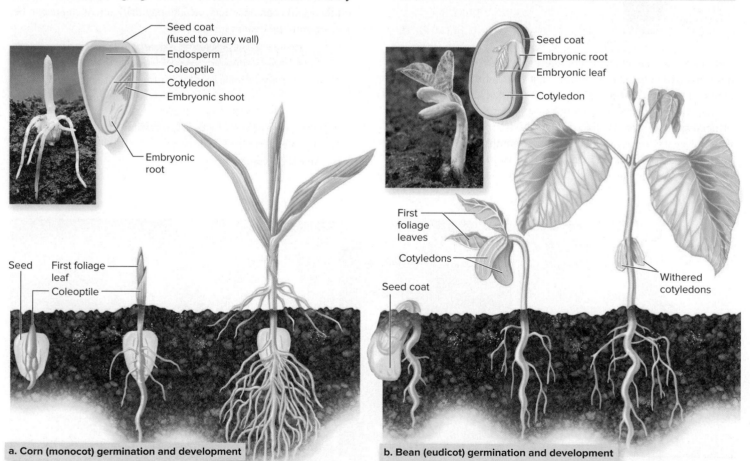

a. Corn (monocot) germination and development

- Seed coat (fused to ovary wall)
- Endosperm
- Coleoptile
- Cotyledon
- Embryonic shoot
- Embryonic root
- Seed
- First foliage leaf
- Coleoptile

b. Bean (eudicot) germination and development

- Seed coat
- Embryonic root
- Embryonic leaf
- Cotyledon
- First foliage leaves
- Cotyledons
- Seed coat
- Withered cotyledons

Figure 24.12 **Seed Germination and Early Seedling Development.**   The root emerges first from a germinating seed, and then the shoot begins to elongate. (a) In a monocot such as corn, a sheathlike coleoptile covers the shoot until the first foliage leaves develop. The cotyledon remains belowground. (b) In green beans and some other eudicots, the cotyledons emerge from the ground. They carry out photosynthesis until the first foliage leaves form.
Photos: (a): ©Dwight Kuhn; (b): ©Ed Reschke/Photolibrary/Getty Images

24.4 Hormones Regulate Plant Growth and Development

A plant's responses to environmental stimuli usually seem much more subtle than those of an animal. Plants cannot hide, bite, or flee; instead they must adjust their growth and physiology to external conditions. The rest of this chapter explores some of the ways in which a plant responds to the changing environment as it grows and develops, from seed germination through senescence (aging and death).

The environment affects plant growth in many ways. Shoots grow up, toward light and against gravity; roots grow down. Many plants leaf out in the spring, produce flowers and fruits, then enter dormancy in autumn—all in response to seasonal changes. Other responses are immediate. When the weather is hot, plants reduce transpiration by closing their stomata. A Venus flytrap snaps shut when a fly wanders across a leaf. Plants may even send signals to one another, warning of such dangers as insect infestations.

Chemicals called hormones regulate many aspects of plant growth, flower and fruit development, senescence, and responses to environmental change. A **hormone** is a biochemical synthesized in small quantities in one part of an organism and transported to another location, where it triggers a response from target cells.

Hormones also coordinate animal growth, development, and responses to the environment (see chapter 28). Unlike animals, however, plants do not have glands dedicated to hormone production. Instead, a plant can produce the same hormone at multiple sites throughout the body.

A plant's hormones may either diffuse from cell to cell or move larger distances by entering xylem or phloem. Either way, when a hormone reaches a target cell, it binds to a receptor protein. This interaction begins a cascade of chemical reactions that ultimately change the expression of genes in target cells.

The interactions and responses of plant hormones are extremely complex and difficult to study, for several reasons. First, plants produce hormones in extremely low concentrations. Second, each type of hormone exerts a wide variety of effects. Third, the same hormone can either stimulate or inhibit a process, depending on its concentration and the developmental stage of the plant. Fourth, the functions of plant hormones may overlap—for example, multiple hormones promote cell division and stem elongation.

The "classic five" plant hormones are auxins, cytokinins, gibberellins, ethylene, and abscisic acid (table 24.2). A plant must produce auxins and cytokinins if it is to develop at all. Both of these hormones occur in all major organs of all plants at all times, and biologists have never found a mutant plant lacking either one. Other hormones are required for normal development, but plants can complete their life cycles without them.

A. Auxins and Cytokinins Are Essential for Plant Growth

Auxins (from the Greek meaning "to grow") are hormones that promote cell elongation in stems and fruits but have the opposite effect in roots. Auxins also control plant responses to light and gravity. The most active auxin is indoleacetic acid (IAA), which is produced in shoot tips, embryos, young leaves, flowers, fruits, pollen, and coleoptiles. These hormones act rapidly, spurring noticeable growth in a grass seedling in minutes.

The first plant hormones described were auxins. In the late 1870s, decades before researchers determined the chemical

T A B L E **24.2** The "Classic Five" Plant Hormones: A Summary

Class	Synthesis Site(s)	Mode of Transport	Selected Actions
Auxins	Shoot apical meristem, developing leaves and fruits	Diffusion between parenchyma cells associated with vascular tissue	• Stimulate elongation of cells in stem • Control phototropism, gravitropism, thigmotropism • Stimulate growth of adventitious roots from stem cuttings • Suppress growth of lateral buds in stem (apical dominance)
Cytokinins	Root apical meristem	In xylem	• Stimulate cell division in seeds, roots, young leaves, fruits • Delay leaf senescence • Stimulate cell division in stem's lateral buds when auxin concentrations are low
Gibberellins	Young shoot, developing seeds	In xylem and phloem	• Stimulate cell division and elongation in roots, shoots, young leaves • Break seed dormancy
Ethylene	All parts, especially under stress, aging, or ripening	Diffusion of gas	• Hastens fruit ripening • Stimulates leaf and flower senescence • Stimulates leaf and fruit abscission (shedding) • Participates in thigmotropism
Abscisic acid	Mature leaves, especially in plants under drought or freezing stress	In xylem and phloem	• Inhibits shoot growth and maintains bud dormancy • Induces and maintains seed dormancy • Stimulates closure of stomata

Figure 24.13
Bending Toward Light.
(a) The shoot of an oat seedling normally bends toward light. (b) When a collar blocks the shoot tip, the plant no longer bends toward the light. (c) The plant does bend if the collar is placed beneath the tip.

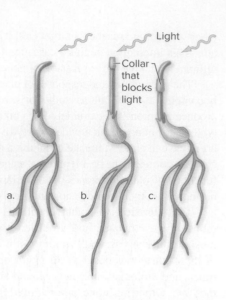

Figure 24.14 Apical Dominance.
Removing a plant's shoot tip (*right*) promotes the growth of lateral buds.
Photos: (both): ©Nigel Cattlin/Alamy

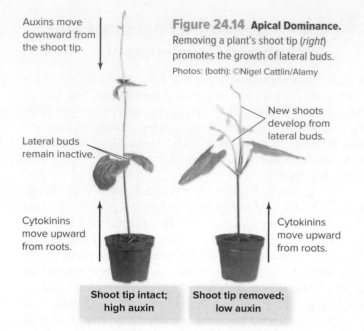

Auxins move downward from the shoot tip.

Lateral buds remain inactive.

New shoots develop from lateral buds.

Cytokinins move upward from roots.

Cytokinins move upward from roots.

Shoot tip intact; high auxin

Shoot tip removed; low auxin

structures of plant hormones, Charles Darwin and his son Francis learned that an "influence" produced at the shoot tip caused plants to grow toward light (figure 24.13). This influence was later discovered to be auxin.

Auxins have commercial uses. These hormones stimulate the growth of adventitious roots from cuttings, which is important in the asexual production of plants. Also, a synthetic compound with auxinlike effects, called 2,4-D (2,4-dichlorophenoxyacetic acid), is a widely used herbicide. The plant cannot completely break down 2,4-D, which accumulates to lethal levels. For reasons that are not completely understood, 2,4-D kills eudicots ("broadleaf weeds") but not grasses, which are monocots. (Learn about other weed killers in chapter 5's Apply It Now box.)

Cytokinins earned their name because they stimulate cytokinesis, or cell division. In flowering plants, most cytokinins affect roots and developing organs such as seeds, fruits, and young leaves. Cytokinins also slow the aging of mature leaves, so these hormones are used to extend the shelf lives of leafy vegetables.

The actions of cytokinins and auxins compete with each other. Cytokinins are more concentrated in the roots, whereas auxins are more concentrated in shoot tips. Cytokinins move upward within the xylem and stimulate lateral bud sprouting. In a counteracting effect called **apical dominance,** the terminal bud of a plant secretes auxins that move downward and suppress the growth of lateral buds. If the shoot tip is removed, the concentration of auxins in lateral buds decreases. Meristem cells in the buds then begin dividing, thanks to the the ever-present cytokinins. Apical dominance explains why gardeners can promote bushier growth by pinching off a plant's shoot tip (figure 24.14).

B. Gibberellins, Ethylene, and Abscisic Acid Influence Plant Development in Many Ways

In 1926, Japanese scientists studying "foolish seedling disease" in rice discovered **gibberellins,** another class of plant hormone that causes shoot elongation (figure 24.15). Although all plants produce gibberellins, this hormone was first discovered in the fungus that causes foolish seedling disease. Affected rice plants grow rapidly, becoming so spindly that they fall over and die. The researchers discovered that a chemical extract of the fungus produced the same symptoms. In 1934, scientists isolated the active compound. We now know of at least 84 naturally occurring gibberellins.

Farmers use gibberellins to stimulate stem elongation and fruit growth in seedless grapes. But gibberellins also have other functions. For example, they trigger seed germination by inducing the production of enzymes that digest starch in the seed. After germination, gibberellins also promote cell division.

Ethylene is a gaseous hormone that ripens fruit in many species. Ethylene released from one overripe apple can hasten the ripening, and eventual spoiling, of others nearby, leading to the expression "one bad apple spoils the bushel." Exposure to ethylene also ripens immature fruits after harvest (figure 24.16). For example, shipping can damage soft, vine-ripened tomatoes. Farmers therefore pick the fruit while it is still hard and green. Ethylene treatment just before distribution to supermarkets yields ripe-looking (if not good-tasting) tomatoes.

All parts of flowering plants synthesize ethylene, especially the shoot apical meristem, nodes, flowers, and ripening fruits. Like other hormones, ethylene has several effects. In most species, it causes petals and leaves to fade and wither. In addition, a damaged plant part produces ethylene, which hastens aging; the plant then sheds the affected part before the problem spreads. This effect was noticed in Germany in 1864, when ethylene in a mixture of gases in street lamps caused nearby trees to lose their leaves.

A fifth plant hormone, **abscisic acid** (abbreviated ABA), counters the growth-stimulating effects of many other hormones. Stresses such as drought and frost stimulate the production of ABA. One immediate effect is to trigger stomata to close, which helps plants conserve water. ABA also inhibits seed germination,

Figure 24.15 Effects of Gibberellins. Gibberellins applied to a bean plant (*right*) stimulate stem elongation. The plant on the left is untreated.
©Custom Medical Stock Photo/Newscom

opposing the effects of gibberellins. Commercial growers apply ABA to inhibit the growth of nursery plants so that shipping is less likely to damage them.

C. Biologists Continue to Discover Additional Plant Hormones

Researchers once recognized only the five types of plant hormones in table 24.2. We now know that plants produce several additional hormones:

- **Brassinosteroids:** These steroid hormones occur throughout the plant but are most abundant in pollen and immature seeds. Plants lacking brassinosteroids are abnormally short, indicating that these hormones are essential for stem elongation.
- **Florigen:** Leaves make the hormone florigen, a protein that induces a shoot apical meristem to produce flowers.
- **Jasmonic acid:** This stress hormone and its volatile relative, methyl jasmonate, not only induce defenses against insects

Figure 24.16 Unripe Fruit. Exposure to ethylene would turn these green tomatoes red.
©Kent Knudson/PhotoLink/Getty Images RF

Apply It **Now**

Plants Fight Back

Ravenous insect larvae—including caterpillars—can quickly devour a plant's leaves. But plants may not be as passive as they seem. Their hidden defenses include a biochemical arsenal that punishes the offending caterpillar, warns the neighbors, and summons the insect's enemies.

Imagine a caterpillar munching on a leaf of a tomato plant. The plant's cells release multiple hormones in response to the injury. One such hormone, systemin, travels in phloem throughout the plant and stimulates the release of another hormone, jasmonic acid. Among other actions, this hormone triggers the synthesis of chemicals that destroy a caterpillar's digestive enzymes.

Plus, jasmonic acid forms a gas that prompts the injured plant's neighbors—some of which may be relatives—to strengthen their own defenses against herbivores. The jasmonic acid message lures caterpillar-killing wasps to the scene, too.

©Steven P. Lynch RF

and pathogens but also signal nearby plants to beef up their own defenses (see this chapter's Apply It Now box).

- **Salicylic acid:** This molecule, familiar to most people as aspirin, helps a plant defend against viruses and other disease-causing agents. When a plant's cells detect an attack, they release salicylic acid, which induces the surrounding cells to die. This so-called hypersensitive response keeps the pathogen from spreading to additional tissue, and the accumulation of salicylic acid makes the entire plant resistant to a wide variety of pathogens.

24.4 MASTERING CONCEPTS

1. What is a hormone?
2. How does a plant hormone exert its effects?
3. List the major classes of plant hormones, and name some of their functions.
4. Give an example of how plant hormones interact.

24.5 Light Is a Powerful Influence on Plant Life

Plants are exquisitely attuned to light. Their lives depend on it, because light is their sole energy source for photosynthesis (see chapter 5). This section illustrates some of the ways that light influences a plant's life.

A response to light requires **photoreceptors,** molecules that detect the wavelength and intensity of light. All photoreceptors have the same basic structure: a protein bound to a pigment molecule that is "tuned to" certain wavelengths. Light absorption triggers a change in the protein, which then stimulates changes in gene expression that alter the plant's appearance or behavior. ⓘ *properties of light,* section 5.2A

Of course, photoreceptors do not act alone. We have already seen that hormones help regulate fundamental functions in plants. A light-activated photoreceptor may alter the action of any hormone by influencing its production, release, or transport, or by changing the expression of receptors in target cells. These complex photoreceptor–hormone interactions remain an active area of plant biology research.

Figure 24.17 Phototropism. This bean seedling shows strong phototropism when the light is off to the side.
©Martin Shields/Science Source

A. Phototropism Is Growth Toward Light

A **tropism** is the orientation of a plant part toward or away from a stimulus such as light, gravity, or touch. All tropisms result from differential growth, in which one side of the responding organ grows faster than the other. **Phototropism,** for example, is growth toward or away from light (figure 24.17).

The plant commonly sold as "lucky bamboo" often has a curled stem, illustrating the effects of phototropism. Farmers grow the plants for a year or more in greenhouses, exposing only one side to light. Periodically rotating each plant directs the stem's growth into a twist.

Phototropism occurs in stems when cells on the shaded side elongate more than cells on the opposite side. Photoreceptors and auxins participate in the response. Photoreceptors absorb light, which causes auxins to migrate to the shaded side of the stem (figure 24.18). The auxin influx causes proteins in the cell membrane to pump protons (H^+) into the cell wall. The protons, in turn, activate enzymes that separate the cellulose fibers, enabling the wall to expand and elongate against turgor pressure. The stem bends toward the light. ⓘ *cell walls,* section 3.6B

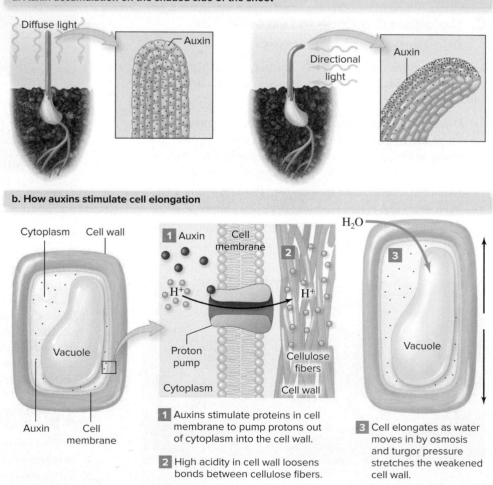

a. Auxin accumulation on the shaded side of the shoot

b. How auxins stimulate cell elongation

1 Auxins stimulate proteins in cell membrane to pump protons out of cytoplasm into the cell wall.

2 High acidity in cell wall loosens bonds between cellulose fibers.

3 Cell elongates as water moves in by osmosis and turgor pressure stretches the weakened cell wall.

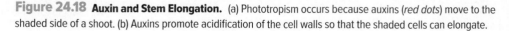

Figure 24.18 Auxin and Stem Elongation. (a) Phototropism occurs because auxins (*red dots*) move to the shaded side of a shoot. (b) Auxins promote acidification of the cell walls so that the shaded cells can elongate.

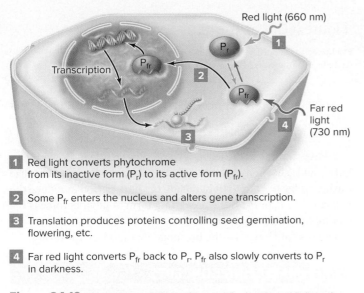

1. Red light converts phytochrome from its inactive form (P_r) to its active form (P_{fr}).

2. Some P_{fr} enters the nucleus and alters gene transcription.

3. Translation produces proteins controlling seed germination, flowering, etc.

4. Far red light converts P_{fr} back to P_r. P_{fr} also slowly converts to P_r in darkness.

Figure 24.19 Two Forms of Phytochrome. P_r absorbs red light and rapidly changes into its biologically active form (P_{fr}), which moves to the cell's nucleus and triggers a variety of responses.

Figure 24.20 Growing Up in the Dark. Seedlings grown in the light (*left*) are green, whereas those grown in the dark (*right*) have pale, elongated stems and tiny leaves.
©Nigel Cattlin/Alamy

B. Phytochrome Regulates Seed Germination, Daily Rhythms, and Flowering

One type of photoreceptor, **phytochrome,** is a light-sensitive protein that exists in two forms (**figure 24.19**). Cells produce phytochrome as P_r, which is inactive. But when P_r absorbs red wavelengths of light (660 nm), it converts nearly instantaneously to P_{fr}, the biologically active form. P_{fr} travels to the cell's nucleus and stimulates the transcription of selected genes. The P_{fr} form of phytochrome can absorb far-red wavelengths of light (730 nm) and revert to P_r; in the dark, P_{fr} also converts slowly to P_r. This back-and-forth transformation of phytochrome influences many important processes.

Seed Germination Although some seeds will germinate in total darkness, others require light. In seeds of lettuce and many weeds, for example, red light stimulates germination, and far-red light inhibits it. Exposure to red light stimulates the conversion of P_r to P_{fr}, which triggers germination. The presence of P_{fr} apparently "informs" the seed that light is available and induces the transcription of genes required for germination. On the other hand, if seeds are buried too deeply in the soil, then P_r is not converted to P_{fr} and germination does not occur.

After germination, phytochrome and another photoreceptor, cryptochrome, control early plant growth. Seedlings grown in the dark have elongated stems, small leaves, a

pale color, and a spindly appearance (**figure 24.20**). The yellowish-white bean sprouts used in Chinese cooking are grown without light. Once exposed to light, normal growth begins: Stem elongation slows, root and leaf development accelerates, and chlorophyll synthesis begins.

Circadian Rhythms **Circadian rhythms** are physiological cycles that repeat daily. In most plants, for example, stomata close at night and reopen in the morning. Some plants, such as the four-o'clock and the evening primrose, open their flowers in late afternoon or at nightfall, when their pollinators are most likely to visit. The prayer plant, *Maranta,* is a houseplant that exhibits "sleep movements." It folds its leaves vertically each night, then moves them to a horizontal position during the day (**figure 24.21**). No one knows the function of sleep movements.

Similar rhythms occur in many protists, fungi, and animals. In all species studied, pigments detect light and transmit signals to a "central oscillator," a set of genes and proteins that keep the clock on track. The details of the central oscillator, well understood in animals, remain unclear in plants.

Figure 24.21 Sleep Movements. The prayer plant exhibits rhythmic sleep movements.
(both): ©Tom McHugh/Science Source

Circadian cycles often continue under laboratory conditions of constant light or dark. They are ingrained. Nevertheless, external conditions such as a change in day length can reset (entrain) the plant's internal clock. Phytochromes absorb light and interact with the oscillator, adjusting the clock to match the new conditions.

Flowering   Plants respond in many ways to changes in **photoperiod,** or day length. In the spring, as days grow longer, buds resume growth and rapidly transform a barren deciduous forest into a leafy canopy. Bud dormancy and leaf abscission are responses to the shorter days that accompany the approach of winter. These seasonal changes illustrate the interactions among photoreceptors, environmental signals, hormones, and the plant's genes.

Photoperiod regulates flowering in some plant species. Traditionally, biologists used the term **long-day plants** for plants that flower when days are longer than a critical length, usually 9 to 16 hours. These plants typically bloom in the spring or early summer and include lettuce, spinach, beets, clover, and irises. Likewise, **short-day plants** produce flowers when days are shorter than some critical length, usually in late summer or fall. Asters, strawberries, poinsettias, potatoes, soybeans, ragweed, and goldenrods are short-day plants. **Day-neutral plants** such as tomatoes flower at maturity, regardless of day length.

However, experiments eventually confirmed that flowering actually requires a defined period of uninterrupted darkness, rather than a certain day length (figure 24.22). Thus, short-day plants are really long-night plants, because they flower only if their uninterrupted dark period exceeds a critical length. Similarly, long-day plants are really short-night plants.

Figure It Out

Duckweed is a short-day plant with a critical photoperiod of 14 hours of daylight. Will duckweed flower when there are 15 hours of daylight?

Answer: No. Duckweed will flower only if nights are longer than 8 hours; it will not bloom if nights are 7 hours long.

Phytochrome in the leaves clearly plays a role in flowering in long- and short-day plants (figure 24.23). Researchers use flashes of far-red or red light during the dark period to induce the formation of P_r or P_{fr}. If the prevailing form of phytochrome is P_{fr}, that molecule travels to the nucleus and indirectly stimulates the production of florigen—the hormone that induces flowering.

24.5   MASTERING CONCEPTS

1. What is the function of photoreceptors?
2. What is auxin's role in phototropism?
3. How does phytochrome help regulate seed germination, circadian rhythms, and flowering?
4. How do long-day plants differ from short-day plants?

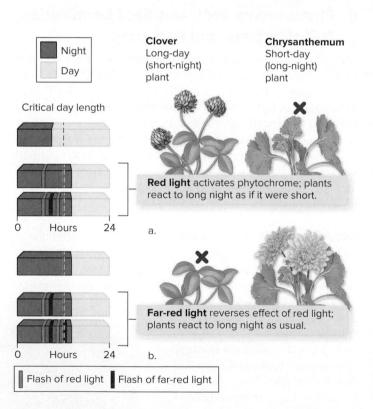

Figure 24.23   Phytochrome and the Last Flash. Interrupting darkness with flashes of light can alter the prevalent form of phytochrome. (a) A long-day (short-night) plant ordinarily does not flower if nights are too long. But a flash of red light converts P_r to P_{fr}, and the plant flowers. A short-day (long-night) plant does not flower; the time of uninterrupted darkness is too short. (b) A flash of far-red light after a flash of red causes P_{fr} to revert to P_r, canceling the effect.

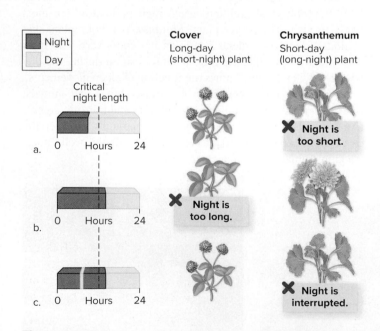

Figure 24.22   Night Length and Flowering. (a) When nights are shorter than a critical length, long-day (short-night) plants produce flowers. (b) When nights exceed a critical length, short-day (long-night) plants flower. (c) If a flash of light interrupts a long night, the short-night plants flower, but the long-night plants do not.

24.6 Plants Respond to Gravity and Touch

Besides light, a developing plant also responds to countless other environmental cues. For example, the more CO_2 in the atmosphere, the lower the density of stomata on leaves. Likewise, a plant in soil with abundant nitrogen produces fewer lateral roots than it would in nutrient-poor soil. Plants can also sense temperature; many require a prolonged cold spell before producing buds or flowers. A warm period in December, before temperatures have really plummeted, does not stimulate apple and cherry trees' buds to "break," but a similar warm-up in late February does induce growth.

Gravity is another important environmental cue. **Gravitropism** is directional growth in response to gravity (figure 24.24). As a seed germinates, its shoot points upward toward light, and its roots grow downward into the soil. Turn the plant sideways, and the stem and roots bend according to the new direction of gravity.

An asymmetrical distribution of auxins plays a role in gravitropism, as in phototropism. As we have already seen, auxin accumulation on one side of a stem causes cell elongation. In roots, however, auxins have the opposite effect: A high concentration of auxins on the "downward" side of a root inhibits cell elongation.

No one knows exactly how gravitropism works, although it is clear that the root cap must be present for roots to respond to gravity. One hypothesis centers on **statoliths,** starch-rich plastids inside root cap cells (see figure 24.24b). These organelles function as gravity detectors, sinking to the bottoms of the cells. Somehow the position of the statoliths tells the cells which direction is down. Turning a root sideways causes the statoliths to

Figure 24.25 Thigmotropism. Tendrils of a wild cucumber plant wrap around the steel wires of a fence.
©Mariëlle Hoefnagels/McGraw-Hill Education

move, redistributing calcium ions and auxins in a way that bends the root downward. ⓘ *plastids,* section 3.4D

What happens when plants are grown in a low-gravity environment, such as the International Space Station? Without gravity, a root cap's statoliths are equally distributed throughout root cells. Astronauts tending "space gardens" have observed roots growing in every direction, even as the stems grow toward the light.

Besides ever-present gravity, a plant also encounters a changing variety of mechanical stimuli, including contact with wind, rain, and animals. Even roots encounter obstacles as they grow. Over time, repeated touching produces shorter, stockier roots and shoots. The mechanism is unknown, but perhaps plants can detect slight changes in cell shape, which somehow induces the expression of genes that regulate cell expansion. Because of the touch response, the cells expand radially instead of lengthening.

The coiling tendrils of twining plants exhibit **thigmotropism,** a directional response to touch (figure 24.25). Specialized epidermal cells detect contact with an object, which induces the tendril to bend. In only 5 to 10 minutes, the tendril completely encircles the object. Auxins and ethylene apparently control thigmotropism.

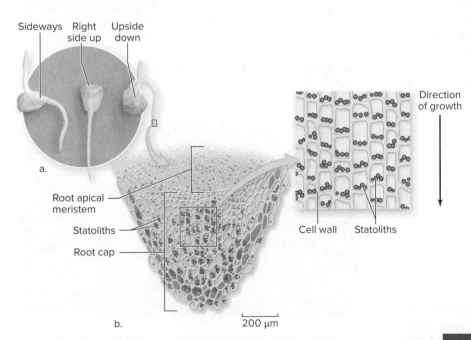

Figure 24.24 Gravitropism. (a) Shoots grow up and roots grow down, whether a seed is oriented sideways, right side up, or upside down. (b) Starch-filled statoliths in the root cap may help the plant detect gravity.
Photo: (a): ©Martin Shields/Science Source

24.6 MASTERING CONCEPTS

1. How do statoliths and auxins participate in gravitropism?
2. How does thigmotropism help some plants climb?

24.7 Plant Parts Die or Become Dormant

A normal part of every plant's life is **senescence,** or aging, when metabolism switches from synthesis to breakdown. Annual plants, whose entire lives span just one growing season, senesce and die after they produce seeds. Biennial plants grow vegetatively in one growing season, enter dormancy, then reproduce and die in the second growing season. Trees and other perennial plants live for more than 2 years, sometimes for centuries. But even their tissues senesce, often seasonally.

The most spectacular example of senescence is the changing colors of the leaves of deciduous trees in autumn. As days begin to shorten, enzymes digest proteins, chlorophyll, and other large molecules in the leaves. Yellow, orange, and red carotenoid pigments, previously masked by chlorophyll, become visible. These newly exposed carotenoids, along with the production of purplish anthocyanin pigments, combine to create the spectacular colors of autumn leaves (see the Burning Question in chapter 5).

Meanwhile, minerals such as nitrogen and phosphorus move out of the leaf for winter storage in the stem and roots. By the time a leaf falls, it is little more than a collection of cell walls and remnants of nutrient-depleted cytoplasm.

The leaf separates from the plant at its **abscission zone,** a specialized layer of cells at the base of the petiole (figure 24.26). This cell layer forms early in development, but abscission does not occur until ethylene stimulates the production of digestive enzymes that degrade the cellulose and pectin that bond the cells together.

Why would a deciduous tree spend energy on leaf production each spring, only to discard its investment each autumn? The reason is that leaves would continuously lose moisture to cold, dry winter winds. Roots cannot absorb water from frozen soil, so the tree could not replace the lost water. The seasonal loss of leaves therefore removes a large surface area from which the plant would otherwise lose water. But what about evergreen trees and shrubs? Their leaves also die, but not all at once. Instead, these plants retain leaves for several years, periodically shedding the oldest ones.

As a deciduous tree sheds its leaves before winter, other plant parts often become **dormant,** entering a state of decreased metabolism. Cells synthesize sugars and amino acids, which function as antifreeze that minimizes cold damage. Growth inhibitors accumulate in buds, transforming them into winter buds covered by thick, protective scales (figure 24.27).

A variety of cues can end a plant's dormancy. In many desert plants, for example, rainfall alone releases the plant from dormancy. Usually, however, dormancy ends with the arrival of spring. Longer days and warmer temperatures trigger growth in the plant's buds. Initially, stored carbohydrates and minerals in the stem and roots provide the raw materials that build new leaves. Soon, however, the fresh foliage produces its own food by photosynthesis, and carbohydrates begin to flow from the leaves to the rest of the plant.

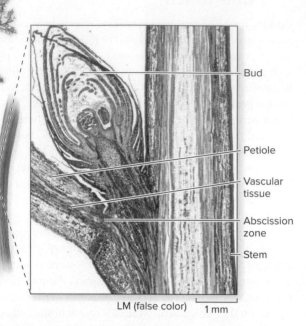

Bud

Petiole

Vascular tissue

Abscission zone

Stem

LM (false color)   1 mm

Figure 24.26 Leaf Abscission. The abscission zone is a region of separation that forms near the base of a leaf's petiole, minimizing the risk of infection and nutrient loss when the leaf is shed.
Photo: ©Steven P. Lynch RF

24.7   MASTERING CONCEPTS

1. How does seed dormancy promote reproductive success? How does seasonal dormancy in deciduous plants promote reproductive success?
2. What are some conditions that can release a plant from dormancy?

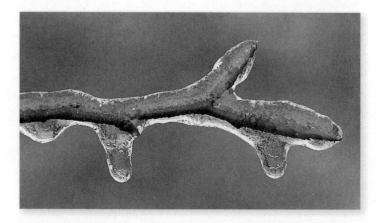

Figure 24.27 Dormancy. Some plants enter a seasonal state of dormancy, which enables them to survive harsh weather. The protective scales give the buds of this species a light brown color.
©L. West/Science Source

INVESTIGATING LIFE

24.8 A Red Hot Chili Pepper Paradox

Chili peppers are famous for their hot, spicy taste. Chilies get their kick from a unique family of chemical compounds called capsaicinoids, including the most famous one: capsaicin. The sensation ranges from the pleasantly mild pimento to the outright painful habanero.

The pungency of chilies is a bit of a paradox. After all, the main function of fruits is to disperse seeds. Many species of flowering plants produce chemicals that make their fruits unpalatable until the seeds are mature. Anyone who has sampled an unripe apple or tomato has experienced this adaptation, which protects the developing seeds from hungry herbivores. Once the seeds mature, however, most fruits lose their defensive chemicals. But chilies retain their pungent arsenal, even after the fruit is mature. Why?

Biologists Joshua Tewksbury, of the University of Washington, and Gary Nabhan, of the University of Arizona, developed a hypothesis to explain this paradox. They realized that some fruit-eating animals disperse seeds intact, but others destroy seeds. Perhaps, they suggested, capsaicinoids deter the harmful seed-destroyers without affecting the beneficial dispersers.

The researchers used field observations and laboratory feeding studies to test their hypothesis. Their study population consisted of wild chiltepin chilies (*Capsicum annuum* var. *glabriusculum*) growing in a canyon in southern Arizona. These shrubby plants produce tiny, round chilies (**figure 24.28**). First, Tewksbury and Nabhan wanted to learn which animals consumed the fruits. After videotaping chili plants for a total of 146 daylight hours, they found that birds called curve-billed thrashers accounted for 72% of all fruits that animals removed from the plants. The researchers also had indirect evidence that small mammals, which are active at night, avoided the chilies.

A laboratory study settled the question. The team captured 5 cactus mice, 5 pack rats, and 10 thrashers from the study site. Each animal was offered three types of fruit: hot chilies from the field sites, nonpungent mutant chilies, and desert hackberries. Although the birds ate all three fruits, the mice and pack rats consumed the hackberries but avoided the chilies (**figure 24.29a**).

According to Tewksbury and Nabhan's hypothesis, the capsaicin in chilies should repel seed-destroying animals without affecting dispersers. Clearly, chilies deter mammals

Figure 24.28 Chilies. The pea-sized fruits of the chiltepin chili plant are very spicy.
©Sierra Vista Herald, Jonathon Shacat/AP Images

but not birds. Do mammals harm chili seeds more than the thrashers do? The researchers set up another laboratory test to answer that question. They fed nonpungent chilies to thrashers, mice, and pack rats; control chilies were never offered to any animal. The team then compared the germination rate of the control seeds with germination rates for seeds that had passed through the digestive systems of thrashers or mammals. Figure 24.29b shows the result. The mice and pack rats destroyed the seeds, which presumably were crushed in the mammals' molars. In contrast, the seeds remained viable after consumption by thrashers.

These experiments help explain the evolutionary forces that select for capsaicin production in mature chili peppers. Interestingly, thrashers and other birds seem insensitive to capsaicin, while most mammals avoid it. The real paradox is how humans transformed this innate mammalian aversion into a worldwide love affair with the chili pepper.

Source: Tewksbury, Joshua J., and Gary P. Nabhan. 2001. Directed deterrence by capsaicin in chillies. *Nature*, vol. 412, pages 403–404.

24.8 MASTERING CONCEPTS

1. What did Tewksbury and Nabhan conclude about the function of capsaicin in mature chili fruits?
2. Explain how the data in figure 24.29 support Tewksbury and Nabhan's hypothesis.

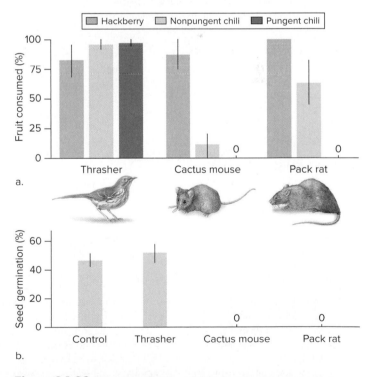

Figure 24.29 Some Like It Hot. (a) Thrashers readily ate pungent chilies; mice and pack rats rejected the spicy fruits. (b) Chili seeds that passed through thrashers remained viable, whereas seeds eaten by mice and pack rats were destroyed. (See appendix B for an explanation of error bars.)

CHAPTER SUMMARY

24.1 Angiosperms Reproduce Asexually and Sexually

A. Asexual Reproduction Yields Clones
- In **asexual reproduction,** identical clones develop from a parent plant.

B. Sexual Reproduction Generates Variability
- **Sexual reproduction** produces genetically variable offspring.
- In **alternation of generations,** the plant's diploid **sporophyte** generation produces **spores** that develop into haploid **gametophytes.** The gametophytes produce **gametes.** These haploid cells fuse at **fertilization** to produce the diploid **zygote,** which develops into a new sporophyte.

24.2 The Angiosperm Life Cycle Includes Flowers, Fruits, and Seeds

A. Flowers Are Reproductive Organs
- **Flowers** are reproductive structures built of whorls of parts attached to a **receptacle.** The **sepals** and **petals** are accessory parts. The **stamens** have pollen-containing **anthers** at their tips. **Carpels** consist of the **ovary** (each of which has one or more **ovules**) and the **style,** topped by the **stigma.**

B. The Pollen Grain and Embryo Sac Are Gametophytes
- Meiosis produces haploid **microspores** (male) and **megaspores** (female).
- Cells in **pollen sacs** undergo meiosis and then mitosis. The resulting **pollen** grain is the male gametophyte.
- Cells in ovules divide meiotically and then mitotically, yielding a female gametophyte, or **embryo sac.** Its cells include the egg and a central cell.

C. Pollination Brings Pollen to the Stigma
- Flowers are usually adapted to either animal or wind **pollination.** In **coevolution,** animal pollinators and flowers select for changes in one another.

D. Double Fertilization Yields Zygote and Endosperm
- Once on a stigma, a pollen grain grows a pollen tube, and its two sperm nuclei move through the tube toward the ovary.

- In **double fertilization,** one sperm nucleus fertilizes the egg to form the diploid zygote; the other fertilizes the central cell's two nuclei, forming the triploid **endosperm.**

E. A Seed Is an Embryo and Its Food Supply Inside a Seed Coat
- A **seed** is an embryo, endosperm (food supply), and **seed coat.** As the embryo grows, one or two **cotyledons** develop.

F. The Fruit Develops from the Ovary
- After fertilization, the ovary (and sometimes other plant parts) develops into a **fruit.**

G. Fruits Protect and Disperse Seeds
- The fruit protects the seeds and aids in dispersal. Animals, wind, and water disperse seeds to new habitats.

24.3 Plant Growth Begins with Seed Germination

- Given water, oxygen, and a favorable temperature, a seed **germinates.** The embryo bursts from the seed coat, and its development resumes.

24.4 Hormones Regulate Plant Growth and Development

- Plants respond to the environment with changes in growth, mediated by the action of **hormones** (figure 24.30).

A. Auxins and Cytokinins Are Essential for Plant Growth
- **Auxins** stimulate cell elongation and are most concentrated at the main shoot tip, which blocks the growth of lateral buds (**apical dominance**).
- **Cytokinins** stimulate cell division in actively developing plant parts, including lateral buds.

B. Gibberellins, Ethylene, and Abscisic Acid Influence Plant Development in Many Ways
- **Gibberellins** stimulate cell division and elongation, and they help break seed dormancy.
- **Ethylene** is a gas that speeds ripening, senescence, and leaf abscission.
- **Abscisic acid** induces dormancy and inhibits shoot growth.

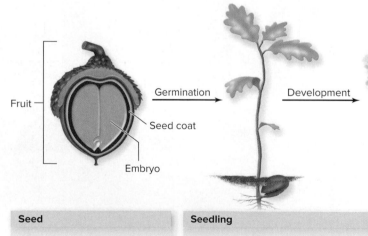

Seed	Seedling	Mature plant
• Abscisic acid maintains seed dormancy until favorable conditions arise. • Gibberellins break seed dormancy. • Cytokinins stimulate cell division in germinating seeds.	• Auxins stimulate stem elongation, suppress lateral bud growth (apical dominance), and control tropisms. • Gibberellins and cytokinins stimulate cell division in shoot and roots. • Cytokinins stimulate lateral bud growth.	• Hormones active in seedling are still active in mature plant. • Abscisic acid inhibits shoot growth and maintains bud dormancy. • Cytokinins delay leaf senescence. • Ethylene hastens fruit ripening and promotes leaf abscission.

Figure 24.30 Hormones and Plant Development: A Summary.

C. Biologists Continue to Discover Additional Plant Hormones

- Brassinosteroids are required for stem elongation. Florigen triggers flowering. Jasmonic acid and salicylic acid stimulate plant defenses.

24.5 Light Is a Powerful Influence on Plant Life

- **Photoreceptors** absorb light energy and influence a plant's growth, development, or other response to the environment.

A. Phototropism Is Growth Toward Light

- A **tropism** is a growth response toward or away from an environmental stimulus. In **phototropism,** a plant grows toward the light.

B. Phytochrome Regulates Seed Germination, Daily Rhythms, and Flowering

- **Phytochrome** controls many plant responses to light, such as flowering.
- Internal biological clocks control daily responses, or **circadian rhythms.**
- Some plants are **day-neutral,** but **short-day (long-night) plants** and **long-day (short-night) plants** use **photoperiod** as a cue in flowering.

24.6 Plants Respond to Gravity and Touch

- **Gravitropism** is growth toward or away from the direction of gravity. Starch-rich **statoliths** in cells help plants detect gravity.
- **Thigmotropism** is growth directed toward or away from a mechanical stimulus such as wind or touch.

24.7 Plant Parts Die or Become Dormant

- **Senescence** is an active and passive cessation of growth. Senescent leaves detach at an **abscission zone.**
- A plant may become **dormant** during cold or dry times.

24.8 Investigating Life: A Red Hot Chili Pepper Paradox

- The capsaicin in chili pepper fruits deters mammals, which would otherwise destroy the chili's seeds. Capsaicin does not affect birds, which eat chili pepper fruits and disperse the seeds intact.

MULTIPLE CHOICE QUESTIONS

1. Although gametes are produced via _____ in plants, genetic variation is primarily the result of _____.
   a. mitosis; meiosis
   b. meiosis; grafting
   c. budding; mitosis
   d. binary fission; asexual reproduction

2. Which of the following is NOT a flower structure typically associated with reproduction?
   a. Anther   c. Sepal
   b. Stigma   d. Carpel

3. What are the products of double fertilization?
   a. Two diploid zygotes
   b. A diploid zygote and a triploid endosperm
   c. A haploid sperm and a diploid zygote
   d. A triploid zygote

4. A fungus causes "foolish seedling" disease in rice. The shoots of infected plants grow so rapidly that they fall over from their own weight. Which plant hormone is the fungus secreting?
   a. Abscisic acid   c. Ethylene
   b. Gibberellin   d. All of the above are possible.

5. Which type of plant tropism is mismatched with its stimulus?
   a. Gravitropism: gravity
   b. Thigmotropism: nutrients
   c. Phototropism: light
   d. All of these are appropriately matched.

Answers to these questions are in appendix A.

WRITE IT OUT

1. Explain how flowers, fruits, and seeds contribute to the reproductive success of angiosperms.

2. Describe the male and female gametophytes of flowering plants.

3. How does an exclusive relationship between a plant and its pollinator benefit each partner? What are the risks of exclusive partnerships?

4. What process begins endosperm formation? How is the endosperm important to plants?

5. From what source does the endosperm receive its nutrients as it develops inside the seed?

6. If fruit production is a measure of fitness, why wouldn't a plant spend all of its energy producing fruits instead of roots and leaves? Why do you think some annual plants die back as they produce fruits?

7. Explain how plants grow toward light. Why is this response adaptive?

8. Develop a hypothesis that explains why it might be adaptive for a plant to flower in response to photoperiod rather than temperature.

9. Chrysanthemums are long-night plants that normally flower in the fall. If you could manipulate photoperiod, what would be the simplest way to prevent mums from blooming (without killing the plants)?

10. How does senescence occur?

PULL IT TOGETHER

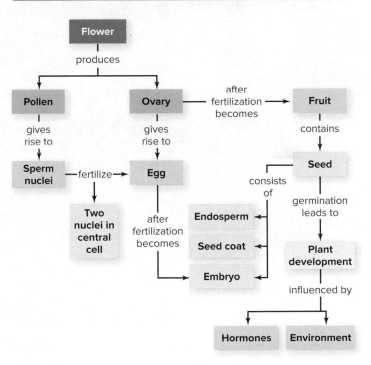

Figure 24.31 Pull It Together: Reproduction and Development of Flowering Plants.

Refer to **figure 24.31** and the chapter content to answer the following questions.

1. Review the Survey the Landscape figure in the chapter introduction, and then add *flowering plants, leaves,* and *sugars* to the Pull It Together concept map.

2. Add the following terms to the concept map: *stamen, anther, carpel, ovule, stigma.*

25 Animal Tissues and Organ Systems

Extreme Endurance Test. Athletes run in a 42-kilometer (26-mile) marathon on the streets of London, pushing their bodies to the limit.
©Daniel Garcia/AFP/Getty Images

LEARN HOW TO LEARN
Flashcard Excellence

While making flashcards, you may be tempted to focus on definitions. For example, after reading this chapter, you might make a flashcard with "simple squamous" on one side and "single layer of flattened cells" on the other. This description is correct, but it won't help you understand the bigger picture. Instead, your flashcards should include realistic questions that cover both the big picture and the small details. Try making flashcards that pose a question, such as "What are the four tissue types in an animal?" or "How do cells, tissues, organs, and organ systems relate to one another?" Write the full answer on the other side; then practice writing the answers on scratch paper until you are sure you know them all.

UNIT 6

Your Body in a Marathon

Imagine that you have registered to run a marathon, and the race is tomorrow morning. To prepare, you've eaten lots of starchy foods today. Your digestive system physically and chemically broke down the meals; sugars and other nutrients then moved from your small intestine into your circulatory system. Your blood delivered the nutrients to your body cells, so your energy reserves are high.

Your pre-race meals are the last step in a preparation process that began long ago. Months of endurance training increased your heart's ability to pump blood throughout your body. You have also felt the discomfort of physical exhaustion many times, and you have practiced paying attention to your body's changing condition as you run. Being mindful of tension in your face and shoulders helps you relax when you are tired, minimizing wasted energy. All of these experiences have physically and mentally prepared you for your race. You are ready.

Fast-forward to the morning of the race. The starter's gun goes off, and you start to run. Every step requires countless interactions between your nervous system and your muscular system. Neurons send signals to muscle cells, telling them it is time to contract. Each muscle, in turn, pulls on a bone. The brain seamlessly orchestrates these individual movements so your overall stride is smooth and balanced.

When the nervous system senses that your muscle cells have ramped up their respiration rate, neurons signal your breathing rate to increase. More oxygen enters your body, and more carbon dioxide escapes. Blood shuttles these gases between body cells and the lungs at an increasing rate as your pulse quickens.

Meanwhile, your muscle cells consume glucose, so your blood sugar level falls. In response, hormones stimulate your liver and fat cells to release stored nutrients into your blood, which delivers these molecules to the energy-hungry muscle cells. Signals from the nervous system shunt blood away from the digestive system, saving energy for the processes that your body needs to run. Sweat pours out of your skin, releasing excess body heat as it evaporates. At the same time, blood vessels near the skin surface dilate; blood arriving at the skin from the body's warm core radiates additional heat.

Hours after the race began, you finally cross the finish line. Your pulse rate, metabolism, breathing, and core body temperature slowly return to their resting levels. Your digestive system receives blood once again, preparing your body for its postmarathon meal—a welcome reward after a job well done.

SURVEY THE LANDSCAPE
Animal Life

An animal's parts include familiar organs such as skin, bones, muscles, and the brain. These organs are composed of tissues, which are themselves made of specialized cell types. All of these parts interact to maintain homeostasis in the animal body.

For more details, study the Pull It Together feature in the chapter summary.

25.1 Specialized Cells Build Animal Bodies

Everywhere we look, form and function are entwined (figure 25.1). The broad, flat surface of a plant's leaf maximizes its exposure to light. A neuron's many branches permit the cell-to-cell connections that are essential to communication in the nervous system. In birds, fluffy down feathers trap pockets of air and conserve warmth.

Anatomy, the study of an organism's structure, describes the parts that compose the body—that is, its form. **Physiology** is a related discipline that considers how those parts work—their function. Unit 5 described the anatomy and physiology of plants; unit 6 turns to animals.

Biologists describe the animal body in terms of an organizational hierarchy (figure 25.2). Most animals have **tissues,** which are groups of specialized cells that interact and provide a specific function. The inner lining of the stomach, for example, is a tissue that secretes stomach acid. An **organ** consists of two or more interacting tissues that function as a unit. The stomach is an organ that consists of muscle, blood, and nerves in addition to the tissue of its inner lining. (Artificial organs are the topic of this chapter's Burning Question.) Still farther up the organizational hierarchy are **organ systems,** which consist of two or more organs that are

Figure 25.1 Form and Function. A mole is a small mammal that spends much of its life in underground tunnels. The broad front paws, sharp claws, tiny eyes, and sensitive whiskers are among this burrowing animal's adaptations.
©image100/Corbis RF

physically or functionally joined. The human digestive system includes not only the stomach but also the small intestine, large intestine, and other organs. ⓘ *levels of organization,* section 1.1A

Burning Question

Can biologists build artificial organs?

Medical technology can help replace body parts damaged by disease or injury. Transplantable organs—taken from living donors or cadavers—include corneas, pancreases, kidneys, skin, livers, lungs, bone marrow, parts of the digestive tract, and hearts. Surgeons have even transplanted entire hands and faces.

Unfortunately, the demand for transplantable organs far exceeds the supply, and many people die while on the waiting list for a lifesaving transplant. Artificial organs, grown in the lab, may one day offer an attractive option (although they are not yet available for most patients).

Suppose a person has a defective urinary bladder. The first step in producing a replacement organ is to harvest healthy bladder cells and allow them to divide in culture. Biologists then "sow" the cells onto a biodegradable polymer (figure 25.A). This scaffold disintegrates after surgeons implant the replacement bladder into a patient. A similar technique might one day produce engineered windpipes, ears, blood vessels, kidney tubules, livers, or cartilage.

Whether an organ comes from a person or a lab bench, one challenge in transplant medicine is preventing the recipient's immune system from rejecting the foreign tissue. Biologists are therefore trying to build artificial organs out of a patient's own cells or out of materials that are "invisible" to the body's defenses. In addition, physicians can prescribe drugs that suppress the immune system, but this strategy increases the risk of deadly infection.

Submit your burning question to
Marielle.Hoefnagels@mheducation.com

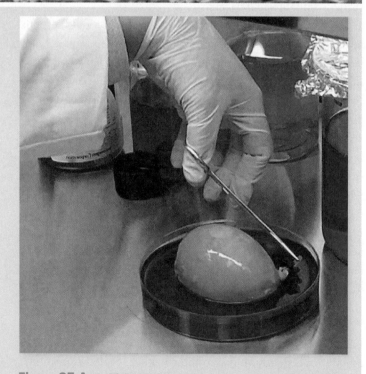

Figure 25.A Artificial Bladder. A researcher dips a bladder-shaped biodegradable mold into a solution seeded with human bladder cells.
©Brian Walker/AP Images

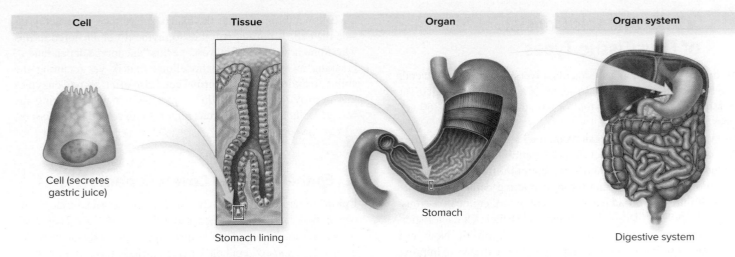

Cell	Tissue	Organ	Organ system

Cell (secretes gastric juice)

Stomach lining

Stomach

Digestive system

Figure 25.2 Organizational Hierarchy Within the Body. Cells make up tissues, which build the organs that form organ systems. In this example, a cell that secretes gastric juice is one specialized cell type in the epithelial tissue that lines the stomach. This organ, in turn, is one of many that make up the digestive system.

The human body contains dozens of organs composed of trillions of specialized cells. To understand the origin of this complexity, turn back the clock to the beginning of your life. Your father's sperm fertilized your mother's egg. Hours later, that zygote divided and became two cells. Cell division continued, and soon a tiny embryo began to take shape.

In these earliest stages of development, your cells were unspecialized and appeared pretty much alike. Then changes occurred. Some cells started expressing new combinations of genes, first developing into the three primary germ layers and later giving rise to all of your body's tissues and organs (see chapter 35). Ectoderm, for example, developed into your skin and nervous system. Endoderm differentiated into your digestive tract, liver, and lungs. Mesoderm gave rise to your muscles, bones, reproductive organs, and several other structures.
ⓘ *primary germ layers,* section 21.1C

Everyone is familiar with the overall form of the human body. Other animal bodies have wildly different shapes, from the flattened tapeworm to the squishy squid to the armored lobster to the scaly snake. But all of these animals have organs that carry out the same basic functions as our own: They sense their environment, acquire food and oxygen, eliminate wastes, protect themselves from injury and disease, and reproduce. Although this unit describes some notable adaptations in other animals, the focus is mainly on humans.

This chapter introduces the basic parts that build animal bodies. Chapters 26 through 35 then consider the organ systems one at a time. Throughout this unit, you will encounter anatomical terms that describe the position of structures in relation to the rest of the body (**figure 25.3**). For example, *dorsal* is a general term that means "toward the spine." The opposite of dorsal is *ventral,* which means "toward the belly." Learning these and the other terms in figure 25.3 should make it easier to remember the arrangement of some of the organs in the animal body.

25.1 MASTERING CONCEPTS

1. What is the difference between anatomy and physiology?
2. What is the relationship among cells, tissues, organs, and organ systems?
3. Trace the body's early development from one fertilized egg to a many-celled organism.

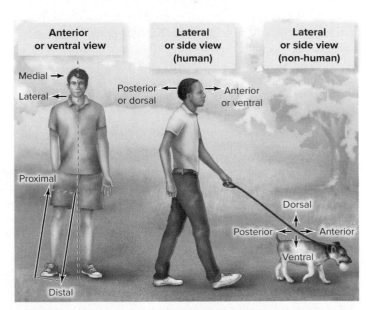

Lateral: away from the midline
Medial: toward the midline
Anterior: toward the front
Posterior: toward the back
Proximal: toward the point of attachment
Distal: away from the point of attachment
Dorsal: toward the spine
Ventral: toward the belly

Figure 25.3 Anatomical Terms. These terms describe the positions of anatomical structures relative to the rest of the animal body.

25.2 Animals Consist of Four Tissue Types

Animal bodies contain a spectacular diversity of specialized cells (see figure 3.8 to review the structure of a basic animal cell). The vertebrate body, including that of humans, has at least 260 cell types.

When people first began examining the microscopic structure of animal bodies, they noticed patterns: Some cell types tended to occur together and share common functions. Eventually, biologists realized that these interacting cells form tissues that fall into four broad categories: epithelial, connective, muscle, and nervous. Table 25.1 summarizes their characteristics, functions, and locations; this chapter's Apply It Now box describes how plastic surgeons reposition these tissues to improve a person's appearance or repair damage.

All animal tissues have a feature in common: The cells are embedded in a nonliving **extracellular matrix** consisting of a ground substance and (usually) fibers. The ground substance, in turn, is a mixture of water, proteins, carbohydrates, and lipids; in bone, the ground substance also includes minerals. Depending on the ingredients, the ground substance may be solid (as in bone), liquid (as in blood), or rubbery (as in cartilage). In most tissues, the extracellular matrix also contains fibers. The most abundant fiber is a strong, flexible protein called collagen. Elastin protein fibers add resiliency.

The extracellular matrix may take up very little space in a tissue, or it may dominate the tissue's volume and properties. In epithelial and muscle tissues, for example, cells are tightly packed, and the matrix is minimal. In connective tissue, in contrast, the matrix often occupies more volume than the cells and defines the properties of the tissue. Nervous tissue has a watery, indistinct extracellular matrix.

Interestingly, most normal body cells cannot survive or replicate when removed from the extracellular matrix. Proteins in the cell membrane bind to the matrix and to the cytoskeleton inside the cell. A normal cell fails to divide if not anchored in this way. Somehow, cancer cells escape this "anchorage dependence," breaking away from the extracellular matrix yet retaining the ability to divide. These abnormal cells also often secrete enzymes that destroy the fibers of the extracellular matrix, clearing the way for cells from a cancerous tumor to invade adjacent tissues. ⓘ *cancer,* section 8.6

A. Epithelial Tissue Covers Surfaces

Epithelial tissues coat the body's internal and external surfaces with one or more layers of tightly packed cells (figure 25.4). They cover organs and line the inside of hollow organs and body cavities. The diverse functions of epithelial tissues include protection, nutrient absorption along the intestinal tract, and gas diffusion in the lungs. These tissues also form **glands,** organs that secrete substances into ducts or into the bloodstream. Glands release breast milk, sweat, saliva, tears, mucus, hormones, enzymes, and many other important secretions.

Epithelial tissues always have a "free" surface that is exposed either to the outside or to a space within the body. On the opposite side, epithelium is anchored to underlying tissues by a layer of extracellular matrix called the basement membrane. (Despite the name, this structure is not the same as the plasma membrane that surrounds all cells.)

Connections called tight junctions often join adjacent epithelial cells into leak-proof sheets. The only way materials can pass through these cell layers is by passing through the cell membranes. The tightly knit structure of epithelial tissue is closely tied to its function as a border between the body's tissues and an open space. ⓘ *animal cell junctions,* section 3.6A

Epithelial tissues are classified partly by the shapes of their cells: squamous (flattened), cuboidal (cube-shaped), or columnar (tall and thin). The number of cell layers is also important.

TABLE **25.1**   **Animal Tissue Types: A Summary**

Tissue Type	Description	Functions	Selected Locations	Embryonic Origin
Epithelial	Single or multiple layer of flattened, cube-shaped, or columnar cells	Cover interior and exterior surfaces of organs; protection; secretion; absorption	Glands; inner linings of blood vessels, lungs, kidney tubules, digestive tract; skin	Endoderm, ectoderm, or mesoderm
Connective	Cells scattered in prominent extracellular matrix	Support, adhesion, insulation, attachment, transportation	Tendons, ligaments, cartilage, bone, blood, fat deposits	Mesoderm
Muscle	Elongated cells that contract when stimulated	Movement	Skeletal muscles, heart, arteries, digestive tract	Mesoderm
Nervous	Cells that transmit electrochemical impulses	Rapid communication among cells	Brain, spinal cord, nerves	Ectoderm

Simple epithelial tissues consist of a single layer of cells, whereas stratified epithelial tissues are made of multiple cell layers. Pseudostratified epithelium contains a single layer of columnar cells, but it appears stratified because of the staggered arrangement of nuclei.

About 90% of human cancers arise in epithelial tissues. Such a cancer is called a carcinoma. The most common carcinomas include cancers of the skin, breast, lung, prostate, and colon.

B. Most Connective Tissues Bind Other Tissues Together

The most widespread tissue type in a vertebrate's body is **connective tissue,** which consists of cells that are embedded within the extracellular matrix rather than being attached to one another. Connective tissues fill spaces, attach epithelium to other tissues, protect and cushion organs, and provide both flexible and

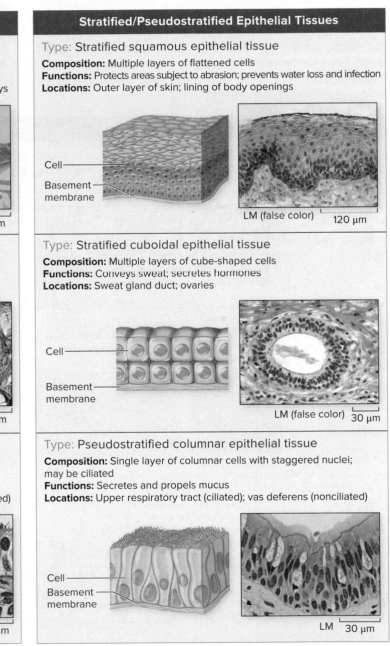

Simple Epithelial Tissues

Type: Simple squamous epithelial tissue
Composition: Single layer of flattened cells
Function: Allows substances to pass by diffusion and osmosis
Locations: Lining of blood vessels; alveoli of lungs; glomeruli of kidneys

Cell
Basement membrane
LM 30 µm

Type: Simple cuboidal epithelial tissue
Composition: Single layer of cube-shaped cells
Functions: Secretes and absorbs substances
Locations: Glands; lining of kidney tubules

Cell
Basement membrane
LM (false color) 20 µm

Type: Simple columnar epithelial tissue
Composition: Single layer of column-shaped cells; may be ciliated
Functions: Secretes and absorbs substances; sweeps egg/embryo along uterine tube
Locations: Lining of digestive tract; bronchi (ciliated); uterine tubes (ciliated)

Cell
Basement membrane
LM (false color) 10 µm

Stratified/Pseudostratified Epithelial Tissues

Type: Stratified squamous epithelial tissue
Composition: Multiple layers of flattened cells
Functions: Protects areas subject to abrasion; prevents water loss and infection
Locations: Outer layer of skin; lining of body openings

Cell
Basement membrane
LM (false color) 120 µm

Type: Stratified cuboidal epithelial tissue
Composition: Multiple layers of cube-shaped cells
Functions: Conveys sweat; secretes hormones
Locations: Sweat gland duct; ovaries

Cell
Basement membrane
LM (false color) 30 µm

Type: Pseudostratified columnar epithelial tissue
Composition: Single layer of columnar cells with staggered nuclei; may be ciliated
Functions: Secretes and propels mucus
Locations: Upper respiratory tract (ciliated); vas deferens (nonciliated)

Cell
Basement membrane
LM 30 µm

Figure 25.4 Epithelial Tissues. Epithelial tissues are composed of tightly packed cells in single or multiple layers that rest atop a basement membrane. These tissues cover body surfaces and line hollow organs and body cavities.

Photos: (simple squamous, simple cuboidal, pseudostratified columnar): ©Ed Reschke/Photolibrary/Getty Images; (stratified squamous, simple columnar): ©Victor P. Eroschenko RF; (stratified cuboidal): ©Alvin Telser/Science Source

firm structural support. Unlike epithelial tissues, connective tissues never coat any body surface.

Connective tissues are extremely variable in both structure and function (**figure 25.5**). **Loose connective tissue** binds other tissues together and fills the space between organs, **dense connective tissue** builds ligaments and tendons, and **adipose tissue** stores energy as fat. **Blood** is a liquid connective tissue that circulates within the body. **Cartilage** and **bone tissue** are connective tissues that make up the vertebrate skeleton. A close look at figure 25.5 reveals that in all connective tissues except adipose tissue, the extracellular matrix occupies more volume than do the cells.

Many cell types occur in connective tissues. In loose connective tissue, dense connective tissue, and adipose tissue, cells called fibroblasts produce and secrete the extracellular matrix. Adipocytes (fat cells) also occur singly or in small clusters in loose and dense connective tissue.

Cartilage, bone, and blood contain still other cell types. Chondrocytes and osteocytes, for example, are cells that secrete the extracellular matrix in cartilage and bone. Blood contains red blood cells, white blood cells, and cell fragments called platelets, all traveling in a liquid ground substance called plasma. The extracellular matrix of blood is unique in two ways: It is a fluid and it lacks protein fibers.

C. Muscle Tissue Provides Movement

Muscle tissue consists of cells that contract (become shorter) when stimulated. Contraction occurs when protein filaments composed of actin and myosin slide past one another inside the cell. The heat generated by muscle contraction is important in body temperature regulation.

The most familiar function of muscle tissue is to move other tissues and organs. Muscle cells attach to soft tissue or bone;

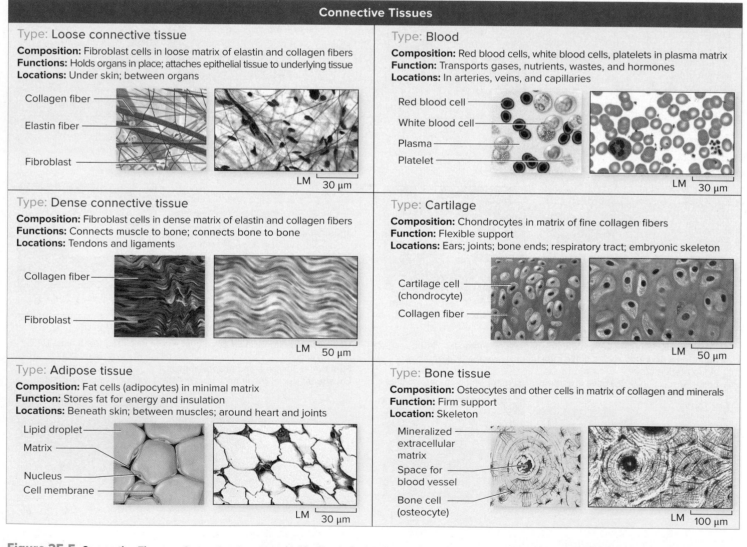

Connective Tissues

Type: Loose connective tissue
Composition: Fibroblast cells in loose matrix of elastin and collagen fibers
Functions: Holds organs in place; attaches epithelial tissue to underlying tissue
Locations: Under skin; between organs
Collagen fiber
Elastin fiber
Fibroblast
LM 30 µm

Type: Blood
Composition: Red blood cells, white blood cells, platelets in plasma matrix
Function: Transports gases, nutrients, wastes, and hormones
Locations: In arteries, veins, and capillaries
Red blood cell
White blood cell
Plasma
Platelet
LM 30 µm

Type: Dense connective tissue
Composition: Fibroblast cells in dense matrix of elastin and collagen fibers
Functions: Connects muscle to bone; connects bone to bone
Locations: Tendons and ligaments
Collagen fiber
Fibroblast
LM 50 µm

Type: Cartilage
Composition: Chondrocytes in matrix of fine collagen fibers
Function: Flexible support
Locations: Ears; joints; bone ends; respiratory tract; embryonic skeleton
Cartilage cell (chondrocyte)
Collagen fiber
LM 50 µm

Type: Adipose tissue
Composition: Fat cells (adipocytes) in minimal matrix
Function: Stores fat for energy and insulation
Locations: Beneath skin; between muscles; around heart and joints
Lipid droplet
Matrix
Nucleus
Cell membrane
LM 30 µm

Type: Bone tissue
Composition: Osteocytes and other cells in matrix of collagen and minerals
Function: Firm support
Location: Skeleton
Mineralized extracellular matrix
Space for blood vessel
Bone cell (osteocyte)
LM 100 µm

Figure 25.5 Connective Tissues. Connective tissues are highly diverse, but they all consist of cells scattered within an extracellular matrix. In most connective tissues, the matrix occupies more space than the cells themselves.

Photos: (loose connective, blood): ©Al Telser/McGraw-Hill Education; (dense connective, adipose, bone): ©Dennis Strete/McGraw-Hill Education; (cartilage): ©Chuck Brown/Science Source

when the cells contract, the body part moves. Digestion, the elimination of wastes, blood circulation, and the motion of the limbs all rely on muscle contraction.

Animal bodies contain three types of muscle tissue (**figure 25.6**). **Skeletal muscle tissue** consists of long cells called muscle fibers. When viewed with a microscope, this tissue appears striped, or striated, because the protein filaments that fill each cell align in a repeated pattern. Most skeletal muscle attaches to bone and provides voluntary movements that a person can consciously control. **Cardiac muscle tissue,** which occurs only in the heart, is also striated, but the cells are shorter, and their control is involuntary. Cardiac muscle cells are electrically coupled with one another at connections called intercalated discs. The cells contract simultaneously to produce the heartbeat. **Smooth muscle tissue** is not striated, and its contraction is involuntary. This type of muscle pushes food along the intestinal tract, regulates the diameter of blood vessels, and controls the size of the pupil of the eye.

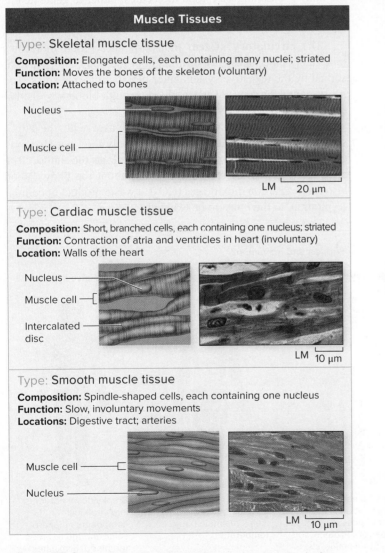

Muscle Tissues

Type: Skeletal muscle tissue

Composition: Elongated cells, each containing many nuclei; striated
Function: Moves the bones of the skeleton (voluntary)
Location: Attached to bones

Nucleus

Muscle cell

LM   20 μm

Type: Cardiac muscle tissue

Composition: Short, branched cells, each containing one nucleus; striated
Function: Contraction of atria and ventricles in heart (involuntary)
Location: Walls of the heart

Nucleus

Muscle cell

Intercalated disc

LM   10 μm

Type: Smooth muscle tissue

Composition: Spindle-shaped cells, each containing one nucleus
Function: Slow, involuntary movements
Locations: Digestive tract; arteries

Muscle cell

Nucleus

LM   10 μm

Figure 25.6 Muscle Tissues. Muscle tissue attaches to other body parts; when the muscle cells contract, the body part moves.

Photos: (skeletal, cardiac): ©Corbis RF; (smooth): ©Dennis Strete/McGraw-Hill Education

D. Nervous Tissue Forms a Rapid Communication Network

Nervous tissue uses electrochemical signals to convey information rapidly within the body. Sensory cells detect stimuli such as the scent of a rose or a prick of its thorn. Other cells then transmit that information along nerves to the central nervous system (brain and spinal cord), which helps you interpret what you experience.

Two main cell types occur in nervous tissue (**figure 25.7**). **Neurons** form communication networks that receive, process, and transmit information. The cell may connect to another neuron at a junction called a synapse, or it may stimulate a muscle or gland. **Neuroglia** (also called glial cells) are cells that support neurons and assist in their functioning. Figure 25.7 shows Schwann cells, neuroglia that form insulating sheaths of myelin around parts of a neuron. As explained in section 26.3C, the myelin sheath speeds the conduction of neural impulses.

Biologists do not yet understand many of the roles of neuroglia in nervous tissue. Once considered to be merely the "glue" that binds neurons together, neuroglia are now known to guide the early development of the nervous system, regulate the concentrations of important chemicals surrounding the neurons, and participate in the formation of synapses.

25.2 MASTERING CONCEPTS

1. If you were given a microscope slide with a slice of tissue on it, how would you classify it into one of the four main tissue types?
2. Suppose you determine that a tissue sample consists of unconnected cells embedded in an extracellular matrix. What criteria would you use to classify it into a specific subtype of tissue?

Nervous Tissue

Type: Nervous tissue

Composition: Neurons, neuroglia
Functions: Detects stimuli; conveys information throughout body
Locations: Brain; spinal cord; nerves

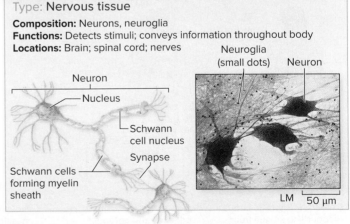

Neuron

Nucleus

Schwann cell nucleus

Schwann cells forming myelin sheath

Synapse

Neuroglia (small dots)

Neuron

LM   50 μm

Figure 25.7 Nervous Tissue. Neurons and several types of neuroglia make up nervous tissue, which specializes in rapid communication. Neuroglia called Schwann cells make up the myelin sheath coating parts of many neurons.

Photo: ©Ed Reschke/Photolibrary/Getty Images

25.3 Organ Systems Are Interconnected

As we have just seen, cells make up the animal body's four basic tissue types. These tissues build organs, such as the heart and lungs; organs interact to form organ systems. This section provides a brief overview of human organ systems, organized by their contributions to the body's function (figure 25.8). Each system may seem distinct, but the function of each one relies on extensive interactions with the others.

A. The Nervous and Endocrine Systems Coordinate Communication

The human **nervous system,** which consists of the brain, spinal cord, and nerves, specializes in rapid communication. Some neurons are sensory receptors that detect stimuli; others relay the sensory input to the spinal cord and brain. Still other neurons carry impulses from the brain or spinal cord to muscles or glands, which contract or secrete products in response.

The **endocrine system** includes glands that secrete hormones, which are communication molecules that affect development, reproduction, mental health, metabolism, and many other functions. Hormones travel within the circulatory system and stimulate a characteristic response in target organs. For example, the pituitary gland in the brain produces and secretes the hormone prolactin into the bloodstream. Prolactin binds to target cells in the breasts of a new mother and promotes milk secretion. Hormones act relatively slowly, but their effects last longer than nerve impulses.

B. The Skeletal and Muscular Systems Support and Move the Body

The **skeletal system** consists of bones, ligaments, and cartilage. Bones protect underlying soft tissues and serve as attachment points for muscles. The marrow within some bones produces the components of blood; bones also store minerals such as calcium.

Individual skeletal muscles are the organs that make up the **muscular system.** When a skeletal muscle contracts, it moves another body part or helps support a person's posture. As noted in section 33.1B, the heat released by contracting skeletal muscles also helps maintain body temperature.

C. The Digestive, Circulatory, and Respiratory Systems Help Acquire Energy

The organs of the **digestive system** dismantle food, absorb the small molecules, and eliminate indigestible wastes. All cells of the body use the digested food molecules, either to generate energy in cellular respiration or as raw materials in maintenance and growth.

The **circulatory system** transports these food molecules (and many other substances) throughout the body. Nutrients absorbed by the digestive system enter blood at the intestines. The heart pumps the nutrient-laden blood through blood vessels that extend to all of the body's cells.

The **respiratory system** exchanges gases with the atmosphere. Cellular respiration requires not only food but also oxygen gas (O_2), which diffuses into blood at the lungs. The circulatory system delivers the O_2 throughout the body. Blood also carries carbon dioxide gas (CO_2), a waste product of cellular respiration, to the lungs to be exhaled.

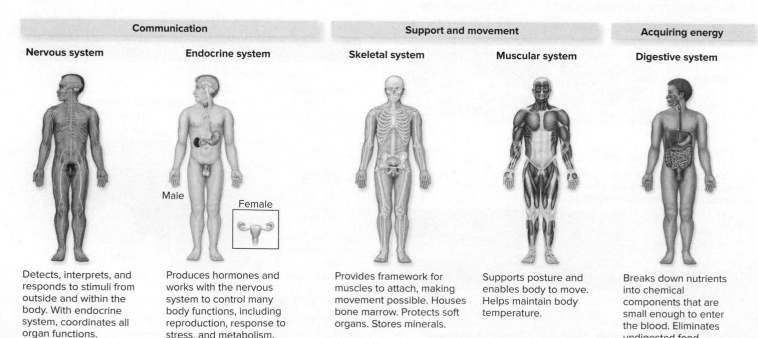

Communication		Support and movement		Acquiring energy
Nervous system	Endocrine system	Skeletal system	Muscular system	Digestive system
Detects, interprets, and responds to stimuli from outside and within the body. With endocrine system, coordinates all organ functions.	Produces hormones and works with the nervous system to control many body functions, including reproduction, response to stress, and metabolism.	Provides framework for muscles to attach, making movement possible. Houses bone marrow. Protects soft organs. Stores minerals.	Supports posture and enables body to move. Helps maintain body temperature.	Breaks down nutrients into chemical components that are small enough to enter the blood. Eliminates undigested food.

Figure 25.8 Human Organ Systems.

D. The Urinary, Integumentary, Immune, and Lymphatic Systems Protect the Body

Cell metabolism generates many waste products in addition to CO_2. These wastes enter the blood, which circulates through the kidneys. These organs are part of the **urinary system,** the organs that remove water-soluble nitrogenous wastes and other toxins from blood and eliminate them in urine. The kidneys also have other protective functions; they adjust the concentrations of many ions, balance the blood's pH, and regulate blood pressure.

One line of physical protection is the **integumentary system,** which consists of skin, associated glands, hair, and nails. Skin is the main component of the integumentary system. This waterproof barrier keeps underlying tissues moist, blocks the entry of microorganisms, and helps maintain body temperature.

The body also fights infection, injury, and cancer. The **immune system** is a huge army of specialized cells, organs, and transport vessels. This complex system launches an attack against cancer cells, viruses, microbes, and other foreign substances. Moreover, the immune system has a "memory" of previous infections. Vaccines build upon this immunological memory by "teaching" the immune system about disease-causing agents the body has never actually encountered.

The **lymphatic system** is a bridge between the immune system and the circulatory system. Lymph originates as fluid that leaks out of blood capillaries and fills the spaces around the body's cells. Lymph capillaries absorb the excess fluid and pass it through the lymph nodes, where immune system cells destroy foreign substances. The fluid then returns to the circulatory system.

E. The Reproductive System Produces the Next Generation

The **reproductive system** consists of organs that produce and transport sperm and egg cells (gametes). Examples include the testes and penis in males and the ovaries and vagina in the female. The female body also can nurture developing offspring in the uterus. Moreover, hormones from the testes and ovaries promote the development of secondary sex characteristics in adults, including the facial hair of a man and the breasts and wide hips of a woman.

The reproductive system illustrates how the organ systems are, in a sense, not separate at all. Consider the uterus, the pear-shaped sac that houses the embryo and fetus. This organ consists mainly of muscle. It also contains nervous tissue, which is why a woman feels cramps when it contracts. Hormones from the endocrine system stimulate these contractions. The entire system is richly supplied with the circulatory system's blood vessels, which also deliver the cells and biochemicals of the immune system.

The remaining chapters in this unit describe the animal organ systems in the order in which they appear in figure 25.8, beginning with communication and ending with reproduction. The exception is the integumentary system, the subject of section 25.5.

25.3 MASTERING CONCEPTS

1. Which organ systems contribute to each of the five general functions of life?
2. What are three examples of interactions between organ systems?

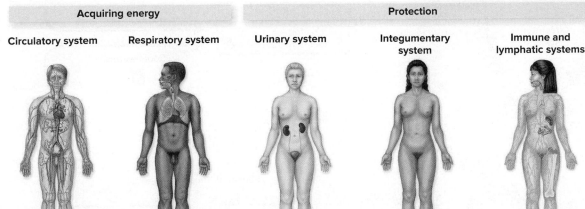

| Acquiring energy | | Protection | | | Reproduction |

| **Circulatory system** | **Respiratory system** | **Urinary system** | **Integumentary system** | **Immune and lymphatic systems** | **Reproductive system**
 Male   Female |

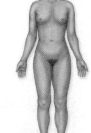
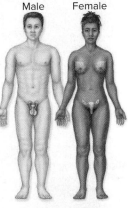

| Vessels carry blood throughout the body, nourishing cells, delivering oxygen, and removing wastes. | Delivers oxygen to blood and removes carbon dioxide. | Excretes nitrogenous wastes and maintains volume and composition of body fluids. | Protects the body, controls temperature, and conserves water. | Protect the body from infection, injury, and cancer. | Manufactures gametes and enables the female to carry and give birth to offspring. |

25.4 Organ System Interactions Promote Homeostasis

So far, this chapter has emphasized cells, tissues, organs, and organ systems. An animal's body, however, consists mostly of water. Some of this moisture makes up the cytoplasm inside every cell. The rest of it forms blood plasma and the **interstitial fluid** that bathes the body's cells. Because interstitial fluid is inside the body but outside the cells, biologists consider it part of the "internal environment." Many organ systems interact to help maintain the correct concentrations of nutrients, salts, hydrogen ions, and dissolved gases in body fluids (**figure 25.9**).

Yet the external environment, which surrounds the body, changes constantly. Temperatures rise and fall; food may be abundant or scarce; water comes and goes. In the midst of this great variability, an animal's body must maintain its internal temperature, its blood pressure, and the chemical composition of its fluids within certain limits. **Homeostasis** is this state of internal stability.

All organisms maintain homeostasis. Without it, a body system may stop functioning, and the organism may die. As just one example, consider what happens if the lungs fill with water. The body can no longer acquire O_2 or dispose of CO_2, yet cells continue to respire. Soon, all available O_2 in the blood is consumed, and CO_2 accumulates to toxic levels. Cells begin to die, and the person will drown unless rescuers arrive quickly.

Many physiological mechanisms that maintain homeostasis use **negative feedback,** in which a change in a condition triggers action that reverses the change. **Figure 25.10** illustrates negative feedback in a familiar situation: maintaining room temperature. When the room gets too warm, the heater turns off. When the temperature is low, the thermostat signals the heater to switch back on.

In all negative feedback systems, **sensors** monitor changes in some parameter. If the value is too high or too low, the sensor activates one or more effectors. The **effector** responds by counteracting the original change. In figure 25.10, the thermostat is the sensor, and the heater is the effector.

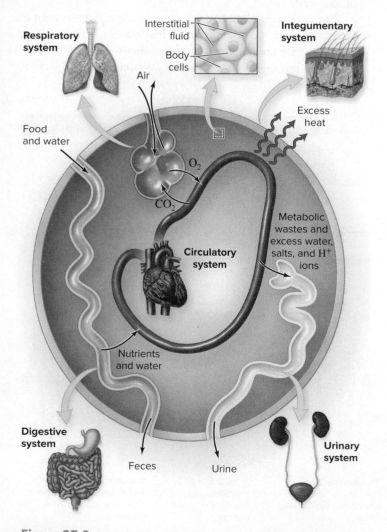

Figure 25.9 **Organ System Interactions.** Organ systems work together to maintain a constant body temperature and optimal concentrations of O_2, CO_2, nutrients, and other substances in the body's fluids.

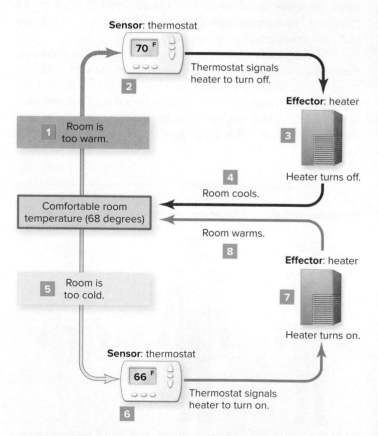

Figure 25.10 **Negative Feedback: An Example.** A negative feedback system maintains room temperature within comfortable limits. (*1*) If the room is too warm, (*2*) the thermostat sends a signal to the heater. (*3*) The heater shuts off, and (*4*) the room cools. (*5*) If the room is too cold, (*6*) the thermostat sends a different signal to the heater, (*7*) which switches on. (*8*) This action warms the room.

Likewise, the body uses sensors and effectors to maintain homeostasis. The "supervisor" that coordinates much of the action is an almond-sized part of the brain called the hypothalamus. If blood pressure rises too high, for example, receptors in the walls of blood vessels signal the hypothalamus to slow the contraction of the heart. The pressure drops. If blood pressure falls too low, the hypothalamus signals the heart to speed up, sending out more blood. ⓘ *blood pressure regulation,* section 30.5B

As you will see throughout this unit, the hypothalamus participates in many negative feedback loops. If oxygen is scarce, the hypothalamus stimulates faster breathing. If the body's temperature deviates from normal, the hypothalamus initiates mechanisms that help the body release or conserve heat. If the blood's salt concentration is too high, hormones from the hypothalamus signal the kidney to release more salt into urine, and so on. ⓘ *breathing control,* section 31.4B; *thermoregulation,* section 33.1; *osmoregulation,* section 33.2

Only a few biological functions demonstrate **positive feedback,** in which the body reacts to a change by amplifying it. Blood clotting and childbirth are examples of positive feedback—once started, they perpetuate their activity. By itself, positive feedback therefore does not maintain homeostasis. Ultimately, however, other controls cut off the positive feedback loop and restore equilibrium. ⓘ *blood clotting,* section 30.2D; *childbirth,* section 35.5E

25.4 | MASTERING CONCEPTS

1. Use figure 25.9 to explain which materials enter and exit the body, and how they do so.
2. What happens to an organism that fails to maintain homeostasis?
3. Distinguish between negative and positive feedback.

25.5 The Integumentary System Regulates Temperature and Conserves Moisture

The integumentary system consists mostly of **skin,** the organ that forms the body's outer surface. Hair, nails, and several types of glands complete the system.

This organ system beautifully illustrates the main themes of this chapter. Not only does skin consist of multiple interacting tissue types, but the integumentary system also protects the body and helps maintain homeostasis in several ways.

The most obvious way that skin helps maintain homeostasis is body temperature regulation. Specialized nerve endings in the skin sense the temperature and convey the information to the hypothalamus. If the temperature is too high, the hypothalamus stimulates blood vessels in the skin to dilate, releasing heat from deeper tissues to the surroundings. Meanwhile, perspiration pours out of sweat glands. When the temperature is too cold, the skin's blood vessels constrict, keeping warm blood away from the body's surface.

The integument contributes to homeostasis in other ways as well. For example, skin conserves water. People who suffer extensive burns lose large amounts of skin and may die of dehydration. Moreover, burn patients are vulnerable to infection because intact skin is the first defense against disease-causing microorganisms. Skin even plays a role in nutrition, because the initial step in vitamin D synthesis occurs when ultraviolet light strikes the skin. Among other functions, vitamin D helps regulate calcium and phosphate concentrations in the blood, which in turn influences bone strength. Inadequate exposure to sunlight or a diet low in vitamin D causes bones to soften and weaken.

Apply It **Now**

Two Faces of Plastic Surgery

The "plastic" in "plastic surgery" has nothing to do with the substance that makes up water bottles. Instead, the word derives from the Greek word *plastikos,* which means "to mold." Plastic surgeons "mold" a person's appearance by moving and reshaping tissues such as fat, bone, and cartilage.

Cosmetic surgery is one field of plastic surgery, and its scope extends far beyond aging movie stars struggling to maintain the illusion of youth. In fact, millions of people undergo cosmetic surgery every year in the United States. Among the most common procedures is a rhinoplasty, commonly called a "nose job," in which a surgeon removes or repositions some of the cartilage and bone of the nose to create a new shape. Liposuction is also popular (**figure 25.B**). The surgeon makes small incisions and uses a surgical vacuum to remove excess fat from the thighs, buttocks, arms, neck, or stomach. Face lifts are common as well. In a typical face lift, a surgeon makes a long incision at the hairline, lifts the skin of the face, and repositions the muscle and connective tissue under the skin. He or she then tightens the skin and trims the excess before reattaching the skin to the face. Other popular procedures include breast augmentation, breast reduction, breast lifts, buttock lifts, tummy tucks, laser skin resurfacing, hair transplants, and collagen injections.

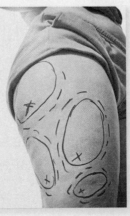

Cosmetic surgery may enhance a healthy person's appearance, but reconstructive plastic surgery has a different goal: to restore the function of damaged body parts. For example, some children are born with a cleft palate (a gap in the bones between the nose and mouth). To repair a cleft palate, a plastic surgeon moves tissue on either side of the gap to close the opening. Reconstructive surgery may also include skin grafts (for burn patients) or breast reconstruction (for women who have lost one or both breasts to cancer). Far from frivolous, these procedures can dramatically improve a patient's ability to function.

Figure 25.B Liposuction Markings.
©image100/Corbis RF

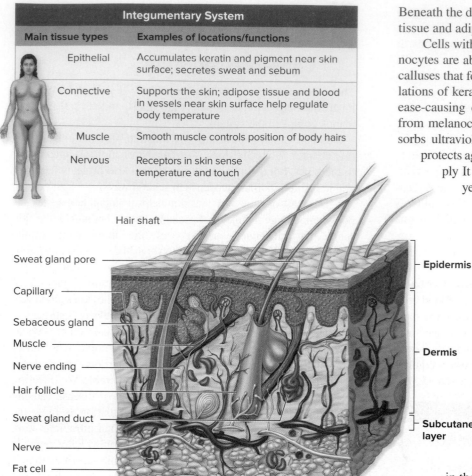

Integumentary System	
Main tissue types	**Examples of locations/functions**
Epithelial	Accumulates keratin and pigment near skin surface; secretes sweat and sebum
Connective	Supports the skin; adipose tissue and blood in vessels near skin surface help regulate body temperature
Muscle	Smooth muscle controls position of body hairs
Nervous	Receptors in skin sense temperature and touch

Hair shaft

Sweat gland pore

Capillary

Sebaceous gland

Muscle

Nerve ending

Hair follicle

Sweat gland duct

Nerve

Fat cell

Epidermis

Dermis

Subcutaneous layer

Figure 25.11 **The Integumentary System.**

Skin is a surprisingly extensive organ. It is only 1 to 2 mm thick on average, although skin on the soles of the feet and in several other places is thicker. Nevertheless, it makes up about 15% by weight of an adult's body.

Human skin has two major layers (figure 25.11): the epidermis and the dermis. The **epidermis** is the outermost layer, and it consists mostly of stratified squamous epithelium. Active cell division replaces dry, dead epidermal cells that are continuously lost to abrasion at the surface. Below the epidermis is the **dermis,** which is composed mostly of collagen but also contains elastin fibers, nerve endings, smooth muscle, sweat and oil glands, and the blood vessels that nourish both skin layers. A thin basement membrane anchors the epidermis to the underlying dermis.

Figure 25.12 **Tattoo.**
To create a tattoo, a needle is used to deposit ink into the dermis of the skin.
©Lee Davenport

Beneath the dermis lies a subcutaneous layer of loose connective tissue and adipose tissue that is not technically part of the skin.

Cells within the epidermis have specialized functions. Keratinocytes are abundant cells that produce the protein keratin. (The calluses that form on the soles of the feet are thick, scaly accumulations of keratin.) The tough, dry keratin protects skin from disease-causing organisms and abrasion. The skin's color derives from melanocytes, cells that produce melanin. This pigment absorbs ultraviolet radiation that can damage DNA and therefore protects against some types of skin cancer (see chapter 8's Apply It Now box). Other epidermal cells detect touch, and yet others help fight infection if the skin is injured.

Hairs also emerge from the epidermis. A hair grows from a follicle anchored in the dermis. Cells at the base of a hair follicle divide, pushing daughter cells up. These cells stiffen with keratin and die to form the hair.

Hair has many functions. A mammal's fur can provide camouflage, as the white fur of an Arctic fox illustrates. Conversely, the striped fur of a skunk acts as a warning. In addition, smooth muscle tissue surrounding follicles can make hair "stand on end" when skin senses cold—this is the basis of goose bumps. In many mammals, erect fur provides insulation that helps retain body heat.

The tissues that make up the dermis also have other functions. Collagen fibers in the dermis support the skin. In addition, humans can generate a wide array of facial expressions, thanks to muscles attached to fibers in the dermis. Nerve endings in the dermis convey signals from sensory receptors to the brain.

Several types of glands originate in the dermis as well. Sweat glands produce perspiration. Mammary glands, which produce milk in female mammals, are derived from sweat glands. Many mammals, including humans, have sebaceous glands, which secrete an oily substance (sebum) that softens and protects the skin and hair.

Injuries that damage the dermis, such as a cut from broken glass, can leave a scar. As the injury heals, the replacement skin contains more collagen fibers than the surrounding undamaged tissue, but it lacks hair follicles and sweat glands. The resulting scar consists of new skin that does not look exactly like the old.

Many people decorate their skin with one or more tattoos. A tattoo looks as if it is on the skin's surface (figure 25.12). If it were, however, it would quickly disappear as epidermal cells sloughed off. Instead, a needle deposits ink directly into the dermis, where the design remains permanently in place.

25.5 MASTERING CONCEPTS

1. How does the integumentary system help the body maintain homeostasis?
2. What tissue types and specialized cells occur in each layer of human skin?

25.6 Vitamins and the Evolution of Human Skin Pigmentation

Human skin color ranges from the very pale to the deeply pigmented (see figure 10.29). How did this variation arise?

Scientists have long understood that the darker the skin, the higher the concentration of melanin. Moreover, indigenous people from the tropics (near the equator) tend to have darker skin than their counterparts from higher latitudes (toward the poles). Tropical areas receive more ultraviolet (UV) radiation than do locations farther north and south. High melanin production is therefore correlated with exposure to UV radiation.

A question arises from these observations: Have differences in UV radiation selected for variation in pigmentation? More specifically, is there a relationship among UV radiation, melanin production, and reproductive success in the human population?

The risk for skin cancer offers one possible answer. The more melanin in the skin, the less UV radiation penetrates below the epidermis, and the lower the risk for cancer. But skin cancer is rarely fatal, and most cases develop after a person's reproductive years. Since skin cancer usually does not interfere with reproductive success, it is probably not a strong selective force in the evolution of skin color.

Oddly enough, nutrition may offer a more convincing explanation for the worldwide distribution of skin pigmentation. The body produces vitamin D when sunlight strikes the skin. This vitamin is crucial to reproductive success; a vitamin D deficiency can cause death or pelvic deformities that make childbirth difficult. One hypothesis that explains the variation in skin pigmentation, then, reflects the body's need for vitamin D. In high latitudes, light skin may maximize the potential to synthesize this essential molecule.

Pennsylvania State University anthropologists Nina Jablonski and George Chaplin tested the strength of the association between vitamin D nutrition and UV radiation. They began by mapping satellite measurements of UV radiation striking Earth's surface. The pair also searched the scientific literature for measurements of skin pigmentation in indigenous peoples anywhere in the world. They then superimposed three zones on their map, corresponding to areas where indigenous people have lightly, moderately, and darkly pigmented skin (figure 25.13).

Jablonski and Chaplin also looked for scientific papers documenting how much UV radiation is needed to synthesize sufficient vitamin D. They compared the results to their map and found a strong association between skin pigmentation and the potential to produce vitamin D. That is, in high latitudes, UV exposure is insufficient to produce vitamin D in people with dark skin. These regions therefore select for light pigmentation.

These results support the hypothesis that vitamin D deficiency becomes an increasingly important selective force at higher latitudes, where sunlight can be scarce. This result leaves open the question of why areas with high UV exposure are associated with dark skin pigmentation. The answer may relate to another vitamin, B_9 (folic acid). UV radiation destroys folic acid in the skin. In men, insufficient folic acid may cause infertility. If a woman has too little folic acid, her children may be born with serious birth defects. Skin pigmentation may therefore reflect a nutritional trade-off between sun-induced loss of folic acid and production of vitamin D.

This study not only provides insight into an interesting question about human evolution but also illustrates two important features of scientific investigation. First, technological advances can offer new ways to test old hypotheses. In this case, satellites enabled Jablonski and Chaplin to peek with unprecedented detail at ultraviolet radiation patterns across the globe. Second, scientists often borrow heavily from the past; dozens of previous studies on skin pigmentation and vitamin D synthesis gave the research team the information they needed to reach their conclusion.

Source: Jablonski, Nina G., and George Chaplin. 2000. The evolution of human skin coloration. *Journal of Human Evolution*, vol. 39, pages 57–106.

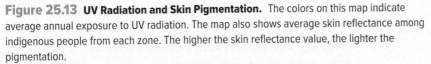

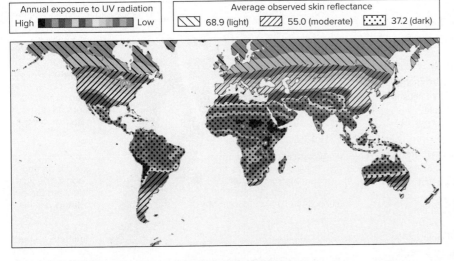

Figure 25.13 UV Radiation and Skin Pigmentation. The colors on this map indicate average annual exposure to UV radiation. The map also shows average skin reflectance among indigenous people from each zone. The higher the skin reflectance value, the lighter the pigmentation.

25.6 MASTERING CONCEPTS

1. Refer to figure 25.13. Explain the overlap between areas colored gray and areas marked with distantly spaced lines. How might vitamin D explain these data?

2. Suggest a specific prediction related to folic acid that follows from the nutritional trade-off hypothesis.

CHAPTER SUMMARY

25.1 Specialized Cells Build Animal Bodies

- **Anatomy** and **physiology** are interacting studies of the structure and function of organisms.
- Specialized cells express different genes. These cells aggregate and function together to form **tissues.** Tissues build **organs,** and interacting organs form **organ systems.**

25.2 Animals Consist of Four Tissue Types

- Animals have four main tissue types.
- Animal tissues consist of cells within an **extracellular matrix** consisting of ground substance and (usually) protein fibers. The matrix may be solid or liquid.

A. Epithelial Tissue Covers Surfaces

- **Epithelial tissue** lines and covers organs; it also forms **glands.** This tissue protects underlying tissue, senses stimuli, and secretes substances.
- Epithelium may be simple (one layer) or stratified (more than one layer), and the cells may be squamous (flat), cuboidal (cube-shaped), or columnar (tall and thin).

B. Most Connective Tissues Bind Other Tissues Together

- **Connective tissues** have diverse structures and functions. Most consist of scattered cells and a prominent extracellular matrix.
- The six major types of connective tissues are **loose connective tissue, dense connective tissue, adipose tissue, blood, cartilage,** and **bone.**

C. Muscle Tissue Provides Movement

- **Muscle tissue** consists of cells that contract when protein filaments slide past one another.
- Three types of muscle tissue are **skeletal, cardiac,** and **smooth muscle.**

D. Nervous Tissue Forms a Rapid Communication Network

- **Neurons** and **neuroglia** make up **nervous tissue.**
- A neuron functions in rapid communication; neuroglia support neurons.

25.3 Organ Systems Are Interconnected

A. The Nervous and Endocrine Systems Coordinate Communication

- The **nervous system** and **endocrine system** coordinate all other organ systems.
- Neurons form networks of cells that communicate rapidly, whereas hormones produced by the endocrine system act more slowly.

B. The Skeletal and Muscular Systems Support and Move the Body

- The bones of the **skeletal system** protect and support the body. Bones also store calcium and other minerals.
- The **muscular system** enables body parts to move and generates body heat.

C. The Digestive, Circulatory, and Respiratory Systems Help Acquire Energy

- The **digestive system** provides nutrients. The **respiratory system** obtains O_2, and the **circulatory system** delivers nutrients and O_2 to tissues.
- The body's cells use O_2 to extract energy from food molecules. The circulatory and respiratory systems eliminate the waste CO_2.

D. The Urinary, Integumentary, Immune, and Lymphatic Systems Protect the Body

- The **urinary system** removes metabolic wastes from the blood and reabsorbs useful substances.
- The **integumentary system** provides a physical barrier between the body and its surroundings.
- The **immune system** protects against foreign substances and cancer.
- The **lymphatic system** connects the circulatory and immune systems, passing the body's fluids through the lymph nodes.

E. The Reproductive System Produces the Next Generation

- The male and female **reproductive systems** are essential for the production of offspring.
- **Figure 25.14** summarizes the human body's organ systems.

25.4 Organ System Interactions Promote Homeostasis

- **Homeostasis** is stability in the internal environment even as external conditions change. Animals may maintain homeostasis in body temperature and in the chemical composition of the blood plasma and the **interstitial fluid.**
- In **negative feedback, sensors** detect a change in the internal environment and activate **effectors** that counteract the change. The overall effect is to restore the parameter to its normal range.
- **Positive feedback** reinforces the effect of a change.

25.5 The Integumentary System Regulates Temperature and Conserves Moisture

- **Skin** helps regulate body temperature, conserves moisture, and contributes to vitamin D production.
- Skin consists of an **epidermis** over a **dermis,** plus specialized structures such as hairs and sweat glands. A basement membrane joins the epidermis to the dermis, and a layer of connective tissue underlies the dermis.
- In the epidermis, keratinocytes accumulate keratin, and melanocytes provide pigment.

25.6 Investigating Life: Vitamins and the Evolution of Human Skin Pigmentation

- Variation in exposure to ultraviolet radiation may select for a range of skin pigmentation, reflecting a balance between folic acid and vitamin D nutrition.

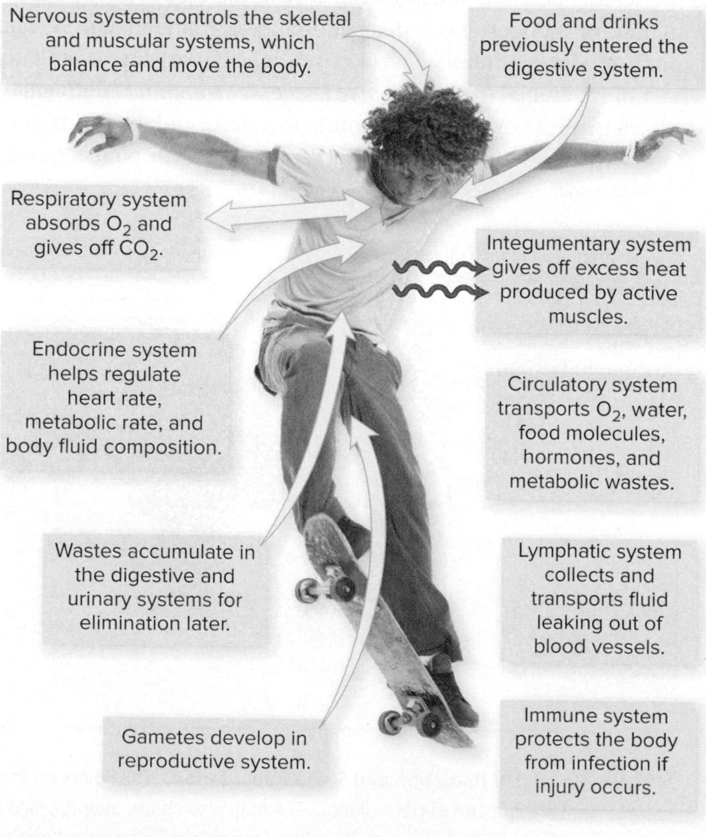

Nervous system controls the skeletal and muscular systems, which balance and move the body.

Food and drinks previously entered the digestive system.

Respiratory system absorbs O_2 and gives off CO_2.

Integumentary system gives off excess heat produced by active muscles.

Endocrine system helps regulate heart rate, metabolic rate, and body fluid composition.

Circulatory system transports O_2, water, food molecules, hormones, and metabolic wastes.

Wastes accumulate in the digestive and urinary systems for elimination later.

Lymphatic system collects and transports fluid leaking out of blood vessels.

Gametes develop in reproductive system.

Immune system protects the body from infection if injury occurs.

Figure 25.14 Organ System Functions: A Summary.

Photo: ©James Woodson/Getty Images RF

MULTIPLE CHOICE QUESTIONS

1. Which of the following represents the correct order of organization of an animal's body?
   a. Cells; organs; organ systems; tissues
   b. Cells; tissues; organ systems; organs
   c. Tissues; cells; organs; organ systems
   d. Cells; tissues; organs; organ systems

2. According to the anatomical terms in figure 25.3, your nose is _____ to your ears.
   a. distal
   b. dorsal
   c. medial
   d. lateral

3. Epithelial tissue consisting of multiple layers of flattened cells is called
   a. simple cuboidal.
   b. simple squamous.
   c. stratified cuboidal.
   d. stratified squamous.

4. Which of the following features do all types of connective tissue share?
   a. Cells are connected firmly to each other.
   b. Each cell contacts an extracellular matrix.
   c. Abundant protein fibers surround cells.
   d. The extracellular matrix is fluid.

5. Smooth muscle is *different* from skeletal muscle because
   a. smooth muscle contraction is involuntary.
   b. skeletal muscle is striated.
   c. smooth muscle contains actin and myosin.
   d. Both a and b are correct.

6. Ovaries produce gametes and hormones; these organs therefore belong to the _____ systems.
   a. immune and integumentary
   b. reproductive and endocrine
   c. circulatory and nervous
   d. urinary and lymphatic

7. Which of the following scenarios does NOT illustrate negative feedback?
   a. In childbirth, contractions stimulate the release of oxytocin, which induces more contractions.
   b. Body temperature climbs so high that a person begins to sweat, which cools the body.
   c. The salt concentration in blood is too high, so the kidneys eliminate salt in urine.
   d. Eating a meal causes a rise in blood sugar, which stimulates body cells to absorb sugar from blood.

8. The inner layer of skin is _____, whereas the outer layer is _____.
   a. epithelial tissue; connective tissue
   b. dermis; epidermis
   c. epidermis; epithelial tissue
   d. epithelial tissue; epidermis

9. How does the integumentary system influence homeostasis?
   a. By preventing water loss
   b. By sensing external temperatures
   c. By preventing infection
   d. All of the above are correct.

Answers to these questions are in appendix A.

WRITE IT OUT

1. Distinguish between the following.
   a. Organs and organ systems
   b. Simple squamous and stratified squamous epithelial tissue
   c. Loose and dense connective tissue
   d. Skeletal and cardiac muscle tissue
   e. Neurons and neuroglia
   f. Negative and positive feedback

2. Use the Internet to research cosmetic surgery. Write a paragraph explaining how a plastic surgeon might manipulate each of the four main tissue types.

3. List and describe six types of connective tissue.

4. Explain the similarities and differences among the three types of muscle tissue.

5. Explain how you use your organ systems (except your reproductive system) while you are exercising at the gym, and provide three hypotheses that might explain why some exercise sessions go better than others.

6. Describe how the circulatory system connects the digestive system with the urinary system.

7. When a person gets cold, he or she may begin to shiver. If the weather is too hot, the heart rate increases and blood vessels dilate, sending more blood to the skin. How does each scenario illustrate homeostasis?

8. Provide nonbiological examples of negative and positive feedback, other than those presented in the chapter.

9. Describe at least one interaction between skin and each of the 10 other organ systems.

PULL IT TOGETHER

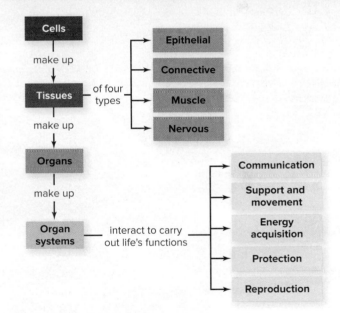

Figure 25.15 Pull It Together: Animal Tissues and Organ Systems.

Refer to figure 25.15 and the chapter content to answer the following questions.

1. What features distinguish the four types of tissue?

2. Add *skin* to the concept map, making at least five connections to existing terms in the map.

3. Review the Survey the Landscape figure in the chapter introduction, and then connect *homeostasis* to the concept map in at least two ways. One connection to *homeostasis* should include a detailed description of how one "life function" (e.g., *energy acquisition*) maintains homeostasis.

26 The Nervous System

Blow to the Head. A kickboxer's head jerks backward after his opponent delivers a kick. Repeated blows to the head can cause long-term brain damage.

©Henrik Sorensen/Riser/Getty Images

LEARN HOW TO LEARN
How to Use a Study Guide

Some professors provide study guides to help students prepare for exams. Used wisely, a study guide can be a valuable tool. One good way to use a study guide is AFTER you have studied the material, when you can go through the guide and make sure you haven't overlooked any important topics. Alternatively, you can cross off topics as you encounter them while you study. No matter which technique you choose, don't just memorize isolated facts. Instead, try to understand how the items on the study guide relate to one another. If you're unclear on the relationships, be sure to ask your instructor.

The Repercussions of Repeated Concussions

Mild traumatic brain injury, also called a concussion, impairs brain function. Concussions can occur when a boxer gets punched, when football players collide, during a traffic accident, or in a fall. People who have recently experienced a concussion may have headaches, trouble concentrating, nausea, and many other symptoms.

Concussions may be mild or severe, depending mainly on how fast the head changes direction. The rotational acceleration of the head—like the motion you make when you nod "yes"—is most important. A layer of fluid typically cushions the brain during minor rotational forces. But when the neck rapidly jerks forward or backward, this protective fluid cannot stop the brain from crashing into the interior of the skull.

The brain consists of millions of neurons, which are cells that communicate via electrochemical signals. For these signals to be properly conducted, each neuron must maintain a specific balance of positively charged ions on the inside and outside of its membrane. Rapid movement of the brain stretches or tears brain cells, disrupting their delicate chemistry. Neuron function becomes impaired.

The immediate result of mild brain trauma may be a brief loss of consciousness, but the consequences may last for hours or days. Surviving neurons must use extra energy to reestablish chemical homeostasis, which leaves less energy for normal brain activity. Doctors therefore advise patients with a concussion to limit physical and mental activity, giving neurons time to recover.

Frequent concussions may also affect brain health in the long term. People who experience multiple concussions are more likely than average to develop depression or neurodegenerative disorders such as Parkinson disease. Moreover, images of athletes' brains show deterioration after numerous traumas. Because football players are especially vulnerable to concussions, these results have initiated a wave of lawsuits between the National Football League and former players who accuse the league of failing to inform and protect them. War veterans also frequently suffer from long-term effects of brain traumas, following repeated exposure to explosions.

Understanding and treating brain injury requires knowledge about the structures and functions of individual neurons, groups of neurons, and the entire nervous system, including each region of the brain. This chapter explores these topics.

SURVEY THE LANDSCAPE
Animal Life

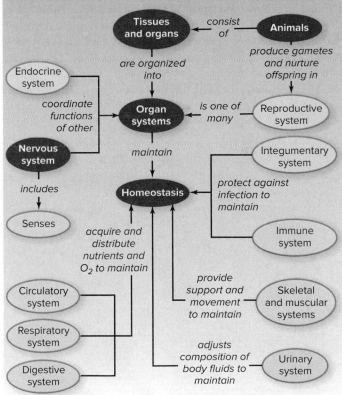

In the nervous system, neurons are cells that communicate with one another and with cells in other organ systems. Along with the endocrine system, the nervous system therefore plays a major role in maintaining homeostasis.

For more details, study the Pull It Together feature in the chapter summary.

26.1 The Nervous System Forms a Rapid Communication Network

Love, happiness, tranquility, sadness, jealousy, rage, fear, and excitement—all of these emotions spring from the cells of the nervous system (figure 26.1). So do language, the sensation of warmth, memories of your childhood, and your perception of pain, color, sound, smell, and taste. The muscles that move when you scratch, chew, blink, or breathe all are controlled by the nervous system, as is the unseen motion that propels food along your digestive tract.

Among the nervous system's most critical functions are the "behind the scenes" activities that keep the body's temperature, ion balance, and other conditions within optimal levels. The negative feedback loops that maintain homeostasis require communication between the sensors (the cells that detect each condition) and the effectors (the muscles and glands that make adjustments). Together, the **nervous system** and the **endocrine system** provide this communication. A major difference between these two organ systems is the speed with which they act. The nervous system's electrochemical impulses travel so rapidly that their effects are essentially instantaneous. The moment you decide to pick up a pencil or type a letter, for example, you can carry out your plan. The endocrine system, the subject of chapter 28, acts much more slowly. Endocrine glands secrete chemical messages called hormones that can take minutes or hours to take effect, but these signals last much longer than do neural impulses. ⓘ *negative feedback,* section 25.4

Nervous tissue includes two basic cell types: interconnected neurons and their associated neuroglia. The **neurons** are the cells

Figure 26.1 Great Fun. Thanks to the nervous system, these roller-coaster passengers can experience a range of sensations and emotions.
©Digital Vision RF

that communicate with one another. The more numerous **neuroglia** (also called glial cells) provide physical support, help maintain homeostasis in the fluid surrounding the neurons, guide neuron growth, and play many other roles that researchers are just beginning to discover. ⓘ *nervous tissue,* section 25.2D

The neurons and neuroglia that make up nervous tissue control mood, appetite, blood pressure, coordination, and the perception of pain and pleasure. Neurons enable animals to sense the environment, screen out unimportant stimuli, move, learn, and remember. Yet despite their diverse functions, all neurons communicate in a similar manner. This chapter explores nervous system diversity, describes how neurons function, and then considers the human nervous system in more detail.

A. Invertebrates Have Nerve Nets, Nerve Ladders, or Nerve Cords

Nervous systems vary greatly in complexity and organization. Depending on the species, a nervous system might be a loose network of relatively few cells or a structure as intricate as the human brain. The more complex the nervous system, the greater the animal's ability to detect multiple stimuli simultaneously and to coordinate responses. Moreover, highly developed nervous systems are associated with complex behaviors and an increased capacity for learning, memory, problem solving, and language.

Figure 26.2 illustrates some examples of animal nervous systems (chapter 21 describes each of the illustrated animals in detail). The simplest are **nerve nets,** which are diffuse networks of neurons in the body walls of hydras, jellyfishes, sea anemones, and other cnidarians. The nerve cells physically touch one another, so that a stimulus at any point on the body spreads over the entire body surface. Nerve nets stimulate muscle cells near the body surface, enabling the animals to move their tentacles or swim.

Cnidarians are relatively simple, radially symmetrical animals. The evolution of bilateral symmetry, however, meant that one end of the body (the head) took in the most sensory information from the environment. Nervous tissue began to accumulate into a brain and sensory structures at the head end, and the nervous system became increasingly centralized.

A flatworm is a simple bilaterally symmetrical animal. Its brain consists of two **ganglia,** or clusters of neurons, at its head end. Two nerve cords run along the body's length; multiple lateral nerves connect the two cords, forming a **nerve ladder.** Paired muscles along the nerve ladder allow the worm to move in a coordinated forward motion. In addition, paired chemical receptors compare the strength of a stimulus on each side of its body.

Segmented worms such as earthworms have a more complex nervous system with a larger brain. A **ventral nerve cord** connected to the brain branches into each segment, so the animal can coordinate its movements. Cells that are sensitive to light, chemicals, and touch occur all over the body surface.

The nervous system of an arthropod (including insects and spiders) consists of a brain, a ventral nerve cord, and organs that

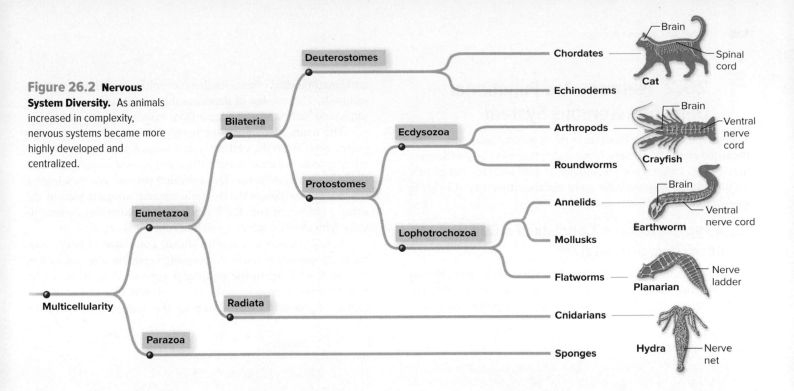

Figure 26.2 Nervous System Diversity. As animals increased in complexity, nervous systems became more highly developed and centralized.

Labels in figure: Deuterostomes, Bilateria, Ecdysozoa, Protostomes, Eumetazoa, Lophotrochozoa, Multicellularity, Radiata, Parazoa

Chordates — Cat — Brain, Spinal cord
Echinoderms
Arthropods — Crayfish — Brain, Ventral nerve cord
Roundworms
Annelids — Earthworm — Brain, Ventral nerve cord
Mollusks
Flatworms — Planarian — Nerve ladder
Cnidarians — Hydra — Nerve net
Sponges

detect light, touch, chemicals, sound, and balance. Many of these animals exhibit complex behaviors, such as the famous "waggle dance" by which honeybees communicate the location of food sources.

Among invertebrates, the octopuses and squids have the most sophisticated nervous systems. These intelligent animals have large brains, keen eyes, and sensitive tentacles. They can even master complex visual tasks, such as opening food jars!

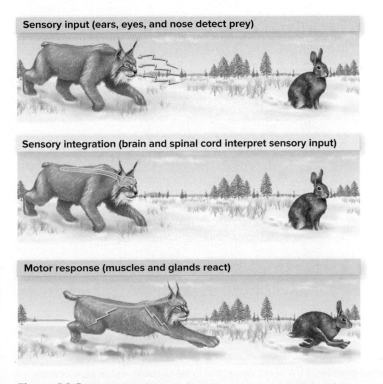

Sensory input (ears, eyes, and nose detect prey)

Sensory integration (brain and spinal cord interpret sensory input)

Motor response (muscles and glands react)

Figure 26.3 Roles of the Nervous System. Sensory organs such as the eyes, ears, and nose receive sensory input. The central nervous system integrates the information and sends signals that initiate appropriate motor responses.

B. Vertebrate Nervous Systems Are Highly Centralized

The trend toward nervous tissue accumulation in the head is most evident in vertebrates. The vertebrate nervous system has two main divisions: the central and peripheral nervous systems. The **central nervous system** consists of the **brain** (inside the skull) and the dorsal, tubular **spinal cord** (see figure 26.2). The main function of these two organs is to integrate sensory information and coordinate the body's response.

The **peripheral nervous system** carries information between the central nervous system and the rest of the body. The sensory pathways of the peripheral nervous system carry information to the spinal cord and brain; the motor pathways transmit nerve impulses from the central nervous system to muscle or gland cells.

To understand how the nervous system regulates virtually all other organ systems, imagine a lynx hunting a hare (**figure 26.3**). Sensory neurons in the peripheral nervous system enable the cat to hear, see, and smell its prey. The lynx's central nervous system interprets this sensory input and decides how to act, and then motor neurons coordinate the skeletal muscles that move the lynx into position to catch the hare. Meanwhile, the cat's heart pumps blood, and its lungs inhale and exhale—all under the control of the central and peripheral nervous systems.

26.1 MASTERING CONCEPTS

1. How are nervous systems adaptive to animals?
2. Describe how nervous systems changed as animal bodies became more complex.
3. Distinguish between the central and peripheral nervous systems in vertebrates.

26.2 Neurons Are Functional Units of a Nervous System

The nervous system's function is rapid communication by electrical and chemical signals. Neurons are the cells that do the communicating, either with one another or with muscles and glands. To understand how neurons carry out their function, it helps to first learn about their structure.

A. A Typical Neuron Consists of a Cell Body, Dendrites, and an Axon

All neurons have the same basic parts (figure 26.4). The enlarged **cell body** contains the nucleus, mitochondria that supply ATP, ribosomes that manufacture proteins, and other organelles. **Dendrites** are short, branched extensions that transmit information toward the cell body. The number of dendrites may range from one to thousands, and each can receive input from many other neurons.

The **axon,** also called the nerve fiber, conducts nerve impulses away from the cell body and toward a muscle, gland, or other neuron. An axon is typically a single long extension that is finely branched at its tip. The axon that permits you to wiggle a big toe, for example, extends about a meter from the base of the spinal cord to the toe. Each tiny terminal extension communicates with another cell at a junction called a synapse.

In many neurons, a **myelin sheath** composed of fatty material coats sections of the axon, speeding nerve impulse conduction (see section 26.3). In the peripheral nervous system, neuroglia called **Schwann cells** form the myelin sheath; other types of neuroglia, **oligodendrocytes,** make up the myelin sheaths in the

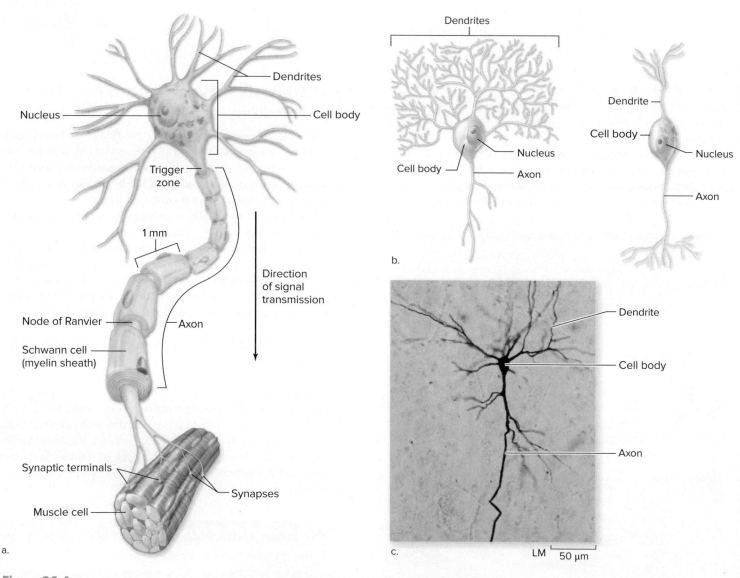

Figure 26.4 **Parts of a Neuron**. (a) A neuron consists of a cell body, one or more dendrites that transmit information to the cell body, and an axon. In many neurons, the axon is encased in a myelin sheath. (b) Neuron shapes vary. A motor neuron with multiple dendrites is on the left, and a sensory neuron with one dendrite is on the right. (c) Several neurons are visible in this micrograph of the brain's outermost layer.

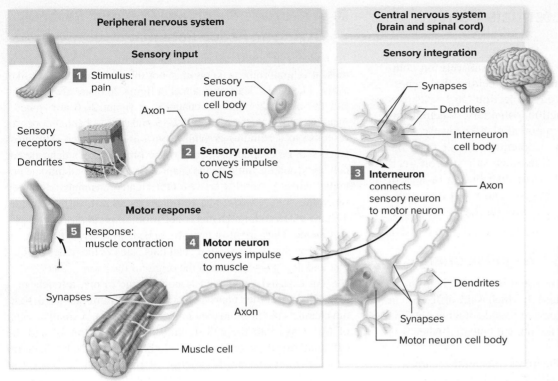

Peripheral nervous system

Sensory input

1 Stimulus: pain

Sensory neuron cell body

Axon

Sensory receptors

Dendrites

2 **Sensory neuron** conveys impulse to CNS

Motor response

5 Response: muscle contraction

4 **Motor neuron** conveys impulse to muscle

Synapses

Axon

Muscle cell

Central nervous system (brain and spinal cord)

Sensory integration

Synapses

Dendrites

Interneuron cell body

3 **Interneuron** connects sensory neuron to motor neuron

Axon

Dendrites

Synapses

Motor neuron cell body

Figure 26.5 **Categories of Neurons.** Sensory neurons transmit information from sensory receptors to the central nervous system. Interneurons connect sensory neurons to motor neurons, which send information from the central nervous system to muscles or glands.

central nervous system. **Nodes of Ranvier** are short regions of exposed axon between sections of the myelin sheath.

To picture the relative sizes of a typical neuron's parts, imagine its cell body is the size of a tennis ball. The axon might then be up to 1.5 kilometers long—the length of about 63 tennis courts—but only a few centimeters thick. The mass of dendrites extending from the cell body would fill an average-size living room.

B. The Nervous System Includes Three Classes of Neurons

Biologists divide neurons into three categories, based on general function (figure 26.5).

- A **sensory neuron** brings information to the central nervous system from the rest of the body. Sensory neurons respond to light, pressure from sound waves, heat, touch, pain, and chemicals detected as odors or taste. The dendrites, cell body, and most of the axon of each sensory neuron lie in the peripheral nervous system, whereas the axon's endings reside in the central nervous system.

- About 90% of all neurons are **interneurons,** which connect one neuron to another within the spinal cord and brain. Large, complex networks of interneurons receive information from sensory neurons, process this information, and generate the messages that the motor neurons carry. Interneurons may also receive signals from other interneurons.

- A **motor neuron** conducts its message from the central nervous system toward an effector (muscle or gland). A motor neuron's cell body and dendrites reside in the central nervous system, but its axon extends into the peripheral nervous system. Thus, motor neurons stimulate muscle cells to contract and stimulate glands to secrete their products into the bloodstream or into a duct. (They are called motor neurons because most lead to muscle cells, not glands.)

Figure 26.5 shows a simplified example of how the three types of neurons work together to coordinate the body's reaction to a painful stimulus. The process begins when a person steps on a tack. Sensory neurons whose dendrites are in the skin of the foot convey the information to the spinal cord. There, the sensory neuron transmits the signal to an interneuron, which in turn synapses on a motor neuron that stimulates muscle contraction in the leg and foot (see section 29.4). The interneuron in the spinal cord also sends action potentials to an area of the brain that interprets the sensation as pain. The signals move so quickly that the foot withdraws at about the same time as the brain perceives the pain.

26.2 **MASTERING CONCEPTS**

1. Describe the parts of a typical neuron.
2. Where is the myelin sheath located?
3. In what direction does a message move in a neuron?
4. What are the functions of each of the three classes of neurons?

26.3 Action Potentials Convey Messages

Your thoughts, sensations, and movements all rely on communication between neurons, as do the "behind the scenes" feedback mechanisms introduced in section 26.1. Neurons send messages by conveying action potentials; a neural impulse is the propagation of action potentials like a wave along an axon. As you will see, action potentials result from the movement of charged particles (ions) across a neuron's cell membrane. This section describes the distribution of ions in neurons, both when the cell is "at rest" and when it is transmitting a neural impulse. ⓘ *ions*, section 2.1B; *cell membrane*, section 3.3

A. A Neuron at Rest Has a Negative Charge

To understand how ions move in a neural impulse, it helps to be familiar with the **membrane potential**, which is the difference in electrical charge between the inside and outside of a neuron. The neuron's **resting potential** is its membrane potential when it is not conducting a neural impulse.

At rest, the interior of a neuron's membrane carries a slightly negative electrical charge relative to the outside because it maintains an unequal distribution of ions across its membrane (**figure 26.6**, step 1). In particular, the concentration of potassium (K^+) is much higher inside the cell than outside, while the reverse is true for sodium (Na^+). One membrane protein that helps maintain this gradient is the **sodium–potassium pump**, which pumps three Na^+ out of the cell for every two K^+ that enter; the energy cost is one ATP molecule per cycle. The sodium–potassium pump operates continuously. ⓘ *sodium–potassium pump*, section 4.5B

The neuron's resting potential reflects a balance of forces on K^+. On one hand, K^+ is more concentrated inside the cell than outside, thanks to the sodium–potassium pump. Membrane proteins called "leakage channels" allow some of this K^+ to diffuse along its concentration gradient and leave the cell. On the other hand, positively charged Na^+ ions outside the cell repel K^+, while large, negatively charged proteins (and other negative ions) inside the cell attract K^+. ⓘ *facilitated diffusion*, section 4.5A

When the opposing forces—the concentration gradient and charge interactions—are equal, the membrane has a net positive charge on the outside and a net negative charge on the inside. This difference in charge is the resting potential, and its magnitude is approximately −70 millivolts (mV). (A volt measures the difference in electrical charge between two points.)

The term *resting potential* is a bit misleading because the neuron consumes a tremendous amount of energy while "at rest." In fact, the nervous system devotes about three quarters of its total energy budget to maintaining the ion gradients that characterize the resting potential. The resulting state of readiness allows each neuron to respond quickly when a stimulus does arrive. The resting potential is therefore like holding back the string on a bow to be constantly ready to shoot an arrow.

B. A Neuron's Membrane Potential Reverses During an Action Potential

To understand how a neuron initiates and sustains action potentials, it is important to know that not all ion channels are alike. Some, like the K^+ leakage channel in figure 26.6, are always open. But the other Na^+ and K^+ channels in figure 26.6 are "gated," meaning they can open and close. Some gated channels open in response to a stimulus arriving from *outside* the cell. Stimulus-gated sodium channels typically occur on a neuron's dendrites or cell body, and they initiate the change in a neuron's membrane potential when a stimulus arrives. Other gated channels open only when the voltage *inside* the cell changes relative to the outside. Voltage-gated sodium channels occur on all parts of the neuron's membrane. They are most densely packed, however, in the parts of the neuron that sustain action potentials, such as the nodes of Ranvier and the "trigger zone" at the origin of the axon.

An external signal—such as a change in pH, a touch, or a signal from another neuron—causes some stimulus-gated sodium channels to open and then immediately close. A small amount of Na^+ leaks into the cell through the open channels, and the cell's interior therefore becomes slightly depolarized. (The term *depolarization* means the interior of the neuron becomes less negative relative to the outside.) This local flow of electrical current is a type of **graded potential** because it weakens with distance from the source of the stimulus. If the stimulus is very weak, the graded potential is small, and the neuron will not "fire."

But if the stimulus is strong enough to cause multiple stimulus-gated channels to open at once, the depolarization may spread from its source to the trigger zone. If enough Na^+ enters to depolarize the trigger zone's membrane to the **threshold potential** (about −50 mV), an action potential begins.

An **action potential** is a brief "all-or-none" depolarization at a patch of an axon's membrane; steps 2, 3, and 4 in figure 26.6 illustrate its events. Once the trigger zone reaches its threshold potential, additional voltage-gated sodium channels in the axon open (figure 26.6, step 2). Driven by both the electrical gradient and the concentration gradient, Na^+ pours into the cell. The axon's membrane now has a positive charge at its interior side. This reversal of the membrane potential lasts only for an instant. Near the peak of the action potential, membrane permeability changes again. Sodium channel gates close, again preventing Na^+ from entering the cell (figure 26.6, step 3). However, the delayed voltage-activated K^+ channels are now open. Repelled by the positively charged ions inside the cell, K^+ pours out.

The loss of K^+ restores the negative charge in the cell's interior (figure 26.6, step 4). The sodium–potassium pump, which operates continuously, returns Na^+ to the membrane's exterior. For a short time, an unusually large number of K^+ channels are open, so the axon briefly hyperpolarizes to a membrane potential less than −70 mV. (*Hyperpolarization* means the interior of the neuron becomes even more negative relative to the outside than it is at rest.) The resting potential is quickly restored, however, once potassium permeability returns to normal. The entire process, from the initial influx of Na^+ to the restoration of the resting potential, takes only 1 to 5 milliseconds to complete.

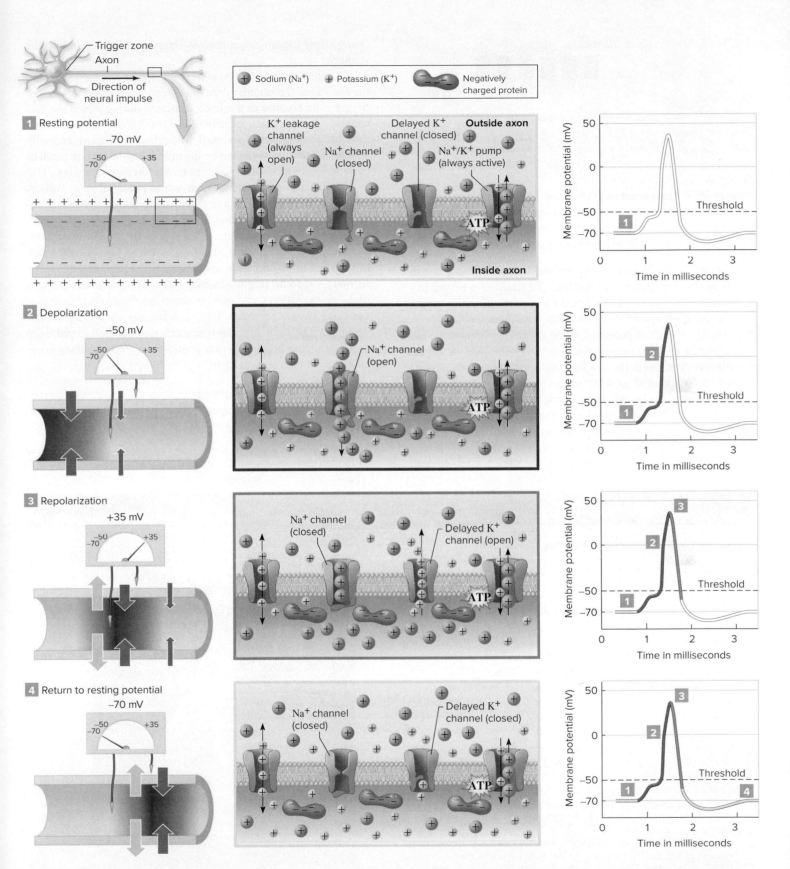

Figure 26.6 Resting Potential and Action Potential. (*1*) At rest, the neuron's membrane potential is about −70 mV. Na⁺ channels are closed, but the Na⁺/K⁺ pump is active and the K⁺ leakage channel is open (as always). (*2*) If the membrane depolarizes to its threshold potential, an action potential begins. The membrane reverses its polarity as Na⁺ pours in through open voltage-gated Na⁺ channels. (*3*) After a split second, Na⁺ channels close. Delayed K⁺ channels open, and K⁺ pours out of the axon. The cell's interior repolarizes, regaining its negative charge. (*4*) The membrane briefly hyperpolarizes as the resting potential is reestablished.

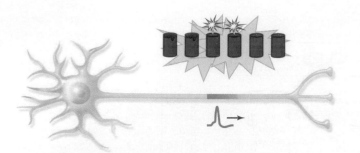

Figure 26.7 Chain Reaction. Lighting one firecracker can set off a chain reaction if the firecrackers are close together, but not if they are far apart. Likewise, only those parts of the neuron with densely packed, voltage-gated Na⁺ channels can sustain action potentials in a chain reaction.

Figure 26.6 shows how an action potential occurs at one small patch of a neuron's membrane. To transmit a neural impulse, however, the electrical signal must move from the trigger zone to the other end of the axon. How does this occur?

During an action potential, some of the Na⁺ ions that rush into the cell diffuse a short distance along the interior of the membrane. As a result, the neighboring patch of the axon reaches its threshold potential as well, triggering a new influx of Na⁺. The resulting chain reaction carries the impulse forward (**figure 26.7**). A neural impulse is therefore similar to people "doing the wave" in a football stadium, when successive groups of spectators stand and then quickly sit. The participants do not change their locations, yet the wave appears to travel around the stadium.

A neural impulse begins at the trigger zone and moves toward the axon's tip. It ordinarily does not spread "backward" toward the trigger zone because of a refractory period during which the Na⁺ channels in each patch of membrane are inactivated and cannot generate another action potential. The refractory period lasts only a couple of milliseconds, but by the time it is over for one patch of membrane, the neural impulse has moved on down the axon.

Unlike a graded potential, an action potential is an "all-or-none" event; that is, either a full-fledged action potential proceeds to completion or none occurs at all. The nervous system therefore detects the strength of a stimulus by measuring the frequency of incoming action potentials, not their size. A light touch to the skin, for example, might produce 10 action potentials per second. A hard hit might generate 100 action potentials per second—along with a more intense sensation. Neurons also distinguish the type of stimulus. For example, we can tell light from sound because light stimulates neurons that transmit impulses to one part of the brain, whereas sound-generated impulses go to another part (see section 26.6).

Figure It Out

If negatively charged chlorine atoms (Cl⁻) move into a neuron, does an action potential become more likely or less likely? Why?

Answer: Less likely; the membrane becomes more polarized.

Burning Question

Do neurons communicate at the speed of light?

Many people assume that neurons work like the metal wiring in a home, conducting electricity at nearly the speed of light (about 300 million meters per second). In fact, biologists once thought that nerve impulses were instantaneous. But experiments eventually demonstrated that a mammal's neurons conduct impulses at a top speed of about 100 m/sec (224 miles per hour). That speed is impressive, and it allows athletes and ordinary people to react quickly to incoming stimuli, but it is just a tiny fraction of the speed of light.

A simple exercise can help you estimate the speed of neural communication in your body. You'll need a partner, a ruler, paper and pencil, and a calculator. Have your partner hold the ruler vertically by the 30 cm mark, and cup your own hand around the 0 cm mark. Your partner should then drop the ruler without saying anything. Grab the ruler as quickly as possible after your partner drops it, and note the centimeter mark where you caught it. Repeat three times; use the average of the three results as your distance measurement (D). Then use the following equation to convert D into a time measurement (T):

$$T = \sqrt{(2 \times D) \div 980}$$

T represents the time (in seconds) it took for receptors in your eye to send neural impulses to the brain, for the brain to process the information and send a neural impulse along a motor neuron to your arm, and for your muscles to close your hand around the ruler. Keep your reaction time in mind as you learn how cells in the nervous system communicate.

©Akihiro Sugimoto/age fotostock RF

Submit your burning question to
Marielle.Hoefnagels@mheducation.com

C. The Myelin Sheath Speeds Impulse Conduction

The greater the diameter of an axon, the faster it conducts an impulse. A squid's "giant axons" are up to 1 mm in diameter. (Much of what biologists know about action potentials comes from studies on these large-diameter nerve fibers.) Axons from vertebrates are a hundredth to a thousandth the diameter of the squid's. Yet even thin vertebrate axons can conduct impulses very rapidly when they are coated with a myelin sheath (figure 26.8a).

Myelin prevents ion flow across the membrane. At first glance it might therefore seem that myelin should prevent the spread of action potentials. But the entire axon is not coated with myelin. Instead, ions can move across the membrane at the nodes of Ranvier. These gaps in the myelin sheath house high densities of voltage-gated sodium channels. When an action potential happens at one node, Na$^+$ entering the axon diffuses to the next node. The incoming Na$^+$ depolarizes the membrane and stimulates the sodium channels to open at the second node, triggering an action potential there. In this way, when a neural impulse travels along the axon, it appears to "jump" from node to node (figure 26.8b).

The neural impulse moves up to 100 times faster when it leaps between nodes than when it spreads along an unmyelinated axon. Not surprisingly, myelinated axons occur in neural pathways where speed is essential, such as those that transmit motor commands to skeletal muscles. Thanks to myelin, a sensory message travels from the toe to the spinal cord in less than 1/100 of a second. Unmyelinated axons occur in pathways where speed is less important, such as in the neurons that trigger the secretion of stomach acid. (This chapter's Burning Question explores the nervous system's speed.)

Illness can result if an axon has too much or not enough myelin. In Tay-Sachs disease, for example, cell membranes accumulate excess lipid and wrap around nerve cells, burying them in fat so that they cannot transmit messages to each other and to muscle cells. An affected child gradually loses the ability to see, hear, and move. In multiple sclerosis, the reverse happens. The immune system destroys oligodendrocytes in the central nervous system, leading to a loss of myelin. Neural messages are therefore weakened or lost. Symptoms of multiple sclerosis include impaired vision and movement, numbness, and tremors.

26.3 | MASTERING CONCEPTS

1. Describe the forces that maintain the distribution of K$^+$ and Na$^+$ across the cell membrane in a neuron at rest.
2. In what way is the term *resting potential* misleading?
3. Differentiate among a graded potential, the threshold potential, and an action potential.
4. How does an axon generate and transmit a neural impulse?
5. What prevents action potentials from spreading in both directions along an axon?
6. How do action potentials indicate stimulus intensity and type?
7. How do myelin and the nodes of Ranvier speed neural impulse transmission along an axon?

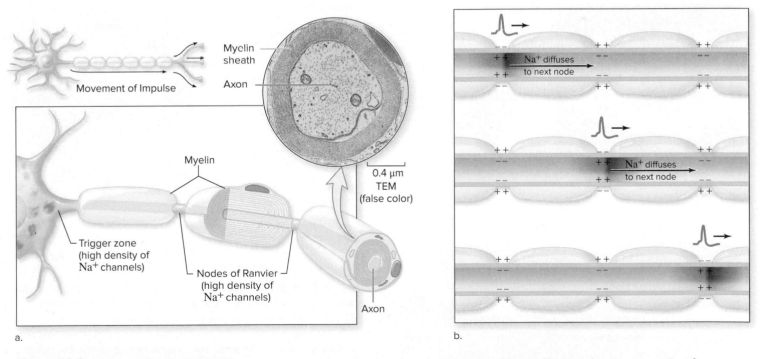

a.   b.

Figure 26.8 The Role of the Myelin Sheath. (a) The myelin sheath on an axon is interrupted by nodes of Ranvier. The inset shows a cross section of a myelinated axon. (b) In myelinated axons, Na$^+$ channels occur only at the nodes of Ranvier. Action potentials appear to "jump" from one node to the next, which speeds impulse transmission along the axon.

Photo: (a): ©Fawcett/Science Source

26.4 Neurotransmitters Pass the Message from Cell to Cell

To form a communication network, a neuron conducting action potentials must convey the information to another cell. Most neurons do not touch each other, so the electrical impulse cannot travel directly from cell to cell. Instead, a neural impulse that reaches the tip of an axon causes the release of a **neurotransmitter,** a chemical signal that travels from a "sending" cell to a "receiving" cell across a tiny space.

A. Neurons Communicate at Synapses

A **synapse** is a specialized junction at which the axon of a neuron communicates with another cell (**figure 26.9**). Every synapse has three main components. First, the **presynaptic cell** is the neuron sending the message. Second, the **postsynaptic cell** receives the message; the receiving cell may be another neuron, a muscle cell, or a gland cell. (The axon of the presynaptic cell usually synapses onto the dendrites or cell body of a postsynaptic neuron, as in figure 26.9a, but a synapse can also form directly on another axon.) Third, the **synaptic cleft** is the space between the two cells.

The axon of a presynaptic cell enlarges at its tip to form a knob-shaped **synaptic terminal** (figure 26.9a and b). Each tiny knob contains many small sacs, or vesicles, that hold

neurotransmitter molecules; dozens of vesicles are visible in figure 26.9c. Immediately opposite the synaptic terminal, receptor proteins in the postsynaptic cell membrane can bind to the neurotransmitters.

The steps in figure 26.9b show how two neurons communicate at a synapse. A neural impulse travels along the membrane of the presynaptic neuron until it reaches the synaptic terminal (step 1). When the membrane of the synaptic terminal depolarizes, voltage-gated ion channels open, and calcium ions (Ca^{2+}) enter the cell (step 2). The calcium influx triggers loaded vesicles to dump their neurotransmitter contents into the synaptic cleft by exocytosis (steps 3 and 4). The neurotransmitter molecules released by the presynaptic cell diffuse across the synaptic cleft and attach to stimulus-gated ion channels on the membrane of the postsynaptic neuron (step 5). The ion channels open. ⓘ *exocytosis,* section 4.5C

The interaction between a neurotransmitter and the ion channel may be excitatory—that is, the membrane of the postsynaptic cell may become depolarized, increasing the probability of an action potential. In figure 26.9, the Na^+ ions entering the postsynaptic cell are having this effect. Conversely, the interaction may be inhibitory. Opening ion channels that admit chloride ions (Cl^-), for example, makes the interior of the cell more negative and reduces the likelihood of an action potential.

The human brain uses at least 100 neurotransmitters. The two most common are the amino acids glutamate and GABA

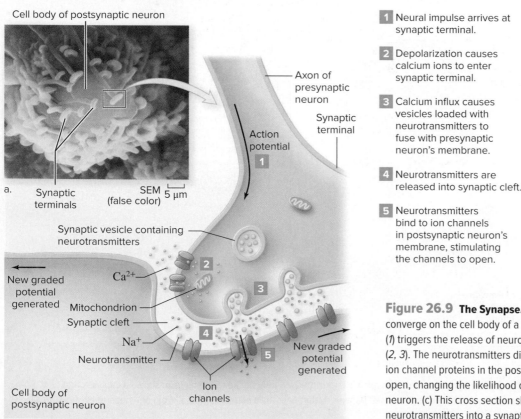

1 Neural impulse arrives at synaptic terminal.

2 Depolarization causes calcium ions to enter synaptic terminal.

3 Calcium influx causes vesicles loaded with neurotransmitters to fuse with presynaptic neuron's membrane.

4 Neurotransmitters are released into synaptic cleft.

5 Neurotransmitters bind to ion channels in postsynaptic neuron's membrane, stimulating the channels to open.

Figure 26.9 The Synapse. (a) Synaptic terminals from many neurons may converge on the cell body of a postsynaptic neuron. (b) An action potential (*1*) triggers the release of neurotransmitters from a synaptic terminal (*2, 3*). The neurotransmitters diffuse across the synaptic cleft (*4*) and bind with ion channel proteins in the postsynaptic neuron's membrane (*5*). The channels open, changing the likelihood of an action potential in the postsynaptic neuron. (c) This cross section shows many synaptic vesicles poised to release neurotransmitters into a synaptic cleft.

Photos: (a): ER Lewis, YY Zeevi and TE Everhart; (c): ©Don W. Fawcett/Science Source

(gamma aminobutyric acid). Other neurotransmitters occur at fewer synapses but still are vital. Serotonin, dopamine, epinephrine, norepinephrine, and acetylcholine are examples.

The same neurotransmitter may excite some neurons but inhibit others. Some receptors, however, are nearly always either excitatory or inhibitory. Most glutamate receptors, for example, are excitatory. Likewise, acetylcholine released from a motor neuron is excitatory, stimulating a muscle cell to contract. In contrast, most GABA receptors are inhibitory.

What happens to the neurotransmitter after it has done its job? If the chemical stayed in the synapse indefinitely, its effect on the receiving cell would be continuous. The nervous system would be bombarded with stimuli. Instead, the neurotransmitter can diffuse away from the synaptic cleft, be destroyed by an enzyme, or be taken back into the presynaptic axon soon after its release, an event called reuptake.

Notice that a synapse is asymmetrical; that is, nerve impulses travel from presynaptic cell to postsynaptic cell and not in the opposite direction. This one-way traffic of information stands in contrast to the cnidarian nerve net, in which nerve impulses travel in all directions (see section 26.1). Unidirectional information flow was a key adaptation that permitted the evolution of dedicated circuits in which one set of neurons communicated with a limited set of postsynaptic cells. Over time, some circuits became associated with specific functions, controlling complex behaviors and forming the specialized sense organs typical of many animals (including vertebrates).

B. A Neuron Integrates Signals from Multiple Synapses

So far, we have considered the events that occur at one synapse. But the human brain consists of billions of neurons, each of which has synaptic connections to a thousand other neurons (see figure 26.9a). As we just saw, some of the synapses are inhibitory, whereas others are excitatory. With so many potentially conflicting signals, how does a neuron determine whether to pass a neural impulse to the next cell in a pathway?

The cell uses a process called **synaptic integration** to evaluate the incoming messages. If the majority of signals reaching the neuron's trigger zone are excitatory, the membrane of the postsynaptic cell is stimulated to generate action potentials of its own. If, on the other hand, inhibitory signals predominate, the postsynaptic cell will not generate action potentials. Synaptic integration, then, is analogous to a voting system.

Clearly, neurotransmission is very much a matter of balance. Too much or too little of a neurotransmitter can cause serious illness; table 26.1 lists a few examples of disorders that are at least partly associated with neurotransmitter imbalances. Moreover, some drugs can alter the functioning of the nervous system by either halting or enhancing the activity of a neurotransmitter. A drug may bind to a receptor on a postsynaptic neuron, blocking a neurotransmitter from binding there. Alternatively, a drug may activate the receptor and trigger an action potential. This chapter's Apply It Now box illustrates how drugs such as nicotine, cocaine, and heroin tamper with neurotransmission.

26.4 MASTERING CONCEPTS

1. Describe the structure of a synapse.
2. What series of events stimulates a presynaptic neuron to release neurotransmitters?
3. What happens to a neurotransmitter after its release?
4. How does synaptic integration determine whether a neuron transmits action potentials?

TABLE **26.1** Disorders Associated with Neurotransmitter Imbalances

Condition	Imbalance of Neurotransmitter in Brain	Symptoms
Alzheimer disease	Deficient acetylcholine (caused by death of acetylcholine-producing cells)	Memory loss, depression, disorientation, dementia, hallucinations, death
Epilepsy	Excess GABA leads to excess norepinephrine and dopamine	Seizures, loss of consciousness
Huntington disease	Deficient GABA	Uncontrollable movements, dementia, behavioral and personality changes, death
Hypersomnia	Excess serotonin	Excessive sleeping
Insomnia	Deficient serotonin	Inability to sleep
Myasthenia gravis	Deficient receptors for acetylcholine at synapse between motor neuron and muscle cell	Progressive muscle weakness
Parkinson disease	Deficient dopamine	Tremors of hands, slowed movements, muscle rigidity
Schizophrenia	Deficient GABA leads to excess dopamine	Inappropriate emotional responses, hallucinations

26.5 The Peripheral Nervous System Consists of Nerve Cells Outside the Central Nervous System

The neurons of the brain and spinal cord interact constantly with those of the peripheral nervous system—the nerve cells outside the central nervous system (figure 26.10a). The peripheral nervous system consists mainly of **nerves,** which are bundles of axons encased in connective tissue. The nerves, in turn, are classified based on where they originate. Twelve pairs of cranial nerves emerge directly from the brain; examples include the optic and olfactory nerves, which transmit information about sights and smells to the brain. The remaining 31 pairs of nerves in the peripheral nervous system are spinal nerves, which emerge from the spinal cord and control most functions from the neck down.

The peripheral nervous system is functionally divided into sensory and motor divisions (figure 26.10b). Sensory pathways carry signals to the central nervous system from sensory receptors in the skin, skeleton, muscles, and other organs. Motor pathways, on the other hand, convey information from the central nervous system to muscles and glands. In most nerves, the sensory and motor nerve fibers form a single cable.

The motor pathways of the peripheral nervous system include the somatic (voluntary) nervous system and the autonomic (involuntary) nervous system. The **somatic nervous system** carries signals from the brain to voluntary skeletal muscles, such as those that enable you to ride a bicycle, shake hands, or talk. The **autonomic nervous system** transmits impulses from the brain to smooth muscle, cardiac muscle, and glands, enabling internal organs to function without conscious awareness.

The autonomic nervous system is further subdivided into the sympathetic and parasympathetic nervous systems (figure 26.11). The **sympathetic nervous system** dominates under stress, including emergencies. When you are startled, you can immediately feel your sympathetic nervous system leap into action as your heart pounds and your breathing rate increases. Neurons of the sympathetic nervous system also slow digestion and boost blood flow toward vital organs like the heart, brain, and muscles necessary for "fight or flight." Still others trigger the adrenal glands to secrete hormones that prolong these effects. The **parasympathetic nervous system** returns body systems to normal during relaxed times ("rest and repose"); heart rate and respiration slow, and digestion resumes. Despite the "fight or flight" and "rest and repose" nicknames, the autonomic nervous system is always active; the two subdivisions maintain homeostasis without conscious thought.

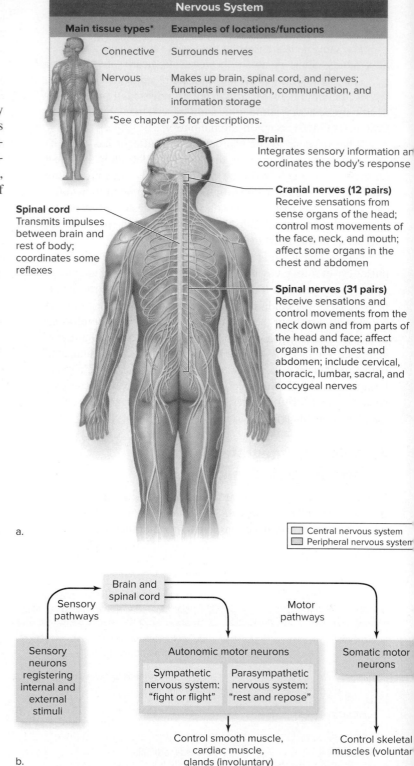

Nervous System	
Main tissue types*	**Examples of locations/functions**
Connective	Surrounds nerves
Nervous	Makes up brain, spinal cord, and nerves; functions in sensation, communication, and information storage

*See chapter 25 for descriptions.

Brain
Integrates sensory information and coordinates the body's response

Cranial nerves (12 pairs)
Receive sensations from sense organs of the head; control most movements of the face, neck, and mouth; affect some organs in the chest and abdomen

Spinal cord
Transmits impulses between brain and rest of body; coordinates some reflexes

Spinal nerves (31 pairs)
Receive sensations and control movements from the neck down and from parts of the head and face; affect organs in the chest and abdomen; include cervical, thoracic, lumbar, sacral, and coccygeal nerves

a.

☐ Central nervous system
☐ Peripheral nervous system

Brain and spinal cord

Sensory pathways

Motor pathways

Sensory neurons registering internal and external stimuli

Autonomic motor neurons

Sympathetic nervous system: "fight or flight"

Parasympathetic nervous system: "rest and repose"

Somatic motor neurons

Control smooth muscle, cardiac muscle, glands (involuntary)

Control skeletal muscles (voluntary)

b.

Figure 26.10 The Human Nervous System. (a) The brain and spinal cord make up the central nervous system, and nerves compose the peripheral nervous system. (b) Sensory pathways of the peripheral nervous system provide input to the central nervous system. The brain and spinal cord, in turn, regulate the motor pathways of the peripheral nervous system.

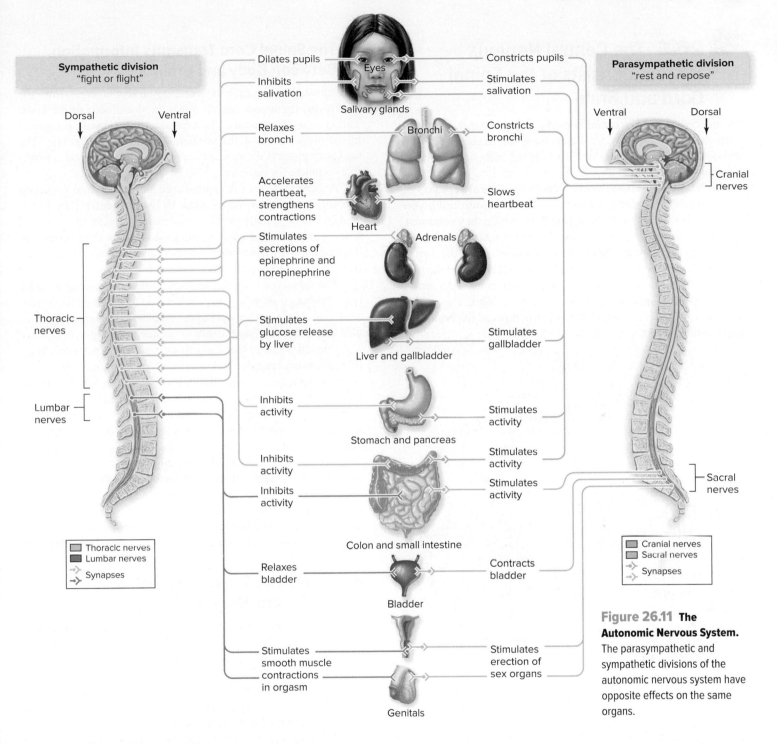

Sympathetic division
"fight or flight"

Dorsal   Ventral

Thoracic
nerves

Lumbar
nerves

☐ Thoracic nerves
☐ Lumbar nerves
⤏ Synapses
⤐ Synapses

Dilates pupils
Eyes
Constricts pupils

Inhibits
salivation
Stimulates
salivation
Salivary glands

Relaxes
bronchi
Bronchi
Constricts
bronchi

Accelerates
heartbeat,
strengthens
contractions
Heart
Slows
heartbeat

Stimulates
secretions of
epinephrine and
norepinephrine
Adrenals

Stimulates
glucose release
by liver
Liver and gallbladder
Stimulates
gallbladder

Inhibits
activity
Stomach and pancreas
Stimulates
activity

Inhibits
activity
Stimulates
activity

Inhibits
activity
Stimulates
activity

Colon and small intestine

Relaxes
bladder
Contracts
bladder
Bladder

Stimulates
smooth muscle
contractions
in orgasm
Stimulates
erection of
sex organs
Genitals

Parasympathetic division
"rest and repose"

Ventral   Dorsal

Cranial
nerves

Sacral
nerves

☐ Cranial nerves
☐ Sacral nerves
⤏ Synapses
⤐ Synapses

Figure 26.11 **The
Autonomic Nervous System.**
The parasympathetic and
sympathetic divisions of the
autonomic nervous system have
opposite effects on the same
organs.

Some illnesses interfere with the function of the peripheral nervous system. In Guillain–Barré syndrome, for example, the immune system attacks and destroys the nerves of the peripheral nervous system. The disease can be life-threatening if it causes paralysis, breathing difficulty, and heart problems. In a person with Bell's palsy, another peripheral nervous system disorder, the cranial nerve that controls the muscles on one side of the face is damaged. The facial paralysis typically strikes suddenly and may either resolve on its own or be permanent. The cause is unknown.

26.5 MASTERING CONCEPTS

1. What is the difference between cranial and spinal nerves?
2. How do the sensory and motor pathways of the peripheral nervous system differ?
3. Describe the relationships among the autonomic, sympathetic, and parasympathetic nervous systems.
4. Cite an example of the opposing actions of the sympathetic and parasympathetic nervous systems.

26.6 The Central Nervous System Consists of the Spinal Cord and Brain

The nerves of the peripheral nervous system spread across the body, but the brain and spinal cord form the largest part of the nervous system. Together, these two organs make up the central nervous system.

Two types of nervous tissue occur in the central nervous system (figure 26.12). **Gray matter** consists of neuron cell bodies and dendrites, along with the synapses by which they communicate with other cells. The outer surface of the brain, and a few inner structures, are composed of gray matter, as is the central core of the spinal cord. Information processing occurs in the gray matter. **White matter** consists of myelinated axons transmitting information throughout the central nervous system. The periphery of the spinal cord and most inner structures of the brain consist of white matter.

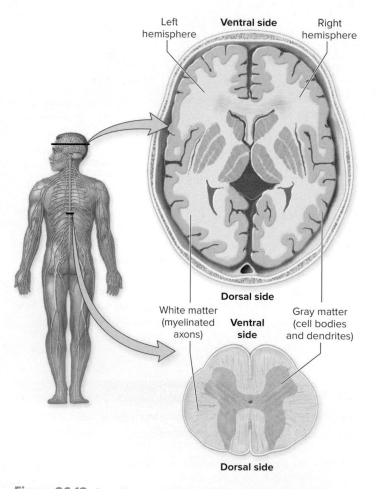

Figure 26.12 Gray Matter and White Matter. Gray matter makes up the exterior of the brain and some internal structures. It also makes up the central core of the spinal cord. Myelin-rich white matter is at the periphery of the spinal cord and forms most of the brain's interior.

A. The Spinal Cord Transmits Information Between Body and Brain

The spinal cord is a tube of neural tissue that emerges from the base of the brain and extends down the dorsal side of the body. This critical component of the central nervous system is encased in the bony armor of the vertebral column, or backbone. The backbone protects the delicate nervous tissue and provides points of attachment for muscles.

A cross section of the spinal cord shows a central H-shaped core of gray matter surrounded by white matter (see figure 26.12). On the dorsal side, the white matter consists of ascending bundles of axons that carry sensory information to the brain. On the ventral side, descending axons transmit motor information from the brain to muscles and glands.

The spinal cord handles reflexes without interacting with the brain (although impulses must be relayed to the brain for awareness to occur). A **reflex** is a rapid, involuntary response to a stimulus that may come from within or outside the body. A **reflex arc** is the neural pathway that controls the reflex; a typical reflex arc links a sensory receptor and an effector such as a muscle.

Consider, for example, the series of events by which you pull your hand away from a painful stimulus such as a thorn (figure 26.13). The arc begins with a sensory neuron whose dendrites reside in your fingertip. When you touch the sharp thorn, an action potential is generated along the neuron's axon. Inside the spinal cord, the axon synapses with an interneuron, which in turn synapses on a motor neuron. The motor neuron's axon exits the spinal cord, and its activation stimulates a skeletal muscle cell to contract. When enough muscle fibers contract, you pull your hand away from the thorn. The interneuron in the spinal cord also sends action potentials to the brain, perhaps prompting you to yell in pain.

B. The Human Brain Is Divided into Several Regions

The human brain weighs, on average, about 1.4 to 1.6 kilograms; it looks and feels like grayish pudding. The brain requires a large and constant energy supply to oversee organ systems and to provide the qualities of "mind"—learning, reasoning, and memory. At any time, brain activity consumes 20% of the body's oxygen and 15% of its blood glucose. Permanent brain damage occurs after just 5 minutes without oxygen.

Anatomically, the brain has three main subdivisions: the hindbrain, the midbrain, and the forebrain (figure 26.14). The **hindbrain** is located toward the lower back of the skull. The **midbrain** is a narrow region that connects the hindbrain with the forebrain. The **forebrain** is the front of the brain. All three subdivisions are obvious early in embryonic development, but the forebrain's rapid growth soon obscures the midbrain and much of the hindbrain.

The midbrain and parts of the hindbrain make up the **brainstem,** the stalklike lower portion of the brain. The brainstem regulates essential survival functions such as breathing and heartbeat. In addition, 10 of the 12 pairs of cranial nerves emerge

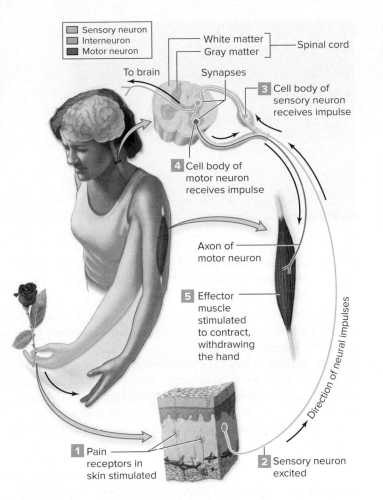

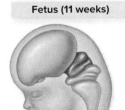

Structure	Selected functions
Hindbrain	
Medulla oblongata	Regulates essential physiological processes such as blood pressure, heartbeat, and breathing
Pons	Connects forebrain with medulla and cerebellum
Cerebellum	Controls posture and balance; coordinates subconscious muscular movements
Midbrain	Relays information about voluntary movements from forebrain to spinal cord
Forebrain	
Thalamus	Processes information and relays it to the cerebrum
Hypothalamus	Homeostatic control of most organs
Cerebrum	
White matter	Transmits information within brain
Gray matter (cerebral cortex)	Sensory, motor, and association areas

Figure 26.13 A Reflex Arc. A reflex arc links a sensory receptor to an effector. (*1*) In response to a painful stimulus, (*2*) the dendrite of a sensory neuron relays an action potential to (*3*) its cell body. In the spinal cord's gray matter, the sensory neuron synapses on multiple neurons, including (*4*) the cell body of a motor neuron. (*5*) The motor neuron, in turn, stimulates muscle cells to contract. The hand withdraws. The brain perceives the stimulus but does not participate in the reflex.

from the brainstem. Among other functions, these nerves control movements of the eyes, face, neck, and mouth along with the senses of taste and hearing.

The brainstem includes two parts of the hindbrain: the medulla oblongata and the pons. The **medulla oblongata** is a continuation of the spinal cord; this region not only regulates breathing, blood pressure, and heart rate but also contains reflex centers for vomiting, coughing, sneezing, defecating, swallowing, and hiccupping. The **pons,** which means "bridge," is the area above the medulla. White matter in this oval mass connects the forebrain to the medulla and to another part of the hindbrain, the cerebellum.

The midbrain is also part of the brainstem. Portions of the midbrain help control consciousness and participate in hearing and eye reflexes. In addition, nerve fibers that control voluntary motor function pass from the forebrain through the brainstem;

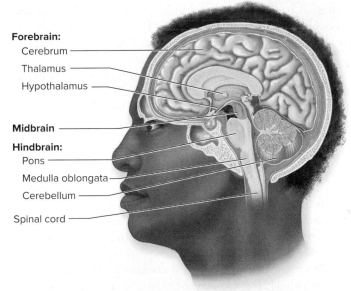

Figure 26.14 The Human Brain. The three major areas of the vertebrate brain are the hindbrain, the midbrain, and the forebrain. They become apparent early in the development of the human embryo.

the death of certain neurons in the midbrain results in the uncontrollable movements of Parkinson disease.

Behind the brainstem is the cerebellum, the largest part of the hindbrain. The neurons of the **cerebellum** ("little brain") refine motor messages and coordinate muscle movements subconsciously. Thanks to the cerebellum, you can complete complex physical skills—like tying your shoes or brushing your teeth—smoothly and rapidly.

By far the largest part of the human brain is the forebrain, which contains structures that participate in complex functions such as learning, memory, language, motivation, and emotion. Three major parts of the forebrain are the thalamus, hypothalamus, and cerebrum.

The **thalamus** is a mass of gray matter located between the midbrain and the cerebrum. This central relay station processes incoming sensory input and sends it to the appropriate part of the cerebrum. The almond-sized **hypothalamus,** which lies below the thalamus, occupies less than 1% of the brain volume, but it plays a vital role in maintaining homeostasis by linking the nervous and endocrine systems. Cells in the hypothalamus are sensitive not only to neural input arriving via the brainstem but also to hormones circulating in the bloodstream. The autonomic nervous system, in turn, relays neural signals from the hypothalamus to involuntary muscles and glands (see figure 26.11). Moreover, hormones produced in the hypothalamus coordinate the production and release of many other hormones from the pituitary gland. All together, neural and hormonal signals from the hypothalamus regulate body temperature, heartbeat, water balance, and blood pressure, along with hunger, thirst, sleep, and sexual arousal. ⓘ *pituitary gland,* section 28.3

The other major region of the forebrain is the **cerebrum,** which controls the qualities of what we consider the "mind"—that is, personality, intelligence, learning, perception, and emotion. In humans, the cerebrum occupies 83% of the brain's volume. It is divided into two **hemispheres,** which gather and process information simultaneously. The cerebral hemispheres work together, interconnected by a thick band of nerve fibers called the corpus callosum.

Each hemisphere controls the opposite side of the body, so damage to the left side of the brain affects the right side of the body (and vice versa). In addition, although each side of the brain participates in most brain functions, some specialization does occur. In most people, for example, parts of the left hemisphere are associated with speech, language skills, mathematical ability, and reasoning, whereas the right hemisphere specializes in spatial, intuitive, musical, and artistic abilities.

The cerebrum consists mostly of white matter—myelinated axons that transmit information within the cerebrum and between the cerebrum and other parts of the brain. But the outer layer of the cerebrum, the **cerebral cortex,** consists of gray matter that processes information. The human cerebral cortex is only a few millimeters thick, but it boasts about 10 billion neurons that form some 60 trillion synapses (figure 26.15). In humans and other large mammals, deep folds enhance the surface area of the cerebral cortex.

Figure 26.15 Technicolor Brain. Special labelling techniques reveal the individual neurons that make up the intricate circuits of the cerebral cortex.
©AFP/Getty Images

Anatomically, the cerebral cortex of each hemisphere is divided into four parts (figure 26.16): the frontal, parietal, temporal, and occipital lobes. The functions of the cerebral cortex, however, overlap across these lobes. Sensory areas receive and interpret messages from sense organs (relayed through the thalamus), as described in chapter 27. Motor areas send impulses to skeletal muscles, which produce voluntary movements. (Damage to the motor regions of a fetus's or an infant's brain can cause cerebral palsy, a disorder of muscle movement.) Association areas analyze, integrate, and interpret information from many brain areas. These are the seats of judgment, problem solving, learning, abstract thought, language, and creativity.

The cerebrum also houses most of the **limbic system,** a loosely defined collection of brain structures that is sometimes called the emotional center of the brain. The thalamus and hypothalamus are part of the limbic system, as are two nearby parts of each temporal lobe: the hippocampus and the amygdala. The **hippocampus** is essential for the formation of long-term memories, whereas the **amygdala** is a center for emotions such as pleasure or fear. The amygdala sends signals to the hypothalamus, which activates the autonomic nervous system and coordinates the physical sensations that accompany strong emotions.

With this perspective on brain anatomy, consider the often-repeated statement that humans use only 10% of their brains. This common notion is a myth. After all, damage to even a tiny part of the brain can have devastating consequences. Moreover, the brain demands a huge amount of energy and oxygen; it does not make sense that the nervous system would waste valuable resources that the body could use in productive ways.

C. Many Brain Regions Participate in Memory Formation

Why is it that you can't remember the name of someone you met a few minutes ago, but you can easily picture your first-grade teacher or your best friend from childhood (figure 26.17)?

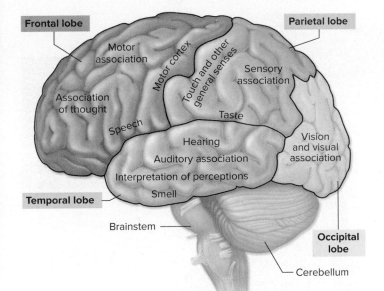

a. The four lobes of the cerebral cortex
(only the left hemisphere is shown)

Division	Function(s)	Brain Region(s)
Sensory	Senses of vision, hearing, smell, taste, and touch	Parietal, occipital, and temporal lobes
Motor	Voluntary movements	Frontal lobe
Association	Judgment, analysis, learning, creativity	Frontal lobe and parts of the parietal, occipital, and temporal lobes

b. Functional divisions of the cerebral cortex

Figure 26.16 **The Cerebral Cortex.** (a) Each hemisphere of the cerebrum is divided into four lobes: the frontal lobe at the front of the head, the temporal lobe above each ear, the occipital lobe at the rear of the head, and the parietal lobe across the top of the head. (b) The functional divisions of the brain (sensory, motor, and association) have been mapped to these lobes.

The answer relates to the difference between short-term and long-term memories. In response to the new acquaintance's name, your brain apparently created a **short-term memory** that remained available only for a few moments. You remember the teacher, however, because you interacted with that person every day for months at a time. This repeated reinforcement allowed your brain to produce a **long-term memory,** which can last a lifetime.

Much of what scientists know about memory comes from research on people with damage to specific parts of the brain. One famous example is a man called Henry Molaison, known in the medical literature by the initials H.M. until his death in 2008. Surgeons removed portions of his temporal lobes and hippocampus in 1953 in an effort to alleviate his severe epilepsy. The surgery accomplished its goal but had an unintended consequence: H.M. was unable to form new memories. Although he could

Figure 26.17 **Memories.** A glance at old photos can bring back memories from times past.
©Gary He/McGraw-Hill Education

recall events that occurred before the surgery, he could not remember what he had eaten for breakfast. Clues from H.M. and other patients suggest that the hippocampus is essential in the formation of long-term memories, but memories are not actually stored there.

No one knows exactly what happens to the brain's neurons and synapses when a new memory forms. For short-term memories, it is possible that neurons in the frontal and parietal lobes connect in a temporary circuit in which the last cell in the series restimulates the first. As long as the pattern of stimulation continues, you remember the thought. When the reverberation ceases, however, so does the memory. On the other hand, long-term memory probably requires stable, permanent changes in neuron structure or function.

Researchers are trying to learn more about memory formation, in part by studying the brains of people with memory loss (amnesia) or with exceptional memories. Practical applications could include drugs that enhance memory in patients with disorders that cause memory loss, including Alzheimer disease. Conversely, pharmaceuticals that selectively erase memories could help people who are struggling in the aftermath of traumatic experiences.

D. Damage to the Central Nervous System Can Be Devastating

The central nervous system is generally well protected from physical injury. The most obvious lines of defense are the bones of the skull and vertebral column, which shield nervous tissue from bumps and blows. Just beneath these bones, **meninges** are layered membranes that jacket the central nervous system. Finally, **cerebrospinal fluid** bathes and cushions the brain and spinal cord. This fluid, made by cells that line ventricles in the brain, further insulates the central nervous system from injury.

Physical damage, however, is not the only threat to the nervous system. Harmful chemicals can damage the brain, as can changes in the concentrations of substances that normally circulate in the bloodstream. The **blood–brain barrier,** which is composed of brain capillaries with specialized walls, helps protect against these chemical threats (figure 26.18). Tight junctions between the tilelike epithelial cells that form the brain's capillaries ensure that only some chemicals can cross into the cerebrospinal fluid that bathes the brain. ⓘ *tight junctions,* section 3.6A

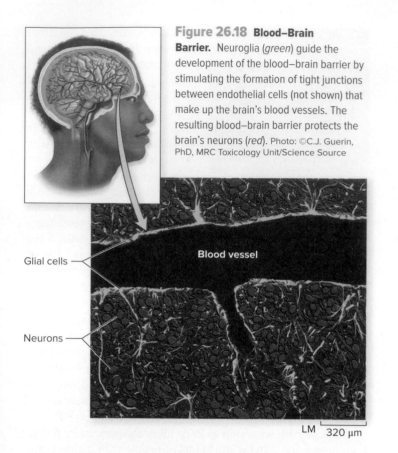

Figure 26.18 Blood–Brain Barrier. Neuroglia (*green*) guide the development of the blood–brain barrier by stimulating the formation of tight junctions between endothelial cells (not shown) that make up the brain's blood vessels. The resulting blood–brain barrier protects the brain's neurons (*red*). Photo: ©C.J. Guerin, PhD, MRC Toxicology Unit/Science Source

Glial cells

Blood vessel

Neurons

LM ├── 320 μm

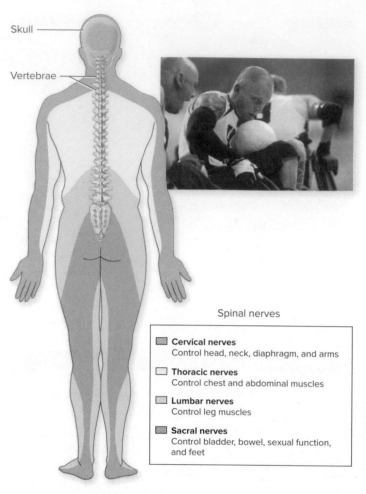

Skull

Vertebrae

Spinal nerves

■ **Cervical nerves**
Control head, neck, diaphragm, and arms

□ **Thoracic nerves**
Control chest and abdominal muscles

■ **Lumbar nerves**
Control leg muscles

■ **Sacral nerves**
Control bladder, bowel, sexual function, and feet

Figure 26.19 Spinal Cord Injuries. A person with a damaged spinal cord may have full or partial paralysis, depending on the site of the injury. The inset shows Mark Zupan, whose neck was broken in a car accident. Although he is a quadriplegic, he retains some use of his arms and became famous as a wheelchair rugby player.

Photo: ©Natalie Behring/Getty Images

The barriers surrounding the central nervous system cannot protect it against severe injuries suffered in car accidents, sports mishaps, and explosions. The chapter opening essay described the dangers of concussions, in which the brain crashes against the interior of the skull. Injuries can also damage or sever the spinal cord, preventing motor impulses from descending from the brain (**figure 26.19**). Depending on the location of the injury, the result may be full or partial paralysis of the arms, torso, and legs.

Brain damage can also result from a more subtle killer: stroke. In a stroke, a burst or blocked blood vessel can interrupt the flow of blood to part of the brain. Deprived of oxygen, some brain cells die, often so many that the stroke is fatal. In other cases, the patient may suffer from temporary or permanent muscle weakness, paralysis, loss of speech, blindness, or other impairment. Often, only one side of the body is affected.

Infectious diseases can also affect the central nervous system. Several viruses and bacteria can cause meningitis, or inflammation of the meninges. Bacterial meningitis is rare but can be fatal. Prions, the infectious proteins described in section 16.6B, also damage the brain. Fungal infections of the central nervous system often afflict people with compromised immune systems.

Many serious illnesses of the central nervous system remain a mystery. Parkinson disease, for example, is a slowly progressing, degenerative disease in which the death of brain cells causes muscle tremors, weakness, slow movement, and loss of coordination. In most cases, no one knows what triggers the cell death.

The death of brain cells also causes Alzheimer disease. The patient, who is usually elderly, suffers from memory loss, confusion, and personality changes. Autopsy reveals tangled neurons and clusters of degenerating nerve endings; neurotransmitter production is also reduced (see table 26.1). Unlike in Parkinson disease, motor function usually remains unaffected.

Another mysterious nervous system disease is amyotrophic lateral sclerosis, also called ALS or Lou Gehrig disease. In ALS, motor neurons in the central nervous system gradually die, so the brain loses control of the body's muscles. The first clues to this disease's onset are often subtle; a person may drop things or suffer from persistent muscle twitches. Eventually, impaired muscle function leaves the patient unable to speak, breathe, swallow, or move, yet the senses and mental function remain unaffected. This devastating disease has no cure.

Apply It **Now**

Drugs and Neurotransmitters

Understanding how neurotransmitters work helps explain the action of some mind-altering illicit and pharmaceutical drugs. The following are some examples, organized by the neurotransmitter affected.

Norepinephrine

Amphetamine drugs are chemically similar to norepinephrine; they bind to norepinephrine receptors and trigger the same changes in the postsynaptic membrane. The resulting enhanced norepinephrine activity heightens alertness and mood. Cocaine, which is chemically related to amphetamine, produces a short-lived feeling of euphoria, in part by blocking reuptake of norepinephrine.

Acetylcholine

Nicotine crosses the blood–brain barrier and reaches the brain within seconds of inhaling from a cigarette. An acetylcholine mimic, nicotine binds to acetylcholine receptor proteins in neuron cell membranes. The nicotine-stimulated neurons signal other brain cells to release dopamine, which provides the pleasurable feelings associated with smoking. Nicotine addiction stems from two sources: seeking the dopamine release and avoiding painful withdrawal symptoms.

Excess acetylcholine accounts for the deadly effects of poisonous nerve gases and some insecticides. These toxic chemicals prevent acetylcholine from breaking down in the synaptic cleft. The resulting buildup of acetylcholine overstimulates skeletal muscles, causing them to contract continuously. The twitching legs of a cockroach sprayed with insecticide demonstrate the effects.

Serotonin

Drugs that increase the amount of norepinephrine or serotonin in a synapse appear to reduce the symptoms of depression. Selective serotonin reuptake inhibitors (SSRIs) block the reuptake of serotonin at the synaptic cleft, causing the neurotransmitter to accumulate (**figure 26.A**).

Endorphins

Humans produce several types of endorphins, molecules that influence mood and perception of pain. Opiate drugs such as morphine, heroin, codeine, and opium are potent painkillers that bind endorphin receptors in the brain. In doing so, they elevate mood and make the pain easier to tolerate.

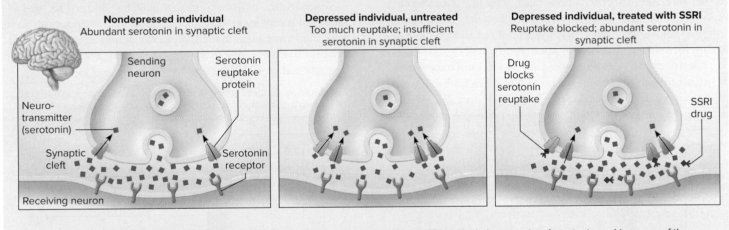

Figure 26.A Anatomy of an Antidepressant. Selective serotonin reuptake inhibitors (SSRIs) block the reuptake of serotonin, making more of the neurotransmitter available in the synaptic cleft. The precise mechanism by which SSRIs relieve depression is not well understood.

Whatever its cause, part of the difficulty in reversing damage to the central nervous system is that mature neurons typically do not divide. The brain and spinal cord therefore cannot simply heal themselves by producing new cells as your skin does after a minor cut. The neurons that survive the damage can, however, form some new connections that compensate for the loss. Therapy can therefore help restore some function to injured tissues. Moreover, stem cells and gene therapy may one day improve the outlook for patients with brain damage or disease. ⓘ *stem cells,* section 11.3A; *gene therapy,* section 11.4D

26.6 MASTERING CONCEPTS

1. Describe the functions of the neurons that form a reflex arc.
2. What are the names, locations, and functions of the main parts of the human brain?
3. Summarize what researchers know and have yet to learn about memory formation and storage.
4. List some structures that protect the central nervous system.
5. To what extent can the nervous system regenerate?

26.7 Scorpion Stings Don't Faze Grasshopper Mice

Of the many types of arthropods that can sting, scorpions are distinctive. The scorpion uses the stinger on its tail to inject venom when threatened. Proteins called neurotoxins (poisons that act on the nervous system) cause most of the sting's worst effects. Some scorpion neurotoxins cause an axon's sodium channels to become stuck in the "open" position. The resulting flood of ions triggers a continuous barrage of action potentials, accompanied by simultaneous sensations of pain, heat, cold, and touch.

Scorpion venom causes intense pain in many mammals, including humans and most rodents. However, grasshopper mice are an exception. These small rodents prey on bark scorpions in the Arizona desert (figure 26.20). When it encounters a scorpion, a grasshopper mouse attempts to bite the scorpion's head. The scorpion fights back, stinging the mouse many times. Oddly, the mouse barely responds, as if it hardly feels the sting.

To verify this observation in the lab, researchers injected scorpion venom or a control solution into the hind paws of typical lab mice and of grasshopper mice. The lab mice licked a venom-injected paw far more vigorously than they did a control paw, an indication of pain. But venom did not induce the paw-licking behavior in grasshopper mice (figure 26.21).

Clearly, the nervous systems of these two types of mice respond differently to scorpion venom, but how? A research team focused on pain receptors in the rodents. Specifically, they used electrodes to measure the number of action potentials in pain receptors as the

Figure 26.20 Dinner Time. A grasshopper mouse encounters its prey, a scorpion.
Photo by Ashlee and Matthew Rowe

animals were injected with a control solution, a low dose of scorpion venom, or a high dose of venom.

The results were clear (figure 26.22). In lab mice, a higher venom dose translated into more action potentials in pain receptors. In grasshopper mice, scorpion venom actually *inhibited* action potentials in pain receptors.

In lab mice and most other mammals, scorpion venom causes sodium channels in pain receptor cells to open. Action potentials from the stimulated neurons soon arrive at the brain, which interprets the message as pain. Grasshopper mice have unique, specialized sodium channels that do not open when scorpion venom binds to them. The brain therefore never receives the message of pain. In fact, the opposite occurs. Scorpion venom actually has a *painkilling* effect on grasshopper mice, and they do not react to painful stimuli following a scorpion sting.

Apparently, a diet rich in scorpions has selected for specialized pain receptors in grasshopper mice. Many generations from now, random changes in the scorpion toxin may again make grasshopper mice vulnerable to the sting. But in this arms race between predator and prey, the predator has taken the lead for now.

Source: Rowe, Ashlee H., and four coauthors. 2013. Voltage-gated sodium channel in grasshopper mice defends against bark scorpion toxin. *Science*, vol. 342, pages 441–446.

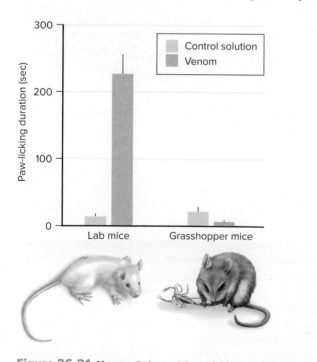

Figure 26.21 Venom, Pain, and Paw-Licking. In lab mice, an injection of scorpion venom stimulated vigorous paw-licking, an indication of pain. Venom had the opposite effect in grasshopper mice. (Error bars reflect standard errors; see appendix B.)

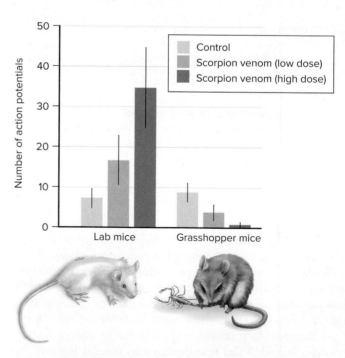

Figure 26.22 Venom Disarmed. The higher the dose of scorpion venom, the more action potentials in a laboratory mouse's pain receptors. In grasshopper mice, the venom reduced the activity of pain receptors. (Error bars reflect standard errors; see appendix B.)

26.7 | MASTERING CONCEPTS

1. The researchers conducted a behavioral experiment and a physiological experiment to learn how grasshopper mice respond to scorpion venom. What was the dependent variable in the behavioral experiment? In the physiological experiment?

2. How does scorpion venom affect grasshopper mice? Why is this response adaptive?

CHAPTER SUMMARY

26.1 The Nervous System Forms a Rapid Communication Network

- The **nervous system** and **endocrine system** work together to coordinate the feedback systems that maintain homeostasis. The two systems differ, however, as summarized in table 26.2.
- Nervous tissue consists of **neurons** and **neuroglia.**

A. Invertebrates Have Nerve Nets, Nerve Ladders, or Nerve Cords

- Invertebrate nervous systems vary widely in complexity. The simplest systems, in cnidarians, are **nerve nets.** A flatworm has collections of neurons called **ganglia** at its head end, with a **nerve ladder** running the length of the body. Earthworm and arthropod nervous systems are more centralized, with **ventral nerve cords** connected to an anterior brain.

B. Vertebrate Nervous Systems Are Highly Centralized

- The vertebrate **central nervous system** consists of the **brain** and **spinal cord.** The **peripheral nervous system** conveys information between the central nervous system and the rest of the body.
- Overall, the nervous system receives sensory information, integrates it, and coordinates a response.

26.2 Neurons Are Functional Units of a Nervous System

A. A Typical Neuron Consists of a Cell Body, Dendrites, and an Axon

- A neuron has a **cell body, dendrites** that receive impulses and transmit them toward the cell body, and an **axon** that conducts impulses away from the cell body.
- Fatty neuroglia called **Schwann cells** or **oligodendrocytes** wrap around portions of some axons to form the **myelin sheath.** The gaps between these insulating cells are **nodes of Ranvier.**

B. The Nervous System Includes Three Classes of Neurons

- A **sensory neuron** carries information toward the central nervous system; an **interneuron** conducts information between two neurons and coordinates responses; a **motor neuron** carries information away from the central nervous system and stimulates an effector (a muscle or gland).

26.3 Action Potentials Convey Messages

A. A Neuron at Rest Has a Negative Charge

- A neuron has a **membrane potential** that changes depending on the cell's activity.

TABLE **26.2** Nervous and Endocrine Systems Compared

Nervous System	Endocrine System
• Uses action potentials and neurotransmitters	• Uses hormones
• Action is instantaneous.	• Action is slow but lasting.
• Signal molecules affect one or a few nearby cells.	• Signal molecules affect many cells throughout the body.

- When not conducting a neural impulse, a neuron's **resting potential** is slightly negative. The **sodium–potassium pump** uses ATP to maintain a chemical gradient in which the K⁺ concentration is much greater inside the cell than outside, and the Na⁺ concentration is greater outside than inside. This concentration gradient, combined with negatively charged proteins within the cell, means the interior of the membrane has a slightly negative charge relative to the outside.

B. A Neuron's Membrane Potential Reverses During an Action Potential

- In one type of **graded potential**, a stimulus causes some Na⁺ to enter the cell, depolarizing the membrane in proportion to the strength of the stimulus. If enough Na⁺ comes in, the membrane may further depolarize to its **threshold potential.**
- When the membrane reaches its threshold potential, an electrical change called an **action potential** begins. Na⁺ and K⁺ quickly redistribute across a small patch of the axon's membrane, triggering a wave of electrochemical changes along the nerve fiber (figure 26.23).

C. The Myelin Sheath Speeds Impulse Conduction

- Myelination increases the speed of neural impulse transmission. The neural impulse rapidly "jumps" from one node of Ranvier to the next.

26.4 Neurotransmitters Pass the Message from Cell to Cell

A. Neurons Communicate at Synapses

- A **synapse** is a junction between a neuron and another cell.
- An action potential reaching the end of an axon causes vesicles in the **synaptic terminals** of the **presynaptic cell** to release **neurotransmitters** into the **synaptic cleft.** These chemicals bind to ion channels on the membrane of the **postsynaptic cell.**

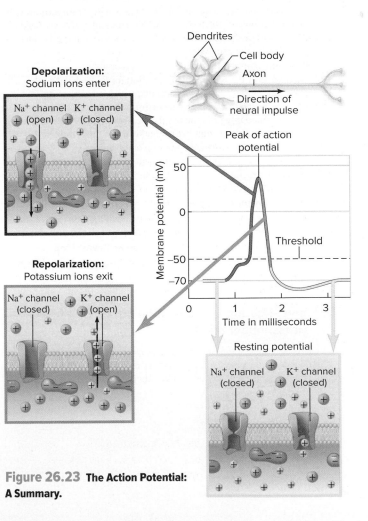

Figure 26.23 The Action Potential: A Summary.

- A neurotransmitter may have an excitatory effect, making an action potential more probable in the postsynaptic cell; an inhibitory interaction has the opposite effect.

B. A Neuron Integrates Signals from Multiple Synapses

- **Synaptic integration** sums excitatory and inhibitory messages, providing fine control of neuron activity.

26.5 The Peripheral Nervous System Consists of Nerve Cells Outside the Central Nervous System

- **Nerves** are bundles of axons that convey information in the peripheral nervous system.
- The peripheral nervous system is divided into the sensory and motor pathways.
- The motor pathways of the peripheral nervous system consist of the **somatic** (voluntary) division and the **autonomic** (involuntary) division.
- Within the autonomic nervous system, the **sympathetic nervous system** controls physical responses to stressful events, and the **parasympathetic nervous system** restores a restful state.

26.6 The Central Nervous System Consists of the Spinal Cord and Brain

A. The Spinal Cord Transmits Information Between Body and Brain

- **White matter** on the periphery of the spinal cord conducts impulses to and from the brain; the central **gray matter** processes information.
- The spinal cord is a reflex center. A **reflex** is a quick, automatic, protective response that travels through a **reflex arc.**

B. The Human Brain Is Divided into Several Regions

- The **brainstem** consists of the midbrain and portions of the hindbrain.
- The **hindbrain** includes three main subdivisions: the **medulla oblongata,** which controls many vital functions; the **cerebellum,** which coordinates movements that do not require conscious awareness; and the **pons,** which bridges the medulla and higher brain regions and connects the cerebellum to the cerebrum.
- The **midbrain** connects the hindbrain and forebrain.
- The major parts of the **forebrain** are the **thalamus,** a relay station between brain regions; the **hypothalamus,** which regulates vital physiological processes; and the **cerebrum,** which processes and transmits information within the brain. The gray matter of the forebrain also houses the **limbic system,** which includes the **amygdala** and **hippocampus** and is involved in emotion and memory.
- The cerebrum's outer layer is the **cerebral cortex,** where information is processed and integrated. Each cerebral **hemisphere** receives sensory input from, and directs motor responses to, the opposite side of the body.

C. Many Brain Regions Participate in Memory Formation

- Biologists have much to learn about memory, but it appears that the brain stores **short-term memories** and **long-term memories** in different ways.

D. Damage to the Central Nervous System Can Be Devastating

- The skull, vertebrae, **meninges, cerebrospinal fluid,** and **blood–brain barrier** all protect the central nervous system.
- Trauma, diseases, and other conditions can damage the nervous system.

26.7 Investigating Life: Scorpion Stings Don't Faze Grasshopper Mice

- Unlike other rodents, grasshopper mice are insensitive to scorpion venom. Researchers have traced this unusual trait to their specialized pain receptors; scorpion venom inhibits their activity.

MULTIPLE CHOICE QUESTIONS

1. Some cells of the central nervous system are located in the
    a. spinal cord.   c. glands.
    b. muscles.   d. Both a and c are correct.

2. What characteristic distinguishes dendrites from axons?
    a. Presence of branches at the tips
    b. Direction of impulse transmission relative to cell body
    c. Whether they direct impulses toward or away from the brain
    d. Presence of sodium channels in the cell membrane

3. The peripheral nervous system _____, whereas the central nervous system _____.
    a. receives sensory stimuli; integrates sensory information
    b. produces sensory information; produces movements
    c. integrates sensory information; receives sensory stimuli
    d. coordinates voluntary actions; coordinates involuntary actions

4. What event triggers an action potential?
    a. Opening of sodium channels
    b. Opening of delayed potassium channels
    c. High concentration of negative ions outside the cell
    d. Activation of the sodium–potassium pump

5. Ordinarily, a neuron's trigger zone is activated before any other part of an axon, so a wave of action potentials occurs as sodium channels open in the direction of the synaptic terminal. What would happen if you artificially stimulated an axon to reach threshold potential midway along its length rather than at the trigger zone?
    a. Action potentials would occur in the direction of the synaptic terminal.
    b. No action potentials would occur.
    c. Action potentials would occur in the direction of the trigger zone.
    d. Both a and c are correct.

6. Which of the following examples of synaptic integration is most likely to lead to an action potential in a postsynaptic cell?
    a. All excitatory and no inhibitory synaptic inputs
    b. An equal mix of excitatory and inhibitory inputs
    c. A majority of inhibitory inputs with only a few excitatory inputs
    d. Both a and b are correct.

7. Which division of the nervous system is responsible for a rapid heartbeat?
    a. Autonomic   c. Parasympathetic
    b. Sympathetic   d. Both a and b are correct.

8. The part of the human brain involved in coordinating muscle movements is the
    a. cerebrum.   c. cerebellum.
    b. medulla oblongata.   d. hypothalamus.

9. Damage to the surface tissue of the spinal cord will most likely affect
    a. information processing within the spinal cord.
    b. the stimulation of sensory organs.
    c. communication between the peripheral nervous system and the brain.
    d. reflex arcs.

10. Underdevelopment of the medulla oblongata could lead to
    a. poor processing of sensory information.
    b. limited language skills.
    c. erratic changes in blood pressure.
    d. unconscious muscle movements.

Answers to these questions are in Appendix A.

WRITE IT OUT

1. Describe some invertebrate nervous systems. Why do animals with simple nervous systems still exist, even after the more complex vertebrate nervous system evolved?

2. How do the nervous and endocrine systems differ in how they communicate?

3. Explain how sensory neurons, interneurons, and motor neurons work together as an insect moves toward a chemical stimulus.

4. Describe the distribution of charges in the membrane of a resting neuron.

5. What is the connection between the threshold potential and an action potential? What happens to sodium and potassium ions as the membrane is depolarized and repolarized? How does the membrane restore its resting potential?

6. Write a nonbiological analogy for resting potential and for depolarization, other than those mentioned in the chapter.

7. What is the function of each of the four membrane proteins shown in figure 26.6? When is each membrane protein channel open?

8. Sketch a synapse; label the axon and synaptic terminal of the presynaptic cell, the postsynaptic cell, the synaptic cleft, the synaptic vesicles, and the receptor proteins.

9. Describe the events that occur at a synapse when a neural impulse arrives at a synaptic terminal of a presynaptic cell.

10. Prescription sleep aids, like Ambien, bind to inhibitory receptors on neurons. Explain how these drugs slow the production of action potentials in affected parts of the brain.

11. What is the relationship between gray matter and white matter in the spinal cord?

12. Cerebral palsy is a nervous system disorder that is often caused by a lack of oxygen during brain development. Impairments in movement, hearing, seeing, and thinking can result. For each of these symptoms, indicate which lobe of the cerebrum was most affected during development.

13. Traumatic brain injury can occur when a person receives a strong blow to the head or when an object enters the brain through the skull. Symptoms can include nausea, loss of sight or hearing, memory loss, and personality changes. Why do symptoms depend strongly on the location and severity of the injury?

14. What is a stroke? Use the Internet to learn the symptoms of stroke.

15. Neuroglia outnumber neurons in the nervous system by about 10 to 1. In addition, neuroglia retain the ability to divide, unlike most neurons. How do these two observations relate to the fact that most brain cancers begin in glial cells?

16. Scientists know little about many common illnesses, including migraines and Alzheimer disease. What ethical considerations make research on these diseases difficult? What are the limitations of using animals as models to study the nervous system?

17. Use the following activity to visualize the ion movements of an action potential: If you're in a classroom with several other students, label half of them as sodium ions and half as potassium ions. Find an object to represent a membrane (e.g., a line in the floor); label four mats or pieces of paper to represent different membrane proteins, and lay them across the line. Choose which side will be the inside of the cell. Act out how the ions move before, during, and after an action potential. Alternatively, if you are working alone or in a small group, use 20 coins of two different types to represent 10 Na⁺ and 10 K⁺ ions. Draw a line on a piece of paper to serve as a membrane, add four types of proteins, and label the inside and outside of the cell. Position the ions before, during, and after an action potential. Then, write a sentence or two about your activity.

PULL IT TOGETHER

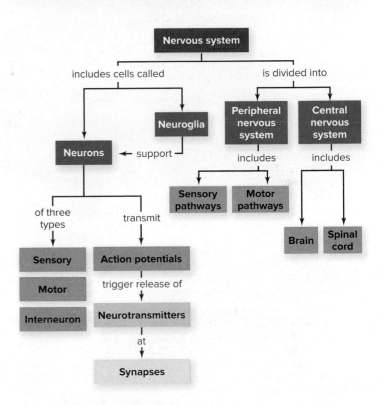

Figure 26.24 Pull It Together: The Nervous System.

Refer to figure 26.24 and the chapter content to answer the following questions.

1. Review the Survey the Landscape figure in the chapter introduction, and then add *homeostasis* to the Pull It Together concept map.

2. Add *axons* and *myelin* to the concept map.

3. What structures are included in the peripheral nervous system?

4. Add the somatic, autonomic, sympathetic, and parasympathetic nervous systems to this concept map.

27 The Senses

Pectines

Eclectic Senses. A snake's forked tongue samples odors; the comblike pectines on a scorpion's "chest" sense textures and chemicals on the ground; and a platypus detects electrical fields with its bill.

(snake): ©Art Wolfe/Science Source; (scorpion): ©Bradley P. Brayfield; (platypus): ©Dave Watts/Alamy

LEARN HOW TO LEARN
Don't Throw That Exam Away!

Whether or not you were satisfied with your last exam, take the time to learn from your mistakes. Mark the questions that you missed and the ones that you got right but were unsure about. Then figure out what went wrong for each question. For example, did you neglect to study the information, thinking it wouldn't be on the test? Did you memorize a term's definition without understanding how it fits with other material? Did you misread the question? After you have finished your analysis, look for patterns and think about what you could have done differently. Then revise your study plan so that you can avoid making the same mistakes in the future.

Different Views on the Same World

Most people know that dogs have extremely sensitive noses and that cats see well in the dark. Here are three lesser-known adaptations that other animals use to sense the world.

The Snake's Forked Tongue

A snake's forked tongue is in constant motion. Each flick represents a sample of the odor molecules left by predators, prey, and potential mates. Inside the mouth, the tips of the forked tongue pass through openings in the palate and contact sensory cells in the snout. The receptors transmit action potentials to the brain, which compares the strength of the signals arriving from each tip. The snake integrates this information, along with the nature of the chemical stimulus, to identify the source. Just as our brains determine the direction of a sound by comparing the stimuli reaching our two ears, the forked tongue allows the snake to sense odors at two points simultaneously.

Scorpion Pectines

A scorpion searches for food and mates by tasting the ground surface with a pair of comb-shaped structures called pectines, which are located on the animal's "chest." Each pecten's flexible spine carries a row of "teeth" adorned with thousands of tiny pegs. A close look at each peg reveals an elongated pore. As the animal walks, it taps its pectines on the ground. Chemicals left by other animals enter the pegs' pores and bind to sensory receptors inside, signaling the nervous system that a meal or mate is nearby. Pectines may also detect the ground's texture, which helps males decide where to deposit sperm packets during mating rituals.

Platypus Electroreception

As a platypus navigates along a riverbed, it nudges rocks aside with its bill to find prey. But the bill doesn't sense its targets by touch alone; it can also detect weak electrical fields. Australian scientists discovered that the sensors are specialized neurons called electroreceptors. The researchers placed platypuses in pools with live and dead batteries and observed that the animals explored only the live batteries, suggesting an attraction to electricity. Closer examination revealed threadlike electroreceptors deep within mucous glands in the bill. The tiniest movement of a shrimp can stimulate these cells to trigger 20 to 50 nerve impulses per second!

The sensory worlds of the snake, scorpion, and platypus may seem strange, but that is only because human sense organs respond to different signals. This chapter focuses on our own senses.

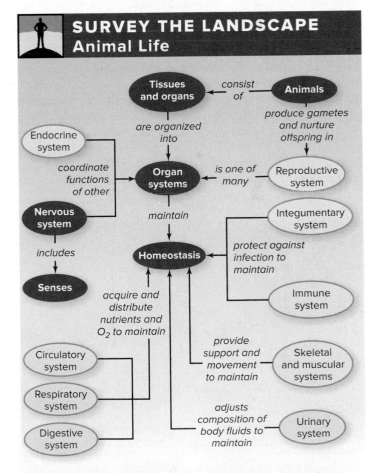

SURVEY THE LANDSCAPE
Animal Life

The senses are the bridge between the environment and an animal's body. They pass along information about the outside world to the body's internal communication networks, which trigger responses that maintain homeostasis.

For more details, study the Pull It Together feature in the chapter summary.

27.1 Diverse Senses Operate by the Same Principles

The human senses paint a complex portrait of our surroundings. Consider, for example, the woman in figure 27.1. The tips of her fingers feel the banjo strings, her eyes see the instrument, and her ears hear the music. Her skin senses the warmth of the sun. She also maintains her balance, thanks to both her ability to feel the position of her limbs and her inner ear's sense of equilibrium. When she pauses for a snack, she will be able to smell and taste her food.

As rich as our own senses are, other animals can detect stimuli that are imperceptible to us. The chapter opening essay described the unusual sensory abilities of the platypus, snake, and scorpion. Mammals have their own keen senses as well. Dogs, for example, have an extremely well-developed sense of smell, which explains why these animals are so useful in sniffing out drugs and other contraband. Bats have an entirely different ability, called echolocation. As a bat flies, it emits high-frequency pulses of sound. The animal's large ears pick up the sound waves that bounce off prey and other objects, and its brain analyzes these echoes to "picture" the surroundings.

An animal's senses are an integral part of its nervous system. As described in chapter 26, the vertebrate nervous system has two main subdivisions: the central and peripheral nervous systems. The central nervous system is composed of the brain and spinal cord. The peripheral nervous system consists of the cranial and spinal nerves that convey information between the central nervous system and the rest of the body.

This chapter turns to the sense organs, which detect stimuli from an animal's own body and its environment. The sensory pathways of the peripheral nervous system send that information to the brain. A **sensation** is the raw input that arrives at the central nervous system. For example, your eyes and hands may inform your brain that a particular object is small, round, red, and smooth. The brain integrates all of this sensory input and consults memories to form a **perception,** or interpretation of the sensations—in this case, of a tomato.

The senses convey vital information about food, danger, mates, and other stimuli. Sensory information also helps animals maintain homeostasis, a state of internal constancy that relies on negative feedback loops. Many of the feedback loops operate without our awareness; for example, we can't directly "feel" our blood pH or thyroid hormone concentration. But we are aware of sights, sounds, smells, and many other stimuli. The central nervous system responds to many types of sensory input by coordinating the actions of muscles and glands, which make adjustments as necessary to maintain homeostasis. ⓘ *negative feedback,* section 25.4

A. Sensory Receptors Respond to Stimuli by Generating Action Potentials

All sense organs ultimately derive their information from **sensory receptor** cells that detect stimuli. The human body includes several types of sensory receptors. **Mechanoreceptors** respond to physical stimuli such as sound or touch. **Thermoreceptors** respond to temperature. **Pain receptors** detect tissue damage, extreme heat and cold, and chemicals released from damaged cells. **Proprioceptors** detect the positions of the limbs, head, and other body parts. **Photoreceptors** respond to light, and **chemoreceptors** detect chemicals.

Each of these cell types "translates" sensory information into the language of the nervous system. **Transduction** is the process by which a sensory receptor converts energy from a stimulus into action potentials. Generally, a stimulus alters the shape of a protein embedded in a sensory receptor's cell membrane, causing the membrane's permeability to ions to change. The resulting movement of ions across the membrane triggers a **receptor potential,** which is a graded potential that occurs in a sensory receptor (figure 27.2). Three of the green lines in figure 27.2 depict receptor potentials that

Figure 27.1 **Sensory Blend.** This woman is experiencing many senses at once.
©RubberBall/Getty Images RF

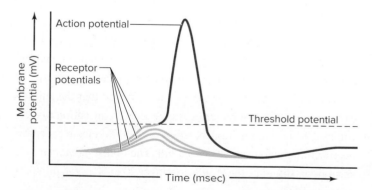

Figure 27.2 **Receptor Potentials.** Green lines in this figure show receptor potentials, three of which do not exceed the threshold potential. The largest receptor potential, however, exceeds the threshold potential, stimulating an action potential (*red line*). The central nervous system detects only stimuli that provoke action potentials.

are below the cell's threshold and therefore do not trigger action potentials; the stimulus remains undetected. But if the receptor potential does exceed the threshold potential, as in the uppermost green line, an action potential occurs in the sensory receptor (red line in the figure). The frequency of action potentials arriving at the brain from specific groups of receptors conveys information about the type and intensity of the stimulus. ⓘ *action potential,* section 26.3

B. Continuous Stimulation May Cause Sensory Adaptation

You may have noticed that your perceptions of some stimuli can change over time. Your first thought when you roll out of bed may be "I smell coffee." But by the time you stand up, pull your clothes on, and wander to the kitchen, you hardly notice the coffee odor anymore. Likewise, the steaming water in a bathtub may seem too hot at first, but it soon becomes tolerable, even pleasant.

These examples illustrate **sensory adaptation,** a phenomenon in which sensations become less noticeable with prolonged exposure to the stimulus. The explanation is that sensory

receptors generate fewer action potentials under constant stimulation. Generally, the response returns only if the intensity of the stimulus changes.

Without sensory adaptation, we would be distinctly aware of the touch of clothing, along with every sight, sound, and odor. Detecting a new stimulus, such as a person speaking, would be difficult. Many receptors therefore adapt quickly. Pain receptors, however, are very slow to adapt. The constant awareness of pain is uncomfortable, but it also alerts us to tissue damage and prompts us to address the source of the pain.

The remainder of the chapter explores the senses in more detail. It first describes the general senses, such as touch and pain, whose receptors are located throughout the body. The chapter then turns to the "special" senses, those that are limited to the head: smell, taste, vision, hearing, and equilibrium. **Figure 27.3** and **table 27.1** summarize the senses.

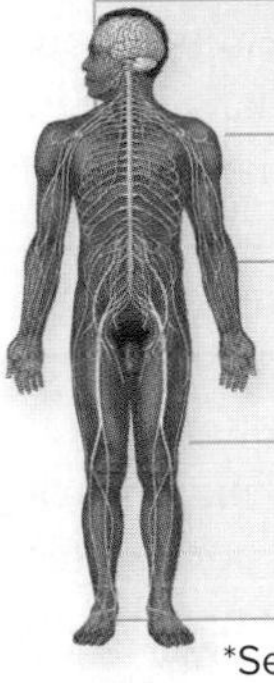

27.1   MASTERING CONCEPTS

1. What role do the senses play in maintaining homeostasis?
2. Distinguish between sensation and perception.
3. What are the major types of sensory receptors?
4. What is a receptor potential?
5. What is sensory adaptation, and how is it beneficial?

Sensory System	
Main tissue types*	**Examples of locations/functions**
Epithelial	Makes up some sensory receptors (e.g., taste cells)
Connective	Makes up part of nose and outer ear and coverings of brain and nerves
Muscle	Skeletal muscle controls opening of eyes and mouth; smooth muscle controls size of iris
Nervous	Makes up the brain, spinal cord, and nerves; functions in sensation, communication, and information storage

*See chapter 25 for descriptions.

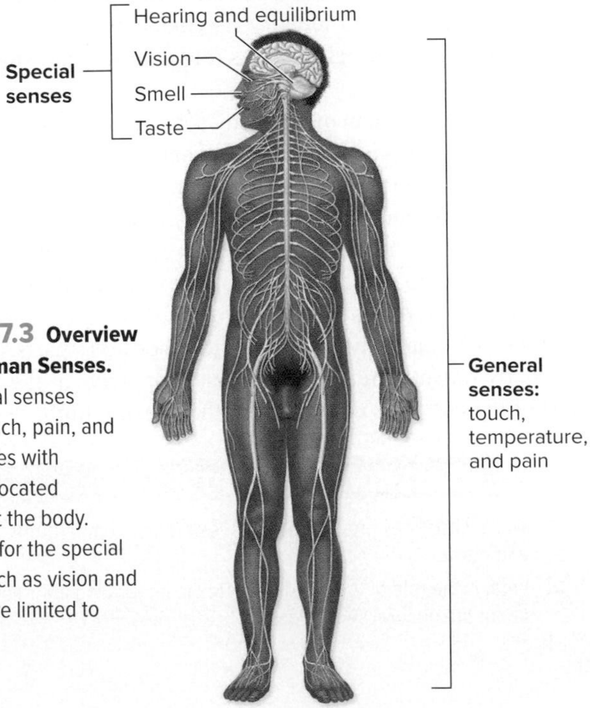

Figure 27.3 Overview of the Human Senses. The general senses include touch, pain, and other senses with receptors located throughout the body. Receptors for the special senses, such as vision and hearing, are limited to the head.

Special senses
- Hearing and equilibrium
- Vision
- Smell
- Taste

General senses: touch, temperature, and pain

TABLE **27.1** **Sense Organs and Receptors: A Summary**

Sense	Sense Organ(s)	Stimulus	Type of Receptor
General senses			
Touch	Skin	Pressure, vibration	Mechanoreceptor
Temperature	Skin	Heat, cold	Thermoreceptor
Pain	Everywhere except the brain	Damage to body tissues	Pain receptor
Position of body parts	Joints, muscles, ligaments	Stretching of muscles and ligaments	Proprioceptor
Special senses			
Smell	Nasal cavity	Airborne molecules	Chemoreceptor
Taste	Mouth and tongue	Dissolved molecules	Chemoreceptor
Vision	Eyes	Light	Photoreceptor
Hearing	Ears	Air pressure waves	Mechanoreceptor
Equilibrium	Ears	Motion of fluid in inner ear	Mechanoreceptor

27.2 The General Senses Detect Touch, Temperature, Pain, and Position

The general senses allow you to detect touch, temperature, or pain with any part of your skin. Each of these senses uses its own types of receptors. Most have their dendrites wrapped (encapsulated) in neuroglia or connective tissue, although some have free (unencapsulated) nerve endings.

The sense of touch comes from several types of mechanoreceptors in the skin (figure 27.4). The receptors that detect light touches, deep pressure, and vibrations each consist of a single encapsulated dendrite. A touch pushes the flexible sides of the capsule inward, generating an action potential in the nerve fiber. Unencapsulated touch receptors include the dendrites that wrap around each hair follicle and sense when the hair bends. The density of touch receptors varies across the body. As a result, the fingertips and tongue are much more sensitive to touch than, say, the skin of the lower back.

Free nerve endings that act as thermoreceptors in the skin respond to temperature. The brain integrates input from many thermoreceptors to determine whether a stimulus is cool, hot, or somewhere in between.

Like thermoreceptors, pain receptors are also free nerve endings, but they detect tissue damage. These neurons respond to the mechanical damage that follows a sharp blow, a cut, or a scrape. Pain receptors also detect extreme heat, extreme cold, noxious chemicals, and substances released from damaged cells.

Nearly every part of the body has pain receptors, except the brain. The brain's inability to feel pain in its own tissues has an important practical application: A patient undergoing brain surgery can remain awake and responsive throughout the procedure. The surgeon can ask the patient to move or respond verbally during surgery, ensuring that critical areas of the brain are not damaged.

Pain is an unpleasant but important response; people who are unable to perceive pain can unknowingly injure themselves. Nevertheless, temporarily suspending the body's pain response with drugs called anesthetics can make some medical treatments tolerable. These drugs work in multiple ways. Local anesthetics such as a dentist's procaine (Novocain) stop pain-sensitive neurons from transmitting action potentials in a limited area of the body. General anesthetics cause a loss of consciousness that prevents the brain from perceiving any pain.

The body also has proprioceptors, which detect the positions of the joints, tendons, ligaments, and muscles. For example, some of the muscles in your neck stretch when you move your head. Encapsulated nerve endings wrap around specialized cells in the muscles, and the dendrites initiate nerve impulses that tell the brain exactly which way your head is facing. These specialized cells are most abundant in body parts with the finest muscle control, such as the hands.

In all of the general senses, input travels along spinal or cranial nerves to the central nervous system. After passing through the thalamus, the information arrives at the cerebral cortex (see figure 26.16). Here, input from multiple receptors is mapped to each body part. We can therefore tell where on the body a sensation is originating and identify its characteristics. For example, we can feel that our right hand is touching the hot, smooth hood of a car.

The brain's role in interpreting sensations can lead to strange errors in perception. Consider the "phantom limb" phenomenon, in which a person who has lost an arm, a leg, or other body part can feel pain and other sensations arising from the limb. Researchers once attributed the pain to damaged nerve endings in the limb's stump, but they now believe that the problem is in the cerebral cortex. Areas that once processed input from the missing limb become partially reassigned to other body parts. The brain evidently misinterprets some signals that arrive at the "rewired" part of the cortex as coming from the missing limb.

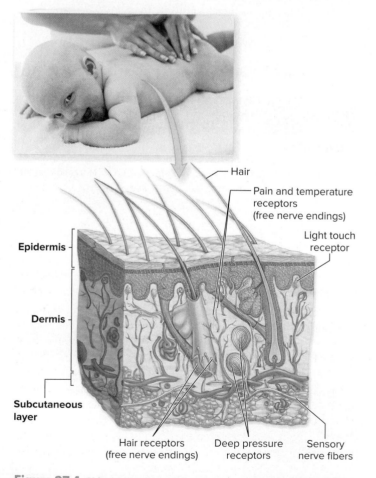

Epidermis

Dermis

Subcutaneous layer

Hair

Pain and temperature receptors (free nerve endings)

Light touch receptor

Hair receptors (free nerve endings)

Deep pressure receptors

Sensory nerve fibers

Figure 27.4 Skin Senses Many Stimuli. Sensory receptors in the skin respond to touch, temperature, and pain.

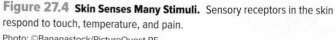
Photo: ©Bananastock/PictureQuest RF

27.2 MASTERING CONCEPTS

1. Which structures provide the senses of touch, temperature, pain, and position?

2. What is the role of the cerebral cortex in integrating information about the general senses?

27.3 The Senses of Smell and Taste Detect Chemicals

The senses of smell (also called **olfaction**) and taste (**gustation**) both depend on the body's ability to detect chemicals (figure 27.5). Animals use these senses to find food, avoid predators, identify suitable habitats, and communicate with one another. Although it is tempting to think of sniffing and tasting as behaviors that are unique to animals, chemoreception is probably the most ancient sense. Bacteria and protists use chemical cues to approach food or move away from danger, so the ability to detect external chemicals must have arisen long before animals evolved.

The two "chemical detection" senses have properties in common. In each case, the stimulus molecule must dissolve in a watery solution, such as saliva or the moist lining of a nasal passage. In addition, the molecule must interact with a chemoreceptor on a sensory cell's membrane.

A. Chemoreceptors in the Nose Detect Odor Molecules

The sense of smell begins at the **nose,** which forms the entrance to the nasal cavity inside the head. Specialized olfactory receptor neurons are located in a patch of epithelium high in the nasal cavity (figure 27.6). The human olfactory epithelium houses about 20 million receptor cells. A bloodhound has 10 times as many, and its olfactory epithelium has dozens of times more surface area than ours. The dog's sense of smell is therefore much more acute than our own.

Each olfactory neuron expresses one type of receptor protein on its cell membrane; each receptor protein, in turn, can bind to a limited set of odorants. A molecule that enters the nose in inhaled air binds to a receptor protein, and the cell then transduces this chemical signal into receptor potentials.

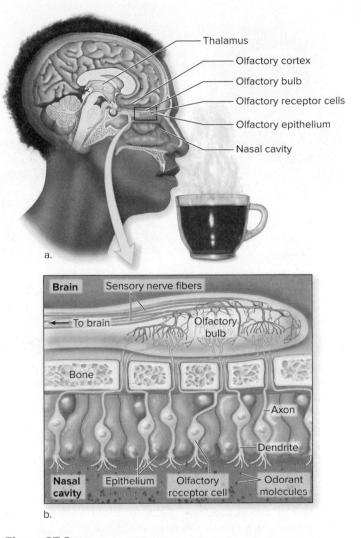

Figure 27.6 The Sense of Smell. (a) Olfaction derives from receptor cells in the nasal cavity. (b) An olfactory receptor cell binds an odorant molecule and transmits neural impulses to cells in the olfactory bulb.

Each olfactory receptor cell synapses with neurons in the brain's olfactory bulb. Unlike with the other senses, information about odors does not first pass through the thalamus. Instead, sensory neurons relay the message directly from the olfactory bulb to the brain's olfactory cortex, which interprets the information from multiple receptors and identifies the odor.

People often associate distinctive odors with vivid memories. A whiff of the perfume Grandma used to wear, for example, may elicit a flood of childhood recollections. The explanation is that the olfactory cortex is embedded in the limbic system, the brain center of memory and emotion. ⓘ *limbic system,* section 26.6B

Many arthropods use chemicals in communication. **Pheromones** are chemical substances that elicit specific responses in other members of the same species. For example, female silk moths release pheromones that attract males, which "smell" the chemical signal from up to several kilometers away. Female scorpions also attract mates with chemicals: Males "taste" female pheromones deposited on sand. The role of pheromones in humans remains an open question; see this chapter's Burning Question.

Figure 27.5 Chemical Senses. (a) Sensory cells in the nose detect the odor of incense, whereas (b) sensory cells in the mouth detect the flavor of spaghetti.

(a): ©Corbis RF; (b): ©White Rock/Getty Images RF

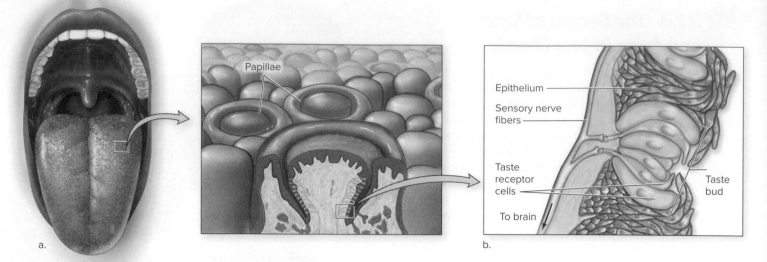

Figure 27.7 The Sense of Taste. (a) Circular papillae scattered on the tongue house taste buds, which contain the taste receptor cells. (b) The receptor cells that make up a taste bud relay signals to sensory neurons, which convey the information to the brain.

B. Chemoreceptors in the Mouth Detect Taste

The nose detects odor molecules in inhaled air, so we can perceive scents originating from near or distant objects. Chemoreceptors in the mouth, however, can taste items only at very close range.

The human mouth contains about 10,000 chemosensory organs called **taste buds** (figure 27.7). The taste buds are most concentrated on the **tongue,** a muscular organ arising from the floor of the mouth. Raised bumps called papillae house numerous taste buds, each containing 50 to 150 chemoreceptor cells that generate action potentials when dissolved food molecules bind to them. At a taste bud's base, the receptors are epithelial cells that synapse at the base of the taste bud onto sensory neurons, which relay information to cranial nerves leading to the medulla and thalamus. From there, input arrives at the gustatory cortex, the region of the cerebral cortex that interprets tastes (see figure 26.16).

The human tasting experience includes five primary sensations: sweet, sour, salty, bitter, and umami. The umami taste, such as the savory flavor of meat, derives from receptors for the amino acid glutamate; the food additive monosodium glutamate (MSG) activates these receptors. Receptors for fatty acids also occur in taste buds, leading some to argue for recognition of a sixth taste (fat). Most taste buds sense all of the primary tastes but respond most strongly to one or two. Our sense of taste depends on the pattern and intensity of activity across all taste neurons.

Food's texture, temperature, and aroma also contribute to its flavor. The intimate relationship between olfaction and taste is especially important. Even flavorful foods taste bland when a stuffy or plugged nose blocks a person's sense of smell.

27.3 MASTERING CONCEPTS

1. How does the brain distinguish one odor from another?
2. How does a taste bud function?

Burning Question

Do humans have pheromones?

Advertisements for "human pheromone" colognes appeal to the desire to attract the opposite sex. Dab some on, they say, and watch your love life blossom. But are there really human pheromones?

Some mammals definitely produce pheromones. A male hamster smeared with vaginal secretions from a female will stimulate sexual advances from another male—but only if the responding male has an intact vomeronasal organ, the pheromone detector.

Studies have also demonstrated that pheromones from human females influence the

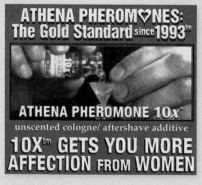

©David Tietz/Editorial Image, LLC

menstrual cycles of other women. However, researchers still know little about how humans detect pheromones. We do have a vomeronasal organ, but no one has ever shown that it is functional. Researchers are working to discover the genes that encode the pheromone-binding receptors in rodents. Comparison with the human genome should provide clues about how human pheromones work and whether the vomeronasal organ plays a role in human life.

Submit your burning question to
Marielle.Hoefnagels@mheducation.com

27.4 Vision Depends on Light-Sensitive Cells

An **eye** is an organ that produces the sense of sight. Animal eyes contain dense concentrations of photoreceptors, the sensory cells that respond to light. A photoreceptor contains a pigment molecule associated with a membrane. **Rhodopsin** is a common light-sensitive pigment. When rhodopsin absorbs light, its shape changes, altering the charge across the membrane and possibly generating an action potential.

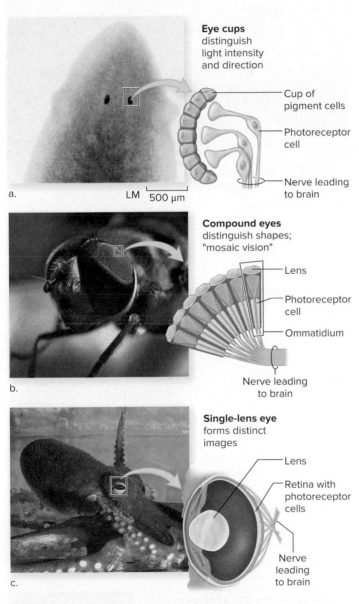

Eye cups distinguish light intensity and direction

— Cup of pigment cells

— Photoreceptor cell

— Nerve leading to brain

a.   LM ⌐500 μm⌐

Compound eyes distinguish shapes; "mosaic vision"

— Lens

— Photoreceptor cell

— Ommatidium

Nerve leading to brain

b.

Single-lens eye forms distinct images

— Lens

— Retina with photoreceptor cells

— Nerve leading to brain

c.

Figure 27.8 Invertebrate Eyes. Invertebrates have several types of eyes, three of which are shown here. (a) Planarian worms have eye cups. (b) Bees and other insects have compound eyes composed of ommatidia. (c) The octopus, a cephalopod, has a single-lens eye.

Photos: (a): ©Kent Wood/Science Source; (b): ©Ingram Publishing/SuperStock RF; (c): ©Corbis RF

A. Invertebrate Eyes Take Many Forms

Invertebrates have several types of eyes (**figure 27.8**). A planarian flatworm's photoreceptor cells, for example, are gathered into two cup-shaped eyes. These simple structures enable the flatworm to detect shadows, which is sufficient for the animal to orient itself in its environment. ⓘ *flatworms,* section 21.4

Most adult insects have paired compound eyes that consist of up to 28,000 closely packed photosensitive units called ommatidia. Each ommatidium contains a lens that transmits light to its own or nearby photoreceptor cells, generating a tiny view of the world. The animal's nervous system then integrates the input from many ommatidia to form a clear image. ⓘ *insects,* section 21.8B

Cephalopods (octopuses and their close relatives) have a single-lens eye that is much like our own. An opening in the eye, the pupil, admits light, which a lens focuses onto photoreceptors at the back of the eye. Action potentials travel along nerves to the brain, where the visual information is interpreted. ⓘ *cephalopods,* section 21.5

B. In the Vertebrate Eye, Light Is Focused on the Retina

Figure 27.9 depicts the vertebrate eye, which is composed of several layers. The **sclera** is the white, outermost layer that protects the inner structures of the eye. Toward the front of the eye, the sclera is modified into the **cornea;** this transparent, curved window bends incoming light rays.

The **choroid** is the layer internal to the sclera. Behind the cornea, the choroid becomes the **iris,** which is the colored part of the eye. The iris regulates the size of the **pupil,** the hole in the middle of the iris. In bright light, the pupil is tiny, shielding the eye from excess stimulation. The pupil grows larger as light becomes dimmer.

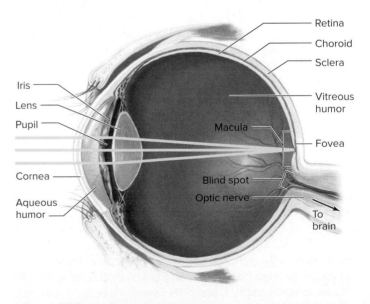

Retina

Choroid

Sclera

Vitreous humor

Macula

Fovea

Iris

Lens

Pupil

Blind spot

Optic nerve

Cornea

To brain

Aqueous humor

Figure 27.9 The Vertebrate Eye. Light passes through the cornea, aqueous humor, pupil, lens, and vitreous humor before striking the retina. Sensory cells in the retina transmit light information to the optic nerve.

A portion of the choroid also thickens into a structure that holds the flexible **lens,** which further bends the incoming light. In a process called visual accommodation, muscles regulate the curvature of the lens to focus on objects at any distance. When a person gazes at a faraway object, the lens is flattened and relaxed. To examine an article closely, however, muscles must pull the lens into a more curved shape.

Blood vessels in the choroid supply nutrients and oxygen to the **retina,** a sheet of photoreceptors that forms the inner-most layer of the eye. Reflection of bright light from the choroid's blood vessels produces the "red eye" effect in photographs. A related phenomenon, called eye shine, occurs in cats and other nocturnal vertebrates. If you aim a flashlight at a cat's eyes in darkness, the light bounces off of a reflective layer behind the retina, called the tapetum. The iridescent tapetum gives photoreceptors another chance to transduce light and helps cats see in one sixth the amount of light required for humans.

The **optic nerve** is a cranial nerve that connects each retina to the brain. The point where the optic nerve exits the retina is called the blind spot because it lacks photoreceptors and there-fore cannot sense light.

Each eyeball also contains fluid that helps bend light rays and focus them on the retina. The watery aqueous humor lies between the cornea and the lens. This fluid cleanses and nour-ishes the cornea and the lens and maintains the shape of the eye-ball. Behind the lens is the vitreous humor, which fills most of the eyeball's volume. This jellylike substance presses the retina against the choroid.

Light rays pass through the cornea, lens, and humors of the eye and are focused on the retina. The Apply It Now box in this section explains how glasses and surgery can improve poor eye-sight. Some vision problems, however, cannot be treated with lenses or surgery. One example is an aging-related disorder called macular degeneration. The macula, a small area near the center of the retina, produces the sharp, central vision required for read-ing. An indentation called the fovea is especially rich in photo-receptors. In macular degeneration, progressive loss of photoreceptors in the macula gradually causes a loss of central vision. In se-vere cases, the center of a person's field of view appears as nothing more than a dark spot. Peripheral vision, however, is unaffected.

C. Signals Travel from the Retina to the Optic Nerve and Brain

Oddly, light has to pass through several layers of cells before reaching the photoreceptors at the back of the retina (**figure 27.10**). The photoreceptors are neurons called rods and cones. **Rod cells,** which are concentrated around the edges of the retina, provide black-and-white vision in dim light and en-able us to see at night. **Cone cells** require more light, but they can detect color; they are concentrated toward the center of the retina. The human eye contains about 125 million rod cells and 7 million cone cells.

The membranes of rods and cones are studded with pig-ments. In a rod cell, the pigment is rhodopsin. The three types of cone cells contain related pigments that absorb short (blue), medium (green), or long (red) wavelengths of light. People who lack one or more types of functional cone cells, due to genetic mutations, are color-blind. Red-green color blindness, caused by mutations on the X chromosome, is more common in males than females. (i) *X-linked disorders,* section 10.7B

A pigment molecule that absorbs light energy changes shape, ultimately triggering receptor potentials in the retina's

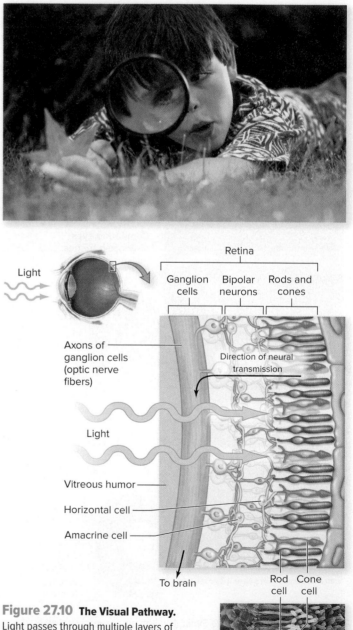

Figure 27.10 The Visual Pathway. Light passes through multiple layers of cells before striking the rods and cones, which transduce light energy into action potentials. Photos: (boy): ©Comstock Images RF; (micrograph): ©Science Photo Library RF/ Getty Images RF

Apply It **Now**

Correcting Vision

Perfect vision requires that the cornea, lens, and eyeball are a certain shape, so that light rays focus precisely on the retina. For those of us whose eyes are not perfectly formed, corrective lenses (eyeglasses and contact lenses) can treat blurry vision by altering the path of light (figure 27.A). A more recent technology for correcting vision problems is laser eye surgery, which vaporizes tiny parts of the cornea, changing the path of light to the retina.

Even people with perfectly shaped eyeballs and corneas usually need reading glasses after the age of about 40. To focus on a very close object, a muscle inside the eye must curve the lens so that it can bend incoming light rays at sharper angles. As we age, the lens becomes less flexible. It therefore becomes difficult for the muscles in the eye to bend the lens enough to clearly focus on nearby objects or printed words. Glasses can help, but laser surgery cannot correct this age-related decline in eyesight.

Sometimes, the cornea becomes clouded or misshapen. Surgeons can replace the defective cornea with one taken from a cadaver. Corneal transplant surgery carries a low risk of immune system rejection because, unlike other transplantable organs, the cornea lacks blood vessels. Another common eye disorder is a cataract, in which the lens of the eye becomes opaque. Cataract surgery is a simple procedure that replaces the clouded lens with a plastic implant.

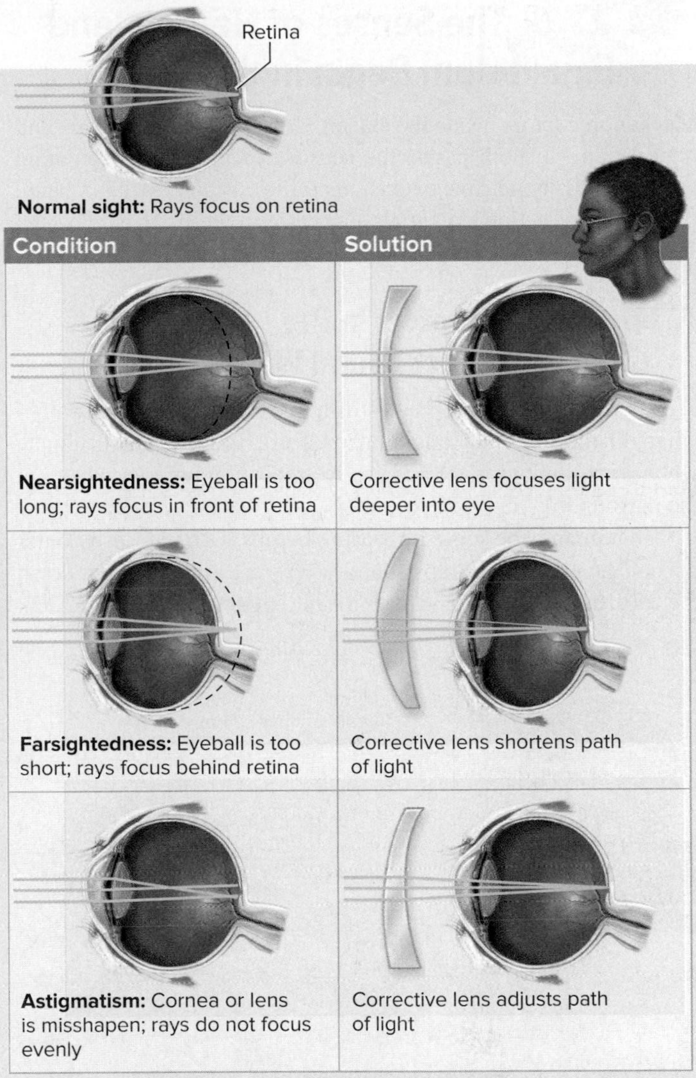

Figure 27.A **Correcting Vision.** If the cornea, lens, and eyeball are perfectly shaped, light rays focus precisely on the retina. Eyeglasses alter the path of light, correcting the blurry vision that occurs when rays focus elsewhere.

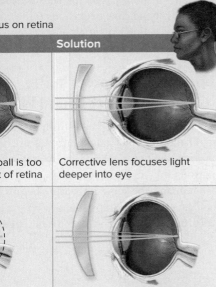

Retina

Normal sight: Rays focus on retina

Condition	Solution
Nearsightedness: Eyeball is too long; rays focus in front of retina	Corrective lens focuses light deeper into eye
Farsightedness: Eyeball is too short; rays focus behind retina	Corrective lens shortens path of light
Astigmatism: Cornea or lens is misshapen; rays do not focus evenly	Corrective lens adjusts path of light

bipolar neurons. These neurons transmit the message to the **ganglion cells,** interneurons that make up the retina's innermost layer. (Horizontal and amacrine cells modify the information along the way.) Ganglion cells are the only cells in the retina that generate action potentials; all others produce only graded potentials.

The axons of the ganglion cells make up the two optic nerves (figure 27.11). Some of these axons criss-cross behind the eyes. The stimulus then passes to the thalamus. From there, the signals go to the primary visual cortex at the rear of the brain for processing and interpretation (see figure 26.16).

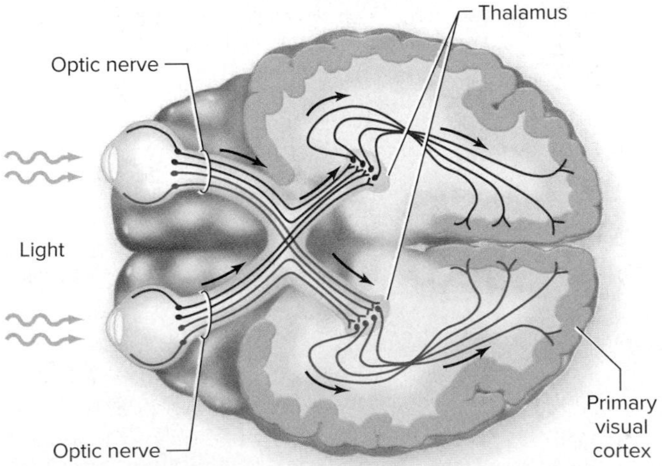

27.4   MASTERING CONCEPTS

1. Describe three types of invertebrate eyes.
2. What are the parts of the vertebrate eye?
3. Trace the pathway of information from retina to brain.

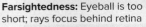

Thalamus

Optic nerve

Light

Optic nerve

Primary visual cortex

Figure 27.11 **From the Eyes to the Brain.** Each optic nerve collects information from one retina and passes it to the thalamus. The signal then passes to the primary visual cortex, which processes and integrates the information.

27.5 The Senses of Hearing and Equilibrium Begin in the Ears

Mechanoreceptors inside the **ear** provide two senses: hearing and equilibrium. In both cases, the sensory receptors are epithelial cells with many hairlike extensions (cilia). When the hairs bend, they provoke action potentials in nearby neurons that relay the signals to the brain.

A. Mechanoreceptors in the Inner Ear Detect Sound Waves

The clatter of a train, the notes of a symphony, a child's wail—what do they have in common? All are sounds that originate when something vibrates and creates repeating pressure waves in the surrounding air.

In humans, the sense of hearing begins with the fleshy outer part of the ear, which traps sound waves and funnels them down the **auditory canal** (ear canal) to the **eardrum** (figure 27.12).

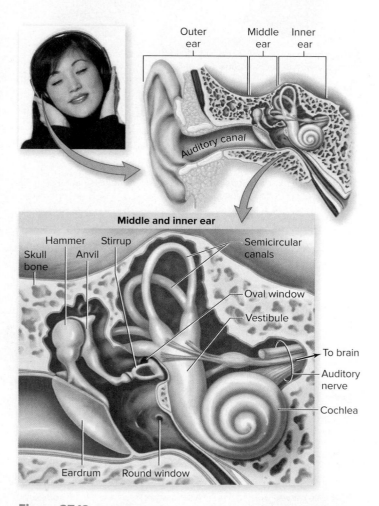

Figure 27.12 The Human Ear. Sound enters the outer ear and vibrates the three bones of the middle ear. The bones, in turn, cause vibrations in the fluid of the pea-sized cochlea in the inner ear.

Photo: ©Rubberball Productions RF

Sound pressure waves in air vibrate the eardrum, which moves three small bones in the middle ear. These bones, called the hammer, anvil, and stirrup, transmit and amplify the incoming sound. When the stirrup moves, it pushes on the **oval window,** a membrane that connects the middle ear with the inner ear. The oval window transfers the vibration to the snail-shaped **cochlea,** where sound is transduced into neural impulses.

The walls of the cochlea, which are made of bone, enclose three fluid-filled ducts (figure 27.13). Two of the ducts, called the vestibular and tympanic canals, form a U-shaped tube with the oval window at one end and the round window at the other. Sound waves trigger back-and-forth motions of this fluid.

Nestled between vestibular and tympanic canals lies the cochlear canal. The **basilar membrane** forms the lower wall of the cochlear canal. Embedded in the basilar membrane are mechanoreceptors called **hair cells,** which initiate the transduction of mechanical energy to receptor potentials. The **tectorial membrane** rests on the hair cells' cilia.

Fluid movement inside the vestibular canal is transmitted to the fluid of the adjacent cochlear canal. (To understand how, imagine a person walking on a covered pool; the footsteps disturb the water below.) As the fluid in the cochlear canal moves, the basilar membrane vibrates as well. The hair cells of the basilar membrane therefore move relative to the overlying tectorial membrane. As their cilia bend, the hair cells depolarize and initiate action potentials in fibers of the **auditory nerve,** a cranial nerve that carries the

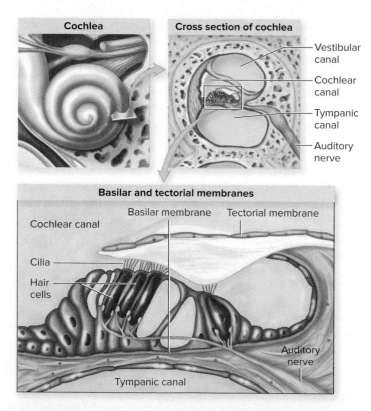

Figure 27.13 The Sense of Hearing. When the fluid inside the cochlea moves, cilia on hair cells bend relative to the tectorial membrane. These hair cells transduce sound waves into action potentials.

impulses to the thalamus. From there, the information passes to the brain's auditory cortex for interpretation (see figure 26.16).

Each sound stimulates a different region of the cochlea (figure 27.14). The high-pitched tinkle of a dinner bell vibrates the wide, rigid region of the basilar membrane at the base of the cochlea; the low-pitched tones of a tuba stimulate the cochlea's tip, deep inside the spiral. The brain interprets the input from different regions of the cochlea as sounds of different pitches. Sound intensity is important as well. Loud sounds cause the basilar membrane to vibrate more than softer sounds, stimulating additional action potentials.

The sense of hearing requires the interaction of many parts of the ear and nervous system. Deafness can occur if any of those components fail to function correctly. This chapter's second Apply It Now box explores the causes and treatments of hearing loss.

B. The Inner Ear Also Provides the Sense of Equilibrium

The sense of equilibrium includes balance and coordination. A part of the inner ear called the **vestibular apparatus** contains the receptors for equilibrium. The vestibular apparatus consists of two pouches (the utricle and saccule) and three semicircular canals (figure 27.15).

Hair cells lining the utricle and saccule detect whether the head is accelerating horizontally (utricle) or vertically (saccule). Both chambers contain a dense, jellylike fluid topped with granules of calcium carbonate. When the body moves forward, backward, up, or down, the movement of the granules lags slightly behind that of the head. The granules therefore indirectly cause cilia on the hair cells to bend, changing the membrane potential in the hair cells. Sensory neurons synapsing on the hair cells send electrical signals along the auditory nerve to the brain.

Motion sickness in a car results from contradictory signals from the vestibular apparatus and the eyes. The inner ear signals the brain that the person is accelerating. At the same time, the eyes focus on stationary objects in the car and indicate that the person is not moving. These mixed signals induce nausea.

The **semicircular canals** are three interconnected, fluid-filled loops that detect whether the head is rotating or tilting.

Their perpendicular orientation ensures that movement in any plane will stimulate one or more of the canals. The enlarged bases of the canals, which open into the utricle, are lined with clusters of hair cells covered by a caplike structure called a cupula. As the head tilts, fluid in a canal shifts. The resulting pressure on the cupula bends the cilia on the hair cells. The stimulated hair cells trigger action potentials in the auditory nerve. The brain interprets the impulses from all three canals in each ear to form a perception of head movements in three dimensions.

Like motion sickness, dizziness also traces its origin to contradictory signals. When you spin rapidly, you set the fluid in your semicircular canals into motion. But that motion does not stop as soon as the spinning ceases. As long as the fluid continues to move, the brain senses motion, even if the eyes can see that the spinning has stopped.

27.5 | **MASTERING CONCEPTS**

1. What are the parts of the ear, and how do they transmit sound?
2. How does the vestibular apparatus provide the sense of equilibrium?

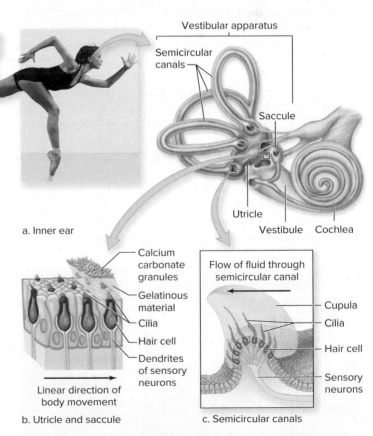

a. Inner ear

b. Utricle and saccule

c. Semicircular canals

Figure 27.15 The Sense of Equilibrium. (a) The vestibular apparatus provides the sense of equilibrium. (b) When the head is accelerating, calcium carbonate granules in the utricle and saccule move, bending the cilia on the hair cells. This motion provokes action potentials. (c) Rotating the head causes fluid to move in one or more semicircular canals, deflecting the cilia on hair cells at the base of each loop.

Photo: (a): ©Digital Vision RF

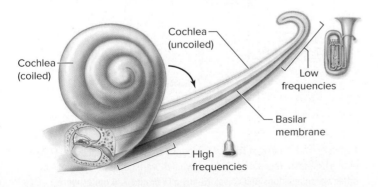

Figure 27.14 The Cochlea, Uncoiled. High-frequency sounds stimulate a different region of the cochlea than do low-frequency sounds.

INVESTIGATING LIFE

27.6 How Do Whales Taste?

Whale-watching is big business, and for good reason: Many humans are fascinated by the size and social behavior of whales. From their complex communication systems to their feeding strategies, whales are intelligent, acrobatic, and fearsome.

Whales belong to a group of mammals called the cetaceans. Biologists divide the 86 or so species of cetaceans into two groups: baleen and toothed whales. The baleen whales, including the enormous blue whale, produce giant oral baleen plates. Huge gulps of ocean water are forced through the baleen, trapping krill and other small prey for the whale to eat. The toothed whales, which include orcas, dolphins, and porpoises, use their teeth to grab fish, seals, and other prey (figure 27.16).

Researchers have known for some time that some whales have a limited sense of taste. For example, behavioral research suggests that bottlenose dolphins have a reduced perception of sweet, umami, and bitter tastes when compared to other mammals. This result is not unique, as some other mammals have their own food-related idiosyncrasies; cats and some otters are indifferent to sweets, and umami sensations are reduced or absent in giant pandas.

DNA sequencing technology has enabled researchers to study flavor perception in unprecedented detail. For example, we now understand that embedded in the cell membrane of each taste receptor cell is a protein or combination of proteins. These proteins are encoded by genes: Combinations of *T1R* (taste receptor type 1) genes encode umami and sweet taste receptors, and *T2R* (taste receptor type 2) genes encode bitter taste receptors. Taste molecules bind to T1R and T2R proteins, producing sensations of umami, sweet, or bitter. (The proteins of sour and salty taste receptor cells are ion channels, and they are encoded by genes other than *T1R* and *T2R*.)

A mutation in any taste-related gene may render a receptor protein nonfunctional, but natural selection often weeds out mutations that reduce the sense of taste. After all, flavors allow animals to detect both nutrients and potentially noxious chemicals in food. In some cases, an animal's feeding behavior may help explain the loss of a taste sensation. Cats are carnivores, so a taste for sweet foods might not help or harm them. Giant pandas eat bamboo, which lacks umami flavor components. Ancient mutations in the panda *T1R* gene would have made little difference in survival.

And what of the cetaceans? A partial dolphin genome released in 2012 revealed that these marine mammals lack intact *T1R* and *T2R* genes. Are these gene mutations unique to dolphins, or do related whale species also lack functional taste receptor genes?

Researchers Ping Feng and Huabin Zhao of Wuhan University in China teamed up with other scientists in China and England to learn more about what flavors whales can taste. The team extracted DNA from preserved tissue samples of seven species of toothed whales (including bottlenose dolphins) and five species of baleen whales. They already knew the sequences of the 3 *T1R* and 10 *T2R* genes of the bottlenose dolphin, based on the partial genome. Using these DNA sequences, the team used the polymerase chain reaction (PCR) to identify and amplify taste receptor genes from the other whale species. Finally, they sequenced each gene they found and searched for disruptive mutations, which either change the reading frame of the gene or cause premature stop codons. ⓘ *PCR,* section 11.2C; *DNA sequencing,* section 11.2B; *mutations,* section 7.7

Figure 27.17 summarizes the results of their analysis. All species of toothed and baleen whales in the study had disruptive

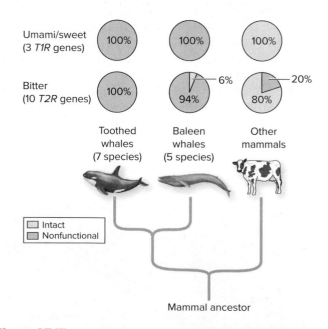

Figure 27.17   Loss of Taste. The genes encoding umami/sweet (*T1R*) and bitter (*T2R*) taste receptor proteins in whales were compared to those of other mammals (dogs and cows). Nearly all *T1R* and *T2R* genes were nonfunctional in whales, whereas most of the corresponding genes in dogs and cows were intact.

Figure 27.16   Salty Snack. A bottlenose dolphin eats an octopus.
©Jeff Rotman/Science Source

mutations in *T1R* genes, indicating that the marine mammals likely cannot taste umami or sweet. Moreover, all *T2R* (bitter) genes were nonfunctional in all seven toothed whale species. One *T2R* gene appears to be intact in three species of baleen whales; however, the researchers suggest that this one bitter taste receptor gene does not substantially confer bitter taste.

The researchers also searched for one sour taste gene in three whale species. In each case, stop codons preceded the end of the gene, strongly indicating that the whales cannot taste sour foods. Using the same method, the team searched for the three genes encoding salty taste receptors. These genes were nearly intact in each whale species, suggesting that the whales can likely taste salt.

Overall, the results indicate that whales have lost four of the five flavor sensations that most other mammals experience; salt is the lone exception. What makes these marine mammals different from their relatives on land? One possible explanation relates to the unique feeding behavior of whales: They typically swallow their food whole. This behavior may have selected against the maintenance of taste receptor genes; over time, the genes have accumulated so many mutations that they now encode nonfunctional receptor proteins. This research also raises new questions: How do whales avoid ingesting toxic or noxious foods without functional bitter taste receptors? Do other whale senses compensate for the loss of taste? Future research may reveal more about sensation in these "tasteless" animals.

Source: Feng, Ping, and five coauthors, including Huabin Zhao. 2014. Massive losses of taste receptor genes in toothed and baleen whales. *Genome Biology and Evolution*, vol. 6, pages 1254–1265.

27.6 MASTERING CONCEPTS

1. How did the researchers locate taste genes in whales? Explain how they knew if a gene was nonfunctional.
2. What data led the researchers to conclude that whales likely cannot taste umami, sweet, bitter, or sour?

Apply It **Now**

Deafness

The sense of hearing is so complex that it is not surprising to know that deafness can take multiple forms. For example, the middle ear may fail to transmit sounds to the inner ear, or the inner ear or auditory nerve may not function, or the brain may not respond to input from the nerve.

What causes hearing loss? Some babies are born deaf because of a genetic mutation, chromosomal abnormality, or prenatal exposure to disease. Other people lose their hearing later because of disease, exposure to loud noise, or injury. Earwax or an ear infection can cause short-term deafness. And nearly everyone suffers some hearing loss later in life as the ear becomes less sensitive to higher frequencies.

Hearing aids can sometimes help treat hearing loss. By amplifying sounds, a hearing aid moves the eardrum more than normal, helping the person hear more clearly. If the middle ear cannot transmit sound, however, a conventional hearing aid is useless. Bone-conduction aids solve this problem by transmitting sound waves to the bones of the skull. The vibrations stimulate the cochlea directly, bypassing the middle ear.

A cochlear implant may restore some hearing to a person who is profoundly deaf (figure 27.B). A surgeon places the device under the skin behind the ear. A microphone in the implant picks up sound; a processor then decomposes it into separate frequency components. Electrodes placed directly in the cochlea stimulate the parts of the auditory nerve corresponding to each frequency. By sending signals directly to the nervous system, cochlear implants compensate for nonfunctioning parts of the middle and inner ear.

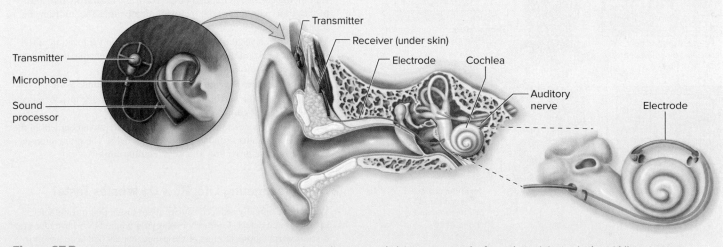

Figure 27.B Cochlear Implant. A cochlear implant stimulates the auditory nerve, helping overcome deafness that originates in the middle or inner ear.

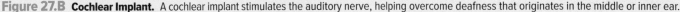

CHAPTER SUMMARY

27.1 Diverse Senses Operate by the Same Principles

- Sense organs send information about internal and external stimuli to the central nervous system. A **sensation** is the raw input received by the central nervous system; a **perception** is the brain's interpretation of the sensation.
- The senses help an animal sense food, mates, and danger; they also help the body maintain homeostasis.

A. Sensory Receptors Respond to Stimuli by Generating Action Potentials

- **Sensory receptors** are sensory neurons or specialized epithelial cells that detect stimuli (figure 27.18).
- Types of sensory receptors include **mechanoreceptors** (pressure, vibration), **thermoreceptors** (heat), **pain receptors** (tissue damage), **proprioceptors** (stretch), **photoreceptors** (light), and **chemoreceptors** (chemicals).
- A sensory receptor selectively responds to a single form of energy and **transduces** it to **receptor potentials,** which change membrane potential in proportion to stimulus strength. If a receptor potential exceeds the cell's threshold, the cell generates action potentials.

B. Continuous Stimulation May Cause Sensory Adaptation

- In **sensory adaptation,** sensory receptors cease to respond to a constant stimulus.

27.2 The General Senses Detect Touch, Temperature, Pain, and Position

- The skin's mechanoreceptors, including encapsulated and free nerve endings, respond to mechanical deflection. Free nerve endings also include thermoreceptors and pain receptors.
- Sensory neurons in muscles help detect the positions of body parts.

27.3 The Senses of Smell and Taste Detect Chemicals

- The senses of **olfaction** and **gustation** detect chemicals dissolved in watery solutions, such as those in the **nose** and mouth.

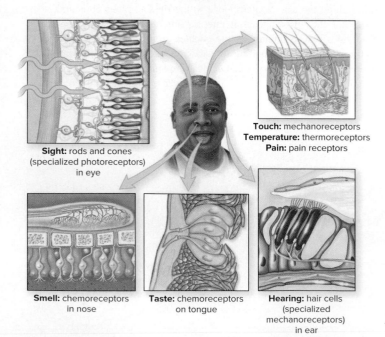

Figure 27.18 **Sensory Receptors: A Summary.**

Touch: mechanoreceptors
Temperature: thermoreceptors
Pain: pain receptors

Sight: rods and cones (specialized photoreceptors) in eye

Smell: chemoreceptors in nose

Taste: chemoreceptors on tongue

Hearing: hair cells (specialized mechanoreceptors) in ear

A. Chemoreceptors in the Nose Detect Odor Molecules

- Odorant molecules bind to receptors in the nose. The brain perceives a smell by evaluating the pattern of signals from olfactory receptor cells.
- **Pheromones** are chemicals that many animals use to communicate with others of the same species.

B. Chemoreceptors in the Mouth Detect Taste

- Humans perceive taste when chemicals stimulate receptors within **taste buds** on the **tongue.**

27.4 Vision Depends on Light-Sensitive Cells

- Photoreceptors in the **eye** contain pigments such as **rhodopsin** associated with membranes.

A. Invertebrate Eyes Take Many Forms

- Invertebrates have eye cups, compound eyes, and single-lens eyes.

B. In the Vertebrate Eye, Light Is Focused on the Retina

- The human eye's outer layer, the **sclera,** forms the transparent **cornea** in the front of the eyeball.
- The next layer, the **choroid,** supplies nutrients to the retina. At the front of the eye, the choroid holds the muscle that controls the shape of the **lens,** which focuses light on the photoreceptors. The **iris** adjusts the amount of light entering the eye by constricting or dilating the **pupil.**
- The innermost eye layer is the **retina,** and the **optic nerve** connects the retina with the brain.

C. Signals Travel from the Retina to the Optic Nerve and Brain

- The retina's photoreceptors are **rod cells,** which provide black-and-white vision in dim light, and **cone cells,** which provide color vision in brighter light.
- Light stimulation alters the pigments embedded in the membranes of rod and cone cells, possibly generating an action potential.
- Photoreceptor cells synapse with bipolar cells that, in turn, synapse with **ganglion cells.** Axons of ganglion cells leave the retina as the optic nerve, which carries information to the thalamus and visual cortex.

27.5 The Senses of Hearing and Equilibrium Begin in the Ears

- Mechanoreceptors bend in response to the motion of fluids in the inner **ear.**

A. Mechanoreceptors in the Inner Ear Detect Sound Waves

- Sound enters the **auditory canal,** vibrating the **eardrum.** Three bones in the middle ear carry these vibrations to the **oval window,** which moves fluid in the **cochlea.** Vibration of the cochlea's **basilar membrane** pushes **hair cells** against the **tectorial membrane.** The **auditory nerve** transmits the impulses to the brain.
- The brain perceives the pitch of a sound based on the location of the moving hair cells in the cochlea. Louder sounds generate more action potentials than softer ones.

B. The Inner Ear Also Provides the Sense of Equilibrium

- In the inner ear, the **vestibular apparatus** includes the utricle, saccule, and **semicircular canals.**
- When the head accelerates, tilts, or rotates, fluid movement within the vestibular apparatus stimulates sensory hair cells. The brain interprets this information, providing a sense of equilibrium.

27.6 Investigating Life: How Do Whales Taste?

- Whales have greatly reduced sensitivity to tastes that most other mammals can detect. Researchers studying whale DNA have identified disruptive mutations in most of the genes that encode taste receptor proteins.

MULTIPLE CHOICE QUESTIONS

1. As you snuggle into bed, you feel the weight of the blankets on your body, but you soon become unaware of the covers. What has happened?
   a. Your skin's touch receptors have become unable to receive information about new stimuli.
   b. Your skin's touch receptors have adapted to the feeling of the blankets.
   c. All of your body's sensory receptors have become unable to receive information about new stimuli.
   d. All of your body's sensory receptors have adapted to the feeling of the blankets.

2. Which type of sensory receptor enables you to feel the position of your legs, even if a table hides your legs from sight?
   a. Thermoreceptor   c. Chemoreceptor
   b. Photoreceptor   d. Proprioceptor

3. In what way are the senses of smell and taste different?
   a. Chemoreceptors detect smell, whereas mechanoreceptors detect taste.
   b. Olfactory receptors bind to chemicals dissolved in gas, whereas taste receptors bind to chemicals dissolved in water.
   c. Smell is processed in the spinal cord, whereas taste is processed in the brain.
   d. We can smell chemicals from distant objects, whereas taste is limited to chemicals at close range.

4. The structures that enable bees to see flowers are
   a. eye cups.   c. ommatidia.
   b. single-lens eyes.   d. maculas.

5. What is the function of hair cells in the cochlea?
   a. Transduce sound waves into neural impulses
   b. Interpret and identify sounds
   c. Funnel sounds to the inner ear
   d. Prevent debris from entering the delicate inner ear

Answers to these questions are in appendix A.

WRITE IT OUT

1. A male moth uses his antennae to detect the concentration of a pheromone on each side of his body, allowing him to fly toward the female that is producing the pheromone. Explain how sensory receptors, peripheral nerves, and the central nervous system interact to allow moths to compare pheromone concentrations in the air surrounding each antenna.

2. What is the role of transduction in the sensory system? How does transduction occur for each of the senses described in this chapter?

3. Explain why evolution has favored slow sensory adaptation to pain stimuli.

4. Try as you might, you cannot tickle yourself. Speculate about why it could be adaptive to respond to surprises but not self-imposed stimuli.

5. How does the nervous system differentiate among odors?

6. Why might cold medicines that dry the nasal cavity make it harder to smell?

7. Explain why some people hold their nose when consuming bad-tasting food or medicine.

8. Suppose you put on glasses belonging to someone who is more farsighted than you. Draw how light passes through the glasses and into your eye. Why will the glasses blur your vision?

9. What are the roles of rods and cones in the sense of sight?

10. In what ways do the cochlea and vestibular apparatus function similarly?

11. In a rare condition called synesthesia, stimulation of one sense causes stimulation of another sense. For example, people with synesthesia have reported seeing bursts of color when stimulated with loud noises. Would you expect synesthesia to be a problem with sensory receptors, peripheral nerves, or the central nervous system? Explain.

PULL IT TOGETHER

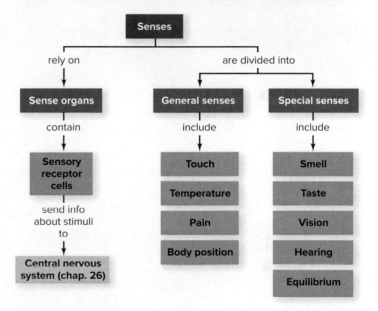

Figure 27.19 Pull It Together: The Senses.

Refer to figure 27.19 and the chapter content to answer the following questions.

1. Review the Survey the Landscape figure in the chapter introduction, and then connect each general and special sense in the Pull It Together concept map to a new term: *homeostasis.*

2. Which sense organs are required for each of the general and special senses?

3. Add *mechanoreceptors, chemoreceptors,* and *photoreceptors* to the concept map.

28 The Endocrine System

Breaking Point. Training the mind and body to relax can help relieve stress when life takes a turn for the worse.

©Juanmonino/E+/Getty Images RF

LEARN HOW TO LEARN
Use Those Office Hours

Most instructors maintain office hours. Do not be afraid to use this valuable resource! Besides getting help with course materials, office hours give you an opportunity to know your professors personally. After all, at some point you may need a letter of recommendation; a letter from a professor who knows you well can carry a lot of weight. If you do decide to visit during office hours, be prepared with specific questions. And if you request a separate appointment, be sure to arrive at the time you have arranged—or let your instructor know you need to cancel.

Reduce Stress for Healthy Living

Maybe midterms are around the corner or your personal relationships are on the rocks. Whatever the cause, stress can promote weight gain, suppress your immune system, and have other serious side effects. How can a mental state—chronic stress—alter your physical health?

This question relates to the endocrine system, the subject of this chapter. Cells in endocrine glands produce hormones, which are chemicals that travel throughout the body in the bloodstream. Stress hormones come from the adrenal glands. For example, when you are frightened, these glands produce epinephrine (commonly called adrenaline), a hormone that makes your heart race.

The adrenal glands also produce hormones in response to long-term stress; the most important of these hormones is cortisol. Physical or mental stress triggers the release of cortisol into the blood, and the body responds in multiple ways. For example, cortisol initiates reactions that prompt liver and fat cells to release stored glucose and fatty acids into the blood. Having these energy sources readily available is useful during stress, when metabolic demands might be higher. The effects of cortisol are therefore adaptive when stress is infrequent and temporary.

Chronic stress, however, maintains high levels of cortisol, which could cause unwanted weight gain. As cells release glucose and fatty acids, appetite increases. Food helps replenish depleted energy reserves, but overeating causes body weight to climb. Interestingly, cortisol's effects vary with the location in the body: Fat is released into the blood at the extremities, whereas fat is stored in the abdomen. Excess "belly fat" is therefore often associated with a high-stress life.

Cortisol also suppresses the immune system, decreasing swelling in tissues damaged by physical stress. But lowered immune function may become harmful over the long term, as the body's ability to fight infection dwindles. Not surprisingly, illness often follows a stressful week.

Behavioral changes can either enhance or suppress the cortisol response. People with chronic stress may have difficulty sleeping or finding time to prepare nutritious meals, further boosting cortisol levels. Intense exercise puts physical stress on the body and temporarily increases cortisol as well. Yoga and other gentle forms of exercise, in contrast, relieve stress, as do deep breathing and stretching.

Understanding how the endocrine system controls your body's chemistry can be empowering. This chapter describes how cortisol and many other hormones participate in the complex internal signals of the endocrine system.

SURVEY THE LANDSCAPE
Animal Life

Hormones are communication molecules produced by the organs of the endocrine system. These molecules circulate throughout the body, altering the activities of the body's other organs. These adjustments help the animal body maintain homeostasis.

For more details, study the Pull It Together feature in the chapter summary.

28.1 The Endocrine System Uses Hormones to Communicate

Animals communicate with one another in many ways, including color displays, sounds, body language, and scents. Likewise, the cells that make up a multicellular organism's body send and receive signals; these cell-to-cell messages coordinate the actions of the body's organ systems.

In flowering plants, hormones orchestrate growth, development, and responses to the environment (see chapter 24). Animal bodies, in contrast, have two main communication systems. The **nervous system,** described in chapter 26, is a network of cells that specialize in sending speedy signals that vanish as quickly as they arrive. The **endocrine system** is the other main communication system within the animal body. As you will see, the endocrine system does not act with the speed of neural impulses, but its chemical messages have something else: staying power.

A. Endocrine Glands Secrete Hormones That Interact with Target Cells

The endocrine system has two main components: glands and hormones. An **endocrine gland** consists of cells that produce and secrete hormones into the bloodstream, which carries the secretions throughout the body. A **hormone** is a biochemical

that travels in the bloodstream and alters the metabolism of one or more cells.

The endocrine system would be ineffective if every hormone acted on every cell in the body. Instead, a limited selection of **target cells** respond to each hormone. Inside or on the surface of each target cell is a receptor protein, which binds to the hormone and initiates the cell's response.

Hormones are analogous to the radio signals that multiple stations simultaneously broadcast into the atmosphere. The receptor proteins, then, are like individual radios. Even when dozens of signals are present, each radio is tuned to one frequency and therefore picks up the signal of just one station. Likewise, each receptor binds to one of the many hormones that may be circulating in the bloodstream. Moreover, just as one house may contain many radios, each tuned to a different station, one target cell may also have receptors for many hormones, each of which initiates a unique response.

Many organs produce hormones. The main endocrine organs in vertebrates are the hypothalamus, pituitary gland, pineal gland, thyroid gland, parathyroid glands, adrenal glands, pancreas, ovaries, and testes (**figure 28.1**). The heart, kidneys, liver, stomach,

Figure 28.1 **Human Endocrine Glands.** The endocrine system includes several glands containing specialized cells that secrete hormones. Additional hormone-secreting cells are scattered among the other organ systems. The hormones circulate throughout the body in blood vessels, which are not shown in this figure.

Endocrine System	
Main tissue types*	**Examples of locations/functions**
Epithelial	Makes up the bulk of most glands and secretes many types of hormones
Connective	Blood circulates hormones throughout the body
Nervous	Parts of the brain secrete some hormones and control release of others; some neurons secrete hormones

*See chapter 25 for descriptions.

Hypothalamus (shown in green) Produces hormones that stimulate or inhibit the release of hormones from the pituitary gland

Thyroid gland Releases thyroid hormones, which regulate metabolism

Pineal gland (shown in blue) Produces melatonin, which helps regulate sleep-wake cycles

Pituitary gland (shown in orange) Produces numerous hormones that affect target tissues directly or stimulate other endocrine glands

Parathyroid glands (behind thyroid) Secrete parathyroid hormone, which helps regulate blood calcium

Adrenal glands Produce hormones that regulate kidney function and contribute to the body's stress response

Pancreas Releases hormones that regulate blood glucose levels

Ovaries (in female) Produce estrogen and progesterone, which mediate monthly changes in the uterine lining and promote secondary sex characteristics

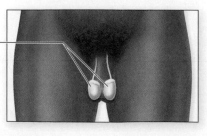

Testes (in male) Produce testosterone, which promotes sperm maturation and secondary sex characteristics

small intestine, and placenta also contain scattered hormone-secreting cells. Together, these organs release dozens of hormones that simultaneously regulate every aspect of our lives, from conception through death.

To illustrate the power of the endocrine system, consider one stage of life that famously involves hormones: puberty. During this period, hormones transform a child's body into that of an adult. Females develop enlarged breasts and wider hips, males acquire a deeper voice and more muscular physique, and new body hair sprouts in both sexes. The same hormones also affect mood, emotions, and feelings of sexual attraction.

Hormones figure prominently in the lives of other animals, too (figure 28.2). For example, a caterpillar undergoes a dramatic metamorphosis as it develops into a butterfly, as does a tadpole transforming into an adult frog.

The endocrine system's effects are not always so extreme, but they are nonetheless present throughout life. Some types of chemical contaminants disrupt these delicate signals. The consequences can be serious and long-lasting, as described in this chapter's Burning Question box.

Because hormones are so powerful, an animal's body strictly regulates the levels of these molecules in the bloodstream. This tight control often occurs by negative feedback interactions. Recall from chapter 25 that in negative feedback, a change in a condition triggers action that reverses the change. Feedback loops ensure that endocrine glands adjust the secretion of all hormones as required to maintain homeostasis. ⓘ *negative feedback,* section 25.4

B. The Nervous and Endocrine Systems Work Together

Although the nervous and endocrine systems both specialize in communication, they differ in many ways. First, neurons use action potentials and neurotransmitters to send messages, whereas the endocrine system employs hormones. Second, each neuron influences only a few cells at a time, whereas hormones circulate throughout the body in the blood and may affect many different cells. Third, the endocrine system communicates much more slowly than the nervous system. A nervous impulse is virtually instantaneous, and its effects disappear as soon as the stimulus vanishes. Hormones take minutes, hours, or even days to exert their effects, which are generally more prolonged.

Despite these differences, the nervous and endocrine systems are tightly integrated—so much so that some biologists refer to them together as the "neuroendocrine system." The most obvious connection is a physical one: the hypothalamus. This region of the brain is clearly part of the central nervous system. Yet the hypothalamus contains neurons called neurosecretory cells that release hormones directly into the bloodstream. Moreover, the hypothalamus directly or indirectly controls the action of many endocrine glands.

The nervous and endocrine systems also share chemical links. For example, some chemicals, such as epinephrine and norepinephrine, can act as neurotransmitters (if released from a neuron) or hormones (if released from an endocrine gland). In addition, neurotransmitters and hormones share some target cells.

Animal physiologists are still learning how hormones and neurons interact to oversee growth and development, influence appetite, regulate the concentrations of vital nutrients in the blood, and ready the body to confront stress. This chapter therefore cannot paint a complete picture of endocrine action. Instead, the objective is to explain some of the best understood hormonal effects—beginning in section 28.2, which describes how target cells respond to hormones.

28.1 MASTERING CONCEPTS

1. What is the overall function of the endocrine system?
2. Describe the relationships among endocrine glands, hormones, and target cells.
3. What is the role of negative feedback in the endocrine system?
4. How do the nervous and endocrine systems differ?
5. Describe how the nervous and endocrine systems interact.

Figure 28.2
Metamorphosis.
Hormones control the transformation of (a) a caterpillar into a butterfly and (b) a tadpole into an adult frog.

(a, both): ©Ken Cavanagh/McGraw-Hill Education; (b): ©Robert Clay/Alamy RF

a.

b.

28.2 Hormones Stimulate Responses in Target Cells

Just as a key fits a lock, each hormone affects only target cells bearing specific receptor molecules. The term *target cells* is a little misleading, because it implies that hormones somehow travel straight from their source to a limited set of cells. In reality, the blood circulating throughout the body contains many hormones at once. Each hormone's target cells are simply those with the corresponding receptors.

This section describes how the interaction between a hormone and its receptor initiates the target cell's response. In general, receptors for water-soluble hormones are on the surface of the target cell. In contrast, lipid-soluble hormones typically interact with receptors inside cells, either in the cytoplasm or in the nucleus.

A. Water-Soluble Hormones Trigger Second Messenger Systems

Most water-soluble hormones are proteins or **peptide hormones** (short chains of amino acids). Insulin, the protein involved in sugar metabolism, is one example. Some water-soluble hormones, the amines, are derived from a single amino acid. This group includes epinephrine, norepinephrine, and dopamine. (i) *amino acids,* section 2.5C

Water-soluble hormones cannot pass readily through the cell membrane. Instead, they bind to receptors on the surface of target cells (**figure 28.3a**). This hormone–receptor interaction triggers a chain reaction called signal transduction. The receptor protein contacts a neighboring protein (the "G protein"). The G protein, in turn, stimulates an enzyme called adenyl cyclase to convert ATP to another molecule, cyclic AMP (cAMP). This product is called a **second messenger,** and it is the molecule that actually provokes the cell's response. Typically, cAMP activates enzymes that produce the hormone's effects. This entire cascade of reactions therefore converts the external "message"—the arrival of the peptide hormone at the outer membrane—into a signal that acts inside the cell. (i) *cell membrane,* section 3.3; *ATP,* section 4.3

In general, water-soluble hormones act rapidly, within minutes of their release. Target cells respond quickly because all of the participating biochemicals are already in place when the hormone binds to the receptor.

Burning Question

Are plastics dangerous?

Every day, humans release thousands of types of pesticides, cosmetics, fragrances, medications, and other products into the air, soil, and water. The environment therefore teems with chemicals that are or may be endocrine disruptors—substances that alter hormonal signaling, often by mimicking a natural hormone. Other endocrine disruptors block the action of hormones, and still others stimulate or inhibit the activity of the glands that produce the hormones in the first place. These chemicals are in our water and food, and they accumulate in the fatty tissues of our bodies.

Many scientists and health professionals are also concerned that some plastics release harmful chemicals that may alter human development. Much of the controversy over plastics centers on a chemical called bisphenol A (BPA). Manufacturers use BPA to make shatterproof polycarbonate bottles, the linings of food cans, the sealants used in dentistry, and many other items. BPA is so common that everyone on Earth has this chemical in his or her tissues. Not only does BPA accumulate over a person's lifetime, but it can also pass from mother to fetus.

Why the concern over BPA? Many researchers have reported that at low doses, BPA replicates the effect of the sex hormone estrogen (see section 28.5), causing reproductive problems and developmental abnormalities in laboratory animals. The U.S. Food and Drug Administration has cited other studies indicating that BPA is safe. The debate is unlikely to end in the near future. Testing for long-term effects of endocrine disruptors in humans is extremely difficult. Besides the ethical issues surrounding human experimentation, other complicating factors include the impossibility of finding BPA-free control subjects, the many stages of development at which endocrine disruptors can act, developmental differences between the sexes, and potential interactions between BPA and other endocrine disruptors.

©Corbis RF

Regardless of the outcome of the BPA debate, other endocrine disruptors have apparently altered the development and reproduction of wild animals, including snails, fishes, frogs, alligators, and polar bears. Limiting the release of harmful chemicals into the environment will help slow the effect of endocrine disruptors on ecosystems. And from a consumer perspective, you can educate yourself about the endocrine-disrupting chemicals you may be ingesting when sipping from a plastic bottle or nibbling on your favorite snack. After all, the widespread effects of these chemicals on wild animals suggest that humans may be at risk, too.

Submit your burning question to
Marielle.Hoefnagels@mheducation.com

B. Lipid-Soluble Hormones Directly Alter Gene Expression

Some hormones are lipid-soluble. The most familiar are the **steroid hormones,** such as testosterone, estrogen, and progesterone. The body synthesizes these and other steroid hormones from cholesterol, which is one reason humans need at least some cholesterol in their diets. Two other lipid-soluble hormones, the thyroid hormones, are derived from a single hydrophobic amino acid (tyrosine). ⓘ *lipids,* section 2.5E

Lipid-soluble hormones easily cross the cell membrane (figure 28.3b); no second messenger is involved. Once inside the cell, the hormone may enter the nucleus and bind to a receptor associated with DNA, triggering the production of proteins that carry out the target cell's response. Alternatively, the hormone may bind to a receptor in the cytoplasm, and the two molecules may travel together to the nucleus. Either way, response time is much slower than for peptide hormones, because the cell must produce new proteins before the hormone takes effect. ⓘ *protein synthesis,* section 7.3

Note that a target cell's sensitivity to either water- or lipid-soluble hormones may change over time. For example, chronically high levels of a hormone may prompt a target cell to destroy some of its receptors; the cell subsequently becomes less sensitive to the hormone. On the other hand, the presence of a hormone may prompt target cells to produce more receptors, boosting sensitivity to the hormone. Adjusting receptor numbers in either direction helps cells maintain homeostasis.

28.2 | MASTERING CONCEPTS

1. How does a hormone affect some cells but not others?
2. Describe the locations of the receptors that bind to water- and lipid-soluble hormones.
3. What is the role of second messengers in hormone action?
4. Why do steroid hormones act relatively slowly?

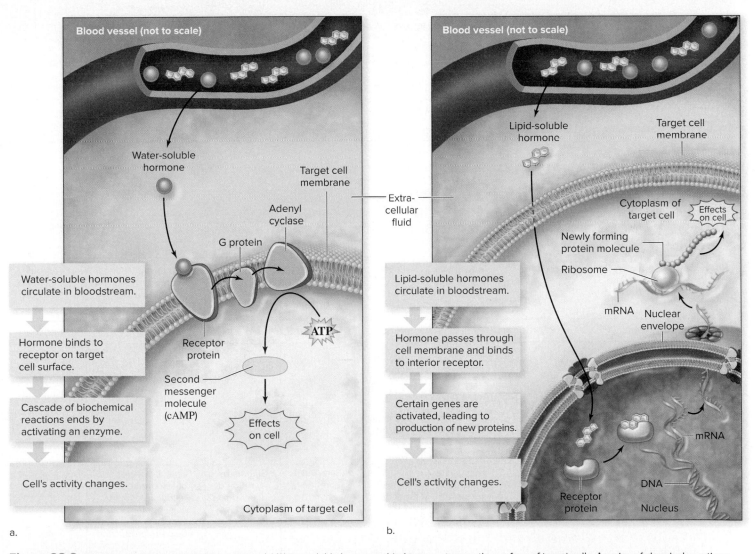

a.   b.

Figure 28.3 Target Cell Responses to Hormones. (a) Water-soluble hormones bind to receptors on the surface of target cells. A series of chemical reactions initiates the target cell's response. (b) Lipid-soluble hormones pass through cell membranes and bind to receptors in the cytoplasm or nucleus. The target cell responds by altering the expression of one or more genes.

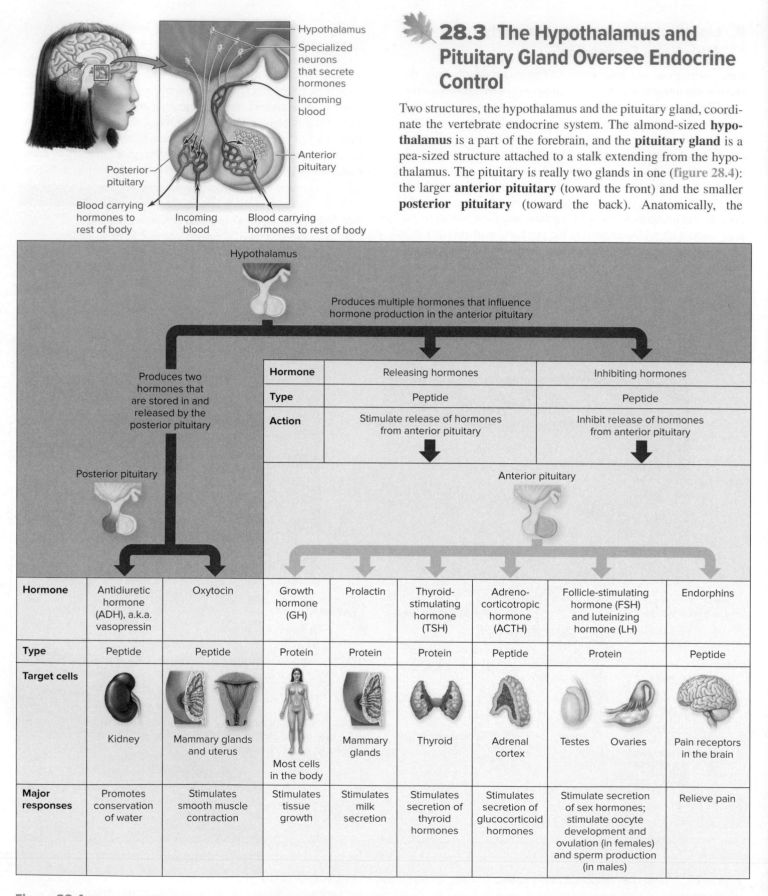

28.3 The Hypothalamus and Pituitary Gland Oversee Endocrine Control

Two structures, the hypothalamus and the pituitary gland, coordinate the vertebrate endocrine system. The almond-sized **hypothalamus** is a part of the forebrain, and the **pituitary gland** is a pea-sized structure attached to a stalk extending from the hypothalamus. The pituitary is really two glands in one (**figure 28.4**): the larger **anterior pituitary** (toward the front) and the smaller **posterior pituitary** (toward the back). Anatomically, the

Hypothalamus

Specialized neurons that secrete hormones

Incoming blood

Anterior pituitary

Posterior pituitary

Blood carrying hormones to rest of body

Incoming blood

Blood carrying hormones to rest of body

Hypothalamus

Produces multiple hormones that influence hormone production in the anterior pituitary

Produces two hormones that are stored in and released by the posterior pituitary

Posterior pituitary

Anterior pituitary

Hormone	Releasing hormones	Inhibiting hormones
Type	Peptide	Peptide
Action	Stimulate release of hormones from anterior pituitary	Inhibit release of hormones from anterior pituitary

Hormone	Antidiuretic hormone (ADH), a.k.a. vasopressin	Oxytocin	Growth hormone (GH)	Prolactin	Thyroid-stimulating hormone (TSH)	Adreno-corticotropic hormone (ACTH)	Follicle-stimulating hormone (FSH) and luteinizing hormone (LH)	Endorphins
Type	Peptide	Peptide	Protein	Protein	Protein	Peptide	Protein	Peptide
Target cells	Kidney	Mammary glands and uterus	Most cells in the body	Mammary glands	Thyroid	Adrenal cortex	Testes Ovaries	Pain receptors in the brain
Major responses	Promotes conservation of water	Stimulates smooth muscle contraction	Stimulates tissue growth	Stimulates milk secretion	Stimulates secretion of thyroid hormones	Stimulates secretion of glucocorticoid hormones	Stimulate secretion of sex hormones; stimulate oocyte development and ovulation (in females) and sperm production (in males)	Relieve pain

Figure 28.4 Hormones of the Hypothalamus and Pituitary: A Summary. The hypothalamus and the pituitary gland secrete hormones that help coordinate the actions of the other endocrine glands.

posterior pituitary is a continuation of the hypothalamus, whereas the anterior pituitary consists of endocrine cells.

Information about body temperature, body fluid composition, and many other stimuli travels from the body's sensory neurons to the hypothalamus. In response, specialized neurons extending from the hypothalamus secrete hormones into both parts of the pituitary gland, forming a direct link between the nervous and endocrine systems (see the anatomical diagram in figure 28.4. As you will see, the overall effect is to maintain homeostasis.

The table in figure 28.4 summarizes the roles of the hypothalamus, posterior pituitary, and anterior pituitary. Note that the posterior pituitary does not synthesize hormones, but it does store and release two hormones that the hypothalamus produces. The hypothalamus controls the anterior lobe of the pituitary, too, but in a different way—by secreting hormones that reach the anterior pituitary through a specialized system of blood vessels.

A. The Posterior Pituitary Stores and Releases Two Hormones

One of the two hormones produced by the hypothalamus and released by the posterior pituitary is **antidiuretic hormone (ADH),** also called vasopressin. If cells in the hypothalamus detect that the body's fluids are too concentrated, the posterior pituitary releases more ADH. This hormone stimulates kidney cells to return water to the blood (rather than eliminating the water in urine). The body's fluids become more dilute. Once balance is restored, ADH production slows. Chapter 33 explains kidney function in more detail.

Oxytocin is the other posterior pituitary hormone. When a baby suckles, sensory neurons in the mother's nipple relay the information to the brain, which stimulates the release of oxytocin. The hormone causes cells in the breast to contract, squeezing the milk through ducts leading to the nipple. Oxytocin also triggers muscle contraction in the uterus, which pushes a baby out during labor. Physicians use synthetic oxytocin to induce labor or accelerate contractions in a woman who is giving birth.
(i) *childbirth,* section 35.5E

ADH and oxytocin also act on the brain, playing a role in bonding, affection, and social recognition in at least some species (see section 28.6). These hormones may also participate in human social attachment and in disorders such as autism.

B. The Anterior Pituitary Produces and Secretes Six Hormones

One of the six hormones that the anterior pituitary gland produces is **growth hormone (GH),** also called somatotropin. This hormone promotes growth and development in all tissues by increasing protein synthesis and cell division rates. Levels of GH peak in the preteen years and help spark adolescent growth spurts. A severe deficiency of GH during childhood leads to pituitary dwarfism (extremely short stature); at the other extreme, a child with too much GH becomes a pituitary giant (figure 28.5).

Figure 28.5 Growth Hormone Abnormality. A pituitary giant poses with his father and young brother.
©AP Images

In an adult, GH does not affect height because the long bones of the body are no longer growing. However, excess GH can cause acromegaly, a thickening of the bones in the hands and face.

Prolactin is an anterior pituitary hormone that stimulates milk production in a woman's breasts after she gives birth. In males and in women who are not nursing an infant, a hormone from the hypothalamus suppresses prolactin synthesis. In nursing mothers, however, a suckling infant triggers nerve impulses that overcome this inhibition.

Four other anterior pituitary hormones all influence hormone secretion by other endocrine glands. **Thyroid-stimulating hormone (TSH)** prompts the thyroid gland to release hormones, whereas **adrenocorticotropic hormone (ACTH)** stimulates hormone release from parts of the adrenal glands. The remaining two anterior pituitary hormones stimulate hormone release from the ovaries and testes: **follicle-stimulating hormone (FSH)** and **luteinizing hormone (LH).** Sections 28.4 and 28.5 describe the roles of these four hormones in more detail.

The anterior pituitary also produces **endorphins,** which are natural painkillers that bind to receptors on target cells in the brain. Usually, however, endorphins are not detectable in the blood, so their status as hormones is questionable.

28.3 MASTERING CONCEPTS

1. How does the hypothalamus interact with the posterior and anterior pituitary glands?
2. List the names and functions of the hormones released by the posterior and anterior pituitary glands.

28.4 Hormones from Many Glands Regulate Metabolism

The thyroid gland, parathyroid glands, adrenal glands, and pancreas secrete hormones that influence metabolism (**figure 28.6**). Hormones from the anterior pituitary control many, but not all, of the activities of these glands (see figure 28.4).

A. The Thyroid Gland Sets the Metabolic Pace

The **thyroid gland** is a two-lobed structure in the neck. The lobes secrete two thyroid hormones, **thyroxine** and **triiodothyronine,** that increase the rate of metabolism in target cells. Under thyroid stimulation, the lungs exchange gases faster, the small intestine absorbs nutrients more readily, and fat levels in cells and in blood plasma decline.

The thyroid hormones illustrate how the hypothalamus and pituitary interact in negative feedback loops (**figure 28.7**). When blood levels of thyroid hormones are low, the hypothalamus secretes thyrotropin-releasing hormone (TRH), which stimulates the anterior pituitary to increase production of thyroid-stimulating hormone (TSH). In response, epithelial cells in the thyroid secrete thyroxine and triiodothyronine. In the opposite situation, TRH secretion slows, so the thyroid glands reduce their production of hormones.

One disorder that affects the thyroid gland is hypothyroidism, a condition in which the thyroid does not release enough hormones. The metabolic rate slows and weight increases. Synthetic hormones can treat many cases of hypothyroidism. In the

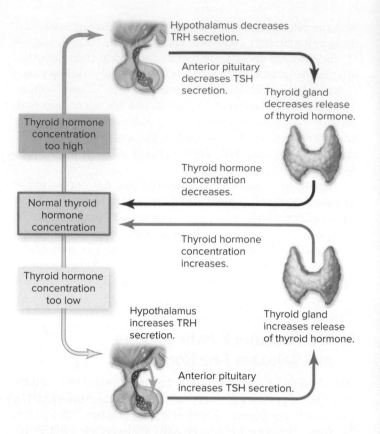

Figure 28.7 Thyroid Hormone Regulation. A negative feedback loop maintains the proper concentration of thyroid hormones in blood.

Source	Thyroid		Parathyroid	Adrenal medulla	Adrenal cortex		Pancreas		Pineal gland
Hormone	Thyroxine, triiodothyronine	Calcitonin	Parathyroid hormone (PTH)	Epinephrine, norepinephrine	Mineralo-corticoids	Gluco-corticoids	Insulin	Glucagon	Melatonin
Type	Amine	Peptide	Protein	Amine	Steroid	Steroid	Protein	Peptide	Amine
Target cells	All tissues	Bone	Bone, digestive organs, kidneys	Blood vessels	Kidney	All tissues	All tissues	Liver, adipose tissue	Other endocrine glands
Major responses	Increase metabolic rate	Increases rate of calcium deposition	Releases calcium from bone, increases calcium absorption in digestive organs and kidneys	Raise blood pressure, constrict blood vessels, slow digestion	Maintain blood volume and electrolyte balance	Increase glucose levels in blood and brain	Increases uptake of glucose	Stimulates breakdown of glycogen into glucose and of fats into fatty acids	Regulates effects of light–dark cycles

Figure 28.6 Hormones That Regulate Metabolism: A Summary. Hormones from several endocrine glands simultaneously influence many metabolic processes.

past, the most common cause of hypothyroidism was iodine deficiency. Both thyroid hormones contain iodine; a deficiency of this essential element causes a goiter, or swollen thyroid gland. Today, iodine-deficient goiter is rare in nations where iodine is added to table salt.

An overactive thyroid causes hyperthyroidism. This disorder is associated with hyperactivity, an elevated heart rate, a high metabolic rate, and weight loss. Graves disease is the most common type of hyperthyroidism. Both former President George H. W. Bush and his wife, Barbara, have this disorder.

Scattered cells throughout the thyroid gland produce a third hormone, **calcitonin,** which decreases blood calcium level by increasing the deposition of calcium in bone. Levels of calcitonin greatly increase during pregnancy and milk production, when a woman's body is under calcium stress. The overall physiological importance of calcitonin in adult humans, however, is usually minimal.

B. The Parathyroid Glands Control Calcium Level

The **parathyroid glands** are four small groups of cells embedded in the back of the thyroid gland. **Parathyroid hormone (PTH)** increases calcium levels in blood and tissue fluid by releasing calcium from bones and by enhancing calcium absorption at the digestive tract and kidneys (see figure 29.9). PTH action therefore opposes that of calcitonin.

Calcium is vital to muscle contraction, neurotransmitter release, blood clotting, bone formation, and the activities of many enzymes. Underactivity of the parathyroids can therefore be fatal. Excess PTH can also be harmful if calcium leaves bones faster

than it accumulates. This condition, called osteoporosis, is most common in women who have reached menopause (cessation of menstrual periods). The estrogen decrease that accompanies menopause makes bone-forming cells more sensitive to PTH, which depletes bone mass. ⓘ *osteoporosis,* section 29.3C

C. The Adrenal Glands Coordinate the Body's Stress Responses

The paired, walnut-sized **adrenal glands** sit on top of the kidneys (*ad-* = near; *renal* = kidney). Each adrenal gland has two regions, which are controlled in different ways and secrete different hormones (**figure 28.8**). The **adrenal medulla,** the inner portion, releases its hormones when stimulated by the sympathetic nervous system. The **adrenal cortex** is the outer portion, and it is under endocrine control.

The adrenal medulla's hormones, **epinephrine** (adrenaline) and **norepinephrine** (noradrenaline), help the body respond to exercise, trauma, fear, excitement, and other short-term "fight or flight" stresses. These water-soluble hormones cause heart rate and blood pressure to climb. In addition, the airway increases in diameter, making breathing easier. The metabolic rate increases, while digestion and other "nonessential" processes slow.

Epinephrine can save the lives of people with severe allergic reactions to bee stings or specific foods. Moments after contacting the allergen, a massive immune system reaction causes the airway to constrict. People with known allergies may therefore carry a self-injectable dose of epinephrine. The epinephrine temporarily reverses the allergic reaction, but symptoms may recur. Anyone experiencing a life-threatening allergic reaction should therefore seek emergency medical help, whether or not epinephrine is available. ⓘ *allergies,* section 34.5C

Unlike the adrenal medulla, the adrenal cortex secretes steroid hormones under the influence of ACTH. These hormones include mineralocorticoids, glucocorticoids, and even a small amount of testosterone. The **mineralocorticoids** maintain blood volume and salt balance. One example, aldosterone, stimulates the kidneys to return sodium ions and water to the blood. This action conserves water and increases blood pressure, which is especially important in compensating for fluid loss from severe bleeding. ⓘ *aldosterone,* section 33.5F

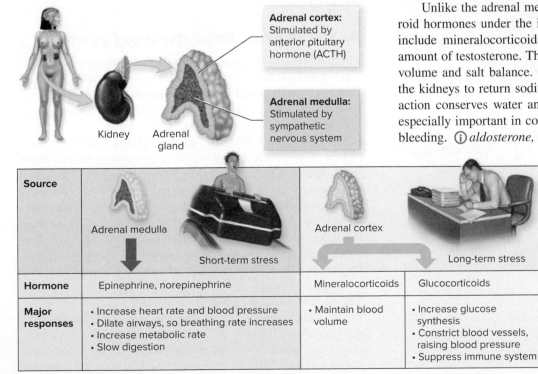

Adrenal cortex: Stimulated by anterior pituitary hormone (ACTH)

Adrenal medulla: Stimulated by sympathetic nervous system

Kidney   Adrenal gland

Source	Adrenal medulla		Adrenal cortex	
	Short-term stress		Long-term stress	
Hormone	Epinephrine, norepinephrine		Mineralocorticoids	Glucocorticoids
Major responses	• Increase heart rate and blood pressure • Dilate airways, so breathing rate increases • Increase metabolic rate • Slow digestion		• Maintain blood volume	• Increase glucose synthesis • Constrict blood vessels, raising blood pressure • Suppress immune system

Figure 28.8 Hormones of the Adrenal Glands: A Summary. The adrenal medulla secretes epinephrine and norepinephrine, which help the body respond to short-term stresses. Mineralocorticoids and glucocorticoids from the adrenal cortex enable the body to survive prolonged stress. The adrenal cortex also secretes small amounts of sex hormones (not shown).

Glucocorticoids are essential in the body's response to prolonged stress (see the chapter opening essay). Cortisol is the most important glucocorticoid. This hormone mobilizes energy reserves by stimulating the production of glucose from amino acids. Glucocorticoids also indirectly constrict blood vessels, which slows blood loss and prevents inflammation after an injury. These same effects, however, also account for the unhealthy consequences of chronic stress. Narrowed blood vessels can lead to heart attacks, and the suppressed immune system leaves a person vulnerable to illness. (i) *immunodeficiencies*, section 34.5B

Prednisone, like other synthetic glucocorticoids, is an anti-inflammatory drug that mimics cortisol's effects. This drug can treat arthritis, allergic reactions, and asthma, but it also suppresses the immune system. In addition, with long-term use of the drug, the adrenal cortex may stop producing its own glucocorticoids. Abruptly stopping treatment may therefore cause a "steroid withdrawal" condition, with symptoms including fatigue, low blood pressure, and nausea. In severe cases, the patient may go into shock, which can be fatal.

Figure It Out

If the anterior pituitary doesn't produce enough ACTH, will the level of cortisol in the blood rise, fall, or stay the same? [*Hint*: Consult figure 28.4.]

Answer: It will fall.

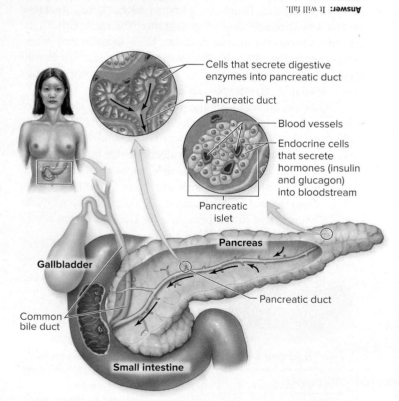

Figure 28.9 The Pancreas. This multifunctional organ has roles in the digestive and endocrine systems. The inset at upper left shows cells that secrete digestive enzymes; the one on its right shows endocrine cells that secrete insulin and glucagon.

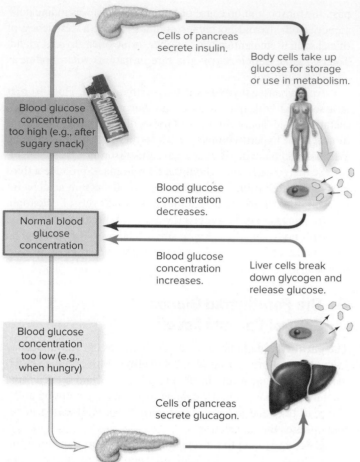

Figure 28.10 Blood Glucose Regulation. In a negative feedback loop, insulin and glucagon control the concentration of glucose in blood.

D. The Pancreas Regulates Blood Glucose

The **pancreas** is an elongated gland, about the size of a hand, located beneath the stomach and attached to the small intestine. Clusters of cells called **pancreatic islets** (also called islets of Langerhans) secrete insulin and glucagon, two hormones that regulate the body's use of glucose (**figure 28.9**).

Insulin and glucagon regulate blood glucose levels (**figure 28.10**). After a meal rich in carbohydrates, glucose enters the circulation at the small intestine. The resulting rise in blood sugar triggers cells in the pancreas to secrete **insulin,** which stimulates cells throughout the body to absorb glucose from the bloodstream. As cells take up sugar, the blood glucose concentration declines, and insulin secretion slows. If blood sugar dips too low, however, other pancreatic cells secrete **glucagon,** which stimulates target cells in the liver to release stored glucose into the bloodstream.

Too Much Glucose in Blood: Diabetes
Failure to regulate blood sugar can be deadly. In **diabetes,** glucose accumulates to dangerously high levels in the bloodstream. Diabetes is a paradox: Sugar pours out of the body in urine, yet the body's cells starve for lack of glucose. Symptoms include frequent urination,

excessive thirst, extreme hunger, blurred vision, weakness, fatigue, irritability, nausea, and weight loss. If the illness remains untreated, complications may include kidney failure, blindness, or a loss of sensation in the hands and feet. Severe diabetes can eventually result in coma and death.

The accumulation of blood sugar can occur for multiple reasons, but two forms of diabetes are most common. In type 1 diabetes, the pancreas fails to produce insulin, so the body's cells never receive the signal to "open the door" and admit glucose. In type 2 diabetes, the body's cells fail to absorb glucose even when insulin is present; this condition is called insulin resistance. In type 2 diabetes, then, insulin "rings the doorbell" but the cell never opens the door.

Fifteen percent of affected individuals have type 1 diabetes, which usually begins in childhood or early adulthood. Typically, the underlying cause is an autoimmune attack on cells of the pancreas, which therefore cannot produce insulin. Type 1 diabetes is sometimes also called insulin-dependent diabetes because insulin injections can replace the missing hormone (**figure 28.11**). ⓘ *autoimmune disorders,* section 34.5A

Type 2 diabetes is much more common. Although it usually begins in adulthood, the incidence of type 2 diabetes among adults and children is rising in developed countries (including the United States). This disease is strongly associated with obesity (**figure 28.12**); nearly all type 2 diabetes patients are overweight. The cause–effect relationship, however, is unclear.

Medicines can help lower blood glucose levels, but the best strategies to prevent and treat type 2 diabetes are to be physically active, reduce calorie intake, and choose healthy foods. For obese patients, gastric bypass surgery may also help. This procedure helps people cut calories by reducing the size of the stomach. But as a bonus, many gastric bypass patients enjoy an immediate reduction in diabetic symptoms, perhaps due to shifting hormones. ⓘ *healthy diet,* section 32.4

Not Enough Glucose in Blood: Hypoglycemia

The opposite of diabetes is **hypoglycemia,** in which excess insulin production or insufficient carbohydrate intake causes low blood sugar. A person with this condition feels weak, sweaty, anxious,

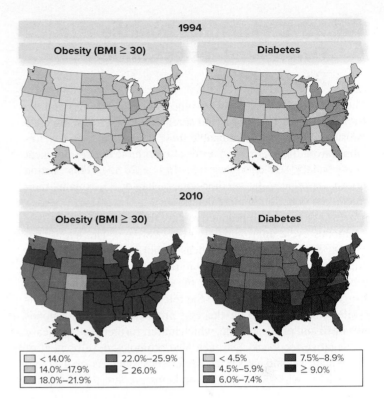

Figure 28.12 Diabetes and Obesity. Data from the Centers for Disease Control and Prevention show that the prevalence of diabetes in adults 18 years and older has increased along with obesity since 1994. (Figure does not include data after 2010 because of changes in survey methods starting in 2011.)

and shaky; in severe cases, hypoglycemia can cause seizures or loss of consciousness.

E. The Pineal Gland Secretes Melatonin

The **pineal gland,** a small brain structure near the hypothalamus, produces the hormone **melatonin.** Darkness stimulates melatonin synthesis, whereas exposing the eye to light inhibits melatonin production. The amount of melatonin therefore "tells" the other cells of the body how much light the eyes are receiving. This interaction, in turn, sets the stage for the regulation of sleep–wake cycles and other circadian rhythms.

A form of depression called seasonal affective disorder (SAD) may be linked to abnormal melatonin secretion. Exposure to additional daylight (or full-spectrum lightbulbs) can elevate mood. Because melatonin levels fall as we age, some people believe that taking extra melatonin might delay aging. However, additional evidence is needed to support this claim.

28.4 MASTERING CONCEPTS

1. Describe the functions of each of the thyroid's hormones.
2. What is the function of parathyroid hormone (PTH)?
3. Compare the hormones of the adrenal cortex and the medulla.
4. Describe the opposing roles of insulin and glucagon.
5. How do darkness and light affect melatonin secretion?

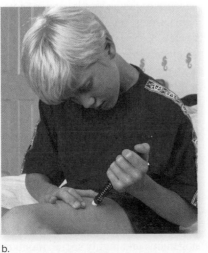

Figure 28.11 Type I Diabetes. (a) Blood glucose meter. (b) A diabetic boy injects himself with insulin.

(a): ©Piotr Adamowicz/Getty Images RF; (b): ©Saturn Stills/Science Source

a.

b.

28.5 Hormones from the Ovaries and Testes Control Reproduction

The reproductive organs include the **ovaries** in females and the **testes** in males. These egg- and sperm-producing organs secrete the steroid hormones that enable these gametes to mature. Hormones from the ovaries and testes also promote the development of secondary sex characteristics (**figure 28.13**). This section briefly introduces the sex hormones; chapter 35 explains their role in reproduction in more detail.

In a woman of reproductive age, the levels of several sex hormones cycle approximately every 28 days. The hypothalamus produces gonadotropin-releasing hormone (GnRH), which stimulates the anterior pituitary to release FSH and LH into the bloodstream. At target cells in the ovary, these two hormones trigger the events that lead to the release of an egg cell. Meanwhile, cells surrounding the egg produce the sex hormones **estrogen** and **progesterone,** which inhibit further GnRH release by negative feedback. Estrogen also promotes development of the female secondary sex characteristics, such as breasts and wider hips, whereas progesterone helps prepare the uterus for pregnancy.

In males, FSH stimulates the early stages of sperm formation in the testes. Sperm production is completed under the influence of LH, which also prompts cells in the testes to release the sex hormone **testosterone.** This hormone stimulates the formation of male structures in the embryo and promotes later development of male secondary sex characteristics, including facial hair, deepening of the voice, and increased muscle growth (see the Apply It Now box).

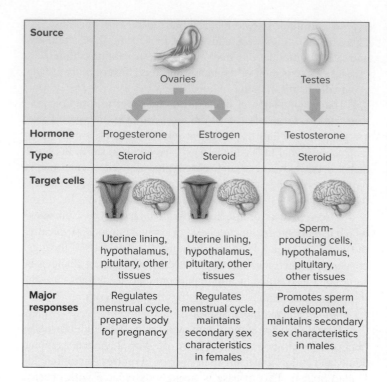

Source	Ovaries		Testes
Hormone	Progesterone	Estrogen	Testosterone
Type	Steroid	Steroid	Steroid
Target cells	Uterine lining, hypothalamus, pituitary, other tissues	Uterine lining, hypothalamus, pituitary, other tissues	Sperm-producing cells, hypothalamus, pituitary, other tissues
Major responses	Regulates menstrual cycle, prepares body for pregnancy	Regulates menstrual cycle, maintains secondary sex characteristics in females	Promotes sperm development, maintains secondary sex characteristics in males

Figure 28.13 **Hormones of the Ovaries and Testes: A Summary.** Hormones produced in the ovaries and testes coordinate reproduction and the development of secondary sex characteristics.

28.5 MASTERING CONCEPTS

1. Which organs contain target cells for FSH and LH?
2. What are the functions of estrogen, progesterone, and testosterone?

Apply It **Now**

Anabolic Steroids in Sports

The anabolic steroids that regularly make headlines in the sporting news are synthetic forms of testosterone. Although these drugs have a legitimate place in medicine, and they can be legally obtained with a prescription, some athletes abuse anabolic steroids as a shortcut to greater muscle mass.

Steroid users may improve strength and performance in the short term, but the drugs are harmful in the long run. In males, the body mistakes synthetic steroids for the natural hormone and lowers its own production of testosterone, causing infertility once use of the drug stops. Impotence, shrunken testicles, and the growth of breast tissue are other possible side effects. Females who abuse steroids may develop a masculine physique, a deeper voice, and facial or body hair. In adolescents, steroids hasten adulthood, stunting growth and causing early hair loss. Finally, research suggests that high doses of steroids may cause psychological side effects such as aggression, mood swings, and irritability. For all of these reasons, health professionals strongly advise against the use of illegal steroids.

INVESTIGATING LIFE

28.6 Addicted to Affection

The sexual behavior of animals fascinates many people, perhaps because of the insight it can lend into human relationships. In our own species, sexual attraction and feelings of love are often intertwined. Love is difficult to study in humans (and impossible to study in other animals), but scientists can examine patterns of sexual behavior in many species other than our own.

Some animals are faithful to their sexual partners for life, whereas others are much more promiscuous. One of the rarest and least understood social behaviors is monogamy. To qualify as monogamous, an animal must mate exclusively with one partner, live with its mate, help with care of the young, and defend the family against intruders. Only a tiny fraction (about 3%) of mammal species are monogamous.

Two closely related species of snub-nosed rodents have given scientists the opportunity to investigate the biological basis of monogamy. Whereas prairie voles (*Microtus ochrogaster*) are monogamous and highly social, montane voles (*Microtus montanus*) are promiscuous and solitary. What accounts for the difference in lifestyle? Antidiuretic hormone (abbreviated ADH and also called vasopressin) apparently plays a role. In the 1990s, researchers learned that if male prairie voles were given ADH, they rapidly formed social attachments, even with females they had not mated with. When ADH's effects were blocked, pair bonds did not form.

One logical explanation for the difference in sexual behavior between the two species is that the monogamous prairie voles have naturally higher levels of ADH than do the promiscuous montane voles. Yet this is not the case; both vole species have similar levels of the pair-bonding hormone. The location of receptors for ADH in the brain, however, does differ between the two species. In monogamous prairie voles, ADH receptors occur in the same brain region where addictive drugs act, and males seem to derive feelings of reward from being with their mates and young. ADH receptors in promiscuous montane voles are located in a different brain area.

A team of researchers, led by Lauren Pitkow and Larry Young from Emory University in Atlanta, wondered what would happen to prairie voles with extra ADH receptors in the brain areas associated with pleasure and rewards. Would the animals form social attachments even more readily? To find out, the researchers inserted the gene encoding the ADH receptor into a virus. They injected the modified virus into the reward areas of the brains of prairie voles, effectively increasing the number of ADH receptors (**figure 28.14**, step 1). Two groups of control voles also received injections. One control group received a different gene (one not related to ADH) in the reward area of the brain. Voles in the second control group received the ADH receptor gene but in a different brain area.

Afterward, each male spent 17 hours in a cage with an adult female that was not sexually receptive (figure 28.14, step 2). The voles then underwent "partner-preference" tests (figure 28.14, step 3). The researchers placed each male into a choice chamber, where he was free to spend as much time as he wanted with his female "partner" or with a different female that was a stranger to him. Male prairie voles do not ordinarily pair-bond with females after less than 24 hours together, unless they have mated. Nevertheless, the voles with the extra ADH receptors in the reward region of the brain spent much more time in contact with their partners than did either group of control voles. The extra ADH receptors evidently made the male prairie voles especially likely to form pair bonds.

This study is interesting because it explicitly links genes, brain chemistry, and social behavior. It also raises the startling possibility that a genetically modified virus can transmit genes that increase social attachment, at least in prairie voles. Could such a "love bug" infect humans, too? Researchers already know that the location of ADH receptors in the brain may play a role in human social behaviors; examples include not only sexual fidelity but also autism, a disorder in which individuals have difficulty forming social attachments. Researchers now are investigating primate ADH receptors to learn more about the biochemistry of attachment in our closest relatives.

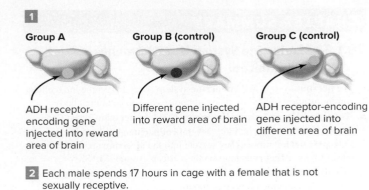

1

Group A	Group B (control)	Group C (control)

ADH receptor-encoding gene injected into reward area of brain

Different gene injected into reward area of brain

ADH receptor-encoding gene injected into different area of brain

2 Each male spends 17 hours in cage with a female that is not sexually receptive.

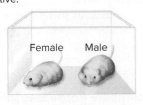

Female   Male

3 In a 3-hour partner-preference test, the male can choose to spend time with his previous companion or a stranger.

Previous female companion   Male   Female stranger

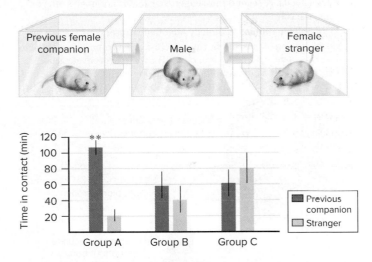

Figure 28.14 More Receptors, More Bonding. Prairie voles injected with ADH receptor genes in the reward center of the brain spent significantly more time with a female partner than with a stranger. Control voles were less likely to bond with their partners. (Asterisks indicate a statistically significant difference within a group; see appendix B.)

Source: Pitkow, Lauren, and five coauthors (including Larry Young). 2001. Facilitation of affiliation and pair-bond formation by vasopressin receptor gene transfer into the ventral forebrain of a monogamous vole. *Journal of Neuroscience,* vol. 21, no. 18, pages 7392–7396.

28.6 MASTERING CONCEPTS

1. How did the researchers test the hypothesis that the number of ADH receptors in the brain's reward area influences pair-bonding?

2. Could the researchers have drawn the same conclusions if they had omitted one of the control groups? Why or why not?

CHAPTER SUMMARY

28.1 The Endocrine System Uses Hormones to Communicate

- The **nervous system** and **endocrine system** specialize in the intercellular communication needed to maintain homeostasis in an animal body.

A. Endocrine Glands Secrete Hormones That Interact with Target Cells

- The endocrine system includes several **endocrine glands** and scattered cells, plus the **hormones** they secrete into the bloodstream. Hormones interact with **target cells** to exert their effects (**figure 28.15**).
- Negative feedback loops ensure that the levels of a hormone in the bloodstream are not too high or too low.

B. The Nervous and Endocrine Systems Work Together

- The nervous system acts faster and more locally than the endocrine system.
- The hypothalamus physically connects the nervous and endocrine systems. In addition, the two communication systems share many messenger molecules and target cells.

28.2 Hormones Stimulate Responses in Target Cells

A. Water-Soluble Hormones Trigger Second Messenger Systems

- **Peptide hormones** are water-soluble and bind to the surface receptors of target cells. A **second messenger** triggers the hormone's effect inside the cell.

B. Lipid-Soluble Hormones Directly Alter Gene Expression

- Most lipid-soluble **steroid hormones** cross cell membranes and bind to receptors in the cytoplasm or nucleus. The receptors activate genes, which direct the synthesis of proteins that provide the cell's response.

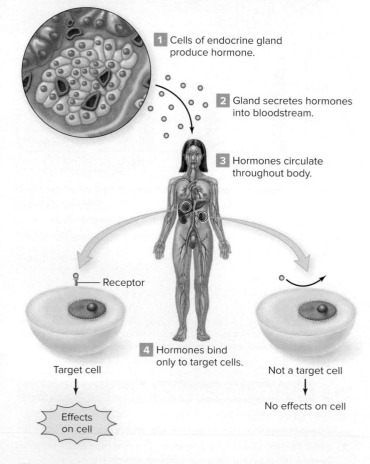

1 Cells of endocrine gland produce hormone.

2 Gland secretes hormones into bloodstream.

3 Hormones circulate throughout body.

Receptor

Target cell

Not a target cell

4 Hormones bind only to target cells.

Effects on cell

No effects on cell

Figure 28.15 Hormones and Target Cells: A Summary.

28.3 The Hypothalamus and Pituitary Gland Oversee Endocrine Control

- Neurons from the **hypothalamus** influence the release of hormones from the **posterior pituitary** and **anterior pituitary** glands.

A. The Posterior Pituitary Stores and Releases Two Hormones

- Two hormones produced by the hypothalamus are released by the posterior pituitary. **Antidiuretic hormone (ADH)** regulates body fluid composition; **oxytocin** stimulates muscle contractions in the uterus and milk ducts.

B. The Anterior Pituitary Produces and Secretes Six Hormones

- **Growth hormone (GH)** stimulates cell division, protein synthesis, and growth throughout the body.
- **Prolactin** stimulates milk production.
- **Thyroid-stimulating hormone (TSH)** prompts the thyroid gland to release hormones.
- **Adrenocorticotropic hormone (ACTH)** stimulates the adrenal cortex to release hormones.
- **Follicle-stimulating hormone (FSH)** and **luteinizing hormone (LH)** stimulate hormone release from the ovaries and testes.
- **Endorphins** are natural painkillers with target cells in the brain.

28.4 Hormones from Many Glands Regulate Metabolism

A. The Thyroid Gland Sets the Metabolic Pace

- **Thyroxine** and **triiodothyronine** from the **thyroid gland** speed metabolism. **Calcitonin** lowers the level of calcium in the blood.

B. The Parathyroid Glands Control Calcium Level

- The **parathyroid glands** secrete **parathyroid hormone (PTH),** which increases the blood calcium concentration.

C. The Adrenal Glands Coordinate the Body's Stress Responses

- The **adrenal gland** has an inner portion, the **adrenal medulla,** which secretes **epinephrine** and **norepinephrine.** These hormones ready the body to cope with a short-term emergency. The **adrenal cortex** secretes **mineralocorticoids** and **glucocorticoids,** which mobilize energy reserves during stress and maintain blood volume and blood composition.

D. The Pancreas Regulates Blood Glucose

- The **pancreatic islets** of the **pancreas** secrete **insulin** and **glucagon,** two hormones that regulate blood glucose levels.
- In **diabetes** mellitus, blood sugar concentrations rise to dangerous levels. Type 1 diabetes occurs when the pancreas fails to produce insulin; in type 2 diabetes, the body's cells do not respond to insulin.
- **Hypoglycemia** is low blood sugar, which is caused by excess insulin or insufficient carbohydrate intake.

E. The Pineal Gland Secretes Melatonin

- The **pineal gland** secretes a hormone, **melatonin,** that may regulate how other glands respond to light–dark cycles.

28.5 Hormones from the Ovaries and Testes Control Reproduction

- In females, FSH and LH stimulate the **ovaries** to secrete **estrogen** and **progesterone,** hormones that stimulate the development of secondary sex characteristics and control the menstrual cycle.
- In males, FSH and LH stimulate the **testes** to secrete **testosterone,** which stimulates sperm cell production and the development of secondary sex characteristics.

28.6 Investigating Life: Addicted to Affection

- Researchers have traced pair-bonding behavior to a receptor that binds ADH in the pleasure-seeking area of the prairie vole's brain.

MULTIPLE CHOICE QUESTIONS

1. The effect of a water-soluble peptide hormone such as insulin is generally quicker than that of a steroid hormone such as estrogen because
   a. peptide hormones exert changes using molecules already present in the target cell.
   b. steroid hormones trigger the synthesis of new proteins.
   c. steroid hormones cannot pass through the cell's plasma membrane.
   d. Both a and b are correct.

2. The parathyroid gland releases hormones when
   a. blood sugar is too high.
   b. bone growth is too slow.
   c. blood calcium levels are too low.
   d. urine is too dilute.

3. Which of the following glands releases hormones when the thyroid hormone concentration in the blood is too low?
   a. Hypothalamus   c. Parathyroid gland
   b. Adrenal gland   d. Pineal gland

4. Which hormone is lipid-soluble and helps conserve water?
   a. ADH
   b. Aldosterone
   c. Estrogen
   d. All of the above are correct.

5. The body receives a series of stress-inducing stimuli throughout the day. In response, glucocorticoids are released from the
   a. adrenal cortex.
   b. anterior pituitary.
   c. hypothalamus.
   d. All of the above are correct.

6. Secretion of melatonin is regulated by
   a. light.   c. stress.
   b. temperature.   d. glucose.

7. To increase male fertility, it would be logical to develop a drug that boosts hormone synthesis at any of the following structures except the
   a. hypothalamus.
   b. thyroid.
   c. anterior pituitary.
   d. testes.

Answers to these questions are in appendix A.

WRITE IT OUT

1. How does the endocrine system interact with the circulatory system?

2. An endocrine disruptor is a molecule that either mimics or blocks the activity of a hormone. Propose a way to test the hypothesis that microwaving foods in plastic containers releases endocrine disruptors.

3. Write a paragraph describing the events that occur from the time an endocrine gland releases a steroid hormone to the time the hormone exerts its effects on a target cell.

4. Sketch the mechanisms of water-soluble and lipid-soluble hormone function.

5. Give two examples of hormones counteracting the effects of one another.

6. Alcohol and caffeine inhibit the effects of antidiuretic hormone. Explain why drinking beer or coffee increases the frequency of urination. If your urine appears dilute (is light yellow) after drinking beer or coffee, does that indicate that you are well hydrated?

7. Some professional baseball players use human growth hormone (a banned substance) to aid in fast recovery after difficult workouts. How would GH help speed a player's recovery? Is GH use the same as anabolic steroid use?

8. Which hormone(s) match each of the following descriptions?
   a. Produced by a woman who is breast feeding
   b. Causes fatigue if too little is present
   c. Causes a decrease in blood glucose level

9. Why can a stressful lifestyle lead to heart attacks? Which hormone is released in response to long-term stress?

10. How might insulin-producing stem cells transplanted to the pancreas help people with type 1 diabetes? Would the same treatment help people with type 2 diabetes?

11. In healthy adults, the concentration of glucose in blood is approximately 80 to 110 milligrams per deciliter (mg/dl). After a carbohydrate-rich meal, however, the concentration may spike to 140 mg/dl. Describe the hormonal action that returns blood glucose to normal.

12. Identify the target cells and effects of estrogen, progesterone, and testosterone.

PULL IT TOGETHER

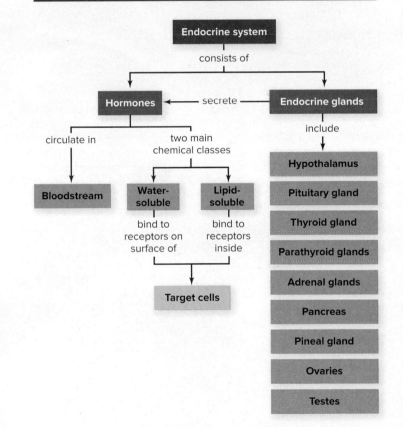

Figure 28.16 **Pull It Together: The Endocrine System.**

Refer to figure 28.16 and the chapter content to answer the following questions.

1. Review the Survey the Landscape figure in the chapter introduction, and then add at least two other organ systems to the Pull It Together concept map. Connect an endocrine gland to each new term, and explain how hormones released from the gland affect the organ system in a way that maintains homeostasis.

2. Connect each hormone discussed in this chapter to the gland that produces it and to either the "Water-soluble" or "Lipid-soluble" box.

3. Describe the relationships among the hypothalamus, pituitary, and other endocrine glands.

29 The Skeletal and Muscular Systems

Helping Hand. Patrick Kane, who lost the fingers of his left hand to illness when he was a baby, shows off his battery-operated prosthetic hand. All five fingers bend independently at the joints; a mobile app lets him program his hand with a variety of grip positions.

©Jeff J Mitchell/Getty Images

LEARN HOW TO LEARN
Make Appointments with Yourself

If you prefer to study alone but often find yourself putting off your solo study sessions, try making recurring "appointments" with yourself. That is, use your calendar to block off time that you dedicate to studying each day. You can reread chapters after class, quiz yourself on course materials, make concept maps, or work on homework assignments. Keep those appointments throughout the semester, so that you never get behind.

Prosthetic Limbs Are Better Than Ever

We use our arms and legs so often that it's hard to imagine life without them. Luckily, people with missing limbs have options. Careful studies of the structure and function of the skeletal and muscular systems, coupled with advances in materials science, have enabled bioengineers to develop sophisticated limb replacements.

Modern prosthetic limbs are much more realistic in both structure and function than the old "wooden legs" of the past. In fact, some artificial limbs might even work better than their natural counterparts, so prosthetics must be evaluated before amputee athletes are allowed to compete in the Olympics.

What makes modern artificial limbs so effective? Engineers can shape strong, lightweight plastics and carbon fiber into prosthetics that absorb weight stresses and mimic the other functions of bones and joints. Also, the socket where the prosthesis connects with the body is custom-made for a secure and comfortable fit.

In the past, a person could control a prosthetic limb by operating cables or switches accessible by a healthy limb. Being able to move an artificial limb just by thinking about it would be a huge improvement. As described in chapter 26, the nervous system coordinates movements by transmitting electrical signals to limb muscles. When an amputee thinks about moving a missing limb, however, these neural impulses reach dead ends. A prosthetic that could "listen" to signals in the brain would simulate the natural communication between nerves and muscles.

One step in that direction is a surgical procedure in which the nerves that previously stimulated an amputated muscle are relocated to an intact muscle in the chest. Afterward, when the patient thinks about moving the amputated arm, the chest muscle contracts instead. Electrodes attached to the chest muscle transmit the electrical signal to a prosthetic arm, causing it to move.

One drawback of this approach is a limited range of motion; another is that the chest muscles also contract each time the patient moves the prosthetic arm. Prosthetics are therefore being developed that will respond directly to electrical signals in the brain. In laboratory studies, monkeys have already used thought-powered artificial limbs, and human trials are under way.

These astonishing feats of prosthetic engineering begin with a basic understanding of the skeletal and muscular systems. This chapter first explores animal skeletal systems, and then explains how muscles interact with bones to produce movements.

SURVEY THE LANDSCAPE
Animal Life

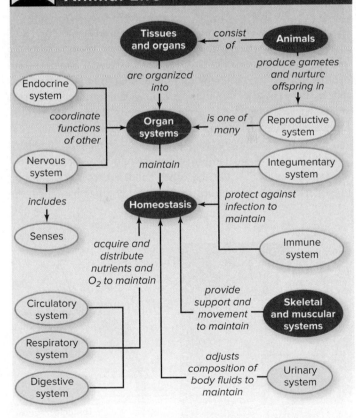

Most animals can flee from danger or approach food, mates, or a source of warmth. All of these movements require the coordinated actions of the skeletal and muscular systems.

For more details, study the Pull It Together feature in the chapter summary.

29.1 Skeletons Take Many Forms

Ask a child what sets animals apart from other organisms, and he or she will likely answer "movement." This response is technically wrong—many bacteria, archaea, protists, fungi, and plants have swimming, creeping, or gliding cells. Yet the child correctly recognizes that animal movements are unmatched in their drama, versatility, and power.

The ability to hop, dig, fly, slither, scuttle, or swim comes from two closely allied, interacting organ systems: the **muscular system** and the **skeletal system,** which together function under the direction of the nervous system. In most animals, organs called **muscles** provide motion; the **skeleton** adds a firm supporting structure that muscles pull against. The skeleton also gives shape to an animal's body and protects internal organs.

The simplest type of skeleton is a **hydrostatic skeleton** (*hydro-* means water), which consists of fluid constrained within a layer of flexible tissue. Many of the invertebrate animals described in chapter 21 have hydrostatic skeletons. The bell of a jellyfish, for example, consists mostly of a gelatinous substance constrained between two tissue layers (**figure 29.1**). To swim, the animal rhythmically contracts the muscles acting on this hydrostatic skeleton, forcibly ejecting water from its body. Other animals with hydrostatic skeletons move by alternately contracting and relaxing muscles surrounding a fluid-filled coelom. ⓘ *coelom,* section 21.1D

The most common type of skeleton is an **exoskeleton** (*exo-* means outside), which acts as a "suit of armor" that protects the animal from the outside (**figure 29.2**). Internal muscles pull against the exoskeleton, enabling the animal to move. The structure of the exoskeleton varies. A clam has a hard, calcium-rich shell, whereas insects and other arthropods have a lightweight, jointed exoskeleton made of chitin. ⓘ *chitin,* section 2.5B

Exoskeletons have advantages and disadvantages. The hard covering protects soft internal organs and provides excellent leverage for muscles. On the other hand, a growing arthropod must periodically molt; until its new skeleton has hardened, the

Figure 29.2 Exoskeleton. Arthropods such as crabs have tough, jointed exoskeletons that consist of chitin. Muscles attach to the exoskeleton's inner surface.
©Photodisc/Getty Images RF

animal is vulnerable to predators. Moreover, a large animal would be unable to support the weight of an exoskeleton; for this reason, exoskeletons occur only in small animals.

An **endoskeleton** (*endo-* means inner) is an internal support structure. Sea stars and other echinoderms, for example, produce rigid, calcium-rich spines and internal plates. Vertebrate animals have endoskeletons made of cartilage or bone (**figure 29.3**). Sharks and rays, for example, are fishes with cartilage skeletons. Other fishes, and all land vertebrates, have skeletons composed primarily of bone.

Like exoskeletons, endoskeletons represent an evolutionary trade-off. On the plus side, an internal skeleton grows with the animal, eliminating molting. Also, an endoskeleton consumes less of an organism's total body mass, so it can support animals as large as a whale. One disadvantage, however, is that the endoskeleton does not protect soft tissues at the body surface.

The vertebrate skeleton's capacity to change over evolutionary time is striking. Each vertebrate species has a distinctive skeleton, yet all are composed of the same types of cells and have similar arrangements. The characteristics of each species' muscles and skeleton reflect common ancestry and the selective forces in its environment. ⓘ *homologous structures,* section 13.4A

29.1 MASTERING CONCEPTS

1. How do the skeletal and muscular systems interact?
2. Describe similarities and differences among the three main types of skeletons.

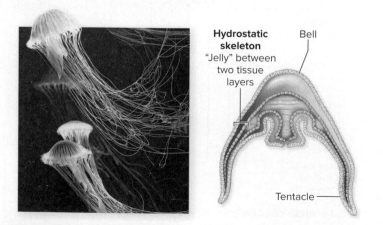

Figure 29.1 Hydrostatic Skeleton. When a jellyfish contracts muscles surrounding its bell, water shoots out of its body. The animal moves forward.
Photo: ©Gabriel Bouys/AFP/Getty Images

Hydrostatic skeleton
"Jelly" between two tissue layers
Bell
Tentacle

Figure 29.3 Endoskeleton. Vertebrates have an endoskeleton composed of cartilage or bone. The bones are clearly visible in this X-ray of a fish.
©Jim Wehtje/Getty Images RF

29.2 The Vertebrate Skeleton Features a Central Backbone

Bones, the organs that compose the vertebrate skeleton, are grouped into two categories (figure 29.4). The **axial skeleton,** so named because it is located in the longitudinal central axis of the body, consists of the bones of the head, vertebral column, and rib cage. The **appendicular skeleton** consists of the appendages (limbs) and the bones that support them.

The axial skeleton shields soft body parts. The skull, which protects the brain and many of the sense organs, consists of hard, dense bones that fit together like puzzle pieces. All of the head bones are attached with immovable joints, except for the lower jaw and the middle ear. These movable jaw and ear bones enable us to chew food, speak, and hear.

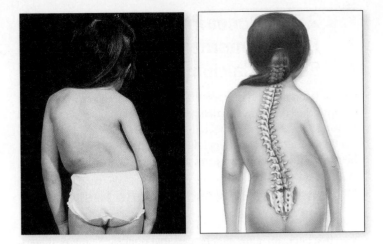

Figure 29.5 Scoliosis. This young girl's spine curves to the side. Spinal surgery or a specially designed brace may help straighten the spine.
Photo: ©Southern Illinois University/Science Source

Figure 29.4 The Human Skeleton. The axial skeleton in humans includes the bones of the head, vertebral column, and rib cage. The bones that compose and support the limbs constitute the appendicular skeleton.

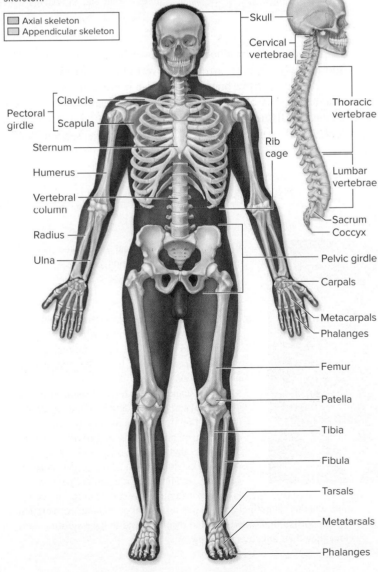

- Axial skeleton
- Appendicular skeleton

Skull
Cervical vertebrae
Clavicle
Pectoral girdle
Scapula
Sternum
Thoracic vertebrae
Rib cage
Humerus
Lumbar vertebrae
Vertebral column
Radius
Sacrum
Coccyx
Ulna
Pelvic girdle
Carpals
Metacarpals
Phalanges
Femur
Patella
Tibia
Fibula
Tarsals
Metatarsals
Phalanges

The **vertebral column** supports and protects the spinal cord. A human vertebral column typically consists of 33 vertebrae, separated by cartilage disks that cushion shocks and enhance flexibility. Scoliosis, in which the vertebral column curves to the side, is a disorder of the axial skeleton (figure 29.5).

Attached to the human vertebral column are 12 pairs of ribs, which protect the heart and lungs. Flexible cartilage in the rib cage allows the chest to expand during breathing.

In the appendicular skeleton, the **pectoral girdle** connects the forelimbs to the axial skeleton; it includes the collarbones (clavicles) and shoulder blades (scapulas). Likewise, the **pelvic girdle** attaches the hindlimb bones to the axial skeleton. The hipbones join the sacrum in the rear and meet each other in front, creating a bowl-like pelvic cavity that protects organs in the lower abdomen.

This chapter's Apply It Now box illustrates how bones reveal clues that are useful to people in several professions.

29.2 MASTERING CONCEPTS

1. What are the components of the axial and appendicular skeletons?
2. What are the locations of the pectoral and pelvic girdles?

Skeletal System		
Main tissue types*	**Examples of locations/functions**	
Connective	Makes up bone, cartilage, tendons, ligaments, marrow of vertebrate skeleton	
Muscle	Skeletal muscle connects to movable bones, enabling voluntary movements	
Nervous	Senses body position and controls muscles	

*See chapter 25 for descriptions.

29.3 Bones Provide Support, Protect Internal Organs, and Supply Calcium

The skeleton not only supports and protects the body but also has several other functions, which may at first glance seem unrelated (table 29.1). Bones connected to muscles provide movement, and bone minerals supply calcium and phosphorus to the rest of the body. Blood cells also form at the marrow inside bones.

A. Bones Consist Mostly of Bone Tissue and Cartilage

A glance back at figure 29.4 reveals that bones take many shapes. Long bones make up the arms and legs, whereas the wrists and ankles consist mainly of short bones. Flat bones include the ribs and skull. Vertebrae are irregularly shaped.

No matter what their shape, bones are lightweight and strong because they are porous, not solid (figure 29.6). The weight of long bones is further reduced by the **marrow cavity,** a space occupying the center of the shaft. The term *marrow cavity* falsely implies that bone marrow exists only in that location. In fact, marrow occurs not only in the central bone shaft but also in flat bones and in the porous ends of long bones.

Bones contain two types of soft, spongy marrow. When a baby is born, all marrow is **red bone marrow,** which is a nursery for blood cells and platelets. In adults, **yellow bone marrow** replaces the red marrow in the marrow cavity. The fatty yellow

TABLE 29.1 Functions of the Vertebrate Endoskeleton: A Summary

Function	Explanation
Support	The skeleton is a framework that supports an animal's body against gravity. It largely determines the body's shape.
Movement	The vertebrate skeleton is a system of muscle-operated levers. Typically, the two ends of a skeletal (voluntary) muscle attach to different bones that connect in a structure called a joint. When the muscle contracts, one bone moves.
Protection of internal structures	The backbone surrounds and shields the spinal cord, the skull protects the brain, and ribs protect the heart and lungs.
Production of blood cells	Many bones, such as the long bones of the arm and leg, contain and protect red marrow, a tissue that produces red blood cells, white blood cells, and platelets.
Mineral storage	The skeleton stores calcium and phosphorus.

marrow does not produce blood; however, if blood cells are in short supply, yellow marrow can revert to red marrow.

Besides marrow, bones also contain nerves and blood vessels. But the majority of the vertebrate skeleton consists of two types of connective tissue: bone and cartilage. Figure 29.6 offers a closer look at both of these tissues. (i) *connective tissue,* section 25.2B

Apply It **Now**

Bony Evidence of Murder, Illness, and Evolution

Skeletons sometimes provide useful clues to past events. Hard, mineral-rich bones and teeth remain intact long after a corpse's soft body parts decay (figure 29.A). These durable remains can help solve crimes, lend insight into human history, and shed light on evolution.

Detectives can use bones to identify the sex of a decomposed murder victim. This technique relies on the differences between male and female skeletons. Most obviously, the average male is larger than the average female. In addition, the front of the female pelvis is broader and larger than the male's, and it has a wider bottom opening that accommodates the birth of a baby. These same features allow anthropologists to determine the sex of ancient human fossils.

Bones can also reveal events and illnesses unique to each person's life. Features of the skull and ribs give clues about a person's age at death. Healed breaks may indicate accidents or abuse. Egypt's King Tut, for

Figure 29.A Old Bones. An archaeologist inspects a grave from the ninth century.
©Viktor Chlad/isifa/Getty Images

example, suffered a severe leg break shortly before he died. Crooked joints may be evidence of arthritis, and patterns of bone thickenings tell whether a person spent his or her life in hard physical labor.

The shapes and sizes of fossilized bones also reveal some of the details of human evolution. Section 15.4 explains how the skeletons and teeth of primate fossils provide clues to brain size, diet, and posture in our ancestors. Animal skeletons also tell the larger story of vertebrate evolution. For example, paleontologists can examine skeletal features to determine whether an extinct animal was terrestrial or aquatic. Air is much less supportive than water, so land dwellers tend to have sturdier skeletons than their aquatic relatives.

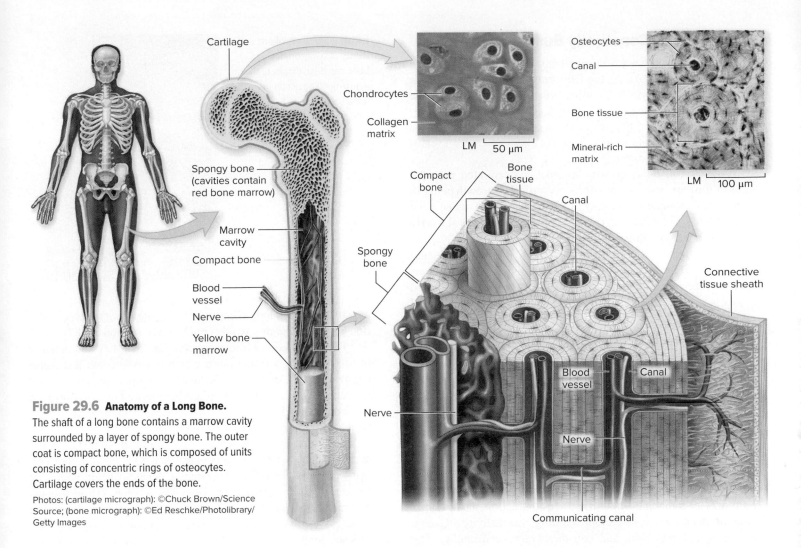

Cartilage

Chondrocytes

Collagen matrix

LM   50 μm

Osteocytes

Canal

Bone tissue

Mineral-rich matrix

LM   100 μm

Spongy bone (cavities contain red bone marrow)

Marrow cavity

Compact bone

Blood vessel

Nerve

Yellow bone marrow

Compact bone

Bone tissue

Spongy bone

Canal

Nerve

Connective tissue sheath

Blood vessel

Canal

Nerve

Communicating canal

Figure 29.6 Anatomy of a Long Bone.
The shaft of a long bone contains a marrow cavity surrounded by a layer of spongy bone. The outer coat is compact bone, which is composed of units consisting of concentric rings of osteocytes. Cartilage covers the ends of the bone.

Photos: (cartilage micrograph): ©Chuck Brown/Science Source; (bone micrograph): ©Ed Reschke/Photolibrary/ Getty Images

Bone tissue consists of cells suspended in a hard extracellular matrix. An **osteocyte** is a bone cell that is embedded in the mineral-rich matrix that it has produced. Each osteocyte inhabits a small space joined to others by narrow passageways. The matrix itself consists mainly of collagen and minerals. Collagen is a protein that gives the bone flexibility, elasticity, and strength. The hardness and rigidity of bone come from the minerals, primarily calcium and phosphate, that coat the collagen fibers.

In **compact bone tissue,** which is hard and dense, osteocytes and the surrounding matrix are organized into concentric rings, each surrounding a central canal. The canals contain nervous tissue and a blood supply. Nutrients and oxygen pass from the blood, and from osteocyte to osteocyte, through a "bucket brigade" of up to 15 cells. Other canals connect the entire labyrinth to the outer surface of the bone and to the marrow cavity.

Spongy bone tissue is much lighter than compact bone, thanks to a web of hard, bony struts that enclose large, marrow-filled spaces. Unlike compact bone, spongy bone tissue does not consist of concentric rings of cells. Instead, the osteocytes secrete layers of matrix as rods and plates that develop along a bone's lines of stress. Each cell acquires nutrients and oxygen directly from the nearby bone marrow.

Bones include both compact and spongy bone tissue, with compact bone concentrated in areas that experience the most mechanical stress. Long bones consist mostly of compact bone, although their bulbous tips contain spongy bone. Likewise, in short bones, compact bone covers the spongy bone. In most flat bones, spongy bone is sandwiched between two layers of compact bone.

Cartilage is the other main connective tissue in the skeleton (see figure 29.6). This rubbery material covers the ends of bones. Cells called chondrocytes secrete the extracellular matrix of collagen and another protein, elastin. Strong networks of collagen fibers resist breakage and stretching, even when bearing great weight. Elastin provides flexibility.

The protein network in cartilage holds a great deal of water, making cartilage an excellent shock absorber. But cartilage lacks a blood supply. As the body moves, water within cartilage cleanses the tissue and bathes it with dissolved nutrients from nearby blood vessels. Nevertheless, the absence of a dedicated blood supply means that injured cartilage is slow to heal.

B. Bones Are Constantly Built and Degraded

Most of the bones of a developing embryo originate as cartilage "models" (figure 29.7). As the fetus grows, each model's matrix hardens with calcium salts, cutting off diffusion to the cartilage cells. The cartilage matrix therefore degenerates, and capillaries penetrate the degenerating areas. Bone cells migrate in via these new blood vessels and secrete bone matrix. These cells eventually become mature osteocytes.

After birth, bone growth becomes concentrated near the ends of the long bones in thin disks of cartilage ("growth plates"). The bones continue to elongate until the late teens, when bone tissue begins to replace the cartilage growth plates. By the early twenties, bone growth is complete.

Even after a person stops growing, bone is continually being remodeled. Bones become thicker and stronger with strenuous exercise such as weight lifting. On the other hand, less-used bones lose mass, thanks to cells that slowly degrade the mineral matrix at the bone surface. For example, astronauts lose bone density if they are in a prolonged weightless environment because their bodies don't have to work as hard as they do against Earth's gravity.

Moreover, broken bones can repair themselves (figure 29.8). Cells near the site of the fracture produce new bone, so that after several weeks, the injury is all but healed.

C. Bones Help Regulate Calcium Homeostasis

As bones grow and become stronger, they absorb calcium from the blood. But bones also act as a reservoir for calcium throughout life. This mineral is vital for muscle contraction, the transmission of neural impulses, blood clotting, the activity of some enzymes, cell adhesion, and cell membrane permeability. The body therefore maintains calcium homeostasis by constantly shuttling calcium between blood and bone. Hormones from the thyroid and parathyroid glands control this exchange in a

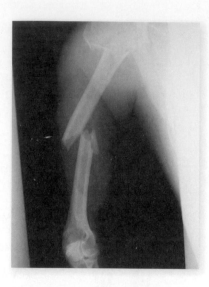

Figure 29.8 Broken Bone. A badly broken arm bone can heal, thanks to the matrix-building action of bone cells.
©Corbis RF

negative feedback loop (figure 29.9). ⓘ *negative feedback,* section 25.4; *thyroid and parathyroid glands,* section 28.4

Bones sometimes lose more calcium than they add, leading to **osteoporosis,** a condition in which bones become less dense (figure 29.10). An astronaut's "disuse osteoporosis," mentioned in section 29.3B, is one example. Much more familiar, however, is the age-related osteoporosis that causes shrinking stature, back pain, and frequent fractures in the elderly.

Both men and women can suffer from osteoporosis, but the disorder is most common in females. The bone mass of the average woman is about 30% less than a man's to begin with, and women live longer than men. In young women, estrogen helps maintain bone density by inducing the death of bone-destroying cells. As estrogen levels drop after menopause, however, bone loss accelerates. To prevent osteoporosis, women should exercise regularly and take calcium supplements daily; after all, high blood calcium prompts bones to accumulate calcium (see figure 29.9). Several drugs can also slow or reverse bone loss.

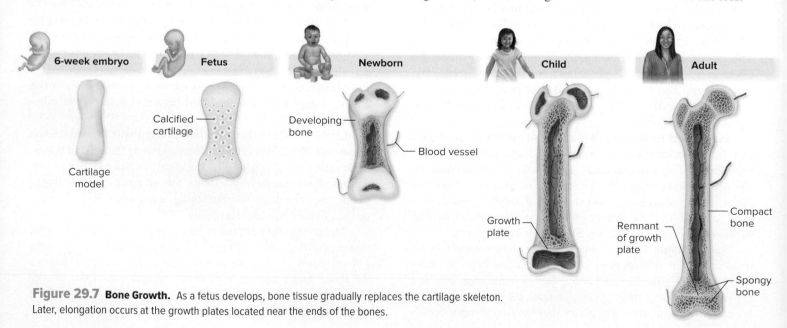

6-week embryo   Fetus   Newborn   Child   Adult

Cartilage model

Calcified cartilage

Developing bone

Blood vessel

Growth plate

Remnant of growth plate

Compact bone

Spongy bone

Figure 29.7 Bone Growth. As a fetus develops, bone tissue gradually replaces the cartilage skeleton. Later, elongation occurs at the growth plates located near the ends of the bones.

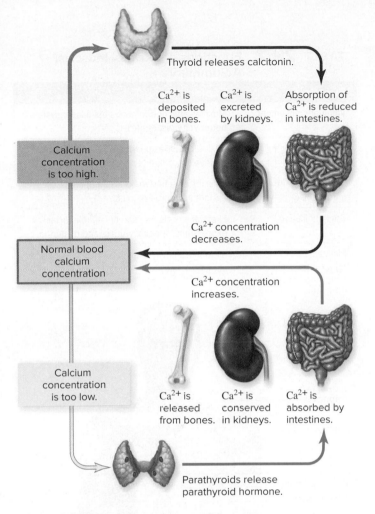

Thyroid releases calcitonin.

Ca^{2+} is deposited in bones.

Ca^{2+} is excreted by kidneys.

Absorption of Ca^{2+} is reduced in intestines.

Calcium concentration is too high.

Ca^{2+} concentration decreases.

Normal blood calcium concentration

Ca^{2+} concentration increases.

Calcium concentration is too low.

Ca^{2+} is released from bones.

Ca^{2+} is conserved in kidneys.

Ca^{2+} is absorbed by intestines.

Parathyroids release parathyroid hormone.

Figure 29.9 **Regulation of Blood Calcium Level.** Bones play a critical role in maintaining calcium homeostasis.

D. Bone Meets Bone at a Joint

A **joint** is an area where two bones meet. Many joints are freely movable, such as those of the knees, hips, elbows, fingers, and toes (**figure 29.11**). These joints consist of movable bones joined by a fluid-filled capsule of fibrous connective tissue. Together, the lubricating fluid and slippery cartilage allow bones to move against each other in a nearly friction-free environment.

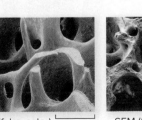
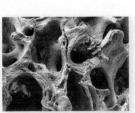

SEM (false color) 1 mm   SEM (false color) 1 mm

Figure 29.10 **Osteoporosis.** In osteoporosis, calcium loss weakens bones. The bone on the left is normal; osteoporosis has leached away part of the bone on the right.

Photos: (healthy): ©Prof. P.M. Motta/Science Source; (diseased): ©Dee Breger/Science Source

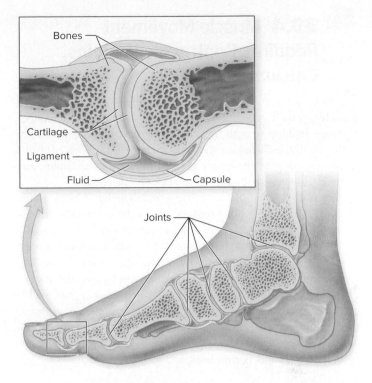

Bones

Cartilage

Ligament

Fluid   Capsule

Joints

Figure 29.11 **Movable Joints.** The bones of the foot are connected by joints that enable the foot and toes to move. The inset shows a fluid-filled capsule of fibrous connective tissue surrounding a movable joint in the big toe.

Two types of connective tissue—tendons and ligaments—help stabilize movable joints. **Tendons** are tough bands of connective tissue that attach bone to muscle; **ligaments** are similar structures that attach bone to bone. A strain is an injury to a muscle or tendon, whereas a sprain is a stretched or torn ligament. A torn anterior cruciate ligament (ACL) is a common type of knee sprain, especially in sports such as basketball and volleyball. The ACL is one of two ligaments that criss-cross at the knee, connecting the thighbone to the shinbone. Surgical reconstruction of the ACL enables many injured athletes to return to their sports.

Arthritis is a common disorder of joints. A very severe form, rheumatoid arthritis, is an inflammation of the joint membranes, usually in the hands and feet. In the more common osteoarthritis, joint cartilage wears away. As the bone is exposed, small bumps of new bone begin to form. Osteoarthritis usually reveals itself as stiffness and soreness after age 40.

29.3 MASTERING CONCEPTS

1. Describe the organization and functions of bone tissue and cartilage.
2. What are the differences between spongy bone and compact bone?
3. How are bones remodeled and repaired throughout life?
4. How do bones participate in calcium homeostasis?
5. What are the relationships among joints, tendons, and ligaments?

29.4 Muscle Movement Requires Contractile Proteins, Calcium, and ATP

The human muscular system includes more than 600 **skeletal muscles,** which generate voluntary movements. (This number does not include smooth muscle and cardiac muscle, which are not typically included in the muscular system.) Figure 29.12 identifies a few of the major skeletal muscles in a human, and table 29.2 lists some functions of skeletal muscles.

Pairs of muscles often work together to generate body movements. Figure 29.13, for example, shows the biceps and triceps muscles. Tendons attach both of these muscles to the bones of the shoulder and lower arm. Each contracting muscle can pull a bone in one direction but cannot push the bone the opposite way. The elbow can bend and straighten because the biceps and triceps operate in opposite directions. That is, when you contract the biceps, the triceps relaxes and the arm bends at the elbow joint. Conversely, the arm extends when the triceps contracts and the biceps relaxes. Many other skeletal muscles occur in similar antagonistic pairs that permit back-and-forth movements.

A. Actin and Myosin Filaments Fill Muscle Cells

Picture a softball player swinging a bat, an action that requires the contraction of many skeletal muscles. Each muscle moves a different body part, yet all are organized in essentially the same way. Figure 29.14 illustrates several levels of muscle anatomy, zooming from the whole organ to the microscopic scale.

The left side of figure 29.14 shows a whole muscle, an organ that consists of multiple tissue types. Connective tissue, for example, makes up tendons and the sheath that wraps around each muscle like a banana peel. Blood vessels in the muscle deliver nutrients and oxygen while removing wastes. Nerves transmit information to and from the central nervous system (chapter 26).

Figure 29.12 The Human Muscular System. The human body has more than 600 skeletal muscles, a few of which are identified here.

Muscular System	
Main tissue types*	**Examples of locations/functions**
Connective	Makes up tendons that attach muscles to bone; surrounds muscle cells, bundles, and whole muscles
Muscle	Connects to bones and soft tissue, enabling movement of body parts
Nervous	Senses body position and controls muscles

*See chapter 25 for descriptions.

TABLE 29.2 Functions of Skeletal Muscles: A Summary

Function	Explanation
Voluntary movement	Muscles attached to bones build lever systems under voluntary control.
Control of body openings	Skeletal muscles provide voluntary control of the eyelids, mouth, and anus.
Maintain posture and joint stability	Muscles attached to bones keep the body upright and stabilize joints.
Communication	Skeletal muscle movements enable facial expressions, speech, writing, and gesturing.
Maintain body temperature	Metabolic activity in skeletal muscle generates abundant heat.

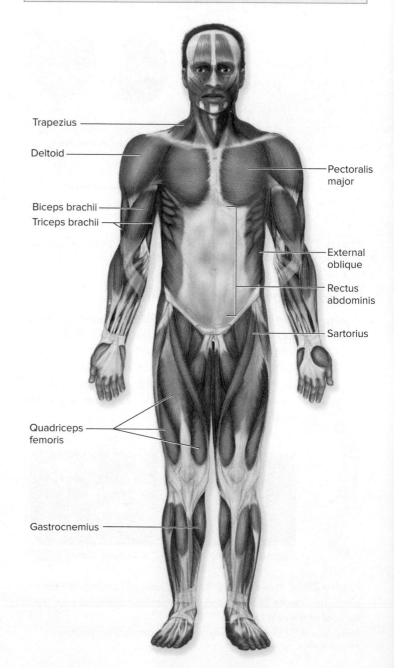

Trapezius

Deltoid

Biceps brachii

Triceps brachii

Pectoralis major

External oblique

Rectus abdominis

Sartorius

Quadriceps femoris

Gastrocnemius

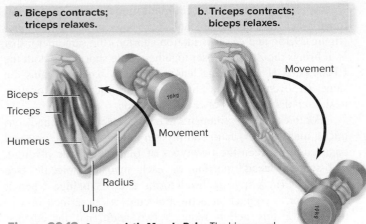

a. Biceps contracts; triceps relaxes.

b. Triceps contracts; biceps relaxes.

Biceps
Triceps
Humerus
Radius
Ulna
Movement
Movement

Figure 29.13 **Antagonistic Muscle Pair.** The biceps and triceps muscles work together to move the lower arm in opposite directions. (a) Contracting the biceps bends the elbow. (b) When the triceps contracts, the arm straightens.

The bulk of the muscle, however, consists of skeletal muscle tissue. The cross sections in the center of figure 29.14 show that muscle tissue is composed of parallel bundles of **muscle fibers,** which are individual muscle cells. The right half of the figure focuses on one muscle cell. Each cell contains multiple nuclei and other organelles. But most of the cell's volume is occupied by hundreds of thousands of cylindrical **myofibrils,** bundles of parallel protein filaments running the length of the cell.

The inset at the top of figure 29.14 depicts the proteins that compose each myofibril. A **thick filament** is made of a protein called **myosin.** A **thin filament** consists primarily of two entwined strands of another protein, **actin,** along with two regulatory proteins called troponin and tropomyosin. Myosin must bind to actin for a muscle to contract and produce movements. In a resting muscle, however, troponin holds tropomyosin over the binding sites on actin. This arrangement keeps the muscle relaxed until the nervous system delivers the signal to contract.

B. Sliding Filaments Are the Basis of Muscle Fiber Contraction

Skeletal muscle tissue appears striped, or striated, because of the alternating arrangement of thick and thin filaments (**figure 29.15a**). These striations divide each myofibril into many functional units, called **sarcomeres.**

Each sarcomere has characteristic lines and bands (figure 29.15b). Z lines form the boundaries of each sarcomere; they are membranes to which thin (actin) filaments attach. Regions called A bands represent the length of the thick (myosin) filaments, including areas that overlap with actin; I bands contain *only* actin; and H zones contain *only* myosin.

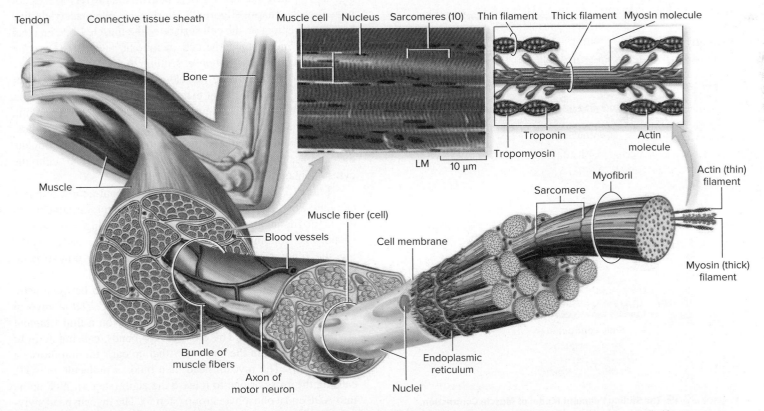

Tendon
Connective tissue sheath
Bone
Muscle
Bundle of muscle fibers
Axon of motor neuron
Blood vessels
Muscle cell   Nucleus   Sarcomeres (10)
Muscle fiber (cell)
Cell membrane
Nuclei
Endoplasmic reticulum
Thin filament   Thick filament   Myosin molecule
Troponin
Tropomyosin
Actin molecule
Sarcomere
Myofibril
Actin (thin) filament
Myosin (thick) filament
LM   10 μm

Figure 29.14 **Skeletal Muscle Organization.** A muscle is an organ enclosed in connective tissue, nourished by blood vessels, and controlled by nerves. Tendons attach skeletal muscles to bones. Bundles of muscle fibers make up most of the muscle's volume; each muscle fiber is a single cell with many nuclei. Most of the cell's volume is occupied by myofibrils, which are, in turn, composed of filaments of the proteins actin and myosin.

Photo: (micrograph): ©Al Telser/McGraw-Hill Education

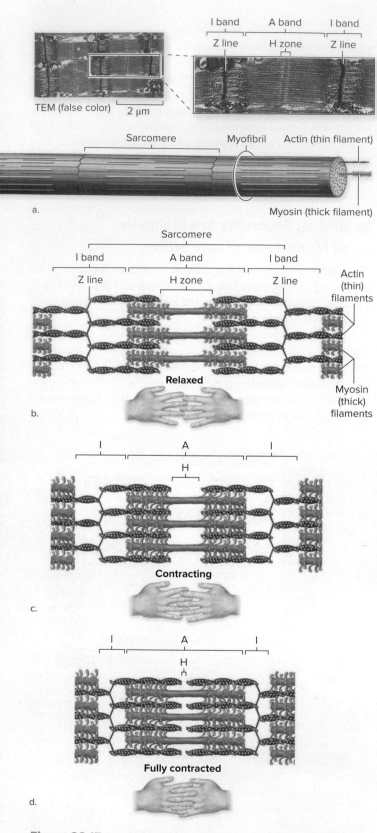

Figure 29.15 **The Sliding Filament Model of Muscle Contraction.**
(a) Myofibrils are divided into functional units called sarcomeres. (b–d) During muscle cell contraction, thick and thin filaments slide past one another, decreasing the length of each sarcomere.

Photos: (a, both): ©Biology Pics/Science Source

According to the **sliding filament model,** a muscle cell contracts when the thin filaments slide between the thick ones (figure 29.15c and d). This motion shortens the sarcomere; that is, it brings the Z lines closer together while shortening both the I bands and H zones. The length of the A bands remains the same. The overall effect is a little like fitting your fingers together to shorten the distance between your hands.

For thick and thin filaments to move past each other and contract a muscle fiber, actin and myosin must touch. The physical connection between the two types of filaments is the pivoting, club-shaped "head" portion of each myosin molecule (see figure 29.14). A myosin head forms a **cross bridge** when it swings out to contact an actin molecule. As described in section 29.4C, this action requires two additional ingredients: calcium and ATP.

C. Motor Neurons Stimulate Muscle Fiber Contraction

Muscles do not contract at random; rather, they wait for electrical stimulation from the central nervous system. A **motor neuron** delivers the signal to contract at a specialized synapse between the neuron and a muscle cell (**figure 29.16**).

When the central nervous system sends the signal to contract, action potentials occur in the motor neuron's axon. These signals stimulate the release of a **neurotransmitter,** acetylcholine, at each synapse (figure 29.16, step 1). When acetylcholine binds to receptors on the cell surface of the muscle fiber, an electrical wave races along the cell membrane (step 2). ⓘ *action potential,* section 26.3; *synapse,* section 26.4A

The membrane of the muscle cell forms tunnel-like structures, called T tubules, that extend deep into the cell's interior. The T tubules touch the muscle cell's endoplasmic reticulum, which surrounds the myofibrils. The electrical waves in the muscle fiber's membrane spread through the T tubules and stimulate the endoplasmic reticulum to release calcium ions (Ca^{2+}) into the cytosol (figure 29.16, step 3).

Figure 29.17 illustrates why muscle contraction requires calcium. The Ca^{2+} ions bind to the troponin proteins attached to the actin filaments (step 1). This binding changes troponin's shape. The tropomyosin proteins therefore move aside, exposing the binding sites on actin. The muscle is now free to contract.

Besides calcium ions, the sliding interaction between actin and myosin also requires energy in the form of ATP. A myosin head attaches to an exposed actin monomer on a thin filament (figure 29.17, step 2). The cross bridge bends, causing actin to slide past myosin in the same way that an oar's motion moves a boat (step 3). The myosin head then binds a molecule of ATP, causing the cross bridge to release the actin (step 4). ATP splits into ADP and a phosphate group (step 5). The myosin head swivels back to its original position, ready to contact an actin monomer farther down the thin filament. ⓘ *ATP,* section 4.3

This sliding repeats about a hundred times per second on each of the hundreds of myosin molecules of a thick filament.

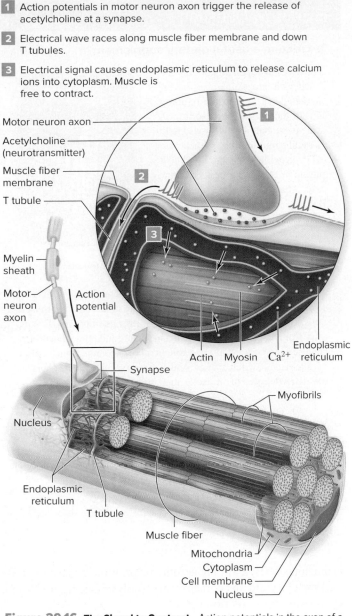

1 Action potentials in motor neuron axon trigger the release of acetylcholine at a synapse.

2 Electrical wave races along muscle fiber membrane and down T tubules.

3 Electrical signal causes endoplasmic reticulum to release calcium ions into cytoplasm. Muscle is free to contract.

Motor neuron axon

Acetylcholine (neurotransmitter)

Muscle fiber membrane

T tubule

Myelin sheath

Motor neuron axon

Action potential

Actin   Myosin Ca^{2+} Endoplasmic reticulum

Synapse

Myofibrils

Nucleus

Endoplasmic reticulum

T tubule

Muscle fiber

Mitochondria

Cytoplasm

Cell membrane

Nucleus

Figure 29.16 The Signal to Contract. Action potentials in the axon of a motor neuron stimulate the release of acetylcholine at a synapse with a muscle fiber. The resulting electrical changes in the muscle cell membrane travel along T tubules to the interior of the muscle cell, leading to the release of calcium ions from the endoplasmic reticulum. Figure 29.17 illustrates the role of calcium ions in muscle contraction.

A skeletal muscle contracts quickly and forcefully due to the efforts of many thousands of "rowers."

How does a muscle *stop* contracting? Shortly after the release of Ca^{2+}, the ions move back into the endoplasmic reticulum by active transport. Troponin and tropomyosin return to their original positions. Even if ATP remains available, actin can no longer interact with myosin. The sarcomere relaxes until the next neural impulse brings a fresh influx of Ca^{2+}. ⓘ *active transport, section 4.5B*

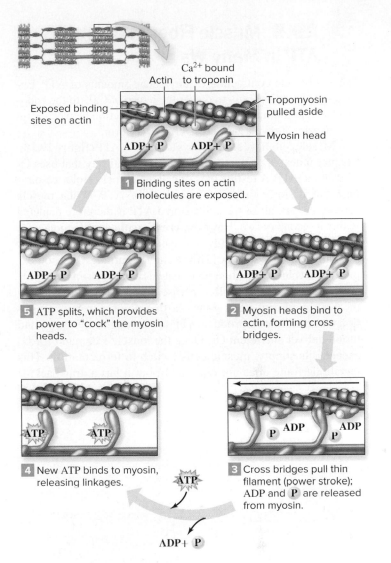

Ca^{2+} bound to troponin

Actin

Exposed binding sites on actin

Tropomyosin pulled aside

Myosin head

ADP+ P   ADP+ P

1 Binding sites on actin molecules are exposed.

ADP+ P   ADP+ P

5 ATP splits, which provides power to "cock" the myosin heads.

ADP+ P   ADP+ P

2 Myosin heads bind to actin, forming cross bridges.

ATP   ATP

4 New ATP binds to myosin, releasing linkages.

ATP

P ADP   P ADP

3 Cross bridges pull thin filament (power stroke); ADP and P are released from myosin.

ADP+ P

Figure 29.17 Calcium and ATP in Muscle Contraction. Calcium ions change the interaction between troponin and tropomyosin, allowing myosin to bind to actin. ATP provides the energy required for myosin filaments to slide past actin filaments.

Some pathogens interfere with the neural signals that stimulate muscle contraction. For example, the bacterium *Clostridium botulinum* produces botulinum toxin (commonly known as Botox). This poison blocks the release of acetylcholine from motor neurons; affected muscles never receive the signal to contract. Injecting tiny amounts of Botox around the eyes and forehead temporarily paralyzes some facial muscles. This treatment reduces the appearance of wrinkles but also causes parts of the face to become expressionless and "frozen."

29.4 MASTERING CONCEPTS

1. What is an antagonistic pair of muscles?
2. Describe how sliding filaments shorten a sarcomere.
3. How do ATP, motor neurons, and calcium ions participate in muscle contraction?

29.5 Muscle Fibers Generate ATP in Many Ways

Skeletal muscle contraction requires huge amounts of ATP. Energy is required not only to power the return of Ca^{2+} to the endoplasmic reticulum but also to break the connection between actin and myosin, allowing new cross bridges to form.

Muscle cells have several ways to produce ATP (figure 29.18). Chapter 6 described aerobic respiration, a pathway that uses O_2 to generate ATP. A resting muscle cell undergoes aerobic respiration, but it stores only small amounts of ATP. When muscle activity begins, all of the cell's stored ATP is therefore depleted within a second or two. However, **creatine phosphate** molecules rapidly donate high-energy phosphates to ADP, temporarily restoring the ATP supply. (This section's Burning Question explores the value of creatine phosphate as a dietary supplement.)

Less than a minute after intense exercise starts, the supply of creatine phosphate is gone. At that point, aerobic respiration can continue to produce ATP for as long as the lungs and blood deliver sufficient O_2. Once the muscle's demand for O_2 exceeds its supply, muscle cells switch to fermentation. This metabolic route does not require O_2, but it has a drawback: It

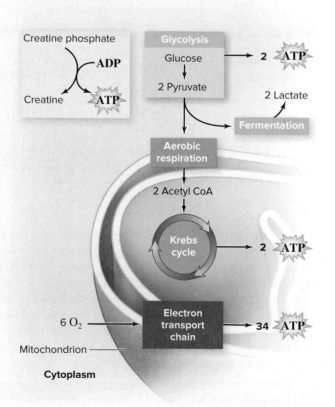

Figure 29.18 Energy Sources. An active muscle cell can use creatine phosphate to replenish its ATP supply for a short time. The cell can also generate ATP in aerobic respiration as long as O_2 is available. If the demand for O_2 exceeds the supply, the cell begins producing ATP by fermentation, which is far less efficient but does not require O_2.

Burning Question

Is creatine a useful dietary supplement?

You may have seen jars of creatine powder on nutrition store shelves, marketed as a muscle-building aid. In theory, an increase in creatine phosphate levels should help skeletal muscle cells generate ATP, providing an energy boost during brief, intense bouts of exercise. After all, creatine phosphate donates its P to ADP, quickly regenerating ATP soon after muscle activity starts. But does this supplement really work?

©Alan Mather/Alamy

Although the idea seems logical, people differ in their response to creatine supplements. For some, taking creatine powder does increase the amount of creatine phosphate inside skeletal muscle cells. But not everyone's athletic performance improves. Results vary from no effect to small gains in sprints and other short-term, intense exercises. Endurance athletes show no gains from creatine supplements, which makes sense: Creatine phosphate plays its short-lived role in muscle metabolism soon after exercise begins.

As a note of caution, the long-term consumption of creatine powder may be harmful. Much of the extra creatine ends up in urine, indicating stress on the kidneys. The possibility of kidney toxicity requires further study.

Submit your burning question to Marielle.Hoefnagels@mheducation.com

is much less efficient than aerobic respiration, generating only the low ATP yield that comes from glycolysis (see figure 29.18).
ⓘ *fermentation,* section 6.8B

Intense exercise may lead to a period of **oxygen debt,** during which the body requires extra O_2 to restore resting levels of ATP and to recharge the proteins that carry oxygen in blood and muscle. Heavy breathing for several minutes after intense muscle activity is a sign of oxygen debt.

Shortly after death, muscles can no longer generate any ATP at all. One consequence is rigor mortis, the stiffening of muscles that occurs within a few hours after death. Without ATP, the cross bridges cannot release from actin. The muscles remain in a stiff position for the next couple of days, until the protein filaments begin to decay.

29.5   MASTERING CONCEPTS

1. Describe the role of creatine phosphate in muscle metabolism.
2. What happens when a muscle cell cannot generate enough ATP by aerobic respiration?

29.6 Many Muscle Fibers Combine to Form One Muscle

This chapter has so far described individual muscle fibers, each of which is either contracted or relaxed. Muscle fibers, however, form bundles that build whole muscles (see figure 29.14); these organs may contract a lot, a little, or not at all. How do the "all-or-none" activities of individual muscle fibers contribute to the more nuanced behavior of whole muscles?

A. Each Muscle May Contract with Variable Force

External electrical stimulation of an isolated muscle fiber produces a **twitch,** a single rapid cycle of contraction and relaxation. Each twitch yields one jerky movement. Yet in the body, the nervous system interacts with muscles to produce smooth, coordinated movements. Two integrated mechanisms maintain this fine control of whole-muscle movement.

The first mechanism operates at the level of the individual muscle cell. A muscle cell contracts whenever a motor neuron stimulates it, but Ca^{2+} is quickly cleared from the cytoplasm of the muscle fiber after each action potential. For a muscle cell to contract to its full range, repeated action potentials must sustain Ca^{2+} availability. A high rate of action potentials, then, ensures smooth, prolonged contractions of individual muscle cells.

The second mechanism that fine-tunes muscle movements involves groups of muscle fibers. Each motor neuron's axon branches at its tip, with each branch leading to a different muscle fiber. Together, a motor neuron and its muscle fibers make up a **motor unit** (**figure 29.19**). The motor unit in figure 29.19, for example, includes three muscle fibers. When a neural impulse arrives, all of the fibers in a motor unit contract at the same time.

Within one muscle, motor units vary in size from tens to hundreds of muscle fibers per motor neuron. A motor neuron that controls only a few muscle fibers produces fine, small-scale responses, such as the eye movements required for reading. A motor unit consisting of hundreds of muscle cells produces large, coarse movements, such as those required for throwing a ball. The more motor units activated, the stronger the force of contraction. In this way, by recruiting different combinations of motor units, the same hand can both grip a hammer and pick up a tiny nail.

B. Muscles Contain Slow- and Fast-Twitch Fibers

Most skeletal muscles contain fibers of two main types, distinguished by the duration of each twitch (**figure 29.20**). **Slow-twitch fibers** have a relatively small diameter and produce twitches of

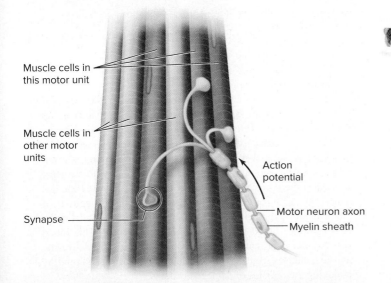

Figure 29.19 Motor Unit. One motor neuron may stimulate multiple muscle fibers simultaneously; the motor unit shown here includes three muscle cells.

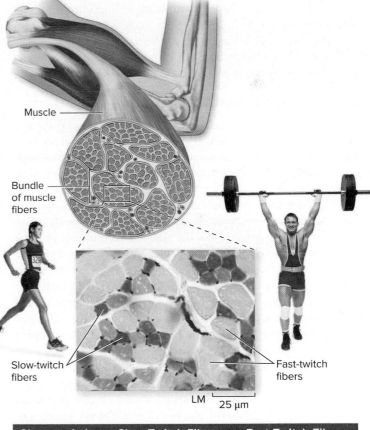

Characteristic	Slow-Twitch Fibers	Fast-Twitch Fibers
Metabolism	Aerobic	Anaerobic
Energy use	Slow, steady	Quick, explosive
Endurance	High	Low

Figure 29.20 Slow-Twitch and Fast-Twitch Fibers. Each muscle contains a mix of slow-twitch and fast-twitch fibers. Special stains reveal mitochondrial activity in each cell type. The red slow-twitch fibers in the photo contain more mitochondria and therefore sustain ATP production for a longer duration than do the yellow fast-twitch fibers.

Photos: (woman): ©RubberBall/Getty Images RF; (tissue): ©Biophoto Associates/Science Source; (man): ©Jack Mann/Photodisc/Getty Images RF

relatively long duration (up to 100 milliseconds long). Abundant capillaries deliver oxygen-rich blood. In turn, ample myoglobin (the red pigment that stores O_2 inside muscle cells) supports the aerobic respiration that regenerates ATP in the fibers' plentiful mitochondria. High-endurance, slow-twitch muscle fibers predominate in body parts that are active for extended periods, such as the flight muscles ("dark meat") of ducks and geese or the back muscles that maintain our upright posture.

Fast-twitch fibers, in contrast, are larger-diameter cells that split ATP quickly in short-duration twitches (as short as 7.5 milliseconds). Anaerobic pathways fuel the short bouts of rapid, powerful contraction that are characteristic of fast-twitch fibers. Muscles dominated by these fibers appear white because they have few capillaries and a lower content of myoglobin. The white breast muscle of a domesticated chicken, for example, can power barnyard flapping for a short time but cannot support sustained, long-distance flight.

The proportion of slow-twitch to fast-twitch fibers affects athletic performance. People with a high proportion of slow-twitch fibers excel at endurance sports, such as long-distance biking, running, and swimming. Athletes who have a higher proportion of fast-twitch fibers perform best at short, fast events, such as weight lifting, hurling the shot put, and sprinting.

C. Exercise Strengthens Muscles

Regular exercise strengthens the muscular system. During the few months after a runner begins training, leg muscles noticeably enlarge. This increase in muscle mass comes from the growth of individual muscle cells rather than from an increase in their number. Exercise-induced muscle growth is more pronounced in a weight lifter, because the resistance of the weights greatly stresses the muscles. Anabolic steroids boost muscle growth by activating the genes encoding muscle proteins, but the potential side effects of illicit steroid use are serious (see the Apply It Now box in chapter 28).

A trained athlete's muscle fibers also use energy more efficiently than those of an inactive person. The athlete's cells contain more enzymes and mitochondria, so his or her muscles can withstand more exertion before fermentation begins. The athlete's muscles also receive more blood and store more glycogen than those of an untrained person.

Like bones, muscles can degenerate from lack of use. After just two days of inactivity, mitochondrial enzyme activity drops in skeletal muscle cells. After a week without exercise, aerobic respiration efficiency falls by 50%. The number of small blood vessels surrounding muscle fibers declines, lowering the body's ability to deliver O_2 to the muscle. Glycogen reserves fall, and the breakdown of lactate occurs less efficiently (see section 6.8). After a few months of inactivity, the benefits of regular exercise all but disappear.

Athletic ability aside, exercise is often followed by sore muscles and joint pain. A hot tub can offer some relief; this chapter's second Burning Question explains why this remedy works.

29.6 | MASTERING CONCEPTS

1. How can one muscle make small and large movements?

2. How do slow- and fast-twitch muscle fibers differ?

3. How does exercise strengthen muscles?

INVESTIGATING LIFE

29.7 Did a Myosin Gene Mutation Make Humans Brainier?

The old admonition not to "bite off more than you can chew" seems to apply especially well to humans. Our chewing muscles are considerably smaller than those of most primates, including chimpanzees and gorillas. We favor soft foods such as bread or cheese, and we would have a hard time chewing the bark, stems, and seeds that some of our primate relatives savor. These differences have spurred researchers to investigate the evolution of the muscles that connect the lower jaw to the other bones of the skull.

Part of the evidence that has helped scientists understand the evolution of jaw muscles came from an unexpected source. It all began with the Human Genome Project, which has allowed researchers to comb through human DNA sequences in search of particular genes. Hansell H. Stedman and associates from the University of Pennsylvania and the Children's Hospital of Philadelphia hoped to catalog every myosin gene in the human genome. Myosin is not just one protein; it is a family of proteins

Burning Question

Why does heat soothe sore muscles and joints?

©liquidlibrary/PictureQuest RF

Few experiences are as relaxing as slipping into a bubbling hot tub. Muscle tension subsides, and aches seem to melt away. Why does heat have this effect?

Conventional wisdom says that heat causes muscles to relax, but this answer is misleading. Technically, a muscle relaxes when actin and myosin filaments slide past one another in a way that makes the sarcomeres longer. Heat does not trigger this sliding action, so it cannot make muscles relax in this sense.

Instead, most of heat's soothing effects on muscles and joints are indirect. First, heat causes the proteins that make up tendons and ligaments to become more fluid and stretchier. Loosening up this connective tissue not only relieves tension on the muscles but also makes stiff joints more mobile. Second, warmth helps to relieve pain in sore, overworked muscles and joints. Heat stimulates thermoreceptors in the skin, which may override impulses from pain receptors. Third, heat may reduce the activity of motor neurons that trigger painful muscle spasms. And finally, heat increases blood flow near the body's surface. This blood speeds healing of damaged muscles in two ways: by delivering nutrients and oxygen that cells require to make repairs, and by removing the remains of damaged cells.

Submit your burning question to
Marielle.Hoefnagels@mheducation.com

encoded by at least 40 closely related genes. Mutations in many of the myosin-encoding genes can lead to loss of muscle function and other serious disorders. Stedman's research was part of an effort to better understand diseases such as muscular dystrophy. ⓘ *Human Genome Project,* section 11.2B

During a search of chromosome 7, however, the team stumbled on an inactive gene that encoded a nonfunctional myosin protein. At first, they thought the inactive gene they had found represented a quirk in the sequences they were searching. They therefore searched the genomes of six widely dispersed human populations originating in locations ranging from Africa to Iceland. All of the human groups had the mutated gene. The team also compared the gene to a homologous DNA sequence in seven species of nonhuman primates, including gorillas and chimpanzees. The results were clear: The researchers had discovered a human myosin gene containing mutations that are not present in nonhuman primates (**figure 29.21**).

Next, the researchers discovered that in both humans and macaque monkeys, only two muscles express the gene, and both participate in the up-and-down jaw movements required for chewing. Because the protein is nonfunctional in humans but functional in macaques, this finding suggested that an ancient genetic mutation made our chewing muscles small and weak, at least compared with those of our close relatives.

That result prompted Stedman's research team to learn more about the chewing apparatus in fossil primates. Bones bear marks indicating where muscles were once attached. Until about 2 million years ago (MYA), primates had large skull bones and robust chewing muscles. These large complexes occurred in the ancestral hominines *Australopithecus* and *Paranthropus,* as well as in contemporary primates such as macaques and gorillas (**figure 29.22**). But a more delicate chewing apparatus appeared in early humans (*Homo erectus/ergaster*) between 1.8 and 2.0 MYA.

The researchers used a "molecular clock" technique to estimate that the mutation in the myosin gene occurred approximately 2.4 MYA—before the migration of *Homo* out of Africa. The timing coincides with a significant trend in human evolution: increasing brain size. Stedman and his colleagues argue that the myosin mutation may have changed the muscles in a way that released a constraint on the size of the brain. Enhanced brain power may have eventually led to the spread of culture, profoundly changing the course of human evolution. ⓘ *molecular clock,* section 13.6B

a. Macaque   b. Gorilla   c. Human

Figure 29.22 Primate Comparison. The temporalis muscle is one of the four main chewing muscles. (a) In macaques and (b) in gorillas, the temporalis attachment area (*red*) occupies most of the side of the skull. (c) In humans, the corresponding area is much smaller, resulting in a much weaker chewing apparatus.

Of course, one mutated myosin gene is, by itself, not likely to have set in motion the entire course of human history. Other changes in the skeletal and nervous systems must also have occurred to spur the evolution of the brain.

The story of this myosin gene illustrates how a routine study can have unexpected results; what began as a simple search through the human genome ended up influencing the study of human evolution. This study also points to the connections among different areas of biology. Stedman and his colleagues combined their knowledge of the muscular and skeletal systems with observations of ancient fossils, gene mutations, and protein functions. Their results give us something to chew on as we contemplate the evolutionary history of our own species.

Source: Stedman, Hansell, Benjamin W. Kozyak, Anthony Nelson, and 7 coauthors. 2004. Myosin gene mutation correlates with anatomical changes in the human lineage. *Nature,* vol. 428, pages 415–418.

29.7 MASTERING CONCEPTS

1. Summarize the hypothesized relationship between the myosin gene mutation and brain size in humans and other primates.

2. Describe the data in figure 29.21. What conclusion did the researchers draw from these data?

Figure 29.21 Myosin Mutation. The DNA sequences for a small portion of the myosin gene are shown for nonhuman primates (upper seven sequences) and humans (lower six sequences). The fragments start at nucleotide 26 in the gene. Shaded boxes highlight differences between the two sets of genes. The mutations in the human genes cause the encoded protein to be truncated (shortened) and nonfunctional.

	30	40	50	60	70	80
Nonhuman						
Woolly monkey	CCCTCCACAGCACTGTACCCCATTTTGTCCGCTGTATTGTGCCCAATGAGTTTAAGCAGTCAG
Pigtail macaque	CCCTCCACAGCACTGTACCCCATTTTGTCCGCTGTATTGTGCCCAATGAGTTTAAGCAGTCAG
Rhesus	CCCTCCACAGCACTGTACCCCATTTTGTCCGCTGTATTGTGCCCAATGAGTTTAAGCAGTCAG
Orangutan	CCCTCCACAGCACTGTACCCCATTTTGTCCGCTGTATTGTGCCCAATGAGTTTAAGCAGTCAG
Gorilla	CCCTCCACAGCACTGTACCCCATTTTGTCCGCTGTATTGTGCCCAATGAGTTTAAGCAGTCAG
Bonobo	CCCTCCACAGCACTGTACCCCATTTTGTCCGCTGTATTGTGCCCAATGAGTTTAAGCAGTCAG
Chimpanzee	CCCTCCACAGCACTGTACCCCATTTTGTCCGCTGTATTGTGCCCAATGAGTTTAAGCAGTCAG
Human						
Africa (pygmy)	CCCTCCATAGC--CGCACCCCATTTTGTCCGCTGTATTATCCCAATGAGTTTAAGCAATCGG
Spain (Basque)	CCCTCCATAGC--CGCACCCCATTTTGTCCGCTGTATTATCCCAATGAGTTTAAGCAATCGG
Iceland	CCCTCCATAGC--CGCACCCCATTTTGTCCGCTGTATTATCCCAATGAGTTTAAGCAATCGG
Japan	CCCTCCATAGC--CGCACCCCATTTTGTCCGCTGTATTATCCCAATGAGTTTAAGCAATCGG
Russia	CCCTCCATAGC--CGCACCCCATTTTGTCCGCTGTATTATCCCAATGAGTTTAAGCAATCGG
South America	CCCTCCATAGC--CGCACCCCATTTTGTCCGCTGTATTATCCCAATGAGTTTAAGCAATCGG

CHAPTER SUMMARY

29.1 Skeletons Take Many Forms

- The **muscular system** and **skeletal system** enable an animal to move.
- An animal's **skeleton** supports its body. **Muscles** act on the skeleton to provide motion (figure 29.23).
- A **hydrostatic skeleton** requires constrained fluid; an **exoskeleton** is on the organism's exterior; and an **endoskeleton** forms inside the body.

29.2 The Vertebrate Skeleton Features a Central Backbone

- The **axial skeleton** consists of the bones of the head, **vertebral column,** and rib cage.
- The **appendicular skeleton** includes the limbs and the bones (**pectoral girdle** and **pelvic girdle**) that attach them to the axial skeleton.

29.3 Bones Provide Support, Protect Internal Organs, and Supply Calcium

- **Bones** are strong and lightweight. Long bones have a **marrow cavity** that contains **red bone marrow** or **yellow bone marrow.**

A. Bones Consist Mostly of Bone Tissue and Cartilage

- Bone tissue derives its strength from collagen and its hardness from minerals. In **compact bone tissue,** cells called **osteocytes** occur in concentric rings. **Spongy bone tissue** has many spaces separated by hard struts. Cartilage is an excellent shock absorber.

B. Bones Are Constantly Built and Degraded

- Bone-degrading and bone-building cells have opposing activities.
- Exercise strengthens bones; conversely, bones weaken with disuse.

C. Bones Help Regulate Calcium Homeostasis

- Hormones control the exchange of calcium between blood and bones. **Osteoporosis** results when bones lose more calcium than they replace.

D. Bone Meets Bone at a Joint

- **Joints** attach bones to each other. Some joints are immovable.
- Freely moving joints consist of cartilage and a connective tissue capsule containing lubricating fluid. **Ligaments** and **tendons** stabilize the joint.

29.4 Muscle Movement Requires Contractile Proteins, Calcium, and ATP

- Many **skeletal muscles** form antagonistic pairs.

A. Actin and Myosin Filaments Fill Muscle Cells

- Skeletal muscle fibers are elongated, multinucleate, striated, and voluntary.
- A skeletal **muscle fiber** contains **myofibrils** composed of **thick filaments** (**myosin**) and **thin filaments** (primarily **actin**).

B. Sliding Filaments Are the Basis of Muscle Fiber Contraction

- A myofibril is a chain of contractile units called **sarcomeres.**
- According to the **sliding filament model,** muscle contraction occurs when thick and thin filaments move past one another. Myosin heads swing out to form **cross bridges** that pull on actin filaments.

C. Motor Neurons Stimulate Muscle Fiber Contraction

- Action potentials in a **motor neuron** trigger **neurotransmitter** (acetylcholine) release into the synapse with each muscle fiber.
  - The arrival of acetylcholine ultimately stimulates Ca^{2+} ions from the endoplasmic reticulum to flood the sarcomere. Calcium binds to troponin, which moves tropomyosin and allows actin to bind to myosin. The cross bridge bends, sliding actin past myosin. ATP binding releases actin from myosin and repositions the myosin head.
  - After the impulse, Ca^{2+} returns to the endoplasmic reticulum; tropomyosin again prevents myosin from binding actin. The muscle relaxes.

29.5 Muscle Fibers Generate ATP in Many Ways

- The energy that powers muscle contraction comes first from stored ATP, then from **creatine phosphate** stored in muscle cells, then from aerobic respiration, and finally from fermentation.
- **Oxygen debt** is a temporary deficiency of O_2 after intense exercise.

29.6 Many Muscle Fibers Combine to Form One Muscle

A. Each Muscle May Contract with Variable Force

- A single stimulation causes a muscle cell to contract and relax (a **twitch**). Prolonged muscle contraction requires repeated action potentials.
- The more **motor units** stimulated, the greater the contraction of the muscle.

B. Muscles Contain Slow- and Fast-Twitch Fibers

- **Slow-twitch fibers** use ATP slowly and regenerate it by aerobic respiration. **Fast-twitch fibers** use ATP quickly and use mostly anaerobic pathways to replenish it.

C. Exercise Strengthens Muscles

- Exercise thickens muscle cells and helps them use energy efficiently.

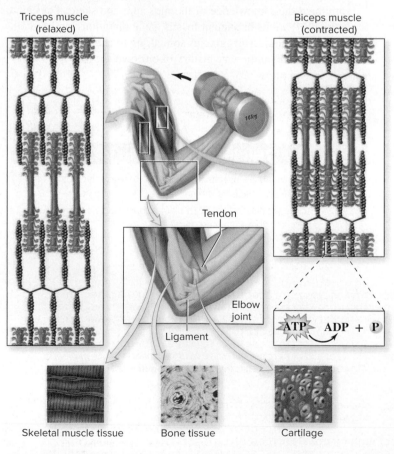

Triceps muscle (relaxed)

Biceps muscle (contracted)

Tendon

Elbow joint

Ligament

ATP → ADP + P

Skeletal muscle tissue   Bone tissue   Cartilage

Figure 29.23 Muscles Move Bones: A Summary.

29.7 Investigating Life: Did a Myosin Gene Mutation Make Humans Brainier?

- In humans, a mutated myosin gene is expressed only in the muscles required for chewing. Nonhuman primates lack the mutation.
- Smaller chewing muscles may have increased brain capacity in humans.

MULTIPLE CHOICE QUESTIONS

1. The axial skeleton is to the appendicular skeleton as
   a. a tree's branches are to its trunk.
   b. a car's body is to its wheels.
   c. a cell's cytoplasm is to its nucleus.
   d. a finger is to a toe.

2. Bone matrix is composed of ____, which give bones flexibility and strength, and ____, which give bones rigidity.
   a. cartilage cells; bone cells
   b. minerals; collagen proteins
   c. bone cells; cartilage cells
   d. collagen proteins; minerals

3. The function of a ligament is to connect
   a. cartilage to bone.
   b. bone to bone.
   c. bone to muscle.
   d. muscle to muscle.

4. Which of the following is arranged in order from smallest to largest?
   a. Motor unit < sarcomere < muscle cell < actin subunit
   b. Muscle cell < actin subunit < sarcomere < motor unit
   c. Actin subunit < sarcomere < muscle cell < motor unit
   d. Sarcomere < motor unit < actin subunit < muscle cell

5. In the following list of events required for muscle contraction, which occurs *third*?
   a. Calcium binds to troponin.
   b. Myosin head binds to actin.
   c. Cross bridges pull thin filament.
   d. Acetylcholine is released from the motor neuron.

6. Within the first few seconds of a 5-minute race, stored ATP is depleted. How do muscles obtain energy for the rest of the race?
   a. High-energy phosphates are transferred from glycogen to ADP.
   b. Creatine phosphate restores the ATP supply at first; then aerobic respiration generates ATP as long as O_2 is available.
   c. Actin filaments donate the high-energy phosphates that myosin filaments require.
   d. Fermentation begins producing new ATP as soon as stored ATP is depleted.

Answers to these questions are in appendix A.

WRITE IT OUT

1. Distinguish among a hydrostatic skeleton, an exoskeleton, and an endoskeleton. What are the advantages and disadvantages of each type of skeleton? Give an example of an animal with each type.

2. Use the Internet to research bone marrow transplants. What do patients who receive these transplants typically have in common?

3. Suppose a young boy severely fractures his femur. Describe how the injury could affect his growth.

4. Bones typically become stronger with exercise. However, some athletes develop stress fractures from overexercising. Why might light exercise strengthen bones but intense exercise cause fractures?

5. Design an experiment to test whether changes in the atmosphere (such as an incoming thunderstorm) cause joint pain. Then, use the Internet to learn whether researchers have found evidence to support a connection between weather and joint pain.

6. How do antagonistic muscle pairs move bones? Give an example of such a pair.

7. Describe four muscle proteins and their functions.

8. Explain how multiple muscle twitches combine to allow you to lift heavy objects.

9. How might your muscles lengthen when you stretch? Use *sarcomere, myosin, actin,* and *tendon* in your answer.

10. Write the sequence of events that leads to a muscle contraction, starting with "Action potentials occur in the axon of a motor neuron."

11. Why do endurance sports require a high proportion of slow-twitch muscle fibers, whereas power sports require more fast-twitch muscle fibers?

PULL IT TOGETHER

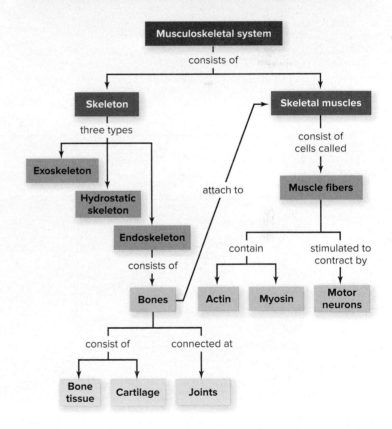

Figure 29.24 Pull It Together: The Skeletal and Muscular Systems.

Refer to figure 29.24 and the chapter content to answer the following questions.

1. Using the Survey the Landscape figure in the chapter introduction and the Pull It Together concept map, explain some ways that the musculoskeletal system maintains homeostasis.

2. How do bones help maintain blood calcium concentrations?

3. Add *exercise* to the concept map in at least three different places.

30 The Circulatory System

Raw Material for Artificial Blood? A technician holds a bag of hemoglobin purified from human blood. The hemoglobin is being tested for use in a blood substitute.

©Philippe Plailly/Science Source

LEARN HOW TO LEARN
Studying in Groups

Study groups can offer a great way to learn from other students, but they can also dissolve into social events that accomplish little real work. Of course, your choice of study partners makes a huge difference; try to pick people who are at least as serious as you are about learning. To stay focused, plan activities that are well suited for groups. For example, you can agree on a list of vocabulary words and take turns adding them to a group concept map. You can also write exam questions for your study partners to answer, or you can simply explain the material to each other in your own words. Focus on what you need to learn, and your study sessions should be productive.

New Ways to Replace Hearts and Blood

Together with blood, the heart is central to vertebrate life. The heart is the circulatory system's only pump; if it fails to do its job, death can come quickly. Blood is the fluid that delivers life-giving nutrients and removes wastes throughout the body. Excessive blood loss therefore also spells death. Can science find replacements for either hearts or blood?

Current treatment options for heart failure include surgery, drugs, and lifestyle changes, but in the future, physicians may have a new tool: stem cells. Scientists may one day coax stem cells to divide and produce daughter cells that differentiate into exactly what is needed to heal damaged tissue. The burgeoning field of tissue engineering, described in chapter 25, has already yielded artificial bladders. No one has yet created a human heart using the same technology, but research looks promising. In one study, scientists "seeded" a mixture of cells onto a scaffold made from a rat's heart. The cells divided and formed a heart-shaped organ that could beat and pump blood. ⓘ *stem cells,* section 11.3A

In addition to the heart, blood is the other half of the circulatory equation. Blood can easily be transferred between people, but transfusions carry a risk: Donated blood may transmit diseases to recipients. Artificial blood would solve this problem. Some blood substitutes are based on hemoglobin, the oxygen-toting protein in red blood cells. Hemoglobin is extracted, stabilized, and mixed with a sterile salt solution to create artificial blood. A cow hemoglobin preparation saved the life of a young woman whose immune system was attacking her own blood, maintaining her circulation for several days until the illness subsided.

Red blood cell replacements may also come from stem cells. Bone marrow, for example, contains stem cells that give rise to all blood cells. A bone marrow transplant replaces a patient's own marrow with that of a healthy person. The transplant procedure entails many risks, but if all goes well, stem cells in the donor's marrow churn out healthy cells in the recipient's body. Such transplants have saved the lives of people with sickle cell disease, an illness in which red blood cells are misshapen.

Someday, stem cells reared in the laboratory may give us a safe, limitless supply of human blood. But stem cell research is a young science, and researchers still have much to learn before the medical applications become widespread. Success will require a basic understanding of how the body works, including the circulatory system—the subject of this chapter.

SURVEY THE LANDSCAPE
Animal Life

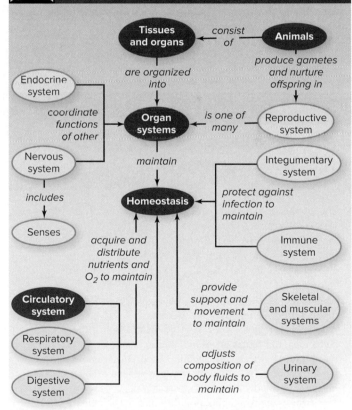

Blood vessels form the body's highways, carrying hormones, water, food molecules, and oxygen throughout the body. Meanwhile, cells produce carbon dioxide and other wastes; blood totes these metabolic byproducts to their disposal sites.

For more details, study the Pull It Together feature in the chapter summary.

30.1 Circulatory Systems Deliver Nutrients and Remove Wastes

Watch a crime drama on TV, and it won't be long until a gunshot wound leaves someone lying in a pool of blood. Unless help arrives immediately, life quickly fades—a vivid reminder of blood's importance.

This bright red fluid is the most visible and familiar part of the **circulatory system,** which consists of blood (or a comparable fluid), a network of vessels that contain the blood, and a heart. The **heart** is a pump that keeps the blood moving through these vessels; animals may have one or more hearts. Overall, the circulatory system's function is to transport materials throughout the body.

Many substances, including glucose and oxygen gas (O_2), travel throughout the circulatory system. Without these resources, the body's cells could not carry out aerobic respiration, which generates the ATP required for life (see chapter 6). The circulatory system also carries off wastes such as carbon dioxide (CO_2).

The circulatory system has extensive connections with organ systems that exchange materials with the environment. For example, blood vessels acquire O_2 and unload CO_2 at gills, lungs, or other organs of the respiratory system (see chapter 31). Nutrients enter the circulatory system at blood vessels near the intestines, which form part of the digestive system (see chapter 32). Blood also circulates through the kidneys, which eliminate many water-soluble metabolic wastes (see chapter 33).

A. Circulatory Systems Are Open or Closed

Some types of animals lack a dedicated circulatory system. Flatworms, for example, use their incomplete digestive tracts not only to absorb nutrients but also for gas exchange. ⓘ *flatworms,* section 21.4

Most animals, however, do have a circulatory system, which may be open or closed (**figure 30.1**). In an **open circulatory system,** fluid is pumped through short vessels that lead to open spaces in the body cavity. There, the fluid can exchange materials with the body's cells before flowing back into the heart through pores. Animals with open circulatory systems include arthropods (such as insects) and most mollusks. ⓘ *mollusks,* section 21.5; *arthropods,* section 21.8

In a **closed circulatory system,** blood remains within vessels that exchange materials with the fluid surrounding the body's cells. Examples of animals with closed circulatory systems include vertebrates, annelids, and some mollusks such as squids and octopuses. ⓘ *annelids,* section 21.6

Both types of circulatory systems have advantages. Open circulatory systems require fewer vessels, and the blood moves under low pressure. The energetic costs of circulation are therefore relatively low. But closed circulatory systems tend to be more efficient than open systems. In a closed circulatory system, blood flows at higher pressure, so nutrient delivery and waste removal can occur more rapidly. Moreover, the vessels of a closed system can direct blood flow toward or away from specific areas

---

Burning Question

What causes bruises?

Anyone who has bumped into a piece of heavy furniture knows that minor injuries often leave bruises. The collision breaks blood vessels, which leak blood into the surrounding tissues. Blood that collects near the skin surface produces the familiar discoloration of a bruise. The color of a bruise indicates its progression through the healing process. Recent bruises are reddish-blue. As the hemoglobin (the oxygen-containing molecules in blood) break down, the bruise turns bluish-purple. Meanwhile, white blood cells remove decayed blood products. A yellow-brown component of blood, called bilirubin, is usually last to disappear.

©Ingram Publishing RF

Bruises are painful because they indicate an injury to the underlying tissues. Touching the bruise stimulates the pain receptors that sense the damage.

Submit your burning question to
Marielle.Hoefnagels@mheducation.com

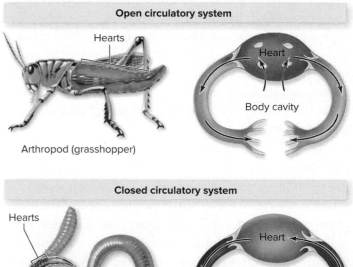

Open circulatory system

Hearts

Heart

Body cavity

Arthropod (grasshopper)

Closed circulatory system

Hearts

Heart

Annelid (earthworm)

Figure 30.1 Open and Closed Circulatory Systems. In an open circulatory system, fluid leaves vessels and bathes cells directly. In a closed circulatory system, blood is confined within vessels.

of the body, depending on metabolic demands. (The Burning Question in this section explains what happens when blood vessels under the skin break.)

B. Vertebrate Circulatory Systems Have Become Increasingly Complex

As vertebrates adapted to life on land, their circulatory systems evolved. One prominent trend was that blood flowing to the organs of gas exchange (gills or lungs) became increasingly separated from blood flowing to the rest of the body (figure 30.2). (Figure 31.4 shows corresponding changes in vertebrate respiratory systems.)

Among the vertebrates, fishes and tadpoles have the simplest circulatory systems (see figure 30.2a). A fish's heart has just two chambers: an **atrium** where blood enters, and a **ventricle** from which blood exits. The heart pumps blood through the gills to pick up O_2 and unload CO_2. The blood then circulates to the rest of the body before returning to the heart.

Other vertebrates divide the circulatory system into two interrelated circuits. In the **pulmonary circulation,** blood exchanges gases at the lungs and returns to the heart; in the **systemic circulation,** blood circulates throughout the rest of the body and back to the heart.

In these animals, the heart may have three or four chambers. The three-chambered heart of a frog, for example, has one undivided ventricle and two atria (see figure 30.2b). The left atrium receives O_2-rich blood from the lungs, and the right atrium receives O_2-depleted blood from the rest of the body. Blood from the two atria mixes in the ventricle, which pumps the blood throughout the body. In turtles, snakes, and lizards, the ventricle is partially divided, an adaptation that increases the separation of oxygenated and deoxygenated blood.

Birds and mammals independently evolved four-chambered hearts with two atria and two ventricles, completing the separation of O_2-rich from O_2-depleted blood (see figure 30.2c). Birds and mammals are endotherms, meaning they spend enormous amounts of energy maintaining a constant body temperature. Their cells therefore consume much higher quantities of nutrients and O_2 than do those of ectotherms, which have variable body temperatures. The four-chambered heart supports these high metabolic rates by maximizing the amount of O_2 reaching tissues. (Interestingly, the ectothermic crocodiles also have four-chambered hearts, reflecting their shared ancestry with birds.) ⓘ *ectotherms and endotherms,* section 33.1A

30.1 MASTERING CONCEPTS

1. What are the components of a circulatory system?
2. Distinguish between open and closed circulatory systems.
3. Describe the circulatory systems of fishes, nonavian reptiles, and mammals. What is the advantage of separating the pulmonary and systemic circulatory pathways?

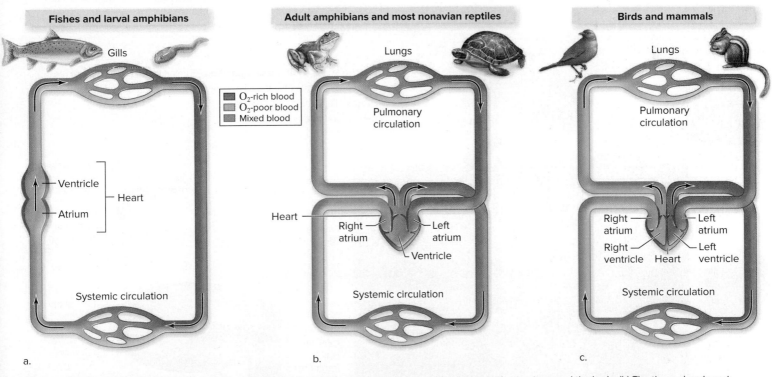

■	O_2-rich blood
■	O_2-poor blood
■	Mixed blood

Figure 30.2 Vertebrate Circulatory Systems. (a) A fish's two-chambered heart pumps blood in a single circuit around the body. (b) The three-chambered heart of a frog or turtle has two atria and one ventricle. Blood from the pulmonary and systemic circuits may mix in the ventricle. (c) A bird or mammal has a four-chambered heart, maximizing the separation of the pulmonary and systemic circuits.

30.2 Blood Is a Complex Mixture

The human circulatory system transports blood, a complex liquid tissue with many functions (table 30.1). Blood not only carries gases, nutrients, hormones, and wastes but also participates in immune reactions and helps maintain homeostasis in several ways.

Blood is a connective tissue that consists of several types of cells and cell fragments (platelets), all suspended in a liquid extracellular matrix called plasma (**figure 30.3**). The cell types are diverse: A milliliter of blood normally contains about 5 million red blood cells, 7000 white blood cells, and 250,000 platelets. This section describes the functions of each component of blood; the Burning Question box explains how donations of whole blood and plasma can save lives. ⓘ *connective tissue,* section 25.2B

A. Plasma Carries Many Dissolved Substances

Plasma is the liquid matrix of blood. This fluid, which makes up more than half of the blood's volume, is 90% to 92% water.

Besides water, more than 70 types of dissolved proteins make up the largest component of plasma. These proteins have many functions. For example, antibodies participate in the body's immune response; high- and low-density lipoproteins transport cholesterol; and clotting factors help stop bleeding following an injury (see section 30.2D). ⓘ *antibodies,* section 34.3C; *cholesterol,* section 2.5E

TABLE 30.1 Functions of Blood: A Summary

Function	Explanation
Gas exchange	Carries O_2 from lungs to tissues; carries CO_2 to the lungs to be exhaled
Nutrient transport	Carries nutrients absorbed by the digestive system throughout the body
Waste transport	Carries urea (a waste product of protein metabolism) to the kidneys for excretion in urine
Hormone transport	Carries hormones secreted by endocrine glands
Formation of interstitial fluid	Blood plasma leaking out of capillaries becomes interstitial fluid that surrounds cells.
Maintenance of homeostasis (temperature, water, pH)	Absorbs heat and dissipates it at the body's surface; regulates cells' water content; buffers in blood help maintain pH of interstitial fluid and lymph.
Protection	Blood clots plug damaged vessels; white blood cells destroy foreign particles and participate in inflammation.

About 1% of plasma consists of dissolved salts, hormones, metabolic wastes, CO_2, nutrients, and vitamins. The concentrations of these dissolved molecules are low, but they are critical. For example, blood usually contains about 0.1% glucose; if the concentration falls to 0.06%, the body begins to convulse.

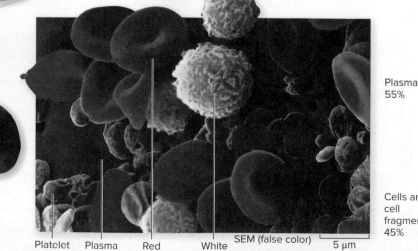

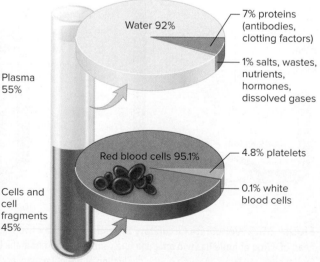

Figure 30.3 Blood Composition. Human blood is a mixture of red blood cells, platelets, and white blood cells suspended in a liquid extracellular matrix called plasma.

Photo: ©National Cancer Institute/Getty Images

Hemoglobin

Platelet   Plasma   Red blood cell   White blood cell   SEM (false color)   5 μm

Plasma 55%

Cells and cell fragments 45%

Water 92%

7% proteins (antibodies, clotting factors)

1% salts, wastes, nutrients, hormones, dissolved gases

Red blood cells 95.1%

4.8% platelets

0.1% white blood cells

B. Red Blood Cells Transport Oxygen

Mature **red blood cells,** or erythrocytes, are saucer-shaped disks that participate in the exchange of O_2 and CO_2. As they fill with the pigment **hemoglobin**—the protein that carries O_2—the red blood cells of humans and some other vertebrates lose their nuclei, ribosomes, and mitochondria. This adaptation maximizes the space available for hemoglobin but also means that the mature cells cannot divide. They generate ATP by fermentation, a pathway that does not require mitochondria. ⓘ *fermentation,* section 6.8B

Tucked into each hemoglobin molecule are four iron atoms, each of which can combine with one O_2 molecule picked up in the lungs (see figure 31.12). Oxygen-saturated hemoglobin is bright red; without bound O_2, hemoglobin has a deeper red color.

Red blood cells originate from stem cells in red bone marrow at a rate of 2 million to 3 million per second. Mature red blood cells leave the bone marrow and enter the circulation. During its life of about 120 days, each red blood cell pounds against artery walls and squeezes through tiny capillaries. Eventually, the spleen destroys the cell and recycles most of its components. ⓘ *bone marrow,* section 29.3A

A person's blood type derives from carbohydrates and other molecules embedded in the outer membranes of red blood cells. Biologists have identified at least 26 human blood group systems, 2 of which are familiar to most people: ABO and Rh. In the ABO blood group system, external carbohydrate "markers" called A and B determine an individual's blood type: A, B, AB, or O. That is, a person's cells may express only marker A (type A blood), only marker B (type B), both A and B (type AB), or neither A nor B (type O). Figure 10.18 explains the genetic basis of ABO blood types.

Knowing a person's blood type is important in blood transfusions because the immune system reacts to "foreign" molecules that are not already present in the blood. Antibodies produced against incompatible blood types cause **agglutination,** a reaction in which the cells clump together (**figure 30.4**). Agglutination following a transfusion of incompatible blood can be fatal. For this reason, a person with type A blood cannot receive a transfusion of type B or type AB blood; his or her plasma contains antibodies that will react against any blood carrying marker B. For people with blood type AB, however, neither marker A nor marker B is "foreign." Type O blood reacts against all blood types except O.

Rhesus (Rh) typing yields the "+" and "−" designations appended to ABO blood groups, such as "A positive" or "O negative." People who express the Rh marker are Rh+, whereas those who lack it are Rh−. Agglutination caused by Rh incompatibility is most important when an Rh− woman becomes pregnant with an Rh+ fetus (see section 34.5D).

C. White Blood Cells Fight Infection

Blood also contains five types of **white blood cells,** or leukocytes (see figure 34.2). These immune system cells are larger than red blood cells, retain their nuclei, and lack hemoglobin.

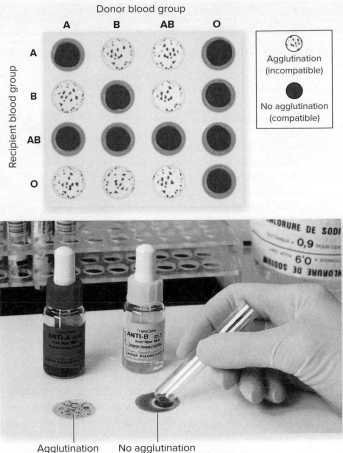

Figure 30.4 ABO Blood Groups. In this chart, the agglutination (clumping) reactions reveal which blood types are incompatible with one another. The photo shows the results of a blood typing test on a sample that contains marker A (*left*) but not marker B (*right*).

Photo: ©Jean Claude Revy - ISM/Phototake

White blood cells originate from stem cells in red bone marrow. Although some enter the bloodstream, most either wander in body tissues or settle in the lymphatic system. These cells participate in many immune responses. Some secrete signaling molecules that provoke inflammation, whereas others destroy microbes or produce antibodies. Chapter 34 explains the interactions of white blood cells in more detail.

White blood cell numbers that are too high or too low can indicate illness. For example, **leukemias** are cancers in which bone marrow overproduces white blood cells. The abnormal white cells form at the expense of red blood cells, so when the patient's "white cell count" rises, the "red cell count" falls. Thus, leukemia also causes anemia. Having too few white blood cells, on the other hand, leaves the body vulnerable to deadly infections. HIV destroys T cells (a type of white blood cell), causing AIDS. Likewise, exposure to radiation or toxic chemicals can severely damage bone marrow, killing many white blood cells. Death occurs rapidly from rampant infection. ⓘ *cancer,* section 8.6; *HIV,* section 16.4B

Burning Question

What is the difference between donating whole blood and donating plasma?

A person who "gives blood" donates 450 to 500 milliliters (about a pint) of blood to a nonprofit blood bank. After being screened for disease-causing agents, the blood may go to patients who need transfusions following trauma or surgery. More commonly, however, the blood is separated into its components. A single blood donation therefore can help several patients.

A person may also donate plasma. In this process, whole blood is removed from a donor's body; then a machine separates out the plasma. The red blood cells and other components are returned

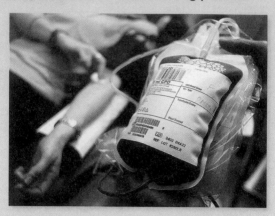

©Toby Melville/Reuters/Corbis

to the donor. The plasma center sells the fluid to pharmaceutical companies, which use it to manufacture treatments for hemophilia, hepatitis, and other diseases. A person can donate plasma more frequently than whole blood, because it takes only about a day or two to replenish the fluid lost in plasma donation. Whole blood takes longer to replace.

Submit your burning question to
Marielle.Hoefnagels@mheducation.com

D. Blood Clotting Requires Platelets and Plasma Proteins

Platelets are small, colorless cell fragments that participate in blood clotting. A platelet originates as part of a huge cell containing rows of vesicles that divide the cytoplasm into distinct regions, like a sheet of stamps. The vesicles enlarge and join together, "shedding" fragments that become platelets.

In a healthy circulatory system, platelets travel freely within the vessels. Sometimes, however, a wound nicks a blood vessel, or the blood vessel's inner lining becomes obstructed. Platelets "catch" on the obstacle and form a clump that temporarily plugs the leak. The platelets attract plasma proteins called clotting factors, which participate in reactions that ultimately produce a web of protein threads. These threads trap red blood cells and platelets, forming a **blood clot**—a plug of solidified blood (**figure 30.5**). The entire cascade includes several positive

feedback loops; once initiated, blood clot formation therefore occurs rapidly. ⓘ *positive feedback,* section 25.4

Blood that clots too slowly can lead to severe blood loss. Hemophilias, for example, are inherited bleeding disorders caused by absent or abnormal clotting factors. Deficiencies of vitamin C or K can also slow clotting and wound healing. Blood that clots too readily is also extremely dangerous. For example, platelets may snag on rough spots in blood vessel linings, producing a clot that may stay in place or travel in the bloodstream to another location. The obstruction may be deadly if it blocks blood flow to the lungs, brain, or heart.

30.2 MASTERING CONCEPTS

1. What are the components of blood?
2. What are the functions of white blood cells?
3. Describe the process of blood clotting.

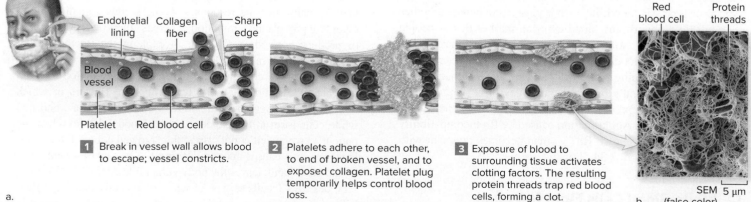

1 Break in vessel wall allows blood to escape; vessel constricts.

2 Platelets adhere to each other, to end of broken vessel, and to exposed collagen. Platelet plug temporarily helps control blood loss.

3 Exposure of blood to surrounding tissue activates clotting factors. The resulting protein threads trap red blood cells, forming a clot.

SEM 5 μm
b. (false color)

Figure 30.5 Blood Clotting. (a) A cut blood vessel immediately constricts. Platelets aggregate at the injured site. Proteins called clotting factors participate in a cascade of reactions, producing a meshwork of protein threads. (b) Red blood cells are trapped by protein threads in a clot.

Photo: (b): ©Steve Gschmeissner/Getty Images RF

30.3 Blood Circulates Through the Heart and Blood Vessels

The plasma, cells, and platelets that make up blood circulate throughout the body in an elaborate system of blood vessels, thanks to the relentless pumping of the heart (**figure 30.6**). The **cardiovascular system** is this entire transportation network (*cardio-* refers to the heart, *vascular* to the vessels).

Figure 30.6 shows the largest of the body's blood vessels, which are classified by size and the direction of blood flow. **Arteries** are large vessels that conduct blood away from the heart; the left half of the figure lists some of the body's major arteries. These branch into **arterioles,** smaller vessels that then diverge into a network of **capillaries,** the body's tiniest blood vessels.

As described in section 30.5, water and dissolved substances diffuse between each capillary and the **interstitial fluid,** the liquid that bathes the body's cells. The interstitial fluid, in turn, exchanges materials with the tissue cells.

To complete the circuit, capillaries empty into slightly larger vessels, called **venules,** which unite to form the **veins** that carry blood back to the heart. The right half of figure 30.6 lists some major veins.

Notice that in figure 30.6, vessels carrying oxygen-rich blood are red, and those carrying oxygen-poor blood are blue. This convention, coupled with the bluish appearance of blood vessels under lightly pigmented skin, has led to the misconception that the blood in veins is actually blue. In fact, blood is always red: It is bright red when oxygenated, and somewhat darker red when deoxygenated. Veins only appear blue because of the way that light of various wavelengths interacts with the skin. As for arteries, these blood vessels tend to be located in deeper tissues, far from the skin's surface. If arteries were visible through skin, they would appear blue, too.

30.3 MASTERING CONCEPTS

1. What is the cardiovascular system?
2. Describe the relationship among arteries, veins, arterioles, venules, and capillaries.

Figure 30.6 **Human Circulatory System.**

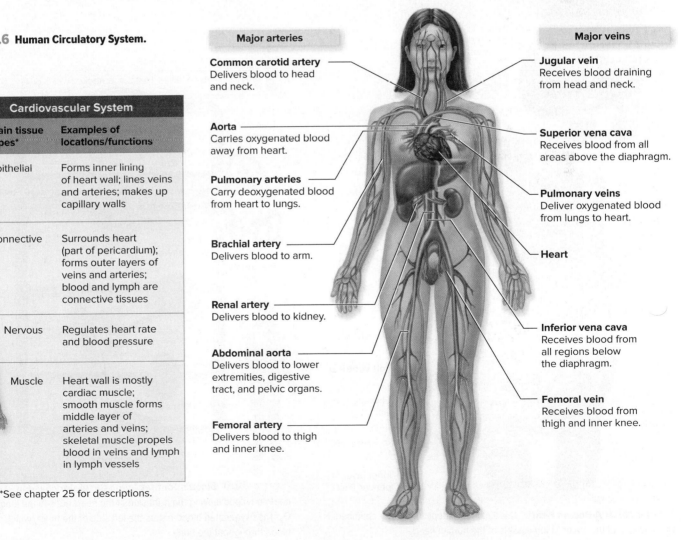

Cardiovascular System	
Main tissue types*	**Examples of locations/functions**
Epithelial	Forms inner lining of heart wall; lines veins and arteries; makes up capillary walls
Connective	Surrounds heart (part of pericardium); forms outer layers of veins and arteries; blood and lymph are connective tissues
Nervous	Regulates heart rate and blood pressure
Muscle	Heart wall is mostly cardiac muscle; smooth muscle forms middle layer of arteries and veins; skeletal muscle propels blood in veins and lymph in lymph vessels

*See chapter 25 for descriptions.

Major arteries

Common carotid artery
Delivers blood to head and neck.

Aorta
Carries oxygenated blood away from heart.

Pulmonary arteries
Carry deoxygenated blood from heart to lungs.

Brachial artery
Delivers blood to arm.

Renal artery
Delivers blood to kidney.

Abdominal aorta
Delivers blood to lower extremities, digestive tract, and pelvic organs.

Femoral artery
Delivers blood to thigh and inner knee.

Major veins

Jugular vein
Receives blood draining from head and neck.

Superior vena cava
Receives blood from all areas above the diaphragm.

Pulmonary veins
Deliver oxygenated blood from lungs to heart.

Heart

Inferior vena cava
Receives blood from all regions below the diaphragm.

Femoral vein
Receives blood from thigh and inner knee.

30.4 The Human Heart Is a Muscular Pump

Each day, the human heart sends a volume equal to more than 7000 liters of blood through the body, and it beats more than 2.5 billion times in a lifetime. This section explores the structure and function of the heart.

A. The Heart Has Four Chambers

Figure 30.7 illustrates the fist-sized human heart. The **pericardium** ("around the heart") is a tough connective tissue sac that encloses the heart and anchors it to surrounding tissues. This protective structure consists of two tissue layers; figure 30.7 shows only the inner layer. Thanks to lubricating fluid between the two layers of the pericardium, the heart is free to move, even during vigorous beating.

The heart's wall consists mostly of the **myocardium,** a thick layer of muscle. Contraction of the **cardiac muscle tissue** that makes up the myocardium provides the force that propels blood through arteries and arterioles. The innermost lining of the heart (and of all blood vessels) consists of **endothelium,** a one-cell-thick layer of simple squamous epithelium. ⓘ *epithelial tissue,* section 25.2A; *cardiac muscle,* section 25.2C

The human heart has four chambers: two upper atria and two lower ventricles. The atria are "primer pumps" that send blood to the ventricles, which pump the blood to the lungs or the rest of the body. Four heart valves ensure that blood moves in one direction. The two **atrioventricular valves (AV valves)** prevent blood from moving back into the atrium when a ventricle contracts; the two **semilunar valves** prevent backflow into the ventricles from the arteries leaving the heart.

B. The Right and Left Halves of the Heart Deliver Blood Along Different Paths

A schematic view of the circulatory system shows the pathway of blood as it travels to and from the heart (**figure 30.8**). The two largest veins in the body, the superior vena cava and the inferior

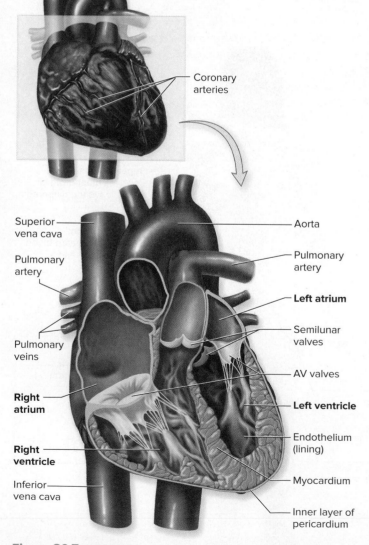

Figure 30.7 A Human Heart. This illustration depicts the four chambers, the valves, and the major blood vessels of the human heart.

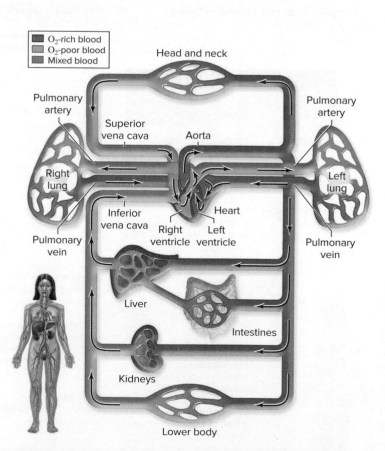

Figure 30.8 Blood's Journey in the Circulatory System. Oxygen-depleted blood leaving the right side of the heart goes to the lungs to pick up O_2. The oxygenated blood enters the left side of the heart, which pumps the blood throughout the body.

vena cava, deliver blood from the systemic circulation to the right atrium. From there, blood passes into the right ventricle and through the **pulmonary arteries** to the lungs, where blood picks up O_2 and unloads CO_2. The **pulmonary veins** carry oxygen-rich blood from the lungs to the left atrium of the heart, completing the pulmonary circuit.

The blood then flows from the left atrium into the left ventricle, the most powerful heart chamber. The massive force of contraction of the left ventricle sends blood into the **aorta,** the largest artery in the body. The blood then circulates throughout the body before returning to the veins that deliver blood to the right side of the heart. The systemic circuit is complete.

In some locations, circulating blood detours from the usual route of capillary to venule to vein. In a **portal system,** blood passes from capillaries into a vein that drains into a second set of capillaries before returning to the heart. For example, in the hepatic portal system, veins leaving the intestines diverge into a capillary bed in the liver, which breaks down toxins. (The hepatic portal vein is the blood vessel "bridge" connecting the intestines and liver in figure 30.8.) From there, veins return the blood to the heart. Another portal system delivers hormones from the hypothalamus to target cells in the anterior pituitary (see figure 28.4).

How does the heart muscle receive its blood supply? Blood does not seep from the heart's chambers directly to the myocardium. Instead, the **coronary arteries**—two vessels that branch off from the aorta—supply blood to the heart muscle (see figure 30.7). A vein entering the right atrium returns blood that has been circulating within the walls of the heart. Blockage of a coronary artery is the most common cause of a heart attack.

C. Cardiac Muscle Cells Produce the Heartbeat

A **cardiac cycle,** or a single beat of the heart, consists of the events that occur with each contraction and relaxation of the heart muscle.

Each heartbeat requires the forceful contraction of cardiac muscle in the myocardium. The sliding filament model of muscle contraction described in chapter 29 applies to cardiac muscle, just as it does to skeletal muscle. Unlike skeletal muscle, however, cardiac muscle does not require stimulation from motor neurons to contract.

Instead, cardiac muscle is "self-excitable"; many cardiac muscle cells contract in unison without input from the central nervous system. Cardiac muscle cells are interconnected, forming an almost netlike pattern (see figure 25.6). Wherever individual cardiac muscle cells meet, gap junctions spread synchronized waves of action potentials from cell to cell. ⓘ *gap junctions,* section 3.6A

Figure 30.9a shows how electrical activity in the heart triggers cardiac muscle contraction. The signal begins at the **sinoatrial (SA) node,** or **pacemaker,** a region of specialized cardiac muscle cells in the upper wall of the right atrium. The SA node sets the tempo of the beat (normally about 75 beats per minute). Each time the cells of the SA node fire, they stimulate the cardiac cells of the atria to depolarize and contract. This action pushes blood from the atria to the ventricles.

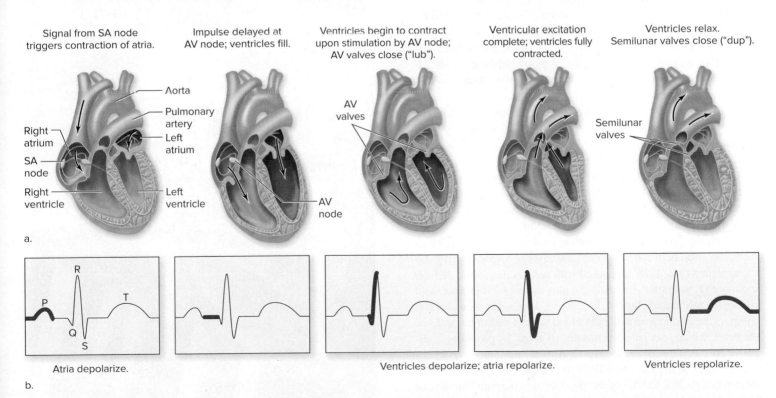

a.

b.

Figure 30.9 Heartbeat. (a) Signals from the sinoatrial (SA) node trigger contraction of the atria and spread to the AV node. After a brief delay, the AV node stimulates contraction of the ventricles. As the heart's valves close, they produce the "lub-dup" sound of the heartbeat. (b) An electrocardiogram (ECG) records the heart's electrical changes. The P wave and QRS complex correspond to the depolarization (contraction) of the atria and ventricles, respectively. The repolarization (relaxation) of the atria is "hidden" in the QRS complex, and the T wave corresponds to the repolarization of the ventricles.

The electrical impulses then race across the atrial wall to the **atrioventricular (AV) node,** a "relay station" in the wall of the lower right atrium. After a brief delay, which gives the ventricles time to fill, the AV node conducts electrical stimulation throughout the ventricle walls. The cardiac cells of the ventricles depolarize, contracting in unison. As a result, blood from the ventricles moves into the aorta and pulmonary artery.

Each depolarization of the atria and ventricles causes electrical changes that are detectable by electrodes placed on the skin. A device called an electrocardiograph amplifies and records these electrical changes, producing a chart called an electrocardiogram (ECG). Figure 30.9b shows a normal ECG pattern. An abnormal waveform may indicate damage to the heart muscle, ion imbalances, and other problems that can reduce heart function.

The familiar "lub-dup" sound of the beating heart comes from the two sets of heart valves closing, preventing blood from flowing backward during each contraction (see figure 30.9b). A heart murmur is a variation on the normal "lub-dup" sound, and it often reflects abnormally functioning valves.

D. Exercise Strengthens the Heart

After you circle the bases in a softball game, you may notice that your heart is beating faster than normal. The explanation for your elevated heart rate relates to the activity of your skeletal muscles, which require lots of ATP. Regenerating that ATP by aerobic respiration requires O_2; as we have already seen, one function of blood is to deliver this essential gas to the body's cells.

As you exercise, your heart meets the increased demand for O_2 by increasing its **cardiac output,** a measure of the volume of blood that the heart pumps each minute. Cardiac output is a function of the heart rate and the volume of blood pumped per stroke. Elevating the heart rate therefore quickly boosts cardiac output during an exercise session.

Just as regular workouts make skeletal muscles stronger, regular exercise also strengthens cardiac muscle. A stronger heart muscle, in turn, boosts the stroke volume. An active person can therefore pump the same amount of blood at 50 beats per minute as a sedentary person's heart pumps at 75 beats per minute. Former cyclist Miguel Indurain provides an extreme example of cardiovascular fitness; at the peak of his career, he reportedly had a resting heart rate of about 28 beats per minute.

Exercise provides several other cardiovascular benefits as well. The number of red blood cells increases in response to regular exercise, and these cells are packed with more hemoglobin, delivering more O_2 to tissues. Exercise can also lower blood pressure and reduce the amount of cholesterol in blood. Moreover, regular activity spurs the development of extra blood vessels within the walls of the heart, which may help prevent a heart attack by providing alternative pathways for blood to flow to the heart muscle.

To achieve the most benefit from exercise, the heart rate must be elevated to 70% to 85% of its "theoretical maximum" for at least half an hour three times a week. One way to calculate your theoretical maximum is to subtract your age from 220 (table 30.2). If you are 18 years old, your theoretical maximum is 202 beats per minute; 70% to 85% of this value is 141 to 172 beats per minute. Tennis,

skating, skiing, racquetball, vigorous dancing, hockey, basketball, biking, or brisk walking can elevate your heart rate to this level.

For some people, however, too much exercise may actually harm the heart. For example, years of training for marathons or other extreme endurance events may stiffen cardiac muscle and artery walls, boosting the chance of arrhythmias and other heart conditions (see the Apply It Now in section 30.5). However, scientists warn that this connection remains tentative and that extreme endurance athletes are generally at low risk for poor cardiovascular health. For now, the benefits of endurance training—even for long races—far outweigh the costs.

30.4   MASTERING CONCEPTS

1. Trace the pathway of an O_2 molecule from the lungs to a respiring cell at the tip of your finger.
2. Why is the heart sometimes called "two hearts that beat in unison"?
3. How does a heartbeat originate and spread?
4. How does exercise affect the circulatory system?

TABLE **30.2**   Target Heart Rates by Age

Age	Theoretical Maximum Heart Rate (beats per minute)	Target Heart Rate During Exercise (beats per minute)
20	200	140–170
25	195	137–166
30	190	133–162
40	180	126–153
50	170	119–145
60	160	112–136
70	150	105–128

©Masterfile RF

30.5 Blood Vessels Form the Circulation Pathway

As the heart's ventricles contract, they push blood to the lungs and the rest of the body. This section describes the system of vessels through which blood travels as it delivers nutrients and removes wastes.

A. Arteries, Capillaries, and Veins Have Different Structures

As we have already seen, arteries carry blood away from the heart, whereas veins return blood to the heart. Despite these opposite functions, the walls of arteries and veins share some similarities (figure 30.10). The outermost layer is a sheath of connective tissue. The middle layer is made mostly of **smooth muscle tissue,** and endothelium forms the innermost layer. (i) *smooth muscle tissue,* section 25.2C

One feature that characterizes arteries is the thick layer of smooth muscle. The muscular walls of major arteries can withstand the high-pressure surges of blood leaving the heart. Farther from the heart, as arteries branch into arterioles, their walls become thinner, and the outermost layer of connective tissue may taper away. Arterioles do retain a layer of smooth muscle that helps regulate blood pressure; section 30.5B describes how this occurs.

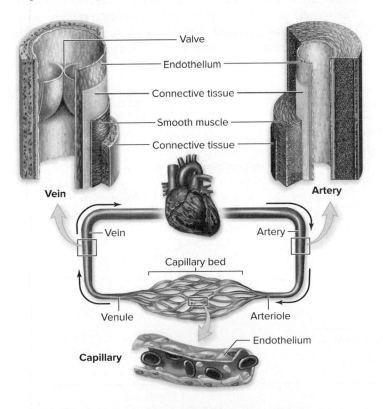

Figure 30.10 Types of Blood Vessels. The walls of arteries and veins consist of connective tissue, smooth muscle, and endothelium. Arteries, which are subject to the highest blood pressure, are much more muscular than veins. In some veins, valves keep blood moving toward the heart (against the force of gravity). The capillary wall consists only of endothelium.

Arterioles branch into **capillary beds,** networks of tiny blood vessels that connect an arteriole and a venule (figure 30.11). Capillaries are tiny but very numerous, providing extensive surface area where materials are exchanged with the interstitial fluid. Because their walls consist of a single layer of endothelial cells, nutrients and gases easily diffuse into and out of capillaries. (i) *diffusion,* section 4.5A

From the capillary beds, blood flows into venules, which converge into veins. These vessels receive blood at low pressure.

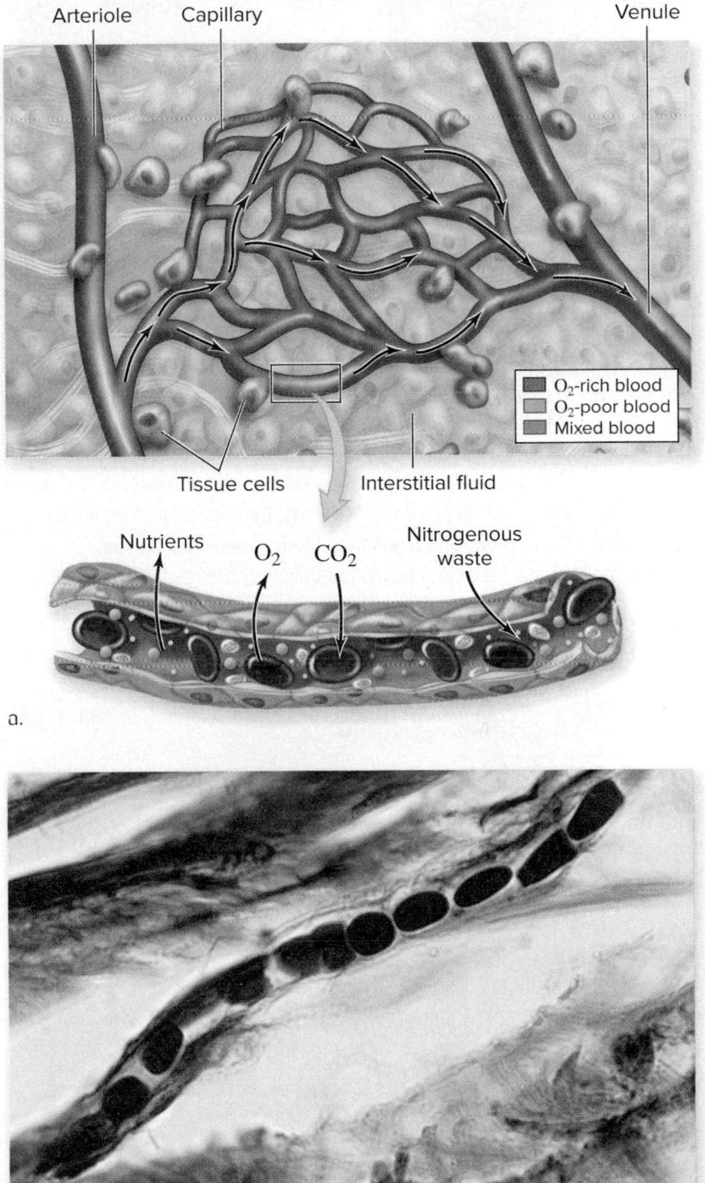

Figure 30.11 Capillaries. (a) A capillary bed is a network of tiny vessels that lies between an arteriole and a venule. At capillaries of the systemic circulation, O_2 and nutrients leave the bloodstream, and CO_2 and other wastes enter the circulation for later disposal at the lungs and kidneys. (b) Capillaries are so small that red blood cells move through them in single file.

(b): ©Ed Reschke/Peter Arnold/Getty Images

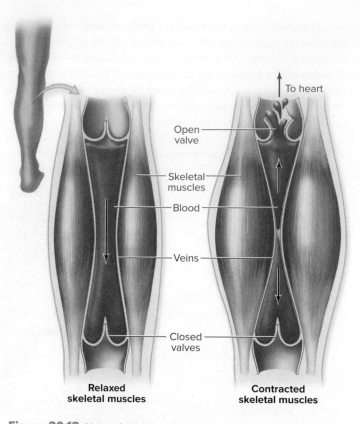

Figure 30.12 Valves in Veins. When skeletal muscles are relaxed, valves prevent blood from flowing backward in veins. Contracted skeletal muscles squeeze the veins, propelling blood through the open valves. In this illustration, the veins appear much larger than they are relative to real muscles.

The smooth muscle layer in their walls is much reduced or even absent (see figure 30.10); in fact, unlike an artery, a vein collapses when empty.

If pressure in veins is so low, what propels blood back to the heart, against the force of gravity? In many veins, flaps called venous valves keep blood flowing in one direction (**figure 30.12**). These valves are especially numerous in the legs. As skeletal muscles in the leg contract, they squeeze veins and propel blood through the open valves in the only direction it can move: toward the heart.

Varicose veins result in part from faulty venous valves. If the flaps of a valve do not completely block the downward flow of blood, the vein can no longer counter the force of gravity acting on the blood. As a result, blood accumulates in the veins of the lower legs. The walls of these distended blood vessels form prominent bulges under the skin.

B. Blood Pressure and Velocity Differ Among Vessel Types

A routine doctor's office visit always includes a blood pressure reading, a good indication of overall cardiovascular health. **Blood pressure** is the force that blood exerts on vessel walls. As the heart drives blood through the vessels, you can feel your blood pressure as a "pulse."

A device called a sphygmomanometer measures blood pressure in an artery in the upper arm, close to the heart (**figure 30.13**). The **systolic pressure,** or upper number in a blood pressure reading, reflects the powerful contraction of the ventricles. The **diastolic pressure,** or low point, occurs when the ventricles relax.

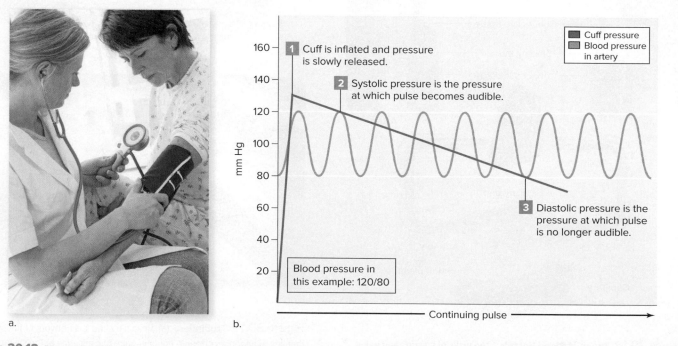

Figure 30.13 Measuring Blood Pressure. (a) The cuff of a sphygmomanometer is inflated until circulation to the lower arm is cut off. Then, as the cuff slowly deflates, the stethoscope detects the sound of returning blood flow. (b) The value on the gauge when thumping is first audible is the systolic blood pressure; this sound is the blood rushing through the arteries past the deflating cuff. The pressure reading when the sound ends is the diastolic blood pressure.

(a): ©BSIP/UIG/Getty Images

Blood pressure readings are in units of "millimeters of mercury," abbreviated mm Hg, because older sphygmomanometers measured the distance over which blood pressure could push a column of mercury. A typical blood pressure reading for a young adult is 110 mm Hg for the systolic pressure and 70 mm Hg for the diastolic pressure, expressed as "110 over 70" (written 110/70). "Normal" blood pressure, however, varies with age, sex, race, and other factors.

Blood pressure decreases with distance from the heart; that is, blood in arteries has the highest pressure, followed by capillaries and then veins (**figure 30.14**). Blood velocity, however, is lowest in the capillaries. The reason is that the total cross-sectional area of capillaries is much greater than that of the arteries or veins. Just as the velocity of a river slows as the water spreads out over a delta, so does the flow of blood slow as it is divided among countless tiny capillaries. This leisurely flow of blood allows adequate time for nutrients and wastes to diffuse across the capillary walls.

Past the capillaries, venules converge into veins. The total cross-sectional area of these blood vessels is again smaller than that of the capillaries. The resulting reduction in cross-sectional area helps speed blood flow back to the heart. To understand why, picture water flowing out of a hose. If you put your thumb over the nozzle, you reduce the area of the opening. What happens? The velocity of water through the nozzle increases.

Overall, a person's blood pressure reflects many factors, including blood vessel diameter, heart rate, and blood volume. The body regulates blood pressure over the long term by raising or lowering the volume of blood. Chapter 33 describes how the kidneys adjust the blood's volume by controlling the amount of fluid excreted in urine.

In the short term, blood vessel diameter and heart rate are under constant regulation by negative feedback (**figure 30.15**). Pressure receptors within the walls of major arteries detect blood pressure and pass that information to the medulla, in the brainstem. The medulla, via the autonomic nervous system, adjusts both heart rate and the diameter of arterioles to maintain homeostasis. ⓘ *negative feedback,* section 25.4; *autonomic nervous system,* section 26.5

The role of the arterioles deserves special mention. **Vasoconstriction** is the narrowing of blood vessels that results from the contraction of smooth muscle in arteriole walls. When arteriole diameter decreases, blood pressure rises. The opposite effect, **vasodilation,** is the widening of blood vessels that occurs when the same muscles relax. Altering arteriole diameter allows the body to increase blood delivery to regions that need it most. During physical activity, for example, skeletal muscles receive additional blood at the expense of organs not in immediate use, such as those in the digestive tract.

Blood pressure that is too low or too high can cause health problems. Hypotension, which is blood pressure that is significantly lower than normal, may cause fainting. At the opposite end

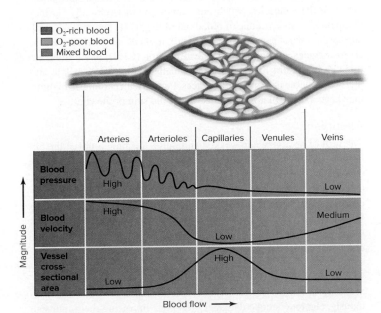

Figure 30.14 Blood Pressure and Velocity. Blood pressure drops with increasing distance from the heart. Blood moves rapidly in the aorta, but its velocity declines steadily as it moves through arteries, arterioles, and capillaries. Blood regains some of its velocity as venules converge into veins.

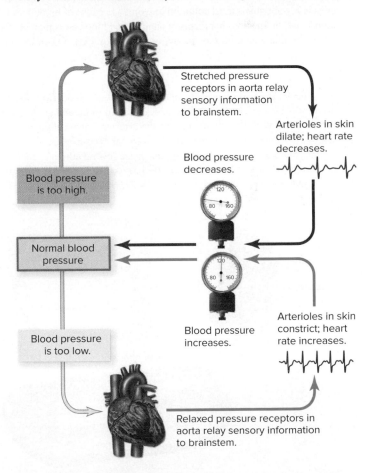

Figure 30.15 Regulation of Blood Pressure. A negative feedback loop regulates blood pressure in the short term. Stretch-sensitive neurons detect pressure in major arteries. If blood pressure climbs too high, signals from the central nervous system cause the blood vessels to dilate and the heart rate to slow. If blood pressure is too low, vessels constrict and the heart beats faster.

of the spectrum, consistently elevated blood pressure, or hypertension, may severely damage the circulatory system and other organs. High blood pressure affects 15% to 20% of adults residing in industrialized nations.

High blood pressure is just one example of illnesses that can affect the human circulatory system; this chapter's Apply It Now box considers several others.

Apply It **Now**

The Unhealthy Circulatory System

Anemias

Anemias are a collection of more than 400 disorders resulting from a decrease in the oxygen-carrying capacity of blood. One symptom is fatigue, reflecting a shortage of O_2 at the body's cells. Some types of anemia are inherited; a mutation in the gene encoding hemoglobin, for example, can cause an inherited form of anemia called sickle cell disease. Other types of anemia are related to diet, especially an iron deficiency, which can prevent cells from producing hemoglobin. Iron-deficiency anemia is most common in women because of blood loss in menstruation. In still other forms of anemia, red blood cells may be too small, be manufactured too slowly, or die too quickly. ⓘ *sickle cell disease,* section 7.7A

Atherosclerosis

Fatty deposits inside the walls of arteries reduce blood flow to the brain or heart muscle (**figure 30.A**). A diet high in fat and cholesterol is associated with this "hardening of the arteries," also called atherosclerosis (*athero-* is from the Greek word for "paste," and *sclerosis* means "hardness"). Atherosclerosis can cause several ailments, including strokes (blocked blood flow to the brain), heart attacks, arrhythmia, and aneurysms. ⓘ *bad and good cholesterol,* section 2.5E; *stroke,* section 26.6D

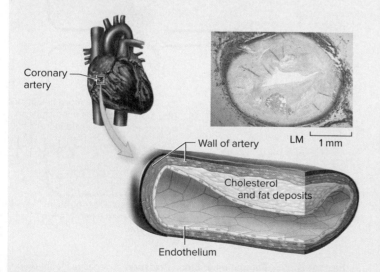

Figure 30.A   Atherosclerosis. Deposits of cholesterol and other fatty materials beneath the inner lining of an artery cause atherosclerosis, or "hardening of the arteries." Atherosclerosis in a coronary artery may cause a heart attack. Photo: ©R Roseman/CMSP/Getty Images

30.5   **MASTERING CONCEPTS**

1. Compare and contrast the structures of arteries, capillaries, and veins.
2. Explain why blood pressure is highest in the arteries and lowest in the veins.
3. How is the regulation of blood pressure an example of negative feedback?

Heart Attack

Blocked blood flow in a coronary artery prevents oxygen delivery to part of the heart muscle (see figure 30.7). Starved for oxygen, muscle cells die. This is a heart attack, and it may come on suddenly. A common treatment for a blocked coronary artery is a bypass operation. A surgeon creates a bridge around the blockage by sewing pieces of blood vessel taken from the patient's chest or leg onto the blocked artery. The procedure's name often includes the number of arteries repaired. A quadruple bypass operation, for example, bridges four obstructed arteries.

Arrhythmia

An arrhythmia is an abnormal heartbeat—that is, the heart beats too slowly, too fast, or irregularly. In mild cases, an arrhythmia may cause fluttering or racing that lasts a few seconds. Such conditions normally do not require treatment. A heart that beats too slowly is dangerous because the body's tissues may not receive enough blood. An electronic pacemaker implanted under the skin is a common treatment for a slow heartbeat. Another serious condition is a ventricular fibrillation, in which the ventricles twitch but cannot contract normally, so they pump only a small volume of blood out of the heart. The resulting sudden cardiac arrest can cause death within minutes. First aid for cardiac arrest includes cardiopulmonary resuscitation (CPR); in addition, external defibrillators that shock an erratically beating heart back to normal are available in many public places. People with chronic arrhythmia may have a defibrillator implanted under the skin of the chest.

Aneurysm

The wall of an artery can weaken and bulge, forming a pulsating, enlarging sac called an aneurysm. Aneurysms are dangerous because they may rupture without warning. Depending on the location and volume of bleeding, a burst aneurysm can be fatal; common locations include the aorta and the arteries of the brain. Aneurysms may result from atherosclerosis, a congenitally weakened area of an arterial wall, trauma, infection, persistently high blood pressure, or an inherited disorder such as Marfan syndrome.

The Effects of Smoking on Cardiovascular Health

Smoking is the most common preventable cause of death. Most famously, cigarette smoke damages the lungs, impairing their ability to deliver O_2 to blood (and, of course, increasing the chance of lung cancer). Tobacco's effects on the circulatory system are less familiar to most people. Nicotine increases heart rate and blood pressure, damages blood vessels, and stimulates the formation of blood clots, increasing the risk of stroke.

30.6 The Lymphatic System Maintains Circulation and Protects Against Infection

Capillary walls are typically rather porous. Red blood cells remain confined to capillaries, but many other components of blood can pass between the endothelial cells that make up capillary walls. How does the fluid that leaks out of these blood vessels return to the circulatory system? The answer is that the **lymphatic system** collects the fluid, removes bacteria, debris, and cancer cells, and returns the liquid to the blood (**figure 30.16**). The lymphatic system is therefore a bridge between the circulatory and immune systems.

Lymph, the colorless fluid of the lymphatic system, travels within **lymph capillaries**—tiny, dead-end vessels that absorb interstitial fluid from the spaces between cells. The composition of lymph is therefore similar to that of blood plasma, minus the proteins that are too large to leave blood capillaries.

The cells of a lymph capillary, however, are not joined as tightly together as those of a blood capillary. Lymph capillaries therefore also admit bacteria, viruses, cancer cells, and other large particles in body tissues. These capillaries then converge into larger lymph vessels that eventually empty into veins in the chest, where the fluid returns to the blood.

Along the way, lymph passes through **lymph nodes,** kidney-shaped organs that contain millions of white blood cells. These infection-fighting cells intercept and destroy cellular debris, cancer cells, and bacteria in the lymph flow. White blood cells can also migrate from lymph nodes into the lymph, which transports them to the blood.

White blood cells also occur in other organs of the lymphatic system. For example, one of the spleen's many functions is to produce, store, and release white blood cells. The thymus is an organ in which specialized white blood cells called T cells develop; these cells distinguish body cells from foreign cells.

Lymph flow is sluggish, because the lymphatic system has no pump. Instead, contractions of surrounding skeletal muscles and valves in lymph vessels help move the fluid. If lymph flow stops, excess fluid accumulates in the body's tissues and causes swelling (edema). This condition commonly occurs in immobile people. Parasites can also cause extreme edema. In an infectious disease called elephantiasis, nematode worms block the flow of lymph and cause grotesquely swollen tissues (see figure 21.17).

Along with the bloodstream, the lymphatic system can carry cancer cells that break off from tumors, seeding new tumors elsewhere. A biopsy of cancerous tissue therefore often includes a sample of nearby lymph nodes. If these lymph nodes are cancer-free, abnormal cells may not have begun to invade other tissues, improving the chance of successful treatment.

30.6 | MASTERING CONCEPTS

1. Where does lymph come from?
2. List the functions of each part of the lymphatic system.
3. How does lymph travel within lymph capillaries?

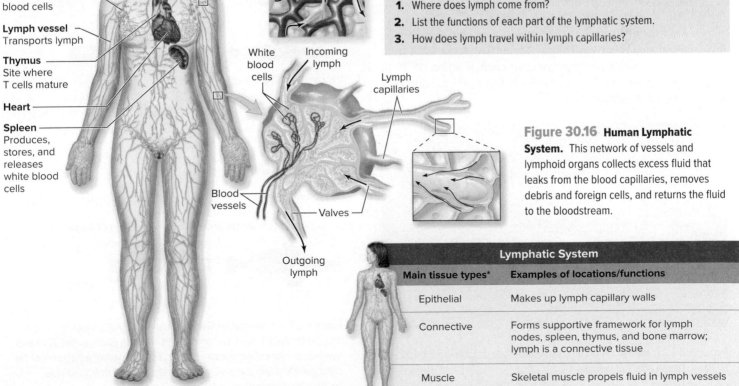

Lymph node
Contains white blood cells

Lymph vessel
Transports lymph

Thymus
Site where T cells mature

Heart

Spleen
Produces, stores, and releases white blood cells

Blood capillary   Lymph capillary
Interstitial fluid (between cells)

White blood cells   Incoming lymph
Lymph capillaries

Blood vessels
Valves

Outgoing lymph

Figure 30.16 Human Lymphatic System. This network of vessels and lymphoid organs collects excess fluid that leaks from the blood capillaries, removes debris and foreign cells, and returns the fluid to the bloodstream.

Lymphatic System	
Main tissue types*	Examples of locations/functions
Epithelial	Makes up lymph capillary walls
Connective	Forms supportive framework for lymph nodes, spleen, thymus, and bone marrow; lymph is a connective tissue
Muscle	Skeletal muscle propels fluid in lymph vessels

*See chapter 25 for descriptions.

INVESTIGATING LIFE

30.7 In (Extremely) Cold Blood

Sometimes, an obscure discovery can turn into a big story. So it was with a report that appeared in the journal *Nature* in 1954, when a researcher named J. T. Ruud confirmed the existence of a fish with colorless blood. The animal in Ruud's report was an icefish that lives deep in the extremely cold ocean waters surrounding Antarctica (**figure 30.17**). The blood of this fish is a ghostly white. Microscopic examination of the blood revealed white blood cells, but few if any red blood cells and never any hemoglobin. The story represented a biological curiosity because it contradicted the commonly held idea that all vertebrate life requires hemoglobin.

Since Ruud's time, researchers have described 16 species of icefishes, none of which has hemoglobin. Interestingly, icefishes have red-blooded relatives that share their frigid habitat. The "family tree" of these fishes is well understood, based on studies of ribosomal RNA, mitochondrial DNA, and many other traits. Could these relationships help researchers answer questions about the genetic origin of the icefishes' colorless blood?

Biologist Thomas J. Near, of Yale University, collaborated with Northeastern University's Sandra K. Parker and H. William Detrich III in an effort to learn more about icefish blood. They knew that decades of previous research had already established some basic facts about hemoglobin. For example, this protein consists of four polypeptides called globins (see figure 31.12); two of the chains are designated alpha, and the other two are called beta. Also, the genes encoding both alpha and beta globins had already been well studied in many vertebrates.

Near, Parker, and Detrich collected 44 specimens representing 33 fish species around the Antarctic. All 16 species of icefishes were included in the study, plus 17 species representing their red-blooded relatives. The researchers extracted DNA from the fishes' cells and used the polymerase chain reaction (PCR) to amplify the stretch of DNA where the globin genes are normally located. ⓘ *PCR,* section 11.2C

Figure 30.17 Icefish. A scuba diver approaches an icefish in the Southern Ocean near Antarctica.
©Rick Price/Corbis

After sequencing this stretch of DNA for all of the specimens, the team discovered that the fishes fell into three categories (**figure 30.18**). One group consisted of the red-blooded fishes with functioning genes encoding both alpha and beta globins. The second category included 15 of the 16 icefish species. All of these fishes had just a small fragment of the alpha globin gene; the beta globin gene was missing entirely. The third category included just one icefish species, *Neopagetopsis ionah,* which had a unique gene arrangement. In this animal's DNA, both the alpha and beta genes were mutated in a way that leaves the fish unable to produce hemoglobin. ⓘ *DNA sequencing,* section 11.2B

The genetic analysis clearly explained why icefishes have colorless blood: Unlike their red-blooded relatives, icefishes simply cannot produce hemoglobin.

But a surprising result emerged when the researchers mapped the gene configurations onto their previously existing phylogenetic tree (**figure 30.19**). The icefish with the unique globin genes, *N. ionah,* was not a "missing link" that gave rise to the other icefishes, as one might expect from its globin gene configuration. Instead, *N. ionah* was right in the middle of the family tree.

Why doesn't *N. ionah* have the same alpha globin gene remnant as every other icefish? The researchers proposed two possible explanations. Identical deletion mutations in the alpha and

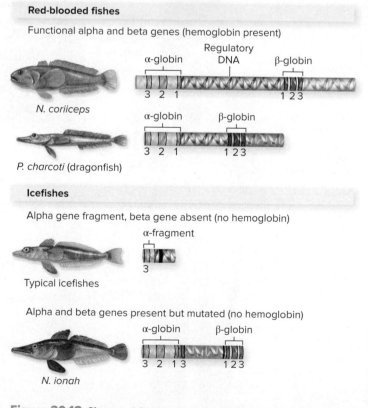

Figure 30.18 Shattered Genes. Analysis of DNA sequences revealed that red-blooded fishes have intact alpha and beta globin genes. No icefish, however, produces hemoglobin. Most icefishes have just a fragment of the alpha gene and lack the beta gene entirely. The icefish *N. ionah* has nonfunctional versions of both genes.

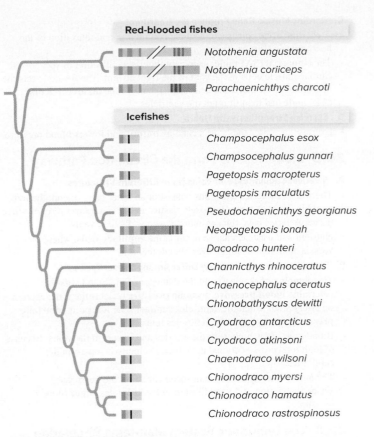

Red-blooded fishes

Notothenia angustata
Notothenia coriiceps
Parachaenichthys charcoti

Icefishes

Champsocephalus esox
Champsocephalus gunnari
Pagetopsis macropterus
Pagetopsis maculatus
Pseudochaenichthys georgianus
Neopagetopsis ionah
Dacodraco hunteri
Channicthys rhinoceratus
Chaenocephalus aceratus
Chionobathyscus dewitti
Cryodraco antarcticus
Cryodraco atkinsoni
Chaenodraco wilsoni
Chionodraco myersi
Chionodraco hamatus
Chionodraco rastrospinosus

Figure 30.19 Icefish Family Tree. The icefish with the unique globin gene cluster, *N. ionah,* is not ancestral to all other icefishes. (Only three of the 17 red-blooded species studied are shown.)

beta globin genes could have occurred four separate times in the 8 million years since icefishes diverged from their red-blooded relatives. The alternative explanation, which is perhaps more likely, is that *N. ionah*'s ancestors acquired the unusual hybrid hemoglobin gene by interbreeding with red-blooded relatives.

The story of the Antarctic icefishes reminds us that scientific inquiry is a journey. What began as a report of an unusual fish in 1954 has blossomed into an in-depth study of the evolution of hemoglobin. And although we now know exactly why icefishes have colorless blood, a new question has emerged: What accounts for *N. ionah*'s unusual globin genes? The answer may lie with future biologists—people whose desire to understand evolution is "in their blood."

Source: Near, Thomas J., Sandra K. Parker, and H. William Detrich III. 2006. A genomic fossil reveals key steps in hemoglobin loss by the Antarctic icefishes. *Molecular Biology and Evolution*, vol. 23, no. 11, pages 2008–2016.

30.7 MASTERING CONCEPTS

1. How did researchers use DNA evidence to determine why icefishes have colorless blood?
2. Which component of blood is likely to transport most of the oxygen in an icefish?

CHAPTER SUMMARY

30.1 Circulatory Systems Deliver Nutrients and Remove Wastes

- A **circulatory system** consists of blood or a similar fluid, a network of vessels, and a **heart** that pumps the fluid to the body cells. The fluid delivers nutrients and O_2, removes metabolic wastes, and transports other substances.

A. Circulatory Systems Are Open or Closed

- In an **open circulatory system,** fluid bathes tissues directly in open spaces before returning to the heart.
- In a **closed circulatory system,** such as that of vertebrates, the heart pumps blood through a continuous system of vessels.

B. Vertebrate Circulatory Systems Have Become Increasingly Complex

- A fish has a two-chambered heart, with an **atrium** that receives blood and a **ventricle** that pumps blood out.
- In other vertebrates, the **pulmonary circulation** delivers oxygen-depleted blood to the lungs, and the **systemic circulation** conveys freshly oxygenated blood to the rest of the body.
- Most land vertebrates have a three- or four-chambered heart.

30.2 Blood Is a Complex Mixture

- Human **blood** is a mixture of water, proteins and other dissolved substances, cells, and cell fragments (table 30.3).

A. Plasma Carries Many Dissolved Substances

- **Plasma** is the fluid component of blood; it transports all other blood components.

B. Red Blood Cells Transport Oxygen

- **Red blood cells** contain abundant **hemoglobin,** a pigment that binds O_2 molecules. Like other blood cells, these cells originate in red bone marrow.
- Surface markers on red blood cells react with antibodies in **agglutination** reactions that reveal a person's blood type.

C. White Blood Cells Fight Infection

- **White blood cells** provoke inflammation, destroy infectious organisms, and secrete antibodies. **Leukemia** is a type of cancer in which bone marrow produces too many white blood cells.

D. Blood Clotting Requires Platelets and Plasma Proteins

- **Platelets** are cell fragments that collect near a wound. Damaged tissue activates plasma proteins that trigger the formation of a network of fibers, trapping additional platelets and perpetuating **blood clot** formation.

30.3 Blood Circulates Through the Heart and Blood Vessels

- The heart drives blood through the vessels of the human **cardiovascular system.**
- **Arteries** are blood vessels that carry blood away from the heart. Arteries branch into smaller **arterioles,** which lead to tiny capillaries.

TABLE **30.3** Components of Blood: A Summary

Component	Function(s)
Plasma	Liquid component of blood; exchanges water and many dissolved substances with interstitial fluid
Red blood cells (erythrocytes)	Carry O_2
White blood cells (leukocytes)	Destroy foreign substances, initiate inflammation
Platelets	Cell fragments that initiate clotting

- **Capillaries** exchange materials with the **interstitial fluid** surrounding the body's cells.
- Capillaries empty into **venules,** which converge into the **veins** that return blood to the heart.

30.4 The Human Heart Is a Muscular Pump

A. The Heart Has Four Chambers

- A sac of connective tissue, the **pericardium,** surrounds the heart. **Cardiac muscle tissue** makes up most of the **myocardium,** the thickest portion of the heart wall. **Endothelium** lines the inside of the heart and all of the body's blood vessels.
- The heart has two atria that receive blood and two ventricles that propel blood throughout the body. The heart's two **atrioventricular (AV) valves** and two **semilunar valves** ensure one-way blood flow.

B. The Right and Left Halves of the Heart Deliver Blood Along Different Paths

- **Pulmonary arteries** and **pulmonary veins** transport blood between the right side of the heart and the lungs.
- Blood exits the left side of the heart at the **aorta,** the artery that carries blood toward the rest of the body.
- Blood typically flows from artery to arteriole, capillary bed, venule, vein, and back to the heart. In a **portal system,** however, blood passes from a capillary bed to a vein to a second set of capillaries before returning to the heart.
- **Coronary arteries** supply blood to the heart muscle itself.

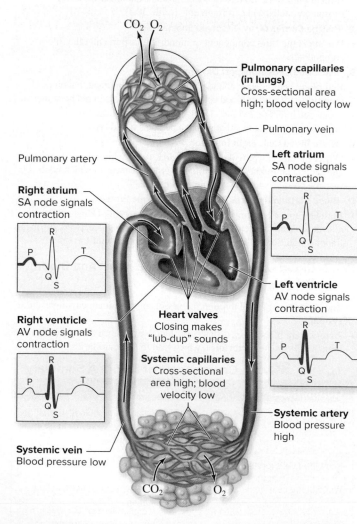

Figure 30.20 The Cardiovascular System: A Summary.

C. Cardiac Muscle Cells Produce the Heartbeat

- A **cardiac cycle** consists of a single contraction and relaxation of the heart muscle (figure 30.20).
- The **sinoatrial (SA) node,** or **pacemaker,** is a collection of specialized cardiac muscle cells in the wall of the right atrium. The SA node sets the heart rate. From there, the heartbeat spreads to the **atrioventricular (AV) node** and then through the ventricles.

D. Exercise Strengthens the Heart

- Exercise increases the heart's **cardiac output** and lowers blood pressure.

30.5 Blood Vessels Form the Circulation Pathway

A. Arteries, Capillaries, and Veins Have Different Structures

- The walls of arteries and veins consist of an inner layer of endothelium, a middle layer of **smooth muscle tissue,** and an outer layer of connective tissue. Arteries have thicker, more elastic walls than veins.
- Nutrient and waste exchange occur at the **capillary beds,** where blood vessels consist of a single layer of endothelium.

B. Blood Pressure and Velocity Differ Among Vessel Types

- The pumping of the heart and the diameter of the blood vessels determine **blood pressure. Systolic pressure** reflects the force exerted on artery walls when the ventricles contract. The low point, **diastolic pressure,** occurs when the ventricles relax.
- Blood pressure is highest in the arteries and lowest in the veins. Because of high total cross-sectional area, blood velocity is lowest in the capillaries.
- The autonomic nervous system speeds or slows the heart rate. **Vasoconstriction** and **vasodilation** in the arterioles adjust blood pressure.

30.6 The Lymphatic System Maintains Circulation and Protects Against Infection

- The **lymphatic system** includes a network of **lymph capillaries** that collect fluid from the body's tissues.
- **Lymph nodes** cleanse the resulting **lymph,** which subsequently returns to the bloodstream.

30.7 Investigating Life: In (Extremely) Cold Blood

- Antarctic icefishes are unique among vertebrates in that they lack hemoglobin. Researchers have traced this unusual characteristic to major deletions in the alpha and beta hemoglobin genes.

MULTIPLE CHOICE QUESTIONS

1. What is the advantage of a four-chambered heart?
   a. It uses the least energy of any type of heart.
   b. It maximizes the amount of O_2 reaching tissues.
   c. It enhances the mixing of blood from the pulmonary and systemic circulation.
   d. Both a and c are correct.

2. What component of blood is matched correctly with one of its functions?
   a. Red blood cell: Initiates clotting
   b. White blood cell: Initiates inflammation
   c. Platelet: Carries oxygen and carbon dioxide
   d. Plasma: Destroys foreign substances

3. Which of the following blood transfusions would be successful?
   a. Transfusing type B blood into a person with type AB blood
   b. Transfusing type AB blood into a person with type A blood
   c. Transfusing type A blood into a person with type O blood
   d. Transfusing type B blood into a person with type A blood

4. Which chamber of the human heart collects the oxygenated blood from the lungs?
   a. Left atrium
   b. Left ventricle
   c. Right atrium
   d. Right ventricle

5. How would the sinoatrial node of an athlete at rest differ from that of a sedentary person?
   a. It would establish a higher rate of contraction.
   b. It would not be any different.
   c. It would establish a lower rate of contraction.
   d. It would trigger a stronger contraction.

6. Rank the following blood vessels from highest blood pressure to lowest blood pressure. Which type of vessel is *third* in your list?
   a. A capillary
   b. An artery
   c. A vein
   d. An arteriole

7. How might the body maintain homeostasis if blood pressure rises too high?
   a. By boosting the volume of blood
   b. By increasing the heart rate
   c. By dilating the blood vessels near the skin
   d. By raising the diastolic pressure

8. How is lymph related to blood plasma?
   a. Both contain white blood cells.
   b. Both are fluids that are propelled by the heart.
   c. Both are types of interstitial fluid.
   d. Both a and c are correct.

Answers to these questions are in appendix A.

WRITE IT OUT

1. How are open and closed circulatory systems similar? How are they different?

2. Some athletes turn to blood doping to gain an unfair competitive advantage. For example, they may take supplements of erythropoietin (EPO), a hormone that stimulates red blood cell production. Why would increasing the number of circulating red blood cells help an athlete? What might be the dangers of having too many red blood cells?

3. People infected with HIV eventually develop a diminished white blood cell count. When examining a blood sample under the microscope, would the blood cell composition of an AIDS patient look different from that of an unaffected individual? Why or why not? [*Hint:* Consult figure 30.3.]

4. Referring to figure 30.4, speculate about which blood type is considered the "universal donor" and which is called the "universal recipient." Explain your answer.

5. One effect of aspirin is to prevent platelets from sticking together. Why do some people take low doses of aspirin to help prevent a heart attack?

6. Describe the path of blood through the heart's chambers and valves, and through the pulmonary and systemic circulations.

7. Describe the events that occur during one cardiac cycle.

8. What is the function of heart valves? Why might a person with a leaky heart valve feel short of breath?

9. Make a chart that compares systemic arteries, capillaries, and systemic veins. Consider the following properties: structure; amount of smooth muscle; presence of valves; cross-sectional area; blood pressure; blood velocity; direction of blood flow relative to the heart; O_2 content of blood.

10. Endothelial cells lining the heart and blood vessels are the only cells to receive nutrients and O_2 directly from blood. How do the rest of the body's cells receive these resources and dispose of wastes?

11. Explain why a doctor listens at your inner elbow with a stethoscope when measuring your blood pressure.

12. What change in blood vessel diameter would raise blood pressure?

13. The carotid artery extends from the heart to the head. Some of the body's blood pressure receptors are located in the carotid sinus, where the carotid artery passes through the neck. If you press lightly on the carotid sinus, what do you predict should happen to your heart rate? What if you press lightly on a spot just *below* the carotid sinus? [*Hint:* Figure 30.15 may help you answer this question.]

14. Where does lymph originate? What propels lymph through the lymph vessels? How is the lymphatic system connected with the circulatory system? The immune system?

15. Describe the interactions between the circulatory system and the respiratory, immune, digestive, and endocrine systems.

16. Name three ways that the circulatory system helps maintain homeostasis.

17. How is the human cardiovascular system similar to and different from the vascular tissue in a plant?

18. Use the Internet to learn more about disorders of the cardiovascular or lymphatic system. Choose one to investigate in more detail. What causes the disease you chose? Who is affected, and what are the consequences? Are there ways to prevent, treat, or cure the disease?

PULL IT TOGETHER

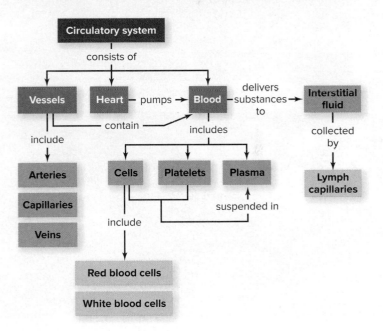

Figure 30.21 **Pull It Together: The Circulatory System.**

Refer to figure 30.21 and the chapter content to answer the following questions.

1. Review the Survey the Landscape figure in the chapter introduction, and then add *homeostasis, respiratory system,* and *digestive system* to the Pull It Together concept map.

2. How do pulmonary and systemic circulation fit into this concept map?

3. Connect *red blood cells* and *white blood cells* with *lymph capillaries* and the three types of blood vessels, using phrases that indicate whether each cell type typically occurs in each vessel.

31 The Respiratory System

Holding Time. Magician David Blaine is among the world's elite breath-holders. He is shown here during a performance in the Philippines.

©Mark Fredesjed Cristino/Demotix/Corbis

LEARN HOW TO LEARN
How to Use a Tutor

Your school may provide tutoring sessions for your class, or perhaps you have hired a private tutor. How can you make the most of this resource? First, meet regularly with your tutor for an hour or two each week; don't wait until just before an exam. Second, if possible, tell your tutor what you want to work on before each session, so that he or she can prepare. Third, take your textbook, class notes, and questions to your tutoring session. Fourth, be realistic. Your tutor can discuss difficult concepts and help you practice with the material, but don't expect him or her to simply give you the answers to your homework.

Twenty-One Minutes . . . and Counting

Suppose your relaxing day at the pool turns competitive when your friend challenges you to a breath-holding contest. You agree, take a deep breath, and dive in. While you're under water, oxygen gas (O_2) in the air you just breathed crosses from the lungs into the blood. Red blood cells carry O_2 to body cells, which use it along with glucose to produce ATP in aerobic respiration. Meanwhile, respiring cells release carbon dioxide (CO_2) into the blood, which travels back to the lungs. Soon, the urge to breathe becomes too strong, and you surface to see if you've won.

Some people take this friendly game to an extreme. Competitive breath-holding tests the limits of the human body. Intense training of the body and the mind results in extraordinary accomplishments—the world record for underwater breath-holding is an astonishing 21 minutes! Many of us can't even last a minute without taking a breath.

How do they do it? A deep understanding of both circulation and respiration helps elite breath-holders tip the competitive balance in their favor. First, they breathe pure O_2 for several minutes before submerging, packing their blood with O_2 reserves. Second, the competitions occur in cold water. The chilly surroundings initiate an O_2-conserving response called the "diving reflex." The heart rate slows, and blood is shunted away from the limbs and toward the heart and brain.

However, breath-holding is not just fun and games. Even during the diving reflex, muscle cells in the extremities continue to produce ATP, releasing CO_2 that is carried away only slowly by the sluggish blood flow. CO_2 accumulation at these tissues produces painful muscle cramps and acidifies the blood.

Recreational breath-holders may use misguided strategies to get an edge on the competition. Rather than breathing pure O_2 before going underwater, as a pro would do, people attempting long breath-holds at a pool might prepare by hyperventilating—that is, by taking several quick, deep breaths. Although the intention is to maximize the amount of O_2 entering the body, rapid breathing has the opposite effect. Only minimal extra O_2 enters the blood, but CO_2 is quickly expelled. The consequence may be disastrous. Because the brain uses the concentration of CO_2 in blood as a guide for adjusting breathing rate, artificially lowering CO_2 before the contest may allow the diver to fight the urge to breathe for too long. Thanks to the depletion of O_2, he or she may then lose consciousness underwater and drown.

No one knows whether competitive breath-holders risk long-term brain damage or other health consequences. It is clear, however, that attempting underwater breath-holding without proper monitoring can result in fatal blackouts—a stark reminder of our unyielding need for oxygen.

SURVEY THE LANDSCAPE
Animal Life

Cellular respiration consumes O_2 as it produces ATP. The lungs—the largest organs of the respiratory system—acquire this essential gas. They also dispose of CO_2, the gaseous waste product of respiration.

For more details, study the Pull It Together feature in the chapter summary.

31.1 Gases Diffuse Across Respiratory Surfaces

Each of us breathes some 20,000 times a day. Most of the time, you inhale and exhale without thinking—unless you happen to be using your breath to fog a mirror, spin a pinwheel, play the trumpet, or make soap bubbles (figure 31.1). Fun aside, breathing is obviously a vital function; if a person is deprived of air, death can occur within minutes.

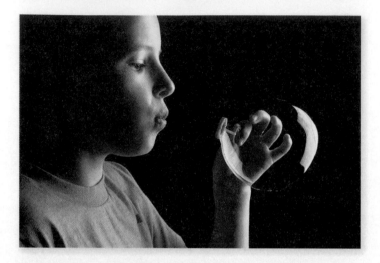

Figure 31.1 Blowing Bubbles. This girl is using air exhaled from her lungs to make a soap bubble.
©Photodisc/Getty Images RF

Breathing is so automatic that it is easy to forget why we do it. Cells use ATP to power protein synthesis, movement, DNA replication, cell division, growth, reproduction, and countless other activities that require energy. As described in chapter 6, animal cells generate ATP in **aerobic respiration,** which consumes oxygen gas (O_2) and generates carbon dioxide gas (CO_2) as a waste product. In most animals, the **respiratory system** is the organ system that does the breathing; it works with the circulatory system (the topic of chapter 30) to deliver O_2 to cells and to eliminate the CO_2 waste. Without gas exchange, cells die. ⓘ *ATP,* section 4.3

The term *respiration* also has two additional meanings, besides its use in aerobic cellular respiration. One is breathing (ventilation), the physical movement of air into and out of the body. Respiration can also mean the act of exchanging gases. External respiration is gas exchange between an animal's body and its environment; internal respiration is gas exchange between tissue cells and the bloodstream.

An animal's **respiratory surface** is the area of its body where external respiration occurs. In humans and other terrestrial vertebrates, the respiratory surface is in the lungs, but other animals use different surfaces (figure 31.2). All of these structures share three characteristics. First, the respiratory surface must come into contact with air or water. Second, each respiratory surface consists of a moist membrane across which O_2 and CO_2 diffuse. Third, its surface area must be relatively large. Not only must the respiratory surface meet the animal's demand for O_2, but it must also eliminate CO_2 fast enough to keep this waste gas from accumulating to toxic levels. ⓘ *diffusion,* section 4.5A

Figure 31.2 Respiratory Surfaces.
Gas exchange may occur (a) across the body surface or via specialized respiratory organs such as (b) tracheae, (c) gills, or (d) lungs.

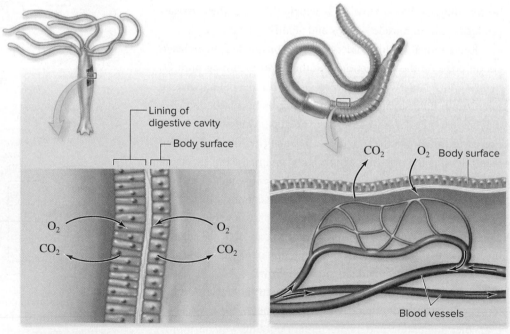

Lining of digestive cavity
Body surface
O_2
CO_2
O_2
CO_2
CO_2 O_2 Body surface
Blood vessels

a. **Body surface:** Many invertebrates and amphibians; gases diffuse across the body's surface.

As you examine figure 31.2, note that either air or water can be a source of O_2 and a "dumping ground" for CO_2. In general, air offers two advantages over water. Air has a higher concentration of O_2 than does water; in addition, air is lighter than water. Less energy is therefore required to move air across a respiratory surface than to move an equal volume of water. However, the requirement that the respiratory surface remain moist puts air-breathing animals at a disadvantage: A respiratory surface exposed to air may dry out, rendering it useless.

This section describes a variety of respiratory surfaces, ranging from the relatively unspecialized to the complex. It may be helpful to refer back to figure 31.2 as you read through the rest of this section. In each case, note that O_2 diffuses from where it is more concentrated (in air or water) to where it is less concentrated (inside the animal body); CO_2 follows its own concentration gradient in the opposite direction. For both gases, maximizing the concentration gradient also maximizes the rate of gas exchange.

A. Some Invertebrates Exchange Gases Across the Body Wall or in Internal Tubules

In animals that are very small or very flat (or both), all cells are close to the external environment. Gas exchange occurs across the moist body wall, as gases simply diffuse into and out of each cell. For example, cnidarians such as *Hydra* and sea anemones have thin, extended body parts with a large surface area for gas exchange. Likewise, flatworms are thin enough for gas exchange and distribution to occur without specialized respiratory or circulatory systems. ⓘ *cnidarians,* section 21.3; *flatworms,* section 21.4

In larger animals, the body's external surface area is insufficient to exchange gases with cells in deeper tissues. These animals often have a circulatory system that transports gases between interior cells and the skin. In an earthworm, for example, blood vessels connect the body's surface with deeper tissue layers. O_2 from the surrounding air diffuses across the skin and into the blood vessels; CO_2 moves in the opposite direction, out of the blood and into the atmosphere. ⓘ *annelids,* section 21.6

In insects and many other arthropods, the circulatory system plays only a minor role in the transportation of gases. These animals have **tracheae,** which are internal, air-filled tubes that connect to the atmosphere through openings, called spiracles, along each side of the abdomen. Inside the animal, tracheae branch into tiny tubules that extend around individual cells. Tracheae therefore bring the outside environment close to every cell of the body. ⓘ *insects,* section 21.8B

Chitin, the material that makes up the exoskeleton, strengthens the walls of the trachcae, but these tubules could not withstand the crushing weight of a very large arthropod. Tracheal breathing may therefore be one factor that constrains insect size; the largest insects are only about 15 centimeters long, and most are much smaller.

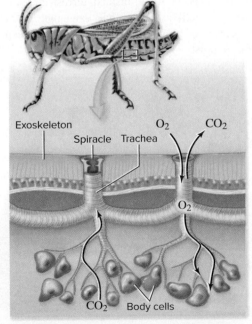

b. **Tracheae:** Many arthropods, including insects; gases diffuse between tracheae and body cells without first entering capillaries.

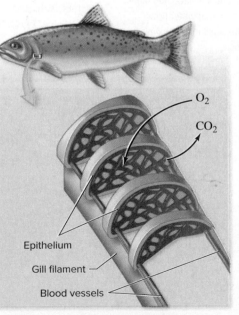

c. **Gills:** Fishes, amphibian larvae, and many invertebrates; gases are exchanged with water at blood vessels covered by a thin layer of epithelium.

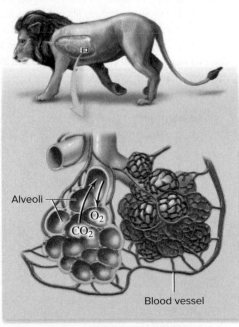

d. **Lungs:** Terrestrial vertebrates; blood vessels exchange gases with air across a moist, thin layer of epithelium inside the body.

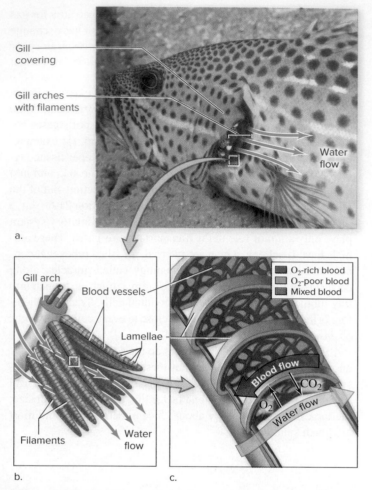

Gill covering

Gill arches with filaments

Water flow

a.

Gill arch

Blood vessels

Lamellae

Filaments

Water flow

b.

O₂-rich blood
O₂-poor blood
Mixed blood

Blood flow

CO₂

O₂

Water flow

c.

Figure 31.3 Fish Gills.   (a) Water flows across the feathery gills of a fish. (b) Each gill arch supports many filaments. (c) A filament is divided into many platelike lamellae that house blood vessels. The direction of water flow across the lamellae opposes that of blood flow in the capillaries, maximizing the exchange of O₂ and CO₂ between water and blood.

Photo: (a): ©Reinhard Dirscherl/Alamy

B. Gills Exchange Gases with Water

Most aquatic organisms have **gills,** highly folded structures containing blood vessels that exchange gases with water across a thin layer of epithelium. Gills occur in some invertebrates, such as mollusks, and in vertebrates such as fishes and amphibians.

A bony fish's gills work by a mechanism called **countercurrent exchange,** in which two adjacent fluids flow in opposite directions and exchange materials with each other (chapter 33 describes other examples of countercurrent exchange). Examine the gills illustrated in **figure 31.3.** Each gill arch is fringed with many elongated filaments, each consisting of platelike lamellae. The lamellae house dense networks of capillaries, the tiniest of blood vessels. In figure 31.3c, water is flowing from left to right across the gill membrane. Blood, however, flows through the capillaries in the opposite direction. This arrangement maximizes gas exchange because the concentration gradient favors O₂ diffusion from water to blood along the entire length of the capillary bed. Meanwhile, CO₂ diffuses from blood to water. ⓘ *fishes,* section 21.13; *capillaries,* section 30.5

The aquatic larvae of amphibians have gills that extend into water from stalks on the side of the head. During metamorphosis, the animal changes into an air-breathing adult, and the gills typically disappear as lungs develop. Some salamanders, however, retain external gills through adulthood. ⓘ *amphibians,* section 21.14

C. Terrestrial Vertebrates Exchange Gases in Lungs

As vertebrates ventured onto the land hundreds of millions of years ago, gills became useless; these delicate, feathery structures simply dry up when exposed to air. Instead, terrestrial habitats selected for **lungs,** saclike organs that exchange gases (**figure 31.4**). The location of lungs inside the body helps keep the respiratory surfaces moist.

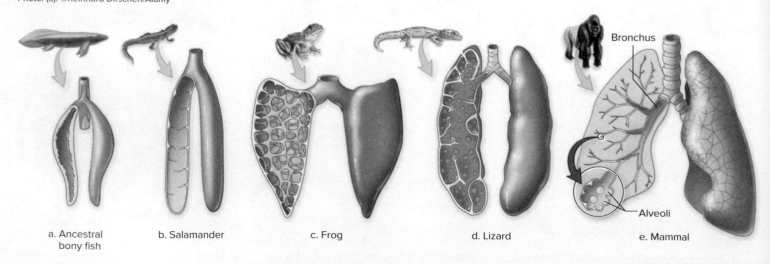

Bronchus

Alveoli

a. Ancestral bony fish

b. Salamander

c. Frog

d. Lizard

e. Mammal

Figure 31.4 Lung Evolution.   (a) Early bony fishes had paired air sacs with capillary networks in their smooth walls. (b) Lungs in salamanders have a few partitions. (c) Lungs become more extensively subdivided in frogs and toads. (d) The lizard lung is even more clearly compartmentalized. (e) The mammalian lung has the most surface area of all, thanks to millions of microscopic air sacs called alveoli.

Simple lungs were already present in the earliest bony fishes. These sacs allowed fishes to gulp air, supplementing gas exchange through gills when water levels became low (see figure 31.4a). The swim bladder that controls buoyancy in most modern fishes is derived from these structures. But in the ancestors of terrestrial vertebrates, the paired sacs developed into much more elaborate organs. ⓘ *lungfishes,* section 21.13B

A look at the lungs of modern vertebrates gives an idea of how these organs have increased in complexity. The lungs of salamanders are little more than air-filled pouches. The lungs of their amphibian relatives, the frogs and toads, are somewhat partitioned, an adaptation that increases the surface area available for external respiration (see figure 31.4b and c). Amphibians also exchange gases through moist skin and the lining of the mouth. The lungs of lizards are much more extensively subdivided (see figure 31.4d), reflecting the many adaptations of these animals to dry land. ⓘ *reptiles,* section 21.15

The mammalian lung's surface area is truly immense; millions of tiny air sacs sit within baskets of capillaries (see figure 31.4e). Mammals have a high metabolic rate, both because they maintain a constant body temperature and because they tend to be very active. The mammalian lung provides the enormous respiratory surface necessary to support this high metabolic rate. (Figure 30.2 shows corresponding changes in vertebrate circulatory systems.) ⓘ *mammals,* section 21.16

Lungs did, however, introduce a new challenge: moving air in and out to renew the O_2 supply and expel CO_2. Vertebrates accomplish this task in a variety of ways (figure 31.5). For example, frogs force air into their lungs under positive pressure and then expel it by contracting muscles in the body wall. In contrast, birds have a respiratory system through which air flows in one direction. O_2-rich air therefore never mixes with O_2-depleted air; moreover, O_2 remains in the respiratory system even during exhalation. Thanks to this arrangement, a bird's respiratory system is extremely efficient, delivering the O_2 required to support the enormous metabolic demands of flight. ⓘ *birds,* section 21.15B

The lungs of mammals, including humans, move air in yet another way: A muscular diaphragm expands the chest area, so air flows passively into the lungs under negative pressure. Air is expelled as the diaphragm relaxes (see section 31.3).

31.1 MASTERING CONCEPTS

1. What is the main function of the respiratory system?
2. What is the relationship between the circulatory and respiratory systems?
3. Differentiate between aerobic cellular respiration, internal respiration, and external respiration.
4. An earthworm prefers moist soil. Why does this animal die if it dries out on a sidewalk?
5. How do an animal's size, activity level, and environment influence the structure and function of its respiratory surface?

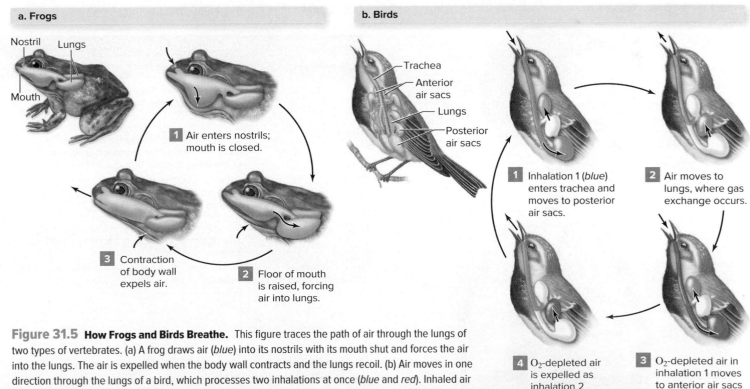

a. Frogs

Nostril Lungs

Mouth

1 Air enters nostrils; mouth is closed.

3 Contraction of body wall expels air.

2 Floor of mouth is raised, forcing air into lungs.

b. Birds

Trachea

Anterior air sacs

Lungs

Posterior air sacs

1 Inhalation 1 (*blue*) enters trachea and moves to posterior air sacs.

2 Air moves to lungs, where gas exchange occurs.

3 O_2-depleted air in inhalation 1 moves to anterior air sacs as inhalation 2 (*red*) enters trachea.

4 O_2-depleted air is expelled as inhalation 2 moves to lungs.

Figure 31.5 How Frogs and Birds Breathe. This figure traces the path of air through the lungs of two types of vertebrates. (a) A frog draws air (*blue*) into its nostrils with its mouth shut and forces the air into the lungs. The air is expelled when the body wall contracts and the lungs recoil. (b) Air moves in one direction through the lungs of a bird, which processes two inhalations at once (*blue* and *red*). Inhaled air enters the trachea and goes first to the posterior air sacs, then passes through the lungs and anterior air sacs before being released. Compare these animals to the human in figure 31.9.

31.2 The Human Respiratory System Delivers Air to the Lungs

The human respiratory system is a continuous network of tubules that delivers O_2 to the circulatory system and unloads CO_2 into the lungs. **Figure 31.6** presents an overview of the system, and **table 31.1** summarizes its functions.

A. The Nose, Pharynx, and Larynx Form the Upper Respiratory Tract

The **nose,** which forms the external entrance to the nasal cavity, functions in breathing, immunity, and the sense of smell. Stiff hairs at the entrance of each nostril keep dust and other large particles out. If a large particle is inhaled, a sensory cell in the nose may signal the brain to orchestrate a sneeze, which forcefully ejects the object. ⓘ *sense of smell,* section 27.3A

Epithelial tissue in the nose secretes a sticky mucus that catches most airborne bacteria and dust particles that get past the hairs. Enzymes in the mucus destroy some of the would-be invaders, and immune system cells under the epithelial layer await any disease-causing organisms that penetrate the mucus.

The nasal cavity also adjusts the temperature and humidity of incoming air. Blood vessels lining the nasal cavity release heat, and mucus contributes moisture to the air. This function ensures that the respiratory surface of the lungs remains moist.

The back of the nose and mouth leads into the **pharynx,** or throat. Like the mouth, the pharynx is part of the digestive and respiratory systems, since both swallowed food and inhaled air pass through the pharynx.

TABLE 31.1 Functions of the Human Respiratory System: A Summary

Function	Explanation
Gas exchange	Lungs exchange oxygen (O_2) and carbon dioxide (CO_2) with blood.
Olfaction (sense of smell)	Breathing moves air across the nose's olfactory epithelium, which detects odors.
Production of sounds, including speech	Movement of air across the vocal cords in the larynx produces sounds.
Maintenance of homeostasis (blood pH)	Breathing volume and rate determine the concentration of CO_2 in blood, which affects blood pH.

Just below and in front of the pharynx is the **larynx,** or Adam's apple, a boxlike structure that produces the voice. Stretched over the larynx are the **vocal cords,** two elastic bands of tissue that vibrate as air from the lungs passes through a slit-like opening called the **glottis.** Vibrations of the vocal cords produce the sounds of speech. A male's voice becomes deeper during puberty because the vocal cords grow longer and thicker. The cords therefore vibrate more slowly during speech, producing lower-frequency sounds that we perceive as a deeper voice.

Another function of the larynx is to direct chewed food toward the esophagus—the tube leading from the mouth to the stomach—and away from the respiratory system. During swallowing, a cartilage flap called the **epiglottis** covers the glottis so that food enters the esophagus, not the lungs.

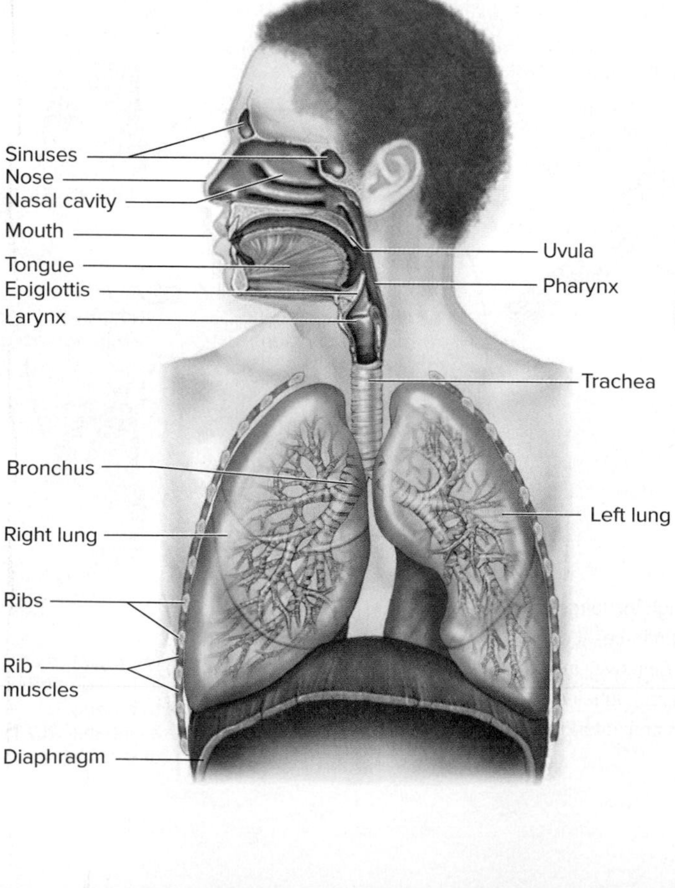

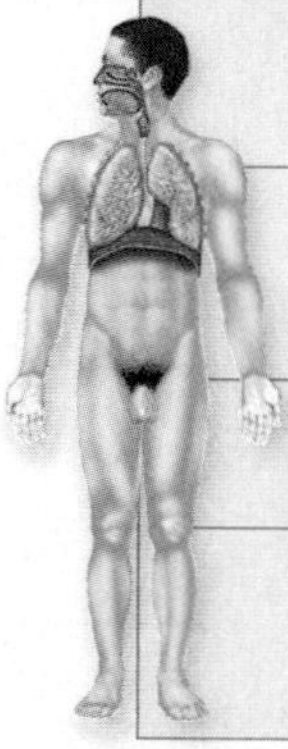

Respiratory System	
Main tissue types*	**Examples of locations/functions**
Epithelial	Enables diffusion across walls of alveoli and capillaries; secretes mucus along respiratory tract.
Connective	Blood (a connective tissue) exchanges gases with lungs; cartilage makes up part of the nose, trachea, bronchi, and larynx.
Nervous	Autonomic nervous system controls smooth muscle in bronchi.
Muscle	Smooth muscle in lungs regulates airflow to alveoli; skeletal muscle in diaphragm expands lungs.

*See chapter 25 for descriptions.

Sinuses
Nose
Nasal cavity
Mouth
Tongue
Epiglottis
Larynx
Uvula
Pharynx
Trachea
Bronchus
Right lung
Left lung
Ribs
Rib muscles
Diaphragm

Figure 31.6 The Human Respiratory System. Inhaled air passes through the mouth and trachea, which divides into two bronchi. The airways branch into increasingly narrow tubes ending in tiny alveoli, where gas exchange occurs (see figure 31.8).

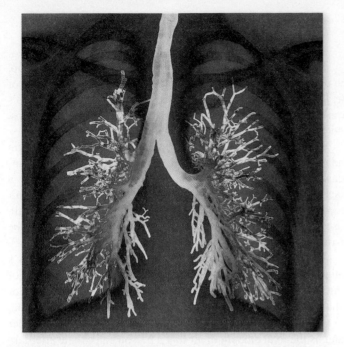

Figure 31.7 The Bronchial Tree. The respiratory passages of the lungs form a complex branching pattern. This is an X-ray image called a bronchogram.
©Innerspace Imaging/Science Source

The entire upper respiratory tract is lined with epithelium that secretes mucus. Dust and other inhaled particles trapped in the mucus are swept out by waving cilia. Coughing brings the mucus up, to be either spit out or swallowed.

Figure 31.8 Alveoli. Each human lung contains some 300 million alveoli, which are tiny air sacs that give the lungs a spongy texture. Gas exchange occurs at the lush capillary network that surrounds each cluster of alveoli.

B. The Lower Respiratory Tract Consists of the Trachea and Lungs

The **trachea,** or windpipe, is a tube just beneath the larynx. C-shaped rings of cartilage hold the trachea open and accommodate the expansion of the esophagus during swallowing. You can feel these rings in the front of your neck, near the lower portion of your throat. Cilia and mucus coat the trachea's inside surface, trapping debris and moistening the incoming air.

The trachea branches into two **bronchi,** one leading to each lung. The bronchi branch repeatedly, each branch decreasing in diameter and wall thickness (figure 31.7). **Bronchioles** ("little bronchi") are the finest branches. The bronchioles have no cartilage, but their walls contain smooth muscle. The autonomic nervous system controls the contraction of these muscles, adjusting airflow in response to metabolic demands. ⓘ *autonomic nervous system,* section 26.5

Each bronchiole narrows into several alveolar ducts, and each duct opens into a grapelike cluster of alveoli, where gas exchange occurs (figure 31.8). Each **alveolus** is a tiny sac with a wall of epithelial tissue that is one cell layer thick. A vast network of capillaries surrounds each cluster of alveoli. Oxygen and CO_2 diffuse through the thin walls of the alveoli and the neighboring capillaries. The interface between the alveoli and the capillaries is the respiratory surface in humans, and it is enormous. The total surface area of the alveoli in a pair of lungs is about 150 square meters, or more than half the area of a tennis court!

31.2 MASTERING CONCEPTS

1. List the components of the upper and lower respiratory tracts.
2. How does the nose function in the respiratory system?
3. Which structure prevents swallowed food from entering the respiratory system?
4. Describe the relationships among the trachea, bronchi, bronchioles, and alveoli.

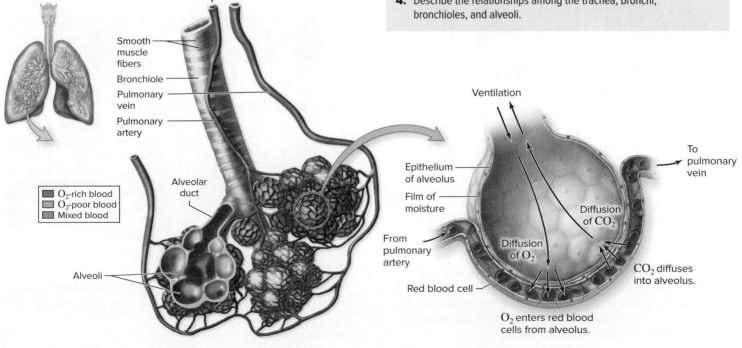

Blood flow

Smooth muscle fibers
Bronchiole
Pulmonary vein
Pulmonary artery

O_2-rich blood
O_2-poor blood
Mixed blood

Alveolar duct

Alveoli

Ventilation

Epithelium of alveolus
Film of moisture

To pulmonary vein

Diffusion of CO_2

From pulmonary artery

Diffusion of O_2

CO_2 diffuses into alveolus.

Red blood cell

O_2 enters red blood cells from alveolus.

31.3 Breathing Requires Pressure Changes in the Lungs

Pay attention to your own breathing for a moment. Each **respiratory cycle** consists of one inhalation and one exhalation (figure 31.9). Each time you **inhale,** air moves into the lungs; when you **exhale,** air flows out of the lungs.

What drives this back-and-forth motion? The answer is that air flows from areas of high pressure to areas of low pressure. Therefore, air enters the body when the pressure inside the lungs is lower than the pressure outside the body. Conversely, air moves out when the pressure in the lungs is greater than the atmospheric pressure.

The body generates these pressure changes by altering the volume of the chest cavity. As a person inhales, skeletal muscles of the rib cage and diaphragm contract, expanding and elongating the chest cavity. The resulting increase in volume lowers the air pressure within the space between the lungs and the outer wall of the chest. The lungs therefore expand, lowering pressure in the alveoli. Air rushes in. Inhalation requires energy because muscle contraction uses ATP. ⓘ *muscle contraction,* section 29.4

The muscles of the rib cage and the diaphragm then relax. As a result, the rib cage falls to its former position, the diaphragm rests up in the chest cavity again, and the elastic tissues of the lung recoil. The pressure in the lungs now exceeds atmospheric pressure, so air flows out. At rest, exhalation is passive—that is, it requires only muscle relaxation, not contraction—and therefore does not require ATP.

Hiccups briefly interrupt the respiratory cycle. In a hiccup, the diaphragm contracts unexpectedly, causing a sharp intake of air; the "hic" sound occurs as the epiglottis closes. Chapter 13's Apply It Now box explores the evolutionary origin of hiccups, which have no known function.

Illness or injury may prevent contraction of the diaphragm and rib muscles, so breathing stops. Mechanical ventilators compensate for this loss of function (figure 31.10). A ventilator is a machine that blows air into a patient's respiratory tract. The air reaches the lungs through a tube inserted into the patient's mouth, nose, or trachea (through an incision). In most cases, the tube disrupts the larynx, so a patient on a ventilator cannot speak.

Medical professionals can test lung function by having a patient blow into a device called a spirometer. One measure of lung capacity is the **tidal volume,** or the amount of air inhaled or exhaled during a quiet breath taken at rest. In a young adult male, the tidal volume is about 500 mL. **Vital capacity,** on the other hand, is the total amount of air that a person can exhale after taking the deepest possible breath, about 4700 mL in a young man.

As we age, vital capacity declines. Illnesses can also interfere with a person's lung function by reducing the elasticity of the lungs, obstructing the airways, or weakening the muscles of the chest. This chapter's Apply It Now box describes some breathing disorders.

31.3   MASTERING CONCEPTS

1. What is the relationship between the volume of the chest cavity and the air pressure in the lungs?
2. Describe the events of one respiratory cycle.
3. What is the difference between tidal volume and vital capacity? What does each measurement indicate about lung function?

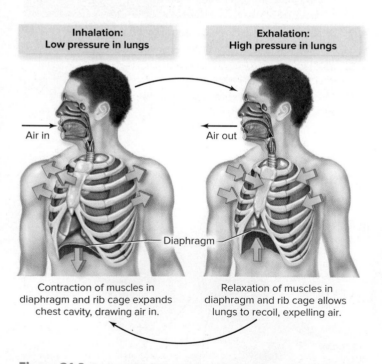

Figure 31.9 How We Breathe. When we inhale, the muscles of the diaphragm and rib cage contract, expanding the chest cavity. The chest cavity therefore has lower air pressure than the atmosphere, so air moves into the lungs. When we exhale, the diaphragm relaxes and the rib cage lowers, reversing the process and pushing air out of the lungs.

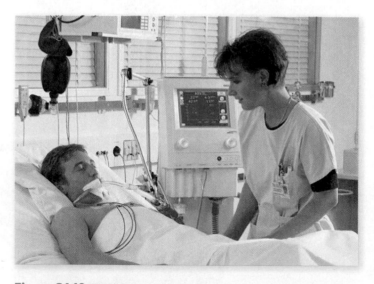

Figure 31.10 Ventilator. A patient with impaired lung function may need a ventilator, a machine that blows air into the respiratory system through a tube inserted into the trachea.
©PRNewsFoto/AP Images

Apply It Now

The Unhealthy Respiratory System

Asthma

During an asthma attack, spasms occur in the smooth muscle lining the lung's bronchi, slowing airflow and causing wheezing. An allergy to pollen, dog or cat dander (skin particles), or dust mites triggers most asthma attacks. Inhalant drugs that treat asthma usually relax the bronchial muscles.

Lung Cancer

Lung cancer is the most common cancer worldwide, and 85% of cases occur in smokers. Other risk factors for lung cancer include exposure to asbestos or radioactivity. Cigarette smoke, asbestos, and radioactive substances contain chemicals that mutate DNA in lung cells; these altered cells may develop into tumors that obstruct the airway. Patients with lung cancer experience chest pain, chronic coughing, and shortness of breath. Moreover, cancerous cells may break away from the tumor and spread throughout the body. ⓘ *cancer,* section 8.6

Apnea

Apnea is the cessation of breathing. It can be voluntary for a short time, as when you hold your breath. Premature infants may stop breathing because the part of the brain that regulates breathing is not fully developed. Overweight adults are especially susceptible to sleep apnea. The fleshy folds of the throat sag into the airway, causing breathing to stop for a short time. The person reflexively clears the throat, and breathing resumes. Because sleep is interrupted many times a night, people with sleep apnea may be extremely tired during the day.

Pneumonia

Pneumonia is an inflammation of the alveoli, usually resulting from an infection with bacteria, viruses, or fungi. Mucus and white blood cells accumulate in inflamed alveoli, impeding gas exchange. Symptoms include coughing, often accompanied by green or yellow sputum; fever; chest pain; and shortness of breath.

Common Cold

Viruses that cause colds infect cells in the upper respiratory tract (see chapter 16). A day or so after infection, the immune system's efforts to get rid of the invading viruses cause the typical symptoms of a cold: coughs, runny nose, and sneezes. There is no cure. Fortunately, colds are usually not a serious health problem, although clogged sinuses can be vulnerable to bacterial infections.

Tuberculosis

Infection with the bacterium *Mycobacterium tuberculosis* causes tuberculosis (also called TB). Symptoms of advanced TB include painful coughs, bloody sputum, fever, and weight loss. When a patient with an active TB infection coughs or sneezes, he or she expels bacteria-laden droplets that nearby people can inhale. Once in the lungs, the bacteria replicate inside immune system cells. Other immune system cells clump around the infected cells. This response keeps the bacteria from spreading, but these clusters also block airways and provide a place for the bacteria to become dormant. The resulting latent infection may later reemerge as full-blown (active) TB. Tuberculosis is especially important today because of the emergence of *Mycobacterium* strains that resist antibiotic treatment.

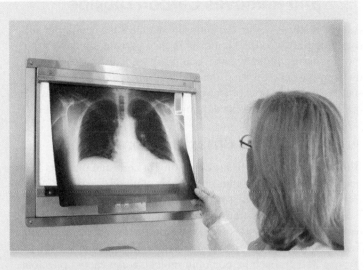

©Echo/Cultura/Getty Images RF

COPD (Emphysema and Chronic Bronchitis)

Emphysema, a term derived from the Greek word for "inflate," is an abnormal accumulation of air in the lungs. Long-term exposure to cigarette smoke and other irritants causes a loss of elasticity in lung tissues. The walls of the alveoli tear, impeding airflow and reducing the surface area for gas exchange. The patient experiences shortness of breath, an expanded chest, and hyperventilation. Emphysema is often accompanied by chronic bronchitis—that is, inflammation of the bronchi. Together, these two illnesses are called chronic obstructive pulmonary disease (COPD).

Cystic Fibrosis

Cystic fibrosis is among the most common inherited diseases. A faulty transport protein in the membranes of epithelial cells causes sticky mucus to accumulate in the lungs. The thick mucus prevents cilia from beating and could cause a chronic cough. Bacteria also thrive in the warm, moist mucus, causing persistent infections.

The Effects of Smoking on the Respiratory System

Besides increasing the risk of heart attack, stroke, and lung cancer, tobacco use causes other problems as well. The very first inhalation of cigarette smoke slows the beating of cilia. Over time, the cilia become paralyzed, and they eventually vanish. Without cilia to remove mucus, coughing alone must clear particles from the airways. Smoking also causes excess mucus production, which favors the reproduction of disease-causing microorganisms. Smokers are therefore especially susceptible to respiratory infections.

The smoker's cough leads to emphysema and chronic bronchitis. As the linings of the bronchioles thicken and become less elastic, they can no longer absorb the pressure changes that accompany coughing. A cough can therefore rupture the delicate walls of alveoli, causing a worsening cough, fatigue, wheezing, and impaired breathing. Meanwhile, cancers of the mouth, larynx, esophagus, and jaw can develop and spread.

Nicotine's addictive properties make quitting difficult, but it pays to stop smoking. In the short term, cilia may reappear, and the thickening of alveolar walls can reverse. In the long term, going tobacco-free greatly cuts the risk of heart disease, stroke, cancer, and chronic respiratory problems.

31.4 Blood Delivers Oxygen and Removes Carbon Dioxide

As we breathe, O_2 from the atmosphere enters the capillaries surrounding the alveoli; CO_2 moves in the opposite direction, from the capillaries to the alveoli. As a result, the air we exhale has a different gas composition than inhaled air (figure 31.11). This section describes how blood transports O_2 and CO_2, and how the brain uses blood gas concentrations to set the breathing rate.

A. Blood Carries Gases in Several Forms

Blood is a connective tissue consisting of cells and platelets suspended in **plasma,** the liquid component of blood (see figure 30.3). Cells and plasma interact in different ways to carry O_2 and CO_2.

Red blood cells transport at least 99% of O_2 in blood; the rest is dissolved in plasma. Red blood cells are packed with **hemoglobin,** an iron-rich pigment protein that binds with O_2 (figure 31.12). Each hemoglobin molecule can carry up to four O_2 molecules. The hemoglobin in a red blood cell becomes almost completely saturated with O_2 after spending only a second or two in the alveolar capillaries.

Not surprisingly, illness or death results when hemoglobin cannot bind O_2. Carbon monoxide (CO), for example, is a colorless, odorless gas in cigarette smoke and in exhaust from car engines, kerosene heaters, wood stoves, and home furnaces. CO binds to hemoglobin more readily than O_2 does, and it is less likely to leave the hemoglobin molecule. When 30% of the hemoglobin molecules carry CO instead of O_2, a person loses consciousness and may go into a coma or even die.

Hemoglobin also carries some CO_2, but it does not bind this gas as readily as it binds to O_2. Instead, an enzyme in red blood cells converts most CO_2 to bicarbonate ions (HCO_3^-). The following sequence of chemical reactions produces bicarbonate (the two double arrows indicate that the reactions are reversible):

$$CO_2 + H_2O \rightleftharpoons H_2CO_3 \text{ (carbonic acid)}$$
$$H_2CO_3 \rightleftharpoons H^+ \text{ (hydrogen ions)} + HCO_3^- \text{ (bicarbonate)}$$

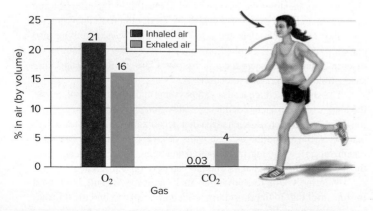
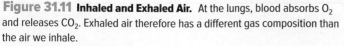

Figure 31.11 Inhaled and Exhaled Air. At the lungs, blood absorbs O_2 and releases CO_2. Exhaled air therefore has a different gas composition than the air we inhale.

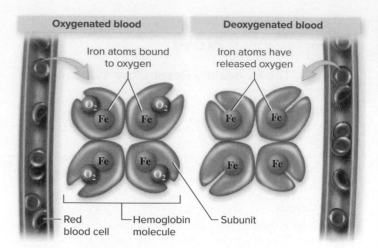

Figure 31.12 Hemoglobin. A red blood cell contains millions of hemoglobin proteins, each of which consists of four subunits. In oxygenated blood, the iron (Fe) atom in each subunit binds to one O_2. Hemoglobin unloads O_2 at body tissues, and deoxygenated blood returns to the heart.

Most of the bicarbonate ions subsequently diffuse into plasma. At the lungs, the ions return to red blood cells and the reactions reverse, liberating CO_2 into the air to be exhaled.

Carbonic acid and bicarbonate dissolved in plasma are an important part of blood's ability to buffer pH changes and maintain homeostasis. Ordinarily, the respiratory and excretory systems work together to maintain blood pH at about 7.4. Some metabolic processes, however, release H^+ into the blood; an example is the production of lactic acid in strenuously exercising muscles. If blood pH declines, excess H^+ reacts with HCO_3^- and forms H_2CO_3, raising the pH. ⓘ *pH, section 2.4A*

Gas exchange in the alveoli and at the body's other tissues relies on simple diffusion (figure 31.13). In external respiration, which occurs at the lungs, O_2 diffuses down its concentration gradient from the alveoli into the blood. At the same time, CO_2 diffuses from the blood to the air in the lungs. The heart then pumps the freshly oxygenated blood to the rest of the body. In internal respiration, O_2 diffuses from blood to the tissue fluid and then into the body's respiring cells, which have the lowest O_2 concentration. CO_2 diffuses in the opposite direction. ⓘ *diffusion, section 4.5A*

B. Blood Gas Levels Help Regulate the Breathing Rate

The heart has an internal pacemaker, so it can beat without stimulation from the nervous system. In contrast, breathing involves coordinated contractions of multiple skeletal muscles (see section 31.3). The brain's breathing control centers in the pons and medulla send out the neural signals that control these muscles. ⓘ *brainstem, section 26.6B*

The medulla helps maintain homeostasis in blood gas concentrations by regulating the depth and rate of breathing; figure 31.14 shows this negative feedback loop. Because CO_2 is

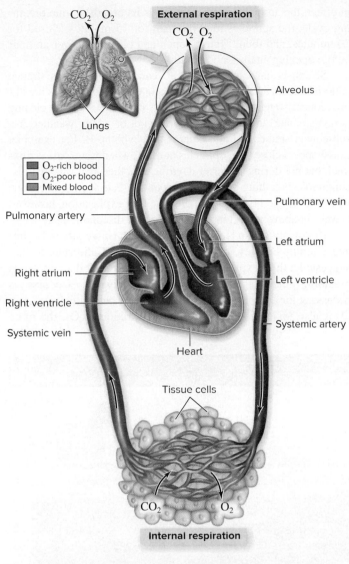

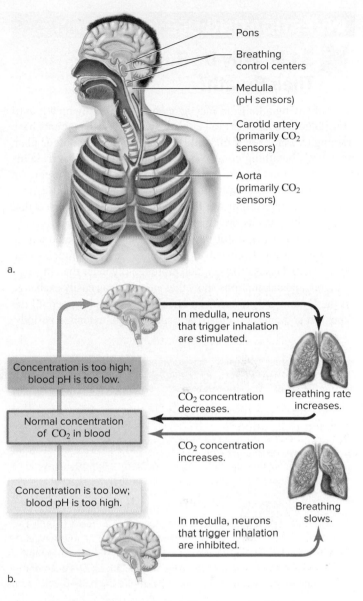

a.

b.

Figure 31.13 Gas Exchange at the Lungs and Tissues. CO_2 diffuses out of blood and into the alveoli of the lungs. O_2 moves in the opposite direction. In the rest of the body, O_2 diffuses out of the blood and into tissues, while CO_2 moves into the bloodstream.

Figure 31.14 Breathing Control. (a) Sensors send information about blood pH and CO_2 levels to the brain. The medulla integrates the signals and regulates the contraction of rib and diaphragm muscles. (b) The control of breathing is an example of a negative feedback loop.

a byproduct of aerobic respiration, monitoring blood CO_2 levels is a good way to determine how quickly cells use oxygen. Chemoreceptors in the body's largest arteries communicate blood CO_2 levels to the medulla. In addition, when the CO_2 level rises, carbonic acid accumulates in blood; the resulting decline in blood pH stimulates chemoreceptors in the medulla. Small increases in the CO_2 concentration of the air we breathe can trigger a tremendous surge in breathing rate. For example, an increase to 0.5% CO_2 in the air (from the normal 0.03%) makes us breathe 10 times faster. ⓘ *negative feedback*, section 25.4

Oxygen level is less important than CO_2 level in regulating breathing. In fact, the concentration of O_2 in blood affects breathing rate only if it falls dangerously low. (This chapter's Burning Question discusses mountaintop breathing.) Sometimes this O_2-regulating system fails, especially in the very young.

Sudden infant death syndrome, in which a baby dies while asleep, may occur when receptors fail to detect low oxygen levels in arterial blood.

31.4 MASTERING CONCEPTS

1. What protein in a red blood cell carries oxygen?
2. In what forms does blood transport O_2 and CO_2?
3. Describe the diffusion gradients for O_2 and CO_2 in the lungs and in the rest of the body.
4. Describe how the brain's breathing control centers use blood CO_2 concentration to alter the breathing rate.

INVESTIGATING LIFE

31.5 Why Do Bugs Hold Their Breath?

A child having a tantrum may scream, "I'll hold my breath until I turn blue!" in an effort to get his way. Luckily, the parents have nothing to fear; even if the child does begin to carry out his plan, his brain's breathing control centers will eventually override his effort to hold his breath. In insects, the situation is different. An insect can hold its breath much longer than we can, and its parents might actually approve: Evidence suggests that an insect that periodically holds its breath may live a long, healthy life.

An insect's respiratory system consists of air-filled tracheae extending inward from valvelike openings called spiracles (see figure 31.2). Gases diffuse much more rapidly in air than in water or blood, and calculations show that an insect can easily exchange all the O_2 and CO_2 it needs by keeping its spiracles open all the time. Yet some insects close their spiracles for extended periods;

in effect, the animal periodically holds its breath. In one breathing cycle, the spiracles may be open for 15 minutes, closed for 20 minutes, and then "flutter" open and closed for over an hour before opening again for 15 minutes.

Scientists have speculated for years about how this strange pattern benefits an insect. Some thought that discontinuous breathing reduces water loss, much as plants conserve water by closing pores (stomata) in their leaves during hot or windy weather. Yet subsequent studies did not support this hypothesis. For example, grasshoppers close their spiracles at night, when water loss is minimal, but not during the drier daytime. Another idea was that discontinuous breathing helps underground insects such as ants cope with low O_2 and high CO_2 conditions. That explanation, however, leaves out aboveground insects that breathe discontinuously.

Biologists Stefan Hetz (from Humboldt University in Berlin) and Timothy Bradley (from the University of California, Irvine) suggested a third explanation. They proposed that discontinuous breathing guards against *too much* O_2. Oxygen, of course, is necessary for aerobic respiration; without it, an animal dies. But O_2 is also toxic. The higher the concentration of O_2, the more

Burning Question

How does the body respond to high elevations?

The human respiratory system functions best near sea level. At high elevations, air density and oxygen availability fall gradually, so that at about 3000 meters above sea level, an individual inhales a third less O_2 with each breath. Each year, 100,000 mountain climbers and high-altitude exercisers experience altitude sickness (table 31.A). More than 180 climbers have died attempting to reach Mount Everest's 8850-meter summit, often from the effects of low oxygen. At high altitudes, the body's effort to get more O_2, by increasing breathing and heart rate, cannot keep pace with declining O_2 concentration. Altitude sickness can be prevented by giving the body time to adjust

©PhotoLink/Getty Images RF

TABLE 31.A   The Effects of High Altitude

Altitude (meters)	Condition	Symptoms
1800	Acute mountain sickness	Headache, weakness, nausea, poor sleep, shortness of breath
2700	High-altitude pulmonary edema (fluid accumulation in lungs)	Severe shortness of breath, cough, gurgle in chest, stupor, weakness; person can drown in accumulated fluid in lungs
4000	High-altitude cerebral edema (fluid accumulation in brain)	Severe headache, vomiting, altered mental status, loss of coordination, hallucinations, coma, death

during the ascent, and it can be cured by descending when symptoms first appear.

Less dramatically, the body also reacts to the low O_2 supply at high altitudes by increasing red blood cell production. When cells in the kidneys do not receive enough O_2, they trigger production of the hormone erythropoietin (also called EPO), which stimulates red blood cell production. Synthetic erythropoietin is a treatment for anemia (a reduction in the O_2-carrying capacity of blood), but some athletes inject themselves with EPO to gain a performance edge. This illicit practice is a form of blood doping, which improves an athlete's stamina but is difficult for sports authorities to detect. Doping with EPO can be dangerous because it thickens the blood and therefore strains the heart.

Submit your burning question to
Marielle.Hoefnagels@mheducation.com

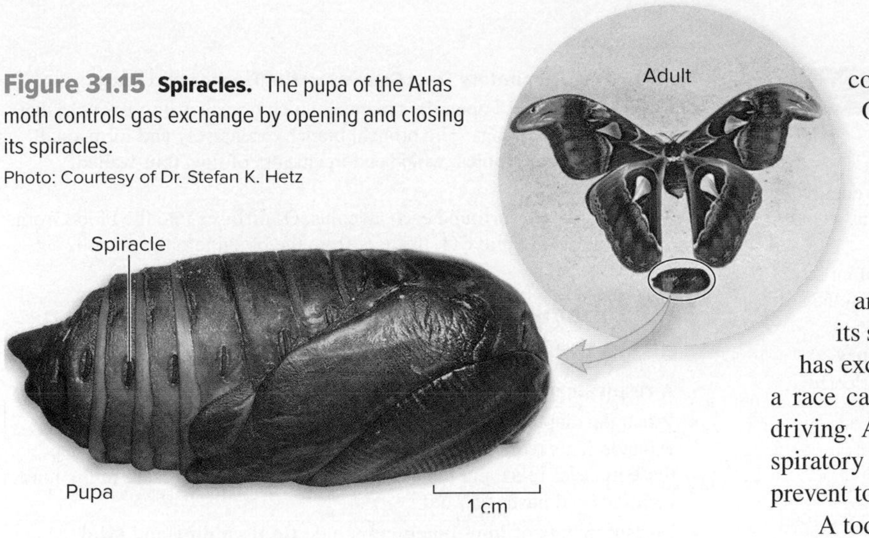

Figure 31.15 Spiracles. The pupa of the Atlas moth controls gas exchange by opening and closing its spiracles.

Photo: Courtesy of Dr. Stefan K. Hetz

Spiracle

Adult

Pupa

1 cm

harmful free radicals produced in the mitochondria. The more free radicals, the more damage to cells; in fact, studies have shown that insects experimentally exposed to the highest concentrations of O_2 live the shortest time.

Hetz and Bradley thought that discontinuous breathing might help the insect's body maintain a constant, safe level of O_2. To test this idea experimentally, they studied pupae of the Atlas moth (*Attacus atlas*; figure 31.15). The pupa is the stage of the life cycle during which the animal transforms from a caterpillar into an adult moth. Hetz and Bradley poked tiny oxygen sensors into spiracles of the Atlas moth pupa and measured the concentration of O_2 inside the tracheae. At the same time, they changed the external concentration of O_2. The results of the experiment supported the "oxygen-guarding" hypothesis: The internal oxygen

concentration stayed the same regardless of how much O_2 the insects were exposed to (figure 31.16).

The oxygen-guarding hypothesis explains one puzzling aspect of discontinuous breathing: It happens only in resting insects. During periods of high metabolic activity, the O_2 concentration inside the tracheae remains low because cells consume O_2 as quickly as it arrives. An active insect therefore has no reason to close its spiracles. At rest, however, the insect respiratory system has excess capacity; this situation is somewhat comparable to a race car's engine, which is much too powerful for everyday driving. An inactive insect compensates for its supercharged respiratory system by opening its spiracles only when necessary to prevent toxic buildup of CO_2.

A toddler who refuses to breathe does nothing but annoy his parents. In contrast, a logical conclusion of Hetz and Bradley's work is that discontinuous breathing helps insects live longer by guarding against excess O_2 exposure. Animals that live longer may have more mating opportunities, which may mean greater reproductive success—and an explanation for how natural selection maintains the behavior in insect populations.

Source: Hetz, Stefan K., and Timothy J. Bradley. 2005. Insects breathe discontinuously to avoid oxygen toxicity. *Nature*, vol. 433, pages 516–519.

31.5 MASTERING CONCEPTS

1. What does this study suggest is the function of discontinuous breathing in the Atlas moth pupa?
2. Compare the amount of time that the spiracles were open in pupae at each O_2 concentration.

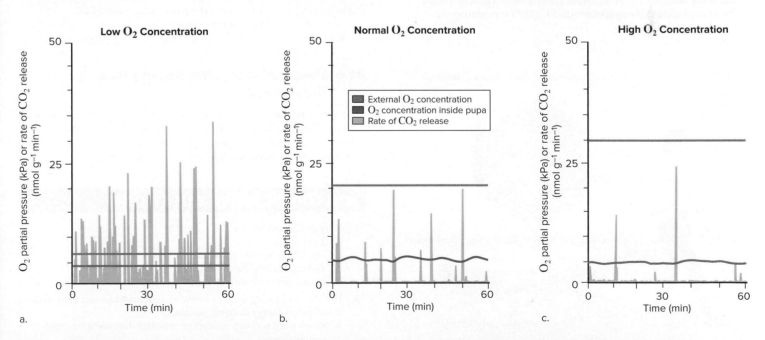

Figure 31.16 The Benefit of Discontinuous Breathing. The concentration of O_2 inside an Atlas moth (*blue lines*) pupa is nearly constant, regardless of whether the external concentration of O_2 (*pink lines*) is (a) low, (b) normal, or (c) high. The researchers made these measurements during the flutter phase, in which spiracles rapidly open and close; CO_2 release (*green bars*) indicates open spiracles. Notice that spiracles open most frequently when O_2 is in extremely short supply.

CHAPTER SUMMARY

31.1 Gases Diffuse Across Respiratory Surfaces

- Cells use oxygen gas (O_2) in **aerobic respiration** to release the energy in food and store the energy in ATP. Carbon dioxide (CO_2) forms as a byproduct of respiration and must be eliminated from the body.
- **Respiratory systems** exchange O_2 and CO_2 with air or water, often in conjunction with a circulatory system that transports gases within the body (**figure 31.17**).
- O_2 and CO_2 are exchanged by diffusion across a moist **respiratory surface.** Body size, metabolic requirements, and habitat have selected for a variety of respiratory surfaces.

A. Some Invertebrates Exchange Gases Across the Body Wall or in Internal Tubules

- Cnidarians, earthworms, and other invertebrates exchange O_2 and CO_2 directly across the body surface.
- Most terrestrial arthropods bring the atmosphere into contact with almost every cell through a highly branched system of **tracheae.**

B. Gills Exchange Gases with Water

- Many aquatic animals exchange gases across **gill** membranes enclosing networks of capillaries. In bony fishes, water flows over the gills in the direction opposite that of blood flow, an arrangement called **countercurrent exchange.**

C. Terrestrial Vertebrates Exchange Gases in Lungs

- Vertebrate **lungs** contain a moist internal respiratory surface. Amphibians have the simplest lungs; birds and mammals have the most complex.

31.2 The Human Respiratory System Delivers Air to the Lungs

A. The Nose, Pharynx, and Larynx Form the Upper Respiratory Tract

- The **nose** purifies, warms, and moisturizes inhaled air. The air then flows through the **pharynx** and **larynx.**
- **Vocal cords** stretched over the larynx produce the voice as air passes through the **glottis.** The **epiglottis** prevents food from entering the trachea through the glottis.

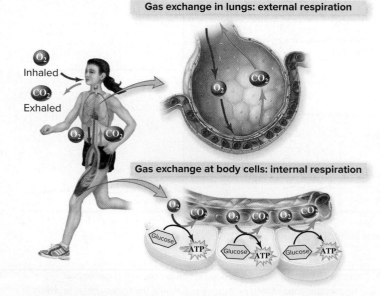

Figure 31.17 Gas Exchange: A Summary.

B. The Lower Respiratory Tract Consists of the Trachea and Lungs

- Cartilage rings hold open the **trachea,** which branches into **bronchi** that deliver air to the lungs. The bronchi branch extensively and form small air tubules, **bronchioles,** which end in clusters of tiny, thin-walled, saclike **alveoli.**
- Many capillaries surround each alveolus. O_2 diffuses into the blood from the alveolar air, while CO_2 diffuses from the blood into the alveoli.

31.3 Breathing Requires Pressure Changes in the Lungs

- A **respiratory cycle** consists of one **inhalation** and one **exhalation.**
- When the diaphragm and rib cage muscles contract, the chest cavity expands. This reduces air pressure in the lungs, drawing air in. When these muscles relax and the chest cavity shrinks, the pressure in the lungs increases and pushes air out.
- Measurements of lung function include **tidal volume** and **vital capacity.**

31.4 Blood Delivers Oxygen and Removes Carbon Dioxide

A. Blood Carries Gases in Several Forms

- Gases are transported in **blood.**
- Almost all oxygen transported to cells is bound to **hemoglobin** in **red blood cells.** Carbon monoxide (CO) poisoning occurs when CO prevents O_2 from binding hemoglobin.
- Some CO_2 in the blood is bound to hemoglobin or dissolved in **plasma.** Most CO_2, however, is converted to bicarbonate ions that form in red blood cells and subsequently diffuse into plasma.
- In the lungs, hemoglobin binds O_2 and releases CO_2 for exhalation. At the body's other tissues, the concentration of O_2 is low, so O_2 diffuses out of the blood and into the respiring cells. CO_2 moves in the opposite direction along its concentration gradient.

B. Blood Gas Levels Help Regulate the Breathing Rate

- The brain typically uses blood pH (an indirect measure of CO_2 concentration) to adjust the breathing rate.
- Only critically low O_2 levels trigger a change in the breathing rate.

31.5 Investigating Life: Why Do Bugs Hold Their Breath?

- At rest, some insects breathe discontinuously by alternately opening and closing the spiracles that connect their tracheae to the atmosphere.
- Experimental evidence suggests that discontinuous breathing helps the insects maintain a safe, low level of O_2 inside their bodies.

MULTIPLE CHOICE QUESTIONS

1. Why do humans breathe?
   a. To eliminate CO_2
   b. To support aerobic respiration in mitochondria
   c. To keep the respiratory surface dry
   d. Both a and b are correct.

2. How did respiratory systems evolve as vertebrates moved away from water and as energy demands increased?
   a. Respiratory surfaces in lungs became less dependent on moist membranes.
   b. The surface area for gas exchange within lungs became greater.
   c. Tracheae became much more numerous.
   d. Both b and c are correct.

3. Air entering the human respiratory system encounters several structures. Which of the following lists is arranged in the order that incoming air passes through them?
   a. Pharynx, larynx, trachea, bronchus
   b. Larynx, pharynx, bronchus, trachea
   c. Trachea, pharynx, larynx, bronchus
   d. Bronchus, trachea, pharynx, larynx

4. Air is pulled into the human respiratory tract mainly because of volume changes in the
   a. nose.   c. chest cavity.
   b. pharynx.   d. trachea.

5. Air flows from areas of _____ to areas of _____.
   a. high O_2 concentration; low O_2 concentration
   b. low O_2 concentration; high O_2 concentration
   c. high pressure; low pressure
   d. low pressure; high pressure

6. At an active muscle, red blood cells drop off _____ and pick up _____.
   a. hemoglobin; bicarbonate   c. hemoglobin; carbon dioxide gas
   b. oxygen gas; bicarbonate   d. oxygen gas; carbon dioxide gas

7. How does the concentration of CO_2 in an expanding bubblegum bubble compare to the concentration of CO_2 in the surrounding air?
   a. The concentration of CO_2 is higher in the bubble.
   b. The concentration of CO_2 is higher in the surrounding air.
   c. The concentration of CO_2 is equal in the bubble and the surrounding air.
   d. More information is necessary because the concentration of CO_2 depends on the concentration of exhaled O_2.

8. Breathing rate is mostly determined by
   a. the concentration of O_2 in the air.
   b. the concentration of O_2 in the blood.
   c. the concentration of CO_2 in the blood.
   d. the amount of hemoglobin in the blood.

Answers to these questions are in appendix A.

WRITE IT OUT

1. Write a paragraph comparing four types of respiratory surfaces (body surface, tracheae, gills, lungs). Your paragraph should describe each surface, list which animals have each, and say whether the surface can function in air, water, or both.

2. Explain why efficient respiratory and circulatory systems evolved as animals' energy demands increased.

3. Describe how lungs differ among the main groups of terrestrial vertebrates.

4. Compare figure 31.4 with figure 30.2. How might the evolution of lungs have selected for new adaptations in the circulatory system? How might changes in the heart have selected for more efficient lungs?

5. How is air cleaned, warmed, and humidified before it reaches the lungs?

6. Trace the path of an O_2 molecule from a person's nose to a red blood cell at an alveolar capillary.

7. A person can choke if a hard candy or other small object obstructs the airway. In drowning, a person's lungs fill with water. Explain how each of these events can cause death. In most circumstances, how does the body react to prevent either of these disasters from occurring?

8. How is the branching pattern of the airways similar to the branching pattern in the blood vessels? What characteristic of airways and blood vessels increases as the structures branch?

9. How does the pressure in the lungs compare to the pressure in the atmosphere during inhalation? During exhalation?

10. Jumping into cold water may initiate the "diving reflex," during which blood is shunted toward the head and away from the limbs. Under these conditions, why might you be able to hold your breath for an extended period?

11. How does blood transport O_2 to cells?

12. How does blood transport most CO_2? In what other way is CO_2 transported?

13. People who suffer from claustrophobia are afraid of being enclosed in small areas. Some claustrophobes fear that they will "use all of the air" in the space and suffocate. Why is it impossible to use all of the air in a space? What does happen to the air in an enclosed space as you respire? Are the changes in the air dangerous?

14. Why is carbon monoxide (CO) so dangerous?

15. A track athlete runs 100 meters while holding his breath. Predict the gas composition of the breath he exhales as he crosses the finish line. How might it compare to the gas composition of a normally exhaled breath?

16. A broken rib can sometimes puncture a lung, causing it to collapse. A collapsed lung leaks air into the chest cavity. Why might it be difficult for a person to fully expand a collapsed lung?

17. How does the brain establish the breathing rhythm?

18. Search the Internet for hypotheses about why humans and other animals yawn; then design an experiment to test one of the hypotheses.

19. The concentration of O_2 in the atmosphere declines with increasing elevation. Why do you think the times of endurance events at the 1968 Olympics, held in Mexico City (elevation: 2200 m), were relatively slow?

20. How are the lungs similar to the stomata of plants, and how are they different?

PULL IT TOGETHER

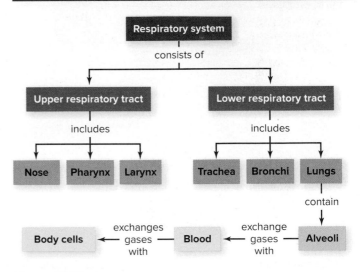

Figure 31.18 Pull It Together: The Respiratory System.

Refer to figure 31.18 and the chapter content to answer the following questions.

1. Using the Survey the Landscape figure in the chapter introduction and figure 31.18, describe how the respiratory system helps an animal to maintain homeostasis.

2. Add O_2 and CO_2 to the concept map; connect these terms with *body cells, blood,* and *alveoli.*

3. Add terms to the concept map to explain the pressure changes that occur during inhalation and exhalation.

32

Digestion and Nutrition

SEM (false color)

Good Bacteria. Yogurt's tart flavor comes from acids that bacteria produce as they ferment milk. The inset shows cells of *Lactobacillus casei*, "probiotic" bacteria that inhabit yogurt and the human gut.

(woman): ©Britt Erlanson/age fotostock; (bacteria): ©Steve Gschmeissner/Science Photo Library/Corbis RF

LEARN HOW TO LEARN
Avoid Distractions

Despite your best intentions, constant distractions may take you away from your studies. Friends, music, TV, social media, text messages, video games, and online shopping all offer attractive diversions. How can you stay focused? One answer is to find your own place to study where no one can find you. Turn your phone off for a few hours; the world will get along without you while you study. And if you must use your computer, create a separate user account with settings that prevent you from visiting favorite websites during study time.

Your Microscopic Partners in Digestion

Food and water are among the familiar necessities of life. Digestive organs break down foods and absorb nutrients, which the body uses to generate energy and build its tissues. The small and large intestines, located near the end of the digestive tract, absorb most of the nutrients and water in food.

Researchers have long known that the intestines do not act alone. Instead, they are home to trillions of microbes, including bacteria and microscopic fungi. These microbes produce the bad-smelling gases that we experience as flatulence. They also participate in digestion, produce essential vitamins, and help break down drugs. Without these inhabitants, harmful species are more likely to colonize our intestines and cause illness. ⓘ *normal microbiota*, section 17.4B

Not just any microbes will do; the optimal mix of species is vital. In fact, researchers have discovered that certain digestive problems can be treated by simply changing the intestinal community. For example, yogurt and other foods may include probiotics, which are live microbes that may benefit the digestive tract. But severe digestive abnormalities warrant more drastic measures. For patients with stubborn intestinal problems, physicians sometimes order a fecal transplant. That is, they transplant a blended, filtered fecal sample from a healthy donor into a patient's intestines. This simple procedure, called bacteriotherapy, can save lives by restoring a healthy microbial community.

The gut's microbes may also be linked to disorders ranging from autism to obesity to heart attacks. Scientists analyzing the gut of autistic children found types of bacteria that are not present in non-autistic children. It is not yet known whether autism promotes a shift in intestinal microbes or if the microbes contribute to autism. Moreover, some microscopic gut inhabitants may increase the risk of obesity. Germ-free mice are much less likely to become obese than are mice with microbe-rich intestines, even on a diet high in fat and simple sugars. Some day, physicians may be able to manipulate microbe communities in the gut to help patients lose weight. Intestinal bacteria may even form a link between red meat and heart attack risk in humans. Frequent beef-eaters—but not vegetarians—harbor microbes that react to a meaty diet by releasing chemicals that can cause heart disease.

Some scientists consider intestinal microbes a collective metabolic organ. The more traditionally recognized organs of the human digestive system—the mouth, esophagus, stomach, intestines, and several accessory organs—are the subjects of this chapter.

LEARNING OUTLINE

32.1 Digestive Systems Derive Nutrients from Food

32.2 Animal Digestive Tracts Take Many Forms

32.3 The Human Digestive System Consists of Several Organs

32.4 A Healthy Diet Includes Essential Nutrients and the Right Number of Calories

32.5 Investigating Life: The Cost of a Sweet Tooth

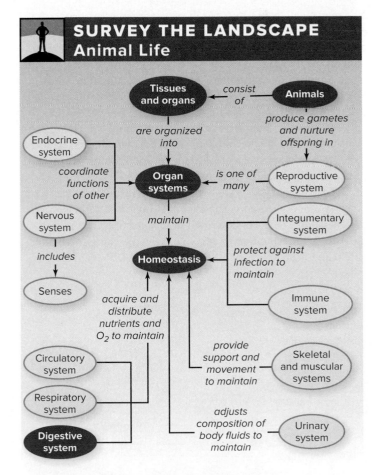

SURVEY THE LANDSCAPE
Animal Life

Animals require food as a source of both energy and raw materials. The digestive system acquires chunks of unprocessed food, converts it into small molecules that the body's cells can use, and delivers them to the blood for distribution throughout the body.

For more details, study the Pull It Together feature in the chapter summary.

32.1 Digestive Systems Derive Nutrients from Food

Is it true that "you are what you eat"? In some ways, the answer is yes. After all, the atoms and molecules that make up your body came from food that you ate (or that your mother ate before you were born). But in other ways, the answer is no. The woman in figure 32.1 may enjoy eating fruit, yet she looks nothing like a mango. Clearly, food is not incorporated whole into her body, even though atoms and molecules derived from food compose her and every other animal.

The explanation for this apparent contradiction lies in the **digestive system,** the organs that ingest food, break it down, absorb the small molecules, and eliminate undigested wastes. As this section describes, some of the molecules absorbed from food do become part of the animal body, but others are used to generate the energy needed for life.

A. Animals Eat to Obtain Energy and Building Blocks

You may recall that an **autotroph** uses inorganic raw materials to assemble its own food; the energy source is often sunlight. Plants and algae are examples of autotrophs. Unlike these organisms, animals need to eat; that is, they are heterotrophs. A **heterotroph** is an organism that must consume food—organic molecules—to obtain carbon and energy. Fungi and many other microbes are heterotrophs, too.

An animal's food is its source of **nutrients,** which are substances required for metabolism, growth, maintenance, and repair. The six main types of nutrients in food are carbohydrates, proteins, lipids, water, vitamins, and minerals.

These nutrients contain two important resources. One is the potential energy stored in the chemical bonds of carbohydrates, proteins, and lipids. As described in chapter 6, cells use each of these fuels in respiration to generate ATP, the energy-rich molecule that powers most cellular activities. ⓘ *ATP, section 4.3*

Figure 32.1
Delicious.
Eating provides the raw materials and energy required for life.
©Stockbyte/PunchStock RF

The second resource in food is the chemical building blocks that make up the animal's body. Simple sugars, fatty acids, amino acids, nucleotides, water, vitamins, and minerals are the raw materials that build, repair, and maintain all parts of the body, from blood to bone (see chapter 2). To the extent that your body incorporates these materials, you really are what you eat.

B. How Much Food Does an Animal Need?

An animal's metabolic rate largely determines its need for food. Endotherms use a lot of energy to maintain a constant body temperature; these animals, such as birds and mammals, have relatively high metabolic rates. In contrast, an ectotherm's body temperature fluctuates with the environment. Examples of ectotherms include invertebrate animals, fishes, amphibians, and nonavian reptiles such as lizards, snakes, and crocodiles. Thanks to this difference in metabolic rate, a sparrow or mouse weighing 25 grams requires much more food than does a 25-gram lizard. ⓘ *endotherms and ectotherms, section 33.1*

Body size also influences the need for food. In general, the larger the animal, the more it needs to eat. When corrected for body size, however, the smallest animals typically have the highest metabolic rates. A hummingbird, for example, has a much higher surface area relative to its body mass than does an elephant. Because the tiny bird rapidly loses heat to the environment, it must consume a lot of food to maintain a constant body temperature. A hummingbird therefore eats its own weight in food every day; an elephant takes 3 months to do the same.

Another factor that affects metabolic rate is an animal's physiological state. Growth and reproduction require more energy and nutrients than simply maintaining the adult body. Baby animals therefore often have ravenous appetites, and a new parent may spend much of its time finding food to fuel its offspring's rapid growth.

C. Animals Process Food in Four Stages

The overall process by which animals use food has four major steps (figure 32.2). First, **ingestion** is the entrance of food into the digestive tract. The second stage, **digestion,** is the physical and chemical breakdown of food. In mammals, this process begins with chewing, which tears food into small pieces mixed with saliva. Chewing therefore softens food and increases the surface area exposed to digestive enzymes. In chemical digestion, enzymes split large nutrient molecules into their smaller components. In the third stage, **absorption,** the nutrients enter the cells lining the digestive tract and move into the bloodstream to be transported throughout the body (see chapter 30). Fourth, in **elimination,** the animal's body expels (egests) undigested food. **Feces** are the solid wastes that leave the digestive tract. ⓘ *enzymes, section 4.4*

One potential source of confusion is the distinction between feces and urine, both of which are animal waste products. Feces are composed partly of undigested food that never enters the

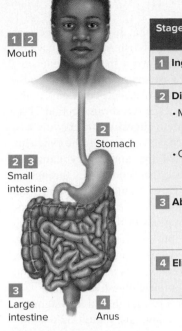

Stage	Description	Location(s) in Human Body
1 Ingestion	Food enters digestive tract	Mouth
2 Digestion		
• Mechanical	Food is physically broken down into small particles	Mouth, stomach
• Chemical	Digestive enzymes break food molecules into small subunits	Mouth, stomach, small intestine
3 Absorption	Water and digested food enter bloodstream from digestive tract	Small intestine (food and water), large intestine (water)
4 Elimination	Undigested food exits digestive tract in feces	Anus

Figure 32.2 Food Processing. The stages by which animals acquire and use food are ingestion, digestion, absorption, and elimination.

body's cells. Urine, on the other hand, is a watery fluid containing dissolved nitrogen and other metabolic wastes produced by the body's cells. Chapter 33 describes how the kidneys produce urine.

D. Animal Diets and Feeding Strategies Vary Greatly

Biologists divide animals into categories based on what they eat and how they eat it (**figure 32.3**). **Herbivores,** such as cows and rabbits, eat only plants. Eagles, cats, wolves, and other **carnivores** hunt other animals for food. **Detritivores** consume decomposing organic matter; dung beetles and earthworms illustrate this diet. Finally, **omnivores** eat a broad variety of foods, including plants and animals. Humans are omnivores, as are raccoons and chickens.

Many animals rely on one or a few kinds of food. Insectivores such as anteaters, bats, flycatchers, praying mantises, and most spiders eat only insects. A piscivore eats fish, and a frugivore eats fruits. The giant panda is a folivore (leaf-eater). Because it eats only bamboo, the giant panda can survive only where that plant thrives. Animals with flexible diets, such as raccoons, can live in a broader range of habitats.

Animals also differ in how they acquire food. Humans and most other animals are **bulk feeders,** ingesting large chunks of food. **Fluid feeders,** in contrast, drink their food: A mosquito takes a "blood meal" after piercing human skin with its mouthparts. **Suspension feeders** are animals that strain particles from water. Sponges are suspension feeders that pass water through their porous bodies, filtering out the suspended organic matter. Corals, clams, and mussels, on the other hand, have feeding

appendages that extend into the water column. Suspension feeding is adaptive in animals that do not move about much, because water carries the food to them. Nevertheless, some suspension feeders, such as baleen whales and basking sharks, are active swimmers.

Still other animals, called **substrate feeders,** live in their food and eat it from the inside. A female parasitic fly, for example, may lay eggs inside a live grasshopper. When the eggs hatch, the larvae feed on the host's tissues. A **deposit feeder** such as an earthworm is also a type of substrate feeder. These animals eat organic matter in soil or other sediments.

Regardless of whether an animal eats plants, animals, or decomposing organic matter, note that all food ultimately comes from autotrophs. Chapter 38 explores this ecological principle in more detail.

32.1 MASTERING CONCEPTS

1. What are two reasons that animals must eat?
2. Explain three factors that affect an animal's metabolic rate.
3. What are the four stages of food use in animals, and where in the human digestive tract does each stage occur?
4. How does the mechanical breakdown of food speed chemical digestion?
5. Define the terms *herbivore, carnivore, detritivore,* and *omnivore.*

a.

b.

c.   d.

Figure 32.3 A Selection of Animal Diets. (a) A giant panda munches on bamboo. (b) This leopard is eating its kill. (c) A mosquito takes a blood meal. (d) The basking shark filters food from water.

32.2 Animal Digestive Tracts Take Many Forms

Because heterotrophs and their food consist of the same types of chemicals, digestive enzymes could just as easily attack an animal's body as its food. Digestion therefore occurs within specialized compartments that are protected from enzyme action.

These compartments may be located inside or outside of cells (figure 32.4). Intracellular digestion occurs entirely inside a cell. For example, a protist such as *Paramecium* engulfs nutrients by phagocytosis and encloses the food in a food vacuole. A loaded food vacuole fuses with another sac containing digestive enzymes that break down nutrient molecules. Sponges are the only animals that rely solely on intracellular digestion. ⓘ *phagocytosis*, section 4.5C; *sponges*, section 21.2

More complex animals use extracellular digestion, releasing hydrolytic enzymes into a digestive cavity connected with the outside world. The enzymes dismantle large food particles; cells lining the cavity then absorb the products of digestion. Food remains outside the body's cells until it is digested and absorbed. Extracellular digestion eases waste removal because indigestible components of food never enter the cells. Instead, the digestive tract simply ejects the waste.

The cavity in which extracellular digestion occurs may have one or two openings (figure 32.5). An **incomplete digestive tract** has only one opening: a mouth that both ingests food and ejects wastes. The animal must digest food and eliminate the residue before the next meal can begin. This two-way traffic limits the potential for specialized compartments that might store, digest, or absorb nutrients. Cnidarians such as jellyfish and *Hydra* have incomplete digestive tracts, as do flatworms. In these organisms, the digestive tract is also called a **gastrovascular cavity** because it doubles as a circulatory system that distributes nutrients to the body cells. ⓘ *cnidarians*, section 21.3; *flatworms*, section 21.4

Most animals have a **complete digestive tract** with two openings; the mouth is the entrance, and the **anus** is the exit. This tubelike digestive cavity is called the **alimentary canal** or **gastrointestinal (GI) tract.** Notice that food passes through in one direction, so the animal can digest one meal at the same time that it acquires its next one. Another advantage of the two-opening digestive system is that regions of the tube can develop specialized areas that break food into smaller particles, digest it, absorb the nutrients into the bloodstream, and eliminate wastes. A complete digestive tract therefore extracts nutrients from food more efficiently than does an incomplete digestive tract.

Diet and lifestyle differences select for digestive system adaptations in all animals, including mammals. **Figure 32.6** shows the digestive tracts of two herbivores and a carnivore. An herbivore's diet is rich in hard-to-digest cellulose from the cell walls of plants. The long digestive tract allows extra time for digestion. The diet of a carnivore, on the other hand, consists mostly or entirely of highly digestible meat. The overall length of the digestive tract is therefore short.

The elk in figure 32.6a is a ruminant, which is an herbivore with a complex, four-chambered organ that specializes in the digestion of grass. Saliva mixed with chewed grass enters the

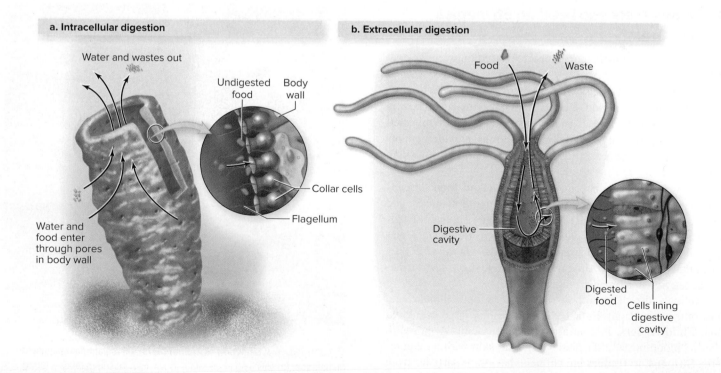

a. Intracellular digestion

Water and wastes out

Undigested food   Body wall

Collar cells

Flagellum

Water and food enter through pores in body wall

b. Extracellular digestion

Food   Waste

Digestive cavity

Digested food   Cells lining digestive cavity

Figure 32.4 Intracellular and Extracellular Digestion. (a) A sponge uses intracellular digestion, which occurs entirely inside cells. (b) Extracellular digestion occurs in a digestive cavity containing enzymes that break down food. Cells lining the digestive cavity absorb the nutrients.

a. Incomplete digestive tract: one opening

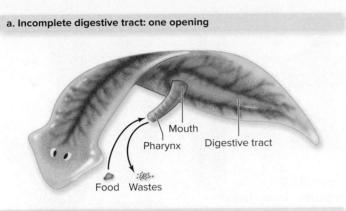

b. Complete digestive tract: two openings

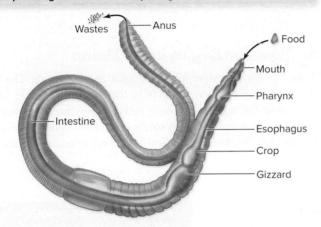

Figure 32.5 Incomplete and Complete Digestive Tracts. (a) In flatworms and other animals with an incomplete digestive tract, the mouth acquires food and eliminates wastes. (b) A complete digestive tract has two openings: the mouth and the anus.

first and largest chamber, the rumen, where fermenting microorganisms break the plant matter down into balls of cud. The animal regurgitates the cud into its mouth; chewing the cud breaks the food down further. When the animal swallows again, the food bypasses the rumen, continuing digestion in the remaining chambers. Cows, sheep, deer, and goats are familiar ruminants. ⓘ *fermentation,* section 6.8B

Figure 32.6 also depicts another structure whose size varies with diet: the pouchlike **cecum,** which forms the entrance to the large intestine. In herbivores, the cecum is large, and it houses bacteria that ferment plant matter. Carnivores have a small or absent cecum. The cecum is medium-sized in omnivores, reflecting a diet based partly on plants. (Section 32.5 describes how diet influences the digestive tracts of tadpoles.)

32.2 MASTERING CONCEPTS

1. Distinguish between intracellular and extracellular digestion.
2. How is an incomplete digestive tract different from a complete digestive tract?
3. Compare and contrast the digestive systems of an elk and a wolf.

a. Ruminant herbivore

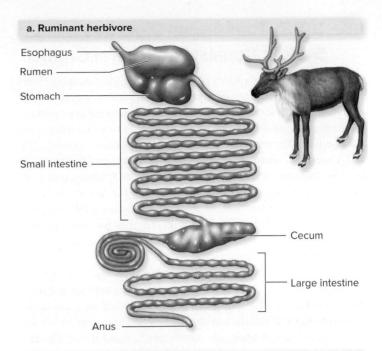

b. Nonruminant herbivore

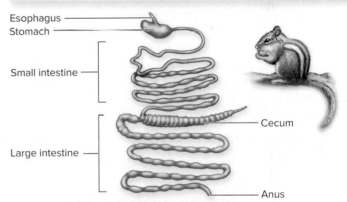

c. Carnivore

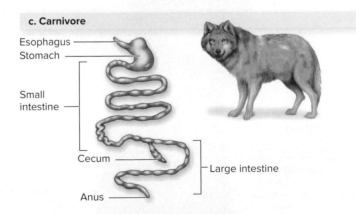

Figure 32.6 Digestive System Adaptations. (a) In ruminants, adaptations to a grassy diet include a microbe-rich rumen, a large cecum, and a long digestive tract. (b) The intestines and cecum of nonruminant herbivores such as rabbits and rodents also harbor microbes that break down cellulose in plants. (c) The protein-rich diet of carnivores is easy to digest, so the digestive system is much shorter and has a reduced cecum.

32.3 The Human Digestive System Consists of Several Organs

The human digestive system consists of the gastrointestinal tract and accessory structures (figure 32.7). The salivary glands and pancreas are accessory structures that produce digestive enzymes; the liver and gallbladder produce and store bile, which assists in fat digestion. The teeth and tongue are also accessory organs.

Layers of smooth muscle underlying the entire digestive tract undergo **peristalsis,** or rhythmic waves of contraction that propel food in one direction (figure 32.8). The autonomic nervous system stimulates these involuntary smooth muscle contractions. Peristalsis also mixes food with enzymes to form a liquid. ⓘ *smooth muscle,* section 25.2C; *autonomic nervous system,* section 26.5

Muscles also control the openings between digestive organs. **Sphincters** are muscular rings that can contract to block the passage of materials. The sphincters at the mouth and anus are composed of skeletal muscle and are under voluntary control, so we can decide when to open our mouth or eliminate feces. The remaining sphincters within the digestive tract, however, are composed of involuntary smooth muscle.

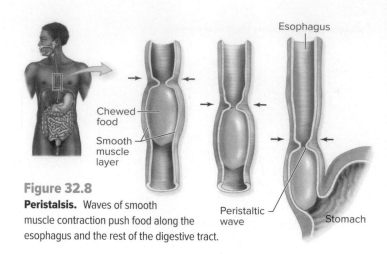

Figure 32.8
Peristalsis. Waves of smooth muscle contraction push food along the esophagus and the rest of the digestive tract.

A. Digestion Begins in the Mouth

A tour of the digestive system begins at the mouth. The taste of food triggers salivary glands in the mouth to secrete saliva. This fluid contains an enzyme, salivary amylase, that starts to break down starch (a polysaccharide) into smaller subunits. Meanwhile, the **teeth**—mineral-hardened structures embedded

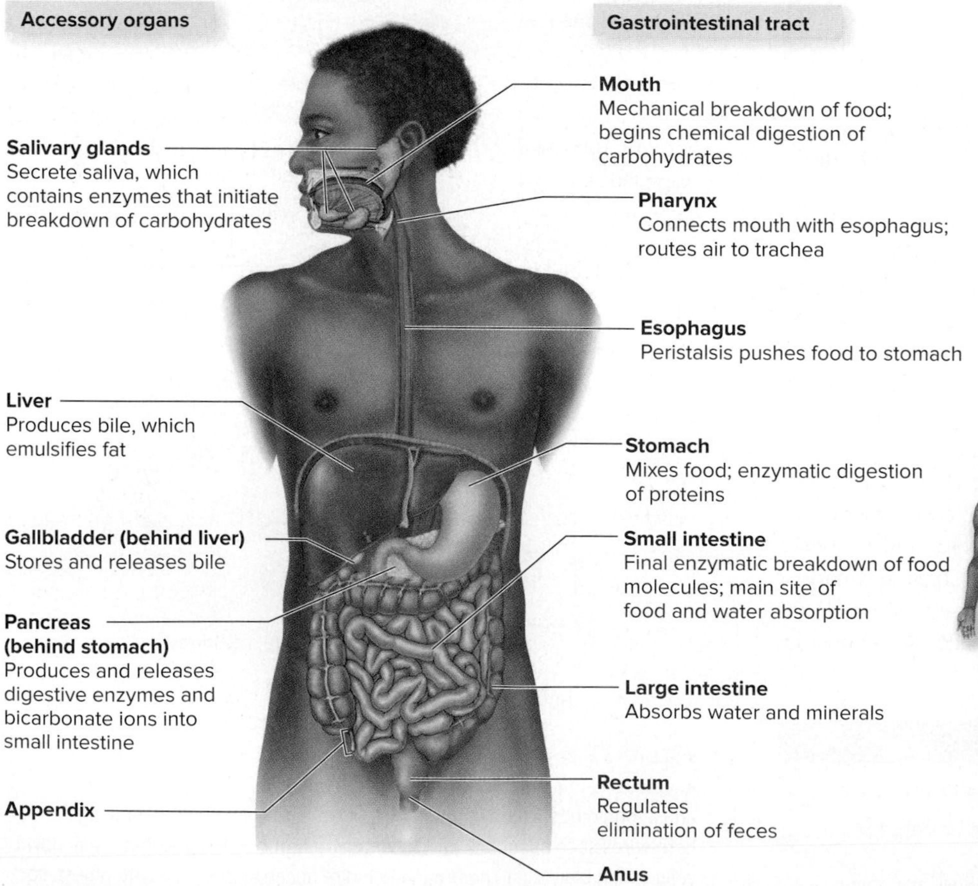

Accessory organs

Salivary glands
Secrete saliva, which contains enzymes that initiate breakdown of carbohydrates

Liver
Produces bile, which emulsifies fat

Gallbladder (behind liver)
Stores and releases bile

Pancreas (behind stomach)
Produces and releases digestive enzymes and bicarbonate ions into small intestine

Appendix

Gastrointestinal tract

Mouth
Mechanical breakdown of food; begins chemical digestion of carbohydrates

Pharynx
Connects mouth with esophagus; routes air to trachea

Esophagus
Peristalsis pushes food to stomach

Stomach
Mixes food; enzymatic digestion of proteins

Small intestine
Final enzymatic breakdown of food molecules; main site of food and water absorption

Large intestine
Absorbs water and minerals

Rectum
Regulates elimination of feces

Anus

Digestive System	
Main tissue types*	**Examples of locations/functions**
Epithelial	Secretes hormones, enzymes, and mucus into digestive tract; absorbs products of digestion; protects mouth, esophagus, and anal canal from pathogens and abrasion.
Connective	Blood (a connective tissue) transports nutrients from the digestive system to all parts of the body; supports esophagus, liver, and digestive lining.
Muscle	Smooth muscle moves food along digestive tract and aids in mechanical digestion; skeletal and smooth muscle control mouth, tongue, esophagus, and anal canal.
Nervous	Stretch receptors signal presence of food in stomach; nerves regulate activity of digestive organs.

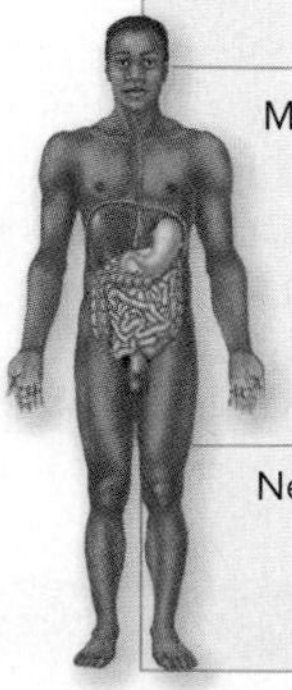

*See chapter 25 for descriptions.

Figure 32.7 The Human Digestive System. Food breaks down as it moves along the digestive tract. Accessory organs aid in digestion.

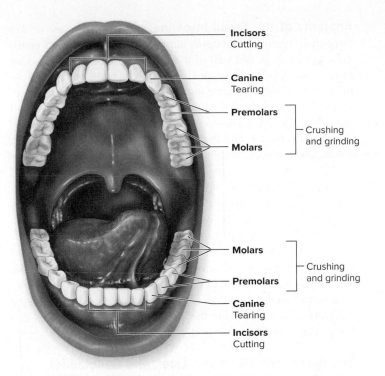

Figure 32.9 The Mouth. Mechanical digestion begins with the teeth, which grasp, cut, crush, and grind food. Salivary glands (such as those under the tongue) contribute fluid and some enzymes.

in the jaws—grasp and chew the food (**figure 32.9**). Chewing is a form of mechanical digestion: Water and mucus in saliva aid the teeth as they tear food into small pieces, increasing the surface area available for chemical digestion. The muscular **tongue** at the floor of the mouth mixes the food with saliva and pushes it to the back of the mouth to be swallowed.

The chewed mass of food passes first through the **pharynx,** or throat, the tube that also conducts air to the trachea. During swallowing, the **epiglottis** temporarily covers the opening to the trachea so that food enters the digestive tract instead of the lungs. From the pharynx, swallowed food and liquids pass to the **esophagus,** a muscular tube leading to the stomach. Food does not merely slide down the esophagus under the influence of gravity; instead, contracting muscles push it along in a wave of peristalsis.

B. The Stomach Stores, Digests, and Churns Food

The **stomach** is a muscular bag that receives food from the esophagus (**figure 32.10**). The stomach is about the size of a large sausage when empty, but when very full, it can expand to hold as much as 3 or 4 liters of food. Ridges in the stomach's lining can unfold like the pleats of an accordion to accommodate a large meal.

The stomach absorbs very few nutrients, but it can absorb some water and ions, a few drugs (e.g., aspirin), and, like the rest of the digestive tract, alcohol. As a result, we feel alcohol's intoxicating effects quickly.

The stomach's main function, however, is to continue the mechanical and chemical digestion of food. Swallowed chunks of food break into smaller pieces as waves of peristalsis churn the stomach's contents. At the same time, the stomach lining produces **gastric juice,** a mixture of water, mucus, salts, hydrochloric acid, and enzymes.

This gastric juice comes from mucous, chief, and parietal cells housed in gastric pits in the epithelium of the stomach. Mucous cells near the entrance to each pit produce mucus. Chief cells secrete a protein that becomes **pepsin,** an enzyme that digests proteins. Parietal cells release hydrochloric acid upon stimulation by a hormone called gastrin. Thanks to this acid, the pH of the gastric juice is low, about 1.5 or 2. The acidity denatures the proteins in food, kills most disease-causing organisms, and activates pepsin so that protein digestion can begin. ⓘ *epithelial tissue,* section 25.2A; *pH,* section 2.4A

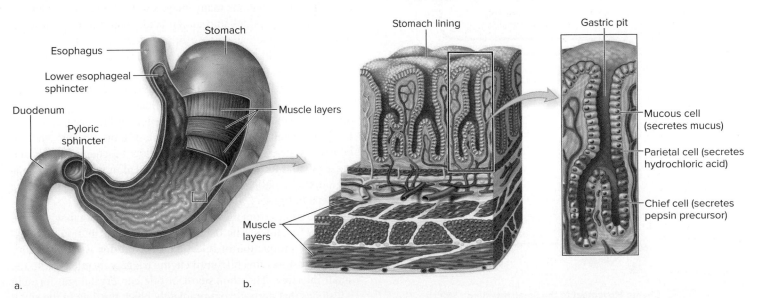

Figure 32.10 The Stomach. (a) Layers of muscle enable the stomach to move food, mix it with gastric juices, and break it into smaller pieces. Two sphincters control the entrance and exit of food. (b) The stomach lining contains cells that secrete mucus and gastric juice.

If gastric juice breaks down protein in food, how does the stomach keep from digesting itself? First, the stomach produces little gastric juice until food is present. Second, mucus coats and protects the stomach lining. Tight junctions between cells in the stomach lining also prevent gastric juice from seeping through to the tissues below. ⓘ *animal cell junctions,* section 3.6A

Chyme is the semifluid mixture of food and gastric juice in the stomach. Small amounts of chyme periodically squirt through the pyloric sphincter that links the stomach and the upper part of the small intestine (the duodenum). A negative feedback loop controls the opening and closing of the pyloric sphincter (**figure 32.11**). ⓘ *negative feedback,* section 25.4

C. The Small Intestine Digests and Absorbs Nutrients

The **small intestine** is a tubular organ that completes digestion and absorbs nutrients and water. Although narrow in comparison with the large intestine, the small intestine's 3- to 7-meter length makes it the longest organ in the digestive system.

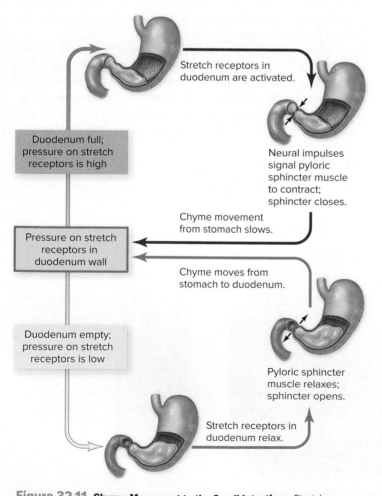

Stretch receptors in duodenum are activated.

Duodenum full; pressure on stretch receptors is high

Neural impulses signal pyloric sphincter muscle to contract; sphincter closes.

Chyme movement from stomach slows.

Pressure on stretch receptors in duodenum wall

Chyme moves from stomach to duodenum.

Duodenum empty; pressure on stretch receptors is low

Pyloric sphincter muscle relaxes; sphincter opens.

Stretch receptors in duodenum relax.

Figure 32.11 Chyme Movement to the Small Intestine. Stretch receptors signal the pyloric sphincter to open or close, depending on the quantity of food in the duodenum.

Anatomy of the Small Intestine   The small intestine has three main regions. The duodenum makes up the first 25 centimeters. Glands in the wall of the duodenum secrete mucus that protects and lubricates the small intestine. In addition, ducts from the pancreas and liver open into the duodenum. The next two regions, the jejunum and the ileum, form the majority of the small intestine. The 2.5-meter-long jejunum absorbs most carbohydrates and proteins, and the 3.5-meter-long ileum absorbs most water, fats, vitamins, and minerals.

Close examination of the hills and valleys along the small intestine's lining reveals millions of **intestinal villi,** tiny finger-like projections that absorb nutrients (**figure 32.12**). The epithelial cells on the surface of each villus bristle with hundreds of **microvilli,** extensions of the cell membrane. Villi and microvilli increase the surface area of the small intestine at least 600 times, allowing for the efficient extraction of nutrients from food. The capillaries that snake throughout each villus take up the newly absorbed nutrients and water, then empty into veins that carry the nutrient-laden blood to the liver. Also inside each villus is a lymph capillary that receives digested fats.

The Role of the Pancreas, Liver, and Gallbladder   The small intestine absorbs water, minerals, free amino acids, cholesterol, and vitamins without further digestion. Most molecules in food, however, require additional processing.

Digestive enzymes released by cells lining the small intestine act on short polysaccharides and disaccharides to release simple sugars, which the small intestine immediately absorbs. These enzymes are called carbohydrases. People who lack one such enzyme, lactase, cannot digest milk sugar; this section's Burning Question discusses this condition.

Most of the digestive enzymes in the small intestine, however, come from the **pancreas,** an accessory organ (see figure 32.7) that sends about a liter of pancreatic "juice" to the duodenum each day. This fluid contains many enzymes. Trypsin and chymotrypsin break polypeptides into amino acids; pancreatic amylase digests starch; pancreatic lipase breaks down fats; and nucleases split nucleic acids such as DNA into nucleotides. In addition to these enzymes, pancreatic juice also contains alkaline sodium bicarbonate, which neutralizes the acid from the stomach.

Fats present an interesting challenge to the digestive system. Lipase enzymes are water-soluble, but fats are not. Therefore, lipase can act only at the surface of a fat droplet, where it contacts water. **Bile** is a greenish-yellow biochemical that disperses the fat into tiny globules suspended in water. The resulting mixture, called an emulsion, increases the surface area exposed to lipase.

Bile comes from the **liver,** a large accessory organ with more than 200 functions. The liver's only *direct* contribution to digestion is the production of bile. The **gallbladder** is an accessory organ that stores this bile until chyme triggers its release into the small intestine. The cholesterol in bile can crystallize, forming gallstones that partially or completely block the duct to the small intestine. Gallstones are very painful and may require removal of the gallbladder. A person can survive without this organ because

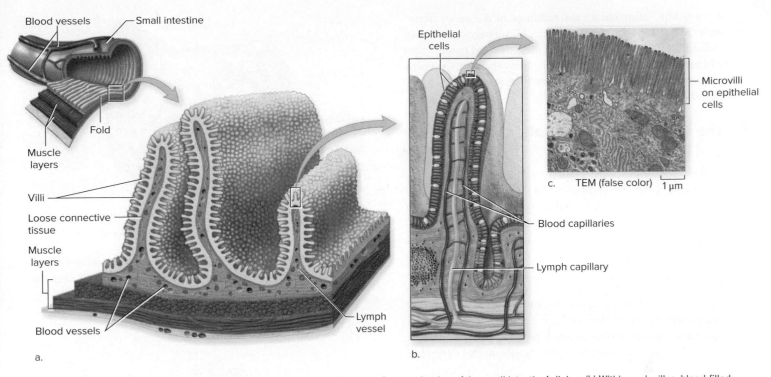

Figure 32.12 **The Lining of the Small Intestine.** (a) Fingerlike villi project from each ridge of the small intestine's lining. (b) Within each villus, blood-filled capillaries absorb digested carbohydrates and proteins, and lymph capillaries absorb digested fats. Epithelial cells coat each villus. (c) Microvilli extending from the epithelial cells add tremendous surface area for absorption.

(c): Courtesy of David H. Alpers, M.D.

surgeons simply redirect the flow of bile from the liver to the small intestine. ⓘ *cholesterol, section 2.5E*

The liver's other functions include detoxifying alcohol and other harmful substances in the blood, storing glycogen or releasing glucose as needed, producing fat and releasing it to adipose tissue, storing iron and fat-soluble vitamins, and producing blood-clotting proteins. The liver receives nutrient-laden blood from the intestines via the hepatic portal vein (see figure 30.8). The blood passes through the liver's extensive capillary beds, which remove bacteria and toxins. The liver also gets "first dibs"

on the nutrients in the blood before it is pumped to the rest of the body. In a condition called cirrhosis, scar tissue blocks this vital blood flow through the liver.

Like the stomach, the small intestine protects itself against self-digestion by producing digestive biochemicals only when food is present. In addition, mucus protects the intestinal wall from digestive juices and neutralizes stomach acid. Nevertheless, many cells of the intestinal lining die in the caustic soup. Rapid division of the small intestine's epithelial cells compensates for the loss, replacing the lining every 36 hours.

Burning Question

What's lactose intolerance?

Lactose, or milk sugar, is a disaccharide in milk. In an infant's small intestine, the lactase enzyme breaks down lactose. Most people stop producing lactase after infancy; after all, milk is not typically part of an adult mammal's diet. If a person without lactase consumes milk, bacteria in the large intestine ferment the undigested sugar, producing abdominal pain, gas, diarrhea, bloating, and cramps.

People with lactose intolerance can prevent these symptoms simply by avoiding fresh milk. Instead, they can choose fermented dairy products such as yogurt, buttermilk, and cheese; bacteria have already broken down the

lactose in those foods. Taking lactase tablets can also prevent symptoms.

Interestingly, people with roots in northern Europe and a few other locations continue to produce lactase as adults, thanks to a long-ago genetic mutation that was adaptive in dairy-herding regions of the world.

Submit your burning question to
Marielle.Hoefnagels@mheducation.com

©Burke/Triolo Productions/Getty Images RF

Figure 32.13 summarizes the locations of the major digestive enzymes, along with their products. Ultimately, as we have seen, the small intestine absorbs these nutrients and passes them to the bloodstream. Cells throughout the body then use the nutrients to generate energy and to build new proteins, carbohydrates, lipids, and nucleic acids (see chapter 2).

D. The Large Intestine Completes Nutrient and Water Absorption

The material remaining in the small intestine moves next into the **large intestine,** which extends from the ileum to the anus while forming a "frame" around the small intestine (**figure 32.14**). At 1.5 meters long, the large intestine is much shorter than the small intestine, but its diameter is greater (about 6.5 centimeters). Its main functions are to absorb water and salts and to eliminate the remaining wastes as feces.

The start of the large intestine is the pouchlike cecum. Dangling from the cecum is the appendix, a thin, worm-shaped tube. Trapped bacteria or undigested food can cause the appendix to become irritated, inflamed, and infected, producing severe pain. A burst appendix can spill its contents into the abdominal cavity and

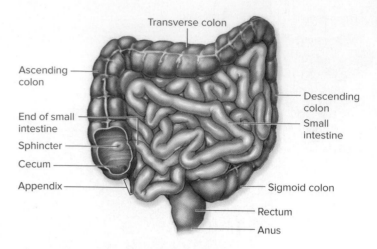

Figure 32.14 The Large Intestine. The large intestine receives chyme from the small intestine. The lining of the intestine absorbs water, salts, and minerals; whatever is left is eliminated as feces.

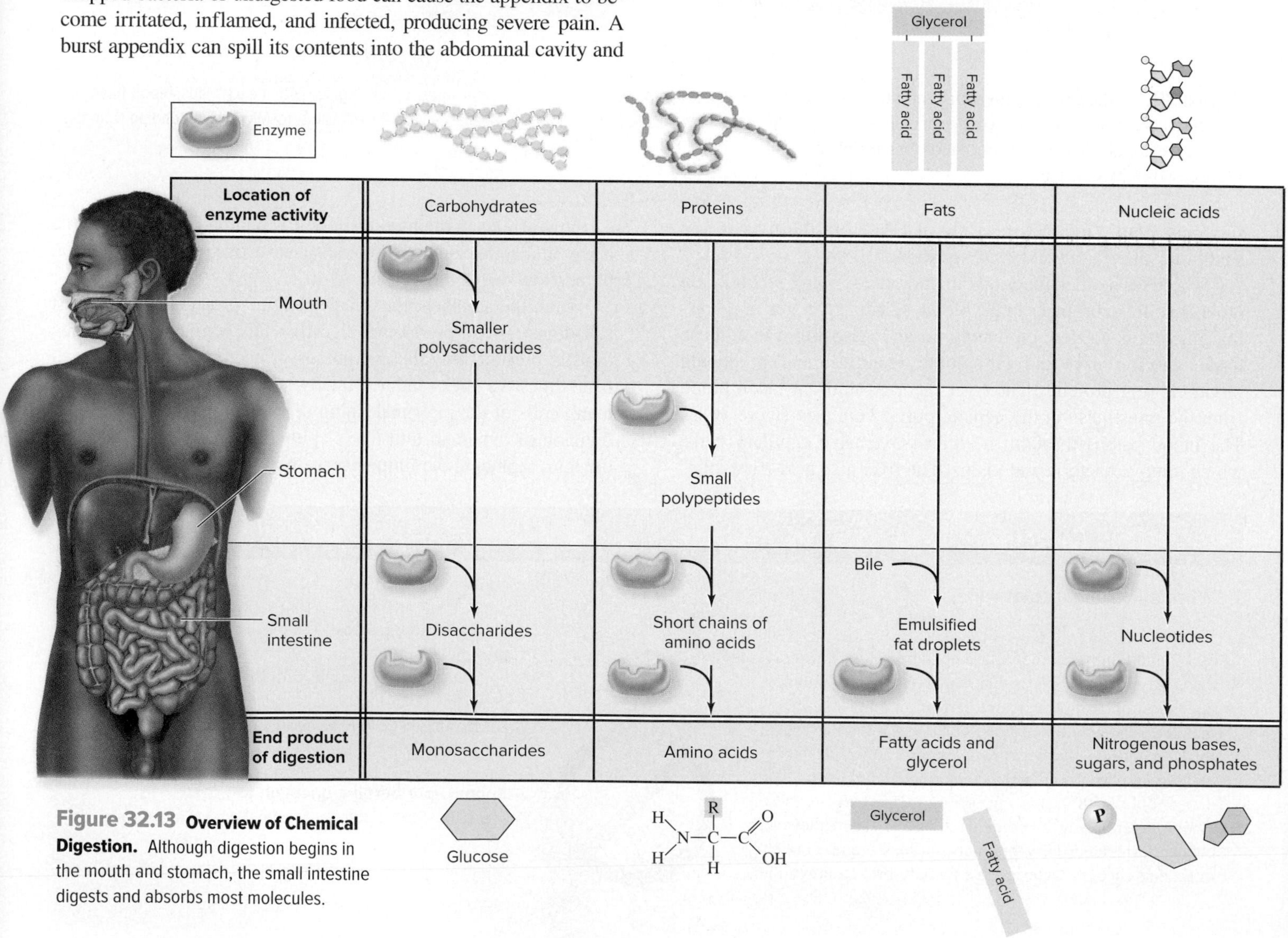

Location of enzyme activity	Carbohydrates	Proteins	Fats	Nucleic acids
Mouth	Smaller polysaccharides			
Stomach		Small polypeptides		
Small intestine	Disaccharides	Short chains of amino acids	Bile → Emulsified fat droplets	Nucleotides
End product of digestion	Monosaccharides	Amino acids	Fatty acids and glycerol	Nitrogenous bases, sugars, and phosphates

Enzyme

Glycerol Fatty acid Fatty acid Fatty acid

Figure 32.13 Overview of Chemical Digestion. Although digestion begins in the mouth and stomach, the small intestine digests and absorbs most molecules.

Glucose

Glycerol

Fatty acid

Apply It **Now**

The Unhealthy Digestive System

The entire length of the human digestive tract is subject to numerous disorders. Some are a nuisance or easily treated, whereas others can be deadly. A few are listed here.

Tooth Decay

Bacteria living in the mouth secrete acids that eat through a tooth's surface, causing cavities. This decayed area can extend to the interior of the tooth, eventually killing the tooth's nerve and blood supply.

Acid Reflux

Gastric juice, normally confined to the stomach, sometimes emerges through the esophageal sphincter and burns the esophagus. This painful condition is commonly known as "heartburn." Antacids can neutralize the acidity.

Vomiting

When a person vomits, the medulla (in the brainstem) coordinates several simultaneous events. The muscles of the diaphragm and abdomen contract, while the sphincter at the entrance of the stomach relaxes. These actions force chyme out of the stomach and up the esophagus. Alcohol, bacterial toxins from spoiled foods, and excessive eating can trigger queasiness and vomiting.

Hepatitis

Hepatitis literally means inflammation of the liver. Many conditions can cause hepatitis, including alcohol abuse, some pharmaceutical drugs, and eating poisonous mushrooms. Some forms of hepatitis are contagious; five viruses (hepatitis A–E) cause most cases of infectious hepatitis.

Diarrhea

If the intestines fail to absorb as much water as they should, the feces become loose and watery. The risk of dehydration is high. Many food- and waterborne viruses, bacteria, and protists cause diarrhea, as can treatment with some antibiotics. Diarrhea is a major cause of death in underdeveloped countries.

©Michael Matisse/Getty Images RF

Constipation

A constipated person defecates less frequently than three times a week, and the feces are hard, dry, and difficult to eliminate. Causes of constipation include an obstructed large intestine, loss of peristalsis, dehydration, starvation, and anxiety. A high-fiber diet prevents constipation by easing the movement of food through the digestive system.

Colon (Colorectal) Cancer

Cancerous tumors may arise in the rectum, colon, or appendix. This is among the most common cancer types, and it is a leading cause of death worldwide. Periodic colon cancer screening leads to early detection, which dramatically improves the survival rate.

Hemorrhoids

These swollen, distended veins protrude into the rectum or anus, causing painful defecation and bleeding. Chronic constipation aggravates hemorrhoids, as do pregnancy and obesity.

---

spread the infection. This chapter's Apply It Now box describes some other examples of illnesses affecting the gastrointestinal tract.

The colon forms the majority of the large intestine. Here, the large intestine absorbs most of the remaining water, ions, and minerals from chyme. Veins carry blood from vessels surrounding the large intestine to the liver.

The remnants of digestion consist mostly of bacteria, undigested fiber, and intestinal cells. These materials, plus smaller amounts of other substances, collect in the rectum as feces. When the rectum is full, the feces are eliminated through the anus.

Our partnership with our intestinal microbes deserves special mention (see the chapter opening essay). Trillions of bacteria, representing about 500 different species, are normal inhabitants of the large intestine. These microscopic residents produce the characteristic foul-smelling odors of intestinal gas and feces, but they also provide many benefits. Most notably, they help prevent infection by disease-causing microorganisms—an interesting application of ecology's competitive exclusion principle. They also decompose cellulose and some other nutrients, produce B vitamins and vitamin K, and break down some

drugs. Antibiotics often kill these normal bacteria and allow other microbes to take over. This disruption causes the diarrhea that may accompany treatment with antibiotics. ⓘ *competitive exclusion,* section 38.1A

Interestingly, a baby is born with a microbe-free digestive tract. The infant begins acquiring its microbiota with its first meal of milk or formula. Bacteria from the mother's skin and the environment gradually colonize the baby's intestines, establishing populations that will persist throughout life.

32.3 MASTERING CONCEPTS

1. Explain the action and importance of peristalsis and sphincters in digestion.
2. Describe the functions of saliva, teeth, and the tongue.
3. How does food move from the mouth to the stomach?
4. Describe the digestion that occurs in the stomach.
5. What are the structure and function of the small intestine?
6. How do the pancreas, liver, and gallbladder aid digestion?
7. Describe the events that occur in the large intestine.

32.4 A Healthy Diet Includes Essential Nutrients and the Right Number of Calories

Healthy eating has two main components. First, the diet must include all of the nutrients necessary to sustain life. A second consideration is the calorie content of food, which must balance a person's metabolic rate and activity level.

A. A Varied Diet Is Essential to Good Health

Nutrients fall into two categories. **Macronutrients** are required in large amounts. Water is a macronutrient; all living cells require water as a solvent and as a participant in many reactions. Organisms use three other macronutrients—carbohydrates, proteins, and lipids—to build cells and to generate ATP. Despite the many nonfat foods on grocery store shelves, the diet must include *all* of these nutrients, including moderate amounts of fat. For adults, a typical 2000-Calorie diet might include 50 grams of protein, 65 grams of fat, and 300 grams of carbohydrates.

Eating a variety of proteins and fats is important because the body requires their components—amino acids and fatty acids—for many functions. An amino acid or fatty acid is called "essential" if the body cannot produce it on its own; these molecules must come from the diet. A "complete protein" is a food that contains all nine essential amino acids. Meats are often complete proteins, as are mixtures of grains (e.g., rice) and legumes (e.g., beans). In humans, only two fatty acids are essential: one omega-3 fatty acid and one omega-6 fatty acid (the "omega" number refers to the position of the double bond in the unsaturated fatty acid tails). Dietitians recommend consuming omega-3 and omega-6 fatty acids in equal amounts. Many modern foods, such as vegetable oils, contain abundant omega-6 fatty acids; canola oil and fish contain omega-3 fatty acids. ⓘ *amino acids,* section 2.5C; *fatty acids,* section 2.5E

TABLE **32.1**   Selected Minerals in the Human Diet

Element	Food Sources	Function(s)	Selected Deficiency Symptoms
Bulk Minerals			
Calcium (Ca)	Milk products, green leafy vegetables	Electrolyte,* bone and tooth structure, blood clotting, hormone release, nerve transmission, muscle contraction	Muscle cramps and twitches, weakened bones, heart malfunction
Chlorine (Cl)	Table salt, meat, fish, eggs, poultry, milk	Electrolyte, part of stomach acid	Muscle cramps, nausea, weakness
Magnesium (Mg)	Green leafy vegetables, beans, fruits, peanuts, whole grains	Muscle contraction, nucleic acid synthesis, enzyme activity	Tremors, muscle spasms, loss of appetite, nausea
Phosphorus (P)	Meat, fish, eggs, poultry, whole grains	Bone and tooth structure; part of DNA, ATP, and cell membranes	Weakness, mineral loss from bones
Potassium (K)	Fruits, potatoes, meat, fish, eggs, poultry, milk	Electrolyte, nerve transmission, muscle contraction, nucleic acid synthesis	Weakness, loss of appetite, muscle cramps, confusion, heart arrhythmia
Sodium (Na)	Table salt, meat, fish, eggs, poultry, milk	Electrolyte, nerve transmission, muscle contraction	Muscle cramps, nausea, weakness
Sulfur (S)	Meat, fish, eggs, poultry	Hair, skin, and nail structure; blood clotting; energy transfer	Brittle hair and nails, stunted growth
Trace Minerals (Partial List)			
Cobalt (Co)	Meat, eggs, dairy products	Part of vitamin B_{12}	Weakness, fatigue, nerve degeneration
Copper (Cu)	Organ meats, nuts, shellfish, beans	Part of many enzymes, iron metabolism in red blood cells	Anemia, low white blood cell counts
Fluorine (F)	Water (in some areas)	Mineralization of bones and teeth	Tooth decay
Iodine (I)	Seafood, iodized salt	Part of thyroid hormone	Goiter (enlarged thyroid gland)
Iron (Fe)	Meat, liver, fish, shellfish, egg yolk, peas, beans, dried fruit, whole grains	Part of hemoglobin and myoglobin, part of some enzymes	Anemia, learning deficits in children
Selenium (Se)	Meat, milk, grains, onions	Part of some enzymes, heart function	Weakened heart, degeneration of cartilage
Zinc (Zn)	Meat, fish, egg yolk, milk, nuts, some whole grains	Part of some proteins, regulation of gene expression	Stunted growth, diarrhea, impaired immune function

*An electrolyte is an ion that helps regulate osmotic balance.

Unlike macronutrients, **micronutrients** are required in very small amounts; recommended daily intakes are measured in milligrams or micrograms. Two examples are minerals and vitamins, neither of which are used as fuel because they do not contain calories. Instead, these micronutrients participate in many aspects of cell metabolism. Tables 32.1 and 32.2 provide details on the sources and functions of minerals and vitamins. Many processed foods are fortified with vitamins and minerals, so deficiencies in developed countries are rare.

The best way to acquire all required nutrients is to eat a varied diet. (Food storage and preparation affect a food's nutrient content, as described in this section's Burning Question.) The U.S. government's food guidelines emphasize whole grains, fresh vegetables, and low-fat dairy products, along with fruits and limited amounts of meat and fat. The Harvard School of Public Health suggests a somewhat different diet that minimizes dairy products, red meat, and starchy processed grains. Whole grains

and vegetable oils, along with abundant vegetables, make up the base of this pyramid.

The indigestible components of food help maintain good health, too. Dietary fiber, for example, is composed of cellulose from plant cell walls. Humans do not produce cellulose-digesting enzymes, so fiber contributes only bulk—not nutrients—to food. This increased mass eases movement of the food through the digestive tract, so cancer-causing ingredients in food contact the walls of the intestines for a shorter period. People who consume abundant fiber in their food therefore have a lower incidence of colorectal cancer. A high-fiber diet also reduces blood cholesterol and helps regulate blood sugar. ⓘ *cellulose,* section 2.5B

A balanced diet delivers many long-term health benefits, including a reduced risk of type 2 diabetes, cancer, osteoporosis, high blood pressure, and heart disease. Fortunately, the labels of packaged foods list not only ingredients but also the

TABLE **32.2** Vitamins in the Human Diet

Vitamin	Food Sources	Function(s)	Selected Deficiency Symptoms
Water-Soluble Vitamins			
B complex vitamins			
Thiamine (vitamin B_1)	Pork, beans, peas, nuts, whole grains	Growth, fertility, digestion, nerve cell function, milk production	Beriberi (confusion, loss of appetite, muscle weakness, heart failure)
Riboflavin (vitamin B_2)	Liver, leafy vegetables, dairy products, whole grains	Energy use	Cracked skin at corner of mouth
Niacin (vitamin B_3)	Liver, meat, peas, beans, whole grains, fish	Growth, energy use	Pellagra (diarrhea, dementia, dermatitis)
Pantothenic acid (vitamin B_5)*	Liver, eggs, peas, potatoes, peanuts	Growth, cell maintenance, energy use	Headache, fatigue, poor muscle control, nausea, cramps
Pyridoxine (vitamin B_6)*	Red meat, liver, corn, potatoes, whole grains, green vegetables	Protein metabolism, production of neurotransmitters	Mouth sores, dizziness, nausea, weight loss, neurological disorders
Biotin (vitamin B_7)*	Meat, milk, eggs	Metabolism	Skin disorders, muscle pain, insomnia, depression
Folic acid (vitamin B_9)	Liver, navy beans, dark green vegetables	Manufacture of red blood cells, metabolism	Weakness, fatigue, diarrhea, neural tube defects in fetus
Cobalamin (vitamin B_{12})	Meat, organ meats, fish, shellfish, milk	Manufacture of red blood cells, maintenance of myelin sheath	Weakness, fatigue, nerve degeneration
Ascorbic acid (vitamin C)	Citrus fruits, tomatoes, peppers, strawberries, cabbage	Antioxidant, production of connective tissue and neurotransmitters	Scurvy (weakness, gum bleeding, weight loss)
Fat-Soluble Vitamins			
Retinol (vitamin A)	Liver, dairy products, egg yolk, vegetables, fruit	Night vision, new cell growth	Blindness, impaired immune function
Cholecalciferol (vitamin D)	Fish liver oil, milk, egg yolk	Bone formation	Skeletal deformation (rickets)
Tocopherol (vitamin E)*	Vegetable oil, nuts, beans	Antioxidant	Anemia in premature infants
Vitamin K*	Liver, egg yolk, green vegetables	Blood clotting	Internal bleeding

*Deficiencies in these vitamins are rare in humans, but they have been observed in experimental animals.

nutrient content in each serving (figure 32.15). Some labels may have additional information; for example, chapter 2's Burning Question explains the distinction between "natural" and "organic."

B. Body Weight Reflects Food Intake and Activity Level

Nutritional labels also list a food's caloric content. This information is determined by burning food in a bomb calorimeter, a chamber immersed in water and designed to measure heat output. Energy released from the food raises the water temperature: 1 food **Calorie** (also called a kilocalorie) is the energy needed to raise 1 kilogram of water from 14.5°C to 15.5°C under controlled conditions.

Most young adults require 2000 to 2400 Calories per day, depending on sex and level of physical activity. Bomb calorimetry studies have shown that 1 g of carbohydrate yields 4 Calories, 1 g of protein yields 4 Calories, and 1 g of fat yields 9 Calories. Although the body cannot extract all of the potential energy in food, these values help explain the link between fatty foods and weight gain; excess dietary fat supplies more energy than most people can use.

Interestingly, the amount of energy that food delivers to the body depends in part on how it is prepared. Chopping food into tiny pieces—say, in a blended smoothie—makes its calories more easily accessible to the digestive system. So does heating it up. Cooked foods are often softer than their raw counterparts and therefore require less mechanical energy (and fewer calories) to chew and digest. Gram for gram, processed foods such as refined flour also deliver more calories than do whole, unprocessed foods. Enzymes in the small intestine easily digest carbohydrates

Figure 32.15 Nutrition Information. The packaging of processed foods includes a standard nutrition label that indicates the calorie and nutrient content in each serving.
©Jill Braaten/McGraw-Hill Education

Burning Question

How can I maximize the nutrient content of my food?

In making everyday food choices, it is tempting to ignore fruits and vegetables in favor of cheaper, more convenient processed foods. Eating fresh foods, however, is correlated with good health. Several strategies can help maximize the nutrients delivered in every bite.

In general, fruits and vegetables are at their peak nutrient level immediately after harvest, regardless of whether the plant was organically or conventionally grown. That's because once a plant is removed from the ground, vitamins begin to break down upon exposure to heat, light, oxygen, and other stressors. Buying local produce decreases the time between harvest and consumption. As a result, local fruits and vegetables typically have higher nutrient content than do those shipped from distant sources. Canning and freezing also deplete many vitamins, but once processing is complete, the food's nutrient content changes little. Therefore, by the time it is consumed, canned or frozen produce may have vitamin levels similar to those in unprocessed fruits and vegetables picked weeks ago.

Food storage and preparation techniques may also affect nutrient content. Cool temperatures and high humidity typically delay vitamin breakdown in harvested plants. Cutting the produce into small pieces exposes more surfaces to the vitamin-damaging conditions, so chopping should be delayed until just before food is prepared. Finally, boiling vegetables leaches water-soluble vitamins into the hot water; steaming the food minimizes this loss.

©Roy Hsu/Getty Images RF

Fortunately, healthy eating does not require us to obsess about complicated rules, "forbidden foods," "miracle foods," and calories. Instead, food writer Michael Pollan recommends this simple strategy for healthy eating: "Eat food, not too much, mostly plants." Regardless of how you select or prepare them, a diet rich in fruits and vegetables is one simple key to good health.

Submit your burning question to
Marielle.Hoefnagels@mheducation.com

TABLE **32.3** Beverage Nutrition

Drink (per 240 mL; 8 oz)	Sugar (g)	Fat (g)	Alcohol* (g)	Calories**
Water, black coffee, tea (unsweetened)	0	0	0	0
Sports drink	14	0	0	55
Light beer	3	0	8	70
Soda	26	0	0	95
Lowfat milk (1%)	13	3	0	100
Orange juice	24	0	0	115
Craft beer	9	0	10	120
Wine	2	0	24	190
Blended coffee drink	30	15	0	205

*Alcohol contains 7 Calories per gram.
**Calories may not exactly reflect values in table because numbers may be rounded.

and protein in a processed food. Conversely, raw, unprocessed foods often contain fiber that speeds passage through the digestive tract, leaving many macronutrients undigested.

Many people overlook the abundant calories in some drinks, even though a gram of sugar in a beverage contains as much energy as a gram of sugar in fruit or a candy bar. Table 32.3 lists nutritional information for a selection of popular drinks.

No matter what we eat or drink, taking in more calories than we expend causes weight gain; those who consume fewer calories than they expend lose weight and may even starve.

What constitutes a "healthful" weight? The most common measure is the body mass index, or BMI (figure 32.16). To calculate BMI, divide a person's weight (in kilograms) by his or her squared height (in meters): BMI = weight/(height)². Alternatively, multiply weight (in pounds) by 704.5, and divide by the square of the height (in inches).

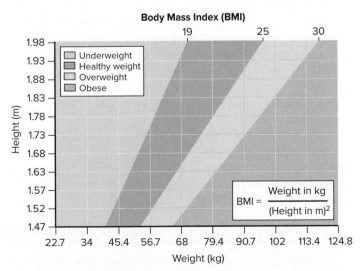

Body Mass Index (BMI)

$$BMI = \frac{\text{Weight in kg}}{(\text{Height in m})^2}$$

Source: U.S. Department of Agriculture: Dietary Guidelines for Americans

Figure 32.16 Body Mass Index. This chart is a quick substitute for a BMI calculation; the intersection of a person's height (in meters) and weight (in kilograms) indicates the BMI range.

Many health professionals consider a person whose BMI is less than 19 to be underweight. A BMI between 19 and 25 is healthy; an overweight person has a BMI greater than 25; and a BMI greater than 30 denotes **obesity.** Morbid obesity is defined as a BMI greater than 40. One limitation of BMI is that it cannot account for many of the details that affect health. An extremely muscular person, for example, will have a high BMI because muscle is denser than fat, yet he or she would not be considered overweight.

Thanks in part to these limitations, many health professionals rely on other indicators of healthy weight, including abdomen circumference. People whose fat accumulates at the waistline ("apples") are more susceptible to health problems such as insulin resistance than are "pears," who are bigger around the hips.

Figure It Out

Consider the following nutritional facts—bacon cheeseburger: 23 g fat, 25 g protein, 2 g carbohydrates; large fries: 25 g fat, 6 g protein, 63 g carbohydrates; large soda: 86 g carbohydrates. How many Calories are in this meal?

Answer: 1160 Calories

C. Starvation: Too Few Calories to Meet the Body's Needs

A diet is inadequate if it either contains too few calories to sustain life or fails to provide an essential nutrient (figure 32.17). In some areas of the world, famine is a constant condition, and millions of people starve to death every year. Hunger strikes, inhumane treatment of prisoners, and eating disorders can also cause starvation.

A healthy human can survive for 50 to 70 days without food—much longer than without air or water. The exact timeline varies from person to person, but the starving human body essentially digests itself. After only a day without food, reserves of sugar and glycogen are dwindling, and the body begins extracting energy from stored fat and muscle protein. Gradually, metabolism slows, blood pressure drops, the pulse slows, and chill sets in. Skin becomes dry and hair falls out as the proteins that form these structures are digested. When the body dismantles the immune system's antibody proteins, protection against infection declines. Mouth sores and anemia develop, the heart beats irregularly, and bones begin to degenerate. Near the end, the starving human is blind, deaf, and emaciated.

Anorexia nervosa, or self-imposed starvation, is refusal to maintain normal body weight. The condition affects about 1 in 250 adolescents, more than 90% of whom are female. The sufferer's body image is distorted; although she may be extremely thin, she perceives herself as overweight. A person with anorexia eats barely enough to survive, losing as much as 25% of her original body weight. She may further lose weight by vomiting, taking laxatives and diuretics, or exercising intensely. About 15% to 21% of people with anorexia die from the disease.

Bulimia is another eating disorder that mainly affects females. Rather than avoiding food, a person with **bulimia** eats large quantities and then intentionally vomits or uses laxatives

Figure 32.17 Malnutrition and Starvation. The malnourished boy on the left has a swollen belly, a sign of a severe protein deficiency. The girl on the right suffers from anorexia nervosa, or self-imposed starvation.

(boy): ©Brennan Linsley/AP Images; (girl): ©Bubbles Photolibrary/Alamy

Figure 32.18 Hormone Deficiency. In a normal mouse, leptin secreted by fat cells affects target cells in the hypothalamus, helping the brain regulate appetite and metabolism. The obese mouse on the left cannot produce leptin and therefore has a ravenous appetite.

©Chicago Tribune/Newscom

shortly afterward, a pattern called "binge and purge." A person with bulimia may or may not be underweight.

Fortunately, eating disorders are often treatable under the combined care of a physician, dietician, and psychologist. These health professionals can help the patient return to a normal body weight, develop healthy eating habits, and address underlying psychological problems.

D. Obesity: More Calories Than the Body Needs

Obesity is increasingly common in the United States (see figure 28.12), and the health consequences can be serious. People who accumulate fat around their waists are susceptible to type 2 diabetes, high blood pressure, and atherosclerosis. High body weight is also correlated with heart disease, acid reflux, urinary incontinence, low back pain, stroke, sleep disorders, and other health problems. In addition, obese people also face higher risk of cancers of the colon, breast, and uterus. ⓘ *diabetes,* section 28.4D

Excess body weight accumulates when a person consumes more calories in food and beverages than he or she expends. For most people, the main culprits are an inactive lifestyle coupled with a diet loaded with sugar and fat. Oversized portions at restaurants contribute to the problem; consuming too few fresh fruits and vegetables does, too.

Nevertheless, many genes contribute to appetite, digestion, and metabolic rate, so a person's family history plays at least a small part in the risk for obesity. Some of these genes encode hormones, which interact in complex and poorly understood ways to maintain the balance between food intake and energy expenditure. Two weight-related hormones are leptin and ghrelin. Stored

fat (adipose tissue) releases leptin into the bloodstream. At the hypothalamus, leptin interacts with receptors and triggers a signal cascade that inhibits food intake and increases metabolic activity. This action explains why leptin-deficient mice become extremely obese (**figure 32.18**). Leptin deficiency, however, is very rare in humans; in fact, in most overweight people, leptin is abundant, but the hypothalamus is somehow leptin-resistant.

Whereas leptin is an indicator of relatively long-term fat storage, ghrelin helps to stimulate hunger in the short term. Cells in the stomach release ghrelin into the bloodstream before a meal, and ghrelin production slows after eating. This hormone interacts with receptors in the hypothalamus and elsewhere.

Research into leptin, ghrelin, and other appetite-related hormones may someday yield new treatments for obesity. In the meantime, the concern over expanding waistlines has fueled demand for low-calorie artificial sweeteners and fats (see the Apply It Now box in section 2.5). Many people also follow fad diets that ban some foods entirely or overemphasize others, often based on shaky (or no) science.

The most healthful way to lose weight, however, is to exercise and reduce calorie intake while maintaining a balanced diet. For people who have difficulty losing weight in this way, stomach-reduction surgery and drugs that either reduce appetite or block fat absorption offer other options.

32.4 MASTERING CONCEPTS

1. Which nutrients are macronutrients and which are micronutrients?
2. Describe the relationship of body weight to calorie intake and energy expenditure.
3. What is body mass index?
4. What are some of the causes and effects of obesity?

INVESTIGATING LIFE

32.5 The Cost of a Sweet Tooth

Most children know that eating candy causes tooth decay. Sweets fuel the growth of oral bacteria, which release tooth-eroding acids as they metabolize sugar. The result? Cavities.

Bacteria also form clumps called plaque on the tooth surface (figure 32.19). Plaque, in turn, hardens into tartar that causes gingivitis—an inflammation of the gums. Untreated gingivitis may develop into periodontitis, with its accompanying damage to the soft tissues and bones supporting the teeth.

Daily brushing and flossing remove bacteria from the teeth, minimizing the effects of harmful microbes in the mouth. These and other modern practices—including regular visits to the dentist—keep our mouths healthy.

The human diet has not always been as sugary as it is now. Thousands of years ago, during the Mesolithic (mid-Stone Age) period, people were hunters and gatherers. Sweet foods would have been a rare luxury. Neolithic people were the first farmers, with innovations that included the domestication of wheat, barley, and other grains. These new crops signaled a major dietary shift toward starch. Human diets subsequently remained pretty much unchanged until the Industrial Revolution (about 150 years ago), when refined sugars became commonplace.

How did the bacteria colonizing the mouth change over this time? Australian researchers Christina Adler and Alan Cooper teamed up with colleagues in Australia and Europe to find out. The team scraped tartar from the teeth of 34 skeletons from Mesolithic, Neolithic, Bronze Age, and Medieval times; they also collected plaque and tartar from living people in Australia. The scientists then extracted DNA from the tartar and plaque samples. Finally, they used the polymerase chain reaction (PCR) to detect and amplify DNA sequences specific to bacteria. ⓘ *PCR, section 11.2C*

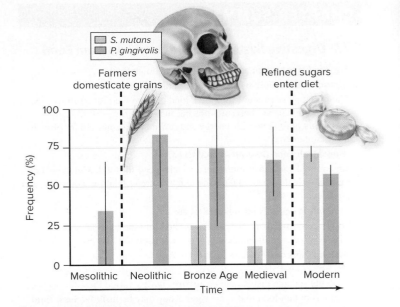

Figure 32.20 The Changing Oral Ecosystem. Researchers collected tartar from the teeth of modern humans and of skeletons ranging from 7500 years old (Mesolithic) to 400 years old (Medieval). The samples revealed that *Streptococcus mutans*, a bacterium associated with tooth decay, has become more common as the human diet has become more sugary. The incidence of another bacterium, *Porphyromonas gingivalis*, has remained stable. (Error bars are a measure of uncertainty in the data; see appendix B.)

The researchers were especially interested in two species of bacteria that cause dental problems today: *Streptococcus mutans* (cavities) and *Porphyromonas gingivalis* (periodontal disease). They used a PCR primer unique to each species to determine if each dental sample contained the bacteria. Their analysis revealed that the incidence of *P. gingivalis* has remained relatively steady since Mesolithic times, but *S. mutans* has become more common along with the availability of sugar (figure 32.20).

This story is an ecological one, as it documents a shift in the bacterial communities occupying the human mouth. But it also raises evolutionary questions. Why hasn't natural selection optimized the human oral ecosystem? That is, why can't our bodies do a better job fighting off the "bad" bacteria that cause tooth decay? Some scientists say our diet has shifted faster than our defenses have evolved, just as today's high rates of diabetes and heart disease reflect a switch to a sedentary lifestyle. In both cases, it seems that our rapid cultural shifts have outstripped the pace of evolution. ⓘ *ecological succession, section 38.2*

Source: Adler, Christina J., and 11 coauthors, including Alan Cooper. 2013. Sequencing ancient calcified dental plaque shows changes in oral microbiota with dietary shifts of the Neolithic and Industrial revolutions. *Nature Genetics*, vol. 45, pages 450–455.

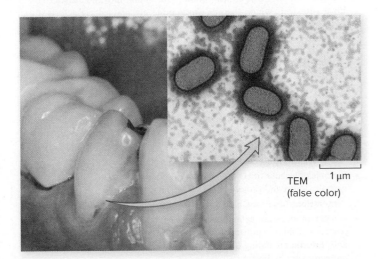

Figure 32.19 Mouth Bacteria. *Porphyromonas gingivalis* bacteria form the plaque that causes gingivitis.

(teeth): ©Backyard Productions/Alamy RF; (bacteria): ©CAMR/A. Barry Dowsett/Science Source

32.5 | MASTERING CONCEPTS

1. How did the researchers analyze the bacterial communities of ancient and modern mouths?

2. What are the differences and similarities between the diversity of mouth bacteria in modern humans versus Mesolithic humans?

CHAPTER SUMMARY

32.1 Digestive Systems Derive Nutrients from Food

- The **digestive system** acquires and breaks down food.

A. Animals Eat to Obtain Energy and Building Blocks

- Plants and algae produce their own food and are therefore **autotrophs,** whereas animals are **heterotrophs** because they consume food. **Nutrients** in food provide energy and raw materials needed for growth and maintenance.

B. How Much Food Does an Animal Need?

- An animal's metabolic rate determines its need for food. Metabolic rate, in turn, reflects the animal's body temperature regulation, body size, and physiological state.

C. Animals Process Food in Four Stages

- Food is **ingested, digested,** and **absorbed** into the bloodstream (**figure 32.21**); indigestible wastes are **eliminated** as **feces.**

D. Animal Diets and Feeding Strategies Vary Greatly

- **Herbivores** eat plants, **carnivores** eat meat, **detritivores** consume decaying organic matter, and **omnivores** have a varied diet.
- Animals obtain food in diverse ways. They may be **bulk feeders, fluid feeders, suspension feeders,** or **substrate feeders.** A **deposit feeder** is a type of substrate feeder.

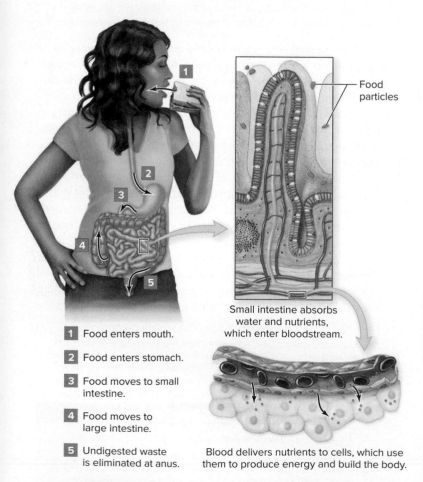

1 Food enters mouth.

2 Food enters stomach.

3 Food moves to small intestine.

4 Food moves to large intestine.

5 Undigested waste is eliminated at anus.

Small intestine absorbs water and nutrients, which enter bloodstream.

Blood delivers nutrients to cells, which use them to produce energy and build the body.

Figure 32.21 The Digestive System: A Summary.

32.2 Animal Digestive Tracts Take Many Forms

- Protists and sponges have intracellular digestion. Their cells engulf food and digest it in food vacuoles.
- Other animals use extracellular digestion, which occurs in a cavity within the body. An **incomplete digestive tract** (also called a **gastrovascular cavity**) has one opening. A **complete digestive tract,** or **alimentary canal (gastrointestinal tract),** has two openings. Food enters through the mouth and is digested and absorbed; undigested material leaves through the **anus.**
- The length of the digestive tract and size of the **cecum** are adaptations to specific diets. In **ruminants,** bacteria inhabiting the rumen help break down hard-to-digest plant matter.

32.3 The Human Digestive System Consists of Several Organs

- Waves of contraction called **peristalsis** move food along the digestive tract. Muscular **sphincters** control movement from one compartment to another.

A. Digestion Begins in the Mouth

- In the mouth, **teeth** break food into small pieces. Salivary glands produce saliva, which moistens food and begins starch digestion.
- With the help of the **tongue,** swallowed food moves past the **pharynx** and through the **esophagus** to the stomach. The **epiglottis** prevents food from entering the trachea.

B. The Stomach Stores, Digests, and Churns Food

- The **stomach** stores food, mixes it with **gastric juice,** and churns it into liquefied **chyme.** Hydrochloric acid in the gastric juice kills microorganisms and denatures proteins. The protein-splitting enzyme **pepsin** begins protein digestion.

C. The Small Intestine Digests and Absorbs Nutrients

- The **small intestine** is the main site of digestion and nutrient absorption. **Intestinal villi** absorb the products of digestion; **microvilli** on each villus provide tremendous surface area.
- The **pancreas** supplies bicarbonate to the small intestine, along with digestive enzymes that break down carbohydrates, polypeptides, lipids, and nucleic acids.
- The **liver** produces **bile,** which emulsifies fat; the **gallbladder** stores the bile and releases it to the small intestine.

D. The Large Intestine Completes Nutrient and Water Absorption

- Material remaining after absorption in the small intestine passes to the **large intestine,** which absorbs water, minerals, and salts. Bacteria digest the remaining nutrients and produce useful vitamins that are then absorbed. Feces exit the body through the anus.

32.4 A Healthy Diet Includes Essential Nutrients and the Right Number of Calories

A. A Varied Diet Is Essential to Good Health

- Metabolism, growth, maintenance, and repair of body tissues all require nutrients from food. **Macronutrients** include carbohydrates, proteins, fats, and water, whereas **micronutrients** include vitamins and minerals.

B. Body Weight Reflects Food Intake and Activity Level

- **Calories** measure the energy stored in food. Fat has 9 Calories per gram; carbohydrates and protein have 4 Calories per gram.

C. Starvation: Too Few Calories to Meet the Body's Needs

- If a person does not eat enough over a long period, the body uses reserves of fat and protein to fuel essential processes. **Anorexia nervosa** and **bulimia** are eating disorders that may reduce calorie intake to dangerously low levels.

D. Obesity: More Calories Than the Body Needs

- A person who eats more calories than he or she expends will gain weight. **Obesity** is associated with many health problems.

32.5 Investigating Life: The Cost of a Sweet Tooth

- Researchers scraped tartar from the teeth of modern humans and of people who lived in Mesolithic, Neolithic, Bronze Age, and Medieval times. They found that the bacterial community in tartar has shifted as our diets have increasingly emphasized starch and refined sugar.

MULTIPLE CHOICE QUESTIONS

1. Which of the following animals uses the most energy per gram of body weight?
   a. Horse
   b. Eagle
   c. Mouse
   d. Python

2. Which digestive organ uses both mechanical and chemical digestion?
   a. Mouth
   b. Stomach
   c. Intestines
   d. Both a and b are correct.

3. At which stage do nutrients enter an organism's bloodstream?
   a. Ingestion
   b. Digestion
   c. Absorption
   d. Elimination

4. Which of the following is NOT true regarding starvation?
   a. Starvation occurs when the body receives fewer calories than it uses.
   b. A healthy human can live without food for up to 70 days.
   c. A starving body uses energy from proteins before it uses energy from fat.
   d. Anorexia nervosa is self-imposed starvation.

5. The protein you eat is mostly
   a. incorporated into your body without modification.
   b. eliminated in feces.
   c. converted to fat before digestion.
   d. dismantled into individual amino acids.

6. A person's body mass index is calculated based on
   a. caloric intake.
   b. height.
   c. weight.
   d. Both b and c are correct.

Answers to these questions are in appendix A.

WRITE IT OUT

1. What are the two main functions of food in an animal's body?

2. On a per-kilogram basis, why does a small mammal such as a shrew require so much more energy than does an elephant or other large mammal?

3. Biologists estimate that carnivores assimilate 90% of the mass in their diet; in contrast, most herbivores assimilate only 30% to 60% of their food. How do these differences in assimilation efficiency relate to the structure of their digestive systems?

4. Trace the movement of food in the digestive tract from mouth to anus.

5. What are the digestive products of carbohydrates, proteins, and fats? Where in the digestive system is each macromolecule digested?

6. Orlistat (Alli) is a weight-loss drug that inhibits the activity of lipases in the small intestine. Why would this be more effective than a drug that blocks the absorption of proteins or carbohydrates? Given that four essential vitamins are fat-soluble, what might be a side effect of blocking fat absorption?

7. Write down the foods you ate today, and use nutrition labels to determine how many Calories you consumed. Then, use calculators on the Internet to estimate how many Calories you require each day based on your age, weight, gender, and activity level. Did you consume more or fewer Calories than you needed? If you repeated these eating habits for several weeks, would you gain or lose weight?

8. Calculate your body mass index using the formula in the text. How could you change your BMI?

9. Fructose and glucose are both monosaccharides, but the body metabolizes these sugars differently. For example, glucose stimulates insulin release from the pancreas (see section 28.4); fructose does not. Moreover, insulin stimulates leptin release. Finally, fructose is more likely than glucose to be converted to fat. Use this information to propose an explanation for the correlation between the skyrocketing consumption of high fructose corn syrup since 1970 and the rise in obesity during the same period.

10. How do the circulatory and muscular systems interact with the digestive system?

11. Design an experiment to test the hypothesis that intestinal bacteria are essential to nutrient absorption in mice.

12. Since chewing gum cannot be digested, many children believe that a swallowed piece will remain in the body for 7 years. Propose an alternative prediction for how long gum might remain in the body, and explain your reasoning.

PULL IT TOGETHER

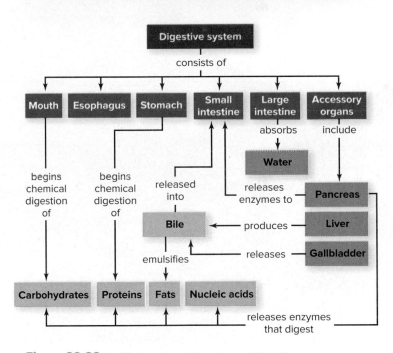

Figure 32.22 Pull It Together: Digestion and Nutrition.

Refer to figure 32.22 and the chapter content to answer the following questions.

1. Review the Survey the Landscape figure in the chapter introduction and the Pull It Together concept map. How do the digestive system and the circulatory system work together to maintain homeostasis?

2. Add the terms *ingestion, digestion, absorption, elimination, chyme, bacteria,* and *peristalsis* to this concept map.

3. How would removing the gallbladder affect the digestion of fats?

33 Regulation of Temperature and Body Fluids

Sea Lizards. Glands on a marine iguana's head eliminate excess salt ingested with food. The salt accumulates as a crust on the lizard's forehead.

©Doug Allan/The Image Bank/Getty Images

LEARN HOW TO LEARN
Pay Attention in Class

It happens to everyone occasionally: Your mind begins to wander while you are sitting in class, so you doodle, check your phone, or doze off. Before you know it, class is over, and you got nothing out of it. How can you keep from wasting your class time this way? One strategy is to get plenty of sleep and eat well so that your mind stays active instead of drifting off. Another is to prepare for class in advance, since getting lost can be an excuse for drifting off. When you get to class, sit near the front, listen carefully, and take good notes. Finally, a friendly reminder can't hurt; make a small PAY ATTENTION sign to put on your desk, where you can always see it.

A Day in the Life of a Marine Iguana

Many environments are hostile to cell survival—they are too hot, too cold, too salty, or not salty enough. Humans cope with some of these extremes in familiar ways; for example, we sweat when we are hot and shiver when we are cold. We can eat salty pretzels, but only to the extent that our kidneys can dispose of the excess salt. Other animals, however, have radically different behavioral and physiological adaptations to both temperature and salt.

Consider the marine iguana lizard. These reptiles spend much of their time regulating their body temperatures. Iguanas and other nonavian reptiles were once called "cold-blooded" (as opposed to "warm-blooded" birds and mammals). But this term is misleading. An iguana's body temperature is not necessarily low; rather, its temperature fluctuates as the lizard exchanges heat with the environment.

Marine iguanas on the Galápagos Islands begin their days basking in the rising sun, draped on boulders and hardened lava, sunning their backs and sides. After about an hour, they turn, raise their bodies, and aim their undersides at the sun.

By midmorning, the air temperature is rising rapidly. Iguanas cannot sweat as humans do; instead, they must escape the intense heat. They lift their bodies by extending their short legs, which removes their bellies from the hot rocks and allows breezes to fan them. By noon, these push-ups are insufficient to stay cool. The iguanas retreat to the shade of rock crevices.

The animals are hungry by midday. The iguanas hang off rocks to reach seaweed by the shore, or they dive into the ocean and eat green algae on the ocean floor. The water is so much colder than the sun-baked rocks that the lizards' body temperatures rapidly drop. Arteries near their body surfaces constrict, helping conserve heat. Even so, the water is too cold for them to stay in for more than a few minutes.

After feeding, the iguanas stretch out on the rocks again, warming sufficiently to digest their meal. They continue basking as the day ends, absorbing enough heat to sustain them through the cooler night temperatures until a new day begins. Meanwhile, the iguana's body must cope with the high salt content of its diet. A gland in the lizard's head rids the animal's body of extra salt. Cells lining the salt gland secrete a fluid that travels in branching tubules, which empty through the nostrils. This fluid carries the salts with it, creating crusts on the lizard's head.

Marine iguanas, and all other animals, maintain homeostasis in many ways. This chapter describes how animals regulate their temperatures and control the composition of their body fluids.

SURVEY THE LANDSCAPE
Animal Life

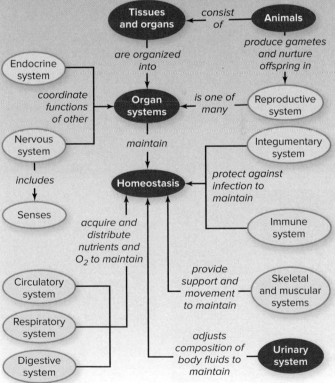

The outside temperature can be hot, mild, or cold; food may be salty or bland; beverages may be acidic or alkaline. Many adaptations help the body maintain homeostasis, even as the outside world changes.

For more details, study the Pull It Together feature in the chapter summary.

665

33.1 Animals Regulate Their Internal Temperature

Animals live nearly everywhere on Earth (figure 33.1). The salty Antarctic Ocean contrasts sharply with the perpetual humidity of a tropical rain forest and the dry, scorching home of a roadrunner. These habitats select for different ways of regulating body temperature, conserving water, and disposing of wastes. In hot, dry areas, an animal must conserve water and keep its body cool enough for cells to function. At the other extreme, frigid salt water surrounds an animal living in the Antarctic Ocean. An animal in this habitat must continually pump excess salts out of its body and generate enough body heat to survive the cold.

Most animals live in more moderate environments, but each species has adaptations that enable it to maintain homeostasis. This chapter begins by describing how animals regulate body temperature and the concentrations of salt and water in their tissues. It moves next to the types of nitrogenous wastes that animals produce, then concludes with the structure and function of the human urinary system.

Figure 33.1 Different Habitats. Penguins thrive on Antarctic ice, parrots inhabit the tropics, and roadrunners live in the desert. The adaptations of each bird reflect its diet, the temperature of its habitat, and many other selective forces.

(penguins): ©Mint Images Limited/Alamy RF; (parrots): ©IT Stock/PunchStock RF; (road runner): U.S. Fish & Wildlife Service/Gary Karamer

A. Heat Gains and Losses Determine an Animal's Body Temperature

Whether an animal lives in Antarctica or the Amazonian rain forest, its body temperature must remain within certain limits. Part of the reason is that extreme temperatures alter biological molecules. Excessive heat can ruin a protein's three-dimensional shape and disrupt its function. Extreme cold solidifies lipids and therefore inhibits the function of biological membranes. In addition, if the temperature inside a cell deviates from an animal's customary body temperature, enzymes function less efficiently and vital biochemical reactions proceed too slowly to sustain life. ⓘ *protein shape,* section 2.5C; *lipids,* section 2.5E; *enzymes,* section 4.4

Thermoregulation is the control of body temperature, and it requires the ability to balance heat gained from and lost to the environment. An animal can also maintain homeostasis by controlling how much heat it produces. When cells generate ATP in aerobic respiration (see chapter 6), they also produce metabolic heat. The more active the animal, the higher its metabolic rate and the more heat it produces.

The main source of an animal's body heat may be external or internal (figure 33.2). An **ectotherm** lacks an internal temperature-regulating mechanism. It thermoregulates by moving to areas where it can gain or lose heat, so its temperature varies with external conditions. The vast majority of animals are ectotherms, including all invertebrates, plus fishes, amphibians, and nonavian reptiles such as the iguana described in this chapter's opening essay.

An **endotherm** regulates its body temperature internally. Most endotherms maintain a relatively constant body temperature by balancing heat generated in metabolism (especially in the muscles) with heat lost to the environment. Mammals and birds are endotherms; insulation in the form of fat, feathers, or fur helps retain their body heat.

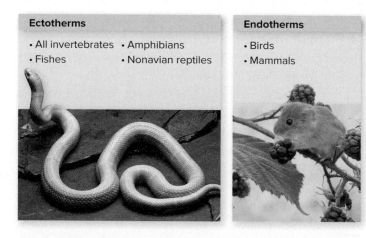

Ectotherms		Endotherms
• All invertebrates • Amphibians		• Birds
• Fishes • Nonavian reptiles		• Mammals

Figure 33.2 Ectotherm and Endotherm. An ectotherm such as a snake alters its behavior to manage the exchange of heat between its body and the environment. In contrast, the mouse is an endotherm; its metabolism generates most of its body heat.

Photos: (snake): ©IT Stock/PunchStock RF; (mouse): ©Janette Hill/Alamy RF

Some animals combine features of both ecto- and endothermy. In some of these "heterotherms," body temperature fluctuates with the animal's activity level. A bat, for example, maintains a high temperature while flying, but its temperature drops while the animal roosts. In other heterotherms, an animal's limbs are much colder than the rest of its body. A penguin's feet are nearly as cold as ice, while the bird's core body temperature is much higher.

Both ectothermy and endothermy have advantages and disadvantages. An ectotherm uses much less energy, and therefore requires less food and O_2, than an endotherm. However, an ectothermic animal must be able to seek or escape environmental heat. An injured iguana that could not squeeze into a crevice to avoid the broiling noonday sun would cook to death. Ectotherms also become sluggish when the temperature is low, which can make it hard for them to escape from predators.

On the other hand, an endotherm maintains its body temperature even in cold weather or in the middle of the night. But this internal constancy comes at a cost. The metabolic rate of an endotherm is generally five times that of an ectotherm of similar size and body temperature. Endotherms therefore require much more food than do ectotherms.

Figure It Out

As the sun sets and the external temperature gets colder, what will happen to a frog's body temperature? What will happen to that of a rabbit?

Answer: The frog's temperature will go down; the rabbit's will stay the same.

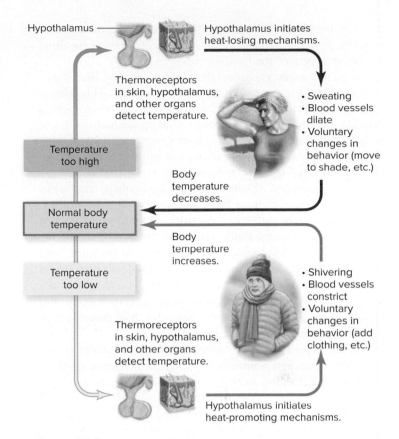

Figure 33.3 Thermoregulation in Humans. Thermoreceptors signal the hypothalamus to trigger the responses that keep body temperature within a certain range.

B. Several Adaptations Help an Animal to Adjust Its Temperature

The hypothalamus detects blood temperature, receives information from thermoreceptors in the skin and other organs, and controls many of the negative feedback loops that maintain homeostasis (**figure 33.3**). These responses may be physiological or behavioral. Sweating is a physiological reaction to heat, but a person who is too hot might also move into the shade—a behavioral response. Thermoregulation therefore depends on the interactions of the musculoskeletal, nervous, endocrine, respiratory, integumentary, and circulatory systems. ⓘ *negative feedback,* section 25.4

One adaptation to extreme cold is a **countercurrent exchange** system, in which two adjacent currents flow in opposite directions and exchange heat with each other (**figure 33.4**). Recall from chapter 30 that arteries carry blood away from the heart, whereas veins transport blood in the opposite direction. The heart is located in the warm core of the body, so arterial blood is warmer than blood in veins near the body's surface. In the limbs of the wolf in figure 33.4, arteries and veins are adjacent to each other.

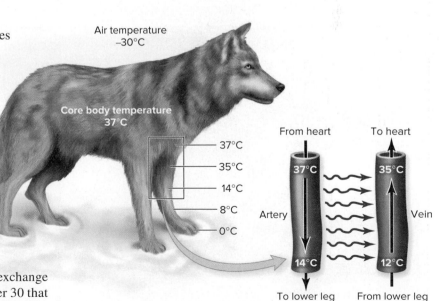

Figure 33.4 Countercurrent Heat Exchange. In the wolf's leg, arteries carrying warm blood lie near veins carrying cooler blood in the opposite direction. Heat moves along its gradient from artery to vein, rather than being lost at the extremities.

Because the blood flows in opposite directions in the two vessels, the gradient favors heat transfer from artery to vein along the entire length of the limb.

Countercurrent exchange therefore conserves heat rather than allowing it to escape through the extremities. This adaptation not only enables Arctic mammals such as wolves to hunt in extreme cold but also allows penguins to spend hours in frigid water. (Note that animals use countercurrent exchange in other ways as well, including gas exchange at gills and solute movement in kidneys.) ⓘ *gills,* section 31.1B

Many birds and mammals, including humans, also make other physiological changes that conserve heat. In cold weather, blood vessels in the extremities constrict, retaining more blood in the warmer core of the body. The animal may also shiver; the contraction of skeletal muscle generates heat. At the same time, muscles in the skin cause feathers or fur to stand on end, trapping an insulating air layer next to the skin. (In humans, this hair-raising response is useless because we have so little body hair. Nevertheless, goose bumps form when the hair muscles contract.) ⓘ *vestigial structures,* section 13.4B

Endotherms are famous for behaviors that help them to conserve heat; birds and mammals may migrate to warmer climates, hibernate, tuck the legs or wings close to the body, or huddle together (**figure 33.5a**). Likewise, ectotherms have a repertoire of body-warming behaviors; they may seek sunlight, sprawl on warm rocks or roadways, and build insulated burrows. Physiological adaptations that conserve heat, such as narrowing the diameter of blood vessels near the body surface, occur in ectotherms as well.

So far, the focus has been on conserving heat, but animals also must maintain homeostasis when the environment is too hot (figure 33.5b). Evaporative cooling from the skin or respiratory surfaces is one way to lower body temperature. For example, humans sweat to cool off, whereas a panting dog allows water to evaporate from the moist lining of its mouth. Likewise, an owl flutters loose skin under its throat to move air over moist surfaces in the mouth.

In addition, when the environment is warm, blood vessels in the extremities dilate and allow more blood to approach the relatively cool body surface. Tiny veins in the face and scalp also reroute blood cooled near the body's surface toward the brain. This adaptation explains why vigorous exercise causes the face of a light-skinned person to turn red.

Behavioral strategies can also help both ectotherms and endotherms to cool off. Many animals escape the sun's heat by swimming, covering themselves with cool mud, or retreating to the shade. Some burrow underground and emerge only at night; others extend their wings to promote cooling. Humans shed extra layers of clothing, and we consume cold food and drinks. We also swim or fan ourselves to increase heat loss.

33.1 MASTERING CONCEPTS

1. Differentiate between ectotherms and endotherms.
2. What are the advantages and disadvantages of endothermy and ectothermy?
3. What are some physiological and behavioral adaptations to extreme cold and extreme heat?

a.   b.

Figure 33.5 Too Cold or Too Hot. (a) Snow monkeys (Japanese macaques) are adapted to cold winters. Their thick fur retains body heat, as does their huddling behavior. (b) This panting dog loses excess heat through its mouth, while the man is sweating. (He has also employed a behavioral strategy that helps him cool off: He has shed his excess clothing.)

(a): ©Akira Kaede/Digital Vision/Getty Images RF; (b): ©Dynamic Graphics/PictureQuest RF

33.2 Animals Regulate Water and Ions in Body Fluids

Besides regulating its temperature, an animal must also maintain homeostasis in the composition of its body fluids. The balance of solutes and water between a cell's interior and its surroundings is critical to life. Cells must retain water even when conditions are dry, yet too much water is damaging. Sodium, chlorine, hydrogen, and other ions are vital to life, but not in excess. ⓘ *ions,* section 2.1B; *water,* section 2.3; *essential minerals,* section 32.4A

In most habitats, therefore, organisms must **osmoregulate;** that is, they control the concentration of ions in their body fluids as the environment changes. Osmoregulation requires managing the gain and loss of water, ions, or both.

A brief review of how water and ions move across membranes will help explain how osmoregulation works. Osmosis is the diffusion of water across a semipermeable membrane; the net direction of water movement is toward the side with the higher concentration of dissolved solutes. If a cell's environment is saltier than the cell itself, water moves out of the cell. In the opposite situation, water moves into the cell. ⓘ *osmosis,* section 4.5A

Unlike water, most ions cross membranes via protein channels. Osmoregulation often requires cells to move ions against their concentration gradient. This process, called active transport, requires energy in the form of ATP. Water may follow the ions by osmosis; as described in section 33.5, our kidneys exploit this mechanism to conserve water during the production of urine.

Bony fishes that live in the ocean and in fresh water face opposite challenges in osmoregulation (**figure 33.6**). Ocean water is much saltier than a fish's cells, so the animal loses water by osmosis, mostly at the gills. The fish therefore drinks seawater, produces little urine, and uses active transport at the gills to get rid of excess salts. In contrast, fresh water is much more dilute than the cells of a fish. A freshwater fish therefore constantly takes in water at its gills and through its skin by osmosis, while losing salts to its surroundings. Its kidneys shed excess water in dilute urine, and the gills take up ions from water by active transport.

Land animals use a combination of strategies to obtain and conserve water (**figure 33.7**). Whereas humans ingest most of their water in food and drink, desert kangaroo rats derive most of their water as a byproduct of metabolism, especially cellular respiration. Animals lose water through evaporation from lungs and the skin surface, in feces, and in urine.

33.2 MASTERING CONCEPTS

1. What is osmoregulation, and why is it important?
2. Describe osmoregulation in marine and freshwater fishes.
3. How do land animals gain and lose water?

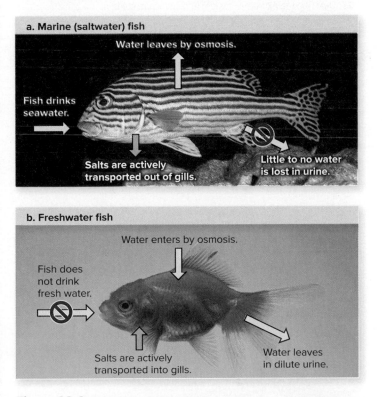

Figure 33.6 Osmoregulation in Fishes. (a) Seawater contains a higher solute concentration than the cells of a marine fish. The animal therefore constantly loses water by osmosis and pumps ions out by active transport. (b) Fresh water contains few dissolved solutes; a fish in that habitat therefore constantly gains water by osmosis and must use active transport to acquire essential ions.

Photos: (both): ©Brand X Pictures/PunchStock RF

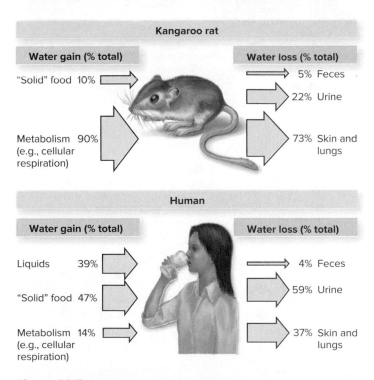

Figure 33.7 Water Gain and Loss. The desert-dwelling kangaroo rat gets all of its water from its food and its metabolism; it loses little in feces and urine. In contrast, a human gets most water from food and drink, losing most of it through urine.

33.3 Nitrogenous Wastes Include Ammonia, Urea, and Uric Acid

Animals produce two main categories of waste. One type is feces, which contain undigested food that passes through the digestive tract without ever entering the body's cells (see chapter 32). The other category is waste produced by the body's cells; **excretion** is the elimination of these metabolic wastes. For example, as described in chapter 31, the respiratory system excretes CO_2, a byproduct of aerobic cellular respiration.

An animal's body also must excrete nitrogen-containing (nitrogenous) wastes, which cells produce during the breakdown of proteins (**figure 33.8**). Proteins are composed of amino acids, which can enter the energy-generating pathways described in chapter 6. During this process, amino groups ($-NH_2$) are stripped from amino acids. Each amino group picks up a hydrogen ion and becomes **ammonia,** NH_3.

Excreting ammonia directly into the environment consumes little energy, but this waste product is very toxic. Animals that excrete ammonia—aquatic invertebrates, most bony fishes, tadpoles, and salamanders—live in habitats with abundant water

that carries the waste away. In these animals, ammonia simply diffuses into the water from blood that flows through the gills.

Land animals expend energy to change ammonia to **urea** or **uric acid.** Both of these substances store higher concentrations of nitrogen than ammonia does. Moreover, urea and uric acid are less toxic than ammonia, so they require less dilution in water. Excreting either of these wastes therefore helps conserve water.

Mammals, adult frogs and toads, turtles, and cartilaginous fishes (sharks and rays) form urea. Land tortoises, lizards, and birds change most of their nitrogenous wastes to uric acid. In birds, uric acid mixes with undigested food to form the familiar "bird dropping."

Insects and spiders also produce uric acid. The excretory system of these animals consists of structures called Malpighian tubules that empty into the gut (**figure 33.9**). The cells lining the Malpighian tubules transport uric acid and ions into the tubules. Water follows by osmosis. Fluid in the Malpighian tubules enters the intestine, where cells lining the rectum reabsorb most of the ions and water. The uric acid mixes in the intestine with undigested food, which the insect eliminates through its anus.

33.3 MASTERING CONCEPTS

1. Why do land animals convert ammonia to urea or uric acid?
2. Describe the excretory system of insects and spiders.

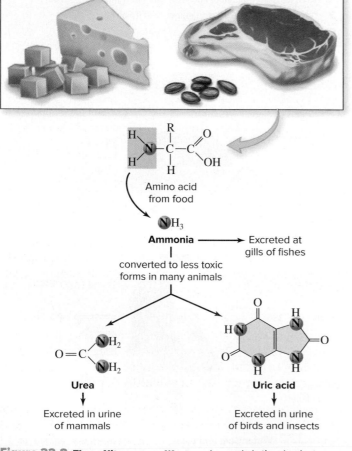

Figure 33.8 Three Nitrogenous Wastes. Ammonia is the simplest byproduct of protein breakdown, but it is toxic. Urea and uric acid are more costly to produce but much less toxic.

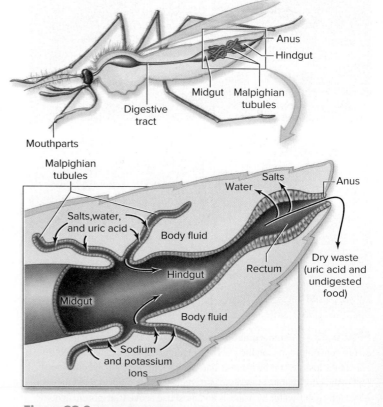

Figure 33.9 Insect Excretion. Insect excretory organs called Malpighian tubules collect uric acid to be eliminated with other wastes.

33.4 The Urinary System Produces, Stores, and Eliminates Urine

The human **urinary system** filters blood, eliminates nitrogenous wastes, and helps maintain the ion concentration of body fluids. The paired **kidneys** are the major excretory organs in the urinary system (**figure 33.10**). Located near the rear wall of the abdomen, each kidney is about the size of an adult fist and weighs about 230 grams.

As the kidneys cleanse blood, a liquid waste called **urine** forms; section 33.5 describes this process in detail. Besides eliminating urea and other toxic substances, kidneys have other functions as well. These organs conserve water, salts, glucose, amino acids, and other valuable nutrients. They also help regulate blood pH.

One other function of the kidneys is to regulate the volume of blood. Look back at the water balance in figure 33.7. What happens if we drink too many fluids or if we lose too much moisture in sweat and breath? Rather than swelling like a balloon or drying up like a leaf, the body adjusts: The kidneys maintain the volume of blood by controlling the amount of water lost in urine. This function is important because blood volume is one factor that influences blood pressure. ⓘ *blood pressure,* section 30.5B

The urine from each kidney drains into a **ureter,** a narrow, muscular tube about 28 centimeters long. Waves of smooth muscle contraction, called peristalsis, squeeze the fluid along the two ureters and squirt it into the **urinary bladder,** a saclike, muscular organ that collects urine. ⓘ *peristalsis,* section 32.3

The **urethra** is the tube that connects the bladder with the outside of the body. In females, the urethra opens between the clitoris and vagina. In males, the urethra extends the length of the penis. The urethra also carries semen in males (see chapter 35). The term *urogenital tract* reflects the intimate connection between the urinary and reproductive systems.

Two sphincters guard the bladder's exit; both must relax before urine can leave the body. An involuntary spinal reflex controls the innermost sphincter, which is made of smooth muscle. The outer sphincter consists of skeletal muscle. Its relaxation is under voluntary control in most people over 2 years old.

The adult bladder can hold about 600 milliliters of urine. As little as 300 mL of accumulating urine stimulates stretch receptors in the bladder. These sensory neurons send impulses to the spinal cord and brain, generating a strong urge to urinate. We can suppress this urge for a short time. Eventually, however, the brain's cerebral cortex directs the sphincters to relax, and bladder muscle contractions force urine out of the body.

Weakened sphincter muscles, overactive bladder muscles, and nerve damage are among the underlying causes of urinary incontinence—the loss of bladder control. Urinary tract infections, pregnancy, prostate enlargement, spinal cord injuries, and many other conditions are associated with incontinence.

Adrenal glands
Produce hormones that regulate blood pressure

Kidneys
Maintain homeostasis in blood composition

Renal artery
Delivers blood to kidney

Renal vein
Drains blood from kidney

Ureters
Convey urine to bladder

Urinary bladder
Stores urine before elimination

Urethra
Conveys urine out of the body

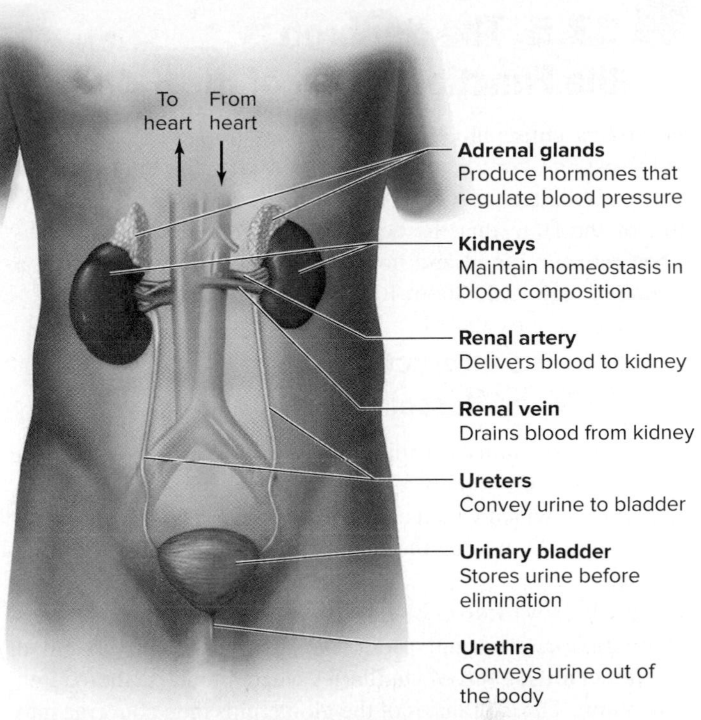

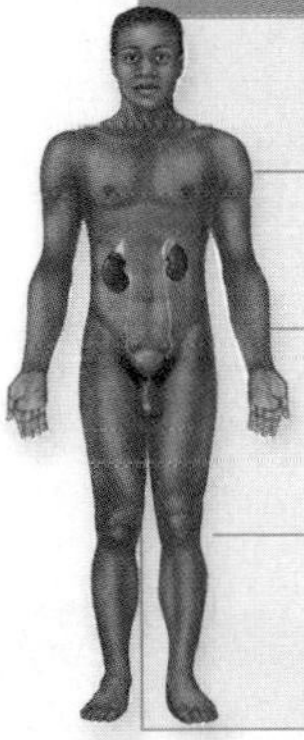

Urinary System	
Main tissue types*	**Examples of locations/functions**
Epithelial	Enables diffusion between nephron and blood; lines ureters and bladder.
Connective	Blood (which kidneys filter) is a connective tissue.
Muscle	Smooth muscle controls flow of blood to and from nephrons; smooth and skeletal muscle sphincters control urine release.
Nervous	Sensory cells in hypothalamus coordinate negative feedback loops that maintain salt concentration in body fluids.

*See chapter 25 for descriptions.

Figure 33.10 The Human Urinary System. The human urinary system includes the kidneys, ureters, urinary bladder, and urethra. This generalized depiction omits the differences between male and female organs. Figures 35.4 and 35.8 show the sex differences in more detail.

Treatment may include strengthening the muscles that control urination. Behavioral changes, alone or combined with medication, may also help. Surgical procedures to treat incontinence include the implantation of an artificial sphincter.

33.4 MASTERING CONCEPTS

1. List the organs that make up the human urinary system. What is the function of each?
2. What is the role of sphincters in the urinary system?

🍁 33.5 The Nephron Is the Functional Unit of the Kidney

The body's entire blood supply courses through the kidney's blood vessels every 5 minutes. At that rate, the equivalent of 1600 to 2000 liters of blood passes through the kidneys each day. Most of the fluid that the nephrons process, however, is reabsorbed into the blood and not released in urine. As a result, a person produces only about 1.5 liters of urine daily.

A. Nephrons Interact Closely with Blood Vessels

Each kidney contains 1.3 million tubular **nephrons**—the functional units of the kidney. As illustrated in **figure 33.11**, each nephron is entwined with a network of capillaries. Our tour of the kidney begins by tracing the flow of blood into and out of these vessels.

Each kidney receives blood via a renal artery, which branches into multiple arteries and arterioles. An arteriole delivers blood to a **glomerulus,** a tuft of capillaries where blood is filtered into the nephron. The capillaries of the glomerulus then converge into another arteriole, which leads to the **peritubular capillaries** that snake around part of each nephron. These blood vessels empty into a venule, which joins the renal vein carrying cleansed blood out of the kidney and (ultimately) to the heart.

Now examine the anatomy of the nephrons in figure 33.11. Each nephron consists of two main parts: a glomerular capsule and a renal tubule. In the kidney's outer portion, or cortex, the glomerular (Bowman's) capsule receives fluid from the glomerulus. From there, the solution travels along the **renal tubule,** a winding passageway consisting of three functional regions: the proximal convoluted tubule, the nephron loop, and the distal convoluted tubule. The **proximal convoluted tubule** leads from the glomerular capsule to the hairpin-shaped **nephron loop** (also called the loop of Henle). The descending limb of the nephron loop dips into the renal medulla toward the kidney's center, and the ascending limb returns to the **distal convoluted tubule** in the cortex. The entire nephron, when stretched out, is about 3 to 4 centimeters long.

A **collecting duct** receives the fluid from several nephrons. Urine from many collecting ducts accumulates in the funnel-like renal pelvis before entering the ureter and urinary bladder and moving out of the body through the urethra.

B. Urine Formation Includes Filtration, Reabsorption, and Secretion

The chemical composition of urine reflects three processes (**figure 33.12**):

1. **Filtration:** Water and dissolved substances are filtered out of the blood at the glomerular capsule.
2. **Reabsorption:** Useful materials such as salts, water, glucose, and amino acids return from the nephron and collecting duct to the blood.
3. **Secretion:** Toxic substances and drug residues are secreted into the nephron to be eliminated in urine, as are hydrogen ions (H^+) and other surplus ions.

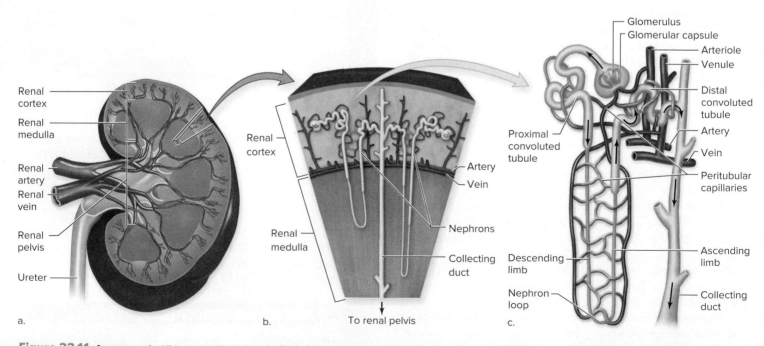

Figure 33.11 **Anatomy of a Kidney.** (a) The kidney is divided into an outer cortex and an inner medulla. Blood arrives via the renal artery. (b) The nephron is the functional unit of the kidney. Each nephron begins in the cortex, descends into the medulla, and returns to the cortex. Collecting ducts carry away the urine produced by each nephron. (c) Nephrons are intimately associated with blood vessels.

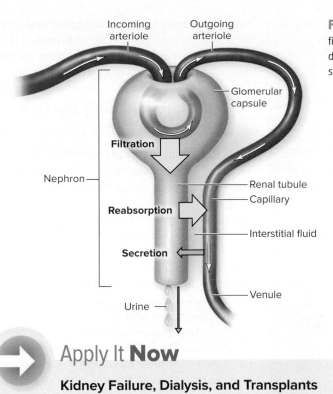

Figure 33.12 **Overview of Urine Formation.** Urine forms from three processes: filtration, reabsorption, and secretion. Arrows represent the flow of water and dissolved substances; each arrow's size reflects the magnitude of the flow. This stylized, schematic view does not reflect nephron anatomy.

When the nephrons fail to do their job, nitrogenous wastes and other toxins accumulate in the blood to harmful levels; in addition, a person may lose too much water and become dehydrated. Without treatment, kidney failure can be fatal (see this chapter's Apply It Now box).

The remainder of this section describes the events that occur at the nephron. **Figure 33.13** summarizes the exchange of materials between blood vessels and a nephron; refer to this figure frequently as you go along.

Apply It **Now**

Kidney Failure, Dialysis, and Transplants

Kidneys can fail for many reasons. The most common causes are diabetes and high blood pressure, which together account for about two thirds of all cases. Other causes are chronic inflammation that causes progressive loss of nephrons, polycystic kidney disease (the accumulation of cysts in the kidneys), scarring from untreated kidney or urinary tract infections, and obstructed urine flow.

A person with diabetes has too much sugar in his or her blood. Over time, high blood sugar damages small blood vessels in the kidneys (and throughout the body). As a result, diabetes reduces kidney function. Likewise, high blood pressure also damages the kidneys' blood vessels. Kidney damage, in turn, may cause the blood to retain excess fluid, driving blood pressure even higher and setting into motion a dangerous cycle.

One treatment option for kidney patients is dialysis (figure 33.A). The dialysis machine pumps blood out of the patient's body and past a semipermeable membrane. Wastes and toxins, along with water, diffuse across the membrane, but blood cells do not. The cleaned blood then circulates back to the patient's body. The procedure requires hours a day, several times a week.

Dialysis membranes cannot replace all of the kidney's functions. For example, nephrons selectively recycle useful components such as glucose and salts to the blood. The dialysis machine cannot do this, although a technician can adjust the concentrations of dissolved compounds removed from the blood.

Transplantation is a second option (figure 33.B). The transplanted kidney may come from a cadaver, or a living person may donate one healthy kidney to a recipient. Surgeons connect the new kidney to the recipient's blood supply and attach its ureter to the bladder, usually leaving the old kidneys in place.

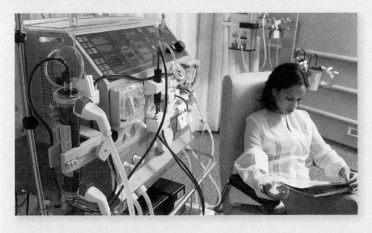

Figure 33.A **Kidney Dialysis.** A dialysis machine can take over some, but not all, of a healthy kidney's functions.
©AJPhoto/Science Source

Figure 33.B **Three Is Not a Crowd.** Surgeons usually insert a new kidney and ureter in a cavity low in the abdomen, leaving the patient's own kidneys in place.

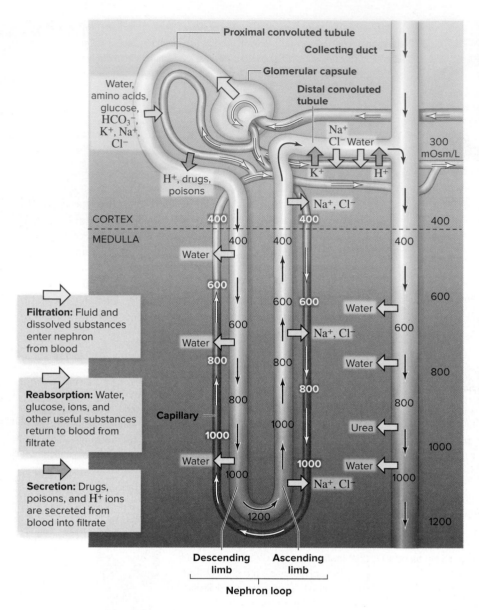

Figure 33.13 **Nephron Function.** This schematic view shows the locations of filtration, reabsorption, and secretion along the nephron and collecting duct. Water moves by osmosis between the nephron and adjacent blood vessels; depending on the region, ions and other substances cross membranes by diffusion or active transport. (Numbers indicate milliosmoles per liter, abbreviated mOsm/L, a measure of ion concentration.)

C. The Glomerular Capsule Filters Blood

Like a baseball mitt catching a ball, the glomerular capsule surrounds the glomerulus. The function of the glomerulus, however, is more like that of a coffee filter: Its pores allow water, urea, glucose, salts, amino acids, and creatinine (a byproduct of muscle contraction) to pass into the glomerular capsule, but large structures such as plasma proteins, blood cells, and platelets remain in the bloodstream. ⓘ *blood composition,* section 30.2

Blood pressure drives substances out of the glomerulus, forcing fluid and small dissolved molecules across the capillary walls and into the nephron. The liquid entering the glomerular capsule,

called the filtrate, is the product of the first step in urine manufacture. The 1600 to 2000 liters of blood per day that pass through the kidneys produce approximately 180 liters of glomerular filtrate.

D. Reabsorption and Secretion Occur in the Renal Tubule

The Proximal Convoluted Tubule Most selective reabsorption occurs at the proximal convoluted tubule. All along the tubule, specialized cells transport sodium ions (Na^+), glucose, amino acids, and other useful solutes into the interstitial fluid surrounding the tubule. Water follows by osmosis. Overall, blood vessels along the proximal convoluted tubule reabsorb almost two thirds of the salts and water that were present in the original filtrate, along with nearly all of the glucose and amino acids.

Secretion also occurs in the proximal convoluted tubule. The antibiotic penicillin, for example, enters the filtrate by secretion. Patients must take penicillin several times a day to compensate for this loss.

Reabsorption and secretion work together to maintain the pH of blood between 7 and 8; any variance from this range is deadly. Acids (H^+-releasing compounds) produced in metabolism lower blood pH. To prevent the pH from dipping too low, H^+ is secreted into the proximal convoluted tubule; bicarbonate ions (HCO_3^-), which raise pH, are reabsorbed into the blood. ⓘ *pH,* section 2.4A

The Nephron Loop Next, the filtrate moves into the nephron loop, which makes a hairpin turn into the medulla. Notice in figure 33.13 that the medulla's solute concentration becomes progressively higher with depth; this gradient is critical to the nephron's ability to conserve water and produce concentrated urine. Also, note that the filtrate in the nephron loop flows in the opposite direction of the blood in the surrounding capillaries—another example of countercurrent exchange.

As the descending limb of the nephron loop dips into the medulla, the solute concentration in the surrounding fluid is higher than that of the filtrate inside. Water therefore leaves by osmosis along the entire length of the descending limb. The cells of the descending limb, however, are impermeable to ions and urea. As water leaves the descending limb, the solute concentration inside the tubule gradually rises until it reaches its maximum at the bottom of the loop.

The filtrate then flows around the bend and into the ascending limb. Here, the filtrate is more concentrated than in the surrounding capillaries, but the wall of the ascending limb is impermeable to urea and water. Therefore, water cannot reenter the filtrate. Ions, however, can leave. Along most of the ascending limb, Na⁺ and Cl⁻ diffuse from the filtrate into the capillaries, which have a lower salt concentration. Toward the end of the ascending limb, however, Na⁺ and other ions move out of the filtrate by active transport. Since water cannot follow, the filtrate that enters the distal convoluted tubule is less concentrated than the tissues and blood in the surrounding cortex.

The Distal Convoluted Tubule   At the distal convoluted tubule, Na⁺ and Cl⁻ ions continue to move out of the filtrate and into the blood by active transport. The walls of the distal convoluted tubule are again permeable to water, which moves by osmosis into the capillaries. At the same time, excess K⁺ and H⁺ in the blood may be secreted into the distal convoluted tubule and collecting duct. As described in section 33.5F, this fine-tuning of urine composition is under delicate hormonal control.

Once the filtrate has passed through the distal convoluted tubule, 97% of the water that was in the original glomerular filtrate has been reabsorbed, and little salt remains.

E. The Collecting Duct Conserves More Water

As the collecting duct descends into the salty medulla, much of the remaining water in the filtrate leaves by osmosis. Some urea also diffuses out of the collecting duct and contributes to the high solute concentration surrounding the nephron loop deep in the medulla.

After reabsorption and secretion, the filtrate is urine. Urine contains water, urea, a small amount of uric acid, creatinine, and several types of ions. Nearly all of the glucose and amino acids present in the original filtrate, however, return to the blood. Nevertheless, the exact chemical composition of urine can vary, as described in this chapter's Burning Question.

F. Hormones Regulate Kidney Function

Kidney function adjusts continuously to maintain homeostasis. For example, if water is scarce, we produce small amounts of concentrated urine. When we drink too much water, urine is abundant and dilute.

Hormones help make these adjustments (**figure 33.14**). When we are dehydrated, osmoreceptor cells in the hypothalamus send impulses to the posterior pituitary gland, which secretes a peptide hormone called **antidiuretic hormone (ADH),** or vasopressin. ADH triggers the reabsorption of more water at the distal convoluted tubule and collecting duct, so the urine becomes more concentrated. Conversely, if blood plasma is too dilute, ADH production stops, and more water is eliminated in the urine. ⓘ *posterior pituitary,* section 28.3A

A diuretic is a substance that increases the volume of urine. The ethanol in alcoholic beverages is one example. Alcohol stimulates urine production partly by reducing ADH secretion, thereby decreasing water reabsorption. By increasing water loss to

urine, alcoholic beverages actually intensify thirst; dehydration also causes the discomfort of a hangover.

Hormones from the adrenal glands also act on the kidneys to regulate blood pressure; these glands are illustrated in figure 33.10. For example, when blood pressure and blood volume dip too low, the outer portion of each adrenal gland releases the steroid hormone **aldosterone.** This mineralocorticoid stimulates the production of sodium channels in the distal convoluted tubule and collecting duct. Water follows Na⁺ ions by osmosis, so blood pressure rises. ⓘ *adrenal glands,* section 28.4C

33.5 | MASTERING CONCEPTS

1. What three processes occur in urine formation?
2. Select a point along the descending limb of the nephron loop in figure 33.13. Compare the solute concentration at that point and in the adjacent interstitial fluid; explain why water moves in the direction it does.
3. Describe the roles of antidiuretic hormone and aldosterone.

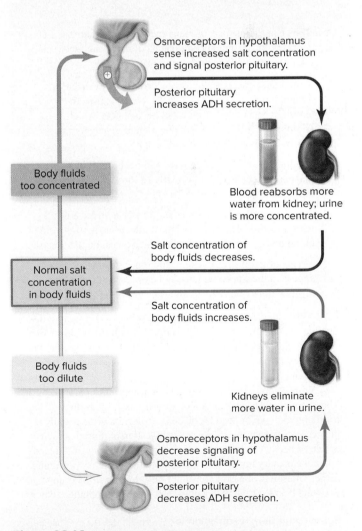

Figure 33.14 Osmoregulation in Humans. A feedback loop regulates the amount of water that blood reabsorbs from the kidneys.

Burning Question

What can urine reveal about health and diet?

©Corbis RF

Urinalysis is a routine part of many medical examinations. Laboratory tests of the chemical components of urine can reveal many health problems:

- More than a trace of glucose may be a sign of diabetes, a high-carbohydrate diet, or stress. Stress causes the adrenal glands to release excess epinephrine, which stimulates the liver to break down more glycogen into glucose. ⓘ *diabetes,* section 28.4D

- Albumin may be a sign of damaged nephrons, since this plasma protein does not normally fit through the pores of intact glomerular capsules.

- Together, pus and an absence of glucose indicate a urinary tract infection. The pus consists of infection-fighting white blood cells along with bacteria, which consume the glucose.

- Traces of marijuana, cocaine, and other substances may appear in the urine. Athletes and employees of some organizations routinely submit urine samples for drug testing.

Although urine is usually odorless, some foods impart a distinctive aroma. Asparagus, for example, contains a sulfur-rich molecule called mercaptan. Many people produce an enzyme that breaks down mercaptan, and the products of the reaction have a strong odor.

What about urine's color—why is it normally pale yellow? The pigment that lends its color to urine is a byproduct of the liver's breakdown of dead blood cells. Colorless urine usually indicates excessive water intake or the ingestion of diuretics such as coffee or beer. A reddish tinge may suggest anything from bleeding in the urinary tract to beet consumption to mercury poisoning. Either vitamin C or carrot consumption can color urine orange. It is always best to check with a physician when urine changes color.

Submit your burning question to
Marielle.Hoefnagels@mheducation.com

INVESTIGATING LIFE

33.6 Sniffing Out the Origin of Fur and Feathers

While visiting the zoo and appreciating the beauty of peacocks, parrots, and flamingos, have you wondered about the origin of their beautiful plumage (**figure 33.15**)? Have you ever admired the fur coat of a giraffe, jaguar, or red fox and asked yourself how hair evolved? Evolutionary biologists have been working to answer the same questions.

Fossil and DNA evidence clearly indicates that mammals and birds arose from two different lineages of reptiles. Existing nonavian reptiles are ectotherms, whereas both birds and mammals are endotherms. Because the ancestors of birds were not closely related to mammalian ancestors, researchers surmise that endothermy evolved independently in these two animal lineages.

It makes sense that feathers and fur evolved at the same time as the elevated metabolic rates associated with birds and mammals. After all, these integumentary adaptations provide insulation that helps the animals maintain their body temperature. To test this hypothesis of simultaneous evolution, scientists would need to examine fossils from several points along the mammal and bird lineages, including animals that lived before and after the evolution of endothermy, fur, and feathers.

This task is easier said than done. Both fur and feathers occasionally show up in fossils, but determining the metabolic style of an extinct animal is difficult. How can we know whether an animal that no longer exists was an endotherm or an ectotherm?

John A. Ruben and his colleagues at Oregon State University believe they have hit on the ideal indicator of endothermy: the inside of the nose. Endothermy requires a high metabolic rate,

Figure 33.15 Pretty Bird. Feathers have many functions, including insulation, flight, and communication. Scientists are working to determine how these structures evolved.

©Pat Pendarvis

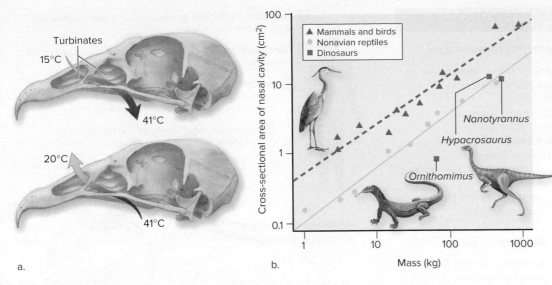

Figure 33.16 **The Nose Knows.**
(a) Turbinates direct air flow within the nasal cavity of endotherms, conserving both heat and water. (b) The nasal cavities of existing endotherms (ranging in size from herons to African cape buffaloes) have a higher cross-sectional area than those of ectotherms of equal size. Dinosaurs had nasal cavities similar in size to those of existing ectotherms.

which in turn means a huge demand for O_2 to fuel respiration. Mammals and birds therefore have high breathing rates. Along with this greater volume of air moving into and out of the lungs, however, comes a higher potential loss of both heat and water from an animal's body. Endotherms minimize this loss with the help of structures called turbinates that direct the flow of air within the nasal cavity (**figure 33.16a**).

Turbinates are internal ridges made of bone or cartilage covered with epithelium. These structures warm and humidify inhaled air before it enters the lungs; they also cool and remove water vapor from air as it is exhaled. The nasal cavities of more than 99% of all existing birds and mammals contain turbinates; no known ectotherms have them. Ruben and his colleagues reasoned that turbinates are therefore the next best thing to direct evidence of endothermy in extinct animals. They might be useful for approximating when endothermy first appeared in the ancestors of mammals and birds.

The researchers began by looking for evidence of bony turbinates in numerous fossils of extinct reptiles thought to be ancestors of mammals, including therapsids. They found turbinates in therapsids that lived during the Permian period, some 250 million years ago. Yet the first evidence of insulating fur appears only in fossils of true mammals, which did not arise until millions of years later, during the Triassic period. The researchers therefore concluded that the reptilian ancestors of mammals developed endothermy long before the origin of fur. ⓘ *geologic time scale*, section 13.1

But birds posed a problem. Dinosaurs with feathers clearly lived 150 MYA (see the opening essay for chapter 13). Did these animals have turbinates? It is hard to know, because the turbinates may have been made of cartilage, a soft tissue that does not leave fossils. Nevertheless, Ruben's team suggested an indirect way to detect whether turbinates were present. The researchers predicted that, in general, endotherms should have broader nasal cavities than ectotherms, both to accommodate the presence of turbinates and to allow for a high breathing rate.

To make sure this was the case, Ruben's research team measured the cross-sectional areas of the nasal cavities of 21 living species of birds, mammals, and nonavian reptiles. In every case, the cross-sectional area of the nasal cavity was larger for an endotherm than for an ectotherm of equal size (figure 33.16b). The consistent, strong relationship suggested that measuring the nasal cavity in a fossil should be a good way to learn if an extinct animal was an endotherm or an ectotherm.

The researchers used imaging technologies to measure nasal cavities inside the fossilized skulls of three dinosaur species that lived about 70 MYA and are closely related to modern birds. The results indicate that the reptiles probably were ectotherms. Endothermy must therefore have arisen after that time, yet feathers already existed at least 150 MYA.

Fur and feathers are adaptations that help mammals and birds stay warm, so it is easy to assume they evolved hand-in-hand with endothermy. Ruben's team paired old-fashioned comparative anatomy with modern technology to reject this assumption. Endothermy was evidently present in mammalian ancestors before there was fur, and feathers apparently existed long before birds became endothermic. Thanks to the efforts of this research team, we now have part of the answer to the puzzle—and it is right in front of our nose.

Sources: Ruben, John A., Willem J. Hillenius, Nicholas R. Geist, et al. 1996. The metabolic status of some late Cretaceous dinosaurs. *Science,* vol. 273, pages 1204–1207.

Ruben, John A., and Terry D. Jones. 2000. Selective factors associated with the origin of fur and feathers. *American Zoologist,* vol. 40, pages 585–596.

33.6   MASTERING CONCEPTS

1. How did Ruben and his colleagues use turbinates and nasal cavities to draw their conclusions?

2. Suppose researchers discover a fossil of a 100-kilogram dinosaur with a nasal cavity greater than 10 square centimeters. According to figure 33.16, how would that finding affect Ruben's conclusions?

CHAPTER SUMMARY

33.1 Animals Regulate Their Internal Temperature

A. Heat Gains and Losses Determine an Animal's Body Temperature
- Animals regulate their body temperatures (**thermoregulate**) with physiological and behavioral adaptations.
- **Ectotherms** use the environment to regulate body temperature. **Endotherms** use internal metabolism to generate heat.

B. Several Adaptations Help an Animal to Adjust Its Temperature
- Adaptations to cold include **countercurrent exchange** systems, insulation, constriction of blood vessels near the body surface, shivering, and hibernation.
- Adaptations to heat include evaporative cooling, dilation of blood vessels near the body surface, and behaviors that help cool the body.

33.2 Animals Regulate Water and Ions in Body Fluids
- **Osmoregulation** is the control of ion concentrations in body fluids. Depending on its habitat, an animal may need to conserve or eliminate water and ions.

33.3 Nitrogenous Wastes Include Ammonia, Urea, and Uric Acid
- Animals **excrete** metabolic wastes. Most nitrogenous wastes come from protein breakdown and include **ammonia, urea,** and **uric acid.**
- Ammonia is highly toxic; its excretion requires abundant water. Urea and uric acid require more energy to produce but are less toxic.

33.4 The Urinary System Produces, Stores, and Eliminates Urine
- The human **urinary system** excretes nitrogenous wastes (mostly urea) and regulates water and electrolyte levels.
- The **kidneys** produce **urine.** Each kidney drains into a **ureter,** which delivers urine to the **urinary bladder** for temporary storage. Urine leaves the body through the **urethra.**

33.5 The Nephron Is the Functional Unit of the Kidney

A. Nephrons Interact Closely with Blood Vessels
- The **nephron** receives filtered blood from the **glomerulus** and exchanges materials with the **peritubular capillaries.**
- The two main regions of the nephron are the **glomerular capsule** and the **renal tubule.** The renal tubule, in turn, includes the **proximal convoluted tubule, nephron loop,** and **distal convoluted tubule.**
- Fluid moves from the nephron into a **collecting duct.**

B. Urine Formation Includes Filtration, Reabsorption, and Secretion
- The nephron **filters** blood and **secretes** wastes and other substances into the filtrate, while blood capillaries **reabsorb** useful components (figure 33.17).

C. The Glomerular Capsule Filters Blood
- Blood pressure forces some components of blood from the glomerulus into the glomerular capsule.

D. Reabsorption and Secretion Occur in the Renal Tubule
- The proximal convoluted tubule returns glucose, amino acids, water, ions, and other solutes to the blood; H$^+$ and HCO$_3^-$ adjustments in this region help maintain blood pH.
- The nephron loop dips into the medulla of the kidney, then returns to the cortex. A concentration gradient between the nephron loop and surrounding fluid drives water out of the filtrate in the descending limb.

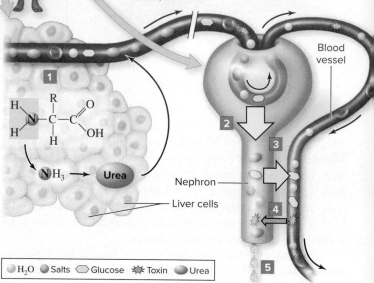

1 Liver cells metabolize amino acids and release urea into the bloodstream.

2 **Filtration (⇨):** Urea, glucose, salts, and other solutes are filtered from the blood at the nephrons.

3 **Reabsorption (⇨):** Water and some other substances are reabsorbed into the bloodstream. Hormones (aldosterone and ADH) regulate this process.

4 **Secretion (⇨):** Toxins are secreted into the nephron.

5 Urine is a mixture that includes water, urea, toxins, and salts.

Blood vessel

Nephron

Liver cells

● H$_2$O   ● Salts   ◯ Glucose   ✳ Toxin   ● Urea

Figure 33.17 Nephron Function: A Summary.

In the ascending limb, salt ions leave the filtrate by diffusion or active transport.
- At the distal convoluted tubule, additional water and salts are reabsorbed into the blood. K$^+$ and H$^+$ may be secreted into the filtrate.

E. The Collecting Duct Conserves More Water
- As the filtrate travels through the collecting duct, water leaves by osmosis.
- Collecting ducts deliver urine to the funnel-like renal pelvis, which drains the fluid into the ureter.

F. Hormones Regulate Kidney Function
- **Antidiuretic hormone (ADH),** secreted by the posterior pituitary gland, regulates water reabsorption from the distal convoluted tubule. ADH increases the permeability of the distal convoluted tubule and the collecting duct, so more water is reabsorbed and urine is more concentrated.
- The adrenal glands release **aldosterone** in response to either low Na$^+$ concentration in the plasma or low blood pressure. Aldosterone causes additional Na$^+$ to be reabsorbed into the blood; water follows by osmosis, raising blood pressure.

33.6 Investigating Life: Sniffing Out the Origin of Fur and Feathers
- Researchers have tested the hypothesis that fur and feathers evolved at the same time as endothermy. Their findings suggest that whereas endothermy preceded fur in the ancestors of mammals, feathers came first in birds.

MULTIPLE CHOICE QUESTIONS

1. Which animal's body temperature would drop the fastest in a cold environment?
   a. A whale
   b. An Arctic bird
   c. A tropical bird
   d. A lizard

2. Whereas a human is likely to _____, a snake _____.
   a. thermoregulate; does not thermoregulate
   b. use the environment to thermoregulate; thermoregulates internally
   c. use positive feedback to thermoregulate; uses negative feedback to thermoregulate
   d. thermoregulate internally; uses the environment to thermoregulate

3. Why do endotherms require more energy than ectotherms?
   a. Because they need more energy to move between hot and cold environments
   b. Because they use metabolic energy to maintain an internal temperature
   c. Because they lack insulation that retains body heat
   d. Both b and c are correct.

4. In a freshwater fish,
   a. water moves into the fish by osmosis.
   b. ions diffuse into the fish.
   c. urine has a high solute concentration.
   d. cells pump ions out of the fish.

5. Complete this analogy: Urine is to _____ as feces is to undigested food.
   a. unabsorbed drinks
   b. metabolic wastes
   c. old blood cells
   d. All of the above are correct.

6. Which of the following functions of the kidney is INCORRECTLY paired with its description?
   a. Filtration: Fluid containing dissolved substances moves into the nephron
   b. Reabsorption: Water returns to the blood
   c. Secretion: Hydrogen ions move into the nephron
   d. All of these are correctly paired.

7. Water moves out of the filtrate of the descending limb of the nephron loop because
   a. there is a high concentration of solutes in the surrounding fluid.
   b. the cells of the descending limb are permeable to ions.
   c. the difference in the diameter of the arterioles creates a pressure gradient.
   d. there is a high concentration of solutes in the descending limb.

8. Which of the following components of glomerular filtrate is NOT normally found in urine?
   a. Amino acids
   b. Water
   c. Urea
   d. Na^+

9. You may feel the urge to urinate after drinking an alcoholic beverage because
   a. alcohol travels quickly through the digestive tract.
   b. the digestive tract does not absorb alcohol.
   c. alcohol is easily reabsorbed at nephrons.
   d. alcohol decreases the reabsorption of water at nephrons.

Answers to these questions are in appendix A.

WRITE IT OUT

1. Birds and insects frequently collect nectar from plants. Birds are endothermic, and insects are ectothermic. Do you think that a given amount of nectar can support a greater mass of insects or of birds? Explain.

2. Explorers of Antarctica must eat thousands more Calories per day than people exploring the tropics. Explain this observation.

3. Imagine you are adrift at sea. If you drink seawater, you will dehydrate much faster than if you have access to fresh water. Explain.

4. What are the three types of nitrogenous wastes? Which animals are most likely to excrete each one, and why?

5. Explain why exercise increases the urea concentration in urine.

6. Shortly after you drink a large glass of water, you will feel the urge to urinate. Explain this observation. Begin by tracing the path of the water, starting at the glomerulus and ending with the arrival of urine in the bladder.

7. How does the kidney reduce the volume of urine to a small fraction of the volume of filtrate that enters the glomerular capsule?

8. How could chronic high blood pressure damage the kidneys? How could very low blood pressure impair kidney function?

9. The fluids deep in the medulla of the kidney have a much higher solute concentration than those in the cortex. Where do the solutes in the medulla come from? What is the role of that high solute concentration in the functioning of the nephron loop?

10. Diuretic drugs prevent ions from being reabsorbed from the nephron. Explain why a diuretic drug might lower blood pressure. Why do some patients have low levels of salts in their blood while taking a diuretic?

PULL IT TOGETHER

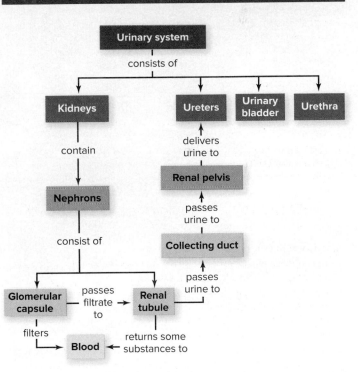

Figure 33.18 Pull It Together: The Urinary System.

Refer to figure 33.18 and the chapter content to answer the following questions.

1. Review the Survey the Landscape figure in the chapter introduction and the Pull It Together concept map. How do the urinary system and the circulatory system work together to maintain homeostasis?

2. Add the terms *glomerulus, proximal convoluted tubule, nephron loop,* and *distal convoluted tubule* to this concept map.

3. Add *water, salts, toxins,* and *urine* to the concept map.

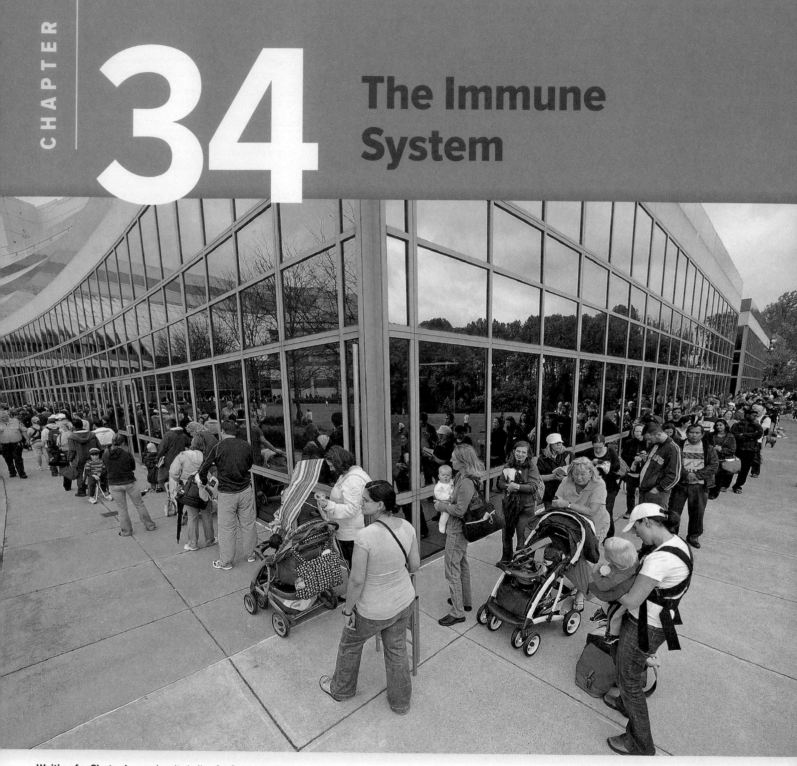

CHAPTER

34 The Immune System

Waiting for Shots. A crowd waits in line for flu vaccinations.

©Tracy A. Woodward/The Washington Post/Getty Images

LEARN HOW TO LEARN
Find a Good Listener

For many complex topics, you may struggle to know how well you really understand what is going on. One tip is to try explaining what you think you know to somebody else. Choose a subject that takes a few minutes to explain. As you describe the topic in your own words, your partner should ask follow-up questions and note where your explanation is vague. Those insights should help draw your attention to important details that you overlooked.

Vaccine Myths and Realities

A child entering day care or kindergarten must already be vaccinated against a host of diseases. Targeted illnesses include diphtheria, tetanus, whooping cough (pertussis), hepatitis A and B, *Haemophilus influenzae* type B, measles, mumps, rubella, polio, rotavirus, and chickenpox. Each vaccine relies on the immune system's amazing ability to recognize and destroy foreign particles, including viruses and bacteria. A vaccine containing weakened measles viruses, for example, "trains" the immune system to recognize measles. Then, if the person later encounters the full-strength virus, the immune system is primed to pounce.

Vaccines have saved millions of lives and prevented enormous suffering since Edward Jenner developed the first smallpox vaccine in 1796. For example, the Centers for Disease Control and Prevention estimates that smallpox, diphtheria, measles, polio, and rubella once killed nearly 650,000 people a year in the United States. Many more people suffered from lingering harm such as paralysis, deafness, blindness, and intellectual disabilities. Ever since routine childhood vaccinations started, however, the overall incidence of these five diseases has declined by 99% in the United States, and mortality has plunged below 100 per year.

Nevertheless, some people refuse to vaccinate their children. They cite a variety of arguments. One is that vaccines can have harmful side effects. It is true that no vaccine is 100% safe, and some children do have medical conditions that preclude vaccines. But for a healthy child, the risk of a deadly reaction to a vaccine is much smaller than the risk of death from infectious disease.

A second objection stems from the suspicion that vaccines cause autism, sudden infant death syndrome, diabetes, asthma, and other illnesses. This argument is based primarily on anecdotal evidence from parents who noticed signs of illness shortly after a child received a vaccination. The events may occur at about the same time, but that does not mean that one causes the other. In fact, numerous epidemiological studies have found no connection between vaccines and autism.

A third argument against vaccination is that the shots are unnecessary, given how rare these diseases are in countries with high vaccination rates. However, global air travel means that diseases easily spread from one part of the world to another. When people refuse to vaccinate their children, they increase the chance that many now-rare diseases will flare up again.

Vaccines represent a tangible benefit of scientific research on the immune system. This complex set of cells, tissues, and organs is the subject of this chapter.

SURVEY THE LANDSCAPE
Animal Life

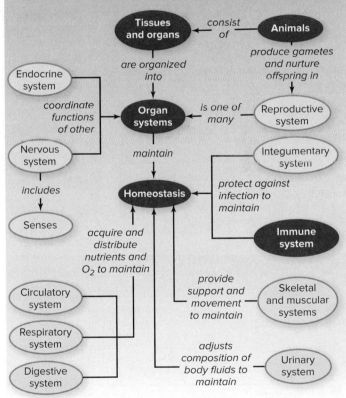

The environment teems with microbes and other organisms that compete for access to the resources inside an animal's body. The immune system fights these pathogens in many ways.

For more details, study the Pull It Together feature in the chapter summary.

34.1 Many Cells, Tissues, and Organs Defend the Body

Disease-causing agents—**pathogens**—are nearly everywhere. Viruses, bacteria, protists, fungi, and worms are in water, food, soil, and air. These pathogens can enter our bodies whenever we eat, drink, breathe, or interact with people and other animals. Yet we are not constantly sick. The explanation is that the **immune system** enables the body to recognize its own cells and to defend against infections, cancer, and foreign substances. The vertebrate immune system consists of a network of cells, defensive chemicals, and fluids that permeate the body (**figure 34.1**).

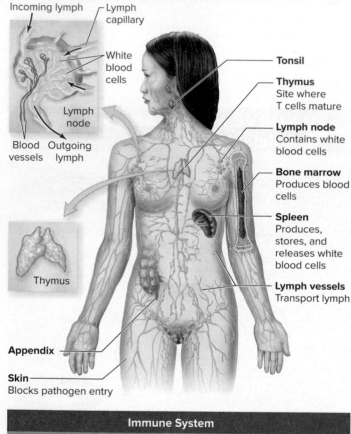

Figure 34.1 The Human Immune System.

Incoming lymph — Lymph capillary
White blood cells
Lymph node
Blood vessels   Outgoing lymph
Thymus

Tonsil

Thymus
Site where T cells mature

Lymph node
Contains white blood cells

Bone marrow
Produces blood cells

Spleen
Produces, stores, and releases white blood cells

Lymph vessels
Transport lymph

Appendix

Skin
Blocks pathogen entry

Immune System	
Main tissue types*	**Examples of locations/functions**
Epithelial	Thymus, spleen, and tonsils consist partly of epithelial tissue; lines lymphatic and blood vessels; lymphoid tissue lies beneath epithelial tissues of the digestive, respiratory, and urinary tracts, guarding potential points of entry for pathogens.
Connective	Immune system cells and chemicals circulate in blood and lymph, which are connective tissues; bone marrow is connective tissue that produces lymphocytes; thymus, spleen, and lymph nodes consist partly of connective tissue.

*See chapter 25 for descriptions.

A. White Blood Cells Play Major Roles in the Immune System

Blood is critical to immune function. Plasma, the liquid matrix of blood, carries defensive proteins called antibodies. In addition, infection-fighting **white blood cells** are suspended in blood plasma and occupy the interstitial fluid between cells. Stem cells in **red bone marrow,** the spongy tissue inside bones, give rise to white blood cells. ⓘ *bone marrow, section 29.3A; composition of blood, section 30.2*

White blood cells fall into five categories and play many roles in the body's defenses (**figure 34.2**). Two of the five types of white blood cells function primarily as **phagocytes,** which are scavenger cells that travel in the bloodstream or wander through body tissues, engulfing bacteria and debris. **Neutrophils,** the most abundant white blood cells, are phagocytes. In addition, white blood cells called **monocytes** give rise to **macrophages,** which also function as phagocytes. Some types of macrophages wander throughout the body; others remain in just one tissue. As described in sections 34.2 and 34.3, macrophages that consume foreign particles play important roles in initiating the body's defenses. ⓘ *phagocytosis, section 4.5C*

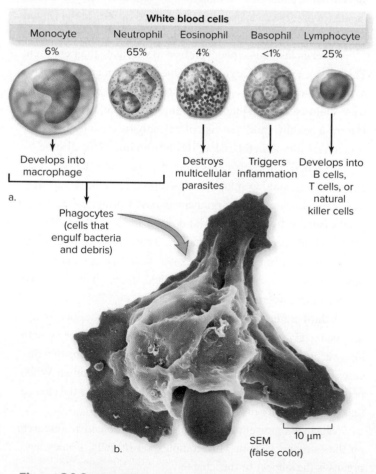

White blood cells				
Monocyte	Neutrophil	Eosinophil	Basophil	Lymphocyte
6%	65%	4%	<1%	25%
Develops into macrophage		Destroys multicellular parasites	Triggers inflammation	Develops into B cells, T cells, or natural killer cells

a.

Phagocytes (cells that engulf bacteria and debris)

SEM
(false color)
10 μm

b.

Figure 34.2 White Blood Cells. (a) Human blood contains five types of white blood cells. (b) A phagocyte (*purple*) engulfs a yeast cell (*red*).

(b): ©SPL/Science Source

Eosinophils, a third type of white blood cell, produce a potent cocktail of chemicals and enzymes that defend mainly against multicellular parasites such as worms.

Basophils are white blood cells that release chemical signals that trigger inflammation and allergies. Their close relatives, mast cells, share similar functions. Like basophils, mast cells originate in red bone marrow. But mast cells do not circulate in blood. Rather, they settle in tissues, especially those near the skin, digestive tract, and respiratory system.

The fifth category of white blood cell, the lymphocytes, includes several cell types. B cells are lymphocytes that mature in red bone marrow, then migrate to lymphoid tissues and into the blood. Lymphocytes called T cells also originate in red bone marrow but mature in the thymus, a small immune organ in the chest (*T* is for *thymus*). From there, T cells migrate throughout the body. Together, B cells and T cells coordinate the body's responses to specific pathogens, as described in section 34.3. Another type of lymphocyte, the natural killer cell, attacks cancerous or virus-infected cells.

B. The Lymphatic System Produces and Transports Many Immune System Cells

We have already seen that the lymphatic system helps regulate the volume of tissue fluid (see chapter 30) and absorbs fat from the small intestine (see chapter 32). A third role, described in this chapter, is to help defend the body against pathogens.

The lymphatic system transports lymph, a colorless fluid that forms when lymph capillaries absorb interstitial fluid from the spaces between cells. Interstitial fluid originates as plasma that has leaked out of blood vessels; the lymphatic system "recycles" this fluid and returns it to the bloodstream. Lymph also contains white blood cells, and it may pick up bacteria, viruses, and cancer cells as it travels in the body.

Other components of the lymphatic system include the lymphoid organs that produce, accumulate, or aid in the circulation of lymphocytes. Red bone marrow and the thymus are two examples of lymphoid organs. In addition, the spleen is a large lymphoid organ containing masses of lymphocytes and macrophages that destroy pathogens in the blood. Surgeons sometimes remove the spleen if it ruptures or becomes cancerous. Although people can live without a spleen, a major side effect is weakened immunity and increased susceptibility to infections.

A lymph node is one of the many small, bean-shaped lymphoid organs located along the lymph vessels. Inside each lymph node, millions of white blood cells engulf dead cells and pathogens circulating in lymph. Lymph nodes also release B and T cells to lymph, which carries them to the blood. When you have an infection, lymph nodes in the neck, armpits, or groin may become swollen and tender as they accumulate extra white blood cells. Many people call these enlarged lymph nodes "swollen glands."

Scattered concentrations of lymphoid tissues also guard the mucous membranes where pathogens may enter the body. Examples include the tonsils (near the throat), the appendix, and patches of lymphoid tissue in the small intestine.

C. The Immune System Has Two Main Subdivisions

Biologists divide the human immune system into two parts: innate defenses and adaptive immunity (figure 34.3). Together, innate defenses and adaptive immunity interact in highly coordinated ways to defend the body against pathogens.

Innate defenses provide a broad defense against any infectious agent. *Innate* refers to the fact that these defenses are always present and ready to function, as described in section 34.2. In adaptive immunity, the body's immune cells not only recognize specific parts of a pathogen but also "remember" previous encounters. Section 34.3 describes adaptive immunity in detail.

34.1 MASTERING CONCEPTS

1. List the cell types that participate in the body's defenses, along with some of their functions.
2. List and describe the components of the lymphatic system.
3. How does the immune system interact with the circulatory system?
4. What are the two subdivisions of the immune system?

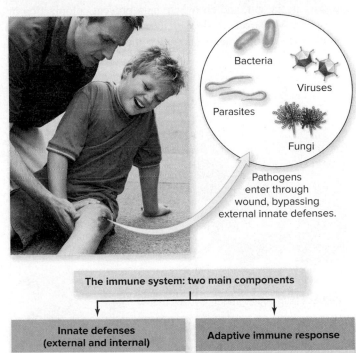

The immune system: two main components

Innate defenses (external and internal)	Adaptive immune response
• Prevent entry of pathogens (external) or attack pathogens that enter the body (internal) • Always active; immediate response • Nonspecific	• Delayed response • Response is strongest and fastest for previously encountered pathogens • Specific to particular pathogens

Figure 34.3 Overview of Body Defenses. The immune system's two arms—innate and adaptive responses—protect the body against disease. Innate defenses are nonspecific but always active, whereas the adaptive immune response creates a "memory" of specific pathogens.

Photo: ©Digital Vision RF

34.2 Innate Defenses Are Nonspecific and Act Early

The body's innate defenses include many components (figure 34.4). This arm of the immune system is called "nonspecific" because it acts against any type of invader.

A. External Barriers Form the First Line of Defense

Physical barriers block pathogens and foreign substances from entering the body. Unpunctured skin is the most extensive and obvious wall; note that the bacteria and other microbes in figure 34.3 are entering the boy's body at his skinned knee. Other physical barriers include mucus that traps inhaled dust particles in the nose; wax in the ears; tears that wash irritants from the eyes and contain an antimicrobial substance called lysozyme; and cilia that sweep bacteria out of the respiratory system. In addition, a bath of strong acid destroys most microbes that reach the stomach. ⓘ *skin,* section 25.5; *stomach,* section 32.3B

An often underappreciated component of this first line of defense is the body's normal microbiota. Resident microbes on the skin, in the gut, and elsewhere help prevent colonization by pathogens. Moreover, studies of lab animals reared in special microbe-free environments show that the immune system does not develop correctly without stimulation from our microscopic companions. ⓘ *normal microbiota,* section 17.4B

B. Internal Innate Defenses Destroy Invaders

A large collection of defensive cells and molecules awaits any microbe that manages to breach the body's external barriers.

White Blood Cells   White blood cells play many roles in the body's innate defenses. Eosinophils secrete substances that destroy parasites. Basophils and mast cells provoke inflammation, attracting additional white blood cells. Natural killer cells destroy cancerous or virus-infected cells. Meanwhile, neutrophils and macrophages consume pathogens by phagocytosis and promote fever. And as described in section 34.3, macrophages play a critical role in activating the body's adaptive immune response.

Inflammation   **Inflammation** is an immediate, localized reaction to an injury or to any pathogen that breaches the body's barriers. The body's familiar response to a minor cut illustrates the effects of inflammation: The area surrounding the wound becomes red, warm, swollen, and painful. Overall, this nonspecific defense recruits immune system cells, helps clear debris, and creates an environment hostile to microorganisms.

Figure 34.5 illustrates the events of inflammation in response to a minor injury. Damaged cells release substances that provoke basophils and mast cells in the skin's dermis to release **histamine,** a biochemical that dilates (widens) blood vessels and causes them to become "leakier"—that is, more permeable to fluids and white blood cells.

As capillaries near the injury become dilated, additional blood arrives, turning the area warm and red. Blood plasma, which carries antimicrobial substances, leaks out of the blood vessels. This fluid dilutes the toxins secreted by bacteria and causes localized swelling. Pressure on the swollen tissues, coupled with chemical signals released from the injured cells, stimulates pain receptors in the skin.

Meanwhile, macrophages and neutrophils squeeze through the blood vessel walls and move into the area, engulfing and destroying bacteria and damaged cells. Pus may accumulate; this whitish fluid contains white blood cells, bacteria, and debris from dead cells.

In medical terminology, the suffix *-itis* indicates inflammation. For example, dermatitis (a rash) signifies inflamed skin, often resulting from contact with an irritant such as poison ivy. Appendicitis is inflammation of the appendix, usually caused by bacterial infection. Aspirin and ibuprofen reduce pain and swelling by blocking the enzymes required for inflammation to occur.

Inflammation may be acute or chronic. Acute inflammation is an adaptation that prevents infection after an injury. The effects usually last only a few days or less, as illustrated by the short-term discomfort of a minor burn or "paper cut." Chronic

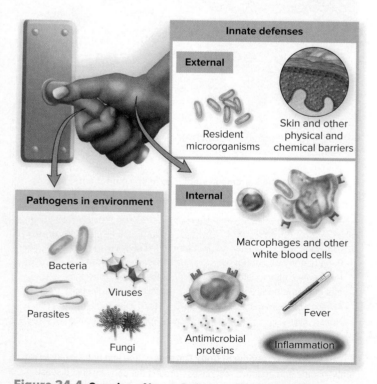

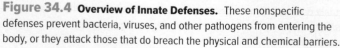

Figure 34.4 Overview of Innate Defenses. These nonspecific defenses prevent bacteria, viruses, and other pathogens from entering the body, or they attack those that do breach the physical and chemical barriers.

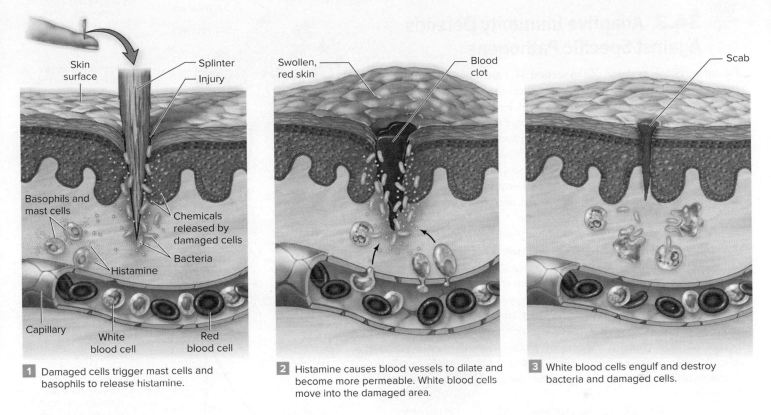

Skin surface — Splinter — Injury

Swollen, red skin — Blood clot

Scab

Basophils and mast cells — Chemicals released by damaged cells — Bacteria — Histamine

Capillary — White blood cell — Red blood cell

1 Damaged cells trigger mast cells and basophils to release histamine.

2 Histamine causes blood vessels to dilate and become more permeable. White blood cells move into the damaged area.

3 White blood cells engulf and destroy bacteria and damaged cells.

Figure 34.5 Inflammation. Immediately after a splinter pierces the skin, chemicals released from damaged cells trigger the inflammatory response.

inflammation, on the other hand, is a prolonged response that may last for months or years. The persistent presence of pathogens or toxins can cause any tissue in the body to become chronically inflamed; genetic mutations may also play a role. Medical problems associated with chronic inflammation include rheumatoid arthritis, atherosclerosis, gum disease, diabetes, Alzheimer disease, depression, cancer, Crohn's disease, celiac disease (gluten intolerance), and many other serious illnesses.

Complement Proteins and Cytokines Many antimicrobial biochemicals participate in the innate defenses. For example, the 25 or so types of **complement** proteins all help to destroy pathogens in the body. When activated, some trigger a chain reaction that punctures bacterial cell membranes. Others cause mast cells to release histamine, and still others attract phagocytes.

Other chemical defenses include **cytokines,** messenger proteins that bind to immune cells and promote cell division, activate defenses, or otherwise alter their activity. For example, cells infected by viruses release **interferons,** which are cytokines that "sound an alarm" to alert other components of the immune system to the infection. White blood cells release **interleukins,** the largest group of cytokines. Their name comes from their role in communicating (*inter-*) between leukocytes, or white blood cells (-*leukins*). One type of interleukin,

produced by macrophages, activates B cells and T cells, provokes inflammation, and causes fever.

Fever Cytokines travel throughout the body in the bloodstream. At the hypothalamus, they can trigger a temporary increase in the set point of the body's thermostat. **Fever,** a rise in the body's temperature, is therefore a common reaction to infection. Although the shivering and chills that accompany fever feel uncomfortable, a mild fever can help fight infection in several ways. A higher body temperature directly inhibits some bacteria and viruses. Fever also counters microbial growth indirectly, because an elevated body temperature reduces the iron level in the blood. Bacteria and fungi require more iron as the temperature rises, so a fever stops the replication of these pathogens. Phagocytes also attack more vigorously when the temperature climbs. ⓘ *hypothalamus,* section 28.3

34.2 MASTERING CONCEPTS

1. List the categories of innate defenses.
2. Describe the physical barriers to infection.
3. How do white blood cells contribute to innate defenses?
4. How can inflammation be both helpful and harmful?
5. How is fever protective?

34.3 Adaptive Immunity Defends Against Specific Pathogens

The innate defenses described in section 34.2 are broad-spectrum weapons. Adaptive immune responses, on the other hand, act against individual targets. Two classes of lymphocytes, B cells and T cells, provide the ammunition in these precision defenses.

The target in an adaptive immune response is an **antigen,** which is any molecule that stimulates an immune reaction by B and T cells. Most antigens are carbohydrates or proteins. Examples include parts of a bacterial cell wall or a virus, proteins on the surface of a mold spore or pollen grain, and unique molecules on the surface of a cancer cell.

The word *antigen* (short for *anti*body-*gen*erating) reflects a crucial part of adaptive immunity: the production of **antibodies,** which are Y-shaped proteins that recognize specific antigens. As described later in this section, each B and T cell is genetically programmed to produce receptors that recognize and bind to only one target antigen. But because every foreign particle contains dozens of molecules that can act as antigens, many sets of lymphocytes respond to invasion by one pathogen.

Note that the cells of the immune system can respond to antigens from pathogens that are "floating" in the body's fluids, but they cannot see *inside* any cell. They can, however, respond to molecules on the surface of a cell. The immune system therefore relies on specialized proteins that display antigens on cell surfaces. These proteins are encoded by a cluster of genes called the **major histocompatibility complex (MHC).** As you will soon see, MHC proteins bound to antigens play a major role in distinguishing between healthy body cells, cancer cells, virus-infected cells, pathogens, and other foreign cells.

A. Macrophages Trigger Both Cell-Mediated and Humoral Immunity

One of the first cell types to respond to infection is the macrophage, which both participates in innate defenses and triggers adaptive immunity. If a macrophage encounters a bacterium or other foreign substance in the body, it engulfs the invader, dismantles it, and links each antigen to an MHC protein on the macrophage surface (figure 34.6, steps 1–2).

A macrophage displaying an antigen on its surface travels in lymph to a lymph node, where the cell encounters collections of T and B cells (figure 34.6, step 3). **Helper T cells** are "master cells" of the immune system because they initiate and coordinate the adaptive immune response. (HIV, the virus that causes AIDS, deals a deadly blow to the immune system by killing helper T cells; section 34.5 describes this illness in more detail.)

When an antigen-presenting macrophage meets a helper T cell with receptors specific to the antigen it is displaying, the two cells bind (figure 34.6, step 4). The activated helper T cell

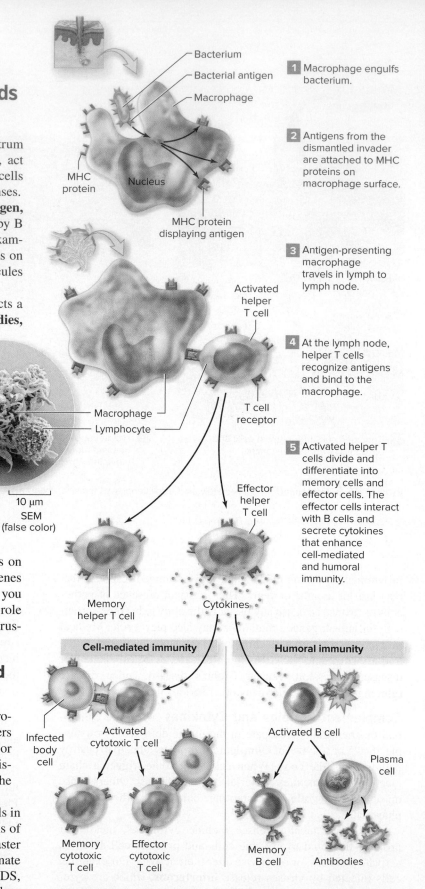

1 Macrophage engulfs bacterium.

2 Antigens from the dismantled invader are attached to MHC proteins on macrophage surface.

3 Antigen-presenting macrophage travels in lymph to lymph node.

4 At the lymph node, helper T cells recognize antigens and bind to the macrophage.

5 Activated helper T cells divide and differentiate into memory cells and effector cells. The effector cells interact with B cells and secrete cytokines that enhance cell-mediated and humoral immunity.

Figure 34.6 Adaptive Immunity: A Summary. A macrophage displays bacterial antigens on its surface. At a lymph node, a helper T cell binds to the macrophage. The activated helper T cell divides, producing effector T cells that help activate and enhance the cell-mediated and humoral responses.

Photo: ©Dr. Olivier Schwartz, Institute Pasteur/Science Source

immediately divides into many identical copies. Some of the copies differentiate into **memory cells,** which remain in the body long after the initial infection subsides. (As you will see, memory cells may be either T cells or B cells.) In general, memory cells launch a quick immune response upon subsequent exposure to the antigen; that is, they "remember" antigens the immune system has already encountered. The other cell copies are "effectors" that act immediately, initiating the cell-mediated and humoral arms of the adaptive immune system (step 5).

In **cell-mediated immunity,** defensive cells kill body cells that are defective or have already been infected by a pathogen. **Humoral immunity,** on the other hand, relies primarily on secreted antibodies (the term *humoral* refers to substances that circulate in body fluids). The rest of this section describes these defenses in more detail.

B. Cytotoxic T Cells Provide Cell-Mediated Immunity

Cytotoxic T cells provide cell-mediated immunity by physically binding to and destroying "suspicious" cells—that is, those that are cancerous, damaged, foreign to the body, or infected with viruses or bacteria. Activation of cell-mediated immunity requires a cytotoxic T cell to bind to a cell presenting an antigen (this requirement is one difference between cytotoxic T cells and the natural killer cells participating in innate immunity). Once activated, a cytotoxic T cell divides, and the resulting cells differentiate into memory cells and effector cells. Cytokines from helper T cells enhance the rate of cell division.

Figure 34.7 shows how a cytotoxic T cell kills a cancer cell. After binding to an antigen on the surface of the cancer cell, the cytotoxic T cell releases perforin proteins that poke holes in the cancer cell's membrane. The cancer cell soon dies.

Cytotoxic T cells also recognize and destroy cells infected with viruses or bacteria. By destroying a damaged or infected cell before it can replicate, cytotoxic T cells can stop a potential problem in its tracks. Moreover, the memory cells, which linger long after the infection subsides, differentiate immediately into cell-killing "machines" if the body encounters the same problem again.

Cell-mediated immunity is one factor that complicates organ transplants. The body perceives any foreign object, including a donated kidney, heart, or skin graft, as something to be destroyed. Cytotoxic T cells bind to and destroy the transplanted cells, provoking tissue rejection. To minimize this risk, physicians use "tissue matching" techniques to ensure that an organ donor's MHC proteins closely match the corresponding proteins in the transplant recipient. Immune-suppressing drugs can further reduce the risk of rejection, but the cost is increased vulnerability to cancer and infectious disease (see section 34.5).

C. B Cells Direct the Humoral Immune Response

The humoral immune response includes millions of different B cells, each producing a unique antibody. Before learning how B cells operate, it is important to understand what antibodies are.

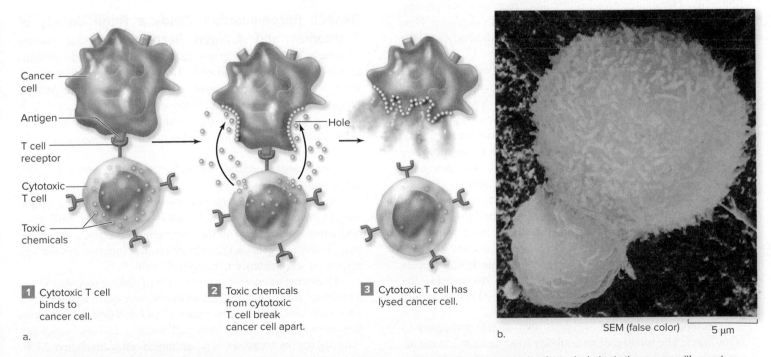

Cancer cell

Antigen

T cell receptor

Cytotoxic T cell

Toxic chemicals

Hole

1 Cytotoxic T cell binds to cancer cell.

2 Toxic chemicals from cytotoxic T cell break cancer cell apart.

3 Cytotoxic T cell has lysed cancer cell.

a.

b.

SEM (false color)   5 μm

Figure 34.7 Cytotoxic T Cells. (a) An effector cytotoxic T cell binds to a cancer cell (step 1) and secretes proteins that poke holes in the cancer cell's membrane (step 2). The cancer cell dies as its membrane disintegrates (step 3). (b) A small cytotoxic T cell attacks a large cancer cell.

(b): ©Dr. Andrejs Liepins/Science Source

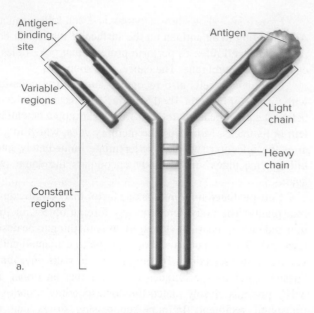

<table>
<tr><td>**Antibody Class**</td><td>**Location**</td><td>**Sample Functions**</td></tr>
<tr><td>IgA</td><td>Lining of digestive, respiratory, and urogenital tracts; also in milk, saliva, and tears</td><td>Protects points of entry into the body</td></tr>
<tr><td>IgD</td><td>B cell membranes</td><td>Antigen receptor; regulates B cell activation</td></tr>
<tr><td>IgE</td><td>Tonsils, skin, mucous membranes</td><td>Stimulates mast cells and basophils to release histamine in inflammatory and allergic responses</td></tr>
<tr><td>IgG</td><td>Circulate in blood plasma; can cross placenta to confer immunity on fetus</td><td>Predominant antibody secreted in secondary immune response</td></tr>
<tr><td>IgM</td><td>B cell membranes; circulate in blood plasma</td><td>Antigen receptor; predominant antibody secreted in primary immune response</td></tr>
</table>

b.

Figure 34.8 **Antibody Structure.** (a) The simplest antibody molecule consists of four polypeptide chains, two long ("heavy") and two short ("light"). Each chain has constant and variable regions; the variable portions form the antigen-binding sites. (b) The five types of antibodies have distinct locations and functions.

Antibodies Are Defensive Proteins Antibodies (also called immunoglobulins) are the main weapons of humoral immunity. These large proteins circulate freely in blood plasma, lymph, and interstitial fluid (see this section's Apply It Now box). Their function is to attack pathogens in the body's fluids, not inside infected cells.

The simplest antibody molecule consists of four polypeptides: two identical light chains and two identical heavy chains (figure 34.8). Together, the four chains form a shape like the letter Y. Each chain has constant and variable regions. The **constant regions** have amino acid sequences that are very similar in all antibody molecules, but the **variable regions** differ a great deal among antibodies. These variable regions determine the specific target antigen to which an antibody binds.

As illustrated in figure 34.8, biologists divide antibodies into five classes, designated IgA, IgD, IgE, IgG, and IgM, based on the structures of their heavy chains (the "Ig" stands for immunoglobulin). Each class has unique chemical properties and predominates in different circumstances. For example, IgE antibodies are responsible for allergies (see section 34.5), whereas IgG is the predominant class of antibody circulating in blood after exposure to a pathogen.

Antibodies are potent weapons that attack pathogens in many ways. The binding of an antibody to an antigen can inactivate a microbe or neutralize its toxins. Antibodies can cause pathogens to clump, making them more apparent to macrophages. They can coat viruses, preventing them from contacting target cells. Antibodies also activate complement proteins, which destroy microorganisms.

Genetic Recombination Yields a Huge Variety of Antibodies and Antigen Receptors Of the human genome's 25,000 or so genes, fewer than 250 encode proteins that specifically bind to antigens. How can one person's lymphocytes produce enough unique antibody proteins and antigen receptor proteins to defeat millions of potential pathogens? As it turns out, generating these diverse molecules is a little like using the limited number of words in a language to compose an infinite variety of stories.

The genes that encode antibodies contain hundreds of small DNA segments that are rearranged in developing lymphocytes. The result: countless lineages of cells that each produce a unique antigen receptor and antibody. Most lymphocytes will never encounter a pathogen with the corresponding antigen, but a few will. Thanks to genetic recombination, the immune system can respond to even newly emerging pathogens.

Producing an enormous assortment of antigen receptors poses a problem: Some of them will no doubt correspond to the body's own molecules. In a process called **clonal deletion,** lymphocytes that recognize the body's own cell surfaces and molecules are "weeded out" by apoptosis, or programmed cell death (figure 34.9). This process, which begins before birth, helps prevent self-immunity. The developing immune system somehow also learns not to attack antigens in food. ⓘ *apoptosis,* section 8.7

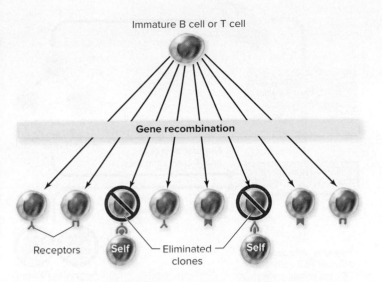

Figure 34.9 **Clonal Deletion.** As lymphocytes develop in a fetus, they are tested against proteins and polysaccharides on the body's own cell surfaces. Clones that match self antigens are eliminated.

Immature B cell or T cell

Gene recombination

Receptors

Self   Eliminated clones   Self

Figure It Out

If a person's DNA can recombine to encode 2700 unique light chains and 11,000 unique heavy chains, how many unique antibodies can be produced?

Answer: 2700 × 11,000 = 29,700,000

Activated B Cells Produce a Surge of Identical Antibodies

On the surface of each B cell is a receptor, which is a version of the antibody that the cell will produce. Until a B cell encounters the antigen it is genetically programmed to recognize, it remains dormant. But when an antigen binds to this surface antibody, the B cell begins to activate. An effector helper T cell binds to the B cell and secretes cytokines that stimulate cell division, completing the activation.

In a process called **clonal selection,** a stimulated B cell divides rapidly, generating an army of memory cells and plasma cells that are clones of the original B cell (figure 34.10). The **plasma cells** immediately secrete huge numbers of antibodies—thousands of molecules each second. At first exposure to the pathogen, B cells produce mostly IgM antibodies. Later, some

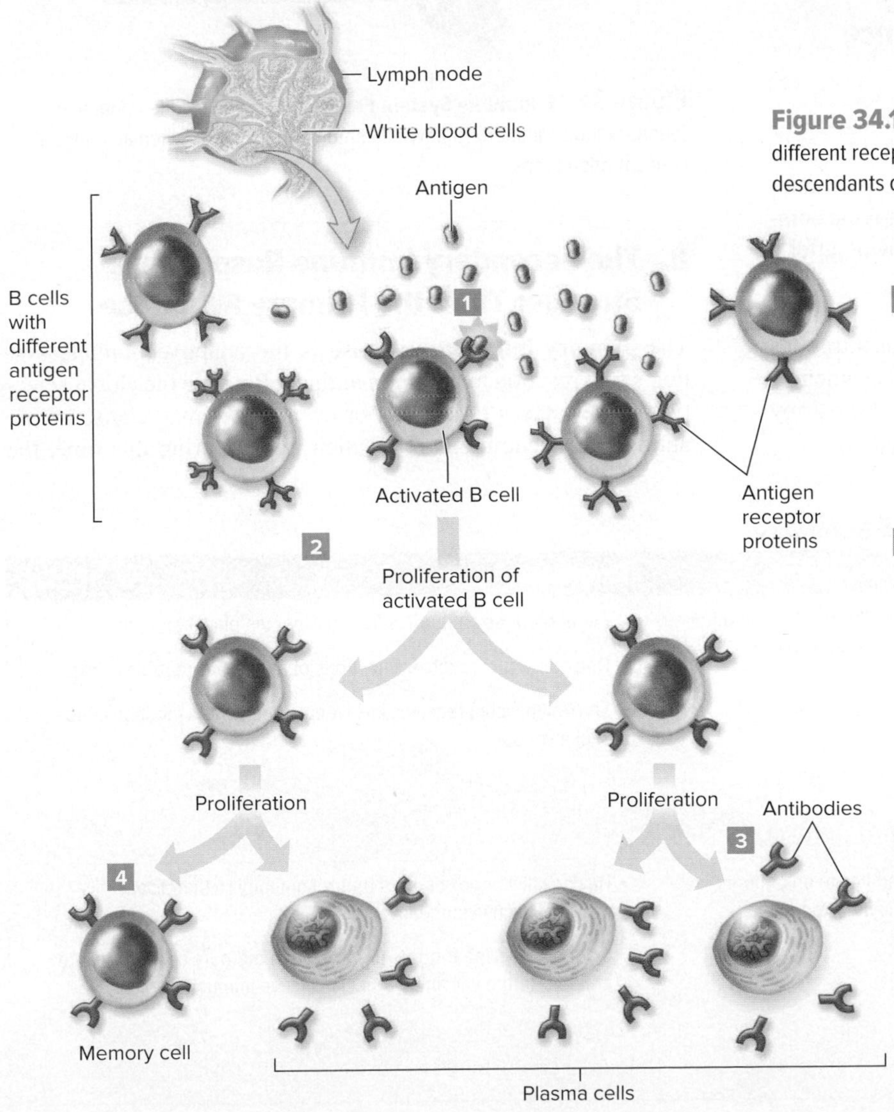

Figure 34.10 **Clonal Selection.** The body produces B cells with many different receptors, but only the cell that binds an antigen proliferates; its descendants develop into memory cells or plasma cells.

Lymph node

White blood cells

Antigen

B cells with different antigen receptor proteins

Activated B cell

Antigen receptor proteins

1 B cell becomes activated when antigen binds to antibody on its surface. Effector helper T cell (not shown) binds to activated B cell and secretes cytokines that stimulate B cell to divide.

2
Proliferation of activated B cell

2 Activated B cell divides rapidly, generating memory cells and plasma cells.

Proliferation   Proliferation   Antibodies

3 Plasma cells produce antibody molecules.

4   **3**

4 Memory cells "remember" exposure to this antigen for long-term immunity.

Memory cell

Plasma cells

daughter cells of the activated B cell switch to IgG or another class of antibodies specific to the same antigen.

The body's immune response is complicated by the fact that each invading virus, bacterium, or other cell has many different antigens on its surface. The immune response therefore involves multiple antibodies, each made by a specific B cell and its clones.

Humoral Immunity Is Passive or Active The humoral immune response is divided into two categories: passive and active (table 34.1). In **passive immunity,** a person receives intact antibodies from another individual. For example, an infant acquires antibodies from its mother in breast milk. Administering antivenom to a victim of a snakebite also illustrates passive immunity.

Active immunity results from the body's own production of antibodies after exposure to antigens in the environment. A person who is bitten by a tick, for example, may begin producing antibodies against the bacteria that cause Lyme disease. Immunity following an illness is called "natural" active immunity. Vaccines, the topic of section 34.4, stimulate "artificial" active immunity because memory cells form in the absence of illness.

D. The Immune Response Turns Off Once the Threat Is Gone

Turning off an immune response once an infection has been halted is as important as turning it on because powerful immune biochemicals can also attack the body's healthy tissues. Immunologists continue to learn more about the precise cytokine combinations that signal the immune system to "back down" after a threat is removed.

Figure 34.11 shows a simplified version of the negative feedback that regulates lymphocyte and antibody concentrations in blood. Once a threat is removed, regulatory T cells somehow reduce the number of dividing B cells and T cells, and the army of plasma cells shrinks until only the memory cells remain.

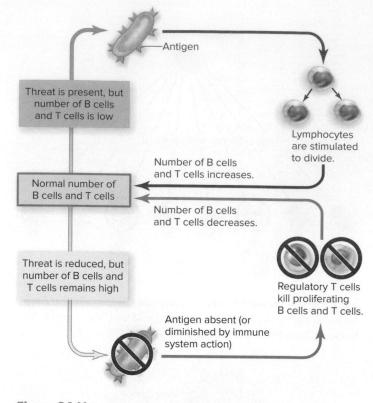

Figure 34.11 Immune System Feedback. This simplified negative feedback loop summarizes how an immune attack stops when an antigen's concentration drops.

E. The Secondary Immune Response Is Stronger Than the Primary Response

The **primary immune response** is the adaptive immune system's first reaction to a foreign antigen. Because the clonal selection process takes time, days or even weeks may elapse before antibody concentrations reach their peak. During this time, the

TABLE **34.1** **Ways to Acquire Immunity: A Summary**

Type	Description	Examples
Passive immunity	One individual acquires antibodies from another individual; temporary (no memory cells are produced)	• Fetus acquires antibodies from mother via placenta or milk. • Dog bite victim receives injections of antibodies to rabies virus. • Snakebite victim receives intravenous antivenom (antibodies to snake venom).
Active immunity	Individual produces antibodies to an antigen; long-lasting (memory cells are produced)	• Having chickenpox confers future immunity to that disease ("natural" active immunity). • Influenza vaccine triggers production of memory cells specific to antigens in the vaccine ("artificial" active immunity).

Apply It **Now**

HIV Tests

Viral infections can be hard to diagnose. Unlike most bacteria, which can be identified based on their growth in petri dishes, viruses cannot reproduce in pure culture. Instead, diagnostic tests may rely on the detection of virus-specific antibodies in a person's saliva, blood, or urine.

The oral swab test is one tool that determines whether a person is producing antibodies against HIV, the virus that causes AIDS. A technician collects saliva by soaking a cotton swab between a person's cheek and gums, then puts the fluid into the wells of a test plate containing a special solution. The mixture in the test plate contains antigens from HIV, plus a substance that changes color when these antigens bind to antibodies from the saliva (figure 34.A). The results are available in about 20 minutes. If the test is positive for antibodies, then other tools can confirm the presence of HIV proteins.

For most people, 2 to 3 months elapse between HIV infection and the production of detectable antibodies. Although antibody tests for HIV are both accurate and sensitive, people who have recently been exposed to HIV may have "false-negative" test results. To be safe, a person who suspects recent exposure to HIV should be tested both immediately and a few months later.

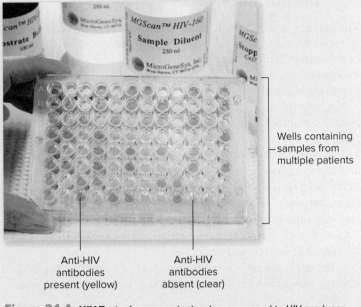

Wells containing samples from multiple patients

Anti-HIV antibodies present (yellow)

Anti-HIV antibodies absent (clear)

Figure 34.A HIV Test. A person who has been exposed to HIV produces antibodies, as revealed by a color change on this test plate.
©Hank Morgan/Science Source

---

pathogen can cause severe damage or death. If a person survives, however, the memory B cells and memory T cells leave a lasting impression—that is, immunological memory.

Thanks to memory cells, the **secondary immune response**—the immune system's reaction the next time it detects the same foreign antigen—is much stronger than the primary response (figure 34.12). Memory B cells transform into rapidly

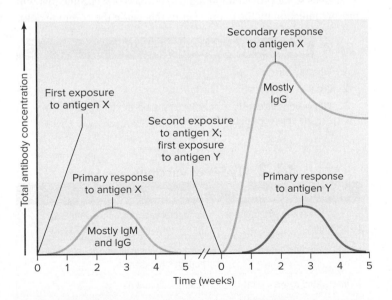

Figure 34.12 Primary and Secondary Immune Responses. The primary immune response leaves memory cells that stimulate a strong secondary immune response following subsequent exposure to the same antigen.

dividing plasma cells. Within hours, billions of antigen-specific antibodies are circulating throughout the host body, destroying the pathogen before it takes hold. Usually, there is no hint that a second infection ever occurred. As described in section 34.4, vaccines create this immunological memory without risking an initial infection.

Figure 34.12 illustrates another difference between the primary and secondary immune responses: the predominant classes of antibodies. During the initial stages of a primary response, IgM predominates, but some B cells quickly switch to producing IgG antibodies. In the secondary response, IgG antibodies dominate. The levels of these antibodies in blood may therefore be useful in diagnosing illness. High concentrations of IgM typically indicate a new infection, while high IgG is consistent with chronic infection.

34.3 MASTERING CONCEPTS

1. What is the relationship between antigens and antibodies?
2. What are the two subdivisions of adaptive immunity, and which cell types participate in each?
3. In your own words, write a paragraph describing the events of adaptive immunity, beginning with a pathogen entering a host's body and ending with the production of memory cells.
4. Describe the structure and function of an antibody.
5. What happens if an immune reaction persists after a pathogen is eliminated?
6. Explain the difference between the primary and secondary immune responses.

34.4 Vaccines Jump-Start Immunity

The immune system is remarkably effective at keeping pathogens and cancer cells from taking over our bodies. Nevertheless, humans do suffer from many incurable infectious diseases caused by viruses and other pathogens. Besides sanitation, the best way to prevent many of these illnesses is vaccination.

A **vaccine** is a substance that stimulates active immunity against a pathogen without actually causing illness. The vaccine consists partly of antigens that "teach" the recipient's immune system to recognize a pathogen. Once taken into the body, the antigens stimulate a primary immune response. Memory cells linger after this initial exposure, ensuring that a subsequent encounter with the real pathogen triggers the rapid secondary immune response (see the Burning Question in this section). Vaccination programs therefore reduce both the incidence and the spread of infectious disease.

The use of vaccines in medicine dates to the late 18th century, when an incurable viral disease called smallpox was ravaging the human population. The fatality rate among infected people was about 25%, and about two thirds of the survivors were left horribly scarred and sometimes blind. But in 1796, a British country physician named Edward Jenner invented a vaccine against smallpox. The preparation contained vaccinia viruses, which cause a mild infection called cowpox. The two viruses—smallpox and vaccinia—are related to each other, and they share some antigens. Exposure to the vaccinia virus therefore confers immunity to smallpox.

This vaccine became the centerpiece of a worldwide smallpox eradication campaign, which began in 1967. By 1980, the World Health Organization had declared that "smallpox is dead." The declaration heralded a milestone in medicine—the eradication of a disease. Today, all known stocks of smallpox virus reside in two labs, one in the United States and the other in Russia. Only the threat of bioterrorism maintains the demand for the smallpox vaccine among some military personnel and "first responders."

Scientists have developed vaccines against many other pathogens as well. The antigens in the vaccines take several different forms (**table 34.2**). Measles and mumps vaccines, for example, contain weakened viruses. Others, such as the vaccine against hepatitis A, contain inactivated viruses that cannot cause an infection. Diphtheria and tetanus vaccines incorporate an inactivated toxin that bacterial pathogens produce. Still others, such as the hepatitis B vaccine, incorporate only a part of the pathogen's surface. The "cervical cancer" vaccine contains proteins identical to those on the human papillomavirus, a pathogen that is strongly correlated with cervical cancer.

Although vaccines have saved countless lives since Jenner's time, they cannot prevent all infectious diseases. For example, researchers have been unable to develop a vaccine against HIV, the rapidly evolving virus that causes AIDS. Influenza viruses also mutate rapidly, so each vaccine is effective for only one flu season. And it has so far proved impossible to develop one vaccine that will prevent infection by the many viruses that cause the common cold.

Burning Question

Why do we need multiple doses of some vaccines?

Vaccines stimulate the immune system to produce memory cells that "remember" their exposure to the antigens in the vaccine. These memory cells can last for decades, triggering the production of antibodies when the real pathogen comes along. Yet most childhood vaccines require a series of shots, spaced out over months or years. And the tetanus and diphtheria vaccines require booster shots at least every 10 years. Why isn't one dose enough?

The answer has to do with the number of memory cells that the body produces after exposure to the vaccine. After just one dose, the number of memory cells may be too small to launch an effective immune response against a pathogen. Repeated doses ensure a stronger immune response because they trigger the formation of more memory cells.

The vaccine's formulation therefore helps determine the need for booster shots. A vaccine that consists of active viruses, such as the one Edward Jenner used against smallpox, causes a mild infection that lasts for about a week. This lengthy exposure to viral antigens stimulates a robust immune response, so booster shots are not usually necessary. On the other hand, when a vaccine contains toxins or inactivated viruses, no infection occurs. The body's exposure to the antigens is therefore relatively brief. In that case, repeated shots help boost the number of memory cells over time.

Submit your burning question to
Marielle.Hoefnagels@mheducation.com

©TRBfoto/Getty Images RF

34.4 MASTERING CONCEPTS

1. What is a vaccine?
2. List the main types of vaccine formulations.
3. Why haven't scientists been able to develop vaccines against HIV and the common cold?

TABLE **34.2** Types of Vaccines

Vaccine Formulation	Examples
Live, weakened (attenuated) pathogen	Polio (oral vaccine), influenza (nasal spray), measles, mumps, chickenpox
Inactivated pathogen	Polio (injectable), influenza (injectable), hepatitis A
Inactivated toxins	Tetanus, diphtheria
Subunits of pathogens	Whooping cough (pertussis), hepatitis B, human papillomavirus, Lyme disease (experimental)

34.5 Several Disorders Affect the Immune System

The immune system may turn against the body's own cells, or it may fail to respond to disease-causing organisms. In addition, harmless substances sometimes trigger an immune response.

A. Autoimmune Disorders Are Devastating and Mysterious

Ideally, the immune system does not attack the body's own cells; as a person develops, lymphocytes corresponding to molecules already present in the body should be eliminated by clonal deletion. In an **autoimmune disorder,** however, the immune system attacks the body's "self antigens." The resulting damage to tissues and organs may be severe (table 34.3).

Several processes typically help the body avoid autoimmunity. The immune system deletes certain self-responding lymphocytes in the red bone marrow and thymus; other mechanisms teach mature lymphocytes not to respond to self antigens. And the immune system simply ignores some self antigens for unknown reasons. Overall, immunologists still have much to learn about how self antigens induce tolerance. Their discoveries may lead to better treatments for autoimmunity.

B. Immunodeficiencies Lead to Opportunistic Infections

An **immunodeficiency** is a condition in which the immune system lacks one or more essential components. A weakened immune system leaves a person vulnerable to **opportunistic pathogens** and cancers that do not normally affect people with healthy immune systems (figure 34.13). Viruses such as HIV can weaken the immune system, as can some inherited diseases and pharmaceutical drugs such as prednisone. ⓘ *prednisone,* section 28.4C

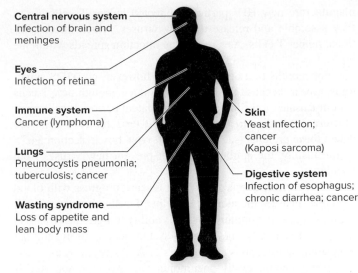

Figure 34.13 Opportunistic Illnesses. People with AIDS and other immunodeficiencies are vulnerable to a wide variety of illnesses that are rare in people with intact immune systems.

Human immunodeficiency virus (HIV) causes acquired immune deficiency syndrome (AIDS). A person can acquire HIV by sexual contact or by using contaminated needles when injecting drugs. (Routine screening of the blood supply has virtually eliminated blood transfusions as an exposure route.) A mother can also transmit HIV to her baby, either during delivery or in breast milk.

Helper T cells are HIV's main target (see figure 16.5). Viral RNA enters a cell, where the enzyme reverse transcriptase transcribes the RNA to DNA. The freshly synthesized DNA inserts itself into the T cell's chromosome. From their new location in the nucleus, the viral genes encode all of the proteins needed to

TABLE 34.3 Examples of Autoimmune Disorders

Disorder	Symptoms	Target(s) of Antibody Attack
Glomerulonephritis	Lower back pain, kidney damage	Kidney cell antigens that resemble *Streptococcus* antigens
Graves disease	Restlessness, weight loss, irritability, increased heart rate and blood pressure	Thyroid gland
Type 1 diabetes	Thirst, hunger, weakness, emaciation	Pancreas
Myasthenia gravis	Muscle weakness, difficulty speaking and swallowing	Skeletal muscle cells (receptors for neurotransmitters)
Rheumatic heart disease	Weakness, shortness of breath	Heart valve cell antigens that resemble *Streptococcus* antigens
Rheumatoid arthritis	Joint pain and deformity	Cells lining joints
Scleroderma	Thick, hard, pigmented skin patches	Connective tissue cells
Systemic lupus erythematosus	Red rash on face, prolonged fever, weakness, kidney damage	DNA, neurons, blood cells

manufacture new HIV particles. Infected helper T cells die as they assemble and release the new viruses, which infect additional helper T cells. As a result, the infection spreads within the body.

For months to a decade or more, however, no AIDS symptoms appear, because the body can produce enough new T cells to compensate for the loss. During this latent period, B cells manufacture antibodies to the virus; rapid tests for HIV exposure detect these proteins (see the Apply It Now box in section 34.3). Unfortunately, the antibodies do not prevent new viruses from forming.

HIV-positive people track their disease progress with blood tests that measure the number of helper T cells. As helper T cell counts decline, the immune system's ability to fight the virus also weakens. Eventually, the immune system fails entirely, and the opportunistic infections and cancers of AIDS begin. Kaposi sarcoma and pneumocystis pneumonia are two common AIDS-associated illnesses in North America, but AIDS patients are susceptible to many other infectious diseases as well.

AIDS is a consequence of a viral infection, but immune deficiency can also be inherited. Each year, a few children are born defenseless against infection due to **severe combined immunodeficiency (SCID),** a disorder in which neither T cells nor B cells function. Decades ago, the parents of a child with SCID had one option: try to isolate the youngster from all possible infectious diseases. Today, SCID has a cure. Most children born with SCID receive bone marrow transplants before they are 3 months old, replacing their defective cells with marrow from a healthy donor. In addition, gene therapy has been used to replace faulty genes in some SCID patients. ⓘ *gene therapy,* section 11.4D

Immunodeficiency is also common in organ transplant recipients. To avoid rejection of a donated organ, transplant recipients must take immune-suppressing drugs for the rest of their lives. Like other people with immunodeficiencies, these patients are vulnerable to opportunistic infections.

C. Allergies Misdirect the Immune Response

In an **allergy,** the immune system is overly sensitive, launching an exaggerated attack on a harmless substance (see **figure 34.14** and the Burning Question in this section). Common **allergens,** or antigens that trigger an allergy, include foods, dust mites, pollen, fur, and oils in the leaves of plants such as poison ivy. The allergens activate B cells to produce IgE antibodies. A first exposure to the allergen initiates a step called sensitization, in which IgE antibodies bind to mast cells. On subsequent exposure, the allergens bind to the molecules attached to the mast cells, causing them to explosively release histamine and other allergy mediators.

The symptoms of an allergic response depend on where in the body the mast cells release mediators. Many mast cells are in the skin, respiratory passages, and digestive tract, so

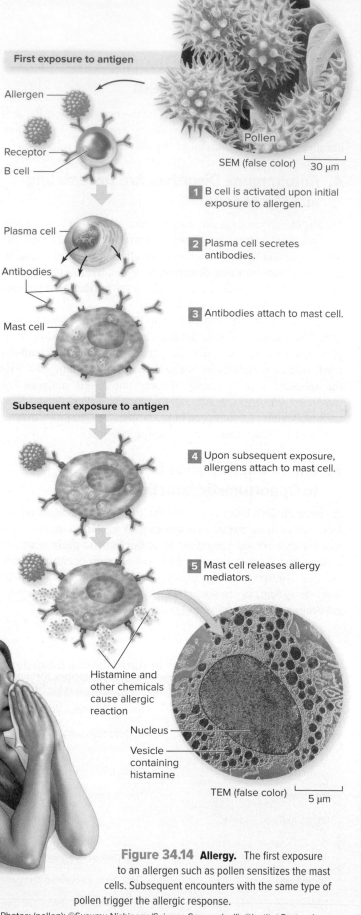

First exposure to antigen

Allergen
Receptor
B cell

Pollen
SEM (false color) 30 μm

1 B cell is activated upon initial exposure to allergen.

Plasma cell

2 Plasma cell secretes antibodies.

Antibodies

3 Antibodies attach to mast cell.

Mast cell

Subsequent exposure to antigen

4 Upon subsequent exposure, allergens attach to mast cell.

5 Mast cell releases allergy mediators.

Histamine and other chemicals cause allergic reaction

Nucleus

Vesicle containing histamine

TEM (false color) 5 μm

Figure 34.14 Allergy. The first exposure to an allergen such as pollen sensitizes the mast cells. Subsequent encounters with the same type of pollen trigger the allergic response.

Photos: (pollen): ©Susumu Nishinaga/Science Source; (cell): ©Institut Pasteur/Phototake

allergies often affect these organs. The result: hives, runny nose, watery eyes, asthma, nausea, vomiting, and diarrhea. Antihistamine drugs relieve these symptoms by preventing the release of histamine or by keeping it from binding to target cells.

Some individuals react to allergens with **anaphylactic shock,** a rapid, widespread, and potentially life-threatening reaction in which mast cells release allergy mediators throughout the body. The person may at first feel an inexplicable apprehension. Then, suddenly, the entire body itches and erupts in hives. Histamine causes blood vessels to dilate, lowering blood pressure. As blood rushes to the skin, not enough of it reaches the brain, and the person becomes dizzy and may lose consciousness. Breathing becomes difficult as the airways in the lungs become constricted. Meanwhile, the face, tongue, and larynx begin to swell. Unless the person receives an injection of epinephrine and sometimes an incision into the trachea to restore breathing, death can come within minutes. ⓘ *epinephrine,* section 28.4C

Anaphylactic shock most often results from an allergy to penicillin, insect stings, or foods. Peanut allergy, for example, affects 6% of the U.S. population and is on the rise. The allergens are proteins that enter the bloodstream undigested. The fact that the initial allergic reaction to peanuts occurs at an average age of 14 months, typically after eating peanut butter, suggests that sensitization occurs even earlier, during breast feeding or before birth.

Early exposure to microorganisms and viruses may be crucial to the development of the immune system. A growing body of evidence has led to the "hygiene hypothesis," which suggests that excessive cleanliness has contributed to recent increases in the incidence of asthma and some allergies (see section 34.6). Apparently, ultraclean surroundings decrease stimulation of the immune system early in life.

D. A Pregnant Woman's Immune System May Attack Her Fetus

Since the immune system rejects foreign cells, it may seem surprising that a woman's body does not destroy her fetus. In general, the female immune response dampens during pregnancy; full immune function returns after the woman gives birth.

One possible source of problems, however, traces to an antigen called the Rhesus (Rh) factor. Some people produce this protein on the surfaces of their red blood cells. A person can be Rh-positive (Rh^+) or Rh-negative (Rh^-). If your blood type is positive (such as "A positive"), the Rh antigen is present; if your blood is Rh-negative (such as "O negative"), your cells lack the Rh antigen. ⓘ *blood types,* section 30.2B

Suppose an Rh^- woman becomes pregnant with an Rh^+ baby. When the baby is born, some of its cells may enter the mother's bloodstream. Her immune system therefore produces antibodies to the Rh antigen. In all subsequent Rh^+ pregnancies, these antibodies can cross the placenta and destroy the fetus's blood. A transfusion of Rh^- blood at birth can save the newborn's life, but this is rarely necessary. Instead, Rh^- women receive an injection of a drug, Rho(D) immune globulin, which prevents the immune response to the Rh antigen.

34.5 MASTERING CONCEPTS

1. How might autoimmunity arise?
2. Which immune system cells does HIV attack, and what is the consequence?
3. Which cells and biochemicals participate in an allergic reaction?
4. How is the Rh factor important in determining whether a pregnant woman's immune system attacks her fetus?

Burning Question

Can people be allergic to meat?

Imagine eating a hamburger and then a few hours later, breaking out in itchy hives. For nearly 4000 people in the United States, this scenario has become a reality, and allergic reactions to meat have become more common in the last few years. Scientists determined that people develop the meat allergy after being bitten by a Lone Star tick (figure 34.B), which is found mostly in the southeastern United States. A biting tick's saliva enters the victim's blood, where the immune system develops antibodies against the many chemicals contained in the saliva. One of those chemicals is a sugar called alpha-gal, which also occurs in meat. When the person subsequently eats meat, antibodies bind to alpha-gal sugars and initiate an immune response. Mast cells release histamine, causing itchiness and dilating blood vessels in the skin. Blood plasma leaking out of the

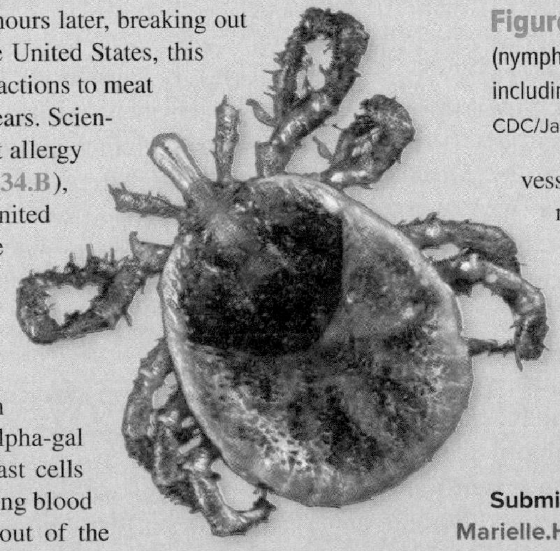

Figure 34.B Lone Star Tick. Immature ticks (nymphs) such as this one transmit many diseases, including an allergy to meat.
CDC/James Gathany

vessels produces hives. In rare cases, the immune reaction is strong enough to put the victim's life at risk.

After developing the allergy, people must avoid meat. Researchers advise using caution when exploring wooded areas in the Lone Star tick's range. To help prevent tick bites, wear clothing that covers the body and use bug spray containing DEET.

Submit your burning question to
Marielle.Hoefnagels@mheducation.com

INVESTIGATING LIFE

34.6 The Hidden Cost of Hygiene

Healthy immune function requires a delicate balance. On the one hand, the immune system must be strong enough to protect the body from dangerous pathogens. On the other hand, if the immune response is too strong, we run the risk of overreacting to harmless substances—as in allergies—or launching an autoimmune attack against our own cells and tissues.

This fine balance reflects our evolutionary history. For millions of years, the human immune system has coevolved with countless bacteria, viruses, and parasitic worms. Many of these hidden residents produce substances that suppress our immune systems, an adaptation that allows them to "fly under the radar" and maintain long-term, chronic infections. Now, however, thanks to improved sanitation and easy access to vaccines and antibiotics, children in developed nations suffer far less from infectious disease than their counterparts in developing countries. At the same time, people in developed countries have a relatively high incidence of allergies, asthma, and autoimmune disorders.

According to an idea called the hygiene hypothesis, eliminating pathogens from everyday life has removed a critical check on the immune system. If so, then people with chronic infections should have dampened immune systems, along with reduced incidence of allergies. Is this the case?

A research team led by Maria Yazdanbakhsh at Leiden University in the Netherlands investigated this question. Along with colleagues from several nations, Yazdanbakhsh studied the incidence of allergies and chronic worm infection among schoolchildren in the west African nation of Gabon. The team predicted that children infected with worms should have a lower incidence of allergies than their uninfected schoolmates.

The study population consisted of 513 children, aged 5 to 14 years. The researchers tested each child for allergies to several substances, including dust mite extract. The team scratched each potential allergen into the skin of each child's forearm, then looked for telltale signs of an allergic reaction (**figure 34.15**). The team also checked for infection with several types of parasitic worms, including a liver fluke. ⓘ *flatworms*, section 21.4

Of the 513 schoolchildren, 57 tested positive for dust mite allergies. Forty-six of these allergic children were worm-free; only 11 were infected with flukes (**figure 34.16**). Statistical analysis of the data indicated that harboring worms significantly lowers the risk for allergies.

A follow-up project revealed that flukes induce white blood cells to release an interleukin that dampens the overall immune system and makes the host's body more hospitable for the worms. This finding suggests the intriguing possibility that parasitic worms might have a role to play in the fight against allergies. Learning which chemicals in worms suppress the immune response may lead to new drugs to treat immune disorders.

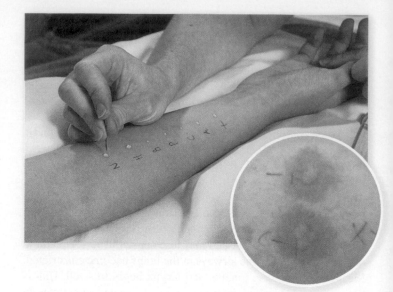

Figure 34.15 Skin Test. Raised welts indicate an allergic reaction to a substance scratched into the skin.

(scratch testing): ©Paul Rapson/Science Source; (welts): ©Joseph Songco/Alamy

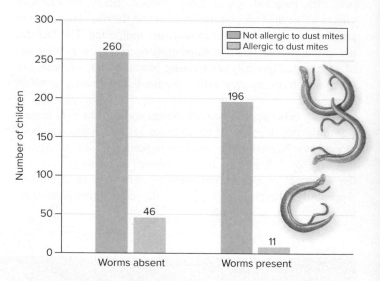

Figure 34.16 Protective Worms? Of the 306 children without parasitic worms, dust mite allergies appeared in 46 (about 15%). In contrast, of the 207 schoolchildren infected with worms, only 11 (about 5%) also had allergies.

In considering these results, it is important not to conclude that hygiene is harmful. No one is suggesting that we return to the days before sanitation and clean water or that we deny these resources to people who lack them now. At the same time, it is interesting to know that allergies and autoimmune diseases may be the price we pay to live a cleaner life.

Source: van den Biggelaar, Anita H. J. and six coauthors, including Maria Yazdanbakhsh. 2000. Decreased atopy in children infected with *Schistosoma haematobium:* A role for parasite-induced interleukin-10. *Lancet,* vol. 356, pages 1723–1727.

34.6   MASTERING CONCEPTS

1. How did researchers document the inverse relationship between worm infection and allergies?
2. Use figure 34.16 to describe the worm–allergy relationship.

CHAPTER SUMMARY

34.1 Many Cells, Tissues, and Organs Defend the Body

- The **immune system** protects the body against **pathogens,** cancer cells, and foreign substances.

A. White Blood Cells Play Major Roles in the Immune System

- Five types of **white blood cells**—**neutrophils, monocytes, eosinophils, basophils,** and **lymphocytes**—participate in immune responses, as do several types of noncirculating cells, such as **mast cells.**
- Neutrophils and **macrophages** (which arise from monocytes) are **phagocytes,** cells that engulf and destroy bacteria and debris.
- **B cells** and **T cells** are lymphocytes that mature in the **red bone marrow** and the **thymus,** respectively. **Natural killer cells** are also lymphocytes.

B. The Lymphatic System Produces and Transports Many Immune System Cells

- The **lymphatic system** plays a crucial role in the immune response. The vessels of the lymphatic system collect and distribute a fluid called **lymph.**
- Besides the thymus, other lymphoid organs include the **spleen** and **lymph nodes.** Immune cells are also concentrated in the tonsils, appendix, and digestive tract.

C. The Immune System Has Two Main Subdivisions

- **Innate defenses** provide broad protection against all pathogens, whereas **adaptive immunity** is directed against specific pathogens (**figure 34.17**). Only adaptive immunity produces immunological memory.

34.2 Innate Defenses Are Nonspecific and Act Early

A. External Barriers Form the First Line of Defense

- Intact skin, mucous membranes, tears, earwax, cilia, and beneficial microbes block pathogens.

B. Internal Innate Defenses Destroy Invaders

- White blood cells destroy bacteria and promote inflammation; natural killer cells destroy cancerous or virus-infected cells; macrophages consume pathogens, promote fever, and activate the immune response.
- Basophils and mast cells trigger **inflammation,** which is an immediate reaction to injury. These cells release **histamine,** a biochemical that causes blood vessels to dilate.
- Redness, warmth, swelling, and pain are associated with inflammation.
- **Complement** proteins interact in a cascade that ends with the destruction of bacterial cells.
- **Cytokines** are antimicrobial molecules that interact with immune system cells and trigger fever. **Interferons** and **interleukins** are examples of cytokines.
- The elevated body temperature of a mild **fever** helps discourage microbial replication.

34.3 Adaptive Immunity Defends Against Specific Pathogens

- Adaptive immunity is directed against specific **antigens.**

A. Macrophages Trigger Both Cell-Mediated and Humoral Immunity

- A macrophage that engulfs a pathogen links antigens from the microbe to specialized proteins on its cell surface. These proteins are encoded by genes of the **major histocompatibility complex (MHC).**
- A **helper T cell** binding to the antigen-presenting cell initiates the **cell-mediated** and **humoral** components of the adaptive immune response. Activated helper T cells divide and differentiate into **memory cells** and into effector cells that help activate cytotoxic T cells and B cells.

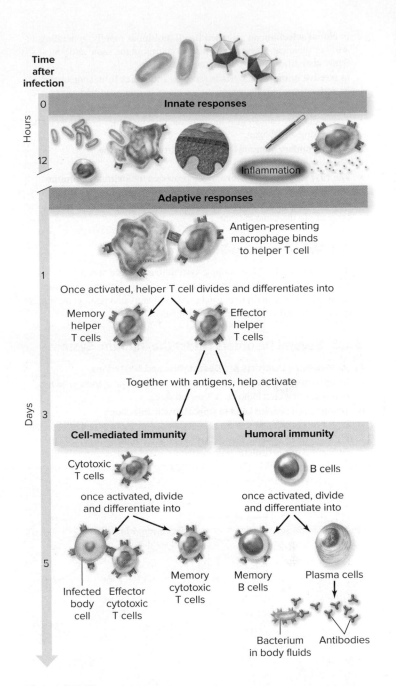

Figure 34.17 **Innate and Adaptive Immunity: A Summary.**

B. Cytotoxic T Cells Provide Cell-Mediated Immunity

- **Cytotoxic T cells** kill cells that are cancerous, damaged, or infected with viruses or bacteria. Memory cytotoxic T cells contribute to long-term immunity.

C. B Cells Direct the Humoral Immune Response

- An **antibody** is a Y-shaped protein composed of two heavy and two light polypeptide chains. Each chain has a **constant region** and a **variable region.** Antibodies bind antigens and form complexes that attract other immune system components.
- Antibodies and antigen receptors are incredibly diverse because DNA segments shuffle during early lymphocyte development. **Clonal deletion** subsequently eliminates lymphocytes corresponding to self antigens.

- In **clonal selection,** an activated B cell multiplies rapidly, generating an army of identical **plasma cells** that all churn out the same antibody. Some also differentiate into memory cells.
- In **passive immunity,** a person receives antibodies from someone else. In **active immunity,** a person makes his or her own antibodies.

D. The Immune Response Turns Off Once the Threat Is Gone
- Tissue damage can occur if the immune response does not dampen after eliminating a pathogen.

E. The Secondary Immune Response Is Stronger Than the Primary Response
- The first encounter with an antigen provokes the **primary immune response,** which is relatively slow (figure 34.18). Its legacy is memory cells that greatly speed the **secondary immune response** on subsequent exposure to the same antigen.

34.4 Vaccines Jump-Start Immunity

- A **vaccine** "teaches" the immune system to recognize specific components of a pathogen, bypassing the primary immune response.
- Some vaccines contain live, weakened, or inactivated pathogens; others incorporate toxins or subunits of pathogens.

34.5 Several Disorders Affect the Immune System

A. Autoimmune Disorders Are Devastating and Mysterious
- An **autoimmune disorder** occurs when the immune system produces antibodies that attack the body's own tissues.

B. Immunodeficiencies Lead to Opportunistic Infections
- An **immunodeficiency** is the absence of one or more essential elements of the immune system. These disorders leave patients vulnerable to cancer and **opportunistic pathogens.**

- The **human immunodeficiency virus (HIV)** kills helper T cells, causing AIDS.
- **Severe combined immunodeficiency (SCID)** is an inherited disease. Drugs that prevent organ transplant rejection also weaken the immune system.

C. Allergies Misdirect the Immune Response
- An **allergy** is an immune reaction to a harmless substance. An **allergen** triggers the production of antibodies, which bind mast cells. On subsequent exposure, these mast cells release allergy mediators such as histamine.
- **Anaphylactic shock** is a life-threatening allergic reaction.

D. A Pregnant Woman's Immune System May Attack Her Fetus
- The immune system of a woman whose cells lack the Rh factor can reject an Rh-positive fetus. A drug can prevent this problem.

34.6 Investigating Life: The Hidden Cost of Hygiene

- According to the hygiene hypothesis, reduced exposure to parasitic worms and other pathogens is associated with an increase in allergies.
- Infection with parasitic worms induces immune cells to release an interleukin that suppresses allergies.

MULTIPLE CHOICE QUESTIONS

1. What may enter lymph capillaries?
   a. Bacteria
   b. Interstitial fluid
   c. Cancer cells
   d. All of the above are correct.

2. Histamine acts on the _____, causing redness and swelling.
   a. white blood cells
   b. cells lining blood vessels
   c. outermost layer of skin
   d. red blood cells

3. Ibuprofen dampens the immune system's inflammation response. What might be a short-term consequence of taking ibuprofen?
   a. Blood vessels may become less permeable.
   b. Antibodies may become less active.
   c. Cytotoxic T cells may begin to divide.
   d. All of the above are correct.

4. The innate immune response is characterized by its
   a. rapid response to invading pathogens.
   b. ability to "remember" pathogens it has already encountered.
   c. ability to produce antibodies.
   d. Both b and c are correct.

5. How does an effector cytotoxic T cell kill an infected body cell?
   a. After binding to an activated helper T cell, the cytotoxic T cell releases antibodies.
   b. After binding to an antigen presented on the surface of the infected cell, the cytotoxic T cell releases toxic proteins.
   c. After binding to the pathogen, the cytotoxic T cell releases cytokines that kill all nearby cells.
   d. After binding to an antibody, the cytotoxic T cell migrates to infected tissues and uses the antibody to kill the cells.

6. Upon exposure to _____ and binding to _____, a B cell divides and differentiates into plasma cells and memory cells.
   a. cytokines released from helper T cells; an antigen from an invader
   b. histamine; an infected cell
   c. interleukins released from macrophages; a complement protein
   d. toxins released from bacteria; an activated helper T cell

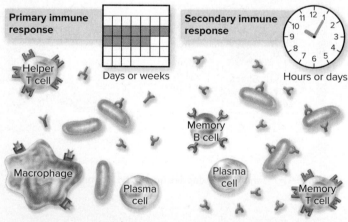

Primary immune response — Days or weeks

Secondary immune response — Hours or days

The primary immune response requires several players. It could take days or weeks for the immune system to produce high numbers of the specific antibodies that target the pathogen's antigens. After the infection is gone, memory cells linger.

The immune reaction to the second encounter of a pathogen is fast and strong. Memory B cells divide into plasma cells that produce antigen-specific antibodies. Memory T cells also divide and differentiate into infection-fighting cells.

Figure 34.18 Primary and Secondary Immune Response: A Summary.

7. In what process is clonal selection important?
   a. Complement function   c. Inflammation
   b. B cell activation   d. Phagocytosis

8. Why is the secondary immune response so much stronger than the primary response?
   a. Because high concentrations of antibodies are already present
   b. Because the phagocytes present the antigens more rapidly
   c. Because memory B cells can rapidly convert to plasma cells
   d. Because protein synthesis occurs more quickly in memory cells

9. How do vaccines prevent infectious disease?
   a. By killing bacteria and viruses
   b. By boosting innate immunity
   c. By triggering a primary immune response
   d. By passive immunity

10. HIV causes immunodeficiency by attacking
    a. memory B cells.   c. plasma cells.
    b. helper T cells.   d. cytotoxic T cells.

Answers to these questions are in appendix A.

WRITE IT OUT

1. Explain the observation that lymphoid tissues are scattered in the skin, lungs, stomach, and intestines.

2. Explain why a scraped knee increases the chance that pathogens will trigger an adaptive immune response.

3. Since fever has protective effects, should we avoid taking fever-reducing medications when ill? Use the Internet to research the consequences of overmedicating a fever and the risk of allowing fever to rise too high.

4. Dead phagocytes are one component of pus. Why is pus a sure sign of infection?

5. As a treatment for bladder cancer, physicians may place a solution containing bovine tuberculosis bacteria into the patient's bladder. The bacteria bind to the bladder wall. Scientists do not fully understand why this treatment causes the patient's immune system to launch an attack against the cancer cells. Using your knowledge of how cytotoxic T cells are activated, write a prediction for how bovine tuberculosis bacteria might help a patient's immune system fight cancer.

6. The environment contains an enormous variety of potential antigens, but the immune system launches an attack only on those that actually enter the body. How does this specificity save energy?

7. What do a plasma cell and a memory cell descended from the same B cell have in common, and how do they differ?

8. Briefly explain the function of each innate and adaptive defense listed in figure 34.17.

9. Which do you think would be more dangerous, a deficiency of T cells or a deficiency of B cells? Explain your reasoning.

10. Compare and contrast how a bacterial population becomes resistant to antibiotics and how a person becomes immune to infections by a particular species of bacteria.

11. Influenza viruses mutate rapidly, whereas the chickenpox virus does not. Why are people encouraged to receive vaccinations against influenza every year, whereas immunity to chickenpox lasts for decades?

12. How do AIDS, SCID, and allergies each relate to the function of the immune system?

13. Explain the difference between clonal deletion and clonal selection; a natural killer cell and a cytotoxic T cell; antibodies and antigens; cell-mediated and humoral immunity; an autoimmune disorder and an immunodeficiency.

14. Humans (and all other organisms) are in an evolutionary battle with a wide variety of pathogens. How does natural selection favor (a) an immune system that adjusts to a changing variety of pathogens and (b) pathogens that evade the immune system?

PULL IT TOGETHER

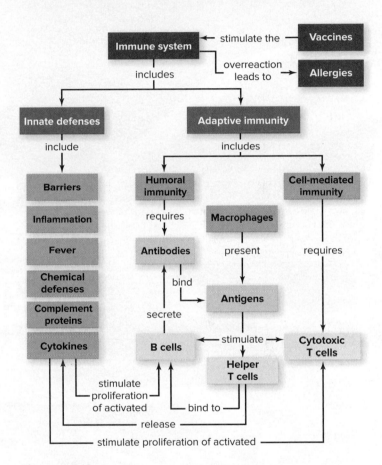

Figure 34.19 Pull It Together: The Immune System.

Refer to figure 34.19 and the chapter content to answer the following questions.

1. Review the Survey the Landscape figure in the chapter introduction and the Pull It Together concept map. How do the actions of the immune system help an animal maintain homeostasis?

2. Add *memory B cells, plasma cells, memory cytotoxic T cells, primary immune response,* and *secondary immune response* to the concept map.

3. Circle the immune system components that a vaccine activates.

4. How do lymph and lymph nodes fit into this concept map?

5. Where else might macrophages fit into this concept map?

35 Animal Reproduction and Development

Runner. South African athlete Caster Semenya was subjected to testing that raised biological and ethical questions about what defines a person's sex.

©Franck Fife/AFP/Getty Images

LEARN HOW TO LEARN
Don't Waste Old Exams

If you are lucky, your instructor may make old exams available to your class. If so, it is usually a bad idea to simply look up and memorize the answer to each question. Instead, use the old exam as a chance to test yourself before it really counts. Put away your notes and textbook, and set up a mock exam. Answer each question without "cheating"; then check how many you got right. Use the questions you got wrong—or that you guessed right—as a guide to what you should study more.

The Sex Spectrum

Many people believe that there are just two kinds of people: males and females. But the truth is considerably more interesting. After all, the development of the genitals depends on interactions among multiple genes, hormones, and embryonic tissues. Any deviation from this complex sequence of events can lead to unexpected results.

For example, some children are born with an intersex condition, in which the anatomy does not match the "standard" male or female. Intersex conditions may affect people who are genetically male (XY) or female (XX) or who have sex chromosome abnormalities.

ⓘ *sex chromosome abnormalities,* section 9.7B

How does biology explain intersex conditions? One cause is androgen insensitivity. In this case, the cells of a child conceived as a boy cannot bind to masculinizing hormones, such as testosterone and other androgens. The genitals may then be ambiguous or appear female. Klinefelter syndrome (XXY), Turner syndrome (XO), and 5-alpha reductase deficiency are also associated with intersex conditions.

Sex determination has implications in sports. In 2009, the International Association of Athletics Federations (IAAF) questioned Caster Semenya's sex due to her "masculine" build and her easy victory in the women's 800-meter run at the World Athletic Championships. She underwent multiple complicated tests to reveal her sex. Months later, the IAAF agreed to let Semenya keep her medal without announcing the results of the tests. However, she subsequently competed in other athletic events for women and won a gold medal in the 2016 Olympics.

Other athletes did not have the same outcome. Following the 2006 Asian Games, runner Santhi Soundarajan was required to return her silver medal after a test revealed that she was genetically male. She was unaware of her condition prior to competing. The public exposure of the devastating news was psychologically taxing on Soundarajan, who reportedly attempted suicide.

Stories such as Soundarajan's raise concerns about whether sex testing is fair to women, especially those whose development does not conform exactly to the female norm. They also stimulate conversations about how much our genes define us. Clearly, gender cannot be considered as only a matter of X and Y.

Regardless of their sex, where do babies come from in the first place? This chapter introduces reproductive anatomy and explains the events leading to fertilization. Section 35.5 then explores how a fertilized egg divides and develops into a fetus. Along the way we will explain contraception, list and describe sexually transmitted diseases, and introduce a few other reproductive perils.

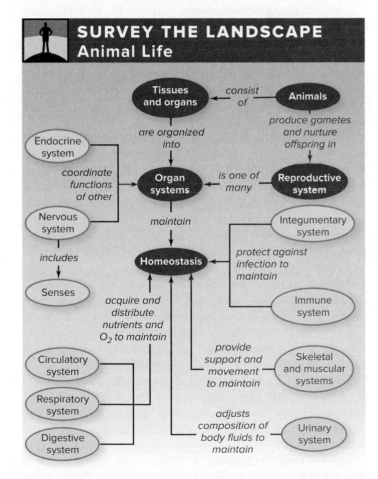

SURVEY THE LANDSCAPE
Animal Life

Reproductive success is the cornerstone of natural selection. The male and female reproductive systems produce the sperm and egg cells that give rise to the next generation.

For more details, study the Pull It Together feature in the chapter summary.

35.1 Animal Development Begins with Reproduction

A monarch butterfly emerges from its chrysalis; a baby bird hatches from an egg; a kitten becomes a full-grown cat. All of these familiar examples illustrate growth and development. Together, reproduction and development are shared features of all multicellular life. Chapter 24 described how flowering plants reproduce and grow; this chapter picks up the topic for animals.

A. Reproduction Is Asexual or Sexual

Like plants, animals may reproduce asexually or sexually (or both). In **asexual reproduction,** the offspring contain genetic information from only one parent. Aside from mutations that occur during replication, all offspring are identical to the parent and to one another. Asexually reproducing animals include sponges, corals, aphids, and some types of lizards. In general, asexual reproduction is advantageous in environments that do not change much over time.

Sexual reproduction requires two parents, each of which contributes half the DNA in each offspring. In many species, sexual reproduction entails high energy costs for attracting mates, copulating, and defending against rivals (see section 36.6). Nevertheless, the benefits of genetic diversity apparently outweigh these costs, especially in a changing environment. Sexual reproduction is therefore extremely common among animals. ⓘ *why sex?* section 9.1

In organisms that reproduce sexually, haploid **gametes** are the sex cells that carry the genetic information from each parent. The gametes—sperm cells from males and egg cells from females—are the products of meiosis, a specialized type of cell division. In meiosis, a diploid cell containing two sets of chromosomes divides into four haploid cells, each containing just one chromosome set. **Fertilization** is the union of two gametes; the product of fertilization is the **zygote,** the diploid first cell of the new offspring.

Sperm and egg come together in a variety of ways. In **external fertilization,** males and females release gametes into the same environment, and fertilization occurs outside the body (figure 35.1a). This strategy is especially common in aquatic animals. Salmon, for example, spawn in streams. Females lay eggs in gravelly nests, and then males shed sperm over them. Other animals with external fertilization include sponges, corals, sea urchins, some crustaceans (such as American lobsters), and some amphibians. Unique "recognition" proteins on the surfaces of the gametes ensure that sperm cells fertilize egg cells of the correct species.

In **internal fertilization,** a male deposits sperm inside a female's body, where fertilization occurs (figure 35.1b). Land animals, including mammals, nonavian reptiles, and birds, commonly use internal fertilization. After copulation, the female may lay hard-shelled, fertilized eggs that provide both nutrition

a.

b.

Figure 35.1 External and Internal Fertilization. (a) A sea urchin releases sperm cells into the water. Meanwhile, females release eggs; fertilization is external. (b) A male black-necked stilt mates with a female. Their offspring will develop inside the female's body until she lays three or four hard-shelled eggs in a nest.

(a): ©Andrew J. Martinez/Science Source; (b): ©Bill Gozansky/Alamy

and protection to developing offspring. A chicken egg is a familiar example. Alternatively, the female may bear live young, as humans and other mammals do.

B. Gene Expression Dictates Animal Development

No matter what the reproductive strategy, development of sexually reproducing animals begins with the zygote. That first cell begins to divide soon after fertilization is complete. As the embryonic animal grows, cells divide and die in coordinated ways to shape the body's distinctive form and function. Developmental biologists study the stages of the animal's growth as cells specialize and interact to form tissues, organs, and organ systems (figure 35.2). The similarities and differences among developing animals can yield important clues to evolution, as illustrated in figure 13.15.

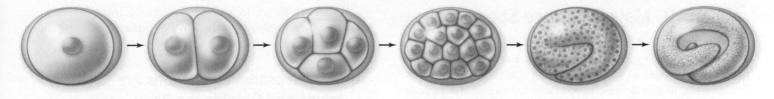

Figure 35.2 Tiny Worm. The life of a nematode worm begins as a single fertilized egg cell. Researchers understand nematode development in detail, describing and mapping the fates of all cells produced from the original zygote.

One key to animal development is **differentiation,** the process by which cells acquire their specialized functions by activating unique combinations of genes in the skin, brain, eyes, heart, and other organs. Another essential process is **pattern formation,** in which genes determine the overall shape and structure of the animal's body, such as the number of segments or the placement of limbs. ⓘ *control of gene expression,* section 7.6

Differentiation and pattern formation involve complex interactions between the DNA inside cells and external signals such as hormones. Many of the molecules that influence animal development in general, and human development in particular, remain unknown. But researchers have found important clues in animals as small as fruit flies. For example, female flies pack a protein called bicoid into one side of each developing egg cell. Bicoid instructs the embryo where to develop a head end. Normally, a high concentration of bicoid protein in the front end of the embryo causes the tissues of the head to differentiate. At the rear of the animal, higher concentrations of another protein, nanos, specify tail formation. When the gene encoding bicoid is mutated in the mother, the embryo develops two rear ends!

These protein gradients are signals that stimulate cells to produce yet other proteins, which ultimately regulate the formation of each structure. Section 7.7 described the importance of homeotic genes in establishing the correct placement of a developing animal's parts. Scientists first discovered homeotic genes by studying mutant flies with legs growing out of their heads. Since that time, additional studies have verified that homeotic genes orchestrate development in humans and all other animal species.

C. Development Is Indirect or Direct

Although many details of animal development remain undiscovered, clear patterns do emerge on a broader scale. For example, biologists distinguish between indirect and direct development. An animal that undergoes **indirect development** spends the early part of its life as a **larva,** an immature stage that looks different from the adult. A caterpillar, for example, is a larva that looks nothing like its butterfly parents; likewise, a tadpole resembles a fish, not the adult frog or salamander that it will grow up to be. Caterpillars, tadpoles, and other larvae often spend most of their time eating and growing. Then, during a process called **metamorphosis,** the larva matures into an adult (see figures 21.8 and 28.2).

Humans and many other familiar animals undergo **direct development:** An infant resembles a smaller version of its parents. The hatching turtle in figure 35.3, for example, is a miniature version of the adult.

This chapter combines reproduction and development, starting with the reproductive anatomy of human males and females. Before you begin, you may find it helpful to review mitosis (chapter 8), meiosis (chapter 9), and the basics of hormone function (chapter 28). The second half of the chapter describes where babies come from—that is, how a fertilized egg grows into a fully formed infant.

35.1 MASTERING CONCEPTS

1. What are the advantages and disadvantages of asexual and sexual reproduction? Of internal and external fertilization?
2. How do genes participate in differentiation and pattern formation?
3. Differentiate between indirect and direct development.

Direct
development

Figure 35.3 From Hatchling to Adult. Like other animals that undergo direct development, a newly hatched tortoise is a miniature version of the adult.

(hatchling): ©Daniel Heuclin/Science Source; (adult): ©Juniors Bildarchiv GmbH/ Alamy RF

35.2 Males Produce Sperm Cells

Both the male and female **reproductive systems** consist of the organs that produce and transport gametes. Each system includes paired **gonads** (testes or ovaries), which contain the **germ cells** that give rise to gametes. Both reproductive systems also include tubes that transport the gametes.

In both sexes, hormones control reproduction and the development of the **secondary sex characteristics,** features that distinguish the sexes but do not participate directly in reproduction. Examples include enlarged breasts and menstruation in adult females and facial hair and deep voices in adult males.

Although the male and female reproductive systems share some similarities, there are also obvious differences. This section details the features and processes that are unique to males.

A. Male Reproductive Organs Are Inside and Outside the Body

Figure 35.4 illustrates the human male reproductive system. The paired **testes** (singular: testis) are the male gonads. The testes lie in a sac called the **scrotum.** Their location outside of the abdominal cavity allows the testes to maintain a temperature about 3°C cooler than the rest of the body, which is necessary for sperm to develop properly. Muscles surrounding each testis can

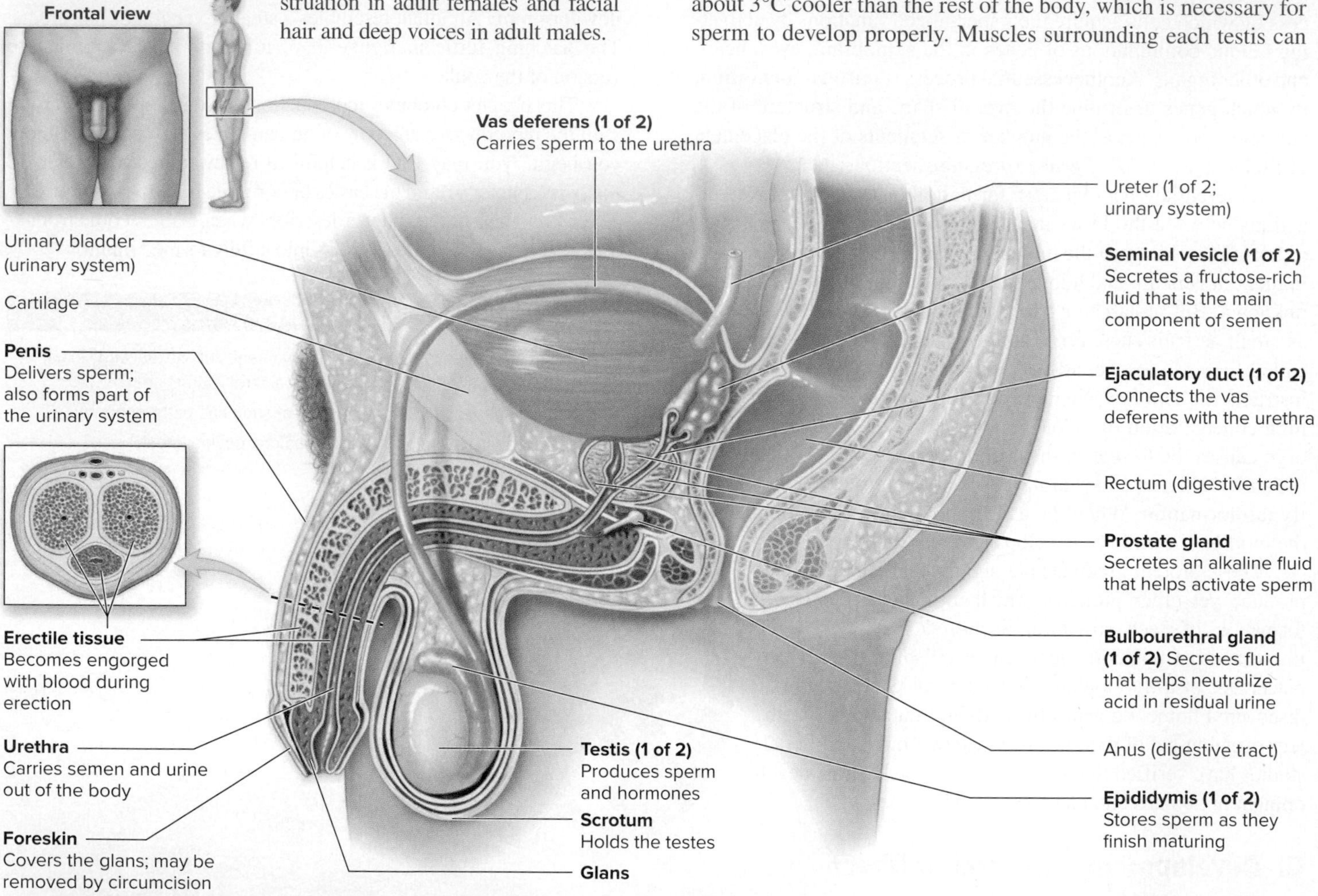

Frontal view

Vas deferens (1 of 2)
Carries sperm to the urethra

Urinary bladder
(urinary system)

Cartilage

Penis
Delivers sperm;
also forms part of
the urinary system

Erectile tissue
Becomes engorged
with blood during
erection

Urethra
Carries semen and urine
out of the body

Foreskin
Covers the glans; may be
removed by circumcision

Ureter (1 of 2;
urinary system)

Seminal vesicle (1 of 2)
Secretes a fructose-rich
fluid that is the main
component of semen

Ejaculatory duct (1 of 2)
Connects the vas
deferens with the urethra

Rectum (digestive tract)

Prostate gland
Secretes an alkaline fluid
that helps activate sperm

**Bulbourethral gland
(1 of 2)** Secretes fluid
that helps neutralize
acid in residual urine

Anus (digestive tract)

Epididymis (1 of 2)
Stores sperm as they
finish maturing

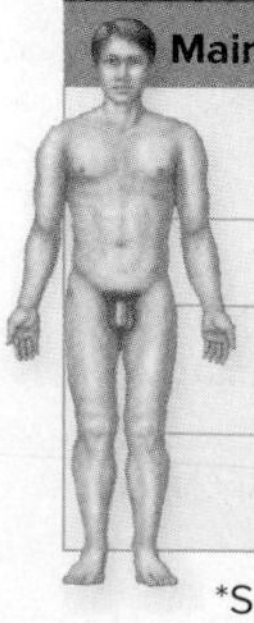

Testis (1 of 2)
Produces sperm
and hormones

Scrotum
Holds the testes

Glans

Reproductive System (Male)		
Main tissue types*	**Examples of locations/functions**	
Epithelial	Lines ducts of reproductive tract; produces sperm cells in seminiferous tubules; produces secretions in accessory glands.	
Connective	Makes up walls of testes; makes up erectile cylinders in penis.	
Nervous	Penis contains sensory nerve fibers and nerve endings; hypothalamus secretes hormones that affect the anterior pituitary.	
Muscle	Smooth muscle surrounds ducts of reproductive tract, propelling sperm out of the body.	

*See chapter 25 for descriptions.

Figure 35.4 The Human Male Reproductive System. The paired testes manufacture sperm cells, which travel through a series of ducts before exiting the body via the urethra in the penis. Various glands add secretions to the semen.

bring the scrotum closer to the body, conserving warmth when the temperature is too cold. If conditions are too warm, the scrotum descends away from the body. Moreover, a network of tiny veins surrounding each testis helps regulate its temperature by exchanging heat with nearby arteries in a countercurrent arrangement. ⓘ *countercurrent exchange,* section 33.1B

A maze of small ducts carries developing sperm to the left or right **epididymis,** a tightly coiled tube that receives and stores sperm from one testis. Each epididymis opens into a **vas deferens,** a duct that travels upward out of the scrotum, bends behind the bladder, and connects with the left or right **ejaculatory duct.** The two ejaculatory ducts empty into the **urethra,** the tube that extends the length of the cylindrical **penis** and carries both urine and semen out of the body. Although these two fluids share the urethra, a healthy male cannot urinate when sexually aroused because a ring of smooth muscle temporarily seals the exit from the urinary bladder.

Semen, the fluid that carries sperm cells, includes secretions from several accessory glands. The two **seminal vesicles,** one of which opens into each vas deferens, secrete most of the fluid in semen. The secretions include fructose (a sugar that supplies energy) and prostaglandins. Prostaglandins are hormonelike lipids that may stimulate contractions in the female reproductive tract, helping to propel sperm. In addition, the single, walnut-sized **prostate gland** wraps around part of the urethra and contributes a thin, milky, alkaline fluid that activates the sperm to swim. The two **bulbourethral glands,** each about the size of a pea, attach to the urethra where it passes through the body wall. These glands secrete alkaline mucus, which coats the urethra before sperm are released.

During sexual arousal, the penis becomes erect, enabling it to penetrate the vagina and deposit semen in the female reproductive tract. At the peak of sexual stimulation, a pleasurable sensation called **orgasm** occurs, accompanied by rhythmic muscular contractions that eject the semen through the urethra and out the penis. **Ejaculation** is the discharge of semen from the penis. One human ejaculation typically delivers more than 100 million sperm cells.

The organs of the male reproductive system may become cancerous. Prostate cancer is the second most common type of cancer in men (behind lung cancer). The resulting prostate enlargement constricts the urethra and may interfere with urination and ejaculation. Prostate cancer usually strikes men older than 50; noncancerous (benign) prostate enlargement affects many older men as well. Testicular cancer, which usually occurs in men younger than 40, is much rarer than prostate cancer. Mutated cells in the testes divide out of control, forming lumps that may be detected in a self-examination. ⓘ *cancer,* section 8.6

B. Spermatogenesis Yields Sperm Cells

Spermatogenesis, the production of sperm, begins when a male reaches puberty and continues throughout life.

Figure 35.5 illustrates the internal anatomy of a testis. Each testis contains about 200 tightly coiled, 50-centimeter-long **seminiferous tubules,** which produce the sperm cells. Large **Sertoli cells** span the width of the seminiferous tubule's wall; these "nurse cells" surround, support, and nourish developing sperm cells. **Interstitial cells,** which fill the spaces between the seminiferous tubules, secrete male sex hormones.

Sperm production begins with diploid germ cells called **spermatogonia** that reside within the wall of a seminiferous tubule (see figure 35.5b). When a spermatogonium divides mitotically, one daughter cell remains in the tubule wall and acts as a stem cell, continually giving rise to cells that become sperm. The other cell becomes a diploid **primary spermatocyte** that moves closer to the tubule's lumen (central cavity).

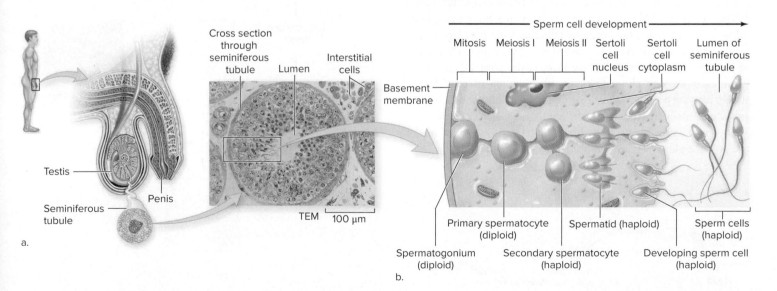

Figure 35.5 Sperm Production. (a) Anatomy of a testis. (b) In the walls of the seminiferous tubules, diploid spermatogonia divide mitotically. Some of the daughter cells undergo meiosis, giving rise to four haploid secondary spermatocytes that differentiate into spermatids. The spermatids, in turn, mature into sperm cells.

Photo: (a): Larry Johnson, Dept. of Veterinary Anatomy and Public Health

In the wall of the seminiferous tubule, the primary spermatocyte undergoes meiosis I, yielding two haploid **secondary spermatocytes.** These cells undergo meiosis II, forming four round, haploid cells called **spermatids** that each contain 23 chromosomes. (Figure 9.16 summarizes how meiosis produces sperm.)

As the spermatids move into the lumen of the seminiferous tubule, they complete their differentiation into **spermatozoa,** or mature sperm cells. They separate into individual cells and develop flagella. They also lose much of their cytoplasm, acquire a streamlined shape, and package their DNA into a distinct head (figure 35.6). Mitochondria just below the head region generate the ATP the sperm needs to move toward an egg cell. The caplike **acrosome** covers the head and releases enzymes that will help the sperm penetrate the egg cell. The entire process, from spermatogonium to sperm cell, takes 74 days in humans. ⓘ *ATP,* section 4.3

Figure It Out

A single ejaculate may contain 240 million sperm cells. How many primary spermatocytes must enter meiosis to produce this many sperm cells, and how many secondary spermatocytes are produced along the way?

Answer: 60 million primary spermatocytes; 120 million secondary spermatocytes

C. Hormones Influence Male Reproductive Function

Hormones play a critical role in male reproduction (figure 35.7). In the brain, the hypothalamus secretes **gonadotropin-releasing hormone (GnRH).** This peptide hormone travels in the bloodstream to the anterior pituitary, where it stimulates the release of two other peptide hormones: **follicle-stimulating hormone (FSH)** and **luteinizing hormone (LH).** Blood carries FSH and LH throughout the body. ⓘ *peptide hormones,* section 28.2A

LH signals interstitial cells in the testes to release the steroid hormone **testosterone** and other male sex hormones (androgens). In the presence of FSH, testosterone affects the body in multiple ways. In adolescents, the hormone stimulates the development of secondary sex characteristics. The testes and penis begin to enlarge at puberty, and hair grows on the face, in the armpits, and at the groin. Testosterone also stimulates the secretion of growth hormone, causing a growth spurt that increases height, boosts muscle mass, and deepens the voice. In adults, testosterone stimulates sperm production, sustains the libido, and controls the activity of the prostate gland. ⓘ *steroid hormones,* section 28.2B

Negative feedback loops maintain homeostasis in the concentrations of these hormones. Negative feedback also explains one consequence of abusing anabolic steroids: infertility or low sperm counts. High concentrations of synthetic steroids cause the testes to produce less testosterone; without testosterone, sperm do not form. ⓘ *negative feedback,* section 25.4

35.2   MASTERING CONCEPTS

1. What are the relationships among gonads, germ cells, gametes, and the zygote?
2. Describe the role of each part of the male reproductive system.
3. What are the stages of spermatogenesis?
4. How do hormones regulate sperm production?

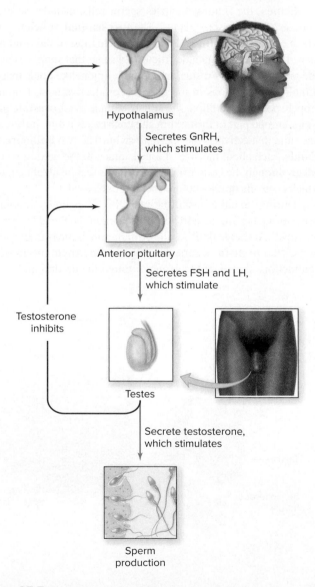

Figure 35.7 **Male Reproductive Hormones.** GnRH, FSH, LH, and testosterone interact in a negative feedback loop to regulate male reproductive function.

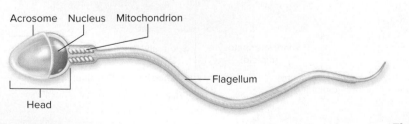

Acrosome   Nucleus   Mitochondrion

Flagellum

Head

Figure 35.6 **Human Sperm.** The DNA in a sperm cell is in the nucleus, which enters the egg cell. The long flagellum propels the sperm.

🍁 35.3 Females Produce Egg Cells

Egg cell production in females is somewhat more complicated than is sperm formation in males, in at least two ways. First, in females, meiosis begins before birth, pauses, and resumes at sexual maturity. Meiosis does not complete until after a sperm cell fertilizes the egg cell. Second, egg cell production is cyclical, under the control of several interacting hormones whose levels fluctuate monthly during a woman's reproductive years. Keep these differences in mind while reading this section.

A. Female Reproductive Organs Are Inside the Body

Female sex cells develop within the **ovaries,** which are paired gonads in the abdomen (**figure 35.8**). Ovaries produce both egg cells and sex hormones. They do not contain ducts comparable to the seminiferous tubules of the male's testes. Instead, within each ovary of a newborn female are about a million oocytes, the cells that give rise to mature egg cells. Nourishing **follicle cells** surround each oocyte.

Approximately once a month, beginning at puberty, one ovary releases the single most mature oocyte. Beating cilia sweep the mature oocyte into the fingerlike projections of one of the two **uterine tubes** (also called Fallopian tubes or oviducts). If sperm are present, fertilization occurs in a uterine tube. The tube carries the oocyte or zygote into a muscular, saclike organ, the **uterus.** During pregnancy, the fetus develops inside the uterus, also called the womb. The **endometrium,** or inner lining of the uterus, has a rich blood supply that is important in both menstruation and pregnancy.

The **cervix** is the necklike narrowing at the lower end of the uterus. The cervix opens into the **vagina,** the tube that leads

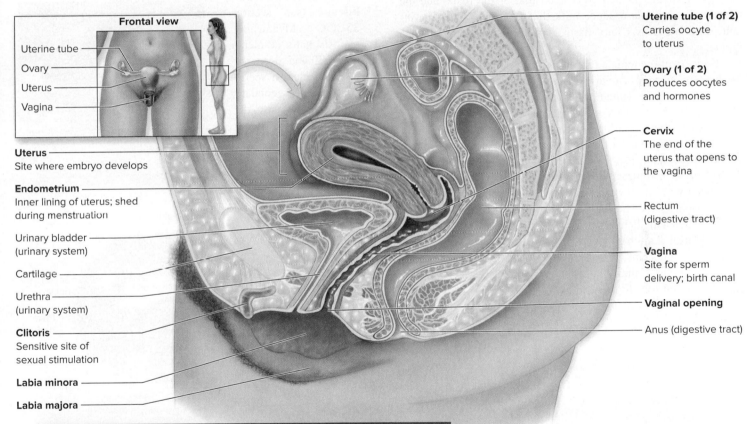

Frontal view

Uterine tube
Ovary
Uterus
Vagina

Uterus
Site where embryo develops

Endometrium
Inner lining of uterus; shed during menstruation

Urinary bladder
(urinary system)

Cartilage

Urethra
(urinary system)

Clitoris
Sensitive site of sexual stimulation

Labia minora

Labia majora

Uterine tube (1 of 2)
Carries oocyte to uterus

Ovary (1 of 2)
Produces oocytes and hormones

Cervix
The end of the uterus that opens to the vagina

Rectum
(digestive tract)

Vagina
Site for sperm delivery; birth canal

Vaginal opening

Anus (digestive tract)

Reproductive System (Female)		
Main tissue types*	**Examples of locations/functions**	
Epithelial	Lines uterus, uterine tubes, and vagina; produces oocytes in ovaries; forms external surface of umbilical cord.	
Connective	Makes up walls of ovaries, uterus, and vagina.	
Nervous	Clitoris contains sensory nerve fibers and nerve endings; hypothalamus secretes hormones that affect the anterior pituitary.	
Muscle	Smooth muscle surrounds uterine tubes, uterus, and vagina.	

*See chapter 25 for descriptions.

Figure 35.8 The Human Female Reproductive System.
One of the two ovaries releases an oocyte each month. This egg cell enters a nearby uterine tube. If a sperm cell fertilizes the oocyte, the offspring develops in the uterus and is delivered through the vagina.

outside the body. The vagina receives the penis during intercourse, and it is the birth canal. Like many other areas of the body, the vagina harbors a community of resident microorganisms. These bacteria lower the pH of the vagina, which helps prevent colonization by harmful bacteria and the yeast *Candida albicans.* Taking antibiotics can disrupt this microbial community and create an opportunity for *Candida* to overgrow, causing a vaginal yeast infection. (i) *normal microbiota,* section 17.4B

Two pairs of fleshy folds protect the vaginal opening on the outside: the labia majora (major lips) and the thinner, underlying flaps of tissue called labia minora (minor lips). The **clitoris** is a 2-centimeter-long structure at the upper junction of both pairs of labia. Rubbing the clitoris stimulates females to experience orgasm. Together, the labia, clitoris, and vaginal opening constitute the **vulva,** or external female genitalia.

The female secondary sex characteristics include the wider and shallower shape of the pelvis, the accumulation of fat around the hips, a higher-pitched voice than that of the male, and the breasts. The **breasts** produce milk that nourishes a nursing infant; each breast has fatty tissue, collagen, milk ducts, and a nipple.

Cancers of the female reproductive system often develop in the breasts, cervix, or ovaries. Breast cancer is the most common cancer type in women. The abnormally dividing cells may originate in the breast's milk-forming tissues or in the milk ducts. Some, but not all, forms of breast cancer have a strong heritable component. A family history is also the leading risk factor for ovarian cancer, which usually starts in the outer lining of the ovary. In contrast, nearly all cases of cervical cancer are associated with the sexually transmitted human papillomavirus (see section 35.4). The Pap test, in which a medical professional uses a microscope to look for abnormal cervical cells, is an important early-detection tool for cervical cancer. (i) *vaccines,* section 34.4

B. Oogenesis Yields Egg Cells

The making of an egg cell—**oogenesis**—begins with an **oogonium,** a diploid germ cell containing 46 chromosomes (**figure 35.9**). Each oogonium grows, accumulates cytoplasm, replicates its DNA, and divides mitotically, becoming two **primary oocytes.** The subsequent divisions of meiosis partition the cytoplasm unequally, so that oogenesis (unlike spermatogenesis) produces cells of different sizes. By the end of meiosis I, the primary oocyte has divided into a small, haploid **polar body** and a larger, haploid **secondary oocyte.**

Ovulation is the release of a secondary oocyte from its follicle. As the follicle ruptures, the egg cell emerges from the ovary's surface; fingerlike projections of the uterine tube move across the ovary and usher the egg into the tube. Following ovulation, the now-ruptured follicle transforms into a gland called a **corpus luteum** (see section 35.3C). Meanwhile, meiosis halts at metaphase II and does not resume unless a sperm contacts the secondary oocyte. In that case, the secondary oocyte again

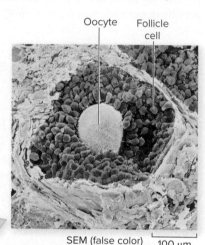

Oocyte   Follicle cell

SEM (false color)   100 μm

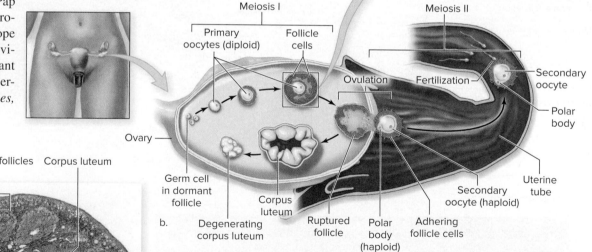

Meiosis I

Primary oocytes (diploid)   Follicle cells   Meiosis II

Ovulation   Fertilization   Secondary oocyte

Ovary   Polar body

Germ cell in dormant follicle   Corpus luteum   Secondary oocyte (haploid)   Uterine tube

b.   Degenerating corpus luteum   Ruptured follicle   Polar body (haploid)   Adhering follicle cells

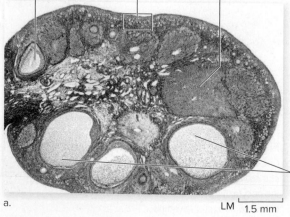

Maturing follicle   Dormant follicles   Corpus luteum

Mature follicles

a.   LM   1.5 mm

Figure 35.9 Egg Cell Production. (a) A cross section of a cat's ovary shows follicles at several stages of development. (b) Follicles contain diploid oogonia, which divide mitotically to produce primary oocytes. These cells undergo meiosis. Every month between puberty and menopause, the most mature follicle ruptures and a haploid secondary oocyte bursts out of the ovary, an event called ovulation. The secondary oocyte completes meiosis II only if fertilized by a sperm cell.

Photos: (a): ©Victor P. Eroschenko RF; (b): ©Prof. P.M. Motta, G. Macchiarelli, S.A. Nottola/Science Source

divides unequally to produce a small, additional polar body and the mature egg cell (or ovum), which contains 23 chromosomes and a large amount of cytoplasm. The polar body produced in meiosis I may divide into two additional polar bodies, or it may decompose.

The egg cell, in receiving most of the cytoplasm, contains all of the biochemicals and organelles that the zygote will use until its own DNA begins to function. The polar bodies normally play no further role in development. Rarely, however, sperm can fertilize polar bodies, and a mass of tissue that does not resemble an embryo grows until the woman's body rejects it. A fertilized polar body accounts for about 1 in 100 miscarriages.

From puberty to menopause (when menstruation stops entirely), monthly hormonal cues prompt an ovary to release one secondary oocyte into a uterine tube. If a sperm penetrates the oocyte membrane, meiosis in the oocyte completes, and the two nuclei combine to form the diploid zygote. If the secondary oocyte is not fertilized, it leaves the body with the endometrium in the menstrual flow.

In some ways, oogenesis is similar to spermatogenesis (table 35.1). Each process starts with a diploid germ cell (spermatogonium or oogonium), which eventually gives rise to the haploid gametes. Also, both testes and ovaries contain gametes in various stages of development. Of course, the two processes also differ. For example, spermatogenesis gives rise to four equal-sized sperm cells, whereas one oogonium yields one functional egg cell and three smaller polar bodies.

Also, the timetable for oogenesis differs greatly from that of spermatogenesis. A male takes about 74 days to produce a sperm cell. In contrast, oogenesis stretches from before birth until after puberty. The ovaries of a 3-month-old female fetus contain 2 million or more primary oocytes. From then on, the oocytes slowly degenerate. At birth, a million primary oocytes are present, their development arrested in prophase I. Only about 400,000 remain by the time of puberty, after which one or a few oocytes complete meiosis I each month. These secondary oocytes stop meiosis again, this time at metaphase II. Meiosis is completed only if fertilization occurs.

C. Hormones Influence Female Reproductive Function

Hormones control the timing of events in the female reproductive system (figure 35.10). The hypothalamus, anterior pituitary, and ovaries are the primary sources of these hormones.

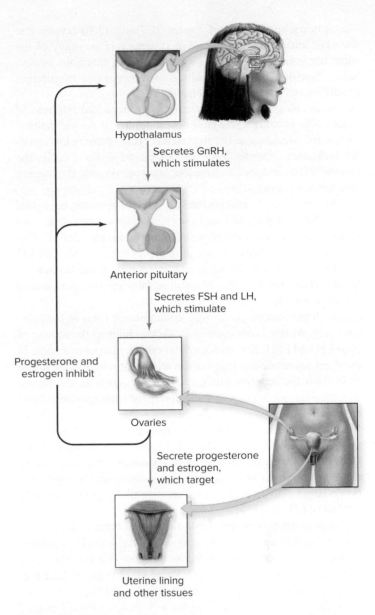

Figure 35.10 Female Reproductive Hormones. Hormones from the hypothalamus regulate the activities of the ovaries and the uterine lining. Progesterone and estrogen, in turn, regulate hormone release from the hypothalamus and anterior pituitary in a negative feedback loop.

TABLE **35.1** **Spermatogenesis and Oogenesis Compared**

Diploid Starting Cell	Product of Mitosis (Diploid)	Products of Meiosis I (Haploid)	Products of Meiosis II (Haploid)
Spermatogonium in seminiferous tubule	Primary spermatocyte	Two secondary spermatocytes	Four equal-sized spermatids
Oogonium in ovary	Primary oocyte	One large secondary oocyte + one small polar body	One large egg cell + three small polar bodies

A quick comparison of figures 35.7 and 35.10 reveals that the male and female reproductive systems rely on many of the same hormones. Table 35.2 summarizes their functions in both sexes. Females, however, produce these hormones in different quantities and on a different schedule.

In females, hormonal fluctuations produce two interrelated cycles. The **ovarian cycle** controls the timing of oocyte maturation in the ovaries, and the **menstrual cycle** prepares the uterus for pregnancy. Figure 35.11 tracks changes in the follicle, the uterine lining, and the levels of four hormones during the ovarian and menstrual cycles.

Menstruation begins on the first day of the menstrual cycle. Low blood levels of two sex hormones, **estrogen** and **progesterone,** signal the hypothalamus to secrete GnRH. This hormone prompts the anterior pituitary to release FSH and LH into the bloodstream. In the ovaries, receptors on the surfaces of follicle cells bind to FSH, stimulating follicles to mature and to release estrogen.

Estrogen has two seemingly contradictory roles in the ovarian cycle. At low concentrations, estrogen inhibits the release of both LH and FSH. But at around the cycle's midpoint, a spike in estrogen accumulation triggers the release of additional LH and FSH from the anterior pituitary. This LH surge in the bloodstream triggers ovulation and transforms the ruptured follicle into a corpus luteum.

The corpus luteum, in turn, secretes progesterone and estrogen, which have multiple effects. These two hormones act together to promote the thickening of the endometrium, preparing the uterus for possible pregnancy. Progesterone and estrogen also act on target cells in the hypothalamus, inhibiting the production of GnRH, LH, and FSH.

If pregnancy does not occur, the corpus luteum degenerates into an inactive scar. Over the next several days, levels of progesterone and estrogen gradually decline (section 35.5 describes what happens if pregnancy does occur). The reduced levels of these hormones no longer maintain the endometrium, which then exits the body through the cervix and vagina as menstrual flow. Lowered progesterone and estrogen levels also release their inhibition of GnRH, LH, and FSH in the brain, and the cycle begins anew.

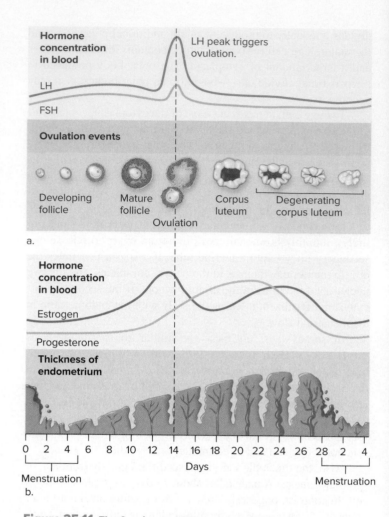

Figure 35.11 The Ovarian and Menstrual Cycles. (a) In the ovarian cycle, LH and FSH coordinate follicle maturation and release from the ovary. (b) Estrogen and progesterone signal the endometrium to thicken, then disintegrate, during each menstrual cycle.

TABLE **35.2**   The Roles of GnRH, LH, and FSH in Human Males and Females

Hormone	Source	Target Cells	Function(s) in Males	Function(s) in Females
GnRH (gonadotropin-releasing hormone)	Hypothalamus	Anterior pituitary	Stimulates release of LH and FSH (continuously)	Stimulates release of LH and FSH (cyclically)
LH (luteinizing hormone)	Anterior pituitary	Testes, ovaries	Stimulates production of testosterone	Stimulates production of estrogen; triggers ovulation midway through menstrual cycle; stimulates development of corpus luteum, which secretes progesterone
FSH (follicle-stimulating hormone)	Anterior pituitary	Testes, ovaries	Enhances production of testosterone-binding protein in Sertoli cells; required for spermatogenesis	Stimulates follicle maturation

D. Hormonal Fluctuations Can Cause Discomfort

Fluctuating concentrations of hormones trigger a variety of conditions that are unique to women. One example is premenstrual syndrome (PMS), a collection of symptoms that appears in the second half of the menstrual cycle. In the days before her menstrual period begins, a woman may experience headache, breast tenderness, cramping, depression, irritability, or dozens of other signs of PMS. The cause of each symptom is unknown, but the correlation with hormonal changes is clear. Over-the-counter drugs provide relief from many symptoms.

Menstrual cramps, which may occur with or without PMS, arise from smooth muscle contractions in the uterus. The function of these painful contractions is to help menstrual fluid exit the uterus. But cramps may also be related to endometriosis or another underlying problem. In endometriosis, cells from the endometrium travel outside the uterus, triggering inflammation and pain. Untreated endometriosis can cause infertility.

As a woman nears the end of her reproductive years, her hormonal fluctuations cease. Ironically, the estrogen withdrawal that accompanies menopause can cause symptoms as well. For example, hot flashes affect most women going through menopause. A hot flash, which typically lasts a few minutes, is a temporary sensation of heat in the face and upper body, coupled with flushed skin and sweating. Researchers speculate that the withdrawal of estrogen somehow affects the temperature set point of the hypothalamus.

35.3 | MASTERING CONCEPTS

1. Describe the role of each part of the female reproductive system.
2. What are the stages of oogenesis?
3. How do hormones regulate the ovarian and menstrual cycles?

35.4 Reproductive Health Considers Contraception and Disease

Birth control, or **contraception,** is the use of devices or practices that work "against conception"; that is, they prevent the union of sperm and egg. Table 35.3 summarizes several methods and estimates the pregnancy rate based on typical (not perfect) use. For comparison, 85 out of 100 fertile women will become pregnant within 1 year if they are sexually active and use no birth control at all. (This section's Burning Question explains how to know when conception is likely.)

Burning Question

When can conception occur?

The short answer to this question is that conception can occur as soon as an ovary releases an egg. Predicting ovulation and avoiding sex near that time can therefore be an effective method of birth control.

Monitoring body clues can help a woman determine when an egg is likely to be released. For example, body temperature (measured upon waking) is elevated shortly after ovulation (figure 35.A). Other clues, such as the quantity and quality of cervical fluid, as well as the position of the cervix, may also help pinpoint ovulation.

Although it may seem logical to predict ovulation by simply counting days after the start of menstruation, this method is less effective. Cycle lengths may vary, so ovulation is not necessarily on day 14.

Having sex several days before or after ovulation may lead to conception. Each egg is viable for about 24 hours, but sperm can survive inside a female for up to 5 days. Also, the ovaries may release one or more additional eggs over a couple of days after initial ovulation.

Submit your burning question to
Marielle.Hoefnagels@mheducation.com

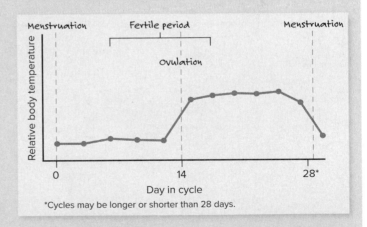

Figure 35.A Thermal Shift. Careful, daily record-keeping is essential to monitoring fertility. Body temperature—measured upon waking each morning—increases after an egg is released. However, the fertile period begins a few days before ovulation. Waiting to have sex until a few days after the thermal shift reduces the chance of conception.

TABLE **35.3** Birth Control Methods

Method	Area(s) Targeted	Mechanism	Advantages	Disadvantages	Pregnancies per 100 Women per Year**
Abstinence	N/A	No intercourse; sperm never encounter egg cell	No cost	Difficult to do	0
Vasectomy*	1	Cuts each vas deferens, so sperm cells never reach urethra	Permanent; does not interrupt spontaneity	Requires minor surgery; difficult to reverse	<1
Withdrawal*	2	Removal of penis from vagina before ejaculation	No cost	Not recommended for sexually inexperienced men; requires great self-control; sperm may leak from penis before withdrawal; sperm spilled on vulva may cause pregnancy	18
Latex or polyurethane condom	2	Worn over penis or inserted into vagina; keeps sperm out of vagina	Protects against sexually transmitted diseases	Disrupts spontaneity; reduces sensation	15
Cervical cap used with spermicide*	3	Kills sperm and blocks cervix	Inexpensive; can be kept in for 24 hours	May slip out of place; must be fitted; less effective for women who have already given birth vaginally	20
Intrauterine device (IUD)*	4	Prevents implantation of preembryo	Does not interrupt spontaneity	Severe menstrual cramps; risk of infection	<1
Tubal ligation*	5	Cuts uterine tubes, so oocytes never reach uterus	Permanent; does not interrupt spontaneity	Requires surgery; risk of infection; difficult to reverse	<1
Fertility awareness method*	6	No intercourse during fertile times, as inferred from body temperature and other clues (see this section's Burning Question)	No cost	Requires careful record-keeping	20
Hormonal contraception (estrogen and/or progesterone)*	4, 6	Prevents ovulation and implantation	Does not interrupt spontaneity; easy to use	Menstrual changes; weight gain; headaches	2–9

*Does not protect against sexually transmitted diseases (STDs).
**For some methods, the number of pregnancies may be much lower if the method is used perfectly every time.

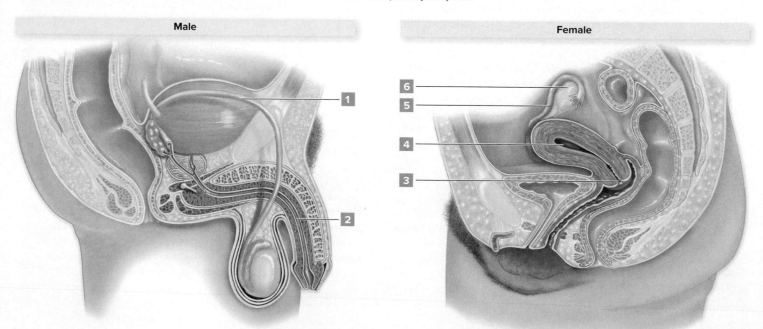

Male

Female

Besides abstinence, the most effective contraceptives are surgical; next are those that adjust hormone concentrations in the woman's body. Birth control pills, patches, vaginal rings, injections, and implants all contain a synthetic form of progesterone. If used correctly, each of these methods prevents ovulation and therefore precludes fertilization. Other methods kill sperm, block the meeting of sperm and oocyte, or prevent a developing embryo from implanting in the lining of the uterus.

Only latex and polyurethane condoms, however, simultaneously prevent pregnancy and protect against **sexually transmitted diseases (STDs),** which spread to new hosts during sexual contact. Vaginal intercourse, oral sex, and anal sex all can provide direct, person-to-person transmission for a wide variety of disease-causing agents. Interestingly, humans are not the only ones to suffer from STDs. Other animal species have STDs of their own; so do plants, which can pass viroids in pollen. ⓘ *viroids,* section 16.6A

Viruses, bacteria, protists, and fungi all can cause STDs; table 35.4 lists a sampling of some of the most common ones, along with their treatments. In the United States, the most common STD is genital warts, which can be caused by more than 40 strains of the human papillomavirus (HPV). Many infections remain symptomless, but others trigger the growth of visible bumps on the male or female genitals. Some strains of HPV that do not cause warts can create a much bigger problem: cancer. Nearly all cervical cancer tumors test positive for DNA from HPV, and the same virus has also been linked to cancers of the mouth and throat in males. If administered before exposure to the virus, the HPV vaccine (commonly called the "cervical cancer vaccine") can help prevent these cancers in both sexes.

Another STD that infects millions of people each year is trichomoniasis, caused by a protist called *Trichomonas vaginalis* (see figure 18.15b). Males can transmit this organism, usually without visible symptoms. Infected women may notice unusual vaginal discharge along with pain, irritation, and itching.

Chlamydia, a third common STD, is caused by the bacterium *Chlamydia trachomatis.* Typical symptoms include discharge from the penis or vagina; an infected person may also experience pain or a burning sensation when urinating. Chlamydia, however, is sometimes called a "silent disease" because some 75% of infected women and 50% of infected men remain symptomless.

Sexually transmitted diseases are especially relevant to young adults, who account for 50% of new infections but only 25% of the sexually active population. Females age 15 to 19, for example, lead the nation in infection rates for both chlamydia and gonorrhea.

Infections with HPV, trichomoniasis, chlamydia, and other agents listed in table 35.4 often remain invisible, but that does not mean they are harmless. For example, in addition to the risk of cancer associated with HPV, untreated chlamydia infections in women can spread to the uterine tubes. Pelvic inflammatory disease, one possible complication, can cause pain and infertility. Moreover, syphilis, gonorrhea, and chlamydia can pass to infants during childbirth.

TABLE 35.4 Examples of Sexually Transmitted Diseases

Disease	Agent	Treatment
Viruses		
HIV/AIDS	Human immunodeficiency virus (HIV)	Combination of drugs that reduce viral replication
Genital warts	Human papillomavirus (HPV)	Removal of warts
Cervical cancer	Human papillomavirus (HPV)	Surgery, radiation, chemotherapy
Genital herpes	Herpes simplex virus	Medications that reduce outbreak frequency and duration
Hepatitis B	Hepatitis B virus	None
Bacteria		
Chlamydia	*Chlamydia trachomatis*	Antibiotics
Gonorrhea	*Neisseria gonorrhoeae*	Antibiotics
Syphilis	*Treponema pallidum*	Antibiotics
Protists		
Trichomoniasis	*Trichomonas vaginalis*	Antiprotozoan drugs
Fungi		
Yeast infection	*Candida albicans*	Antifungal drugs

Most STDs are also associated with an elevated risk for acquiring and transmitting HIV. One explanation is that many STDs cause sores through which viral particles can enter or leave the body. A second reason is that any infection can cause inflammation and other defensive reactions. As described in chapter 34, these responses attract the types of white blood cells that HIV infects. Efforts to prevent STDs in general can therefore also help reduce infections by HIV in particular.

The best way to prevent sexually transmitted diseases is to abstain from sex entirely. The second best way is to develop a long-term, monogamous relationship with a partner who has recently been tested for STDs and is therefore known to be disease-free. The third best option is to properly use a latex or polyurethane condom throughout a sexual encounter. In addition, vaccines can protect against some sexually transmitted viruses, including HPV and hepatitis B.

35.4 MASTERING CONCEPTS

1. Describe how three birth control methods work.
2. List and describe three common STDs.
3. Describe two reasons that a symptomless infection with an STD can be harmful.

35.5 The Human Infant Begins Life as a Zygote

So far, this chapter has described gamete production in the male and female reproductive systems. This section now turns to the development of a baby, which occurs inside the female's body (see this section's Burning Question for an interesting exception). As you will see, fertilization produces the zygote; this single cell divides many times as it develops into a preembryo, an embryo, a fetus, and a newborn baby (table 35.5).

A. Fertilization Initiates Pregnancy

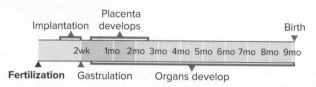

After intercourse, sperm cells swim toward the oocyte. Of the 100 million sperm that begin the journey in the vagina, only about 200 approach the egg cell in a uterine tube.

Those that make it must penetrate two layers to contact the ovum. An outer layer of follicle cells surrounds a thin, clear "jelly layer" of proteins and carbohydrates that encases the oocyte (figure 35.12). On contact with the follicle cells

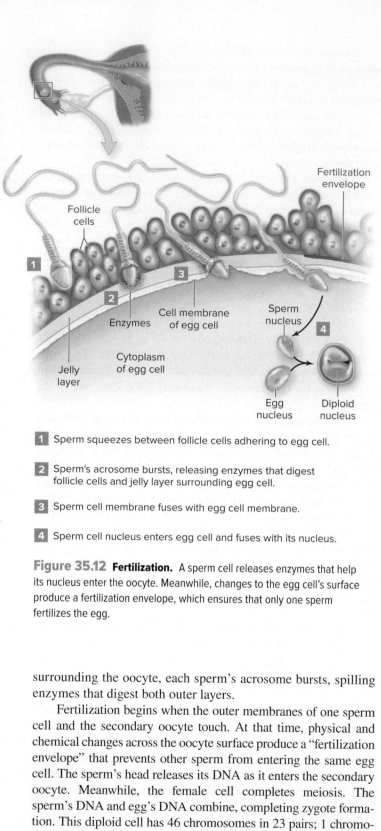

1 Sperm squeezes between follicle cells adhering to egg cell.

2 Sperm's acrosome bursts, releasing enzymes that digest follicle cells and jelly layer surrounding egg cell.

3 Sperm cell membrane fuses with egg cell membrane.

4 Sperm cell nucleus enters egg cell and fuses with its nucleus.

Figure 35.12 Fertilization. A sperm cell releases enzymes that help its nucleus enter the oocyte. Meanwhile, changes to the egg cell's surface produce a fertilization envelope, which ensures that only one sperm fertilizes the egg.

TABLE 35.5	Stages in Development from Fertilization to Birth: A Summary	
Name		**Description**
Zygote		Fertilized egg cell
Preembryonic stage		
Morula		Solid ball of 16 or more cells
Blastocyst		Hollow sphere consisting of outer layer (trophoblast) and inner cell mass; implants into endometrium beginning about 1 week after fertilization
Gastrula		Structure consisting of three germ layers (endoderm, mesoderm, ectoderm)
Embryonic stage		Three germ layers differentiate into organ systems
Fetal stage		Organ systems continue to develop and become functional

surrounding the oocyte, each sperm's acrosome bursts, spilling enzymes that digest both outer layers.

Fertilization begins when the outer membranes of one sperm cell and the secondary oocyte touch. At that time, physical and chemical changes across the oocyte surface produce a "fertilization envelope" that prevents other sperm from entering the same egg cell. The sperm's head releases its DNA as it enters the secondary oocyte. Meanwhile, the female cell completes meiosis. The sperm's DNA and egg's DNA combine, completing zygote formation. This diploid cell has 46 chromosomes in 23 pairs; 1 chromosome of each pair comes from each parent. (i) *homologous chromosomes,* section 9.2

Occasionally, a woman ovulates two or more egg cells at once, and a different sperm fertilizes each one. If all the zygotes complete development, the result is twins, triplets, quadruplets, or even higher-order multiple births. Because each zygote is

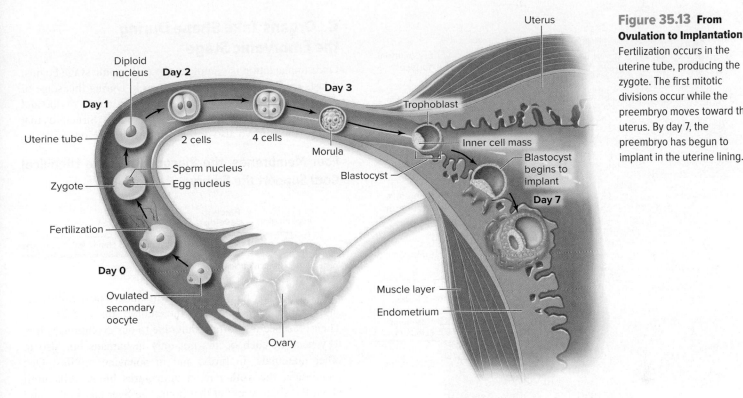

Figure 35.13 From Ovulation to Implantation. Fertilization occurs in the uterine tube, producing the zygote. The first mitotic divisions occur while the preembryo moves toward the uterus. By day 7, the preembryo has begun to implant in the uterine lining.

derived from a separate sperm and egg cell, the siblings will not be genetically identical, and they may be of different sexes. At the opposite end of the spectrum are couples that have trouble conceiving any child at all; the Apply It Now box, "Assisted Reproductive Technologies," explains how technology can help. ⓘ *multiple births,* section 9.5C

B. The Preembryonic Stage Ends When Implantation Is Complete

The first 2 weeks of prenatal development have a variety of names, but we will call this period the **preembryonic stage.**

About 36 hours after fertilization, the zygote divides for the first time, beginning a period of rapid mitotic cell division called **cleavage** (**figure 35.13**). (Recent research suggests that many preembryos have mutations and chromosomal abnormalities that halt development during cleavage.) Cleavage results in a **morula,** a solid ball of 16 or more cells. *Morula* is Latin for mulberry, a small fruit that the preembryo resembles. Three days after fertilization, the morula is still within the uterine tube but moving toward the uterus.

Surprisingly, the morula is about the same size as the zygote. The reason is that the initial cleavage divisions split the zygote—an unusually large cell—into smaller and smaller cells. Once the

preembryo consists of cells that are "normal"-sized, however, cell size stabilizes.

The mass of cells reaches the uterus 3 to 6 days after fertilization. The preembryo then hollows out, its center filling with fluid that seeps in from the uterus. This fluid-filled ball of cells is called a **blastocyst.** The blastocyst's outer layer, the trophoblast, will eventually form the fetal portion of the placenta. The cells inside the blastocyst form the **inner cell mass,** the cells that will develop into the embryo itself. (The inner cell mass is also the source of embryonic stem cells.) ⓘ *stem cells,* section 11.3A

In a process called **implantation,** the blastocyst becomes embedded in the lining of the uterus (see figure 35.13). Implantation begins within a week of fertilization. The trophoblast not only secretes digestive enzymes that eat through the outer layer of the uterine lining but also sends projections into the uterine lining. Until the placenta forms, the preembryo obtains nutrients from these digested endometrial cells.

The trophoblast cells now secrete **human chorionic gonadotropin (hCG),** the hormone that is the basis of pregnancy tests. For a while, hCG maintains progesterone production at the corpus luteum; progesterone prevents menstruation and further ovulation. In this way, the blastocyst helps to ensure its own survival; if the uterine lining were shed, the blastocyst would leave the woman's body, too. Levels of hCG peak about 10 weeks after fertilization.

During the second week of development, the blastocyst completes implantation. A space called the amniotic cavity forms within a sac called the amnion, which lies between the inner cell

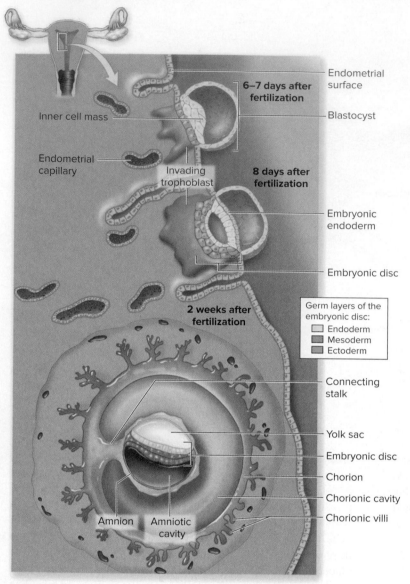

Figure 35.14 Implantation and Gastrulation. The preembryo implants into the endometrium and develops three germ layers by 2 weeks after fertilization.

mass and the trophoblast (figure 35.14). The inner cell mass flattens and forms the **embryonic disc,** which will develop into the embryo.

As the preembryo continues to develop, one layer of the embryonic disc becomes **ectoderm,** and another layer becomes **endoderm.** Soon, a middle **mesoderm** layer forms from the ectoderm. The **gastrula** is the resulting three-layered structure. The formation of the gastrula is called gastrulation, and it begins at about 2 weeks postfertilization. By now, the woman's urine contains enough hCG for an at-home pregnancy test to detect. ⓘ *germ layers,* section 21.1C

The preembryo may split during the first 2 weeks of development, forming identical twins. Depending on when the split occurs, the twins may or may not share the same amnion and placenta; the later the split, the more structures the twins will share. If a preembryo splits after day 12 of pregnancy, the twins are unlikely to separate completely, and they may be conjoined.

C. Organs Take Shape During the Embryonic Stage

Once implantation is complete, the **embryonic stage** begins; it lasts until the end of the eighth week. During this stage of development, cells of the three layers continue to divide and differentiate, forming all of the body's organs. Structures that support the embryo also develop during this period.

Four Membranes, the Placenta, and the Umbilical Cord Support the Embryo

Four thin layers of tissue, called extraembryonic membranes, support, protect, and nourish the embryo (figure 35.15a,b). Their presence is an important clue to our evolutionary history because each occurs not only in humans but also in other mammals, in birds, and in nonavian reptiles. One membrane, the **yolk sac,** manufactures blood cells until about the sixth week; at that point, the liver takes over, and the yolk sac starts to shrink. In addition, parts of the yolk sac develop into the intestines and germ cells. By the third week, an outpouching of the yolk sac forms the **allantois,** another extraembryonic membrane. It, too, manufactures blood cells, and it gives rise to blood vessels in the umbilical cord. The **amnion** is the transparent sac that contains the amniotic fluid. This fluid cushions the embryo, maintains a constant temperature and pressure, and protects the embryo if the woman falls. Cells collected from amniotic fluid are used in many prenatal medical tests. ⓘ *amnion,* section 21.10B

Some cells of the trophoblast develop into the **chorion,** the outermost extraembryonic membrane. **Chorionic villi** are fingerlike projections from the chorion that extend into the uterine lining. These structures establish the beginnings of the **placenta,** which will connect the developing embryo with its mother's uterus (figure 35.15c). The placenta begins to form in the embryonic stage, but it is not completely functional until about 10 weeks after fertilization (early in the fetal stage). Arteries and veins in the **umbilical cord** connect the fetus to the placenta. Umbilical cord blood is a rich source of stem cells used to treat an ever-expanding list of disorders.

The fully developed placenta consists of pools of the mother's blood surrounding the chorionic villi, which contain fetal blood vessels. Many substances are exchanged at the placenta, although the cells of the chorionic villi block most large proteins and red blood cells. Carbon dioxide, urea, and other wastes from the fetus pass to the mother's blood. Glucose, amino acids, oxygen, ions, and some antibodies from the mother move in the opposite direction, nourishing and protecting the offspring. At the same time, harmful substances can also cross from mother to fetus; examples include some disease-causing agents and drugs such as cocaine, alcohol, and nicotine.

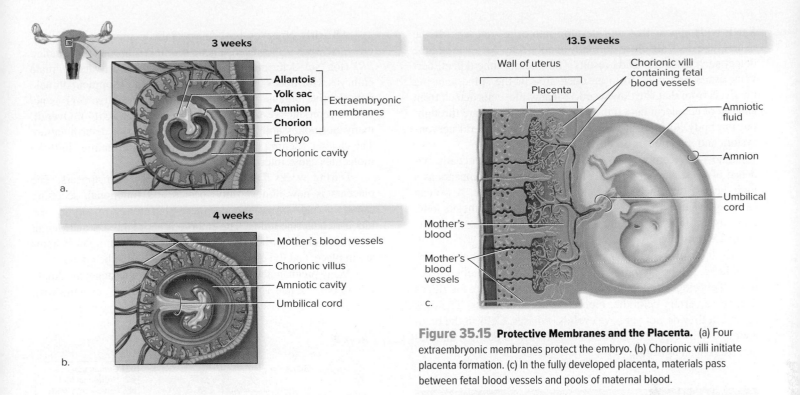

3 weeks

Allantois ┐
Yolk sac │ Extraembryonic
Amnion ├ membranes
Chorion ┘
Embryo
Chorionic cavity

a.

4 weeks

Mother's blood vessels
Chorionic villus
Amniotic cavity
Umbilical cord

b.

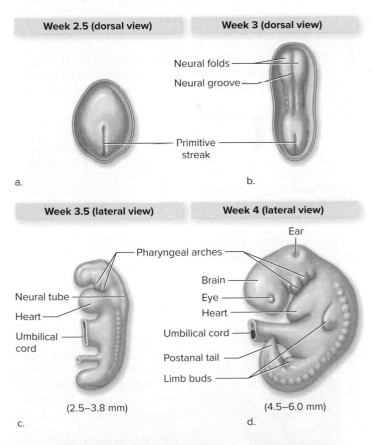

13.5 weeks

Wall of uterus
Chorionic villi containing fetal blood vessels
Placenta
Amniotic fluid
Amnion
Mother's blood
Mother's blood vessels
Umbilical cord

c.

Figure 35.15 Protective Membranes and the Placenta. (a) Four extraembryonic membranes protect the embryo. (b) Chorionic villi initiate placenta formation. (c) In the fully developed placenta, materials pass between fetal blood vessels and pools of maternal blood.

Organ Formation Begins in the Third Week of Development

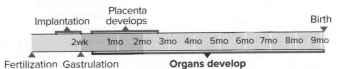

Implantation   Placenta develops   Birth

2wk 1mo 2mo 3mo 4mo 5mo 6mo 7mo 8mo 9mo

Fertilization Gastrulation   **Organs develop**

As the extraembryonic membranes, placenta, and umbilical cord develop, so does the embryo itself. Beginning during the third week of prenatal development, distinct organs form. Over several months, ectoderm cells develop into the nervous system, sense organs, outer skin layers, hair, nails, and skin glands. Splits in the mesoderm form the coelom, which becomes the chest and abdominal cavities. Mesoderm cells develop into bone, muscle, blood, the inner skin layer, and the reproductive organs. Endoderm cells form the organs of the digestive and respiratory systems. ⓘ *coelom*, section 21.1D

Prior to organ development, a furrow called the **primitive streak** appears along the back of the embryonic disc, forming a longitudinal axis around which other structures organize as they develop (**figure 35.16**). For example, the primitive streak gives rise to the **notochord**, a flexible, rodlike structure that forms the basic framework of the vertebral column. The notochord induces ectoderm to differentiate into the central nervous system. The neural groove in figure 35.16 is the first sign of this process. The neural groove folds into a hollow **neural tube,** which eventually develops into the brain and spinal cord. A bulge containing the heart also appears in the third week and begins to beat at around day 22. ⓘ *notochord*, section 21.10A; *central nervous system*, section 26.6

The fourth week of the embryonic period is a time of rapid growth and differentiation. Blood cells begin to form and to fill developing blood vessels. Immature lungs and kidneys appear.

Week 2.5 (dorsal view)

Primitive streak

a.

Week 3 (dorsal view)

Neural folds
Neural groove
Primitive streak

b.

Week 3.5 (lateral view)

Pharyngeal arches
Neural tube
Heart
Umbilical cord

(2.5–3.8 mm)

c.

Week 4 (lateral view)

Ear
Brain
Eye
Heart
Umbilical cord
Postanal tail
Limb buds

(4.5–6.0 mm)

d.

Figure 35.16 Early Organ Formation. (a) The primitive streak is the embryo's longitudinal axis. (b) Ectoderm folds around the neural groove to form the neural tube. (c) The heart becomes prominent during week 3. (d) The head, tail, and four limb buds are clearly visible in week 4.

Small buds appear that will develop into arms and legs. If the neural tube does not close normally at about day 28, a neural tube defect such as spina bifida results. (In a neural tube defect, nervous tissue protrudes from an open area of the spine, causing paralysis from that site downward.) Meanwhile, cells detach from an area of ectoderm called the neural crest and migrate throughout the body, forming the foundation of the peripheral nervous system and many other organs.

The 4-week embryo has a distinct head and jaw and early evidence of eyes, ears, and nose. The digestive system appears as a long, hollow tube that will develop into the intestines. A woman carrying this embryo, which is now only about 6 millimeters long, may suspect that she is pregnant because her menstrual period is about 2 weeks late.

By the fifth week, the embryo's head appears disproportionately large. Limbs extending from the body end in platelike structures. Tiny ridges run down the plates, and by week 6, the ridges deepen as certain cells die, molding fingers and toes. The eyes open, but they do not yet have eyelids or irises. Cells in the brain are rapidly differentiating. The embryo is now about 1.3 centimeters from head to rump.

Burning Question

Can males become pregnant?

Human males do not have the reproductive organs necessary to become pregnant. In the future, a man who wishes to become pregnant may elect to receive a uterus transplant—a procedure that has already been successful in women lacking a uterus. A male recipient would also receive a vagina, have pelvic reconstruction, and take many artificial hormones.

Unlike in humans, the males of sea horses and some other bony fishes normally bear the offspring. In these unusual animals, the leisurely courtship ritual lasts several days, during which the male and female swim and "dance" together and hold one another's tail. The female produces eggs, which she deposits into a brood pouch on the male's abdomen. He then releases sperm, fertilizing the eggs.

Male seahorse

Baby emerging from brood pouch

©Dr. Paul Zahl/Science Source

The young sea horses develop in the wall of the male's brood pouch for 2 to 4 weeks. In the meantime, the female does not pursue other mates, and she visits the male each day. At the end of the pregnancy, he gives birth to dozens or hundreds of miniature sea horses, which receive no further parental care. He can then accept more eggs, beginning the process anew.

Submit your burning question to Marielle.Hoefnagels@ mheducation.com

All early embryos have unspecialized reproductive structures. At week 7, however, a gene on the Y chromosome, called *SRY* (for "sex-determining region of the Y"), is activated in male embryos. Hormones then begin to stimulate development of male reproductive organs and glands. Without an active *SRY* gene, female reproductive structures develop (**figure 35.17**). Overall, many hormones and proteins participate in sex determination. The chapter opening essay describes how abnormalities in these molecules cause intersex conditions.

During weeks 7 and 8, a cartilage skeleton appears. The placenta is now almost fully formed and functional, secreting estrogen and progesterone; these hormones maintain the blood-rich uterine lining. The embryo is about the size and weight of a paper clip. By the end of the eighth week, all organ systems are in place, and the embryo is now considered to be a fetus.

Many pregnancies end during the embryonic stage; the Apply It Now box, "When a Pregnancy Ends Before Birth," explains why.

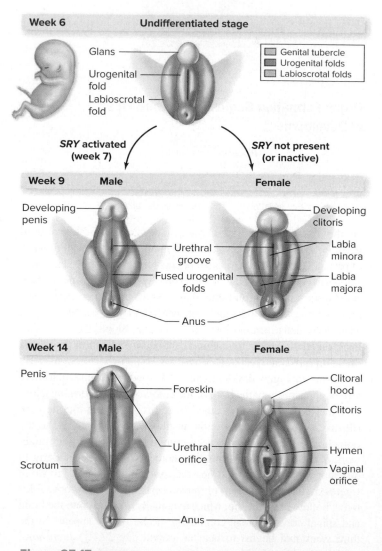

Figure 35.17 Development of Male and Female Genitals. Until about week 7 of development, embryos have undifferentiated genitals. If the *SRY* gene on the Y chromosome is activated, development continues as a male. Otherwise, female genitals develop.

Apply It **Now**

When a Pregnancy Ends Before Birth

Many conditions can cause miscarriage (also called a spontaneous abortion), in which an undeveloped fetus separates from the uterus. The woman expels the fetus and placenta (if present) from her body, ending the pregnancy. A miscarriage may occur any time before 7 months of development, when a fetus can survive outside the womb. But most occur early in a pregnancy, often reflecting a chromosomal abnormality in the embryo or fetus. If a miscarriage occurs later, during the second trimester, the problem usually originates with the woman, not the fetus. Diabetes, high blood pressure, hormonal abnormalities, infectious diseases, uterine problems, or drug abuse may be to blame, but often the cause of a miscarriage is unclear. ⓘ *nondisjunction,* section 9.7B

Just as a miscarriage is the loss of a pregnancy before the fetus is viable, a stillbirth occurs if a baby dies in the uterus after about the 24th week (near the end of the second trimester). The same conditions that contribute to miscarriage are also associated with stillbirths.

An induced abortion is the deliberate termination of a pregnancy. A combination of drugs can cause an abortion within the first 9 weeks of pregnancy. Alternatively, a surgical abortion may be an option later in the pregnancy (typically through week 24). In a surgical abortion, a medical professional may use a manual syringe or an electric pump to remove the embryo or fetus from the uterus. Abortions in the third trimester are rare, in part because they pose higher risks of serious complications such as heavy bleeding or infection.

D. Organ Systems Become Functional in the Fetal Stage

The third stage of prenatal development is the **fetal stage,** which lasts from the beginning of the ninth week through the full 38 weeks of development. The fetal stage therefore begins about two thirds of the way through the first trimester (3-month period) of pregnancy and ends 9 months after fertilization. During this time, the fetus grows considerably (**figure 35.18**). Organs begin to function and interact, forming organ systems.

As the first trimester progresses, the body proportions of the fetus begin to appear more like those of a newborn. Bone begins to form and will eventually replace most of the cartilage, which is softer. Soon, as the nerves and muscles begin to coordinate their actions, the fetus will move its arms and legs. Physical differences between the sexes are usually detectable by ultrasound after the 12th week. From this time, the fetus sucks its thumb, kicks, and makes fists and faces, and baby teeth begin to form in the gums. More than half of all pregnant women experience the nausea and vomiting of morning sickness in their first trimester.

During the second trimester, body proportions become even more like those of a newborn. By the fourth month, the fetus has hair, eyebrows, eyelashes, nipples, and nails. Bone continues to replace the cartilage skeleton. The fetus's muscle movements become stronger, and the woman may begin to feel a slight fluttering in her abdomen. By the end of the fifth month, the fetus curls into the classic head-to-knees "fetal position." As the second trimester ends, the woman feels distinct kicks and jabs. The fetus is now about 30 centimeters long.

In the final trimester, fetal brain cells rapidly connect into networks, and organs differentiate further and grow. The bones become fully developed, and a layer of fat develops beneath the skin. The digestive and respiratory systems mature last, which is why infants born prematurely often have difficulty digesting milk and breathing. The fetus may move vigorously, causing back pain and frequent urination in the woman as it presses against her bladder. About 266 days (38 weeks) after a single sperm burrowed into an oocyte, a baby is ready to be born.

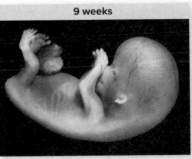

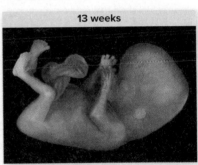

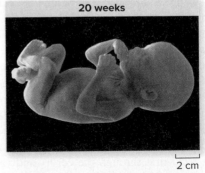

Figure 35.18 The Fetal Stage. These photographs show the development of a fetus from 9 weeks to 20 weeks after fertilization.

Photos: (9 weeks, 13 weeks): ©Science Source; (20 weeks): ©James Stevenson/Science Source

Apply It **Now**

Assisted Reproductive Technologies

What is something that you desperately want but cannot have? Money? A new car? Better health? Less stress? One more chance to talk to a loved one who has passed away? For millions of couples, the answer is "a baby." About 12% of women of child-bearing age in the United States have trouble conceiving a child.

To make a baby, sperm and egg must meet and merge; this process ordinarily occurs in the woman's reproductive tract. Abnormal gametes or blockages that impede this meeting of cells can result in infertility (the inability to conceive). Assisted reproductive technologies such as those described below can sometimes help.

Artificial Insemination

In artificial insemination, a doctor places donated sperm in a woman's reproductive tract. A woman might seek artificial insemination if her partner is infertile or carries an allele for an inherited illness. Alternatively, she may want to be a single parent or to raise a child with a partner other than the biological father.

Surrogate Motherhood

If a man produces healthy sperm but his partner's uterus is absent or cannot maintain a pregnancy, a surrogate may carry the pregnancy. The sur-

©Nancy R. Cohen/Getty Images RF

rogate may undergo artificial insemination; in that case, she would be the genetic mother of the child. Alternatively, a surrogate may receive a fertilized egg and be genetically unrelated to the resulting child.

In Vitro Fertilization

In *in vitro* fertilization (IVF), sperm meets oocyte outside the woman's body. A woman might undergo IVF if her ovaries and uterus work but her uterine tubes are blocked, or if she is using donated eggs instead of her own. A physician combines a few of a woman's eggs, chemicals that mimic those in the female reproductive tract, and sperm in a dish. The fertilized eggs divide two or three times and are introduced into the oocyte donor's (or a surrogate's) uterus. If all goes well, a pregnancy begins—often with twins or triplets.

Oocyte Donation

College newspapers often run ads recruiting healthy young women to become egg donors. In exchange for a payment, the donor receives several daily injections of a superovulation drug, has her blood checked for hormones, and then undergoes minor surgery to collect the oocytes. The prospective father's sperm fertilize the donor's oocytes *in vitro,* and the resulting embryos are placed in a woman's uterus.

E. Muscle Contractions in the Uterus Drive Childbirth

Implantation · Placenta develops · Birth
Fertilization · Gastrulation · 2wk · 1mo 2mo 3mo 4mo 5mo 6mo 7mo 8mo 9mo · Organs develop

Labor refers to the strenuous work a woman performs as she gives birth. The process typically occurs in three stages (figure 35.19). During the first stage of labor, the fetus presses down and ruptures the amniotic sac, causing an abrupt leakage of amniotic fluid ("water breaking"). Labor may also begin with a discharge of blood and mucus from the vagina, or a woman may feel mild contractions in her lower abdomen about every 20 minutes.

As labor proceeds, hormones prompt the smooth muscle that makes up the wall of the uterus to contract with increasing frequency and intensity. The cervix dilates (opens) a little more each time the baby's head presses against it. By the end of the first stage of labor, the cervix has stretched open to about 10 centimeters.

The second stage of labor is delivery, during which the baby typically descends head-first through the cervix and vagina. In the third (and last) stage of labor, the uterus expels the placenta.

The events of childbirth illustrate positive feedback because the process reinforces itself (as opposed to negative feedback, in which a process counteracts an existing condition). At the onset of labor, stretched sensory receptors at the cervix relay the message to the hypothalamus, which triggers release of the hormone oxytocin from the posterior pituitary. Oxytocin travels in the bloodstream and stimulates muscles in the uterus to contract, pushing the baby out (for this reason, physicians sometimes use synthetic oxytocin to induce labor). As the baby emerges, the cervix stretches farther and stimulates more oxytocin production, which intensifies the contractions, and so on. After the baby is born, other hormonal changes stop the cycle. *positive feedback,* section 25.4

Not all births go according to plan. For example, about 3% of babies have a "breech presentation," in which the baby's feet, knees, or buttocks appear first instead of the head. Breech deliveries are more difficult than head-first deliveries and account for some cesarean sections ("C-sections"). In this procedure, a surgeon removes the child from the uterus through an incision in the abdomen. Nearly one third of all babies in the United States are delivered by cesarean section, either by choice or because the life of the mother or the baby is at risk.

Sometimes, a pregnancy ends with a premature birth. An infant is premature if it is born before completing 35 weeks (about 8 months) of gestation. A fetus born before 22 weeks of gestation is not viable, but babies born after that time may survive. Premature infants are at risk for many health problems, so they typically

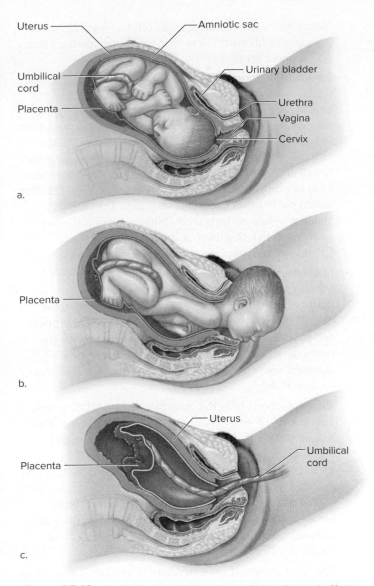

a.

b.

c.

Figure 35.19 Childbirth. (a) In the first stage of labor, the cervix dilates and the amniotic sac breaks. (b) In the second stage, the baby is pushed out of the birth canal. (c) The final stage is delivery of the placenta.

spend their first weeks or months in incubators at a neonatal intensive care unit. The incubators control the temperature and protect the infants from infection.

35.5 MASTERING CONCEPTS

1. What are the events of fertilization?
2. What are the relationships among the zygote, morula, blastocyst, inner cell mass, and gastrula?
3. What is implantation, and when does it occur?
4. Which supportive structures develop during the embryonic period? What are their functions?
5. When do sex differences appear, and what triggers them?
6. What are the events of the second and third trimesters?
7. Which events make up the three stages of labor?

35.6 Birth Defects Have Many Causes

The birth of a live, healthy baby seems against the odds, considering the complexity of human development. Of every 100 egg cells exposed to sperm, 84 are fertilized. Of these, 69 implant in the uterus, 42 survive a week or longer, 37 survive 6 weeks or longer, and only 31 are born alive. Of those that do not survive, about half have chromosomal abnormalities too severe to maintain life.

About 97% of newborns are apparently normal. In the remaining cases, genetic abnormalities, vitamin deficiencies, toxins, or viruses disrupt prenatal development and cause a **birth defect**—any abnormality that occurs during development and causes death or disability in the child.

Some birth defects result from a chromosomal abnormality or a faulty gene that acts during prenatal development. An extra copy of chromosome 21, for example, is the most common cause of Down syndrome. Faulty genes cause Tay-Sachs disease, phenylketonuria, Rett syndrome, and other disorders. ⓘ *trisomy 21, section 9.7B; Rett syndrome, section 10.7C*

Teratogens are substances that cause birth defects. People encounter teratogens everywhere, including the workplace. Women who work with textile dyes, lead, some photographic chemicals, semiconductor materials, mercury, and cadmium face increased risk of miscarriage and birth defects in their children. Biologists have described several other types of teratogens:

- **Alcohol:** A pregnant woman who consumes alcohol risks fetal alcohol syndrome. A child with fetal alcohol syndrome has a small head, misshapen eyes, and a flat face and nose. The child has impaired intellect, ranging from minor to severe learning disabilities. Because everyone metabolizes alcohol slightly differently, physicians advise all pregnant women to avoid alcohol.

- **Cigarettes:** Smoking during pregnancy increases the risk of miscarriage, stillbirth, and prematurity. Carbon monoxide crosses the placenta and robs rapidly growing fetal tissues of oxygen. Other chemicals in cigarette smoke prevent nutrients from reaching the fetus. The placentas of women who smoke lack important growth factors, thus lowering birth weight.

- **Illicit drugs:** A pregnant woman's use of cocaine, heroin, methamphetamine, and other illicit drugs is associated with poor fetal growth and low birth weight. In addition, cocaine increases the risk of urinary tract defects. However, it is often difficult to isolate the effect of each drug on fetal health because drug-addicted women also tend to have poor nutrition and poor health overall.

- **Excess or insufficient vitamins:** The acne medicine isotretinoin (Accutane), derived from vitamin A, causes miscarriages and defects of the heart, nervous system, and face. Vitamin deficiencies can also cause birth defects.

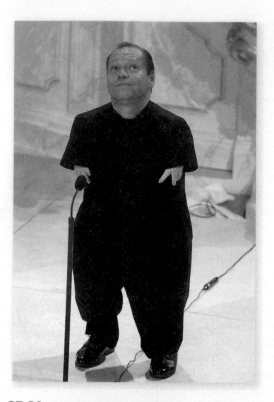

Figure 35.20 **Effects of Thalidomide.** Grammy award-winning baritone Thomas Quasthoff has unusually short limb bones, and his hands resemble flippers. His mother took thalidomide to combat morning sickness early in pregnancy.
©Peer Grimm/AP Images

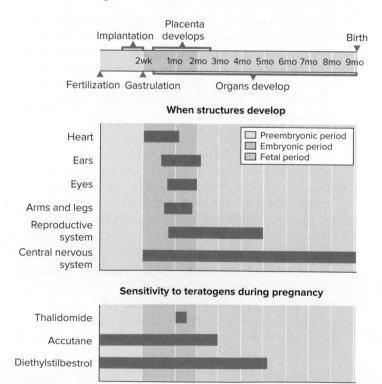

Figure 35.21 **Critical Periods of Development.** Body parts differ in their critical periods. The type of birth defect resulting from each drug depends on which structures are developing at the time of the exposure.

Neural tube defects such as spina bifida, for example, are associated with insufficient folic acid in the mother's diet. ⓘ *vitamins,* section 32.4A

- **Prescription drugs:** During the late 1950s, a drug called thalidomide was prescribed as a sedative and as a treatment for morning sickness in pregnant women. The drug caused severe limb shortening and other serious birth defects in the children of these women (**figure 35.20**). From the 1940s through the 1960s, many pregnant women with a history of miscarriages were prescribed a synthetic form of estrogen, called diethylstilbestrol (DES). The drug turned out to be an endocrine disruptor. The "DES children"—and perhaps their children's children—face an elevated risk of cancer of the reproductive tract. ⓘ *endocrine disruptors,* section 28.2

- **Starvation:** Inadequate nutrition during pregnancy increases the incidence of miscarriage and damages the placenta, causing low birth weight, short stature, tooth decay, delayed sexual development, learning disabilities, and possibly intellectual disabilities. Starvation greatly raises the risk of a child developing obesity, type 2 diabetes, and other conditions later in life. ⓘ *starvation,* section 32.4C

- **Viral infection:** HIV, the virus that causes AIDS, infects 15% to 30% of infants born to HIV-positive women. This risk drops sharply if an infected woman takes antiviral drugs while pregnant. Fetuses infected with HIV are at risk for low birth weight, prematurity, and stillbirth. In addition, the virus that causes rubella (German measles) causes birth defects such as deafness, cataracts, and heart disease; herpes simplex viruses can infect the fetal nervous system and cause neurological disabilities such as intellectual disabilities or cerebral palsy.

The "critical period" of a structure is the time during which its development is most susceptible to damage by a teratogen (**figure 35.21**). About two thirds of birth defects stem from a disruption during the embryonic period, because developing organs are especially sensitive to damage.

The longest bar in figure 35.21 corresponds to the development of the central nervous system—that is, the brain and spinal cord. The brain is vulnerable throughout prenatal development, as well as during the first 2 years of life. Because of the brain's long critical period, many birth defect syndromes include intellectual disabilities. The continuing sensitivity of the brain after birth explains why toddlers who accidentally ingest lead-based paint suffer impaired learning.

35.6   MASTERING CONCEPTS

1. What is a birth defect?
2. What are some examples of genetic conditions and teratogens that cause birth defects?
3. Suppose that a woman uses illicit drugs for 2 weeks before finding out that she is nearly 3 months pregnant. What structures may be affected in the developing fetus?

INVESTIGATING LIFE

35.7 The Ultimate Sacrifice

Of all the surprising behaviors in the animal kingdom, among the most puzzling is sexual cannibalism. A cannibal, of course, is an organism that eats other members of its species. Sexual cannibalism is a special case in which a female eats a male before, during, or after copulation.

Sexual cannibalism occurs in arthropods, including praying mantises, scorpions, and some groups of spiders. The Australian redback spider, *Latrodectus hasselti,* however, offers a particularly interesting example. This arachnid, which is a relative of the North American black widow, has a bizarre mating ritual in which the male volunteers his body for the female to consume.

The male redback spider has two mating structures, called palps, on his head. The male fills each palp with sperm that he has deposited on a web. He courts a female; if she accepts him, he inserts a palp into one of her two sperm storage organs and begins delivering sperm. A few seconds later, with the palp still inserted, he does something unusual: He offers up his abdomen for her to eat. Using the palp as a sort of pivot, he somersaults his body onto her fangs. Observations of spiders in the field suggest that, most of the time, the female eats the male.

The most surprising feature of the redback spider's behavior is that the male appears to cooperate in his own demise. The female does not position the male's body into her jaws; rather, he moves his own body toward her mouth. And unlike other arthropods, which often try to avoid being eaten during mating, the male redback spider does not struggle or try to escape.

This behavior is puzzling: By allowing his mate to kill him, he forfeits all opportunities to mate again in the future. Natural selection should weed out behaviors that decrease reproductive success. Yet the male's behavior is consistent and predictable.

Scientists have proposed many hypotheses to explain sexual cannibalism, not only in the redback spider but also in other species. One possibility is that the female uses the nutrients from her meal to help support the growth of her eggs and offspring, increasing both her and her mate's reproductive success. Or perhaps she simply mistakes the male for prey.

These explanations may apply to other species, but not to the Australian redback spider. After all, the female is much larger than the male: The average female is about 1 centimeter long and weighs some 256 milligrams; the male's body, in contrast, is only 3 to 4 millimeters long and weighs just over 4 milligrams. Moreover, laboratory studies suggest that consuming one tiny male neither boosts the mass of a female's egg sac nor increases the number of eggs. Researchers have therefore ruled out a nutritional explanation for sexual cannibalism in this species.

How, then, does the male redback spider's strange behavior increase his fitness? Evolutionary biologist Maydianne Andrade investigated this question while she was a graduate student at the University of Toronto at Mississauga. Andrade knew that a female Australian redback spider may mate more than once, setting up a "paternity competition" between the males. Andrade's experiment simulated this competition and allowed her to determine how many of each female's offspring were fathered by each mate.

Andrade arranged two consecutive matings between virgin females and either normal or sterile virgin males (**figure 35.22**). The order in which the matings occurred—normal male first or sterile male first—was randomized. Andrade carefully observed how long each pair copulated and noted if each male was cannibalized. After the second mating, Andrade counted the number of hatched and unhatched eggs in each female's egg sac. Andrade assumed that all of the unhatched eggs were "fertilized" by sterile males. These counts enabled Andrade to estimate the proportion of the offspring fathered by each male. In particular, she was curious whether a male that mates second and is cannibalized has greater reproductive success than a male that survives mating.

Overall, copulation lasted from 6 to 31 minutes, but Andrade noticed an interesting pattern. A male that was cannibalized spent about 25 minutes in copulation, more than twice as long as one that survived the mating (about 11 minutes). While the female occupied herself with her meal, the dying male evidently put the

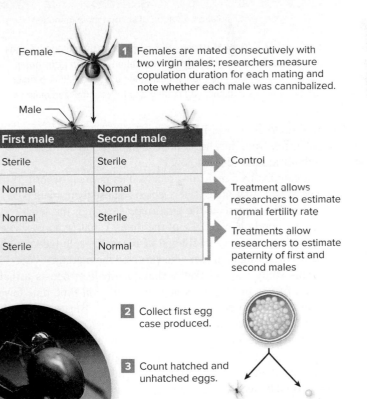

First male	Second male	
Sterile	Sterile	→ Control
Normal	Normal	→ Treatment allows researchers to estimate normal fertility rate
Normal	Sterile	⎫ Treatments allow researchers to estimate paternity of first and second males
Sterile	Normal	⎭

1 Females are mated consecutively with two virgin males; researchers measure copulation duration for each mating and note whether each male was cannibalized.

2 Collect first egg case produced.

3 Count hatched and unhatched eggs.

Hatched: fertilized by normal male   **Unhatched:** fertilized by sterile male

4 Plot paternity of second male against copulation duration (see figure 35.23).

Figure 35.22 Paternity Testing. Researchers mated female Australian redback spiders with two males (normal or sterile) to determine how mating duration and cannibalism affect the probability of fathering offspring. The inset photo shows a male on the abdomen of a female, illustrating the size difference between the sexes.

Photo: ©Tim Wimborne/Reuters/Corbis

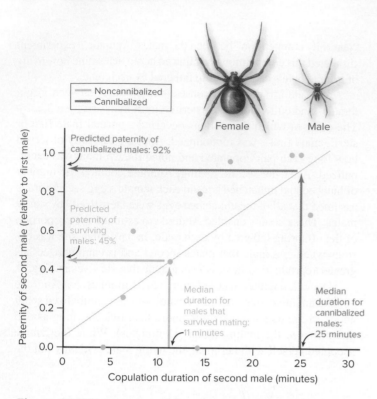

Figure 35.23 **Time Matters.** This graph plots the percent of eggs fertilized by the second male against copulation duration (*gray dots*). Andrade fit the gray line to the data; she then used the line to predict paternity for males that survived mating and for cannibalized males.

extra time to good use: Longer copulation predicted a big boost in paternity, from 45% for a noncannibalized male to 92% for a cannibalized male (figure 35.23).

A cannibalized male also received an indirect benefit. If the female ate her first mate, she rejected a subsequent male 67% of the time. In contrast, if the first male survived, the female rejected a second male just 4% of the time. A male who was eaten therefore had a higher likelihood of being the sole father to the female's offspring than one who survived the mating.

This study showed that sexual cannibalism boosts a male spider's fitness in two ways: longer copulation time plus fewer mating opportunities for subsequent males. But wouldn't the male be even better off by escaping and pursuing other females? The answer appears to be no. In Australian redback spiders, mating opportunities for males are exceedingly rare. Even if a male mates and then manages to find a second female, his palps are likely too damaged to mate again. A copulating male can make the best of his good fortune by offering up his own body—all of it—as the ultimate sacrifice to the next generation.

Source: Andrade, Maydianne C. B. 1996. Sexual selection for male sacrifice in the Australian redback spider. *Science,* vol. 271, pages 70–72.

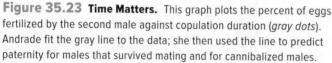

35.7   MASTERING CONCEPTS

1. Explain why sexual cannibalism represents an evolutionary paradox.
2. Use figure 35.23 to predict the proportion of offspring that a male redback spider will sire if he is the female's second mate and copulates for 20 minutes.

CHAPTER SUMMARY

35.1 Animal Development Begins with Reproduction

A. Reproduction Is Asexual or Sexual
- **Asexual reproduction** does not require a partner and yields identical offspring.
- In **sexual reproduction,** two haploid **gametes** unite and form a **zygote,** the first cell of a new offspring. **Fertilization** may occur in the environment (**external fertilization**) or inside an animal's body (**internal fertilization**).

B. Gene Expression Dictates Animal Development
- Development requires **differentiation,** the formation of specialized cells. In **pattern formation,** the animal takes on its overall shape and structure.

C. Development Is Indirect or Direct
- Animals that undergo **indirect development** have a **larva** stage that does not resemble the adult. **Metamorphosis** transforms the larva into an adult.
- In **direct development,** young animals resemble miniature adults.

35.2 Males Produce Sperm Cells

- The **reproductive systems** of both males and females include **gonads,** which house the **germ cells** that give rise to gametes.
- Other sex organs deliver or nurture the gametes. Males and females have different **secondary sex characteristics,** traits that do not directly participate in reproduction.

A. Male Reproductive Organs Are Inside and Outside the Body
- Developing sperm originate in **seminiferous tubules** within the paired **testes.** Also inside the testes are **Sertoli cells** that surround the seminiferous tubules and **interstitial cells** that secrete hormones. A pouch called the **scrotum** contains the testes.
- Sperm travel through the **epididymis, vas deferens,** and **ejaculatory duct,** and they exit the body with **semen** through the **urethra** (within the **penis**) during **orgasm** and **ejaculation.**
- The **prostate gland, seminal vesicles,** and **bulbourethral glands** add secretions to semen.

B. Spermatogenesis Yields Sperm Cells
- **Spermatogenesis** begins with diploid **spermatogonia,** which divide mitotically to yield a stem cell and a **primary spermatocyte.** The first meiotic division produces two haploid **secondary spermatocytes.** In meiosis II, the secondary spermatocytes divide, yielding four haploid **spermatids.** The spermatids develop into **spermatozoa,** or mature sperm cells.
- A mature sperm cell has a flagellum and a head that contains the chromosomes. A caplike **acrosome** covers the head.

C. Hormones Influence Male Reproductive Function
- **Gonadotropin-releasing hormone (GnRH)** from the hypothalamus stimulates the anterior pituitary gland to release **follicle-stimulating hormone (FSH)** and **luteinizing hormone (LH).** In males, these hormones affect the testes, triggering the release of **testosterone** necessary for sperm formation and the development of secondary sex characteristics.

35.3 Females Produce Egg Cells

A. Female Reproductive Organs Are Inside the Body
- Egg cells (and their nourishing **follicle cells**) originate in the **ovaries.** Each month after puberty, one ovary releases an egg cell into a **uterine tube,** which leads to the **uterus.** A blood-rich **endometrium** lines the inside of the uterus. The **cervix** leads to the **vagina,** which connects the uterus with the outside of the body.
- The **vulva,** or external genitalia, consists of the labia, **clitoris,** and vaginal opening. **Breasts** deliver milk to infants.

B. Oogenesis Yields Egg Cells
- In **oogenesis, oogonia** divide mitotically to form two **primary oocytes.** In meiosis I, the cytoplasm of the primary oocyte divides unevenly as it splits into one large haploid **secondary oocyte** and a much smaller haploid **polar body.** In meiosis II, the secondary oocyte again divides unequally, yielding the large haploid **ovum** and another small polar body.

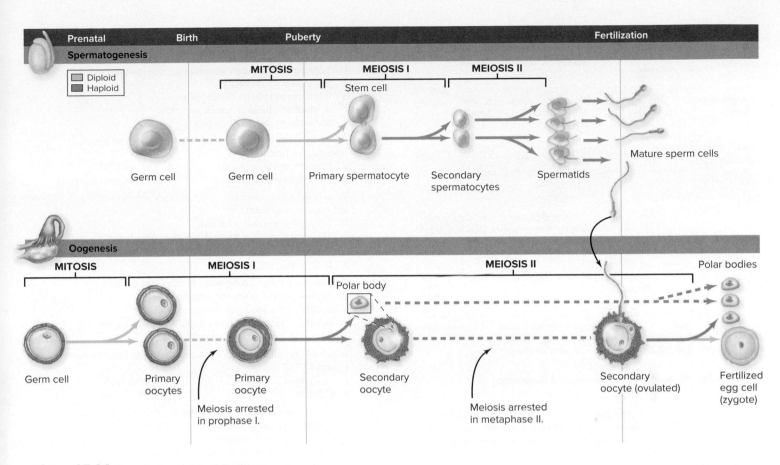

Figure 35.24 **Gamete Formation: A Summary.**

- Meiosis in the female begins before birth and completes at fertilization.
- Figure 35.24 compares gamete formation in males and females. The steps of sperm and oocyte formation are similar, and the products of each process are haploid. But the male and female gametes look different, and they form according to different timetables.

C. Hormones Influence Female Reproductive Function

- The ovaries secrete **estrogen** and **progesterone,** hormones that stimulate development of female sexual characteristics. GnRH, FSH, LH, estrogen, and progesterone control the **ovarian cycle** and the **menstrual cycle.**
- After **ovulation,** the ruptured follicle develops into the **corpus luteum,** which secretes hormones that prepare the body for pregnancy. If pregnancy does not occur, the endometrium is shed in menstrual flow.

D. Hormonal Fluctuations Can Cause Discomfort

- Premenstrual syndrome, menstrual cramps, endometriosis, and hot flashes are associated with changing hormone concentrations.

35.4 Reproductive Health Considers Contraception and Disease

- **Contraception** is the use of behaviors, barriers, hormones, spermicidal chemicals, or other devices that prevent pregnancy.
- Pathogens that cause **sexually transmitted diseases (STDs)** spread via sexual contact. Viruses and bacteria cause most STDs, which may cause infertility.

35.5 The Human Infant Begins Life as a Zygote

A. Fertilization Initiates Pregnancy

- A male produces sperm cells and delivers them to the female's body, where fertilization occurs (figure 35.25).

- In fertilization, a sperm cell burrows through the two layers surrounding a secondary oocyte. The two united cells constitute the diploid zygote.

B. The Preembryonic Stage Ends When Implantation Is Complete

- The **preembryonic stage** lasts from fertilization until the end of the second week of development.
- After fertilization, **cleavage** divisions produce the **morula.** Between days 3 and 6, the morula arrives at the uterus and hollows out, forming a **blastocyst.** An outer layer of cells (the trophoblast) and an **inner cell mass** form. Cells of the trophoblast secrete **human chorionic gonadotropin (hCG),** which prevents menstruation. **Implantation** occurs between days 6 and 14.

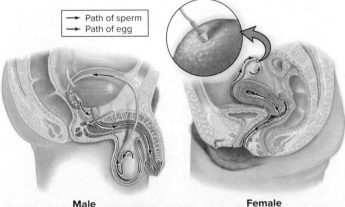

Figure 35.25 **Path to Fertilization: A Summary.**

Name		Duration
Zygote		About 24 hours after ovulation
Preembryonic stage	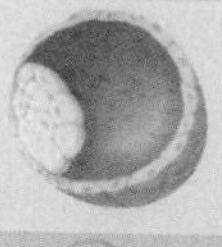	From first cell division until implantation (about 2 weeks)
Embryonic stage		From implantation until about 8 weeks after fertilization
Fetal stage	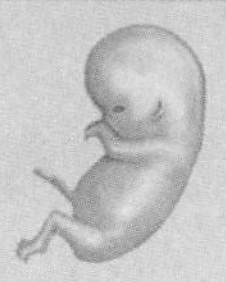	From end of eighth week until birth (end of first trimester, plus second and third trimesters)

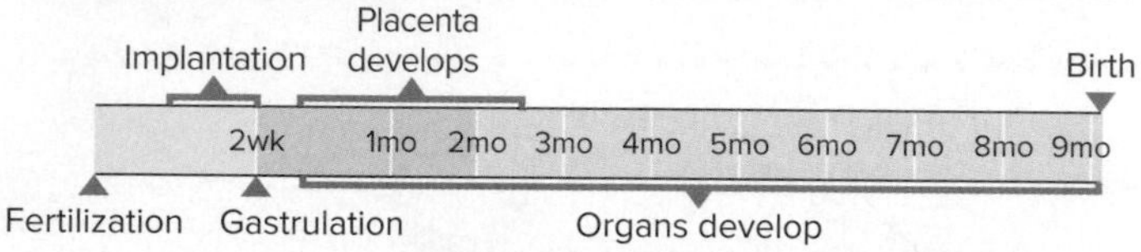

Figure 35.26 **Stages in Development: A Summary.**

- During the second week, the amniotic cavity forms as the inner cell mass flattens, forming the **embryonic disc. Ectoderm** and **endoderm** form, and then **mesoderm** appears, establishing the three primary germ layers of the **gastrula.**

C. Organs Take Shape During the Embryonic Stage

- The **embryonic stage** lasts from the second week through the eighth week of development.
- **Chorionic villi** extending from the **chorion** start to develop into the **placenta.** The **yolk sac, allantois,** and **umbilical cord** form as the **amnion** swells with fluid.
- Organs form throughout the embryonic period. The **primitive streak** forms the longitudinal axis around which new structures appear, including the **notochord, neural tube,** arm and leg buds, heart, facial structures, skin specializations, sex organs, and skeleton.

D. Organ Systems Become Functional in the Fetal Stage

- Structures continue to elaborate during the **fetal stage,** which lasts from the ninth week of development until the baby is ready for birth.
- Figure 35.26 summarizes the stages of development.

E. Muscle Contractions in the Uterus Drive Childbirth

- Labor begins as the fetus presses against the cervix. In a **positive feedback** loop, cervical stretching triggers the release of oxytocin, which stimulates uterine contractions that push the baby farther out, causing even more stretching. Eventually, the mother expels baby and placenta.

35.6 Birth Defects Have Many Causes

- Genetic abnormalities, dietary deficiency, or exposure to **teratogens** such as chemicals or viruses can cause **birth defects.**
- Each body structure has a different critical period, which is the time in development during which it is especially vulnerable to teratogens.

35.7 Investigating Life: The Ultimate Sacrifice

- During copulation, a male Australian redback spider offers his body for the female to eat.
- Males that are cannibalized copulate longer and therefore father more offspring than those that survive mating.

MULTIPLE CHOICE QUESTIONS

1. Unchanging environments are advantageous to organisms
   a. that reproduce asexually.
   b. that undergo indirect development.
   c. in which fertilization occurs internally.
   d. that undergo pattern formation.

2. Which hormones play central roles in both male and female reproductive function?
   a. Estrogen and testosterone
   b. Progesterone and LH
   c. LH and FSH
   d. Testosterone and FSH

3. What is the relationship between a primary spermatocyte and a spermatogonium?
   a. They are genetically identical.
   b. The spermatocyte is haploid and the spermatogonium is diploid.
   c. The spermatocyte has a flagellum; the spermatogonium does not.
   d. Both b and c are correct.

4. Why can't a fertilized polar body develop into a fetus?
   a. Because it has too few chromosomes
   b. Because it has too many chromosomes
   c. Because it lacks sufficient cytoplasm and organelles
   d. Because it carries excess flagella

5. Which of the following is a birth control method that protects against STDs?
   a. Vasectomy
   b. Condom
   c. Withdrawal
   d. HPV vaccine

6. The acrosome's enzymes allow a sperm cell to penetrate the
   a. proteins and carbohydrates encasing the oocyte.
   b. follicle cells surrounding the oocyte.
   c. plasma membrane of the secondary oocyte.
   d. Both a and b are correct.

7. What is the correct order of structures to develop after fertilization?
   a. Zygote, gastrula, blastocyst, morula, fetus
   b. Zygote, fetus, gastrula, morula, blastocyst
   c. Fetus, zygote, morula, blastocyst, gastrula
   d. Zygote, morula, blastocyst, gastrula, fetus

8. What part of the blastocyst develops into the embryo?
   a. Yolk sac
   b. Placenta
   c. Polar body
   d. Inner cell mass

9. A baby born four weeks before its due date is most likely to have an underdeveloped
   a. central nervous system.
   b. heart.
   c. reproductive system.
   d. All of the above are correct.

10. A chemical that can cause a birth defect is a
    a. toxin.
    b. trisomy.
    c. mutagen.
    d. teratogen.

Answers to these questions are in appendix A.

WRITE IT OUT

1. What is differentiation? How do protein gradients cause cells to differentiate?

2. Some newborn mammals can walk and carry out other life functions independently. Human babies, on the other hand, are born helpless. Speculate about the trade-offs in these two reproductive strategies. What selective forces might limit the stage of development at which humans are born?

3. How are the human male and female reproductive tracts similar, and how are they different? How are the structures of the testis and ovary similar and different?

4. Briefly explain the contradictory roles of estrogen in the human female reproductive cycle.

5. What are the functions of LH and FSH in males and females?

6. How are the timetables different for oogenesis and spermatogenesis in humans?

7. In many species, a female may mate with multiple males, increasing her chance of conception. Speculate about how this behavior may select for males with faster sperm or more numerous sperm per ejaculate.

8. Point mutations usually occur during interphase of mitosis, but most chromosomal abnormalities arise during meiosis. Given the differences between gamete production in males and females, why is it reasonable to predict that more point mutations occur during sperm production and more chromosomal abnormalities appear in egg cells?

9. Is each of the following cell types haploid or diploid? How does each cell type relate to the others?
   a. An oogonium
   b. A primary spermatocyte
   c. A spermatid
   d. A secondary oocyte
   e. A polar body derived from a primary oocyte

10. Write a paragraph describing the path of sperm to egg, starting with sperm in the epididymis. *Hint:* See figure 35.25.

11. Use the Internet to learn more about sexually transmitted diseases. Choose one to study in detail. What type of infectious agent causes the disease? What are the symptoms and long-term consequences of infection? Is a treatment available? Who is most affected?

12. What would happen if two sperm fertilized the same egg cell? If two sperm fertilized two egg cells?

13. A zygote is a single cell whose weight is immeasurably small, yet after countless cell divisions it develops into a newborn weighing approximately 3.5 kilograms. While in the uterus, the developing infant obtains the atoms that make up its ever-increasing mass from its mother. Where does the mother obtain these atoms?

14. After ovulation, LH transforms the ruptured follicle into the corpus luteum. As LH levels decline, the corpus luteum degenerates. However, if fertilization occurs, the preembryo starts producing a hormone called hCG, which is similar in structure to LH and prevents the corpus luteum from degenerating. Why is it adaptive for the preembryo to maintain the corpus luteum?

15. For each of the following pairs of phrases, indicate if the first is greater than the second, if the second is greater than the first, or if the two are equal.
   a. The number of cells in a morula; the number of cells in a gastrula
   b. The thickness of the endometrium after ovulation; the thickness of the endometrium after menstruation
   c. The number of eggs released from an ovary during each ovarian cycle; the number of sperm released during ejaculation
   d. The number of chromosomes in an egg cell; the number of chromosomes in a sperm cell

16. What is the source of nutrients for a developing preembryo? For an embryo? For a fetus?

17. Provide a general explanation for why men have nipples. *Hint:* Figure 35.17 may help.

18. Consult a website that describes and illustrates fetal development. What technology do you think would be necessary to enable a fetus born in the fourth month to survive in a laboratory setting?

19. During labor, a baby may remain in the birth canal for hours. How does the baby not suffocate during this time?

20. Most physicians clamp and cut the umbilical cord immediately after birth, separating the baby from the nutrients and respiratory gases it receives from the placenta; other physicians wait to cut the cord for several minutes after birth, allowing the remaining cord blood to enter the baby's body. Design an experiment to test whether immediate clamping or delayed clamping of the umbilical cord is a better practice for healthy babies.

21. Use the Internet to research the rate of cesarean sections (surgical removal of the baby) in the United States. Why might the incidence be so high?

22. What is the significance of the critical period in determining the type and severity of a birth defect?

23. What kinds of studies and information would be necessary to determine whether exposure to a potential teratogen can cause birth defects a year later? How would such an analysis differ if it were a man or a woman who was exposed?

PULL IT TOGETHER

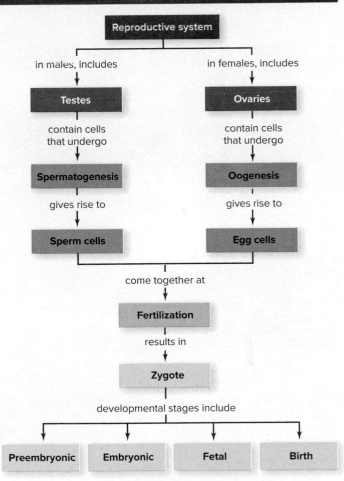

Figure 35.27 Pull It Together: Animal Reproduction and Development.

Refer to figure 35.27 and the chapter content to answer the following questions.

1. Review the Survey the Landscape figure in the chapter introduction and the Pull It Together concept map. Explain why an evolutionary biologist might say that the reproductive system is "selfish" compared to the body's other organ systems.

2. Add ovulation to the concept map. What hormonal changes typically accompany ovulation? Add those hormones to the concept map as well.

3. Write into the concept map the approximate time after fertilization that each of the following structures is present: preembryo, embryo, fetus, newborn.

4. Add the terms *placenta, follicle, polar body, gonad, gametes, meiosis,* and *mitosis* to this concept map.

36 Animal Behavior

Stotting. A springbok leaps straight up into the air with its legs stiffly extended and its back arched, a behavior called stotting. No one knows why springboks stot.

©Gerald Hinde/Gallo Images/Getty Images RF

LEARN HOW TO LEARN
What's Your Learning Style?

Students differ in how they prefer to receive information: Some love to hear lectures, others like reading, and still others thrive on hands-on activities in lab. You may find it helpful to search for a learning styles inventory online to discover more about your own preferences. At the very least, the advice on the website may alert you to study techniques that you have not tried before.

UNIT 7

Risky Business

It is a hot morning in the African bush, and a herd of springboks grazes on shoots of new grass. The animals are alert as they feed. Their ears swivel and their tails twitch away flies. Every now and then, one pauses, looks up, and scans the horizon. Death crouches in the tall grasses in the form of a hungry cheetah. The wind shifts, and one springbok jerks up its head. Then it does something that looks awfully odd. Instead of running away, the springbok leaps into the air. Its nose points to the ground as it bounces up and down on stiff legs. In a flash it races away, but the odd bouncing, called stotting, seems to be contagious. As they race away, other springboks also stot, sometimes in midstride. The cheetah charges, but it is a half-hearted attempt. The spotted cat stops short and watches the herd disappear.

Mobbing is another risky animal behavior, and you can see it closer to home. Terns (a type of gull) will mob any animal that intrudes into their nesting colony, from cats to foxes to humans. When a peregrine falcon sails over a colony of terns, for example, the birds fly up from their nests and attack. Screaming, they dive-bomb the falcon. This truly is risky business. The peregrine is such an agile flier that it is perfectly capable of rolling over on its back in midair and striking at a mobber or grabbing it in lethal talons.

Other birds also mob predators. If a crow spies a great horned owl or red-tailed hawk, it will send a loud alarm call. The sound draws other crows and often smaller birds; sometimes a huge mixed flock is recruited. Giving their different alarm calls, the birds settle on tree branches near the predator. Some fly straight at it, veering away only at the last second. Eventually, the predator flies away, out of the nesting territories of the birds.

Stotting and mobbing are spectacular behaviors, but if you watch any animal long enough, something interesting is bound to happen. It might spin a web, curl into a ball, start barking, groom itself, hide, jump into the air, yawn, or swish its tail. Even the most mundane behavior can prompt interesting questions: Why does the animal do that? How does it do that? Was it born knowing how to perform the behavior, or did it learn by imitating other animals?

In pondering such questions, the mind may wander to some of the strange things we do, like kissing, playing basketball, or whistling. How did those behaviors ever start, and why do they persist? The answers to many such questions remain elusive, but this chapter explores some of what we do know about a wide spectrum of animal behaviors, both simple and complex.

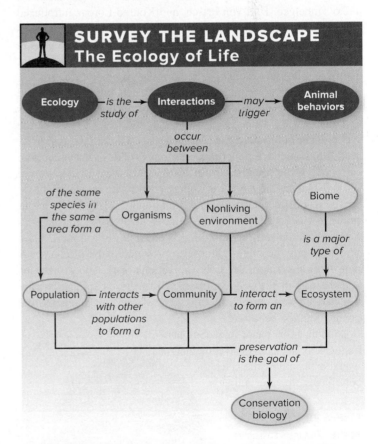

SURVEY THE LANDSCAPE
The Ecology of Life

Animal behaviors include a huge variety of interactions within an animal's own species, with predators and prey of other species, and with the nonliving environment. The animal's physiology explains how these behaviors happen, and natural selection explains why animals do what they do.

For more details, study the Pull It Together feature in the chapter summary.

36.1 Animal Behaviors Have Proximate and Ultimate Causes

Animal behavior is an ideal topic to bridge the animal physiology and ecology units of this textbook. An animal's behavior represents the integration of its muscular, nervous, endocrine, reproductive, and other physiological systems. At the same time, the animal interacts with other members of its own species and with the plants and other organisms in its habitat. These interactions form part of the subject of ecology.

The study of animal behavior has evolved throughout human history. To early humans, the behaviors of animals must have been familiar. After all, a hunter's knowledge of an animal's habits might mean the difference between survival and starvation, and dangerous animals might lurk around the bend of any trail. Nowadays, few people have firsthand experience with wild animals, although many own pets (see this chapter's Apply It Now box).

As popular knowledge of animal behavior has moved to the periphery, the science of animal behavior has developed and blossomed into several related disciplines, including behavioral ecology and evolutionary psychology. In the 1930s, naturalists Niko Tinbergen, Karl von Frisch, and Konrad Lorenz introduced the term *ethology* for the scientific study of animal behavior, especially in the natural environment. An ethologist must understand what animals do, beginning with firsthand observations that do not interfere with the animals. The emphasis in ethological studies, however, is on experiments. Karl von Frisch, for example, used experiments to help decode the meaning of the honeybees' famous "waggle dance." Tinbergen, von Frisch, and Lorenz received the Nobel Prize in Physiology or Medicine in 1973 for their research.

Animal behavior encompasses many fields of research. If an observer of the dung beetle in figure 36.1 were to ask several biologists what the animal is doing and why, he or she might receive very different answers. A functional morphologist might point to the muscles and exoskeleton involved in the beetle's head-down ball-rolling motion. A physiologist might explain how the body mobilizes glucose to provide a burst of energy. A neurobiologist might consider the actions of the motor neurons and the role of the beetle's brain. All of these explanations are directed at **proximate** causes, or the "how" mechanisms that bring about the behavior.

On the other hand, an evolutionary biologist or a behavioral ecologist would explain how the behavior contributes to an animal's

Figure 36.1 Dung Beetle. This insect assumes a distinctive head-down posture as it rolls a ball of dung to its nest.
©James Hager/Robert Harding World Imagery/Getty Images

Proximate: "How?" (Physiology)

Ultimate: "Why?" (Evolution)

fitness and how it might have evolved by natural selection. These latter explanations are directed at **ultimate** causes, which explore the evolutionary basis ("why") of a behavior—that is, how the behavior promotes the animal's survival or reproductive success. For example, the dung beetle will stow its prize underground. It will lay eggs inside the ball, which provides a protected habitat and guaranteed food supply to fuel the growth and development of the beetle's offspring. ⓘ *fitness,* section 12.3D

This chapter begins with approaches at the proximate level of causation—descriptions of innate and learned behavior patterns—followed by the interplay of genes and environment in behavior. The chapter then leads into ultimate causation, focusing on the role of behavior in survival and reproduction. Although we treat these topics separately, a recent trend in animal behavior research is to combine both levels of causation in the same study. Teams including geneticists, physiologists, ecologists, and evolutionary biologists can work together to arrive at a much more complete understanding of the causes of a behavior, from its control at the cellular and molecular level to its evolutionary history.

36.1 | MASTERING CONCEPTS

1. What is ethology?
2. Distinguish between the proximate and ultimate causes of an animal's behavior.

Apply It **Now**

Puppy Love

If you own and love a pet, you may wonder how the animal's experience of your relationship compares with your own. The cat may jump in your lap, curl up, and purr. The dog may greet you exuberantly when you come home. Do these behaviors mean that your pet loves you?

However tempting it is to project human emotions onto animals—that is, to be *anthropomorphic*—it is unscientific to do so. Scientists have no way to measure objectively the love that humans feel, let alone the emotions that other animals may have. Your cat may seem to enjoy her time in your lap, but is she simply seeking a soft, warm, comfortable place to rest? Your dog may seem happy to see you, but perhaps he is simply behaving in ways that he has learned will earn attention or treats.

This question of animal emotions illustrates the sometimes frustrating reality that science cannot answer every question that interests us. Our inability to peer into the animal mind adds mystery to our relationships with our nonhuman companions. Perhaps this mystery helps make pet ownership one of the world's most popular hobbies.

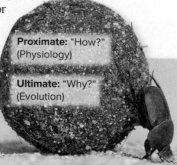
©Getty Images/Digital Vision RF

36.2 Animal Behaviors Combine Innate and Learned Components

Biologists traditionally classify animal behaviors based on the way they are acquired and how changeable they are. All animals are born with the ability to perform some behaviors; other behaviors must be learned.

A. Innate Behaviors Do Not Require Experience

The ethological approach to animal behavior has stressed the importance of **innate,** or instinctive, behavior patterns. An animal can perform innate behaviors at birth; learning is not involved. These behaviors tend to be stereotypic; that is, they are performed in a predictable way each time.

Two simple examples of innate behaviors are reflexes and taxes (singular: taxis). A **reflex** is an instantaneous, automatic response to a stimulus. If you touch a hot pan on the stove, you quickly withdraw your hand without even thinking about it. In fact, the reflex circuitry is located in the spinal cord and does not involve the brain at all (see figure 26.13); only the sensation of pain, which comes later, requires the brain. Another innate behavior is a **taxis,** a movement toward or away from a stimulus. An earthworm, for example, tends to crawl toward the high humidity of moist soil, and a cockroach scurries away from light.

A more complex behavior sequence is a **fixed action pattern (FAP),** a motor response that is initiated by an environmental stimulus and that continues to completion, even if the stimulus is withdrawn. For example, Tinbergen studied fish called three-spined sticklebacks. In this species, breeding males have bright red undersides and defend territories in which females lay eggs. The sight of an intruding male elicits an aggressive response consisting of lunges and bites.

Tinbergen focused on the specific stimuli that triggered the aggressive threat display (**figure 36.2**). By presenting male sticklebacks with a series of models, Tinbergen determined that many shapes would elicit the threat display, as long as the object was red on the bottom. Accurate clay models of sticklebacks without

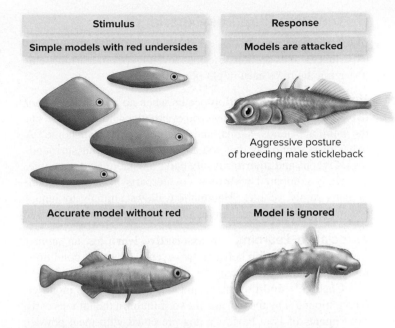

Figure 36.2 Color Matters. Male sticklebacks act aggressively toward clay models that look nothing like fish, as long as the undersides are red. They do not, however, display aggression toward a realistic model of a fish lacking a red belly.

the color red, however, were not effective. In male sticklebacks, red serves as the sign stimulus that releases the threatening FAP. Apparently, a male stickleback is more or less "programmed" to threaten any intruding breeding male. This behavior not only helps ensure that he is the one that fertilizes eggs laid in his territory but also helps protect the eggs from cannibalism.

Behavioral geneticists have delved into the proximate causes of some FAPs. A favorite model species is the fruit fly. One fruit fly gene with powerful behavioral effects is named *fruitless* (*fru,* for short). Normal *fru* function is required for proper development of the motor neurons that innervate muscles involved in courtship. Mutations in *fru* disrupt the complex, six-part FAP that leads to copulation (**figure 36.3**). One mutant form of *fru,* for example, causes males to court other males, moving in conga lines several flies long.

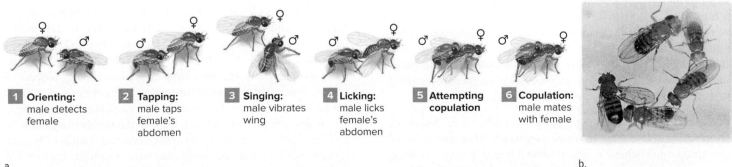

1 **Orienting:** male detects female

2 **Tapping:** male taps female's abdomen

3 **Singing:** male vibrates wing

4 **Licking:** male licks female's abdomen

5 **Attempting copulation**

6 **Copulation:** male mates with female

a.

b.

Figure 36.3 Fruit Fly Courtship. (a) Sequence of normal courtship behaviors. (b) A line of courting males with mutated *fru* genes.

(b): Courtesy of Professor Daisuke Yamamoto, Tohoku University/JST-ERATO project

B. Learning Requires Experience

In **learning,** an animal alters its behaviors as an outcome of its experiences. This section describes several types of learning, emphasizing the roles each might play in survival.

Habituation   **Habituation** occurs when an animal learns *not* to respond to a stimulus. Have you ever noticed how you tune out the familiar sounds, sights, and smells of your own home? A behaviorist would say that you are habituated to your surroundings. Pigeons and squirrels in city parks are habituated to people, especially compared with their counterparts in habitats where humans rarely venture. Habituation allows animals to ignore normal stimuli while remaining alert to the out-of-the-ordinary.

Associative Learning   In **associative learning,** an animal learns the relationship between two events. Classical conditioning and operant conditioning are forms of associative learning.

   Classical conditioning occurs when a behavior is modified ("conditioned") by the pairing of two stimuli. In the famous early experiments of Ivan Pavlov, a dog presented with meat powder would invariably salivate. If Pavlov repeatedly accompanied the meat stimulus with the sound of a bell, the dog would soon salivate upon hearing the bell, even with no meat present. The dog formed an association between the two stimuli. Classical conditioning plays an important role in the real world as well. For example, an animal may become sick shortly after eating a noxious food. If the animal associates the two events, it may avoid that food in the future.

   Another type of associative learning is **operant conditioning,** in which an animal learns to associate a behavior with its consequences. For example, a bird may learn that pressing a bar or pecking a target results in a reward or punishment. This type of learning is involved in most animal training: A dog learns to associate a command ("Sit!") with the behavioral response that earns a food reward. Operant conditioning also undoubtedly plays a role in the lives of animals in the wild.

Imprinting   **Imprinting** is a kind of rapid learning that occurs during a restricted time early in an animal's life without obvious reinforcement. The animal retains the memory throughout life. In the 1930s, Lorenz found that during a "sensitive period" of their lives, goslings would imprint on the first moving object they encountered. In one experiment, goslings raised in an incubator accepted Lorenz as "mother" and followed him wherever he went (**figure 36.4a**). Animals that have not been properly imprinted, however, may have trouble recognizing their own species when it comes time to mate. Biologists who raise endangered species in captivity therefore may use hand puppets resembling parent birds to feed the young (figure 36.4b).

Observational Learning   Some animals can learn by **observational learning,** in which they watch what others do and then imitate the behavior. Observational learning saves the time, energy, and risk involved in trial-and-error learning. Sometimes a learned behavior pattern spreads through a group and persists as a tradition, as when one young female Japanese snow monkey

a.

b.

Figure 36.4 Imprinting. (a) Konrad Lorenz and his flock of greylag geese. (b) A hand puppet modeled on an adult condor is used to feed an endangered condor chick.

(a): ©Nina Leen//Time Life Pictures/Getty Images; (b): ©Corbis

began washing sweet potatoes in the ocean. Although older monkeys were slow to adopt the washing behavior, the younger animals apparently learned by observing their companions. Over several years, the practice spread throughout the entire group.

Animal Cognition   Some animals, especially many mammals and birds, have cognitive abilities that extend to reasoning, problem solving, tool use, and symbolic communication. Much of the research in this area suggests that animals use a combination of operant conditioning, observational learning, and insight to develop these abilities.

   Although tool use was once thought to be a hallmark of humans, nonhuman animals also learn to use objects as aids in obtaining food. Tool-using animals include woodpecker finches and crows that use sticks to probe for insects, sea otters that use rocks to open shellfish, and chimpanzees that modify objects to collect termites. In lab experiments, chimpanzees even figured out how to stack boxes to reach bananas suspended high above.

   A few types of animals can learn to use words or symbols to communicate. The great apes cannot make the sounds of human speech, but they can communicate using symbols. One of the early students was a chimp named Washoe that learned about 250 words in American Sign Language. Washoe could string words together into phrases and even invented new combinations of signs. Other chimps that were trained on computer keyboards learned to communicate with one another about such matters as the location of hidden food.

The use of words is not limited to primates; an African grey parrot named Alex learned to communicate by speech. Alex had an English vocabulary of more than 150 words, could count, and could answer questions about an object's shape, color, and material (**figure 36.5**). It's not clear that animals actually use such abilities in the wild, however, and researchers disagree about whether nonhuman animals can learn true language.

C. Genes and Environment Interact to Determine Behavior

Biologists once placed innate and learned behavior patterns at opposite ends of a spectrum (formerly called "nature versus nurture"). Genes were assumed to control innate behavior, whereas experience in the environment governed learning. We now know that reality is considerably more complicated: Innate behavior patterns do not occur without interaction with the environment, and learning has a strong genetic component. ⓘ *is there a gay gene?* section 7.6

Animal behavior therefore does not follow a strict "innate or learned" dichotomy. Every animal begins life with a range of possible behavior patterns, some of which have more genetic input than others. The social, biological, and physical contexts, plus the experiences that an animal acquires as it matures, help determine how it actually behaves at any time.

The interplay of genes and environment is well illustrated by studies of the development of bird song. White-crowned sparrows are widespread across North America and have been a favorite subject of animal behaviorists. Under natural conditions, a young male hears the songs of his father and other neighbors during his first year. A species-typical song is crucial to a male's ability to defend a territory and attract a mate. During the first few months of life, the young male instinctively begins to sing. At first, he produces a "subsong" consisting of variable notes that sound little like the typical full song. During the year, he adds elements to the song that are more like those used by adults, so

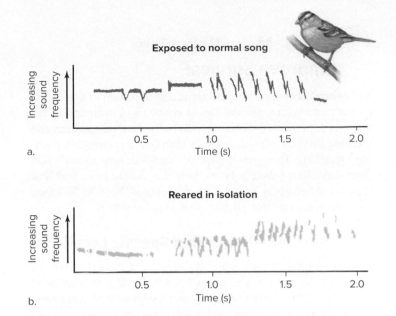

Exposed to normal song

a.

Reared in isolation

b.

Figure 36.6 **Song Learning.** (a) A male white-crowned sparrow that hears a normal song while young can perform the song correctly. (b) A bird reared in isolation cannot sing normally.

that by the first breeding season, his song has crystallized to sound much like his father's (**figure 36.6a**).

Unlike a wild bird, a white-crowned sparrow reared in isolation in the laboratory develops an abnormal song (figure 36.6b). If it hears taped or live adult male songs between 10 and 50 days after hatching, however, it develops a normal song at breeding time, demonstrating a sensitive period for song learning. When hearing the song of a different species instead, the young birds end up no better than those that heard nothing. Therefore, some sort of genetic template guides the bird to learn the correct song, but only if exposed to it at a certain time in development. Moreover, birds that were deafened after hearing normal song during the sensitive period sang abnormally as adults, indicating that they need to hear themselves sing to produce normal song.

Research on another bird species, the zebra finch, has revealed some of the neural basis of song learning. In these birds, exposure to song activates a gene called *zenk* in the brain, but *only* if it is zebra finch song. Several areas in the brain serve as song centers. One such region adds new neurons as a song is being learned. Thus, as we discover more about how a behavior pattern actually emerges, the traditional dichotomy of "innate versus learned" seems like a gross oversimplification.

Figure 36.5 **Talkative Bird.** Scientist Irene Pepperberg used Alex, an African grey parrot, in studies of cognition.
©Rick Friedman/Corbis

36.2 MASTERING CONCEPTS

1. What is the traditional distinction between innate and learned behaviors?
2. Describe three examples of innate animal behaviors.
3. Name four types of learning.
4. What are some examples of animal cognition?
5. How does song learning in birds illustrate how genes and experience interact to determine an animal's behavior?

36.3 Many Behaviors Improve Survival

Behavioral ecology is a relatively recent approach to animal behavior that explores the ecological context and evolution of behavioral patterns. Most behavior can be linked to adaptations that increase survival, reproduction, or both (see this chapter's Burning Question). This section explores survival: how animals find their way around their habitats, how they locate food, and how they avoid becoming food for another animal. Section 36.4 then takes up behaviors that promote reproductive success.

A. Some Animals Can Find Specific Locations

In many animal species, members of one sex disperse from the place of birth and move to a new habitat. Once there, they need to be able to find their way to resources such as food and mates. Many animals also return regularly to a home refuge.

Consider the desert ant, an insect that feeds on dead arthropods in the barren salt pans of the Sahara Desert in Tunisia. For their size, these ants travel some of the longest distances of any animal. They follow a meandering path as they search hundreds of meters from their nest, but once they find a prey item, they take a direct route back home (**figure 36.7**). Likewise, many species of birds migrate thousands of miles each year, returning to the same nesting location where they bred the previous year.

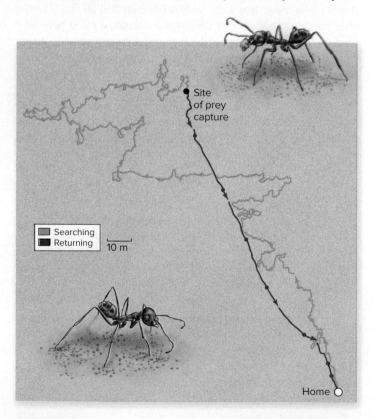

Figure 36.7 Shortcut. A foraging desert ant takes a meandering path as it leaves its nest in search of food but takes a much more direct route upon returning home.

How do animals perform such feats? Researchers have identified several types of cues.

Sun, Star, or Magnetic Orientation   Some animals can orient themselves by maintaining a constant angle to a celestial object, such as the sun. This strategy may also require an internal clock if the object's angle changes during the day.

Night migrants use the moon or the star patterns to maintain their heading during their nocturnal flights. Indigo buntings judge their location by referring to the position of the north star, which always appears stationary in the sky. A young bunting does not develop this ability, however, unless it is exposed to the northern sky before beginning its migratory flight in autumn.

Bats, salmon, and homing pigeons may use magnetic fields to orient themselves, but this sense is difficult to isolate and study experimentally. African mole rats, however, provide a well-documented example. These rodents live in underground colonies and are nearly blind. They dig the longest tunnels of any mammal, more than 200 meters, and they typically orient their tunnels in a southerly direction. Researchers confirmed that the mole rats detect magnetic cues by allowing the animals to build nests in a light-proof arena surrounded by electric coils that shifted the magnetic field. The mole rats altered the orientation of their nests in proportion to the shift in the magnetic field (**figure 36.8**). ⓘ *magnetic bacteria,* section 3.7

Path Integration   Some animals can estimate their present position by calculating how fast and in what direction they have traveled from their previous position. The desert ant in figure 36.7 keeps track of its turns, accelerations, decelerations, and number of steps taken in each direction. The ant integrates this information to monitor its whereabouts. Once the animal is near its home, it uses visual and olfactory landmarks to find the exact nest location.

Piloting   In piloting, an animal uses distant objects as landmarks when finding its way, just as you might use a distant mountain as a landmark when on a hike. Landmarks are also useful for locating objects in familiar areas close to home, as when a squirrel retrieves a food cache. Ethologist Niko Tinbergen showed that beewolf wasps use landmarks to find their nest when returning from a foraging trip. Homing pigeons also use landmarks when in the vicinity of their loft. Piloting requires the animal to know enough about its landscape to move toward familiar scenes; this strategy therefore does not work in unfamiliar terrain.

True Navigation   Navigation requires that an animal in unfamiliar terrain can understand its current position. Imagine that researchers displace an animal from its home and prevent it from collecting information about its location along the way. If the animal uses true navigation, it will return home, even without relying on any stimulus from the destination. The animal must therefore use other information to know where it is and how to get where it wants to go. Only a few animal species, including the homing pigeon, sea turtle, and spiny lobster, are known to use true navigation; researchers debate how these animals do it.

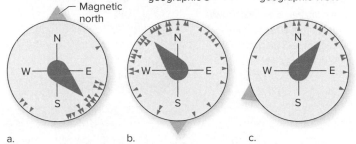

| Positions of the nests in the local magnetic field | Positions of the nests with magnetic north turned to geographic S | Positions of the nests with magnetic north turned to geographic WSW |

a.   b.   c.

Figure 36.8 Magnetic Field Detection. African mole rats use magnetic fields to orient their nests. In each of these diagrams, the green triangle indicates magnetic north, blue triangles represent nests, and the blue "pointer" shows the average nest position. (a) Local (unmanipulated) magnetic field. (b) Magnetic field reversed. (c) Magnetic north shifted toward the southwest.

Photo: ©Tom McHugh/Science Source

Figure 36.9 Tool User. A chimpanzee uses a rock to crack palm nuts.
©Dr Clive Bromhall/Oxford Scientific/Getty Images

B. Animals Balance the Energy Content and Costs of Acquiring Food

Animals spend much of their time **foraging,** or searching for and collecting food. Their foraging techniques vary greatly. Some animals, such as the desert ant, wander in search of dead insects. Spiders spin webs that ensnare prey. Packs of African wild dogs work cooperatively to hunt large, dangerous prey such as zebras. Some species of fish have modified fin spines, which they use to entice prey; the angler fish attracts passing animals by waving a lure near its mouth. And as we have already seen, a variety of bird and mammal species use tools to collect food (**figure 36.9**).

Burning Question

Do lemmings really commit mass suicide?

A vintage Disney documentary, *White Wilderness,* purports to show the "mass suicide" behavior that makes lemmings famous. Masses of lemmings scurry toward the sea, while the narrator describes the scene. "They've become victims of an obsession, a one-track thought: Move on, move on, keep moving on," he says. Then, as the rodents approach the cliff's edge, the narrator intones, "This is the last chance to turn back. Yet over they go, casting themselves bodily into space." Dozens of lemmings hurtle over the cliff, splash into the water, swim a short distance, and drown.

This evocative imagery solidified the popular notion that lemmings commit mass suicide when their populations grow too large. In reality, however, the mass suicide of lemmings is a myth. These rodents have populations that fluctuate wildly, peaking every 3 to 4 years. As the population reaches its height, lemmings begin to disperse to new habitats. Sometimes, an obstacle such as a lake or river blocks their path. Lemmings can swim, but if the distance is too great, they will drown before reaching the shore.

From an evolutionary viewpoint, mass suicide makes no sense. How can a population inherit the desire to kill itself? Even if such a trait were to arise in some lemmings, the ones lacking the self-destructive desire would presumably have the greatest reproductive success. Over many generations, a genetic tendency toward mass suicide should gradually disappear.

And what of the *White Wilderness* footage? Researchers who investigated the history of the film discovered that the entire lemming sequence was faked. The filmmakers purchased the lemmings from schoolchildren in Manitoba and transported the animals to landlocked Alberta, Canada. The crew then forced the lemmings into the water to create footage of the supposed mass suicide.

Submit your burning question to
Marielle.Hoefnagels@mheducation.com

©Jim Cartier/Science Source

For many animals, learning plays an important role in finding food. They may form a **search image,** a collection of cues that enable a predator to detect inconspicuous prey. In laboratory studies, blue jays were trained to peck a target for a reward when shown a slide of a moth hidden on a tree trunk; they received no reward if they pecked when no moth was present. The blue jays greatly improved their performance on this task over time, but not if more than one species of moth was included in the trials. This result is consistent with the formation of a search image.

Food contains the energy that an animal needs to live, but finding, acquiring, and handling food costs energy. Over the long term, an animal cannot spend more energy collecting food than it gains by consuming it. Behavioral ecologists have developed models to predict which foraging decisions maximize an animal's fitness.

Optimal foraging theory predicts that an animal's food-finding strategy should maximize the amount of energy (measured in calories) collected per unit time. For example, if given the choice between a plate of celery and an equal-sized plate of chocolate cake, a person seeking to maximize energy gain would choose the cake. But if the cake were broken into tiny crumbs and scattered over a wide area, the additional foraging costs for the cake might make celery the better deal.

In predicting an animal's optimal foraging strategy, it is important to keep in mind that each animal has a limited ability to remember a previous visit to a site or how profitable it was. A bumblebee, for example, moves from flower to flower as it forages for nectar and pollen. How does it decide which flower to visit next? Models that account for the bee's limited brain power predict that the insect should follow a simple rule: visit the closest flower other than the one the bee last visited. That is pretty much what bees actually do.

Once an animal secures its food, it may incur considerable handling costs before it can actually eat. Consider, for example, the northwestern crows that search along the shore at low tide for snails called whelks. When they spot a whelk, they pick it up, fly over a rocky area, swoop upward, and drop the snail on the rocks below. If the shell breaks open, the crow eats the contents. If not, the bird repeats the process. To forage optimally, the crows should minimize the total amount of energy they spend dropping each whelk.

In an effort to find out how close these crows get to their optimal behavior, researchers dropped whelks of different sizes from platforms of known height. They found that smaller whelks needed to be dropped from much greater heights than large whelks. So it was no surprise that crows prefer larger whelks, which also contain more food than smaller ones. But how high should the crow fly to get the most food for the least effort? The platform tests revealed that drops from 5 meters minimized the total effort (figure 36.10). Drops from higher than 5 m weren't much more likely to break the shell but were more costly to the crow.

When the researchers analyzed their observations of the crows, they found that the average height from which the birds

Height of Drop (m)	Average Number of Drops Required to Break Shell	Total Flight Height (Number of Drops × Height per Drop)
2	55	110
3	13	39
5	6	30
7	5	35
15	4	60

Figure 36.10 Optimal Foraging in Crows. Researchers repeatedly dropped whelks from a platform until the shell broke. The height that required the least total effort was 5 meters. Crows actually dropped snails from an average of 5.23 m, close to the most efficient height.

actually dropped whelks was 5.23 m, close to the 5-m height predicted from the platform tests. Thus, in this instance, crows seemed to be foraging optimally; they were maximizing food intake while minimizing costs.

C. Avoiding Predation Is Another Key to Survival

Animals have many morphological and behavioral adaptations that reduce their chances of becoming someone else's meal (figure 36.11). Body markings often come into play. Many animals have camouflage, which helps animals become less visible by matching their background. Others have shapes and markings that allow them to masquerade as something inedible, such as a twig or leaf. Still others frighten would-be predators by looking and acting like predators themselves. At least one type of caterpillar, for example, resembles a snake head. Particularly common are fake eyespots, such as those on the underwings of moths; when displayed suddenly, the eyespots trigger innate fright responses in predators.

Some animals protect themselves with poisons or other weapons, such as the quills of a porcupine or the noxious spray of a skunk. Instead of hiding, many such species have distinctive warning colors or patterns that make them conspicuous to potential predators. Examples include the black-and-white stripes on a skunk, the orange-and-black patterns on the wings of a monarch butterfly, and the brightly colored body of the red velvet ant. The latter is actually a type of wasp, and its nickname ("cow killer") suggests that its sting is particularly ferocious. ⓘ *warning coloration,* section 38.1C

In addition to weapons, behaviors can also deter predators. Individuals may use **distraction displays** that divert a predator's

a.

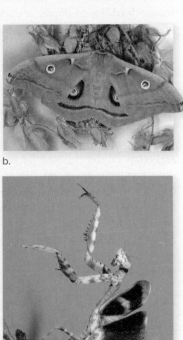

b.

c.

d.

e.

Figure 36.11 Insect Defenses. (a) The wings of the brimstone butterfly look like leaves. (b) The eyespots on this moth startle a would-be predator. (c) The rear end of a swallowtail caterpillar looks like a snake's head. (d) This defensive display makes a praying mantis look larger than it really is. (e) This insect, commonly called a "red velvet ant," is actually a wasp, and it can deliver a painful sting.

(a): ©Stefano Stefani/Getty Images RF; (b): ©Steven P. Lynch/McGraw-Hill Education; (c): ©Anna Pomaville/iStock/Getty Images RF; (d): ©WILDLIFE GmbH/Alamy; (e): ©James H. Robinson/Science Source

Figure 36.12
Meerkat Sentries.
These meerkats are on the lookout for predators. When a sentry barks the alarm call, the entire community dashes underground to safety.
©Roy Toft/National Geographic/Getty Images

attention away from an occupied nest or den. One example is the broken-wing display of killdeers, which lures the predator away from young.

Gazelles have a conspicuous jumping display, called stotting, that deters predators (see the chapter opening photo). Stotting seems to signal "I see you, I'm in good shape, and I can outrun you." Other functions for stotting are also possible, however, such as distracting the predator from a fawn or simply attempting to get a better view of the predator in tall vegetation.

Group living can also offer a defense. An advantage of living in groups is the "many eyes" effect; that is, many individuals are more likely to spot a predator than one individual searching alone (figure 36.12). Groups can also form a protective circle around vulnerable young, or they can "mob" predators.

Perhaps the simplest advantage of being in a group is the dilution effect. By getting into a large group, one's chances of being picked off by a predator are reduced. For example, many species of fish quickly form schools in the presence of a predator; the greater the threat, the tighter the school. Such groups have been called **selfish herds:** Individuals behaving selfishly try to position themselves so that as many of their companions as possible are between them and the predator. In a school of fish, the result is an ever-tightening group.

36.3 MASTERING CONCEPTS

1. What are four mechanisms by which animals find their way to a specific location?
2. Describe how optimal foraging theory explains an animal's foraging decisions.
3. What are some behaviors that help animals avoid predation?

36.4 Many Behaviors Promote Reproductive Success

Section 36.3 demonstrated that behavioral adaptations are important for an animal's survival. This section turns to reproductive behaviors. Most reproductive activities fall into the category of **social behaviors,** which are interactions among members of the same species. Besides courtship and mating, other social behaviors include interactions among animals that live in groups (the subject of section 36.5). A field of study called **sociobiology** attempts to understand social behavior in the context of an animal's fitness.

A. Courtship Sets the Stage for Mating

Many species have elaborate courtship behaviors, often involving a long chain of actions leading to copulation; recall the fruit flies illustrated in figure 36.3. Birds are especially well known for their courtship displays and dances (figure 36.13). If you have seen the movie *March of the Penguins,* you might recall the calls, postures, promenades, and bows that occur when pair bonds are formed or when mates reunite after a long absence.

Courtship rituals have at least three functions. First, these displays are generally species-specific, so they help to ensure that an animal is directing its mating attempts at the correct species. Matings between closely related species often yield no offspring or offspring of reduced fitness. ⓘ *reproductive barriers,* section 14.2

A second function of courtship is to coordinate the physiological and hormonal states of the participants. One of the best-studied examples is the ring dove. If nesting materials are available, the male courts the female using a "bow-coo" display. This behavior stimulates the female's pituitary gland to release a hormone that stimulates cells in her ovaries to secrete estrogen. Within a day or two, both birds begin nest construction. The doves soon copulate. The presence of a nest stimulates progesterone synthesis, followed by egg laying. Both sexes incubate the eggs, a process that stimulates the production of yet another hormone, which stimulates the regurgitation of food for the hatched young. ⓘ *reproductive hormones,* section 28.5

In many mammals, courtship is necessary to trigger ovulation (egg production) in the female. If you have ever observed cats mating, you may have wondered about all the yowling. The male not-so-gently bites the back of the female's neck prior to copulation. The bite, together with withdrawal of the male's barbed penis, leads to a surge of hormones that trigger ovulation.

A third function of courtship is to provide each participant with information about the suitability of the other as a mate. A female bird, for example, may evaluate the courtship rituals of many males before selecting a mate.

B. Sexual Selection Leads to Differences Between the Sexes

Females of many species seem especially picky when it comes to mate choice. Recall from chapter 35 that females produce relatively small numbers of large gametes. A female's egg cells contain yolk and other substances used for the growing embryo. Males, on the other hand, produce many tiny sperm that each contain little more than a haploid set of chromosomes. This disparity in gamete size is thought to set the stage for the tendency of females to be the more choosy sex, given their large investment in a relatively small number of eggs. Males, with their small investment in a large number of relatively cheap sperm, tend to be less discriminating and to compete more strongly than females for access to mates.

Charles Darwin noticed this tendency, and he argued that traits that improved an individual's chances of getting a mate could be selected for independently from selection for survival traits. He coined the term **sexual selection** for the form of natural selection that results from variation in the ability to obtain mates (see section 12.6). Sexual selection can lead to **sexual dimorphism,** a situation in which the two sexes look very different. Many sexually selected traits seem to be of no value to survival; they may even be harmful.

Some sexually selected traits result from intrasexual selection, which occurs when males compete with one another for access to females. Northern elephant seals offer a spectacular example (figure 36.14). Males are several times larger than females, have an elongated snout that they use to make loud roaring sounds, and have tough skin around their necks that protects them from the blows of other males. The strongest males win access to the large numbers of females hauled out on the beach. As a result, reproductive success varies tremendously. Fewer than one third of males get to copulate at all in a given season; those that do mate may sire more than 50 pups. Male–male competition therefore maintains the sexual dimorphism among elephant seals.

Figure 36.13 Courtship Behaviors.
(a) Atlantic puffins rub their beaks together as a prelude to mating. (b) "Sky pointing" is one step in the elaborate courtship dance of Laysan albatrosses.

(a): ©Mark Smith/Flickr Open/Getty Images RF; (b): ©Paul & Paveena Mckenzie/Oxford Scientific/Getty Images

a.

b.

Figure 36.14 Northern Elephant Seals. These two large males are fighting to establish dominance. The winner mates with the females in the territory.

©Roberta Olenick/All Canada Photos/Corbis

Antlers are also subject to intrasexual selection. Mammals of the deer family grow antlers before each breeding season. These structures, which typically occur only on males, are made entirely of bone. They are therefore costly to produce. Males use their antlers in fights that establish dominance in bachelor herds. As with the elephant seals, winners gain access to females for breeding; the antlers are discarded at the end of the breeding season.

Intersexual selection results in traits that are attractive to members of the opposite sex. One spectacular example is the peacock (**figure 36.15**). In contrast to the normal-sized, drab tail feathers of the female, the male has a "train" consisting of hugely elongated feathers with iridescent spots. The male spreads his

Figure 36.15 Spectacular Feathers. The remarkable train of the peacock is a sexually selected trait.

(tail fanned): ©Corbis/SuperStock RF; (tail down): ©Sara Venter/Alamy RF

train in the presence of a female. Feathers with more spots attract more females, possibly explaining this example of "runaway selection."

Sexually selected traits can be very costly, to the point where they compromise survival. Why should a female prefer a mate that may look (or sound) attractive if he is actually less likely to survive? Part of the answer is that some traits are so costly that they may be reliable indicators of fitness. For instance, the brightness of a rooster's comb is an honest indicator of the male's parasite load: Males with the brightest combs have the fewest parasites, and females prefer them as mates. It is also possible that an extravagant trait such as a peacock's tail serves as a handicap: A male that can afford such a tail must be fit indeed!

C. Animals Differ in Mating Systems and Degrees of Parental Care

Animals display almost every imaginable mating system. Some species, such as corals, simply shed eggs and sperm into the environment. At the opposite extreme are some birds that form pairs for life.

Species that are strongly sexually dimorphic tend to have polygamous mating systems. **Polygamy** is a general term that encompasses all forms of multiple mating, in which either males or females have multiple sexual partners. Polygyny is a form of polygamy in which one male has exclusive sexual access to more than one female. In polygynous animals, the male is not likely to help care for the young. The dominant male elephant seal, for example, mates with many females and provides no care for his offspring. Males are even known to trample pups from the previous year's matings as they attempt to copulate with females.

On the other hand, species in which males and females resemble each other tend to have monogamous mating systems. **Monogamy** means that both male and female have one sexual partner. In monogamous species, both parents typically provide care for offspring. Most bird species are monogamous, with both parents needed to feed and defend the vulnerable nestlings. Some small mammals are also monogamous, as described in section 28.6.

In birds and mammals, males rarely care for their young alone. Indeed, extensive paternal care is relatively rare in mammals; after all, males cannot produce milk. Nevertheless, males can protect the group and may even supply food to the mother and young. In contrast, among many species of amphibians and fish, the male parent is the one that typically tends the nest and eggs. (A male sea horse even becomes pregnant, as described in section 35.5's Burning Question.)

Why should birds and mammals behave differently than fishes and amphibians? The answer likely involves confidence of paternity. Birds and mammals have internal fertilization, which means that some time elapses between copulation and the arrival of fertilized eggs or newborns. The female can take additional mates during that interval, so each male's confidence of paternity is relatively low. But animals with internal fertilization may increase their confidence of paternity in many ways. Some males guard the female before and after copulation (see section 36.6). Others insert a plug after insemination, reducing the chance that another male will follow. Some spiders enhance their probability of paternity by offering their own bodies as a "nuptial gift" (see section 35.7). And some dragonflies even have a penis with a special "sperm scoop" that can remove a previous male's sperm!

Most species of fish and amphibians, on the other hand, have external fertilization. In jawfish, for instance, a male fertilizes a female's eggs and immediately scoops them into his mouth, where they remain until they hatch (figure 36.16). Because the male's confidence of paternity is high, he enhances his reproductive success by investing his time and energy protecting the egg mass.

Figure 36.16 Quite a Mouthful. A male yellowhead jawfish protects the eggs he has fertilized by keeping them in his mouth until they hatch.
©UIG via Getty Images

D. Human Reproductive Choices May Reflect Natural Selection

How do the reproductive behaviors of nonhuman animals compare to those of humans? The range of human sexual behavior obviously varies greatly from person to person, but it is possible to use what we know about other animals to make some general predictions. Evolutionary psychologists and other researchers can then test these predictions by gathering evidence about human behavior.

First, consider the disparity in reproductive investment between human females and males. As in other mammals, human females make a larger investment in gamete production than do males. A pregnant woman also undergoes 9 months of gestation, followed by milk production for up to several years. We might therefore expect females to be the choosier sex, while men might be less discriminating when it comes to choice in a mate.

What do the data reveal? One survey of college undergraduates asked how many sexual partners they would like to have across different time spans. The result: Men wanted many more partners than did women (figure 36.17a). These subjects were also asked how likely they would be to have sexual

intercourse with a person after having known him or her for periods ranging from 1 hour to 5 years. The sexes were similar at 5 years, but men were much more inclined than women to have sex at the shorter time spans (figure 36.17b). Other studies suggest that men most highly value physical attractiveness in a potential mate. Women most highly value a male's resources (wealth), although his looks are also important.

These types of surveys are subject to the criticism that the preferences simply reflect cultural norms and do not necessarily stem from underlying selection pressures. Nevertheless, the results do seem to hold up in a wide range of cultures. In addition, other types of studies quantify human behavior and fitness. In a number of cultures, both a woman's and a man's reproductive success are positively correlated with the husband's socioeconomic status. This finding is consistent with the value women place on wealth in choosing a mate.

A second prediction arises from the degree of sexual dimorphism in humans. Although sexual dimorphism in our species is nothing like that of elephant seals, it is still substantial. Males of most races are taller, heavier, and more muscular than females,

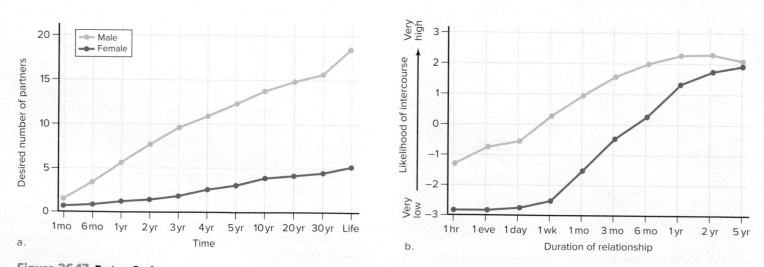

Figure 36.17 Partner Preferences. (a) On average, men would prefer to have more sexual partners over a lifetime than would females. (b) On average, males are more likely than females to agree to sexual intercourse after having known a potential partner for a shorter time.

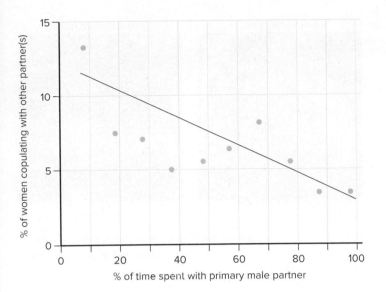

Figure 36.18 Mate Guarding in Humans? In one study, the more time a female spent with her primary male partner, the less likely she was to have copulated with other males.

suggesting the importance of male–male competition in the past. Given the association between mating system and degree of dimorphism, we might expect a certain degree of polygyny in humans. In fact, anthropologists have reported that some 85% of known societies worldwide have either occasional or frequent polygyny, typically where men of high status have more than one wife.

Third, like all other mammals, humans have internal fertilization. Therefore the female, but not the male, has high confidence of parentage. Researchers working in the United Kingdom have found that the percentage of women copulating with males other than their primary partner increased the more time the primary partner spent away from the woman, thus lowering his confidence of paternity of any resulting offspring (figure 36.18). Thus, we might expect mate guarding or other tactics so that males can increase their certainty of paternity prior to investing in offspring. Indeed, evidence of mate guarding by men comes in many forms. In medieval times, chastity belts prevented women from having intercourse while their husbands were away. And in many cultures, adultery by a woman is considered a punishable offense against her husband.

36.4 MASTERING CONCEPTS

1. List three functions of animal courtship rituals.
2. Describe examples of intrasexual selection and of intersexual selection.
3. What is the relationship between sexual dimorphism and mating systems?
4. Explain how confidence of paternity influences the degree of parental care by a male.
5. How do observations of reproductive behavior in nonhuman animals enable researchers to make predictions about humans?

36.5 Social Behaviors Often Occur in Groups

Lots of animals live in flocks, herds, prides, schools, or other types of groups. But not all groups are considered societies. A society is a group of individuals of the same species that is organized in a cooperative manner, extending beyond sexual and parental behavior. This section describes some of the social behaviors that occur in groups.

A. Group Living Has Costs and Benefits

Animals profit in several ways by living in groups. One advantage is in finding food. For example, groups can share information about the location of food. One striking example is the dance language of honeybees, in which foragers return to the hive and communicate the location of food sources (figure 36.19). In addition, groups can cooperate in the capture of prey not available to a lone individual.

Another advantage of group living is protection. Animals can huddle together to conserve body heat, providing protection from cold, wind, or other environmental conditions. (Anonymous

Figure 36.19 Waggle Dance. Bees observing a waggle dance interpret the dancer's movements relative to the angle of the sun. If the food is toward the sun (1), the straight run of the dance points directly up. If the food source lies at a 45-degree angle relative to the sun (2), the straight run of the dance is 45 degrees relative to vertical. If the food source is away from the sun (3), the straight run points directly down.

aggregations, however, might not fit the biological definition of a society.) Similarly, being part of a large group helps protect against predators, as with schools of fish. Groups of individuals are also more likely to spot predators, allowing others to spend more time foraging or in other activities.

Living in groups, however, does have costs. Predators may be more likely to detect groups of animals than dispersed individuals. Also, members of the same species are likely to compete for the same resources, leading to an increase in aggressive behavior. This aggression may interfere with the rearing of young, as when a new cadre of males takes over a lion pride and kills the cubs sired by the previous males.

Additionally, group living may promote the spread of diseases and parasites. For example, in colonies of cliff swallows, the number of blood-sucking bugs on each nestling increases as a function of colony size. Infested chicks grow much more slowly than do uninfested chicks.

B. Dominance Hierarchies and Territoriality Reduce Competition

One way group members reduce the cost of competing with one another is by establishing dominance hierarchies. A **dominance hierarchy** is a social order with dominant and submissive members. Individuals learn their place in the "pecking order" from previous interactions, reducing the time, energy, and risk of fighting. In general, higher-ranking individuals have higher fitness than do lower-ranking individuals. Those at the bottom of the hierarchy presumably fare better than they would if they were on their own, and their status may improve with time.

For example, rhesus monkeys live in social groups of 20 to 50 individuals. Adult females have a linear dominance order that is family-based: Offspring rank just below their mother in the hierarchy and above the next adult female and her family. Once established, the hierarchy is maintained by threat and submission displays that rarely lead to costly fights (figure 36.20). Males have a separate linear order that is based on seniority in the group. Males leave the natal group at puberty and eventually join another group, where they start out at the bottom and work their way up as more senior males die or leave.

Another way to reduce competition is to occupy a **territory,** which is space that an animal defends against intruders. The key to territory formation is economic defendability, such that the benefits (exclusive access to resources) outweigh the costs (energy expenditure, risk of injury, and loss of feeding or mating opportunities). Note that territoriality is not exclusively associated with group living; solitary animals such as bobcats establish territories as well.

Olfactory or vocal signals often establish the boundaries of a territory. For example, hyena clans mark their borders with latrine areas, whereas male gibbons signal theirs with loud whoop-gobble calls. Within breeding colonies of gulls, each pair defends a small area around the nest from marauding neighbors.

Territorial animals spend much time and energy patrolling their boundaries, especially in newly established territories

Submissive grin   Open-mouthed threat

a.   b.

Figure 36.20 Maintaining Dominance. Rhesus monkeys display (a) submission to monkeys that occupy higher positions and (b) threats to those that are lower in the hierarchy.

Photos: (a, b): Courtesy of Stephen H. Vessey

(figure 36.21). Eventually, neighbors need only occasional reminders to keep out. If the owner should leave or die, however, neighbors or immigrants will quickly move in.

C. Kin Selection and Reciprocal Altruism Explain Some Acts of Cooperation

Group living incurs both costs and benefits, and it is easy to see how social behavior might be maintained in a population if individuals in groups have higher fitness than those living alone. But social living often involves cooperation with group mates, and cooperation can be costly. Occasionally, a social behavior even seems to be **altruistic;** that is, the behavior seems to lower one animal's fitness for the good of others.

The definition of *altruism* does not include parental behavior, in which parents risk their own lives for the sake of their offspring. After all, parents increase their own fitness by helping their offspring survive. Instead, altruism *reduces* an individual's fitness. Behavioral ecologists have therefore reasoned that altruistic individuals should be selected against in favor of selfish individuals. But if that is the case, how can altruistic behaviors be maintained?

Figure 36.21 Keep Out. Two lions fight over territory in Kenya.
©imageBROKER/Alamy

Part of the answer may lie in the distinction between direct fitness and indirect fitness. An individual's own offspring account for its direct fitness. But what about other relatives that are not direct descendants? It's true that we directly pass genes on to only our own offspring, but we share genes through common descent with all our close relatives. In fact, we share half our genes with our full siblings and with our own offspring. We share one quarter of our genes with our full nieces and nephews, and one eighth of our genes with first cousins. Another way to look at it is that on average, we share as many genes with one of our own offspring as we do with two of our nieces and nephews.

An individual's **inclusive fitness** is the sum of direct and indirect fitness. Individuals, then, can be expected to behave so as to maximize their inclusive fitness. The most familiar way to maximize fitness is to help one's own offspring to survive and reproduce, but another possibility is to maximize indirect fitness.

In **kin selection,** an individual reduces its own direct fitness but assists the survival and reproduction of nondescendant relatives. For instance, Belding's ground squirrels put themselves at risk by giving alarm calls when they spot a predator such as a weasel or a coyote (figure 36.22). Females with relatives in the vicinity are most likely to call. Males, which disperse each year, are unlikely to have relatives in the group and rarely sound alarm calls.

Researchers have rigorously tested the role of kin selection in the Florida scrub jay, where some birds forgo breeding and help their parents raise siblings. Why? Part of the answer may lie in the observation that this species occupies a limited habitat in which all available territories may be filled. Stay-at-home helpers may increase their direct fitness by inheriting their natal territory, should the parents leave or die. In the meantime, they increase their indirect fitness because their parents raise more offspring than they would without help.

Not all altruistic behavior, however, occurs among close relatives. In **reciprocal altruism,** individuals might help others at a cost to themselves if it is likely that they will be paid back at a later time. For reciprocal altruism to work, the individuals must have a high likelihood of encountering one another for possible repayment. For example, vampire bats feeding at night on blood from cattle and horses return to a community roost during the day. At the roost, mothers often regurgitate blood to their offspring, but adults also regurgitate to unrelated individuals that have not fed the night before. The latter, however, only get fed if they have associated with the donor frequently in the past and are thus likely to return the favor in the future.

You may have noticed that neither kin selection nor reciprocal altruism represents "pure" altruism. In the former, the helper increases its inclusive fitness by helping relatives; in the latter, the helper gets repaid, possibly with interest and possibly when it is needed most. It should not be surprising that pure altruism is nonexistent in animals; such traits would lower fitness and be selected out of the population.

Humans may be uniquely prone to altruistic behavior. In laboratory tests, infants at 18 months of age immediately assist an unrelated adult whose hands are full and who needs help picking up a dropped object. Even at the age of 12 months, infants

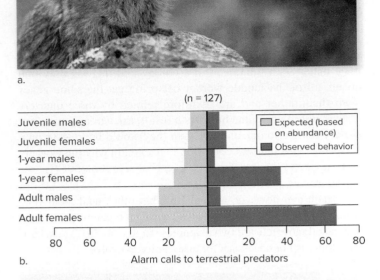

a.

b. **Alarm calls to terrestrial predators**

(n = 127)

Legend: Expected (based on abundance); Observed behavior

Figure 36.22 **Sound the Alarm.** (a) A female ground squirrel gives an alarm call. (b) The bars on the left show the distribution of alarm calls expected if ground squirrels called in proportion to their abundance in the population. Instead, the bars on the right show that adult females and young females made the majority of the calls.

(a): ©Rod Planck/Science Source

will point toward objects that an adult appears to have lost. Starting at age 3, human children become more selective in whom they assist, being more likely to help a child who has previously helped them. Because altruism appears so early in development, across many cultures, and without reward, some researchers think it is innate.

D. Eusocial Animals Have Highly Developed Societies

The epitome of organization is **eusociality,** a social structure that includes extensive division of labor, especially in reproduction. The characteristics of eusociality are cooperative care of young by nonparents; the presence of nonreproductive adults (often called sterile castes) that care for the breeders; and overlap of generations such that offspring assist parents in raising siblings.

In a honeybee colony, the queen does nothing but lay eggs. Her daughters are workers that will never reproduce; instead, they forage and tend the queen and her offspring. Males (drones) contribute little to colony life, flying off and trying to mate with queens that are forming new colonies. Rather than a group of individuals, a bee colony can be considered a single "superorganism." The sterile workers are no different from organs in the body, which help maintain life but can't exist on their own.

Eusociality is relatively common in ants, bees, and wasps, possibly because of their unusual mechanism of sex determination.

Whereas sex chromosomes such as XX and XY determine the sex of many animals, these insects use a method called haplo-diploidy. Males in this order develop from unfertilized eggs and are therefore haploid; females develop from fertilized eggs and are diploid. Since males are already haploid, they produce genetically identical sperm by mitotic cell division. If only one male mates with the queen in a colony and there is only one queen, all of the female worker offspring get the same genes from their father and are therefore related by three quarters instead of the usual one half. As a result, the workers are more closely related to one another than they would be to their own offspring. This unusual condition may explain why the workers give up reproducing for the good of the colony. ⓘ *sex chromosomes,* section 9.2

There are, however, animals that have the usual method of sex determination and have evolved eusociality anyway; termites and naked mole rats are two examples (**figure 36.23**). Thus, haplodiploidy is not necessary for eusociality to evolve.

36.5 MASTERING CONCEPTS

1. What are some benefits and costs of group living?
2. Explain the roles of dominance hierarchies and territoriality in animal societies.
3. Define *altruism,* and explain why evolution should select against it.
4. What is eusociality?

a.

b.

Figure 36.23 Eusocial Animals. (a) A queen termite is surrounded by nonreproductive workers. (b) Pups suckle from a naked mole rat queen.

INVESTIGATING LIFE

36.6 The "Cross-Dressers" of the Reef

The scene on TV is familiar: Two bighorn sheep rear up on their hind legs and bash their heads together with bone-crunching force. Only one will win the prize—the exclusive "right" to mate with a female. Throughout the animal kingdom, however, fighting is not the only form of competition that determines a male's reproductive success. If a female mates with many males in a short period, the competition shifts to a much smaller field: her reproductive tract. Only one sperm will fertilize each egg cell.

Sometimes, numbers decide the winner. Males that produce the most sperm are most likely to fertilize an egg cell, just as the owner of several raffle tickets has a better chance of winning a prize than someone who buys only one. But behavior also often plays a role. In some species, for example, a male may try to block other males from approaching a female he has already mated with; in effect, the male prevents his rivals from buying raffle tickets. Males also do the equivalent of destroying raffle tickets that others have purchased. That is, they remove other males' sperm before depositing their own.

Sperm competition has selected for unusual behavioral adaptations in a squidlike mollusk called the Australian giant cuttlefish (*Sepia apama*). During the winter mating season, these animals congregate by the hundreds of thousands on the reefs of southern Australia. ⓘ *mollusks,* section 21.5

The two cuttlefish sexes look different from each other (**figure 36.24**). A female has shorter arms than a male, and her skin has dark patches on a white background. The male's arms are longer and whiter, and he displays moving patterns of zebra-like stripes on his skin during courtship rituals.

When a female accepts a male's mating attempt, the two animals align head-to-head, and he inserts a sperm packet into a

Figure 36.24 Australian Giant Cuttlefish. In this species, males are substantially larger than females.

pouch near her mouth. He also tries to flush out the sperm packets that other males have deposited. After mating, she retreats to her den and removes eggs, one by one, from her mantle cavity. She passes the eggs through her sperm pouch to be fertilized.

With a sex ratio of at least four males to every female, the competition to father offspring is fierce. What reproductive behaviors has natural selection favored in male cuttlefish?

A study led by Marié-Jose Naud of Australia's Flinders University and Roger Hanlon of the Marine Biological Laboratory at Woods Hole, Massachusetts, showed that mate guarding is one successful strategy. That is, the largest males mate with females and guard them afterward, fighting off rivals to prevent subsequent insemination of their mates. Direct observation, followed by DNA analysis, revealed that males that guarded a female for up to 40 minutes after mating fertilized significantly more eggs than those that guarded for less than 20 minutes (**figure 36.25**).

Nevertheless, smaller males do get to mate. How do they gain access to females? A team of Australian biologists, led by Mark Norman, noticed that some males are female impersonators. A small male can convincingly disguise himself as a female by hiding the arms that reveal his sex and changing his skin color. (The ploy is so realistic that other males, including other female impersonators, often try to mate with the mimic.) Norman's team observed more than 20 examples of sexual mimicry, but they were unable to tell how often mimics slipped past a guard, approached a female, mated, and successfully fertilized eggs.

Hanlon, Naud, and their colleagues began to answer these questions by videotaping matings and analyzing DNA to determine which male fathered each female's egg (**figure 36.26**). They found that impersonators deceived a guarding male and approached a female about 30 times out of 62 attempts. Five of the 30 mimics tried to mate with the female. The guard male interrupted one attempt, and the female rejected another, but three

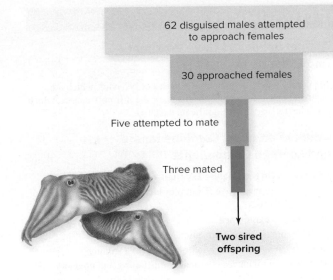

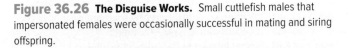

Figure 36.26 The Disguise Works. Small cuttlefish males that impersonated females were occasionally successful in mating and siring offspring.

were successful. DNA analysis showed that two of these three female impersonators fathered the first egg laid after mating. The third did not. Because females store sperm, however, the team could not rule out the possibility that sperm from the third mimic fertilized a subsequent egg.

Even within the same species, there may be more than one path to reproductive success. In polygamous animals such as the cuttlefish, natural selection favors the male with the most successful sperm—and he may not be the largest or the strongest individual. True, the brawniest males can fight off rivals and preserve their own access to a female, but the fierce competition and constant distractions often leave females with additional mating opportunities. A small male has little chance of winning a head-to-head competition with a larger rival, but he does have another tool—deception. Thanks to his changeable skin, even a small cuttlefish has a shot at winning in the great raffle of life.

Sources: Hanlon, Roger T., Marié-Jose Naud, Paul W. Shaw, and Jonathan N. Havenhand. 2005. Transient sexual mimicry leads to fertilization. *Nature,* vol. 433, page 212.

Naud, Marié-Jose, Roger T. Hanlon, Karina C. Hall, et al. 2004. Behavioural and genetic assessment of reproductive success in a spawning aggregation of the Australian giant cuttlefish, *Sepia apama. Animal Behaviour,* vol. 67, pages 1043–1050.

Norman, Mark D., Julian Finn, and Tom Tregenza. 1999. Female impersonation as an alternative reproductive strategy in giant cuttlefish. *Proceedings of the Royal Society of London B.,* vol. 266, pages 1347–1349.

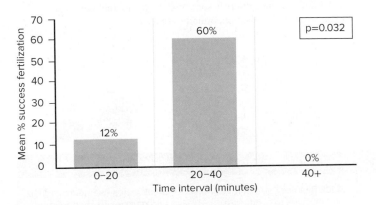

Figure 36.25 Mate Guarding Works. Researchers timed the mate-guarding activities of 22 male cuttlefish and used DNA to determine the paternity of the resulting fertilized eggs. Males that guarded their mates from 20 to 40 minutes ($n = 10$) had greater reproductive success than those that guarded for less than 20 minutes ($n = 8$) or more than 40 minutes ($n = 4$). The p-value indicates that these values are significantly different from one another (see appendix B).

36.6 MASTERING CONCEPTS

1. Why does evolution select for males that mimic females, even though the chance of success is relatively low?

2. Large male cuttlefish guard their mates. Use figure 36.26 to estimate the percent of guarding males that successfully deterred mating attempts by disguised males.

CHAPTER SUMMARY

36.1 Animal Behaviors Have Proximate and Ultimate Causes

- Researchers may study the **proximate** causes of behavior, which are anatomical and physiological mechanisms, or the **ultimate** causes, which explain a behavior's adaptive significance (figure 36.27).

36.2 Animal Behaviors Combine Innate and Learned Components

A. Innate Behaviors Do Not Require Experience

- **Innate** behaviors are instinctive. Examples include a **reflex,** a **taxis,** and a **fixed action pattern.**

B. Learning Requires Experience

- **Learned** behaviors result from an animal's experiences.
- In **habituation,** an animal learns *not* to respond to a stimulus.
- **Associative learning** includes **classical conditioning** and **operant conditioning.**
- **Imprinting** occurs only during a restricted time in an animal's life.
- In **observational learning,** an animal learns by watching what others do.
- Animal cognition reflects multiple types of learning.

C. Genes and Environment Interact to Determine Behavior

- All learned behaviors have a genetic component, and all innate behaviors are triggered by interactions with the environment.

36.3 Many Behaviors Improve Survival

- **Behavioral ecology** is the study of how behaviors contribute to an animal's fitness.

A. Some Animals Can Find Specific Locations

- Animals find locations using the sun, stars, magnetic fields, path integration, piloting, or navigation.

B. Animals Balance the Energy Content and Costs of Acquiring Food

- **Foraging** behaviors are directed at finding and handling food. An animal that hunts visually may develop a **search image** of its prey.
- According to **optimal foraging theory,** an animal's behavior reflects a balance between calories gained from food and energy spent in foraging.

C. Avoiding Predation Is Another Key to Survival

- Adaptations that help animals avoid predation include deceptive markings, noxious weapons, **distraction displays,** and living in groups.
- Each individual in a **selfish herd** looks out for its own self-interest.

36.4 Many Behaviors Promote Reproductive Success

- **Social behaviors** occur between members of the same species. **Sociobiology** studies the evolutionary significance of social behaviors.

A. Courtship Sets the Stage for Mating

- Courtship behaviors function in species recognition, hormone release, and partner evaluation.

B. Sexual Selection Leads to Differences Between the Sexes

- In animals that are **sexually dimorphic,** males look different from females. Sexual dimorphism is a consequence of **sexual selection.**

C. Animals Differ in Mating Systems and Degrees of Parental Care

- Sexually dimorphic species are often **polygamous;** species in which male and female look alike are often **monogamous.**
- In males, parental care is correlated with confidence of paternity.

D. Human Reproductive Choices May Reflect Natural Selection

- Observations of animal reproductive behaviors have enabled researchers to form and test hypotheses about human sexual behaviors.

36.5 Social Behaviors Often Occur in Groups

A. Group Living Has Costs and Benefits

- An animal that lives in a group may be more conspicuous to predators but may also receive information about food and threats to the community.

B. Dominance Hierarchies and Territoriality Reduce Competition

- **Dominance hierarchies** reinforce social status within a group, whereas **territorial** behaviors help an animal retain access to resources.

C. Kin Selection and Reciprocal Altruism Explain Some Acts of Cooperation

- **Altruism** occurs if an animal reduces its own fitness while improving the fitness of an unrelated individual.
- An animal's **inclusive fitness** includes not only its offspring but also its close relatives. Behavioral ecologists propose that **kin selection** explains why animals are most likely to sacrifice themselves for close relatives.
- In **reciprocal altruism,** an animal that shares resources with an unrelated individual receives future benefits.

D. Eusocial Animals Have Highly Developed Societies

- **Eusocial** animals have the most complex societies, often including workers that support the reproduction of only a few members of the society.

36.6 Investigating Life: The "Cross-Dressers" of the Reef

- In Australian giant cuttlefish, some males disguise themselves as females, increasing their opportunities to mate.

Figure 36.27 Proximate and Ultimate Causes of Selected Behaviors.

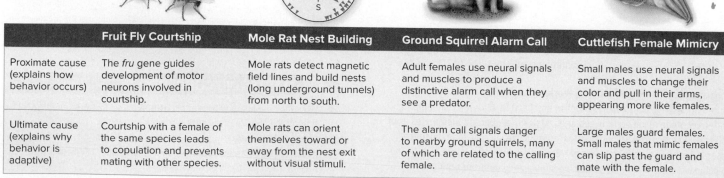

	Fruit Fly Courtship	Mole Rat Nest Building	Ground Squirrel Alarm Call	Cuttlefish Female Mimicry
Proximate cause (explains how behavior occurs)	The *fru* gene guides development of motor neurons involved in courtship.	Mole rats detect magnetic field lines and build nests (long underground tunnels) from north to south.	Adult females use neural signals and muscles to produce a distinctive alarm call when they see a predator.	Small males use neural signals and muscles to change their color and pull in their arms, appearing more like females.
Ultimate cause (explains why behavior is adaptive)	Courtship with a female of the same species leads to copulation and prevents mating with other species.	Mole rats can orient themselves toward or away from the nest exit without visual stimuli.	The alarm call signals danger to nearby ground squirrels, many of which are related to the calling female.	Large males guard females. Small males that mimic females can slip past the guard and mate with the female.

MULTIPLE CHOICE QUESTIONS

1. Which of the following questions is most closely associated with an ultimate cause of a behavior?
   a. How does a squid alter its color to blend in with its environment?
   b. How does a hummingbird's body shape influence its ability to hover?
   c. How do alarm calls affect prairie dog fitness?
   d. Which cues do monarch butterflies use during migration?

2. Although the trains that rumble through your town once bothered you, now you barely even hear them. This example illustrates
   a. habituation.
   b. associative learning.
   c. a reflex.
   d. imprinting.

3. To test optimal foraging theory, you assemble 40 plates of food, scatter them randomly in a field, bury them under a thin layer of dirt, and leave them overnight. Each plate contains one of the following: peanuts in shells, peanuts without shells, carrots, and celery. Based on optimal foraging theory, which food is most likely to be eaten?
   a. Peanuts in shells
   b. Peanuts without shells
   c. Carrots
   d. Celery

4. Male stag beetles have mouthparts that are greatly enlarged compared to those of the females. This difference is likely to result from
   a. distraction displays.
   b. sexual selection.
   c. monogamy.
   d. external fertilization.

5. In most mammal species, a male is _____ to provide parental care because his confidence of paternity is _____.
   a. likely; high
   b. unlikely; high
   c. likely; low
   d. unlikely; low

6. If it were to occur in nature, which of the following would be an example of true altruism—that is, a behavior that reduces an animal's fitness for the good of others?
   a. An adult orangutan protecting an unrelated infant orangutan from human poachers
   b. A ground squirrel putting itself in danger to protect several cousins and siblings
   c. A bird giving food to an unrelated chick when the favor is likely to be returned later
   d. None of these are examples of true altruism.

Answers to these questions are in appendix A.

WRITE IT OUT

1. Classify each of the following descriptions as either a proximate or an ultimate cause of behavior. For all proximate causes, speculate about a possible ultimate cause of the behavior.
   a. A surge of epinephrine initiates a flight response in a gazelle.
   b. Turtles use Earth's magnetic field lines as guides during migration.
   c. A small mammal ignores low-energy food sources that are difficult to obtain, maximizing energy gain from foraging.
   d. A bird distracts a predator near its nest, increasing the nestlings' chance of survival.

2. In what ways is a taxis similar to and different from a tropism in a plant (see chapter 24)? Give an example of a taxis and a tropism.

3. What do nonthreatening odors, repetitive sounds, and constant touch have in common?

4. Explain how biologists identify genes controlling behavior in a fruit fly. Why might it be difficult to identify similar genes in humans?

5. Niko Tinbergen investigated how female digger wasps find their nests. After a wasp emerged, Tinbergen moved the pine cones surrounding the nest to a nearby location. Upon its return, the wasp flew to the pine cones, not the nest. Briefly describe the system that the wasp used to find its nest.

6. For multiple decades, scientists in Russia have selectively bred one group of foxes for aggression and bred a second group for friendliness toward humans. The offspring of aggressive foxes tend to be aggressive, while those of friendly foxes tend to be friendly. What do these results reveal about whether a fox's disposition is influenced by genes or the environment? What type of disposition would you predict for a baby fox captured from the wild?

7. European cuckoos and American cowbirds are "brood parasites" that lay their eggs in the nests of other birds. The chicks are generally raised by "adoptive parents" whose own chicks receive less food and often are killed by the larger intruders. What kinds of behaviors does natural selection favor in the cuckoo and cowbird chicks? In the adoptive parents?

8. Prey animals may use distraction displays when a predator is nearby. What selective pressures do these prey animals exert on predators? Explain why distraction displays may become more convincing through evolutionary time.

9. Suppose you read a story in a popular science magazine about social rodents that kill themselves and their offspring if the group becomes overcrowded. Why might a biologist be skeptical of these findings?

10. In the 1930s, the population geneticist J. B. S. Haldane famously said, "I would lay down my life for two brothers or eight cousins." What was he talking about?

PULL IT TOGETHER

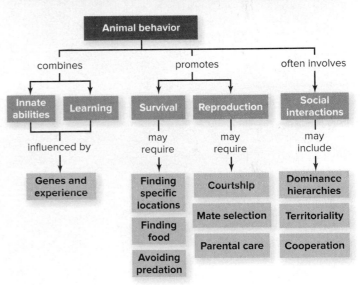

Figure 36.28 **Pull It Together: Animal Behavior.**

Refer to figure 36.28 and the chapter content to answer the following questions.

1. Review the Survey the Landscape in the chapter introduction. Then, classify each of the nine behaviors listed in the Pull It Together concept map as typically triggered (a) by animals of the same species or a different species; (b) only by animals of the same species; (c) only by animals of a different species; (d) by the nonliving environment.

2. Add *foraging, habituation, kin selection, fixed action pattern, eusocial animals, inclusive fitness, imprinting,* and *reciprocal altruism* to this concept map.

3. Select one of the three boxes under "Reproduction" in figure 36.28. For your selected box, provide a specific example of an animal behavior. What are the proximate and ultimate causes of the behavior you described?

Nutria. These large rodents are native to South America. They are invasive in Louisiana, where their large colonies can destroy marshlands.
©blickwinkel/Alamy

LEARN HOW TO LEARN
Expand on What You're Learning

Many students make the mistake of memorizing information without really thinking about what they are learning. This strategy not only makes schoolwork boring but also limits your ability to answer questions. Instead of repeating someone else's words as you study, try elaborating on what you have learned. Here are some ideas: Explain how the topic relates to other course material or to your own life; draw a picture of important processes; discuss the topic with someone else; explain it to yourself out loud or in writing; or think of additional examples or analogies.

Turning "Boom" into "Bust"

Population ecology can be a study of extremes. Hundreds of species teeter on the brink of extinction, but others are booming. For example, deer overpopulation is an enormous problem in many areas. These large animals graze in gardens and cause deadly car crashes. Other out-of-control animal populations include swimming rodents called nutria in Louisiana, brown tree snakes in Guam, and koalas in Australia. In each case, population control means boosting the death rate, cutting the birth rate, or both.

At one time, wolves and other predators kept deer in check, but humans have all but eliminated these top carnivores. Hunting, which boosts the deer population's death rate, can help compensate for the loss. Regions with too many deer may even allow hunters to take does (female deer), which reduces the reproductive rate faster than restricting the hunt to bucks (males).

Hunting also reduces populations of nutria. Fur farmers took the rodents to Louisiana in the 1930s. Some of the animals escaped and thrived in the nearby marshlands, eating plants and destroying thousands of hectares as their population grew. Hunting nutria is vital to protecting the native ecosystem, and the Coastwide Nutria Control Program offers five dollars for each nutria tail. Thanks to this incentive, hundreds of thousands of nutria are harvested each year.

Similarly, efforts to control the brown tree snake focus on increasing the death rate. The reptiles likely arrived on Guam during World War II, hitching a ride on cargo shipped to the island from its native East Asia. Since that time, the snakes have feasted on Guam's native birds and small reptiles. The most common snake control strategy is to capture the pests in traps. Another tactic is leaving dead rats laced with acetaminophen, which destroys the snake's liver. Researchers are also investigating ways to prevent reproduction.

Reducing birth rates is the primary method for controlling koala populations. Koalas are destroying eucalyptus trees and starving to death in some areas of Australia. Surgically sterilizing male koalas is one method of birth control. Scientists are also giving female koalas birth control implants that block the release of an egg.

Animal pests illustrate the importance of studying how populations grow, stabilize, and decline. This chapter introduces the factors that affect populations, including that of our own species.

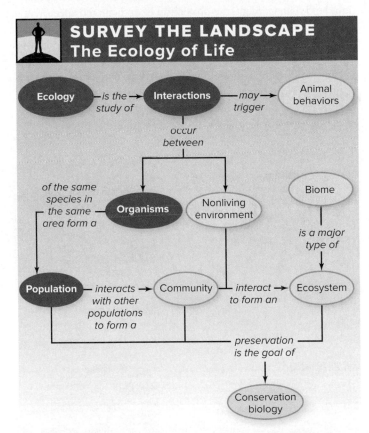

SURVEY THE LANDSCAPE
The Ecology of Life

Within a population, individuals interact as they compete for limited nutrients, energy, and mates. Those that acquire these resources most efficiently are most likely to have high reproductive success.

For more details, study the Pull It Together feature in the chapter summary.

37.1 A Population Consists of Individuals of One Species

Take a moment to think of all the ways you have interacted with the world today. You have certainly breathed air, and you have probably sipped a beverage, eaten one or more meals, put on some clothes, and greeted other people. You may have stepped on the grass or driven a car or played with a pet. Such interactions are part of the science of **ecology,** the study of the relationships that organisms have with each other and with the environment.

Ecologists classify these relationships at several levels (**figure 37.1**). The nutria described in the chapter opening essay represent a **population:** a group of interbreeding organisms of one species occupying a location at the same time. A **community** includes all of the populations, representing multiple species, that interact in a given area. An **ecosystem** is a community plus its nonliving environment, including air, water, minerals, and fire. This chapter describes population ecology, whereas chapter 38 describes community- and ecosystem-level processes.

The study of population ecology is relevant to disease prediction, land management, the protection of endangered species, and many other fields. Typical questions that population ecologists might ask include "Which weather conditions favor reproduction among rodents that transmit human diseases?" "How large a patch of old-growth forest does a breeding pair of spotted owls require?" "How many deer should hunters cull to keep the herd healthy?" "How many humans can Earth support?" To answer such questions, ecologists begin by describing the population.

A. Density and Distribution Patterns Are Static Measures of a Population

The **habitat** is the physical location where the members of a population normally live. The ocean, desert, and rain forest are typical examples, but an organism's habitat might even be another organism. Your body, for example, is home to billions of microbes. ⓘ *normal microbiota,* section 17.4B

Population density is the number of individuals of a species per unit area or unit volume of habitat. Density varies greatly among species. Billions of bacteria can occupy a spoonful of soil, whereas the largest trees may live at densities of one trunk per 10 or more square meters. Because a tiny individual generally requires fewer resources than a large one, it is not surprising that the smallest organisms tend to live at the highest densities. This chapter's Apply It Now box describes some of the ways that biologists estimate population density.

Population distribution patterns describe how individuals are scattered through the habitat space (**figure 37.2**). Organisms may occur in a random pattern if individuals neither strongly attract nor repel one another and the environment is relatively homogeneous (for example, on an exposed rock surface or in the deep shade under a dense forest canopy). More uniform spacing may occur if individuals repel one another, as in the case of strongly territorial animals. Most often, however, a population's distribution is somewhat clumped, with individuals being attracted to one another or to the most favorable patches of habitat.

Population density and distribution measurements provide static "snapshots" of a population at one time. By repeatedly using the same method to measure a population's size over multiple generations, ecologists can determine whether the population is growing, shrinking, or stable. Section 37.2 explains the factors that determine the population's fate.

B. Isolated Subpopulations May Evolve into New Species

The overall area occupied by a population is not necessarily fixed. Habitat destruction can eliminate some or all of a population, as explained in chapter 40. Conversely, individuals in a population can disperse to new habitats. Such range expansion is most obvious for invasive species that spread from their point of introduction. Examples include brown tree snakes, fire ants, zebra mussels, gypsy moths, European starlings, and kudzu. But dispersal also occurs as new habitats become available. For example, glaciers gradually retreat after an ice age, and new volcanic islands occasionally emerge from the sea. Plants, animals, and microbes from nearby areas typically colonize the new habitat.

As organisms disperse to new habitats, they may form local subpopulations. Individuals belonging to a subpopulation are more likely to breed with one another than with members outside the group (see section 40.7). But that does not necessarily mean that each subpopulation is isolated from the others. For example, butterflies may flit from meadow to meadow, or the wind may carry a plant's seeds from one subpopulation to another.

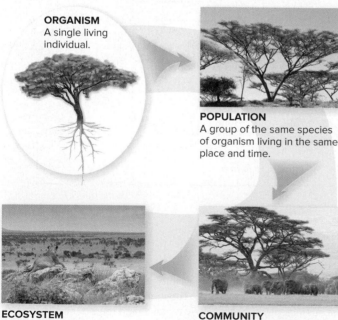

ORGANISM
A single living individual.

POPULATION
A group of the same species of organism living in the same place and time.

ECOSYSTEM
The living and nonliving components of an area.

COMMUNITY
All populations that occupy the same region.

Figure 37.1 From Organism to Ecosystem.

Photos: (population): ©Gregory G. Dimijian, M.D./Science Source; (ecosystem): ©Bas Vermolen/Getty Images; (community): ©Daryl Balfour/Gallo Images/Getty Images

a. Random   b. Uniform   c. Clumped

Figure 37.2 Population Distribution Patterns. (a) These lichens illustrate random spacing in their rocky habitat. (b) These penguins defend the space around their nests, leading to a uniform distribution. (c) The clumped spacing of these palms reflects unevenly distributed resources—especially water.
Photos: (a): ©iStockphoto.com/seraficus RF; (b): ©Rashman/Shutterstock RF; (c): ©PhotoAlto/PunchStock RF

Sometimes, however, a subpopulation becomes physically separated from its parent population, limiting the further exchange of individuals and setting the stage for the evolution of a new species. How might this occur? The isolated subpopulation often encounters a novel combination of selective forces, such as differences in sunlight, moisture, or nutrient availability. Previously rare variants may therefore have the greatest reproductive success in the new habitat. Over many generations, natural selection is likely to shape a genetically distinct subpopulation. If these genetic differences lead to new reproductive barriers, the subpopulation will be unable to interbreed with the parent population. The result: an entirely new species. Chapter 14 describes this process, which is called speciation.

 ## Apply It **Now**

Many Ways to Measure Populations

Collecting long-term data on plant and animal populations can help scientists learn which species are healthy and which are threatened with extinction. Population data can also help ecologists measure the effects of catastrophes such as fires or oil spills. Hunting, trapping, and fishing regulations are based on population estimates, as are decisions on where to build houses, dams, bridges, and pipelines. But how do we know how many individuals of each species inhabit an area?

Simple counts are occasionally possible. For example, aerial photos can reveal the number of newborn seal pups on an island (see figure 37.7). Most population estimates, however, rely on sampling techniques. One common way to estimate the size of a plant population is to count the number of stems in randomly selected locations, such as within a 1-meter square or along a 50-meter line.

A widely used technique to estimate animal populations is called mark–recapture. Suppose researchers want to know how many squirrels inhabit a park. They begin by placing baited nest boxes in trees. Each captured squirrel receives a unique identifying tattoo before being released. Assume the park researchers mark and release 50 squirrels (M = number marked = 50) in a month. During the next month, they capture 25 squirrels, of which 5 bear the tattoo (f = fraction of marked animals in "recapture" sample = 5/25 = 0.20). The number of marked squirrels divided by this fraction estimates the total squirrel population size; in this case, the population is M/f = 50/0.20 = 250.

Researchers also use other ways to estimate the size of an animal population. For example, ecologists may use sound to sample marine

Figure 37.A Remote Camera Snapshot of Mountain Lion.
U.S. National Park Service

mammal populations; audio recording devices attached to a boat or a buoy "listen" for whale or dolphin calls. On land, one possibility is to count the number of droppings within a defined area. Another is to set up animal-triggered camera traps. Bait is placed near a heat detector attached to a camera. When a mountain lion or other animal saunters over to collect the treat, the detector picks up the animal's body heat, and the camera snaps its picture (**figure 37.A**). By setting up multiple camera traps, ecologists can compare the population densities in multiple locations.

37.2 Births and Deaths Help Determine Population Size

Some populations grow, whereas others remain stable. Still others decline, sometimes to extinction. **Population dynamics** is the study of the factors that influence these changes in a population's size (table 37.1).

Births and deaths are the most obvious ways to add to and subtract from a population. If a population adds more individuals than are subtracted, the population grows. If the opposite happens, the population shrinks (and may become extinct). The population size remains unchanged if additions exactly balance subtractions.

Regionally, dispersal to new habitats can also increase or decrease a population; **immigration** is the movement of individuals into a population, and **emigration** occurs when individuals leave. For example, immigration has tremendously increased the human population in the United States over the past 200 or so years. Most Americans are either immigrants or descended from immigrants, and today migration accounts for about half of population growth in this country. Likewise, nonhuman species also disperse to new habitats. They may actively swim, fly, or walk; alternatively, wind or water currents may move individuals into or out of a population.

A. Births Add Individuals to a Population

A population's **birth rate** is the number of new individuals produced per individual in a defined time period. For example, the human birth rate worldwide is about 18.7 births per 1000 people per year.

The number of offspring an individual produces over its lifetime depends on many variables. The number of times it reproduces in its lifetime and the number of offspring per reproductive episode are important, as is the age at first reproduction. All other things being equal, the earlier reproduction begins, the faster the population will grow.

A population's **age structure,** or distribution of age classes, helps determine its birth rate (**figure 37.3**). A population with a large fraction of prereproductive individuals will grow. As these individuals enter their reproductive years and produce offspring, the prereproductive age classes swell further, building a foundation that ensures future growth. Conversely, a population that consists mainly of older individuals will be stable or may even decline. This situation can doom a population of endangered plants if, for example, habitat destruction makes it impossible for seedlings to establish themselves. With few young individuals to replace those that die of old age, the population may go extinct.

TABLE **37.1** Factors Affecting Population Growth: A Summary

Factor	Affected by ...
Additions	
Births	• Number of reproductive episodes per lifetime • Number of offspring per reproductive episode • Age at first reproduction • Population age structure (proportion at reproductive age)
Immigration	• Availability of dispersal mechanism • Availability of suitable habitat
Subtractions	
Deaths	• Availability of nutrients • Predation • Accidents • Genetic/infectious disease
Emigration	• Availability of dispersal mechanism

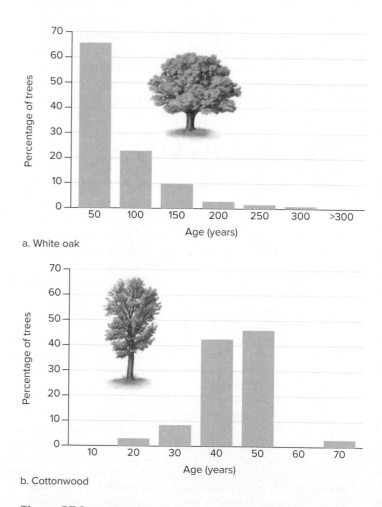

a. White oak

b. Cottonwood

Figure 37.3 **Age Structures.** (a) The white oak population is dominated by younger individuals, indicating high potential for future growth. (b) This population of cottonwoods has few individuals in the youngest age classes. The old trees will likely die soon. Lacking young trees to take their place, the population is probably doomed.

B. Survivorship Curves Show the Probability of Dying at a Given Age

A population's **death rate** is the number of deaths per unit time, scaled by the population size. The causes of death may include accidents, disease, predation, and competition for scarce resources. In the human population, for example, the overall death rate is approximately 7.9 per 1000 people per year; section 37.5 describes human death rates in more detail.

Each individual in a population will eventually die; the only question is when. But species vary tremendously in their patterns of survivorship. In species that produce many offspring but invest little energy or care in each one, the probability of dying before reaching reproductive age is very high. In other species, heavy parental investment in a small number of offspring means that most individuals survive long enough to reproduce.

To help interpret which pattern might apply to a particular species, population biologists developed the **life table,** a chart that shows the probability of surviving to any given age. (Life insurance companies use life tables to compute premiums for clients of different ages.) **Figure 37.4a**, for example, shows a life table for yellow-eyed penguins. The declining number of survivors in each age class reflects the effects of predation, disease, food scarcity, and all other factors that prevent an individual from reaching its theoretical life span.

The values in a life table are often plotted onto a **survivorship curve,** a graph of the proportion of surviving individuals at each age. Figure 37.4b shows a graph of the values from the penguin life table. Note that the *y* axis data are plotted on a logarithmic scale, not a linear one. The log scale makes it easier to see trends along the entire range of values, from 0 to 1000.

The survivorship curve reveals important details about the lives of yellow-eyed penguins. For example, only about 32% of the chicks survive their first year, although the death rate slows thereafter. Moreover, combining life table data with information about fertility at each age can help fill in additional details. For example, reproduction typically begins in year 3 of a yellow-eyed penguin's life. According to the survivorship curve, only about 25% of the penguins that hatch in any given year reach reproductive age. Compare this figure to our own species, in which about 99% of offspring survive long enough to reproduce.

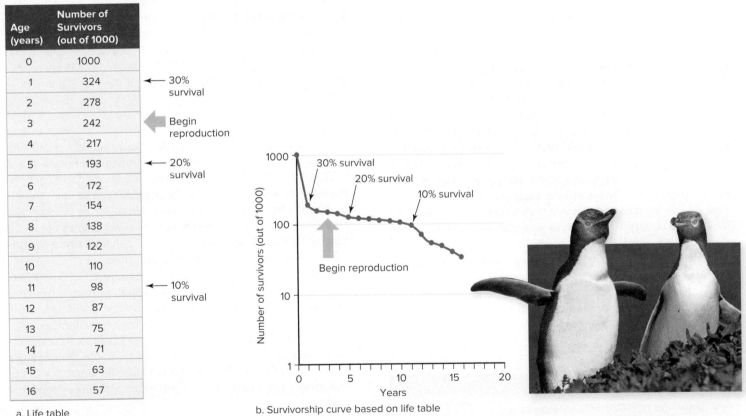

Age (years)	Number of Survivors (out of 1000)
0	1000
1	324
2	278
3	242
4	217
5	193
6	172
7	154
8	138
9	122
10	110
11	98
12	87
13	75
14	71
15	63
16	57

a. Life table

b. Survivorship curve based on life table

Figure 37.4 Penguin Survivorship. (a) This life table lists the number of survivors remaining each year out of a group of 1000 yellow-eyed penguins that hatch at the same time. (b) A survivorship curve is a graph of the data in a life table. The curve reveals that most penguins die before age 1, after which the death rate levels off.

Photo: ©Kevin Schafer/Corbis

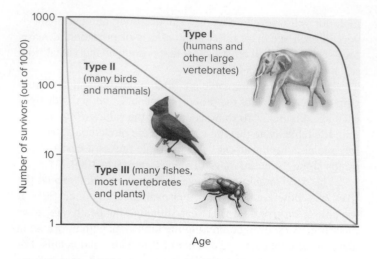

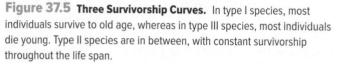

Figure 37.5 Three Survivorship Curves. In type I species, most individuals survive to old age, whereas in type III species, most individuals die young. Type II species are in between, with constant survivorship throughout the life span.

The survivorship curves of many species follow one of three general patterns (figure 37.5). Type I species, such as humans and elephants, invest a great deal of energy and time in each offspring. Most individuals live long enough to reproduce, and the death rate is highest as individuals approach the maximum life span. Type II species, including many birds and mammals, may also provide a great deal of parental care. However, the threats of predation and disease are constant throughout life, and these organisms have an equal probability of dying at any age. The type II line is therefore straight. Type III species, such as many fishes and most invertebrates and plants, may produce many offspring but invest little in each one. Most offspring of type III species therefore die at a very young age. Section 37.4 looks more closely at the evolutionary trade-offs that these survivorship curves reflect.

Of course, these generalized examples do not describe all populations. Many species have survivorship curves that fall between two patterns. The yellow-eyed penguins in figure 37.4, for example, have a survivorship curve that combines features of both type II and type III curves.

37.2 | MASTERING CONCEPTS

1. Under what conditions will a population grow?
2. What factors determine the birth rates and death rates in a population?
3. What is the relationship between a life table and a survivorship curve?
4. Describe the three patterns of survivorship curves.

37.3 Population Growth May Be Exponential or Logistic

Any population will grow if the number of individuals added exceeds the number removed. The **per capita rate of increase,** r, is the difference between the birth rate and the death rate. (Note that this simplified formula for r ignores the effects of immigration and emigration.) For example, a population with 35 births and 10 deaths per 1000 individuals per year is growing at a rate (r) of 25 people per 1000 (0.025/yr), or 2.5% per year. Any population with a positive value of r is growing. On the other hand, a negative r means the population is shrinking. If r equals zero, the population is stable.

But how fast will a population grow, and how large can the population become? Two mathematical models, called exponential and logistic growth, illustrate two simple patterns of population growth.

A. Growth Is Exponential When Resources Are Unlimited

In the simplest model of population growth, **exponential growth,** the number of new individuals is proportional to the size of the population. In exponential growth, the number of individuals added during any time interval (G) depends only on the per capita rate of increase (r) and the initial size of the population (N), as expressed by this equation:

$$G = rN$$

For example, suppose that 100 aquatic animals called rotifers are placed in a tank under ideal growth conditions. If r is 0.22 per day, then how many rotifers are added each day? The answer depends on the size of the population: The more rotifers in the tank, the larger the capacity to add offspring. **Figure 37.6** shows how to calculate the size of each generation and plots the running total on a graph. A characteristic **J-shaped curve** emerges when exponential growth is plotted over time.

Growth resulting from repeated doubling (1, 2, 4, 8, 16, 32, ...), such as in bacteria, is exponential. Species introduced to an area where they are not native may also proliferate exponentially for a time, since they often have no natural population controls. **Figure 37.7** shows another example. Researchers have counted the grey seal pups born on Sable Island, Nova Scotia, since the 1960s. For four decades, the number of pups was proportional to the total population, implying exponential growth.

Figure It Out

Suppose the grey seals in figure 37.7 have an r of 0.13 per year. What size population would produce 41,500 pups in 1 year?

Answer: $G = 41,500$ and $r = 0.13$. N is therefore $41,500 / 0.13 = 319,231$ seals.

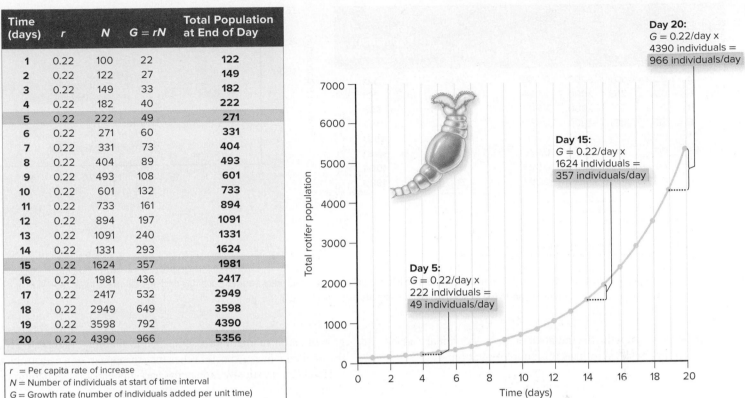

Time (days)	r	N	G = rN	Total Population at End of Day
1	0.22	100	22	122
2	0.22	122	27	149
3	0.22	149	33	182
4	0.22	182	40	222
5	0.22	222	49	271
6	0.22	271	60	331
7	0.22	331	73	404
8	0.22	404	89	493
9	0.22	493	108	601
10	0.22	601	132	733
11	0.22	733	161	894
12	0.22	894	197	1091
13	0.22	1091	240	1331
14	0.22	1331	293	1624
15	0.22	1624	357	1981
16	0.22	1981	436	2417
17	0.22	2417	532	2949
18	0.22	2949	649	3598
19	0.22	3598	792	4390
20	0.22	4390	966	5356

r = Per capita rate of increase
N = Number of individuals at start of time interval
G = Growth rate (number of individuals added per unit time)

Day 20:
G = 0.22/day x 4390 individuals = 966 individuals/day

Day 15:
G = 0.22/day x 1624 individuals = 357 individuals/day

Day 5:
G = 0.22/day x 222 individuals = 49 individuals/day

Figure 37.6 Exponential Population Growth. For a population of rotifers with unlimited resources, the number of individuals added each day (G) is the product of the per capita rate of increase (r) and the number of individuals at the beginning of the interval (N). With each generation, G increases, even though r remains constant. Colored bars in the table represent time points that are labeled in the graph.

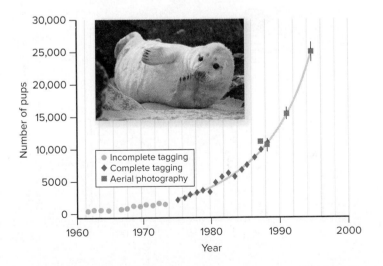

Figure 37.7 Seal Pup Population. Tagging and aerial surveys of the grey seal pups on Sable Island, Nova Scotia, revealed 40 years of exponential population growth. (Growth began to slow after the year 2000.)

Photo: ©Ronald Wittek/Photographer's Choice/Getty Images

B. Population Growth Eventually Slows

Exponential growth may continue for a short time, but it cannot continue indefinitely because some resource is eventually depleted. **Environmental resistance** is the combination of external factors that keep a population from reaching its maximum growth rate. Environmental resistance includes competition, predation, and anything else that reduces birth rates or increases death rates. (Section 37.3C describes these factors in more detail.)

Thanks to environmental resistance, every habitat has a **carrying capacity,** which is the maximum number of individuals that the ecosystem can support indefinitely. This carrying capacity imposes an upper limit on a population's size. How does this limit affect the population's growth rate? According to the **logistic growth** model, the early growth of a population may be exponential, but growth slows to zero as the population's size approaches the habitat's carrying capacity (figure 37.8). The resulting **S-shaped curve** depicts the leveling off of a population in response to environmental resistance.

Day	r	N	K	$\left(\frac{K-N}{K}\right)$	$G = rN\left(\frac{K-N}{K}\right)$	Total Population at End of Day
1	0.22	100	2000	0.950	21	121
6	0.22	253	2000	0.874	49	301
11	0.22	577	2000	0.711	90	668
16	0.22	1087	2000	0.456	109	1197
21	0.22	1578	2000	0.211	73	1651
26	0.22	1851	2000	0.075	30	1881
31	0.22	1954	2000	0.023	10	1964
36	0.22	1986	2000	0.007	3	1989

r = Per capita rate of increase
N = Number of individuals at start of time interval
K = Carrying capacity
G = Growth rate (number of individuals added per unit time)

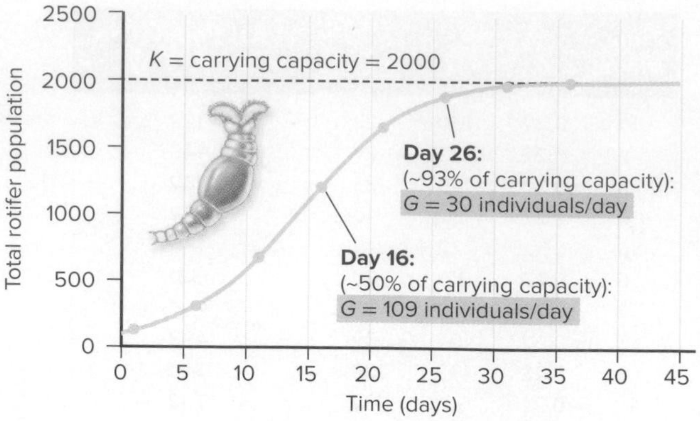

Figure 37.8 Logistic Population Growth. When resources become limited, the number of individuals added each day declines. Numbers for N, (K − N)/K, and G are shown for selected days only. Colored bars in the table represent time points that are labeled in the graph.

In logistic growth, the number of new individuals added after each time interval follows this simple equation:

$$G = rN\left(\frac{K-N}{K}\right)$$

G equals the number of individuals added per unit time; r is the per capita rate of increase; N is the number of individuals in the population at a given time; and K is the carrying capacity. On the right side of the equation, the term rN is the growth rate that would occur if resources were not limiting. The term (K − N)/K, however, accounts for the increasing environmental resistance as the population approaches the carrying capacity. When N is very small relative to K, this expression yields a numerical value near 1, so the population growth rate is high. When N is near K, the value of the expression drops to near 0, and the growth rate is low.

Figure It Out

Suppose that the carrying capacity for grey seals on Sable Island is 1 million and that r = 0.13 per year. Once the population reaches 500,000, how many new pups will be added in the following year?

Answer: (K − N)/K = (1,000,000 − 500,000)/1,000,000 = 0.5. Therefore, G = 0.13 × 500,000 × 0.5 = 32,500.

Logistic curves often describe populations of disease-causing organisms in confined groups of plants or animals, as in a greenhouse or on a feedlot. The S-shaped curve simply indicates that the disease cannot continue to spread after all available hosts are infected. In this case, the carrying capacity is determined largely by the number of susceptible hosts.

The carrying capacity of an ecosystem, however, typically is not fixed. A drought that lasts for a decade may be followed by a year of exceptionally heavy rainfall, causing a flush of new plant growth and a sudden increase in food availability. Alternatively, the food on which a species relies may disappear, or a catastrophic flood can drastically reduce the carrying capacity as habitat is destroyed.

In studying population growth, it is important to understand that some species do not fit neatly into either the exponential or logistic models. In collared lemmings, for example, the population fluctuates on a 4-year cycle (figure 37.9). The origin of such boom-and-bust cycles remains mysterious, but predation by stoats (a type of weasel) appears to be one of the main factors regulating the ups and downs of the lemming population.

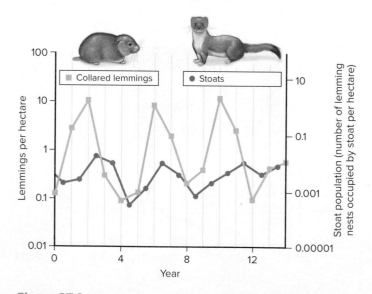

Figure 37.9 Population Cycle. The population of collared lemmings fluctuates regularly over a 4-year period. Research suggests that a major influence on the lemming population is the number of stoats, a type of weasel that preys on the lemmings.

C. Many Conditions Limit Population Size

A combination of factors determines the size of most populations. Consider the population of songbirds in your town. Some lose their lives to cold weather or food shortages, whereas others succumb to infectious disease or the jaws of a cat. These and other limits on the growth of the bird population fall into two general categories: density-dependent and density-independent (**figure 37.10**).

Density-dependent factors are conditions whose growth-limiting effects increase as a population grows. Most density-dependent limits

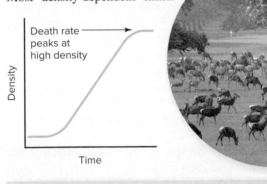

Density-dependent limits

• Competition for resources within or among species	• Infectious disease • Predation

a.

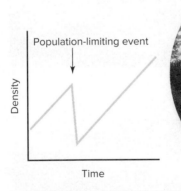

Density-independent limits

• Natural disasters • Industrial accidents	• Habitat destruction

b.

Figure 37.10 Factors Limiting Populations. (a) Density-dependent factors become more important as the habitat becomes crowded. This herd of deer may soon overwhelm its habitat. (b) Density-independent factors such as forest fires limit populations at all densities.

Photos: (a): ©Tim Graham/Getty Images; (b): Courtesy of John McColgan, Alaska Fire Service/Bureau of Land Management

are **biotic**, meaning they result from interactions with living organisms. Within a population, competition for space, nutrients, sunlight, food, mates, and breeding sites is density-dependent. When many individuals compete for limited resources, few may be able to reproduce, and population growth slows or even crashes.

For example, competition for food may finally slow the growth of the invasive brown tree snake on Guam (see the chapter opening essay). After consuming most of the small prey on the island, the snake population is finally showing signs of stress and overcrowding, and reproduction is apparently slowing down.

Other species can also exert density-dependent limits on a population's growth. Two or more species may compete fiercely for the same nutrients and space (see section 38.1A). Infectious disease also takes a toll. Viruses, bacteria, and other microbes spread by direct contact between infected individuals and new hosts. The higher the host population density, the more opportunity for these disease-causing organisms to spread. Likewise, a higher population density may lead to a higher probability of death by predation, as depicted in figure 37.9.

Density-independent factors exert effects that are unrelated to population density. Most density-independent limits are **abiotic**, or nonliving. Natural disasters such as fires, floods, volcanic eruptions, and severe weather are typical density-independent factors. The high winds of a hurricane, for example, may destroy 50% of the birds' nests in a forest, without regard to the density of the bird population. Likewise, a lava flow kills everything in its path. The effects of oil spills and other industrial accidents are density-independent, too. In addition, as described in chapter 40, habitat destruction related to human activities is a density-independent factor that is pushing many species to the brink of extinction.

No matter what combination of factors controls a population's size, it is worth remembering the importance of these limits in natural selection and evolution. Many individuals do not survive long enough to reproduce. Those that do manage to breed have the adaptations that allow them to escape the biotic and abiotic challenges that claim the lives of many of their counterparts. As these "fittest" individuals pass their genes on, the next generation contains a higher proportion of offspring with those adaptations.

37.3 | MASTERING CONCEPTS

1. What is the per capita rate of increase?
2. What conditions support exponential population growth?
3. What is environmental resistance, and what is its relationship to the carrying capacity?
4. Use figures 37.6 and 37.8 to explain why G increases as a population grows exponentially but G decreases as a population under logistic growth approaches the carrying capacity.
5. Distinguish between density-dependent and density-independent factors that limit population size, and give three examples of each.

37.4 Natural Selection Influences Life Histories

As we have already seen, a population's size depends in part on birth rates and death rates. Species differ widely, however, in the timing of these events. Population ecologists therefore find it useful to document a species' **life history,** which includes all events of an organism's life from conception through death. The main focus of life history analysis is the adaptations that influence reproductive success.

Life tables and population growth curves, described in sections 37.2 and 37.3, document two elements of an organism's life history. Other aspects include developmental rate, life span, social behaviors, reproductive timing, mate selection, number and size of the offspring, number of reproductive events, and amount of parental care. This section explores how natural selection shapes some of these life history traits. ⓘ *social behaviors, section 36.5*

A. Organisms Balance Reproduction Against Other Requirements

A species' life history reflects a series of evolutionary trade-offs; after all, supplies of time, energy, and resources are always limited. Just as an investor allocates money among stocks, bonds, and real estate, a juvenile organism divides its efforts among growth, maintenance, and survival. After reaching maturity, another competing demand—reproduction—joins the list.

Reproduction is extremely costly. Many animals, for example, devote time and energy to attracting mates, building and defending nests, and incubating or gestating offspring. Once the young are hatched or born, the parents may feed and protect them. All of these activities limit a parent's ability to feed itself, defend itself, or reproduce again. In plants, the reproductive investment is also substantial. Flowers, fruits, and defensive chemicals cost energy to produce and take away from a plant's photosynthetic area, reducing the ability to capture sunlight.

Besides the total effort allocated to reproduction, the timing is also critical. An organism that delays reproduction for too long may die before producing any offspring at all. On the other hand, reproducing too early diverts energy away from the growth and maintenance that may be crucial to survival.

The solutions to these trade-offs vary tremendously among species. A bacterial cell, for example, may grow for just 20 minutes before dividing asexually. A female winter moth mates once and lays hundreds of fertilized eggs just before she dies. A pigweed plant sheds 100,000 seeds in the one summer of its life. These and many other organisms mature early, produce many offspring in a single reproductive burst, and die. Humans, elephants, and century plants, on the other hand, mature late and produce only a few offspring throughout their long lives.

B. Opportunistic and Equilibrium Life Histories Reflect the Trade-Off Between Quantity and Quality

Although each species is unique, ecologists have discovered that reproductive strategies fall into patterns shaped by natural selection. One prominent trade-off is the balance between offspring quality and quantity.

At one extreme are species that have an **opportunistic life history** (also called an *r*-selected life history), in which individuals tend to be short-lived, reproduce at an early age, and have many offspring that receive little care (**figure 37.11a, b**). The population's growth rate can be very high if conditions are optimal and individuals face little competition for resources; the *r* in "*r*-selected" reflects a large per capita rate of increase. In general, however, each offspring has a very low probability of surviving to reproduce; this pattern is typical of species with type III survivorship curves (see figure 37.5). Weeds, insects, and many other invertebrates typically have opportunistic life histories.

At the other extreme are species with an **equilibrium life history** (also called a *K*-selected life history), in which individuals

Figure 37.11 Opportunistic and Equilibrium Life Histories. (a) Rice plants and (b) tussock moths produce many small offspring and invest little in each one. (c) Each coconut produced by this tree requires a large energy investment. (d) Grizzly bears devote extensive time and resources to rearing a small number of young.

Photos: (a): USDA/Keith Weller; (b): ©Andrew Darrington/Alamy; (c): ©Flat Earth Images RF; (d): ©DLILLC/Corbis RF

tend to be long-lived, to be late-maturing, and to produce a small number of offspring that receive extended parental care (figure 37.11c, d). High parental investment in each offspring means that most live long enough to reproduce. The *K* in "*K*-selected" stands for the carrying capacity; density-dependent factors such as intense competition for resources keep these populations close to the carrying capacity. Many large mammals (organisms with survivorship curves that approximate type I) have equilibrium life histories.

Species vary widely in their life histories; strict adherence to an opportunistic or equilibrium strategy is the exception, not the rule. For example, some species have intermediate life histories with type II survivorship curves. In addition, even populations of large animals with equilibrium life histories fluctuate greatly in response to changes in their environments.

Nevertheless, these two strategies illustrate the important point that natural selection shapes a species' life history characteristics. These traits therefore reflect the competing demands of reproduction and survival. Section 37.6 explores how unique habitats—dark, toxic waters—influence this basic trade-off.

37.4 | MASTERING CONCEPTS

1. Explain why an organism cannot produce large numbers of offspring that receive extended parental care.
2. Distinguish between opportunistic and equilibrium life histories.

37.5 The Human Population Continues to Grow

The human population has grown exponentially in the last 2000 years (**figure 37.12**). So far, we have found ways to escape many of the forces that limit the growth of other animal populations, such as food availability and disease. Yet exponential growth cannot continue indefinitely. This section describes how the principles of population growth apply to the human population.

A. Birth Rates and Death Rates Vary Worldwide

In early 2017, the world's human population was over 7.4 billion. Since Earth's land area is about 150 million square kilometers, the average population density is about 49 people per square kilometer of land area, but the distribution is far from random. For the most part, the highest population densities worldwide occur along the coastlines and in the valleys of major rivers; very few people live on high mountains, in the middle of the world's major deserts, or on Antarctica. Two countries—China and India—account for one third of all humans.

Overall, the human population growth rate is about 1% per year and declining. Demographers project that zero population growth may happen during the 22nd century, but no one is certain. Nor do we know how many people will inhabit Earth when that

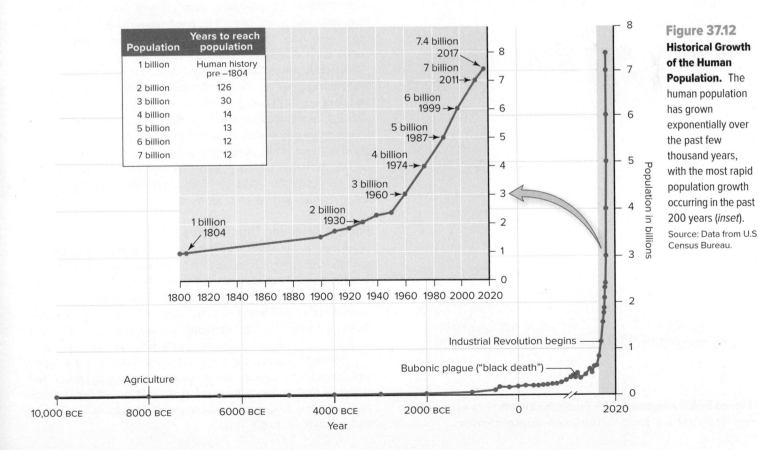

Figure 37.12

Historical Growth of the Human Population. The human population has grown exponentially over the past few thousand years, with the most rapid population growth occurring in the past 200 years (*inset*).

Source: Data from U.S. Census Bureau.

Population	Years to reach population
1 billion	Human history pre –1804
2 billion	126
3 billion	30
4 billion	14
5 billion	13
6 billion	12
7 billion	12

occurs. Clearly, however, less-developed countries are growing at much faster rates than are more-developed countries (**figure 37.13**).

What explains the difference in growth rates? Each country's economic development influences its progress along the **demographic transition,** during which birth rates and death rates shift from high to low (**figure 37.14**). In the first stage of the demographic transition, population growth is minimal because both birth rates and death rates are high. Then, during the second stage, improved living conditions and disease control lower the death rate, but birth rates remain high. This transitional period therefore sees the rapid population growth typical of the world's less-developed countries. During the third stage of the demographic transition, birth rates fall; the difference between birth rates and death rates is once again small. The world's more-developed countries have entered this stage, and a few even have declining populations because death rates exceed birth rates.

Factors Affecting Birth Rates

As described in section 37.2, a population's age structure helps predict its future birth rate. **Figure 37.15** shows the age structures for the world's three most populous countries. In India and many other less-developed countries, a large fraction of the population is entering its reproductive years, suggesting a high potential for future growth. In the United States, as in many other developed countries, the population consists mainly of older individuals. Such populations are stable or declining.

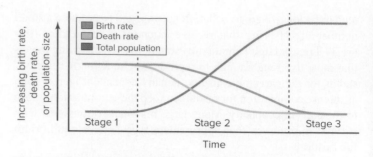

Figure 37.14 **The Demographic Transition.** During stage 1, birth rates and death rates are high, so the population remains small. Population growth is rapid in stage 2, when death rates fall faster than birth rates. In stage 3, birth rates and death rates are both low, and the population stabilizes.

As recently as 1990, China's age structure resembled that of present-day India. But China's population now shows a decline in the youngest age classes, so its future growth rate should decline. Between 1980 and 2016, the Chinese government controlled runaway population growth with drastic measures, limiting families to one child. (Couples are now allowed two children.) Although China remains the world's most populous nation, biologists expect India to take the lead by 2025.

A close look at China's age structure diagram reveals another consequence of that country's one-child policy. Although the numbers of older males and females are approximately equal, the diagram reveals a bias toward male children in younger age classes. According to cultural tradition in China, parents prefer male children. Couples who were allowed just one child may have ended pregnancies with female fetuses in hopes of conceiving a male later. Population biologists are closely studying the social and demographic consequences of this gender shift.

Overall, why do birth rates tend to decline as the demographic transition progresses? One explanation for this trend is the availability of family planning programs, which are relatively inexpensive and have immediate results. Social and economic factors play an important role as well. Educated women are most likely to learn about and use family planning services, have more opportunities outside the home, and may delay marriage and childbearing until after they enter the workforce. Delayed childbearing often means fewer children and therefore slows population growth.

Factors Affecting Death Rates

Besides the birth rate, the death rate is the other major factor influencing a population's growth rate. Overall, average life expectancy has increased steadily throughout history—for example, from 22 during the Roman Empire to 33 in the Middle Ages. Today, life expectancy averages around 68 worldwide; it exceeds 80 in a few of the most-developed countries. This dramatic increase reflects centuries of scientific advances in agriculture, medicine, and public health.

Table 37.2 shows the top five causes of death in developed and developing countries. Heart disease, stroke, lung cancer, and dementia top the list in the developed world. Infectious diseases rank relatively low, thanks in large part to sanitation, antibiotics, and vaccines (**figure 37.16**).

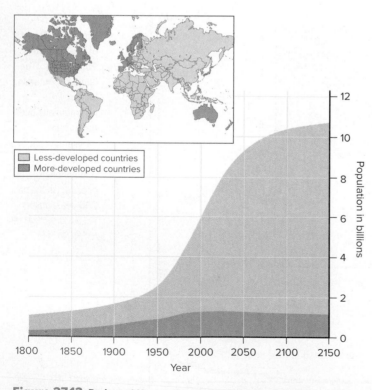

Figure 37.13 **Projected Human Population Growth.** Future population growth will continue to be concentrated in less-developed countries.

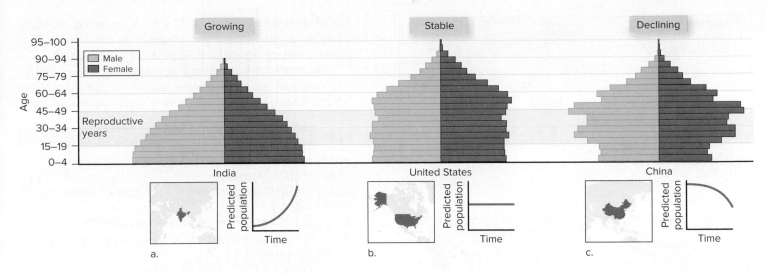

Figure 37.15 **Age Structures for Three Human Populations.** In age structure diagrams, the width of each bar is proportional to the percent of individuals in that age class. (a) India's population is likely to continue to grow because a high proportion of individuals are in prereproductive age classes. (b) The population of the United States is stable, with roughly equal numbers of people of each age group. (c) China's future growth rate should decline because most of its members are in reproductive or postreproductive age classes.

Source: Data from U.S. Census Bureau, International Data Base.

In contrast, deadly diseases such as respiratory infections, diarrhea, and HIV/AIDS are more prominent in developing countries. Crowded conditions facilitate the spread of cholera and other waterborne diseases, especially in areas with limited access to clean drinking water. AIDS has taken an especially high toll in sub-Saharan Africa, where about 4.7% of adults are HIV-positive. ⓘ *HIV,* section 16.4B

Human population projections take into account the factors affecting birth rates and death rates around the world. This chapter's Burning Question offers a peek into the uncertain future.

B. The Ecological Footprint Is an Estimate of Resource Use

The human population cannot continue to grow exponentially because living space and other resources are finite. Worldwide, increasing numbers of people will mean greater pressure on land, water, air, and fossil fuels as people demand more resources and generate more waste.

Ecologists summarize each country's demands on the planet by calculating an ecological footprint. Just as an actual footprint shows the area that a shoe occupies with each step, an **ecological footprint** measures the amount of land area needed to support a country's overall lifestyle. The calculation includes, among other

TABLE **37.2** **Top Five Causes of Death in High- and Low-Income Countries**

Rank	High-Income Countries	Low-Income Countries
1	Heart disease (coronary artery blockage)	Lower respiratory infection (pneumonia, acute bronchitis)
2	Stroke	HIV/AIDS
3	Lung cancer	Diarrhea (e.g., cholera, rotavirus)
4	Alzheimer disease and other dementias	Stroke
5	Lower respiratory infection (pneumonia, acute bronchitis)	Heart disease (coronary artery blockage)

Source: World Health Organization, "The 10 Leading Causes of Death by Country Income Group."

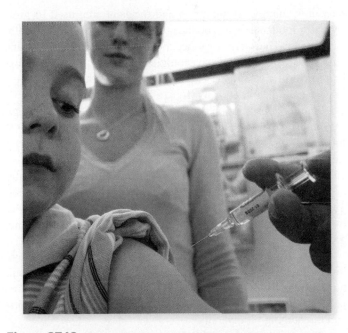

Figure 37.16 **Lifesaver.** Disease prevention, including widespread vaccination, is one factor that has allowed the human population to continue to grow.

©Peter Cade/Iconica/Getty Images

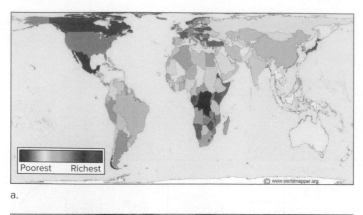

a.

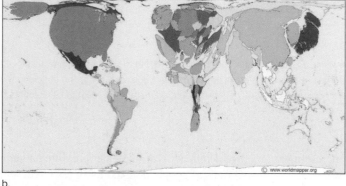

b.

Figure 37.17 **Ecological Footprint.** (a) In a traditional map, each country's size is proportional to its land area. (b) Here, each country's size is proportional to its ecological footprint. Note that the mapmakers grouped countries into 12 regions and assigned each region (not each individual country) a color according to its wealth.

measures, energy consumption and the land area used to grow crops for food and fiber, produce timber, and raise cattle and other animals. The land areas occupied by streets, buildings, and landfills are also part of the ecological footprint. Not surprisingly, the world's wealthiest and most populous countries have the largest ecological footprints (figure 37.17).

Energy consumption accounts for about half of the ecological footprint. The wealthiest countries make up less than 20% of the world's population yet consume more than half of the energy. Less-developed countries, however, will take a larger share of energy supplies as their populations grow and their economies become more industrialized. Since the vast majority of the energy comes from fossil fuels, the result will be increased air pollution and acid rain. Moreover, the accumulation of CO_2 and other greenhouse gases is implicated in global climate change.

Food production is another significant element of the ecological footprint. Overall, both the demand for food and agricultural productivity rise each year. To boost food production, people often expand their farms into forests, destroying habitat and threatening biodiversity.

Because the ecological footprint focuses on land area, it does not include water consumption. Nevertheless, the availability of fresh water has declined worldwide as people have demanded more water for agriculture, industry, and household use. In many poor countries, less than half the population has access to safe

water for drinking and cooking. As a result, waterborne diseases such as cholera periodically surge through the dense populations of India and many African and South American countries.

Chapter 40 further explores deforestation, species extinctions, increased fuel consumption, global climate change, and other problems related to the expanding human population.

37.5 MASTERING CONCEPTS

1. Which parts of the world have the highest and lowest rates of human population growth?

2. How does the demographic transition reflect changes in birth rates and death rates?

3. What factors affect birth rates and death rates worldwide?

4. What are some of the environmental consequences of human population growth?

Burning Question

What will happen to the human population?

No one knows exactly how growth rates will change in the future, so it is impossible to predict when or at what level the human population will level off. In 2015, the United Nations issued three projections for the world's population, assuming high, medium, and low growth rates. The highest projection says the Earth's population will be around 16.6 billion (and still growing) in 2100. The medium estimate shows the population stabilizing at around 11.2 billion by 2100. The low estimate predicts that the population will peak at just over 8.7 billion in 2050 and then it will decline to about 7.3 billion by 2100.

These projections are only as good as their assumptions. Will birth rates in developing countries continue to decline? If so, by how much, and how fast will it happen? Will birth rates decline by the same amount in all countries? Will more-developed countries be willing and able to provide family planning services even as the rural populations of less-developed countries continue to grow? The answers to these questions will determine the future of Earth's human population.

Submit your burning question to
Marielle.Hoefnagels@mheducation.com

©Melanie Stetson Freeman/Christian Science Monitor/The Image Works

INVESTIGATING LIFE

37.6 A Toxic Compromise

Birth and death rates influence a population's overall growth, but natural selection acts on a smaller scale. Evolution selects for life histories that maximize fitness—that is, the number of offspring that live to reproductive age. For example, section 12.8 described how changes in fishing regulations can alter the life histories of commercially fished species.

Harsh environments can also affect life history traits. Fish called Atlantic mollies (*Poecilia mexicana*) are an ideal species for studying the trade-offs of living in a challenging environment. Atlantic mollies live in four distinct habitats within Mexico. The most stressful environments are toxic streams flowing through dark caves (**figure 37.18**). These streams contain a high concentration of poisonous hydrogen sulfide, H_2S (commonly associated with its "rotten egg" odor). Other mollies populate nearby caves containing nontoxic water. Atlantic mollies also live in surface waters, which may be either poisonous or nontoxic. Measuring the life history qualities of the mollies in the four habitats could reveal the evolutionary cost of a stressful environment.

Biologists Rüdiger Riesch and Ingo Schlupp at the University of Oklahoma teamed up with Martin Plath from the University of Frankfurt to learn more about the life history adaptations of Atlantic mollies. They anesthetized and preserved pregnant females collected from each habitat. At the lab, they measured the number and mass of embryos in each female. They also determined the developmental stage of each embryo and estimated what its mass would have been at birth.

Their results were clear: Females in toxic cave streams produced fewer, larger offspring than females in open, nontoxic streams (**figure 37.19**). The offspring of females from habitats with only one environmental stress—that is, only toxic water or only darkness—were intermediate in number and mass. The data suggest that in the harshest environments, females sacrifice quantity in favor of quality.

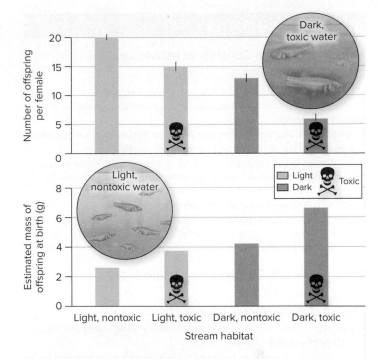

Figure 37.19 Life History Trade-Offs. Dark, toxic water selected for fewer, larger offspring than did light or nontoxic streams. In the top graph, error bars represent standard errors (see appendix B).

Investing large amounts of energy into each offspring is adaptive in environments where resources might be limited. Paradoxically, caves and toxic streams contain plenty of food, courtesy of abundant microbes that use H_2S as an energy source. But this food is not easy to acquire, for two reasons. First, finding food in the dark is more difficult than in a well-lit habitat, even for animals adapted to darkness. Second, food in toxic streams is most abundant in the deeper waters, where oxygen is scarce and mollies can visit only for short periods. Larger offspring are better swimmers; they can therefore find more food and are less likely to starve than smaller offspring.

When Atlantic mollies first encountered caves and toxic streams, their new habitats began selecting for new life history adaptations. A molly adapted to dark, toxic caves would be outcompeted in light, nontoxic water, where other mollies produce more offspring. These life history changes might now prevent interbreeding between molly populations, even though no geographic barrier separates them—a possible case of parapatric speciation in action. ⓘ *parapatric speciation,* section 14.3B

Source: Riesch, Rüdiger, Martin Plath, and Ingo Schlupp. 2010. Toxic hydrogen sulfide and dark caves: Life-history adaptations in a livebearing fish (*Poecilia mexicana,* Poeciliidae). *Ecology,* vol. 91, pages 1494–1505.

37.6 MASTERING CONCEPTS

1. Use figure 37.19 to predict the life history traits of mollies in a toxic stream near a cave opening, where a limited amount of light is available.

2. Explain why mollies in light, nontoxic streams have smaller offspring than do mollies in dark, toxic streams.

Figure 37.18 Dark Water. Some Atlantic mollies inhabit toxic water in caves.

(both): ©Stephen Alvarez/National Geographic Creative/Getty Images

CHAPTER SUMMARY

37.1 A Population Consists of Individuals of One Species

- **Ecology** considers interrelationships between organisms and their environment. It includes interactions at the **population, community,** and **ecosystem** levels.

A. Density and Distribution Patterns Are Static Measures of a Population

- A **habitat** is the location where an individual normally lives.
- **Population density** is a measure of the number of individuals per unit area or volume of habitat, and **population distribution** describes how individuals are distributed within the habitat.

B. Isolated Subpopulations May Evolve into New Species

- Dispersal to new habitats may produce subpopulations that exchange individuals on a limited basis. An isolated subpopulation may evolve into a new species if genetic isolation leads to one or more reproductive barriers.

37.2 Births and Deaths Help Determine Population Size

A. Births Add Individuals to a Population

- The factors that determine a population's size over time are part of the study of **population dynamics.**
- A population grows when more individuals are added through birth or **immigration** than leave due to death or **emigration.**
- The **birth rate** is the number of new individuals produced per capita in a defined time period. A population's birth rate depends on many factors, including the **age structure.**

B. Survivorship Curves Show the Probability of Dying at a Given Age

- The **death rate** reflects the number of deaths per unit time.
- A **life table** shows the number of survivors remaining at each age. **Survivorship curves** fall into three general patterns, reflecting the balance between the number of offspring and the amount of parental investment in each.

37.3 Population Growth May Be Exponential or Logistic

- The difference between the birth rate and the death rate is r, the **per capita rate of increase.**

A. Growth Is Exponential When Resources Are Unlimited

- Population growth that is proportional to the size of the population is **exponential** and produces a **J-shaped curve.**

B. Population Growth Eventually Slows

- In response to **environmental resistance,** the population may stabilize indefinitely at the habitat's **carrying capacity.** A plot of the resulting **logistic growth** produces a characteristic **S-shaped curve.**

C. Many Conditions Limit Population Size

- **Density-dependent factors** such as infectious disease, predation, and competition have the greatest effect on crowded populations. Most such factors are **biotic,** or living.
- **Density-independent factors** such as natural disasters kill the same fraction of the population regardless of the population's density. Most such factors are **abiotic** (nonliving).

37.4 Natural Selection Influences Life Histories

- The **life history** of a species includes all events from birth to death but typically emphasizes the factors that affect reproduction.

A. Organisms Balance Reproduction Against Other Requirements

- Organisms must allocate limited time, energy, and resources among growth, maintenance, survival, and reproduction.

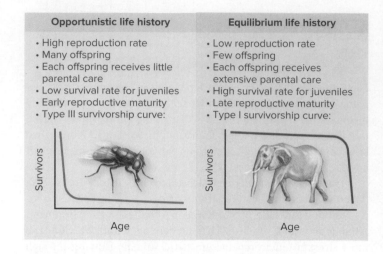

Opportunistic life history	Equilibrium life history
• High reproduction rate • Many offspring • Each offspring receives little parental care • Low survival rate for juveniles • Early reproductive maturity • Type III survivorship curve:	• Low reproduction rate • Few offspring • Each offspring receives extensive parental care • High survival rate for juveniles • Late reproductive maturity • Type I survivorship curve:

Figure 37.20 Opportunistic and Equilibrium Species: A Summary.

B. Opportunistic and Equilibrium Life Histories Reflect the Trade-Off Between Quantity and Quality

- Species with an **opportunistic (r-selected) life history** produce many offspring but expend little energy on each, whereas species with an **equilibrium (K-selected) life history** invest heavily in rearing relatively few young (figure 37.20).

37.5 The Human Population Continues to Grow

A. Birth Rates and Death Rates Vary Worldwide

- Each stage in the **demographic transition** is characterized by a unique combination of birth rates and death rates. In less-developed countries, birth rates are high and death rates are low, producing rapid population growth. As economic development increases, birth rates decline and population growth slows.
- Education, access to contraceptives, and government policies affect birth rates.
- The top causes of death vary around the world; they include heart disease, stroke, cancer, and infectious disease.

B. The Ecological Footprint Is an Estimate of Resource Use

- Countries differ in their **ecological footprint.** Sustained population growth will continue to strain supplies of natural resources such as fossil fuels, farmland, and clean water.

37.6 Investigating Life: A Toxic Compromise

- Researchers have analyzed the life histories of fish living in habitats ranging from dark, toxic caves to nontoxic surface streams.
- Females living in dark, toxic caves invest in fewer, larger offspring than do fish living in a less stressful environment. Food is abundant but hard to acquire in the toxic caves, and the larger juveniles are less likely to starve than are their smaller counterparts.

MULTIPLE CHOICE QUESTIONS

1. Population size increases when
   a. the sum of birth rate and death rate exceeds the sum of immigration and emigration.
   b. the sum of birth rate and immigration exceeds the sum of death rate and emigration.

   c. the sum of birth rate and emigration exceeds the sum of death rate and immigration.
   d. the sum of death rate and immigration exceeds the sum of birth rate and emigration.

2. An age structure for a rat population reveals an equal number of young and adults. How will the population size likely change over time?
   a. The population size will increase.
   b. The population size will decrease.
   c. The population size will remain approximately the same.
   d. The population may go extinct.

3. As a population's size increases toward the ecosystem's carrying capacity,
   a. environmental resistance increases.
   b. the rate of population growth slows.
   c. the death rate increases.
   d. All of the above are correct.

4. An opportunistic life history emphasizes
   a. large offspring size.
   b. high offspring quantity.
   c. late reproduction.
   d. type I survivorship.

5. Which of the following is typical of developed countries such as the United States?
   a. A smaller ecological footprint than less-developed countries
   b. Sustained exponential growth
   c. Low birth rates and low death rates
   d. A shift from opportunistic to equilibrium life history

Answers to these questions are in appendix A.

WRITE IT OUT

1. List some of the ways you have interacted with your surroundings today. Categorize each item on your list as a population-, community-, or ecosystem-level interaction.

2. Chapter 39 describes biomes, the major types of ecosystems. Pick a biome and discuss a community that might be found in that biome. What populations may be included in the community you selected?

3. Describe the difference between population density and distribution. Why aren't organisms always distributed evenly throughout their habitat?

4. Why might an ecologist be interested in studying population dynamics?

5. Calculate the anticipated size of the rotifer population in figure 37.6 after 21 days. Explain your answer.

6. Refer to the logistic model of population growth shown in figure 37.8. What would be the growth rate (*G*) of the population if the population size (*N*) equaled the carrying capacity (*K*)? Explain your answer.

7. Suppose that in a population with an equilibrium life history, 60% of individuals are younger than reproductive age. Is the population likely to grow, remain stable, or decline? Why? Why might this question not apply to a population with an opportunistic life history?

8. Define the following terms: *per capita rate of increase, environmental resistance,* and *carrying capacity.*

9. Use the Internet to find a species that is growing out of control and a species that is endangered. Compare their birth rates, death rates, and other factors that are leading to the differences in the populations' fates. What techniques do researchers use to study each species?

10. Domesticated animals, such as house cats, may disturb ecosystems. Are outdoor cats a density-independent or density-dependent factor limiting the populations of birds? Explain your answer. Then, brainstorm as many ways as you can think of for how home bird feeders might raise or lower bird population sizes.

11. Some animal behaviors seem at odds with survival and reproduction. For example, when food is scarce, a female scorpion may eat her own offspring. In addition, section 35.7 describes a male spider that offers his own body for his mate to consume. Explain each of these behaviors in terms of the trade-offs described in section 37.4A.

12. Distinguish between opportunistic and equilibrium life histories, and give an example of an organism with each type of life history.

13. A species with an opportunistic life history occupies a habitat where conditions fluctuate in 2-year cycles—that is, years with optimal conditions for population growth alternate with suboptimal years. Graph the population size of the species over 6 years; then indicate on the same graph how the population of a species with an equilibrium life history would change over the same period.

14. What is your ecological footprint? To find out, search for the Personal Footprint Calculator on the Global Footprint Network website (or find a similar calculator on the Internet). What can you do to reduce your ecological footprint?

PULL IT TOGETHER

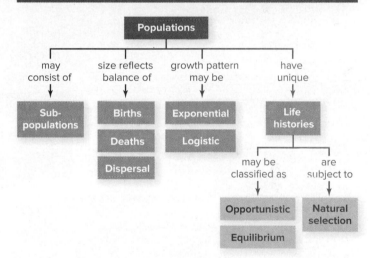

Figure 37.21 Pull It Together: Populations.

Refer to **figure 37.21** and the chapter content to answer the following questions.

1. Review the Survey the Landscape figure in the chapter introduction, and then add *ecosystems, nonliving environment,* and *communities* to the Pull It Together concept map.

2. Add *G* (the number of individuals added during any time interval) to the concept map. Explain how *G* changes throughout exponential and logistic growth.

3. Add the following terms to this concept map: *density-dependent, density-independent, carrying capacity,* and *age structure.*

38

Communities and Ecosystems

Yard Work. Many homeowners spray pesticides on their lawns to kill unwanted plants, animals, and fungi. These chemicals may end up in streams and lakes, altering food webs in unknown ways.

©Huntstock Images/Brand X Pictures/Getty Images RF

LEARN HOW TO LEARN
Perform Your Best on Exams

Last-minute cramming for exams may be a classic college ritual, but it is not usually the best strategy. If you try to memorize everything right before an exam, you may become overwhelmed and find yourself distracted by worries that you'll never learn it all. Instead, work on learning the material as the course goes along. Then, on the night before the exam, get plenty of rest, and don't forget to eat on exam day. If you are too tired or too hungry to think, you won't be able to give the exam your best shot.

Everyday Ecology: Meat and the Perfect Lawn

Humans shape ecology, sometimes to an extreme degree. For example, most of the chicken, turkey, pork, and beef consumed in the United States comes from CAFOs, or concentrated animal feeding operations. These facilities represent "incomplete" ecosystems. They house extremely high densities of animals, even though they lack plants and decomposers. Their food comes from faraway croplands, and the manure is piped out. Spreading these wastes on nearby land can cause nutrient overloads, altering those ecosystems as well. CAFOs may even spur the evolution of antibiotic-resistant bacteria (see section 17.5).

Closer to home, urban and suburban yards may also be out of balance. A conventional lawn or garden may be heavily irrigated, depleting the local drinking water supply. Homeowners may apply chemical fertilizers to boost the nutrient content of the soil, and they may spray pesticides to kill weeds, insects, fungi, and other pests. Some of the fertilizer ends up in nearby streams and lakes—possibly causing an explosion of algal growth that leads to fish deaths. The pesticides may enter local waterways as well. Dead leaves and lawn clippings may go into the trash, along with the nutrients they contain. Starved of organic matter, soils may have few decomposers.

What can you do to promote an environmentally friendly lifestyle? If you can afford their higher prices, consider buying pasture-raised beef, dairy products, pork, poultry, and eggs. Pasture-raised animals live at much lower densities than do those in a CAFO, consuming plants grown on-site and generating much smaller volumes of waste. In the yard, take an evolutionary perspective. Native plants are adapted to the local climate, soil, insects, and other environmental factors. Therefore, they typically require less water, fertilizer, and pesticides than do exotic plants.

Cultivating a diverse mixture of native plants in your garden also helps local animal life. Bees and other pollinators adapted to the region often seek out native plants for food. The plants, in turn, rely on the pollinators for reproduction. Other "green" tips include adding compost to soil instead of relying on chemical fertilizers. Compost provides food for earthworms, insects, and soil microbes, and it poses a much lower threat of nutrient pollution than do chemical fertilizers.

Farm animals, bees, and garden plants form just a few strands in a complex web of interactions that sustain every ecosystem on Earth. This chapter describes the ties that bind species together.

SURVEY THE LANDSCAPE
The Ecology of Life

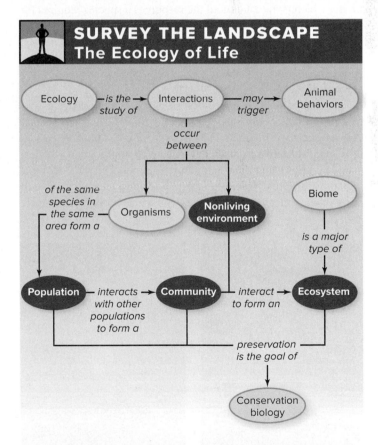

No species has an exclusive hold on its habitat. Multiple species form complex webs of interactions ranging from symbiosis to competition to predation. Along the way, each individual exchanges energy and materials with the nonliving environment.

For more details, study the Pull It Together feature in the chapter summary.

767

38.1 Multiple Species Interact in Communities

Chapter 37 described the ecology of **populations,** which consist of members of the same species that inhabit the same area. Douglas fir trees, for example, form a population in the Pacific Northwest. But no population lives in isolation; this chapter therefore extends the study of ecology to communities and ecosystems.

A **community** is a group of interacting populations. Each of the forests pictured in **figure 38.1** houses a bustling **biotic,** or living, community of photosynthetic organisms, animals, and microbes. Each community is part of an **ecosystem** that includes not only the living organisms but also the **abiotic,** or nonliving, environment. Note that the communities in figure 38.1 have few (if any) species in common, largely because the physical and chemical conditions in the two locations are so different. That is, Douglas fir trees and mushrooms cannot survive in salt water, whereas kelp and damselfish cannot live in soil.

Each species in a community has a characteristic home and way of life. Recall from chapter 37 that a **habitat** is the place where members of a population typically live, such as a forest canopy or the bottom of a river. The habitat is one part of the **niche,** which is the total of all the resources a species requires for its survival, growth, and reproduction. In addition to the habitat, the niche also includes the temperature, light, water availability, salinity, fire, and other abiotic conditions where the species lives. Biotic interactions, such as an organism's place in the food chain, are part of the niche as well.

Communities usually consist of enormous numbers of species. Some are easily visible, whereas others are microscopic. One on one, their interactions may seem simple—a whale eats an otter, or a wasp kills a caterpillar. But an attempt to map all interactions within a community quickly becomes complicated. Individuals of different species compete for limited resources, live in or on one another, eat one another, and try to avoid being eaten.

The first half of this chapter describes community-level interactions, whereas sections 38.3 and 38.4 consider ecosystem-level processes. In studying these topics, keep in mind that population-, community-, and ecosystem-level interactions are the selective forces that shape the evolution of each species. Plants, animals, and all other organisms must be able to defend themselves and to acquire resources for growth, maintenance, and reproduction. The adaptations that characterize each species—the ability to produce thorns or live in salt water or catch a gazelle—ultimately trace their origins to genetic mutations. But these features persist over multiple generations because they have enhanced reproductive success in a dangerous and competitive world.

A. Many Species Compete for the Same Resources

Competition occurs when organisms vie for the same limited resource, such as shelter, nutrients, water, light, or food (**figure 38.2**). Since neither participant obtains all of the resource that it needs, the effects of competition are negative for both.

We saw in section 37.3 that competition within a species—for food, mates, breeding sites, and other resources—can limit a population's growth. Likewise, organisms also compete with other species, especially those that occupy a similar niche.

Competition can help shape the species composition of a community. Consider the **competitive exclusion principle,** which states that two species cannot coexist indefinitely in the same niche. The two species will compete for the limited resources that they both require, such as food, nesting sites, or soil nutrients. According to the competitive exclusion principle, the species that acquires more of the resources will eventually "win." The less successful species dies out. (Competitive exclusion explains how the microbes that normally occupy the human intestinal tract provide such good protection against harmful invaders; see chapter 32.)

Figure 38.1 **Two Communities.** An evergreen forest on land and a kelp forest in the ocean both include populations of multiple species.
(mushroom): ©IT Stock Free/Alamy RF; (kelp): ©David Hall/Science Source

Figure 38.2 **Competition.** A blackbird (*left*) fights over apples with a fieldfare (*right*) in England.
©FLPA/Alamy

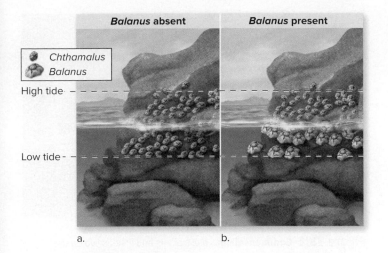

Balanus absent   **Balanus present**

Chthamalus
Balanus

High tide -

Low tide -

a.   b.

Figure 38.3 **Competitive Exclusion.** (a) When the barnacle *Balanus* is absent, *Chthamalus* adults occupy the entire intertidal zone. (b) When the faster-growing *Balanus* is present, however, competition for space limits *Chthamalus* to the upper intertidal zone.

Figure 38.3 illustrates a classic example of competitive exclusion, which affects two species of barnacles living in the intertidal zone along Scotland's shoreline. When the faster-growing species (*Balanus*) is removed from the intertidal zone, its slower-growing competitor (*Chthamalus*) occupies the entire area. What if both species are present? *Balanus* crowds out *Chthamalus* by competitive exclusion, but only in the lower intertidal zone. In the upper region, *Chthamalus* has the advantage: It can tolerate dehydration while the tide is out. *Balanus* cannot survive in that area, even when *Chthamalus* is absent.

Introduced species sometimes displace native species by competitive exclusion. Zebra mussels, for example, are native to the Caspian Sea in Asia. These mollusks were accidentally introduced into the Great Lakes in the 1980s and have since spread to many waterways in the United States and Canada. The tiny filter feeders reproduce rapidly and have crowded out native mussel species, with which they compete for food and oxygen. The effects of the zebra mussel invasion have rippled through the rest of the lake community as well. Zebra mussels have greatly increased water clarity, which has changed the amount of light available to aquatic plant communities. In turn, the altered plant species composition has triggered changes in the community of fishes. ⓘ *invasive species,* section 40.5A

Competitive exclusion, however, is not inevitable; coexistence in overlapping niches is also possible. After all, if interspecific competition reduces fitness, then natural selection should favor organisms that avoid competition. Therefore, another possible outcome of competition is **resource partitioning,** in which multiple species use the same resource in a slightly different way or at a different time. For

example, multiple species of rockhopper penguins live on islands in the southern Indian Ocean, occupying similar niches: The birds all appear similar, and they all eat similar foods. When researchers tracked penguin movements, however, they found that the populations feed in different places and at different times (**figure 38.4**). Their feeding locations and times reduce competition and therefore improve the reproductive success of all populations.

B. Symbiotic Interactions Can Benefit or Harm a Species

In a **symbiosis,** two species share a close (and often lifelong) relationship in which one typically lives in or on the other. The relationship between symbiotic species may take several forms, defined by the effect on each participant.

Mutualistic relationships are symbioses that improve the fitness of both partners. Clownfish living among the tentacles

Figure 38.4 **Resource Partitioning.** Two species of rockhopper penguins form three populations. All have similar diets, but each feeds at a different time or place.
Photo: ©Enrique R. Aguirre Aves/Oxford Scientific/Getty Images

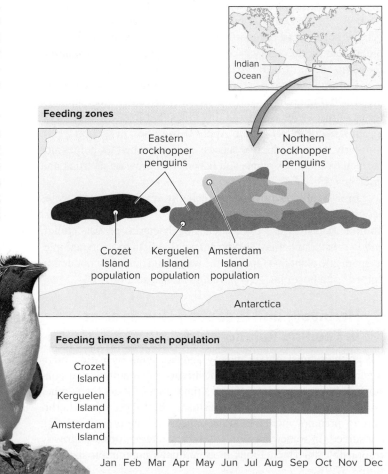

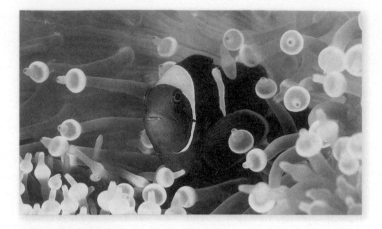

Figure 38.5 Mutualism. A clownfish in Indonesia is safe from predators among the tentacles of a sea anemone; the fish chases away animals that would otherwise nibble on the anemone.
©Reinhard Dirscherl/WaterFrame/Getty Images

Figure 38.6 Commensalism. Moss plants and lichens live on the branches and trunks of trees, enhancing their own sun exposure without harming their hosts.
©Dr. Parvinder Sethi RF

of sea anemones are a classic example (figure 38.5), as are mycorrhizal fungi. In the latter case, the fungus acquires nutrients and water that it shares with its host plant; the plant feeds sugars to its live-in partner. Many of the bacteria in our intestines are also mutualistic; these microbes consume nutrients from our food but also produce vitamins and defend us against disease. Section 22.5 describes a mutualistic relationship in which ants occupy specialized compartments in branches and attack caterpillars that might otherwise eat the tree's leaves. ⓘ *normal microbiota,* section 17.4B; *mycorrhizae,* section 20.7B

Commensalism is a type of symbiosis in which one species benefits but the other is not significantly affected. Most humans, for example, never notice the tiny mites that live, eat, and breed in our hair follicles (see the Apply It Now box in section 21.8). Similarly, the reproductive success of a tree is neither helped nor harmed by the moss plants and lichens that grow on its trunk and branches (figure 38.6).

In a symbiotic relationship called **parasitism,** one species acquires resources at the expense of a living host. The most familiar parasites are disease-causing bacteria, protists, fungi, and worms. Plants may also be parasites. Mistletoe, for example, is a parasitic plant that taps into the water- and nutrient-conducting "pipes" of a host plant. ⓘ *parasitic plants,* section 23.4

C. Herbivory and Predation Link Species in Feeding Relationships

All animals must obtain energy and nutrients by eating other organisms, living or dead. An **herbivore** is an animal that consumes plants; a **predator** is an animal that kills and eats other animals, called **prey.** As in parasitism, the fitness of the herbivore or predator increases at the expense of the organism being consumed. In some cases, predator–prey interactions are directly responsible for fluctuations in an animal's population size (see figure 37.9).

Natural selection favors plant defenses against herbivores, which may eat leaves, roots, stems, flowers, fruits, or seeds. The loss of leaf and root tissue reduces the plant's ability to carry out photosynthesis; consumption of flowers or immature fruits and seeds compromises the plant's reproductive success. Some plant species deter herbivores with thorns, milky sap, or distasteful or poisonous chemicals. The spicy hot chemicals in chili peppers, for example, discourage attack by both fungi and small mammals (see section 24.8). At the same time, many herbivores have adaptations that correspond to the plant's defenses. The caterpillars of monarch butterflies, for example, tolerate the noxious chemicals in milkweed plants.

Likewise, predation exerts strong selective pressure on prey animals, which often have adaptations that help them avoid being eaten (figure 38.7). Camouflage and warning coloration are two examples. An interesting variation on the theme of warning coloration is mimicry, in which different species develop similar appearances. For example, a harmless species of fly may have yellow and black stripes similar to those of a bee. The stripes deter predators even though the fly cannot sting.

Besides camouflage and warning coloration, prey species may also have hard shells, pincers, stingers, or other defensive adaptations (some of which are also useful in capturing their own prey). And as described in chapter 36, prey animals also have a repertoire of defensive behaviors, including stotting, fleeing, fighting, releasing noxious chemicals, or forming a tight group.

Only those predators that can defeat prey defenses will live long enough to reproduce and care for their young. Acute senses, agility, sharp teeth, and claws are common among predators. Camouflage is adaptive in predators as well as prey. Tigers and other big cats, for example, have markings that hide their shape against their surroundings, which helps them sneak up on their prey. Hunting in groups is a behavioral adaptation that helps predators capture large prey (figure 38.8).

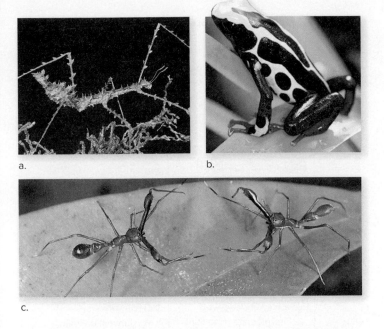

a.

b.

c.

Figure 38.7 Prey Defenses. (a) This insect from Madagascar resembles the leaves in its habitat. (b) Warning coloration advertises a poison dart frog's defenses. (c) These jumping spiders look like ants. Many predators avoid ants, which are aggressive and unpalatable.

(a): ©Kevin Schafer/The Image Bank/Getty Images; (b): ©MedioImages/ SuperStock RF; (c): ©Simon D. Pollard/Science Source

D. Closely Interacting Species May Coevolve

Some connections between species are so strong that the species directly influence one another's evolution. In **coevolution,** a genetic change in one species selects for subsequent changes in the genome of another species. Of course, all interacting species in one community have the potential to influence one another, and they are all "evolving together." These genetic changes are considered coevolution only if scientists can demonstrate that adaptations specifically result from the interactions between the species.

One example of coevolution is the relationship between lodgepole pines and birds called crossbills that eat the trees' seeds (**figure 38.9**). In areas with crossbills, the pine trees produce large seed cones with thick, protective scales. The birds, however, have a corresponding adaptation: Their bills are largest in forest regions where pines have thick cones.

Another example of an evolutionary "arms race" links predators and prey animals. Amphibians called rough-skinned newts produce exceedingly high concentrations of a toxin that binds to sodium channels in a predator's muscles, usually causing paralysis and death. Garter snakes, however, routinely eat the newts. Their muscle cells have modified sodium channels that are resistant to the poison. Snakes with resistant sodium channels have greater reproductive success than those with typical channels; on the other hand, newts are vulnerable to resistant snakes. Natural selection therefore simultaneously favors more potent newt toxins and more resistant snakes.

Flowering plants and insects have also coevolved. As described in chapter 24, a plant may rely on one insect species for pollination, and the insect may eat nectar from only that plant.

38.1 MASTERING CONCEPTS

1. Distinguish among communities, ecosystems, and populations.
2. Name some abiotic and biotic parts of your environment.
3. Distinguish between habitat and niche.
4. What is the competitive exclusion principle?
5. Give examples of three types of symbiotic relationships.
6. Describe adaptations that protect against herbivory and predation.
7. Define *coevolution* and describe an example.

Figure 38.8 Predator Cooperation. By working together, a pack of wolves can bring down an elk that is much larger than the wolves themselves. Sharp teeth and claws aid in prey capture as well.

U.S. National Park Service/Doug Smith

Figure 38.9 Coevolution. In the Rocky Mountains, crossbills eat the seeds of lodgepole pines. The birds select for pine cones with thick, protective scales. In turn, the cones select for birds with larger, stronger bills.

©Craig W. Benkman

38.2 Succession Is a Gradual Change in a Community

From oceans to mountaintops, many species share each habitat. As described in section 39.3, however, communities vary greatly in diversity. Ecologists consider diversity to be a function of two measures, called species richness and species evenness (figure 38.10). One way to measure **species richness** is simply to count the species occupying a habitat. A patch of prairie, for instance, may contain about 100 plant species, whereas an equal-sized area of desert might house only 6 types of plants. In this example, the prairie has greater species richness than the desert.

But two communities with the same species richness may not be equally diverse. **Species evenness,** or relative abundance, describes the proportion of the community that each species occupies. In our patch of prairie, for example, suppose that one type of plant accounts for 90% of the individuals in the community, with 99 species making up the remaining 10%. Because one species has such high relative abundance, that community is less diverse than one in which, say, each of the 100 species makes up 1% of the community.

The numbers and types of species that form each community may seem constant, but that is only because we usually observe them over a relatively short period. **Succession** is a gradual change in a community's species composition. Ecologists define two major types of succession: primary and secondary.

Primary succession occurs in an area where no community previously existed. When a volcano erupts, for example, lava may obliterate existing life, a little like suddenly replacing an intricate painting with a blank canvas. Road cuts and glaciers that scour the landscape also expose virtually lifeless areas on which new communities eventually arise.

Figure 38.11 illustrates primary succession in New England, a region typically occupied by deciduous forest. The process begins with a patch of bare rock. Hardy **pioneer species** such as lichens are first to colonize the area. Lichens release organic acids that erode

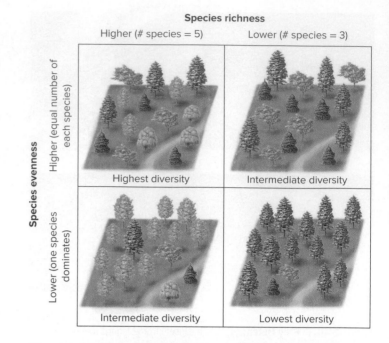

Figure 38.10 **Measures of Community Diversity.** Species diversity combines two components: species richness and species evenness.

the rock, producing crevices where sand and dust accumulate. Decomposing lichens add organic material, eventually forming a thin covering of soil. Then rooted plants such as herbs and grasses invade. Soil continues to form, and larger plants such as shrubs appear. As these new plants take root, a changing variety of birds,

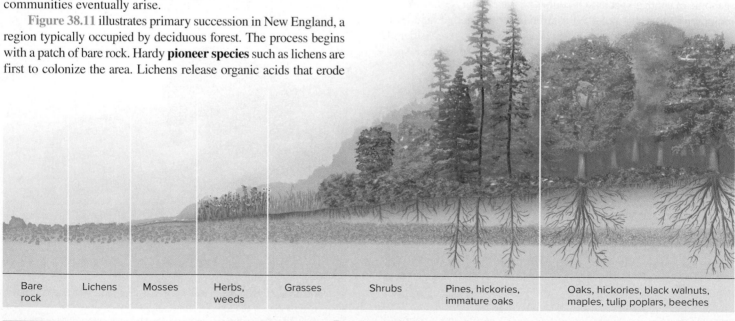

| Bare rock | Lichens | Mosses | Herbs, weeds | Grasses | Shrubs | Pines, hickories, immature oaks | Oaks, hickories, black walnuts, maples, tulip poplars, beeches |

Time (hundreds of years)

Figure 38.11 **Primary Succession.** It takes centuries for a mature forest community to develop on a patch of bare rock. The example shown here includes plant species typical of New England; a region with another soil type and climate would have a different mix of species.

Figure 38.12 Secondary Succession. A forest fire can devastate an existing community. Soon, however, seedlings sprout and absorb the nutrients in the tree's ashes. The forest eventually regrows.
©Jerry Dodrill/Aurora/Getty Images

mammals, and other vertebrates joins the community as well. Next come the young trees. Finally, hundreds of years after lichens first arrived on the bare rock, the soil becomes rich enough to support a stable, mature forest community. ⓘ *lichens,* section 20.7D

In contrast to primary succession, **secondary succession** occurs where a community is disturbed but not destroyed. Because some soil and life remain, secondary succession occurs faster than primary succession. Fires, hurricanes, and agriculture commonly trigger secondary succession (**figure 38.12**).

Primary and secondary succession share a common set of processes. The first plants to arrive are usually opportunistic, with rapid reproduction and efficient dispersal. These early colonists often alter the physical conditions in ways that enable other species to become established. The new arrivals, in turn, continue to change the environment. Some early colonists do not survive the new challenges, further altering the community. When pine trees invade a site, for example, they simultaneously shade out lower-growing plants while attracting species that grow or feed on pines. Later in succession, the dominant species are usually long-lived, late-maturing, equilibrium species that are strong competitors in a stable environment. ⓘ *opportunistic and equilibrium species,* section 37.4B

A century ago, ecologists hypothesized that primary and secondary succession would eventually lead to a so-called **climax community,** which is a community that remains fairly constant. We now know, however, that few (if any) communities ever reach true climax conditions. In the Pacific Northwest, for example, old-growth forests are 500 to 1000 years old, yet they are still changing in their structure and composition. Major disturbances such as fire, disease, and severe storms can leave a mark that lasts for centuries. On a smaller scale, pockets of local disturbance, such as the area affected when a large tree blows over, create a patchy distribution of successional stages across a landscape.

Succession is not limited to land; it occurs in aquatic communities as well. A young lake, for example, is too low in nutrients to support abundant phytoplankton. These lakes are therefore clear and sparkling blue. As a lake ages, however, nutrients accumulate from decaying organisms and sediment. Algae thrive, turning the water green and murky. In time, a lake continues to fill with sediments and transforms into a freshwater wetland, where the soil is permanently or seasonally saturated with water. Wetlands often host spectacularly diverse assemblages of plants and animals that rely on the interface between land and water. Eventually, the wetland fills in completely and becomes dry land.

38.2 MASTERING CONCEPTS

1. How do ecologists measure species diversity in a community?
2. How is natural selection apparent in ecological succession?
3. What processes and events contribute to primary and secondary succession?
4. How do disturbances prevent true climax communities from developing?

Burning Question

Could human life be supported in space or on Mars?

Astronauts can already make months-long visits to space, courtesy of the International Space Station orbiting Earth. Special "transfer vehicles" periodically deliver food, and wastes are loaded into spacecraft that are destroyed in the atmosphere.

During the 10-month journey to the Red Planet, however, a space capsule would have to be a fully self-contained ecosystem, complete with plants (to feed the astronauts) and microbes. Without a way to acquire new resources, the occupants would have to be fanatical about recycling all essential elements. Water would be purified and reused. Human wastes would be composted into a form that onboard plants could use. Luckily, the byproduct of photosynthesis is O_2. But imagine the catastrophe if an unwanted stowaway, such as a plant pathogen, killed all the onboard plants!

In anticipation of a future Mars mission, researchers are working to find the best concentrations of CO_2, minerals, and other raw materials that plants need. Determining which crop plants grow well together is another high priority.

Submit your burning question to Marielle.Hoefnagels@mheducation.com

38.3 Ecosystems Require Continuous Energy Input

Sections 38.1 and 38.2 described the biotic interactions among members of a community. We now turn to ecosystems, which represent a still larger scale of ecology. Ecosystems include everything from polar ice to the open ocean to the tropical rain forest, each with its own unique community and set of abiotic conditions. (This chapter's Burning Question considers the difficulties of building artificial ecosystems.)

To understand ecosystem-level interactions, it is useful to recall from unit 1 that all organisms consist of both matter and energy. One way to remember this is to picture a candy bar's nutrition label, which lists the nutrient and calorie (energy) content of the snack. The food label mirrors the contents of a living cell. Inside each cell are organic molecules such as fats, sugars, and proteins. These molecules consist of carbon, hydrogen, oxygen, nitrogen, and other elements. Moreover, the covalent bonds of organic molecules store potential energy that cells can use to do work. ⓘ *organic molecules,* section 2.5

Both energy and nutrients are critical to the two properties shared by all ecosystems on Earth. First, energy flows through ecosystems in one direction only. All ecosystems therefore rely on a continuous supply of energy from some outside source, usually the sun. Second, the atoms that make up every object in an ecosystem are constantly recycled. This section and the next describe these two properties and their consequences for ecosystem function.

A. Food Webs Depict the Transfer of Energy and Atoms

Many energy and nutrient transfers occur in the context of food chains and food webs. A **food chain** is a linear sequence of feeding relationships: A beetle eats a plant, a bird eats the beetle, and so on (**figure 38.13**). Each organism's **trophic level** describes its position in the food chain.

Trophic levels are defined relative to the ecosystem's energy source. The first trophic level in any food chain is a primary producer. A **primary producer,** or **autotroph** ("self-feeder"), is any organism that can use energy, CO_2, H_2O, and other inorganic substances to produce all the organic material it requires. For most ecosystems, the energy source is sunlight. Plants, algae, cyanobacteria, and some other microorganisms are primary producers that use photosynthesis to trap solar energy in the bonds of organic chemicals such as glucose (see chapter 5).

Nevertheless, a few ecosystems rely on energy sources other than sunlight. For example, some bacteria and archaea are primary producers that can extract energy from inorganic chemicals such as iron or manganese. Countless numbers of these microbes form the bases of complex hydrothermal vent communities that never see the sun. The opening essay for chapter 39 describes these unique ecosystems in more detail.

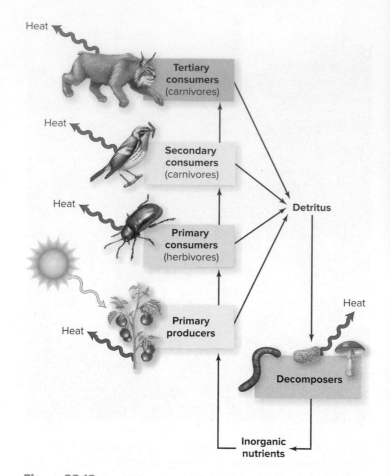

Figure 38.13 Trophic Levels. Producers are at the base of the food chain; consumers occupy the other trophic levels. All organisms contribute detritus (wastes and dead bodies) to the ecosystem. Decomposers are microbes that return the nutrients to their inorganic form, which producers absorb.

All of the other trophic levels consist of **consumers,** or **heterotrophs** ("other eaters"), which obtain energy either from producers or from other consumers. For example, the primary consumers in figure 38.13 are herbivores, which eat the primary producers. Secondary consumers are carnivores (meat-eaters) that eat primary consumers, and tertiary consumers eat secondary consumers.

All organisms leave behind **detritus** consisting of dead tissue and organic wastes such as feces. Scavengers are animals that eat this material. Vultures, crows, raccoons, flies, earthworms, and many other animals are scavengers. **Decomposers,** such as many fungi and bacteria, are microbes that complete the recycling process; they secrete enzymes that digest the remaining organic molecules in detritus. As they do so, they fuel their own growth and reproduction, but they also return carbon, nitrogen, phosphorus, and other inorganic nutrients to the environment. Without these crucial microbes, dead bodies and organic wastes would tie up all useful nutrients, and ecosystems would grind to a halt. This chapter's Apply It Now box describes how we employ decomposers in community wastewater treatment facilities.

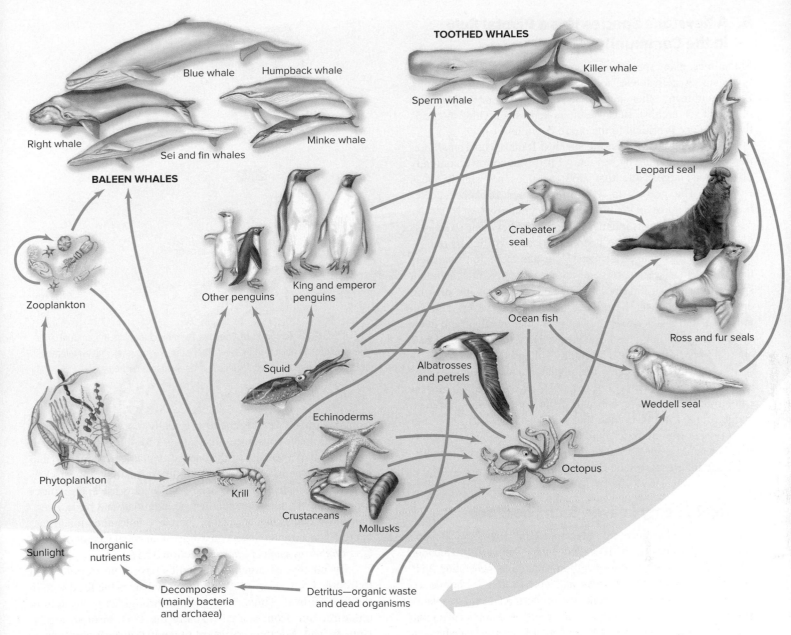

Figure 38.14 Antarctic Web of Life. The interactions among Antarctic residents form a complex network. Note that producers, consumers, and scavengers are all present, and that decomposers release inorganic nutrients that producers can use. For simplicity, heat is not illustrated.

Autotrophs and decomposers have opposite roles in ecosystems. Whereas autotrophs absorb inorganic nutrients and produce organic molecules, decomposers return the elements in those organic molecules to their inorganic form. Both roles are critical to ecosystem function.

Of course, feeding relationships in an ecosystem are more complex than a simple food chain might suggest. A **food web** is a network of interconnected food chains, such as the Antarctic web in figure 38.14. Keep in mind that this diagram is highly simplified. Not shown, for example, are the worms, hagfishes, sharks, and other organisms that feed for months or years on the carcasses of dead whales that sink to the seafloor. Nor does the diagram depict the many transfers between aquatic ecosystems and the land.

Careful examination of a food web diagram reveals that even the fiercest top predator, such as a killer whale, relies on other organisms, many of them microscopic. The same principle applies to the food web in which humans participate. Think of everything you have eaten today: In all likelihood, your meals and snacks have included both plant and animal products. Humans have developed an extremely complex global food chain that includes organisms harvested from land and water all over the world. Some of the connections are surprising. For example, pigs and chickens raised on commercial farms around the world eat millions of tons of fish harvested near South America each year. Conversely, commercially raised catfish eat soybean meal, corn, and rice—all of which grow on land.

B. A Keystone Species Has a Pivotal Role in the Community

Sometimes, many species in a community depend on one type of organism. A kelp forest, for example, houses snails, shrimps, fishes, and many other species. Sea urchins devour the kelp, and otters eat sea urchins. Because otters keep sea urchins in check, their presence is critical to the entire community.

The sea otter is a **keystone species:** It makes up a small portion of the community by weight, yet its influence on community diversity is large. Note that *keystone* does not simply mean *essential.* The grasses in a prairie are not keystone species because they make up the bulk of the community.

How do ecologists identify keystone species? One strategy is to measure what happens to a community after artificially removing each species, one at a time. Using this method, researchers discovered that sea stars are keystone predators that maintain species diversity in tide pools. The sea stars normally prey on diverse invertebrates. Removing sea stars, however, allowed mussels and barnacles to take over, crowding out algae and other invertebrates. Without the keystone predator, the tide pools lost 7 out of 15 species, and the community collapsed.

Many keystone species are predators, but mutualists may also be keystone species. For example, mycorrhizal fungi help coniferous trees acquire nutrients from soil, and they produce underground fruiting bodies that small rodents eat. Owls and other predators hunt these small mammals. The fungi are considered keystone species because their small biomass is disproportionate to their enormous influence on community structure.

C. Heat Energy Leaves Each Food Web

The total amount of energy that is trapped, or "fixed," by all autotrophs in an ecosystem is called gross primary production. Autotrophs use much of this energy in respiration, generating ATP for their own growth, maintenance, and reproduction. As they do so, they lose heat energy (the second law of thermodynamics explains this loss). The remaining energy in the producer level is called **net primary production;** it is the amount of energy available for consumers to eat. ⓘ *laws of thermodynamics,* section 4.1B

Primary production varies widely across ecosystems on land, depending largely on temperature and moisture. The warm, wet tropical rain forests therefore have among the highest rates of net primary production per square meter; deserts have the lowest. In aquatic ecosystems, the availability of inorganic nutrients such as phosphorus is more important. Nutrient-polluted waters therefore often become overgrown with algae (see section 38.4).

Consumers also produce heat energy in every metabolic reaction. Because of these inefficiencies, only a small fraction of the potential energy in one trophic level fuels the growth and reproduction of organisms at the next trophic level. On average, about one tenth of the energy at one trophic level is available to the next rank in the food chain (**figure 38.15**).

The "10% rule" provides a convenient estimate, but it ignores the fact that food quality varies widely. The transfer efficiency

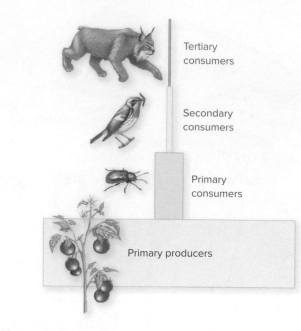

Figure 38.15 Pyramid of Energy. Each block depicts the amount of energy stored in each trophic level. This example assumes that an average of 10% of the energy in any trophic level is available to the next.

from one trophic level to the next actually ranges from about 2% to 30%. Primary consumers that eat hard-to-digest plants convert only a small percentage of the energy available to them into animal tissue, whereas meat is easy to digest. In addition, ectothermic animals such as insects, fishes, and lizards use energy much more efficiently than do endothermic mammals and birds. A trophic level consisting of lizards therefore consumes much less energy than a trophic level consisting of an equal weight of birds. ⓘ *endotherms and ectotherms,* section 33.1A

Eventually, as organic molecules pass from trophic level to trophic level, all of the stored energy leaves the food web in the form of heat. Thus, energy flows through an ecosystem in one direction: from source (usually the sun), through organisms, to heat. For the ecosystem to persist, it must have a continual supply of energy. If the energy source goes away, so does the ecosystem.

A **pyramid of energy** represents each trophic level as a block whose size is directly proportional to the energy stored in biomass per unit time (see figure 38.15). Because every organism loses heat to the environment, the energy pyramid explains why food chains rarely extend beyond four trophic levels. An organism in a still higher trophic level would have to expend tremendous effort just to find the small amount of food available, and that small amount would not be enough to make all that effort pay off.

The loss of energy at each trophic level suggests a way to maximize the benefit we get from crops we grow for food. The most energy available in an ecosystem is at the producer level. The lower we eat on the food chain, the more people we can feed. A person can do this by getting protein from beans, grains, and nuts instead of from meat and dairy (**figure 38.16**).

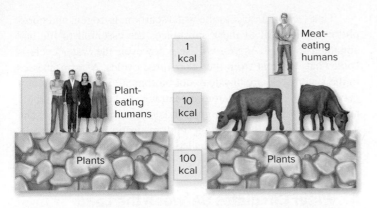

Figure 38.16 The Energetic Cost of Meat. These simplified pyramids show that the energy available in plants can support many more vegetarians than it can meat-eaters.

Figure It Out

Consult the food web in figure 38.14. Assuming the 10% rule is correct, about how many kilograms of krill would it take to support one 40-kilogram emperor penguin?

Answer: 4000 kg

D. Harmful Chemicals May Accumulate in the Highest Trophic Levels

The shape of the energy pyramid has another consequence for ecosystems. In **biomagnification,** a chemical becomes most concentrated in organisms at the highest trophic levels. Biomagnification happens for pollutants and other chemicals that share two characteristics. First, they dissolve in fat. Animals eliminate water-soluble chemicals in their urine but retain fat-soluble chemicals in fatty tissues. Second, chemicals that biomagnify are not readily degraded. A highly degradable chemical would not persist long enough in the environment to ascend food chains.

Mercury is a persistent pollutant that illustrates biomagnification (figure 38.17). Coal-fired power plants and mines release this element into the air and water. Moreover, many household products contain mercury that enters water, air, or soil via sewage treatment plants and incinerators. Imagine mercury entering water from a nearby power plant. Bacteria in the sediments soon convert the mercury into a fat-soluble form called methylmercury. Its concentration in the water is initially low. But methylmercury that enters an organism's body is not eliminated in urine or other watery wastes. As one animal eats another, all of the methylmercury stored in the prey ends up in the predator. Each predator eats many prey, so the mercury accumulates in the predator's tissues. In organisms at the fourth level of the food chain, mercury concentrations may be 100,000 times greater than at the base of the food web. High levels of mercury have prompted health professionals to recommend that people limit their consumption of long-lived, carnivorous fish such as shark and tuna.

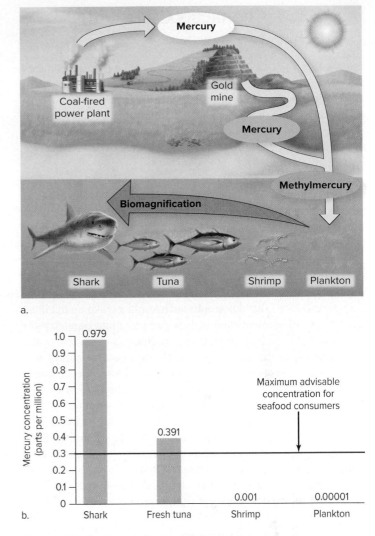

Figure 38.17 Biomagnification. (a) Power plants and gold mines are among the main sources of mercury pollution. Mercury is converted to methylmercury in sediments; after entering the food chain, the methylmercury concentration increases with each successive trophic level. (b) Concentrations of mercury in shark meat, tuna, and shrimp vividly illustrate biomagnification.

38.3 MASTERING CONCEPTS

1. Identify the trophic levels of a food chain.
2. What sources of energy sustain ecosystems?
3. What roles do primary producers and decomposers play in ecosystems?
4. Woodpeckers dig nesting cavities in tree trunks; many other bird species subsequently use abandoned woodpecker cavities. In these forest communities, what term might be used to describe woodpeckers?
5. How efficient is energy transfer between trophic levels in food webs?
6. Draw an energy pyramid for an ecosystem with three levels of consumers.
7. Explain how biomagnification disproportionately affects organisms at the top of a food chain.

38.4 Chemicals Cycle Within Ecosystems

Energy flows in one direction, but all life must use the elements that were present when Earth formed. In **biogeochemical cycles,** interactions of organisms and their environment continuously recycle these elements. If not for this worldwide recycling program, supplies of essential elements would have been depleted as they became bound in the bodies of organisms that lived eons ago.

Whatever the element, all biogeochemical cycles have features in common (figure 38.18). Each element is distributed among four major storage reservoirs: organisms, the atmosphere, water, and rocks and soil. Depending on the reservoir, the element may combine with other elements and form a solid, liquid, or gas.

These "pools" are not isolated; instead, transfers among the pools form the basis of each biogeochemical cycle. Along the way, the elements undergo chemical changes. For example, recall from figure 38.13 that autotrophs such as plants take up the inorganic forms of elements and incorporate them into organic molecules such as lipids, carbohydrates, proteins, and nucleic acids. If an animal eats the plant, the elements may become part of animal tissue. If another animal eats the herbivore, the elements may be incorporated into the predator's body. Eventually, decomposers consume the dead bodies and organic wastes, releasing inorganic forms of the elements into the environment.

These transfers and transformations are the root of the term *biogeochemical cycle*: *bio-* refers to the role of organisms; *geo-* stands for the inorganic components, such as those found in the earth; *-chemical* relates to the chemical transformations; and *cycle* is for the endless transfers from reservoir to reservoir.

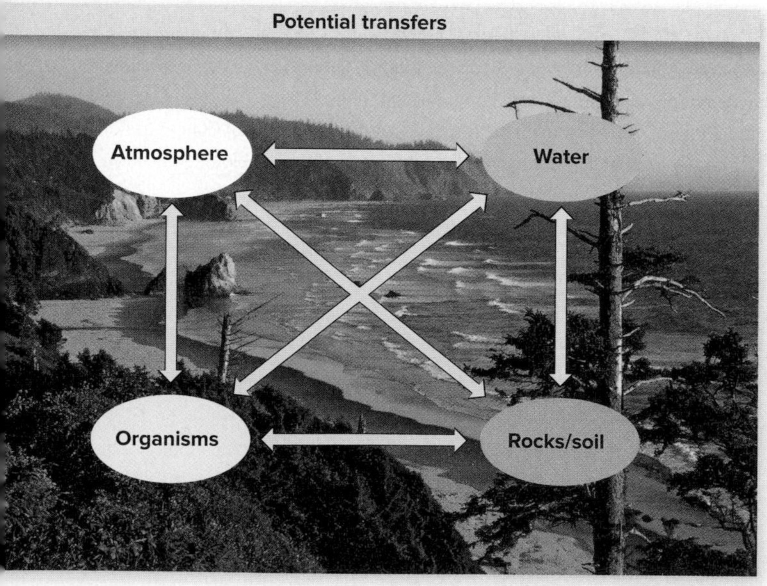

Figure 38.18 Biogeochemical Cycle. Water and inorganic nutrients cycle among four basic storage reservoirs: the atmosphere, water, organisms, and rocks and soil. A coastal ecosystem illustrates all four.
Photo: ©Corbis RF

This section describes the water, carbon, nitrogen, and phosphorus cycles. All of these substances are essential to life and abundant in cells. As you will see, however, the water cycle is somewhat different from the other three cycles. Most processes in the water cycle are physical, not biological. Its status as a biogeochemical cycle is therefore tenuous. Nevertheless, carbon, nitrogen, and phosphorus compounds can dissolve in water, and water movement is important in transporting these elements among the storage pools. Knowledge of the water cycle is therefore essential to understanding the three nutrient cycles.

A. Water Circulates Between the Land and the Atmosphere

Water covers much of Earth's surface, primarily as oceans but also as lakes, rivers, streams, ponds, swamps, snow, and ice (figure 38.19). Water also occurs below the land surface as groundwater. ⓘ *world water resources,* section 39.4

The main processes that transfer water among these major storage compartments are evaporation, precipitation, runoff, and percolation. The sun's heat evaporates water from land and water surfaces (figure 38.19, step 1). Water vapor rises on warm air currents, then cools and forms clouds (step 2). If air currents carry this moisture higher or over cold water, more cooling occurs, and the vapor condenses into water droplets that fall as rain, snow, or other precipitation (step 3). Some of this precipitation falls on land, where it may run along the surface. Streams unite into rivers that lead back to the ocean (step 4), where the sun's energy again heats the surface, continuing the cycle.

Rain and melted snow may also soak (percolate) into the ground, restoring soil moisture and groundwater (step 5). This underground water feeds the springs that support many species. Spring water evaporates or flows into streams, linking groundwater to the overall water cycle.

Although most processes in the water cycle are physical, organisms do participate (step 6); after all, water is essential to life. Plant roots absorb water from soil and release much of it from their leaves in transpiration. The lush plant life of the tropical rain forests draws huge amounts of water from soil and returns it to the atmosphere. In addition, animals drink water and consume it with their food, returning it to the environment through evaporation and urination. ⓘ *transpiration,* section 23.2A

Because trees play such a critical role in the global water cycle, the wholesale destruction of forests may have important consequences. Deforestation reduces the potential for transpiration, an important avenue by which water returns to the atmosphere. This decrease in transpiration may trigger changes in the global water cycle and, by extension, in the global climate. In addition, in an intact forest, rain and melting snow seep slowly through the forest floor and into the soil, helping to recharge groundwater. Logging and burning destroy the spongy organic matter that accumulates on forest floors, so less water percolates into soil and more pours over the land surface. This runoff carries nutrients and sediments from the eroding soil, so it reduces water quality in nearby streams and lakes. ⓘ *deforestation,* section 40.2

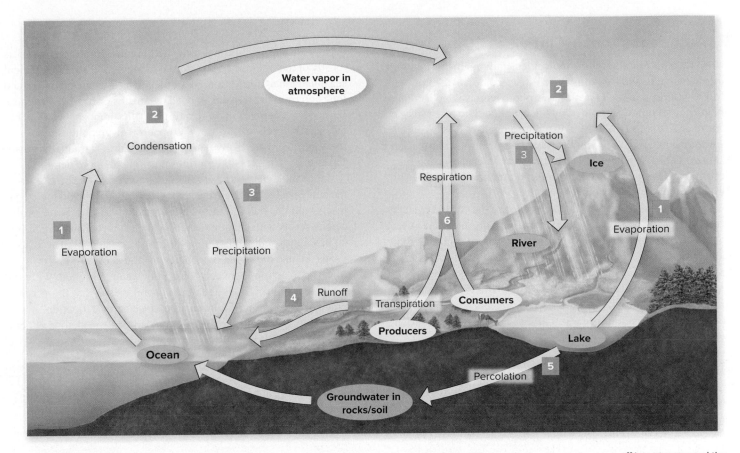

Figure 38.19 The Water Cycle. Water falls to Earth as precipitation. Organisms use some water, and the remainder evaporates, runs off into streams and the ocean, or enters the ground. Transpiration and respiration return water to the environment.

 Apply It **Now**

What Happens After You Flush

Most people probably give little thought to what happens to whatever they flush down the toilet. But the engineers and technicians who run community sewage treatment facilities think about it all the time.

Decades ago, many towns simply released untreated sewage into rivers and oceans. Because human waste harbors disease-causing organisms, this practice posed an obvious threat to public health. It was also harmful for another reason: The raw sewage killed fish and other aquatic animals. The problem was not that human diseases spread to aquatic organisms. Instead, aquatic microbes decayed the feces and other organic wastes in the sewage. As the microbes respired, they used all the available O_2 dissolved in the water (see section 38.4E). Huge numbers of fishes and other aquatic organisms suffocated and died.

Federal law now mandates that communities treat wastewater before releasing it. Treatment plants harness the power of microorganisms to consume the organic matter in sewage before it enters waterways. In trickling filters, for example, sewage-eating bacteria and archaea are given "dream homes"—all the organic matter they can eat, along with plenty of moisture and

©C Squared Studios/Getty Images RF

O_2 (see figure 17.14c). After the microbes have done their job, the treated water contains a very low concentration of organic matter.

The presence of these microscopic workers explains why communities prohibit dumping used motor oil or organic solvents such as paint thinner down the drain. Toxic chemicals can poison the bacteria and archaea that degrade sewage, making water treatment impossible.

Thanks in part to water quality laws, U.S. river ecosystems have largely recovered from the past onslaught of untreated sewage. The most commonly used forms of sewage treatment do not, however, remove all chemicals from the waste stream. In particular, a wide array of household chemicals and pharmaceutical drugs often remain in the water discharged from a sewage treatment plant. The antibiotics in soaps and hand sanitizers are especially common, as are hormones excreted in the urine of women taking birth control pills. Ecologists are still studying the effects of these chemicals on wildlife. In the meantime, experts recommend against flushing medications—or anything other than human waste—down the toilet.

B. Autotrophs Obtain Carbon as CO₂

Carbon is a part of all organic molecules, and organisms continually exchange it with the atmosphere (figure 38.20). Autotrophs absorb atmospheric CO_2 and use photosynthesis to produce organic compounds, which they incorporate into their tissues (figure 38.20, step 1). Cellular respiration releases carbon back to the atmosphere as CO_2 (step 2). Dead organisms and wastes contribute organic carbon to soil or water (step 3). Bacteria and fungi decompose these organic compounds and release CO_2 to the soil, air, and water as they respire (step 4).

Some types of archaea also participate in the carbon cycle by metabolizing organic compounds and releasing methane (CH_4) into the atmosphere. Most of these microbes live in anaerobic habitats such as wetlands, marine sediments, and the intestines of humans, cattle, and other animals.

The exchange of carbon between organisms and the environment is relatively rapid, but more stable pools of carbon also exist. A substantial fraction of soil carbon consists of persistent, decay-resistant organic matter called humus. Limestone consists mostly of the calcium carbonate exoskeletons and shells of ancient sea inhabitants. In addition, fossil fuels such as coal and oil formed long ago from the remains of dead organisms (step 5). When these fuels burn (step 6), carbon returns to the atmosphere as CO_2. Decades of accumulation of CO_2 and other greenhouse gases in the atmosphere are likely responsible for Earth's gradually warming climate. Chapter 40 describes this topic in more detail.

One of the largest reservoirs of carbon is the ocean. CO_2 from the atmosphere dissolves in ocean water. Most of the dissolved gas reacts with the water to form carbonic acid (H_2CO_3). Some of this carbon reacts with calcium to form calcium carbonate, which precipitates into sediments on the ocean floor (step 7). These sediments are one of the major stable repositories of carbon.

The reaction between CO_2 and water, coupled with the accumulation of CO_2 in the atmosphere, means that the ocean is gradually becoming more acidic. Ocean acidification harms coral reefs by dissolving the calcium carbonate skeletons of coral

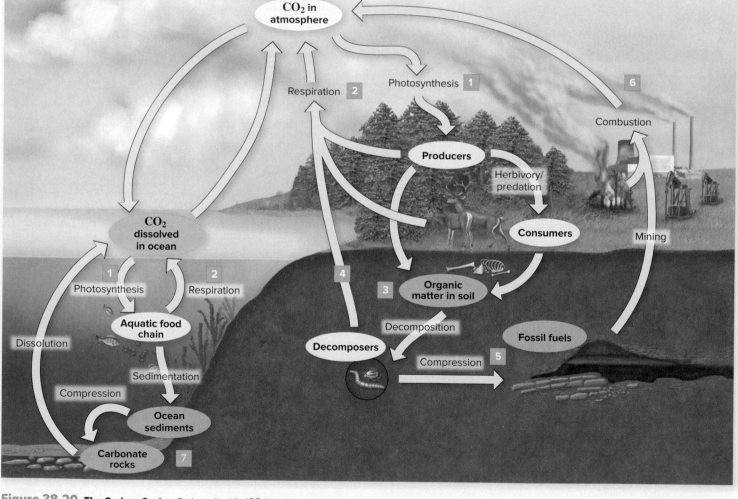

Figure 38.20 The Carbon Cycle. Carbon dioxide (CO_2) in the air and water enters ecosystems through photosynthesis and then passes along food chains. Respiration and combustion return carbon to the abiotic environment. Carbon can be retained for long periods in carbonate rocks and fossil fuels. The archaea that produce methane (CH_4) are omitted from the cycle for simplicity.

animals, joining global climate change as another side effect of greenhouse gas buildup. ⓘ *pH scale,* section 2.4

C. The Nitrogen Cycle Relies on Bacteria

Nitrogen is an essential component of proteins, nucleic acids, and other biochemicals in living cells. **Figure 38.21** depicts the nitrogen cycle.

Although the atmosphere is about 78% nitrogen gas (N_2), most organisms cannot use this form of nitrogen. The nitrogen cycle therefore depends on **nitrogen fixation,** the process by which some bacteria and archaea convert N_2 into ammonium ions, NH_4^+ (figure 38.21, step 1). Examples of nitrogen-fixing bacteria include *Rhizobium,* which lives in nodules on the roots of legume plants such as beans, peas, and clover (see figure 17.12 and figure 23.4). Many farmers alternate nonlegume crops, such as corn, with legumes to enrich the soil with biologically fixed nitrogen. In addition, farmers often boost plant growth by applying nitrogen fertilizers to their fields (step 2). Fertilizer production relies on an industrial-scale form of nitrogen fixation.

Nitrogen can also occur in the form of nitrate (NO_3^-). In a process called **nitrification,** bacteria and archaea convert ammonium to nitrate (step 3). This process occurs both in soil and in the ocean. In addition, the combustion of fossil fuels such as coal, oil, and natural gas releases NO_2 gas and other nitrogen oxides into the atmosphere (step 4). These compounds dissolve in precipitation; in soil, they are converted to nitrate. Excess nitrogen deposition from the atmosphere may be altering some low-nitrogen ecosystems, such as bogs where carnivorous plants thrive. ⓘ *carnivorous plants,* section 23.5

Plants and other autotrophs can absorb either ammonium or nitrate and incorporate it into the organic molecules that make up their own bodies (step 5). Consumers then acquire the nitrogen by eating the producers (step 6), and so on up the food chain. Decomposers release some ammonia when they decay the dead bodies and wastes (step 7).

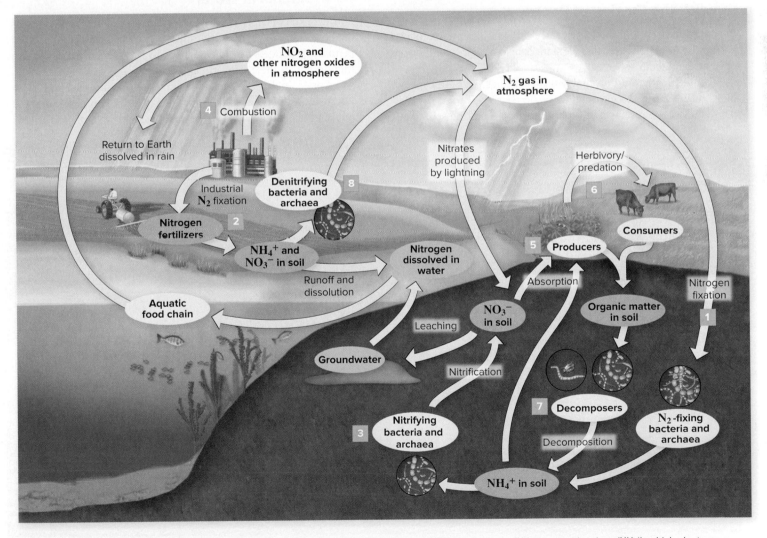

Figure 38.21 The Nitrogen Cycle. Nitrogen-fixing bacteria and archaea convert atmospheric nitrogen gas (N_2) to ammonium ions (NH_4^+), which plants can absorb. Nitrogen returns to the abiotic environment in urine and during the decomposition of organic matter. Bacteria and archaea convert ammonium to nitrate, NO_3^- (another form plants can use). Microbes also convert nitrate to N_2, completing the cycle.

Yet another group of microbes completes the cycle. In **denitrification,** bacteria and archaea return nitrogen to the atmosphere as they convert nitrate to N_2 (step 8). This process, which is a form of anaerobic respiration, occurs where O_2 is scarce, such as wetlands, water-saturated soils, groundwater, and ocean sediments. ⓘ *anaerobic respiration,* section 6.8A

Nitrification and denitrification occur in a fish aquarium, a miniature ecosystem with a nitrogen cycle of its own. Fish release toxic ammonia as a waste product, which begins to accumulate soon after fish are placed into a new aquarium. Given enough time, however, populations of nitrifying bacteria grow on the gravel and in the aquarium's filter. These microbes convert ammonia to nitrate. Other microbes carry out denitrification, eliminating the nitrogen as N_2 gas. When starting a new aquarium, one way to jump-start this process is to obtain a used filter or some gravel from an established aquarium. The microbes in the transplanted biofilms will give the new tank a head start in establishing a functioning nitrogen cycle.

D. The Phosphorus Cycle Begins with the Erosion of Rocks

Phosphorus occurs in nucleic acids, ATP, and membrane phospholipids; in vertebrates, this element is also a major component of bones and teeth.

Unlike in the carbon and nitrogen cycles, the atmosphere plays little role in the phosphorus cycle (**figure 38.22**). Instead, the main storage reservoirs for phosphorus are marine sediments

and rocks. As phosphate-rich rocks erode, they gradually release phosphate ions (PO_4^{-3}) into water (figure 38.22, step 1).

Many soils, however, are relatively low in phosphorus; a deficiency of this element often limits plant growth. Humans therefore mine phosphate rocks to produce plant fertilizers (step 2). Animal waste also contains abundant phosphorus; some people harvest the guano (droppings) of birds and bats for use as fertilizer. As described in section 38.4E, however, too much phosphorus can damage an ecosystem.

Autotrophs absorb phosphorus, often with the help of mycorrhizal fungi (step 3). Consumers move the element throughout the food web (step 4), and decomposers eventually return inorganic phosphates to soil and water (step 5). Much of the phosphate, however, joins the sediments raining down onto the ocean floor (step 6). After many millions of years, geological uplift returns some of this underwater sedimentary rock to the land (step 7).

E. Excess Nitrogen and Phosphorus Cause Problems in Water

Nitrogen and phosphorus are essential to life; in fact, deficiencies of these nutrients typically limit the growth of algae and other primary producers in aquatic ecosystems. Extra nutrients can therefore alter the ecological balance in water.

For example, sewage and fertilizers carry nitrogen and phosphorus into waterways. In a process called **eutrophication,** these nutrients trigger the rapid growth of algae and other phytoplankton

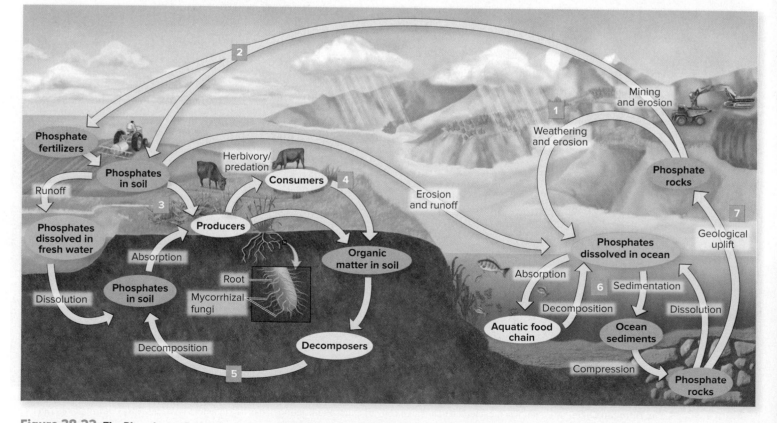

Figure 38.22 **The Phosphorus Cycle.** As phosphate-rich rocks erode, they release phosphorus that plants can absorb and pass to the rest of the food chain. Decomposers return phosphorus to the abiotic environment. Fertilizers have increased phosphorus availability to both terrestrial and aquatic organisms.

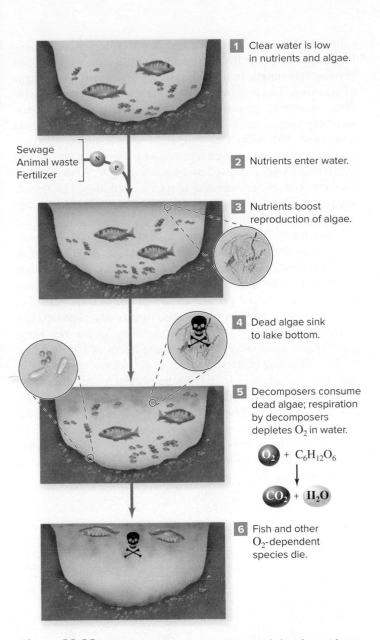

1 Clear water is low in nutrients and algae.

Sewage
Animal waste
Fertilizer

2 Nutrients enter water.

3 Nutrients boost reproduction of algae.

4 Dead algae sink to lake bottom.

5 Decomposers consume dead algae; respiration by decomposers depletes O_2 in water.

$$O_2 + C_6H_{12}O_6$$
$$\downarrow$$
$$CO_2 + H_2O$$

6 Fish and other O_2-dependent species die.

Figure 38.23 Eutrophication. Excess nitrogen and phosphorus trigger algae blooms. After the algae die, decomposers consume their bodies and deplete dissolved oxygen in the water.

in the water (**figure 38.23**). At first, photosynthesis by these producers releases O_2 into the upper water column. But when they die, their bodies sink from the surface to deeper waters, where microbes decompose their dead bodies. Cellular respiration by the decomposers depletes dissolved O_2 in the water. Without O_2, most fish and other animals cannot survive. Eutrophication is therefore associated not only with algae blooms but also with massive fish kills.

Moreover, some algae release toxins. Breathing the air near an algae bloom may therefore cause itchy eyes and a sore throat. In severe cases, touching or ingesting the water may cause serious illness or even death. ⓘ *harmful algae blooms,* section 18.2B

F. Terrestrial and Aquatic Ecosystems Are Linked in Surprising Ways

Sometimes, nutrients in an ecosystem can come from unexpected sources. Consider, for example, a mountain stream in Alaska. The eggs of Alaska salmon hatch in the streambed, and over the next year or so, the hatchlings develop into juveniles as they make their way to the mouths of their home rivers. As they transform into adults, their bodies adjust to seawater, and the fish swim into the Pacific Ocean. They spend as many as 4 years in the ocean, eating crustaceans and small fish. During this portion of their life cycle, the salmon (and their prey) ultimately rely on nutrients that well up from the ocean bottom.

Eventually, the fully grown salmon readjust to fresh water and swim upstream, back to the same waters where they hatched. As they make their way to their home streams to spawn, their bodies carry nutrients from the ocean to the heart of Alaska (**figure 38.24**). Bears and eagles feast on the salmon, and decomposers consume the remains. The nutrients they release support the growth of algae, which form the foundation of the stream's food chain. Moreover, the bears and eagles may carry their prey away from the stream. The salmon's nutrients—originally from the ocean—end up fertilizing faraway grasses and trees. The adult salmon therefore form a link between the biogeochemical cycles of the ocean and the land.

38.4 MASTERING CONCEPTS

1. What features do biogeochemical cycles share?
2. Describe the main abiotic reservoirs for water, carbon, nitrogen, and phosphorus.
3. What unique roles do bacteria and archaea play in the nitrogen cycle?
4. How can nutrient pollution lead to O_2-depleted water?
5. Describe how a terrestrial ecosystem can interact with a faraway aquatic ecosystem.

Figure 38.24 Fish Feast. Sockeye salmon contain nutrients from the sea; a brown bear carries these nutrients from water to land.

(fish): ©Jeff Mondragon/Alamy; (bear): ©Digital Vision/Getty Images RF

INVESTIGATING LIFE

38.5 Two Kingdoms and a Virus Team Up to Beat the Heat

In his 1733 poem entitled *On Poetry, a Rhapsody,* Jonathan Swift wrote:

> *So, naturalists observe, a flea*
> *Has smaller fleas that on him prey;*
> *And these have smaller still to bite 'em;*
> *And so proceed ad infinitum.*

Swift's observation, though not literally true, showed amazing insight into the relationships between the merely small and the truly microscopic. We now know that symbiotic arrangements can cross all phylogenetic lines. Symbioses exist between animals and plants (ants and acacias), plants and fungi (mycorrhizae), protists and animals (protozoa in termite guts), and bacteria or protists and fungi (lichens). Mutualism is a particularly interesting form of symbiosis, in which both species benefit from the deal. It is not that each species is advocating the success of the other's genome. Rather, if an alliance gives both parties a selective advantage, then *vive la différence!*

Researchers at the Noble Foundation in Ardmore, Oklahoma, have discovered an unusual three-way symbiosis and documented its survival benefit to the partners. Biologists Luis Márquez and Marilyn Roossinck knew of a type of panic grass (*Dichanthelium lanuginosum*) that endured the scorching geothermal soils of Yellowstone National Park. Previous studies had shown that the grass benefited from a fungus growing in its roots. The fungus (*Curvularia protuberate*) is an endophyte, an organism that lives between a plant's cells without causing disease. By itself, neither plant nor fungus could grow at temperatures above 38°C (100°F). The grass–fungus relationship somehow allows the plant to survive in soil temperatures up to 65°C (about 150°F). As long as the grass survives, so does the fungus, but each is doomed without the other.

By itself, a plant–endophyte relationship is not unusual; section 20.8 described endophytes that help cacao plants fight off other fungi. But Márquez and his team dug a bit deeper. They knew that many fungi are themselves infected with viruses. For example, a mycovirus (a virus that infects fungi) helps shape the biology of the fungus responsible for a disease of chestnut trees. The team wondered whether the fungus within the grass might also have a partnership with a virus.

The first thing to do was to see if a mycovirus was present. Mycoviruses usually have genomes made of double-stranded RNA, which is distinctive from the fungus's own double-stranded DNA. The researchers extracted nucleic acids from the fungal cells and found telltale fragments of double-stranded RNA within the fungal tissue. ⓘ *nucleic acids,* section 2.5D

Next, the team wanted to see if the presence of the mycovirus within the fungal endophytes helped plants survive in hot soils. They found a colony of the fungus that contained very low amounts of viral RNA. By repeatedly drying, freezing, and thawing this fungus in the laboratory, they "cured" the fungus of its virus. The researchers then set up an experiment with three treatment groups: grass with the wild-type (virus-infected) fungus, grass without any fungus at all, and grass inoculated with virus-free fungus. They submitted all three groups of plants to a tough regimen of 65°C soil temperatures for 10 hours a day, followed by 14 hours at 37°C. After 14 days, they checked to see how the plants were doing. The results were clear: Only the plants with the virally infected fungus survived the ordeal (figure 38.25).

The results seemed promising, but the team had to be certain it was the mycovirus that did the trick. After all, some unseen difference between the two fungi (something other than the virus) may have accounted for the heat tolerance. They therefore inserted a special genetic marker in one of their virus-free fungus samples and grew this marked fungus on a plate next to a fungus that contained the mycovirus. They let the two fungi grow across each other and took samples from the entwined hyphae. They moved these samples to new plates, and in one of the 35 subcultures they found that the marked fungus had picked up the virus. They inoculated grass with this newly infected fungus and turned up the heat. Sure enough, these plants survived, too (figure 38.26).

Finally, the scientists decided to see if the mycovirus–fungus partnership could work its magic in tomato plants, a distant relative of grass. They repeated the treatments and

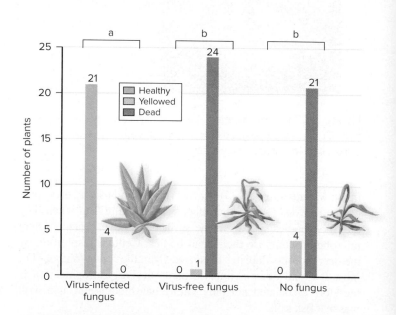

Figure 38.25 Beating the Heat. Plants that were inoculated with the normal virus-infected fungus survived the heat. Not so for plants infected with virus-free fungi or for nonsymbiotic plants without fungal endophytes. (See appendix B for a brief discussion of the notation that represents statistically significant differences between treatments.)

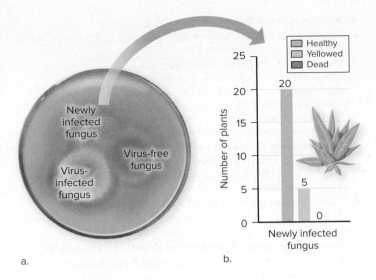

Figure 38.26 Reinfecting a "Cured" Fungus. (a) The researchers grew virus-infected (wild-type) and virus-free fungi on agar in a petri dish. In the zone where the two colonies grew together, previously virus-free fungi reacquired the virus. (b) Plants that were colonized by the newly infected fungi were heat-tolerant once more.

found that the plant–fungus–virus trio survived the heat, but the others did not.

Nearly 300 years after Jonathan Swift penned his whimsical poem, biologists are coming to realize that symbiotic relationships are the rule rather than the exception. Over hundreds of millions of years of shared evolutionary history, the struggle for existence has generated some unusual working relationships. The grass–fungus–virus partnership apparently allows all of the partners to colonize habitats that were previously unavailable. No one knows how the mycovirus-infected endophytes help their hosts beat the heat; perhaps they scavenge harmful oxygen free radicals produced by the heat-stressed plants. Whatever the mechanism, however, the relationship appears to be deeply "rooted" in evolutionary history.

Source: Márquez, Luis M., Regina S. Redman, Russell J. Rodriguez, and Marilyn J. Roossinck. 2007. A virus in a fungus in a plant: Three-way symbiosis required for thermal tolerance. *Science,* vol. 315, pages 513–515.

38.5 MASTERING CONCEPTS

1. Use figure 38.25 to describe the relationships among panic grass, the fungus, and the virus.

2. The fungus–virus partnership helped young grass plants survive at very high temperatures. What would be the benefits of inoculating all of our food crop plants with the fungus–virus team? What else would you need to know before recommending that strategy? Can you think of any possible drawbacks?

CHAPTER SUMMARY

38.1 Multiple Species Interact in Communities

- **Communities** consist of coexisting **populations** of multiple species.
- An **ecosystem** includes a **biotic** community plus its **abiotic** environment.
- Each species in a community has a place where it normally lives (**habitat**) and a set of resources necessary for its life (**niche**).
- Within a community, species interact in many ways (table 38.1).

A. Many Species Compete for the Same Resources
- Populations that share a habitat often **compete** for limited resources. Competition reduces the fitness of both species.
- According to the **competitive exclusion principle,** two species cannot indefinitely occupy exactly the same niche.
- In **resource partitioning,** competition between multiple species with similar niches restricts each species to a subset of available resources.

B. Symbiotic Interactions Can Benefit or Harm a Species
- **Symbiotic** relationships include **mutualism** (both species benefit), **commensalism** (one species benefits, whereas the other is unaffected), and **parasitism** (one species benefits, but the host is harmed).

C. Herbivory and Predation Link Species in Feeding Relationships
- **Herbivory** is an interaction in which a consumer eats a plant; a **predator** is an animal that kills and eats another animal (its **prey**).
- Plants and prey animals have defenses against herbivores and predators; in addition, animals have adaptations that help them capture food.

D. Closely Interacting Species May Coevolve
- In **coevolution,** the interaction between species is so strong that genetic changes in one population select for genetic changes in the other.

38.2 Succession Is a Gradual Change in a Community

- The diversity of a community depends on **species richness** (the number of species) and **species evenness** (relative abundance).
- As species interact with one another and their physical habitats, they change the composition of the community. This process is called **succession.**
- **Primary succession** occurs in a previously unoccupied area, beginning with **pioneer species** that allow soil to develop, paving the way for additional organisms to thrive.
- **Secondary succession** is more rapid than primary succession because soil does not have to build anew.
- Succession may lead toward a stable **climax community,** but true long-term stability is rare. Pockets of local disturbance mean that most communities are a patchwork of successional stages.

TABLE 38.1 Species Interactions: A Summary

Interaction	Effects on Species 1	Effects on Species 2
Competition	–	–
Symbiosis		
Mutualism	+	+
Commensalism	+	0
Parasitism	+	–
Herbivory	+	–
Predation	+	–

38.3 Ecosystems Require Continuous Energy Input

A. Food Webs Depict the Transfer of Energy and Atoms

- An organism's **trophic level** depends on the number of steps between it and the ultimate source of energy in the ecosystem.
- At the base of each **food chain** are **primary producers (autotrophs)** that harness energy from the sun or inorganic chemicals.
- **Consumers (heterotrophs)** include primary consumers (herbivores) that eat primary producers. Secondary, tertiary, and higher-order consumers eat other consumers. **Decomposers** break down **detritus** (nonliving organic material) into inorganic nutrients.
- Interconnected food chains form **food webs.**

B. A Keystone Species Has a Pivotal Role in the Community

- A **keystone species** makes up a small proportion of a community's biomass but has a large influence on the community's composition.

C. Heat Energy Leaves Each Food Web

- The total amount of energy that producers convert to chemical energy is gross primary production. After subtracting energy for maintenance and growth, the energy remaining in producers is **net primary production.**
- Food chains rarely extend beyond four trophic levels because only a small percentage of the energy in one trophic level transfers to the next level. Most energy is lost to the surroundings as heat.
- A **pyramid of energy** depicts the amount of energy at each trophic level.

D. Harmful Chemicals May Accumulate in the Highest Trophic Levels

- **Biomagnification** concentrates nondegradable, fat-soluble chemicals in the highest-order consumers.

38.4 Chemicals Cycle Within Ecosystems

- **Biogeochemical cycles** are geological and chemical processes that recycle chemicals essential to life (**figure 38.27**), including transfers among organisms, the atmosphere, water, and rocks and soil.

A. Water Circulates Between the Land and the Atmosphere

- The water cycle describes the movement of water in air and on or below Earth's surface. Plants and other organisms participate in the water cycle, too.

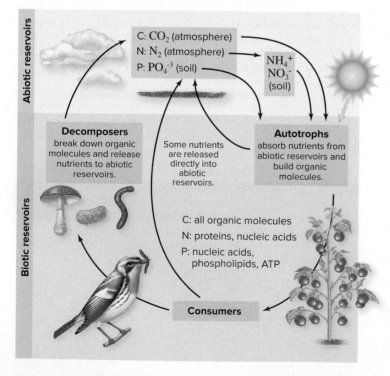

Figure 38.27   Biogeochemical Cycles: A Summary.

B. Autotrophs Obtain Carbon as CO_2

- Autotrophs use CO_2 to produce organic molecules. Cellular respiration and burning fossil fuels release CO_2.

C. The Nitrogen Cycle Relies on Bacteria

- **Nitrogen-fixing** microbes convert atmospheric nitrogen (N_2) to ammonium (NH_4^+), which plants can incorporate into their tissues. Decomposers convert the nitrogen in dead organisms back to ammonium. **Nitrification** converts the ammonium to nitrate (NO_3^-), whereas **denitrification** converts nitrate to nitrogen gas.

D. The Phosphorus Cycle Begins with the Erosion of Rocks

- Rocks release phosphorus as they erode. Autotrophs incorporate the phosphates into organic molecules; decomposers return inorganic phosphates to the environment.

E. Excess Nitrogen and Phosphorus Cause Problems in Water

- During **eutrophication,** excess nutrients cause algae blooms. Microbes decompose dead algae, depleting oxygen in the water.

F. Terrestrial and Aquatic Ecosystems Are Linked in Surprising Ways

- Salmon carry nutrients from the ocean to inland ecosystems.

38.5 Investigating Life: Two Kingdoms and a Virus Team Up to Beat the Heat

- A three-way symbiosis involving a grass plant, a fungus, and a virus enables plants to survive extremely high temperatures.

MULTIPLE CHOICE QUESTIONS

1. Which of the following is an example of an ecological community?
   a. The many types of microbes living in a human intestine
   b. An ant colony
   c. The people living in your neighborhood
   d. The cells of a platypus

2. Cyanobacteria are photosynthetic bacteria that live in water. Two species, isolated from the same habitat, use different wavelengths of light in photosynthesis. This is an example of
   a. competitive exclusion.   c. resource partitioning.
   b. commensalism.   d. biomagnification.

3. When researchers experimentally excluded stream insects called caddis flies from submerged ceramic tiles, algae growth on the tiles was much higher than when caddis flies were present. This experiment tested the effects of _____ on community structure.
   a. predation   c. herbivory
   b. camouflage   d. mutualism

4. Some types of crabs live in clumps of coral. The crab defends its home, protecting the coral from sea stars and other predators. The interaction between crabs and corals is an example of
   a. resource partitioning.
   b. competitive exclusion.
   c. commensalism.
   d. mutualism.

5. A large tree falls over in an old-growth forest, allowing light to reach a formerly shaded area. A few weeks later, what types of rooted plants will dominate the open patch?
   a. Lichens and mosses
   b. Herbs and weeds
   c. Large shrubs
   d. Oak trees

6. When you eat a carrot, you are acting as a(n)
   a. heterotroph and primary consumer.
   b. autotroph and herbivore.
   c. heterotroph and secondary consumer.
   d. autotroph and primary producer.

7. In a prairie, which trophic level should have the highest biomass?
   a. Predatory birds
   b. Grasses
   c. Seed-eating rodents
   d. Soil fungi that decompose organic matter

8. Refer to the food web in figure 38.14. If krill were to disappear from the ecosystem, what species would most likely become more abundant?
   a. Phytoplankton
   b. Squid
   c. Penguins
   d. Toothed whales

9. Which statement best explains why persistent, fat-soluble chemicals such as methylmercury accumulate in the highest trophic levels?
   a. Animals at the highest trophic levels eat the lowest-quality food.
   b. The amount of biomass increases in each trophic level, and large organisms accumulate the most methylmercury.
   c. Large organisms are often herbivores, so they would consume methylmercury directly.
   d. Methylmercury does not leave an animal's body; each predator therefore consumes the mercury contained in many prey.

10. Which cycle relies the least on decomposers?
    a. Carbon cycle
    b. Water cycle
    c. Phosphorus cycle
    d. Nitrogen cycle

Answers to these questions are in appendix A.

WRITE IT OUT

1. How does a community differ from an ecosystem?

2. Researcher G. F. Gause tested how two species of paramecia (one-celled protists) responded to food availability. When grown in separate dishes, each population's size was proportional to the amount of food (the more food, the larger the population). However, when both were grown in the same dish, only one species survived. What conclusions about competition can you draw from these two results?

3. In an example of mimicry, a harmless jumping spider physically resembles an aggressive species of ant. Explain why this type of mimicry can exist only if the spiders are less abundant than the ants.

4. Suppose a plot of forest is cleared of trees in anticipation of a new shopping mall. However, after the bulldozers are gone, the company runs out of money, and the land sits undisturbed for many years. Describe the events that may occur in the years following the damage to the forest. What are these community changes called?

5. Imagine that you could build a covered enclosure around a small ecosystem, blocking out all light and preventing gas exchange with the environment. How would the total amount of organic material, available energy, and nutrients in the ecosystem change over time?

6. Search the Internet for "ecosystem services." What does the term mean? List some examples of ecosystem services and reflect on what services might be most important to you.

7. Australian researchers removed a parasitic plant called mistletoe from a forest and observed the consequences. During the 3-year experiment, over 30% of the insect and bird species disappeared from the mistletoe-free forest. Another forest in which no mistletoe was removed saw no change in species diversity. What might the biologists conclude from these data?

8. In a eutrophic lake, algae are abundant and dissolved oxygen levels are low. Predict how the pyramid of energy might appear for an ecosystem under these conditions.

9. Mountain yellow-legged frogs live in the Sierra Nevada mountains. Their tadpoles mainly eat algae. One predator of adult frogs is a garter snake, which is eaten by bullfrogs. Recently, a chytrid fungus has killed many adult mountain yellow-legged frogs. How might this change affect the algae, garter snakes, and bullfrogs?

10. Officials in the northeastern United States are proposing to serve dogfish—a type of shark—in schools, prisons, and homeless shelters. The fish are cheap and abundant, and their populations interfere with efforts to catch more valuable fish such as cod. However, some people argue that serving dogfish to schoolchildren is risky; use figure 38.17 to explain this argument.

11. Imagine that you could follow a single water molecule, carbon atom, nitrogen atom, and phosphorus atom from one animal into two abiotic reservoirs. Which atom or molecule is most likely to remain near the animal? Explain your answer.

12. Suppose a friend says, "I hate germs! I wish we could kill all bacteria!" What would happen if your friend didn't have bacteria in her body? What would happen to nutrient cycles without bacteria?

PULL IT TOGETHER

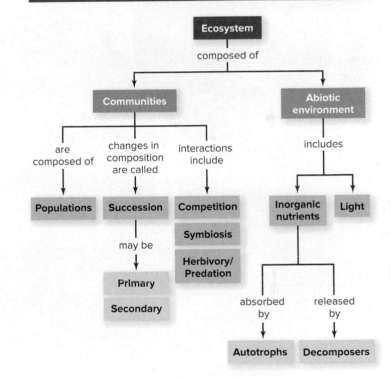

Figure 38.28 Pull It Together: Communities and Ecosystems.

Refer to figure 38.28 and the chapter content to answer the following questions.

1. Review the definitions of *ecology, populations, communities,* and *ecosystems* using the Survey the Landscape figure in the chapter introduction and the Pull It Together figure. Then, classify each of the following as a question relating to populations, communities, or ecosystems:
   a. How do gophers and moles compete for resources?
   b. How do monkeys establish dominance within a group?
   c. How does pollution affect crop plants?

2. Where do mutualism, commensalism, and parasitism fit into this concept map?

3. Make another concept map that shows nutrient cycling on land. Include *producers, consumers, decomposers, carbon, nitrogen, phosphorus, atmosphere,* and *soil;* you may also add other concepts.

39 Biomes

Return from the Bottom of the Sea. The submersible *Alvin* completes a research mission in the Atlantic Ocean. The inset shows deep ocean fish and giant tube worms near a hydrothermal vent.

©Perry Thorsvik/National Geographic Stock; (inset): ©Dr. Ken MacDonald/Science Source

LEARN HOW TO LEARN
Use Your Campus Resources

Many first-time college students struggle to keep up with assignments outside of class. Time management can be part of the problem, but poor reading and writing skills may also share the blame. Most campuses offer free resources that can help. For example, look for seminars and workshops aimed at improving reading comprehension. Writing centers are also common; the staff should include consultants trained to help you be a more proficient writer.

Deep-Sea Hydrothermal Vents Host Unique Ecosystems

Oceans cover most of Earth's surface, yet these vast ecosystems largely remain unexplored. Humans can visit the seafloor for only short periods, and then only while confined inside a pressure-resistant vessel such as the submersible *Alvin* (see the chapter opening photo).

Alvin played a significant role in the history of biology. In 1977, it carried a group of geologists to the sea bottom near the Galápagos Islands, directly above cracks in Earth's crust called deep-sea hydrothermal vents. At the time, no one thought life could exist in these areas of intense pressure and temperature extremes where Earth's crust is born. But the researchers were astounded to see a thriving ecosystem. Tube-like worms waved, anemones clung, crabs crawled, and shrimp grazed. Most of these animals were unknown to science. Since that time, many more vent ecosystems—including hundreds of new species—have been discovered at hydrothermal vents worldwide.

The environment at a hydrothermal vent could hardly be more different from our own familiar surroundings. The pressure is 300 times greater than at Earth's surface, and the water temperature can exceed 400°C. No animals can survive in the hottest water, but tube worms can live at 100°C (the boiling point of water at Earth's surface). Moreover, the pH can be as low as 2.8, about as acidic as orange juice. Oxygen can be scarce, and the hydrogen sulfide that spews from the Earth would be toxic to most organisms. Researchers continue to explore the enzymes and other molecules that allow vent communities to thrive in conditions that would kill most organisms. One motivation for this research is that hydrothermal vents may hold clues to life's origins. (i) *origin of life,* section 15.1

Hydrothermal vents also had another surprise in store. Biologists once thought that life at the inky blackness of the deep-sea ocean floor was supported only by a gentle "snow" of organic material drifting down from the sunlit surface. Hydrothermal vent communities, however, are self-supporting, courtesy of bacteria and archaea that tap energy from inorganic chemicals instead of sunlight. (i) *chemotrophs,* section 17.2B

The discovery of these ecosystems has also helped to dispel the long-held idea that life exists only on the thin layer between bedrock and Earth's atmosphere. Nevertheless, life on and near Earth's surface is by far more familiar. This chapter describes the forests, grasslands, waters, and other ecosystems that are essential to our own survival.

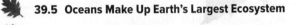
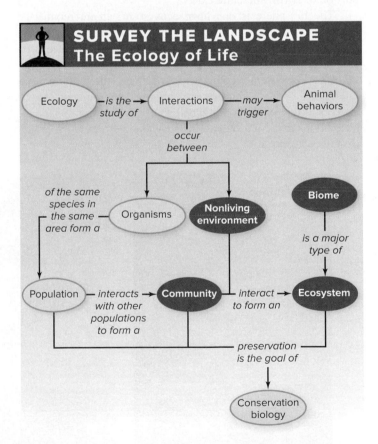

SURVEY THE LANDSCAPE
The Ecology of Life

Each location on Earth may be unique, but the world's ecosystems can nevertheless be grouped into broad categories called biomes. On land, climate is the most important factor determining the distribution of forests, deserts, prairies, and other biomes. In water, temperature and the availability of light and nutrients are key.

For more details, study the Pull It Together feature in the chapter summary.

39.1 The Physical Environment Determines Where Life Exists

Picture the following landscapes in your mind: a prairie, the seashore, the Sahara Desert, a jungle, and the top of Mount Everest. These and all other locations on Earth form a patchwork of unique habitats, each with its own set of conditions. Fire regularly ravages the prairie but not the beach; water is scarce in the desert but not in the jungle. The species that are native to each location have adaptations that correspond to these conditions. The same basic evolutionary process—natural selection—has produced unique populations and communities of organisms in nearly every possible habitat (figure 39.1).

Indeed, life abounds almost everywhere on Earth, even in places once thought to be much too harsh to support it (see the Burning Question in section 17.3). Scientists have discovered life in Arctic ice, salt flats, hot springs, mines that plunge miles below Earth's surface, and hydrothermal vents (see the chapter opening essay). All of these areas are part of the **biosphere,** the portion of Earth where life exists.

The biosphere is one huge **ecosystem,** an interconnected community of organisms and their physical environment. Recall from chapter 38 that community ecologists study the **biotic** interactions among species, such as competition, predation, and mutualism. Ecosystem ecologists incorporate the community's interaction with its **abiotic,** or nonliving, environment. As we have already seen, both biotic and abiotic interactions shape the adaptations that contribute to the survival and reproductive success of each species.

Ecologists divide the biosphere into **biomes,** which are major types of ecosystems characterized by a particular climate and a distinctive group of species. Forests, deserts, and grasslands are examples of terrestrial biomes. Lakes, streams, and oceans are water-based ecosystems. Although it is convenient to classify each ecosystem as belonging to one biome or another for convenience, keep in mind that no ecosystem exists in isolation. Water, air, sediments, and organisms travel freely from one part of the biosphere to another.

Gardeners know that every species is adapted to a limited set of conditions; no "superorganism" can exist everywhere (see section 12.3's Burning Question). A plant adapted to bright sunlight will not thrive on the shady side of the house, nor will a cactus survive in a pond. Fish need water, and earthworms cannot live on a sandy beach. This chapter begins by describing some of the "rules" that determine how distinctive communities develop in each biome, beginning with abiotic features.

Many abiotic factors determine the limits of each species' distribution. The ultimate abiotic factor is an energy source, since no ecosystem can exist without one (see section 39.3). **Primary producers,** also called autotrophs, are organisms that can harvest this energy source as they build organic molecules. These primary producers provide the energy, nutrients, and habitats that support animals, fungi, and the rest of the food web.

Sunlight is the energy source for most ecosystems. On land, plants are the dominant primary producers. In water, however, most photosynthesis occurs courtesy of **phytoplankton:** microscopic, free-floating, photosynthetic organisms such as cyanobacteria and algae. A few ecosystems, such as deep-sea hydrothermal vents, are based on chemical energy, not sunlight. At these vents,

a.

b.

c.

d.

Figure 39.1 Habitat Variety. (a) A reef shark swims near a sea fan, a type of coral. (b) This young manzanita tree lives on a rocky slope on Santa Rosa Island in California. (c) A saguaro cactus stands tall in the Arizona desert. Its massive stem absorbs and stores water during rare rainstorms. (d) Broad, flat lily pads soak up sunshine on temperate and tropical ponds around the world.

the producers are microbes that extract energy from hydrogen sulfide and other inorganic chemicals.

Besides sunlight, the major abiotic factors that determine the numbers and types of plants on land are temperature and moisture. All organisms are adapted to a limited temperature range; trees, for example, cannot live where the temperature is too low. In addition, all life requires water. The plants that grow where water is abundant, such as in the tropical rain forest, have very different adaptations than the vegetation that characterizes a desert ecosystem.

Nutrient availability is another crucial abiotic factor that often determines an ecosystem's productivity. On land, soil provides essential mineral elements such as nitrogen and phosphorus. In aquatic ecosystems, both nutrients and sunlight are often scarce, especially with increasing depth and distance from the shoreline.

Other abiotic factors may also be important in some ecosystems. For example, the amount of dissolved oxygen influences the types of animals that can live in water. Aquatic plants and phytoplankton release oxygen into the water column as they carry out photosynthesis; oxygen also dissolves into water at the interface with the atmosphere. But respiration by aquatic organisms can deplete oxygen in water. Some microbes can tolerate water with little or no oxygen, but most fishes and other animals will suffocate and die.

Another important abiotic factor in water and on land is salinity. Many organisms are adapted to seawater or salty soils, but others are not. The problem of life in a salty habitat lies mostly in the difficulty of extracting water from a saline solution. This is why human castaways adrift on the ocean can die of thirst, even while surrounded by water. Likewise, saline soils prevent agriculture in many parts of the world, because most crop plants are not salt-tolerant.

Fire is an essential abiotic condition in some terrestrial biomes. In grasslands, for example, periodic fires kill trees that might otherwise take over. In coniferous forests, many adult trees die in fires, but their cones open only after prolonged exposure to heat. The seeds germinate after the fire, and the young trees thrive with little competition for sunlight or nutrients.

The rest of this chapter explains the distribution of the main biomes on land and in water. As you read this material, remember that the biomes we see today have not been in place forever. Over hundreds of millions of years, the continents have moved, sea levels have risen and fallen, and climates have shifted (see this chapter's Burning Question). The central United States, for example, was once under the sea, which explains why fossils of marine animals are abundant in land-locked states such as Oklahoma. Likewise, 375 million years ago, the landmass that now includes the islands of Arctic Canada was once very near the equator. Long-buried fossils tell the tales of these tremendous ecosystem shifts. ⓘ *continental drift,* section 13.3A

39.1 MASTERING CONCEPTS

1. What are the relationships among ecosystems, communities, biomes, and the biosphere?

2. What abiotic conditions influence the distribution of species on land and in water?

Burning Question

How do we know what Earth's climate was like in the past?

Earth's climate has undoubtedly changed throughout its long history. The amount of energy from the sun has fluctuated, as have the amount of sunlight reflected into space and the amount of heat trapped by the atmosphere. Fossils reveal some clues to our climatological history, but they cannot tell the whole story. Fortunately, scientists have additional ways of knowing about the climates of the past.

For example, coral animals produce annual growth rings in their skeletons, providing records about water temperature and salinity that may span a few hundred years. Likewise, tree rings can reveal the temperature and rainfall in a particular region over the past few thousand years. ⓘ *corals,* section 21.3; *tree rings,* section 22.4D

But what about records spanning a longer period? Here, climate scientists turn to a nonliving source of information: ice cores (figure 39.A). Scientists have drilled into the thick ice sheets

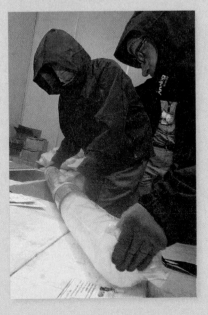

Figure 39.A **Ice Core.** Researchers inspect an ice core from a glacier.
©Romeo Gacad/AFP/Getty Images

covering Greenland and Antarctica, collecting core samples spanning hundreds of thousands of years. Thanks to seasonal changes in the amount of snowfall, the cores have annual layers that researchers can count to measure time. The thickness of each layer indicates the amount and type of precipitation for each year. Moreover, bubbles in the cores contain air samples trapped when the snow was compressed into ice. Analyzing these bubbles reveals a continuous record of changes in the gas composition of the atmosphere. Other data, ranging from oxygen isotopes to pollen, yield valuable clues about the prevailing temperature at the time the ice formed.

Submit your burning question to
Marielle.Hoefnagels@mheducation.com

39.2 Earth Has Diverse Climates

Earth has a wide variety of climates, from the year-round warmth and moisture of the tropics to the perpetually chilly poles. Why does each part of the planet have a different climate? The answer relates to the curvature of the planet and to the tilt of its axis (**figure 39.2**).

At the equator, the sun is overhead (or nearly so) all year; equatorial regions therefore receive the most intense sunlight, and the temperature is warm year-round. Thanks to Earth's curvature, however, the sun's rays hit other parts of the surface at a slant. Because the same amount of sunlight is distributed over a larger area, the average temperature falls with distance from the equator. **Figure 39.3** shows the resulting broad temperature bands from the equator to the poles.

The tilt of Earth's axis accounts for seasonal temperature changes in nonequatorial regions (see figure 39.2). From March through September, the northern hemisphere tilts toward the sun and experiences the warm temperatures of spring and summer. During the rest of the year, cooler temperatures prevail as the northern hemisphere tilts away from the sun. The seasons are the opposite in the southern hemisphere.

Equatorial regions receive not only the most light but also the most precipitation. When sunlight heats the air over the equator, the air rises, expands, and cools. Because cool air cannot hold as much moisture as warm air, the excess water vapor condenses, forming the clouds that pour rain over the tropics.

Air that rises near the equator also travels north and south (**figure 39.4**). As the air cools at higher latitudes, its density

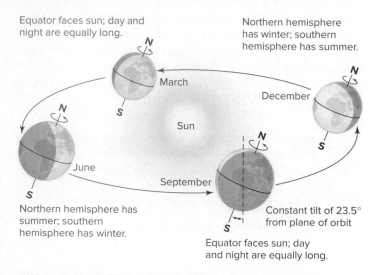

Figure 39.2 Earth's Seasons. The tilt of Earth's axis produces distinct seasons in the northern and southern hemispheres as Earth travels around the sun.

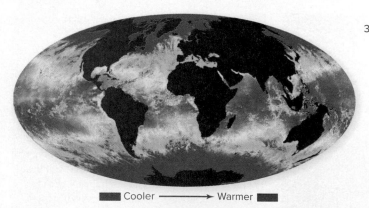

Figure 39.3 From Warm to Cold. The colored bands on this map show that Earth's surface is warmest at the equator and coldest at the poles.
MODIS Oceans Group, NASA Goddard Space Flight Center

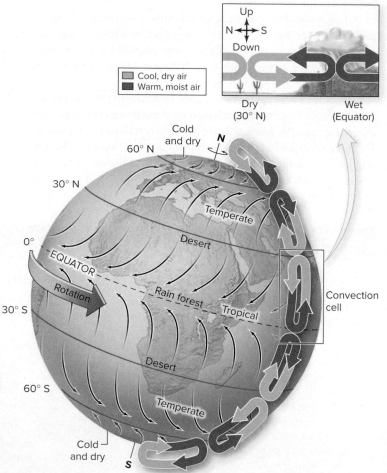

Figure 39.4 Patterns of Air Circulation and Moisture. In each of Earth's six convection cells, air cools as it rises and releases its moisture as rain. Conversely, descending air masses pick up warmth and absorb moisture from the land.

increases, and it sinks back down to Earth at about 30° North and South latitude. Here the warming air absorbs moisture from the land, creating the vast deserts of Asia, Africa, the Americas, and Australia. Some of the air continues toward the poles, rising and cooling at about 60° North and South latitude, bringing the rains that support temperate (midlatitude) forests in these areas. The air rises, and some again continues toward the poles, where precipitation is quite low. The rest returns to the equator, where the air heats up again, and the cycle begins anew.

A cycle of heating and cooling, rising and falling air is called a convection cell. The planet has six such convection cells (three north of the equator and three south). As figure 39.4 illustrates, Earth's major winds correspond to these convection cells. (From Earth, the winds appear deflected toward the east and west because the planet rotates beneath them.) Together, these winds power major ocean currents, including the Gulf Stream along the east coast of North America.

Ocean currents, in turn, influence coastal climates, in part because they transport cold and warm water around the globe (figure 39.5). For example, bands of cold water flow along the west coast of North America, while currents along the east coast are relatively warm. In addition, large water bodies heat up and cool down much more slowly than does the land. Coastal regions therefore often have milder climates than do inland areas at the same latitude. On a hot summer day, beachgoers notice this effect as they enjoy cooling breezes from the sea. Conversely, during the winter, the ocean releases stored heat. Changes in these currents, such as those that occur during an El Niño event, can therefore trigger ecological upheaval.

Mountain ranges also influence climate, in two ways. First, the top of a mountain is generally cooler than its base. Second,

a.

b.

Figure 39.6 Rain Shadow. (a) Precipitation falls on the windward side of a mountain range, leaving the other side with a dry climate. (b) The Andes Mountains create a massive rain shadow; note the difference in vegetation between wet Chile and dry Argentina in this satellite image.

Photo: (b): Jacques Descloitres, MODIS Land Rapid Response Team, NASA/GSFC

mountains often block wind and moisture-laden clouds on their upwind side. The **rain shadow** on the downwind side of the mountain has a much drier climate (figure 39.6).

39.2   MASTERING CONCEPTS

1. Explain this statement: "If Earth's axis were not tilted, there would be no seasons."
2. Moving north and south from the equator, what are the major climatic regions of the world?
3. How do prevailing winds, ocean currents, and mountain ranges affect climate?

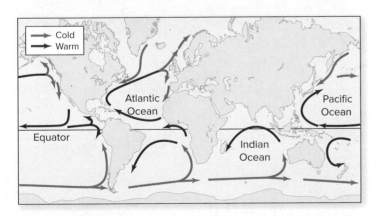

Figure 39.5 Ocean Currents. Earth's prevailing winds produce the major ocean currents, which redistribute water and nutrients throughout the oceans.

39.3 Terrestrial Biomes Range from the Lush Tropics to the Frozen Poles

Earth's climatic zones give rise to huge bands of characteristic types of vegetation, which correspond to the terrestrial biomes (figure 39.7). Temperature and moisture are the main factors that determine the dominant plants in each location (figure 39.8). The overall pattern of vegetation, in turn, influences which microorganisms and animals can live in a biome.

Soils form the framework of terrestrial biomes because they directly support plant life. Although soil may seem like "just dirt," it is actually a complex mixture of rock fragments, organic matter, and microbes (see figure 23.3). Climate influences soil development in many ways. Heavy rain may leach nutrients from surface layers and deposit them in deeper layers, or it may remove them entirely from the soil. In addition, in a warm, moist climate, rapid decomposition may leave little organic material in the soil. In cold, damp areas, on the other hand, undecomposed peat may accumulate in the soil.

The following sections consider some of the world's major terrestrial biomes. The map in figure 39.7 shows the original range of each biome. It is important to remember, however, that humans have drastically reduced many natural biomes, replacing them with farmland, suburban housing, and cities. In addition, as described in chapter 40, human activities threaten much of the native habitat that remains.

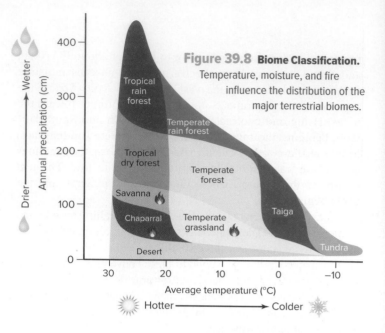

Figure 39.8 Biome Classification. Temperature, moisture, and fire influence the distribution of the major terrestrial biomes.

In studying these biomes, pay attention to the unique communities that occupy each one. Notice, also, that the number of species occupying each biome generally declines with distance from the equator. No one understands the full explanation for this phenomenon. Perhaps the high primary production near the equator provides the most opportunities for specialized consumers to evolve. Perhaps year-round favorable conditions at the equator drive few species to extinction. Or perhaps some other mechanism is at work.

✓ ■ Tropical and subtropical rain forests
■ Tropical and subtropical dry broadleaf forests
■ Tropical and subtropical coniferous forests
✓ □ Temperate deciduous forests
✓ ■ Temperate coniferous forests
✓ ■ Boreal forests/taiga

✓ □ Tropical and subtropical grasslands, savannas, and shrublands
✓ □ Temperate grasslands, savannas, and shrublands
■ Flooded grasslands and savannas
■ Montane grasslands and shrublands
✓ ■ Tundra

✓ ■ Mediterranean shrublands (chaparral)
✓ □ Deserts and dry shrublands
□ Water bodies
✓ □ Rock and ice

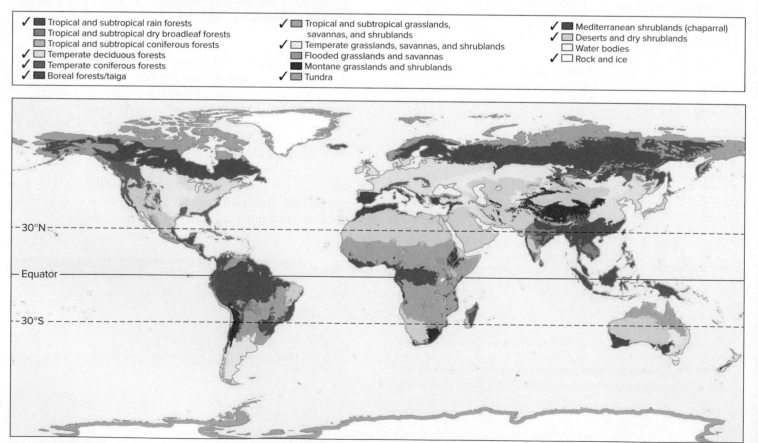

Figure 39.7 Earth's Major Terrestrial Biomes. The biomes that are marked with a check are described in this chapter.

A. Towering Trees Dominate the Forests

Forests supply many of the resources that we use every day: lumber, paper, furniture, and foods such as wild mushrooms and nuts. The trees in the forest also provide wildlife habitat and protect soil from erosion. Moreover, like all plants, trees absorb CO_2 from the atmosphere and use it in photosynthesis, a process that also releases O_2. The wood of a living tree is an especially important long-term "carbon storehouse" that helps offset CO_2 released when humans burn fossil fuels. ⓘ *carbon cycle,* section 38.4B

Tropical Rain Forests The **tropical rain forests** encircle the equator in Africa, Southeast Asia, and Central and South America; their location ensures that the climate is almost constantly warm and moist. These forests are home to a stunning diversity of species connected by intricate networks of relationships.

The warm, wet equatorial climate means favorable conditions for plant growth year-round (**figure 39.9**). A 10-square-kilometer area of tropical rain forest is likely to house 750 tree species, including broadleaf evergreen trees with tall, straight trunks that form the forest canopy. Vegetation in the shade beneath the canopy includes climbing vines and epiphytes, which are small plants that grow on the branches, bark, or leaves of another plant. Tree saplings and countless smaller species grow in the deep shade of the forest floor. These understory plants often have large leaves that maximize the capture of scarce light.

Animal life is similarly diverse. That same forest patch might contain 60 species of amphibians, 100 species of lizards and snakes, 125 species of mammals, 400 species of birds, and thousands or even millions of insect species. The herbivores eat leaves, fruits, and other plant parts. Large cats, birds of prey, snakes, and other carnivores consume the herbivores. Many of the animals have similar adaptations to life in the trees: bright colors that maximize visibility, coupled with loud calls that echo throughout the forest canopy.

Despite the lush plant growth, tropical rain forest soils are usually nutrient-poor and low in organic matter. The heat and humidity speed decomposition, but the mycorrhizal roots of giant trees absorb the nutrients efficiently. Animals recycle nutrients, too. Leaf-cutter ants, for example, cultivate gardens of fungi on decomposing leaf fragments.

Worldwide, people are logging and burning tropical rain forests to make room for crops and domesticated animals. As described in chapter 40, tropical rain forest destruction threatens indigenous people and global water and carbon cycles. As plants near extinction, we lose potential medicines and the ancestors of many of our most important domesticated plants. These older varieties are a valuable resource to plant breeders, who are always searching for disease-resistance genes to breed into modern crop plants.

Figure 39.9 **Tropical Rain Forest.** Plants in the tropical rain forest form distinct layers, from the tallest trees emerging from the canopy to the tiniest residents of the shady forest floor. This rain forest is in southern Thailand.
Photo: ©Muzhik/Shutterstock RF

Temperate Deciduous and Coniferous Forests The world's temperate forests occupy large areas between 30° and 60° North latitude. **Temperate deciduous forests** are dominated by trees that shed their foliage in autumn, whereas **temperate coniferous forests** contain mostly evergreen conifers that lose only a few leaves at a time (**figure 39.10**). These forests once covered parts of Asia, western Europe, North America, South America, Australia, and New Zealand, but logging, agriculture, and urbanization have decimated most of the world's native temperate forests. ⓘ *deforestation,* section 40.2

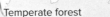

a.   b.

Figure 39.10 **Temperate Forests.** (a) Trees that lose their leaves each autumn dominate temperate deciduous forests such as this one in Pennsylvania. (b) Temperate coniferous forests have evergreen trees such as these in Wyoming's Grand Teton National Park.
Photos: (a): ©Digital Archive Japan/Alamy RF; (b): ©Comstock/PunchStock RF

Deciduous trees occur where summers are warm, winters are cold, and precipitation is approximately constant throughout the year. Usually, one or two tree species predominate, as in oak–hickory or beech–maple forests. Shade-tolerant shrubs grow beneath the towering trees. Below them, small flowering plants grow in early spring, when light penetrates the leafless tree canopy. Herbivores include seed- and nut-eating mice and birds, white-tailed deer, and gray squirrels. Red foxes and snakes are common carnivores. Raccoons eat a wide variety of foods, including insect grubs, acorns, frogs, and bird eggs. Many of these animals adjust to the seasons by putting on fat in the summer and hibernating in the winter, or they may store caches of seeds and nuts that sustain them when food is scarce. Still others migrate to warmer areas for the winter.

Mild winters, cool summers, and abundant rain and fog favor the temperate coniferous forest (sometimes called the temperate rain forest). Most trees in the temperate coniferous forest are evergreens such as spruce, pine, fir, and hemlock, all of which have waxy, needlelike leaves adapted to year-round photosynthesis. Understory shrubs include alder and hazelnut. Herbivores include birds, deer, and squirrels. Snakes, bobcats, weasels, and coyotes hunt the herbivores, and black bears are omnivores.

The temperate coniferous forests include the world's largest trees, the giant sequoias, which dominate the ancient redwood forests along the west coast of North America. Scientists studying the redwood forest canopy began climbing redwood trees in the 1990s and were astonished to find "forests in the air." Organic matter had accumulated into thick soils high on the branches of the giant redwoods. Berry bushes, ferns, and entire trees grew in the soil. The researchers also found insects, other invertebrates, chipmunks, birds, and even salamanders in the aerial ecosystem.

Taiga (Boreal Forests)
North of the temperate zone in the northern hemisphere lies the cold, snowy **taiga** (figure 39.11). This biome is also called the boreal forest or the northern coniferous forest. The long, harsh winter can last more than 6 months, so the growing season is short. Moisture can be scarce in winter, when water may remain frozen for months.

Soils in the taiga are cold, damp, acidic, and nutrient-poor. The low temperature and acidic pH slow decomposition, and nutrients tend to stay in the leaf litter above the soil, rather than entering the topsoil. Spruce, fir, pine, and tamarack (larch) are the dominant trees. The needles of these evergreens resist water loss and can carry out photosynthesis whenever the weather is warm enough to support it. Mycorrhizal fungi help the plants maximize nutrient uptake from the leaf litter. Seed- or leaf-eating herbivores include the woodland caribou, porcupines, red squirrels, chipmunks, snowshoe hares, and moose. As in the deciduous forest, many of these animals hibernate or store

Figure 39.11 **Taiga.** Woodland caribou forage for food in the Canadian boreal forest.
Photo: ©Thomas Kitchin/Science Source

food for the winter. Carnivores include lynx, gray wolves, and wolverines. Migratory birds visit during the short growing season.

Caribou, reindeer, berries, and fungi have provided food to humans and other animals for millennia. Logging, oil exploration, hunting, and trapping, however, are rapidly depleting the boreal forests.

B. Grasslands Occur in Tropical and Temperate Regions

Vast seas of grasses sustain huge herds of large, grazing animals such as bison and zebras. But the rich soils that support the grasslands also make these prime areas for agriculture. Grasslands are therefore endangered around the world.

Tropical Savannas
Tropical **savannas** are grasslands with scattered trees or shrubs and bands of woody vegetation along streams (figure 39.12). The weather is warm year-round, with distinct wet and dry seasons. Perennial grasses dominate the savanna along with patches of drought- and fire-resistant trees and shrubs such as palms, acacias, and baobabs. These plants have deep roots, thick bark, and trunks that store water.

Figure 39.12 Tropical Savanna. A herd of wildebeest roams the savanna in Kenya's Masai Mara National Reserve.

Photo: ©Arthur Morris/Corbis

The Australian savanna is home to many birds and kangaroos, whereas the grassland in Africa features herds of grazers such as zebras, giraffes, wildebeest, gazelles, and elephants. Lions, cheetahs, wild dogs, birds of prey, and hyenas prey on the herbivores, and vultures and other scavengers eat the leftovers. During the dry season, great herds of animals migrate enormous distances in search of water. In many tropical savannas, termites are major detritivores, and their huge nest mounds dot the landscape. It was from the African savanna that our own ancestors evolved, emerged, and spread. ⓘ *human evolution,* section 15.4

Widespread overgrazing by domesticated animals threatens to turn large areas of tropical savanna to desert. Hunting and poaching also endanger some populations of large animals. ⓘ *expanding deserts,* section 40.2

Temperate Grasslands The **temperate grasslands** are also known as the prairies of North America, the steppes of Russia, and the pampas of Argentina (**figure 39.13**). The climate is moderately moist, with hot summers and cold winters. These ecosystems have few if any trees, partly because annual rainfall is often not sufficient to support them. Grazing and fire also suppress tree growth. The tips of a tree's branches, where growth occurs, are easily destroyed by fire or herbivores. In contrast, the perennial buds of grasses lie protected below the soil surface and resprout soon after the fire passes.

Wind-pollinated grasses dominate this biome, although sunflowers, thistles, and other flowering plants are also common. Bison, elk, and pronghorn antelope, whose teeth and digestive systems are adapted to a grassy diet, were originally the large, grazing herbivores of North America; hunting has caused their numbers to dwindle. Other herbivores include prairie chickens, insects such as grasshoppers, and rodents such as prairie dogs and mice. Some of these small animals burrow into the soil to hide from predators, whereas others have camouflage. Coyotes, bobcats, snakes, and birds of prey feed on the herbivores.

The deep, black prairie soils of the North American grasslands are famously fertile and rich in organic matter. Because these soils are ideal for cultivation, only small patches remain of the vast grasslands that once occupied three continents. Farmland has replaced prairie, with wheat and corn taking the place of diverse grasses. Most prairie remnants are too small to sustain the herds of large herbivores that once roamed the plains.

C. Whether Hot or Cold, All Deserts Are Dry

All **deserts** are dry, receiving less than 20 centimeters of rainfall per year. They ring the globe at 30° North and South latitude, and additional deserts occur in the rain shadows of tall mountains. Sparse desert life means soils are low in organic matter.

Although deserts have a reputation for being hot, the temperature can vary dramatically. In a hot desert such as the Sonoran, which spans parts of Arizona and Mexico, few clouds filter the sun's strongest rays, and the days can be scorchingly hot. In China's cold Gobi Desert, in contrast, the average annual temperature is below freezing.

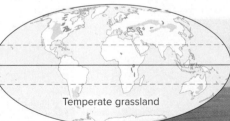

Figure 39.13 Temperate Grassland. Bison once dominated the North American prairie, a temperate grassland.

Photo: ©Photodisc/Getty Images RF

Parts of the Sahara are so dry that they are nearly devoid of life. Other desert habitats, however, are species-rich (figure 39.14). Desert plants often have long taproots, quick life cycles that exploit the brief rainy periods, fleshy stems or leaves that store water, and spines or toxins that guard against thirsty herbivores. Many use CAM photosynthesis, a water-saving variation on the photosynthetic pathway. ⓘ *CAM photosynthesis,* section 5.6

Desert herbivores include jackrabbits and kangaroo rats, which eat seeds and leaves. Snakes and cougars hunt the herbivores. Most desert animals burrow or seek shelter during the day, then become active when the sun goes down. Standing water is scarce, so most water comes from the animal's food.

Humans are drawn to the warm climates of the desert southwest of the United States. Urban development, however, changes the desert ecosystem drastically by diverting huge amounts of water from faraway rivers. Even so, deserts are among the few biomes that are expanding worldwide, as unsustainable agriculture eats away at forests and savannas.

Figure 39.14 Desert. Saguaro cacti and many other desert plants have water-conserving adaptations. This is the Sonoran Desert in the southwestern United States. Photo: ©Ed Reschke

D. Fire- and Drought-Adapted Plants Dominate Mediterranean Shrublands (Chaparral)

Despite the name, **Mediterranean shrublands** occur not only around the Mediterranean Sea but also in other small areas along the west coasts of North and South America, Australia, and South Africa. Summers are hot and dry; winters are mild and moist. As the name suggests, the dominant vegetation is shrubby plants (figure 39.15), including poison oak, manzanita, and scrub oak. The plants have thick bark and small, leathery, evergreen leaves with thick cuticles and hairs that slow moisture loss during the dry summers. Herbivores include jackrabbits, mule deer, and birds and rodents that forage for seeds under the shrub canopy; some of their predators include coyotes, foxes, snakes, and hawks.

Mediterranean shrublands are especially susceptible to fires because the vegetation dries out during the summer. The sandy soils retain little water. The fire-adapted plants resprout from underground parts or produce seeds that germinate only after the heat of a fire. People once burned chaparral to make room for grazing livestock. Since then, large human communities have moved onto the shrublands, which nevertheless remain at great risk for fire. Burning the plant cover makes the soil susceptible to erosion, raising the risk of mud slides and further property damage.

Figure 39.15 Mediterranean Shrubland. Dry summers, wet winters, and fire shape this chaparral ecosystem in California.
Photo: ©Andrew Brown; Ecoscene/Corbis

E. Tundras Occupy High Latitudes and High Elevations

The **tundra** is a low-temperature biome. The largest region of tundra occurs in a band across the northern parts of Asia, Europe, and North America. The Antarctic continent also has a small amount of tundra. In addition, at middle latitudes, alpine tundra occupies high mountaintops between the tree line and areas of permanent snow and ice cover.

Snow covers the Arctic and Antarctic tundra during the bitterly cold and dark winter. Temperatures venture above freezing for a few months each year, and summer sunlight is intense. Because cold temperatures slow decomposition, tundra soils are rich in organic matter. Below the top layer of soil is a zone called **permafrost,** where the ground remains frozen year-round. Permafrost limits rooting depth, which prevents the establishment of large plants (figure 39.16). The shallow tundra soil supports reindeer lichens, mosses, dwarf shrubs, and low-growing perennial plants such as sedges, grasses, and broadleaf herbs.

Penguins, seals, and other vertebrates visit the Antarctic tundra, but the Arctic tundra has much more diverse animal life. Inhabitants include caribou, musk oxen, reindeer, lemmings, hares, foxes, and wolverines, all of which have thick, warm fur. Polar bears sometimes visit coastal areas of the Arctic tundra to den. In the summer, migratory birds raise their young and feed on the insects that flourish in the tundra and its ponds.

The shallow soil, short growing season, and slow decomposition make the tundra a very fragile environment that recovers slowly from disturbance. In recent decades, oil drilling in the Arctic tundra has significantly increased the human presence.

F. Polar Ice Caps Are Cold and Dry

Of all of the world's terrestrial biomes, **polar ice** at the North and South Poles is the least explored (figure 39.17). Although the two poles seem superficially similar, they differ from each other. The northern ice cap is a relatively thin ice layer that covers the Arctic Ocean. Antarctica, the southernmost continent, is a landmass covered with a thick layer of ice. Both ice caps interact extensively with water and therefore share characteristics of both terrestrial and aquatic biomes.

Life is difficult at both poles. Both Antarctica and the Arctic ice cap are extremely cold, dry, and windy year-round. The primary producers in ice and the surrounding ocean are phytoplankton. The light passing through the ice is dim, even in the summer. Yet these phytoplankton support a unique food web consisting of bacteria, archaea, worms, crustaceans, and icefishes (see section 30.7). All of these organisms have antifreeze chemicals that prevent deadly ice crystals from forming in their cells.

Larger animals that exploit the polar food web have thick layers of insulating fat, coupled with fur or feathers. Examples include polar bears, seals, whales, and birds that use ice for migration, hunting, and protection. On Antarctica, vertebrates include penguins and seals. Whales and squid glide through the waters surrounding Antarctica as well.

Arctic and Antarctic tundra

Figure 39.16 Arctic Tundra. In the Alaskan tundra, lichens and small plants dominate. Permafrost prevents the growth of large plants.
Photo: ©Michael DeYoung/Corbis

Scientists are increasingly concerned that human-induced global climate change will continue to cause the polar ice caps to melt. In addition to altering the ocean ecosystem and driving rare species to extinction, the resulting rise in sea level could flood coastal cities. ⓘ *global climate change,* section 40.4

39.3	**MASTERING CONCEPTS**

1. How do climate and soil composition determine the characteristics of terrestrial biomes?
2. Infer one adaptation of plants and one adaptation of animals to the abiotic conditions in any four biomes.
3. Which biomes are supported by fire and grazing?

Polar ice

Figure 39.17 Polar Ice. These emperor penguins live in the frozen desert of Antarctica, Earth's coldest continent. Photo: ©Corbis RF

39.4 Freshwater Biomes Include Lakes, Ponds, and Streams

Although terrestrial biomes are most familiar to us, aquatic ecosystems occupy much more space. In fact, water covers about 71% of Earth's surface. This water moves continuously among the ocean, atmosphere, land surface, and groundwater, providing connections among all biomes. ⓘ *water cycle, section 38.4A*

As illustrated in **figure 39.18**, most of Earth's water is in the ocean. Only about 3% of all water has a low enough salt content to be considered "fresh." Glaciers and the great polar ice sheets of Greenland and Antarctica tie up more than two thirds of this fresh water, and most of the rest is in groundwater, soil, and the atmosphere. Lakes and rivers on Earth's surface contain only about 0.009% of the world's water supply.

Groundwater, lakes, and rivers make up only a tiny sliver of the global water "pie," but that sliver is vital to humans for drinking water and irrigation. Most other terrestrial species rely on this fresh water as well.

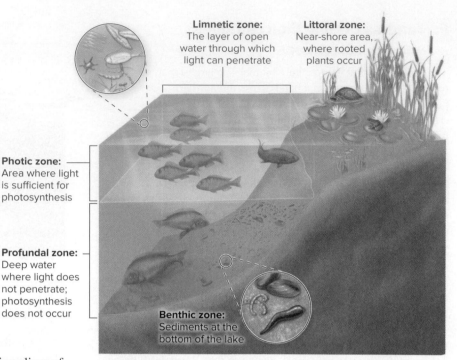

Limnetic zone: The layer of open water through which light can penetrate

Littoral zone: Near-shore area, where rooted plants occur

Photic zone: Area where light is sufficient for photosynthesis

Profundal zone: Deep water where light does not penetrate; photosynthesis does not occur

Benthic zone: Sediments at the bottom of the lake

Figure 39.19 Zones of a Lake. Distance from the shore and the availability of light define the zones of a lake. Each zone houses a unique assortment of species.

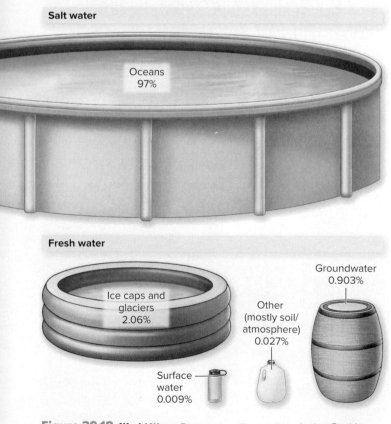

Salt water

Oceans 97%

Fresh water

Ice caps and glaciers 2.06%

Other (mostly soil/atmosphere) 0.027%

Groundwater 0.903%

Surface water 0.009%

Figure 39.18 World Water Resources. This analogy depicts Earth's water supply in a series of containers, ranging from a large aboveground swimming pool (the oceans) to a small bottle (surface water).

A. Lakes and Ponds Contain Standing Water

Standing water includes lakes and ponds, which differ from one another mainly in size and depth. A lake is generally larger and deeper than a pond.

Light penetrates the regions of a lake to differing degrees, creating zones with characteristic groups of organisms (**figure 39.19**). The lake's **photic zone,** where light is sufficient for photosynthesis, is subdivided into littoral and limnetic zones. The **littoral zone** is the shallow shoreline region where rooted plants occur. Productivity is high, thanks to abundant nutrients from the shore. Phytoplankton and plants such as cattails thrive here, providing food and shelter for myriad invertebrates, fishes, amphibians, and other animals. The **limnetic zone** is the layer of open water where light penetrates. Phytoplankton are the dominant producers in the limnetic zone.

In contrast to the photic zone, the **profundal zone** is the deep region of water where light does not penetrate. Scavengers and decomposers, both here and in the **benthic zone** (the sediment at the lake bottom), rely on a gentle rain of organic material from above to supply both energy and nutrients.

Lakes change over time (**figure 39.20**). Younger lakes are often deep, steep-sided, and low in nutrients. These lakes are **oligotrophic,** which means they are nutrient-poor and therefore low in productivity. They are clear and sparkling blue, because phytoplankton aren't abundant enough to cloud the water.

As a lake ages, however, nutrients accumulate from decaying organisms and sediment. Such a lake is **eutrophic,** which means

a.

b.

Figure 39.20 Nutrients Make a Difference. (a) Oregon's Crater Lake is oligotrophic. It formed about 7000 years ago, and few nutrients have made their way into the crystal-clear water. (b) Algae are accumulating in the shallow, nutrient-rich waters of this eutrophic pond.

(a): ©ML Sinibaldi/Corbis; (b): ©Pat Watson/McGraw-Hill Education

it is nutrient-rich and high in productivity. The rich algal growth turns the water green and murky. Not surprisingly, nutrient-rich urban wastewater and farm runoff carrying phosphate-rich fertilizers can speed this process. The nutrients promote the excessive growth of algae, which sink to the lake bottom after they die. As the dead algae decompose, deep waters are rapidly depleted of oxygen. Many fish and other animals die. In addition, microbes respond by switching to anaerobic metabolism; the byproducts are gases with unpleasant, "rotten egg" odors. (i) *eutrophication,* section 38.4E

A eutrophic lake eventually fills completely with sediments and transforms into a freshwater **wetland**—an area where soil is permanently or seasonally saturated with water. These nutrient-rich swamps, bogs, or marshes host diverse species with adaptations to both land and water.

B. Streams Carry Running Water

Streams include brooks, creeks, and rivers that carry water and sediment from all portions of the land toward the ocean (or an interior basin such as the Great Salt Lake). Along the way, streams provide moisture and habitat to aquatic and terrestrial organisms.

Rivers are the largest streams. They change as they flow toward the ocean (**figure 39.21**). At the headwaters, the water is relatively clear, and the stream channel is narrow. Where the current is swift, turbulence mixes air with water, so the water is rich in oxygen. As the river flows toward the ocean, it continues to pick up sediment and nutrients from the channel. The river widens as small streams draining additional land areas contribute more water. As the land flattens, the current slows. The river is now murky, restricting photosynthesis to the banks and water surface. As a result, the oxygen content is low relative to the river upstream.

Rivers depend on runoff from the land for water and nutrients. Dead leaves and other organic material fall into the river and add to the nutrients. On the other hand, rivers also return nutrients to the land. Many rivers flood each year, swelling with meltwater and spring runoff and spreading nutrient-rich silt onto floodplains. When a river approaches the ocean, its current slows and deposits fine, rich soil that forms new delta lands at the mouth of the river.

39.4 MASTERING CONCEPTS

1. Describe the types of organisms that live in each zone of a lake or pond.
2. Contrast oligotrophic and eutrophic lakes.
3. How does a river change from its headwaters to its mouth?

Figure 39.21 A River Changes Along Its Course. A narrow, swift stream in the mountains becomes a slow-moving river as it accumulates water and sediments along its journey to the ocean.

39.5 Oceans Make Up Earth's Largest Ecosystem

The oceans, covering 70% of Earth's surface and running 11 kilometers deep in places, form the world's largest biome. Most photosynthesis on Earth occurs in the vast oceans, contributing enormous amounts of oxygen to the atmosphere. Moreover, oceans absorb so much heat from the sun that they help stabilize Earth's climate.

Food webs in the ocean need nutrients and sunlight. Supplies of both of these resources—and therefore primary production—are highest in shallow waters near the coasts. In deep water away from the coasts, however, both energy and nutrients can be scarce.

A. Land Meets Sea at the Coast

For humans, the world's coasts are a vital source of food, transportation, recreation, and waste disposal. Most of the world's largest cities are located along coastlines, and human populations in those cities are expected to increase in the future. As the human population expands, concerns about coastal ecosystems will likely increase. Several types of ecosystems border shorelines.

Estuaries   An **estuary** is an area where the fresh water of a river meets the salty ocean (**figure 39.22a**). When the tide is out, the water may not be much saltier than water in the river. The returning tide, however, may make the water nearly as salty as the sea. Organisms that can withstand these extremes in salinity receive nutrients from both the river and the tides. Estuaries therefore house some of the world's most productive ecosystems.

In the open water of an estuary, phytoplankton account for most of the productivity. In addition, salt-tolerant grasses and other flowering plants dominate the salt marshes that often occur along the fringes of an estuary. Together, these producers support many animals. For example, more than half of the commercially important fish and shellfish species spend some part of their life cycle in an estuary. Migratory waterfowl feed and nest here as well. Moreover, almost half of an estuary's photosynthetic products go out with the tide and nourish coastal communities.

Intertidal Zone   Along coastlines, the **intertidal zone** is the area between the high tide and low tide marks. This region of constant change is alternately exposed and covered with water as the tide rises and falls.

Species-rich mangrove forests occupy intertidal areas in the tropics (figure 39.22b). An entirely different intertidal zone is a sandy beach, which features long strips of bare sand that make for beautiful vistas and all-day exposure to the sun. Constantly shifting sands mean that few producers can take root on the beach, but ocean water delivers a constant supply of dead organisms and other nutrients. Crabs feed on the organic debris and then burrow into the sand to escape the pounding waves, while shorebirds forage for worms and other small invertebrates.

Some intertidal zones are rocky, not sandy (figure 39.22c). Here, organisms often attach to rocks, preventing wave action from carrying them away. Holdfasts attach large marine algae (seaweeds) to rocks. Suction and sticky threads fasten mussels to rocks. Sea anemones, sea urchins, sea stars, and snails live in pools of water that form between rocks.

Coral Reefs   Colorful and highly productive coral reef ecosystems border tropical coastlines where the water is clear and sediment-free. **Coral reefs** are vast underwater structures of calcium carbonate built by coral animals. The living coral is but a thin layer atop the remains of ancestors. A coral reef, then, is at the same time an immense graveyard and a thriving ecosystem.

The tissues of the coral animals house symbiotic algae that are essential for the coral's—and the ecosystem's—survival. The sun penetrates the clear, shallow water, allowing photosynthesis to occur, and constant wave action brings in additional nutrients.

a.   b.   c.   d.

Figure 39.22 Coastal Ecosystems. (a) An estuary on the coast of Mozambique. (b) Mangrove trees on Okinawa, Japan. (c) Rocky intertidal zone along the Oregon coast. (d) Coral reef in the Red Sea.

(a): ©John Warburton-Lee Photography/Alamy; (b): ©Flickr Open/Getty Images RF; (c): ©Craig Tuttle/Getty Images RF; (d): ©Digital Vision/Getty Images RF

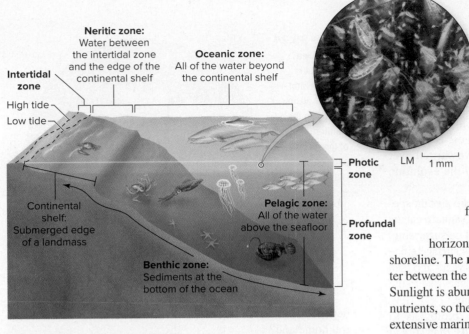

Neritic zone: Water between the intertidal zone and the edge of the continental shelf

Oceanic zone: All of the water beyond the continental shelf

Intertidal zone

High tide

Low tide

Continental shelf: Submerged edge of a landmass

Pelagic zone: All of the water above the seafloor

Benthic zone: Sediments at the bottom of the ocean

Photic zone

Profundal zone

LM   1 mm

Figure 39.23 Zones of the Ocean. The intertidal, neritic, oceanic, and benthic zones of the ocean each harbor unique types of organisms. The inset shows the sea surface microlayer, which brims with phytoplankton, zooplankton, tiny animals, eggs, larvae, and nutrients.

Photo: ©Andy Mann/National Geographic/Getty Images

The nooks and crannies of a coral reef provide food and habitat for a huge variety of organisms (figure 39.22d). The Great Barrier Reef of Australia, for example, is composed of some 400 species of coral and supports more than 1500 species of fishes, 400 of sponges, and 4000 of mollusks. Other residents include algae, snails, sea stars, sea urchins, sea turtles, and countless types of microorganisms. This remarkably rich biodiversity has prompted some people to call coral reefs "the rain forests of the sea."

B. The Open Ocean Remains Mysterious

The oceans cover most of Earth's surface, but we know less about marine life than we do about biodiversity in a single tree in a tropical rain forest. Not only is this ecosystem vast, but it also houses populations that are sometimes small, usually very dispersed, and nearly always difficult for us to observe. Biologists have explored only 5% of the ocean floor and 1% of the huge volume of water above.

Marine biologists divide the ocean surface into horizontal zones (**figure 39.23**). The intertidal zone is the shoreline. The **neritic zone** is the area of relatively shallow seawater between the intertidal zone and the edge of the continental shelf. Sunlight is abundant here, and sediments from the coast contribute nutrients, so the neritic zone supports high primary production and extensive marine food webs. The great kelp forests that fringe many cool-water coastal areas are neritic ecosystems (see figure 38.1). The **oceanic zone** is the deep water beyond the continental shelf.

The ocean can also be subdivided according to depth. The **pelagic zone** consists of all of the water above the seafloor. The upper layer of the pelagic zone, the photic zone, is the only area where photosynthesis can occur. At the very top of the pelagic zone, the sea surface houses phytoplankton and the zooplankton that feed on them (see figure 39.23, inset). Large sea animals such as fishes and whales scoop up vast quantities of krill and other zooplankton.

Below the photic zone is the profundal zone, where light is too dim for photosynthesis. Nevertheless, the bodies and wastes of top-dwelling organisms provide a continual rain of nutrients to the species below, including great numbers and varieties of jellyfishes, fishes, whales, dolphins, mollusks, echinoderms, crustaceans, and organisms yet to be discovered. The unusual communities that occupy hydrothermal vents add biodiversity to the benthic zone (see the chapter opening essay).

The most productive ocean environments arise in zones of **upwelling,** where deep currents force cold, nutrient-rich lower layers of water to move upward. The influx of nutrients causes phytoplankton to "bloom," and with this widening of the food web base, many ocean populations grow. Upwelling generally occurs on the western side of continents, such as along the coasts of southern California, South America, parts of Africa, and the Antarctic. A climatic event called El Niño periodically shifts these ocean currents, greatly disrupting coastal ecosystems (see this chapter's Apply It Now box).

39.5 | MASTERING CONCEPTS

1. Describe some adaptations that characterize life in estuaries, intertidal zones, and coral reefs.
2. List and define the major zones of the ocean.
3. How is upwelling important to ocean ecosystems?

INVESTIGATING LIFE

39.6 There's No Place Like Home

Today, Earth's biomes include great bands of forests, deserts, savannas, and grasslands. Other biomes, such as tundra and Mediterranean shrublands, occupy smaller ranges. But as we saw in section 39.1, continents and climates have shifted over time. These slow changes have directly affected the evolution of plants.

Clearly, ecological constraints determine where a plant species can and cannot grow. After all, finding a giant saguaro cactus in a tropical rain forest or a fern in a desert would be surprising. Yet plants have spread across nearly all of Earth's surface since green algae first colonized land some 450 million years ago (see section 19.1). These aquatic ancestors eventually gave rise to the hundreds of thousands of plant species we see today.

This observation has inspired evolutionary biologists to ask questions. As plants produce spores or seeds that spread into new habitats, how likely are they to survive, thrive, reproduce, and even give rise to new species? Does the answer

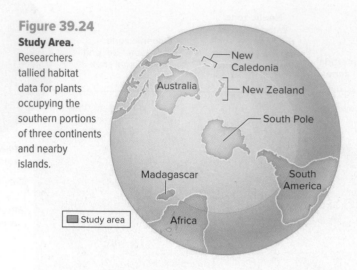

Figure 39.24

Study Area. Researchers tallied habitat data for plants occupying the southern portions of three continents and nearby islands.

□ Study area

depend on the similarity between the new habitat and the ancestral homeland?

Small-scale studies hinted that species rarely shift from one biome to a different one, but the questions demanded larger-scale research. A group of Australian scientists led by Michael Crisp, with collaborators across the world, tackled the challenge. The researchers focused on plants occupying the southern hemisphere. Their study area included large parts of South America, Africa, Madagascar, Australia, New Zealand, and associated islands (**figure 39.24**). The researchers chose this area partly because multiple biomes occupy each continent and the nearby islands, making it possible to detect species shifts from one habitat to another.

Rather than studying the plants directly, the researchers used published data from 89 peer-reviewed papers. They were particularly interested in the distribution of vascular plants in their study area, along with molecular data indicating the evolutionary relationships among all of the species. All together, their data set included around 11,000 plant species, which they sorted into seven biomes (**figure 39.25**). Some of the biomes are subtypes of biomes described in this chapter; others are combinations of biomes. ⓘ *vascular plant diversity*, section 19.1A

The researchers used the current range of each species, plus its evolutionary relationship with its close relatives, to learn which biome the ancestors of each species most likely inhabited. For example, if species X occurs only in the grassland, but all of its closely related sister species occupy the savanna, the researchers would infer that the ancestors of species X originated in the savanna. They would then score species X as having undergone a biome shift.

Overall, the analysis showed that such biome shifts did occur, but only rarely. That is, an average of 96% of sampled plant species occupy the same biome that their ancestor inhabited (**figure 39.26**). The researchers then dug a little deeper into the relatively few cases for which a plant species occupied a different biome from its ancestors. Not surprisingly, they found that biome similarity affected the likelihood of species shifts. For

Apply It Now

El Niño Years

Winds normally blow from east to west at the equator. As a result, warm water gets trapped deep in the western Pacific. During an El Niño event, however, the easterly winds go slack, allowing warm water to drift across the ocean from west to east. The warm water releases heat into the atmosphere, increasing Earth's average temperature (figure 39.B).

El Niño may trigger unusual weather worldwide: monsoons in the central Pacific, typhoons in Hawaii, torrential rains in South America, blizzards in the Rockies, warm winters in the northeastern United States, flooding in southern California, and severe droughts in southern Africa. Moreover, as warm water moves east, it prevents cold, nutrient-rich water below from rising to the surface ("upwelling"). The ecological consequences are grim. Fishes move to deeper, cooler waters; the animals that normally eat them may starve.

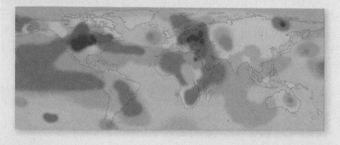

Figure 39.B **El Niño.** The El Niño event of 2015 helped fuel an unusually hot year. Red areas are warmer than average; blue areas are cooler.

Figure 39.25

Biomes Considered in Study	Description	Image	Number of Species Sampled
Sclerophyll	Many shrubs; nutrient-poor soils; similar to Mediterranean shrubland		7250
Arid	Some shrubs and trees, few grasses; dry		1683
Wet forest	Many trees; abundant precipitation		1005
Temperate grassland	Many grasses; cold winters; climate varies but is not arid		504
Savanna	Many grasses, few shrubs and trees; summer rain and winter drought		242
Alpine	Shrubs and grasses above the tree line		186
Bog	Permanently wet soil that lacks surface water		84

Figure 39.25 Biome Descriptions. The researchers classified ecosystems in the study area into seven biomes.

Photos: (sclerophyll): ©DLILLC/Corbis RF; (arid): ©Paul Edmondson/Getty Images RF; (wet forest): ©Photo 24/Getty Images RF; (temperate grassland): ©DLILLC/Corbis RF; (savanna): ©Natphotos/Getty Images RF; (alpine): U.S. Geological Survey/P. Carrara; (bog): ©Steven P. Lynch RF

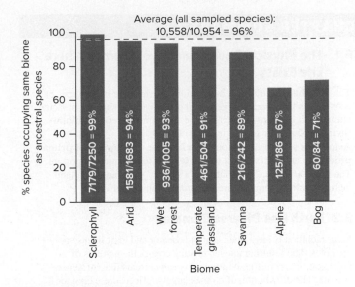

Average (all sampled species):
10,558/10,954 = 96%

% species occupying same biome as ancestral species

Sclerophyll	Arid	Wet forest	Temperate grassland	Savanna	Alpine	Bog
7179/7250 = 99%	1581/1683 = 94%	936/1005 = 93%	461/504 = 91%	216/242 = 89%	125/186 = 67%	60/84 = 71%

Biome

Figure 39.26 Conservative Plants. Plant species in the study area were unlikely to move to new biomes. The ratio within each bar represents the proportion of species sampled in each biome that had ancestors in the same biome. For example, of the 7250 species currently in sclerophyll, 7179 (99%) had ancestors in sclerophyll.

suggest that natural selection typically does not tolerate dramatic shifts in ecology.

Besides shedding light on the past evolution of plants, this type of research also has important implications today. For example, as you will see in chapter 40, invasive plants are altering many ecosystems. If plants "prefer" their ancestral biome types, then an invasive species is most likely to become established in an ecosystem that resembles its home range. Understanding which invasive plants pose the greatest risk to each biome can help conservation biologists focus their eradication efforts.

On an even wider scale, this finding is one of many that highlight a possible consequence of global climate change (see section 40.4). Although deserts are becoming larger, many other biomes are shrinking as the human population grows and global climate change accelerates. Are species that live in a contracting biome likely to adapt to the changing climate? If existing species rarely diversify and colonize new biomes, then natural selection is unlikely to meet the pace of the changing environment. Many species may go extinct, with unpredictable consequences to global ecology.

Source: Crisp, Michael, and 9 coauthors. 2009. Phylogenetic biome conservatism on a global scale. *Nature,* vol. 458, pages 754–756.

39.6 MASTERING CONCEPTS

1. According to figure 39.26, which biome has the highest proportion of plants that have ancestors occupying the same biome? Explain what this result means.

2. Explain how past species shifts can help researchers predict future species survival under global climate change.

example, 95 of the 102 plant species that entered the arid biome had ancestors in the sclerophyll biome, which is ecologically similar but somewhat wetter. The scientists never documented a shift between dissimilar biomes, such as from grassland to a bog or from savanna to an alpine biome. These striking results

CHAPTER SUMMARY

39.1 The Physical Environment Determines Where Life Exists

- The **biosphere** is subdivided into **biomes,** which are major types of **ecosystems** that occupy large geographic areas and share a characteristic climate and group of species. Each ecosystem incorporates **biotic** (living) and **abiotic** (nonliving) interactions.
- Autotrophs support each biome. On land, the most important **primary producers** are plants; in water, **phytoplankton** are most important.
- The major abiotic factors that limit a species' distribution on land include sunlight, temperature, moisture, salinity, and fire.

39.2 Earth Has Diverse Climates

- Solar radiation is most intense at the equator and least intense at the poles. The resulting uneven heating creates the patterns of precipitation, prevailing winds, and ocean currents that influence climate. The distribution of biomes largely reflects these climatic zones (figure 39.27).
- A mountain range can affect local climate by producing a **rain shadow.**

39.3 Terrestrial Biomes Range from the Lush Tropics to the Frozen Poles

A. Towering Trees Dominate the Forests

- The **tropical rain forest** is warm and wet, with diverse life. Nutrients cycle rapidly, leaving soils relatively poor.
- In **temperate deciduous forests,** rainfall is evenly distributed throughout the year, summers are warm, and winters are cold. Their soils are fertile. **Temperate coniferous forests** usually have somewhat poorer soils, more rainfall, milder winters, and cooler summers.
- The **taiga** is a very cold northern (boreal) coniferous forest. Conifers conserve moisture during the long winters, when freezing temperatures mean liquid water is scarce.

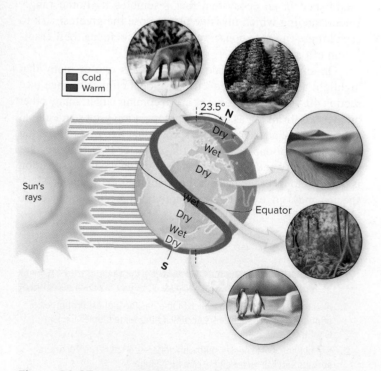

Figure 39.27 Climate Patterns: A Summary.

B. Grasslands Occur in Tropical and Temperate Regions

- Tropical **savannas** have alternating dry and wet seasons and are dominated by grasses, with sparse shrubs and woody vegetation. Migrating herds of herbivores graze on the plants.
- Fire, grazing, and seasonal drought keep **temperate grasslands** free of trees.

C. Whether Hot or Cold, All Deserts Are Dry

- **Deserts** are dry; plants in desert biomes have adaptations that help them obtain and store water. Many desert animals seek shelter during the day and become active at night.

D. Fire- and Drought-Adapted Plants Dominate Mediterranean Shrublands (Chaparral)

- **Mediterranean shrublands** such as California's chaparral, have dry summers and moist, mild winters. Many of the plants are fire-adapted.

E. Tundras Occupy High Latitudes and High Elevations

- **Tundra** can occur in the Arctic, in the Antarctic, or high on mountaintops in temperate zones.
- Arctic and Antarctic tundras have very cold, long winters. A layer of frozen soil called **permafrost** lies beneath the surface and prevents the growth of trees.

F. Polar Ice Caps Are Cold and Dry

- **Polar ice** forms at Earth's poles, where the climate is extremely cold and dry. Phytoplankton in melted ice support a food web that includes zooplankton and many types of vertebrates.

39.4 Freshwater Biomes Include Lakes, Ponds, and Streams

A. Lakes and Ponds Contain Standing Water

- In a lake's **photic zone,** light is sufficient for photosynthesis. The **littoral zone** of a lake is the shallow area where rooted plants occur, whereas phytoplankton are the primary producers in the **limnetic zone,** the upper layer of open water.
- The **profundal zone** is the dark, deeper layer below the photic zone, and the lake bottom is the **benthic zone.** Nutrients arriving from the upper layers support life in the profundal and benthic zones.
- Young, deep **oligotrophic** lakes are clear blue, with few nutrients to support algae. Over time, however, nutrients gradually accumulate, and algae tint the water green. The lake becomes a productive, or **eutrophic,** lake. As sediments accumulate, the lake transforms into a **wetland.**

B. Streams Carry Running Water

- In rivers, organisms are adapted to local currents. Near the headwaters, the channel is narrow, and the current is swift. As the river accumulates water and sediments, the current slows, and the channel widens.

39.5 Oceans Make Up Earth's Largest Ecosystem

A. Land Meets Sea at the Coast

- An **estuary** is a highly productive area where rivers empty into oceans, and life is adapted to fluctuating salinity. Residents of the rocky **intertidal zone** cling tightly to the substrate as the tide ebbs and flows. In tropical regions, **coral reefs** support many thousands of species.

B. The Open Ocean Remains Mysterious

- The region of ocean near the shore is the **neritic zone.** Open water is the **oceanic zone.** Vertical subdivisions of the ocean include the **pelagic zone** (open water above the ocean floor) and the benthic zone (the bottom). The most productive areas are in the neritic zones, where **upwelling** occurs.

39.6 Investigating Life: There's No Place Like Home

- Plants tend to remain in the same biome as their ancestors. As climate change causes biomes to shift, many plants may not adapt to dramatic shifts in ecology.

MULTIPLE CHOICE QUESTIONS

1. What is a major primary producer in many aquatic biomes?
   a. Underwater plants
   b. The sun
   c. Phytoplankton
   d. Insect larvae

2. Why are the poles colder than the equator?
   a. Because the poles receive solar radiation that is less intense
   b. Because the poles receive less rainfall than the equator
   c. Because ocean currents bring warm water to the equator
   d. Because the tilt of the Earth points the poles away from the sun

3. What location most likely has a dry climate?
   a. The east side of a mountain
   b. The upwind side of a mountain
   c. The west side of a mountain
   d. The downwind side of a mountain

4. A biome with high average temperature and low to moderate annual rainfall is a
   a. tropical savanna.
   b. tropical rain forest.
   c. temperate deciduous forest.
   d. taiga.

5. Which of the following correctly ranks Earth's water sources from largest to smallest?
   a. Surface water (fresh) > ocean > groundwater > ice
   b. Ocean > ice > groundwater > surface water (fresh)
   c. Groundwater > ocean > surface water (fresh) > ice
   d. Ocean > surface water (fresh) > ice > groundwater

6. A worm that lives in lake sediment is in the ___ zone.
   a. limnetic
   b. profundal
   c. benthic
   d. littoral

7. Why do estuaries have such high primary production?
   a. The changing salinity of the water
   b. High nutrient input from the land
   c. The absence of consumers that eat producers
   d. All of the above are correct.

Answers to these questions are in appendix A.

WRITE IT OUT

1. How does the fact that Earth is a sphere tilted on its axis influence the distribution of life?

2. Explain why sunlight is most intense at the equator.

3. Tropical rain forests have nutrient-poor soils, and people who clear the forests for farming often abandon the land after a few years. However, the same soil previously supported diverse, abundant rain forest life. Explain this paradox.

4. Explain why fire-adapted plants are common in Mediterranean shrublands. Why might fire-adapted plants be much less common in (or absent from) rain forests?

5. List adaptations of desert and polar ice animals to the climate of their respective biomes.

6. Why might it be unlikely for a rain forest plant in central Africa to have descendants that colonize the nearby savanna?

7. Polar bears live on the ice cap near the North Pole. Their numbers are dwindling, apparently because of both pollution and global climate change. List some specific ways that the melting of the Arctic ice cap might affect polar bear populations.

8. Find where you live on the map in figure 39.7. In which major terrestrial biome is your home? Then, use the Internet to learn more about the ecosystem(s) near your home. In what ways do your surroundings differ from the biome description in section 39.3? In what ways are they the same?

9. Nuisance aquatic plants such as hydrilla can disrupt the ecology of the littoral zone of a lake. Two of the most common ways to control nuisance aquatic plants are herbicides (chemicals that kill plants) and biological control (introducing fungi or animals that consume the plants). How might each strategy help or harm the lake ecosystem?

10. Use the clues provided to determine which biome houses each of the following four fish. Yellowfin tuna swim in deep salt water; young sea bass occupy areas where fresh and salt water mix; brook trout require clear, cool, oxygen-rich fresh water; and catfish prefer warm, quiet fresh water with a slow current.

11. Make a concept map depicting the relationships among the zones of the ocean. What is the main energy source in each zone?

12. In a eutrophic lake, photosynthesis by phytoplankton does not maintain deep-water oxygen levels. Why not?

13. The biomes described in this chapter do not include those that humans create, such as cities, villages, croplands, rangelands, and tree farms. How are these biomes similar to and different from the biomes in this chapter?

14. Hundreds of millions of years ago, Earth's landmasses were joined into one supercontinent, Pangaea. If Pangaea had never broken up, do you think there would be more biodiversity, less biodiversity, or the same amount of biodiversity as today? Explain your answer.

PULL IT TOGETHER

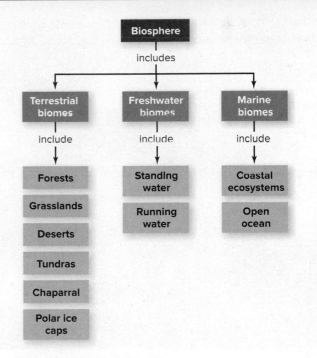

Figure 39.28 Pull It Together: Biomes.

Refer to figure 39.28 and the chapter content to answer the following questions.

1. Review the Survey the Landscape figure in the chapter introduction. What major factors determine how scientists classify each part of the biosphere into a biome?

2. What types of forests occur on Earth, and what combination of conditions favors each type?

3. What types of terrestrial biomes are likely to have standing and running water? Make connections between terrestrial and aquatic biomes in the concept map.

40 Preserving Biodiversity

The Everglades. Saw grass, trees, and countless other species make up the biodiversity of the Everglades National Park in Florida. Pollution, water diversion projects, and invasive species have devastated the entire Everglades ecosystem.

©Raul Touzon/National Geographic/Getty Images

LEARN HOW TO LEARN
Make Your Own Review Sheet

If you are facing a big exam, how can you make sense of everything you have learned? One way is to make your own review sheet. The best strategy will depend on what your instructor expects you to know, but here are a few ideas to try: Make lists; draw concept maps that link ideas within and between chapters; draw diagrams that illustrate important processes; and write mini-essays that explain the main points in each chapter's learning outline.

The Endangered Everglades

The world has no shortage of degraded ecosystems, yet the Florida Everglades region is remarkable for the scope of its destruction. Before 1900, much of Florida was a continuous waterway. This vast area included estuaries, saw grass plains, mangrove swamps, and tropical hardwood forests. The Everglades was home to marsh grasses, cypress trees, egrets, eagles, herons, panthers, and alligators. Now, many of these species are in trouble.

Many factors have contributed to the Everglades catastrophe. First, humans have dramatically altered the flow of water since early in the 20th century. Engineers have built canals, levees, and pumps to prevent flooding and to direct water to farms and cities. Meanwhile, the human population in the area has skyrocketed, demanding not only land but also fresh water.

Changing the path of water has had dramatic consequences. In addition to robbing wetland species of water, canals that direct water toward the sea alter the salt balance of estuaries, too. During torrential rains, for example, the influx of fresh water robs mangroves of salt water, which leads to declines in populations of sea grasses and various animals. Wading bird populations have plummeted to 10% of their levels a century ago. Dozens of species have become endangered.

Second, nutrients from agriculture have played a major role in the decline of the Everglades. Nitrogen and phosphorus from fertilizer and dairy operations wash into the water, causing massive algae blooms. Native saw grass cannot compete with plants that tolerate the nutrient pollution. ⓘ *eutrophication,* section 38.4E

Third, native Everglades species must now compete with exotic invaders. The most famous of these is the Burmese python, a snake imported from southeast Asia for the pet trade. These massive snakes have no natural predators, and they devour huge numbers of other animals. Green iguanas, monitor lizards, wild boars, and domesticated cats are among the other troublesome introduced animals. Moreover, invasive plants such as Australian pines, hydrilla, and water hyacinths crowd out native species.

Efforts have been underway to undo the damage. The Kissimmee River Restoration Project, for example, will gradually restore flow to this major river's channel and the surrounding floodplain wetlands. The project is just one part of the much larger Comprehensive Everglades Restoration Plan. With a price tag estimated at $9.5 billion, the CERP is "the world's largest ecosystem restoration effort." Monitoring its progress will require a small army of conservation biologists. This chapter describes some of the many challenges scientists such as these face worldwide.

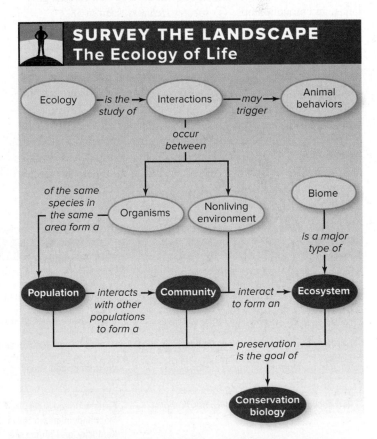

SURVEY THE LANDSCAPE
The Ecology of Life

Human activities are associated with a dramatic reduction in biodiversity, but conservation efforts at the population, community, and ecosystem levels can help reverse the decline.

For more details, study the Pull It Together feature in the chapter summary.

40.1 Earth's Biodiversity Is Dwindling

For more than 3 billion years, evolution has produced an extraordinary diversity of life, both obvious and unseen; unit 4 provided an overview of Earth's inhabitants. Humans simply cannot live without these other species (figure 40.1). We use other organisms for food, shelter, energy, clothing, and drugs. Microbes carry out indispensable tasks, including digesting food in our intestines, decaying organic matter in soil, fixing nitrogen, and producing oxygen (O_2). Plants and microbes absorb carbon dioxide (CO_2) and purify the air, soil, and water. Wetland plants reduce the severity of floods. Insects pollinate our crops. The remains of species that lived millions of years ago provide the fossil fuels that sustain our economies. The list goes on and on.

Clearly, our existence as a species depends on **biodiversity**—the variety of life on Earth. Biologists measure biodiversity at three levels: genetic, species, and ecosystem. Genetic diversity is the amount of variation that exists within a species. This aspect of biodiversity is essential for populations to adapt to changing conditions. The next level, species diversity, accounts for the number of species that occupy the biosphere. Finally, ecosystem diversity means the variety of ecosystems on Earth, such as deserts, rain forests, grasslands, and mountaintops. ⓘ *species richness,* section 38.2

One way to monitor species biodiversity is to count how many species are at risk of extinction. **Extinction** means that the last individual of a species has perished. (Note that conservation biologists sometimes distinguish between species that are totally extinct and those that are extinct in the wild. The dodo is extinct, whereas a bird called the Guam rail exists in captivity but is extinct in its natural habitat.) An **endangered species** has a high risk of extinction in the near future, and a **vulnerable species** is likely to become extinct in the more distant future. The International Union for Conservation of Nature (IUCN) combines endangered and vulnerable species into one umbrella category ("threatened").

a.

b.

Figure 40.1 We Need Other Species. (a) A leaf-nosed bat captures an insect; bats consume many animals that humans consider pests. (b) A water buffalo is not only a work animal but also a source of dairy products. Moreover, the animal's dung fertilizes the rice fields that feed many people.
(a): ©Dr. Merlin D. Tuttle/Science Source; (b): ©dionfeleo bastian1/RooM/Getty Images

The data suggest that Earth is in the midst of a biodiversity crisis (table 40.1, figure 40.2). The current extinction rate of vertebrates is some 100 to 1000 times the "background" species extinction rate, which estimates how quickly species disappeared

TABLE **40.1** A Few of the World's Endangered Species

Species	Former Range	Threat(s)
Green pitcher plant (*Sarracenia oreophila*)	Alabama, Tennessee, Georgia	Habitat loss, overharvesting for commercial trade
Asian slipper orchids (genus *Paphiopedilum*)	Southeast Asia, parts of India and China	Habitat loss, overharvesting for commercial trade
American burying beetle (*Nicrophorus americanus*)	Arkansas, Kansas, Oklahoma, Nebraska, South Dakota	Habitat loss
Northern bluefin tuna (*Thunnus thynnus*)	Atlantic Ocean, Mediterranean and Black Seas	Overfishing
California condor (*Gymnogyps californianus*)	Western United States	Habitat loss, shooting, lead poisoning, toxic substances in environment
Red-cockaded woodpecker (*Picoides borealis*)	Eastern United States from Florida to New Jersey and Maryland, inland to Texas, Oklahoma, Missouri, Kentucky, and Tennessee	Habitat loss
Guam rail (bird) (*Gallirallus owstoni*)	Guam	Invasive species (brown tree snake; see the opening essay for chapter 37)
Sumatran rhino (*Dicerorhinus sumatrensis*)	Indonesia, Malaysia	Poaching

before human intervention. According to the IUCN, threatened groups include amphibians, mammals, reptiles (including birds), bony fishes, corals, plants, and many other types of organisms. Overall, the current biodiversity crisis is comparable to the five mass extinctions that have occurred in the past 500 million years (see figure 13.2).

Conservation biologists study the preservation of biodiversity at all levels. These scientists try to determine why species disappear, and they develop strategies for maintaining diversity. This final chapter focuses first on the main causes of the loss of biodiversity: habitat destruction and degradation, nonnative species, and overexploitation. Each of these threats to biodiversity is intimately related to the ever-expanding human population, and the problems will only become worse as our population continues to grow. Nevertheless, the chapter ends on a hopeful note, with some ways people can help counteract the biodiversity crisis. ⓘ *human population growth,* section 37.5

Figure 40.3 Habitat Loss. Large cities such as San Francisco replace native habitat with pavement and buildings, altering land and water.
©Corbis RF

40.1 MASTERING CONCEPTS

1. What is the value of diversity to humans and to ecosystems as a whole?
2. Describe the relationships among the three levels of biodiversity. Why is each level important?
3. Differentiate among extinct, endangered, and vulnerable species.
4. What is conservation biology?

40.2 Many Human Activities Destroy Habitats

Habitat destruction is the primary threat to biodiversity. Humans have altered more than 50% of the land, replacing prairies, wetlands, and forests with farms, rangeland, and cities (figure 40.3). The link to biodiversity is obvious: Destroying a habitat makes it difficult or impossible for its occupants to survive and reproduce, so populations crash.

One form of habitat destruction is **deforestation,** the removal of all trees from a forested area. Like all plants, trees absorb CO_2 from the atmosphere and use it in photosynthesis, a process that also releases O_2. The wood of a living tree is an important long-term "carbon sink" that helps offset CO_2 released when humans burn fossil fuels (see section 40.4). Trees also play a critical role in the water cycle, returning water to the atmosphere via transpiration. ⓘ *water cycle,* section 38.4A

Forests harbor a tremendous diversity of organisms that provide wildlife habitat and supply many of the resources that we use every day; deforestation eliminates these resources. In addition, forest destruction removes the upper layer of organic matter on the forest floor, so rainwater enters streams rather than soaking into the soil. As the soil erodes away, waterways become choked with nutrients and sediments.

All of the world's forest biomes are under threat. Logging, mining, and exploration for oil and gas are rapidly depleting the northern coniferous forest. In addition, nearly all of the world's native temperate forests are already gone. In North America, for example, people have cleared the land to create farmland, obtain fuel, and make room for cities. Today, less than 1% of the original temperate forest survives.

Closer to the equator, people are logging and burning tropical rain forests to make room for crops and domesticated animals

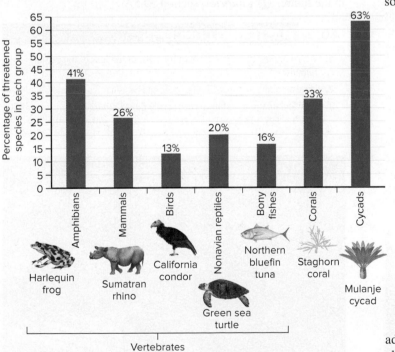

Figure 40.2 Biodiversity Crisis. Species throughout the tree of life are in peril, but some groups are harder hit than others. Each bar represents the percentage of known species that are considered threatened; one example from each group is shown.
Source: Data from IUCN Red List.

(figure 40.4). When trees burn, they release stored carbon into the atmosphere, contributing to the greenhouse effect. Ironically, the same soils that support the lush rain forest produce poor crop yields. The tropical climate explains this apparent contradiction. Warm temperatures promote rapid decomposition of organic matter, and heavy rains deplete soil nutrients. Once native plants give way to crops or grazing animals, the nutrient-poor soils harden into a cementlike crust. Species disappear and food webs topple, threatening biodiversity in the entire region.

Prairies and other temperate grasslands have also disappeared as the human population has expanded (figure 40.5). Grassland soils are among the most fertile in the world, and their rolling hills are ideal for cultivation. Fields of corn, soybeans, wheat, and other crops have replaced nearly all of the North American prairie. ⓘ *temperate grasslands,* section 39.3B

Forests and grasslands are shrinking worldwide, but deserts are expanding as unsustainable agriculture eats away at forests and savannas. For example, widespread drought and overgrazing by domesticated animals threaten to turn large areas of tropical savanna to desert. But human activities can also destroy native deserts. Many people are drawn to the warm climate of the southwestern United States. Desert cities such as Phoenix, Las Vegas, and Palm Springs demand huge amounts of water for household use, irrigation, and recreation. The water comes from faraway rivers, changing the desert ecosystem and reducing the river's flow.

Freshwater habitats in general are vulnerable to destruction. Damming, for example, completely alters river ecosystems (figure 40.6). Deep reservoirs replace waterfalls, rapids, and wetlands, where birds and many other species breed. Areas that were once seasonally flooded become dry. Water temperature, oxygen content, and nutrient levels all change, triggering shifts in food webs above and below the dam. Dams also disrupt the migration of fishes and other aquatic animals.

Another threat to freshwater biodiversity is alterations to a river's path. Along the banks of the Mississippi River, for example, levees built to prevent flooding alter the pattern of sediment deposition. The levees confine the river to its channel, so the water's flow rate increases. Fast-moving water, in turn, erodes nutrient-rich sediments. These sediments, which once spread over floodplains during periodic floods, now remain in the river. Once the

Figure 40.5 **Plowed.** Farmland has replaced the native Iowa prairie.
USDA Natural Resources Conservation Service/Lynn Betts

nutrients arrive at the Gulf of Mexico they stimulate the growth of algae, causing the additional problems described in section 40.3.

Coastlines are also suffering from habitat destruction. Many fishes and invertebrates spend part of their lives in estuaries, and diverse algae and flowering plants support the food web. Yet humans have drained and filled estuaries for urbanization, housing, tourism, dredging, mining, and agriculture. These activities affect life in the oceans, too. The loss of coastal habitats can threaten populations of marine animals such as bluefin tuna, grouper, and cod. These problems are only expected to get worse. Most of the world's largest cities are located along coastlines, and human populations in those cities will probably continue to rise in the future. ⓘ *estuaries,* section 39.5A

40.2   MASTERING CONCEPTS

1. Which human activities account for most of the loss of terrestrial habitat?
2. How do dams and levees alter river ecosystems?
3. Why is damage to estuaries especially devastating?

Figure 40.4 **Coming Through.** A road into the rain forest brings new human settlement and agriculture.
©Wayne Lawler; Ecoscene/Corbis

Figure 40.6 **Big Dam.** Dams such as this one in Turkey eliminate streamside habitat and disrupt the migration of fishes and other animals.
©Ed Kashi/Corbis

40.3 Pollution Degrades Habitats

Pollution is any chemical, physical, or biological change in the environment that harms living organisms. Pollution degrades the quality of air, water, and land, threatening biodiversity worldwide.

A. Water Pollution Threatens Aquatic Life

A diverse array of pollutants affects rivers, lakes, and groundwater (table 40.2). For example, mining operations often release inorganic pollutants such as heavy metals or cyanide into the water, whereas shipping accidents and leaking oil wells add petroleum.

Raw sewage can also be a major pollutant. In addition to carrying disease-causing organisms, sewage also contains organic matter and nutrients such as nitrogen and phosphorus. When released into waterways, organic matter fuels the growth of bacteria, whose respiration depletes the water of oxygen. Fish and other organisms die. Meanwhile, in a process called eutrophication, nutrients from the sewage fertilize phytoplankton in the water (see figure 38.23). The resulting algae blooms are unsightly. Moreover, when the algae die, the microbes that decompose their dead bodies further deplete dissolved oxygen in the water.

Fertilizer and animal wastes that enter waterways also cause eutrophication. For example, nutrients from agricultural and urban lands drained by the Mississippi River find their way into the Gulf of Mexico (figure 40.7). There, the nutrients fuel algae blooms. On a local scale, toxins from the algae may kill fishes, manatees, and other sea life. A larger-scale problem, however, is the oxygen-depleted zone that forms each summer near the seafloor off the coast of Louisiana. The seasonal lack of oxygen in this "dead zone" kills many animals, disrupting not only the Gulf's food web but also its economy: Commercially important fish and shrimp cannot live in oxygen-depleted water. ⓘ *harmful algae bloom,* section 18.2B

Some pollutants seem deceptively harmless. Sediments, for example, reduce photosynthesis by blocking light penetration

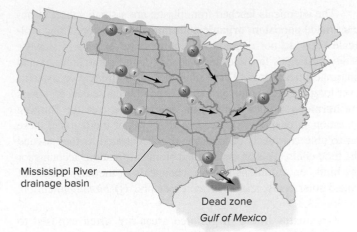

Figure 40.7 **Dead Zone.** Nutrient-rich runoff enters the major river systems draining into the Mississippi River. The combined waters then pour into the Gulf of Mexico. The result is a zone of seasonal oxygen depletion off the coast of Louisiana.

into water. Even heat can be a pollutant. Hot water discharged from power plants reduces the ability of a river to carry dissolved oxygen, harming fishes and other aquatic organisms.

Toxic chemicals and trash also pollute the open ocean. For example, ocean currents have concentrated millions of tons of plastic into two huge, interconnected "garbage patches" in the northern Pacific (figure 40.8). Floating just below the water surface are tiny pellets, called "nurdles," that form when plastic debris disintegrates. Fishes, sea turtles, and sea birds mistake the plastic for their natural food. The pellets can lodge in intestines and kill the animals outright, or chemicals from the plastic may accumulate in their tissues.

TABLE **40.2**	Examples of Chemical Water Pollutants
Organic	**Inorganic**
• Sewage	• Chloride ions
• Detergents	• Heavy metals (mercury, lead, chromium, zinc, nickel, copper, cadmium)
• Pesticides	• Nitrogen from fertilizer
• Pharmaceutical drugs	• Phosphorus from fertilizer and sewage
• Cosmetics	• Cyanide
• Antibacterial soaps	• Selenium
• Petroleum	
• Persistent organic pollutants (e.g., PCBs, PAHs)	

Figure 40.8 **Death by Plastic.** This young albatross died after eating plastic garbage in the ocean. The inset map shows where plastic is accumulating in the northern Pacific Ocean.

Photo: NOAA National Marine Sanctuaries/Claire Fackler

The chemicals leached from plastic are a small subset of human-made **persistent organic pollutants,** carbon-containing molecules that do not degrade (or that degrade extremely slowly). Many pesticides fall into this category, as do some solvents and pharmaceutical drugs. These substances contaminate ecosystems over long periods. Some of these compounds cause cancer; others are hormone mimics that disrupt reproduction (see the Burning Question in chapter 28). Because they do not biodegrade, these fat-soluble chemicals become more concentrated as they ascend the food chain. This process, called biomagnification, accounts for the high concentrations of toxic chemicals in the fatty tissues of tunas, polar bears, and other top predators. ⓘ *biomagnification,* section 38.3D

Organisms living in polluted areas are often exposed to several pollutants at once (figure 40.9). A waterway may contain not only garbage but also raw sewage, crude oil, gasoline, lead and other heavy metals, pesticides, and countless other toxic substances spilled from nearby factories. The toxic soup threatens not only human health but also the entire aquatic food chain.

B. Air Pollution Causes Many Types of Damage

Smog is a type of air pollution that forms a visible haze in the lower atmosphere (figure 40.10). Industrial smog occurs in urban and industrial regions where power plants, factories, and households burn coal and oil. The resulting smoke and sulfur dioxide (SO_2) may form a dark haze. Photochemical smog forms when nitrogen oxides and emissions from vehicle tailpipes undergo chemical reactions in the presence of light, producing ozone (O_3) and other harmful chemicals that injure plants and cause severe respiratory problems in humans. Warm, sunny areas with heavy automobile traffic have the most photochemical smog, but winds may carry the pollutants to sparsely populated areas. ⓘ *nitrogen cycle,* section 38.4C

Air also carries suspended **particulates,** tiny bits of matter that float in the air. Examples include road dust, volcanic ash, soot from partially burned fossil fuels, mold spores, pollen, and acidic particles. The damage they cause extends beyond the occasional need to dust off bookshelves and window sills. Most harmful are particles that are 2.5 μm in diameter or smaller. Not only do they become trapped deep within the lungs, but the heavy metals and toxic organic compounds in these particles make them especially likely to trigger inflammation, shortness of breath, asthma, or even cancer.

Other forms of air pollution are less visible but perhaps more harmful than smog. One example is **acid deposition:** acidic rain, snow, fog, dew, or dry particles. Because the atmosphere contains CO_2 and water, all rainfall includes some carbonic acid (H_2CO_3) and is therefore slightly acidic, with a pH around 5.6. The combustion of fossil fuels, however, releases sulfur and nitrogen oxides (SO_2 and NO_2) into the atmosphere. These compounds react with water, forming sulfuric acid and nitric acid. The acids return to the Earth as acid deposition. ⓘ *pH scale,* section 2.4

Coal-burning power plants release the most sulfur and nitrogen oxides, although emissions of these pollutants have declined since the 1980s. In the United States, winds carry airborne acids hundreds of miles east and northeast of the power plants in the Midwest. As a result, rainfall in parts of the eastern United States has a pH of about 4.7. Acid deposition also affects western North America, Europe, the former Soviet Union, China, and India.

Figure 40.9 Water Pollution. A child wades in garbage-strewn water in Borneo. Heavily polluted waters are likely to be fouled with many contaminants.
©Digital Vision/PunchStock RF

Figure 40.10 Smog. Air pollution plagues many cities. This is Santiago, Chile.
©Reuters/Corbis

Most lakes have a pH between 6 and 8; acid deposition can lower it to 5 or less. The acid leaches toxic metals such as aluminum or mercury from soils and sediments, causing fish eggs to die or yield deformed offspring. Lake-clogging algae replace aquatic flowering plants. Organisms that feed on the doomed species must seek alternative food sources or starve, which disrupts or topples food webs. Eventually, lake life dwindles to a few species that can tolerate increasingly acidic conditions.

Acid deposition alters forests, too. As soil pH drops, aluminum ions released from soil enter roots and stunt tree growth. Affected trees become less able to resist infection or to survive harsh weather. As a result, acid deposition is thinning high-elevation forests throughout Europe and on the U.S. coast from New England to South Carolina (**figure 40.11**).

Chemicals that destroy ozone can also be extremely harmful air pollutants. Ozone is an atmospheric molecule with two faces. As we have already seen, ozone in photochemical smog at Earth's surface is harmful. In the upper atmosphere, however, the stratospheric **ozone layer** blocks damaging ultraviolet (UV) radiation from the sun. Located about 10 to 50 kilometers above Earth's surface, the ozone layer forms when UV radiation from the sun reacts with oxygen gas (O_2). UV radiation damages biological molecules such as DNA, causing genetic mutations. The stratospheric ozone layer thus protects organisms from much of the harmful radiation that would otherwise strike Earth.

In the past several decades, the ozone layer has thinned over parts of Asia, Europe, North America, Australia, and New Zealand, and a "hole" has formed over Antarctica (**figure 40.12**). As the ozone layer thins, UV radiation increases at Earth's surface. In humans, exposure to short-wavelength UV radiation can cause skin cancer or cataracts. Ozone depletion may also indirectly contribute to species extinctions. For example, UV radiation can harm the phytoplankton that support aquatic food webs. UV radiation also damages the embryos and larvae of amphibians, perhaps contributing to the continued decline of their populations.

What is damaging the ozone layer? The main culprits are chlorine, fluorine, and bromine gases, some of which enter the atmosphere as persistent, human-made chlorofluorocarbon (CFC) compounds. These compounds were once used in refrigerants such as Freon, as propellants in aerosol cans, and to produce

Figure 40.11 Acid Deposition. Rainfall with a pH of about 4 severely damaged this forest in the Czech Republic.
©Oliver Strewe/Stone/Getty Images

foamed plastics. They can persist for decades in the upper atmosphere, catalyzing chemical reactions that break down ozone. An international treaty signed in 1987, the Montreal Protocol, banned the use of CFCs. Experts estimate that at mid-latitudes, the ozone layer should recover by 2050; healing the hole over Antarctica might take some 25 years longer.

40.3 MASTERING CONCEPTS

1. How do toxic chemicals, nutrients, sediments, and heat affect aquatic ecosystems?
2. What are major sources of industrial smog, photochemical smog, particulates, and acid deposition?
3. What effects do smog, particulates, acid deposition, and the thinning ozone layer have on life?

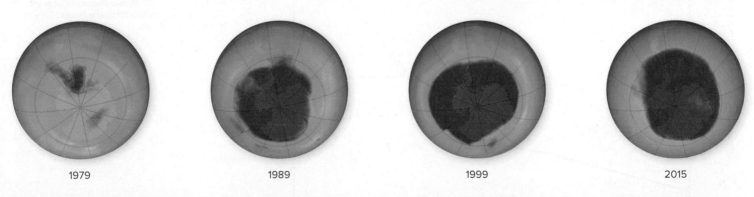

1979	1989	1999	2015

Figure 40.12 Antarctic Ozone Hole. Satellite images reveal the ozone hole over Antarctica from 1979 to 2015. Each image was taken on September 15, when the ozone hole nears its maximum size. In these images, purple and blue represent thinned areas of the ozone layer; yellow and orange areas have more ozone.
(all): NASA Ozone Watch

40.4 Global Climate Change Alters and Shifts Habitats

We now turn to air pollutants with the potential to do the most harm of all: greenhouse gases. In the past, scientists debated whether human activities could actually change something as complex as Earth's overall climate. Now, the scientific consensus is clear: We can and do.

A. Greenhouse Gases Warm Earth's Surface

CO_2 is a colorless, odorless gas present in the atmosphere at a concentration of about 400 parts per million. Although it makes up a tiny fraction of the atmosphere, CO_2 is one of several gases that contribute to the **greenhouse effect,** an increase in surface temperature caused by heat-trapping gases in Earth's atmosphere. As illustrated in **figure 40.13**, sunlight passes through the atmosphere and reaches Earth's surface. Some of the energy is reflected, but some is absorbed and reradiated as heat. Greenhouse gases block the escape of this heat from the atmosphere, just as transparent panes of glass trap heat inside a greenhouse.

Carbon dioxide is one greenhouse gas; others include methane, nitrous oxide (N_2O), and CFCs (see the Burning Question in this section). Livestock and landfills release methane, and nitrogen fertilizers are a major source of N_2O. These other gases actually trap heat much more efficiently than does CO_2, but because they are less abundant, they contribute only half as much to the greenhouse effect.

In a sense, the greenhouse effect supports life, because Earth's average temperature would be much lower without its blanket of greenhouse gases. And greenhouse gases have many

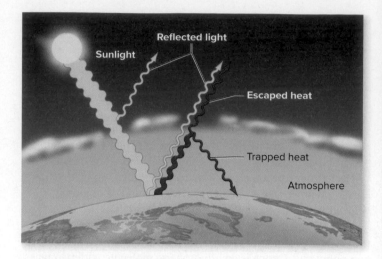

Figure 40.13 **The Greenhouse Effect.** Solar radiation heats Earth's surface. Some of this heat energy is reradiated to the atmosphere, but some is trapped near the surface by CO_2 and other greenhouse gases.

sources—including volcanoes—that have nothing to do with our actions.

But CO_2 has been steadily accumulating in the atmosphere for the past century, largely due to human activities (**figure 40.14**). The primary culprit is the use of fossil fuels such as coal, oil, and gas. Tropical deforestation and other combustion activities also add a share. All together, human activities release some 38 billion metric tons of CO_2 into the atmosphere each year. Photosynthesis temporarily removes some of this carbon from the atmosphere and incorporates it into plant tissues. Overall, however, more CO_2 is added than is removed. ⓘ *carbon cycle,* section 38.4B

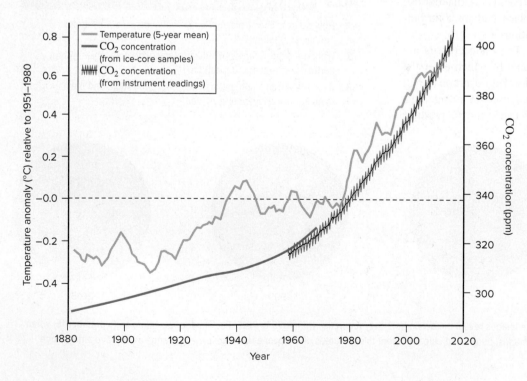

Figure 40.14 **CO_2 and Global Average Temperature.** CO_2 continues to accumulate in the atmosphere; the low point of each year's CO_2 concentration reflects peak photosynthesis in the northern hemisphere. As CO_2 accumulates, the average global temperature is also increasing.

Sources: NOAA Earth System Research Laboratory; NASA Goddard Institute for Space Studies Surface Temperature Analysis.

This accumulation of CO_2—along with climbing levels of other greenhouse gases—was accompanied by an increase in average global temperatures in the 20th century (see figure 40.14). Climate models predict that these trends will continue unless we take immediate action. The United Nations Climate Change Conference in 2015 laid the groundwork for change, with 196 nations agreeing to hold the increase in global temperature below 2°C and to balance greenhouse gas emissions with removals before 2100.

Earth's average temperature is rising overall, but some areas will become warmer and others will become cooler. The commonly used phrase "global warming" is therefore somewhat misleading. **Global climate change** is a more accurate term for past and future changes in Earth's weather patterns.

B. Global Climate Change Has Severe Consequences

Earth's gradual warming that has already occurred has been associated with many measurable effects, including the shrinking of alpine glaciers and polar ice sheets (figure 40.15). This process is self-reinforcing. Melting sea ice exposes seawater, which absorbs more heat than ice. The water becomes slightly warmer, which further accelerates the melt, exposing more seawater, and so on.

When floating sea ice melts, sea level does not rise (just as ice melting in a full glass of lemonade does not cause the liquid to overflow). However, the loss of ice over land—such as the ice covering Greenland and Antarctica—has contributed to a rise in sea level. This change could eventually alter the ocean ecosystem and flood coastal cities.

Weather conditions in tropical and temperate regions are also changing, thanks in part to a rise in sea surface temperature. In the tropics, warmer water means higher winds and more evaporation, which increase the amount of rainfall in a severe storm. In dry areas, on the other hand, global climate change may mean more intense and longer droughts, fewer cold snaps, and more heat waves.

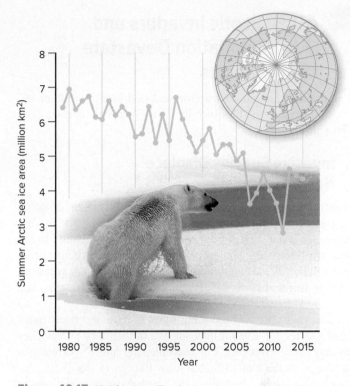

Figure 40.15 Melting Ice. The Arctic polar ice cap is shrinking, owing to higher temperatures at the North Pole.

(graph): Data from NASA; (bear): USGS/Canadian Coast Guard, photo by Patrick Kelley

These changes kill some organisms outright, whereas others become stressed and vulnerable to disease. Still others may migrate to higher latitudes or higher elevations (see section 40.7). At the southern ends of their ranges, where temperatures are rising, some species have become locally extinct. In addition, scientists are tracking events known to occur at the same time each year. In the United Kingdom, butterflies are emerging and amphibians are mating a few days earlier than usual; in North America, many plants are flowering and birds are migrating earlier.

Continued climate change will affect not only wild organisms but also agriculture and public health. Growing seasons in temperate areas are lengthening, and the southern United States may become too dry to sustain many traditional crops. Drought-related water shortages may affect more than a billion people. Tropical diseases such as malaria may move into temperate areas.

Ocean life is also vulnerable to CO_2 accumulation. One problem is that CO_2 buildup causes ocean water to become more acidic. The lowered pH causes the calcium carbonate shells of oysters, clams, and other mollusks to dissolve, along with the exoskeletons of coral animals. The loss of these organisms disrupts the entire ocean food web.

Burning Question

What does the ozone hole have to do with global climate change?

Many people confuse global climate change and the ozone hole. These two problems are largely separate, but they do share two common threads. First, the chlorofluorocarbon gases that deplete the ozone layer are also greenhouse gases, contributing to a warmer atmosphere. Second, the greenhouse effect may cause the hole in the ozone layer to grow. A thick, heat-trapping "blanket" of greenhouse gases in the troposphere (the lowest part of the atmosphere) means less heat reaches the stratosphere, where the ozone layer is. A cooler stratosphere, in turn, extends the time that stratospheric clouds blanket the polar regions in winter. These clouds of ice and nitric acid speed the chemical reactions that deplete stratospheric ozone.

Submit your burning question to
Marielle.Hoefnagels@mheducation.com

40.4 MASTERING CONCEPTS

1. Why is CO_2 accumulating in Earth's atmosphere?
2. Describe how and why Earth's climate changed during the past century.
3. How does global climate change threaten biodiversity?

40.5 Exotic Invaders and Overexploitation Devastate Many Species

In addition to habitat destruction and pollution, two other important threats to biodiversity are invasive species and overexploitation.

A. Invasive Species Displace Native Organisms

An introduced species (also called a nonnative, alien, or exotic species) is one that humans bring to an area where it did not previously occur. When people move from one location to another, we often bring along our pets, crops, livestock, and ornamental plants. We also unintentionally carry microbes, parasites, and stowaways such as rodents and insects on ships, cars, and planes.

This transport may seem harmless at first, and many introduced species die. Even if they survive in their new homes, they may not cause problems. For example, the house sparrow was introduced to the United States from Europe in the 1850s. Although it has spread throughout the North American continent, it has not caused obvious ecological problems. In addition, at least 5000 nonnative plant species live in U.S. ecosystems, introduced from agriculture and urbanization. Most have done no apparent harm.

To be considered **invasive,** an introduced species must begin breeding in its new location and spread widely from the original point of introduction. A species is likely to meet this criterion if its new habitat is similar to its old home, only without natural predators. In addition, according to some definitions, the species must harm the environment, human health, or the economy. Of every 100 species introduced, only 1 persists to take over a niche. Nonetheless, the Global Invasive Species Database lists 498 types of invasive plants, animals, and microorganisms in the United States alone; many more invasive species occur worldwide.

Some invasive species are staggeringly destructive (figure 40.16). The opening essay in chapter 37 describes the problems caused by nutria and brown tree snakes, and section 38.1 refers to the invasion of zebra mussels in the Great Lakes. Other examples of invasive species in North America include these:

- Hydrilla is an aquatic plant that forms dense mats on water surfaces. These mats block light, alter food webs, and reduce recreational use of lakes and rivers.
- Kudzu is a fast-growing plant that smothers native plants in the southeastern United States (see figure 40.16a).
- Fungi have all but eradicated American chestnut and American elm trees.
- Caterpillars of the gypsy moth devour the foliage of hundreds of species of hardwood trees in North America.

a.

b.

Figure 40.16 Invasive Species. (a) An invasive population of kudzu in Virginia blankets every surface in sight. (b) European starlings form enormous, destructive flocks in North America.
(a) ©Steve St. John/Getty Images RF; (b): ©Alex Saberi/National Geographic/Getty Images

- The larvae of an Asian beetle called the emerald ash borer destroy the bark of ash trees. Since they arrived in 2002, these beetles have killed millions of ash trees in the midwestern and eastern United States and Canada.
- European starlings are birds that were released in New York City's Central Park in 1890. Huge flocks of starlings now reside all across North America, fouling cities with their droppings and devouring crops (see figure 40.16b).
- The marine toad is a voracious omnivore that competes with and preys on native amphibians in Florida.
- Asian carp are large, fast-breeding fish that were originally introduced to catfish ponds in the South but have since spread along the Mississippi River toward the Great Lakes. They eat algae voraciously, outcompeting other plankton-eaters for food.

When an invasion does occur, the harm may be ecological and economic. A nonnative species not only changes the composition of a community but also may carry diseases that spread to native species. The economic costs include everything from the purchase of herbicides that kill invasive weeds, to the loss of grain eaten by hungry birds and rodents, to declining tax revenue when invasive aquatic plants interfere with boating and recreation.

B. Overexploitation Can Drive Species to Extinction

Another cause of species extinction is **overexploitation:** harvesting the members of a species faster than they can reproduce (figure 40.17). The market for exotic pets, for example, is harming populations of many species of mammals, birds, snakes, lizards, amphibians, and fishes (see chapter 21's Apply It Now box).

Many of the most famous examples of species extinctions result from overhunting of terrestrial animals. The dodo, for example, was a flightless bird that once lived on the Indian Ocean island of Mauritius. In the late 17th century, humans hunted the dodo for food while introducing other species to the island. The dodo soon went extinct. In the United States, commercial-scale hunting nearly drove the American bison to extinction in the 1800s. The passenger pigeon and Carolina parakeet did go extinct in the 1900s, victims of overhunting and habitat destruction.

The best illustration of widespread overexploitation is the recent collapse of ocean fisheries (see section 12.8). Since the 1950s, some 90% of the world's large, predatory ocean fishes

Figure 40.18 Bycatch. Nontarget animals such as this sea turtle sometimes get caught in nets meant for other species. Fishermen often discard these "bycatch" animals, dead or alive.
©Jeffrey Rotman/Science Source

have disappeared, including tuna, flounder, halibut, swordfish, and cod. Superefficient fishing boats harvest the adults faster than the fishes can reproduce. Moreover, fishing pressure has shifted to other species as predatory fishes have vanished. Fishing equipment scrapes and scours the seafloor, destroying habitat for many other species. In addition, many marine mammals, seabirds, sea turtles, and nontarget fish species are killed accidentally when they are caught up as "bycatch" in the nets set for the target species (figure 40.18). Even farmed seafood may contribute to the problem: Ocean fishes are fed to farmed shrimp and salmon, depleting marine food webs.

Improved management practices can help an overharvested population recover its numbers. Consider, for example, the Chesapeake Bay blue crab. Thanks to overharvesting, pollution, and habitat loss, the federal government declared the crab fishery a disaster in 2008. The state of Virginia had already stopped issuing new commercial crabbing licenses in the 1990s, but in 2009, the state began buying back existing licenses as well. The overall goal of this approach, combined with other regulations, was to reduce pressure on the dwindling blue crab population. According to a 2012 report, the harvest rate is now sustainable, and the population has begun to rebound.

Figure 40.17 Unsustainable Harvest. Customs officials in Lhasa, the capital of Tibet, inspect confiscated tiger, leopard, and otter skins. Selling these skins is illegal, yet poaching remains one of the main threats to mammal populations in Asia.
©China Photos/Getty Images

40.5 MASTERING CONCEPTS

1. What features characterize an invasive species?
2. How do invasive species disrupt ecosystems?
3. List examples of species declines caused by overexploitation.

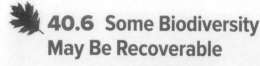

40.6 Some Biodiversity May Be Recoverable

As the human population continues to grow, pressure on natural resources will only increase. One key to reversing environmental decline will therefore be to slow the growth of the human population. In addition, although some species are gone forever, humans may have the power to undo some of our past mistakes. For example, thanks in part to the Endangered Species Act of 1973, some species that faced extinction, such as the bald eagle, have recovered (figure 40.19).

A. Protecting and Restoring Habitat Saves Many Species at Once

One important conservation tool is to set aside parks, wildlife refuges, and other natural areas and to protect them from destruction, invasive species, and hunting. Preserving critical habitat is a good conservation tool because it saves not just one endangered species but also the many other species that share its habitat. For example, the red-cockaded woodpecker is native to the southeastern United States, where it builds its nest in a cavity that it excavates high on the trunk of a mature pine tree. Although most of its preferred habitat has been logged, private property owners and the government have cooperated to save some of the remaining habitat. Other animals that use the nesting cavities, including birds, mammals, snakes, amphibians, and insects, also benefit from the woodpecker recovery plan.

Reversing habitat destruction is a second conservation tool. Major restoration projects, such as the Everglades plan described in this chapter's opening essay, show that recovery, although costly and difficult, may yet reverse some species declines. On a smaller scale, we can also help species bypass degraded habitats by supplying wildlife corridors through housing developments or

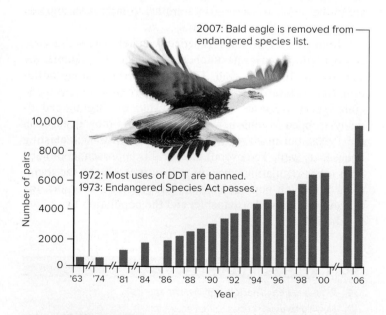

Figure 40.19 Good News for Bald Eagles. Bald eagles were near extinction in the 1960s, thanks to habitat loss, hunting, and exposure to the pesticide DDT. In the 1970s, new laws stimulated a steady recovery for the species.

Figure 40.20 Taking the High Road. Fish use "ladders" such as this one to migrate over dams. ©David R. Frazier Photolibrary, Inc.

building "fish ladders" over dams (figure 40.20). Bioremediation, a strategy that uses plants and microbes to clean up polluted soils, may also play a role in some cases.

B. Some Conservation Tools Target Individual Species

One straightforward way to save a species is to protect its individuals. For example, it is illegal to collect endangered carnivorous plants such as Venus flytraps and pitcher plants in the wild. In the ocean, northern and southern right whales were nearly hunted to extinction for their blubber in the 1800s. They remain endangered, but it is now illegal to kill them. Likewise, the catastrophic decline of Atlantic cod in the past few decades prompted the closure of some fisheries off Newfoundland's coast, along with strict quotas for the overall catch. Whether conservation efforts are in time to save the cod fishery remains to be seen.

Predator control programs can also help. For example, rats, weasels, dogs, and other predators transported by European settlers endanger the great spotted kiwi, a flightless bird native to New Zealand. Removing these predators from nature preserves has made life much easier for the endangered kiwis. Conversely, restoring prey populations can help endangered predators. The number of Iberian lynx in Spain and Portugal tripled after conservation biologists reestablished nearby populations of rabbits, the lynx's primary food source.

Figure 40.21 Black Rhino. Ecotourism may provide some hope for the survival of the endangered black rhinoceros.
©Michele Burgess/Alamy

Figure 40.22 Cataloging Insect Diversity. This researcher is using a lamp and a white sheet to attract and study nocturnal insects in a tropical rain forest in French Guiana.
©Patrick Landmann/Science Source

In areas where wildlife poaching is a problem, changing the local economic incentives can be a powerful conservation tool. Poaching is profitable if a dead animal is worth more money than a live one in its natural habitat. But ecotourism can turn this economic calculation on its head. By attracting visitors who pay to see endangered animals, conservation may bring in more money than poaching. A black rhino conservation program in Namibia, for example, hires former poachers as armed guards who protect the animals from hunters. Meanwhile, guides lead tourists hoping to spot a rhino in the wild (figure 40.21).

Protective laws and predator control programs may help halt the decline of a threatened species, but sometimes it is possible to boost reproduction as well. A captive breeding program can be very useful. Biologists can capture adults from the wild, allow them to reproduce in captivity, then nurture and protect the young until they are old enough to return to their native habitat. The California condor, red wolf, and black-footed ferret are notable examples. But this solution does not work for species whose habitat is gone (submerged after dam construction, for example) or still under the same pressures that threatened the species in the first place.

The biotechnology revolution plays a role in yet another approach to species conservation. In one project, researchers are using DNA to identify bison whose genes are uncontaminated with those of domesticated cattle. The "purest" bison are set aside as the best candidates for reestablishing wild bison herds. In the future, it may even be possible to recover extinct species using DNA extracted from preserved specimens. Scientists who are sequencing DNA from a frozen baby mammoth, for example, may one day be able to produce a live mammoth by using a cloning technique similar to the one used to make Dolly the sheep. One possible approach would be to replace the DNA in a fertilized elephant egg with mammoth DNA and then implant the resulting embryo into an elephant's uterus. After gestation, a woolly mammoth would be born—some 10,000 years after its species went extinct. ⓘ *cloning,* section 11.3B

C. Conserving Biodiversity Involves Scientists and Ordinary Citizens

Regardless of which tools are used to preserve biodiversity, all conservation efforts require a scientific approach. To get a true measure of Earth's biodiversity, taxonomists must catalog all organisms, not just vertebrates and plants (figure 40.22). Evolutionary biologists must continue to analyze the relationships among all species. Preserving biodiversity also requires an understanding of which species need help, whether current conservation efforts are working, and the consequences to ecosystems as species disappear.

But not every important question has a scientific answer. Are the only species worth saving the photogenic ones, such as giant pandas? Or do we also commit to saving the worms, algae, bacteria, and fungi so essential to global ecology? How much money should we spend on conservation? Should developed countries help poor nations with their efforts? How do we balance the need for conservation with the need for economic growth? Which of the tangled threads that tie all life together should we sacrifice to other interests?

It would be very difficult to halt life on Earth completely, short of a global catastrophe such as a meteor collision or a nuclear holocaust. Just the presence of life, however, does not guarantee that the surviving species will have the diversity that humans value. It is safest to try to protect the remaining resources for the future, while maintaining a reasonable standard of living for all people. Scientists and politicians, as well as ordinary citizens, share this heavy burden (see the Apply It Now box). Part of the solution lies within you and how you choose to live (see this chapter's second Burning Question). Do whatever you can to preserve the diversity of life, for in diversity lies resiliency and the future of life on Earth.

40.6 MASTERING CONCEPTS

1. What is the relationship between human population growth and conservation biology?
2. List and describe the tools that conservation biologists use to preserve biodiversity.
3. How can scientists, governments, and ordinary citizens work together for conservation?

Apply It **Now**

Environmental Legislation

Legal training, political savvy, and scientific information about human impacts on other species can combine to stimulate legislation that helps protect the environment. In 1970, President Richard Nixon created the Environmental Protection Agency by executive order. Since that time, Congress has passed several laws to combat some of the worst environmental problems in the United States. Following are a few major pieces of environmental legislation:

©Brand X Pictures/Punch-Stock RF

- The *Endangered Species Act* of 1973 requires that the U.S. Secretary of the Interior identify threatened and endangered species. The overall goals are to prevent extinction and to help endangered species recover their numbers. Since the act was implemented, nearly 700 species of vertebrate and invertebrate animals and nearly 900 species of plants and lichens have been classified as threatened or endangered. Only a few dozen species have been removed from the list because they have either recovered or become extinct, or because new information revealed that their populations are larger than had been thought.

- The *Clean Air Act,* passed in 1970 and amended several times since, sets minimum air quality standards for many types of air pollutants. Since 1970, emissions of nitrogen and sulfur oxides, lead, carbon monoxide, particulates, and other pollutants have declined, leading to significant improvements in regional air quality.

- Among other provisions, the *Clean Water Act* of 1972 required nearly every city to build and maintain a sewage treatment plant, drastically reducing the discharge of raw sewage into rivers and lakes. The 1987 *Water Quality Act* regulates water pollution from industry, agricultural runoff, overflow from sewage treatment plants during storms, and runoff from city streets. Many of the nation's surface waters have recovered from past unregulated discharge of phosphorus, other nutrients, and toxic chemicals.

INVESTIGATING LIFE

40.7 Up, Up, and Away

Both historians and scientists benefit from looking into the past. Imagine a French historian's delight when she discovers a long-lost photograph of Paris from the early 1900s. She may compare the photograph with modern pictures taken from the same vantage point, noting changes in the buildings, signs, clothing, cars, and other objects.

Similarly, scientists use snapshots from the past to understand the causes and effects of climate change. Photographs may play a role, but other types of information can also reveal useful clues about ongoing changes. Animal and plant specimens collected years ago for museum collections, for example, can offer glimpses into past biodiversity. If these relics still contain DNA, researchers can directly compare the genes of past populations with those of today. Changes in genetic diversity, in turn, reveal how climate change affects evolution.

Biologists at the University of California, Berkeley, seized one such opportunity to compare past and present. A century ago, in 1915 and 1916, naturalists from the university's Museum of Vertebrate Zoology traveled throughout California, documenting exactly where they trapped and collected small mammals and other organisms. Sites throughout Yosemite National Park were among the sampling locations, and the animals they caught are still preserved in the museum's collections.

In the 100 years since those surveys occurred, the average temperature at Yosemite has climbed by about 3°C. A research team led by Emily Rubidge and Craig Moritz used the museum's preserved mammals to investigate how elevated temperatures might affect two closely related species—alpine chipmunks and lodgepole chipmunks—inhabiting the park (figure 40.23).

They began by conducting a new survey. Over a 5-year period they used live traps to capture chipmunks at all of the original collection sites. Lodgepole chipmunks still inhabited all of the original locations, but alpine chipmunks were missing at some sites they once occupied. When repeated attempts to find this species at an original site failed, the researchers sampled nearby higher-elevation areas instead. The modern survey revealed that the lowest elevation where alpine chipmunks live is now about 500 meters higher than it was a century ago.

Figure 40.23 **Two Chipmunks.** These small mammals occupy Yosemite National Park and other parts of California's Sierra Nevada mountains.

(alpine): ©Missy Leone; (lodgepole): ©Tom McHugh/Science Source

The researchers reasoned that the alpine chipmunks responded to the gradually warming climate by moving to higher elevations, where temperatures are cooler. Populations that had once overlapped were increasingly fragmented into isolated subpopulations, each occupying its own mountaintop. The team predicted that this isolation would change the genetic diversity of alpine chipmunks compared to their still-widespread relatives, the lodgepole chipmunks (figure 40.24). Specifically, thanks to the "bottleneck effect," the isolated alpine chipmunk subpopulations should become increasingly different from one another. And with less crossbreeding among neighboring groups, overall genetic diversity should decrease. (i) *bottleneck effect,* section 12.7B

To test their prediction, Moritz and his colleagues compared DNA from museum specimens with modern chipmunk DNA. They removed a small skin sample from each preserved chipmunk and extracted each animal's DNA. In all, they collected genetic material from 147 century-old museum specimens (88 alpine chipmunks and 59 lodgepole chipmunks). For comparison, they also collected DNA from 146 alpine chipmunks and 115 lodgepole chipmunks they caught when they revisited the original sampling locations.

The next step was to amplify the genetic material using the polymerase chain reaction and to sequence seven carefully selected regions of each chipmunk's genome. A software program used these sequences, coupled with the location at which each chipmunk had been caught, to calculate the genetic diversity in each chipmunk population. (i) *DNA sequencing,* section 11.2B; *polymerase chain reaction,* section 11.2C

Comparing the genes of modern and historical populations strongly supported the authors' prediction that groups of alpine chipmunks had become genetically isolated (figure 40.25). That is, modern populations of alpine chipmunks were much more different from each other than were historical populations of the same species. In contrast, modern and historical DNA from lodgepole chipmunks revealed few differences among populations.

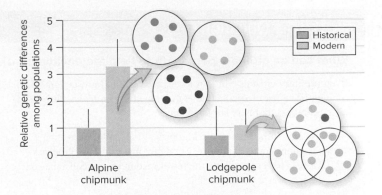

Figure 40.25 Alpine Isolation. Populations of alpine chipmunks have diverged genetically relative to lodgepole chipmunks. Numbers on the *y* axis represent differences among modern chipmunk populations, relative to the differences in the historical samples. Inset circles illustrate the degree of overlap among subpopulations. Bars represent 95% confidence intervals; see appendix B.

The researchers also determined that increasing isolation among modern alpine chipmunk populations has led to a significant loss of genetic diversity in the species. Of the seven regions of nucleotides analyzed in this study, the researchers detected an average of nine alleles per region in historical DNA samples. In contrast, the researchers detected an average of only seven alleles per region in the modern DNA samples.

The outcome of this study does not bode well for alpine chipmunks. As Yosemite's climate continues to warm, alpine chipmunks are likely to retreat to still higher elevations, leading to further erosion of genetic diversity. A population with low genetic variation, in turn, has less potential to adapt to a changing environment. The consequence of climate change on alpine chipmunks is therefore two-pronged: Increasing temperatures cause lower-elevation populations to disappear, which decreases the genetic variation of higher-elevation populations and boosts their chance of extinction. The same trend is likely to endanger many other species, especially those in mountain habitats.

This study is among the first to connect global climate change directly with microevolution. Researchers have seldom been able to document the consequences of climate change with such precision. Just as a historian values a long-lost photo, this study highlights the importance of historical climate records, species ranges, and preserved specimens when studying how ecosystems react to a changing environment.

Source: Rubidge, Emily M., and five coauthors, including Craig Moritz. 2012. Climate-induced range contraction drives genetic erosion in an alpine mammal. *Nature Climate Change*, vol. 2, pages 285–288.

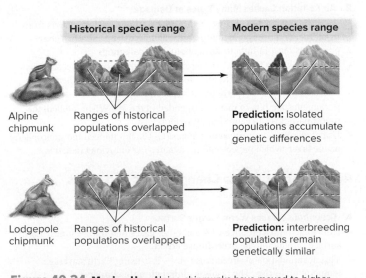

Figure 40.24 Moving Up. Alpine chipmunks have moved to higher elevations over the past century, as temperatures in Yosemite have climbed. Researchers predicted that habitat fragmentation would increase genetic isolation among alpine chipmunk populations.

40.7 MASTERING CONCEPTS

1. Use figure 40.24 to explain the genetic isolation of modern alpine chipmunk populations (see figure 40.25).

2. How might the isolation of modern alpine chipmunk subpopulations have led to reduced genetic diversity in the species?

Burning Question

What can an ordinary person do to help the environment?

Even ordinary citizens can join forces to clean and preserve the environment. The list of small actions that together can make a big difference is endless, but here are a few ideas:

- Use less stuff. Manufacturing, packaging, and transporting consumer goods uses energy and raw materials and produces waste. The less you buy, the fewer resources you consume and the less waste you discard.

- Use reusable canvas shopping bags to carry whatever you do buy.

- Choose foods, lumber, and other products that reflect sustainable practices. Buying organic food, for example, reduces the use of pesticides in agriculture. Shade-grown coffee plantations provide habitat for a wide variety of tropical plants and animals. Before purchasing seafood, consult a Seafood Watch guide for your area to learn which fish species are threatened. Lumber and paper certified by the nonprofit Forest Stewardship Council has been harvested and produced in an environmentally responsible way.

- Conserve energy. Replace conventional lightbulbs with compact fluorescent bulbs, recycle, carpool, drive a fuel-efficient car, ride a bicycle, or turn down the thermostat. Pouring filtered tap water into a reusable bottle rather than buying bottled water not only saves energy but also reduces landfill waste.

- Check with your electric company to see whether you can select renewable energy sources, such as wind or solar energy.

- Eat less meat. Farm animals and their manure emit copious greenhouse gases—especially methane—into the atmosphere.

©Photodisc/Getty Images RF

- Pay attention to what you discard and pour down the drain. Pharmaceutical drugs, petroleum products, and harsh chemicals can end up in waterways and harm ecosystems.

- Encourage your local governments to set aside land for parks.

- Write to state and federal lawmakers to ask them to support legislation that can help protect the environment. The house.gov and senate.gov websites have contact information for your House and Senate representatives.

- If you garden, avoid invasive species. Instead, choose native plant species that attract wildlife.

- Don't buy rare or exotic species as pets. Collecting these animals harms native wildlife populations.

- Donate time or money to groups that save critical habitat.

Submit your burning question to
Marielle.Hoefnagels@mheducation.com

CHAPTER SUMMARY

40.1 Earth's Biodiversity Is Dwindling

- **Biodiversity** means the variety of life on Earth, and it includes diversity at the genetic, species, and ecosystem levels.
- Increasing numbers of species are threatened with **extinction** or are **endangered** or **vulnerable. Conservation biologists** study and attempt to preserve biodiversity.
- Figure 40.26 summarizes the major causes of species extinctions.

40.2 Many Human Activities Destroy Habitats

- Agriculture, logging, and urbanization contribute to **deforestation** in tropical and temperate regions. Prairies and other grasslands also have been destroyed, especially for agriculture.
- Drought and overgrazing are expanding Earth's deserts.
- Dams and levees alter the species that live in and near rivers.
- Preserving estuaries is important because they are breeding grounds for many species.

40.3 Pollution Degrades Habitats

- **Pollution** is any environmental change that harms living organisms.

A. Water Pollution Threatens Aquatic Life

- Excessive nutrient levels cause eutrophication. Sediments and heat also pollute aquatic ecosystems.

- Water and sediments can be contaminated by a mixture of toxic substances such as **persistent organic pollutants,** heavy metals, spilled oil, and plastics.

B. Air Pollution Causes Many Types of Damage

- Air pollutants include heavy metals, **particulates,** and emissions from fossil fuel combustion in automobiles and industries. Some of these pollutants react in light to form photochemical **smog.**
- **Acid deposition** forms when nitrogen and sulfur oxides react with water in the upper atmosphere to form nitric and sulfuric acids. These acids return to Earth as dry particles or in precipitation.
- Particulates are bits of matter suspended in the air. When inhaled, the tiniest particles can cause lung problems.
- Use of chlorofluorocarbon compounds (CFCs) has thinned the stratospheric **ozone layer,** which protects life from damaging ultraviolet radiation.

40.4 Global Climate Change Alters and Shifts Habitats

A. Greenhouse Gases Warm Earth's Surface

- Human activities produce CO_2 and other gases that trap heat near Earth's surface, producing the **greenhouse effect.**
- This accumulation of greenhouse gases is causing Earth's average temperature to rise. The result is **global climate change.**

B. Global Climate Change Has Severe Consequences

- As the temperature increases, polar ice is melting, sea level is rising, coral reefs are declining, and species ranges are shifting.

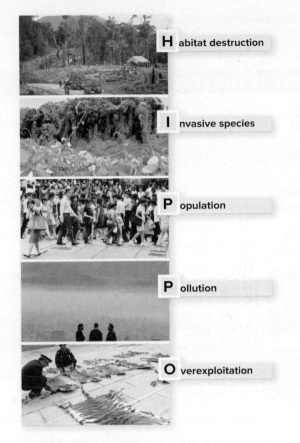

H abitat destruction

I nvasive species

P opulation

P ollution

O verexploitation

Figure 40.26 Threats to Biodiversity: A Summary.

Photos: (destruction): ©Wayne Lawler; Ecoscene/Corbis; (invasion): ©Steve St. John/Getty Images RF; (population): ©Rodrigo A Torres/Glow Images RF; (pollution): ©Reuters/Corbis; (overexploitation): ©China Photos/Getty Images

40.5 Exotic Invaders and Overexploitation Devastate Many Species

A. Invasive Species Displace Native Organisms
- **Invasive species** consume or outcompete native organisms.

B. Overexploitation Can Drive Species to Extinction
- **Overexploitation** means individuals are harvested faster than they can reproduce. High fishing pressure endangers global fisheries.

40.6 Some Biodiversity May Be Recoverable

A. Protecting and Restoring Habitat Saves Many Species at Once
- Protected reserves, habitat restoration, and wildlife corridors are conservation tools that protect multiple species simultaneously.

B. Some Conservation Tools Target Individual Species
- Harvest management, predator exclusion, economic incentives, captive breeding, and biotechnology can help save one species at a time.

C. Conserving Biodiversity Involves Scientists and Ordinary Citizens
- Conservation biologists can monitor biodiversity trends and recommend strategies for saving threatened species, but everyone can choose actions that preserve or deplete biodiversity.

40.7 Investigating Life: Up, Up, and Away

- As temperatures in Yosemite National Park have climbed, alpine chipmunks have moved to higher elevations. The resulting fragmentation of the gene pool is associated with increased genetic differences among the isolated subpopulations and lower genetic diversity overall.

MULTIPLE CHOICE QUESTIONS

1. Which of the following is NOT one of the main causes of today's biodiversity crisis?
   a. Natural disasters, such as earthquakes
   b. Habitat destruction and degradation
   c. Overexploitation
   d. Introduction of nonnative species

2. Which human activity has been the most harmful to biodiversity in tropical forests?
   a. Housing   c. Transportation
   b. Agriculture   d. Tourism

3. What is the connection between agriculture in the midwestern United States and the Gulf of Mexico's "dead zone"?
   a. Pesticides from farmlands are killing ocean life.
   b. Nutrient enrichment causes oxygen depletion in Gulf waters.
   c. River sediments block out the light needed for photosynthesis in the Gulf.
   d. Farmlands use up all of the nitrogen in the water, so the Gulf waters are starved for nutrients.

4. What has been the result of the 1987 Montreal Protocol, which banned the use of CFCs?
   a. It has slowed the expansion of the ozone hole.
   b. It is accelerating the destruction of the ozone layer.
   c. It is slowly helping to restore the ozone layer.
   d. It has caused the ozone hole to disappear.

5. Why is deforestation associated with global climate change?
   a. Because the loss of forest animals reduces the CO_2 released into the atmosphere
   b. Because the loss of trees reduces the amount of photosynthesis occurring on the planet
   c. Because the loss of forest makes more land available for agriculture
   d. Because the loss of tree root systems leads to erosion

6. Which strategy would be most likely to restore overexploited ocean species with the least potential for harm to the ecosystem?
   a. Eating only farmed seafood
   b. Dumping nutrients into the ocean to increase primary production
   c. Passing regulations that limit harvests
   d. Adding invasive species to increase competition in aquatic food webs

Answers to these questions are in appendix A.

WRITE IT OUT

1. List the main threats to biodiversity worldwide.

2. How does human population growth contribute to each of the main factors causing species extinctions?

3. When trees are removed from an area, patches or strips of untouched trees often intersperse the deforested land. How is the abiotic environment on the edge of these strips or patches different than before the area was disturbed? What changes in vegetation would you expect to see in the next few years? How might animals be affected by forest fragmentation?

4. Nanoparticles are tiny bits of metal that are used in sunscreens, as a wastewater treatment, and for many other purposes. Recent evidence suggests that nanoparticles are toxic to phytoplankton, the primary producers at the base of many aquatic food chains. Phytoplankton use the energy in sunlight to produce organic matter, and they consume CO_2 and release O_2. Predict some possible consequences to biodiversity if nanoparticles become a more common pollutant.

5. What are examples of pollutants in air and in water? Which of these pollutants eventually reach land?

6. Suppose you throw a small piece of plastic in the garbage. List three places where the plastic might be found months later.

7. Use the Internet to research ways to make homes more energy efficient. How does reducing your monthly energy bill relate to the conservation of biodiversity?

8. In what ways is the greenhouse effect both beneficial and detrimental?

9. Explain how habitat destruction, the increasing human population, and pollution contribute to climate change.

10. DNA evidence recently confirmed the existence of a "pizzly bear," the offspring of a polar bear and a grizzly bear. Scientists hypothesize that some polar bears are staying on mainland because of the warming climate, so polar bears are encountering grizzlies more often than in the past. Pizzly bears may be less fit than either polar bears or grizzlies, so some people are advocating that they be killed. Make an argument for or against this strategy. If human actions are contributing to the breeding behavior, do we have an ethical obligation to intervene? Do you think polar bears have a better chance at avoiding extinction if humans eliminate their hybrid offspring? Would evidence that polar bears and grizzlies had interbred in the past change your answers?

11. People often move plants from one part of the world to another. Sometimes, an introduced plant species can become invasive, taking over native plant populations. The U.S. Department of Agriculture manages the National Invasive Species Information Center, whose website maintains a list of invasive plants. Which plant species are considered invasive in your home state? Why are those species harmful? Should invasive plants be eradicated? How?

12. One way to combat invasive species is to kill the invaders. In Hawaii, officials shoot feral cats, goats, and pigs. In Australia, the government fought zebra mussels by adding chlorine and copper to a bay, killing everything living in the water. Do you think that these approaches are reasonable? Suggest alternative strategies.

13. In the southeastern United States, several species of freshwater mussels are extinct or threatened because of habitat destruction. In the past, they were also harvested for the button trade. How would a population ecologist (see chapter 37) approach the problem of species recovery for these animals?

14. Phytoremediation is the use of plants to treat environmental problems. Search the Internet for applications of phytoremediation. What are the benefits of phytoremediation? If you were trying to discover plants suitable for use in phytoremediation, what qualities would you look for? Can you foresee any problems with phytoremediation?

15. Refer back to section 12.7, which describes the bottleneck effect. With this information in mind, why might recovery be difficult for species, such as cheetahs, that are nearly extinct?

PULL IT TOGETHER

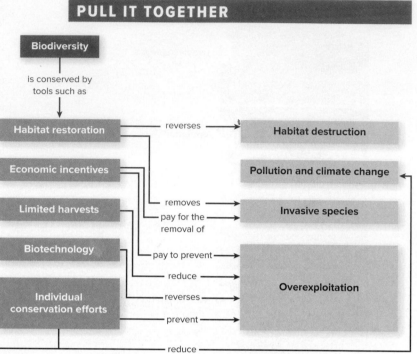

Figure 40.27 **Pull It Together: Preserving Biodiversity.**

Refer to figure 40.27 and the chapter content to answer the following questions.

1. Review the Survey the Landscape figure in the chapter introduction. Add the terms *population, community,* or *ecosystem* along the right edge of figure 40.27. For each threat to biodiversity, connect to one of the new boxes. Each new connecting phrase should explain how the threat affects biodiversity at the selected level of biological organization. For example, one connecting phrase might explain how invasive species affect communities.

2. Name three ways you can alter your lifestyle in a way that promotes conservation practices.

3. Give examples of government actions that threaten biodiversity and examples of government actions that preserve biodiversity.

Appendix A | *Answers to Multiple Choice Questions*

Answers to *Mastering Concepts, Write it Out,* and *Pull It Together* questions can be found following each question in SmartBook™.

Chapter 1
1. b 2. d 3. a 4. a 5. d 6. d 7. b 8. b
9. c 10. d

Chapter 2
1. b 2. c 3. c 4. c 5. a 6. a

Chapter 3
1. c 2. d 3. d 4. a 5. b 6. b 7. c

Chapter 4
1. b 2. d 3. a 4. a 5. d 6. b 7. c

Chapter 5
1. c 2. c 3. b 4. a 5. c 6. a 7. b 8. d

Chapter 6
1. a 2. d 3. b 4. a 5. c 6. a 7. a 8. c

Chapter 7
1. c 2. a 3. c 4. a 5. d 6. a 7. d 8. a
9. b 10. d

Chapter 8
1. c 2. b 3. c 4. a 5. a

Chapter 9
1. c 2. a 3. c 4. d 5. c 6. c

Chapter 10
1. c 2. b 3. d 4. c 5. b 6. b 7. b 8. a
9. b 10. c

Chapter 11
1. c 2. a 3. b 4. c 5. a 6. d 7. b

Chapter 12
1. b 2. b 3. c 4. a 5. c 6. a

Chapter 13
1. d 2. b 3. c 4. b 5. a 6. c 7. a

Chapter 14
1. c 2. d 3. b 4. d 5. c

Chapter 15
1. b 2. d 3. a 4. b 5. d 6. b 7. c 8. c 9. d 10. a

Chapter 16
1. a 2. d 3. b 4. a 5. d 6. c 7. b 8. d 9. c

Chapter 17
1. c 2. b 3. d 4. c 5. c 6. a 7. a 8. a

Chapter 18
1. a 2. d 3. d 4. b 5. c

Chapter 19
1. d 2. a 3. d 4. d 5. b 6. c 7. a 8. d 9. b 10. a

Chapter 20
1. d 2. a 3. c 4. b 5. c 6. c 7. d 8. c 9. b 10. a

Chapter 21
1. a 2. c 3. a 4. d 5. d 6. b 7. a 8. b 9. d 10. b
11. c 12. d 13. a 14. d 15. c

Chapter 22
1. c 2. b 3. b 4. d 5. a 6. c 7. a 8. d 9. d 10. a

Chapter 23
1. b 2. d 3. d 4. d 5. a 6. b 7. b 8. c

Chapter 24
1. a 2. c 3. b 4. b 5. b

Chapter 25
1. d 2. c 3. d 4. b 5. d 6. b 7. a 8. b 9. d

Chapter 26
1. a 2. b 3. a 4. a 5. d 6. a 7. d 8. c 9. c 10. c

Chapter 27
1. b 2. d 3. d 4. c 5. a

Chapter 28
1. d 2. c 3. a 4. b 5. a 6. a 7. b

Chapter 29
1. b 2. d 3. b 4. c 5. b 6. b

Chapter 30
1. b 2. b 3. a 4. a 5. c 6. a 7. c 8. a

Chapter 31
1. d 2. b 3. a 4. c 5. c 6. d 7. a 8. c

Chapter 32
1. c 2. d 3. c 4. c 5. d 6. d

Chapter 33
1. d 2. d 3. b 4. a 5. b 6. d 7. a 8. a 9. d

Chapter 34
1. d 2. b 3. a 4. a 5. b 6. a 7. b 8. c
9. c 10. b

Chapter 35
1. a 2. c 3. a 4. c 5. b 6. d 7. d 8. d 9. a 10. d

Chapter 36
1. c 2. a 3. b 4. b 5. d 6. a

Chapter 37
1. b 2. c 3. d 4. b 5. c

Chapter 38
1. a 2. c 3. c 4. d 5. b 6. a 7. b 8. a 9. d 10. b

Chapter 39
1. c 2. a 3. d 4. a 5. b 6. c 7. b

Chapter 40
1. a 2. b 3. b 4. c 5. b 6. c

Appendix B | *A Brief Guide to Statistical Significance*

Experiments often yield numerical data, such as the height of a plant or the incidence of illness in vaccinated children (see, for example, figure 1.12). But how are we to know whether an observed difference between two samples is "real"? For example, if we do find that 100 vaccinated children become sick slightly less often than 100 unvaccinated ones, how can we make sure that this outcome does not simply reflect random variation between samples of 100 children?

A statistical analysis can help. The dictionary definition of *statistics* is "the science that deals with the collection, analysis, and interpretation of numerical data, often using probability theory." Note that the analysis is grounded in probability theory, a branch of mathematics that deals with random events. A statistical test is therefore a mathematical tool that assesses variation, with the goal of determining whether any observed differences between treatments can be explained by the variation that random events produce.

Researchers use many types of statistical tests, depending on the type of data collected and the design of the experiment. A

description of these tests is beyond the scope of this appendix. For now, it is enough to understand that in each statistical test, the researcher computes a value (the "test statistic") that takes into account the sample size and the variability in the data. The researcher then determines the likelihood that the observed test statistic could be explained by chance alone.

An imaginary experiment will help you understand the role of variability in accepting or rejecting a hypothesis. Suppose that you have two friends, Pat and Kris, both of whom play softball. Pat claims to be able to hit a ball farther than Kris, but Kris disagrees. You therefore set up a test of the null hypothesis, which is that Pat and Kris can hit the ball equally far. You ask both of your friends to hit the ball one time, and Pat's ball does go farther. But Kris wants to re-do the test. This time, Kris's ball goes farther. Evidently, two hits apiece is not sufficient for you to settle the matter.

You therefore decide to improve the experiment (figure B.1). This time, each player gets to hit 10 balls, and you use a tape measure to determine how far each ball traveled from home plate. Figure B.2 shows two possible outcomes of the contest. In each

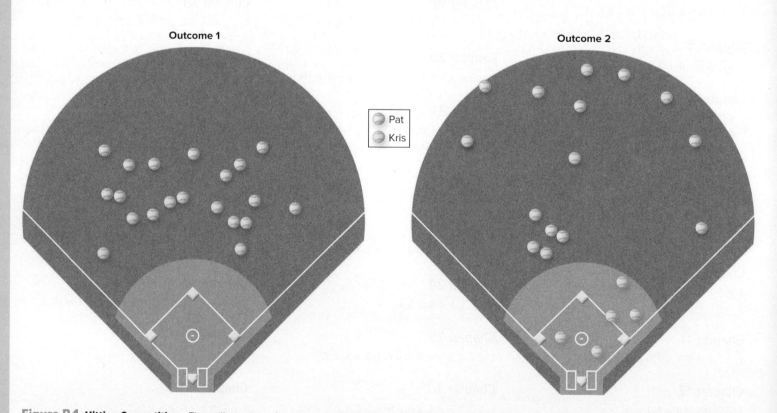

Figure B.1 Hitting Competition. These illustrations show two possible outcomes in a hitting contest between Pat and Kris. Note that the batting distances in Outcome 1 were much less variable than they were in Outcome 2.

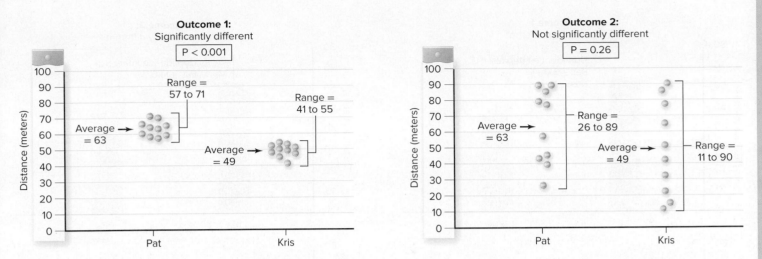

Figure B.2 Statistical Significance. Pat's 10 hits traveled an average of 63 meters, whereas Kris's average was 49 meters. In Outcome 1, this difference is considered highly significant at P < 0.001; in Outcome 2, however, the variability in hitting distances means that the difference between the two batters is not statistically significant.

scenario, Pat's average distance is 63 meters, compared with 49 meters for Kris. Pat therefore appears to be the better hitter.

But look more closely at the data. In Outcome 1, the hitting distances are much less variable than they are in Outcome 2. How does this variability influence our conclusions about Pat and Kris?

The answer lies in statistical tests. The goal of each statistical test is to calculate the probability that an observed outcome would arise *if the null hypothesis were true*. This probability is called the P value. The lower the P value, the greater the chance that the difference between two treatments is "real."

Generally, biologists accept a P value of 0.05 or smaller as being statistically significant. In other words, a difference between two treatments is statistically significant when the probability that we would observe that result by chance alone is 5% or less. Moreover, a P value of 0.01 or smaller is considered highly significant. Note that the meaning of the word *significant* is different from its use in everyday language. Ordinarily, "significant" means "important." Not so in statistics, where "significant" simply means "likely to be true," and "highly significant" means "very likely to be true."

Returning to our experiment, the null hypothesis is that Pat and Kris are equally good hitters. The amount of variability in

Outcome 1 was small, and the calculated value of our test statistic suggests that we can reject this null hypothesis with a P value of <0.001. In other words, there is a 99.999% chance that Pat really is a better hitter than Kris and that our results are not simply due to chance (see figure B.2). In Outcome 2, however, the hitting distances were much more variable, and the calculated P value is 0.26. We therefore cannot reject the null hypothesis with confidence. After all, given these data, the chance that Pat really is the better batter is only 74%. In science, a 26% chance of incorrectly rejecting the null hypothesis is unacceptably high.

Scientists often include information about statistical analyses along with their data (**figure B.3**). The most common technique is to graph the average for each treatment and then add error bars that reflect the amount of variability in the data. The longer the error bar, the greater the amount of variability and the less confident we are in the accuracy of our result.

An error bar usually indicates either the standard error or the 95% confidence interval; you may wish to consult a statistics reference to learn more about each. Because these two measures have slightly different meanings, researchers should always note the type of error bar depicted on a graph.

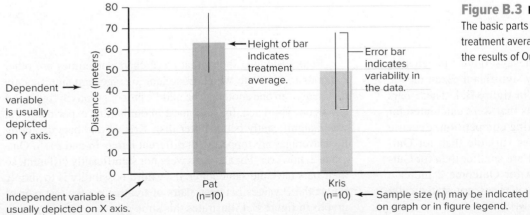

Figure B.3 Representing Statistics on a Graph. The basic parts of a bar graph include the axes, treatment averages, and error bars. This graph shows the results of Outcome 2.

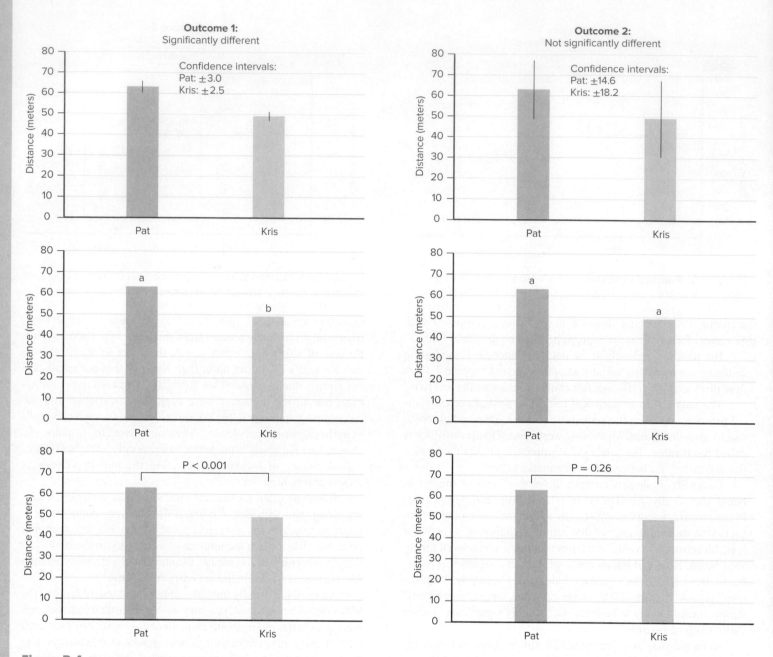

Figure B.4 Three Ways to Represent Statistics. Error bars, letters, and P values are all ways to indicate whether the difference between two treatments is statistically significant.

Figure B.4 shows three common ways to indicate whether an observed difference is statistically significant. Examine the error bars on the top pair of graphs in figure B.4. These bars indicate the 95% confidence intervals that were calculated for the two possible outcomes in our hitting competition. Because the data in Outcome 1 are much less variable than for Outcome 2, the error bars for Outcome 1 are smaller than for Outcome 2. The overlapping error bars for Outcome 2 indicate that the difference between Pat and Kris is not statistically significant.

Figure B.4 also shows that researchers sometimes use other methods to illustrate which treatment averages are significantly different from one another. The middle pair of graphs in figure B.4 shows one approach. In Outcome 1 of our batting experiment, Pat was a significantly better hitter than Kris, so the bars depicting their averages are topped with different letters (*a* and *b*). In Outcome 2, however, Pat and Kris were not significantly different, so their bars have the same letter, *a*. A third possibility is to simply report the P values between pairs of treatments; the last pair of graphs in figure B.4 illustrates this strategy.

Appendix C | *Units of Measurement*

Metric Prefixes

Symbol	Prefix	Multiplier
G	giga	One billion = 1,000,000,000 = 10^9
M	mega	One million = 1,000,000 = 10^6
k	kilo	One thousand = 1,000 = 10^3
h	hecto	One hundred = 100 = 10^2
da	deka (or deca)	Ten = 10 = 10^1
d	deci	One-tenth = 0.1 = 10^{-1}
c	centi	One-hundredth = 0.01 = 10^{-2}
m	milli	One-thousandth = 0.001 = 10^{-3}
µ	micro	One-millionth = 0.000001 = 10^{-6}
n	nano	One-billionth = 0.000000001 = 10^{-9}

Metric Units and Conversions

	Metric Unit	Metric to English Conversion	English to Metric Conversion
Length	1 meter (m)	1 km = 0.62 mile	1 mile = 1.609 km
		1 m = 1.09 yards = 39.37 inches	1 yard = 0.914 m
		1 cm = 0.394 inch	1 foot = 0.305 m = 30.5 cm
		1 mm = 0.039 inch	1 inch = 2.54 cm
Mass	1 gram (g)	1 t = 1.102 tons (U.S.)	1 ton (U.S.) = 0.907 t
	1 metric ton (t) = 1,000,000 g = 1,000 kg	1 kg = 2.205 pounds	1 pound = 0.4536 kg
		1 g = 0.0353 ounce	1 ounce = 28.35 g
Volume (liquids)	1 liter (L)	1 L = 1.06 quarts	1 gallon = 3.79 L
		1 mL = 0.034 fluid ounce	1 quart = 0.95 L
			1 pint = 0.47 L
			1 fluid ounce = 29.57 mL
Temperature	Degrees Celsius (°C)	°C = (°F − 32)/1.8	°F = (°C × 1.8) + 32
Energy and Power	1 joule (J)	1 J = 0.239 calorie	1 calorie = 4.186 J
		1 kJ = 0.239 kilocalorie ("food Calorie")	1 kilocalorie ("food Calorie") = 4,186 J
Time	1 second (sec)		

Appendix D | *Periodic Table of the Elements*

Columns: Groups 1–18. Members of the same group have the same number of valence electrons

Representative elements: Valence shell always gains or loses the same number of electrons (if any)

Representative elements: Valence shell always gains or loses the same number of electrons (if any)

Key

| 1 |
| Hydrogen |
| **H** |
| 1.0079 |

Atomic number
Element
Symbol
Atomic mass

Transition elements: Valence shell may gain or lose variable numbers of electrons

Rows: Periods 1–7. Members of the same period have the same number of orbitals

Period

Period	1	2	3	4	5	6	7	8	9	10	11	12	13	14	15	16	17	18
1	1 Hydrogen **H** 1.0079																	2 Helium **He** 4.0026
2	3 Lithium **Li** 6.941	4 Beryllium **Be** 9.0122											5 Boron **B** 10.811	6 Carbon **C** 12.012	7 Nitrogen **N** 14.0067	8 Oxygen **O** 15.9994	9 Fluorine **F** 18.9984	10 Neon **Ne** 20.179
3	11 Sodium **Na** 22.989	12 Magnesium **Mg** 24.305											13 Aluminum **Al** 26.9815	14 Silicon **Si** 28.086	15 Phosphorus **P** 30.9738	16 Sulfur **S** 32.064	17 Chlorine **Cl** 35.453	18 Argon **Ar** 39.948
4	19 Potassium **K** 39.098	20 Calcium **Ca** 40.08	21 Scandium **Sc** 44.956	22 Titanium **Ti** 47.867	23 Vanadium **V** 50.942	24 Chromium **Cr** 51.996	25 Manganese **Mn** 54.938	26 Iron **Fe** 55.847	27 Cobalt **Co** 58.933	28 Nickel **Ni** 58.693	29 Copper **Cu** 63.546	30 Zinc **Zn** 65.39	31 Gallium **Ga** 69.723	32 Germanium **Ge** 72.59	33 Arsenic **As** 74.922	34 Selenium **Se** 78.96	35 Bromine **Br** 79.904	36 Krypton **Kr** 83.80
5	37 Rubidium **Rb** 85.468	38 Strontium **Sr** 87.62	39 Yttrium **Y** 88.905	40 Zirconium **Zr** 91.22	41 Niobium **Nb** 92.906	42 Molybdenum **Mo** 95.94	43 Technetium **Tc** (98)	44 Ruthenium **Ru** 101.07	45 Rhodium **Rh** 102.905	46 Palladium **Pd** 106.4	47 Silver **Ag** 107.868	48 Cadmium **Cd** 112.41	49 Indium **In** 114.82	50 Tin **Sn** 118.71	51 Antimony **Sb** 121.76	52 Tellurium **Te** 127.60	53 Iodine **I** 126.904	54 Xenon **Xe** 131.29
6	55 Cesium **Cs** 132.905	56 Barium **Ba** 137.327	*57 Lanthanum **La** 138.91	72 Hafnium **Hf** 178.49	73 Tantalum **Ta** 180.948	74 Tungsten **W** 183.84	75 Rhenium **Re** 186.2	76 Osmium **Os** 190.2	77 Iridium **Ir** 192.2	78 Platinum **Pt** 195.08	79 Gold **Au** 196.967	80 Mercury **Hg** 200.59	81 Thalium **Tl** 204.38	82 Lead **Pb** 207.19	83 Bismuth **Bi** 208.980	84 Polonium **Po** (209)	85 Astatine **At** (210)	86 Radon **Rn** (222)
7	87 Francium **Fr** (223)	88 Radium **Ra** (226)	**89 Actinium **Ac** (227)	104 Rutherfordium **Rf** 261	105 Dubnium **Db** 262	106 Seaborgium **Sg** 266	107 Bohrium **Bh** 264	108 Hassium **Hs** 269	109 Meitnerium **Mt** 268	110 Darmstadtium **Ds** (271)	111 Roentgenium **Rg** (272)	112 Copernicium **Cn** (277)	113 Ununtrium/ Nihonium **Uut/Nh** (284)	114 Flerovium **Fl** (289)	115 Ununpentium/ Moscovium **Uup/Mc** (288)	116 Livermorium **Lv** (293)	117 Ununseptium/ Tennessine **Uus/Ts** (294)	118 Ununoctium/ Oganesson **Uuo/Og** (294)

Inner Transition Elements

*Lanthanides — 6

58 Cerium **Ce** 140.12	59 Praseodymium **Pr** 140.907	60 Neodymium **Nd** 144.24	61 Promethium **Pm** (145)	62 Samarium **Sm** 150.35	63 Europium **Eu** 151.96	64 Gadolinium **Gd** 157.25	65 Terbium **Tb** 158.925	66 Dysprosium **Dy** 162.50	67 Holmium **Ho** 164.930	68 Erbium **Er** 167.26	69 Thulium **Tm** 168.934	70 Ytterbium **Yb** 173.04	71 Lutetium **Lu** 174.97

**Actinides — 7

90 Thorium **Th** 232.038	91 Protactinium **Pa** (231)	92 Uranium **U** 238.03	93 Neptunium **Np** (237)	94 Plutonium **Pu** (244)	95 Americium **Am** (243)	96 Curium **Cm** (247)	97 Berkelium **Bk** (247)	98 Californium **Cf** (251)	99 Einsteinium **Es** (252)	100 Fermium **Fm** 257.095	101 Mendelevium **Md** (258)	102 Nobelium **No** (259)	103 Lawrencium **Lr** (262)

Metals: Lose electrons in redox reactions
Nonmetals: Gain electrons in redox reactions (except for group 18)
Semi-metals (metalloids): Intermediate properties

Appendix E | *Amino Acid Structures*

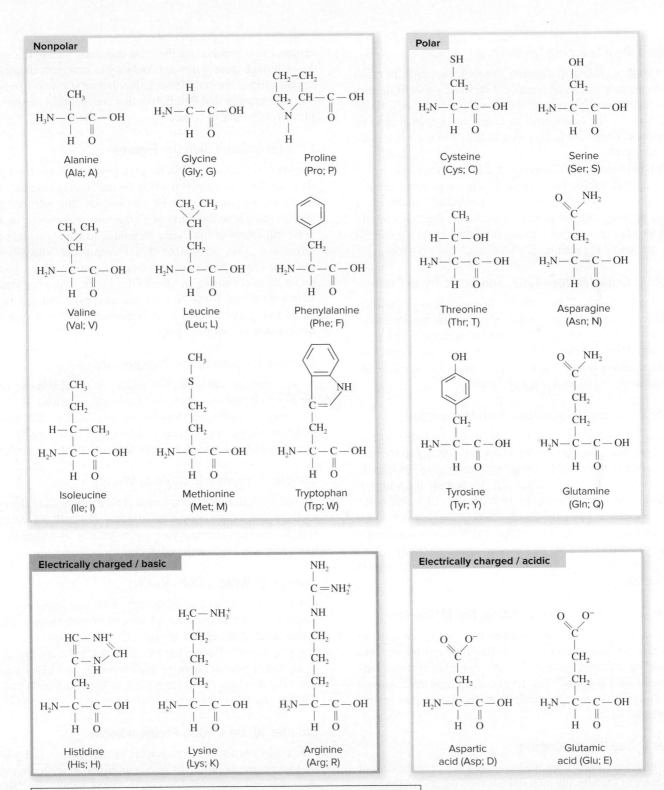

Nonpolar

Alanine
(Ala; A)

Glycine
(Gly; G)

Proline
(Pro; P)

Valine
(Val; V)

Leucine
(Leu; L)

Phenylalanine
(Phe; F)

Isoleucine
(Ile; I)

Methionine
(Met; M)

Tryptophan
(Trp; W)

Polar

Cysteine
(Cys; C)

Serine
(Ser; S)

Threonine
(Thr; T)

Asparagine
(Asn; N)

Tyrosine
(Tyr; Y)

Glutamine
(Gln; Q)

Electrically charged / basic

Histidine
(His; H)

Lysine
(Lys; K)

Arginine
(Arg; R)

Electrically charged / acidic

Aspartic
acid (Asp; D)

Glutamic
acid (Glu; E)

Note: Amino acids are arranged according to each R group's chemical properties.

Appendix F | *Learn How to Learn*

Chapter 1: Real Learning Takes Time

You got good at basketball, running, dancing, art, music, or video games by putting in lots of practice. Likewise, you will need to commit time to your biology course if you hope to do well. To get started, look for the Learn How to Learn tip in each chapter of this textbook. Each hint is designed to help you use your study time productively.

With practice, you'll discover that all concepts in biology are connected. The Survey the Landscape figure in every chapter highlights each chapter's place in the "landscape" of the entire unit. Use it, along with the more detailed Pull It Together concept map in the chapter summary, to see how each chapter's content fits into the unit's big picture.

Chapter 2: Organize Your Time, and Don't Try to Cram

Get a calendar and study the syllabus for every class you are taking. Write each due date in your calendar. Include homework assignments, quizzes, and exams, and add new dates as you learn them. Then, block out time well before each due date to work on each task. Success comes much more easily if you take a steady pace instead of waiting until the last minute.

Chapter 3: Interpreting Images from Microscopes

Any photo taken through a microscope should include information that can help you interpret what you see. First, read the caption and labels so that you know what you are looking at—usually an organ, a tissue, or an individual cell. Then study the scale bar and estimate the size of the image. (For a review of metric units, consult appendix C.) Finally, check whether a light microscope (LM), scanning electron microscope (SEM), or transmission electron microscope (TEM) was used to create the image. Note that stains and false colors are often added to emphasize the most important features.

Chapter 4: Focus on Understanding, Not Memorizing

When you are learning the language of biology, be sure to concentrate on how each new term fits with the others. Are you studying multiple components of a complex system? Different steps in a process? The levels of a hierarchy? As you study, always make sure you understand how each part relates to the whole.

Chapter 5: See What's Coming

Start by reviewing the Survey the Landscape figure at the start of each chapter to see how the material fits with the rest of the unit. Then check out the Learning Outline. Each heading is a complete sentence that summarizes the most important idea of the section. Read through these statements before you start each chapter. You can also flip to the end of the chapter before you start to read; the chapter summary and Pull It Together concept map can provide a preview of what's to come.

Chapter 6: Don't Skip the Figures

As you read the narrative in the text, pay attention to the figures; they are there to help you learn. Some figures summarize the narrative, making it easier for you see the "big picture." Other illustrations show the parts of a structure or the steps in a process; still others summarize a technique or help you classify information. Also, remember that students use illustrations in different ways. Once you encounter a figure's callout, you may prefer to stop reading to absorb the entire figure, or you may switch back and forth between the narrative and the figure's parts. Being attentive to your preferences will help you to be more systematic as you study.

Chapter 7: Pause at the Checkpoints

As you read, get out a piece of paper and see if you can answer the Figure It Out and Mastering Concepts questions. If not, you may want to study a bit more before you move on. Each section builds on the material that came before, and mastering one chunk at a time will make it much easier to learn whatever comes next.

Chapter 8: Explain It, Right or Wrong

As you work through the multiple choice questions at the end of each chapter, make sure you can explain why each correct choice is right. You can also test your understanding by taking the time to explain why each of the other choices is wrong.

Chapter 9: Write It Out—Really!

Get out a pen and a piece of scratch paper, and answer the open-ended Write It Out questions at the end of each chapter. This tip applies even if the exams in your class are multiple choice. Putting pen to paper (as opposed to just saying the answer in your head) forces you to organize your thoughts and helps you discover the difference between what you know and what you only THINK you know.

Chapter 10: Be a Good Problem Solver

This chapter is about the principles of inheritance, and you will find many sample problems within its pages; in addition, interactive problems are available online. Need help? The "How to Solve a Genetics Problem" guide at the end of this chapter shows a systematic, step-by-step approach to answering the most common types of questions. Keep using the guide until you feel comfortable solving any problem type.

Chapter 11: Vary Your Study Plan for Healthy Learning

Your study sessions may become stale if you do the same things over and over. After all, it is difficult to focus after watching countless animations or listening to hours of podcasts. Instead, try switching between strategies that are passive (watching and listening) and those that are active (drawing and writing). You might watch a video on Punnett squares and then try to draw one yourself. Or you could listen to a podcast about cloning and then write a paragraph describing the process in your own words. Keeping variety in your study plan will help you stay engaged.

Chapter 12: Practice Your Recall

Here's an old-fashioned study tip that still works. When you finish reading a passage, close the book and write what you remember—in your own words. In this chapter, for example, you will learn about several forces of evolutionary change. After you read about them, can you list and describe them without peeking at your book? Try it and find out!

Chapter 13: Make a Chart

One way to organize the information in a chapter is to make a summary chart or matrix. The chart's contents will depend on the chapter. For this chapter, for example, you might write the following headings along the top of a piece of paper: "Type of Evidence," "Definition," "How It Works," "What It Tells Us," and "Example." Then you would list the lines of evidence for evolution along the left edge of the chart. Start filling in your chart, using the book at first to find the information you need. Later, you should be able to re-create your chart from memory.

Chapter 14: Don't Neglect the Boxes

You may be tempted to skip the chapter opening essays and boxed readings because they're not "required." Read them anyway. The contents should help you remember and visualize the material you are trying to learn. And who knows? You may even find them interesting.

Chapter 15: Write Your Own Test Questions

Have you ever tried putting yourself in your instructor's place by writing your own multiple choice test questions? It's a great way to pull the pieces of a chapter together. The easiest questions to write are based on definitions and vocabulary, but those will not always be the most useful. Try to think of questions that integrate multiple ideas or that apply the concepts in a chapter. Write 10 questions, and then let a classmate answer them. You'll probably both learn something new.

Chapter 16: Take the Best Possible Notes

Some students take notes only on what they consider "important" during a lecture. Others write down words but not diagrams, or they write what's on the board but not what the instructor is saying. All of these strategies risk losing vital information and connections between ideas that could help in later learning. Instead, write down as much as you can during lecture, including sketches of the diagrams and notes on what the instructor is telling you about the main ideas. It will be much easier to study later if you have a complete picture of what happened in every class.

Chapter 17: Skipping Class?

Attending lectures is important, but you may need to skip class once in a while. How will you find out what you missed? If your instructor does not provide complete lecture notes, you may be able to copy them from a friend. Whenever you borrow someone else's notes, it's a good idea to compare them with the assigned reading to make sure the notes are complete and accurate. You might also want to check with the instructor if you have lingering questions about what you missed.

Chapter 18: What's the Point of Rewriting Your Notes?

Your notes are your record of what happened in class, so why should you rewrite them after the lecture is over? One answer is that the abbreviations and shorthand that make perfect sense while you take notes will become increasingly mysterious as time goes by. Rewriting the information in complete sentences not only reinforces learning but also makes your notes much easier to study before an exam.

Chapter 19: Take Notes on Your Reading

Many classes have reading assignments. Taking notes as you read should help you not only retain information but also identify what you don't understand. Before you take notes, skim through the assigned pages once; otherwise, you may have trouble distinguishing between main points and minor details. Then read them again. This time, pause after each section and write the most important ideas in your own words. What if you can't remember the material or don't understand it well enough to summarize the passage? Read it again, and if that doesn't work, ask for help with whatever isn't clear.

Chapter 20: Use All Your Resources

Whether you are using a print book or an ebook, Connect and other websites offer a wealth of online quizzes, animations, and other resources that can help you learn biology. As you study, pay attention to which concepts are sinking in and which are more difficult to understand. Take regular breaks from reading and try a digital learning resource associated with your book. For example, practice questions can help you learn the material and test your understanding. Also, check for animations that take you through complex processes one step at a time. Sometimes the motion of an animation can help you understand what's happening more easily than studying a static image.

Chapter 21: Think While You Search the Internet

Some assignments may require you to use the Internet. But the Internet is full of misinformation, so you must evaluate every site you visit. Collaborative sites such as Wikipedia may be unreliable because anyone can change any article. For other sites, ask the following questions: Are you looking at someone's personal page? Is there an educational, governmental, nonprofit, or commercial sponsor? Is the author reputable? Does the page contain opinions or facts? Are facts backed up with documentation? Answering these questions will help ensure that the sites you use are credible.

Chapter 22: Bite-Sized Pieces

Many students think they need to read a whole chapter in one sitting. Instead, try working through one topic at a time. Read just one section of the chapter, compare it to your class notes, and test yourself on what you have learned. Think of each chapter as a meal: You eat a sandwich one bite at a time, so why not tackle biology the same way?

Chapter 23: A Quick Once-Over

Unless your instructor requires you to read your textbook in detail before class, try a quick preview. Read the chapter outline to identify the main ideas; then look at the figures and the key terms in the narrative. Previewing a chapter should help you follow the lecture, because you will already know the main ideas. In addition, note taking will be easier if you recognize new vocabulary words from your quick once-over. Return to your book for an in-depth reading after class to help nail down the details.

Chapter 24: Know Yourself

Setting aside time to study is one important ingredient for academic success; another is paying attention to your work habits throughout the day and night. Are you most alert in the morning, afternoon, or evening? Block off time to study during periods when you are at your best. Your study time will be much more productive if you are not fighting to stay awake.

Chapter 25: Flashcard Excellence

While making flashcards, you may be tempted to focus on definitions. For example, after reading this chapter, you might make a flashcard with "simple squamous" on one side and "single layer of flattened cells" on the other. This description is correct, but it won't help you understand the bigger picture. Instead, your flashcards should include realistic questions that cover both the big picture and the small details. Try making flashcards that pose a question, such as "What are the four tissue types in an animal?" or "How do cells, tissues, organs, and organ systems relate to one another?" Write the full answer on the other side; then practice writing the answers on scratch paper until you are sure you know them all.

Chapter 26: How to Use a Study Guide

Some professors provide study guides to help students prepare for exams. Used wisely, a study guide can be a valuable tool. One good way to use a study guide is AFTER you have studied the material, when you can go through the guide and make sure you haven't overlooked any important topics. Alternatively, you can cross off topics as you encounter them while you study. No matter which technique you choose, don't just memorize isolated facts. Instead, try to understand how the items on the study guide relate to one another. If you're unclear on the relationships, be sure to ask your instructor.

Chapter 27: Don't Throw That Exam Away!

Whether or not you were satisfied with your last exam, take the time to learn from your mistakes. Mark the questions that you missed and the ones that you got right but were unsure about. Then figure out what went wrong for each question. For example, did you neglect to study the information, thinking it wouldn't be on the test? Did you memorize a term's definition without understanding how it fits with other material? Did you misread the question? After you have finished your analysis, look for patterns and think about what you could have done differently. Then revise your study plan so that you can avoid making the same mistakes in the future.

Chapter 28: Use Those Office Hours

Most instructors maintain office hours. Do not be afraid to use this valuable resource! Besides getting help with course materials, office hours give you an opportunity to know your professors personally. After all, at some point you may need a letter of recommendation; a letter from a professor who knows you well can carry a lot of weight. If you do decide to visit during office hours, be prepared with specific questions. And if you request a separate appointment, be sure to arrive at the time you have arranged—or let your instructor know you need to cancel.

Chapter 29: Make Appointments with Yourself

If you prefer to study alone but often find yourself putting off your solo study sessions, try making recurring "appointments" with yourself. That is, use your calendar to block off time that you dedicate to studying each day. You can reread chapters after class, quiz yourself on course materials, make concept maps, or work on homework assignments. Keep those appointments throughout the semester, so that you never get behind.

Chapter 30: Studying in Groups

Study groups can offer a great way to learn from other students, but they can also dissolve into social events that accomplish little real work. Of course, your choice of study partners makes a huge difference; try to pick people who are at least as serious as you are about learning. To stay focused, plan activities that are well suited for groups. For example, you can agree on a list of vocabulary words and take turns adding them to a group concept map. You can also write exam questions for your study partners to answer, or you can simply explain the material to each other in your own words. Focus on what you need to learn, and your study sessions should be productive.

Chapter 31: How to Use a Tutor

Your school may provide tutoring sessions for your class, or perhaps you have hired a private tutor. How can you make the most of this resource? First, meet regularly with your tutor for an hour or two each week; don't wait until just before an exam. Second, if possible, tell your tutor what you want to work on before each session, so that he or she can prepare. Third, take your textbook, class notes, and questions to your tutoring session. Fourth, be realistic. Your tutor can discuss difficult concepts and help you practice with the material, but don't expect him or her to simply give you the answers to your homework.

Chapter 32: Avoid Distractions

Despite your best intentions, constant distractions may take you away from your studies. Friends, music, TV, social media, text messages, video games, and online shopping all offer attractive diversions. How can you stay focused? One answer is to find your own place to study where no one can find you. Turn your phone off for a few hours; the world will get along without you while you study. And if you must use your computer, create a separate user account with settings that prevent you from visiting favorite websites during study time.

Chapter 33: Pay Attention in Class

It happens to everyone occasionally: Your mind begins to wander while you are sitting in class, so you doodle, check your phone, or doze off. Before you know it, class is over, and you got nothing out of it. How can you keep from wasting your class time this way? One strategy is to get plenty of sleep and eat well so that your mind stays active instead of drifting off. Another is to prepare for class in advance, since getting lost can be an excuse for drifting off. When you get to class, sit near the front, listen carefully, and take good notes. Finally, a friendly reminder can't hurt; make a small PAY ATTENTION sign to put on your desk, where you can always see it.

Chapter 34: Find a Good Listener

For many complex topics, you may struggle to know how well you really understand what is going on. One tip is to try explaining what you think you know to somebody else. Choose a subject that takes a few minutes to explain. As you describe the topic in your own words, your partner should ask follow-up questions and note where your explanation is vague. Those insights should help draw your attention to important details that you overlooked.

Chapter 35: Don't Waste Old Exams

If you are lucky, your instructor may make old exams available to your class. If so, it is usually a bad idea to simply look up and memorize the answer to each question. Instead, use the old exam as a chance to test yourself before it really counts. Put away your notes and textbook, and set up a mock exam. Answer each question without "cheating"; then check how many you got right. Use the questions you got wrong—or that you guessed right—as a guide to what you should study more.

Chapter 36: What's Your Learning Style?

Students differ in how they prefer to receive information: Some love to hear lectures, others like reading, and still others thrive on hands-on activities in lab. You may find it helpful to search for a learning styles inventory online to discover more about your own preferences. At the very least, the advice on the website may alert you to study techniques that you have not tried before.

Chapter 37: Expand on What You're Learning

Many students make the mistake of memorizing information without really thinking about what they are learning. This strategy not only makes schoolwork boring but also limits your ability to answer questions. Instead of repeating someone else's words as you study, try elaborating on what you have learned. Here are some ideas: Explain how the topic relates to other course material or to your own life; draw a picture of important processes; discuss the topic with someone else; explain it to yourself out loud or in writing; or think of additional examples or analogies.

Chapter 38: Perform Your Best on Exams

Last-minute cramming for exams may be a classic college ritual, but it is not usually the best strategy. If you try to memorize everything right before an exam, you may become overwhelmed and find yourself distracted by worries that you'll never learn it all. Instead, work on learning the material as the course goes along. Then, on the night before the exam, get plenty of rest, and don't forget to cat on cxam day. If you are too tired or too hungry to think, you won't be able to give the exam your best shot.

Chapter 39: Use Your Campus Resources

Many first-time college students struggle to keep up with assignments outside of class. Time management can be part of the problem, but poor reading and writing skills may also share the blame. Most campuses offer free resources that can help. For example, look for seminars and workshops aimed at improving reading comprehension. Writing centers are also common; the staff should include consultants trained to help you be a more proficient writer.

Chapter 40: Make Your Own Review Sheet

If you are facing a big exam, how can you make sense of everything you have learned? One way is to make your own review sheet. The best strategy will depend on what your instructor expects you to know, but here are a few ideas to try: Make lists; draw concept maps that link ideas within and between chapters; draw diagrams that illustrate important processes; and write mini-essays that explain the main points in each chapter's learning outline.

Glossary

3 prime (3′): the end of a nucleic acid strand terminating at the third carbon of a sugar molecule

5 prime (5′): the end of a nucleic acid strand terminating at the fifth carbon of a sugar molecule

A

abiotic: nonliving

abscisic acid: plant hormone that inhibits shoot growth, maintains bud dormancy, induces seed dormancy, and stimulates stomatal closing

abscission zone: specialized layer of cells from which leaf petiole detaches from plant

absolute dating: determining the age of a fossil in years

absorption: the process of taking in and incorporating nutrients

accessory pigment: photosynthetic pigment other than chlorophyll *a* that extends the range of light wavelengths useful in photosynthesis

acetyl CoA: molecule that enters the Krebs cycle in cellular respiration; product of partial oxidation of pyruvate

acid: a molecule that releases hydrogen ions into a solution

acid deposition: low pH precipitation or dry particles that form when air pollutants react with water in the upper atmosphere

acrosome: protrusion covering the head of a sperm cell, containing enzymes that enable the sperm to penetrate layers around the oocyte

actin: protein that forms thin filaments in muscle cells; also part of cytoskeleton

action potential: depolarization that propagates along the cell membrane of a neuron

activation energy: energy required for a chemical reaction to begin

active immunity: immunity generated when an individual produces his or her own antibodies

active site: the part of an enzyme to which substrates bind

active transport: movement of a substance across a membrane against its concentration gradient, using a carrier protein and energy from ATP

adaptation: inherited trait that permits an organism to survive and reproduce

adaptive immunity: defense system that recognizes and remembers specific antigens

adaptive radiation: divergence of multiple new species from a single ancestral type in a relatively short time

adhesion: the tendency of water to hydrogen bond to other compounds

adipose tissue: type of connective tissue consisting of fat cells embedded in a minimal matrix

adrenal cortex: outer portion of an adrenal gland; secretes steroid hormones such as mineralocorticoids and glucocorticoids in response to long-term stress

adrenal gland: one of two endocrine glands atop the kidneys; secretes stress-related hormones

adrenal medulla: inner portion of an adrenal gland; secretes epinephrine and norepinephrine in response to short-term stress

adrenocorticotropic hormone (ACTH): hormone produced in the anterior pituitary; stimulates secretion of hormones from adrenal cortex

adult stem cell: cell that can give rise to a limited subset of cells in the body

adventitious root: root arising from stem or leaf

aerobic respiration: complete oxidation of glucose to CO_2 in the presence of O_2, producing ATP

age structure: distribution of age classes in a population

agglutination: clumping together of cells

alcoholic fermentation: metabolic pathway in which NADH reduces the pyruvate from glycolysis, producing ethanol and CO_2

aldosterone: mineralocorticoid hormone produced in the adrenal cortex; stimulates water conservation at the kidneys

alga (pl. algae): aquatic, photosynthetic protist

alimentary canal: two-opening digestive tract; also called gastrointestinal (GI) tract

alkaline: having a pH greater than 7

allantois: extraembryonic membrane that forms as an outpouching of the yolk sac; manufactures blood cells and gives rise to umbilical cord's blood vessels

allele: one of two or more alternative forms of a gene

allele frequency: number of copies of one allele, divided by the number of alleles in a population

allergen: antigen that triggers an allergic reaction

allergy: exaggerated immune response to a harmless substance

allopatric speciation: formation of new species after a physical barrier separates a population into groups that cannot interbreed

alternation of generations: the sexual life cycle of plants and many green algae, which alternates between a diploid sporophyte stage and a haploid gametophyte stage

altruism: behavior that lowers an animal's fitness for the good of others

alveolus (pl. alveoli): microscopic air sac where gas exchange occurs in mammalian lungs

amino acid: an organic molecule consisting of a central carbon atom bonded to a hydrogen atom, an amino group, a carboxyl group, and an R group

amino group: a nitrogen atom single-bonded to two hydrogen atoms

ammonia: nitrogenous waste (NH_3) generated by breakdown of amino acids and excreted by many fishes

amnion: extraembryonic membrane that contains amniotic fluid

amniote: vertebrate in which protective membranes surround the embryo (amnion, chorion, and allantois); reptiles and mammals

amniotic egg: reptile or monotreme egg containing fluid and nutrients within membranes that protect the embryo

amoeboid protozoan: unicellular protist that produces pseudopodia

amphibian: type of tetrapod vertebrate that may live on land but reproduces in water

amygdala: forebrain structure involved in emotions

anaerobic respiration: cellular respiration using an electron acceptor other than O_2

analogous: similar because of convergent evolution (instead of common ancestry)

anaphase: stage of mitosis in which the spindle pulls sister chromatids toward opposite poles of the cell

anaphase I: anaphase of meiosis I, when spindle fibers pull homologous chromosomes toward opposite poles of the cell

anaphase II: anaphase of meiosis II, when centromeres split and spindle fibers pull sister chromatids toward opposite poles of the cell

anaphylactic shock: rapid, widespread, severe allergic reaction

anatomy: the study of an organism's structure

ancestral character: characteristic already present in the ancestor of a group being studied

anchoring (or adhering) junction: connection between two adjacent animal cells that anchors intermediate filaments in a single spot on the cell membrane

aneuploid cell: cell with too many or too few chromosomes, but not involving a full extra or missing set

angiogenesis: the production of new blood vessels

angiosperm: a seed plant that produces flowers and fruits; includes monocots and eudicots

Animalia: kingdom containing multicellular eukaryotes that are heterotrophs by ingestion

annelid: segmented worm; phylum Annelida

anorexia nervosa: eating disorder characterized by refusal to maintain normal body weight

antenna pigment: photosynthetic pigment that passes photon energy to the reaction center of a photosystem

anterior pituitary: the front part of the pituitary gland

anther: pollen-producing structure at tip of stamen

antibody: defensive Y-shaped protein that binds to a specific antigen

anticodon: a three-base portion of a tRNA molecule; the anticodon is complementary to one codon

antidiuretic hormone (ADH): hormone released from the posterior pituitary; promotes water conservation; also called vasopressin

antigen: molecule that elicits an immune reaction by B and T cells

anus: exit from a complete digestive tract

aorta: the largest artery leaving the heart

apical dominance: the suppression of growth of lateral buds by the intact terminal bud of a plant

apical meristem: meristem at tip of root or shoot

apicomplexan: nonmotile protist with cell containing an apical complex; obligate animal parasite

apoptosis: programmed cell death that is a normal part of development

appendicular skeleton: the limb bones and the bones that support them in the vertebrate skeleton

arachnid: type of chelicerate arthropod with eight legs

Archaea: one of two domains of prokaryotes

arteriole: small artery

artery: vessel that carries blood away from the heart

arthropod: segmented animal with an exoskeleton and jointed appendages; phylum Arthropoda

artificial selection: selective breeding strategy in which a human allows only organisms with desired traits to reproduce

ascomycete: fungus that produces sexual spores in a sac called an ascus

ascospore: haploid sexual spore produced in an ascus

ascus (pl. asci): in ascomycetes, a saclike structure in which ascospores are produced

asexual reproduction: form of reproduction in which offspring arise from only one parent

associative learning: type of learning in which an animal connects two paired stimuli (classical conditioning) or pairs a behavior with its consequences (operant conditioning)

atom: a particle of matter; composed of protons, neutrons, and electrons

atomic number: the number of protons in an atom's nucleus

atomic weight: the average mass of all atoms of an element

ATP (adenosine triphosphate): a molecule whose high-energy phosphate bonds power many biological processes

ATP synthase: enzyme complex that admits protons through a membrane, where they trigger phosphorylation of ADP to ATP

atrioventricular (AV) node: specialized cardiac muscle cells that delay the heartbeat, giving the ventricles time to fill

atrioventricular (AV) valve: flap of heart tissue that prevents the flow of blood from a ventricle to an atrium

atrium: heart chamber that receives blood

auditory canal: ear canal; funnels sounds from the outer ear to the eardrum

auditory nerve: nerve fibers that connect the cochlea and vestibular apparatus to the brain

autoimmune disorder: immune reaction to antigens associated with the body's own cells

autonomic nervous system: in the peripheral nervous system, motor pathways that lead to smooth muscle, cardiac muscle, and glands

autosomal dominant: inheritance pattern of a dominant allele on an autosome

autosomal recessive: inheritance pattern of a recessive allele on an autosome

autosome: a nonsex chromosome

autotroph: organism that produces organic molecules by acquiring carbon from inorganic sources; primary producer

auxin: plant hormone that promotes cell elongation in stems and fruits, controls tropisms, stimulates growth of adventitious roots, and suppresses growth of lateral buds

axial skeleton: the central axis of a vertebrate skeleton; consists of the bones of the head, vertebral column, and rib cage

axillary bud: undeveloped shoot in the angle between stem and petiole; also called a lateral bud

axon (nerve fiber): extension of a neuron that transmits messages away from the cell body and toward another cell

B

B cell: type of lymphocyte that matures in red bone marrow; if activated, B cells secrete antibodies

bacillus (pl. bacilli): rod-shaped prokaryote

background extinction rate: steady, gradual loss of species through natural competition or loss of genetic diversity

Bacteria: one of two domains of prokaryotes

bacteriophage (phage): a virus that infects bacteria

balanced polymorphism: condition in which multiple alleles persist indefinitely in a population

bark: tissues outside the vascular cambium

base: a molecule that either releases hydroxide ions into a solution or removes hydrogen ions from it

basidiomycete: fungus that produces sexual spores on a basidium

basidiospore: haploid sexual spore produced on a basidium

basidium (pl. basidia): in basidiomycetes, a club-shaped structure that produces basidiospores

basilar membrane: the lower wall of the cochlear canal; vibrates in response to sound

basophil: type of circulating white blood cell that triggers inflammation and allergy

behavioral ecology: study of the ecological context and the evolution of behavioral patterns

benign tumor: mass of abnormal cells that does not have the potential to spread

benthic zone: sediment at the bottom of an ocean or a lake

bilateral symmetry: body form in which only one plane divides the animal into mirror image halves

bile: digestive biochemical that emulsifies fats

binary fission: type of asexual reproduction in which a prokaryotic cell divides into two identical cells

biodiversity: the variety of life on Earth

biogeochemical cycle: geological and biological processes that recycle elements vital to life

biogeography: the study of the distribution patterns of species across the planet

biological species: a population, or group of populations, whose members can interbreed and produce fertile offspring

biomagnification: tendency of certain chemicals to accumulate in the highest trophic levels

biome: one of the world's several major types of ecosystems, characterized by a distinctive climate and group of species

biosphere: part of Earth where life can exist

biotic: living

bird: type of reptile with feathers and wings

birth defect: abnormality that causes death or disability in a newborn

birth rate: the number of new individuals produced per individual per unit time

bivalve: type of mollusk with a two-part, hinged shell

blade: flattened part of a leaf

blastocyst: preembryonic stage consisting of a fluid-filled ball of cells; develops into the embryo and part of the placenta

blastula: stage of early animal embryonic development; a sphere of cells surrounding a fluid-filled cavity

blood: type of connective tissue consisting of cells and platelets suspended in a liquid matrix

blood clot: plug of solidified blood

blood pressure: force that blood exerts against blood vessel walls

blood–brain barrier: close-knit cells that form capillaries in the brain, limiting the substances that can enter

bone: organ consisting of bone tissue, cartilage, and other tissues

bone tissue: connective tissue consisting of osteocytes and other cells embedded in a mineralized matrix

bony fish: type of vertebrate chordate with a bony skeleton, a swim bladder, and paired fins

bottleneck: sudden reduction in the size of a population

brain: a distinct concentration of nervous tissue, often encased within a skull, at the anterior end of an animal

brainstem: the midbrain plus two parts of the hindbrain (the pons and medulla oblongata)

breast: milk-producing organ in female mammals

bronchiole: small, branched airway that connects bronchi to alveoli

bronchus (pl. bronchi): one of two large tubes that branch from the trachea

brown alga: multicellular, photosynthetic aquatic protist with swimming spores and brownish accessory pigments

bryophyte: plant that lacks vascular tissue; includes liverworts, hornworts, and mosses

buffer: weak acid/base pair that resists changes in pH

bulbourethral gland: small gland near the male urethra that secretes mucus

bulimia: eating disorder in which a person eats large quantities and then intentionally vomits or uses laxatives shortly afterward

bulk element: an element that an organism requires in large amounts

bulk feeder: animal that ingests large pieces of food

bundle-sheath cell: thick-walled plant cell surrounding veins; site of Calvin cycle in C_4 plants

C

C_3 pathway: the Calvin cycle

C_4 pathway: a carbon fixation pathway in which CO_2 combines with a three-carbon molecule to form a four-carbon compound

caecilian: type of amphibian that lacks limbs

calcitonin: thyroid hormone that decreases blood calcium levels

calorie: the energy required to raise the temperature of 1 gram of water by 1°C under standard conditions

Calorie: a measure of the energy content in food; equal to 1000 calories or 1 kilocalorie

Calvin cycle: in photosynthesis, metabolic pathway in which CO_2 is fixed and incorporated into a three-carbon carbohydrate

CAM pathway: carbon fixation that occurs at night; CO_2 is later released for use in the Calvin cycle during the day

cancer: class of diseases characterized by uncontrolled division of cells that invade or spread to other tissues

capillary: tiny vessel that connects an arteriole with a venule

capillary bed: network of capillaries where the circulatory system and body tissues exchange materials

capsid: protein coat of a virus

carbohydrate: compound containing carbon, hydrogen, and oxygen in a ratio 1:2:1

carbon fixation: the initial incorporation of carbon from CO_2 into an organic compound

carbon reactions: the reactions of photosynthesis that use ATP and NADPH to synthesize carbohydrates from carbon dioxide

carboxyl group: a carbon atom double-bonded to an oxygen and single-bonded to a hydroxyl group

cardiac cycle: sequence of contraction and relaxation that makes up the heartbeat

cardiac muscle tissue: involuntary muscle tissue composed of branched, striated, single-nucleated contractile cells

cardiac output: the volume of blood that the heart pumps each minute

cardiovascular system: circulatory system

carnivore: animal that eats animals

carpel: female part at a flower's center; consists of an ovary, a style, and a stigma

carrying capacity: maximum number of individuals that an ecosystem can support indefinitely

cartilage: type of connective tissue consisting of cells surrounded by a rubbery collagen matrix

cartilaginous fish: type of vertebrate chordate with a cartilage skeleton and paired fins

Casparian strip: waxy barrier in root endodermis

catastrophism: the idea that brief upheavals such as floods, volcanic eruptions, and earthquakes are responsible for most geological formations

cecum: the entrance to the large intestine

cell: smallest unit of life that can function independently

cell body: enlarged portion of a neuron that contains most of the organelles

cell cycle: sequence of events that occur in an actively dividing cell

cell membrane: the boundary of a cell, consisting of proteins embedded in a phospholipid bilayer

cell plate: in plants, the materials that begin to form the wall that divides two cells

cell theory: the ideas that all living matter consists of cells, cells are the structural and functional units of life, and all cells come from preexisting cells

cell wall: a rigid boundary surrounding cells of many prokaryotes, protists, plants, and fungi

cell-mediated immunity: branch of adaptive immune system in which defensive cells kill invaders by direct cell–cell contact

cellular slime mold: protist in which feeding stage consists of individual cells that come together as a multicellular "slug" when food runs out

centipede: type of mandibulate arthropod with one pair of legs per repeating subunit

central nervous system: brain and the spinal cord

centromere: small section of a chromosome where sister chromatids attach to each other

centrosome: part of the cell that organizes microtubules

cephalization: development of sensory structures and a brain at the head end of an animal

cephalopod: type of marine mollusk with arms connected to its head

cerebellum: area of the hindbrain that coordinates subconscious muscular responses

cerebral cortex: outer layer of the cerebrum; participates in perception, voluntary movement, language, thought, and other functions

cerebrospinal fluid: fluid that bathes and cushions the central nervous system

cerebrum: region of the forebrain that controls intelligence, learning, perception, and emotion

cervix: lower, narrow part of the uterus

chaparral: Mediterranean shrubland

charophyte: type of green algae thought to be most closely related to terrestrial plants

chelicerate: arthropod with clawlike mouthparts (chelicerae); horseshoe crabs and arachnids

chemical bond: attractive force that holds atoms together

chemical reaction: interaction in which bonds break and new bonds form

chemiosmotic phosphorylation: reactions that produce ATP using ATP synthase and the potential energy of a proton gradient

chemoreceptor: sensory receptor that responds to chemicals

chemotroph: organism that derives energy by oxidizing inorganic or organic chemicals

chiton: type of marine mollusk with eight flat, overlapping shells

chlorophyll a: green pigment that plants, algae, and cyanobacteria use to harness the energy in sunlight

chloroplast: organelle housing the reactions of photosynthesis in eukaryotes

chordate: animal that at sometime during its development has a notochord, a hollow nerve cord, pharyngeal slits or pouches, and a postanal tail; phylum Chordata

chorion: outermost extraembryonic membrane; helps establish the placenta

chorionic villus (pl. villi): fingerlike projection extending from the chorion to the uterine lining; site of exchange between a woman's blood and a fetus

choroid: middle layer of the eyeball, between the sclera and the retina

chromatid: one of two identical DNA molecules that make up a replicated chromosome

chromatin: collective term for all of the DNA and its associated proteins in the nucleus of a eukaryotic cell

chromosome: a continuous molecule of DNA wrapped around protein in the nucleus of a eukaryotic cell; also, the genetic material of a prokaryotic cell

chyme: semifluid mass of food and gastric juice that moves from the stomach to the small intestine

chytridiomycete (chytrid): microscopic fungus that produces motile cells

ciliate: protist with cilia-covered cell surface

cilium (pl. cilia): one of many short, movable protein projections extending from a cell

circadian rhythm: physiological cycle that repeats daily

circulatory system: organ system consisting of the heart and vessels that distribute blood (or a comparable fluid) throughout the body

clade: monophyletic group of organisms consisting of a common ancestor and all of its descendants

cladistics: phylogenetic system that defines groups by distinguishing between ancestral and derived characters

cladogram: treelike diagram built using shared derived characteristics

classical conditioning: type of associative learning in which an animal pairs two stimuli

cleavage: period of rapid cell division following fertilization

cleavage furrow: in dividing animal cells, the indentation that begins the process of cytokinesis

climax community: community that persists indefinitely if left undisturbed

clitoris: small, highly sensitive female sexual organ at the junction of the labia

clonal deletion: elimination of lymphocytes with receptors for self antigens

clonal selection: rapid division of a stimulated B cell, generating memory B cells and plasma cells that are clones of the original B cell

cloning vector: a self-replicating structure that carries DNA from cell to cell

closed circulatory system: circulatory system in which blood remains confined to vessels

club moss: type of seedless vascular plant with scales or needles and club-shaped reproductive structures

cnidarian: animal with radial symmetry, two germ layers, a jellylike interior, and cnidocytes; phylum Cnidaria

cnidocyte: cell in cnidarians that can fire a toxic barb for predation or defense

coccus (pl. cocci): spherical prokaryote

cochlea: spiral-shaped part of the inner ear, where vibrations are translated into nerve impulses

codominance: mode of inheritance in which two alleles are fully expressed in the heterozygote

codon: a triplet of mRNA bases that specifies a particular amino acid

coelacanth: rare type of lobe-finned fish; oldest existing jawed vertebrate

coelom: fluid-filled animal body cavity that forms completely within mesoderm

coevolution: genetic change in one species selects for subsequent change in another species

cofactor: inorganic or organic substance required for activity of an enzyme

cohesion: the attraction of water molecules to one another

cohesion–tension theory: theory that explains how water moves under tension in xylem

collecting duct: tubule in the kidney into which nephrons drain urine; site of final adjustments to urine composition

collenchyma: elongated living plant cells with thick, elastic cell walls

commensalism: type of symbiosis in which one member increases its fitness without affecting the other member

community: group of interacting populations that inhabit the same region

compact bone tissue: solid, hard bone tissue consisting of tightly packed concentric rings of bone cells in a mineralized matrix

companion cell: specialized parenchyma cell that transfers materials into and out of a sieve tube element

competition: struggle between organisms for the same limited resource

competitive exclusion principle: the idea that two or more species cannot indefinitely occupy the same niche

competitive inhibition: change in an enzyme's activity occurring when an inhibitor binds to the active site, competing with the enzyme's normal substrate

complement: group of proteins that help destroy pathogens

complementary: in DNA and RNA, the precise pairing of purines (A and G) to pyrimidines (C, T, and U)

complete digestive tract: digestive tract through which food passes in one direction from mouth to anus

compound: a molecule including different elements

compound leaf: leaf that is divided into leaflets

concentration gradient: difference in solute concentrations between two adjacent regions

cone: a pollen- or ovule-bearing structure in many gymnosperms

cone cell: photoreceptor cell in the retina that detects colors

conidium (pl. conidia): asexually produced fungal spore

conifer: type of gymnosperm with cones and evergreen leaves that are needlelike or scalelike

conjugation: a form of horizontal gene transfer in which one cell receives DNA via direct contact with another cell

connective tissue: animal tissue consisting of widely spaced cells in a distinctive extracellular matrix

conservation biology: study of the preservation of biodiversity

constant region: amino acid sequence that is the same for all antibodies

consumer: organism that eats other organisms; heterotroph

contact inhibition: property of most noncancerous eukaryotic cells; inhibits cell division when cells contact one another

contraception: use of devices or practices that prevent pregnancy

control: untreated group used as a basis for comparison with a treated group in an experiment

convergent evolution: the evolution of similar adaptations in organisms that do not share the same evolutionary lineage

coral reef: underwater deposit of calcium carbonate formed by colonies of coral animals

cork cambium: lateral meristem that produces cork cells and parenchyma in woody plant

cornea: in the eye, a modified portion of the sclera that forms a transparent, curved window that admits light

coronary artery: artery that provides blood to the heart muscle

corpus luteum: gland formed from a ruptured ovarian follicle that has recently released an oocyte

cortex: ground tissue between epidermis and vascular tissue in roots and stems

cotyledon: seed leaf in angiosperms

countercurrent exchange: arrangement in which two adjacent currents flow in opposite directions and exchange materials or heat

coupled reactions: two simultaneous chemical reactions, one of which provides the energy that drives the other

covalent bond: type of chemical bond in which two atoms share electrons

craniate: animal with a cranium

cranium: part of the skull that encloses the brain

creatine phosphate: molecule stored in muscle fibers that can donate its high-energy phosphate to ADP, regenerating ATP

crista (pl. cristae): fold of the inner mitochondrial membrane along which many of the reactions of cellular respiration occur

crocodilian: type of reptile that lives in or near water; its elongated snout has eyes on top

cross bridge: connection that forms between a myosin head and actin during muscle contraction

crossing over: exchange of genetic material between homologous chromosomes during prophase I of meiosis

crustacean: type of mandibulate arthropod with two pairs of antennae

culture: the knowledge, beliefs, and behaviors that humans transmit from generation to generation

cuticle: waterproof layer covering the aerial epidermis of a plant

cycad: type of gymnosperm with palmlike leaves and large cones

cytokine: messenger protein synthesized in immune cells that influences the activity of other immune cells

cytokinesis: distribution of cytoplasm into daughter cells in cell division

cytokinin: plant hormone that stimulates cell division, delays leaf senescence, and promotes growth of lateral buds

cytoplasm: the watery mixture that occupies much of a cell's volume. In eukaryotic cells, it consists of all materials, including organelles, between the nuclear envelope and the cell membrane

cytoskeleton: framework of protein rods and tubules in eukaryotic cells

cytosol: the fluid portion of the cytoplasm

cytotoxic T cell: lymphocyte that participates in adaptive immunity by binding to and killing cancerous, damaged, foreign, or infected cells

D

day-neutral plant: plant that does not rely on photoperiod to stimulate flowering

death rate: the number of deaths per individual per unit time

decomposer: organism that releases inorganic nutrients as it externally digests wastes and dead organic matter

deforestation: removal of tree cover from a previously forested area

dehydration synthesis: formation of a covalent bond between two molecules by loss of water

deletion: loss of one or more genes from a chromosome

deletion mutation: removal of one or more nucleotides from a gene

demographic transition: phase of economic development during which birth rates and death rates shift from high to low

denaturation: modification of a protein's shape so that its function is destroyed

dendrite: thin neuron branch that receives neural messages and transmits information to the cell body

denitrification: conversion of nitrate to N_2

dense connective tissue: type of connective tissue with fibroblasts and dense collagen tracts

density-dependent factor: population-limiting condition whose effects increase as a population's density increases

density-independent factor: population-limiting condition that acts irrespective of population density

deoxyribonucleic acid (DNA): genetic material consisting of a double strand of nucleotides, each containing the sugar deoxyribose

dependent variable: response that may be under the influence of an independent variable

deposit feeder: type of substrate feeder that strains partially decayed organic matter from sediment

derived character: characteristic not found in the ancestor of a group being studied

dermal tissue: tissue covering a plant's surface

dermis: layer of connective tissue that lies beneath the epidermis in vertebrate skin

descent with modification: Darwin's term for evolution, describing gradual change from an ancestral type of organism

desert: type of terrestrial biome with very low precipitation and variable temperature; plants are adapted to dry conditions

determinate growth: growth that halts at maturity

detritivore: animal that eats decomposing organic matter

detritus: feces and dead organic matter

deuterostome: clade of bilaterally symmetrical animals in which the first opening in the gastrula develops into the anus

diabetes: disease associated with elevated blood sugar levels

diastolic pressure: lower number in a blood pressure reading; reflects relaxation of the ventricles

diatom: photosynthetic aquatic protist with two-part silica wall

differentiation: process by which cells acquire specialized functions

diffusion: movement of a substance from a region where it is highly concentrated to an area where it is less concentrated

digestion: the physical and chemical breakdown of food

digestive system: organ system consisting of the intestines and other organs that ingest and dismantle food, absorb nutrients, and eliminate wastes

dihybrid cross: mating between two individuals that are heterozygous for two genes

dikaryotic: containing two genetically distinct haploid nuclei

dinoflagellate: unicellular, aquatic protist with two flagella of unequal length; many have cellulose plates

diploid cell: cell containing two full sets of chromosomes, one from each parent; also called $2n$

direct development: development pattern in which a juvenile animal resembles an adult

directional selection: form of natural selection in which one extreme phenotype is fittest, and the environment selects against the others

disaccharide: a simple sugar that consists of two bonded monosaccharides

disruptive selection: form of natural selection in which the two extreme phenotypes are fittest

distal convoluted tubule: region of the renal tubule that connects the nephron loop to the collecting duct; site of reabsorption and secretion

distraction display: behavior that distracts a predator away from an animal's offspring

DNA microarray: collection of short DNA fragments of known sequence placed in defined spots on a small square of glass

DNA polymerase: enzyme that adds new DNA nucleotides and corrects mismatched base pairs in DNA replication

DNA probe: labeled, single-stranded fragment of DNA used to reveal the presence of a complementary DNA sequence

DNA profiling: the use of variable parts of the genome to detect genetic differences between individuals

DNA technology: practical application of knowledge about DNA

domain: broadest (most inclusive) taxonomic category

dominance hierarchy: social order with dominant and submissive members

dominant allele: allele that is expressed whenever it is present

dormancy: state of decreased metabolism

dorsal, hollow nerve cord: tubular nerve cord that forms dorsal to the notochord; one of the four characteristics of chordates

double fertilization: in angiosperms, process by which one sperm nucleus fertilizes the egg and another fertilizes the two haploid nuclei in the embryo sac's central cell

duplication: chromosomal abnormality that produces multiple copies of part of a chromosome

E

ear: sense organ of hearing and equilibrium

eardrum: structure that transmits sound from air to the middle ear

earthworm: type of annelid with a saddlelike thickening near its head; ingests soil

echinoderm: unsegmented deuterostome with a five-part body plan, radial symmetry in adults, and a spiny outer covering; phylum Echinodermata

ecological footprint: measure of the land area needed to support an individual or a population

ecology: study of relationships among organisms and between organisms and the environment

ecosystem: a community and its nonliving environment

ectoderm: outermost germ layer in an animal embryo; develops into skin and nervous system

ectotherm: animal that lacks an internal mechanism that keeps its temperature within a narrow range; invertebrates, fishes, amphibians, and nonavian reptiles

effector: any structure that responds to a stimulus

ejaculation: discharge of semen through the penis

ejaculatory duct: tube that deposits sperm into the urethra

electromagnetic spectrum: range of naturally occurring radiation

electron: a negatively charged particle that orbits the atom's nucleus

electron transport chain: membrane-bound molecular complex that shuttles electrons to slowly extract their energy

electronegativity: an atom's tendency to attract electrons

electrophoresis: technique used to sort DNA fragments by size

element: a pure substance consisting of atoms containing a characteristic number of protons

elimination: the expulsion of waste from the body

embryo sac: female gametophyte in angiosperms

embryonic disc: in the preembryo, a flattened, two-layered mass of cells that develops into the embryo

embryonic stage: stage of human development lasting from the end of the second week until the end of the eighth week of gestation

embryonic stem cell: stem cell that can give rise to all types of cells in the body

emergent property: quality that results from interactions of a system's components

emigration: movement of individuals out of a population

endangered species: species facing a high risk of extinction in the near future

endergonic reaction: a chemical reaction that requires a net input of energy

endocrine gland: concentration of hormone-producing cells in an animal

endocrine system: organ system consisting of glands and cells that secrete hormones

endocytosis: form of transport in which a membrane engulfs substances to bring them into a cell

endoderm: innermost germ layer in an animal embryo; develops into lining of gut and other internal organs

endodermis: the innermost cell layer of root cortex

endomembrane system: eukaryotic organelles that exchange materials in transport vesicles

endometrium: inner uterine lining that is shed during menstruation and supports an embryo during pregnancy

endophyte: fungus that colonizes a plant without triggering disease symptoms

endoplasmic reticulum (ER): interconnected, membranous tubules and sacs in a eukaryotic cell

endorphin: pain-killing peptide produced in the anterior pituitary

endoskeleton: skeleton on the inside of an animal

endosperm: triploid tissue that stores food for the embryo in an angiosperm seed

endospore: dormant, thick-walled structure that enables some bacteria to survive harsh conditions

endosymbiont theory: the idea that mitochondria and chloroplasts originated as free-living bacteria engulfed by other cells

endothelium: layer of epithelial tissue that lines blood vessels and the heart

endotherm: animal that maintains its body temperature by using heat generated from its own metabolism; birds and mammals

energy: the ability to do work

energy shell: group of electron orbitals that share the same energy level

enhancer: DNA sequence that helps regulate gene expression and lies outside the promoter

entropy: randomness or disorder

envelope: layer of phospholipids and proteins surrounding the protein coat of some viruses

environmental resistance: combination of factors that limit population growth

enzyme: an organic molecule that catalyzes a chemical reaction without being consumed

eosinophil: type of white blood cell that attacks multicellular parasites

epidermis (animal): the outermost layer of skin

epidermis (plant): single layer of cells covering the leaves, stem, and roots

epididymis: tube that receives and stores sperm from one testis

epigenetics: The study of changes in gene expression that do not involve changes in the DNA sequence

epiglottis: cartilage that covers the glottis, routing food to the digestive tract during swallowing

epinephrine (adrenaline): hormone secreted by the adrenal medulla; readies the body for short-term stress; also can act as a neurotransmitter

epistasis: one gene masks the phenotype associated with another gene

epithelial tissue (epithelium): animal tissue consisting of tightly packed cells that form linings, coverings, and glands

equilibrium (K-selected) life history: reproductive strategy characterized by long-lived, late-maturing individuals that produce few offspring, with each receiving heavy parental investment

erosion: wearing away of soil by water and wind

esophagus: muscular tube that leads from the pharynx to the stomach

essential nutrient: substance vital for an organism's metabolism, growth, and reproduction

estrogen: steroid sex hormone produced in ovaries of female vertebrates

estuary: area where fresh water in a river meets salty water of an ocean

ethylene: plant hormone that ripens fruit, stimulates senescence, stimulates leaf and fruit shedding, and participates in thigmotropism

eudicot: type of angiosperm; embryo has two cotyledons and pollen grains have three or more pores

euglenoid: unicellular, flagellated protist with elongated cell

Eukarya: domain containing eukaryotes

eukaryote: organism composed of one or more cells containing a nucleus and other membrane-bounded organelles

eumetazoan: animal with true tissues

eusociality: social organization reflecting extensive division of labor in reproduction

eutrophic: nutrient-rich

eutrophication: addition of nutrients to a body of water, triggering an algae bloom

evaporation: the conversion of a liquid to a vapor (gas)

evolution: descent with modification; change in allele frequencies in a population over time

excretion: elimination of metabolic wastes

exergonic reaction: an energy-releasing chemical reaction

exhalation: movement of air out of the lungs

exocytosis: form of transport in which vesicles fuse with the cell membrane to carry materials out of a cell

exon: portion of an mRNA that is translated after introns are removed

exoskeleton: skeleton on the outside of an animal

expanding repeat mutation: type of mutation in which the number of copies of a three- or four-nucleotide sequence increases over several generations

experiment: a test of a hypothesis under controlled conditions

exponential growth: population growth pattern in which the growth rate is proportional to the size of the population

external fertilization: release of gametes by males and females into the same environment; fertilization occurs outside the body

extinction: the death of the last individual of a species

extracellular matrix: nonliving substances that surround animal cells; includes ground substance and fibers

eye: organ that detects light and produces the sense of sight

F

F_1 (first filial) generation: the offspring of the P generation in a genetic cross

F_2 (second filial) generation: the offspring of the F_1 generation in a genetic cross

facilitated diffusion: form of passive transport in which a substance moves down its concentration gradient with the aid of a transport protein

facultative anaerobe: organism that can live with or without O_2

$FADH_2$: coenzyme that carries electrons in respiration

fast-twitch fiber: large-diameter muscle cell that produces twitches of short duration

fatty acid: long-chain hydrocarbon terminating with a carboxyl group

feather: in birds, an epidermal outgrowth composed of keratin

feces: solid wastes that leave the digestive tract

fermentation: metabolic pathway in which NADH from glycolysis reduces pyruvate

fertilization: the union of two gametes

fetal stage: stage of human development lasting from the beginning of the ninth week of gestation through birth

fever: rise in the body's temperature

fiber: elongated sclerenchyma cell

fibrous root system: branching root system arising from stem

filtration: removal of water and solutes from the blood, as occurs at the glomerulus

first law of thermodynamics: energy cannot be created or destroyed, just converted from one form to another

fish: vertebrate animal with fins and external gills

fitness: an organism's contribution to the next generation's gene pool

fixed action pattern (FAP): type of innate behavior; stereotyped sequence of events that is performed to completion once initiated

flagellated protozoan: unicellular heterotrophic protist with one or more flagella

flagellum (pl. flagella): a long, whiplike appendage a cell uses for motility

flatworm: unsegmented worm lacking a coelom; phylum Platyhelminthes

flower: the reproductive structure in angiosperms that produces pollen and eggs

fluid feeder: animal that drinks its food

fluid mosaic: two-dimensional structure of movable phospholipids and proteins that form biological membranes

fluke: type of parasitic flatworm

follicle cell: nourishing cell surrounding an oocyte

follicle-stimulating hormone (FSH): hormone produced in the anterior pituitary; stimulates secretion of sex hormones in both sexes

food chain: linear series of organisms in which those in one level eat those at a lower level

food web: network of interconnecting food chains

foot (mollusk): ventral muscular structure that provides movement in mollusks

foraging: animal behavior related to finding and collecting food

foraminiferan: amoeboid protozoan with a calcium carbonate shell

forebrain: front part of the vertebrate brain

fossil: any evidence of an organism from more than 10,000 years ago

founder effect: genetic drift that occurs when a small, nonrepresentative group of individuals leaves their ancestral population and begins a new settlement

frameshift mutation: type of mutation in which nucleotides are added or deleted by any number other than a multiple of three, altering the reading frame

free-living flatworm: planarian or marine flatworm

frog: type of amphibian with a large mouth and no neck or tail

fruit: seed-containing structure in angiosperms

fruiting body: organ that produces sexual spores in a fungus

Fungi: kingdom containing mostly multicellular eukaryotes that are heterotrophs by external digestion

G

G$_0$ phase: resting phase of the cell cycle in which the cell continues to function but does not divide

G$_1$ phase: gap stage of interphase in which the cell grows and carries out its basic functions

G$_2$ phase: gap stage in interphase in which the cell makes its final preparations for division

gallbladder: organ that stores bile from the liver and releases it into the small intestine

gamete: a sex cell; sperm or egg cell

gametophyte: haploid, gamete-producing stage of the plant life cycle

ganglion: cluster of neurons

ganglion cell: interneuron in the retina that generates action potentials

gap junction: connection between two adjacent animal cells that allows cytoplasm to flow between them

gastric juice: mixture of water, mucus, salts, hydrochloric acid, and enzymes produced at the stomach lining

gastrointestinal (GI) tract: two-opening digestive tract; also called alimentary canal

gastropod: type of mollusk with a broad, flat "stomach-foot"

gastrovascular cavity: digestive chamber with a single opening

gastrula: early animal embryo consisting of two (cnidarians) or three (other animals) tissue layers

gene: sequence of DNA that codes for a specific protein or RNA molecule

gene flow: the movement of alleles between populations

gene pool: all of the genes and their alleles in a population

gene therapy: treatment that supplements a faulty gene in a cell with a functioning version of the gene

genetic code: correspondence between specific nucleotide sequences and amino acids

genetic drift: change in allele frequencies that occurs purely by chance

genome: all the genetic material in an organism

genotype: an individual's combination of alleles for a particular gene

genotype frequency: number of individuals of one genotype, divided by the number of individuals in the population

genus: taxonomic category that groups closely related species

geologic timescale: a division of Earth's history into eons, eras, periods, and epochs defined by major geological or biological events

germ cell: specialized cell that gives rise to gametes

germination: resumption of growth after seed dormancy is broken

germline mutation: a DNA sequence change that occurs in the cells that give rise to gametes

gibberellin: plant hormone that breaks seed dormancy and stimulates cell division in roots, shoots, and leaves

gill: highly folded respiratory surface containing blood vessels that exchange gases in water

ginkgo: type of gymnosperm with fan-shaped leaves

gland: organ that secretes substances into the bloodstream or into a duct

global climate change: long-term changes in Earth's weather patterns

glomeromycete: mycorrhiza-forming fungus lacking sexual spores

glomerular (Bowman's) capsule: cup-shaped site of filtration at one end of a nephron

glomerulus: ball of capillaries containing blood to be filtered at a nephron

glottis: slitlike opening between the vocal cords

glucagon: pancreatic hormone that raises blood sugar level by stimulating liver cells to break down glycogen into glucose

glucocorticoid: hormone secreted by the adrenal cortex; among other functions, boosts glucose levels in blood and brain in response to long-term stress

glycerol: a three-carbon alcohol that forms the backbone of triglycerides and phospholipids

glycocalyx: sticky layer composed of proteins and/or polysaccharides that surrounds some prokaryotic cell walls; slime layer or capsule

glycolysis: a metabolic pathway occurring in all cells; one molecule of glucose splits into two molecules of pyruvate

gnetophyte: type of gymnosperm with some characteristics resembling flowering plants

golden alga: photosynthetic aquatic protist with two flagella and yellowish accessory pigments

Golgi apparatus: a system of flat, stacked, membrane-bounded sacs that packages cell products for export

gonad: gland that manufactures hormones and gametes in animals; ovary or testis

gonadotropin-releasing hormone (GnRH): reproductive hormone produced in the hypothalamus in both sexes; stimulates the release of FSH and LH from the anterior pituitary

graded potential: a local flow of electrical current in a neuron; the flow weakens with distance from the source of the stimulus

gradualism: theory that proposes that evolutionary change occurs gradually, in a series of small steps

Gram stain: technique for classifying bacteria into two main groups based on cell wall structure

granum (pl. grana): a stack of flattened thylakoid discs in a chloroplast

gravitropism: directional growth response to gravity

gray matter: nervous tissue in the central nervous system; consists mostly of neuron cell bodies, dendrites, and synapses

green alga: photosynthetic protist that has pigments, starch, and cell walls similar to those of land plants

greenhouse effect: increase in surface temperature caused by carbon dioxide and other heat-trapping atmospheric gases

ground tissue: plant tissue that makes up most of the primary plant body; composed mostly of parenchyma

growth factor: a protein that signals a cell to divide

growth hormone (GH): hormone produced in the anterior pituitary; stimulates tissue growth

guard cells: pair of cells flanking a stoma

gustation: the sense of taste

gymnosperm: a plant with seeds that are not enclosed in a fruit; includes conifers, *Ginkgo*, gnetophytes, and cycads

H

habitat: physical place where an organism normally lives

habituation: type of learning in which an animal learns not to respond to irrelevant stimuli

hagfish: type of invertebrate chordate with a cranium but no jaws

hair cell: mechanoreceptor that initiates sound transduction in the cochlea

half-life: the time it takes for half the atoms in a sample of a radioactive substance to decay

haploid cell: cell containing one set of chromosomes; also called *n*

Hardy–Weinberg equilibrium: situation in which allele frequencies and genotype frequencies do not change from one generation to the next

heart: muscular organ that pumps blood (or a comparable fluid) throughout the body

heartwood: dark-colored, nonfunctioning secondary xylem in woody plant

helper T cell: lymphocyte that coordinates activities of other immune system cells

hemisphere: one of two halves of the cerebrum

hemoglobin: pigment that carries oxygen in red blood cells

herbaceous plant: plant with green, soft stem at maturity

herbivore: animal that eats plants

heterotroph: organism that obtains carbon and energy by eating another organism; consumer

heterozygote advantage: condition in which a heterozygote has greater fitness than homozygotes, maintaining balanced polymorphism in a population

heterozygous: possessing two different alleles for a particular gene

hindbrain: lower, posterior portion of the vertebrate brain

hippocampus: forebrain structure involved in memory formation

histamine: biochemical that dilates blood vessels and increases their permeability; involved in inflammation and allergies

homeostasis: a state of internal constancy in the presence of changing external conditions

homeotic: describes any gene that, when mutated, leads to organisms with structures in the wrong places

hominid: any of the "great apes" (orangutans, gorillas, chimpanzees, and humans)

hominin: humans and their extinct ancestors on the human branch of the primate evolutionary tree

hominoid: any lesser or great ape, including humans

homologous: similar in structure or position because of common ancestry

homologous pair: two chromosomes that look alike and have the same sequence of genes

homozygous: possessing two identical alleles for a particular gene

horizontal gene transfer: transfer of genetic information from one cell to another cell that is not its descendant

hormone: biochemical synthesized in small quantities in one place and transported to another

hornwort: type of bryophyte with a tapered, hornlike sporophyte

horseshoe crab: type of chelicerate arthropod with a hard, horseshoe-shaped exoskeleton

horsetail: type of seedless vascular plant with abrasive stems and leaves

host range: the kinds of organisms or cells that a virus can infect

human chorionic gonadotropin (hCG): hormone secreted by tissue of the placenta; indirectly prevents menstruation

human immunodeficiency virus (HIV): virus that causes acquired immune deficiency syndrome (AIDS)

humoral immunity: branch of adaptive immune system in which B cells secrete antibodies in response to a foreign antigen

humus: chemically complex, jellylike organic substance in soil

hybrid (genetics): producing a mix of offspring for one or more traits; heterozygous

hybrid (speciation): the offspring of individuals from two different species

hydrogen bond: weak chemical bond between opposite partial charges on two molecules or within one large molecule

hydrolysis: splitting a molecule by adding water

hydrophilic: attracted to water

hydrophobic: repelled by water

hydrostatic skeleton: skeleton consisting of constrained fluid in a closed body compartment

hypertonic: describes a solution in which the solute concentration is higher than on the other side of a selectively permeable membrane

hypha (pl. hyphae): a fungal filament; the basic structural unit of a multicellular fungus

hypoglycemia: low blood sugar caused by excess insulin or insufficient carbohydrate intake

hypothalamus: small forebrain structure beneath the thalamus that controls homeostasis and links the nervous and endocrine systems

hypothesis: a testable, tentative explanation based on prior knowledge

hypotonic: describes a solution in which the solute concentration is lower than on the other side of a selectively permeable membrane

I

immigration: movement of individuals into a population

immune system: organ system consisting of specialized cells that defend the body against infections, cancer, and foreign substances

immunodeficiency: condition in which the immune system lacks one or more components

impact theory: idea that mass extinctions were caused by impacts of extraterrestrial origin

implantation: embedding of the blastocyst into the uterine lining

imprinting: type of learning that occurs during a sensitive period early in life and occurs without obvious reinforcement

inclusive fitness: the sum of an individual's direct and indirect fitness

incomplete digestive tract: digestive tract with one opening that takes in food and ejects wastes

incomplete dominance: mode of inheritance in which a heterozygote's phenotype is intermediate between the phenotypes of the two homozygotes

independent variable: a factor that is hypothesized to influence a dependent variable

indeterminate growth: growth that persists indefinitely

indirect development: development of a juvenile animal into an adult while passing through intervening larval stages

inflammation: immediate, localized reaction to an injury or any pathogen that breaches the body's barriers

ingestion: the act of taking food into the digestive tract

inhalation: movement of air into the lungs

inheritance of acquired characteristics: the idea that an organism can inherit the traits that its parent acquired during its lifetime

innate: describing an instinctive behavior that occurs the same way each time, independently of experience

innate defense: cell or substance that provides generalized protection against all infectious agents

inner cell mass: cells in the blastocyst that develop into the embryo

insect: type of mandibulate arthropod with one pair of antennae, a three-part body, six legs, and wings

insertion mutation: addition of one or more nucleotides to a gene

insulin: pancreatic hormone that lowers blood sugar level by stimulating body cells to take up glucose from the blood

integumentary system: organ system consisting of skin and its outgrowths, which sense the external environment and protect the body

intercalary meristem: meristem between the nodes of a mature stem

interferon: type of cytokine released by a virus-infected cell

interleukin: type of cytokine involved in communication between white blood cells

intermediate filament: component of the cytoskeleton; intermediate in size between a microtubule and a microfilament

intermembrane compartment: the space between a mitochondrion's two membranes

internal fertilization: use of a copulatory organ to deposit sperm inside a female's body

interneuron: neuron that connects one neuron to another in the central nervous system

internode: stem area between the points of leaf attachment

interphase: stage preceding mitosis or meiosis, when the cell carries out its functions, replicates its DNA, and grows

intersexual selection: choice of mates by one sex from among competing members of the opposite sex

interstitial cell: testosterone-secreting endocrine cell in a testis

interstitial fluid: liquid that bathes cells in an animal body

intertidal zone: region along a coastline between the high and low tide marks

intestinal villus (pl. villi): tiny projection on the inner lining of the small intestine

intrasexual selection: competition between members of the same sex for access to the opposite sex

intron: portion of an mRNA molecule that is removed before translation

invasive species: introduced species that establishes a breeding population in a new location and spreads widely from the original point of introduction

inversion: abnormality in which a portion of a chromosome flips and reinserts itself

invertebrate: animal without a backbone

ion: an atom or group of atoms that has lost or gained electrons, giving it an electrical charge

ionic bond: attraction between oppositely charged ions

iris: colored part of the eye; regulates the size of the pupil

isotonic: condition in which a solute concentration is the same on both sides of a selectively permeable membrane

isotope: any of the forms of an element, each having a different number of neutrons in the nucleus

J

jaws: bones that frame the entrance to the mouth

joint: area where two bones meet

J-shaped curve: plot of exponential growth over time

K

karyotype: a size-ordered chart of the chromosomes in a cell

keystone species: species whose effect on community structure is disproportionate to its biomass

kidney: excretory organ in the vertebrate urinary system

kilocalorie (kcal): a measure of the energy content in food; equal to one thousand calories or one food Calorie

kin selection: the sacrifice of one individual's fitness for the sake of the genes it shares with related animals

kinetic energy: energy of motion

kinetochore: protein that attaches a chromosome to the spindle in cell division

kingdom: taxonomic category below domain

Krebs cycle: stage in cellular respiration that completely oxidizes the products of glycolysis

L

lac **operon:** in *E. coli,* three lactose-degrading genes plus the promoter and operator that control their transcription

lactic acid fermentation: metabolic pathway in which NADH from glycolysis reduces pyruvate, producing lactic acid

lamprey: type of vertebrate chordate with a cartilage skeleton and cranium but no jaws

lancelet: type of invertebrate chordate with an elongated body and no heart

large intestine: part of the digestive tract that connects the small intestine to the anus; absorbs water and salts and eliminates wastes

larva: immature stage of animal development; does not resemble the adult of the species

larynx: boxlike structure in front of the pharynx

latent: describes an infection in which viral genetic material in a host cell is not expressed

latent infection: for viruses that infect animal cells, an infection in which viral genetic material is replicated along with that of a dividing host cell, but new viruses are not produced

lateral line: network of canals that extends along the sides of fishes and houses receptor organs that detect vibrations

lateral meristem: meristem whose daughter cells thicken a root or stem

law of independent assortment: Mendel's law stating that during gamete formation, the segregation of the alleles for one gene does not influence the segregation of the alleles for another gene

law of segregation: Mendel's law stating that the two alleles of each gene are packaged into separate gametes

leaf: flattened organ that carries out photosynthesis

learning: alteration of an animal's behavior to reflect experience

leech: type of annelid with a saddlelike thickening near its head; typically lives in fresh water and eats other animals

lens: structure in the eye that bends incoming light

leukemia: cancer in which bone marrow overproduces white blood cells

lichen: association of a fungus and a green alga or cyanobacterium

life history: the events of an organism's life, especially those that are related to reproduction

life table: chart that shows the probability of surviving to any given age

ligament: band of fibrous connective tissue that connects bone to bone across a joint

ligase: enzyme that catalyzes formation of covalent bonds in the DNA sugar–phosphate backbone

light reactions: photosynthetic reactions that harvest light energy and store it in molecules of ATP or NADPH

lignin: tough, complex molecule that strengthens the walls of some plant cells

limbic system: collection of forebrain structures involved in emotion and memory

limnetic zone: layer of open water in a lake or pond where light penetrates

linkage group: group of genes that tend to be inherited together because they are on the same chromosome

linkage map: diagram of gene order and spacing on a chromosome, based on crossover frequencies

linked genes: genes on the same chromosome

lipid: hydrophobic organic molecule consisting mostly of carbon and hydrogen

littoral zone: shallow region along a shoreline where rooted plants occur

liver: organ that produces bile; detoxifies blood; stores iron and some vitamins; synthesizes blood proteins; and stores and releases glucose

liverwort: type of bryophyte with upright or flattened gametophytes

lizard: type of reptile with legs, external ear openings, and movable eyelids

lobe-finned fish: type of bony fish with fleshy fins; closely related to tetrapods

locus: physical location of a gene on a chromosome

logistic growth: population growth pattern in which the growth rate levels off as environmental resistance grows

long-day (short-night) plant: plant that flowers when dark periods are shorter than a critical length

long-term memory: memory that can last from hours to a lifetime

loose connective tissue: type of connective tissue with widely spaced fibroblasts and a loose network of fibers

lung: saclike structure where gas exchange occurs in air-breathing vertebrates

lungfish: type of lobe-finned fish with lungs

luteinizing hormone (LH): hormone produced in the anterior pituitary; stimulates secretion of sex hormones in both sexes

lymph: colorless fluid that originates when interstitial fluid enters lymph capillaries

lymph capillary: dead-end vessel that collects and distributes lymph

lymph node: one of many lymphoid organs located along lymph capillaries; contains white blood cells and fights infection

lymphatic system: organ system consisting of lymphoid organs and lymph vessels that recover excess tissue fluid and aid in immunity

lymphocyte: type of white blood cell; T cell, or B cell, or natural killer cell

lysogenic infection: type of infection in which the genetic material of a bacteriophage is replicated along with the host cell's chromosome

lysosome: organelle in a eukaryotic cell that buds from the Golgi apparatus and enzymatically dismantles molecules, bacteria, and worn-out cell parts

lytic infection: type of infection in which a bacteriophage enters a cell, replicates, and causes the host cell to burst (lyse) as it releases the new viruses

M

macroevolution: large-scale evolutionary change

macronutrient: nutrient required in large amounts

macrophage: type of phagocyte that also helps initiate the adaptive immune response

major histocompatibility complex (MHC): cluster of genes that encode proteins that display antigens on cell surfaces

malignant tumor: mass of abnormal cells that has the potential to invade adjacent tissues and spread throughout the body

mammal: type of tetrapod vertebrate with hair and mammary glands; embryo is enclosed in an amnion

mammary gland: milk-producing gland in mammals

mandibulate: arthropod with jawlike mouthparts (mandibles); crustaceans, insects, centipedes, and millipedes

mantle: dorsal fold of tissue that secretes a shell in most mollusks

marrow cavity: space in a bone shaft that contains marrow

marsupial: type of mammal that bears live young, which typically complete development in the mother's pouch

marsupium: pouch in which the immature young of many marsupial mammals nurse and develop

mass extinction: the disappearance of many species over relatively short expanses of time

mass number: the total number of protons and neutrons in an atom's nucleus

mast cell: immune system cell that triggers inflammation and allergy; settles in tissues rather than circulating in the blood

matrix: the inner compartment of a mitochondrion

matter: substance that takes up space and is made of atoms

mechanoreceptor: sensory receptor sensitive to physical deflection

Mediterranean shrubland (chaparral): type of terrestrial biome with rainy, mild winters and dry, hot summers; dominant plants have thick bark and leathery leaves

medulla oblongata: part of the brainstem that controls many involuntary vital functions

medusa: free-swimming form of a cnidarian

megaspore: in seed plants, spore that gives rise to female gametophyte

meiosis: division of genetic material that halves the chromosome number and yields genetically variable nuclei

melatonin: hormone produced in the pineal gland; regulates effects of light–dark cycles

membrane potential: difference in electrical charge across a neuron's cell membrane

memory cell: lymphocyte produced in an initial infection; launches a rapid immune response upon subsequent exposure to an antigen

meninges: membranes that cover and protect the central nervous system

menstrual cycle: hormonal cycle that prepares the uterus for pregnancy

meristem: localized region of active cell division in a plant

mesoderm: embryonic germ layer between ectoderm and endoderm in an animal embryo; develops into muscles, the skeleton, and the circulatory system

mesophyll: photosynthetic tissue in a leaf's interior

messenger RNA (mRNA): a molecule of RNA that encodes a protein

metabolism: the biochemical reactions of a cell

metamorphosis: developmental process in which an animal changes drastically in body form between the juvenile and the adult

metaphase: stage of mitosis in which chromosomes are aligned down the center of a cell

metaphase I: metaphase of meiosis I, when homologous chromosome pairs align down the center of a cell

metaphase II: metaphase of meiosis II, when chromosomes line up down the center of a cell

metastasis: spreading of cancer

microevolution: relatively short-term changes in allele frequencies within a population or species

microfilament: component of the cytoskeleton; made of the protein actin

micronutrient: nutrient required in small amounts

microspore: in seed plants, spore that gives rise to male gametophyte

microtubule: component of the cytoskeleton; made of subunits of tubulin protein

microvillus (pl. microvilli): extension of the cell membrane that increases the surface area of an epithelial cell of a villus

midbrain: part of the brain between the forebrain and hindbrain

millipede: type of mandibulate arthropod with two pairs of legs per repeating subunit

mineral: essential element other than C, H, O, or N

mineralocorticoid: hormone secreted by the adrenal cortex; maintains blood volume and salt balance

mitochondrion (pl. mitochondria): organelle that houses the reactions of cellular respiration in eukaryotes

mitosis: division of genetic material that yields two genetically identical nuclei

modern evolutionary synthesis: the idea that genetic mutations create the variation upon which natural selection acts

molecular clock: application of the rate at which DNA mutates to estimate when two types of organisms diverged from a shared ancestor

molecule: two or more atoms joined by chemical bonds

mollusk: unsegmented animal with a soft body, mantle, muscular foot, and visceral mass; phylum Mollusca

monocot: type of angiosperm; embryo has one cotyledon and pollen grains have one pore

monocyte: type of white blood cell that gives rise to macrophages

monogamy: mating system in which males and females have one sexual partner

monohybrid cross: mating between two individuals that are heterozygous for the same gene

monomer: a single unit of a polymeric molecule

monophyletic: describes a group of organisms consisting of a common ancestor and all of its descendants

monosaccharide: a sugar that is one five- or six-carbon unit

monotreme: type of mammal that lays eggs

morula: preembryonic stage consisting of a solid ball of cells

moss: type of bryophyte with leaflike gametophytes topped with stalklike sporophytes

motor neuron: neuron that transmits a message from the central nervous system toward a muscle or gland

motor unit: a motor neuron and all the muscle fibers it contacts

muscle: organ that powers movements in animals by contracting; consists of muscle tissue and other tissue types

muscle fiber: muscle cell

muscle tissue: animal tissue consisting of contractile cells that provide motion

muscular system: organ system consisting of skeletal muscles whose contractions form the basis of movement and posture

mutagen: any external agent that causes a mutation

mutant: an allele, genotype, or phenotype that is not the most common in a population or that has been altered from the "typical" (wild-type) condition

mutation: a change in a DNA sequence

mutualism: type of symbiosis that improves the fitness of both partners

mycelium: assemblage of hyphae that forms an individual fungus

mycorrhiza: mutually beneficial association of a fungus and the roots of a plant

myelin sheath: fatty material that insulates some nerve fibers in vertebrates, speeding nerve impulse transmission

myocardium: thick, muscular layer of the heart wall

myofibril: cylindrical subunit of a muscle fiber, consisting of parallel protein filaments

myosin: protein that forms thick filaments in muscle cells

N

NADH: coenzyme that carries electrons in glycolysis and other reactions of cellular respiration

NADPH: coenzyme that carries electrons in photosynthesis

natural killer cell: type of lymphocyte that participates in innate defenses by destroying cancerous or virus-infected cells

natural selection: differential reproduction of organisms based on inherited traits

negative feedback: regulatory mechanism in which a change in a condition triggers action that reverses the change

nephron: functional unit of the kidney; adjusts composition of blood and produces urine

nephron loop: hairpin-shaped region of the renal tubule where water and ions are reabsorbed into blood

neritic zone: region of an ocean from the coast to the edge of the continental shelf

nerve: bundle of nerve fibers (axons) bound together in a sheath of connective tissue

nerve ladder: nervous system consisting of two nerve cords connected by transverse nerves

nerve net: diffuse network of neurons in cnidarians

nervous system: organ system consisting of nerves, the brain, and other structures that specialize in rapid communication

nervous tissue: tissue type whose cells (neurons and neuroglia) form a communication network

net primary production: energy available to consumers in a food chain, after cellular respiration and heat loss by producers

neural tube: embryonic precursor of the central nervous system

neuroglia: type of cell in nervous tissue; supports and assists neurons

neuron: type of cell in nervous tissue; receives, processes, and transmits information via electrochemical signals

neurotransmitter: chemical passed from a neuron to receptors on another neuron or on a muscle or gland cell

neutral (solution): neither acidic nor basic

neutron: a particle in an atom's nucleus that is electrically neutral

neutrophil: the most abundant type of white blood cell; functions as a phagocyte

niche: all resources a species uses for survival, growth, and reproduction

nitrification: conversion of ammonium to nitrate

nitrogen fixation: conversion of N_2 to NH_4^+, a form plants can use

nitrogenous base: a nitrogen-containing compound that forms part of a nucleotide

node: point at which leaves attach to a stem

node of Ranvier: short region of exposed axon between two sections of myelin sheath

nodule: root growth housing nitrogen-fixing bacteria

noncompetitive inhibition: change in an enzyme's shape occurring when an inhibitor binds to a site other than the active site

nondisjunction: failure of chromosomes to separate at anaphase I or anaphase II of meiosis

nonpolar covalent bond: a covalent bond in which atoms share electrons equally

norepinephrine (noradrenaline): hormone secreted by the adrenal medulla; readies the body for short-term stress; also can act as a neurotransmitter

nose: organ that forms the entrance to the nasal cavity inside the head; functions in breathing and olfaction

notochord: flexible rod extending along a chordate's back; one of the four characteristics of chordates

nuclear envelope: the two membranes bounding a cell's nucleus

nuclear pore: a hole in the nuclear envelope

nucleic acid: a long polymer of nucleotides; DNA or RNA

nucleoid: the part of a prokaryotic cell where the DNA is located

nucleolus: a structure within the nucleus where components of ribosomes are assembled

nucleosome: the basic unit of chromatin; consists of DNA wrapped around eight histone proteins

nucleotide: building block of a nucleic acid, consisting of a phosphate group, a nitrogenous base, and a five-carbon sugar

nucleus (atom): central part of an atom

nucleus (cell): the membrane-bounded sac that contains DNA in a eukaryotic cell

nutrient: any substance that an organism uses for metabolism, growth, maintenance, and repair of its tissues

O

obesity: condition characterized by an unhealthy amount of body fat; body mass index greater than 30

obligate aerobe: organism that requires O_2 for generating ATP

obligate anaerobe: organism that must live in the absence of O_2

observational learning: type of learning in which an animal imitates the behavior of others

oceanic zone: open sea beyond the continental shelf

olfaction: the sense of smell

oligodendrocyte: type of neuroglia that forms a myelin sheath around some axons in the central nervous system

oligosaccharide: intermediate-length carbohydrate consisting of 3 to 100 monosaccharides

oligotrophic: low in nutrients

omnivore: animal that eats many types of food, including plants and animals

oncogene: gene that normally stimulates cell division but when overexpressed leads to cancer

oogenesis: the production of egg cells

oogonium: diploid germ cell in an ovary; divides mitotically to yield two primary oocytes

open circulatory system: circulatory system in which fluid circulates freely through the body cavity (is not confined to vessels)

operant conditioning: type of associative learning in which an animal connects a behavior with its consequences

operator: in an operon, the DNA sequence between the promoter and the protein-encoding regions

operon: group of related bacterial genes plus a promoter and operator that control the transcription of the entire group at once.

opportunistic (*r*-selected) life history: reproductive strategy characterized by short-lived, early-maturing individuals that have many offspring, with each receiving little parental investment

opportunistic pathogen: infectious agent that normally does not cause disease in a person with a healthy immune system

optic nerve: nerve fibers that connect the retina to the brain

optimal foraging theory: theory that predicts an animal should maximize energy collection per unit time

orbital: volume of space where a particular electron is likely to be

organ: two or more tissues that interact and function as an integrated unit

organ system: two or more physically or functionally linked organs

organelle: compartment of a eukaryotic cell that performs a specialized function

organic molecule: compound containing both carbon and hydrogen

organism: a single living individual

orgasm: pleasurable sensation, accompanied by involuntary muscle contractions, associated with sexual activity

osmoregulation: control of an animal's ion concentration

osmosis: simple diffusion of water through a selectively permeable membrane

osteocyte: mature bone cell surrounded by matrix

osteoporosis: condition in which bones become less dense

ostracoderm: extinct jawless fish

outgroup: basis for comparison in a cladistics analysis

oval window: membrane between the middle ear and the inner ear

ovarian cycle: hormonal cycle that controls the timing of oocyte maturation in the ovaries

ovary (animal): female gonad; organ that produces egg cells and hormones

ovary (plant): base of a carpel in a flower; encloses one or more ovules

overexploitation: harvesting members of a species faster than they can reproduce

ovulation: release of an oocyte from an ovarian follicle

ovule: egg-bearing structure that develops into a seed in gymnosperms and angiosperms

oxidation: the loss of one or more electrons by a participant in a chemical reaction

oxidation–reduction (redox) reaction: chemical reaction in which one reactant is oxidized and another is reduced

oxygen debt: after vigorous exercise, a period in which the body requires extra oxygen to restore ATP and creatine phosphate to muscle and to recharge oxygen-carrying proteins

oxytocin: hormone released from the posterior pituitary; stimulates muscle contraction in breasts and uterus

ozone layer: atmospheric zone rich in ozone gas (O_3), which absorbs the sun's ultraviolet radiation

P

P (parental) generation: the first, true-breeding generation in a genetic cross

pacemaker: specialized cardiac muscle cells that set the pace of the heartbeat; the sinoatrial (SA) node

pain receptor: sensory receptor that detects mechanical damage, temperature extremes, or chemicals released from damaged cells

paleontology: the study of fossil remains or other clues to past life

pancreas: gland between the spleen and the small intestine; produces hormones, digestive enzymes, and bicarbonate

pancreatic islet (islet of Langerhans): cluster of cells in the pancreas that secretes insulin and glucagon

parapatric speciation: formation of new species when part of a population enters a habitat bordering the parent species' range, and the two groups become reproductively isolated

paraphyletic: describes a group of organisms that contains a common ancestor and some, but not all, of its descendants

parasitism: type of symbiosis in which one member increases its fitness at the expense of a living host

parasympathetic nervous system: part of the autonomic nervous system; dominates during relaxed times and opposes the sympathetic nervous system

parathyroid gland: one of four small groups of cells behind the thyroid gland; secretes parathyroid hormone

parathyroid hormone (PTH): hormone produced in the parathyroid gland; stimulates increase in blood calcium level

parazoan: animal without true tissues; a sponge

parenchyma: unspecialized plant cells making up majority of ground tissue

parental chromatid: chromatid containing genetic information from only one parent

parsimonious: in cladistics, describes the evolutionary tree that requires the fewest steps to construct from a set of observations

particulate: small piece of matter suspended in air

passive immunity: immunity generated when one individual receives antibodies from another individual

passive transport: movement of a solute across a membrane without the direct expenditure of energy

pathogen: disease-causing agent

pattern formation: developmental process that establishes the body's overall shape and structure

pectoral girdle: bones that connect the forelimbs to the axial skeleton

pedigree: chart showing family relationships and phenotypes

peer review: evaluation of scientific results by experts before publication in a journal

pelagic zone: all water above the ocean floor

pelvic girdle: bones that connect the hindlimbs to the axial skeleton

penis: cylindrical male organ of copulation and urination

pepsin: enzyme that begins the digestion of proteins in the stomach

peptide bond: a covalent bond between adjacent amino acids; results from dehydration synthesis

peptide hormone: water-soluble hormone consisting of a short chain of amino acids; typically binds to a receptor on the cell surface

peptidoglycan: material in bacterial cell wall

per capita rate of increase (r): difference between the birth rate and the death rate in a population

perception: the brain's interpretation of a sensation

pericardium: connective tissue sac that encloses the heart

pericycle: outermost layer of root vascular cylinder; produces branch roots

periderm: dermal tissue covering woody plant part; consists of cork, cork cambium, and (in some plants) parenchyma cells

periodic table: chart that lists elements according to their properties

peripheral nervous system: portion of the nervous system that transmits information to and from the central nervous system

peristalsis: waves of muscle contraction that propel food along the digestive tract

peritubular capillaries: blood vessels that surround and exchange materials with a nephron

permafrost: permanently frozen ground in tundra

peroxisome: membrane-bounded sac that houses enzymes that break down fatty acids and dispose of toxic chemicals

persistent organic pollutant: carbon-based chemical pollutants that remain in ecosystems for long periods

petal: flower part interior to sepals

petiole: stalk that supports a leaf blade

pH scale: a measurement of how acidic or basic a solution is

phagocyte: cell that engulfs and digests foreign material and cell debris

phagocytosis: form of endocytosis in which the cell engulfs a large particle

pharyngeal slit (or pouch): opening in the pharynx of a chordate embryo; one of the four characteristics of chordates

pharynx: tube just behind the oral and nasal cavities; the throat

phenotype: observable characteristic of an organism

pheromone: chemical an organism releases that elicits a response in another member of the species

phloem: vascular tissue that transports sugars and other dissolved organic substances in plants

phloem sap: solution of water, minerals, sucrose, and other biochemicals in phloem

phospholipid: molecule consisting of glycerol, two hydrophobic fatty acids, and a hydrophilic phosphate group

phospholipid bilayer: double layer of phospholipids that forms in water; forms the majority of a cell's membranes

phosphorylation: the addition of a phosphate to a molecule

photic zone: region in a water body where light is sufficient for photosynthesis

photon: a packet of light or other electromagnetic radiation

photoperiod: day length

photoreceptor: molecule or cell that detects quality and quantity of light

photorespiration: a metabolic pathway in which rubisco reacts with O_2 instead of CO_2, counteracting photosynthesis

photosynthesis: biochemical reactions that enable organisms to harness sunlight energy to manufacture organic molecules

photosystem: cluster of pigment molecules and proteins in a chloroplast's thylakoid membrane

phototroph: organism that derives energy from sunlight

phototropism: directional growth response to unidirectional light

phylogenetics: field of study that attempts to explain the evolutionary relationships among species

phylogeny: graphical depiction of evolutionary relationships among species

physiology: the study of the functions of organisms and their parts

phytochrome: a type of photoreceptor in plants

phytoplankton: microscopic photosynthetic organisms that drift in water

pilus (pl. pili): short projection made of protein on a prokaryotic cell

pineal gland: small gland in the brain that secretes melatonin

pioneer species: the first species to colonize an area devoid of life

pith: ground tissue inside a ring of vascular bundles in roots and stems

pituitary gland: pea-sized endocrine gland attached to the hypothalamus; secretes hormones that influence other endocrine glands

placebo: inert substance used as an experimental control

placenta: structure that connects the developing fetus to the maternal circulation in placental mammals

placental mammal: type of mammal in which the developing fetus is nourished by a placenta

placoderm: extinct armored fishes with jaws

Plantae: kingdom consisting of multicellular, eukaryotic autotrophs

plasma: watery, protein-rich fluid that forms the matrix of blood

plasma cell: B cell that secretes large quantities of one antibody

plasmid: small circle of double-stranded DNA separate from a cell's chromosome

plasmodesma (pl. plasmodesmata): connection that allows cytoplasm to flow between adjacent plant cells

plasmodial slime mold: protist in which the feeding stage consists of a plasmodium containing multiple nuclei

plate tectonics: theory that Earth's surface consists of several plates that move in response to forces acting deep within the planet

platelet: cell fragment that orchestrates clotting in blood

pleiotropy: multiple phenotypic effects of one genotype

polar body: small, haploid byproduct of female meiosis; typically cannot be fertilized

polar covalent bond: a covalent bond in which electrons are attracted more to one atom's nucleus than to the other

polar ice: type of terrestrial biome that is cold, dry, and windy year-round; phytoplankton in ice and surrounding water are the primary producers

pollen: male gametophyte in seed plants (gymnosperms and angiosperms)

pollen sac: pollen-producing cavity in anther

pollination: transfer of pollen to female reproductive part

pollution: physical, chemical, or biological change in the environment that harms organisms

polychaete: type of marine annelid with paddle-shaped appendages

polygamy: mating system in which males or females have multiple sexual partners

polygenic: caused by more than one gene; polygenic traits are typically expressed as a continuum of possible phenotypes

polymer: a long molecule composed of similar subunits (monomers)

polymerase chain reaction (PCR): biotechnology tool that rapidly produces millions of copies of a DNA sequence of interest

polyp: sessile form of a cnidarian

polypeptide: a long polymer of amino acids; it is called a protein once it folds into its functional shape

polyphyletic: describes a group of organisms that excludes the most recent common ancestor of all members of the group

polyploid cell: cell with extra chromosome sets

polysaccharide: carbohydrate consisting of hundreds of monosaccharides

pons: oval mass in the brainstem where white matter connects the forebrain to the medulla and cerebellum

population: interbreeding members of the same species occupying the same region

population density: number of individuals of a species per unit area or volume of habitat

population distribution: pattern in which individuals are scattered throughout a habitat

population dynamics: study of the factors that influence changes in a population's size

portal system: arrangement of blood vessels in which a capillary bed drains into a vein that drains into another capillary bed

positive feedback: a process that reinforces an existing condition

postanal tail: muscular tail that extends past the anus; one of the four characteristics of chordates

posterior pituitary: the back part of the pituitary gland

postsynaptic cell: neuron, muscle cell, or gland cell that receives a message at a synapse

postzygotic reproductive isolation: separation of species due to selection against hybrid offspring

potential energy: stored energy available to do work

prebiotic simulation: experiment that attempts to re-create the conditions on early Earth that gave rise to the first cell

predator: animal that kills and eats other animals

prediction: anticipated outcome of the test of a hypothesis

preembryonic stage: first 2 weeks of human development

preimplantation genetic diagnosis (PGD): use of DNA probes to detect genetic illness in an embryo before implanting it into a uterus

pressure flow theory: theory that explains how phloem sap moves under pressure from source to sink

presynaptic cell: neuron that releases neurotransmitters into a synaptic cleft

prey: animal that a predator kills and eats

prezygotic reproductive isolation: separation of species due to factors that prevent the formation of a zygote

primary growth: cell division in apical meristems; lengthens shoot and root tips

primary immune response: immune system's response to its first encounter with a foreign antigen

primary oocyte: diploid cell arising from an oogonium; undergoes meiosis I and yields a haploid polar body and a haploid secondary oocyte

primary producer: species forming the base of a food web; autotroph

primary spermatocyte: diploid cell arising from a spermatogonium; undergoes meiosis I and yields two haploid secondary spermatocytes

primary structure: the amino acid sequence of a protein

primary succession: series of changes in a community's species composition in an area previously devoid of life

primate: mammal with opposable thumbs, eyes in front of the skull, a relatively large brain, and flat nails instead of claws; includes prosimians, simians, and hominoids

primitive streak: furrow along the back of the embryonic disc in the third week of human development; longitudinal axis around which later structures develop

principle of superposition: the idea that lower rock layers are older than those above them

prion: infectious particle composed of protein

producer (autotroph): organism that uses inorganic sources of energy and carbon

product: the result of a chemical reaction

product rule: the chance of two independent events occurring equals the product of the individual chances of each event

profundal zone: deep region of a lake or ocean where light does not penetrate

progenote: collection of nucleic acid, protein, and lipids that was the forerunner to cells

progesterone: steroid sex hormone produced in ovaries of female vertebrates

prokaryote: a cell that lacks a nucleus and other membrane-bounded organelles; bacteria and archaea

prolactin: hormone produced in the anterior pituitary; stimulates milk secretion

prometaphase: stage of mitosis just before metaphase, when the nuclear membrane breaks up and spindle fibers attach to kinetochores

promoter: a control sequence at the start of a gene; attracts RNA polymerase and (in eukaryotes) transcription factors

prophage: DNA of a lysogenic bacteriophage that is inserted into a host cell's chromosome

prophase: stage of mitosis when chromosomes condense and the mitotic spindle begins to form

prophase I: prophase of meiosis I, when chromosomes condense and become visible, and crossing over occurs

prophase II: prophase of meiosis II, when chromosomes condense and become visible

proprioceptor: sensory receptor that detects the position of body parts

prosimian: type of primate; a lemur, aye-aye, loris, tarsier, or bush baby

prostate gland: male structure between the bladder and the penis; produces a milky, alkaline fluid that activates sperm

protein: a polymer consisting of amino acids and folded into its functional three-dimensional shape

protist: eukaryotic organism that is not a plant, fungus, or animal

proton: a particle in an atom's nucleus carrying a positive charge

protostome: clade of bilaterally symmetrical animals in which the first opening in the gastrula typically develops into the mouth

protozoan: unicellular protist that is heterotrophic and (usually) motile

proximal convoluted tubule: region of the renal tubule between the glomerular capsule and the nephron loop; site of reabsorption and secretion

proximate: describing the mechanistic causes of behavior

pseudocoelom: fluid-filled animal body cavity lined by endoderm and mesoderm

pseudogene: a DNA sequence that is very similar to that of a gene; a pseudogene is transcribed but its mRNA is not translated

pulmonary artery: artery that leads from the right ventricle to the lungs

pulmonary circulation: blood circulation between the heart and lungs

pulmonary vein: vein that leads from the lungs to the left atrium

punctuated equilibrium: theory that life's history has been characterized by bursts of rapid evolution interrupting long periods of little change

Punnett square: diagram that uses the genotypes of the parents to reveal the possible results of a genetic cross

pupil: opening in the iris that admits light into the eye

pyramid of energy: diagram depicting energy stored at each trophic level at a given time

pyruvate: the three-carbon product of glycolysis

Q

quaternary structure: the shape arising from interactions between multiple polypeptide subunits of the same protein

R

R group: an amino acid side chain

radial symmetry: body form in which any plane passing through the body from the mouth to the opposite end divides the body into mirror images

radioactive isotope: atom that emits particles or rays as its nucleus disintegrates

radiolarian: amoeboid protozoan with a silica shell

radiometric dating: a type of absolute dating that uses known rates of radioactive decay to date fossils

radula: a chitinous, tonguelike strap in many mollusks

rain shadow: downwind side of a mountain, with a drier climate than the upwind side

ray: parenchyma cells extending from center of woody stem or root

ray-finned fish: type of bony fish with fan-shaped fins

reabsorption: return of useful substances to the blood at the renal tubule and collecting duct

reactant: a starting material in a chemical reaction

reaction center: a molecule of chlorophyll a (and associated proteins) that participates in the light reactions of photosynthesis

receptacle: attachment point for flower parts

receptor potential: localized change in membrane potential (a graded potential) in a sensory receptor

recessive allele: allele whose expression is masked if a dominant allele is present

reciprocal altruism: altruistic act performed in anticipation of payback in the future

recombinant chromatid: chromatid containing genetic information from both parents as a result of crossing over

recombinant DNA: genetic material spliced together from multiple sources

red alga: multicellular, photosynthetic marine protist with red or blue accessory pigments

red bone marrow: marrow that gives rise to blood cells and platelets

red blood cell: disk-shaped blood cell that contains hemoglobin

reduction: the gain of one or more electrons by a participant in a chemical reaction

reflex: an instantaneous, automatic response to a stimulus

reflex arc: neural pathway that controls a reflex, often linking a sensory receptor directly to an effector

relative dating: placing a fossil into a sequence of events without assigning it a specific age

renal tubule: portion of a nephron that adjusts filtrate composition; consists of the nephron loop and the proximal and distal convoluted tubules

repressor: in an operon, a protein that binds to the operator and prevents transcription

reproductive system: organ system consisting of organs that produce and transport gametes and that may nurture developing offspring

reptile: type of tetrapod vertebrate with scaly or feathery skin; embryo is enclosed in an amniotic egg

reservoir (virus): a host species that serves as a continual source of viral infection for other species

resource partitioning: use of the same resource in different ways or at different times by multiple species

respiratory cycle: one inhalation followed by one exhalation

respiratory surface: part of an animal's body that exchanges gases with the environment

respiratory system: organ system consisting of the lungs or comparable organs that acquire O_2 and release CO_2

resting potential: membrane potential for a neuron not conducting a nerve impulse

restriction enzyme: enzyme that cuts double-stranded DNA at a specific base sequence

retina: sheet of photoreceptors that forms the innermost layer of the eye

reverse transcriptase: enzyme that uses RNA as a template to construct a DNA molecule

rhodopsin: pigment that transduces light into an electrochemical signal in photoreceptors

ribosomal RNA (rRNA): a molecule of RNA that, along with proteins, forms a ribosome

ribosome: a structure built of RNA and protein where mRNA anchors during protein synthesis

ribulose bisphosphate (RuBP): the five-carbon molecule that reacts with CO_2 in the Calvin cycle

ribonucleic acid (RNA): nucleic acid typically consisting of a single strand of nucleotides, each containing the sugar ribose

RNA polymerase: enzyme that uses a DNA template to produce a molecule of RNA

RNA world: the idea that independently replicating RNA molecules were precursors to life

rod cell: photoreceptor in the retina that provides black-and-white vision

root: belowground part of most plants

root cap: cells that protect root apical meristem from abrasion

root hair: epidermal outgrowth that increases root surface area

rough endoplasmic reticulum: ribosome-studded portion of the ER where secreted proteins are synthesized

roundworm: unsegmented worm with a pseudocoelom; phylum Nematoda

rubisco: enzyme that adds CO_2 to ribulose bisphosphate in the carbon reactions of photosynthesis

ruminant: herbivore with a four-chambered organ specializing in grass digestion

S

S phase: the synthesis phase of interphase, when DNA replicates

S-shaped curve: plot of logistic growth over time

salamander: type of amphibian that superficially resembles a lizard

sample size: number of subjects in each experimental group

sapwood: light-colored, functioning secondary xylem in woody plant

sarcomere: one of many repeated units in a myofibril of a muscle cell

saturated fatty acid: a fatty acid with single bonds between all carbon atoms

savanna: type of terrestrial biome with wet and dry seasons and warm weather year-round; grassland with scattered trees

Schwann cell: type of neuroglia that forms a myelin sheath around some axons in the peripheral nervous system

scientific method: a systematic approach to understanding the natural world based on evidence and testable hypotheses

sclera: the outermost layer of the eye; the white of the eye

sclereid: relatively short sclerenchyma cell

sclerenchyma: rigid plant cells that support mature plant parts

scrotum: the sac containing the testes

search image: a collection of cues that enable a predator to detect inconspicuous prey

second law of thermodynamics: every reaction loses some energy to the surroundings as heat; entropy always increases

second messenger: molecule that translates a stimulus at the cell's exterior into an effect inside the cell

secondary growth: cell division in lateral meristems; increases girth of shoots and roots

secondary immune response: immune system's response to subsequent encounters with a foreign antigen

secondary oocyte: haploid cell that undergoes meiosis II and yields a haploid polar body and a haploid egg cell

secondary sex characteristic: trait that distinguishes the sexes but does not participate directly in reproduction

secondary spermatocyte: haploid cell that undergoes meiosis II and yields two haploid spermatids

secondary structure: a "substructure" within a protein, resulting from hydrogen bonds between parts of the peptide backbone

secondary succession: series of changes in a community's species composition following a disturbance

secretion: addition of substances to the fluid in a renal tubule

seed: in gymnosperms and angiosperms, a plant embryo packaged with a food supply inside a tough outer coat

seed coat: protective outer layer of seed

seedless vascular plant: plant with vascular tissue but not seeds; includes true ferns, club mosses, whisk ferns, and horsetails

segmentation: division of an animal body into repeated subunits

selective permeability: the property that enables a membrane to admit some substances and exclude others

selfish herd: group in which each animal tries to move as close to the center as possible

semen: fluid that carries sperm cells out of the body

semicircular canal: one of three perpendicular fluid-filled structures in the inner ear; provides information on the position of the head

semilunar valve: flap of heart tissue that prevents the flow of blood from an artery to a ventricle

seminal vesicle: structure that contributes fluid, fructose, and prostaglandins to semen

seminiferous tubule: tubule within a testis where sperm form and mature

senescence: aging

sensation: raw sensory input that reaches the central nervous system

sensor: in homeostatic responses, a structure that monitors changes in a parameter

sensory adaptation: lessening of sensation with prolonged exposure to a stimulus

sensory neuron: neuron that transmits information from a stimulated body part to the central nervous system

sensory receptor: cell that detects stimulus information

sepal: part of the outermost whorl of a flower

Sertoli cell: cell that supports developing sperm cells within a seminiferous tubule

severe combined immunodeficiency (SCID): inherited immune system disorder in which neither T cells nor B cells function

sex chromosome: a chromosome that carries genes that determine sex

sex pilus: appendage that transfers DNA from one cell to another in conjugation

sex-linked: describes genes or traits on the X or Y chromosome

sexual dimorphism: difference in appearance between males and females

sexual reproduction: the combination of genetic material from two individuals to create a third individual

sexual selection: type of natural selection resulting from variation in the ability to obtain mates

sexually transmitted disease (STD): illness caused by a pathogen that spreads during sexual contact

shoot: aboveground part of a plant

short-day (long-night) plant: plant that flowers when uninterrupted dark periods are longer than a critical length

short tandem repeat (STR): short DNA sequences that vary in length among individuals in a population

short-term memory: memory only available for a few moments

sieve plate: porous area at end of sieve tube element

sieve tube element: conducting cell that makes up sieve tube in phloem

signal transduction: conversion of an external signal into a response inside a cell

simian: type of primate; a monkey

simple diffusion: form of passive transport in which a substance moves down its concentration gradient without the use of a transport protein

simple leaf: leaf with undivided blade

sink: plant part that receives sugars in phloem

sinoatrial (SA) node: specialized cardiac muscle cells that set the pace of the heartbeat; the pacemaker

skeletal muscle tissue: voluntary muscle tissue consisting of long, unbranched, striated cells with multiple nuclei

skeletal muscle: organ that generates voluntary movements between pairs of bones; composed of bundles of skeletal muscle tissue and other tissue types

skeletal system: organ system consisting of bones, ligaments, and cartilage that support body structures and attach to muscles

skeleton: structure that supports an animal's body

skin: the outer surface of the body

sliding filament model: sliding of actin and myosin past each other to shorten a muscle cell

slow-twitch fiber: small-diameter muscle fiber that produces twitches of long duration

small intestine: part of the digestive tract that connects the stomach with the large intestine; site of most chemical digestion and absorption

smog: type of air pollution that forms a visible haze in the lower atmosphere

smooth endoplasmic reticulum: portion of the ER that produces lipids and detoxifies poisons

smooth muscle tissue: involuntary muscle tissue consisting of nonstriated, spindle-shaped cells

snake: type of reptile that lacks legs and has a forked tongue

social behavior: interaction between members of the same species

sociobiology: study of social behavior in the context of an animal's fitness

sodium–potassium pump: protein that uses energy from ATP to transport Na^+ out of cells and K^+ into cells

soil: rock and mineral particles mixed with organic matter, air, and water

solute: a chemical that dissolves in a solvent, forming a solution

solution: a mixture of a solute dissolved in a solvent

solvent: a chemical in which other substances dissolve, forming a solution

somatic cell: body cell that does not give rise to gametes

somatic cell nuclear transfer: technique used to clone a mammal from an adult cell

somatic mutation: a DNA sequence change that occurs in nonsex cells

somatic nervous system: in the peripheral nervous system, motor pathways carrying signals to skeletal (voluntary) muscles

source: plant part that produces or releases sugar

speciation: formation of new species

species: a distinct type of organism

species evenness: measure of biodiversity; the relative abundance of the species in a community

species richness: measure of biodiversity; number of species in a community

spermatid: one of four haploid cells produced in meiosis II of spermatogenesis

spermatogenesis: the production of sperm

spermatogonium: diploid germ cell in the wall of a seminiferous tubule; divides mitotically to yield a stem cell and a primary spermatocyte

spermatozoon (pl. spermatozoa): mature sperm cell

sphincter: muscular ring that contracts to close an opening

spinal cord: tube of nervous tissue that extends through the vertebral column

spindle: a structure of microtubules that aligns and separates chromosomes in mitosis or meiosis

spirillum (pl. spirilla): spiral-shaped prokaryote

spleen: lymphoid organ that produces and stores lymphocytes and destroys worn-out red blood cells

sponge: simple, asymmetrical animal lacking true tissues and gastrulation; phylum Porifera

spongy bone tissue: bone tissue with large spaces between a web of bony struts

spore (fungus): microscopic reproductive cell, produced sexually or asexually

spore (plant): haploid product of meiosis in the sporophyte; develops mitotically into the gametophyte

sporophyte: diploid, spore-producing stage of the plant life cycle

stabilizing selection: form of natural selection in which extreme phenotypes are less fit than the optimal intermediate phenotype

stamen: male flower part interior to petals; consists of a filament and anthers

standardized variable: any factor held constant for all subjects in an experiment

statistically significant: unlikely to be attributed to chance

statolith: starch-containing plastid in root cap cell that functions as a gravity detector

stem: part of a plant that supports leaves

stem cell: undifferentiated cell that divides to give rise to additional stem cells and cells that specialize

steroid hormone: a lipid-soluble hormone that can freely diffuse through a cell membrane and bind to a receptor inside the cell

steroid: lipid consisting of four interconnected carbon rings

stigma: in angiosperms, pollen-receiving tip of style

stoma (pl. stomata): pore in a plant's epidermis through which gases are exchanged with the atmosphere

stomach: J-shaped compartment that mechanically and chemically breaks down food received from the esophagus

stroma: the fluid inner region of the chloroplast

style: in angiosperms, the stalklike upper part of a carpel

substitution mutation: replacement of one nucleotide in a gene with another

substrate feeder: animal that lives in its food and eats it from the inside

substrate-level phosphorylation: ATP formation occurring when an enzyme transfers a phosphate group from a high-energy donor molecule to ADP

succession: change in the species composition of a community over time

survivorship curve: graph of the proportion of individuals that survive to a particular age

suspension feeder: animal that eats by straining particles out of water

symbiosis: a close and often lifelong ecological relationship in which one species typically lives in or on another

sympathetic nervous system: part of the autonomic nervous system; mobilizes the body to respond quickly to environmental stimuli and opposes the parasympathetic nervous system

sympatric speciation: formation of a new species within the boundaries of a parent species

synapse: junction at which a neuron communicates with another cell

synapsid: vertebrate with a single opening behind each eye orbit; mammals and their immediate ancestors

synaptic cleft: space into which neurotransmitters are released between two cells at a synapse

synaptic integration: a neuron's overall response to incoming excitatory and inhibitory neural messages

synaptic terminal: enlarged tip of an axon; contains synaptic vesicles filled with neurotransmitters

synovial joint: joint between two freely movable bones connected by a fluid-filled capsule of fibrous connective tissue

systematics: field of study that includes taxonomy and phylogenetics

systemic circulation: blood circulation between the heart and the rest of the body, except the lungs

systolic pressure: upper number in a blood pressure reading; reflects contraction of the ventricles

T

T cell: type of lymphocyte that matures in the thymus; may become a helper T cell or a cytotoxic T cell

taiga (boreal forest): type of terrestrial biome with long, dry winters and cool summers with moderate precipitation; coniferous trees dominate

tapeworm: type of parasitic flatworm

taproot system: large central root and its lateral branches

target cell: cell that expresses receptors for a particular hormone

taste bud: cluster of cells that detect chemicals in food

taxis: directed movement toward or away from a stimulus

taxon: a group of organisms at any rank in the taxonomic hierarchy

taxonomy: the science of describing, naming, and classifying organisms

technology: the practical application of scientific knowledge

tectorial membrane: membrane in contact with hair cells of the cochlea

telomerase: enzyme that extends telomeres, enabling cells to divide continuously

telomere: noncoding DNA at the tip of a eukaryotic chromosome

telophase: stage of mitosis in which chromosomes arrive at opposite poles and nuclear envelopes form

telophase I: telophase of meiosis I, when homologs arrive at opposite poles

telophase II: telophase of meiosis II, when chromosomes arrive at opposite poles and nuclear envelopes form

temperate coniferous forest: type of terrestrial biome with cool summers, mild winters, and moderate to high precipitation all year; coniferous trees dominate

temperate deciduous forest: type of terrestrial biome with warm summers, cold winters, and moderate precipitation all year; deciduous trees dominate

temperate grassland: type of terrestrial biome with hot summers, cold winters, and low to moderate precipitation; grasses dominate; grazing, fire, and drought restrict tree growth

template strand: the strand in a DNA double helix that is transcribed

tendon: band of fibrous connective tissue that attaches a muscle to a bone

teratogen: substance that causes birth defects

terminal bud: undeveloped shoot at a stem's tip

terminator: sequence in DNA that signals where the gene's coding region ends

territory: a space that an animal defends against intruders

tertiary structure: the overall shape of a polypeptide, resulting mostly from interactions between amino acid R groups and water

test cross: a mating of an individual of unknown genotype to a homozygous recessive individual to reveal the unknown genotype

testis (pl. testes): male gonad; organ that produces sperm and hormones

testosterone: steroid sex hormone produced primarily in the testes of male vertebrates

tetanus: maximal muscle contraction caused by continual stimulation

tetrapod: vertebrate with four limbs

thalamus: forebrain structure that relays sensory input to the cerebrum

theory: well-supported scientific explanation

thermoreceptor: sensory receptor that responds to temperature

thermoregulation: control of an animal's body temperature

thick filament: in muscle cells, filament composed of myosin

thigmotropism: directional growth response to touch

thin filament: in muscle cells, filament composed of actin

threshold potential: potential to which a neuron's membrane must be depolarized to trigger an action potential

thylakoid: disclike structure that makes up the inner membrane of a chloroplast

thylakoid space: the inner compartment of the thylakoid

thymus: lymphoid organ in the upper chest where T cells learn to distinguish foreign antigens from self antigens

thyroid gland: gland in the neck that secretes two thyroid hormones (thyroxine and triiodothyronine) and calcitonin

thyroid-stimulating hormone (TSH): hormone produced in the anterior pituitary; stimulates secretion of thyroid hormones

thyroxine: one of two thyroid hormones that increase the rate of cellular metabolism

tidal volume: volume of air inhaled or exhaled during a normal breath

tight junction: connection between two adjacent animal cells that prevents fluid from flowing past the cells

tissue: group of cells that interact and provide a specific function

tongue: muscular structure on the floor of the mouth; mixes food and aids in swallowing

tooth: mineral-hardened structure embedded in the jaw; grasps and chews food

topsoil: uppermost soil layer

trace element: an element that an organism requires in small amounts

trachea (pl. tracheae): in vertebrates, the respiratory tube that connects the larynx to the bronchi; the "windpipe." In invertebrates, a branched tubule that brings air in close contact with cells, facilitating gas exchange

tracheid: long, narrow conducting cell in xylem

transcription: production of RNA using DNA as a template

transcription factor: in a eukaryotic cell, a protein that binds a gene's promoter and regulates transcription

transduction (gene transfer): transfer of DNA into a cell via a virus

transduction (sensory function): process by which a sensory receptor converts energy from a stimulus into action potentials

transfer RNA (tRNA): a molecule of RNA that binds an amino acid at one site and an mRNA codon at its anticodon site

transformation: type of horizontal gene transfer in which an organism takes up naked DNA without cell-to-cell contact

trans fat: unsaturated fat with straight fatty acid tails

transgenic: containing DNA from multiple species

translation: assembly of an amino acid chain according to the sequence of nucleotides in mRNA

translocation: exchange of genetic material between nonhomologous chromosomes

transpiration: evaporation of water from a leaf

transposable element (transposon): DNA sequence that can move within a genome

triglyceride: lipid consisting of one glycerol bonded to three fatty acids

triiodothyronine: one of two thyroid hormones that increase the rate of cellular metabolism

trilobite: extinct type of arthropod with three body lobes

trophic level: an organism's position along a food chain

tropical rain forest: type of terrestrial biome with year-round high temperatures and precipitation; abundant, diverse plant species include large trees

tropism: orientation toward or away from a stimulus

true-breeding: always producing offspring identical to the parent for one or more traits; homozygous

true fern: type of seedless vascular plant with fronds bearing sporangia on their undersides

trypanosome: flagellated protist that causes human diseases transmitted by biting insects

tuatara: rare type of reptile resembling a lizard

tumor: abnormal mass of tissue resulting from cells dividing out of control

tumor suppressor gene: gene that normally prevents cell division but when inactivated or suppressed causes cancer

tundra: type of terrestrial biome with low temperature and short growing season; lichens, mosses, grasses, and shrubs dominate

tunicate: type of invertebrate chordate with a tunic covering its saclike body

turgor pressure: the force of water pressing against the cell wall

turtle: type of reptile with a shell made of bony plates

twitch: contraction and relaxation of a muscle cell following a single stimulation

U

ultimate: describing the evolutionary origin or adaptive advantage of an animal's behavior

ultraviolet (UV) radiation: portion of the electromagnetic spectrum with wavelengths shorter than 400 nm

umbilical cord: ropelike structure containing blood vessels that connect an embryo or fetus with the placenta

uniformitarianism: the idea that today's gradual geological processes (e.g., erosion and sedimentation) have also occurred in the past, producing changes in Earth over time

unsaturated fatty acid: a fatty acid with at least one double bond between carbon atoms

upwelling: upward movement of cold, nutrient-rich lower layers of a body of water

urea: nitrogenous waste derived from ammonia and excreted by mammals, amphibians, and some reptiles and fishes

ureter: muscular tube that transports urine from the kidney to the bladder

urethra: tube that transports urine (and semen in males) out of the body

uric acid: nitrogenous waste derived from ammonia and excreted by most reptiles (including birds)

urinary bladder: muscular sac where urine collects

urinary system: organ system consisting of the kidneys or comparable organs that maintain body fluid composition

urine: liquid waste produced by kidneys

uterine tube: tube that conducts an oocyte from an ovary to the uterus

uterus: muscular, saclike organ where embryo and fetus develop

V

vaccine: substance that initiates a primary immune response to a pathogen without actually causing a disease

vacuole: membrane-bounded storage sac in a cell, especially the large central vacuole in a plant cell

vagina: conduit from the uterus to the outside of the body; receives the penis during intercourse; the birth canal

valence shell: outermost occupied energy shell of an atom

variable: any changeable element in an experiment

variable region: amino acid sequence that is different for every antibody

vas deferens: tube that transports sperm from an epididymis to an ejaculatory duct

vascular bundle: a strand of xylem, phloem, and other tissues in a stem or leaf

vascular cambium: lateral meristem that produces secondary xylem and secondary phloem

vascular tissue: conducting tissue for water, minerals, and organic substances in plants

vasoconstriction: decrease in the diameter of a blood vessel

vasodilation: increase in the diameter of a blood vessel

vegetative plant parts: nonreproductive parts (roots, stems, and leaves)

vein (animal): vessel that returns blood to the heart

vein (plant): vascular bundle inside leaf

ventral nerve cord: nerve cord that runs along the ventral side of an animal

ventricle: heart chamber that pumps blood out of the heart

venule: small vein

vertebra: one unit of the vertebral column composed of bone or cartilage that supports and protects the spinal cord

vertebral column: bone or cartilage that supports and protects the spinal cord

vertebrate: animal with a backbone

vertical gene transfer: passage of DNA from one generation to the next; cell division

vesicle: a membrane-bounded sac that transports materials in a cell

vessel element: short, wide conducting cell in xylem

vestibular apparatus: structure in the inner ear that provides information on the position of the head and changes in velocity

vestigial: having no apparent function in one organism but homologous to a functional structure in another species

viroid: infectious RNA molecule

virus: infectious agent that consists of genetic information enclosed in a protein coat

visceral mass: part of a mollusk that contains the digestive, circulatory, excretory, and reproductive organs

vital capacity: maximal volume of air that can be forced out of the lungs during one breath

vocal cord: elastic tissue band that covers the larynx and vibrates as air passes, producing sound

vulnerable species: species facing a high risk of extinction in the distant future

vulva: external female genitalia; the labia, clitoris, and vaginal opening

W

water mold: filamentous, heterotrophic protist; also called an oomycete

water vascular system: system of canals in echinoderms; functions in locomotion, feeding, sensation, gas exchange, and excretion

wavelength: the distance a photon moves during a complete vibration

wax: lipid consisting of fatty acids connected to alcohol or other molecules

wetland: ecosystem in which soil is saturated with water

whisk fern: type of seedless vascular plant with branched stems but no obvious leaves

white blood cell: one of five types of blood cells that participate in the immune response

white matter: nervous tissue in the central nervous system; consists of myelinated axons

wild-type: the most common allele, genotype, or phenotype

wood: secondary xylem

woody plant: plant with stems and roots made of wood and bark

X

X inactivation: turning off all but one X chromosome in each cell of a mammal (usually female) early in development

X-linked: describes traits controlled by genes on the X chromosome

xylem: vascular tissue that transports water and dissolved minerals in plants, in addition to some hormones

xylem sap: solution of water, dissolved minerals, and hormones in xylem

Y

yeast: unicellular fungus

yellow bone marrow: fatty marrow that replaces red bone marrow as bones age

Y-linked: describes traits controlled by genes on the Y chromosome

yolk sac: extraembryonic membrane that forms beneath the embryonic disc and manufactures blood cells

Z

zoospore: flagellated spore produced by chytrids

zygomycete: fungus that produces zygospores

zygospore: diploid resting spore produced by fusion of haploid cells in zygomycetes

zygote: the fused egg and sperm, which develops into a diploid individual

Index